INSECTS AS NATURAL ENEMIES

INSECTS AS NATURAL ENEMIES

A Practical Perspective

Edited by

Mark A. Jervis

Cardiff School of Biosciences
Cardiff University, Cardiff,
Wales, United Kingdom

A C.I.P. Catalogue record for this book is available from the Library of Congress.

ISBN-10 1-4020-1734-0 (HB) Springer Dordrecht, Berlin, Heidelberg, New York
ISBN-10 1-4020-2625-0 (e-book) Springer Dordrecht, Berlin, Heidelberg, New York
ISBN-13 978-1-4020-1734-6 (HB) Springer Dordrecht, Berlin, Heidelberg, New York
ISBN-13 978-1-4020-2625-6 (e-book) Springer Dordrecht, Berlin, Heidelberg, New York

---

Published by Springer,
P.O. Box 17, 3300 AA Dordrecht, The Netherlands.

Printed on acid-free paper

Cover picture:
Parasitoid wasp (*Aphidius ervi*) attack posture and host aphid (*Microlophium carnosum*) defence posture. Globules of aphid siphuncular secretion are clearly visible.

Printed in the Netherlands

For Julia,
George and William,
and in memory of
my parents

CONTENTS

PREFACE

The past three decades have seen a dramatic increase in practical and theoretical studies on insect natural enemies. The importance and appeal of insect predators, and of parasitoids in particular, as research animals derives from the relative ease with which many species can be cultured and experimented with in the laboratory, the simple life-cycles of most parasitoids, and the increasing demand for biological control of insect pests.

This book – an updated and considerably expanded version of Jervis and Kidd (1996) – is intended to guide enquiring students and research workers to those approaches and techniques that are most appropriate to the study and evaluation of predators and parasitoids. It is neither a practical manual nor a 'recipe book' – most chapters are accounts of major aspects of the biology of natural enemies, punctuated with advice on which experiments or observations to conduct and how, in broad terms, to carry them out. Detailed protocols are usually not given, but guidance is provided, where necessary, on literature that may need to be consulted on particular topics.

I hope that *Insects as Natural Enemies: A Practical Perspective* will successfully both encourage and assist research into the fascinating biology of insect predation and parasitism.

REFERENCE

Jervis, M. and Kidd, N. (1996) *Insect Natural Enemies: Practical Approaches to Their Study and Evaluation*. Chapman and Hall, London, pp. 491.

Contributors' Addresses

Jacques J.M. van Alphen
Section Evolutionary Biology
Institute of Evolutionary and
Ecological Sciences
University of Leiden
P.O. Box 9516
NL-2300 RA Leiden
Email: alphen@rulsfb.leidenuniv.nl

Leo W. Beukeboom,
Evolutionary Genetics
Centre for Ecological and
Evolutionary Studies
University of Groningen
Kerklann 30
P.O. Box 14
NL-9750 AA Haren
The Netherlands
Email: l.w.beukeboom@biol.rug.nl

Robert D.J. Butcher
Department of Biology and Biochemistry
University of Bath
Claverton Campus
Bath BA2 7AY
UK
Email: r.butcher@bath.ac.uk

Michael J.W. Copland
WyeBugs
Wye Campus
Imperial College London
Wye
Ashford TN25 5AH
UK
Email: info@wyebugs.co.uk

Mark D.E. Fellowes
School of Animal and Microbial Sciences
The University of Reading
Whiteknights
Reading RG6 6AJ
UK
Email: m.fellowes@reading.ac.uk

Ian. C.W. Hardy
School of Biosciences
University of Nottingham
Sutton Bonington Campus
Loughborough LE12 5RD
UK
Email: ian.hardy@nottingham.ac.uk

Jeffrey A. Harvey
Department of Multitrophic Interactions
Netherlands Institute of Ecology
Postbox 40
6666 ZG Heteren
The Netherlands
Email: j.harvey@nioo.knaw.nl

George. E. Heimpel
Department of Entomology
University of Minnesota
1980 Folwell Avenue
St. Paul, MN 55108
USA
Email: heimp001@tc.umn.edu

Mark A. Jervis
Cardiff Schoool of Biosciences
Cardiff University
P.O. Box 915
Cardiff CF10 3TL
Wales
UK
Email: jervis@cf.ac.uk

Neil A.C. Kidd
Cardiff Schoool of Biosciences
Cardiff University
P.O. Box 915
Cardiff CF10 3TL
Wales
UK
Email: kiddn@cf.ac.uk

Paul J. Ode
Department of Entomology
North Dakota State University
202 Hultz Hall
1300 Albrecht Boulevard
Fargo
ND 58105-5346
USA
Email: Paul.Ode@ndsu.nodak.edu

Wilf Powell
Division of Plant and Invertebrate Ecology
Rothamsted Research
Harpenden AL5 2JQ
UK
Email: wilf.powell@bbsrc.ac.uk

Michael Siva-Jothy
Department of Animal and Plant Sciences
University of Sheffield
Sheffield S10 2TN
UK
Email: m.siva-jothy@sheffield.ac.uk

Keith Sunderland
Department of Entomological Sciences
Horticulture Research International
Wellesbourne CV35 9EF
UK
Email: keith.sunderland@hri.ac.uk

William O.C. Symondson
Cardiff School of Biosciences
Cardiff University
P.O. Box 915
Cardiff CF10 3TL
Wales
UK
Email: symondson@cf.ac.uk

Bas J. Zwaan
Section Evolutionary Biology
Institute of Evolutionary and
Ecological Sciences
University of Leiden
P.O. Box 9516
NL-2300 RA Leiden
Email: zwaan@rulsfb.leidenuniv.nl

FORAGING BEHAVIOUR 1

M. D. E. Fellowes, J. J. M. van Alphen and M. A. Jervis

1.1 BEHAVIOUR OF INSECT PARASITOIDS AND PREDATORS

In this chapter, we consider practical aspects of the foraging behaviour of insect natural enemies in its widest sense. Initially, most insect natural enemies must locate the habitat where potential victims may be found. Within that habitat, the victims themselves must be discovered. Once a patch of potential targets is identified, the predator or parasitoid must choose its victim. Furthermore, in judging host quality, a parasitoid must decide whether to feed from the host, to oviposit, or to do both. If she does decide to oviposit, then there are questions of sex allocation and offspring number that need to be addressed (Figure 1.1). All of these activities fall under the aegis of 'foraging behaviour'.

Studies of the foraging behaviour of insect natural enemies lie at the heart of much of modern ecology. These studies have taken two broadly defined pathways, where the emphasis is determined by the interest of the researcher (see below). Irrespective of the motivation of the researcher, it is clear that any attempt to understand the foraging behaviour of a predator or parasitoid will greatly benefit from knowledge gleaned from both approaches. This cross-fertilisation of ideas is something we try to emphasise in this chapter.

In addition, we provide a review of the foraging behaviour of insect natural enemies. This is meant to be illustrative, with stress placed on the experiments used to study the behaviour itself. For greater detail on the behaviour of parasitoids one should refer to the complementary volumes of Godfray (1994) and Quicke (1997). The literature on insect predators is much

Which patches to visit

Whether to oviposit and/or feed on a host

What clutch size to produce for each host attacked (and whether to oviposit in an already parasitised host, i.e. superparasitise it)

How much blood to remove from a host when feeding

What sex ratio of progeny to produce

When to leave a patch

Figure 1.1 Foraging decisions: In adopting either the functional or the causal approach to studying predator and parasitoid behaviour, it is useful to consider that foraging insect natural enemies are faced with a number of consecutive or simultaneous decisions. Listed in this figure are some of the decisions that may need to be made by a gregarious host-feeding parasitoid.

more diffuse, but New (1991) provides a good introduction to the behaviour of predators in general, while Dixon (2000) reviews the behaviour of ladybird beetles (Coccinellidae).

Where possible, we deal with insect predators and parasitoids together, although there are some sections (e.g. sex ratio manipulation) where the examples come exclusively from the parasitoid literature, and other sections (e.g. superparasitism/cannibalism) where both are dealt with, but separately. Nevertheless, many of the approaches to studying foraging behaviour, and the theory underpinning it, are similar for both predators and parasitoids.

There are five main parts to this chapter. First, we describe the methodological approaches that underpin studies of the foraging behaviour of insect natural enemies. Second, we discuss how the predators and parasitoids find the habitat patches where potential victims may be found. Third, we move on to consider search modes within a patch, leading to host or prey location. Fourth, we reflect on what occurs after the victim is found, and deal with issues such as host preference and sex allocation behaviour. Finally, we touch briefly upon some of the wider ecological consequences of insect natural enemy foraging behaviour.

1.2 METHODOLOGY

1.2.1 THE CAUSAL APPROACH

Until the late 1970s, parasitoid foraging behaviour was mostly studied from a proximate (i.e. causal or mechanistic) standpoint, with a strong emphasis on identifying which stimuli parasitoids respond to both in finding and in recognising their hosts. Through this approach, fascinating insights into parasitoid foraging behaviour have been gained, and it has been demonstrated that often an intricate tritrophic relationship exists between phytophagous insects, their host plants and parasitoids. We now know the identities of several of the chemical compounds eliciting certain behaviours in parasitoids. Some of the research in this field has been devoted to the application to crops of chemical substances, as a means of manipulating parasitoid behaviour in such a way that parasitism of crop pests is increased. Many parasitoid species display individual plasticity in their responses to different cues. Associative learning (subsection 1.5.2) of odours, colours or shapes related to the host's environment has been described for many parasitoid species (e.g. Dugatkin and Alfieri, 2003; Meiners *et al.*, 2003).

Often, causal questions do not involve elaborate theories. Questions of whether an organism responds to a particular chemical stimulus or not, or whether it reacts more strongly to one stimulus than to another, lead to straightforward experimental designs. It is in the technical aspects of the experiment rather than the underlying theory that the experimenter needs to be creative. However, the study of causation can be extended to ask how information is processed by the central nervous system. One can ask how a sequence of different stimuli influences the behavioural response of the animal, or how responses to the same cue may vary depending on previous experience and the internal state of the animal (Putters and van den Assem, 1988; Morris and Fellowes, 2002).

Two different causal approaches have been adopted in the study of the integrated action of a series of different stimuli on the behaviour of a foraging animal:

1. The formalisation of a hypothesis into a model of how both external information and the internal state of the animal result in behaviour, and the testing, through experiments, of the predictions of the model. Waage (1979) pioneered this approach for parasitoids. Artificial neural network models have also been used to analyse sex allocation behaviour in parasitoids (Putters and Vonk, 1991; Vonk *et al.*, 1991).
2. The statistical analysis of time-series of behaviour to assess how the timing and sequence of events influences the behaviour of the organism. An example of this approach is the analysis of the factors influencing patch time allocation of a parasitoid, using the proportional hazards model (Haccou *et al.*, 1991; Tenhumberg *et al.*, 2001a).

1.2.2 THE FUNCTIONAL APPROACH

The functional approach to the study of parasitoid behaviour is based on Darwinian ideas initially formalised by MacArthur and Pianka (1966) and Emlen (1966). Termed 'natural selection thinking' by Charnov (1982), it asks how natural selection may have moulded the behaviour under study.

Because foraging decisions (Figure 1.1) determine the number of offspring produced, foraging behaviour must be under strong selection pressures. Assuming that natural selection has shaped parasitoid searching and oviposition behaviour in such a way that it maximises the probability of leaving as many healthy offspring as possible, it is possible to predict the 'best' behaviour under given circumstances. In the real world, no 'Darwinian monsters' exist that can produce limitless numbers of offspring at zero cost. Because resources are often limiting and because reproduction incurs a cost (e.g. in materials and energy [see subsection 2.14] and foraging time) to an individual, increasing investment in reproduction must always be traded off against other factors decreasing fitness (e.g. more offspring often means smaller offspring having shorter life-spans). Thus, producing the maximum possible number of offspring may not always be the optimal strategy.

We refer to natural selection thinking as the **functional approach**, because its aim is to define the function of a particular behaviour. To achieve this goal, it is necessary to show that the behaviour contributes more to the animal's fitness than alternative behaviours in the same situation. The foraging behaviour of female parasitoids has a direct influence on both the number and the quality of their offspring, so it is particularly suited for testing optimisation hypotheses.

The functional approach can be applied not only to theoretical problems but also to problems such as the selection, the evaluation and the mass-rearing of natural enemies for biological control.

There are two ways of investigating functional problems in behavioural ecology. One is to predict quantitatively, using optimality models, the 'best' behaviour under given conditions. The other is to take account of the possibility that the optimal behavioural strategy will be dependent on what other individuals, attacking the same host or prey population, are doing. Both approaches can be used to inform practical studies of insect natural enemies, and in a similar fashion, the results of practical studies can be used to construct more realistic models.

Optimality Models

Optimality models are used to predict how an animal should behave so as to maximise its fitness in the long-term. They can be designed by determining:

1. **What decision assumptions apply**, i.e. which of the forager's choices (problems) are to be analysed. Some of the decisions faced by foraging natural enemies are shown in Figure 1.1. Sexually reproducing gregarious parasitoids need to make the simultaneous decision not only of what size of clutch to lay but also of what sex ratio of progeny to produce. The progeny and sex allocation of such parasitoids may be easier to model if the two components are assessed independently; i.e. it is assumed that the female need make only one decision. In a formal model, the decision studied must be expressed as one or more algebraic **decision variables**. In some models of progeny (clutch size) allocation, the decision variable is the number of eggs laid per host, while in most models of patch exploitation the decision variable is patch residence time.
2. **What currency assumptions or optimality criteria apply**, i.e. how the various choices are to be evaluated. A model's currency is the criterion used to compare alternative values of the decision variable (in other words, it is what is taken to be maximised by the animal in the short-term for long-term fitness gain). For example, some foraging models maximise the net rate of energy gain

while foraging, whereas others maximise the fitness of offspring per host attacked.

3. **What constraint assumptions apply**, i.e. what factors limit the animal's choices, and what limits the 'pay-off' that may be obtained. There may be various types of constraint upon foragers; these range from the phylogenetic, through the developmental, physiological and behavioural, to the animal's time-budget. Taking as an example clutch size in parasitoids, and the constraints there may be on a female's behavioural options, an obvious constraint is the female's lifetime pattern of egg production. In a species that develops eggs continuously throughout its life, the optimal clutch size may be larger than the number of eggs a female can possibly produce at any one time. An example of both a behavioural and a time-budget constraint upon the behavioural options of both parasitoids and predators is the inability of the forager to handle and search for prey simultaneously. Here, time spent handling the prey is at the cost of searching for further prey. For a detailed discussion of the elements of foraging models, see Stephens and Krebs (1986) or Cezilly and Benhamou (1996).

Sometimes the investigator knows, either from the existing literature or from personal experience, the best choices of decision assumption, currency assumption or constraint assumption. If it is impossible to decide on these based on existing knowledge, one can build models for each alternative and compare the predictions of each model with the observed behaviour of the parasitoid or predator. In this way, it is possible to gain insight into the nature of the selective forces working on the insect under study (Waage and Godfray, 1985; Mangel, 1989a; Cezilly and Benhamou, 1996).

Early optimality models assumed a static world in which individual parasitoids search for hosts. While these models are useful research tools, they ignored the possibility that for a forager, today's decision may affect tomorrow's internal state which may in turn affect tomorrow's decision, and so on. The internal state of a searching parasitoid changes during adult life: its egg load (the number of mature eggs in the ovaries) and its energy reserves may decrease, and the probability that it will survive to another day decreases. The optimal behavioural strategy will depend on these changes. Likewise, the environment is not static. Bad weather or the start of an unfavourable season can also influence the optimal strategy. Dynamic foraging models have been designed to take into account internal physiological changes and changes in the environment (Chan and Godfray, 1993; Weisser and Houston, 1993, Tenhumberg *et al.*, 2001b).

Implicit in some optimality models is the assumption that the forager is omniscient or capable of calculation, e.g. that a parasitoid wasp has some knowledge of the relative profitability of different patches without actually visiting them (Cook and Hubbard, 1977). Behavioural studies on parasitoids have shown, however, that insects can behave optimally by employing very simple quick 'rule' mechanisms such as the mechanism determining patch time allocation in *Venturia canescens* described in section 1.12 and the males-first mechanism, used by some species in progeny sex allocation (subsection 1.9.5 and Figure 1.15). These mechanisms approximate well the optimal solution in each case.

Evolutionarily Stable Strategies

Almost all parasitoids leave the host *in situ*. Thus, there is always the possibility that other parasitoids may find the same host and also oviposit in it. The optimal behaviour of the first female thus depends on what other parasitoids may do. Likewise, the best time allocation strategy for a parasitoid leaving a patch in which it has parasitised a number of hosts depends both on the probability that other wasps will visit that patch and on the probability that other parasitoids may have already exploited the patches it visits next. For this reason, problems concerning the allocation of patch time, progeny and sex require models in which the evolutionarily stable strategy (ESS; Maynard Smith, 1974) is

calculated. The ESS approach, which is based on game theory, asks what will happen in a population of individuals that play all possible alternative strategies. A strategy is an ESS if, when adopted by most members of a population, it cannot be invaded by the spread of any rare alternative strategy (Maynard Smith, 1972). In seeking an ESS, theoreticians are looking for a strategy that is robust against mutants playing alternative strategies. The ESS, like the optimum in models for single individuals, is calculated using a cost-benefit analysis. We refer the reader to Maynard Smith (1982) and Parker (1984) for descriptions of the classical ESS models and for details of how to calculate the ESS in static models.

Why use Optimality and ESS Models?

Sometimes, experimental tests of optimality and ESS models will produce results not predicted by the models. At other times, only some of the predictions of the theoretical model are confirmed by empirical tests. Rarely is a perfect quantitative fit between model predictions and empirical test results obtained. Irrespective of whether a good fit is obtained, valuable insights are likely to be gained into the behaviour of the insect. Construction of models helps in the precise formulation of hypotheses and quantitative predictions, and allows us to formulate new hypotheses when the predictions of our model are not met. Thus, optimality and ESS models are nothing more or less than research tools.

Ideally, both causal and functional questions should be asked when studying the foraging behaviour of insect parasitoids and predators. In the subsections on superparasitism (1.8) and patch time allocation (1.12), we will show how, by ignoring functional questions, one may hamper the interpretation of data gathered to establish that a certain mechanism is responsible for some type of behaviour. Ignoring causal questions can likewise hamper research aimed at elucidating the function of a behaviour pattern; e.g. research into causal factors can demonstrate the existence of a constraint, not accounted for in a functional model, upon the behaviour of the parasitoid. Both causal and functional approaches are required for a thorough understanding of parasitoid behaviour.

1.2.3 THE COMPARATIVE METHOD

Introduction

Perhaps the approach with the longest pedigree in studying animal behaviour is the **Comparative Method**. With this method, data are collated across species, and a search is made for statistical patterns (Harvey and Pagel, 1991). One advantage of this method is that data are often already available (although often widely scattered) in the literature (e.g. see analyses by Blackburn, 1991a,b; Mayhew and Blackburn, 1999; Jervis *et al.*, 2001, 2003). Until recently, sets of species-average data were usually analysed in much the same way as within-species data. However, the fundamental assumption of most early statistical analyses – that species-comparative data are independent observations (i.e. are independent of each other) – may not hold. Cross-species data may actually be non-independent, because the species are related through phylogeny (i.e. they share an evolutionary history). If this is the case, then sample sizes will be artificially inflated (pseudoreplication), the number of degrees of freedom will be overestimated, and thus spuriously significant results may be obtained (Type I errors). Comparative biologists have become increasingly aware of this pitfall and methods have been developed which use phylogenetic information in conjunction with species-data sets to generate independent values for statistical analysis (Felsenstein, 1985; Harvey and Pagel, 1991; Harvey and Nee, 1997).

The Method of Independent Contrasts

Probably the most commonly employed method involves **'independent comparisons'**, also known as **'independent contrasts'** (originally developed by Felsenstein, 1985: simple examples are given in Harvey and Pagel, 1991; Purvis and Rambaut, 1995; Harvey, 1996). Independent contrasts methods are designed to detect **'correlated evolution'** (the extent to which change in one trait is associated with change in another). They assume that the branches of a phylogeny can be modelled by a Brownian motion process, such that successive changes are independent of one another and that the expected total change summed over many independent changes is zero. The original method for contrast analysis assumes that the lengths of branches in the phylogeny are known, but often they are not, and then have to be assumed to be equal (e.g. see Jervis *et al.*, 2001, 2003). [Branch lengths are estimated by **genetic distances** (divergence times, in relation to the present, estimated from the fossil record or from molecular clocks) or by the number of character changes, determined from a cladistic analysis (see Harvey and Pagel, 1991).]

An independent contrast is obtained from each node in the phylogeny for each measured variable. Imagine that you are studying five species (Figure 1.2a): It is clear that A and B, and X and Y are more closely related to each other than to members of the other clade, and by comparing their values we then only include independent evolutionary trajectories. By calculating an ancestral trait value at node 2 (the mean of X and Y) with the value for species Z, we gain another contrast. Finally, by comparing the mean of node 3 (the mean of the values for node 2 and species Z) with node 1, we gain our final independent contrast. Therefore we gain *four* contrasts from *five* species. If both traits of interest in the comparative analysis are continuous variables, then typically the data are analysed using a linear regression, constrained to pass through the origin (i.e. there should be no intercept in the regression model, Garland *et al.*, 1992).

As we have seen, with a perfectly resolved phylogeny of n species, there are $n-1$ possible contrasts available. This may result in statistical difficulties when data sets are small. While this problem may alleviated by the addition of an extra species to the analysis, a more invidious difficulty is introduced when a phylogeny is poorly resolved. While phylogenies usually consist of bifurcating lineages, for some taxa the phylogeny may not be well resolved, and so it will contain polytomies (trifurcations, etc.) and hence fewer nodes for a given number of species. Thus, the number of contrasts obtained is less than with a fully resolved (bifurcating) tree, reducing the size of the data set and also the statistical power of subsequent analyses. Reconsider the figure with species A, B, X, Y and Z. If we do not know the evolutionary relationships among X, Y and Z, we must assume that they all originated from the same common ancestor (a polytomy; Fig 1.2b). We now only have two contrasts (species A vs B and node 1 vs node 2) and hence no statistical power in the analysis. Such problems can only be overcome by obtaining a well-resolved phylogeny. Unfortunately, such phylogenies are not always available.

Although a well-resolved, published taxonomy (more often than not, based exclusively on morphology) can be used to approximate the true phylogenetic tree, it is important to be aware that some currently accepted taxonomic groupings may not be monophyletic, i.e. they may not contain all the descendants of a common ancestor. Phylogeny-based comparative methods assume groupings to be monophyletic, so using an incorrect phylogeny will seriously undermine the value of the analyses undertaken. In the absence of molecular-based (i.e. DNA) phylogenies, cladistically-based taxonomies are the most suitable taxonomies for comparative studies as they are intended to closely reflect phylogeny.

There are now a number of software packages available that allow users to perform rigorous comparative analyses, many of which are freely available over the internet (http://evolution.genetics.washington.edu/phylip/software.html provides access to many of these packages, and more). The volume by Harvey and Pagel (1991) remains the best introduction to modern comparative approaches.

The Comparative Method can also be used to make predictions concerning the ecology of a species. Hardy *et al.* (1992a), used a phylogenetic tree, based solely on morphological characters (now published incorporating molecular data; Schilthuizen *et al.*, 1998), of the six species of

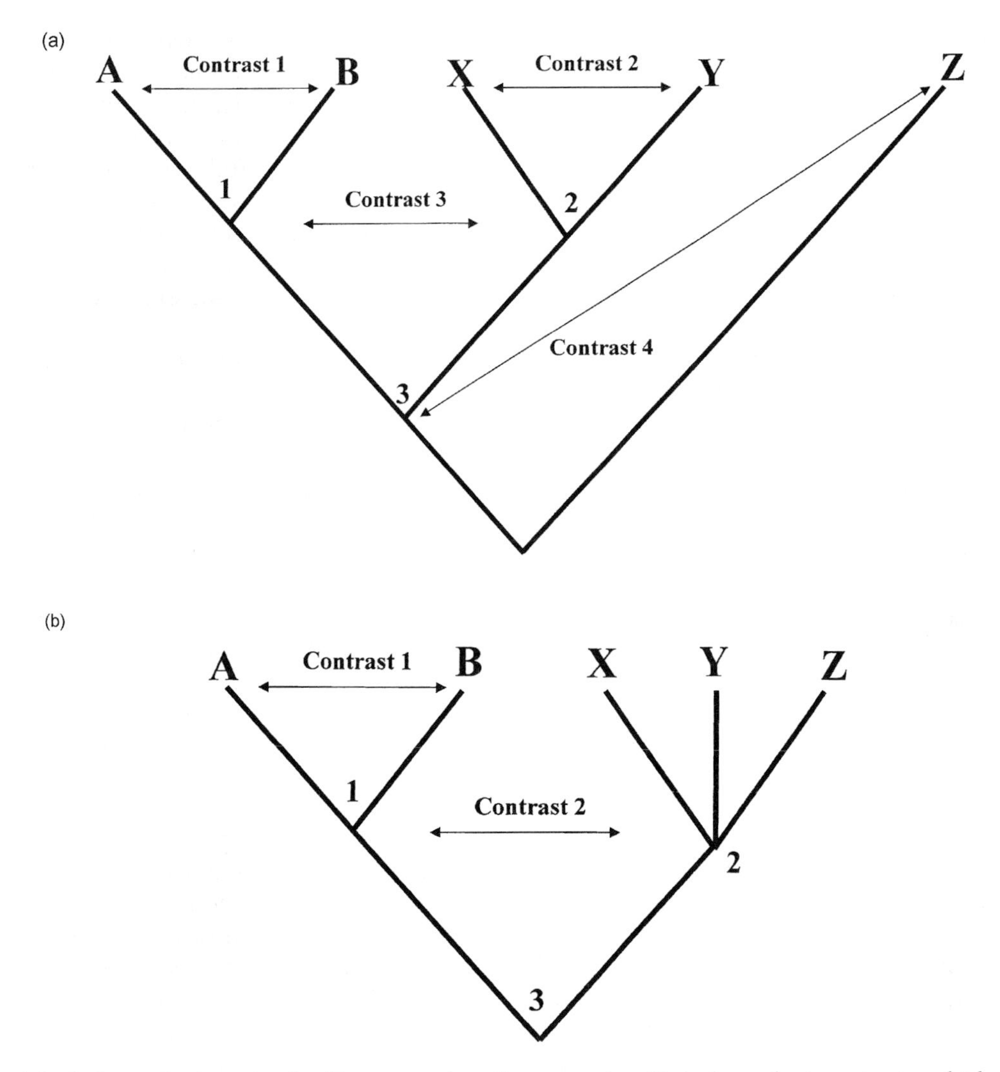

Figure 1.2 Independent contrasts: For n species, there are $(n-1)$ independent contrasts which can be calculated for a fully resolved phylogeny (a). Where the phylogeny is less well resolved, resulting in polytomies, the number of independent contrasts possible is severely diminished (b), reducing the potential power of the analysis. Numerals refer to nodes in the phylogeny. See text for explanation.

Leptopilina occurring in Europe, to predict where in the environment *L. longipes*, a species whose hosts and host habitat were unknown, would be found (Figure 1.3). The five other species are all parasitoids of *Drosophila*. The tree divides initially into two branches. When we examine how the character host habitat choice is distributed over the tree (i.e. the character is 'mapped' onto the tree), it appears that the upper branch of the tree contains the species finding its hosts in fermenting fruits (*L. heterotoma*), while the other branch contains species finding their hosts in fungi and/or decaying plant matter (*L. clavipes*, *L. australis* and *L. fimbriata*). Because *L. longipes* is most closely related to *L. fimbriata*, it was predicted that it is attracted, like its close relative, to decaying plant material. Subsequently, *L. longipes* was trapped with baits

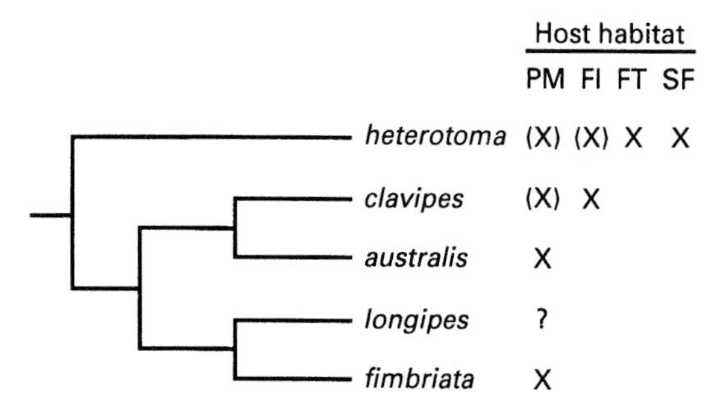

Figure 1.3 Cladogram, based on adult morphology, of the *Leptopilina* species (Hymenoptera: Eucoilidae, parasitoids of *Drosophila*) occurring in northwestern Europe. Microhabitat use is 'mapped' onto the ends of the tree branches. X = principal microhabitat, (X) = microhabitat from which a species has occasionally been recovered. PM = decaying plant material; FI = fungi; FT = fermenting fruit; SF = sap fluxes. Microhabitat use by *L. longipes* was predicted from that species' position on the cladogram.

comprising rotting cucumber containing *Drosophila* larvae, and during fieldwork it was also found on decaying stalks of the umbellifer *Heracleum* and on fungi.

1.3 THE TREATMENT OF PARASITOIDS PRIOR TO THEIR USE IN EXPERIMENTS

1.3.1 REARING

The species and quality of the host a parasitoid is reared on can have a marked influence on their subsequent behaviour, for example through its effect on egg load and life expectancy (subsections 2.7.3. and 2.8.3). While the influence of the host-related phenotypic variation in natural enemy traits on the results of laboratory trials may be mimimised or even avoided by altering the rearing regime, more insidious problems arise when insect populations are reared in mass culture. The first problem is ubiquitous and unavoidable. Natural selection will operate in the controlled environment room as much as anywhere else, changing the genetic composition of the population and, as a result, potentially influencing the behaviour of the population of interest (e.g. Matos *et al.* 2000). There are two approaches to avoiding the complications of such adaptation. The first is to simply measure the traits of interest before significant selection occurs, so ideally few generations will have passed between capture and experiment. Second, if one is interested in using selection experiments to probe the nature of the trait, then we recommended that outbred populations of the species of interest be maintained for at least 10 generations in the laboratory. This allows adaptation to laboratory conditions to occur, and should avoid complications from any inadvertent selection pressures during the experiment.

A more serious problem results from small effective population sizes, leading to genetic drift and inbreeding depression. Testing the variation in an inbred population at best results in an underestimate of the variation in natural populations, and at worst provides a skewed view of the true variation present.

1.3.2 EXPERIENCE

It has been shown for several parasitoid species that an individual's previous experience can modify its behaviour (section 1.5.2). This phenomenon has been observed in all phases of the foraging process and often involves responses to chemical stimuli (Vet and Dicke, 1992). Previous ovipositions in hosts of a certain species can also influence host species selection in choice experiments (van Alphen and Vet, 1986), while the decision to oviposit into an already parasitised host (i.e. superparasitise) also depends on previous experience with unparasitised or parasitised hosts (Visser *et al.*, 1992b; Hubbard *et al.*, 1999).

Thus, when designing experiments, one should always be aware that the previous history of an individual may influence its behaviour. It can affect the results of experiments on patch time allocation, superparasitism and also the results of experiments in which interactions between adult parasitoids are studied. Even storing parasitoids in the absence of either hosts or host-related cues can have an effect. For example, among wasps so treated, older individuals may superparasitise more frequently than younger ones. Visser *et al.* (1990) showed that it even matters whether wasps are stored in a vial singly or with other females prior to conducting an experiment. Such effects have also caused problems when parasitoids and hosts are mass-reared for biological control purposes. Rearing the apple pest *Cydia pomonella* on an artificial diet reduced the ability of the parasitoid *Hyssopus pallidus* to respond to host location cues, as it changed the composition of the kairomones normally found in the host's frass (Gandolfi *et al.*, 2003).

Having said that, we should point out that conditioning parasitoids, by allowing them to search and oviposit for some time before an experiment, can be a sensible practice. Inexperienced parasitoids often show lower encounter rates and are less successful in handling their hosts (Samson-Boshuizen *et al.*, 1974). By allowing parasitoids access to hosts before they are actually used in an experiment, one can often save many hours that would otherwise be wasted in observing parasitoids that are 'unwilling' to search. Often, however, it is advisable to use freshly emerged, inexperienced females, for example in choice experiments, either where different host plants, host instars or host species are offered or where the olfactory responses of parasitoids to different chemicals are studied.

Often, one is interested in the performance of natural populations. These comprise individuals with different experiences and different degrees of experience, so using only inexperienced females in the laboratory gives a distorted view of what happens in nature. One approach is to collect adults from the field for study in the laboratory. A large enough sample should give a reasonable idea of how individuals in the population behave on average. However, one should be aware of the problems of **genotype-by-environment interactions**, where not all genotypes respond to changes in environment in the same way. Ideally, the laboratory conditions will reflect what is likely to be encountered in the field, especially in terms of temperature.

Because experience can influence subsequent behaviour, the results of experiments in which an insect encounters two situations in succession can depend on which situation is encountered first. In such cases, one should take care that in half of the replicates one situation is encountered first, while in the other half the sequence is reversed.

1.3.3 SEX RATIOS

While such effects of experience can have a great influence on the behaviour of insects, more subtle problems ought to be borne in mind. An often overlooked problem in sex ratio studies is the possible presence of *Wolbachia* and other male-killing bacteria in the study organism (see Chapters 3, 5 and 6). For example, Majerus *et al.* (1998) found that almost 50% of females from a Japanese population of the coccinellid *Harmonia axyridis* were attacked by a male-killing bacterium, resulting in a heavily female-biased sex ratio. Those ladybirds from a Mongolian and a Russian population had low ($<2\%$) or no infection. To confirm the presence of bacteria, one can simply 'cure' the experimental individuals by treating them with antibiotics (subsections 3.4.2 and 6.5.2).

Selfish genetic elements (regions of the chromosome that are inherited in a non-Mendelian manner during segregation, resulting in their becoming over-represented in gametes) provide another means of sex ratio distortion. *Nasonia vitripennis* has been found to commonly carry *psr* (parental sex ratio), a selfish genetic element that results in the production of male-only broods by causing fertilised eggs (normally female) to become male. Such distortion of the sex ratio will have a considerable influence on the population ecology of *N. vitripennis* (reviewed in Godfray, 1994) and could potentially influence the outcome of sex ratio studies if present in a laboratory culture.

More often, changes in sex ratios will result from conditional sex allocation (section 1.9.3) or local mate competition (section 1.9.2). Bernal *et al.* (1999) found that two species of *Metaphycus*, parasitoids of scale insects, showed much more female-biased sex ratios if provided with larger hosts. This is likely to result from conditional sex allocation (where female parasitoids preferentially place female offspring in larger hosts). Since these parasitoids may be used as biocontrol agents attacking scale insect pests of citrus trees, Bernal *et al.* suggest that using rearing protocols that maximise the proportion of females would be economically sensible.

1.4 HANDLING BEHAVIOURAL DATA

1.4.1 RECORDING BEHAVIOUR

The equipment used to record insect behaviour has developed rapidly over the past 10 years, driven primarily by advances in computing power. Nevertheless, many (if not most) studies of insect foraging behaviour rely upon direct observation and note-taking. This approach is not without drawbacks, in that it is difficult to avoid bias in recording. The simplest way around this is to use video-recording equipment, so that two independent observers can time and assess the behaviours of interest. A development of such techniques involves "intelligent" video systems, which have a number of advantages. Details of these are given in Chapter 4 (subsection 4.2).

1.4.2 ANALYSING BEHAVIOURAL DATA

Because insects may change their behaviour in response to experiences gained while foraging, and because their internal state (e.g. egg load) changes during the foraging process, one cannot simply add all the events of a certain category occurring during an observation period. In a time-series of events, the different behavioural events are not independent.

The standard statistical methods described in numerous textbooks are in general inappropriate for the analysis of some behavioural data, because the connection between the succession as well as the duration of acts cannot adequately be taken into account. Haccou and Meelis' (1992) book on the statistical analysis of time-series of behavioural events is recommended as a useful introduction to the most appropriate approach.

1.4.3 BEHAVIOURAL RESEARCH IN THE FIELD

Whether behavioural research is aimed at answering fundamental questions or deals with the use of parasitoids and predators in biological control, the ultimate goal of interest is the performance of the insects in the field. The small size of many parasitoids makes observation of their behaviour in the field often difficult or impossible. This applies particularly to the monitoring of the movements of individuals, for example between **patches** (patches can be defined either as units of host/prey spatial distribution or as limited areas in which natural enemies search for hosts/prey; often there is a hierarchy of patches, e.g. tree, branch, leaf, leaf-mine). The movements of larger insects such as ichneumonids and sphecids can be more easily observed. Dispersal of small parasitoids in the field can be studied by placing patches with hosts (e.g. potted, host-bearing plants) and releasing marked adults. By checking the host plants at regular intervals for the presence of marked individuals, it is possible to obtain information on the speed at which the insects move between host plants, on the time they spend searching each patch and on the spatial distribution of parasitoids over the available patches. When hosts are later examined, the aforementioned data can be related to the amount of parasitism in each patch. By using marked parasitoid individuals, one can distinguish between insects released for the experiment and those occurring naturally. Large wasps can be marked with paint on the thorax, using a fine paintbrush. By using different colours or colour combinations one can distinguish between different individuals, or groups. Small wasps can be marked with fluorescent dusts, but this has the disadvantage that one must remove wasps from the experimental plot to detect the dust mark under ultraviolet light. Genetic markers have also been used to monitor parasitoids in the field (Kazmer and Luck, 1995). Other workers have suggested that phenotypically distinguishable mutants may prove useful in studying population dynamics, although there are obvious drawbacks with this approach (Snodgrass, 2002).

Many species, when observed in the field, continue foraging normally. Janssen (1989) used a stereomicroscope mounted on a tripod in the field, and used it to observe the foraging behaviour of parasitoids on patches (sap streams and fermenting fruits) containing *Drosophila* larvae. Casas (1990) also recorded the behaviour of *Sympiesis sericeicornis* while the parasitoid searched for its leaf-miner host on potted apple trees in the field.

Other natural enemy species are easily disturbed when approached, and disturbance can be avoided in some cases by using binoculars (Waage, 1983).

1.5 HOST/PREY LOCATION BEHAVIOUR

1.5.1 INTRODUCTION

With the exception of ambush predators, insect predators and parasitoids employ a heirarchy of behaviours that enable them to locate and choose their prey. These behaviours are generally associated with either:

1. Host/prey habitat location.
2. Host/prey location.

Within each of these levels (which, of course, are really part of a continuum and are only delineated for our convenience), individual parasitoids and predators will generally follow a behaviour pattern that will change in response to cues. However, while such a scheme may allow us to visualise the foraging process, it must be remembered that these behaviours will be influenced by learning (a plastic response to experience) and genetic variation (both within- and between-population variation in responses to cues).

During searching, two important types of cue will influence insect natural enemy behaviour. **Attractant stimuli** induce a change in forager behaviour that results in orientation to areas that either contain hosts or are likely to contain hosts. **Arrestant stimuli** act by eliciting a reduction in the distance or area covered per unit time by the forager within such areas. These stimuli can act at a number of scales, with distinct cues influencing the behaviour of the forager over differing distances.

1.5.2 HOST/PREY HABITAT LOCATION BY PARASITOIDS AND PREDATORS

The literature concerning host habitat location consists largely of papers showing which stimuli (cues) attract parasitoids and predators to the host's habitat (see Vinson, 1985, for a review). Few papers deal with functional aspects of this step in the foraging sequence (but see, e.g. Le Ru and Makosso, 2001; Gohole *et al.*, 2003). The emphasis on causal aspects of host habitat finding reflects the fact that it is much easier to answer qualitative questions, such as which odour acts as an attractant, than it is to answer the question of why one odour should be attractive and another not in terms of the contribution to fitness of the insect natural enemy.

Parasitoids spend a significant proportion of their adult lives searching for places where hosts can potentially be found. They may use visual, acoustic or olfactory cues to locate potential host patches. Certainly, for parasitoids olfactory cues are more important. Often, visual and acoustic cues can guide a parasitoid to its host over a short distance only, in contrast to olfactory cues that can act over much longer distances.

It is difficult to demonstrate the use of visual cues in host habitat location by insect predators and parasitoids, because the use of other, olfactory and acoustic, cues has to be excluded. Van Alphen and Vet (1986) investigated the searching behaviour of *Diaparsis truncatus*, an ichneumonid parasitoid of larvae of the twelve-spotted asparagus beetle, *Crioceris asparagi*. Larvae of the beetle feed inside the green berries of the *Asparagus* plant. It was shown, by placing green-painted wooden beads on *Asparagus* plants, that *D. truncatus* females respond from a distance to the berries of *Asparagus*. The parasitoids landed more often on the slightly larger wooden beads than on the green *Asparagus* berries, which is consistent with the hypothesis that the parasitoids respond to visual cues. Such an approach may be adopted for parasitoids of other insects living in fruits. Colour alone is enough to induce oviposition attempts by the parasitoid *Aphidius ervi* (Battaglia *et al.* 2000). Female *A. ervi* were found to show a strong response to wavelengths similar to those reflected by its primary host, the aphid *Acyrthosiphon pisum*.

Coccinellids are perhaps the best-studied insect predators (Dixon, 2000), and it appears that visual cues can play a large part in longer-distance prey location. Hattingh and Samways (1995) found that the ladybird *Chilocorus nigritus* initially orientated towards a simulated tree line, and then showed a preference for simple ovate leaves over more complex leaf shapes.

In a detailed study, Harmon *et al.* (1998) investigated the role of colour and vision in the close-proximity foraging of the adults of four coccinellid species (*Coccinella septempunctata, Hippodamia convergens, Harmonia axyridis* and *Coleomegilla maculata*). The first three species were more efficient foragers in the light, supporting the conclusion that vision plays an important role in prey capture. When the ladybirds were offered red and green colour morphs of the pea aphid, *Acyrthosiphon pisum, C. septempunctata* preferred the clone whose colour contrasted with the test background. In contrast, *H. axyridis* always preferred red pea aphid clones (Harmon *et al.*, 1998). However, it does appear that coccinellid adults and larvae use different cues in host location, with visual cues proving to be relatively unimportant for the larvae. For example, while both visual and odour cues are important in the host location by adult *Chilocorus nigritus*, the larvae fail to recognise these cues and instead rely upon physical contact (Hattingh and Samways, 1995). Such results illustrate the importance of considering all stages of the life-cycle: adults and larvae will often forage differently.

The importance of visual cues in host location has rarely been considered for parasitoid species. Searching females of the tachinid fly *Exorista japonica* will respond to the colour of a model host, responding to this before responding to tactile, size and odour cues (Tanaka *et al.*, 1999).

Some parasitoids respond to acoustic stimuli produced by the host, and so execute host habitat location and host location in one step, and this appears to be much more common among dipteran parasitoids (Feener and Brown, 1997). Cade (1975), whilst broadcasting the song of the male cricket *Gryllus integer* from a loudspeaker to study the mating behaviour of the crickets in the field, discovered that a tachinid parasitoid (*Euphasiopteryx ochracea*) of the cricket was attracted by the song. Burk (1982) similarly demonstrated this for the tachinid *Ormia lineifrons*. Soper *et al.* (1976), using tape recordings, showed that the sarcophagid parasitoid *Colcondamyia auditrix* finds male cicadas by this means (**phonotaxis**). Phonotaxis by the tachinids *Ormia depleta* and *O. ochracea* has been demonstrated using synthesised male calling songs (Fowler and Kochalka, 1985; Walker, 1993, Adamo *et al.*, 1995). Both Fowler (1987) and Walker (1993) carried out experiments in which the synthesised calls of a range of several host cricket species were simultaneously broadcast in the field. Allen (1998) found that the parasitoid *Homotrixa alleni*, an ormiine fly, locates the bushcricket *Sciarasaga quadrata* by orientating towards calling males. Gravid female flies were most likely to search when calling was maximal, and by using trapped male *S. quadrata*, it was shown that there was a positive correlation between call duration and the number of flies attracted to the bush crickets (Allen, 1998).

Chemical communication, both between insects and between plants and insects, plays a very important role in determining the behaviour of parasitoids and predators. Any chemical conveying information in an interaction between two individuals is termed an **infochemical** (Dicke and Sabelis, 1988). Infochemicals are divided into **pheromones**, which act intraspecifically, and **allelochemicals**, which act interspecifically. Allelochemicals are themselves subdivided into **synomones, kairomones** and **allomones**. A synomone is an allelochemical that evokes in the receiver a response that is adaptively favourable to both the receiver and the emitter; a kairomone is an allelochemical that evokes in the receiver a response that is adaptively favourable only to the receiver, not the emitter; an allomone is a allelochemical that evokes in the receiver a response that is adaptively favourable only to the emitter (Dicke and Sabelis, 1988). The majority of parasitoids and many insect predators respond to volatile kairomones or synomones in the long-distance location of their hosts. These chemicals may originate from:

1. The host itself, e.g. from frass, during moulting, during feeding, sex pheromones and aggregation pheromones, i.e. the chemicals involved are kairomones for the parasitoids;
2. From the host's food plant, i.e. the chemicals involved are synomones for the parasitoids; or
3. From some interaction between host and food plant, e.g. feeding damage, i.e. the chemicals involved are synomones for the parasitoids.

The attraction responses by parasitoids to odours from any source can be studied using various olfactometers and wind tunnels, or by observing the responses of parasitoids to odour sources following release of the insects in the field.

Two types of **airflow olfactometer** are commonly used to study responses to olfactory cues. One is the glass or clear Perspex **Y-tube olfactometer** (Figure 1.4). The insect can be given a choice either between odour-laden air (test) and odour-free but equally moist air (control) or between air laden with one odour and air laden with another odour. Although Y-tube olfactometers have been criticised because odour plumes may mix where the two arms of the olfactometer meet due to turbulence, and that choice is no longer possible once the insect has passed the junction of the tube, impressive results have been obtained. Smoke can be passed through the apparatus to test for unwanted turbulence, but tobacco smoke must be avoided as it is absorbed by the tubing and it can affect the outcome of future experiments. By passing NH_4OH vapour over HCL, a fine smoke of NH_4Cl crystals can be created and the vapour channelled through the Y-tube. After

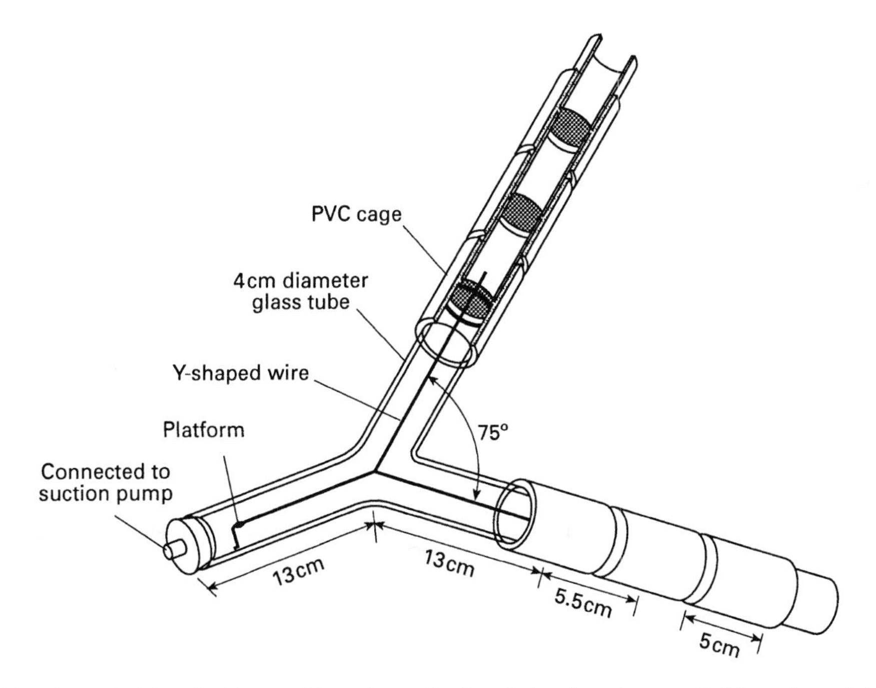

Figure 1.4 Host/prey habitat finding and host/prey finding behaviour in parasitoids and predators: design of the Y-tube air-flow olfactometer used by Sabelis and van der Baan (1983). The Y-shaped wire within the tube cavity provides a walking surface for small predators and parasitoids. For details of operation, see text.

testing, the crystals can easily be washed from the tubing. Turbulence, if detected, can often be reduced by adjusting the flow speed of the air.

With diurnally active insects, a diffuse light source is often required to illuminate the apparatus to encourage the insects to move towards the fork of the tube. This light should not cause the olfactometer to overheat, and so to avoid this a cold-light source (e.g. fibre optics) ought to be used.

To eliminate the effects of any asymmetry in the apparatus, the chambers need to be alternated for each 'run'. It is recommended that parasitoids be tested individually, rather than in batches, because either interference or facilitation may occur between insects and so bias the results. The apparatus should be washed, first with alcohol and then with distilled water, between runs to prevent any response of parasitoids to any trail left by previous individuals. Finally, consideration needs to be given to the possibility of left- and right-handedness in the insects. By analysing the number of left and right turns in the apparatus, it is possible to test, statistically, whether wasps tend to move more to the right or more to the left. The null hypothesis will be that the distribution of turns by parasitoids should be equal in both arms irrespective of the position of the chambers. An additional test of turn preference is to perform several runs when both chambers are empty, although insects may be unwilling to move through the apparatus in the absence of any odour. Some parasitoid species show 'handedness', i.e. a tendency to turn more in one direction than another (J. Pritchard, unpublished).

Even when great care is taken in the design of olfactometer experiments and the analysis of data, the results of olfactometry may be difficult to interpret (Kennedy, 1978). This applies especially to Y-tube olfactometry. The Y-tube, when employing a light source, simultaneously presents test insects with two types of stimulus, light and air current, to which the insect might respond by phototaxis and anemotaxis, but presents the two odours (or odour and non-odour) separately at only one point in the apparatus – the fork, which represents the 'decision point'. Responding by phototaxis and anemotaxis to the common air current, insects might be entrained past the decision point and become behaviourally trapped in the wrong arm (Vet *et al.*, 1983). Another type of airflow olfactometer, designed by Pettersson (1970) to study the responses of aphids, lacks this and other disadvantages. A modification (Figure 1.5) of the **Pettersson olfactometer** by Vet *et al.* (1983) and further developed by others (e.g. Sengonca and Kranz, 2001) has proved quite popular among students of parasitoid behaviour. It is constructed mainly of transparent Perspex, and has a central arena with four arms. Air is drawn out of the arena via a hole in the centre of the bottom plate. Air flows into the arena via four arms. Insects may therefore be exposed to as many as four different odours. Air speed in each arm can be controlled with a valve and an anemometer. Care is taken that air speed is equal in all arms. Before an experiment is performed, an NH_4OH smoke test (see above) can be carried out to test for unwanted turbulence and to show that a clear, straight boundary exists between odour fields. Diffuse light of equal intensity on all four sides of the arena prevents asymmetric attraction of insects to light. Insects are introduced through a hole in the bottom plate by temporarily disconnecting the tube from the air pump. Observations are best made using a video camera placed directly overhead, because a human observer may disturb the insects by his or her movements. Such systems can be further automated by integrating commercially available components, such as CCD (charge-coupled device) cameras, with a computer program that incorporates a positioning and tracking algorithm (Vigneault *et al.*, 1998; see Section 4.2).

The Pettersson olfactometer thus allows an insect to choose between four different odour fields, and repeated choices by the insect are also made possible.

In non-automated versions of the Pettersson olfactometer, the final choice by an insect is usually considered to be made when it enters the narrow tube through which air laden with odour enters the arena. Both because airflow in

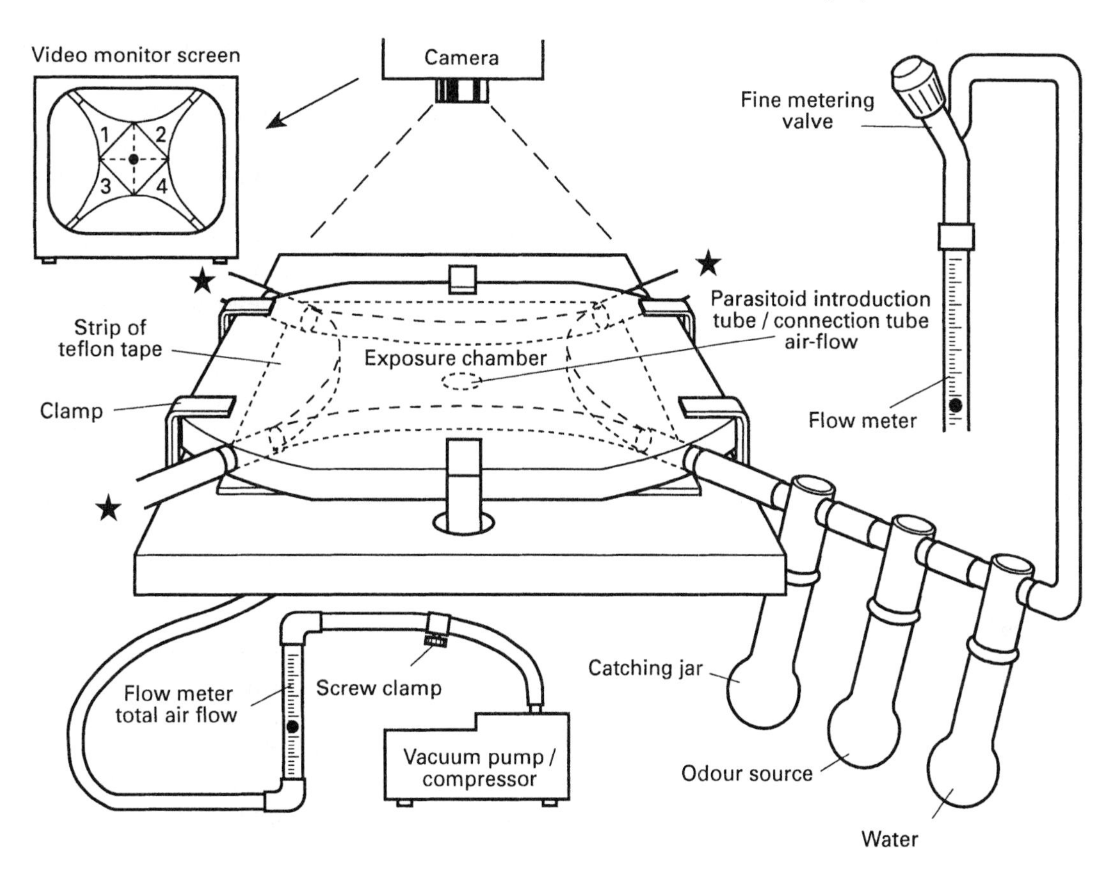

Figure 1.5 Host/prey habitat finding and host/prey finding behaviour in parasitoids and predators: Design of the Pettersson olfactometer, as used by Vet *et al.* (1983). The catching jar is used to collect any insects that move into an outflow tube. For details of operation, see text. Reproduced by permission of Blackwell Publishing.

this narrow region is strong and because many parasitoids have an aversion to entering narrow crevices, some insect species avoid this area and turn without entering.

Other parasitoids react to the odour stimulus by flying vertically upwards. Because flight is impossible in the narrow space between the base and the olfactometer cover, the insects will hit the top plate, and after a number of these aborted flight attempts become so disturbed that they cannot be expected to choose odour fields.

Often, one or more of the odours offered in an olfactometer comprises a mixture of many unidentifiable volatile substances, the concentrations of which in the odour fields are unknown. This does not pose a problem if the responses of an insect to a mixed odour source and a clean air control are compared, because only the test odour is the potential attractant. However, when testing for attraction to two odour sources (e.g. the odours of two different food plants of the host), there may be problems of interpretation. One of the odour sources may be more attractive than the other because the insect responds to one or more substances in that odour source that are lacking in the other. Alternatively, both odour sources may be qualitatively similar but the insect may be differentially attracted because of differences in the concentration of an attractant component of an odour. It also needs to be borne in mind that a combination of a qualitative difference and a quantitative difference may be responsible for differential attraction.

The ultimate solution to the above problem would be to isolate the attractants and test whether differential attraction to odour sources is due either to differences in chemical composition between the sources or differences in concentration of their chemical components.

With any airflow olfactometer it is important to ensure, before carrying out any experiments, that air flows through the apparatus at a constant rate (usually the rate is low). With the Y-tube and Pettersson olfactometers, both of which are **hypobaric systems** (i.e. air is sucked out), a good quality vacuum pump should be used. Flow meters of the correct sensitivity i.e. neither over- nor under-sensitive, should also be employed.

Static-air olfactometers can also be used with predators and parasitoids to measure chemotactic responses to odour gradients. One such olfactometer, used successfully by Vet (1983), is shown in Figure 1.6. The device consists of three chambers. The parasitoid or predator is released into the middle chamber and its subsequent choice of outer chamber containing a test odour recorded. Vet (1983) also recorded the time

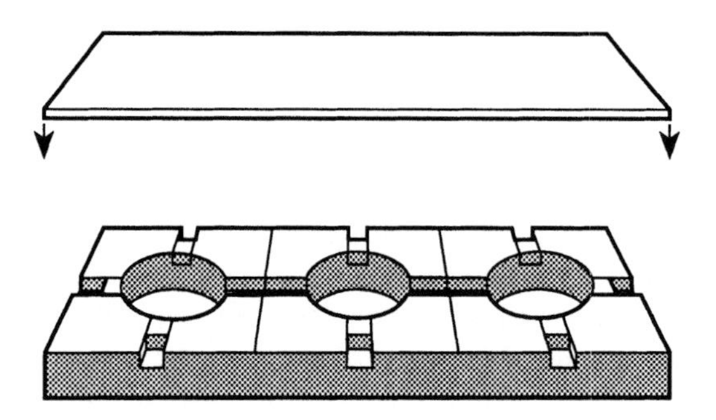

Figure 1.6 Host/prey habitat finding and host/prey finding behaviour in parasitoids and predators: Static-air olfactometer of Vet (1983). The apparatus comprises three perspex blocks glued together and covered by a single glass lid (a single perspex block would also suffice, although it may be more difficult to escavate). The excavations (chambers) in the blocks are connected by corridors. The chambers, measured internally, are 50 mm wide and 16 mm deep; the corridors are 10 mm wide and 5 mm deep. For details of operation, see text. Reproduced by permission of E.J. Brill (Publishers) Ltd.

taken for females to reach an odour source chamber.

As noted above, not all parasitoids can be successfully tested in olfactometers, because they are prevented from flying. Flying parasitoids can be tested in wind tunnels (Figure 1.7), but it is difficult to keep track of the smaller species.

When even a wind tunnel cannot be used, one could try the following:

1. Placing potentially attractive odour sources in an array, either in the field, in a large field cage, or in a large controlled environment chamber;
2. Releasing a large number of adult females and examining the odour sources frequently;
3. Removing each insect that lands on the odour sources.

If more individuals than expected, based on a random distribution, land on a particular odour source, this can be taken as evidence that the odour source is attractive. If different odour sources are offered, it may be possible to rank them in terms of their attractiveness. The problem with this type of experiment is that the number of parasitoids or predators trapped on a particular source is a function both of the number of insects landing on the source and of the time they spend there. Ideally, the test individuals should be caught immediately after arrival on the source, but this is not always possible. Another problem is that the experimental design does not exclude the effect of interactions between individuals, e.g. some parasitoids repel conspecifics or chase them away.

At the University of Leiden, the field release method (involving counting of the numbers of females attracted to uninfested and infested cassava plants) has been used successfully in field experiments with *Apoanagyrus* (= *Epidinocarsis*) *lopezi*. This approach has also been used to compare the attractiveness of different microhabitats containing *Drosophila* larvae to several species of *Leptopilina*. The method also allows a functional analysis of habitat choice. If one knows: (a) the encounter rates with hosts in the different microhabitats, (b) the species

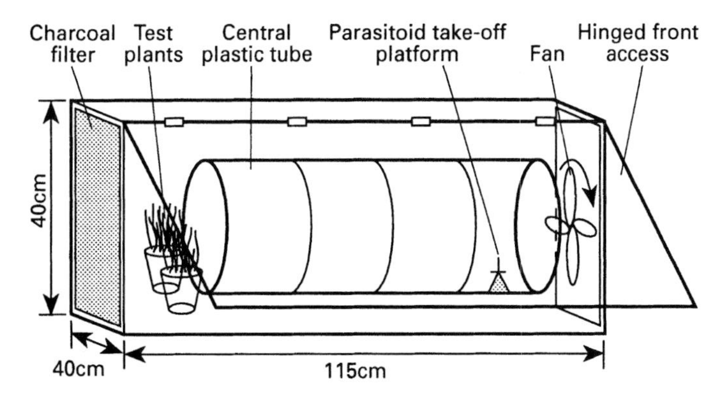

Figure 1.7 Host/prey habitat finding and host/prey finding behaviour in parasitoids and predators: Design of wind tunnel used by Grasswitz and Paine (1993) to study the behaviour of *Lysiphlebus testaceipes* (Hymenoptera: Braconidae), a parasitoid of aphids. The main (rectangular) chamber was constructed of Plexiglass, and the central (cylindrical) test section was constructed of Mylar.

composition of the host larvae in each microhabitat, and (c) the survival rates of parasitoid eggs deposited in each of the host species, one can calculate the relative profitability of each microhabitat for the parasitoid and then predict which ones the parasitoids should visit when given a choice. This approach was employed by Janssen *et al.* (1991). Another approach to studying functional aspects of host habitat location by parasitoids is to consider the reliability and detectability of a cue (Vet and Dicke, 1992). This approach contrasts cues having a high detectability but a low predictive value regarding the presence of hosts, with cues having a low detectability but a high predictive value. Cues with a high detectability are odours emitting from potential host plants. Cues with a high reliability are substances produced by the host plant in reaction to the presence of the host and substances emitted directly from the host. Vet and Dicke (1992) assume that high reliability cues are produced in smaller amounts than general host plant odours. It is unfortunately hard to see how reliability and delectability could be measured in a quantitative way, making the testing of the concept difficult. However, there are now a number of studies (discussed below) that have gone some way to overcoming these problems.

As with all behavioural tests that purport to investigate insect preferences (e.g. for an odour, a host plant, size of host etc.), statistical problems abound. Two will be considered here. First, at which point is the test insect provided with too many choices to make effective comparisons and how does this affect the sample size required? In part, the answer to this depends on whether the experimenter is interested in extremes (insect prefers an odour to the control) or in forming a rank order of preference among odours. Raffa *et al.* (2002) show that an increased number of replicates is required to show the latter, and provide an excellent guide as to how to maximise experimental power. Second, what is the best way to analyse the data? There are several established statistical approaches. Data from two-way trials can usually be analysed using a standard probit or logistic regression approach. However, standard probit models are not suitable for preference assays, and Sakuma (1998) provides an extension of the standard probit method to overcome these problems.

Many studies have shown that parasitoids and insect predators respond to odours produced by the host plants of their potential victims (see Vet, 1999 for a review). In some cases, parasitoids respond to the odour of host-free (undamaged) plants, but frequently herbivore damage is required before a response to plant cues is observed. Wind tunnel trials showed that *Aphidius ervi*, a common parasitoid of the pea aphid, *Acyrthosiphon pisum*, preferred to fly towards aphid-damaged bean plants. However, washing the plants to remove aphid cues did not reduce this preference, indicating that induced plant volatiles were being used as cues (Du *et al.*, 1996). When extracts of the plant volatiles were applied to filter paper and placed in a wind tunnel, a similar effect was seen (Powell *et al.*, 1998). If *A. ervi* females are provided with a choice between volatiles collected from pea aphid-damaged plants or from black bean aphid (*Aphis fabae*)-damaged plants, they are much more likely to fly towards the former (Powell *et al.*, 1998). This is evidence that there are host-specific cues in the plant volatiles (Du *et al.*, 1996; Powell *et al.*, 1998).

Predators also respond to similar cues. In a field experiment, Drukker *et al.* (1995) found that psyllid-infested pear trees attracted significantly more anthocorid predators than uninfested trees. Scutereanu *et al.* (1997) collected volatiles from the headspace (i.e. the air directly above the leaf) of attacked and unattacked trees, and using a Y-tube olfactometer found that the anthocorids preferentially chose the airstream containing volatiles from attacked trees. Using mass spectrometry, they found six volatiles that were significantly more common in the headspace of attacked trees (the monoterpene (E, E)-α-farnesene, the phenolic methyl salicylate, and four green leaf compounds). Only the monoterpene and the phenolic compounds elicited the preference in the bugs (Scutereanu *et al.*, 1997). Methyl salicylate has also been shown to influence the behaviour of other predatory arthropods, including phytoseiid mites (Dicke *et al.*, 1990), and anthocorids are attracted to methyl salicylate produced from mite-infested beans (Dwumfour, 1992). However, not all increases in volatile emission following herbivore attack lead to an increase in natural enemy recruitment. For example, induced secondary defences in cucumber plants correlate with an increase in volatiles, and this deters predatory mites (Agrawal *et al.*, 2002).

It has been suggested that plants might produce these synomones as an indirect defence. By actively recruiting natural enemies of their herbivores, the damage suffered by the plant will be reduced over time (see Vinson, 1999). This hypothesis is not without critics. Van der Meijden and Klinkhamer (2000) point out that plants may not benefit from the presence of koinobiont parasitoids (subsection 1.5.7), since they do not immediately kill the host and damage continues to occur after the host is attacked. More direct benefits accrue to the plant from the recruitment of predators, and parasitoids may simply be subverting this interaction.

Synomone production by a herbivore-infested plant will depend on the feeding mechanism of the herbivore species. Maize plants attacked by the aphid *Rhopalosiphum maidis* (a phloem-feeder) do not increase volatile production, whereas the lepidopteran *Spodoptera littoralis* (a leaf-chewer) and the stem-borer *Ostrinia nubilalis* induced large changes in volatile production (Turlings *et al.*, 1998). However, plants attacked by *O. nubilalis* released a *lower* quantity of volatiles, and these included several unidentified volatiles that were not induced by *S. littoralis* attack, supporting the hypothesis that herbivore-specific volatiles may be produced by infested plants.

Of course, a given herbivore species may not necessarily induce the same production of volatiles on different host plants, and different natural enemies may in turn show different responses to these plant-produced cues. A good example of this is provided by the work of Takabayashi *et al.* (1998) who studied two different tritrophic systems. In the first system, the parasitoid *Cotesia kariyai* was preferentially attracted to plant (corn) volatiles produced by damage from *Pseudaletia separata*. Surprisingly, this attraction was only elicited when the plant was attacked by early larval instars of the host; feeding by late instars did not induce any preference. Plants attacked by younger *P. separata* instars produce higher proportions of terpenoids and indole volatiles. As early instar *P. separata* parasitised by *C. kariyai* consume less leaf material than older larvae, it is beneficial for the plant if the herbivores are attacked when young (Takabayashi *et al.*, 1995). But host age does not always influence parasitoid behaviour. Gouinguene *et al.* (2003) showed that *Microplitis rufiventris*, a parasitoid that cannot successfully attack first instar *Spodoptera litoralis*, could not distinguish among maize plants attacked by different instars of the herbivore.

In the second system, the congeneric parasitoid *C. glomerata* preferred volatiles produced by *Rorippa indica* (Cruciferae) plants infested with its host *Pieris rapae* over clean air. However, the parasitoid's preference is for artificially damaged plants over herbivore-damaged plants, although in both cases the plant releases (Z)-3-hexanol and (E)-2-hexenal. The technique used to artificially damage the plant seems to produce larger amounts of those volatiles. Takabayashi *et al.* (1998) suggest that the parasitoids in these systems use different mechanisms to overcome the reliability-detectability problem. *Cotesia kariyae* responded to volatiles that provide direct evidence of the presence of potential suitable hosts, whereas *C. glomerata* responded to volatiles produced in response to plant damage. *Rorippa indica* has few herbivore species, of which *Pieris brassicae* is one of the dominant, so responding to general damage cues is likely to lead the parasitoid to potential hosts.

There may also be differences in the response elicited by such cues between generalist and specialist predators or parasitoids. Röse *et al.* (1998) found that the specialist parasitoid *Microplitis croceipes* is attracted to insect-damaged cotton plants, whereas artificial damage (i.e. without herbivore kairomones) is enough to attract the generalist *Cotesia marginiventris*. In contrast, Dickens (1999) found that both generalist (*Podisus maculiventris*) and specialist (*Perillus bioculatus*) predators of Colorado potato beetle showed similar responses to the systemic volatiles produced by infested plants.

It is advisable to include at least three treatments when studying the potential role of insect herbivore host plants in natural enemy host location. These treatments are based on the following questions: First, does the predator or parasitoid respond to unattacked plants? Second, is artificial damage enough to generate a response (clipping using a sterilised pair of scissors or a hole-punch)? Third, does herbivore damage induce a response (allowing the herbivores to feed, before removing them and washing away any direct cues which may emanate from the host, such as frass)? In addition, one may ask if there is a synergy between host-plant and host-insect cues. Havill and Raffa (2000) showed that gypsy moth (*Lymantria dispar*) larvae fed on an artificial diet were not attractive to a foraging braconid parasitoid, *Glyptapanteles flavicoxis*, whereas caterpillars that had fed on their main host plant, poplar, were.

As an aside, Rutledge and Wiedenmann (2003) attempted to alter the preference for different host plants in the parasitoid *Cotesia sesamiae*, a braconid parasitoid of stem-borers. *C. sesamiae* preferentially attacks hosts in sorghum, and after four generations of artificial selection (attempting to get the parasitoids to show a preference for cabbage plants), no change was found in foraging behaviour. This suggests that there is little genetic variation (in this species at least) in response to plant cues in parasitoids.

Many species of insect herbivore communicate with conspecifics using infochemicals such as sex pheromones. These provide reliable cues to the presence of potential prey individuals. Pickett *et al.* (1992) identified and synthesized several aphid sex pheromones, and these have proved to be highly attractive to *Praon* spp. parasitoids in field trials (Hardie *et al.*, 1994). However, it appears that other species of aphid parasitoid (*Aphidius ervi* and *A. eadyi*) do not respond to these cues in field-sited pheromone traps. This may result from the behaviour of the parasitoids; these species do not appear to fly towards point sources (Stowe *et al.*, 1995). This may explain why aphids, placed on plants near sex pheromone sources in the field, suffered significantly greater parasitism than those aphids kept in the absence of the odour source (Powell *et al.*, 1998).

Using a combination of four-way olfactometer (see above) and Observer software (subsection 4.2.9), Couty *et al.* (1999) found that *Leptopilina boulardi*, a parasitoid of *Drosophila melanogaster*, was attracted by a combination of the odours of both rotting fruit and a kairomone left by adult *D. melanogaster* on the substrate tested.

Predators also respond to the presence of prey and plant odours. The black bean aphid, *Aphis fabae*, produces a kairomone that attracts *Metasyrphus corollae*, a predatory hover-fly (Shonouda *et al.*, 1998), and the coccinellid *Hippodamia convergens* is attracted by plant cues released when the aphid, *Myzus persicae*, feeds (Hamilton *et al.*, 1999). However, predators differ from parasitoids in that the host location behaviour may differ between the adult and larval stages, and also within the larval instars. Bargen *et al.* (1998) found that first instar larvae of the hover-fly, *Episyrphus balteatus*, were attracted to aphid cues, but not to honeydew. Older larvae did not respond to these volatiles, but aphid extracts, honeydew and sucrose did provide cues.

The olfactory responses of foraging parasitoids and predators may vary with age, nutritional state and experience. It is important to take account of these factors when designing experiments. Ideally, preliminary experiments should be carried out to test for any effects. Synovigenic species (subsection 3.4) may spend the first few days of adult life searching, not for hosts, but for foods such as nectar and honeydew, which supply nutrients for egg development (see Chapter 8). Therefore, when young or starved, they may be unresponsive to host plant and host odours. Some parasitoids may even be repelled very early in adult life by an odour, which later on in life is used in host-finding. *Exeristes ruficollis* responds in this manner to the odour of pine oil (Thorpe and Caudle, 1938). Such responses need not be fixed. In an elegant study, Lewis and Takasu (1990) showed that female *Microplitis croceipes* can learn to recognise different artificial odours associated with food and host sources. Starved individuals showed a preference for the odour associated with the food source, whereas well-fed females preferentially moved towards the host-associated odour.

Furthermore, the ecological context of the interaction may need to be considered. Orr *et al.* (2003) found that the likelihood of a phorid fly parasitoid successfully locating its ant host, *Linepithema humile*, depended on (host) interspecific competitive interactions. Successful host location was more likely when the host was interacting with a species that elicited a chemical, rather than a physical response. Le Ru and Makosso (2001) found that foraging coccinellid predators (*Exochomus flaviventris*) can distinguish between the odours of cassava infested with parasitised and unparasitised mealybugs, preferentially orientating towards the cassava-unparasitised mealybug complex. In a similar study, Van Baaren and Nenon (1996) studied two mealybug parasitoids. Both are monophagous species, with *Apoanagyrus lopezi* attacking the cassava mealybug (*Phenacoccus manihoti*) and *Leptomastix dactylopii* attacking the citrus mealybug (*Planococcus citri*). Both parasitoid species readily responded to the odours of infested plants or unattacked hosts, but not to those produced by parasitised hosts. However, rather than the parasitoids ignoring the odour of parasitised hosts, it may be that parasitised hosts have an additional odour, which acts as a deterrent. Such work strongly suggests that simplistic approaches to tritrophic systems may underplay the importance of other species in altering the pattern of the interaction.

There is one frequently overlooked aspect of experimental design associated with studies of natural enemy responses to odours. Not only do the enemies themselves show both phenotypic and genotypic variation in response to cues, the plants and prey insects themselves show variation in the signal itself. For example, the parasitoid *Diaeretiella rapae* shows different responses to the volatiles released by two near-isogenic strains of *Brassica oleracea*, which differ only in the production of isothiocyanates (Bradburne and Mithen 2000). Such results also hold across cultivars of the same plant species (Gowling and van Emden, 1994).

Genetic variation in kairomone production is also found among aphids. Not only will parasitoids show differential responses to different clones of aphids, the clones themselves will also show variation in response to alarm pheromone (Müller, 1983). It cannot be overemphasised that researchers studying aphid-natural enemy interactions should work with several different clones of aphids. To be pedantic, since aphids within a clone are for all practical purposes genetically identical, the clone is the replicate. Many studies of aphid-natural enemy interactions are essentially performed without replication, since only one clone is used.

There are many studies that show learning in parasitoid wasps. Cues that elicit no response in naïve females can induce a response when they have been experienced in association with host contact (e.g. Fukushima *et al.*, 2002; Meiners *et al.*, 2003). This is known as **associative learning**, defined as a response to a stimulus that usually does not induce a response, after that stimulus has been experienced in combination with another stimulus to which the animal already shows an innate response. The behavioural plasticity allowed by associative learning provides considerable flexibility in parasitoid foraging strategies.

If *Aphidius ervi* females are allowed to experience oviposition on the plant-host complex (here the pea aphid, *Acyrthosiphon pisum*, on broad bean, *Vicia faba*), then they are significantly more likely to orientate towards the plant-host complex than naïve females. Naïve females will orientate towards a source of host volatiles (Du *et al.*, 1996; the innate response), but experienced females will also show an increased response to volatile cues from uninfested plants, which is likely to be an example of associative learning (Guerrieri *et al.*, 1997).

The effect of learning on behaviour may depend on the experience and physiological state of the parasitoid. Female *Leptopilina boulardi* (a eucoilid parasitoid of *Drosophila melanogaster*) will associate odour cues with host presence, increasing ovipositor searching when exposed to the cue (Pérez-Maluf and Kaiser, 1998). This increase in searching behaviour was not associated with mating or prior oviposition experience, although both factors did influence some parameters of host searching. Females with prior oviposition experience showed a higher latency and reduced probing duration, whereas mated females tended to have a reduced latency and increased probing duration (Pérez-Maluf and Kaiser, 1998). Female *L. boulardi* show heritable variation in these responses to learned cues (Pérez-Maluf *et al.*, 1998).

The likelihood of learning appears to be related to the nature of the substrate the parasitoid is searching. Duan and Messing (1999) suggest that parasitoid acceptance of less preferred hosts may be more likely to change with experience, than if the parasitoid is allowed to learn cues associated with preferred host-substrate complexes. If this is the case, then it is possible that associative learning will not be equally likely to be found with all potential host species. Therefore, the absence of learning in one situation may not reflect what will be found with other potential hosts. For example, Morris and Fellowes (2002) found that natal host influenced the likelihood of learning in the pupal parasitoid *Pachycrepoideus vindemmiae.* Females that emerged from *Musca domestica* only showed a preference for that host species after gaining experience attacking it. Experience gained in attacking *Drosophila melanogaster* did not change preference. In a similar manner, wasps that emerged from *D. melanogaster* only showed a preference for that host when allowed to gain experience in attacking *D. melanogaster* pupae; experience gained on *M. domestica* did not alter preference. While such results support Duan and Messing's hypothesis, it needs further testing.

In experiments investigating learning in parasitoids, it is often best to use novel cues, which can be controlled and measured by the investigator. In studies of associative learning, odours such as vanilla and strawberry essence have been successfully used. Iizuka and Takasu (1998) used this approach to show associative learning by the pupal parasitoid *Pimpla luctuosa.* In addition, they found that females ceased attacking dummy hosts which had the previously learned odour after several failed oviposition attempts, which suggests that parasitoids can also learn to ignore cues (Iizuka and Takasu, 1998).

1.5.3 HOST LOCATION BY PARASITOIDS

Inferring Behaviour From Morphology

Perhaps one of the more straightforward means of deducing how a predator or parasitoid may locate its prey is to pay close attention to the insect's morphology. For example, pipunculid flies have extremely well developed compound eyes, and in the females the forward-facing facets are considerably enlarged (Jervis, 1992), so it can be inferred that host-finding in these parasitoids relies on vision (confirmed by Forbes P. Benton, see Waloff and Jervis, 1987).

However, some caution should be used when inferring behaviour from morphology, and ideally, the insect would be studied carefully to confirm that the trait does aid predation or parasitism. Nevertheless, a small amount of basic biology and natural history will provide a huge amount of help in understanding the species of interest.

Genetic Variation in Foraging Behaviour

Wajnberg and Colazza (1998) used a combination of automated recording and statistical techniques to study the foraging behaviour of the parasitoid *Trichogramma brassicae*, and found not only that the searching efficiency of females within a patch determined the number of hosts they encountered, but also that there was significant genetic variation among females in this trait. Whether such genetic variation plays a role in allowing populations to adapt to different habitats is a question that deserves a great deal of attention. Van Nouhuys and Via (1999) studied variation among populations of the parasitoid *Cotesia glomerata* attacking small cabbage white butterfly (*Pieris brassicae*) caterpillars in wild and agricultural habitats. These habitats present very different environments to the foraging parasitoids, as in the agricultural habitat every plant that a parasitoid lands on may carry its host. Wasps that originated from the wild habitat tended to move more between plants, perhaps reflecting the spatial characteristics of wild host populations. Jia *et al.* (2002) found genetic variation in response to herbivore-induced plant volatiles in the predatory mite, *Phytoseiulus persimilis*. Isofemale lines (subsection 3.2.3) showed a negative correlation between the likelihood of patch location and patch residence time, suggesting a trade-off between prey location and reproduction.

It is unfortunate that few studies have considered the importance of genetic variation in natural enemy behaviour, yet studies such as these illustrate that artificial selection techniques may be applied to improve the effectiveness of biological control agents during mass-rearing programmes (but see Rutledge and Wiedenmann, 2003, for a counter-example).

Kairomones

Having arrived in a potential host habitat, a parasitoid begins the next phase in the search for hosts. Often, insects show arrestment in response to contact with kairomones of low volatility deposited by their hosts on the substratum. Materials containing such kairomones (sometimes referred to as 'contact chemicals') have been shown to include host salivary gland or mandibular gland secretions, host frass, and homopteran honeydew and cuticular secretions. Several herbivore species have evolved traits which reduce the build-up of frass near to where they feed, reducing the likelihood of their location by foraging parasitoids. This is a relatively common behaviour in caterpillars dwelling in leaf shelters, who can eject their frass with considerable force, depositing it some distance from the potential host (Weiss, 2003).

Kairomones present on the host itself have also been shown to induce oviposition behaviour by several parasitoid species. For example, the parasitoid *Aphidius ervi* shows strong responses to pea aphid (*Acyrthosiphon pisum*) siphuncle secretions, but only at short very short range or on actual contact, and the presence of these kairomones alone is enough to induce oviposition behaviour (Battaglia *et al.*, 2000). A similar response is shown by the parasitoid *Diadromus pulchellus*. This wasp responds to the presence of soluble polypeptides present in the cocoons of its host, the leek moth *Acrolepiopsis assectella* (Benedet *et al.*, 1999).

Because stronger responses may be found to a kairomone after prior oviposition experience in the presence of the substance, an initial experiment ought to be performed using parasitoids with previous oviposition experience. The next series of experiments would involve comparing the response of parasitoids to patches of potential host habitat, within which hosts have never occurred (e.g. clean host plant leaves), with the response to patches within which hosts have previously occurred for some time. The following changes in behaviour might be observed in the searching insects: a decrease in walking speed, an increase in the rate of turning, a sharper angle of turn at the patch edge, an increase in the number and frequency of ovipositor stabs, an alteration in position of the antennae, an increase in the amount of drumming with the antennae and an increase in the amount of time spent standing still. Video recording equipment, together with the computer software discussed in Chapter 4 (subsection 4.2.9) can be used to record and analyse alterations in these behavioural components. If such equipment is not available, then the insect's path can be traced with a felt-tip pen on a Petri dish lid, and a map measurer then used to measure the distance travelled. If, during the experiment, the path trace is marked at regular (for example, three-second) intervals, alterations in the speed of walking, over short time scales, can be measured. Path tortuosity can be evaluated by measuring the angle between tangents drawn at intervals along the path.

A useful additional analysis that can be carried out involves designating areas of an arena, e.g. the kairomone-treated area and the clean area, and measuring the proportion of the total time available that the insect spends searching each area. If the parasitoid or predator can be shown to have spent a greater proportion of its time in the treated area, then it has been arrested by the kairomone.

Once it has been demonstrated that patches within which hosts have occurred contain a stimulus to which parasitoids respond by arrestment, further experiments can be performed to elucidate the nature of the stimulus. To eliminate the possibility that the arrestment response is to some physical property of the patch (e.g. the texture of the wax secretions left by mealybugs, or depressions caused by feeding larvae), one can attempt to dissolve the putative kairomone either in distilled water, hexane or another suitable solvent, and then apply the solution to a surface, for example a leaf or a glass plate, which has never borne host larvae. If an arrestment response is still observed, it can be concluded that the soluble substance is a kairomone. For a detailed experimental study of the arrestment response in a parasitoid, conducted along these lines, see Waage (1978) (Figure 1.8a,b).

Kairomones provide quantitative, in addition to qualitative information. Several parasitoid species, when presented with several patches of kairomone in different concentrations, have been shown to spend longer periods searching those patches with the higher kairomone concentrations than the patches with the lower concentrations, at least over part of the range of concentrations (Waage, 1978; Galis and van Alphen, 1981; Budenberg, 1990; Hare and Morgan, 2000) (Figure 1.8 shows Waage's experimental design). Because kairomone concentration varies with host density, parasitoids can obtain information concerning the profitability of a patch, even before they encounter hosts. Honeydew produced by the aphid *Brevicoryne brassicae* provides not only a qualitative cue in host location, but also is a source of information on the density of hosts within a patch for the parasitoid *Diaeretiella rapae* (Shaltiel and Ayal, 1998).

1.5.4 RESPONSES TO PARASITOID ODOURS AND PATCH MARKS

Parasitoid odours

Janssen *et al.* (1991) showed in olfactometer experiments that *Leptopilina heterotoma* is

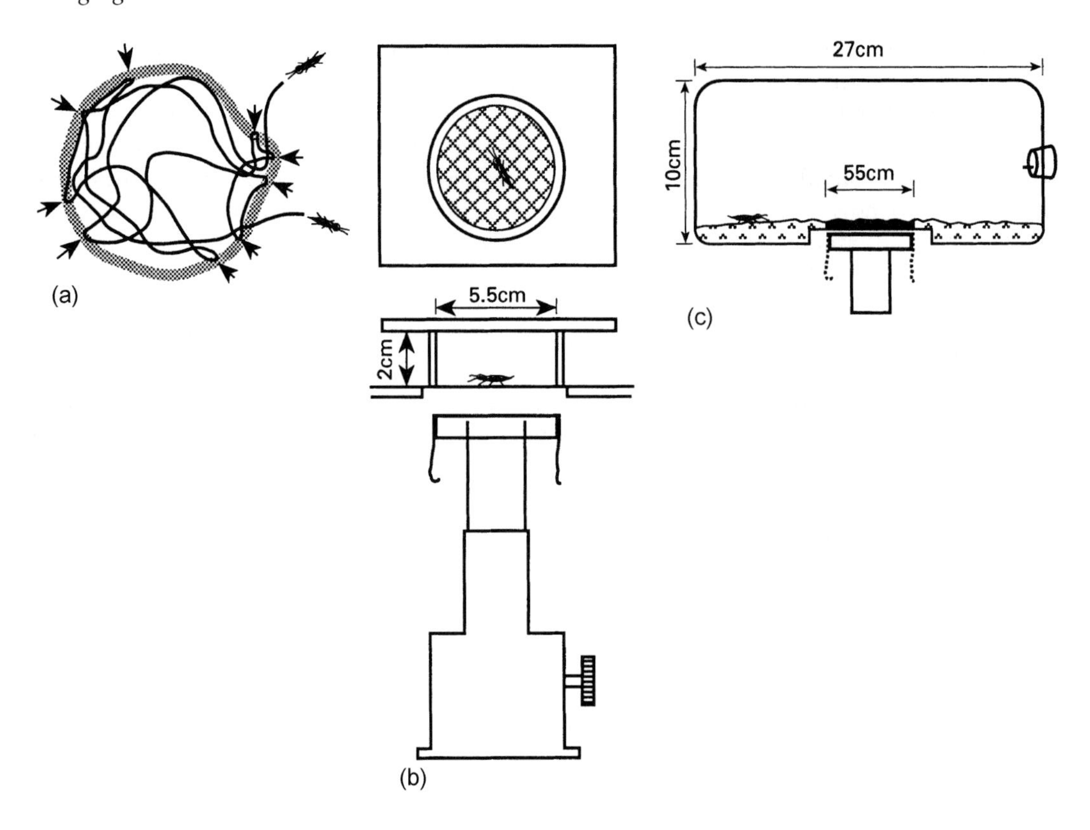

attracted to the odour of stinkhorn fungi containing larvae of *Drosophila phalerata*. When these patches are offered in an olfactometer together with similar patches on which searching females of *L. clavipes* are present, *L. heterotoma* avoids the odour fields of patches containing *L. clavipes* females. The conclusion from these observations is that *L. clavipes* produces an odour whilst searching, which repels its competitor *L. heterotoma*, at least when the latter is presented with the choice between host-containing patches emitting this odour and host-containing patches that lack the odour. Price (1981), suggests that the function of the strong odour emitted by some female ichneumonids, noticed when these insects are handled, likewise signals the insects' presence to other parasitoids. Furthermore, kairomones combined with odours from conspecifics may help parasitoids avoid intraspecific competition. *Venturia canescens* will normally orientate towards host kairomones, but will avoid the odour plumes which contain both host kairomones and the odour of conspecific females (Castelo *et al.*, 2003).

Höller *et al.* (1991) found evidence that foraging primary parasitoids of aphids are repelled by odours produced by adult hyperparasitoids. Furthermore, individuals of the aphid *Sitobion avenae* that have been attacked by a primary parasitoid, *Aphidius ervi*, show differential responses to odours released by a hyperparasitoid, *Alloxysta victrix*. At 120 hours after attack, the aphids are attracted to the volatiles, yet at 160 hours after attack they are repelled by the same cue. Since unattacked aphids show no responses to these cues, Guerra *et al.* (1998) suggest that as behavioural control passes from aphid to parasitoid over time, the adaptive benefits of responding to these cues will also change.

In cases where the odour of a parasitoid repels conspecifics, the substance is a pheromone, whereas in cases where heterospecific competitors are repelled, there is some justification in describing the substance as an allomone. However, because of similar problems to those mentioned when discussing patch marking (below), the use of the term allomone should be avoided here.

It is not known how widespread the use of repellent odours is among insect parasitoids, largely because it has not been studied in a systematic way. Like other infochemicals used by parasitoids, odours produced by adult parasitoids can potentially have a profound effect on patch choice and time allocation by individual wasps and thus on the distribution of parasitoids over a host population.

Figure 1.8 Arrestment and patch time allocation in parasitoids: Jeff Waage's (1978) classic study of *Venturia canescens*. (a) An experiment carried out to investigate arrestment behaviour of *V. canescens* in response to contact with a kairomone. The path of a walking female was observed on a glass plate, upon which 1 ml of ether extract of ten pairs of host (*Plodia interpunctella*; Lepidoptera) manidibular glands had been placed and allowed to evaporate. Stippling denotes the edge of the patch. Upon encountering the patch edge from the outside, a female stops and begins to apply the tips of its antennae rapidly upon the substratum. It then proceeds onto the patch at a reduced walking speed (inverse orthokinesis). Within the patch, the wasp occasionally stops walking and probes the substratum with its ovipositor. When the wasp encounters the patch edge from within the patch, it turn sharply away from the edge. Presumably due to waning of the arrestment response, e.g. through habituation or sensory adaptation to the chemical stimulus, the wasp eventually leaves the patch. (b) Apparatus used to test the hypothesis that the patch edge response of *V. canescens* is to the removal of the chemical stimulus, and not to the patch edge *per se*. The terylene gauze screen was impregnated with host mandibular secretion by confining ten fifth instar host larvae between two sheets of gauze. The lower sheet was then stretched over a Petri dish, as in the figure. By raising and rapidly lowering the contaminated screen, Waage could precisely control when a wasp (in the upper chamber) was 'on' and 'off' the patch. A wasp's movements were traced with a felt-tip pen on the plate glass roof of the chamber (nowadays this could be done using video recording coupled with analysis of movements using computer software, see section 1.4). Over the first centimetre travelled following stimulus removal, most wasps made a reverse turn, which may be considered to be a klinotactic response (i.e. directed), because the turn oriented the wasps back towards the the point from where the stimulus was removed. Thus, Waage concluded that the patch edge response of *V. canescens* is to the removal of the chemical stimulus, not to the patch edge *per se*, i.e. his hypothesis was supported. (c) Apparatus used by Waage to test the effect of kairomone concentration on patch residence time. 'Patches' were made by confining different numbers of host larvae, together with food medium, between terylene gauze sheets for several hours. The larvae were then removed. For each kairomone concentration, the contaminated patch of food medium (minus the hosts) was held over the central part of the floor of the chamber (blackened area). An empty Petri dish was raised beneath the patch (see next experiment). Two arbitrary time intervals (14 s and 60s continuously off a patch) were used as criteria for determining patch-leaving by wasps. Application of either of these criteria indicated that the duration of the first visit to a patch increased markedly with increasing kairomone concentration.

The apparatus was also used to test the effect of ovipositions on patch residence time. A patch of host-contaminated food medium was stretched over the central part of the chamber floor, and at the onset of the experiment a dish containing thirty host larvae was raised beneath the patch. Each wasp was allowed to make an oviposition into a host as soon as she entered the patch. During the resting period following that oviposition, the dish containing host larvae was replaced with an empty one. Oviposition was found to produce a marked increase in the duration of the first patch visit by a wasp. Another experiment was carried out by Waage, which demonstrated that oviposition does not elicit a significant arrestment response in the absence of the kairomone. This experiment employed apparatus (b). A host-contaminated terylene gauze screen, with or without host larvae beneath it, was raised beneath the chamber. A single wasp was exposed either to the chemical stimulus alone for the duration of one bout of probing, or to the chemical stimulus with hosts present for one oviposition of similar duration. The screen was then lowered, so removing the kairomone stimulus, and the time taken for the wasp to leave the chamber floor and then climb onto one of the chamber sides was recorded (this behaviour being interpreted as the cessation of any response elicited by the contact chemical). No significant difference in the amount of time taken to abandon the host area was observed between the treatments with oviposition and those without. From Waage (1978), reproduced by permission of Blackwell Publishing.

Patch-marking

Some parasitoid species are known to leave chemical marks on surfaces they have searched (Galis and van Alphen, 1981; Sheehan *et al.*, 1993). This marking behaviour can have a number of functions. By leaving a scent mark on the substratum, a parasitoid can avoid wasting time and energy in searching already visited areas. A female can also use the frequency with which she encounters marks to determine how well she has searched the patch, and so assist in the decision when to leave the patch. When encountered by conspecific or heterospecific competitors, marks sometimes induce the competitor to leave an area. *Pleolophus basizonus, Orgilus lepidus, Asobara tabida, Microplitis croceipes, Halticoptera rosae* and *H. laevigata* (Price, 1970; Greany and Oatman, 1972; Galis and van Alphen, 1981; Sheehan *et al.*, 1993; Hoffmeister, 2000; Hoffmeister and Gienapp, 2001) mark areas they search, and females spend less time in areas previously searched by conspecifics. In the case of a heterospecific competitor, the marker substance could be termed an allomone. However, leaving the patch may not always be in the interest of the competitor; the competitor may stay and superparasitise the hosts parasitised by the first female (subsection 1.8.4). Thus, the use of the term allomone should be avoided in this context.

The use of patch-marker substances can be demonstrated by offering patches containing kairomone, but not hosts, to a parasitoid. After the parasitoid has left the patch, a second parasitoid is introduced on to the same patch. If the second insect always stays on the patch for a shorter period than the first, the existence of a mark left by the first has been demonstrated.

Predators may also patch-mark. Nakashima *et al.* (2002) provide the only currently known example, showing that the insect predator *Orius sauteri* avoided patches where they had recently foraged, although this behaviour was lost when the predator had not recently fed. The patch marks appear to be relatively short lived (< 1 hour), and may simply prevent the females from foraging in an area previously searched.

Not only insect natural enemies respond to such cues. The prey themselves may also respond to odour cues left by foraging predators or parasitoids. For example, spider mites (*Tetranychus urticae*) will avoid foraging in patches that have previously held predators, and this avoidance is greater if the predators have been feeding on *T. urticae* (Grostal and Dicke 2000). Most studies of predator and parasitoid foraging behaviour assume that such avoidance does not take place.

1.5.5 SEARCH MODES WITHIN A PATCH

While kairomones and other cues can arrest parasitoids and predators in host/prey patches and so increase the probability of encounter, host/prey location is itself likely to be in response to non-chemical, e.g. visual and tactile cues. For example, in coccinellid predators, prey honeydew acts as an arrestant stimulus for adults (van den Meiracker *et al.*, 1990: *Diomus* sp., *Exochomus* sp.; Heidari and Copland, 1993: *Cryptolaemus montrouzieri*), but the prey are located in response to visual cues (Stubbs, 1980: *Coccinella septempunctata*; Heidari and Copland, 1992: *C. montrouzieri*). Stubbs (1980) devised a method for calculating the distance over which prey are detected (Heidari and Copland, 1992, describe a modification of the technique). Coccinellid larvae are arrested by honeydew (Carter and Dixon, 1984), but location of the prey occurs only upon physical contact (*Coccinella septempunctata*). Unlike the adults, the larvae do not use cues to orient themselves towards the prey.

It has been shown for a number of predators that arrestment occurs as a consequence of prey capture (Dixon, 1959a; Marks, 1977; Nakamuta, 1982; Murakami and Tsubaki, 1984; Ettifouri and Ferran, 1993). In this way, the insect's searching activities are concentrated in the immediate vicinity of the previously captured prey, increasing the probability of locating a further prey individual. The adaptive value of such behaviour for predators of insects that have a clumped distribution, such as aphids, is obvious. Predators also show arrestment after capturing a prey individual but failing to feed on it - even a failed encounter with prey is an indication that a clump of prey has been found (Carter and Dixon, 1984). Carter and Dixon (1984) argue that the latter behaviour is particularly important for early instars of coccinellids, since the prey capture efficiency of these instars is relatively low. In final instar larvae of the coccinellid *Harmonia axyridis*, arrestment in response to prey capture occurs only if the predators are provided with the same prey species as they were reared upon, indicating a strong conditioning effect (Ettifouri and Ferran, 1993). Arrestment of the aboreal ponerine ant *Platythyrea modesta* is affected by prey size. Small prey required contact, whereas larger prey, such as grasshoppers, elicit arrestment at a distance (Djieto-Lordon *et al.*, 2001). Following arrestment, the ants attacked without antennation. Small prey species are killed using pressure from the mandibles, whereas larger prey are stung.

Arrestment in the above cases can be studied in the same way as arrestment of natural enemies in response to kairomones, i.e. by analysing the search paths of predators and parasitoids and by measuring the proportion of the total time available spent searching designated unit areas of an arena.

Species of parasitoid attacking the same hosts may differ in the way they search a patch. In parasitoids of concealed anthomyiid, calliphorid, drosophilid, muscid, phorid, sarcophagid and sepsid fly larvae, at least three different **search modes** exist (Vet and van Alphen, 1985). Wasps may either:

1. Probe the microhabitat with their ovipositors until they contact a host larva **(ovipositor search);**
2. Perceive vibrations in the microhabitat caused by movements of the host and use these cues to orient themselves to the host **(vibrotaxis)** which is then probed with the ovipositor; or
3. Drum, with their antennae, the surface of the microhabitat until they contact a host (**antennal search**).

To determine which search mode a parasitoid species uses is easy in the case of ovipositor search or antennal search, where brief observation of a searching female suffices to classify her search mode. However, it can be difficult to prove that vibrotaxis occurs, because of the possibility that the parasitoid locates its hosts by reacting to a gradient in kairomone concentration or some other chemical cue, or to infrared radiation from the host. Therefore, we will consider this search mode in more detail.

Parasitoids have been shown to respond to vibratory stimuli issuing from foraging hosts when they are searching for potential victims. Meyhöfer and colleagues (Meyhöfer *et al.*, 1994; 1997) found that the leaf-miner *Phyllonorycter malella* produces vibrations while feeding, and that the parasitoid *Sympiesis sericeicornis* responds to these cues by increased rates of turning in the vicinity (**vibrokinesis**). There is also some evidence for vibrotaxis, but this is more circumstantial. For example, *Asobara tabida* and *Leptopilina longipes*, two common parasitoids of *Drosophila* species, will fail to locate immobilised hosts (van Alphen and Drijver, 1982; van Dijken and van Alphen, 1998). Indeed, it has been suggested that the rover/sitter polymorphism in larval *Drosophila melanogaster* may be maintained by frequency-dependent selection resulting from the relative proportions of vibrotactic parasitoids within the community of larval parasitoids (Osborne *et al.* 1997).

However, in a valuable review Meyhöfer and Casas (1999) point out some pitfalls in the study of the use of vibratory stimuli by parasitoids searching for hosts. Many of these are associated with experimental design, where the use of immobilised larvae (e.g. by freezing, dipping in hot water, needle insertion) introduces the confounding factors associated with reduced metabolic rate (influencing heat or CO_2 output) and changes in the chemical cues emanating from potential hosts. Unless these confounding factors are controlled for, it is difficult to confirm that changes in parasitoid behaviour are the result of responses to vibratory cues. A second issue they raise is the need to confirm that the host does indeed produce vibratory cues to which the parasitoid can and does respond. Very few studies satisfactorily deal with these issues, although techniques such as laser vibrometry are available to characterise these vibrational signals (Meyhöfer *et al.*, 1994).

Wäckers *et al.* (1998) used laser vibrometry to infer the ability of the pupal parasitoid *Pimpla turionellae* to locate potential hosts. Clearly, the host pupae themselves do not produce vibrations; instead, the parasitoid itself appears to generate vibrations that can then be used, in a manner analogous to SONAR, to locate hosts. The technique used by Wäckers *et al.* was particularly ingenious. By using paper cylinders of differing thickness and a cigarette filter to serve as a 'host', the authors were able to show that as the thickness of the substrate increased, the number of oviposition attempts decreased. They suggest that the parasitoid responded to differences in resonance between hollow and solid sections of substrate, and that increasing thickness of paper reduced the ability of the parasitoid to distinguish between sections.

Vibrations may also be used by potential hosts as a warning that a parasitoid may be about to attack. Bacher *et al.* (1997), again using laser vibrometry, showed that the late instar larvae and pupae of the leaf-miner *Phyllonorycter malella* reacted defensively to certain frequencies produced by oviposition insertion by the parasitoid *Sympiesis sericeicornis*. Such ability to avoid attack may prove to be common among leaf-miners. Using the same system, Djemai *et al.* (2001) used artificial vibrations matched to the frequencies resulting from *Sympiesis sericeicornis* attack, and this elicited the same defensive behaviours in the host. This provides excellent empirical support for the conclusions drawn by the earlier study.

The reason why it is important to determine the search mode of a parasitoid or predator is that different search modes lead to different encounter rates with hosts in the same situation. Thus, a parasitoid using vibrotaxis as a search mode may be more successful in finding hosts when the hosts occur at low densities, while ovipositor search may be more profitable at high host densities. Antennal search results in encounters with larvae on the surface, while ovipositor search can also result in encounters with hosts buried in the host's food medium. However, Broad and Quicke (2000) showed that the use of vibrotaxis is positively correlated with host depth in the substrate, controlling for parasitoid size. This suggests that in substrates where ovipositor searching is time-consuming (e.g. where the host is relatively deep in the substrate), vibrotaxis may be more common than the aforementioned argument suggests.

Often, the searching behaviour of a parasitoid comprises a combination of search modes, as the insect responds to different cues while locating a host. It is therefore not always possible to place the behaviour of a parasitoid in one category.

Predators may employ a combination of search modes. The larvae of the predatory water beetle *Dytiscus verticalis* may either behave as sit-and-wait predators when prey density is high, or hunt actively for prey when prey density is low (Formanowicz, 1982). Such variety is common, and many species that are traditionally considered to be ambush predators (e.g. mantids, see below) frequently actively search for prey.

Pit-dwelling antlion (*Myrmeleon* spp.) larvae provide the classic example of an ambush predator. The larvae excavate funnel-shaped holes in loose sand, and it is that the latter prevents potential prey from escaping. The spatial distribution of the antlion *Myrmeleon immaculatus* reflects that of prey density, minimising the need to move to a new pit location (Crowley and Linton, 1999). The antlion larva waits at the base of the pit, with only its relatively large mandibles projecting from the sand. Once a victim becomes trapped, the larva suddenly grabs its prey and drags it under the sand. This has the advantage of rendering physical defences, such as biting or formic acid, useless (New, 1991). Given that the ambush strategy is risky (i.e. the presence of food is unpredictable) and that manufacturing and maintaining the pit is costly (Lucas, 1985; Hauber, 1999), it is unsurprising that antlions have relatively low metabolic rates (van Zyl *et al.*, 1997). Larvae can survive for relatively long periods without food, albeit at the cost of a long development period.

In contrast to situations where camouflage is critically important, some 'sit-and-wait' predators employ conspicuous colouration, e.g. several species of orb-web spider. The spiny spider, *Gasteracantha fornicata*, has a strikingly coloured yellow and black striped dorsal surface. Spiders which were dyed black captured fewer prey individuals, supporting the hypothesis that bright colours helped attract visually-orienting prey (Hauber, 2002).

1.5.6 HOST RECOGNITION BY PARASITOIDS

Generally, specific (although not necessarily host *species* specific) host-associated stimuli need to be present for the release of oviposition behaviour by parasitoids following location of a prospective host. The role these stimuli play in host recognition has been investigated mainly by means of very simple experiments.

For many parasitoids, host size appears to be important for host recognition. In a classic experiment, Salt (1958) presented female *Trichogramma* with a small globule of mercury – smaller than a host egg – and observed that the parasitoid did not respond to the globule. However, Salt then added minute quantities of mercury to the globule, whereupon a female would mount it, examine it and attempt to pierce it with her ovipositor. When Salt continued adding quantities of mercury to the globule, a globule size was reached where a wasp again did not recognise it as a prospective host.

Host shape can be important in host recognition. A number of workers have placed inanimate objects of various kinds inside either hosts or host cuticles from which the host's body contents have been removed, and have shown that some host shapes are more acceptable than others.

One needs to be cautious in interpreting the results of experiments where hosts or host dummies of various sizes and shapes are presented to parasitoids. If a parasitoid is found to attempt oviposition more often in large dummies than in small ones, or in rounded dummies than in flattened ones, the stimuli involved could be visual, tactile or both. Some investigators have failed to determine precisely which of these stimuli are important. Similar caution needs to be applied to experiments in which dummies of different textures are presented to parasitoids.

As can often be inferred from direct observations on the behavioural interactions of parasitoids and hosts, movement by the host can be important in releasing oviposition behaviour. A simple experiment for investigating the role of host movement in host recognition involves killing hosts, attaching them to cotton or nylon threads, moving both these and similarly attached living hosts before parasitoids, and determining the relative extent to which the dead and living hosts are examined, stabbed, drilled or even oviposited in by the parasitoids.

Kairomones play a very important (although not necessarily exclusive) role in host recognition by parasitoids. Strand and Vinson (1982) showed in an elegant series of experiments how, if glass beads the size of host eggs are uniformly coated with material present in accessory glands of the female host (host eggs normally bear secretions from these glands), and are presented to females of *Telenomus heliothidis* (Scelionidae), the insects will readily attempt to drill the beads with their ovipositors. Female parasitoids, when presented with either clean glass beads or host eggs that had been washed in certain chemicals, were, on the whole, unresponsive. Strand and Vinson (1983) analysed the host accessory gland material and isolated from it (by electrophoresis of the material) proteins, two of which were shown to be more effective in eliciting drilling of glass beads. It cannot be assumed from these findings that *T. heliothidis* will recognise any object as a host provided it is coated in kairomone. Host size and shape are also important criteria for host acceptance by *T. heliothidis*. Similar findings have been reported for several other species (e.g. *Trissolcus brochymenae*: Conti *et al.*, 2003; *Ixodiphagus hookeri*: Takasu *et al.*, 2003). In a number of cases, the active compound has been identified. O-caffeoylserine, produced by the cassava mealybug, elicits host acceptance behaviour in the encyrtid parasitoids *Acerophagus coccois* and *Aenasius vexans* (Calatayud *et al.*, 2001).

Weinbrenner and Völkl (2002) took a different approach to understanding the importance of contact kairomones in host recognition by *Aphidius ervi*. Wet pea aphids were not accepted as hosts, which the authors suggest resulted from the parasitoids being unable to detect the host's kairomones.

A useful approach to studying the role of kairomones would be to take a polyphagous parasitoid species and determine whether the recognition kairomone is different or the same for each of its host species. Van Alphen and Vet (1986) showed that the braconid parasitoid *Asobara tabida* discriminates between the kairomone produced by *Drosophila melanogaster* and that produced by *D. subobscura*.

Acceptance of a prospective host for oviposition also depends upon whether the host is already parasitised. This important aspect of parasitoid behaviour is dealt with later in this chapter (section 1.8).

1.5.7 HOST AND PREY SELECTION

Host Species Selection

Many parasitoid species are either polyphagous or oligophagous. Strictly monophagous species are relatively uncommon. When different potential host species occur in different habitats, a parasitoid 'decides' which host species is to be attacked by virtue of its choice of habitat in which to search. Sometimes, potential host species can be found coexisting in the same patch (e.g. two aphid species living on the same host plant, larvae of different fly species feeding in the same corpse). In these cases, experiments on host species selection are relevant, and can demonstrate whether the parasitoid has a **preference** for either of the species involved. Preference is defined as follows: a parasitoid or predator shows a preference for a particular host/prey type when the proportion of that type oviposited in or eaten is higher than the proportion available in the environment. This is the traditional 'black box' definition (Taylor, 1984), so-called because it does not specify the behavioural mechanisms responsible. For example, a parasitoid may encounter a host individual and accept it, but the host may then escape before the parasitoid has an opportunity to oviposit (likewise, prey may escape from a predator following acceptance). If host types differ in their ability to escape, they will be parasitised to differing extents even though they may be accepted at the same rate. Conversely, they may be accepted at different rates but be parasitised to the same extent. It could be argued that preference, to be more meaningful behaviourally, ought to be defined in terms of the proportion of hosts or prey accepted. However, it may not be possible in experiments to observe and score the number of acceptances (one reason being that the insects do not display obvious acceptance behaviour).

Often (section 1.11 describes a different approach), experiments designed to test for a preference score the number of hosts parasitised or prey fed upon after a certain period of exposure where equal numbers of each species have been offered. There is, however, a problem with this approach: the number of hosts oviposited in/prey eaten depends on the number of encounters with individuals of each species, and the decision to oviposit/feed on the less preferred species may be influenced by how often the female gets the opportunity to oviposit/feed on the preferred species. Encounter rates (section 1.15) may also be unequal for the two host/prey species, due to factors such as differences in size or activity. Therefore, species selection should preferably be investigated in such a way that encounter rates with both species are equal. This requires pilot experiments, with equal numbers of each species offered simultaneously, to calculate the ratio in which both types should be presented so as to equalise encounter rates.

Mathematical formulae used for quantifying preference (whether for species or for stages) are many and varied (Cock, 1978; Chesson, 1978a, 1983; Settle and Wilson, 1990), but the most widely used measure of preference is the following (Sherratt and Harvey, 1993):

$$\frac{E_1}{E_2} = c\frac{N_1}{N_2} \quad (1.1)$$

where n_1 and n_2 represent the numbers of two host/prey types available in the environment, and E_1 and E_2 represent the numbers of the two host/prey types eaten or oviposited in. The parameter c is the **preference index** and can be viewed as a combined measure of preference and encounter probability (section 1.11). A value of c between zero and one indicates a preference for host/prey type 2, whereas a value of c between one and infinity indicates a preference for host/prey type 1. Mathematical formulae used in testing whether preference varies with the relative abundance of the different host/prey types are discussed in later in this chapter (section 1.11).

A rather more sophisticated approach has been suggested by Sakuma (1998), using probit analysis. This method overcomes the problems associated with standard probit analysis (an all-or-nothing approach), taking into account differences in the strength of the stimulus (e.g. number of hosts, quantity of odour cues etc.) The program (available from Masayuki Sakuma, Graduate School of Agriculture, Kyoto University, Kyoto 606-8502, Japan), involves a regression of the probit-transformed number of responses against logged dose (or here, number of hosts). Such an approach would be suitable also for analysing preference data from olfactometer experiments.

Optimal host selection models predict that the acceptance of a less profitable host species depends on the encounter rate with the more profitable host species. The less profitable species should always be ignored if the encounter rate with the more profitable species is above some threshold value, but should be attacked if the encounter rate with the more profitable species is below that threshold value (Charnov, 1976; Stephens and Krebs, 1986). Note that if recognition of prey is not instantaneous, then acceptance of the less profitable host species depends on the encounter rates with both of the host species and on the time it takes for recognition to take place. Often, for the convenience of the researcher, relatively high densities of hosts, resulting in high encounter rates, are offered in laboratory experiments. This will produce a bias towards more selective behaviour. For example, in laboratory experiments with high encounter rates, the *Drosophila* parasitoid *Asobara tabida* is selective when offered the choice between two host species differing in survival probability for its offspring (van Alphen and Janssen, 1982) and avoids superparasitism (van Alphen and Nell, 1982). However, in the field, when encounter rates are equal to or lower than one host per hour, wasps always generalise and superparasitise (Janssen, 1989). If one is interested in knowing the performance of a parasitoid species in the field, where host densities are often very low, one should use host densities equivalent to those occurring in the field. The high densities often offered in the laboratory may allow the researcher to obtain much data over a short period of observation, but the insect's behaviour in such experiments may not be representative of what happens in the field.

To understand the adaptive significance of host preferences the relative profitability of different host species can be assessed, in the first instance, by recording the survival rates of parasitoid progeny in the different hosts. Even if no differences in the probability of parasitoid offspring survival are recorded, one cannot automatically assume that the host species concerned are equally profitable. Handling times may vary with host species, as may the fecundity and other components of the fitness of parasitoid progeny, and ideally, these ought to be measured.

Experiments on prey choice by predators are influenced by prey densities offered in a manner similar to that described above for parasitoids. Because searching activity is influenced by the amount of food in the gut (more precisely, the degree of satiation), a predator's feeding history may determine the outcome of experiments on prey choice (Griffiths, 1982; Sabelis, 1990, but this may not always be the case, e.g. see De Kraker *et al.*, 2001).

So far, we have considered innate host and prey preferences. Preferences may alter with experience (see section 1.11) Preferences may also change with the physiological state of the predator or parasitoid. For example, Sadeghi and Gilbert (1999, 2000b) found that the hoverflies *Episyrphus balteatus* and *Syrphus ribesii* both preferentially attacked pea and rose aphids over nettle aphids, and that the strength of this relationship weakened with time. The influence of host deprivation and egg load on oviposition rates differed between the species.

One crucial, yet almost completely ignored, factor in parasitoid host choice behaviour concerns the presence of genetic variation within a given population. Without this variation, populations will not be able to adapt to changes in the host community. Genetic variation explains the variation among parasitoid populations in host preference. Rolff and Kraaijeveld (2001) found that virulent lines of the parasitoid *Asobara tabida* were more likely to accept *Drosophila melanogaster*, a host species with a strong immune response, than control lines which preferentially attacked the non-resistant species, *Drosophila subobscura*. Host species selection is further discussed in section 1.11.

Host Stage Selection

Parasitoids may encounter different developmental stages of the host within a patch. Those stages potentially vulnerable to attack may differ in their profitability. For **idiobionts** – parasitoids in which the host does not grow beyond the stage attacked and which therefore is a fixed 'parcel' of resource (Figure 1.9a) – small host stages may provide inadequate amounts of resource to permit the successful development of offspring. Even where successful development of idiobiont progeny is possible in small hosts, the offspring are small and therefore oviposition constitutes less of a fitness gain (in parasitoids, body size determines components of fitness such as fecundity, longevity and searching efficiency. Although they exploit a growing amount of host resource, **koinobionts** – those parasitoids that allow their hosts to continue to feed and develop (Figure 1.9b) – also may display a positive relationship between adult body size and host size, although the relationship may not be linear (Sequeira and Mackauer, 1992b; Harvey *et al.*, 1994, 1999). For both idiobionts and koinobionts, smaller hosts may require less time for handling and represent less of a risk of injury resulting from the defensive behaviour of the host (Gross, 1993). For koinobionts (most of which are endoparasitoids), small hosts may present parasitoid progeny with less of a mortality risk from encapsulation (van Alphen and Drijver, 1982: Sagarra and Vincent, 1999; section 2.10). Females of both idiobionts and koinobionts may also gain in fitness from ovipositing in or on older larvae, owing to the fact that under field conditions host mortality resulting from predation and/or intraspecific competition is more severe in early host stages than in late ones (Price, 1975). Thus, it is often the case that parasitoids prefer to attack certain host stages and even avoid or reject other stages for oviposition.

Nevertheless, for several idiobiont parasitoids of fly pupae, there is evidence that females will preferentially avoid older hosts, since these provide fewer resources for their developing offspring (e.g. King, 1998).

The distribution of hosts of different size over the host plant may influence encounter rates and thus host stage selection behaviour. Later instars of mealybugs are often surrounded by earlier instars. Encounter rates with younger instars may be higher, and those with larger ones lower than predicted, based on their densities. Young maggots of calliphorid and drosophilid fly species feed near the surface of the substrate, while older larvae may burrow deeper, possibly out of the reach of parasitoids.

Often, hosts are not passive victims of their parasitoids. Maggots and caterpillars may wriggle, or otherwise defend themselves. Mealybugs may 'flip' the posterior end of their body, or throw droplets of honeydew onto parasitoids attempting to parasitise them. Aphids may drop from the host plant, while leafhoppers and planthoppers may jump away (Gross, 1993, gives a comprehensive review of behavioural defences of hosts against parasitoids; see also subsection 1.20.3). Such behavioural defences of hosts (which are often more effective in later host

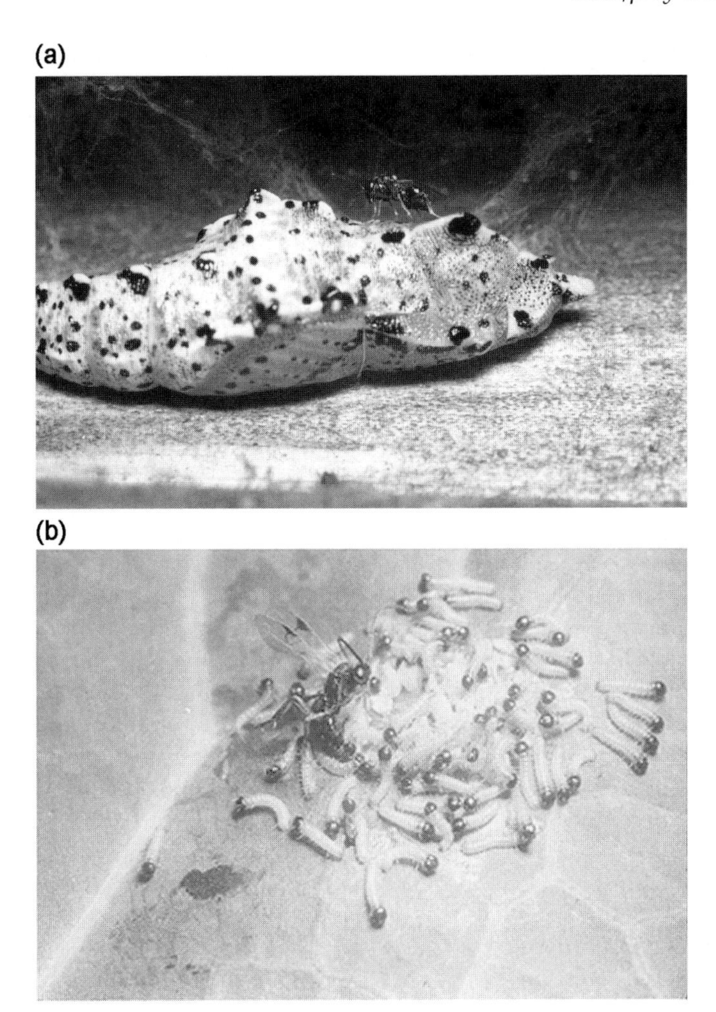

Figure 1.9 Idiobiont and koinobiont parasitoids (both gregarious) of the same host species, *Pieris brassicae* (Lepidoptera: Pieridae): **(a)** *Pteromalus puparum* (Pteromalidae) ovipositing into host's pupal stage. For the parasitoid's progeny, the pupa is a 'fixed' parcel of resource, as it is non-feeding, non-growing stage. This parasitoid is therefore an idiobiont. **(b)** *Cotesia glomerata* (Braconidae) ovipositing into newly hatched host larvae which will continue to feed, grow and develop during parasitoid development. This parasitoid is therefore a koinobiont. Source: Premaphotos, UK.

stages) can cause a problem of data interpretation. Should encounters that do not result in parasitism of the host be scored as acceptances or as rejections? If the parasitoid clearly displays behaviour that is normally associated with host acceptance, such as the turning and stinging shown by encyrtids (Figure 1.10), the encounter should be classified as an acceptance.

The ability of late stage hosts to defend themselves from attack better than early stage hosts may account for a host stage preference. The cost in terms of lost **'opportunity time'** (time that could be spent in more profitable behaviour) when attacks on late stage hosts fail, may outweigh the fitness gain per egg laid (Kouame and Mackauer, 1991). Defence against parasitoids may also incorporate an immune response. It is generally the case that the risk of encapsulation is higher in later compared with earlier host larval instars (section 2.10). This suggests that foraging endoparasitoids should preferentially attack host stages with a weaker immune response, whereas ectoparasitoids, which are not exposed in a similar manner to the host's immune response, should show a host stage/immunity-based preference. This hypothesis requires testing.

Host size selection by parasitoids is not limited to the decision of whether to oviposit or reject the host. It also involves the decision of which sex the offspring ought to be, and, for gregarious parasitoids, how many eggs to lay (Figure 1.1). For practical reasons, we analyse those decisions as isolated steps, but one should bear in mind that they are interrelated, and that it is wise to study host size selection in combination with clutch size and sex allocation decisions. Host size selection in relation to clutch size is discussed further in section 1.6.

Predators are usually less specific in their choice of prey than parasitoids, although some predators show a preference for larger prey or certain instars (Griffiths, 1982; Thompson, 1978; Cock, 1978). Prey size selection in predators may also change with the size of the predator (Griffiths, 1982).

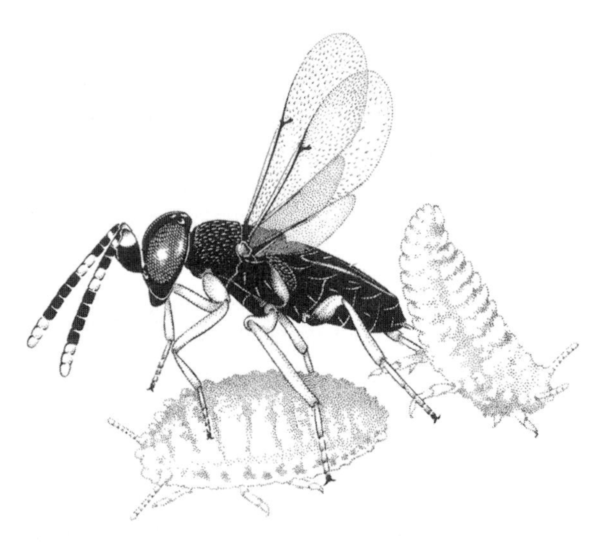

Figure 1.10 Host acceptance behaviour in the encyrtid parasitoid *Apoanagyrus lopezi*: The female examines the host with its antennae. Acceptance is indicated by the wasp turning towards the host to insert its ovipositor. Sometimes, the host escapes whilst the wasp is turning – acceptance therefore does not necessarily lead to oviposition.

Host selection decisions by one female may alter over time during an experiment because these decisions are affected by experience, egg load and stochastic variation in encounter rates. Such changes in decisions are one of the reasons why partial preferences are always found instead of the absolute, i.e. all-or-none preferences predicted by static prey choice models. If one is interested in questions such as how egg load should influence host selection, one should construct dynamic optimisation models as described by Mangel and Clark (1988. see Heimpel *et al.*, 1998, for an example).

1.6 CLUTCH SIZE

Since a host represents a limited amount of resource, and parasitoid offspring have the potential to compete for that resource (sections 2.9 and 2.10), gregarious parasitoids must make an additional decision after accepting a host for oviposition: how many eggs to lay in (or on) a host. Waage and Godfray (1985), Waage (1986) and Godfray (1987a,b) have addressed this question. Here we are mainly concerned with variation in the size of clutches allocated to hosts of a fixed size, although we shall eventually consider host size variation.

Given that the amount of resource a developing parasitoid obtains will determine its fitness, a **fitness function** $f(c)$ can be used to describe the fitness of each offspring in a clutch of size c allocated to hosts of a certain size. The fitness gain to the mother per host attacked is therefore the product of clutch size and the *per capita* fitness function, i.e. $cf(c)$. The value of c where $cf(c)$ is maximised is the parental optimum clutch size, known as the **'Lack clutch size'**, after David Lack (1947) who studied clutch size in birds. Predicted and observed fitness functions for three parasitoid species are shown in Figure 1.11. In each case, the probability of survival to the adult stage is used as the measure of fitness.

Fitness function curves can be constructed as follows:

1. By exposing hosts to individual parasitoids or and examining/dissecting some of these hosts immediately after oviposition to determine clutch size, and rearing parasitoids from the remainder to determine offspring survival (Figure 1.11a,b). If larval mortality arising from resource competition occurs late in development, and dead larvae are not consumed by surviving larvae, one may simply record the numbers of emerged and unemerged offspring (Figure 1.11c);
2. By manipulating parasitoid clutch sizes. This is relatively easy in the case of ectoparasitoids, as different clutch sizes can be obtained simply by adding or removing eggs, manually, from clutches present on the host's body surface (Hardy *et al.*, 1992b; Zaviezo and Mills, 2000). With this technique, however, there is a risk of damaging eggs during manipulation. With endoparasitoids, clutch sizes can be manipulated by interrupting oviposition, by allowing superparasitism to occur, or by exchanging the host for one of a different size after the wasp has examined it but immediately before it has the opportunity to begin ovipositing in it (Klomp and Teerink, 1962). However, a major problem with at least the latter technique is that the parasitoid may alter its sex allocation behaviour. The final egg clutch sex ratio could influence progeny fitness and therefore the optimum clutch size (Waage and Ng, 1984).

Other models predict that the best strategy for a parasitoid is to maximise fitness per unit time rather than per host attacked. If there is a cost in time to laying an egg, it may benefit a female to cease adding more eggs to a host and to allocate the time saved to locating a new host. The fitness gain from leaving hosts and searching for new ones will increase as the travel time between oviposition sites decreases, i.e. as host availability increases. As hosts become more abundant, females should leave each host sooner, i.e. produce *smaller clutch sizes*. Thus, with the maximisation of fitness per unit time models, females maximise fitness per host attacked (i.e. produce Lack clutch sizes) only when hosts are scarce. *Trichogramma minutum* appears to be a species whose strategy is to maximise fitness per unit time. Schmidt and Smith (1987a) presented females with nine host eggs attached by glue to a cardboard base and employed various egg spacing treatments: the eggs were situated with their centres either 2, 3, 4 or 5 millimetres apart on a grid. Clutch size was found to decrease with increased crowding of eggs, i.e. increasing host density per unit area.

There are also models that take into account egg-limitation constraints, i.e. they assume that the parasitoid has a limited number of eggs to lay at any one time. Such a parasitoid is always in a position where available eggs are fewer than potential clutch sites. If eggs are severely limiting, i.e. egg load is much smaller than the number of hosts available (this could be due to the fact that the female has laid most of her eggs, e.g. Heimpel *et al.*, 1998), a female should spread out her eggs between hosts so that the fitness gain *per egg*, rather than per clutch, is maximised. When *per capita* fitness of offspring decreases monotonically (as in Figure 1.11), the optimal clutch size under severe egg-limitation is always one. See Zaviezo and Mills, (2000) for the effects of female life expectancy on optimal clutch size.

What if data and model do not match? As can be seen from Figure 1.11, clutch sizes predicted by optimality models tend to differ from the ones recorded in experiments. This discrepancy may occur because:

1. The wrong category of model has been used. For example, the parasitoid's strategy may be that of maximising fitness per unit time rather than per host attacked;
2. The model does not take account of stochastic variability in certain parameters (Godfray and Ives, 1988);
3. The measure of fitness (e.g. offspring survival) used may be inappropriate.

If fitness is measured instead by total offspring fecundity, a closer fit between model and data may be obtained (Waage and Ng, 1984; Waage and Godfray, 1985). Measuring offspring

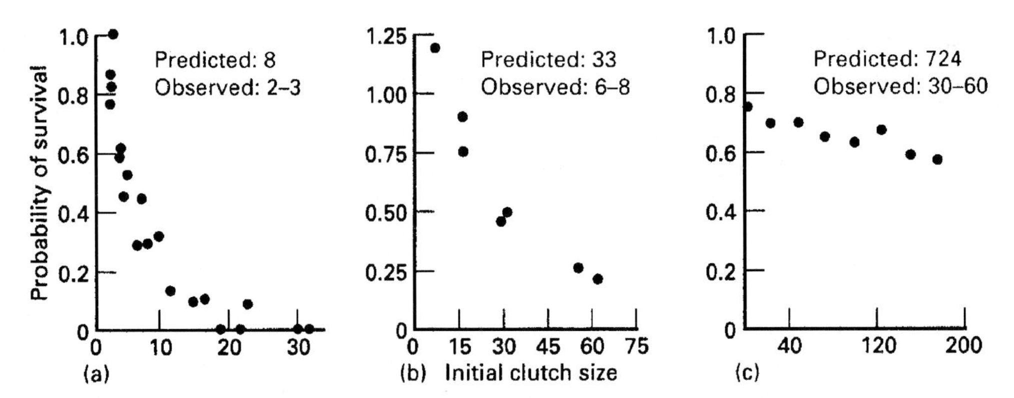

Figure 1.11 Optimal progeny allocation in gregarious parasitoids – clutch size: The *per capita* fitness of offspring as a function of clutch size, estimated by the probability of survival in initial clutches of different sizes (Observed = observed clutch size; Predicted = predicted by calculation of *cf*(*c*) (see text). (a) *Trichogramma evanescens* in eggs of the cabbage moth, *Mamestra brassicae*; (b) *Telenomus farai* in eggs of the bug *Triatoma phyllosoma pallidipennis* (the overestimate of survival in this case is attributable to sampling error); (c) *Dahlbominus fuliginosus* on pupae of the sawfly *Neodiprion lecontei*. Source: Waage and Godfray (1985) and Waage (1986), who used data from Pallewata (1986), Escalante and Rabinovich (1979) and Wilkes (1963). In all three cases, there is a continuous decline in *per capita* fitness with increasing clutch size. For some other gregarious parasitoid species there is evidence of an Allee effect, i.e. an initial rise then a fall in fitness. Such a dome-shaped fitness relationship may prove to be common among gregarious endoparasitoids, because in such parasitoids small larval broods often perish entirely due to their inability either to overcome host physiological defences or to consume all the host tissues (a prequisite in some species for successful pupation and emergence). Reproduced by permission of Blackwell Publishing and Elsevier Science.

fecundity is likely to prove very time-consuming, so an alternative procedure is to measure offspring body size or weight; both of these factors are usually good predictors of fecundity in parasitoids (subsection 2.7.3). Release-recapture experiments with different size classes of parasitoids in the field may provide useful information on size–fitness relationships, although ideally these relationships should be directly measured. Le Masurier (1991) used a combined measure of fitness – the product of progeny survival and the calculated mean egg load at emergence (a measure of lifetime fecundity) of the surviving female progeny. The egg load for each emerging wasp was determined indirectly, from a regression equation relating egg load to head width. Interestingly, le Masurier found that the fitness function curve he constructed for a British population of *Cotesia glomerata* in larvae of *Pieris brassicae* showed no density-dependent effect of clutch size on fitness, and this therefore prevented him from calculating the optimum clutch size for that host. All he could predict was that females should lay at least the maximum number of eggs he recorded in a host.

It has become increasingly apparent that measurements of size-fitness relationships are strongly influenced by environmental variation (Rivero and West, 2002). In addition, laboratory studies tend to underestimate the disadvantages of small body size in parasitoid wasps (Hardy *et al.* 1992b). Therefore the assumption that there is a general size-fitness correlation in parasitoids needs to be treated with a little caution.

A further factor to consider is variation in host size. If a gregarious parasitoid's strategy is that of maximising fitness per host attacked, then the optimal clutch size ought to increase with increasing host size. How do gregarious parasitoids (and solitary parasitoids for that matter) measure host size? Schmidt and Smith (1985) studied host size measurement in *Trichogramma minutum*. Females allocated fewer progeny to host eggs that were partially embedded in the substratum than into host eggs that were fully exposed. Since the eggs were of identical diameter and surface chemistry, it was concluded that the mechanism of host size determination is neither chemosensory nor visual, but is essentially mechanosensory, based on accessible surface area. Schmidt and Smith (1987b) subsequently observed the behaviour of individual *T. minutum*, during the host examination phase, on spherical host eggs of a set size, and recorded: (a) the frequency of and intervals between contacts with the substratum bearing the eggs, and turns made by the wasps, and (b) the number of eggs laid per host. In analysing the data, seven variables were considered: the total number of substratum contacts, the mean interval between such contacts, the interval between the last contact and oviposition, the longest and shortest interval between contacts, the total interval between the first three contacts, and the interval between the first contact with the host and the first contact with the substratum (*initial transit*). Of these, only the duration of the initial transit across the host surface showed a significant linear relationship with the number of eggs deposited. As the duration of the wasp's initial transit increases, more eggs are laid. By interrupting the path of wasps during their initial transit, and thereby reducing their initial transit time, Schmidt and Smith (1987b) succeeded in reducing the number of progeny laid by a female. Schmidt and Smith concluded that wasps are able to alter progeny allocation by measuring short time intervals. Interestingly, the duration of initial transit was found to be the same for both large and small wasps (Schmidt and Smith, 1987b, 1989).

Large-bodied gregarious parasitoids (and solitary parasitoids, for that matter) are likely to measure host size in other ways, for example by determining whether the tips of the antennae reach certain points on the host's body. Such stimuli are thought to be tactile (e.g. King, 1998). Alternately, simple visual examination of the whole host may provide the correct cues.

1.7 HOST-FEEDING

The females of many synovigenic parasitoids (subsections 1.16.2 and 2.3.4) not only parasitise hosts but also feed on them (Jervis, 1998; Jervis and Kidd, 1986, 1999; Heimpel and Collier, 1996; Ueno, 1998a,b, 1999a,b,c). **Host-feeding** supplies the females with materials for continued egg production and for somatic maintenance (Bartlett, 1964; Jervis and Kidd, 1986, 1999). Giron *et al.* (2002) showed that the parasitoid *Eupelmus vuilletti* host fed upon the host's haemolymph. The haemolymph is rich in proteins and various sugars, and it is these sugars that are responsible for the increased longevity of *E. vuilletti* which have been allowed to host feed. In some parasitoid species, host feeding causes the host to die (so-called 'destructive' host-feeding), so rendering it unsuitable for oviposition. Even with those species that remove small quantities of host materials such that the host survives feeding ('non-destructive' host-feeding), the nutritional value of the host for parasitoid offspring may, as a result of feeding, be reduced and the female may lay fewer (gregarious species), or no eggs in it. For example, lepidopteran hosts previously host-fed upon by *Pimpla nipponica* produced fewer and smaller wasps when subsequently parasitised (Ueno, 1997). Thus, while host feeding potentially increases future fitness via subsequently increased egg production, the fitness gain is at the cost of current reproduction.

Most authors have supposed that host feeding has a short-term effect on parasitoid fecundity. However, by using radioactively labelled amino acids, Rivero and Casas (1999) showed that a significant proportion of the resources gained by the aphelinid *Aphytis melinus*, a parasitoid of scale insects, were stored and used gradually throughout the life of the wasp. Such techniques are particularly under-utilised in studies of parasitoid behaviour (see section 2.13).

A general prediction of models of destructive host-feeding behaviour is that the fraction of hosts fed upon by females should increase with decreasing host availability, at least over the upper range of host densities (Figure 1.12; Jervis and Kidd, 1986; Chan and Godfray, 1993) – a prediction borne out by the few empirical studies that have been carried out to date (DeBach, 1943; Bartlett, 1964; Collins *et al.*, 1981; Bai and

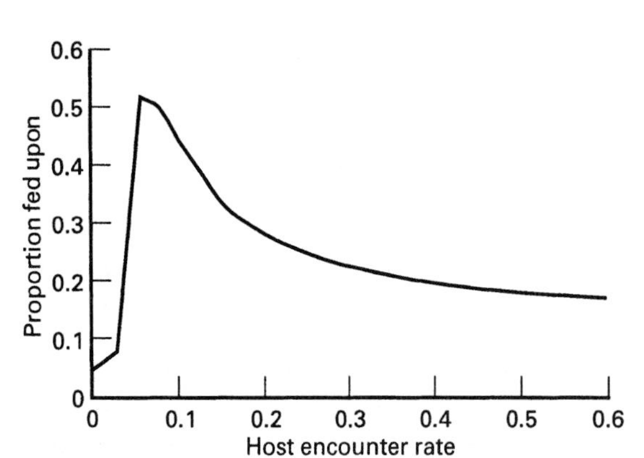

Figure 1.12 Host-feeding by parasitoids. Most models predict that the relationship between the fraction of hosts fed upon and host availability should be dome-shaped, increasing at low levels of host availability and decreasing at moderate to high levels. [The monotonic decline over the mid- to high range is supported empirically. e.g. see Sahragard *et al.* (1991).] The functional explanation for the small fraction of hosts fed upon at low host encounter rates is that the female adopts a 'cutting of losses' tactic – the encounter rate is too low to meet (*via* host-feeding) the wasp's energy requirements, and so the female oviposits in every host encountered (Jervis and Kidd, 1986). Models also predict that host-feeding is more likely when nutrient reserves and/or gut contents are at or below a critical level. Low nutrient levels and and low gut contents presumably warn of the impending risk of starvation and/or egg-limitation. In general, the critical level of nutrient reserves/gut contents depends on the current egg load and *vice versa* (Heimpel and Collier, 1996).

Mackauer, 1990; Sahragard *et al.*, 1991; Thu and Ueno, 2002).

Given that the fitness gains from ovipositing may vary in relation to host stage (subsection 1.5.7), it is likely that the decision either to host-feed or to oviposit also depends on host stage (Kidd and Jervis, 1991; Rosenheim and Rosen, 1992). Indeed, observational and experimental studies of destructively host-feeding parasitoids show in them a tendency to feed preferentially or exclusively on earlier host stages and to oviposit preferentially or exclusively on/in later ones (Kidd and Jervis, 1991; Rosenheim and Rosen, 1992). A similar relationship is likely to apply to different-sized hosts of the same developmental stage. Furthermore, environmental factors such as temperature may influence rates of host-feeding (Urbaneja *et al.*, 2001b). For example, the egg parasitoid *Trichogramma turkestanica* host-feeds on *Ephestia kuehniella* at a greater rate when reared at lower temperatures, although it is not clear why this is so (Hansen and Jensen, 2002).

Models predict that the decision to host-feed versus oviposit depends on the parasitoid's egg load: host feeding is more likely when egg load is low (Chan and Godfray, 1993; McGregor, 1997; see Heimpel and Collier, 1996, and Jervis and Kidd, 1999, for reviews). Rosenheim and Rosen (1992) tested this prediction experimentally using the scale insect parasitoid *Aphytis lingnanensis*. Egg load was manipulated by using wasps of different sizes (egg load being a function of body size) and also by holding parasitoids, prior to their exposure to hosts, at different temperatures (the rate of oöcyte maturation and therefore the rate of accumulation of mature eggs in the ovaries being a function of temperature, subsections 2.3.4 and 2.7.4). Manipulating egg load in this way ensured that previous history of host contact could be eliminated as a possible confounding variable. Alternative methods of manipulating egg load, e.g. depriving parasitoids of hosts or allowing them to oviposit, do not separate the effects of egg load and experience. Rosenheim and Rosen found in their experiments that egg load did not significantly affect the decision to host-feed or oviposit on (small) hosts.

However, although egg load was not directly manipulated, more recent work does support the hypothesis that the likelihood of host feeding is related to egg load (e.g. see Ueno, 1999b, Heimpel and Rosenheim, 1995 and Heimpel et al. 1996).

The decision whether to host feed or oviposit may also depend on the wasp's nutritional state, see Heimpel and Collier (1996).

1.8 HOST DISCRIMINATION

1.8.1 INTRODUCTION

Salt (1932) was the first researcher to clearly demonstrate the ability of a parasitoid to discriminate between hosts that contain the egg of a conspecific and hosts that have not been parasitised, and later (Salt, 1961) he showed that this ability – now known in the literature as **host discrimination** – occurs in the major families of parasitoid Hymenoptera.

It is now understood that females of some parasitoid species are able to discriminate between:

1. Parasitised hosts and unparasitised hosts (numerous published studies have shown this, although the conclusions drawn in some are questionable, see below);
2. Parasitised hosts containing different numbers of eggs (Bakker *et al.*, 1990);
3. Hosts containing an egg of a conspecific from one containing their own egg.

Notwithstanding such sophisticated abilities, **superparasitism** – the laying of an egg in an already parasitised host – is a common phenomenon among insect parasitoids. The occurrence of superparasitism or, expressed statistically, the occurrence of a random egg distribution among hosts, has often led to the erroneous conclusion that a parasitoid is unable to discriminate between parasitised and unparasitised hosts (Hemerik and van der Hoeven, 2003).

Dipteran parasitoids rarely show host discrimination abilities, primarily as the females of many never come into contact with potential hosts, instead often relying on host-seeking larvae (reviewed in Feener and Brown, 1997; but see Lopez *et al.*, 1995).

The effects of superparasitism on progeny development and survival are discussed in Chapter 2 (subsections 2.9.2 and 2.10.2)

1.8.2 INDIRECT METHODS

There are two approaches to determining whether parasitoids are able to discriminate between parasitised and unparasitised hosts. One is to dissect hosts (subsection 2.6) and calculate whether the recorded egg distribution deviates significantly from a Poisson (i.e. random) distribution (Salt, 1961). Van Lenteren *et al.* (1978) have shown that such a procedure is not without pitfalls. They point out that if the method is applied to egg distributions from hosts collected in the field, there is a risk that mixtures of samples with regular (i.e. non-random) egg distributions but different means may add up to produce a random distribution (Figure 1.13). This is one of the reasons why a random egg distribution does not constitute proof of the inability to discriminate. Another problem van Lenteren *et al.* (1978) identified concerning the analysis of egg distributions is that with gregarious parasitoids the distribution of eggs depends not only upon the number of ovipositions but also on the number of eggs laid per oviposition.

There are further problems associated with the use of egg distributions. Van Alphen and Nell (1982) recorded random egg distributions when single females of *Asobara tabida* were placed with 32 hosts for 24 hours. Because not all replicates produced random distributions and because other experiments had unequivocally shown that females of this species are able to discriminate between parasitised and unparasitised hosts, the random egg distributions could not be explained by a lack of discriminative ability.

In van Alphen and Nell's experiments the replicates with a high mean number of eggs had random distributions, whereas replicates with lower means had regular ones. It was therefore concluded that *A. tabida* discriminates between unparasitised and parasitised hosts, but is unable to assess whether one or more eggs are present in a parasitised host. Egg distributions are a mixture of the regularly distributed first eggs laid in hosts and of the randomly distributed supernumerary eggs. At lower means, the contribution of the regular distribution of the first eggs is not masked by the random distribution of the supernumerary eggs, whereas at higher means it is.

Even when egg distributions more regular than a Poisson distribution are found, one cannot establish with certainty that a parasitoid is able to discriminate between parasitised and unparasitised hosts. The recorded egg distribution could result from parasitised hosts having a much

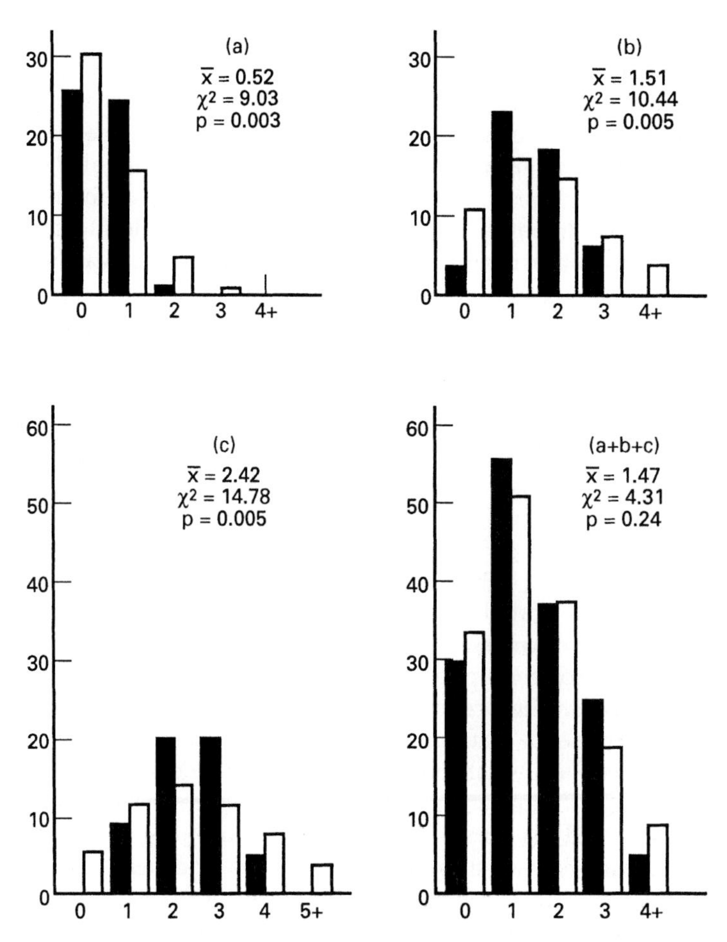

Figure 1.13 Host discrimination by parasitoids: Egg distributions for *Leptopilina heterotoma* (=*Pseudeucoila bochei*) parasitising *Drosophila melanogaster*. Three groups (a,b,c) of around fifty host larvae were presented to female wasps, and the hosts subsequently dissected and examined for wasp eggs. The mean number of eggs recovered per host larva was different in each case. Although in all three cases superparasitism occurred, when the egg distribution was compared with the distribution that would have been obtained had the wasps been ovipositing at random (i.e. a Poisson distribution), the egg distribution was found to be more regular, indicating that the parasitoids discriminate. However, if data from all three distributions are pooled (a + b + c), a distribution is obtained that is indistinguishable from a Poisson distribution – a result that would lead to the erroneous conclusion that the parasitoid species studied cannot discriminate. Source: van Lenteren *et al.* (1978), reproduced by permission of Blackwell Publishing.

lower probability of being encountered, either because they move less than healthy hosts or because they leave the host plant. It is also possible that encounter rates with parasitised hosts are lower because the parasitoid does not re-visit previously searched areas with the same probability, e.g. when it always walks upwards along branches or when it marks areas already visited and avoids re-searching such areas.

The previous examples show that there are major pitfalls associated with using egg distributions to determine whether a parasitoid can discriminate between parasitised and unparasitised hosts. Other components of the behaviour of the parasitoid, or of the behaviour of the hosts, can influence egg distributions. Moreover, a regular egg distribution with a mean number of eggs much greater than one requires more than just an ability to discriminate between parasitised and unparasitised hosts. This has already been illustrated in the above-mentioned example of *A. tabida* where no regular egg distributions are found. The following example illustrates how, in *Leptopilina heterotoma*, different mechanisms are responsible for egg distributions tending to be regular even at a high mean number of eggs per host (Bakker *et al.*, 1972). One explanation for this phenomenon is that *L. heterotoma* is able to discriminate between hosts containing different numbers of eggs. There is, however, a second possible interpretation: when the parasitoid is able to distinguish hosts containing an egg of her own from those containing eggs of conspecifics and avoids ovipositing in the former, regular egg distributions would result. Therefore, it is impossible to decide, based on egg distributions alone, whether a parasitoid is able to assess the number of eggs already present in a host.

Experiments therefore need to be carried out in which a parasitoid female is offered a choice of hosts containing different numbers of eggs, all laid by other (conspecific) females. Bakker *et al.* (1990) offered hosts containing two eggs and hosts containing one egg of other females to individual *L. heterotoma*. The wasps oviposited significantly more often in hosts containing a single egg, thus showing that *L. heterotoma* is indeed able to distinguish between hosts containing different egg numbers. Visser (1992) showed that *L. heterotoma* females are also able to recognise hosts containing their own eggs. Thus, both of the above mechanisms may have contributed to the regular egg distributions found by Bakker *et al.* (1972).

It is thus clear that, by comparing observed egg distributions with those predicted by a Poisson distribution, one can neither conclude that a parasitoid is able to discriminate between parasitised and unparasitised hosts, nor conclude that it lacks this ability. This does not mean a statistical analysis of egg distributions is useless; it is possible to construct models predicting distributions of eggs for parasitoids having different abilities to avoid superparasitism (e.g. discriminating between healthy hosts and parasitised hosts and counting, discriminating but not counting, discriminating between hosts parasitised by self and hosts parasitised by others), and to compare the theoretical egg distributions with distributions recorded in experiments. Bakker *et al.* (1972) and Meelis (1982) adopted this approach when investigating whether wasps are able to assess the number of eggs already laid in a host. These authors assumed that parasitoids search randomly, and that there exists a certain probability that the wasp will lay an egg when it encounters a larva. This probability is 1.0 at the first encounter, but is lower at subsequent encounters. By keeping the probability of oviposition at the subsequent encounters constant, the model could be used to describe superparasitism by *A. tabida*.

1.8.3 DIRECT OBSERVATIONS OF BEHAVIOUR

The other method of determining whether parasitoids are able to discriminate between parasitised and unparasitised hosts involves observing the insects, and recording and comparing encounters resulting in oviposition and rejection of the different host categories. This method provides behavioural evidence that the parasitoid under study rejects parasitised hosts more often than unparasitised hosts. It is, however, wise to use other behavioural criteria in addition to acceptance/encounter ratios; differences in behaviour may also be found in patch time allocation, progeny (i.e. clutch size) and sex allocation (van Alphen *et al.*, 1987).

Because distributions of parasitoid eggs among hosts potentially have an important effect on parasitoid-host population dynamics, one requires a good statistical description of those distributions, for incorporation into population models. We prefer to use the observed behaviour as a basis for a model calculating egg distributions, instead of inferring the underlying behaviour from an analysis of the egg distributions.

An intriguing question is why it took so long before evidence was found of host discrimination by dipteran parasitoids. Host discrimination by hymenopteran parasitoids was discovered long ago, but the phenomenon was only recently described for tachinid flies (Lopez *et al.*, 1995). In field and laboratory experiments, Lopez and colleagues showed that *Myiopharus doryphorae* and *M. aberrans*, both parasitoids of Colorado Beetle (*Leptinotarsa decemlineata*) larvae, almost always reject parasitised larvae, whereas they readily oviposit in unparasitised larvae. While little is known of the host discrimination ability of non-hymenopteran parasitoids, even less is understood of the situation of where it is the parasitoid larva, rather than the ovipositing female, that actively seeks hosts. Larvae of the staphylinid parasitoid *Aleochara bilineata* locate and attack fly pupae. Royer *et al.* (1999) found that these larvae can distinguish hosts that were self-parasitised from those that were attacked by conspecifics, and that this was based on chemical cues. Superparasitism was more common when hosts were scarce, and if given a choice, *A. bilineata* larvae would preferentially attack hosts that contained the related species *A. bipustulata*, rather than conspecifics (Royer *et al.*, 1999).

Edwards and Hopper (1999) took a novel approach to investigate levels of superparasitism by *Macrocentrus cingulum*, a braconid parasitoid of the European corn-borer, *Ostrinia nubilalis*. Since *M. cingulum* is polyembryonic, the number of parasitoid larvae present per host does not reflect the number of females that have attacked that host. By using random amplified polymorphic DNA (RAPD) markers, the authors were able to identify the number of females that had oviposited.

1.8.4 SUPERPARASITISM

Many, if not all, parasitoids are able to discriminate between parasitised and unparasitised hosts, but superparasitism is a common feature in nature (van Alphen and Visser, 1990; Godfray, 1994), posing the question: 'why and when should parasitoids superparasitise?'

Van Lenteren (1976) addressed this question from the standpoint of causation. He assumed that superparasitism was caused by a failure to discriminate. He found that females of *L. heterotoma* inexperienced with unparasitised hosts readily oviposited in already parasitised hosts but avoided ovipositing in parasitised hosts after they had been able to oviposit in unparasitised ones. He concluded from this that parasitoids superparasitise because they are unable to discriminate between parasitised and unparasitised hosts until they have experienced oviposition in unparasitised hosts. A similar conclusion was drawn by Klomp *et al.* (1980) for *Trichogramma embryophagum*.

A functional approach to the problem is to ask whether it is adaptive for a parasitoid always to avoid superparasitism. Van Alphen *et al.* (1987) re-analysed the data of van Lenteren (1976) and Klomp *et al.* (1980), starting with the hypothesis that superparasitism can be adaptive under certain conditions. They reasoned that host discrimination is an ability which the parasitoid can use to decide either to reject a parasitised host or to superparasitise it, depending on the circumstances, i.e. superparasitism is not the result of an inability to discriminate. Van Alphen *et al.* (1987) argued that an inexperienced female arriving on a patch containing only parasitised hosts should superparasitise, because the probability of finding a better patch elsewhere is low. In a similar vein, Sirot *et al.* (1997) showed through modelling that the tendency to superparasitise should vary with egg load and life expectancy.

Van Lenteren's (1976) inexperienced wasps not only rejected parasitised hosts more often than unparasitised ones but also encountered significantly fewer hosts in experiments involving patches containing only parasitised hosts, compared with similar experiments involving patches containing the same density of unparasitised hosts. It was known that *Leptopilina heterotoma* females search by stabbing with the ovipositor, twice per second, in the substrate, and it was possible to measure both the surface area of a host and that of the patches. It was possible therefore to calculate, from the numbers of encounters observed during a 30 minute observation period, that inexperienced wasps spent on average 13.12 minutes searching and handling hosts when introduced on to a patch with parasitised hosts, whereas they spent on average 2.14 minutes when introduced on to patches with unparasitised hosts. It is unclear how such a difference could have escaped the attention of the observer! Van Alphen *et al.* (1987) interpreted the differences in behaviour between inexperienced wasps and experienced wasps as evidence that inexperienced wasps do recognise parasitised hosts. Thus, van Lenteren's (1976) data do not support the conclusion that host discrimination needs to be learnt.

Experiments by van Alphen *et al.* (1987), involving *L. heterotoma* and *Trichogramma evanescens*, confirmed that females inexperienced with unparasitised hosts are, like experienced wasps, already able to discriminate, although inexperienced females superparasitise more frequently. This example shows that alternative hypotheses can be overlooked if one asks only causal questions.

Static and dynamic optimality models as well as ESS models (subsection 1.2.2) have been published (Iwasa *et al.*, 1984; Parker and Courtney, 1984; Charnov and Skinner, 1985; Hubbard *et al.*, 1987; van der Hoeven and Hemerik, 1990; Visser *et al.*, 1990; Field and Keller, 1999; Hemerik *et al.*, 2002), showing that superparasitism is often adaptive. The models predict that oviposition in already parasitised hosts, though resulting in fewer offspring than ovipositions in unparasitised hosts, may still be the better option when either there is no time available to search for and locate unparasitised hosts or when unparasitised hosts are simply not available. By ovipositing into an already parasitised host under such conditions, a female may increase her fitness if there is a finite chance that her progeny will out-compete the other progeny (see subsection 2.10.2). Experimental tests of some of these models have shown that parasitoids behave in such a way that the models' predictions are at least met qualitatively (Hubbard *et al.*, 1987; Visser *et al.*, 1990; van Alphen *et al.*, 1992). For example, Sirot *et al.* (1997) tested predictions that superparasitism by *Venturia canescens* would be less common if females were provided with food, reducing their risk of mortality. As predicted, superparasitism rates were correlated with egg load and previous access to (non-host) food.

Attacking previously parasitized hosts is evidently adaptive if females can kill parasitoid eggs or young larvae present in the host. As yet, there are few good examples of this in the literature. *Encarsia formosa* can kill eggs present in hosts by grabbing them with her ovipositor (Netting and Hunter, 2000). A similar effect is seen with *Haplogonatopus atratus* (Dryinidae), where the female wasp will kill parasitoid larvae present in the host before ovipositing (Yamada and Kitashiro, 2002).

Female parasitoids can often discriminate between hosts that have been self-parasitised from those that have been attacked by a conspecific. *Venturia canescens* females will avoid superparasitising hosts that contain their own progeny, a behaviour mediated by the presence of a marking pheromone (Hubbard *et al.*, 1987). Such ability to discriminate among hosts led to the suggestion that females would increase their inclusive fitness by avoiding hosts that contain kin (Fellowes, 1998), and indeed, female *V. canescens* will avoid attacking hosts containing relatives (Marris *et al.*, 1996). However, *V. canescens* is parthenogenetic, and this may be an example of extended self-recognition, rather than kin discrimination. Ueno (1994) studied the behaviour of *Itoplectis narayanae*, and found that whereas females would avoid parasitising hosts they had previously attacked, there was no difference in their likelihood of attacking hosts that contained kin or unrelated conspecifics.

While in the examples above the females recognise their own odour marks, others distinguish between self-parasitized hosts and those attacked by conspecifics by different means. Ueno and Tanaka (1996) found that *Pimpla nipponica* females do not deposit chemical markers, but instead use visual location cues to avoid self-superparasitising.

Avoidance of self-superparasitism may be one reason that patches are incompletely exploited (e.g. Outreman *et al.*, 2001).

With *Venturia canescens*, the likelihood of avoiding superparasitism increases in the

20 minutes after oviposition if the females have been provided with alternative hosts during the interval, but this does not occur if the female is deprived of other hosts. This suggests that the females can rapidly obtain information on the number of hosts in the patch, and this influences their decision to superparasitise (Hubbard *et al.*, 1999). *Anaphes victus*, a mymarid parasitoid of curculionid beetle eggs, can learn to avoid marked hosts in 4 hours, and are quicker to learn if the mark was made by a close relative (van Baaren and Boivin, 1998).

Just where the oviposition deterrent marker originates is unclear, although it is usually suggested that it originates from the female's Dufour's gland. The pteromalid *Dinarmus basalis* avoids superparasitising hosts that have been attacked over 20 hours previously. Gauthier and Monge (1999) found that the marker originated from the parasitoid egg, and required contact between the egg and the host for at least 4 hours before the deterrent effect became evident. However, with *Leptopilina boulardi* and *Asobara tabida*, parasitoids of drosophilids on fermenting substrates, the mark spreads within the host within about a minute (van Alphen and Nell, 1992).

Experience is often important in determining whether a female will superparasitise a potential host. Naïve *Cotesia flavipes* females will readily attack hosts that contain a conspecific, yet experienced females will reject such hosts. This discrimination is influenced by the presence of a patch-marking odour (Potting *et al.*, 1997). Nufio and Papaj (2001) review patch-marking behaviour in parasitoids.

When parasitoids attack a host that already contains a developing conspecific, the likelihood is that the larvae will fight to the death for ownership of the resource. While it may be expected that older larvae will have a competitive advantage, this does not appear to be the case with *Venturia canescens*, where first instar larvae are more likely to kill older larvae in the same host (Marris and Casperd, 1996). This result appears to explain why the level of superparasitism by *V. canescens* females is higher the longer the period of time that has elapsed since the host was first attached.

1.8.5 MULTIPARASITISM

Multiparasitism (oviposition in a host attacked by heterospecifics) has been less studied than superparasitism. In general, it is thought that the ability to identify hosts attacked by other species is less frequent than discrimination against hosts attacked by conspecifics. There are two main situations where females should discriminate against hosts containing a heterospecific egg or larva. First, competitively inferior species should avoid attacking hosts where a superior competitor has already oviposited (subsection 2.10.2). Second, where the outcome of competition depends upon the time since the host was initially attacked (subsection 2.10.2), the multiparasitising female should be able to detect this factor and incorporate it when making the decision of whether to parasitise or not.

Ueno (1999c) tested this latter prediction, using *Pimpla nipponica* and *Itoplectis naranyae*, two solitary parasitoids of moth larvae. When presented with *Galleria mellonella* larvae, both species preferred attacking previously unattacked hosts when the time since parasitism of the host by the heterospecific parasitoid was over 48 hours. However, if less than 24 hours had passed since the initial attack, then no such preference was shown. How the parasitoids can distinguish the time since the initial parasitism is not known. Bokonon-Ganta *et al.* (1996) found that competitively inferior species do not always avoid ovipositing in hosts previously attacked by a competitor. *Gyranusoidea tebygi*, a parasitoid of the mango mealybug, *Rastrococcus invadens*, will readily accept hosts that have previously been attacked by *Anagyrus mangicola*, although their offspring generally fail to survive.

1.8.6 CANNIBALISM

In many ways, cannibalism by predators can be considered analogous to superparasitism. Cannibalism is a common feature of the behaviour of many predatory insects and is probably a consequence of polyphagy (New, 1991; Dostalkova *et al.*, 2002).

Consuming unrelated conspecifics will have two main benefits (Polis, 1981; Elgar and Crespi, 1992;

Anthony, 2003). First, when resources are scarce, the added nourishment gained will increase the survival chances of the cannibal (e.g. the green lacewing *Chrysoperla carnea*, Duelli, 1981). Second, potential competitors are removed from the patch. When alternative resources are common, then it is unlikely that consuming relatives will be beneficial, but when resources are limiting it may be better to eat kin so that some individuals survive, rather than sacrificing all (Fellowes, 1998).

Cannibalism has been most intensively studied in the Coccinellidae, where some species can complete their larval development on conspecific eggs (Dimitry, 1974). *Adalia bipunctata* will frequently consume conspecifics (Hodek and Honěk, 1996), although adult females and young larvae will avoid their own and sibling eggs, respectively (Agarwala and Dixon, 1993). Males that fathered the eggs do not show any such discrimination. In many non-social insects, such avoidance would be explained by environmental cues, rather than through direct genetic cues (Fellowes, 1998). Joseph *et al.* (1999) investigated these cues using the ladybird *Harmonia axyridis*. Third instar *H. axyridis* larvae avoid cannibalising kin, and when they do cannibalise them, they take longer to attack kin than non-kin. Their results suggest that environmental cues are unimportant, with discrimination linked to genetic differences among the individuals (Joseph *et al.*, 1999).

Given that there will be heterogeneity in habitat quality, it is perhaps unsurprising that there is heritable variation in cannibalistic behaviour in *H. axyridis* (Wagner *et al.*, 1999). When conditions are favourable, cannibalism is maladaptive given that foraging larvae are more likely to encounter kin. However, in unfavourable patches, increased propensity to cannibalism will increase the development rate and survivorship of the cannibal (Wagner *et al.*, 1999).

1.8.7 INTRAGUILD PREDATION

If we consider that cannibalism is analogous to superparasitism, then it is reasonable to compare intraguild predation to multiparasitism. **Intraguild predation** is a combination of predation and competition, and occurs when two predators share a common prey species, but one (or both) of the predators will also attack the other. Such interactions are likely to be common, with many adult predators attacking the eggs and larval stages of other species, as well as their own (see Rosenheim *et al.*, 1995). For example, the anthocorid bug *Orius laevigatus* is frequently used to control the thrips *Frankliniella occidentalis*, a pest of many greenhouse crops. Phytoseiid mites, such as *Neoseiulus cucumeris*, are also used in thrips control. Wittmann and Leather (1997) found that due to intraguild predation by *O. laevigatus* on *N. cucumeris*, the use of both agents together was unlikely to increase the degree of control. However, *O. laevigatus* does not prey upon another predatory mite (*Iphiseius degenerans*), making a pairing much more suitable for *F. occidentalis* control (Wittmann and Leather, 1997).

1.9 SEX ALLOCATION

1.9.1 INTRODUCTION

Haplodiploidy (the production of haploid males from unfertilised eggs and diploid females from fertilised eggs; also known as arrhenotokous parthenogenesis, subsection 3.3.2) allows female wasps to determine the sex of their offspring. Such control of sex allocation has made the parasitic Hymenoptera a favoured subject for behavioural ecologists (Godfray, 1994; Godfray and Shimada, 1999). Fisher (1930) was the first to show that natural selection will favour equal investment in the sexes in a panmictic population. The Hymenoptera, however, frequently show sex ratios that are strongly divergent from equality, and explaining these differences has resulted in robust models that receive strong support from empirical investigations. In general, patterns of sex allocation are influenced by two main factors: the population's mating structure and the environmental conditions experienced. While both will select for the ability to maximise fitness through manipulation of offspring sex ratio, the patterns of sex allocation they influence are quite different.

1.9.2 LOCAL MATE COMPETITION

The first broad pattern that needs to be explained is a cross-species one: sex ratios (usually expressed as the proportion of the progeny that are male) can vary from highly female-biased to equality, or much more rarely, become male-biased. For example, Bernal *et al.* (1998) reported that mated female *Coccophagus semicircularis* produce strongly female-biased sex ratios; the *Drosophila* parasitoid *Leptopilina heterotoma* has a sex ratio near equality, whereas the closely related species *L. boulardi* has a male-biased sex ratio (Fauvergue *et al.*, 1999).

This variation has been explained by the **Theory of Local Mate Competition** (Hamilton, 1967). Under conditions of local mate competition **(LMC)** ovipositing females are predicted to lay an increasingly female-biased offspring sex ratio as the likelihood of sib-mating increases. Imagine a patch where only one female oviposits. With an unbiased sex ratio, the males will compete with each other for matings with their sisters. The ovipositing female can increase her fitness by increasing the number of female offspring in her brood, to the point where only enough males are required to ensure that all females in the brood are mated. Biasing the sex ratio in this manner will have two advantages (Antolin, 1993; Ode *et al.*, 1998). First, competition between sons for access to mates will be lower. Second, the number of available mates for each male will be increased. Conditions suitable for LMC are most likely to be met when patches are discreet (the classic example is that of the pollinating fig wasps where only one female will oviposit within each fig fruit; Hamilton, 1979), when patches are defended by females from attack by other searching females, or when the density of females is low and offspring tend to mate near the emergence site. With increased numbers of females ovipositing in a patch, male offspring will be more likely to compete with each other for access to females. Therefore each female should increase the proportion of males in her brood, until the point is reached where mating is random, and the optimal sex allocation is to produce equal numbers of male and female offspring (Figure 1.14). Hamilton (1967) predicted that the **'unbeatable' sex ratio** is $(n-1)/2n$, where n is the number of females colonising a patch of resource, on which their offspring mate at random.

Fellowes *et al.* (1999a) in a comparative study of the sex ratios of non-pollinating fig wasps (which are primarily inquilines or parasitoids) used the wing morphology of males to distinguish between species that naturally vary in the levels of LMC experienced. Species with wingless males have to mate within the fig fruit, and hence are likely to experience high levels of sib-mating. In contrast, species with winged males will mate outside the fig, often after dispersal, resulting in near-random mating. In some species of non-pollinating fig wasps a proportion of males are winged. These species should experience an intermediate level of LMC (termed *partial* LMC; Hardy, 1994). Theory predicts that sex ratios should follow the rank order of winged males > dimorphic males > wingless males, and a phylogenetically controlled analysis confirmed this pattern.

Within species, similar patterns are also found. Indeed, Salt (1936) noted that the proportion of male *Trichogramma evanescens* offspring emerging from hosts increased when more females oviposited in a patch.

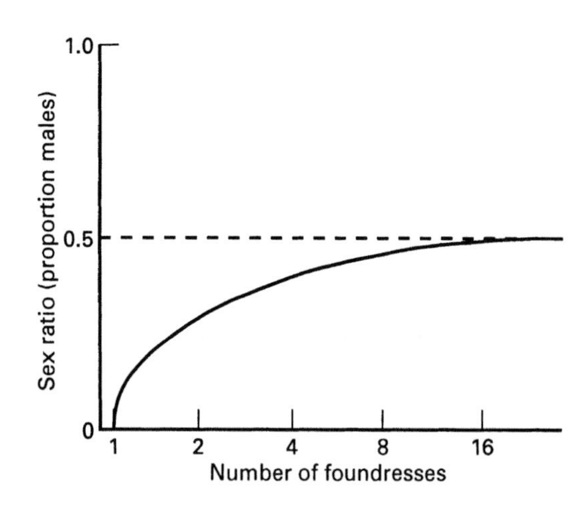

Figure 1.14 Sex allocation by parasitoids: The optimal primary sex ratio in relation to the number of females exploiting a patch (see text for explanation).

This observation was confirmed and developed upon by Waage and Lane (1984), who found that while sex ratio did increase with the number of ovipositing females, the sex ratio was more male-biased than expected. They explained this by proposing that females were less likely to survive superparasitism.

1.9.3 CONDITIONAL SEX ALLOCATION

The second pattern of sex allocation behaviour that needs to be explained is the preferential oviposition of one offspring sex (usually female) in better quality hosts. This was explained by the **Theory of Conditional Sex Allocation** (Charnov, 1979; Charnov *et al.*, 1981). The size of parasitoids (especially idiobionts) is often determined by host size. As there is a positive correlation between parasitoid body size and fitness, larger hosts should be preferred. However, this relationship is much stronger for female parasitoids, so an ovipositing female will maximise her fitness by placing female eggs in better quality hosts (e.g. Morris and Fellowes, 2002). This relationship between size and fitness is often assumed, with few studies investigating the benefits of larger female size in the field (see subsection 2.7.3).

Aphelinus abdominalis is a common parasitoid of several aphid species. In a detailed study, Honěk *et al.* (1998) investigated the sex allocation behaviour of this parasitoid when presented with several potential host species. In all four aphid species (*Macrosiphum euphorbiae*, *Metapolophium dirhodum*, *Sitobion avenae* and *Rhopalosiphum padi*), females preferentially placed male offspring in smaller hosts. If females were provided with small hosts only, then over time the sex ratio became less male-biased. Interestingly, virgin females (i.e. those constrained to produce male progeny) initially favoured small hosts, but over time preferentially attacked larger hosts when provided with a choice.

1.9.4 SEX ALLOCATION AND MASS-REARING

Female-biased sex ratios are clearly the preferable outcome of parasitoid mass-rearing for biocontrol programmes, for economic reasons that mirror the selective pressures of LMC. Thus, the study of sex allocation in insect parasitoids is an important area of study not only for those interested in evolutionary ecology, but also for practitioners of biological control (Luck, 1990). Indeed, such considerations have underpinned several studies of parasitoid sex allocation behaviour. For example, Sagarra and Vincent (1999) suggested that *Anagyrus kamali*, a parasitoid of the Hibiscus Mealybug, should be reared on larger hosts to maximise the production of female progeny. However, one point that is often missed in such studies is that conditional sex allocation is usually a *relative* behaviour. Female offspring will be placed in larger hosts, but if only large hosts are presented then the sex ratio will approach equality.

In a survey of the sex ratios of parasitoids and predators mass-reared for biological control purposes, Heimpel and Lundgren (2000; see also Lundgren and Heimpel, 2003) found that while predators all had a 50:50 sex ratio, a large proportion of the parasitoids had a more male-biased sex ratio than expected. Such work suggests that the producers of biocontrol agents may be able to improve the quality of their product with changed rearing techniques.

1.9.5 DENSITY-DEPENDENT SHIFTS IN SEX RATIO

In a population of wasps adjusting sex allocation as predicted by Hamilton's (1967) model, the proportion of male offspring produced per female will be higher at high wasp densities than at low densities (this prediction is more likely to apply to idiobionts, since females of koinobionts will find it hard to estimate the final size of the host (King, 1989; Kraaijeveld *et al.*, 1999). Thus, individual optimisation does not go hand in hand with maximal female production in a population. This is one reason why the mass-rearing of parasitoids often does not result in desired female-biased sex ratios (see above).

Models like Hamilton's (1967), which predict adaptive shifts in sex allocation in response to the presence of other females, raise the question of how these shifts can be achieved. Waage (1982) was the first to show that simple fixed mechanisms, such as always laying one or more male eggs first (Figure 1.15), can lead to variable sex ratios under different conditions, close to those predicted by the functional models (Waage and Ng, 1984; Waage and Lane, 1984). Rather than counting the number of hosts in a patch and calculating what fraction of her offspring should be sons, the female can lay a son, then lay the number of daughters he can fertilise, then lay another son, and so on. Other mechanisms are also known where the stimulus to change the sequence of sex allocation comes from contacts with marks of conspecifics or with parasitised hosts (Viktorov, 1968; Viktorov and Kochetova, 1971).

To show that females adjust sex ratio in response to the presence of other females, offer patches containing equal numbers of standardised hosts to different densities of parasitoid females (as in a mutual interference experiment, subsection 1.14.3). Primary sex ratio can be determined as described below. If females use a simple **'males first' rule**, a shift in sex ratio can be found with this set-up. Alternatively, one could offer a fixed number of standardised hosts per female in experiments with different numbers of females. In such an experiment, the number of hosts increases with the number of wasps. Where females use a 'males first' rule, no sex ratio adjustment should be found unless, that is, females react to parasitoid odour or marks.

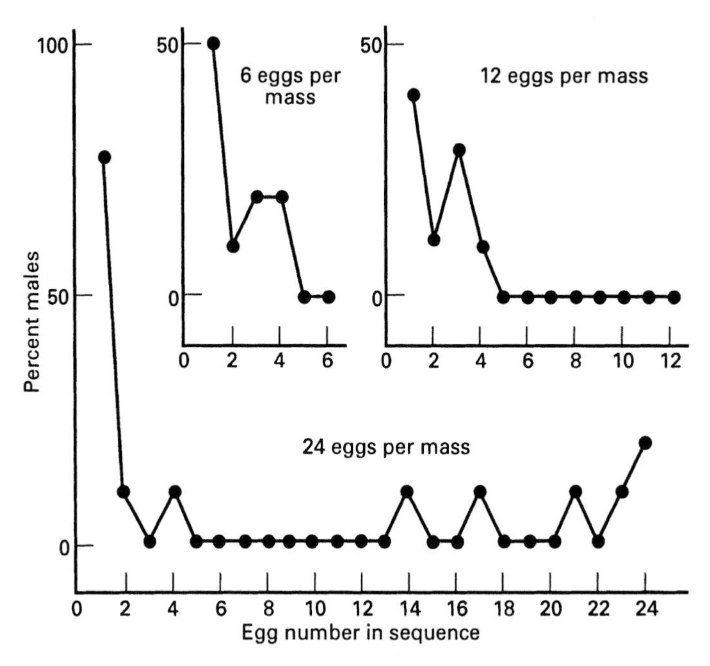

Figure 1.15 Sex allocation by parasitoids: The sequence in which the solitary scelionid parasitoid *Gryon pennsylvanicum* (=*atriscapus*) lays male and female eggs in host egg masses of 6, 12 and 24 eggs. Data are based on observations of mated females and subsequent dissection of host eggs at the time of adult emergence. In all egg masses, males (usually one per mass) are placed in the first few host eggs. By this strategy, every host egg mass, independent of size, is ensured a male wasp offspring. In very large egg masses, e.g. of 24 eggs, a second male is sometimes produced towards the end of the sequence, suggesting that females measure the ratio of males to females and keep it constant for a particular size of egg mass. The males-first strategy alone will also produce an adaptive increase in sex ratio with increased female crowding, since each wasp will lay fewer eggs per egg mass and therefore allocate proportionately more males. Source: Waage (1982), reproduced by permission of Blackwell Publishing.

Strand (1988) adopted a quite different approach in studying density-dependent shifts in sex ratio in *Telenomus heliothidis*. He kept groups of around 200 females together, without hosts, for variable periods of time shortly after emergence and mating, then allowed subsequently isolated females to oviposit in unparasitised hosts (Strand also tested for the effects of subsequent isolation by varying the isolation period). With this method, one can rule out the possibility that females alter sex ratio in response to encounters with already parasitised hosts.

Simple mechanisms such as 'males first' can easily be studied in solitary parasitoids by collecting a sequence of hosts parasitised by an individual female and rearing each of the hosts in separate containers. Investigating such behaviour in gregarious parasitoids is more difficult, but by interrupting oviposition at various points during the laying of egg clutches and rearing the parasitoids, information on the sequence of male and female eggs can be obtained. Sometimes the sequence of male and female eggs can be inferred from the behaviour of the female (see below).

1.9.6 MEASURING PRIMARY SEX RATIOS

Theories of sex allocation deal with the oviposition decisions of female wasps. Tests of these theories require accurate measurement of the sex ratio of the oviposited eggs, i.e. the allocated or **primary sex ratio**. Often, however, due to differential mortality of male and female immatures the sex ratio of emerging parasitoids, the so-called **secondary sex ratio**, does not always reflect the primary sex ratio (e.g. van Baaren *et al.*, 1999). To take account of this problem, the following methods can be used:

1. Wellings *et al.* (1986) designed a statistical method for estimating primary sex ratios from recorded secondary sex ratios. With this method, sex-specific larval mortality of immatures is corrected for;
2. Werren (1980), in laboratory experiments, determined the number of eggs laid in each host, and compared this number with the number of parasitoids emerging from each host in the remainder of a cohort. He found that progeny mortality was negligible and that secondary sex ratio could be taken as an acceptable measure of primary sex ratio, at least as far as his experiments were concerned. The problem with this method of determining how representative secondary sex ratio is of primary sex ratio, is that it requires large samples to minimise sampling error;
3. Van Dijken (1991) dissected eggs laid by *Apoanagyrus lopezi* from the host mealybugs and counted chromosome numbers of cells in metaphase. She was able to determine whether an unfertilised or a fertilised egg had been laid (see subsection 3.4.1 for details of the method);
4. Cole (1981), Suzuki *et al.* (1984) and Strand (1989) discovered for some species that one can actually determine whether a female parasitoid fertilises an egg or not, on the basis of differences in the insect's abdominal movements during oviposition – a feature common to these species is a pause during oviposition of a fertilised (female) egg;
5. Flanders (1950) and Luck *et al.* (1982) used a non-destructive method for *Aphytis* that involves determining the sex of an egg from the position in which it is laid: wasps lay male and female eggs on the host's dorsal surface and ventral surface, respectively.
6. De Menten *et al.* (2003) used fluorescence *in situ* hybridization (FISH) to sex ant eggs. Such approaches are likely to be successful with parasitoid wasps.

1.10 FUNCTIONAL RESPONSES

Understanding how predators and parasitoids respond to changes in prey and host density is critical to gaining a grasp of the interactions between natural enemies and their victims. Solomon (1949) coined the term **functional response** when describing the response shown by *individual* natural enemies to varying host (prey) density; with increasing host availability, each enemy will attack more host individuals.

Several types of functional response are possible (Figure 1.16; see Hassell, 2000 for a detailed review):

Type 1: where there is a rectilinear rise to a maximum (N_x) in the number of prey eaten per predator as prey density increases. The response is described by the following equation:

$$N_a = a'TN \qquad (1.2)$$

where N_a is the number of hosts parasitised or prey eaten, n is the number of hosts or prey provided, T is the total time available for search, and a' is an acceleration constant, the instantaneous attack rate (equation 1.2 applies only when $N < N_x$).

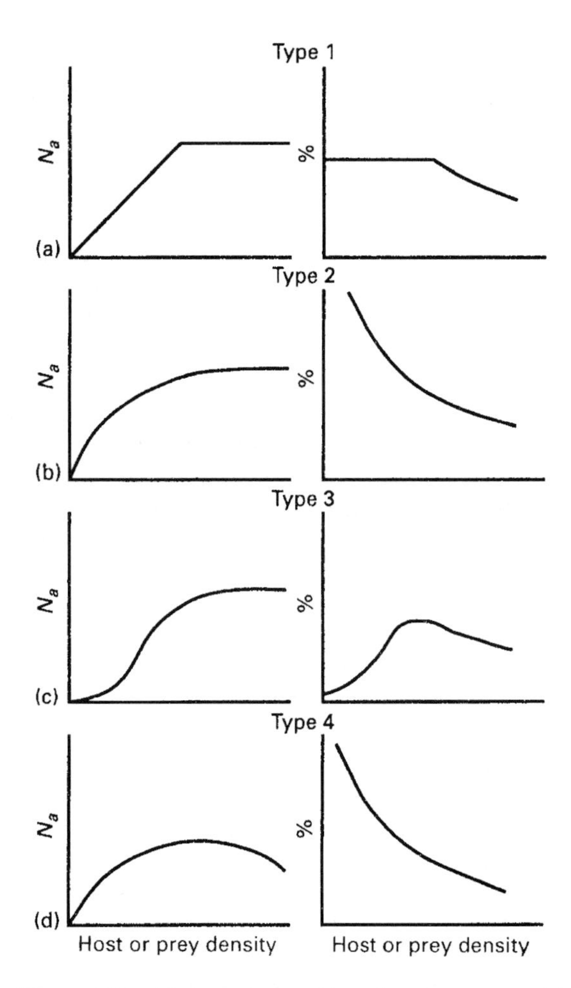

Figure 1.16 The four types of functional response: (a) Type 1; (b) Type 2; (c) Type 3; (d) Type 4. N_a = number of hosts parasitised or numbers of prey eaten; % = percentage of hosts parasitised or eaten.

The Type 1 response is likely to be found when handling times (see below) are negligible and eggs are in limited supply.

Type 2: where the response rises at a constantly decreasing rate towards a maximum value, i.e. the response is curvilinear, in contrast with the Type 1 response. Holling (1959a, b) predicted such a response, reasoning that the acts of quelling, killing, eating and digesting prey are time-consuming activities (collectively called the **handling time**) and reduce the time available for further search, and that with increasing prey density a predator will spend an increasing proportion of its total time available not searching:

$$T_s = T - T_h N_a \qquad (1.3)$$

where T_s is the actual time spent searching, and T_h is the handling time. The type 2 functional response is probably the most commonly reported in parasitoids (Fernandez-Arhex and Corley, 2003).

Type 3: where the response resembles the Type 2 response except that at lower prey densities it accelerates. The response is thus sigmoid.

Type 4: where the response resembles the Type 2 response except that at higher densities it declines, producing a dome-shape.

Sabelis (1992) also recognises a fifth type of response, which is intermediate between the Type 1 and the Type 2. This response appears to be shown by some predatory mites, and will not be discussed further.

The functional response of a predator or parasitoid species is usually measured as follows: individual insects are confined in an arena (e.g. cage), with different numbers of prey or hosts, for a fixed period of time (Figure 1.17). At the end of the experiment, the natural enemies are removed and either the number of prey killed or the number of hosts parasitised (or both, in the case of some host-feeding parasitoids, see below) is counted. Hosts are either dissected or reared until emergence of the parasitoids. From the counts made, a graph can then be plotted relating the number of prey or hosts attacked

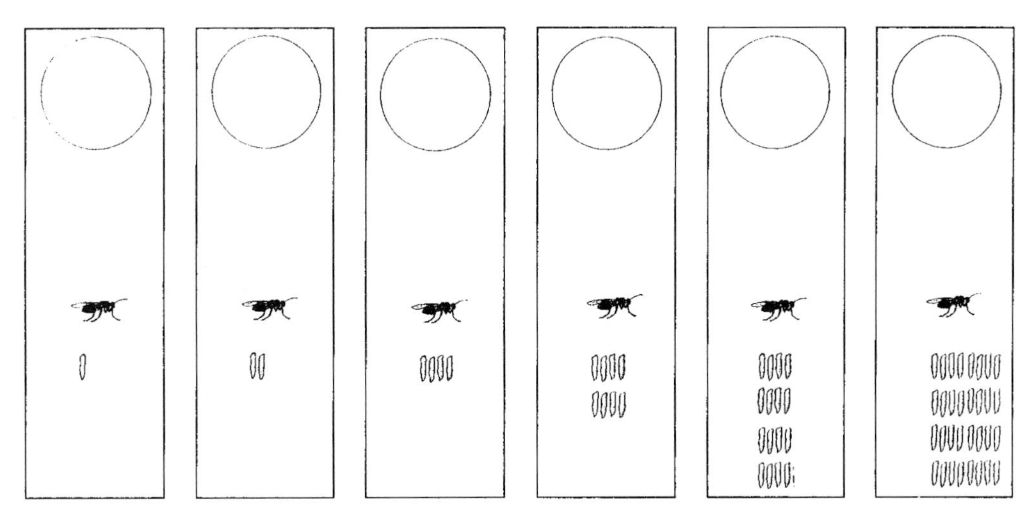

Figure 1.17 Functional responses of parasitoids and predators: Schematic representation of traditional design of a functional response experiment. Circles denote the experimental host patch, and rectangles the experimental arena. See text for discussion.

to the number offered. The plots are then compared with mathematical models (Holling, 1959a,b, 1966; Rogers, 1972; Royama, 1971; Mills, 1982a; Arditi, 1983; Casas *et al.*, 1993; Casas and Hulliger, 1994). With predators, the functional responses of the different larval instars, as well as those of the adults, can be measured. With both predators and parasitoids, functional responses in relation to prey and hosts of different sizes can be measured.

There are two likely reasons why a Type 3 response may be recorded using the aforementioned experimental set-up:

1. As host density decreases at the lower range of host densities, the parasitoid spends an increasing proportion of the total time available in non-searching activities. For example, at lower host densities, *Venturia canescens* spends a greater proportion of its time performing activities such as walking and resting on the sides of the experimental cage. Similar behaviour is probably responsible for the Type 3 response observed in parasitoids and predators that are offered unpreferred prey species: in *Aphidius uzbeckistanicus, Coccinella septempunctata* and *Notonecta glauca,* a Type 3 response was recorded when the unpreferred host and prey species was provided, compared with a Type 2 when the preferred species was provided (Dransfield, 1979; Hassell *et al.*, 1977). The parasitoid under investigation may be a host-feeder, mainly feeding upon hosts rather than ovipositing, at low host densities (section 1.7). [N.B. in host-feeding parasitoids that feed and oviposit on different host individuals, we may distinguish between the following functional responses: that for parasitism alone, that for host-feeding alone, and that for parasitism and feeding combined (i.e. the 'total' functional response (Kidd and Jervis, 1989).] If, as is likely, the handling time for feeding encounters is longer than for oviposition encounters, a Type 3 response for parasitism may result (Collins *et al.*, 1981).
2. Handling times may be shorter at higher host densities. In solitary parasitoids, this is an unlikely cause of a sigmoid functional response, but gregarious parasitoids may decrease clutch size at higher host densities (section 1.6), and so decrease handling time per host. Predators may ingest less food

from each prey item at higher prey densities (Figure 1.18) and so reduce handling times. When extracting food from a prey item becomes increasingly difficult with the time spent feeding on it, predators may optimise the overall rate of food intake by consuming less of each individual prey item when the rate of encounters with prey is high (Charnov, 1976; Cook and Cockrell, 1978). Optimal foraging models predict this behaviour, while a similar prediction can be made based on a causal model relating the amount of food in the gut to the amount eaten from each prey. Some authors have argued that the optimal foraging model can be refuted because there is a causal explanation for the observed behaviour. However, the reader should remember that causal and functional explanations are not mutually exclusive; indeed, they complement each other. When predictions of a causal and a functional model are quantitatively similar, this can be taken as evidence that the mechanism does not constrain optimisation of a behavioural trait.

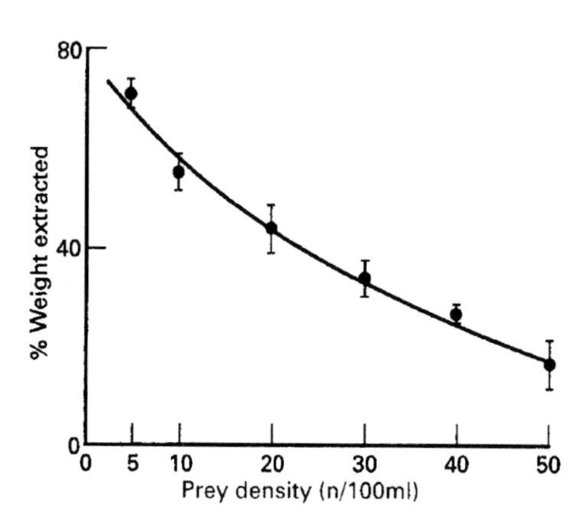

Figure 1.18 The relationship between prey availability and the percentage of the mass of each prey individual consumed by the belostomatid bug *Diplonychus rusticum*: The bug is more 'wasteful', eating proportionately less of each prey (*Chironomus plumosus*) as prey density increases. This effect is predicted by optimal foraging (i.e. functional) and gut-filling (i.e. causal) models, and is shown by a wide variety of predators. Source: Dudgeon (1990).

A Type 4 functional response will occur if:

1. When dealing with prey individuals, other prey individuals interfere with the predator and cause it to abort the attack more frequently at high prey densities than at low densities;
2. The prey have a well-developed group defence reaction that is more effective at high prey densities than at low ones.

The classical functional response experiment assumes there is a homogeneous environment, or at least it does not consider the spatial distribution of prey and hosts. However, most insects are patchily distributed and the spatial distribution of hosts or prey within an experimental arena is likely to vary significantly with the density of the insects. Predators and parasitoids respond to differences in prey and host densities between patches by adjusting the amount of time spent in each patch (subsection 1.12). By allowing the parasitoid, rather than the experimenter, to determine the amount of time it spends in an experimental patch (in a so-called **variable-time experiment**), a different type of functional response may be obtained compared to experiments where the time spent is fixed by the experimenter (so-called **fixed-time experiments**; van Alphen and Galis, 1983; Collins, *et al.*, 1981; Hertlein and Thorarinsson, 1987). Van Lenteren and Bakker (1978) suggest that in fixed-time experiments, some parasitoids are likely to show a Type 2 response, rather than a Type 3, because parasitoids are caused to revisit low-density patches they would otherwise leave. Thus, a Type 2 response may be an artefact of the fixed-time experimental design.

Since the type of functional response found in an experiment depends very much on the experimental design adopted, one should first clearly define what sort of question one wishes to address before measuring a functional response. Often, a functional response is measured to provide insights into the suitability of a parasitoid as a biological control agent. The problem is then how one can use the information generated by the experiments to predict the performance of the parasitoid in the field. The context in which the data will be used is one of population dynamics, and thus relates to the response of the parasitoid *population* to host density. The spatial structure of natural host populations, and the interactions between individual parasitoids in the population, makes it hard to relate the results of experiments on individuals in single-patch, single-parasitoid experiments to processes occurring at the population level.

If single-patch such experiments are, nevertheless, to be carried out, the minimum requirements for experimental design should be as follows: the foraging insect should be observed continuously (in most functional response experiments carried out to date, parasitoid and predator behaviour have not been examined directly at all, never mind continuously!), and a record made of how the parasitoid spends its time in the experimental arena. Parasitoids should be allowed to leave the arena when they decide to leave, so the experiment should be a variable-time one. It may prove difficult for the observer to decide when an experiment should be terminated. A parasitoid may leave the experimental patch for a short period, but then return and continue searching for hosts. Experiments may need to be terminated after the insect has spent an arbitrary period of time outside the patch (Waage, 1979; van Alphen and Galis, 1983), but of course the choice of the period is subjective and it acts as a **censor** in the data (a censor is a factor, other than a decision by the foraging insect, that terminates an experiment, e.g. a decision by the experimenter or an external disturbance, see Haccou *et al.*, 1991). A solution to the problem of when to terminate an experiment is to use an arena containing two patches. Once the insect has left the first patch and arrived in the second one, the experiment can be terminated.

However, there is a drawback to conducting such experiments under artificial conditions in the laboratory. The searching efficiency of the natural enemy will be influenced by the spatial structure of the patch (e.g. variation in plant architecture) and in the age-structure of the victims. As an illustration of this, consider the work of Ives *et al.* (1999). Foraging *Aphidius ervi* will preferentially attack second and third instar pea aphids within a patch, and among patches there is variation in foraging efficiency resulting from variation in plant architecture. Ives *et al.* found that when aphid numbers were low *A. ervi*, a parasitoid species that would normally be considered to exhibit a strong Type 2 functional response, shows a Type 1 response. Foraging experiments should therefore be conducted under a range of scenarios, of which at least some would reflect more natural foraging conditions.

Ideally, functional response experiments should measure encounter rates with concurrently available patches containing different densities of hosts. To do this, a multi-patch experiment needs to be carried out. Such an experiment might show that high density patches are found more easily by the parasitoid, since such patches produce greater quantities of volatile attractants than low density patches (subsection 1.5.3).

Functional response experiments have so far not taken into account the possibility that the response of a parasitoid to patches of different densities depends on whether patches are scarce or common in the habitat. In 'poor' habitats when distances between patches are large and high density patches are scarce, parasitoids should, when exploiting low-density patches, stay longer and parasitize more hosts. Finally, functional response experiments need to take account of the reaction of a parasitoid to the presence of conspecifics.

An alternative to measuring functional responses is to undertake an integrated analysis of all factors affecting patch time allocation in parasitoids (subsection 1.12).

Determining the type of functional response is an important step that needs to be taken by the investigator before he or she attempts to obtain parameter estimates from functional response models. Wrong estimates may be obtained if a model for a Type 2 response is used to estimate parameters from what is in reality a Type 3 response, and *vice versa*. For advice on how to determine the functional response, and for information on curve-fitting routines, see Trexler *et al.* (1998), Casas and Hulliger (1994), Juliano (2001) and Schenk and Bacher (2002).

1.11 SWITCHING BEHAVIOUR

Species and host stage preference by natural enemies has been discussed in subsection 1.5.7. Preference (parameter c in equation 1.1) may not be constant but may vary with the relative abundance of two prey types, in which case if the predator or parasitoid eats or oviposits in disproportionately more of the more abundant type (c increases as N_1/N_2 increases) it is said to display **switching behaviour** (Murdoch, 1969) or **apostatic selection** (Clarke, 1962), the latter term being used by geneticists. Where disproportionately more of the rarer type is accepted (c increases as N_1/N_2 decreases) **negative switching** is said to occur (Chesson, 1984).

(Positive) switching behaviour has aroused the interest of students of population dynamics because it is associated with a Type 3 functional response (to prey type N_1) (subsection 1.10) (Murdoch, 1969; Lawton *et al.*, 1974).

Switching behaviour in parasitoids has been observed by Cornell and Pimentel (1978) in *Nasonia vitripennis*, van Alphen and Vet (1986) in *Asobara tabida*, Chow and Mackauer (1991) in *Aphidius ervi* and *Praon pequodorum*, and probably by Lill (1999), while switching in insect predators has been observed by Lawton *et al.* (1974) in the waterboatman *Notonecta glauca* and the damselfly *Ischnura*. Other examples are given in Sherratt and Harvey (1993).

Switching can be tested for by offering parasitoids combinations of different host species in single-patch experiments. The combined density of the two host species should be kept constant, but the relative abundance of the two species should vary among treatments. If the mechanism causing the switching is to be determined, full records of parasitoid (and host) behaviour ought to be made. As in other host selection experiments (subsection 1.5.7) females should be observed continuously and the number of acceptances and ovipositions scored, to show whether females accept more or fewer individuals of a host type than they successfully parasitize (this possibility is usually ignored by authors).

One can either conduct fixed-time experiments in which depletion of the hosts is prevented by replacing each parasitised host by an unparasitised one of the same species, or allow depletion and terminate experiments when the parasitoid leaves the patch. Bear in mind that switching, like many other aspects of parasitoid and predator behaviour, is likely to be affected by previous experience of the natural enemy (see below).

The resulting data can be analysed using Murdoch's (1969) null or **no-switch model**:

$$P_1 = cF_1/(1 - F_1 + [cF_1]) \qquad (1.4)$$

where F_1 is the proportion of host species 1 in the environment, P_1 is the proportion of species 1 among all the hosts oviposited in, and c (a parameter we have already mentioned in subsection 1.5.7) is a combined measure of preference and encounter probability for species 1, given by:

$$c = \frac{N_2}{N_1}\frac{E_1}{E_2} \qquad (1.5)$$

which is a rearrangement of equation 1.1. In the absence of switching behaviour, c is a constant. c can be estimated in various ways, although it is convenient to estimate it when $N_1 = N_2$. The value of P_1 for any level of availability of species 1 can be estimated by substituting the estimated value of c in equation 1.4, and an expected no-switch curve plotted (Figure 1.19). If the parasitoid species' preference is not constant but alters with changing host availability (or encounter rate) the observed proportion of host species 1 among all the accepted or parasitised hosts will be higher than expected when species 1 is abundant and lower than expected when species 1 is rare.

Elton and Greenwood (Elton and Greenwood, 1970, 1987; Greenwood and Elton, 1979; see Sherratt and Harvey, 1993) provide a model

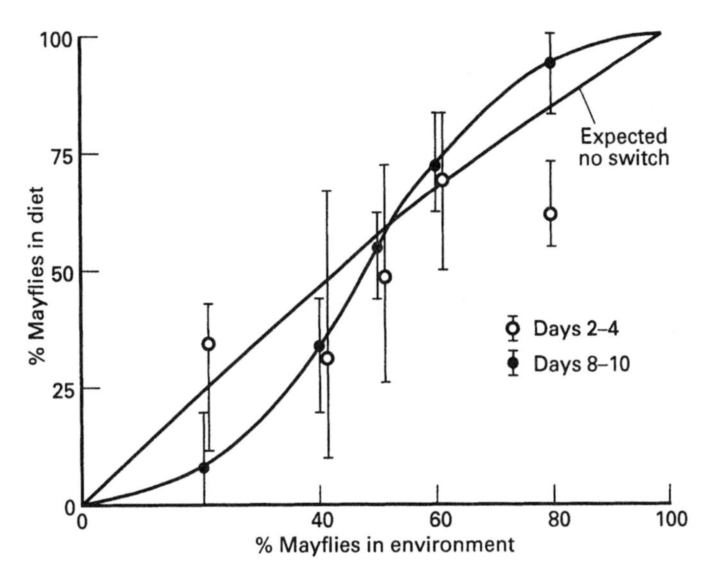

Figure 1.19 Switching in insect natural enemies: The percentage of mayfly larvae in the diet of *Notonecta* as a function of their relative abundance in the habitat. The almost straight line is the 'no-switch' curve. See text for further explanation. Source: Lawton *et al.* (1974).

which can be used for the detection of switching and other forms of frequency-dependent selection, and which includes a measure of the deviation from constant preference as one of the parameters. Another model developed by Manly and colleagues (Manly *et al.*, 1972; Manly, 1972, 1973, 1974; see Sherratt and Harvey, 1993, for a discussion) takes account of prey depletion (this model requires modification before it can be used to take account of depletion of unparasitised hosts). The latter model can also be easily generalised for more than one prey or host type.

Tinbergen (1960) suggested, as a mechanism for switching, that predators form a **search image** of the most abundant prey species, i.e. they experience a perceptual change in the ability to detect a cryptic prey type, and this change does not occur when that type is rare (Lawrence and Allen, 1983; Guilford and Dawkins, 1987 give reappraisals of the evidence for search image formation). However, switching could well result from other behaviour such as active rejection of the less preferred host species as the preferred hosts becomes more abundant– a prediction of optimal prey selection models (subsection 1.5.7). Note that Murdoch's definition of switching is couched in terms of *relative prey* density, whereas optimal foraging models refer to *absolute* densities or encounter rates with prey.

Lawton *et al.* (1974) investigated whether experience with a particular prey species may be a contributory mechanism in the switching behaviour of *Notonecta* presented with *Asellus* and *Cloeon* (in this case, negative switching was recorded over days 2 to 4 of the experiment and positive switching over days 8 to 10 (Figure 1.19). In a separate experiment, they measured the proportion of successful attacks on *Asellus* prey in relation to the proportion of this prey available in the environment during the previous seven days, and found that the more *Asellus* the predator was exposed to, the greater was the proportion of successful attacks recorded. While this strongly suggests that experience with *Asellus* affects the predator's prey capture efficiency, it does not prove conclusively that it does so, since no information was obtained on the encounter rates, and therefore the experience of the insects, during the pre-experimental period. The development of a search image could be ruled out as a mechanism for switching in this predator/prey system since: (a) in the switching test *Notonecta* took different prey species in a random sequence instead of attacking prey in 'runs' (Lawton *et al.*, 1974), and (b) the prey were unlikely to have been cryptic in the experimental tanks used.

Switching behaviour may not necessarily be adaptive. Chow and Mackauer (1991) found that *A. ervi* and *P. pequodorum* switched to the alfalfa aphid when pea aphids and alfalfa aphids were offered to wasps in a 1:3 ratio. However, Chow and Mackauer hypothesise that since alfalfa aphids are more likely than pea aphids to escape from an attacking wasp, a foraging wasp incurs a potentially higher cost in lost 'opportunity time' (subsection 1.5.7) when attacking alfalfa aphids. Furthermore, since it is possible that alfalfa aphids are poorer quality hosts in terms of offspring growth and development, wasps may not derive a fitness gain from switching to alfalfa aphids.

For a comprehensive review of switching and frequency-dependent selection in general, see Sherratt and Harvey (1993).

1.12 PATCH TIME ALLOCATION

1.12.1 INTRODUCTION

One aspect of parasitoid foraging behaviour where the causal approach and the functional approach have traditionally coexisted is patch time allocation. We will consider first which factors affect patch time allocation and second how one can analyse the interplay of the different factors.

1.12.2 FACTORS AFFECTING PATCH TIME ALLOCATION

Patch time allocation is likely to be affected by the following:

1. A parasitoid's previous experience;
2. Its internal state (e.g. egg load, energy reserves);
3. Patch kairomone concentration;
4. Encounters with unparasitised hosts;
5. Encounters with parasitised hosts;
6. The timing of encounters;
7. Encounters with the marks of other parasitoids;
8. Encounters with other parasitoid individuals;
9. Superparasitism;
10. Genetic variation.

Some of these factors can be studied through experiments in which all the other factors are excluded. For example, the effect of kairomone concentration can be investigated without involving hosts at all (subsection 1.5). To eliminate the effects of encounters with other parasitoids and their marks, the experimental design shown in Figure 1.20 can be used. However, it may be impossible with some experiments to separate the effects of different factors. A notorious problem is the analysis of the factors that determine how long a parasitoid will stay on a patch that initially contains only unparasitised hosts. Because the parasitoid oviposits in the unparasitised hosts it encounters, the number of unparasitised hosts decreases while the number of parasitised hosts increases. Thus, with the passage of time, the parasitoid experiences a decreasing encounter rate with unparasitised hosts and an increasing encounter rate with parasitised hosts. Because both the temporal spacing and the sequence of encounters with parasitised

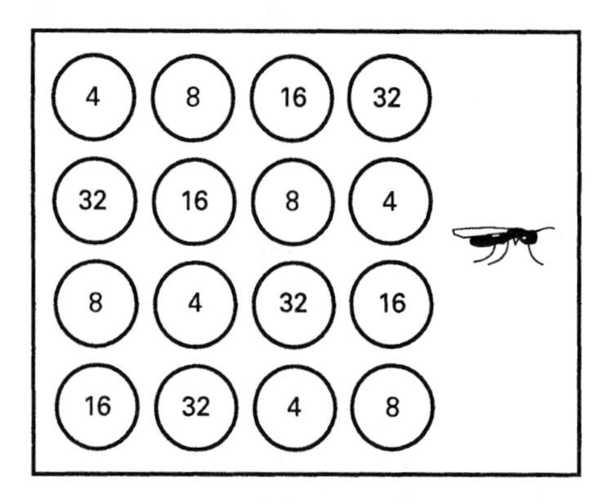

Figure 1.20 Patch time allocation by individual parasitoids and predators: Schematic representation of one suggested experimental design for an experiment for studying patch time allocation. A randomised arrangement of patches (denoted by circles) is used. This experimental design can be used in the study of aggregative responses.

and unparasitised hosts are stochastic in nature, encounter rates with both types of host do not alter in a monotonic, smooth fashion.

In some parasitoid species, encounters with unparasitised hosts have an incremental effect on the time spent in a patch (van Alphen and Galis, 1983; Haccou *et al.*, 1991). This poses the question: 'what effect do encounters with *parasitised* hosts have on patch time allocation, and how does the relative timing of encounters with parasitised and unparasitised hosts influence the period spent in individual patches?'.

1.12.3 ANALYSING THE INTERPLAY OF DIFFERENT FACTORS

Two distinct hypotheses can be formulated about the effect of encounters with parasitised hosts on patch residence times. The functional hypothesis is as follows: given that a parasitoid is able to discriminate between parasitised and unparasitised hosts, encounter rates with both host types provide the parasitoid with information on host density and the degree of exploitation of a patch. This information allows the wasp to determine when to leave the patch, e.g. high encounter rates with parasitised hosts in combination with low encounter rates with unparasitised hosts signal a high level of exploitation of the patch. Because it could be more profitable for the wasp to move on and search for a better quality patch, the insect might decide to leave. Van Lenteren (1976) recognised this as one of the functions of host discrimination and showed, through single patch experiments, that wasps continued to search on patches in which parasitised hosts were immediately replaced by unparasitised ones, whereas wasps allowed to search on similar but unreplenished patches attempted to leave the experimental arena after most of the hosts had been parasitised. The functional hypothesis states that encounters with both unparasitised and parasitised hosts affect patch time, but it does not specify the mechanism involved.

The causal hypothesis formulates explicitly how encounters with parasitised hosts affect patch time. This hypothesis is an extension of a mechanistic model for patch time allocation proposed by Waage (1979) for the parasitoid *Venturia canescens*. Although this model was shown to be an incorrect description of the behaviour of *V. canescens* (Driessen *et al.*, 1995), it is still valuable as a conceptual model, and it can be applied to many other parasitoid species. Waage assumed that a female parasitoid, when entering a patch containing hosts, has a certain motivation level for searching the patch, the level being set by previous experience and kairomone concentration on the patch. If the wasp does not locate and oviposit in hosts, the motivation level will decrease steadily over time to a threshold value whereupon the parasitoid leaves the patch. However, with each oviposition that occurs, an incremental change in motivation occurs. The initial level of motivation, combined with linear decreases of motivation during searching periods and increases in motivation following ovipositions, determines how long the parasitoid will stay in the patch (Figure 1.21). The causal hypothesis assumes there is an additional effect of a rejection of a parasitised host, causing a decrease in motivation level (Figure 1.21b). Like the functional hypothesis, the causal hypothesis predicts shorter patch residence times with increasing patch exploitation, all other things being equal.

A rigorous test of the causal hypothesis ought to demonstrate whether the mechanism by which shorter patch residence times come about is an increase in the tendency to leave the patch after a rejection of a parasitised host. Such a test implies that one is able to assess the relative effects on the motivation to search of ovipositions in unparasitised hosts (i.e. increments), of the time interval between encounters, and of rejections of parasitised hosts (i.e. decrements). To illustrate how difficult it is to determine whether the rejection of parasitised hosts causes a decrease in the motivation to search, we will discuss in some detail the experimental evidence given by van Lenteren (1991). In one experiment, individual females of *Leptopilina heterotoma* were allowed to search on a 1 centimetre diameter patch of yeast containing four unparasitised hosts and 16 parasitised hosts. Each host

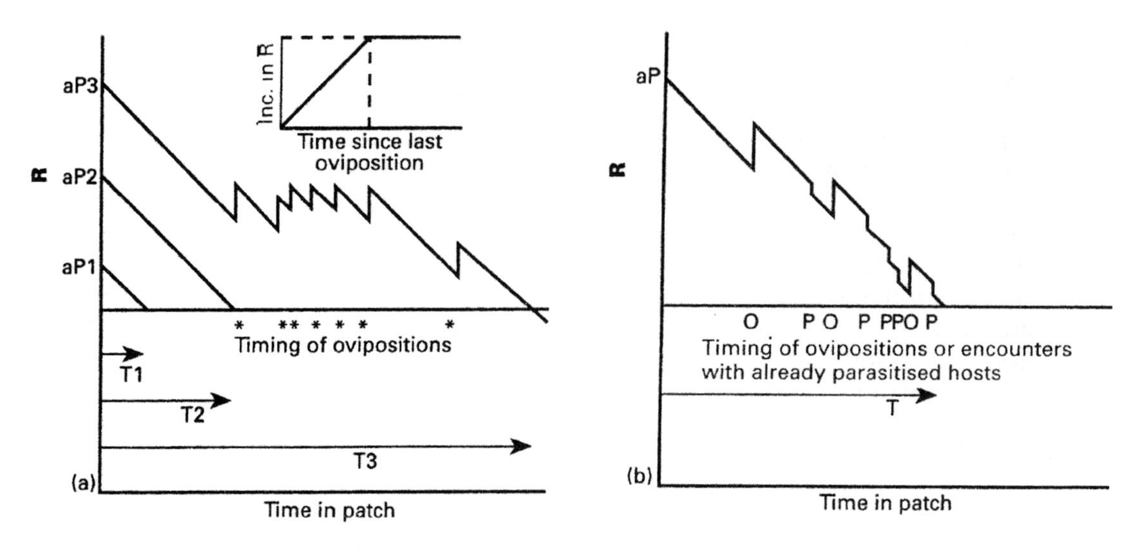

Figure 1.21 Patch time allocation by parasitoids: (a) Waage's (1979) causal model of patch residence time. R = responsiveness of the female parasitoid to the patch edge (a function of the number of hosts in the patch [P1, P2, P3] and a constant a, the quantity of kairomone produced per host). Asterisks denote ovipositions. An oviposition results in an increment of R (inset: the size of the increment depends on linearly on the amount of time that has elapsed since the previous oviposition, and the increment cannot exceed a maximum value). T1, T2 and T3 are the resulting patch residence times for three different cases. (b) Waage's model, modified to incorporate the decremental effect of encounters with parasitised hosts. Symbols as in (a) except that O denotes an oviposition, and P denotes an encounter with an already parasitised host. Source of (a): Waage (1979), reproduced by permission of Blackwell Publishing.

parasitised during the experiments was immediately replaced by an unparasitised one. As a control, single females of *L. heterotoma* searched a similar patch containing only four unparasitised hosts, and any unparasitised hosts parasitised during the experiment were replaced by unparasitised ones.

Wasps stay longer on patches with four unparasitised hosts than on patches with four unparasitised hosts and 16 parasitised hosts. Van Lenteren (1991) argued that because there were no significant differences in *average* time interval between ovipositions in unparasitised hosts in the two treatments, the differences in patch residence times between the treatments can be attributed only to a detrimental effect on patch residence time of encounters with parasitised hosts. In our opinion, however, this conclusion is not supported by van Lenteren's data.

First, consider whether it is at all valid to conclude from the observation that *average* time intervals between ovipositions did not differ between experiment and control, and that there is no difference in the effect of ovipositions on patch residence time between the two treatments. This conclusion would be valid only if the parasitoid itself uses average intervals to assess patch profitability. As Haccou *et al.* (1991) have shown, the effect of an oviposition on the probability of a wasp leaving a patch depends on its timing; hence it is important *when* and *where* the longest intervals occur.

Despite a lack of statistical differences between the *average* values, important differences in interval times between ovipositions could occur between test and control treatments.

An alternative explanation for van Lenteren's (1991) results is that the differences in patch residence time are caused solely by the decrease in motivation over time that results from the extra time spent in rejecting parasitised hosts in the treatment with parasitised hosts. This time could otherwise be spent in ovipositing in unparasitised hosts. Rejection of a parasitised host takes between 2 and 6 seconds (Haccou *et al.* 1991), and with on average 33 rejections in the control treatment; this behaviour may account for an important part of patch residence time. If the decrease in the motivation to search (indicated by the sloping lines in 1.21a and 1.21b) continues during the time spent in rejections, these small decrements may accumulate over time, causing the parasitoid to reach the threshold motivation rate for patch-leaving sooner than when no parasitised hosts are encountered. Intervals between encounters with unparasitised hosts would, on average, be slightly longer in experiments with parasitised hosts than in those without them, as indeed they were: 84 compared with 79 seconds. Although these differences are not significant, the time lost in rejection of parasitised hosts gradually accumulates, and so may be responsible for the ultimate differences in patch residence times.

Clearly, one cannot test the causal hypothesis simply by determining whether patch residence times and search times differ significantly between treatments. What is required is an analysis in which the relative weight of effects of the influencing factors and their timing are estimated from the data and tested statistically. For this reason, Haccou *et al.* (1991) adapted Cox's (1972) proportional hazards model (subsection 1.2.1) to study the problem. They analysed a new set of experimental data using the model. No effect of encounters with parasitised hosts on the probability of patch-leaving was found. If such an effect exists at all, we expect it to be a small one. It might be detected in experiments in which there is a high proportion of encounters with parasitised hosts. Other attempts to determine whether rejections of parasitised hosts increase the probability of patch-leaving, all pre-dating Haccou *et al.* (1991), i.e. van Alphen and Galis (1983), van Alphen and Vet (1986), have, like van Lenteren's (1991) study, similar problems of interpretation. Hence, the first evidence confirming the hypothesis first formulated by Waage (1979) comes from Hemerik *et al.* (1993). They used the proportional hazards model to analyse their experimental results and demonstrated in female *Leptopilina clavipes* that encounters with parasitised hosts decrease the tendency to search the patch.

Finally, the effect of encounters with parasitised hosts may depend on the previous experience of the parasitoid. It is thus possible that encounters with parasitised hosts could also *increase* the tendency to search on a patch, as is the case when they decide to superparasitise (van Alphen *et al.*, 1987).

Studies of patch residence times of insect predators are rare. It is evident that patch residence time may be influenced by the predator's level of satiation, but this is unlikely to be a straightforward relationship. For example, wolf spider (*Schizocosa ocreata*) patch residence time is influenced by hunger, but only in an interaction with spider age and sex (Persons, 1999).

Thiel and Hoffmeister (2004) investigated patch time allocation decisions in relation to habitat quality. They showed that *Asobara tabida* reduces both the patch residence time and the degree of path exploitation when patches increase in abundance in the habitat, as predicted by optimal foraging theory. For a discussion of the proximate mechanisms involved, see their paper.

1.12.4 GENETIC VARIATION IN PATCH TIME ALLOCATION

As discussed above, Haccou *et al.* (1991) used Cox's regression model to investigate the patch-leaving behaviour of parasitoid wasps. Using this technique, Wajnberg *et al.* (1999) studied the behaviour of *Telenomus busseolae*, attacking the eggs of *Sesamia nonagrioides*. Not only did they find that female *T. busseolae* increased their tendency to leave a patch after each successful oviposition attempt, but also that the genotype of the ovipositing female influenced this behaviour. However, most workers consider patch-leaving (and indeed most parasitoid or predator behaviour) rules as a species-specific trait (Driessen *et al.*, 1995; Wajnberg *et al.*, 1999), rather than a variable characteristic among the individuals under study. This is rather short-sighted in many ways, as it assumes: (a) that all populations of a given species will respond in a similar way to different hosts or patches, and (b) that the trait is fixed, whereas it is likely that there is heritable variation for the trait, and that natural selection may change the response found in a population over time.

One of the more convenient methods for studying genetic variation in behaviour involves the use of **isofemale lines**, where the offspring of singly inseminated females are progressively inbred until the majority of loci are homozygous (subsection 3.2.2.) This provides a series of independent lines varying in their genetic background, and the broad-sense heritability (a combination of additive and non-additive components of genetic variation) of a given trait can be estimated (see Chapter 3, subsection 3.5.2). A secondary advantage of founding a series of isofemale lines is that while variation within a line will be lost, a series of independent lines will maintain the heritable variation from within the population of interest, and mixing the lines will reconstitute the majority of the variation. However, it should be borne in mind that there is a risk involved with using isofemale lines. Since loci should be homozygous, sub-lethal deleterious recessive alleles may be expressed, introducing additional variation in the behaviour of interest.

If more detailed information on the narrow-sense (additive) components of genetic variation is required (e.g. so that the short-term response to selection can be predicted), then there are several techniques that are readily available, of which sib-sib and parent-offspring analyses are probably the most convenient (Falconer and MacKay, 1996; Lynch and Walsh, 1998).

1.13 PATCH DEFENCE BEHAVIOUR

When a predatory insect finds a resource, there is a trade-off between allocating time to consuming it or to defending it against competitors (Field and Calbert, 1998). However, the resources utilised by parasitoids are relatively long-lived, and as a result potential hosts in a patch may not yet suitable for attack, or any offspring that have already been invested in are vulnerable to attack themselves. As a result, some parasitoids defend patches of hosts against conspecific and heterospecific intruders, occasionally leading to the death of one of the protagonists (Perez-Lachaud *et al.*, 2002). This patch defence behaviour consists of two components, **resource defence** (where competing females are prevented from gaining access to potential hosts; Waage, 1982) and **maternal care** (where previously parasitised hosts are protected from superparasitism or hyperparasitism; van Alphen and Visser, 1990), and the relative importance of both factors will influence patch defence behaviour (Field and Calbert, 1998). Thus, patch defence can be an alternative competitive strategy to one of allowing conspecifics on the same patch and competing with them through superparasitism (subsection 1.8.4).

Patch defence is only advantageous under a limited set of conditions. The following factors favour defence of patches:

1. Synchronous development of the hosts in the patch;
2. Rapid development of the host to a stage which can no longer be attacked by the parasitoids, or rapid development of the parasitoid offspring to a stage at which they have a competitive advantage in cases of superparasitism;
3. Short travel times between patches. When travel times are long, intruders are likely to be 'reluctant' to lose the contest for the patch. This would prolong fighting and increase the cost of defence;
4. Species of parasitoid in which adult females have a very low probability of finding more than one host or host patch during adult life. This factor allows them to spend a long time guarding their brood, and it has undoubtedly played an important role in the evolution of brood-guarding (e.g. *Goniozus nephantidis*, Hardy and Blackburn, 1991). The females of brood-guarding parasitoids spend a long period of post-reproductive life guarding one host or patch of hosts;
5. Patches should be of a defensible size; Larger patches are harder to defend.

Patch defence was first described for scelionid egg parasitoids (Waage, 1982), which defend small and intermediate-sized host egg masses. However, it is also found in braconids (e.g. *Asobara citri*), ichneumonids (e.g. *Rhyssa persuasoria, Venturia canescens*), and bethylids (e.g. *Goniozus nephantidis*).

In some parasitoids, such as the aforementioned scelionid egg parasitoids, patch defence appears to be a fixed response to an intruder (but see below). However, in other species such as the braconid *Asobara citri*, patch defence and fighting behaviour decrease in frequency with increasing patch size and increasing numbers of intruders, and wasps may switch to competition through superparasitism.

Patch defence can have a pronounced effect on the distribution of adult parasitoids over host patches. It can lead to a regular distribution of parasitoids, and is thus one of the factors reducing aggregation (see below).

Whether patch defence or competition by superparasitism is the better strategy depends on species-specific traits such as the encounter rate with hosts and the handling time. Thus, it is possible that one species attacking a certain host defends hosts or patches against intruders, while another parasitoid species attacking the same host does not. Patch defence is a poorly studied aspect of parasitoid biology and clearly deserves more attention from both pure and applied entomologists.

Perhaps one of the best-studied systems involves the scelionid wasp *Trissolcus basalis*, a parasitoid of the egg masses of pentatomid bugs (Field, 1998; Field and Calbert, 1998, 1999; Field *et al.*, 1997, 1998). Here, if a female finds a patch, she will initially search for, and oviposit in, suitable hosts. Later, she will patrol the patch, still ovipositing until the patch is depleted. Once this occurs, the female will remain on the patch for about 5 hours, before departing, although the length of time spent guarding will depend on patch quality. If two females find a patch, then at first both will exploit it without aggression. However, after a period of time, fighting will be initiated and the females take the roles of "intruder" or "resident", with the resident usually being the female that first arrived. The likelihood of a female *T. basalis* initiating conflict will depend on several factors: the number of potential hosts in the patch and the encounter rate with them, the asymmetry in arrival time and the number of conspecifics encountered, and the number of eggs invested in the patch. The resident will guard the patch, as it is by attacking within the first 3 hours after oviposition by the resident that the intruder will maximise her fitness. During this period, the female will be exposed to a trade-off between exploiting new hosts and guarding the patch. The intruder will regularly attempt to cryptically invade the patch, and eventually will succeed once the resident has left.

Patch defence should preferably be studied in multi-patch experiments. In single-patch experiments, one could easily underestimate the significance of defence behaviour. A classic example of this is the fighting and chasing which occurs when two females of *Venturia canescens* meet whilst searching the same patch. This aggressive behaviour is an important component of mutual interference in laboratory experiments with *V. canescens* (Hassell, 1978; subsection 1.14.3). The function of the fighting and chasing is not easily understood from such experiments because the behaviour leads, on large patches, to a decrease in attack rate for both wasps but not to the permanent exclusion of the intruding wasp. However, field observations of *V. canescens* searching for *Anagasta* larvae feeding on fallen figs suggest that a fig containing a host larva can be successfully defended against intruding competitors; the latter move on to nearby figs following an aggressive encounter (Driessen *et al.*, 1995).

1.14 DISTRIBUTION OF PARASITOIDS OVER A HOST POPULATION

1.14.1 INTRODUCTION

The distribution of parasitoids over a spatially structured (i.e. heterogeneous) host population has attracted considerable attention from theoretical ecologists. Hassell and May (1973) and Murdoch and Oaten (1975), among others, have shown that this is one of the key features affecting stability of parasitoid-host population models (Chapter 7). It is remarkable that the 'boom' in theoretical papers has not been paralleled by a similar surge in empirical studies on parasitoid behaviour.

1.14.2 AGGREGATION

The term **aggregation** is usually used to refer to the host-searching behaviour of parasitoids. Parasitoids may be more attracted to patches of high host density than to patches of low host density or they may show a stronger degree of arrestment in patches of high host density (subsection 1.5.3). Insect ecologists refer to an **aggregative response** of parasitoids and predators, because the aforementioned patch response behaviour leads to the concentration of parasitoids and predators in high-density patches. Latterly, the term aggregation has also been applied to the concentration of parasitoids on patches of low host density or on certain patches irrespective of the number of hosts they contain; this can occur if parasitoids are attracted to some patches in response to stimuli that are either negatively correlated with or independent of host density. In studies of population dynamics the term aggregation has also been used in a statistical sense, in terms of both the variance in parasitoid distribution and the covariance between the distributions of host and parasitoid (Godfray and Pacala, 1992).

Aggregation of adult parasitoids is the result of two different processes:

1. Differences among patches in the probability with which they are discovered by parasitoids. In a heterogeneous environment, it is likely that not every patch has the same probability of being detected, even if all patches are otherwise similar. Patches may also differ in the probability of detection by parasitoids because of differences in host density or other aspects of quality of the patch;
2. The period of time that each parasitoid stays in a patch after discovering it. The number of parasitoids visiting a patch and the period of time they stay there determines the amount of 'search effort' devoted to a patch.

Aggregation can be measured in two main ways:

1. Individual parasitoids can be presented with several patches of different host density, as in studies of patch time allocation (Figure 1.20);
2. Several parasitoids at one time can be presented with several patches of different host density (Figure 1.22).

When measuring aggregation using the second of these experimental designs, one should ideally monitor the behaviour of all parasitoids in all patches and record the time each parasitoid spends in each patch. In laboratory experiments with a modest number of host patches and with the insects continuously observed with video recording equipment, this is possible, but in field experiments such observations are very labour-intensive and often impossible to make. Published field studies on aggregation have therefore relied on periodic observations of the patches (e.g. Waage, 1983; Cronin 2003a).

One problem associated with studying aggregation in the field is deciding upon the spatial scale at which aggregation should be measured. Clumped distributions of hosts may occur at

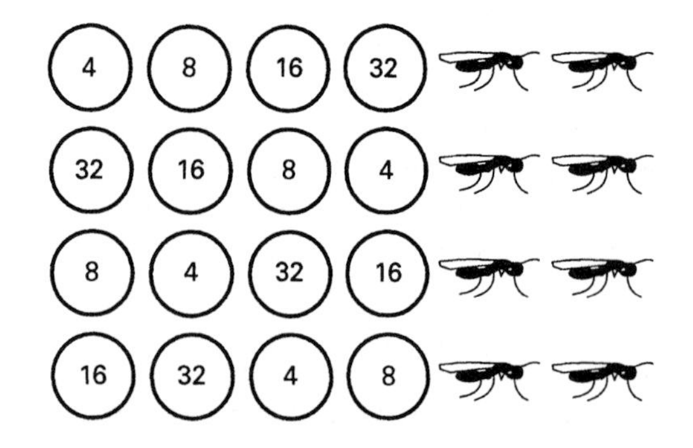

Figure 1.22 Aggregative responses of parasitoids and predators: Schematic representation of the design for an experiment used for detecting and measuring the aggregative response, and which takes account of the effects on interactions between foragers.

different levels of host distribution, and so may aggregation by parasitoids (e.g. Doak 2000). It is often possible, for practical purposes, to define what a patch is. For example, when studying the distribution of parasitoids of the cassava mealybug within a cassava field, cassava plant tips infested with mealybugs are the most relevant foraging units, whereas if one wants to compare biological control between different fields, whole cassava fields can be considered as patches.

The dispersal behaviour of the predator or parasitoid itself may also influence the aggregation pattern of aggregation seen. The likelihood of the minute fairyfly (Mymaridae) egg parasitoid *Anagrus sophiae* laying all of its eggs is correlated with dispersal distance among patches. Females that have dispersed over 250 metres from their natal patch will oviposit all their eggs in that patch (Cronin and Strong, 1999).

Such patterns may also be environment-dependent. Cronin (2003b) found that parasitism of the planthopper *Prokelisia crocea* by the egg parasitoid *Anagrus columbi* depended on the location of the planthopper's host plant, prairie cordgrass. When plant patches were surrounded by other grass species, parasitism rates were lower on the periphery of the patch, whereas if the host plants were surrounded by mudflat, attack rates were even throughout the patch. The dispersal behaviour of the parasitoid also varied, with cordgrass patches surrounded by non-host grasses having a higher likelihood of colonization by the egg parasitoid. It is evident that, ideally, studies of parasitoid aggregations should combine knowledge of host distributions with an understanding of parasitoid dispersal and foraging behaviour.

1.14.3 INTERFERENCE

Before we describe these phenomena, we need to stress that mutual interference, pseudo-interference and indirect mutual interference are concepts that can only be properly understood with reference to mathematical models, in particular those of searching efficiency (Chapter 7). The reader is therefore recommended to consult the literature dealing with modelling of parasitoid-host population dynamics (e.g. Hassell, 2000).

The tendency for some parasitoids and predators to cease searching and to leave the immediate vicinity after an encounter with a conspecific would account for the results of laboratory experiments designed to measure emigration rates in relation to parasitoid density. In these experiments, the proportion of female parasitoids leaving a single, fixed density host patch increased significantly with increasing numbers of parasitoids (see Figure 1.23 for experimental design). It has also been observed that when females encounter either an already parasitised host or a parasitoid mark on the substratum, they move away from the area where the encounter occurred. Any of these behavioural interactions are likely to cause the searching efficiency of a natural enemy in the single patch experiment to be reduced, a phenomenon known as **mutual interference** (Hassell and Varley, 1969; Visser and Driessen, 1991; Lynch, 1998; Kristoffersen *et al.*, 2001; Elliott, 2003).

The study of mutual interference began with Hassell and Varley (1969) who noted an inverse relationship between parasitoid searching efficiency and the density of searching parasitoids:

$$\log_{10} a' = \log_{10} a - m \log_{10} P \qquad (1.6)$$

where P is the density of searching parasitoids; a' is the effective attack rate or area of discovery per generation, $a'P = \log_e$ [initial number of hosts/number of hosts surviving parasitism]; a is the attack rate in the absence of interference. The gradient m is the measure of the extent of mutual interference. Such a relationship is to be expected because, as parasitoid density increases, individual parasitoids will waste an increasing proportion of their searching time in encounters with other conspecifics. Similar patterns are to be found with insect predators, and even among closely related species the importance of mutual interference will vary (Elliott, 2003).

Free *et al.* (1977) argued, using deductive models, that marked parasitoid aggregation (e.g.

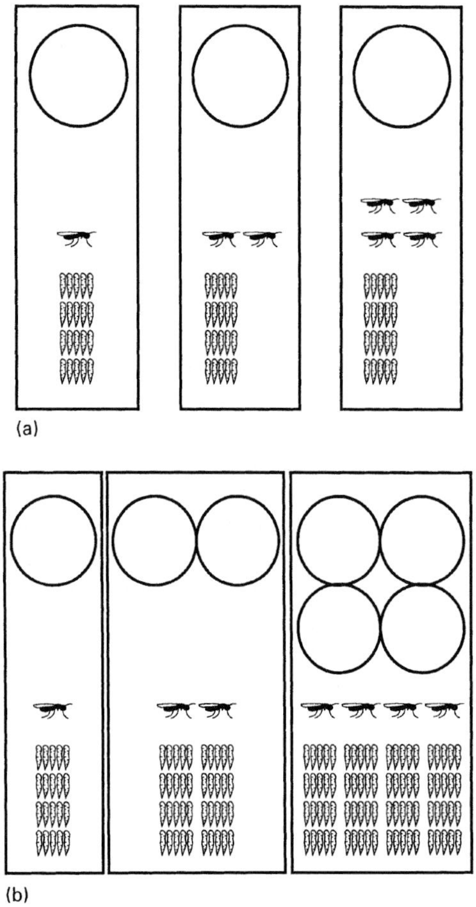

(a)

(b)

Figure 1.23 Mutual interference and pseudo-interference: Schematic representation of the design of two types of experiment for studying interference: (a) the design normally adopted for measuring interference, with a single host patch (denoted by a circle); (b) the design used by Visser *et al.* (1990): either a single parasitoid female searches a single unit patch containing 20 hosts, or two females search a double unit patch containing 40 hosts, or four females search a quadruple unit patch containing 80 hosts. Hassell (1971a,b) used a design similar to that given in Figure 1.22, although the number of foraging parasitoids was also varied. Both Visser *et al.*'s and Hassell's designs takes account of the multipatch context in which interference occurs.

resulting from a strong tendency of parasitoid individuals to spend longer periods of time in higher host density patches, and the consequent differential exploitation of patches) can lead to apparent interference, termed **pseudo-interference**, even if behavioural interference is lacking. As a consequence of parasitoids aggregating in high-density regions (because these are initially the most profitable) a higher proportion of the hosts in the whole area (i.e. experimental cage) is parasitised than would be possible with random search. If parasitoids do not respond (i.e. by dispersal) rapidly to the declining profitability of the high host density (i.e. more heavily exploited patches), then overall searching efficiency will be lower at high parasitoid densities. Thus, pseudo-interference results from 'overaggregation' by the parasitoids (Hassell, 1982a). In a population of optimally foraging parasitoids capable of responding rapidly to exploitation, overall searching efficiency would, at high parasitoid densities, be the same as for random search.

A third form of interference has been identified (Visser and Dreissen, 1991; Visser *et al.*, 1999). **Indirect mutual interference** was first found in the parasitoid *Leptopilina heterotoma*, a generalist parasitoid of drosophilids. Mutual interference is not found in this species, but as a result of superparasitism, searching efficiency is reduced at the population level, but not at the the level of the patch (Visser and Dreissen, 1991).

These different forms of interference can lead to the stabilisation of consumer-victim population interactions, and as a result has proved important in studies linking individual behaviour and population dynamics. In an instructive study, Visser *et al.* (1999) used data collected by Jones (1986; studying *Trybliographa rapae* attacking *Delia radicum*) to explore how the three different forms of interference may influence host-parasitoid population dynamics. In this case, the effect of interference depended on host distribution and the parasitoid's arrival and departure rules. Mutual interference did not appear to be important, but both indirect mutual interference and pseudo-interference reduced parasitoid search rate, their relative importance depending upon host distribution.

Traditionally, parasitoid attack rates (number of hosts parasitised per unit time) are used when considering interference relationships. However, if one is concerned with optimal behaviour, encounter rates should be considered. Visser *et al.* (1990) and van Dijken and van Alphen (1991) went further and calculated the mean number of realised offspring per female parasitoid per unit of patch time as a measure of individual efficiency. ESS models developed by Visser *et al.* (1992a) predict that the presence of other females on a patch reduces this efficiency, even when the number of hosts per female is held constant. This interference is not caused by behavioural encounters that decrease the encounter rate with hosts, but results from the parasitoids staying for longer on patches and superparasitising.

It needs to be stressed that when investigating interference phenomena, the parasitoid densities used and the host spatial distribution pattern must reflect those found in the field. As pointed out by Free *et al.* (1977), few experimenters take account of this requirement.

Another factor to consider is the size of the experimental arena. Jones and Hassell (1988) found *per capita* searching efficiencies to be lower in field cages than in laboratory cages and interference to be more marked in the latter, the volume of which was relatively small. Jones and Hassell attributed the interference to an unnaturally high frequency of encounters between searching parasitoids (*Trybliographa rapae*). The much lower searching efficiency in the field cages was presumably due to the greater opportunities for parasitoids to spend time performing behaviour other than searching in close proximity to hosts.

It should also be noted that not all interactions between foraging parasitoids are necessarily negative. For example, if parasitoids respond to the presence of others by reducing handling time or by avoiding areas previously searched by conspecifics (through patch marking), then 'positive interference' may occur (Visser *et al.*, 1999).

1.15 MEASURING ENCOUNTER RATES

The encounter rate of individual parasitoids with hosts is an important parameter in many optimality models. Because not every encounter will be followed by oviposition, and because not every oviposition will be in an unparasitised host, encounter rate is not equal to the number of hosts parasitised per unit time.

Optimality models divide the time budget of a foraging animal into searching time, recognition time and handling time. Encounter rate is expressed and measured as the number of encounters per unit of searching time, thereby excluding recognition time and handling time. Because encounter rates are not always a linear function of host density, it is necessary to measure them at a range of host densities.

To measure encounter rates, observe a female parasitoid continuously during some time period and make a complete record of her behaviour. From this record, the number of encounters and the net period of time spent searching can be calculated. The encounter rate of a parasitoid searching a patch containing a number of hosts may not be constant over the foraging period for the following reasons:

1. Parasitised hosts are encountered at a lower rate and the number of parasitised hosts increases during the observation period. A lower encounter rate with parasitised hosts may occur because hosts are paralysed by the wasp, and so move less (van Alphen and Galis, 1983).
2. The search effort of the wasp decreases, either in response to contact with its own marker substance (subsection 1.5.4) or because its supply of mature eggs dwindles.

One method of eliminating some of the causes of decreased encounter rate is to replace each parasitised host with an unparasitised one during the course of an experiment. This is not always possible, e.g. sessile hosts such as scale insects cannot easily be removed and replaced. Replacing parasitised hosts may also affect encounter rate: it may disturb the searching wasp and so decrease encounter rate or it may increase encounter rate when the parasitoid is of a species that reacts to host movements and the freshly introduced hosts move more than those already present. Finally, a parasitoid may learn, during the experiment, that the observer is introducing better quality hosts and simply walk towards the forceps or paint brush used to introduce the new host, as has often been observed with alysiine braconid parasitoids.

Therefore, when measuring encounter rates, one should not replace parasitised hosts but instead keep the period of observation short, in order to avoid accumulation of parasitised hosts and marker substance.

Encounter rates can be used to calculate predicted rates of offspring deposited per unit time with a particular optimal foraging model.

Measuring encounter rates using a single patch experimental design will overestimate the encounter rates that would be recorded in a multi-patch, i.e. natural environment, because the time spent in interpatch travel (i.e. transit time) is not accounted for. Since it is often difficult or impossible to measure transit times, the simplest approach is to measure, over a fixed period, the attack rate of a known number of parasitoids foraging in a spatially heterogeneous environment (Waage, 1979; Hassell, 1982a).

1.16 LIFE-HISTORY TRAITS AND FORAGING BEHAVIOUR

1.16.1 INTRODUCTION

Insect parasitoids display an enormous diversity of life-histories (Blackburn, 1991a,b; Godfray, 1994; Quicke, 1997; Mayhew and Blackburn, 1999; Jervis *et al.*, 2001, 2003). Some species have very short development times, can live for only a few days as adults, and emerge with all their eggs ready to be laid, in contrast to other species which develop slowly, can live for several months as adults, and produce new eggs throughout adult life. Some species produce a large number of small eggs, whereas others produce a small number of large eggs. These different life-history traits are associated with differences in searching and host selection behaviour. When designing experiments on parasitoid behaviour, it is important to be aware that this is so. We discuss this in relation to egg production strategy.

1.16.2 EGG-LIMITATION VERSUS TIME-LIMITATION

Parasitoids can be divided into pro-ovigenic and synovigenic species, (subsection 2.3.4). Pro-ovigenic parasitoids emerge with their full potential lifetime complement of mature eggs, whereas synovigenic parasitoids emerge with at most only part of their complement, this fraction varying considerably among synovigenic species (ranging from very nearly one down to zero) (Jervis *et al.*, 2001). These different patterns of egg production can be understood as adaptations to differences in the spatial and temporal distribution patterns of hosts (Jervis *et al.*, 2001; Ellers and Jervis, 2003) – natural selection can be expected to lead to reproductive strategies that approach (but not necessarily attain) a quantitative match between egg supply and the availability of suitable hosts.

Pro-ovigenic species are expected to behave in laboratory experiments in a time-limited manner, because even when large numbers of hosts are offered, these numbers do not exceed the number of mature eggs carried by the parasitoid. On the other hand, synovigenic species are expected to behave in an egg-limited manner, often exhausting their daily egg supply in a few hours when the number of hosts offered to them exceeds the number of mature eggs in their ovaries (this is a gross oversimplification of the difference between pro- and synovigeny; see Ellers *et al.* 2000a, for further details).

Whereas Heimpel and Rosenheim (1998) concluded, from a literature survey of fifteen species, that egg-limitation is common in the field, the results of empirical field studies suggest that only some females may, in reality, experience egg-limitation (Weisser *et al.*, 1997; Ellers *et al.*, 1998; Heimpel *et al.*, 1998; Casas *et al.*, 2000). Thus, it appears that parasitoids have evolved strategies that reduce the risk of egg-limitation. However, concomitant with these would be an increased risk of time-limitation, the risk being heightened by any cost, to life-span, of egg production (Ellers *et al.*, 2000a). Indeed, West and Rivero (2000), using a sex ratio-based method to measure the relative importance of egg- and time-limitation among eight parasitoid species, concluded that on average, most species are at an intermediate position along the egg-/time- limitation continuum, with a bias towards time-limitation.

Stochasticity in host availability is thus likely to play an important role in determining optimal egg loads, as has been argued by Rosenheim *et al.* (1996), Sevenster *et al.* (1998), Ellers *et al.* (2000a) and Ellers and Jervis (2000a). Theoretical studies have shown it to be a major influence, and that the patchy distribution of hosts is a major source of stochasticity (Rosenheim, 1996; Ellers *et al.*, 1998, 2000a). If stochasticity is high, investment is shifted away from life-span to eggs, i.e. towards an optimal egg load that is higher than the expected number of hosts found, and thus a lower incidence of egg-limitation. The ability of synovigenic parasitoids to mature eggs throughout life further reduces the incidence of egg-limitation; furthermore, it also reduces the degree to which individuals are time-limited (i.e. they have a surplus of eggs but not too many) (Ellers *et al.*, 2000a). However, synovigenic females will still experience transient levels of egg-limitation (Heimpel and Rosenheim, 1998; Heimpel *et al.*, 1998; Casas *et al.*, 2000; Rosenheim *et al.*, 2000).

The question of whether time- and egg-limitation, when observed in the laboratory, reflect the field situation or whether they are an artifact of unnaturally high host densities can be addressed by obtaining some measure of oviposition rate under field conditions, and comparing this with the average rate of egg production in parasitoids. The outcome of experiments on aspects of parasitoid biology as diverse as patch time allocation, functional responses, host selection, sex allocation, superparasitism or encounter rates, will all depend critically on whether the experimental conditions place the parasitoid under the constraint of time- or egg-limitation. Either experiments can be run under conditions representing both of these constraints, or an experimental design can be chosen that is relevant to the particular question one is asking. For example, when asking about the performance of a parasitoid immediately following field release, present females in experiments with a superabundance of hosts so that they are egg-limited, but when asking about the performance of the parasitoid after the host population has been suppressed below a damage threshold, present females with low densities of hosts so that the parasitoids are time-limited. If one is asking evolutionary questions, it is advisable to choose a situation (e.g. range of host densities, host spatial distribution pattern) closest to what the wasps experience most often in nature.

1.17 THE COST OF REPRODUCTION

In many studies of time allocation, recognition time and handling time are taken to be the only time costs involved in oviposition. However, as discussed in Chapter 2, a trade-off can exist between reproductive effort and survival. In at least one case, it appears that egg deposition, as opposed to egg production, incurs a survival cost (see subsection 2.8.3).

1.18 AGE-DEPENDENT FORAGING DECISIONS

Although parasitoids may have a longer life expectancy when they lay fewer eggs, they do not live forever. The older they become, the less likely they are to survive to another day. In addition, young adult parasitoids may be more fecund than older females (De Vis *et al.*, 2002; Riddick, 2003). Because of the diminishing probability of survival with increasing age, parasitoids should become less selective and accept more host types for oviposition (Iwasa *et al.*, 1984). For example, young *Lysiphlebus cardui* preferentially attack second and third instar *Aphis fabae*, whereas older wasps show no preference (Weisser, 1994). All other things being equal, older wasps will superparasitise and accept less suitable hosts more readily than younger ones, a prediction that is supported empirically (Roitberg *et al.*, 1992, 1993). One can try to make use of this alteration in behaviour with age in experiments that require parasitoids to oviposit in non-preferred hosts, e.g. parasitised individuals and unpreferred species.

1.19 FORAGING BEHAVIOUR AND TAXONOMY

Taxonomists work primarily with preserved specimens and until recently relied heavily on external morphological characters to describe species (Gauld and Bolton, 1988; Quicke, 1993). This is in most cases a satisfactory state of affairs, because differences in external morphology can often be found, even between closely related species. Sometimes, however, morphologically identical specimens can be collected from populations found in ecologically different situations, e.g. attacking a different host species, occurring on different host plants or in different geographical regions. The question then is whether these populations belong to one species or not – an important question, not only in deciding whether a parasitoid is a specific natural enemy of a target pest, but also because the scientific name of an organism is used in publications.

By comparing the host habitat-finding behaviour and host selection behaviour of different populations, one can establish whether important ecological differences exist between them. Differences in host-habitat finding and/or host species selection can theoretically result in reproductive isolation between the two populations, which occupy different niches by virtue of the differences in their searching behaviour. When interpopulation differences in foraging behaviour are found, one should then determine whether cross-matings are possible. If such matings do not occur either in the laboratory or in the field, it is reasonable to conclude that the populations are 'good' biological species.

Vet *et al.* (1984) discovered *Asobara rufescens* by studying microhabitat location of wasps initially believed to be *A. tabida. Asobara rufescens* had until then gone unrecognised and its populations had been considered conspecific with *A. tabida.* Similarly, van Alphen (1980) discovered a new species of *Tetrastichus*, which attacks the twelve-spotted asparagus beetle, *Crioceris duodecimpunctatum*, by showing that it rejected the eggs of *Crioceris asparagi*, the host *of Tetrastichus coeruleus.* Information about the foraging behaviour of insect natural enemies may therefore prove useful in taxonomy and systematics, although this is complicated by behavioural plasticity (Japyassu and Viera, 2002).

1.20 FORAGING BEHAVIOUR AND HOST RESISTANCE

1.20.1 INTRODUCTION

Not all prey or host individuals are equally worth attacking. It has become increasingly clear that the success rate of natural enemy attack can vary due to host/prey defence. Such resistance may take many forms, but this can be conveniently divided into physiological and behavioural defences.

1.20.2 PHYSIOLOGICAL HOST RESISTANCE

Physiological defences to endoparasitoid attack centre on the innate immune response of insects, which typically involves the parasitoid egg being isolated in a melanised capsule (subsection 2.10.2). The immune response is not, however, the only means by which hosts avoid attack: many herbivorous insects sequester plant secondary chemicals that can be deployed as a means of defence against their natural enemies. *Utetheisa ornatrix*, an arctiid moth, feeds on legumes from which it sequesters pyrrolizidine alkaloids (Eisner *et al.* 2000). These alkaloids are passed onto the eggs, and this acts as a deterrent against the predatory lacewing, *Ceraeochrysa cubana*. However, the amount of alkaloid passed down to the eggs varies, depending on the host plant the parents have been feeding on. The moth's eggs are laid in batches of about twenty, and the lacewing will sample two or three before deciding to accept or reject the batch of eggs. Since the variation in noxious chemicals within a batch is low, sampling a small number will provide a reliable indicator of prey quality. If there is considerable variation among batches in alkaloid concentration, sampling all batches is worthwhile (Eisner *et al.*, 2000).

It has been suggested that such secondary chemicals are more likely to be sequestered by specialist herbivores, than by generalists. This leads to the prediction that generalists should be subject to greater levels of attack by natural enemies, and that levels of attack should reflect the presence of the secondary chemicals in the community. In an elegant experiment, Camara (1997) tested the latter hypothesis under natural conditions. Buckeye butterfly larvae, *Junonia coenia*, were reared on plants that contained iridoid glycosides (*Kickxia elatine* and *Plantago lanceolata*) or an artificial medium lacking the defensive chemicals. In the sites where many plants contained iridoid glycosides, fewer larvae that had been fed on the plants were consumed by predators, whereas no differences in predation were found in sites with lower proportions of the glycoside-containing plants (Camara, 1997). Above, we saw that individuals can vary in the amount of secondary chemicals acquired from host plants; here, it is clear that there is also likely to be substantial variation among populations.

1.20.3 BEHAVIOURAL DEFENCES

Perhaps the classic example of a behavioural defence against predator attack is provided by **aphid dropping behaviour** (Losey and Denno, 1998a,b,c). Here, aphids drop from the plant in response to predator cues, although this may not always be to the benefit of the aphid, as ground predators will often successfully attack the escapees (Losey and Denno, 1998c). Pea aphids (*Acyrthosiphon pisum*) show genetic variation in dropping behaviour, and this behaviour is influenced both by the nutritional quality of the plants the aphid clones are reared on (Fellowes *et al.*, submitted) and by ambient temperature (Stacey and Fellowes, 2002). Aphids can also escape predation by the production of winged morphs in response to the presence of predators (Weisser *et al.*, 1999).

For aposematic prey species (and indeed most species), both physiological and behavioural defences may be intertwined. Tullberg *et al.* (2000) showed that while two species of lygaeid bugs (*Lygaeus equestris* and *Tropidothorax leucopterus*) are unpalatable to birds, the likelihood of being preyed upon was in part determined by the degree of aggregation of the larvae. Fewer attacks occurred when the larvae were in groups, compared to individual larvae.

Irrespective of the means of avoiding attack, it is clear that not all hosts are equal in value to a foraging predator. They vary not only in quality as a resource, but also in terms of the likelihood of their being successfully attacked, and this will vary among individuals, populations and species. Additionally, although much less studied, it is clear that there will also be variation at a similar series of scales in natural enemy 'virulence'.

1.21 INSECT NATURAL ENEMY FORAGING BEHAVIOUR AND COMMUNITY ECOLOGY

The role of insect natural enemy foraging behaviour in determining community interactions has been implicitly rather than explicitly implicated in many aspects of community ecology. These issues are perhaps best illustrated by the work of Müller *et al.* (1999), who studied an aphid-natural enemy (primarily parasitoids) system in Rush Meadow, an abandoned field in south-western England. By producing a quantitative food web of the interacting species, Müller and her colleagues were able to use the web to predict the strength of both direct and indirect interactions within the community. Such webs are immensely time-consuming to develop, but the return, measured in terms of detailed knowledge of the system, is potentially enormous (e.g. Schönrogge and Crawley, 2000; Lewis *et al.*, 2002) (see subsection 6.3.13). This web can be used to convincingly demonstrate the concept of apparent competition, where two species that do not directly compete for resources indirectly compete because of shared natural enemies (subsection 7.3.10; Holt, 1977). Müller and Godfray (1997) tested for apparent competition between the grass aphid, *Rhopalosiphum padi*, and the nettle aphid, *Microlophium carnosum*, mediated by shared natural enemies. They found that foraging ladybirds were attracted in increased numbers to the experimental site, mainly by the presence of grass aphids. However, the increased numbers of coccinellids preferentially attacked the nettle aphids, providing a clear example of apparent competition (Müller and Godfray, 1997).

Such food webs not only illustrate the importance of indirect effects with a community, but also point to the direct influence of insect natural enemies on community structure. Müller and Godfray (1999) studied why two species of aphids (*Aphis jacobaeae* and *Brachycaudus cardui*), common in surrounding areas, were uncommon in Rush Meadow. By excluding predators from artificially inoculated aphid colonies, they found that the aphids were able to colonise the field, but were prevented from doing so by the presence of predators and parasitoids. Clearly, the foraging decisions of insect natural enemies will have a significant influence on the community dynamics of any terrestrial ecosystem.

Such quantitative food webs also illustrate another important facet of insect natural enemy community ecology. While many predators and parasitoids are thought to have an extremely wide host range (the **fundamental niche**), these ignore the effects of host preference or competition, which results in a much narrower range of regularly attacked victims (the **realised niche**).

The importance of the community context in trying to understand the foraging behaviour of an insect predator or parasitoid should not be underestimated. Foraging success will often be influenced by the host plant a potential victim is attacking. For example in *Encarsia formosa* (a parasitoid of whitefly) foraging success is greater on glabrous varieties of cucumber (Hulspas-Jordaan and van Lenteren, 1978). Indirect effects of plants will also be common, as host species reared on poor quality plants will have reduced population growth rates, and parasitoids emerging from such hosts may be smaller and less fecund (Stadler and Mackauer, 1996), while predators may consume more prey individuals for the same return.

From the above examples, it is readily apparent that the foraging behaviour of insect natural enemies is often (within bounds) context-specific, and the range of species attacked will depend on both direct and indirect effects with the community. Moreover, the strength of these interactions may be influenced by the herbivore's host plants. Without an understanding of the natural history of the species of choice, designing laboratory-based systems to assay behaviour is fraught with difficulty.

1.22 CONCLUDING REMARKS

The study of the foraging behaviour of insect predators and especially parasitoids has provided a model system for ecologists for many years. Applied ecologists use such systems in the hope that information obtained from them will inform biological control measures. Population ecologists favour host-parasitoid systems because they offer a relatively tractable system to explore the dynamics of victim-enemy relationships (Godfray and Shimada, 1999).

The past three decades have seen rapid advances in the study of insect natural enemies, particularly as regards parasitoid behaviour. These developments have mainly been in response to advances in ecological theory. The availability of a whole suite of models of parasitoid-host population dynamics that incorporate important behavioural characteristics of parasitoids, combined with the rapid progress in behavioural ecology, has led to the formulation of more precise and quantitative hypotheses. In addition, tools for the analysis of complex time-series of behaviour have become available, allowing us to address problems that previously could not be analysed properly. Behavioural studies of parasitoids have been conducted along two main lines: the functional analysis of behaviour, which has been guided by largely model-based theories on the evolution of animal behaviour, and the causal analysis of behaviour, which has been guided much less by models.

Developments in causal and functional analyses of behaviour have been, for a large part, independent. As we hope to have indicated in this chapter, research can benefit from an increased integration of the study of mechanisms and the study of the function of behaviour.

There is a two-way interaction between theory and empirical research: behavioural and population models can guide us in the design of experiments, while the results of experiments stimulate new theory. Theories of population dynamics and behavioural ecology are now often concerned with the behaviour of parasitoids and predators in a spatially heterogeneous environment, with patchily distributed hosts or prey. For reasons of convenience, behavioural studies on natural enemies have often been conducted in single-patch environments. Empirical results from experimental studies on the behaviour of natural enemies in multi-patch environments, preferably in natural settings, can provide the information needed both to test current theories and to develop new ones.

In hand with this call for a meshing of empirical and theoretical ecology is the continuing need for researchers to have a firm grasp of the natural history of their species. Natural history is difficult, if not impossible, to learn in the lecture theatre or laboratory, and there is no better way of beginning to understand the ecology of any system than spending time with it in the field. Becoming thoroughly acquainted with the study organism(s) must be a prerequisite for any successful research programme.

1.23 ACKNOWLEDGEMENTS

We are very grateful for the helpful comments of Emma Wittmann on earlier drafts, for the advice given by Jacintha Ellers, for the help of Kerry Elliott in preparing the manuscript, and for the valuable contribution made by Ian Hardy and Paul Ode to section 1.2.3.

THE LIFE-CYCLE 2

M.A. Jervis, M.J.W. Copland and J.A. Harvey

2.1 INTRODUCTION

This chapter is concerned with approaches and techniques used in studying those aspects of parasitoid and predator life-cycles that are relevant to the topics covered by other chapters in this book. To illustrate what we mean, consider the female reproductive system of parasitoids, discussed in some detail in section 2.3. As pointed out by Donaldson and Walter (1988), at least some knowledge of its function, in particular of the dynamics of egg production, is crucial to a proper understanding of foraging behaviour in parasitoids. The state of the ovaries may determine: (a) the duration of any pre-oviposition period following eclosion; (b) the rate of oviposition, (c) the frequency and duration of non-ovipositional activities, e.g. host-feeding; and (d) the insect's response to external stimuli, e.g. odours, hosts (Collins and Dixon, 1986) (subsection 1.5.1). Note that egg load (defined in subsection 1.2.2) is now often incorporated into foraging models, as it is becoming increasingly clear that key foraging decisions depend upon the insect's reproductive state (Jervis and Kidd, 1986; Mangel, 1989a; Chan and Godfray, 1993; Heimpel and Rosenheim, 1995). It also follows from the above that a female parasitoid's searching efficiency depends upon the functioning of its reproductive system and this may in turn influence parasitoid and host population processes (Chapter 7).

Comparative studies have provided useful insights into the factors that determine patterns of cross-species variation in the life-history traits of parasitoids and predators. The results of these investigations are only touched upon in this chapter; for further details, see Blackburn (1991a,b), Gilbert and Jervis (1998), Mayhew and Blackburn (1999), Strand (2000), Jervis *et al.* (2001,2003) and Mayhew and Glaizot (2001) on parasitoids, and Dixon (2000, and references cited therein) on predatory coccinellids. Godfray (1994) and Quicke (1997) should also be consulted for information on comparative aspects of parasitoid biology.

Much of the chapter is devoted to methods of recording variation in key life-history traits Investigators should be mindful of the potential for trade-offs to occur between life-history variables, as predicted by general life-history theory (e.g. see Roff, 2002). Examples of phenotypic trade-offs are given in the various sections on fecundity, adult longevity, development and growth and immature survival. Partly because of such trade-offs, caution should be exercised in using individual life-history traits as proxy measures of fitness (see Roitberg *et al.*, 2001). Genetic aspects of trade-offs are discussed in Chapter 3.

2.2 ANATOMICAL STUDIES ON NATURAL ENEMIES

2.2.1 INTRODUCTION

A general introduction to insect structure and function can be found in most standard entomological texts, e.g. Wigglesworth (1972), Chapman (1998), Richards and Davies (1977), Commonwealth Scientific and Industrial Research Organisation (1991). Individual topics are covered in texts such as Snodgrass (1935) on morphology; Engelmann (1970) and Kerkut and Gilbert (1985) on insect reproduction. There are also texts such as Hodek (1973), Gauld and Bolton (1988), Quicke (1997) and McEwen *et al.* (2001), that deal with aspects of the anatomy and morphology of particular taxonomic groups of insect natural enemy. This section is concerned with methods used for investigating the internal anatomy of predators and parasitoids, the emphasis being placed on the female reproductive system.

2.2.2 TECHNIQUES

Dissection

Many insect natural enemies, particularly parasitoids, are so small that routine investigations of their internal anatomy might, at first sight, seem impossible to undertake. One approach to anatomical investigation is to fix and then embed insects in wax or resins, and then to cut, using a microtome, serial sections of the body. This method is, however, technically difficult and there usually arise problems such as distortion (e.g. due to hardness of the cuticle), inadequate fixation and the difficulty of reconstructing sections into a three-dimensional model. A far easier approach is to dissect the insect.

In order to carry out dissection, the following equipment will be required: a stereomicroscope with incident lighting (preferably fibre optics, see below), ordinary or cavity microscope slides, insect saline (e.g. 7.5 g NaCl/litre) and some fine pins. The latter are best securely mounted either in glass tubes (4 mm diameter and approx. 50 mm long) or in matchsticks.

For parasitoid wasps (Trichogrammatidae, Mymaridae and others up to 25 mm long), place one droplet of insect saline on to a microscope slide and place the insect in the droplet. Use insects that have been recently killed either with ether, carbon dioxide, with some other suitable killing agent, or by freezing. Individuals that have been dead for more than an hour at room temperature, and also ones that have been preserved in alcohol, are very difficult to dissect, so storing insects in a deep freeze is highly recommended. When dissecting, ensure that the insect's body is dorsal side up, feet down. With one pin, restrain the insect from floating or otherwise moving in the saline, either by piercing its thorax, or by holding the pin across the female's petiole. With the second pin, make small lateral incisions in the distal part of the gaster, preferably where there is an intersegmental membrane. Place the point of the second pin firmly upon the tip of the insect's gaster and pull the latter gently away from the remainder of the gaster. The abdominal wall should part in the region of the incisions, and the abdominal contents should then spill out into the saline droplet. With a little practice, this technique will permit examination of the entire reproductive system, and also of the mid and hind gut. By carefully noting the positions of all the various organs during dissection, it should be possible to reconstruct the spatial arrangement of the organs and associated structures (Figure 2.1).

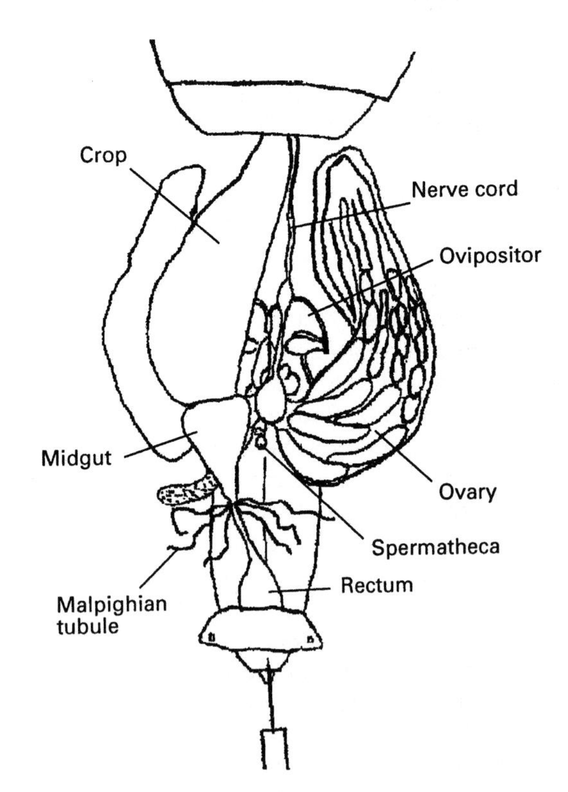

Figure 2.1 Dissection of the gaster of female *Nasonia vitripennis* (Pteromalidae). The point of a micropin is used to pull away the tip of the gaster and reveal the internal organs.

More difficult manipulation may be required in the case of parasitoid wasps with long ovipositors that are housed within the body as a spiral (e.g. Eurytomidae) or extended forward in a 'horn' above the thorax (e.g. *Inostemma* species (Platygastridae)).

Two points need to be borne in mind when using the aforementioned technique. First, the insect must be kept covered in saline solution at all times. If it dries out, it cannot be satisfactorily reconstituted. Second, if water rather than saline is used, some structures may expand and become seriously distorted. Unless a fibre optics system is being used, avoid using an under-stage light source (useful for assisting the examination of some structures) for periods longer than a few minutes only, as the specimen will dry out very quickly.

The above technique can be used for small predators and small dipteran parasitoids, but with large insects such as carabid beetles and hover-flies a small, water-filled, wax-bottomed dish should be used instead of a microscope slide and saline droplet. Gilbert (1986) describes a technique for dissecting adult hover-flies (Syrphidae) (Figure 2.2a) that can also be applied to dipteran parasitoids and predatory beetles and bugs. The insect is placed on its back (on a slide or wax-bottomed dish, dry or under saline) and is secured with an entomological pin inserted through the thorax. Using a second entomological pin, a small tear is made in the intersegmental membrane at the junction of the thorax and abdomen. The end of one arm of a fine forceps is then inserted into this hole and the forceps are then used to grip the first abdominal sternite. Then, using a micropin (preferably one having a point that has been slightly bent near its tip), make lateral incisions in the abdomen, following the line of the pleura to the terminalia. Finally, peel back the abdominal sternites to reveal the internal organs (Figure 2.2b,c). The crop (very large in hover-flies) can be removed in its entirety using forceps, and its contents (pollen and/or nectar) subsequently examined and analysed. The reproductive system can be examined *in situ*, under saline. Carabid beetles are dissected in a similar

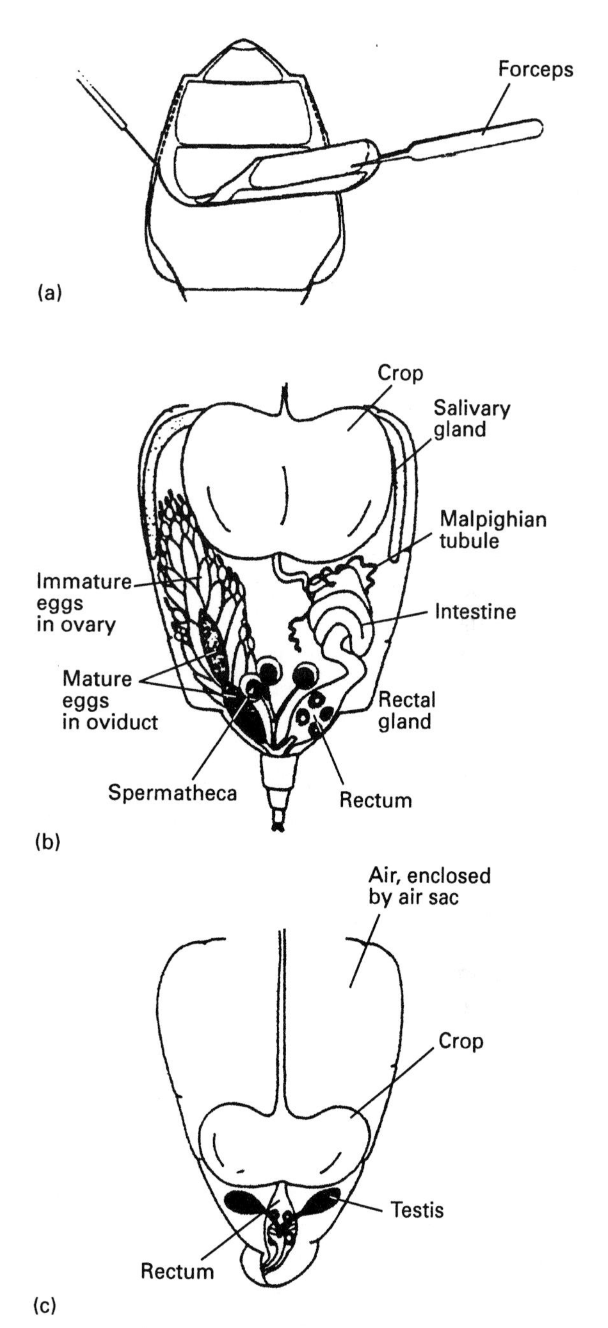

Figure 2.2 Dissections of hover-fly (Syrphidae) abdomen: (a) dissection procedure; (b) internal anatomy of female, (c) internal anatomy of male. Source: Gilbert (1986). Reproduced by permission of Cambridge University Press.

fashion, except that the insect is placed on its front. Figure 2.3 shows the gut of a typical carabid beetle.

It is very difficult to interpret the structure of an insect's reproductive system, or that of other organs, if the structure has been fixed and preserved. If a permanent record of a dissection is needed, the insect's organs are best photographed or drawn as soon as possible. Semi-permanent mounts can be made with water-soluble mountants such as polyvinyl pyrrolidone (Burstone, 1957) or glycerol, but anatomical features are bettter observed in freshly dissected insects. Anatomical features are enhanced by the use of specialist optics such as phase contrast, interference and dark ground illumination, with a transmission compound microscope.

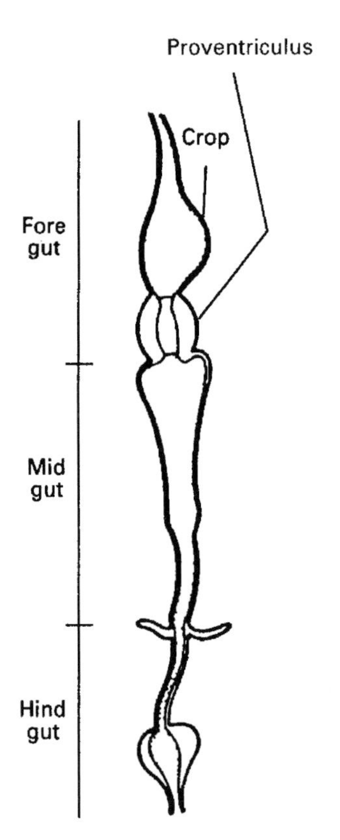

Figure 2.3 The gut of a typical carabid beetle. Source: Forsythe (1987). Reproduced by permission of The Richmond Publishing Co. Ltd.

Microscopy

There is a limit to the information that can be obtained from dissection. Histological and histochemical techniques will reveal the location of lipids, carbohydrates, nucleic acids and many more specific materials in, for example, the reproductive organs (see also section 2.14). Such techniques have been crucial to our understanding of oögenesis in parasitoids (King and Richards, 1969; King *et al.*, 1969b, 1971; Davies *et al.*, 1986). Combined with electron microscopy, they can reveal the detailed structure of secretory tissues, egg oöplasm (e.g. see Le Ralec, 1995), and can demonstrate the effects of diet and temperature on structures such as mitochondria and cell membranes. Davies (1974), for example, showed how in *Nasonia vitripennis* the ultrastructure of flight muscle alters with the age of the adult insect and with variations in adult diet.

2.2.3 OVIPOSITOR AND MALE GENITALIA

The ovipositor of female parasitoids may need to be examined in detail in order to understand the mechanics of oviposition, while the secondary genitalia of male dragonflies may need to be examined in order to study sperm competition (subsection 4.5.2). Light microscopy and scanning electron microscopy (SEM) are usually employed to study these structures. In order to examine whole mounts with light microscopy, clear and stain them following standard protocols, whereas to examine sections, e.g. of ovipositors, embedding, sectioning and staining needs to be carried out; standard protocols (embedding in Spurr's medium and staining, e.g. with Toluidine Blue) were followed, for example, by Austin (1983) and Quicke *et al.* (1992). Greater detail of external morphology can be seen using SEM (e.g. King and Fordy, 1970; Quicke *et al.*, 1992; Jervis, 1992). Specimens of small Hymenoptera and of Diptera are best prepared for SEM by critical-point drying them (Postek *et al.*, 1980), whereas specimens of larger and more hard-bodied insects require only air drying.

Snodgrass (1935) described the basic structure of both male and female insect genitalia, while Scudder (1971) interpreted the structure of the ovipositor in Hymenoptera. For details of ovipositor structure and function in parasitoids, including in some cases the mechanism of egg movement, see Austin and Browning (1981), Austin (1983), Jervis (1992), Field and Austin (1994), Quicke *et al.* (1994), Le Ralec *et al.* (1996), Austin and Field (1997), Kozanec and Belcari (1997), Gerling *et al.* (1998), van Lenteren *et al.* (1998), Rahman *et al.* (1998), Le Lannic and Nenon (1999), Vilhelmsen *et al.* (2001), Heraty and Quicke (2003) and Zacaro and Porter (2003).

Parasitoids, in common with other insects, possess a diversity of sensilla on the ovipositor (King and Fordy, 1970; Gutierrez, 1970; Weseloh, 1972; Hawke *et al.*, 1973; Greany *et al.*, 1977; van Veen, 1981; Jervis, 1992; Kozanec and Belcari, 1997; Consoli *et al.*, 1999). The function (i.e. mechanoreception, chemoreception) of the sensilla can be provisionally inferred from their external morphology, but corroboration needs to be obtained by examining them in detail using transmission electron microscopy, by observing female oviposition behaviour, and by carrying out electrophysiological studies. The role of ovipositor sensilla in host acceptance by parasitoids (section 1.5.5) has long been appreciated.

Except for a few studies such as those by Domenichini (1953), Sanger and King (1971), Teder (1998) and Chiappini and Mazzoni (2000), little work has been done on the functional morphology of male genitalia in either dipteran or hymenopteran parasitoids. The structure and function of the secondary genitalia of male dragonflies are better understood, having been examined by several authors, including Waage (1979a,1984), Artiss (2001) and Cordoba-Aquilar (2002).

2.3 FEMALE REPRODUCTIVE ORGANS

2.3.1 OVARIES

The reproductive organs of hymenopteran (Figures 2.1, 2.4a,b,d, 2.5) and dipteran (Figure 2.4c,e) parasitoids comprise a pair of ovaries which themselves comprise several ovarioles in which the eggs (oöcytes) develop. In parasitoid wasps (King and Richards, 1969) and flies (Coe, 1966) the ovarioles are of the **polytrophic** type. Within each follicle, nurse cells (trophocyte cells: fifteen or more in hymenopteran parasitoids) surround the developing oöcyte, providing it with nutrients (Figure 2.6a). The oöcyte becomes increasingly prominent as it passes down the ovariole. Each oöcyte, together with its associated trophocyte cells, originates from a single cell. It seems that in order to develop eggs as rapidly as possible, the protein production machinery of all the trophocyte cells passes materials into the oöcyte. The follicle cells, which may also pass materials from the haemolymph, secrete the egg shell, i.e. the **chorion**. As the oöcyte matures, the trophocyte cells break down. The follicular epithelium creates a small pore (the **micropyle**) in the chorion, through which the sperm enters to penetrate the egg membrane and effect fertilisation.

To examine the ovarioles of a dissected insect, remove the ovaries (their attachment to the abdominal wall may need to be severed), place them on a microscope slide in a drop of insect saline, and tease the ovarioles apart with micropins. Then gently place a cover-slip over the ovaries. The number of ovarioles can then be counted and their contents viewed.

In both hymenopteran and dipteran parasitoids, the number of ovarioles per ovary varies both interspecifically (Flanders, 1950; Price, 1975; Dowell, 1978; Jervis and Kidd, 1986; Quicke, 1997) and intraspecifically (e.g. van Vianen and van Lenteren, 1986). Many chalcidoid wasps have an average of three ovarioles per ovary (*Encarsia formosa* has an average of eight to ten, depending on the population studied), whereas in ichneumonoid wasps the range of interspecific variation is much wider (Iwata, 1959, 1960, 1962; Cole, 1967; Quicke, 1997). In some species of Ichneumonidae ovariole number alters according to whether the females are of the first or the second field generation, female body size being taken into account, i.e. there is a seasonal dimorphism (Cole, 1967). In predatory coccinellids, as in parasitoids, there is both

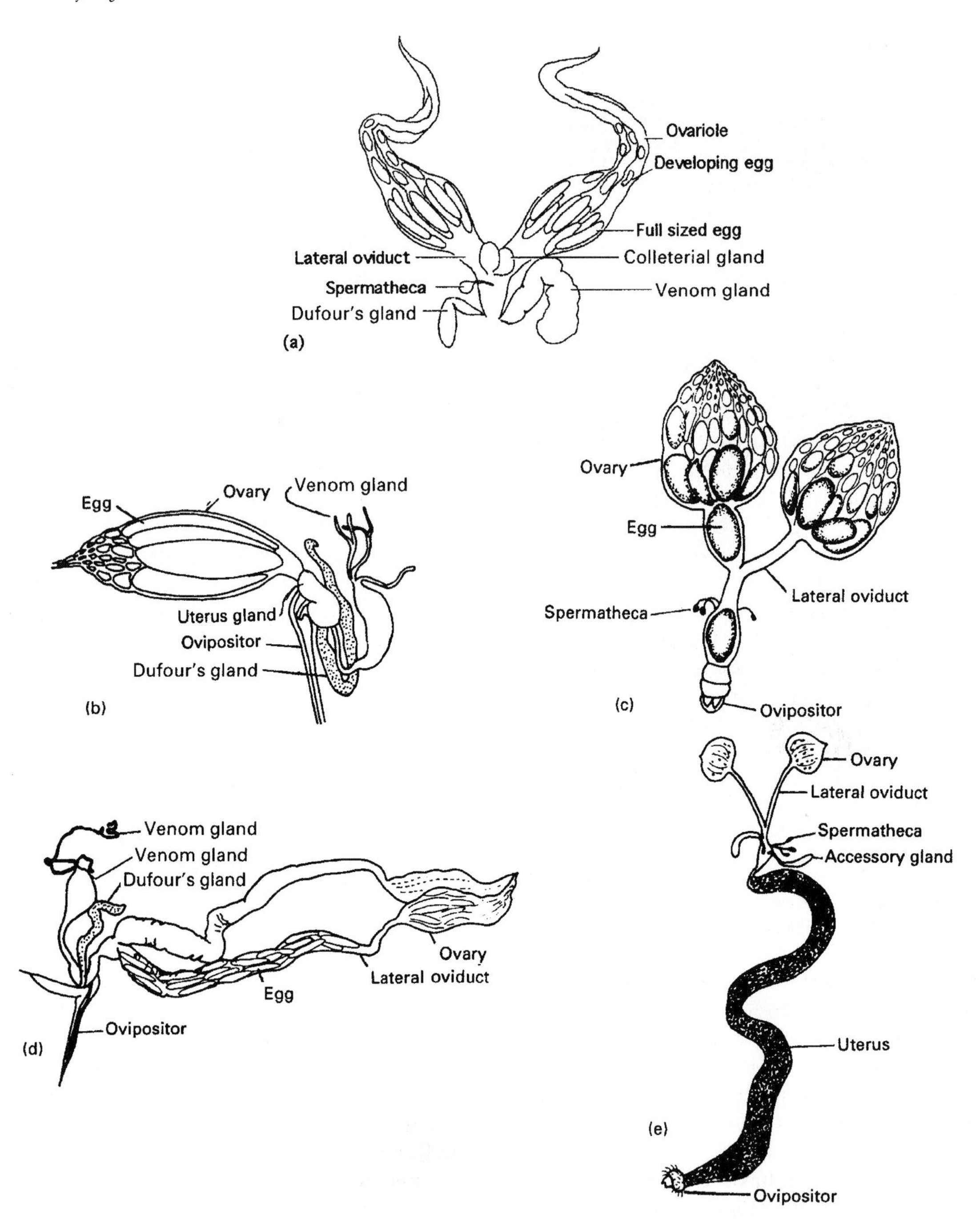

Figure 2.4 The reproductive systems of some parasitoid wasps and flies: (a) gravid female *Coccophagus atratus* (Aphelinidae) 24 hours after emergence (source: Donaldson and Walter, 1988.) (b) *Trachysphyrus albatorius* (Ichneumonidae) (source: Pampel, 1914, in Price, 1975.) (c) *Hyperecteina cinerea* (Tachinidae) (source: Clausen *et al.*, 1927, in Price, 1975.) (d) *Enicospilus americanus* (Ichneumonidae) (source: Price, 1975.) (e) *Leschenaultia exul* (Tachinidae) (source: Bess, 1936 in Price, 1975.). Note: acid gland = poison gland = venom gland. (a) reproduced by permission of Blackwell Publishing; (b), (c), (d) and (e) by permission of Plenum Publishing Corporation.

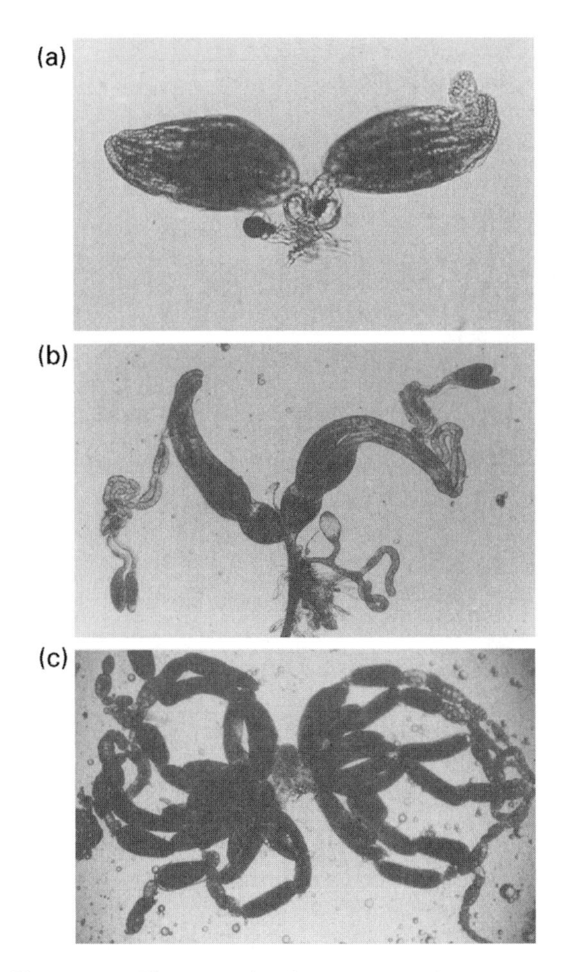

Figure 2.5 The reproductive systems of some parasitoid wasps: (a) *Gonatocerus* sp. (Mymaridae); (b) *Cotesia* sp. (Braconidae); (c) unidentified Eulophidae.

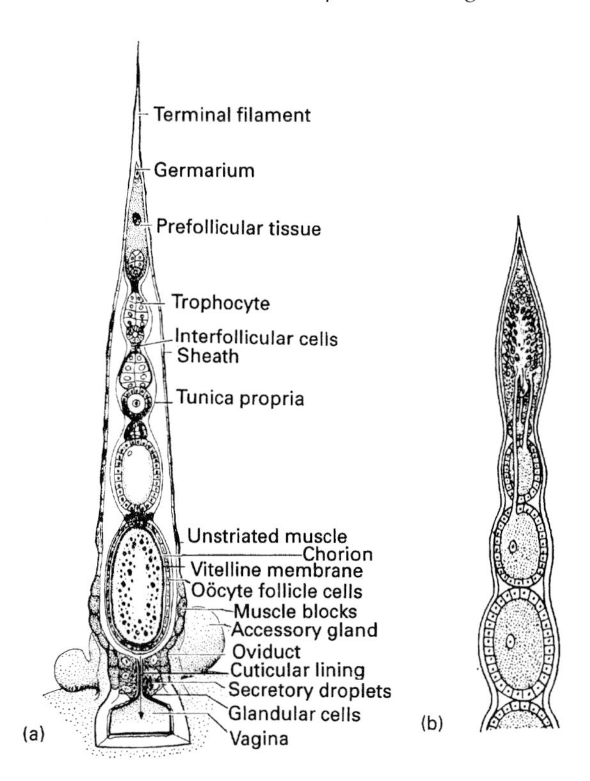

Figure 2.6 Examples of ovariole structure in natural enemies: (a) polytrophic type, in *Nasonia vitripennis* (Pteromalidae) (source: King and Ratcliffe, 1969); (b) telotrophic type as found in coccinellid beetles and heteropteran bugs (source: de Wilde and de Loof, 1973). (a) reproduced by permission of The Zoological Society of London; (b) by permission of Elsevier Science.

intra- and interspecific variability in ovariole number (Iperti, 1966; Stewart *et al.*, 1991). Welch (1993) reviews ovariole number in Staphylinidae.

Predator ovaries fall into several categories. Those of chrysopid lacewings and carabid and gyrinid beetles have polytrophic ovarioles (e.g. Figure 2.7), but coccinellid beetles and predatory heteropteran bugs have **telotrophic** ovarioles (Figure 2.6b). In the latter, the trophocyte cells, instead of accompanying the oöcyte as it moves down the ovariole, remain in the swollen distal end of the ovariole and remain attached to the egg by a lengthening cytoplasmic strand that conveys the nutrients. Telotrophic ovarioles are therefore short, but they are often numerous.

A measure of female reproductive potential can be obtained by counting the total number of oöcytes – mature and immature – within the ovaries and oviducts (subsection 2.7.1). It is a fairly simple procedure to count the number of mature eggs in species that possess enlarged lateral oviducts in which the eggs accumulate (subsection 2.3.2), but care is needed in the case of species that store (albeit for a brief period) some or all of their eggs within the basal part of the ovariole. With practice, it is possible to recognise mature eggs by their slightly opaque appearance resulting from the presence of yolk within (i.e. in anhydropic species, subsection 2.3.4). Immature oöcytes, particularly the smaller ones, are more difficult to count. A stain

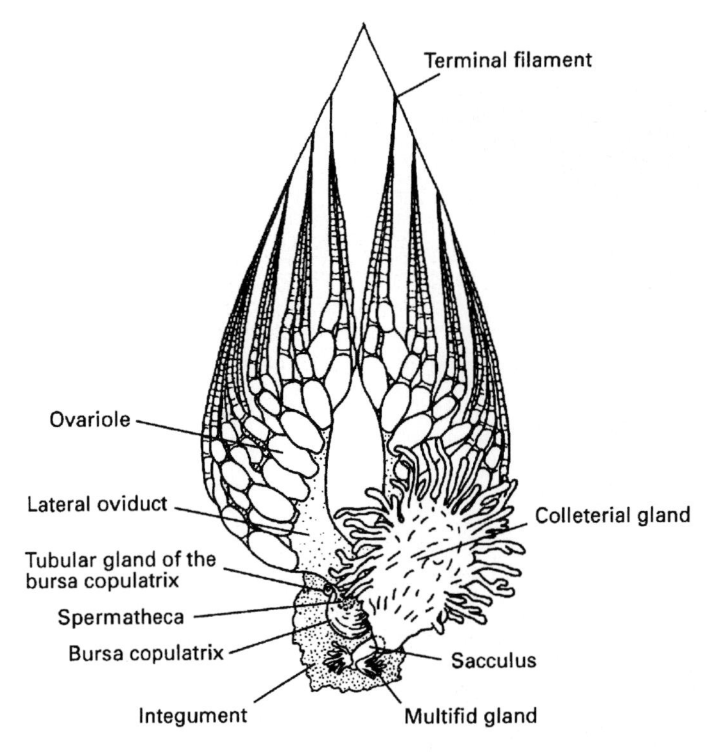

Figure 2.7 Schematic representation of reproductive system in female *Chrysopa septempuncata* (Neuroptera), dorsal aspect. Source: Principi (1949). Reproduced by permission of W. Junk, Publishers.

such as acetocarmine can be used to reveal them more clearly: the stain is taken up by these oöcytes, because they lack a chorion (in mature oöcytes, only the surrounding follicle becomes stained; the follicle is eventually lost prior to the mature egg entering the oviduct).

2.3.2 OVIDUCTS

The ovarioles empty into the lateral oviducts (Figures 2.4, 2.5, 2.8). In most Hymenoptera, each lateral oviduct includes an obvious glandular region – the calyx (Figure 2.8) – which secretes materials onto the egg as it is laid (Rotheram, 1973a,b). In some Braconidae and Ichneumonidae the calyx is the source of polydnaviruses (baculoviruses of the family Polydnaviridae) (Stolz and Vinson, 1979; Stolz, 1981; Cook and Stolz, 1983; Stoltz *et al.*, 1984; Strand *et al.*, 1988; Fleming, 1992). The latter, which replicate in the cells of the calyx, play a role in preventing encapsulation of the parasitoid egg (subsection 2.10.2) and in modifying the host's growth, development, morphology and behaviour (Vinson and Barrass, 1970; Vinson, 1972; Cloutier and Mackauer, 1980; Vinson and Iwantsch, 1980a; Stolz, 1986; Strand *et al.*, 1988). Cheillah and Jones (1990) raised an antibody against the extracted polydnaviral proteins of *Chelonus* sp. and then used it to reveal the location of such proteins in the wasp's reproductive system.

In some synovigenic parasitoid wasps the lateral oviducts can accommodate a small number of eggs, e.g. 9–12 per oviduct in *Coccophagus atratus* (Donaldson and Walter, 1988) (anhydropy, subsection 2.3.4). In others the oviducts are greatly elongated, to form distinctive 'uteri', and can accommodate very large numbers of small eggs (Figures 2.4d and 2.5b) (hydropy, subsection 2.3.4).

The lateral oviducts join to form the common oviduct, a largely muscular structure that in turn becomes confluent with the vagina and (in wasps) the ovipositor stylets. In some tachinid parasitoids, egg storage (and incubation) occurs

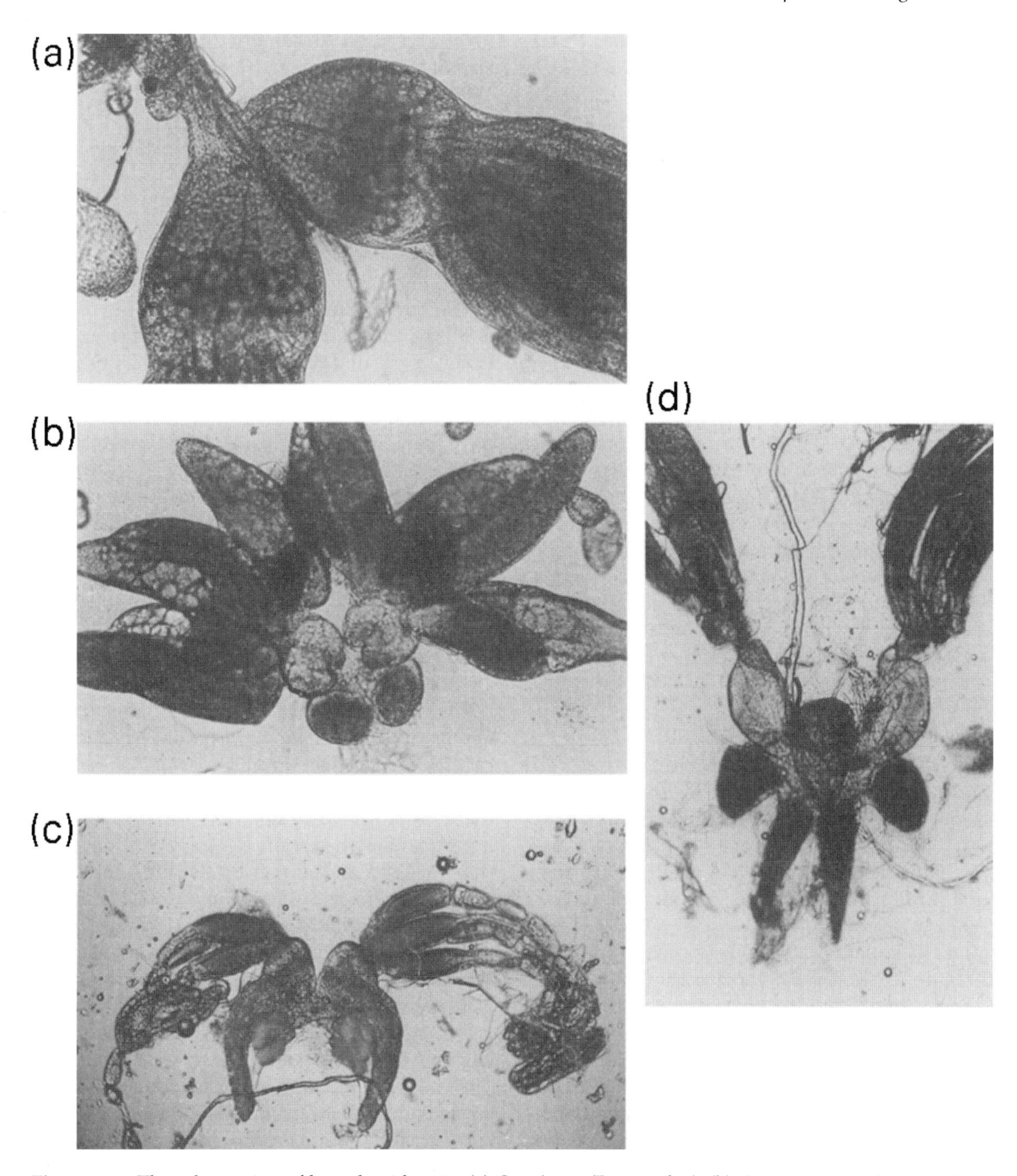

Figure 2.8 The calyx region of lateral oviduct in: (a) *Cotesia* sp. (Braconidae); (b) *Aprostocetus* sp. (Eulophidae) (also showing one pair of colleterial glands); (c) *Torymus* sp. (Torymidae) (also showing two pairs of colleterial glands); (d) *Macroneura vesicularis* (Eupelmidae) (showing calyx lobes, i.e. the very long structures, and two pairs of colleterial glands).

in the common oviduct, e.g. *Cyzenis albicans* (Hassell, 1968). In wasps, forward-pointing spines in the vagina push the egg into the ovipositor at or before oviposition (Austin and Browning, 1981). As it passes down the ovipositor, the egg is squeezed to a small diameter, a process that has been shown to trigger embryonic development (Went and Krause, 1973). Embryonic development of haploid (male) eggs of the ichneumonid parasitoid *Pimpla turionellae* can also be triggered by experimental injection, not involving egg deformation, of calcium ionophore A23187 (Wolf and Wolf, 1988). The chorion of the hymenopteran egg is remarkably flexible, so experiments on the initiation of embryogenesis can be carried out on mature eggs that have been removed from the ovarioles or lateral oviducts of a wasp. The eggs can be manipulated in various ways on a microscope slide, in saline solution, to show, for example, what degree of compression is required to trigger embryogenesis. In the tachinid *Cyzenis albicans* eggs, when laid, contain a fully-formed first-instar larva (Hassell, 1968).

2.3.3 SHAPE, SIZE AND NUMBER OF EGGS

The shape of eggs in parasitoid wasps and flies varies considerably between groups (Iwata, 1959, 1960, 1962; Hagen, 1964a; Quicke, 1997). Egg types found among parasitoid wasps include those with a simple ovoid shape, those that are greatly elongated (Figure 2.9a,b), those with a distinctive stalk at the micropyle end, and those with a double-bodied appearance (Figure 2.9c). For a review of the range of egg types found among parasitoids see Hagen (1964a) and Quicke (1997).

Some eggs (hydropic-type eggs, subsection 2.3.4) characteristically increase greatly in size following deposition in the host's haemocoel. Among Braconidae for example, eggs of Euphorinae expand in volume a thousand times (Ogloblin, 1924; Jackson, 1928), and those of *Praon palitans* (Aphidiinae) over six hundred times (Schlinger and Hall, 1960).

Within a parasitoid wasp species the number and the size of mature oöcytes in the ovaries are, in general, positively correlated with the size of the female (e.g. see O'Neill and Skinner, 1990; Rosenheim and Rosen, 1992; Visser, 1994; but see Fitt, 1990). This observation has important implications for foraging models, since larger females may, theoretically, obtain larger fitness returns per host and also, compared with smaller females, they can utilise a series of hosts in more rapid succession (Skinner, 1985a; O'Neill and Skinner, 1990).

The number of mature oöcytes in the ovaries is a function of the number of ovarioles, which is also correlated with body size within a species (e.g. Banquart and Hemptinne, 2000). Data on oöcyte number, oöcyte size and ovariole number have been gathered for a limited number of species.

In the damselfly *Coenagrion puella*, the carabid beetle *Brachinus lateralis*, and the hover-fly *Episyrphus balteatus*, egg size is not correlated with female size (Banks and Thompson, 1987a; Juliano, 1985; Branquart and Hemptinne, 2000), but it is correlated within species of Gerridae and predatory Coccinellidae (Kaitala, 1991; Dixon, 2000).

The size of a female's eggs may alter during her lifetime. Giron and Casas (2003b) recently demonstrated that *Eupelmus vuilletti* reduces egg provisioning with age: with increasing age, there is a marked decrease in reproductive investment with respect to egg size, and sugar, protein, lipid and energy content. Egg size was a good predictor of offspring fitness, measured as survival of neonate larvae. Wallin *et al.* (1992) showed that in carabid beetles egg size decreases with increasing oviposition rate.

Between parasitoid species, ovariole number is a good predictor of fecundity, as Price (1975) has shown for Ichneumonidae and Tachinidae (Figure 2.10). It remains to be tested whether or not a correlation exists between body size and ovariole number, on a broad, between-species basis.

Blackburn (1991a) and Jervis *et al.* (2003) showed, through comparative analyses, that among parasitoid wasps there is not a positive relationship between adult size and lifetime fecundity (fecundity is defined in subsection 2.7.1),

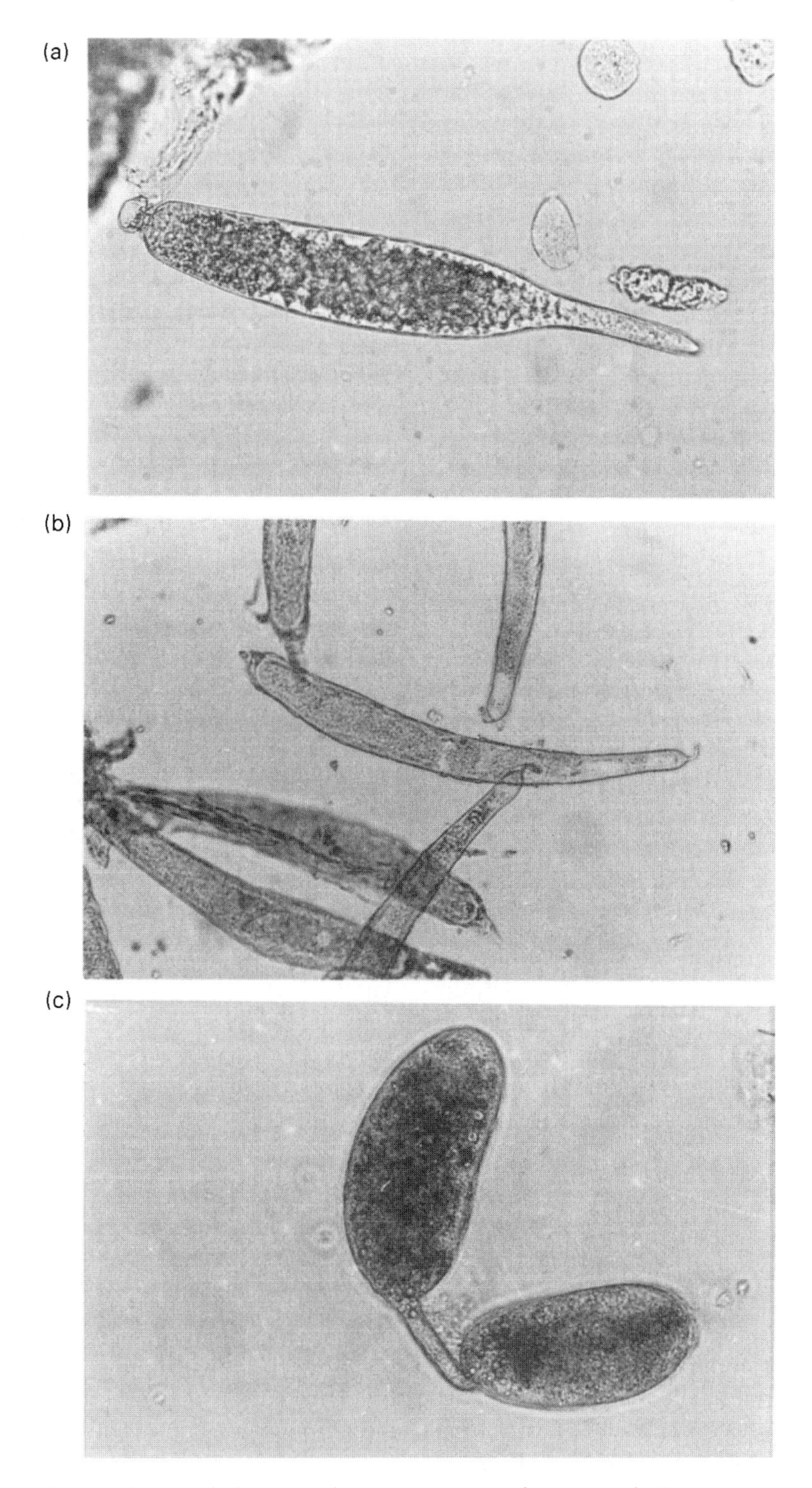

Figure 2.9 Eggs dissected out of the reproductive systems of parasitoid Hymenoptera: (a) unidentified Mymaridae; (b) *Cotesia* sp. (Braconidae); (c) unidentified Encyrtidae.

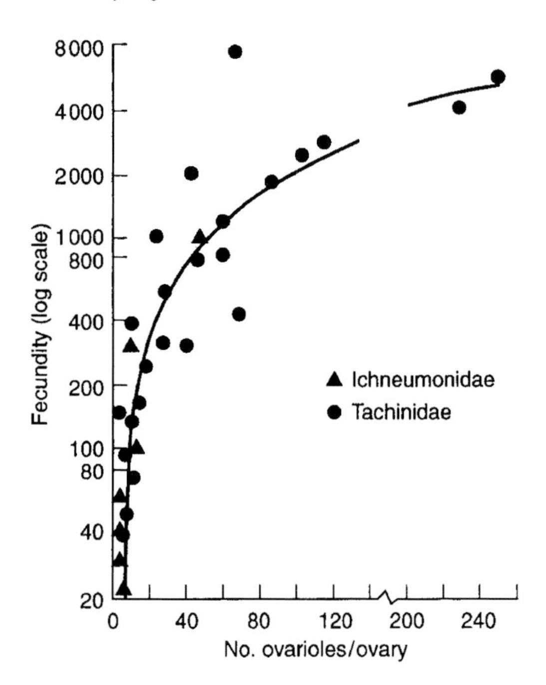

Figure 2.10 The relationship between fecundity (note log scale) and the number of ovarioles per ovary in Ichneumonidae and Tachinidae. Data points represent means for individual species. Source: Price (1975). Reproduced by permission of Plenum Publishing Corporation.

although Blackburn detected one when he controlled for egg size. When adult size is controlled for, species with a high fecundity (the maximum number of eggs reported to have been laid by an individual of a species) tend to have smaller eggs, indicating a trade-off between fecundity and egg size (small eggs require less of a material investment) (see Blackburn's paper for further discussion). Mayhew and Blackburn (1999) showed, also through a comparative analysis, that koinbionts produce smaller eggs than do idiobionts.

The interspecific relationships among predatory Syrphidae and among Coccinellidae with respect to ovariole number, mature oöcyte number, oöcyte size and female body size, and their biological significance, are discussed by Gilbert (1990) and Dixon and Guo (1993). Note that in predatory Coccinellidae large species produce proportionately smaller eggs, relative to their body size, than smaller ones. For a discussion of the adaptive significance of the egg-size-body size relationship in Coccinellidae, see Dixon (2000).

2.3.4 'OVIGENY' AND RELATED TRAITS

Ovigeny Index

Among insects, even among members of the same order, there may be considerable variation in the degree to which the female's lifetime potential egg complement is mature when she emerges into the environment following pupal development. For example, the orders Lepidoptera and Hymenoptera each include at one extreme species that emerge with a fully developed lifetime egg complement, and at the other extreme species that emerge with only immature oöcytes (Flanders, 1950; Dunlap-Pianka *et al.*, 1977; Jervis *et al.*, 2001). The **'ovigeny index'** – the ratio, expressed as a proportion, of the initial mature egg load to the lifetime potential fecundity (subsection 2.71) – was devised by Jervis *et al.* (2001) to quantify variation in the degree of egg development shown by insects both interspecifically and intraspecifically. Ovigeny index = 1 (**'strict pro-ovigeny'** *sensu* Jervis *et al.*, 2001) indicates that all the female's oöcytes are mature upon emergence, whereas ovigeny index = 0 (**'extreme synovigeny'**) denotes emergence with no mature oöcytes. A continuum of ovigeny index exists among parasitoid wasps, ranging from strict pro-ovigeny, through weak then strong synovigeny, to extreme synovigeny (Jervis *et al.* 2001); the same probably applies to parasitoid Diptera and also insect predators as a whole.

The numerator in the calculation of ovigeny index – **initial egg load** (the number of mature, i.e. fully chorionated [layable] eggs in newly emerged females – is in many species easily measured through dissection. Lifetime potential fecundity, the denominator in the calculation of ovigeny index, is measured by adding the number of immature oöcytes (also measured through dissection) to the initial egg load. Alternatively, it can be approximated by measuring the average lifetime realised fecundity (subsection

2.7.1) achieved under conditions of high host abundance (hosts supplied *ad libitum*) and high food availability/quality.

The usefulness of the ovigeny index as a measure of female fitness is discussed in subsection 1.16.2. Ovigeny index can also be used as a simple measure of the allocation of resources to reproduction at the start of adult life (see section 2.14), and thus to seek some of the classic trade-offs predicted by general life-history theory (see Bell and Koufopanou, 1986; Smith, 1991; Stearns, 1992; Roff, 2002). For example, in parasitoid wasps, ovigeny index and life-span are negatively correlated both within species (Jervis *et al.*, 2001, using data in Ellers and van Alphen 1997) and across species (Jervis *et al.*, 2001; Jervis *et al.*, 2003), suggesting that there is a cost, to life-span, of concentrating reproductive effort into early adult life (Jervis *et al.*, 2001). At least within species, the negative correlation is attributable to the differential allocation of capital resources between initial eggs on the one hand, and fat body reserves (which contribute to maintenance metabolism) on the other (Ellers and van Alphen, 1997) (section 2.14). Ovigeny index has also been used to explore the body size-related trade-off between current and future reproduction (Ellers and Jervis, 2003).

Other life-history variables found to be correlates of ovigeny index are: egg resorption capability (associated with a low index), egg type (hydropy is associated with a high index, anhydropy with a low index), and body size (negatively correlated with ovigeny index, both between and within species) (Jervis *et al.*, 2001, 2003; Ellers and Jervis, 2003). Host-feeding species tend to have a low index, as do idiobionts (Jervis *et al.*, 2001). Ovigeny index is hypothesised to be correlated with the degree of resource carry-over (i.e. from pupa to adult) (see section 2.14): an index of 1 indicates that the materials used for lifetime reproduction derive entirely from larval resources, whereas indices of <1 indicate that the materials used for lifetime reproduction derive only partly from carried-over resources, the females relying upon external nutrient inputs to mature their remaining oöcytes). This difference in life-history strategy closely parallels the concept of 'capital' versus 'income' breeding (see Drent and Daan, 1980; Boggs, 1992, 1997a).

For details of the criteria used in deciding whether a species is strictly pro-ovigenic or synovigenic, see Jervis *et al.* (2001). Note that some species categorised by authors as pro-ovigenic are, in reality, weakly synovigenic (see Mills and Kuhlmann, 2000; Jervis *et al.*, 2001)

Autogeny/anautogeny in Synovigenic Insects

Presumably due to there being insufficient resource carry-over from the larval stage, some synovigenic species can mature some eggs without first feeding (i.e are **autogenous**), whereas others must feed (i.e. are **anautogenous**). Hover-fly (Syrphidae) species are synovigenic-autogenous (Gilbert, 1991). The tachinid *Cyzenis albicans* is synovigenic-autogenous (Hassell, 1968). Predatory coccinellids are synovigenic-anautogenous. The green lacewing *Chrysoperla carnea* is anautogenous when reared only on prey, but is autogenous when given a non-prey food, together with prey, during larval life (McEwen *et al.*, 1996).

In an autogenous host-feeding species, the females must consume host blood in order to mature eggs (Jervis and Kidd, 1986).

Hydropy and Anhydropy

Flanders (1942) distinguished between two types of egg in parasitoid wasps, **hydropic** and **anhydropic**, based on the function of the chorion. Hydropic eggs, which are restricted to endoparasitoid species, usually swell to a considerable degree within hours or a few days of being deposited within the host's haemolymph (Schlinger and Hall, 1960). Compared with the mature ovarian eggs, the swollen eggs in euphorine Braconidae are 1000 times larger in terms of volume. The swelling occurs as a result of the uptake, via the thin, permeable chorion, of components of the host's haemolymph (Ferkovich and Dillard, 1987). Anhydropic eggs, which occur among ectoparasitoid as well as endoparasitoid species, have a relatively thick, rigid, impermeable chorion, and any apparent swelling they undergo is slight and mostly the result of the embryo having developed into the first instar larva.

Hydropic eggs contain little yolk, which is mainly comprised of lipids (Le Ralec, 1995). Their oöplasm contains numerous ribosomes and mitochondria, both organelles apparently being derived from the female's trophocytes, via the nutritive pore (King *et al.*, 1971; Le Ralec, 1995). Proteins, rather than being acquired from the host's haemolymph, are synthesised *de novo* within the oöplasm, from amino acids which have been obtained from the host (Ferkovich and Dillard, 1987). The major contribution by the mother to its progeny is thus a protein synthesis apparatus to enable complete embryonic development (Le Ralec, 1995). Anhydropic eggs, by contrast, contain much yolk. Their oöplasm contains numerous lipoid bodies. Proteins, mainly composed of vittelin, are also present, but their character varies among species. In species whose females consume host haemolymph ('host-feed', see subsection 1.7), the protein bodies are typical of insects generally (King and Richards, 1969; Kunkel and Nordin, 1985; Le Ralec, 1995) but in species that do not host-feed they appear to be atypical, although their biochemical composition has yet to be clarified (Le Ralec, 1995). In anhydropic egg-producing species, the mother contributes to its progeny sufficient sources of both energy-rich (lipid) and nitrogen-rich (protein) materials to enable embryonic development to be completed.

It is reasonable to conclude from the above that the greatest degree of parental (female) investment per egg is made by anhydropic egg-producing species. Indeed, Godfray (1994) and Mayhew and Blackburn (1999) assumed the selection pressures for divergence in egg size among parasitoids to be linked to the selection pressures for divergence in egg type (hydropy/anhydropy), with the result that small egg size is associated with hydropic egg production, and large egg size associated with anhydropic egg production. Jervis *et al.* (2001, 2003) therefore took hydropy and anhydropy to be proxy measures of such investment when seeking a link between egg type and the timing of egg production (ovigeny index). In a comparative analysis of over sixty parasitoid wasp species, hydropic egg-producing species were shown to have, on average, a significantly higher ovigeny index (see above) than anhydropic species. Given that Jervis *et al.* (2003) have shown ovigeny index to equate with initial egg load, the aforementioned result accords well with the trade-off, between egg number and egg size across species, predicted for animals generally by life-history theory (Smith and Fretwell, 1974), and established empirically for parasitic (mainly parasitoid) wasps by Berrigan (1991). Therefore, the hydropy/anhydropy distinction would seem to be a valid comparative measure of parental investment per egg. A more convincing case in support of this assumption could be made if egg type and egg volume were shown to be positively correlated. An alternative approach would be to show that hydropy and anhydropy are linked to cross-species variation in body size. The rationale behind the existence of such a relationship is that in parasitoid wasps egg volume and body size are positively correlated, irrespective of the method by which volume is calculated (Berrigan, 1991; Blackburn, 1991a). Ideally, future research into interspecific patterns of maternal egg provisioning should involve measuring allocation per egg in terms of total energy and of the amounts of key nutrients, using the techniques applied by Giron and Casas (2003b) to *Eupelmus vuilletti*.

Egg Resorption

In synovigenic-anhydropic parasitoids, oöcytes, when they become mature, are not immediately discharged into the lateral oviduct. Usually a maximum of only a few (three in *Encarsia formosa;* van Lenteren *et al.*, 1987) mature eggs can be stored per ovariole at any moment in time. These eggs, however, can be retained for only a brief period of time, as they have limited storage life, and space has to be made for other mature oöcytes to enter the lateral oviduct. If a female is deprived of hosts for a sufficiently long period (i.e. host are absent or are otherwise very scarce), she does not jettison such eggs but begins resorbing them, commencing with the oldest (see below). In *Nasonia vitripennis* only the pycnotic residue of the follicle cell nuclei remains after resorption (King and Richards, 1968), although in a few species females may deposit partially resorbed eggs (Flanders, 1950). In some cases, even developing oöcytes may be resorbed (Jervis and Kidd, 1986; van Lenteren *et al.*, 1987 for reviews). By resorbing eggs, the female can use the energy and materials obtained from the eggs to maintain herself and to sustain ovigenesis until hosts are again available. Through egg resorption, eggs are returned to the body of the wasps with only a partial loss of energy and materials, instead of the total loss that would occur if the eggs were jettisoned. Egg resorption is, however, a last-resort survival tactic (Jervis and Kidd, 1986). Egg resorption can be a form of egg-limitation in synovigenic parasitoids, since whilst a female is in the process of resorbing eggs, she may be temporarily incapable of ovipositing even if hosts become available (Jervis and Kidd, 1986, 1999; Heimpel and Rosenheim, 1998).

Eggs that are undergoing resorption can be detected at the proximal ends of the ovarioles by their unusual shape (and sometimes colour in hemipteran bugs) compared with unaffected eggs (Figure 2.11a,b). Because of the partial removal of the chorion, eggs that have recently begun to be resorbed may, unlike unaffected eggs, increase in size when dissected out in water, and will certainly take up stains such as acetocarmine or trypan blue more readily (King and Richards, 1968).

As they are being resorbed, eggs shrink and finally disappear, leaving remnants of the exochorion. The latter are probably voided through the egg canal at the next oviposition, although in some Encyrtidae part of the chorion (the aeroscopic plate) remains in the ovariole or is voided into the haemocoel (Flanders, 1942).

The time of onset of resorption in host-deprived wasps varies, depending on the availability of food. A female *Nasonia vitripennis* that is starved will begin resorbing eggs earlier than a female that is given honey (Edwards, 1954). Heimpel *et al.* (1997a) recorded egg resorption in starved *Aphytis melinus* but not in honey-fed ones over the 36 hour experimental period. In host-deprived, honey-fed females of *Nasonia vitripennis* oöcyte development continues, albeit slowly. Among starved female *Phanerotoma franklini*, some females apparently did not live long enough to resorb eggs, whereas sugar-fed females monitored to natural death began to resorb eggs after around 30 days, and by 40 days had resorbed all of their eggs (Sisterton and Averill, 2002).

The rate of egg resorption can be measured using the chemical colchicine, which stops cell division by interfering with microtubule formation, and therefore halts production of further mature eggs. Rates measured for parasitoids vary from one to several days (Edwards, 1954; Benson, 1973; Bartlett, 1964; Anunciada and Voegele, 1982; van Lenteren *et al.*, 1987). In completely starved *Nasonia vitripennis*, when the terminal oöcyte of one ovariole has begun to be resorbed, it is followed by those in other ovarioles. With continued starvation, the penultimate oöcyte will also start being resorbed, first in one ovariole and then in the others, and so on (King and Richards, 1968).

If a female parasitoid is deprived of hosts for a long enough period for resorption to commence, the number of mature oöcytes in the ovaries (egg load), will depend on both: (a) the rate of oögenesis (which will be much lower in starved females than in females that have access to non-host foods, subsection 2.7.3) and (b) the rate of resorption (King, 1963; van Lenteren *et al.*, 1987).

2.3.5 EGG-LIMITATION

As discussed in Chapters 1 (subsection 1.16.2) and 7 (subsection 7.3.7), the degree to which a parasitoid is egg-limited is an important consideration when studying parasitoid foraging behaviour, from the standpoints of fitness gain and searching efficiency. The size of the parasitoid's mature egg load determines the number of eggs the female can lay at a given

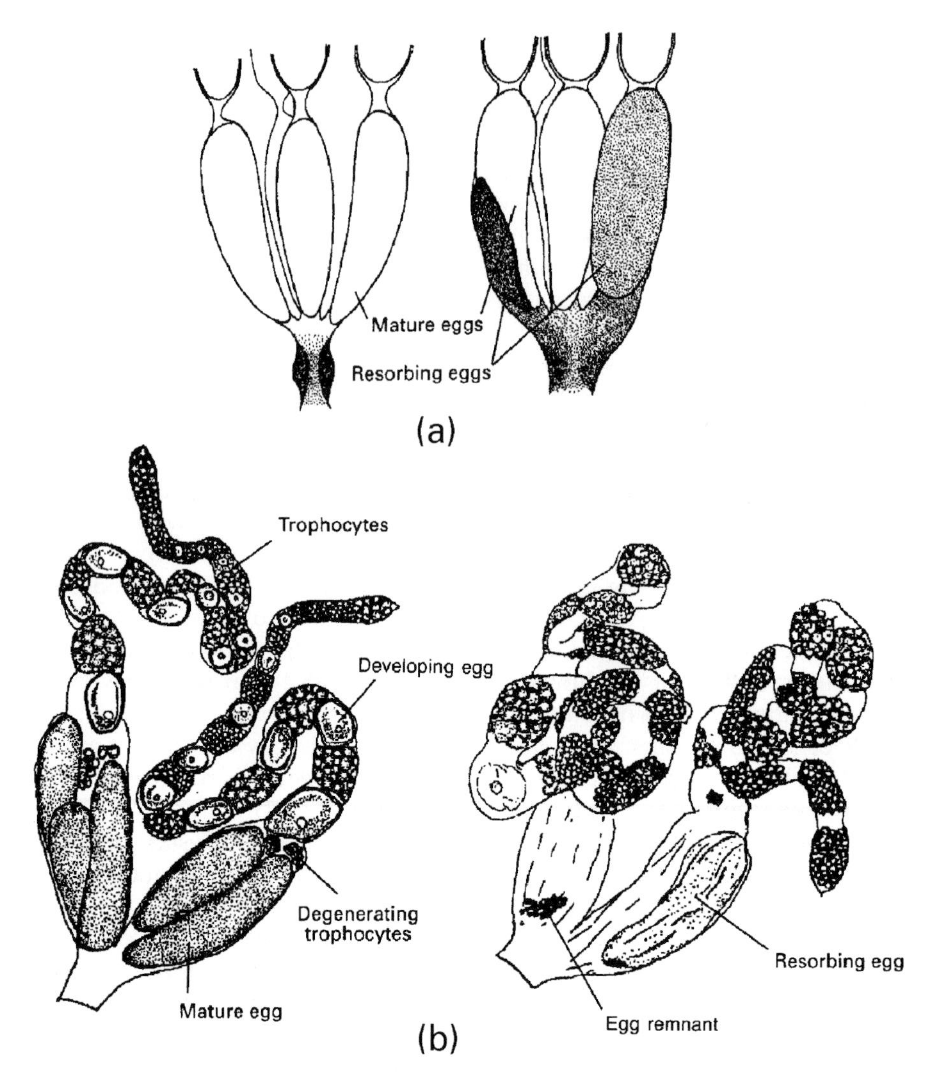

Figure 2.11 Egg resorption in synovigenic-anhydropic parasitoid wasps: (a) *Nasonia vitripennis* (Pteromalidae) (source: King and Richards, 1968.); (b) *Habrobracon hebetor* (Braconidae). (source: Grosch, 1950.) [In both cases, the ovarioles of a non-resorbing female are shown on left, and those of a resorbing female are shown on right.] (a) reproduced by permission of The Zoological Society of London; (b) by permission of The Marine Biological Society, Woods Hole, Massachusetts.

moment in time (Heimpel and Rosenheim, 1996) (Figure 2.12). What, then, sets the upper limit to egg load – is it the rate of ovigenesis or the storage capacity?

If, in a species that is not currently resorbing eggs, not all the ovarioles are found to contain a mature egg at any instant in time when ovigenesis is at its maximum, i.e. there is asynchrony among ovarioles, then the ceiling to egg load is set by the rate of ovigenesis, not storage capacity. On the other hand, if at any instant in time all the ovarioles contain a full-sized egg and the lateral oviducts are also full of eggs, then the ceiling is likely to be set by storage capacity (in which case one must ask: does ovigenesis cease when the maximum storage capacity is reached?). *Coccophagus atratus* apparently belongs to the second category. If females of this

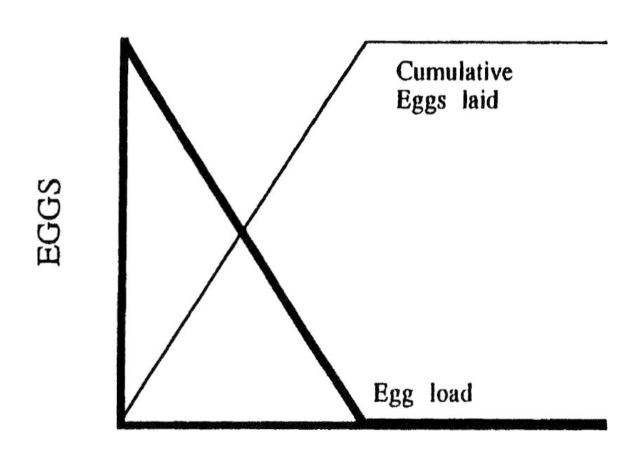

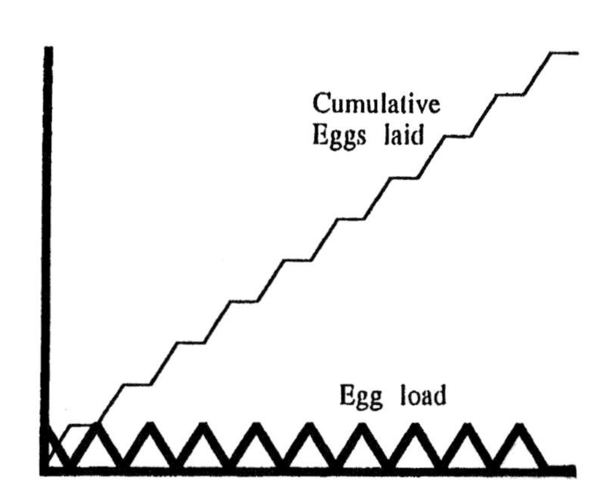

Figure 2.12 The changes in egg load and the cumulative number of eggs laid by a strictly-pro-ovigenic and a strongly synovigenic species in relation to successive oviposition events. From Heimpel and Rosenheim (1998). Reproduced by kind permission of Elsevier Science.

species are withheld from hosts but fed on honey following eclosion and are dissected after varying periods, the egg load is found to increase during the first 24 hours of adult life and thereafter remain constant (Figure 2.13). Since in this species there is no evidence for egg resorption, egg numbers are probably limited by the storage capacity of the ovarioles/ lateral oviducts, with ovigenesis ceasing when there is no room for further eggs (Donaldson and Walter, 1988). It would be interesting to know the frequency with which ovigenesis is switched on and off in females that are foraging under natural conditions.

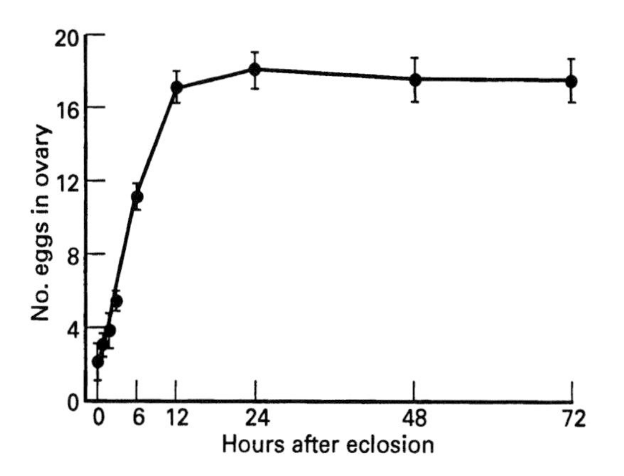

Figure 2.13 The number of full-sized eggs in the ovaries of *Coccophagus atratus* (Aphelinidae), recorded at various intervals after female eclosion. Source: Donaldson and Walter (1988), reproduced by permission of Blackwell Publishing.

To measure the rate of ovigenesis in a synovigenic parasitoid in relation to different treatments, expose each of several large cohorts of standardised (e.g. newly emerged) females to a particular environmental condition, e.g. type of diet, temperature level, and follow the cohorts through until the last females die. Each day, dissect part of each cohort and examine the condition of the ovaries in the females, recording the number of mature eggs. The age-specific and average daily rate of ovigenesis (plotted as an **ovigenesis schedule**) can be compared for the different treatments. A detailed protocol for an investigation of this type, concerned with the effects of different temperatures, may be found in Kajita and van Lenteren (1982).

2.3.6 MOTIVATION TO OVIPOSIT

A number of theoretical models indicate that the motivation to oviposit (and to host-feed) depends upon egg load. How does a parasitoid perceive the size of its egg load? Donaldson and Walter (1988), in a detailed study on ovipositional activity and ovarian dynamics in *Coccophagus atratus*, showed that when females were exposed to an abundance of hosts they deposited eggs within defined bouts of ovipositional activity that were initiated only when the female had accumulated approximately eighteen full-sized eggs (Figure 2.4a). This finding suggests that egg load, possibly perceived via stretch receptors in the lateral oviducts (Collins and Dixon, 1986), affects the motivation to oviposit.

2.3.7 SPERMATHECAL COMPLEX

The spermatheca (Figures 2.1, 2.2, 2.4, 2.7, 2.14 and 2.15) is the sperm storage organ of females. Syrphidae, Tachinidae and Pipunculidae have three (Figure 2.2c; Kozanek and Belcari, 1997), whereas Hymenoptera have only one (Quicke, 1997). In Hymenoptera the spermatheca is situated at or near the confluence of the lateral oviducts. The spermathecal complex comprises a capsule (the storage vessel or 'spermathecal reservoir'), a gland or pair of glands which may help to attract, nourish and possibly activate sperm, and a muscular duct through which sperm are released (or witheld, see section 1.9) as an egg passes along the common oviduct (vagina).

In parasitoid wasps, the spermatheca is noticeably pigmented yellow, dark red or black (a possible adaptation for protecting sperm from the adverse affects of UV light), a useful feature to look out for when dissecting females. Using transmitted light it is usually possible to observe, at high magnifications, the movement of sperm, if sperm are present, within the capsule. To detect such movement, observations must be made within 5 minutes of dissecting the recently killed female. Hardy and Godfray (1990) determined whether or not field-caught parasitoids were virgins, by examining the spermatheca of dissected females. They were able to distinguish between empty spermathecae, those containing living sperm (present as a writhing mass) and those containing dead sperm (inadvertently killed by the dissection process). The spermathecae of Pipunculidae are enclosed within the sclerotised base of the ovipositor, and so are difficult to examine and dissect.

Suggested studies on sperm use, depletion and competition are described in Chapter 4 (section 4.5).

2.3.8 ACCESSORY GLANDS

In many female insects there are obvious glands, occurring as a pair or two pairs of pouches, associated with the anterior end of the common oviduct (vagina), which are termed **accessory** or **colleterial glands** (Figures 2.4a,e, 2.7 and 2.8) (King and Ratcliffe, 1969; Quicke, 1997). It is generally understood that they produce secretions which coat the egg as it is laid. These glands are present in nearly all chalcidoid parasitoids; different families have different numbers and arrangements (King and Copland, 1969; Copland and King, 1971, 1972a,b,c,d; Copland *et al.*, 1973; Copland, 1976), but hardly anything is known about their function. They have been implicated in the formation of feeding-tubes of host-feeding Hymenoptera (Flanders, 1934) but they seem to be equally developed in species that do not host-feed. Some Torymidae have the largest glands, and *Eupelmus urozonus* (Eupelmidae) has both large glands and enormous extensions from the calyx. Noting the condition of the glands in dissected females under various experimental treatments may be instructive as to their function.

2.3.9 DUFOUR'S (ALKALINE) GLAND

The Dufour's or alkaline gland (Figures 2.4a,b,d, 2.15 and 2.16) is well developed in Hymenoptera. It discharges into the anterior common oviduct at the base of the ovipositor. In parasitoids it is the source of the parasitoid marker substances (pheromones) discussed in

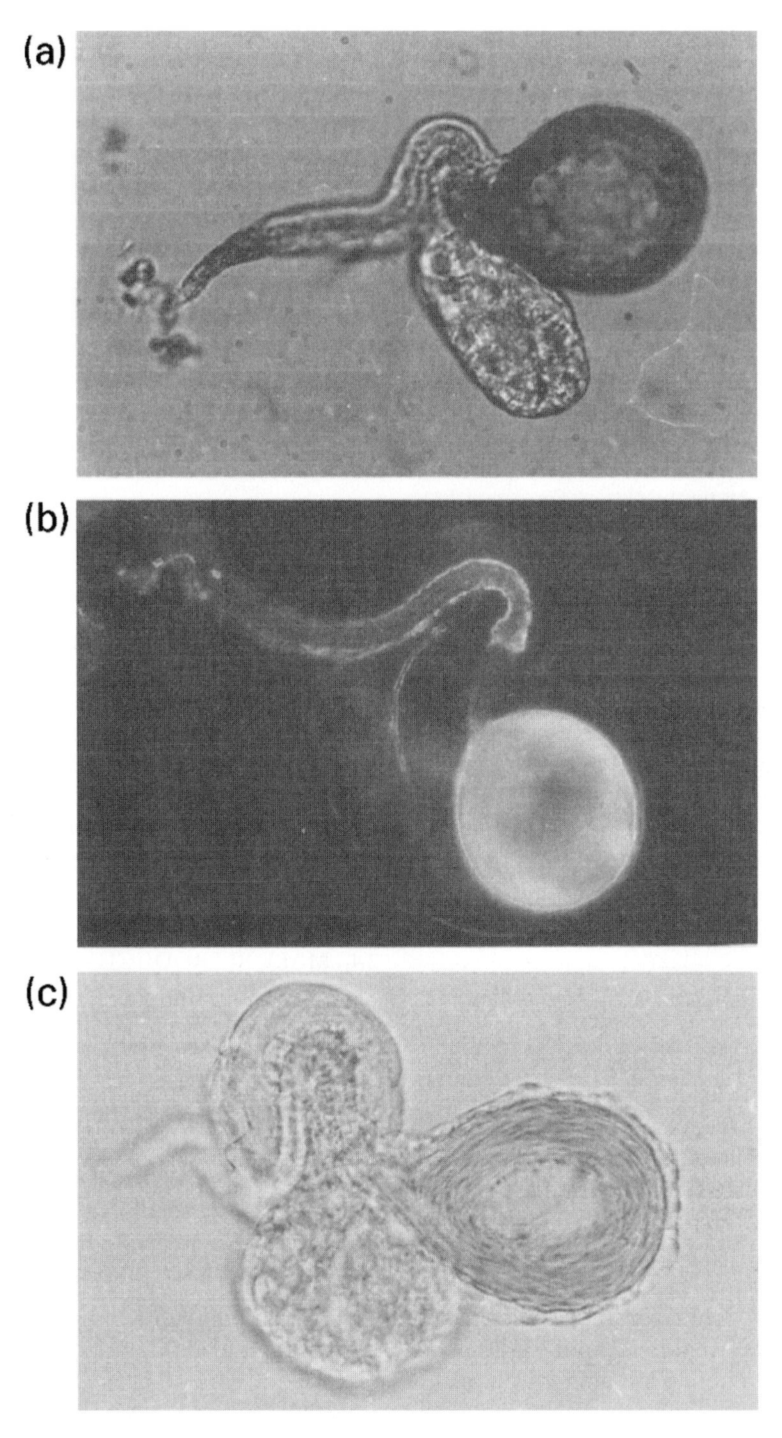

Figure 2.14 The spermatheca in parasitoid wasps: (a) *Aprostocetus* sp. (Eulophidae) (showing pigmented capsule); (b) *Eurytoma* sp. (Eurytomidae) (capsule and gland united, showing the gland's collecting duct); (c) *Nasonia vitripennis* (Pteromalidae) (showing sperm).

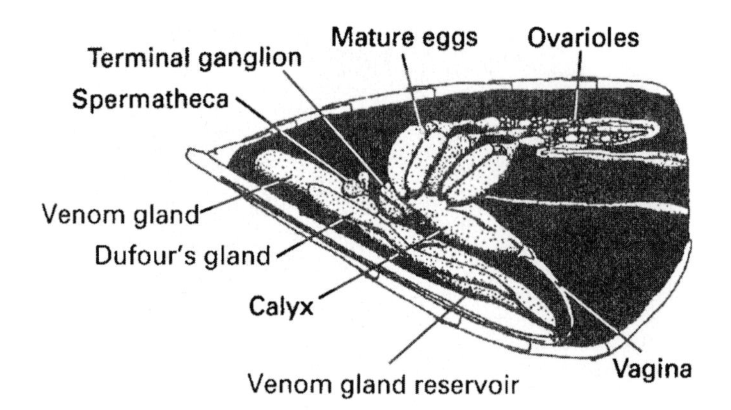

Figure 2.15 The reproductive system of female *Aphelinus* (Aphelinidae), showing the position of the spermatheca, the venom gland, Dufour's gland, the venom gland reservoir and other structures. Source: Copland (1976). Reproduced by permission of Pergamon Press.

subsection 1.5.3 and section 1.18. The Dufour's gland is normally a thin-walled sac containing an oily secretion. It is a long tubular structure in most chalcids but may be extremely small in some braconid wasps, e.g. *Cotesia glomerata*, concealed among the bases of the ovipositor stylets. Gas chromatography can be used to reveal the chemical composition of gland secretion; Marris *et al.* (1996) showed that in *Venturia canescens* there are quantitative between-strain differences in composition, indicating that different genetic lines produce characteristic cocktails of marker pheromone.

2.3.10 VENOM GLAND (ACID GLAND, POISON GLAND)

The venom gland (= **acid gland, poison gland**), like the Dufour's gland, empties into the base of the ovipositor (Figure 2.4a,b,d). It is either a simple structure as in Chalcidoidea (Figure 2.17), a convoluted tubular structure as in Ichneumonidae, or a structure of intermediate complexity as in some Braconidae (Figure 2.17) (see also Quicke, 1997). The venom of some idiobionts (subsection 1.5.6) induces permanent paralysis, arrested development or death in the host, whereas that of koinobionts induces temporary paralysis or no paralysis at all (see Quicke, 1997, for a discussion of these and other effects). Associated with the venom gland is a reservoir that has muscular walls; the reservoir may have additional secretory functions (Robertson, 1968; van Marle and Piek, 1986). The venom gland has been reported to be a source of viruses or virus-like particles. The structure and function of the venom gland system of Hymenoptera has been investigated by several workers (Ratcliffe and King, 1969; Piek, 1986; see also Quicke, 1997, and references contained therein), but there is considerable scope for further investigative work into gland structure and function.

2.4 MALE REPRODUCTIVE SYSTEM

An example of the reproductive system in male Hymenoptera is shown in Figure 2.18. The system comprises a pair of testes and usually a pair of accessory glands. For further details, see Quicke (1997). The possible role of secretions from the latter in parasitoid mating behavior is discussed in subsection 4.10.3.

2.5 SEX RATIO

This aspect of parasitoid and predator biology (including the causes of biased primary and secondary sex ratios), is dealt with in Chapters 1 (section 1.9) and 3 (section 3.4). The role of *Wolbachia* endosymbionts in biasing sex ratios is touched upon in Chapters 3, 4, and 6. Some of the biotic and physical factors discussed elsewhere in this chapter (below) may influence

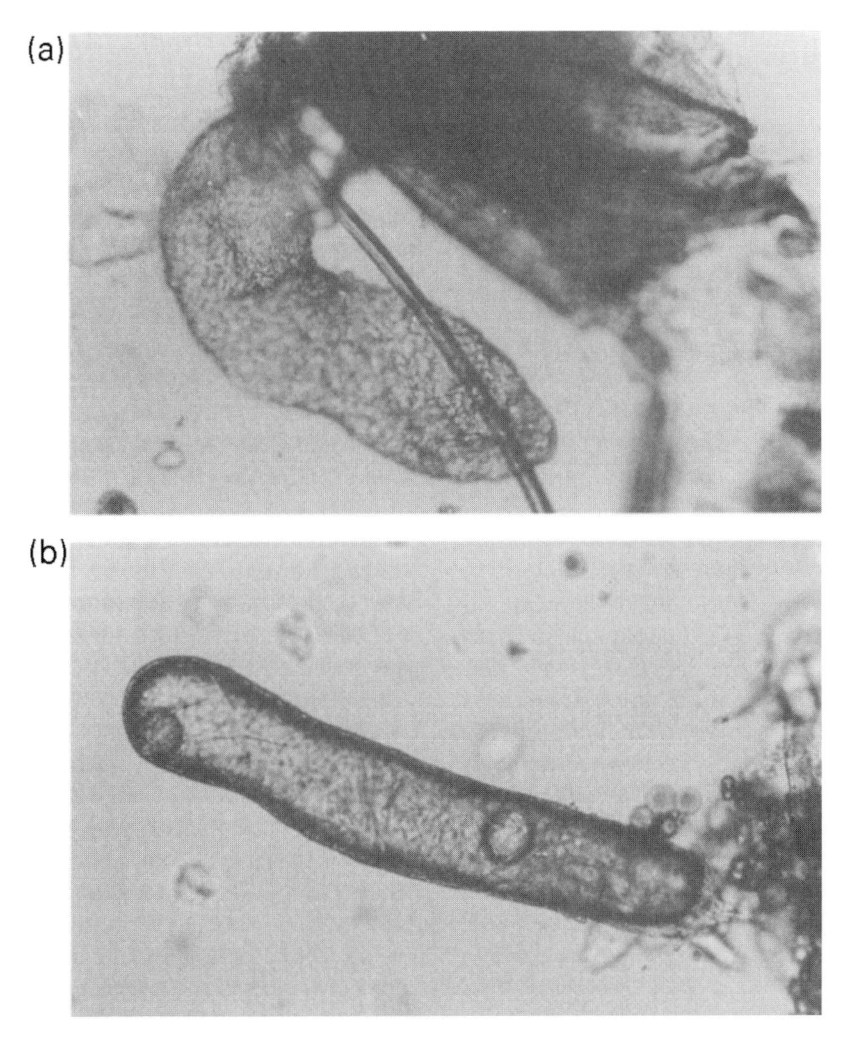

Figure 2.16 Dufour's or alkaline gland in parasitoid wasps: (a) *Eurytoma* sp. (Eurytomidae); (b) *Colastes* sp. (Braconidae).

secondary sex ratio. For a protocol for studying the effects of (constant and variable) temperatures on progeny sex ratio in parasitoids, see Kfir and Luck (1979).

2.6 LOCATING EGGS IN HOSTS

Parasitoid eggs may need to be located in or on hosts for a variety of reasons, including the measurement of fecundity and parasitism (subsection 2.7.3, sections 7.2, 7.3), investigations of parasitoid behaviour (subsection 1.5.6, sections 1.6, 1.8, 1.10) and studies of parasitoid communities (subsections 6.2.9, 6.3.5). The degree of difficulty experienced in locating eggs will depend upon factors such as the relative sizes of the host and the parasitoid egg, the amount of fat body tissue, whether the eggs lie within organs or in the haemocoel, the size of other organs, and the degree of sclerotisation of the host integument (Avilla and Copland, 1987). The eggs of endoparasitoids are generally much more difficult to locate than those of ectoparasitoids.

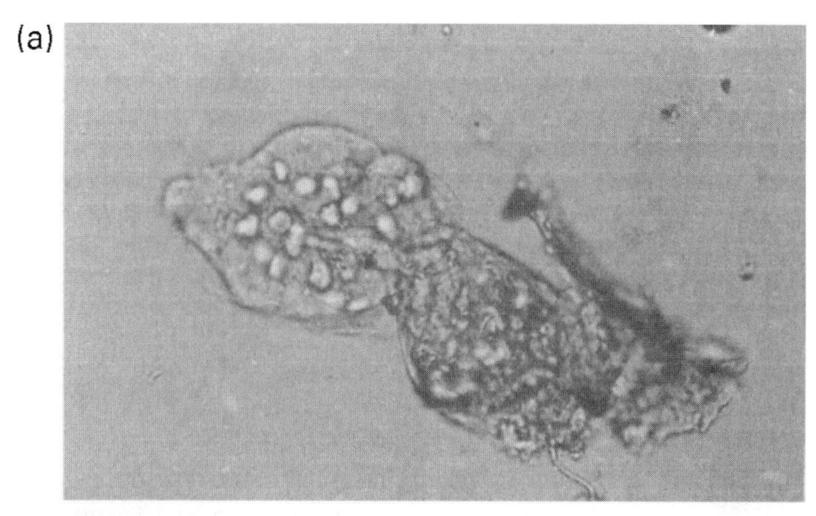

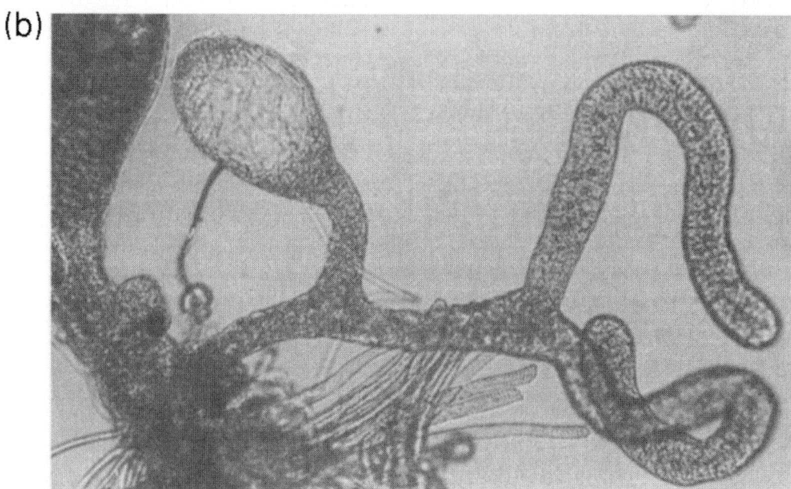

Figure 2.17 The venom gland in parasitoid wasps: (a) unidentified Mymaridae, showing simple gland and reservoir; (b) *Cotesia* sp., showing more complex, (i.e. bifurcate) gland and reservoir.

Preferably, hosts should be killed either: (a) by narcotising them (e.g. using CO_2, ethyl acetate) – in which case they should be dissected shortly afterwards, or (b) by placing them in a deep-freeze - in which case they can remain dissectable for several months. Attempting to locate eggs in hosts that have been preserved in alcohol is likely to prove very difficult indeed.

If endoparasitoid eggs prove difficult to locate, parasitised hosts should be kept alive long enough for the eggs to swell (i.e. in hydropic species) and/or the first instar larvae to form, the parasitoid immature stage in either case becoming more easily visible.

2.7 FECUNDITY

2.7.1 INTRODUCTION

The term **fecundity** refers to an animal's reproductive output, in terms of the total number of eggs produced or laid over a specified period, and should be distinguished from **fertility** which refers to the number of viable progeny

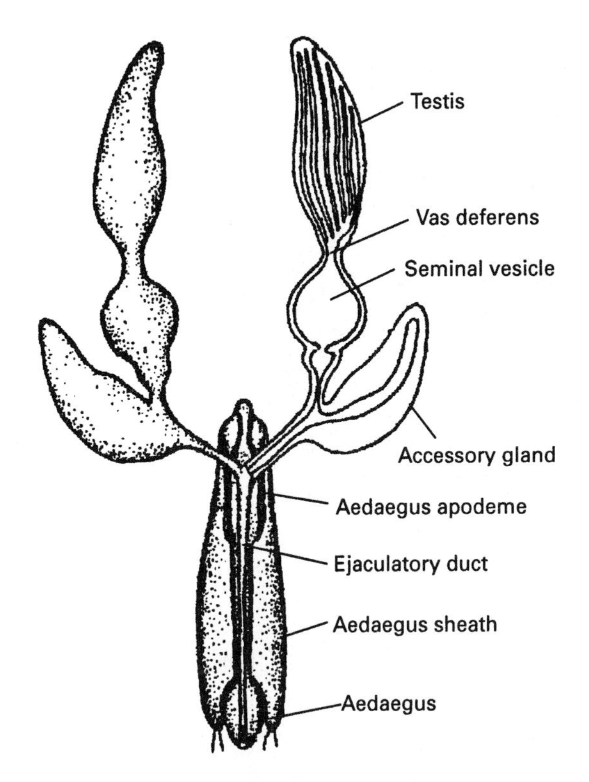

Figure 2.18 Schematic representation of reproductive system in male Chalcidoidea. Source: Sanger and King (1971). Reproduced by permission of The Royal Entomological Society of London.

that ensue. From the standpoint of population dynamics, fertility is the more important parameter, as it is the number of progeny entering the next generation. However, because fertility can be relatively difficult to measure (Barlow, 1961), fecundity measurements are often used instead.

A distinction is drawn between **potential fecundity** and **realised fecundity**. A species' potential fecundity is usually taken to be the maximum number of eggs that can potentially be laid by females. For example, in the laboratory we might take a strictly pro-ovigenic parasitoid (subsection 2.3.4), dissect its ovaries at eclosion and then count the number of eggs (all mature) contained within. This number is the insect's potential lifetime fecundity. Synovigenic parasitoids emerge with some immature eggs, so in these insects potential fecundity is the number of mature eggs (the initial egg load) plus the number of immature eggs.

Potential fecundity can be compared with the number of eggs actually laid over the life-span when excess hosts are provided in the laboratory, i.e. lifetime realised fecundity. The figure for lifetime realised fecundity is likely to fall short of the estimate for lifetime potential fecundity – this applies especially to females whose realised fecundity is measured in the field, where female life-span is likely to be significantly shorter (Leather, 1988).

Fecundity is a variable feature of a species, influenced by a range of intrinsic and extrinsic (physical and biotic) factors. The evaluation of a natural enemy for biological control requires a study of the influence of these factors (and of possible interaction effects between certain factors) on potential and realised fecundity, and if possible, fertility. The data can be used in estimating a species' intrinsic rate of increase which is discussed later in this chapter (section 2.11). Fecundity (potential or realised) is also used as a measure of individual fitness in insects (e.g. see Visser, 1994; Ellers *et al.*, 1998; Roitberg *et al.*, 2001).

When assessing the influence of a particular biotic factor on lifetime realised fecundity, it is important to determine to what extent variation in fecundity can be explained by variation in longevity. For example, take the positive relationship between female size and fecundity. The greater longevity of larger females compared with smaller females could be the sole reason why larger females are more fecund. Females may have the same average daily egg production irrespective of body size, but by living longer, larger females lay more eggs over their life-span (Sandlan, 1979b). For a discussion of fecundity-longevity relationships within and among species of predatory Coccinellidae, see Dixon (2000).

It is possible to obtain measures of realised fecundity without actually counting eggs: Takagi (1985) and Hardy *et al.* (1992b) counted the number of adult offspring produced and took account of the intervening mortality processes, so deriving estimates of the number of eggs originally deposited.

2.7.2 COHORT FECUNDITY SCHEDULES

A (realised) **fecundity schedule** for a parasitoid or predator species can be constructed by taking a cohort of standardised females (standardised in terms of physiological age, size, and oviposition and sexual experience) and exposing them individually to some chosen set of constant environmental conditions from adult emergence until death. The number of eggs laid per female per day is then plotted, giving the **age-specific realised fecundity** of the species (Figure 2.19; see also Figure 2.65). The data obtained from the experiment can also be used to calculate both the lifetime realised fecundity of the species (used by evolutionary ecologists as a measure of fitness, see Roitberg *et al.*, 2001), and the **average daily oviposition rate** (lifetime realised fecundity divided by the average longevity). Using the same data, the **cumulative realised fecundity** of the parasitoids can also be plotted against either female age (Figure 2.20) or cumulative degree-days (Minkenberg 1989) (subsection 2.9.3). It is expressed as the proportion of the highest mean total number of eggs laid by females of any one treatment (e.g. temperature or host density treatment), this total representing the maximal fecundity realisable by females. The usefulness of the cumulative realised fecundity measure is that it tells us to what extent parasitoids achieve their maximum lifetime fecundity/fitness under particular conditions, and allows easier comparison of the effects of different treatments. Using the data from a fecundity schedule, the parameters m_x (age-specific fecundity) and l_x (age-specific survival) can be used in the calculation of the intrinsic rate of increase (r_m) of the parasitoid population (section 2.11). If fecundity schedules are constructed for cohorts held under different host/prey availability regimes, the number of hosts or prey parasitised or eaten can be recorded and the data used to plot age-specific and lifetime

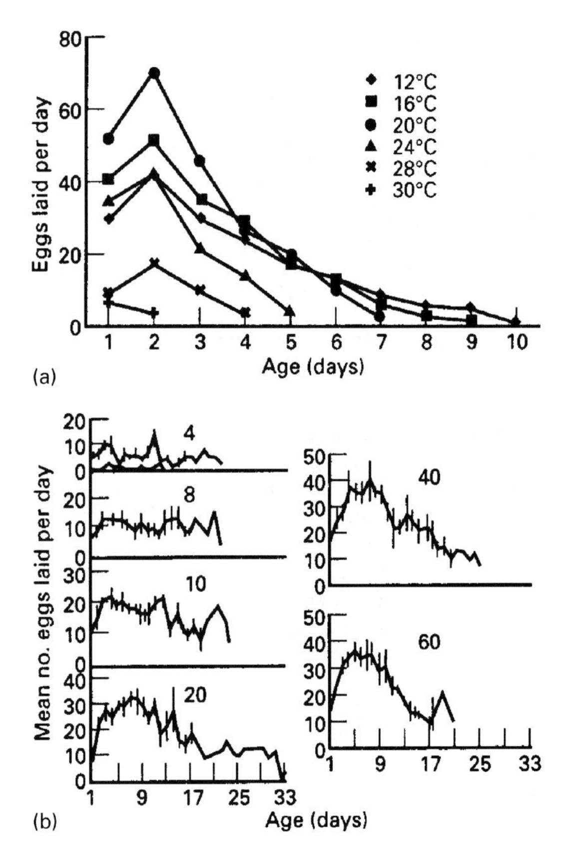

Figure 2.19 The age-specific fecundity schedule for two parasitoid species: (a) *Aphidius matricariae* maintained at different temperatures and at constant host density conditions (source: Hag Ahmed, 1989); (b) *Dicondylus indianus* (Dryinidae) maintained at different host densities (4–60) and constant temperature conditions. The plot of host density 2 treatment is shown along with that of the host density 4 treatment. (Source: Sahragard *et al.* (1991). Reproduced by permission of Blackwell Verlag GmbH.

functional responses (the numbers parasitised or eaten versus the numbers available (section 1.10), as was done by Bellows (1985).

An important consideration when using the aforementioned experimental design is that as time goes on, the data are limited to progressively fewer females. To obtain fecundity data that are statistically meaningful, particularly data for the latter part of adult life, a very large starting density of parasitoid or predator females may be required. This, however, may

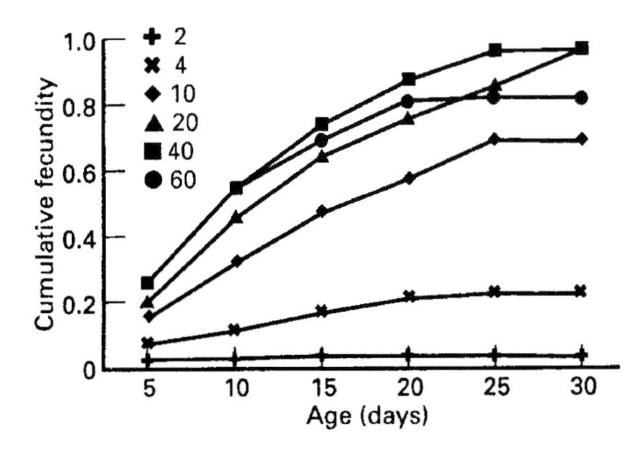

Figure 2.20 The cumulative realised fecundity of the dryinid wasp *Dicondylus indianus*, measured over the lifetime of females, at different levels of host availability. Fecundity is expressed as the proportion of the highest mean total number of eggs laid by females of any one treatment, this total representing the maximal fecundity that could be realised. Source: Sahragard *et al.* (1991). Reproduced by permission of Blackwell Verlag GmbH.

increase the investigator's workload to an unacceptable level.

In most parasitoids and in predators, the realised fecundity schedule (and also the ovigenesis schedule, see subsection 2.3.5) will show a rise in the number of eggs produced or laid per day until a maximum rate of productivity is reached. Thereafter a gradual decrease occurs until reproduction ceases altogether at or shortly before the time of death (see Kindlmann *et al.*, 2001, for a discussion of this **'trangular fecundity function'**) (Figure 2.19) (if there is a period of post-reproductive life, it is usually very short, see Jervis *et al.*, 1994, for exceptions). Fecundity schedules vary between species, depending on the reproductive strategies of the insects, e.g. strict pro-ovigeny and different degrees of synovigeny (subsection 2.3.4). As we shall describe below, environmental factors (temperature, humidity, photoperiod, light quality, light intensity, host or prey availability) modify these patterns in a number of ways, and ideally the role of each factor in influencing the schedule ought to be investigated separately. This, however, may not be practicable, in which case the usual procedure is to expose a predator or parasitoid to an excess of prey or hosts (replenished/replaced daily), at a temperature, a relative humidity, or a light intensity similar to the average recorded in the field (Dransfield, 1979; Bellows, 1985).

2.7.3 EFFECTS OF BIOTIC FACTORS ON FECUNDITY

Host Density (parasitoids)

If fecundity schedules are constructed for a parasitoid species over a range of host densities, females will be found to lay on average more eggs per day at higher host densities than at low ones (Figure 2.21). Also, the lifetime pattern of oviposition, i.e. the shape of the curve, varies with host density. There may be a shift in the fecundity schedule, with wasps concentrating oviposition into the earlier part of adult life (Figure 2.19b). At high host densities, hosts are more readily available for the wasps to attack, whereas at low densities oviposition rates are lower because the wasps have to search a greater area (and probably for a longer period of time), so expending energy that might otherwise be used in ovigenesis (Sahragard *et al.*, 1991).

As far as lifetime fecundity is concerned, the relationship with host density is either a

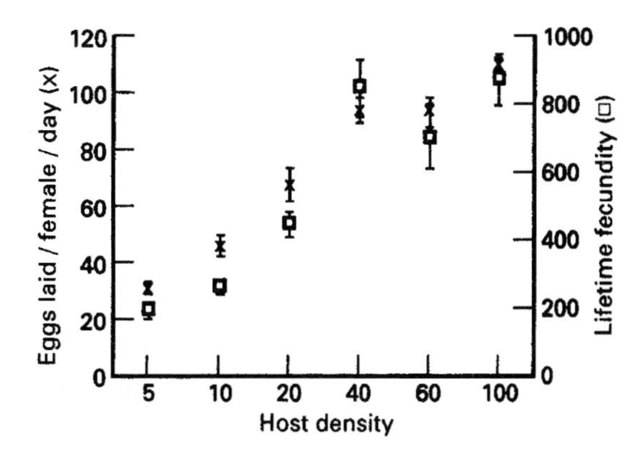

Figure 2.21 The relationship between fecundity (measured as both the mean number of eggs laid per day and the total number of eggs laid over adult life) and host availability in the parasitoid *Aphidius smithi* (Braconidae). Based on data taken from Mackauer (1983).

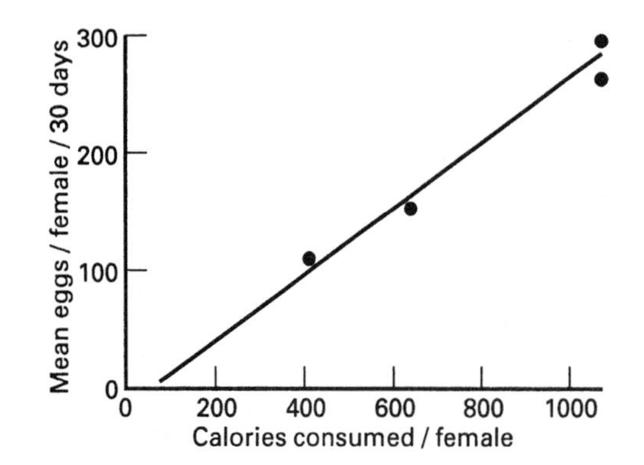

Figure 2.22 Fecundity as a function of ingestion rate in the predatory pentatomid bug *Podisus maculiventris*. Source: Beddington *et al.*, (1976b), who used data from Mukerji and LeRoux (1969). Reproduced by permission of Blackwell Publishing.

curvilinear one, resembling a Type 2 functional response (defined in section 1.10), or a sigmoid one, resembling a Type 3 functional response (Figure 2.23).

A difficulty that may arise when using low host densities is **ovicide**, i.e. the removal of eggs from parasitised hosts, although the number of (ecto)parasitoid species that practice ovicide is considerably smaller than the number of predator species that do so. Predaceous females of chrysopid lacewings are well-known for eating their own eggs in laboratory cultures (Principi and Canard, 1984), as are some coccinellids (Michaud, 2003). Where cannibalism is suspected, video recording techniques may help in determining the number of eggs lost in fecundity experiments.

Food Consumption

Non-predaceous Females

The females of many parasitoid and some predator species (e.g. *Chrysoperla carnea* (Chrysopidae) and adults of all aphidophagous Syrphidae) feed as adults solely on materials such as honeydew, nectar and pollen (Chapter 8), and consume substitute foods such as diluted honey in the laboratory. Females that are either deprived of food or experience a reduced intake (but are given water) lay fewer eggs or no eggs at all. Some non-host/prey foods have a more beneficial affect on fecundity than others (Leius, 1962; Principi and Canard, 1984; Krishnamoorthy, 1984; Heimpel and Jervis, 2004; Jervis *et al.*, 2004).

For an experimental investigation into the effects of adult nutrition on the fecundity schedule of a parasitoid to be ecologically meaningful, the effects of food provision need to be considered in the light of variations in host availability. This is done by taking a cohort of standardised females and providing the insects with one of a range of host densities (see **Host Density**, above) and with a chosen diet for the duration of their lives, the hosts and food being replenished daily. If the effects upon on ovigenesis of combined host deprivation/food provision are to be investigated, then, obviously, hosts are not provided to one set of females. One likely effect of providing food to females is that, at low host densities, females maintain a higher rate of oviposition than they can when deprived of food. As far as the effects of food provision on lifetime fecundity are concerned, it will be necessary to carry out a statistical analysis to show whether or not any improvement in lifetime fecundity brought about by feeding is simply a result of an increase in longevity and not an increase in the daily rate of ovigenesis (subsection 2.8.3).

Predaceous Females

We would expect the fecundity of predaceous females to be strongly influenced by prey availability. This relationship was modelled in a simple way by Beddington *et al.* (1976b) and Hassell (1978). If it is assumed firstly that some of the food assimilated by the female needs to be allocated to maintenance metabolism (and will therefore be unavailable for ovigenesis), and secondly that there is insufficient 'carry over' of food reserves from larval development for the laying of any eggs (i.e. synovigeny-anautogeny), then there will be a threshold prey ingestion rate, c, below which reproduction ceases, but above which there is some positive dependence between fecundity F and ingestion rate I. If it is assumed thirdly that this relationship is linear, then (Beddington *et al.*, 1976b):

$$F = \frac{\lambda}{e}(I - c) \qquad (2.1)$$

where e, λ and c are constants; e is the average biomass per egg. There is empirical support for this model (Mukerji and LeRoux, 1969; Mills, 1981) (Figure 2.22). In Mills' (1981) experiment five feeding levels were used, the daily ration of individual females corresponding to between 1 times and 2 times the average female weight.

To express fecundity in terms of prey density, we first assume ingestion rate to be proportional to the number of prey eaten, N_a, so that:

$$I = kN_a \qquad (2.2)$$

where k is a constant which depends upon the biomass (size) of each prey. Combining equations 2.1 and 2.2 with the simplest functional response model, Holling's (1966) disc equation (subsection 5.3.7), gives:

$$F = \frac{\lambda}{e}\left[\frac{ka'N}{1 + a'T_hN} - c\right] \qquad (2.3)$$

This model predicts that fecundity will rise at a decreasing rate (i.e. will decelerate) towards an upper asymptote as prey density increases, in the manner of the Type 2 functional response (section 1. 10), and also that the curve will be displaced forwards along the prey axis, i.e. away from the origin. There is empirical support for this relationship, both from laboratory studies (Dixon, 1959a; Ives, 1981a; Matsura and Morooka, 1983 (Figure 2.23a,b) and from field studies (Wratten, 1973; Mills, 1982) (Figure 2.24a,b; note that the data in [a] in this figure are expressed as logarithms). Anautogenous, obligate host-feeding parasitoids will have a similar fecundity/host density curve. In autogenous predators however, ovigenesis and oviposition can occur without the female first feeding on prey, so the curve of these insects will not be displaced along the prey axis.

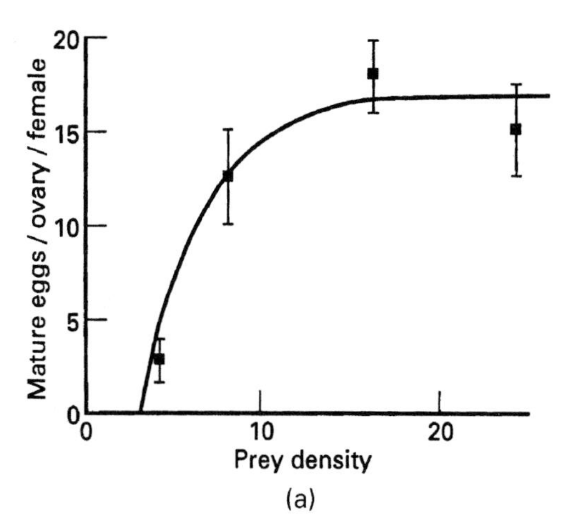

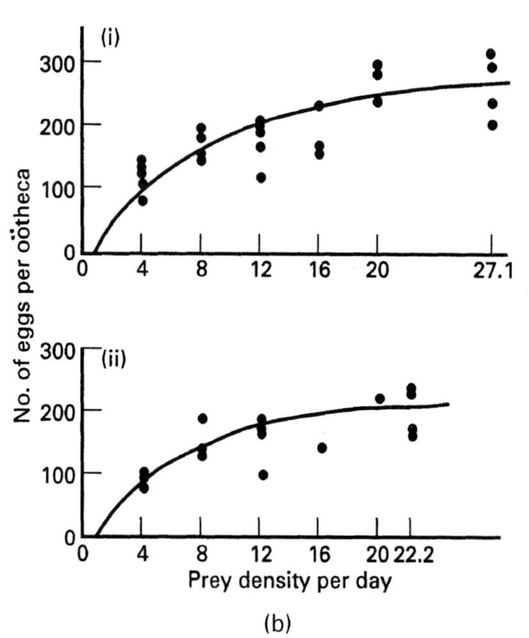

Figure 2.23 Fecundity as a function of prey density: (a) in the coccinellid beetle *Adalia decempunctata* (source: Beddington *et al.*, 1976b, who used data from Dixon, 1959); (b) in the mantid *Paratenodera angustipennis:* (i) first ovipositions, (ii) second ovipositions (oötheca = egg mass). Below the intercept of the curve (fitted by eye) with the prey axis, the insects allocate matter to maintenance processes only (source: Matsura and Morooka, 1983). (a) reproduced by permission of Blackwell Scientific Publications Ltd; (b) by permission of Springer-Verlag.

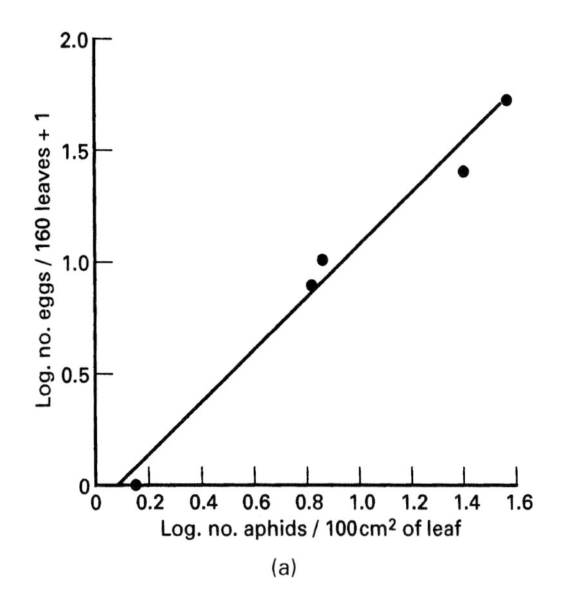

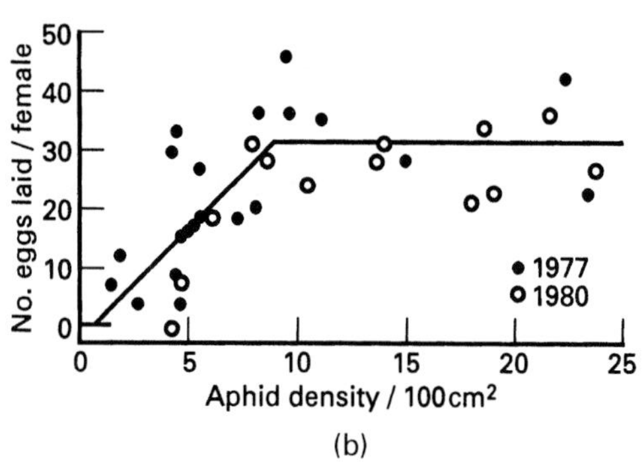

Figure 2.24 Fecundity as a function of prey density: (a) relationship between logarithm of number of eggs laid by the coccinellid *Adalia bipunctata*, and logarithm of density of aphids in the field (data from Wratten, 1973); (b) relationship between number of eggs laid per adult *Adalia bipunctata* and aphid density in the field (source: Mills, 1982b). (a) reproduced by permission of Blackwell Scientific Publishing Ltd; (b) by permission of The Association of Applied Biologists.

In the bug *Anthocoris confusus*, the viability (fertility) of eggs also varies with prey availability (Evans, in Beddington *et al.*, 1976b). This relationship may be due to the female allocating less biomass per developing egg at lower prey densities, i.e. e in equation 2.1 is not a constant (Beddington *et al.*, 1976b).

There are also grounds for questioning the assumption that k in equation 2.2 is a constant (Beddington *et al.*, 1976b). If this assumption is correct, then the relationship between fecundity and the number of prey actually killed will be rectilinear, which is the case for *Coccinella undecimpunctata aegyptiaca* (Figure 2.25). However, as noted in Chapter 1 (section 1.10), when the rate of encounter with prey is high, some predators consume proportionately less of each prey item. This behaviour will alter the shape of the fecundity versus prey killed curve, from rectilinear to curvilinear (Beddington *et al.*, 1976b). The shape of the fecundity versus prey density curve will also be altered, having an earlier 'turnover' point and also being more 'flat-topped' (Beddington *et al.*, 1976b).

Supplying predators with non-prey foods together with prey might lower the ingestion rate threshold, since less of the prey biomass assimilated by the female needs to be allocated to maintenance metabolism. If so, the fecundity-prey density curve of an anautogenous species will be shifted backwards along the prey axis, i.e. towards the origin. The shape of the curve is also likely to be altered.

Prey and Host Quality

Prey quality is likely to affect fecundity, as has been shown for Coccinellidae, Carabidae, Anthocoridae, and host-feeding Aphelinidae (Hariri, 1966; Blackman, 1967; Hodek, 1973;

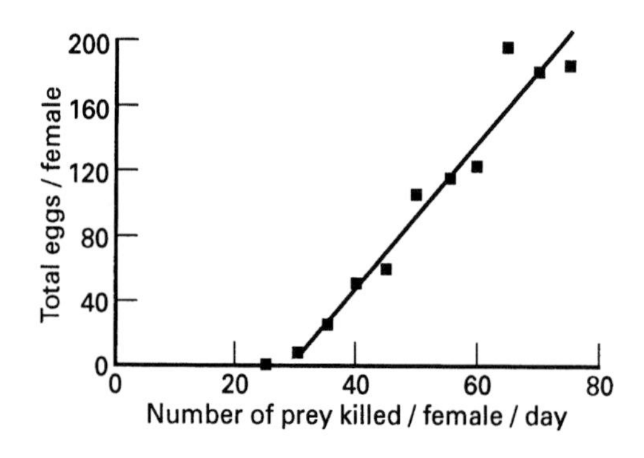

Figure 2.25 The relationship between fecundity and prey consumption rate in *Coccinella undecimpunctata aegyptiaca*. Source: Beddington *et al.* (1976b), who used data from Hodek (1973). Reproduced by permission of Blackwell Publishing.

Wilbert and Lauenstein, 1974; Spieles and Horn, 1998; Evans *et al.*, 1999; Venzon *et al.*, 2002). Some coccinellids and carabids are unable to reproduce at all if confined to a diet of certain prey species (Hodek, 1973; Spieles and Horn, 1998; Evans *et al.*, 1999).

Blackman (1967) found that adults of the coccinellid beetle *Adalia bipunctata* fed on *Aphis fabae* during both larval development and adult life were less than half as fecund as those fed on *Myzus persicae*. Also, their eggs were smaller and less fertile. By carrying out another experiment in which adult beetles were fed on the opposite prey species to that fed upon by the larvae, Blackman tested whether the prey species given to larvae affected the fecundity of the adult. It did not – fecundity depended strongly upon the species fed upon by the adult. However, it is not clear from Blackman's experiments whether he controlled for the effects of prey availability. The results of a study by Hariri (1966) are shown in Figure 2.26.

Evans *et al.* (1999) showed that when two species of predatory Coccinellidae are exposed to limited numbers of their preferred aphid prey, fecundity is enhanced if females are supplied with an additional prey species (a weevil), despite the fact that females given weevils alone cannot produce eggs.

In predators such as coccinellids the pre-oviposition period may be either shortened or prolonged, depending on the prey species fed upon by the female (Hodek, 1973).

Consumption of Food Supplements and Substitutes (predaceous females)

As we have suggested, fecundity is very likely to vary with the availability (and the quality) of plant-derived and other non-host/prey foods (especially so in the case of species having a high requirement for such nutrient input), taken either as supplements (when prey are available) or as substitutes (when prey are absent). Several predators have been shown to have a higher rate of egg production when given non-host foods as a supplement (e.g. see Cocuzza *et al.*, 1997a, Crum *et al.*, 1998), but except for some artificial diets, non-prey foods are a poor substitute for prey materials, in terms of their effects on

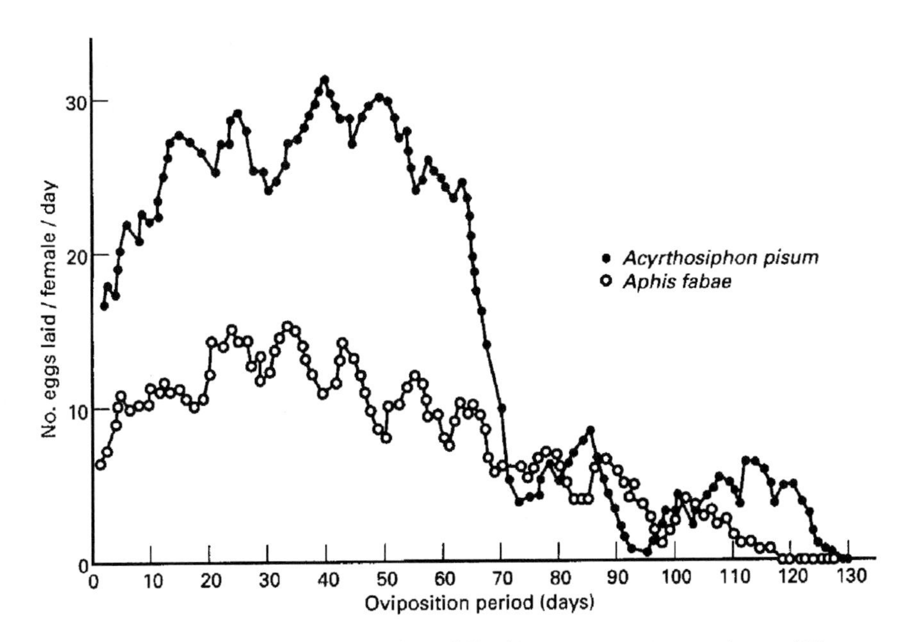

Figure 2.26 Fecundity of the coccinellid beetle *Adalia bipunctata* maintained on different prey species, *Acyrthosiphon pisum* and *Aphis fabae*. Source: Hariri (1966). Reproduced by permission of W. Junk Publishers.

fecundity (e.g. see Cocuzza *et al.*, 1997a; Evans *et al.*, 1999) (this may not apply to predator species whose diet is normally comprised largely of plant materials). In *Aphytis melinus* the benefit, to fecundity, of host-feeding, cannot be realised unless females also feed on sugar (Heimpel *et al.*, 1997a). For further discussion, see section 8.6.

Mutual Interference

Mutual interference between female parasitoids results in a reduction in individual searching efficiency (subsection 1.14.3) which will result in a reduction in the rate of oviposition, i.e fecundity. In the predator *Anthocoris confusus* fecundity declined with increasing adult density, despite the fact that prey density was high at all times and was unlikely to limit egg production through prey exploitation (Evans, 1976) (Figure 2.27). To determine whether mutual interference was a result of confining predators in his experimental cages, Evans (1976) measured fecundity in relation to predator density in females in a large cage within which they were free to move from plant to plant. A significant decrease in fecundity with increasing predator density was still recorded.

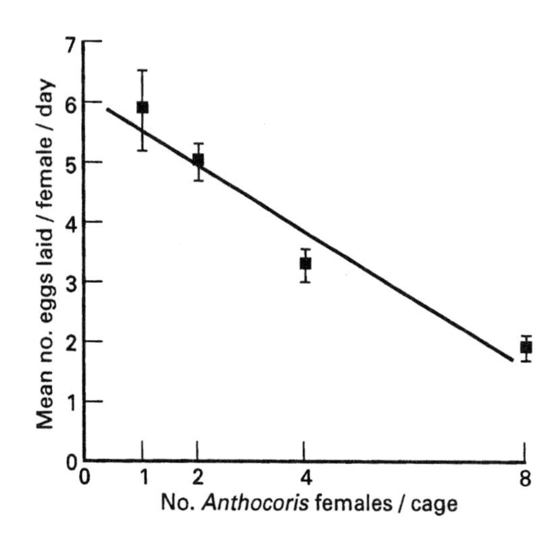

Figure 2.27 The relationship between fecundity and predator density in the predator *Anthocoris confusus*. There was a decline in fecundity despite aphid prey density being high at all times, i.e. the cause of the decline was mutual interference, not exploitation of prey. Source: Evans (1976). Reproduced by permission of Blackwell Publishing.

Mutual interference, and therefore interference-mediated reductions in fecundity, cannot be assumed to occur in all predators. For example, Hattingh and Samways (1990) found no evidence for mutual interference in adults of three species of *Chilocorus* (Coccinellidae). Feeding rate did not decrease and dispersal did not increase with increasing beetle density.

Female Body Size

In the laboratory, lifetime fecundity, and also reproductive correlates such as ovariole number and egg load (the latter usually recorded either at or shortly after eclosion), have been shown to increase with increasing body size within species (Figure 2.28) (e.g. Sandlan, 1979b; Mani and Nagarkatti, 1983; Ernsting and Huyer, 1984; Nealis *et al.*, 1984; Scott and Barlow, 1984; Waage and Ng, 1984; Bellows, 1985; Juliano, 1985; Liu, 1985a; Takagi, 1985; Collins and Dixon, 1986; Opp and Luck, 1986; van Vianen and van Lenteren, 1986; Banks and Thompson, 1987a; Moratorio, 1987; van den Assem *et al.*, 1989; Heinz and Parrella, 1990; O'Neill and Skinner, 1990; le Masurier, 1991; Hardy *et al.*, 1992b; Rosenheim and Rosen, 1992; Sequeira and Mackauer, 1992b; Croft and Copland, 1993; Zheng *et al.*, 1993b; King and King, 1994; Visser, 1994; Weisser *et al.*, 1997; Ellers *et al.*, 1998; Olson and Andow 1998; Taylor *et al.*, 1998; Mills and Kuhlmann, 2000; Harvey *et al.* 2000, 2001; Martínez-Martínez and Bernal, 2002; Pexton and Mayhew, 2002) [Note that there are a few exceptions to this pattern, e.g. see Rotheray and Barbosa, 1984; Bigler *et al.*, 1987; Corrigan and Lashomb, 1990; Visser, 1994; Coombs, 1997; Mills and Kuhlmann, 2000.]

Some of the restricted number of field studies conducted to date have demonstrated a positive intraspecific relationship between body size and fecundity (Visser, 1994; Kazmer and Luck, 1995; Ellers *et al.*, 1998, 2001)

In *Nasonia vitripennis* the slope of the egg load-body size relationship recorded 48 and 72 hours

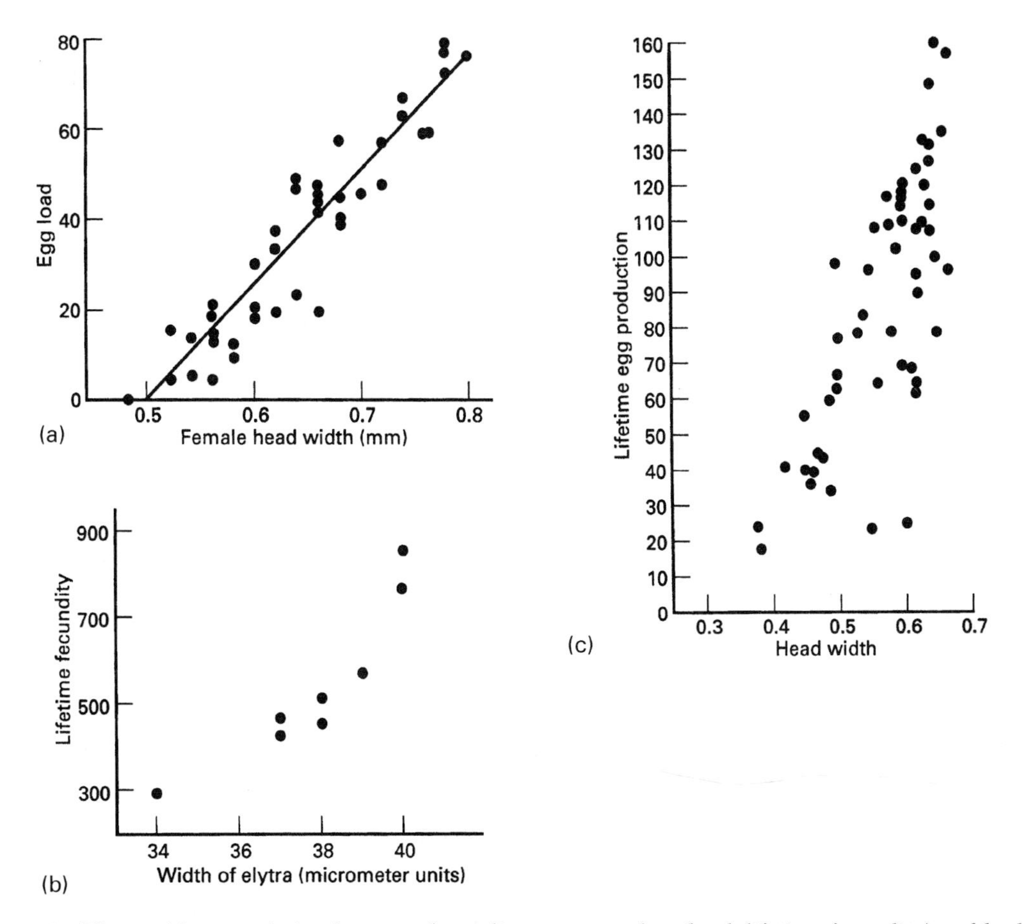

Figure 2.28 The positive correlation between fecundity measures (egg load, lifetime fecundity) and body size in females: (a) egg load in *Nasonia vitripennis* (source: O'Neill and Skinner, 1990); (b) lifetime fecundity in *Notiophilus biguttatus*; elytra width is expressed in micrometer units (100 units = 5.0 mm) (source: Ernsting and Huyer, 1984.); (c) lifetime fecundity in *Lariophagus distinguendus* (Pteromalidae) (source: van den Assem *et al.*, 1989). (a) reproduced by permission of The Zoological Society of London; (b) by permission of Springer-Verlag; (c) by permission of E.J. Brill (Publishers) Ltd.

after emergence was steeper in unfed females than in fed ones (Rivero and West, 2002). This result could explain why at least some researchers have recorded a difference between the size-fecundity plots of field and laboratory populations of a species. Small-sized wasps emerge with smaller fat reserves, and so rely more than large wasps upon obtaining food to fuel ovigenesis (Rivero and West, 2002). Because in the field food can often be limiting (Heimpel and Jervis, 2003), small-sized wasps suffer disproportionately in terms of their realised fecundity.

In some species, the relationship between fecundity and body size correlates over only part of the size range, with fecundity reaching a maximum in insects above a threshold size, e.g. *Aphidius ervi* (Sequeira and Mackauer, 1992b). It is therefore important, in experiments, to provide the complete field range of host sizes to parasitoids, so as to avoid obtaining a misleading impression of the 'true' size-fecundity relationship.

In predators, larger females have a shorter pre-oviposition period than smaller ones (Zheng *et al.*, 1993b), and this may contribute to their higher lifetime fecundity.

Body size is usually measured in terms of the width or length of some body part such as the head, thorax, the hind tibia. Some authors have measured dry body weight. Body size (or mass) is influenced, within species, by:

1. Larval feeding history, i.e. prey availability, host size, host species during development, quality of host diet (note that this includes plant resistance effects, i.e. bottom-up effects), clutch size, superparasitism (Dixon, 1959a; Russel, 1970; Hodek, 1973; Dransfield, 1979; Sandlan, 1979b; Cornelius and Barlow, 1980; Beckage and Riddiford, 1983; Principi and Canard, 1984; Waage and Ng, 1984; Scott and Barlow, 1984; Juliano, 1985; Liu, 1985a; Sato *et al.*, 1986; Eller *et al.*, 1990; Bai and Mackauer, 1992; Zheng *et al.*, 1993a,b; Harvey *et al.*, 1993, 1994; Van Dijk, 1994; Bernal *et al.*, 1999; Harvey *et al.*, 2000; Martínez-Martínez and Bernal, 2002) (Figure 2.29);

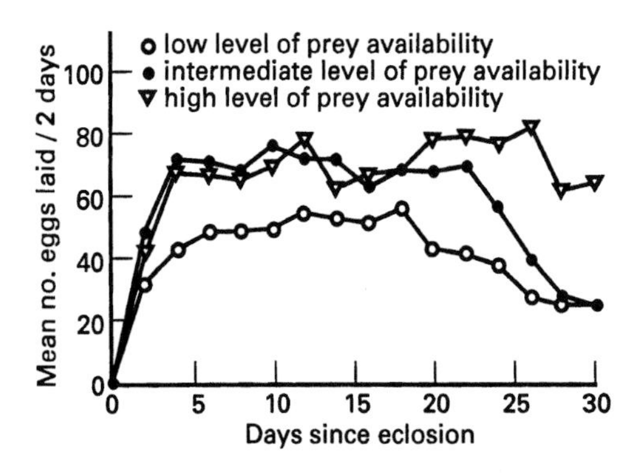

Figure 2.29 The effect of larval feeding history on fecundity in the lacewing *Chrysoperla carnea*. The data points indicate the average number of eggs laid, per 2-day period, of females provided with different levels of prey availability as larvae. Zheng *et al.* (1993b) showed that when lacewing larvae are fed fewer prey than they can potentially consume, they develop into smaller and less fecund adults than when they are given an overabundance of prey. Adults of *C. carnea* are non-predaceous, feeding on nectar, pollen and honeydew, and fecundity is also affected by consumption of these foods. Therefore female fecundity is determined both by larval feeding history and by adult food consumption. Source: Zheng *et al.*, (1993b).

2. The temperature during larval development (Ernsting and Huyer, 1984; Nealis *et al.*, 1984; Van Dijk, 1994) (Figure 2.30).

If an experiment, for whatever purpose, requires females to be of different sizes/fecundities, by far the simplest way of sorting insects according to size is to measure the parasitoids or predators when they are pupae (pupal and adult size being strongly correlated) so avoiding any difficulties and/or harmful side effects associated with handling the adults.

Mating

Female predators and dipteran parasitoids, if they are either unmated or sperm-depleted, lay much smaller numbers of eggs (e.g. very few in coccinellids, see Dixon, 2000, half as many in the bug *Podisus maculiventris*, see De Clercq

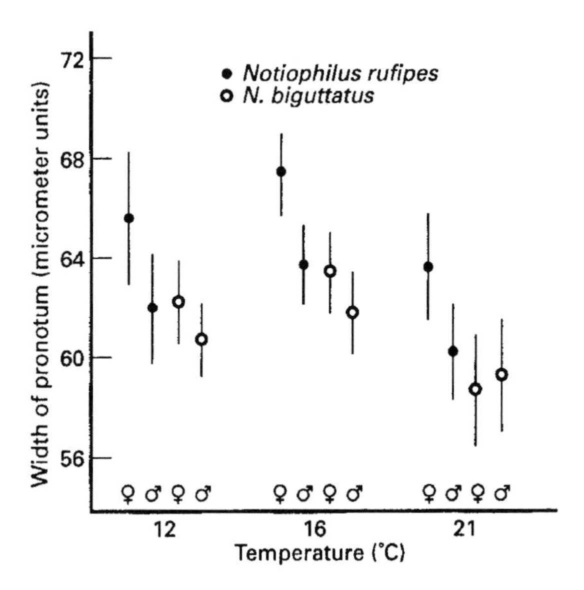

Figure 2.30 The effect of temperature during larval development upon adult size (as measured by pronotum width) in two species of carabid beetle. Means and the corresponding 95% confidence limits are shown, expressed in micrometer units (100 units = 2.5 mm). The data show a decline in adult size at either side of an optimum temperature for total biomass production. The effects upon size are translated into variations in fecundity. Source: Ernsting and Huyer (1984). Reproduced by permission of Springer-Verlag.

and Degheele, 1997) or none at all. Eggs, if laid, are infertile. To achieve their full reproductive potential, females of some species may need to mate several times (Sem'yanov, 1970; Ridley, 1988). By contrast, if a female arrhenotokous hymenopteran parasitoid lacks sperm for whatever reason, she can lay viable (male) eggs, so her fecundity should not be affected by mating. Mating was found not to affect egg load in the braconid wasp *Phanerotoma franklini*, but in this case mating was not confirmed to have occurred in all cases (Sisterton and Averill, 2002).

In experiments aimed at testing for the effects of mating, it is essential to establish that mating really has taken place. Caging females with a male is no guarantee that the insects have either engaged in mating behaviour or have (in the case of females) been inseminated. If an effect of mating upon fecundity is found, the question arises, in the case of females, as to whether oviogenesis has been enhanced because of the nutrient contribution made by the male, in the form of sperm or spermatophore.

Field Predation

Predator-induced mortality of adult parasitoids and predators may cause realised fecundity to be reduced well below the level achieved under laboratory conditions. The extent of the reduction can be estimated by marking and releasing individuals and cohorts of parasitoids and predators, recording predation events (see Heimpel *et al.*, 1997b), and then relating the field survivorship data to the natural enemies' fecundity schedule recorded under optimum laboratory conditions.

2.7.4 EFFECTS OF PHYSICAL FACTORS ON FECUNDITY

Temperature

The rate of egg production, and hence the age-specific and the lifetime fecundity of predators, and parasitoids, will vary in relation to temperature (Force and Messenger, 1964; Hämäläinen *et al.*, 1975; van Lenteren *et al.*, 1987; Braman and Yeargan, 1988; Miura, 1990; Hentz *et al.*, 1998; Ellers *et al.*, 2001) (Figure 2.31). The influence of temperature on the fecundity schedule of a natural enemy species can be investigated by taking cohorts of standardised females, and exposing each of them to one of a range of temperatures for their lifetimes. Females of all the cohorts are exposed to the same conditions of host/prey, food and water availability (hosts and prey need to be replaced daily), humidity and photoperiod etc.. A constant humidity will probably be the most difficult of all these factors to maintain. Temperature may influence the rate of prey consumption (Mills, 1981; Pickup and Thompson, 1990), so temperature-related variation in prey consumption should be looked for.

The effect of temperature upon egg load in a synovigenic insect can be investigated by following the protocol, used for *Aphytis* parasitoids, of Rosenheim and Rosen (1992). Parasitoid pupae are isolated, and adults, when they emerge, are kept with a supply of food (honey), at each of

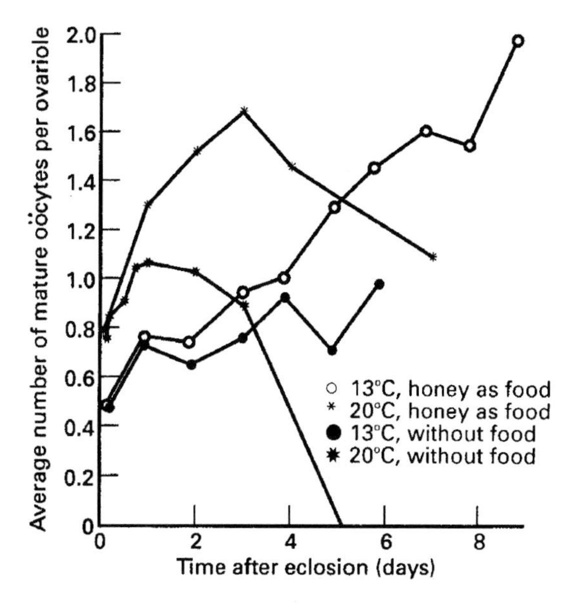

Figure 2.31 The number of mature oöcytes per ovariole in the parasitoid *Encarsia formosa* (Aphelinidae) kept for several days after eclosion without hosts, either without food or on a diet of honey, at two different temperatures. Source: van Lenteren *et al.* (1987). Reproduced by permission of Blackwell Verlag GmbH.

a range of temperatures for 24 hours. The adults are then dissected and the numbers of mature eggs they contain are counted. The results of Rosenheim and Rosen's (1992) study are shown in Figure 2.32, which also shows the influence of body size upon early-life potential fecundity (subsection 2.7.3).

It is generally the case that there is an optimum temperature range outside of which the insect either cannot maintain ovigenesis and oviposition or is unable to do so for long (Force and Messenger, 1964; Greenfield and Karandinos, 1976) (Figures 2.19a and 2.33). Although there is great variation from species to species, the limits to the favourable range for oviposition are often narrower than those for ovigenesis (Bursell, 1964).

Within the optimum range, one effect of higher temperature on the pattern of oviposition is to shift the fecundity schedule, with the ovigenesis/oviposition maximum occurring earlier in life (Siddiqui *et al.*, 1973; Ragusa, 1974; Browning and Oatman, 1981; Miura, 1990).

In the coccinellid *Adalia bipunctata* fecundity increases up to 20°C, correlating well with the increase in food consumption rate. However, above that temperature fecundity declines despite a continued increase in consumption.

Higher temperatures may constrain fecundity through increased metabolic costs i.e. daily maintenance requirement (Mills, 1981; Ellers *et al.*, 2001), although Ives (1981a) found no significant influence of temperature on the maintenance requirement of the two *Coccinella* species he studied.

No attempts appear to have been made to describe mathematically the relationship between oviposition rate and temperature, as has been done with development. Several workers have found that alternating temperatures increase insect fecundity (Messenger, 1964a; Barfield *et al.*, 1977a; Ernsting and Huyer, 1984), and thus it may be invalid to estimate oviposition rates in the field directly from constant temperature data. A similar approach to that used for estimating development based on cyclical temperature regimes might give more meaningful results but has not yet been attempted (subsection 2.9.3).

Some adult predators may be able to maintain maximal levels of ovigenesis through thermoregulation achieved either by thermal preference

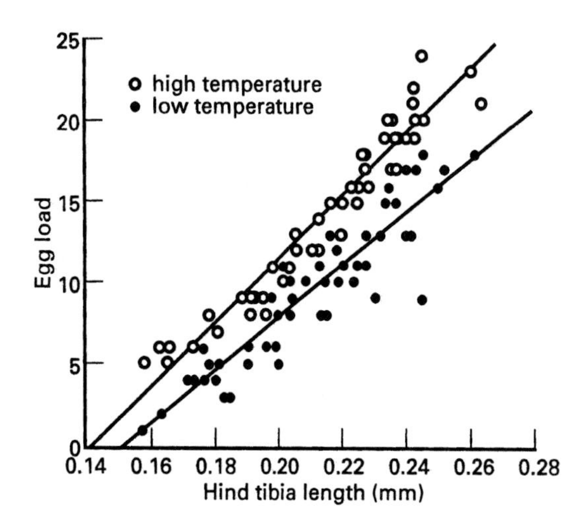

Figure 2.32 The influence on egg load of parasitoid size and the temperature at which females have previously been held from eclosion, in *Aphytis lingnanensis* (Aphelinidae). Source: Rosenheim and Rosen (1992). Reproduced by permission of Blackwell Publishing.

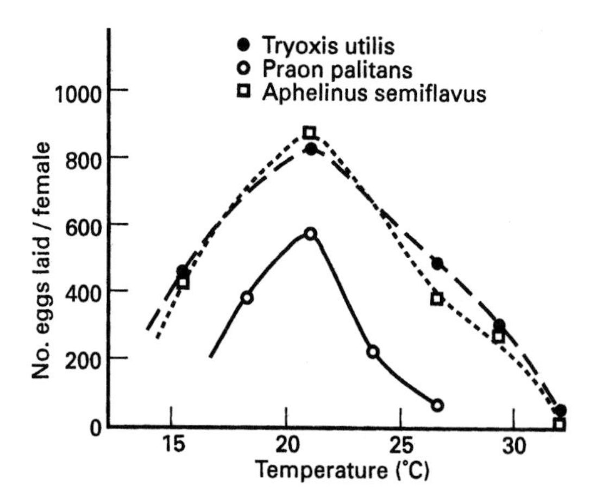

Figure 2.33 Comparison of mean lifetime fecundity of the aphid parasitoids *Praon palitans, Trioxys utilis* (Braconidae) and *Aphelinus semiflavus* (Aphelinidae), over a range of constant temperatures. Source: Force and Messenger (1964). Reproduced by permission of The Ecological Society of America.

behaviour (including basking), by employing physiological mechanisms and by employing physical adaptations such as melanisation of the integument (Dreisig, 1981; Brakefield, 1985; Miller, 1987; Stewart and Dixon, 1989).

Temperature is known to influence the length of pre-oviposition period in parasitoids and predators (e.g. see Stack and Drummond, 1997; Seal *et al.*, 2002).

Acclimation to temperature extremes may be useful in inundative releases, but any benefits could be offset by fitness costs (Scott *et al.*, 1997).

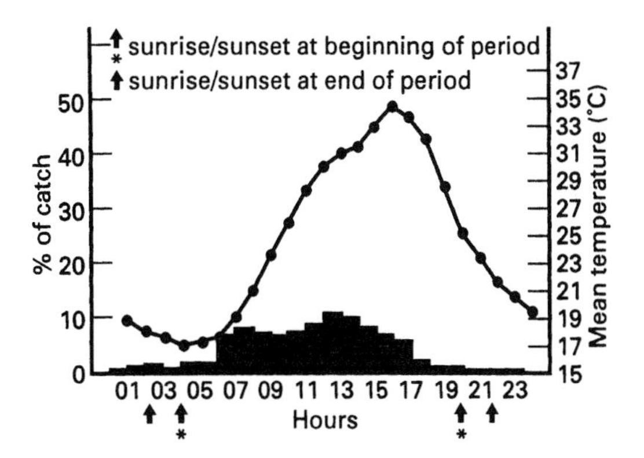

Figure 2.34 Diurnal flight activity patterns in *Encarsia formosa* (Aphelinidae) in the greenhouse (data for May – June). Percentage of the mean daily catch (by air suction trap) of wasps for each hour (histograms) and mean temperature (curve). Source: Ekbom, (1982). Reproduced by permission of Elsevier Science.

Light Intensity and Photoperiod

The intensity, duration and quality of light have an important influence on the biology and behaviour of most insects. High light intensity seems to increase the general activity of diurnal predators and parasitoids. For example, adults of the coccinellid beetle *Cryptolaemus montrouzieri* spend a greater proportion of their time walking and make more attempts to fly in bright light than under dim light conditions (Heidari, 1989). Light quality and intensity may also influence the close-range perception of hosts. Care must therefore be taken in fecundity experiments to provide sufficient light for normal activity, but bear in mind that in the field, bright light conditions are normally associated with increased radiant heat. Laboratory experiments that involve varying light intensity alone will require the radiant heat component of light to be removed, using suitable glass and water filters. Even cold fibre optic lamps used in microscopy can raise the body temperature of dark-coloured insects by at least 2°C above ambient. A thermocouple (see Unwin and Corbet, 1991) inserted into the body of a dead insect will enable the heat absorbed from a light source to be measured and suitable infra-red filters to be devised (Heidari and Copland, 1993).

Most natural enemy species will show strong diurnal peaks of behavioural activity, foraging being mainly confined to the photophase, as in many parasitoids and some carabid beetles (Luff, 1978; Ekbom, 1982; Ruberson *et al.*, 1988) (Figure 2.34). The photophase in fecundity experiments should therefore be the same as that experienced in the field; a continuous light regime may result in a higher fecundity than would be achieved in the field (Lum and Flaherty, 1973). Because of its effects on food consumption, photoperiod length may also influence larval growth and development rates in larval predators (which in turn will influence adult fecundity, subsection 2.7.3) and the rate of ovigenesis.

Weseloh (1986b) showed that the egg load of females of the egg parasitoid *Ooencyrtus kuvanae* (Encyrtidae) kept under long-day conditions increases more rapidly than that of females kept under short-day conditions, and that this is reflected in differences in progeny production. *Anagyrus kamali*, an encyrtid parasitoid of the hibiscus mealybug, is unusual in that its lifetime fecundity is highest under conditions of continuous darkness – this life-history characteristic would help to keep mass-rearing costs to a minimum (Sagarra *et al.*, 2000b). Hentz *et al.* (1998) found no significant effect of photoperiod on fecundity in *Chelonus* sp. near *curvimaculatus*.

Photoperiod and light quality and intensity should also be investigated for their effects on reproductive diapause induction (see 2.12.3), particularly where parasitoids and predators are being employed in artifically lit environments (e.g. see Stack and Drummond, 1997).

Humidity

Decreasing humidity may increase potential and realised fecundity in predators, through an increase in prey consumption by juveniles and females (e.g. see Heidari, 1989), but it may also decrease realised fecundity in predators and parasitoids by reducing searching efficiency and longevity (see below). In fecundity experiments care must therefore be taken to control humidity, so that it is around the field average.

Field Weather Conditions

The influence of field weather conditions upon the realised fecundity of insect natural enemies has rarely been investigated, undoubtedly because of the often immense practical difficulties involved. Weather can affect fecundity in a variety of ways, through its effects on foraging activity (Fink and Vökl, 1995; Weisser *et al.*, 1997), host/prey and non-host/prey food availability and quality, larval growth rate and survival, ovigenesis, and female survival. Weisser *et al.* (1997) estimated the lifetime reproductive success (lifetime realised fecundity) of the parasitoid *Aphidius rosae* in relation to wind and rain conditions by means of simulation modelling. They first developed a simulation model to predict patterns of parasitism of aphid colonies in the field as a function of weather conditions, then they parameterised the model using data from both laboratory and field experiments on parasitoids. Periods of relatively 'good' and relatively 'bad' weather were simulated using real weather data. They showed that only a small proportion of females was able to realise oviposition levels close to the maximum lifetime realised fecundity, as measured in the laboratory.

Barometric pressure is also likely to affect fecundity in insect natural enemies; see Roitberg *et al.* (1993).

2.8 ADULT LONGEVITY

2.8.1 INTRODUCTION

The life-span of an individual insect can be divided into two phases: (a) the development period from hatching of the egg until adult eclosion (section 2.9), and (b) the period of adult life, usually referred to as longevity (Blackburn, 1991a,b). An obligatory or facultative period of dormancy may intervene during the lifetime of an individual to extend either development or adult longevity for a variable period of time (see section 2.12).

Adult longevity may be studied from a variety of standpoints. For evolutionary biologists, it is a component of individual fitness (Waage and Ng, 1984; Hardy *et al.*, 1992b; Roitberg *et al*, 2001; Rivero and West, 2002), the assumption being that: (a) the longer a male can live, the more females he can inseminate, and therefore the more eggs he can fertilise; and (b) the longer a female can live, the more eggs she will lay. In both cases, the proviso 'all else being equal' applies. Adult longevity is also studied from the point of view of population dynamics, because of its relationship to female fecundity, the prey death rate and the predator rate of increase. Most studies on natural enemies measure adult longevity in the laboratory – there is a dearth of studies that measure it under natural conditions. Individual marking techniques that can be used to measure adult survival in the field are discussed in Chapter 6 (subsections 6.2.10, 6.2.11).

Longevity, like fecundity, is a highly variable species characteristic, influenced by a range of physical and biotic factors. The commonest experiments into the effects of these factors involve taking a cohorts of standardised females (subsection 2.7.2) and exposing each of them to one of a range of constant environmental conditions from eclosion until death. The resulting data are then plotted. Mean length of adult life can be plotted against variables such as body size, temperature, humidity, host or prey density, sugar concentration (in diet), and pesticide or other toxin (e.g. Bt, allelochemical) concentration. However, this method of expressing longevity data has major drawbacks (see below).

Evidence for a reproduction-survival trade-off has been found in some predators and parasitoids in relation to prey availability (Ernsting and Isaaks, 1991, Kaitala, 1991; Kopelman and Chabora, 1992; Valicente and O'Neill, 1995; Ellers *et al.*, 2000b). A cross-species trade-off was also observed in the gerrids that Kaitala studied. See Dixon (2000) for a discussion of the reproduction-survival trade-off within and among species of Coccinellidae. A cross-species trade-off between ovigeny index and life-span (subsection 2.3.4) was recorded by Jervis *et al.* (2001, 2003).

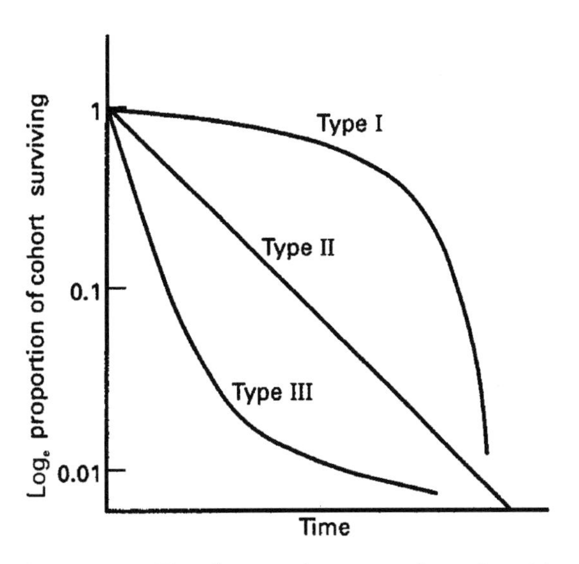

Figure 2.35 The three main types of survivorship curve: Type I – mortality concentrated in the oldest age classes; Type II – constant risk of death; Type III – mortality concentrated in the youngest age classes. Note the logarithmic scale.

2.8.2 SURVIVAL ANALYSIS

Frequently, in the literature, longevity data are presented as the mean length of adult life plus or minus its 95% confidence limit or standard deviation or standard error. However, when statistical comparisons between treatments are made, authors overlook the fact that individual longevity data are rarely normally distributed. For statistical comparisons between treatments to be biologically meaningful, the data are best presented in other ways such as **cohort survivorship curves**, which show the fraction of each cohort surviving at a particular moment in time (Figure 2.35). Such curves fall into 3 categories: **Type I**, in which the risk of death increases with age; **Type II**, in which there is a constant risk of death, i.e. the risk is independent of age; and **Type III**, in which the risk of death decreases with age.

Survival data have been compared by plotting survivorship curves and calculating the time to 50% mortality (LT_{50}) for each treatment and assessing the statistical significance of differences in this quantity. A major difficulty with this approach is that, at a particular point on the time axis, one or more of the curves might comprise few observations. Also, the 50% mortality level is subjective. As pointed out by Crawley (1993), generalised linear interactive modelling techniques (available in the form of the GLIM statistical package, Crawley, 1993) offer one of the best means of analysing survival data. The data can be analysed statistically in terms of **survivorship** (proportion of individuals from the cohort still alive at a particular point in time), the **age at death**, and the **instantaneous risk of death** (also termed the **'age-specific instantaneous death rate'** by biologists or **'hazard function'** by statisticians). Generalised linear interactive modelling can be used to determine which of a variety of available models (exponential, log-normal, Weibull) best describe the observed data. Having decided upon the most appropriate model, the effects of different experimental treatments can then be compared. For details of the procedure, see Crawley (1993, 2002).

The Weibull model has been used to to analyse survival data for parasitoids (Tingle and Copland, 1989; Hardy *et al.*, 1992b). The Weibull frequency distribution was originally considered as a model of human survivorship (Gehan and Siddiqui, 1973) and has commonly been used in engineering as a 'time to failure' model. The Weibull distribution is extremely flexible, possessing either positive or negative skewness, so allowing all three types of survival curve (I, II, III) to be analysed (Cox and Oakes, 1984).

The advantage of using the Weibull model to describe survival curves is that it summarises the information contained in a curve as both a rate parameter and a shape parameter. The fraction (F) of the cohort surviving at time t is given by:

$$F = 1 - \exp(-\{(t/b)^c\}) \qquad (2.4)$$

Statistical packages such as GLIM estimate the most appropriate value of the shape parameter c, and allow the rate (or scale) parameter b to be a linear combination of explanatory variables. Hardy *et al.* (1992b) examined survivorship in females of the bethylid wasp *Goniozus nephantidis*, and found that for each of two treatments a curve based on a Weibull distribution showed some systematic deviation from the observed curve. Having noted a relationship between longevity and body size in females (see below), they allowed the logarithm of the distribution's rate parameter to be a linear function of female size. Incorporation of female size significantly improved the fit, and therefore the explanatory power, of the model in the two treatments.

If you are dealing with a particularly long-lived species, you do not have to wait until all the individuals have died in order to terminate an experiment. The experiment can be terminated earlier, and certain statistical analyses can be used to take account of individuals that die at an unknown time, i.e. insects that are 'censored', statistically speaking (Crawley, 1993). Such analyses can also be used to take account of individuals that are accidentally lost or killed during the experiment.

Siekmann *et al.* (2001) applied Cox's Proportional Hazards Model (which is available in several statistical software packages) to longevity data obtained for *Cotesia rubecula* given a single meal, the concentration and timing of which varied among treatments. Their analysis showed that the risk of sugar-fed females starving to death was reduced by up to 73% in comparison with unfed wasps, depending on sugar concentration and timing, and that wasps need to locate food at least once a day if they are to avoid starving to death.

2.8.3 EFFECT OF BIOTIC FACTORS ON ADULT LONGEVITY

Host and Prey Density

Non-predaceous Females

In some studies host/prey density appears to have had little or no effect upon adult survival in such insects; at least when average longevity is used as the measure of survival (Liu, 1985b; Mackauer, 1983). Visual inspection of survival curves suggests the same, although a survival analysis of the type discussed above needs to be carried out on such data. Nevertheless, there is evidence for an effect of host availability on survival in some parasitoids. In some species, host-deprived females are able to live longer than undeprived ones kept under otherwise identical conditions (e.g. see Tran and Takasu, 2000); presumably they are able to do so because they obtain energy for maintenance from egg resorption, and/or they do not incur the life-span costs of oviposition (see above). A host density-related trade-off between reproduction and adult life-span has been recorded in a few parasitoids. Ellers *et al.* (2000b) exposed *Asobara tabida* to different host density regimes, and found that: (a) the total number of eggs produced (those laid in hosts plus those remaining in the females upon death) correlated with host density, and (b) there was a negative linear correlation between physiological life-span and the number of eggs produced – each egg that was produced decreased life-span by an equal amount. The significance of the shape of the trade-off function in *A. tabida* and in insects generally is discussed by Ellers *et al.* (2000b). Another parasitoid species in which there is a host availability-related trade-off between reproduction and life-span is *Leptopilina boulardi* (Kopelman and Chabora, 1992). Apparently in this species life-span declines in relation to increased oviposition rate, given that the species is pro-ovigenic (this rules out an effect of ovigenesis rate) and females in the different host density treatments produced the same number of progeny (see discussion in Jervis *et al.*, 2001).

Predaceous Females

Most information on predaceous females relates to cases where the longevity of females deprived of hosts or prey (deprived for either the whole or part of an experimental period) is compared with that of undeprived females. As one might expect, longevity is found to be shortest in the deprived females – they cannot satisfy their metabolic requirements for maintenance.

There are few published studies in which the longevity of predaceous females has been related to either availability of prey/hosts or consumption rate. Longevity is positively related to prey consumption rate in ovipositing *Coccinella undecimpunctata* over a wide range of prey densities (Ibrahim, 1955). In non-ovipositing *Thanasimus dubius* (Cleridae) longevity becomes a direct function of prey density only at low levels of prey availability (Turnbow *et al.*, 1978); the probable reason for the lack of a relationship at higher levels of prey/host availability in this cases is that at these levels the predators' maintenance requirements are fully satisfied.

By varying prey availability, Ernsting and Isaaks (1991), Kaitala (1991) and Nakashima and Hirose (1999) obtained evidence for a reproduction-survival trade-off in the predators they studied (see their papers for details).

Prey and Host Quality

In predatory Coccinellidae and Anthocoridae adult longevity may be significantly affected by the prey species fed upon by the adult (Hodek, 1973; Chyzik *et al.*, 1995; Mendes *et al.*, 2002). This also applies to destructively host-feeding parasitoids (Wilbert and Lauenstein, 1974). The host stage fed upon influences longevity in the host-feeding bethylid *Cephalonomia stephanoderis* (Lauzière *et al.*, 2000). With both host/prey species and host/prey stage effects, it is important to establish whether they are atributable to differences in the quantity of prey/host materials ingested or differences in host/prey quality *sensu stricto*.

In parasitoids, host species affects longevity via its effect on body size (see above).

Host-feeding by Parasitoids

Consumption of host blood improves longevity in some host-feeding wasp species (e.g. *Eupelmus vuilletti:* Giron *et al.*, 2002) but not in others (e.g. *Diadromus subtilicornis*: Tran and Takasu, 2000), while in the host-feeding parasitoid wasp *Aphytis melinus*, consumption of host blood positively influences longevity only if sugar-rich food is also taken (Heimpel *et al.*, 1997a) (see Jervis and Kidd, 1986, and Heimpel and Collier, 1996, for reviews). By directly injecting females with the sugars that are abundant in host blood (trehalose, sucrose), Giron *et al.*, (2002) showed that these sugars are solely responsible for the greater longevity of host-fed females.

Non-host and Non-prey Foods

Many studies have shown that, in the absence of hosts or prey, many parasitoids and predators given carbohydrate-rich foods, e.g. diluted honey solutions, live significantly longer than insects that are either starved or given only water (see reviews by Hagen, 1986; Jervis and Kidd, 1986; van Lenteren *et al.*, 1987; Heimpel and Jervis, 2004; see also Chapter 8).

Several studies have also revealed that longevity varies with the quality of food consumed. A simple experiment involves providing predators or parasitoids with one of a range of different diets, e.g. different sugars or combinations of sugars in solution, or even different nectars or honeydews, and comparing the effects of these on survival. However, for investigations of the effects of non-host food consumption on longevity (and fecundity) to have relevance to the field situation (particularly in biological control), they should involve first identifying the natural diet of parasitoids and predators, and then providing insects with the same or very similar foods (Chapter 8).

For details of protocols for determining the effects of biochemical components of non-host foods on longevity, see Finch and Coaker (1969) and Wäckers (2001). The effects of sugar-feeding on carbohydrate and lipid levels in parasitoids have been investigated by Olson *et al.* (2000), Fadamiro and Heimpel (2001) and Casas *et al.* (2003) (see section 2.13 for biochemical techniques).

Body Size

A positive correlation between body size and longevity, at least across the lower range of host body sizes, has been shown for the adults (in some cases males as well as females) of several parasitoid species (e.g. see Sandlan, 1979b; Mani and Nagarkatti, 1983; Waage and Ng, 1984; Bellows, 1985; Hooker *et al.*, 1987; van den Assem *et al.*, 1989; Hohmann *et al.*, 1989; Heinz and Parella, 1990; Hardy *et al.*, 1992b; Harvey *et al.*, 1994; West *et al.*, 1996; Ellers *et al.*, 1998; Fidgen *et al.*, 2000; Rivero and West, 2002; also see review by Visser, 1994). Exceptions include *Goniozus nephantidis* (in which larger females live longer than smaller ones if hosts are provided, but smaller females live slightly longer than larger ones if hosts are not available) (Hardy *et al.*, 1992b), and both *Asobara tabida* and *Nasonia vitripennis* (in which the body size effect does not occur in fed females) (Ellers *et al.*, 1998; Rivero and West, 2002). The size effect upon the longevity of unfed female *A. tabida* and *N. vitripennis* is attributable to the smaller fat reserves of small females: such females can obtain additional energy for maintenance by feeding, but note that females cannot either supplement or replenish their fat body reserves (Ellers *et al.*, 1998; Rivero and West, 2002) (see also section 2.14). Interestingly, in the two *Metaphycus* species studied by Bernal *et al.* (1999) the body size effect occurs in fed females and not in unfed ones; the amounts of lipid reserves were not measured in this case.

Because a size-longevity relationship may not appear to exist for some species (e.g. Takagi, 1985, on *Pteromalus puparum*), it is important, in experiments, to provide parasitoids with a range of host sizes equivalent to that occurring in the field. Few studies have measured longevity in relation to body size under actual field conditions (see West *et al.* 1996).

Blackburn (1991a) showed through a comparative analysis (subsection 1.2.3) of 474 hymenopteran parasitoid species that, across the species within a taxon, there is no correlation between body size and life-span (see Blackburn, 1991a, and Jervis *et al.*, 2003, for discussion).

Sokolovska *et al.*, (2000) found positive correlations between longevity and male and female body size among Odonata, but their meta-analysis has been severely criticised by Thompson and Fincke (2002).

Burkhard *et al.* (2002, used the degree of wing wear and injury to estimate size-specific survivorship in field populations of the predatory fly *Scathophaga stercoraria*, and showed that in females longevity increased with body size in both flight seasons but that in males it increased slightly in the spring and decreased in the autumn. Burkard *et al.* (2002), however, urge caution in applying the method (see their paper for details).

Mating

As discussed in Chapter 4 (subsection 4.5.3) frequent mating may shorten life-span in both females and males. Several laboratory studies have shown that in predatory Coccinellidae unmated females live longer than mated ones; the same also applies to males (Dixon, 2000) (see also Taylor *et al.*, 1998, on a predatory stonefly). However, there is no difference in longevity between mated and virgin females of the predatory bug *Podisus maculiventris* (De Clercq and Degheele, 1997).

In experiments aimed at testing for the effects of mating on longevity, one must establish that mating really has occurred: simply caging females with a male is no guarantee either that the insects have engaged in mating behaviour or have (in the case of females) been inseminated. If a positive effect of mating upon female longevity is found, the question arises as to whether longevity is enhanced due to the nutrient contribution made by the male, in the form of sperm or spermatophore.

2.8.4 EFFECT OF PHYSICAL FACTORS ON ADULT LONGEVITY

Temperature

There will be an optimum range of temperatures outside of which survival is severely reduced (Jackson, 1986; Krishnamoorthy, 1989). In general, and usually in males as well as females, longevity decreases with increasing temperature within the optimum range (Abdelrahman, 1974; Hofsvang and Hågvar, 1975a,b; Sahad, 1982, 1984; Nealis and Fraser, 1988; McDougall and Mills, 1997; Hentz *et al.*, 1998; Tran and Takasu, 2000; Liu and Tsai, 2002; Seal *et al.*, 2002) (e.g. see Figure 2.36), although for some species no more than a trend may be apparent (Barfield *et al.*, 1977a; Cave and Gaylor, 1989; Miura, 1990).

Most experiments designed to demonstrate the effect of temperature on adult longevity involve exposing insects to constant temperatures, which ignores the fact that in nature temperatures will fluctuate during each day, the lowest temperatures occurring at night. Ideally, longevity ought to be studied at temperature extremes that are part of a cyclical regime, but such an approach has rarely been adopted. Ernsting and Isaaks (1988) measured the survival of the carabid *Notiophilus biguttatus*, given excess prey, at a constant 10°C regime compared with a daily fluctuating (20°C day/10°C night) regime. The lower survival of beetles held under the fluctuating regime could simply be explained by the higher average daily temperature at that regime. Minkenberg (1989) incorporated a more realistic fluctuating temperature regime in his experimental design. He exposed the eulophid *Diglyphus isaea* to each of three constant temperatures (15°, 20°, 25°C) and to a fluctuating regime that involved the temperature increasing linearly from 0100 h to 0300 h, decreasing from 1500 h to 1700 h, and being fixed at 22°C from 0300 h to 1500 h and at 18°C from 1700 h to 0100 h. Survival of wasps held under the fluctuating regime (average daily temperature 20.3°C) was much lower than at the constant 20°C regime.

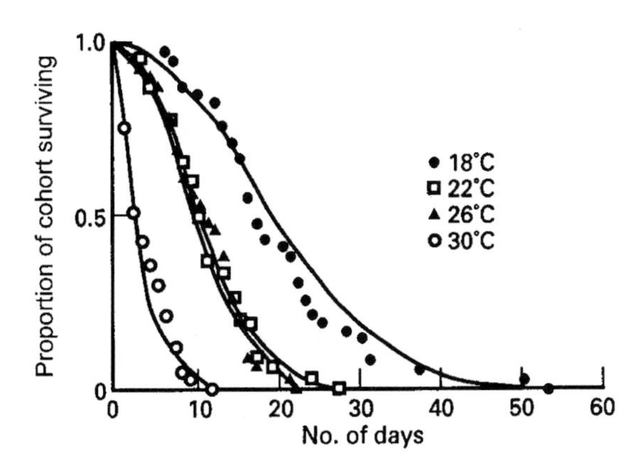

Figure 2.36 Survivorship of *Anagyrus pseudococci* (Encyrtidae) at four different constant temperature regimes. Source: Tingle and Copland (1989). Reproduced by permission of Lavoisier Abonnements.

Some parasitoids and predators overwinter as adults, and may be exposed to near- and/or sub-zero temperatures. Cold tolerance of adult *Bathyplectes curculionis* (Ichneumonidae) was studied by Berberet *et al.* (2002), and that of *Harmonia axyridis* (Coccinellidae) by Watanabe (2002).

Humidity

It is clear from many experimental studies that natural enemy adults have particular humidity requirements for survival (Kfir, 1981; Wysoki *et al.*, 1988; Herard *et al.*, 1988).

Although it would appear to be quite easy to carry out an experiment designed to measure survival at different humidities, there is the problem of maintaining the insects for a sufficiently long period for statistical comparisons to be made. Insects deprived of food are likely to die quite quickly, but if they are provided with honey or sucrose solutions (see above), it may be difficult to separate the effects upon longevity of the water content of the air and that of the food. Similarly, it may be difficult to set up an experiment that incorporates some degree of biological realism in the form of a plant surface, since the latter will be actively transpiring.

Small-bodied insects, because of their high surface area to volume ratio, will be more prone to desiccation at low humidities than large-bodied insects; see Jervis *et al.* (2003) who discuss this, from a comparative perspective, in relation to parasitoid wasps.

Photoperiod

Little is known about the influence of photoperiod on longevity. Given that some predaceous insects are active only during certain periods of the day or night, one might expect longevity to be influenced by photoperiod. In the parasitoid *Ooencyrtus kuvanae* photoperiod experienced upon adult eclosion influences both longevity and the rate of progeny production. Short-day conditions resulted in females producing fewer progeny but living longer. Switching photoperiods after twelve days failed to alter this once it had been established (Weseloh, 1986b). In *Anagyrus kamali* longevity, like lifetime fecundity, is highest under continuous darkness (Sagarra *et al.*, 2000). Hentz *et al.* (1998), however, found no significant effect of photoperiod on longevity in *Chelonus* sp. near *curvimaculatus*.

Note that some entomophagous insects are nocturnal, e.g. certain Ichneumonidae (Gauld and Huddleston, 1976); Vespidae, Pompilidae and Rhopalosomatidae (Gauld and Bolton, 1988).

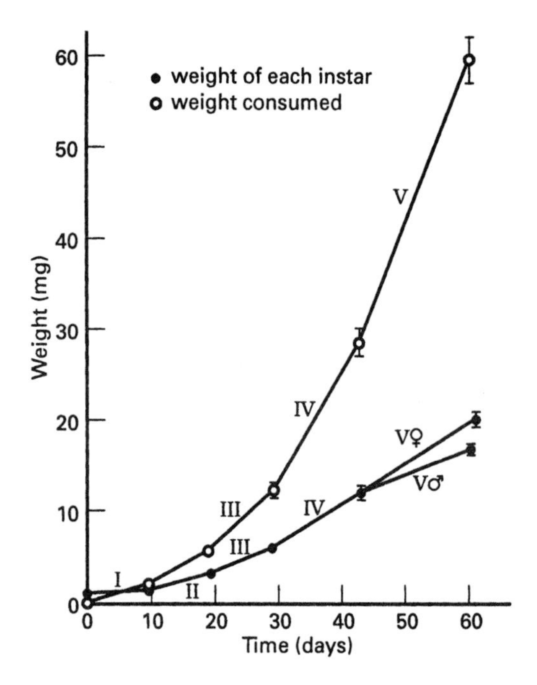

Figure 2.37 Development in the bug *Blepharidopterus angulatus* (Miridae) given excess prey, showing the length of time spent in each larval instar, the body weight at the start of each instar, and the cumulative wet weight of lime aphids consumed up to the start of each instar. Roman numerals denote instars. Source: Glen (1973).

2.9 GROWTH AND DEVELOPMENT OF IMMATURES

2.9.1 INTRODUCTION

Development refers to the morphological, anatomical and physiological changes shown by each individual insect from the time the egg is laid to the time the adult ecloses. **Growth** refers to the increase in biomass of the insect during the period between hatching from the egg and the end of the larval phase of the life-cycle, as shown in Figure 2.37, or between instars. The larval phase in predators comprises long periods of feeding and brief periods of moulting. Typically, biomass increases steadily throughout each instar. At the time of the moult, biomass falls slightly due to the loss both of the exuvium and of some water (which is not immediately replaced, as the insect is not feeding) (Chapman, 1998). In some aquatic insects there is no decrease in biomass at the moult; instead there is an increase due to absorption of water through either the cuticle or the gut; in *Notonecta glauca* this increase is very large (Wigglesworth, 1972).

Field, as opposed to laboratory, measurements of growth and development in predators and parasitoids have been made on few species (e.g. Griffiths, 1980, on ant-lions; Banks and Thompson, 1987b, on damselflies).

For predators the protocols for measuring growth and development in relation to certain physical and biotic factors are relatively straightforward. For example, to study the influence of prey availability, take a series of cohorts of newly hatched larvae and present each insect with one of a range of chosen prey densities (prey of a fixed size), at a constant temperature, humidity, photoperiod, either for the duration of the insect's life or for the duration of one or a few instars only. On each day of larval life, replace the prey. Larval development is measured simply in terms of the period of time between moults or other events (e.g. egg hatch and pupation). Larval growth can be measured as the dry or fresh weight gain, including exuvia weight, or the body size increase (e.g. measured in terms of head width) between instars, although the standard measure of growth rate is, the **mean relative growth rate** (**MRGR**):

$$\text{MRGR} = \frac{l_n(W_f) - l_n(W_i)}{d} \qquad (2.5)$$

where W_i is the initial weight of the insect, W_f is the final weight of the insect, and d is the period of time over which growth is measured. Some workers (Paradise and Stamp, 1990, 1991) have expressed growth rate differently, as the fresh weight gained/instar duration) × the average fresh weight of the predator during the instar.

For parasitoids, the protocols for measuring the influence of physical and biotic factors on growth and development may be rather more complex than for predators. Endoparasitoids are a particular problem, since the sizes and weights of larvae cannot easily be measured and the larvae often cannot easily be assessed as to their stage of development (Mackauer, 1986; Sequeira and Mackauer, 1992b; Harvey *et al.*, 1994) (see below).

A necessary prerequisite for studying many aspects of larval development, particularly instar-related aspects of biology, in predators and parasitoids is the ability to distinguish between the different instars. In some cases, it is relatively easy to tell the instars apart, using features such as mouthpart structure, the degree of wing development, the number and position of prominent setae, spines and other cuticular structures, the structure of the tracheal system and associated spiracles, and body colour patterns. However, in other insects, obvious distinguishing features may be lacking. Morphometric techniques may therefore be required. Thompson (1975, 1978), for example, decided upon the instar of the damselflies larvae he studied (*Ischnura elegans*), by means of both a frequency distribution plot of head widths of randomly field-collected larvae and a regression of modal head width against probable instar number. Even better discrimination between instars was obtained by plotting head width against body length (Thompson, 1978).

2.9.2 EFFECTS OF BIOTIC FACTORS ON GROWTH AND DEVELOPMENT

Food Consumption

Introduction

Predator larvae need to consume several prey individuals during development, and each successive instar will show a maximum rate of growth and development at different levels of prey availability. Generally, with increasing prey density, at least across the low and medium ranges, larval predators consume more prey, develop faster, gain more weight and so attain a higher final size (Dixon, 1959a; Lawton *et al.*, 1980; Scott and Barlow, 1984; Pickup and Thompson, 1990; Zheng *et al.*, 1993a,b; Bommarco, 1998; Dixon, 2000). Where development rate increases non-linearly with prey consumption rate (see below), development rate stops increasing above a certain prey density while growth continues. Growth and development also vary in relation to prey quality.

Food consumption by insects is a subject in its own right, and the associated literature is very large (Waldbauer, 1968; Beddington *et al.*, 1976b; Kogan and Parra, 1981; Scriber and Slansky, 1981; Slanksy and Scriber, 1982,1985; Slansky and Rodriguez, 1987; Karowe and Martin, 1989; Farrar *et al.*, 1989). The approach we are recommending here is that of Beddington *et al.* (1976b), as it provides one of the most useful bases for predicting predator-prey population dynamics (Chapter 7). The various problems inherent in measuring food consumption and utilisation by insects and other arthropods are discussed in Waldbauer (1968), Lawton (1970), Ferran *et al.* (1981) and Pollard (1988).

The basic protocol for studying the effects of prey availability and prey consumption on growth and development in predators has already been outlined. Other measurements can also be taken in order that various **nutritional indices** can be calculated; these measurements are of:

1. The biomass of the prey materials ingested (biomass is best measured in terms of dry weight, since prey remains are likely to lose water before retrieval), The predator's **efficiency of conversion of ingested food into body substance (ECI)** can then be calculated as follows:

$$\text{Conversion efficiency} = \frac{M}{C - D} \times 100 \quad (2.6)$$

   where M is the increase in biomass of the predator, C is the biomass of captured prey, and D is the biomass of the captured prey that is not consumed ($C{-}D$ is therefore the biomass of prey actually ingested). According to Cohen (1984, 1989), predaceous insects with piercing, suctorial mouthparts (e.g. Heteroptera) ought to have higher ECI values than predators with chewing mouthparts, because they obtain a larger proportion of highly digestible materials from their prey (a process assisted by pre-oral digestion) (see also Cohen, 1995, and Cohen and Tang, 1997).The ECI is a measure of *gross* growth efficiency, since biomass losses in the form of faeces and excreta are not accounted for.
2. C, D and the biomass which appears as faeces (F), and the products of nitrogenous excretion (U). The predator's **utilisation efficiency**, i.e. the efficiency with which the prey biomass *captured* is converted into predator biomass, can then be calculated:

$$\text{Utilisation efficiency} = \frac{C - D - F - U}{C} \times 100 \quad (2.7)$$

3. C, D, F and U (as in 1. and 2.). The predator's **assimilation efficiency**, i.e. the efficiency with which the prey biomass consumed is converted into predator biomass, can then be calculated:

$$\text{Assimilation efficiency} = \frac{C - D - F - U}{C - D} \times 100 \quad (2.8)$$

4. C, D, and F (as in 1. and 2.). The predator's **digestive efficiency** (also termed 'approximate digestibility'), i.e. the efficiency with which the prey biomass *ingested is* digested and absorbed, can be calculated:

$$\text{Digestive efficiency} = \frac{C - D - F}{C - D} \times 100 \quad (2.9)$$

Other nutritional indices used in studies of food consumption by insects are discussed by Waldbauer (1968) and Slansky and Scriber (1982, 1985).

Growth Rate

At least some of the food a larva consumes needs to be allocated to maintenance metabolism. Because of this, growth will stop if consumption falls below a certain threshold (this threshold will become higher as the insect grows and its maintenance requirements increase). The energy allocated to growth can be assumed to be a linear function of food intake (Beddington *et al.*, 1976b):

$$G = \delta (I - B) \quad (2.10)$$

where G is the the growth rate (biomass accumulated per unit time, e.g. fresh weight gain, including exuvium weight, divided by the number of days spent in the instar) of each juvenile stage, I is the rate of ingestion of food (biomass of prey consumed per unit time, in comparable units to G, see equation 2.6), and δ and B (the threshold ingestion rate, analogous with parameter c in equation 2.1) are constants. Mills (1981) gives an alternative model.

Figure 2.38 shows the relationship between growth rate and consumption rate in larval *Notonecta;* the relationship conforms to that

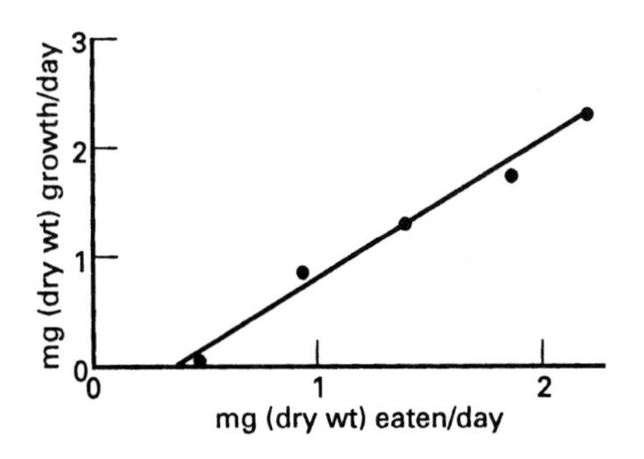

Figure 2.38 Growth rate as a function of ingestion rate in final instar of *Notonecta undulata*. Source: Beddington *et al.*, (1976b), who used data from Toth and Chew (1972). Reproduced by permission of Blackwell Publishing.

predicted by equation (2.10). As can be seen from the intercept of the line with the abscissa, the predator needs to consume a minimum amount of food for any growth to occur.

Should the increase in respiratory rate be *non-linear*, then growth rate will be non-linear and conform to the following model (Beddington *et al.*, 1976b):

$$G = \delta\,(\log_e I - B) \tag{2.11}$$

Development Rate

If W_i is the initial weight (biomass) of an instar (teneral weight), W_f is the final weight achieved, and W is the total weight gain, then $W = W_f - W_i$. The ratio W/G will define the duration, d, of the instar, and development rate, $1/d$, is given by the following linear model (Beddington *et al.*, 1976b):

$$\frac{1}{d} = \frac{\delta}{W}\,(I - B) \tag{2.12}$$

If it is assumed, for simplicity's sake, that W remains a constant, then (Beddington *et al.*, 1976b):

$$\frac{1}{d} = \alpha\,(I - B) \tag{2.13}$$

where α and B are constants. Equation 2.13 still predicts a simple, linear relationship between development rate and consumption rate.

As pointed out by Beddington *et al.* (1976b), equation (2.12) ignores the fact that the larvae of some predators may, under conditions of food scarcity, moult to the next instar at significantly lower body weights than when food is abundant. W_i, W_f and W are therefore functions of consumption rate and thus of prey availability – weight gain in each instar cannot be assumed to be constant. Figure 2.39 shows how, in the damselfly *Ischnura elegans*, larvae fed at low prey densities moulted to smaller individuals, i.e. they moulted earlier than better fed larvae, having gained less weight. Mills (1981) demonstrated, through a regression analysis of the relationship between W and consumption rate and teneral weight in *Adalia bipunctata*, a significant dependence in both cases; consumption rate explained 47–75% of the variance.

Thus, the relationship in some predators is more complex than that described by equation 2.13. Lawton *et al.* (1980) provide the following non-linear model:

$$\frac{1}{d} = \alpha\,(\log_e N_a - B) \tag{2.14}$$

where N_a is the number of hosts fed upon. An alternative non-linear model is provided by

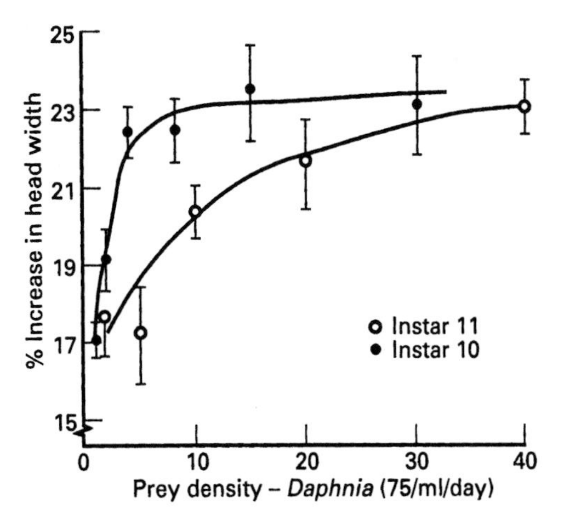

Figure 2.39 The effect of prey density on the percentage increase in head width at the moult in *Ischnura elegans*. Source: Lawton *et al.*, (1980). Reproduced by permission of Blackwell Publishing.

Mills (1981). Both models describe a decelerating curve for the relationship between development rate and consumption rate. Curves of this type were obtained in the laboratory for both *Ischnura elegans* (Figure 2.40) and *Adalia bipunctata* (Figure 2.41) Lawton *et al.* (1980) gave, as well as a dependence of W on consumption rate, three other reasons to account for non-linearity in the case of *Ischnura*:

1. Variation in k (equation 2.2) with prey availability. In *Ischnura* k declined with prey availability (Figure 2.42), the predators wasting proportionately more of each of the prey they kill at higher densities (adaptive behaviour in many predators, section 1.10; Cook and Cockrell, 1978; Giller, 1980; Sih, 1980; Kruse, 1983; Bailey, 1986; Dudgeon, 1990, although some predators may go to the extreme of not consuming any part of the prey – see Yasuda, 1995). However, as Lawton *et al.* (1980) point out, a decline in utilisation efficiency in *Ischnura* cannot be the sole reason for the non-linear dependence of development rate upon prey consumption rate. If it is, daily growth rates plotted against prey biomass assimilated ($C-[D+F]$) ought to be linear (equation (2.10) – they are not (Figure 2.43).
2. A decrease in assimilation efficiency with increasing consumption rate. Lawton (1970) had suggested that this can occur with over-feeding at high levels of prey availability, causing defaecation to take place before digestion is complete. Lawton *et al.* (1980) investigated whether assimilation efficiency varied with prey availability. Since it does not do so in *Ischnura* (Figure 2.42), this hypothesised effect could not account for the non-linear dependence of development rate on consumption rate.
3. A non-linear increase in respiratory rates with increasing consumption rate (see equation 2.11). Lawton *et al.* (1980) concluded that this effect, together with the variation in k and W, accounted for the observed relationship in Figure 2.40. Circumstantial evidence to support the conclusion regarding change in respiratory rates comes from Lawton

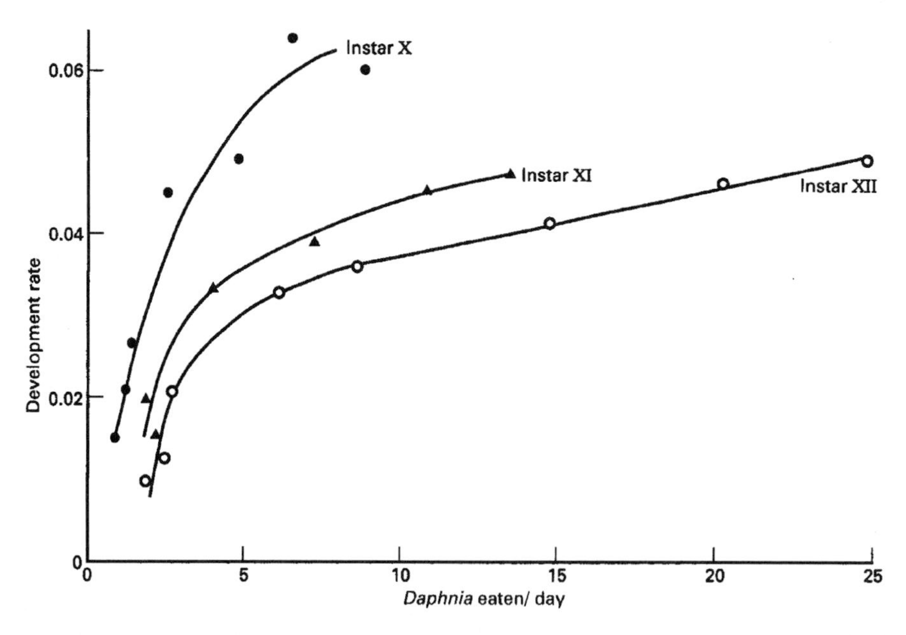

Figure 2.40 Development rates in *Ischnura elegans* (Odonata: Zygoptera) larvae as a function of the number of prey killed per day. Only larvae that successfully completed their development in each stage were included in the calculations. Source: Lawton *et al.*, (1980). Reproduced by permission of Blackwell Publishing.

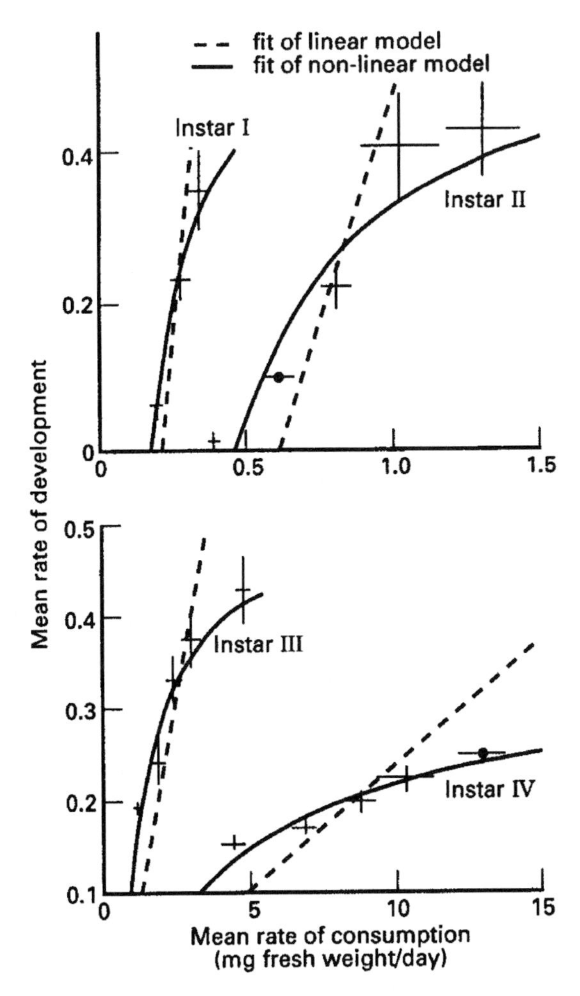

Figure 2.41 The relationship between mean development rate and consumption rate for the four larval instars of *Adalia bipunctata* (Coccinellidae). Indicated is the fit of linear and non-linear models of development (see text). Source: Mills (1981). Reproduced by permission of Blackwell Publishing.

et al.'s behavioural observations: larvae held at high prey densities frequently engage in more waving of the gills than other larvae, suggesting that they are under oxygen stress. Respirometric methods would need to be employed to establish whether respiratory rates do indeed alter.

To obtain the relationship between development rate and prey availability, both equation (2.13) and equation (2.2) can be incorporated

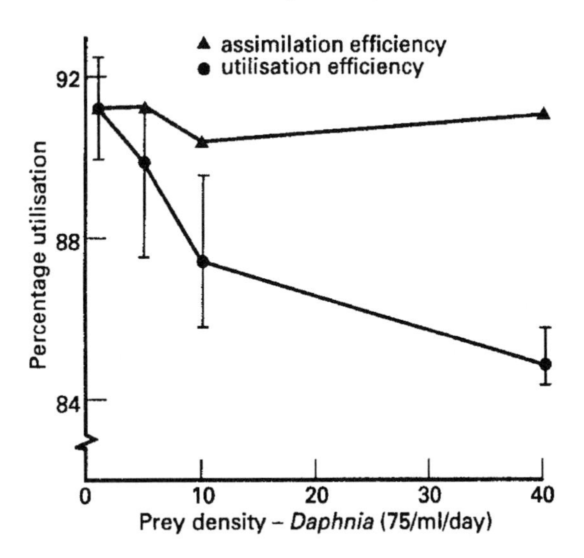

Figure 2.42 The relationships between assimilation and utilisation efficiencies and prey density in the eleventh instar of *Ischnura elegans* (Odonata: Zygoptera). Utilisation efficiency clearly declines with increased prey density. Source: Lawton *et al.*, 1980). Reproduced by permission of Blackwell Publishing.

into the simple functional response model (Holling's disc equation, subsection 7.3.7) (Beddington *et al.*, 1976b):

$$\frac{1}{d} = \alpha\left(\frac{ka'NT}{1 + a'T_hN} - B\right) \qquad (2.15)$$

Equation 2.15 describes a decelerating curve, like a Type 2 functional response (section 1.10). As pointed out by Beddington *et al.* (1976b), the curve is unlikely to go through the origin. Unless the weight at which a species is able to moult to the next instar is very flexible, the effect of B will be to displace the curve along the prey axis. Put another way, there will be a threshold prey density (and therefore consumption rate) below which growth and development cannot take place. Examples of this are shown in Figure 2.44. In those species that consume proportionately less of each prey item when encounter rates, i.e. levels of prey availability, are high (k declines) the curve will be somewhat different in shape: flatter-topped, with an earlier 'turnover' point (Beddington *et al.*, 1976b; also see Yasuda, 1995).

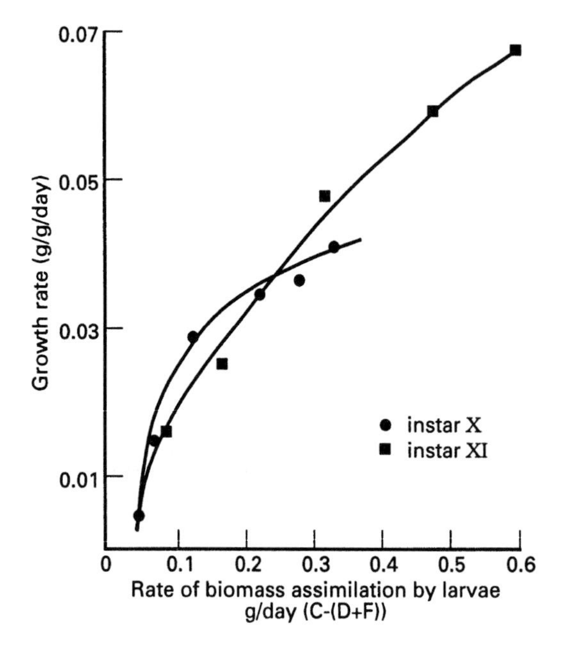

Figure 2.43 The effect of daily rate of biomass assimilation on growth rate for instars X and XI of *Ischnura elegans* (Odonata: Zygoptera). Growth rate is measured as g/g/d increase in weight and is calculated by dividing weight gained during the instar by instar duration. These figures were corrected for the initial weight of the larvae. Wet weights were used for initial and final weights. Only larvae that successfuly completed their development in each instar were used in the calculations. Source: Lawton *et al.*, (1980). Reproduced by permission of Blackwell Publishing.

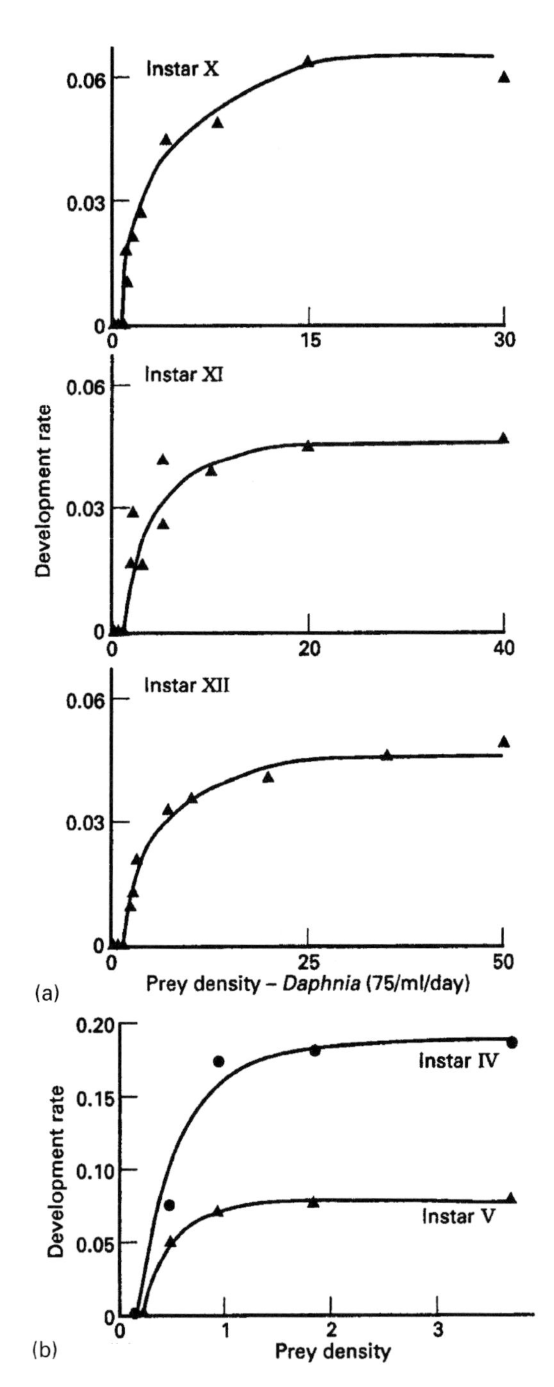

Figure 2.44 Development rates as a function of prey density in different instars of: (a) *Ischnura elegans* (Odonata: Zygoptera) (source: Lawton *et al.*, 1980); (b) *Notonecta undulata*. (source: Beddington *et al.*, 1976b, who used data from Toth and Chew, 1972). Reproduced by permission of Blackwell Publishing.

Variation in growth and development between and within instars

Figure 2.37 shows both the cumulative increase in prey biomass consumed and the increase in weight of nymphs of the bug *Blepharidopterus angulatus* as they develop. Later instars account for most of the total consumption and growth that occurs. In the green lacewing *Chrysoperla carnea* the third (final) instar accounts for 80.5–82.8% of the total consumption and 80.45–85.6% of the total growth that occurs (data from Zheng *et al.*, 1993a).

Figure 2.44 shows that the development rate versus prey availability curves differ between instars. As pointed out by Beddington *et al.*

(1976b), this is to be expected from the between-instar differences that exist with respect to: (i) attack rate (a') and handling time (T_h), both of which are parameters in the functional response model, subsection 7.3.7); (ii) metabolic rate, which will increase with instar by a certain power of the body weight – this affects B in equation 2.14; and (iii) the constants α and k (Beddington *et al.*, 1976b).

Examination of the growth rate versus consumption rate plots for *Adalia bipunctata* (Figure 2.45) reveals that the slope (which represents conversion efficiency) decreases as the insects pass through the instars. This change in the slope is partly attributable to increased metabolic costs in later instars, as can be seen from the intercepts with the y-axis, representing basal respiratory rates. However, the main cause is likely to be a decline in digestive efficiency, since compared with earlier instars, later instars of *Adalia* consume a greater proportion of each prey item, i.e. k increases with instar (Mills, 1981). To understand the relationship between the proportion of each prey consumed and digestive efficiency, consider the surface area/volume ratio difference between food boluses of different sizes. A larger bolus will have proportionately less of its surface area exposed to digestive fluids than a smaller bolus.

Conversion efficiency can also vary with consumption rate *within* an instar. Third instar larvae of *Chrysoperla carnea* provided with low prey densities have, as expected, a reduced consumption rate compared with third instar larvae given high prey densities, but they have a higher conversion efficiency (Zheng *et al.*, 1993a). A similar difference in conversion efficiency is shown by the early instars of the bug *Blepharidopterus angulatus* (Glen, 1973). Two possible reasons for this effect in the case of *C. carnea* were put forward by Zheng *et al.* (1993a):

1. Digestive efficiency is increased, due to the smaller quantities of prey being ingested by larvae given low prey densities;
2. Third instar larvae, like some spiders, reduce their metabolism in response to prey scarcity.

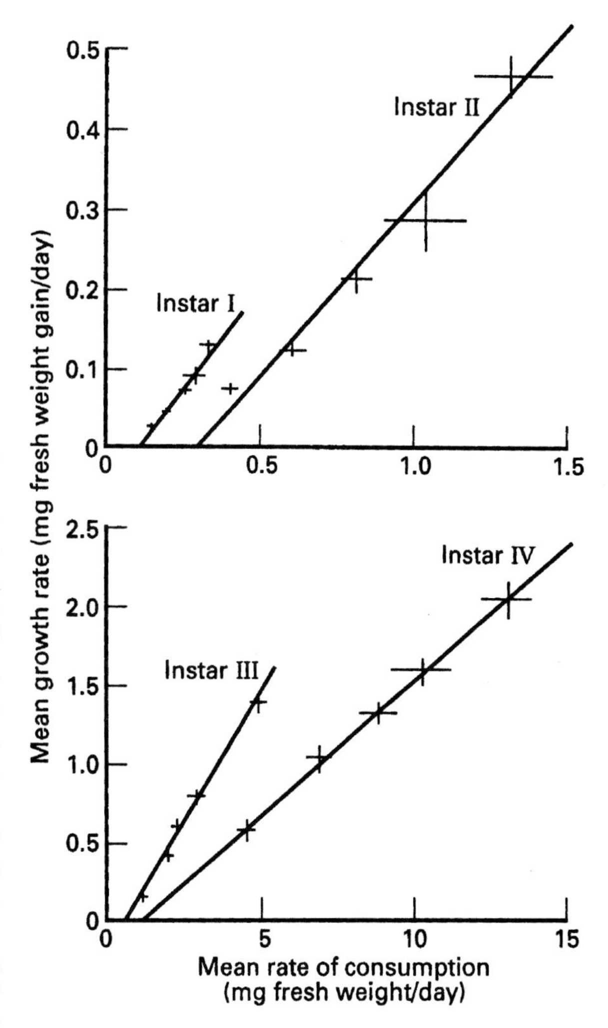

Figure 2.45 The linear dependence of average growth rates on food consumption rate for the four larval instars of the coccinellid beetle *Adalia bipunctata*. Note that the slope of the relationship, representing the gross food conversion efficiency, decreases as the insect develops. This is partly due to increased metabolic costs, as can be observed from the y-axis intercepts representing basal respiratory rates, but it is mainly due to a decline in digestive efficiency with instar. Source: Mills, (1981). Reproduced by permission of Blackwell Publishing.

Feeding History

Can predators recover from the deleterious effects upon growth and development brought about in previous instars by prey scarcity?

To answer this question, a cohort of larvae can be exposed to high levels of prey availability throughout two instars, e.g. the third and the fourth in a coccinellid, and another cohort can be exposed to a prey availability regime that alters from low to high between these two instars. The fourth instar insects from the two regimes can then be compared with respect to weight gain and instar duration. In this experiment consumption rate and the various nutritional efficiencies should be measured, to determine whether the compensatory effects shown by the test cohort are a result of changes in one or more of these factors within the later instar.

A similar experiment to the above was carried out by Paradise and Stamp (1991) on the mantid *Tenodera sinensis*. These authors showed that: (a) first and second instar mantids given a small quantity of prey attained a smaller size and spent more time in those instars than mantids provided with as much prey as they could eat, but that (b) in two out of three cohorts, mantids reared during the first instar on a poor diet recovered during the second instar when they were switched to a higher diet, gaining as much weight as, and spending less time in that instar than, those given a high diet throughout. The larvae of the later instar compensated for poor feeding in the earlier instar by having a higher consumption rate.

Zheng *et al.* (1993a) conducted a similar experiment with the green lacewing, *Chrysoperla carnea*, but over the entire larval development period. Larvae were either provided with a large quantity of prey over all three instars (HHH regime), or they were given a low quantity over the first two instars and a large quantity during the third (LLH regime). No significant difference in the duration of the third instar was found between larvae in the two regimes, but the overall duration of development from eclosion to pupation was significantly longer in the LLH larvae, i.e. recovery in development rate was partial. The dry weight gain of third instar larvae was not significantly different in the two treatments, and the same applied to the overall weight gain over the whole of larval development, i.e. recovery in growth was complete. Third instar larvae in the LLH regime consumed as many prey as those in the HHH regime, and the same applied to larvae over the whole of their development.

Limited recovery from suboptimal feeding conditions can, at least in the laboratory, be achieved in some Odonata (*Lestes sponsa*) by the larva passing through an additional instar. However, instar number is constrained and an increase in any linear dimension is limited to around 25–30% (D.J. Thompson, personal communication).

Can predators with higher growth rates in one instar maintain the advantage through subsequent instars? To answer this question, the aforementioned experimental design can be reversed, so that in the test cohort the prey availability regime alters from high to low. Experiments carried out by Fox and Murdoch (1978) on the backswimmer *Notonecta hoffmani* show that larvae can maintain a growth advantage during larval development.

Non-prey Foods

As with the fecundity versus prey density relationship, two effects of providing non-prey foods together with prey might be to lower the prey ingestion rate threshold, thus shifting the development rate versus prey availability curve nearer to the origin, and to alter the shape of the curve. Predator larvae may require a lower minimum number of prey items in order to develop at all, and they may develop more rapidly at and above this minimum.

That development rate is increased by provision of non-prey foods is demonstrated by experiments conducted on larvae of the lacewing *Chrysoperla carnea* (McEwen *et al.*, 1993a). At the three test prey densities offered to the predators during development, larvae given an artificial honeydew with prey required significantly fewer prey, developed significantly more rapidly, and attained a significantly higher adult weight than larvae given water with prey. Some predators can complete larval development when prey are absent, if certain non-prey foods are available, e.g. the bug *Orius insidiosus* (Anthocoridae) (Kiman and Yeargan, 1985), and the coccinellid *Coleomegilla maculata* (Smith, 1961, 1965). Predators such as the bug *Blepharidopterus angulatus* cannot complete development on a diet of honeydew alone, but nymphs that are switched from honeydew to a diet of aphids after the third instar can complete development (Glen, 1973).

Prey Species

Larval growth and development might be expected to vary in relation to prey species. Examples of studies demonstrating this effect in coccinellids include those of Blackman (1967) and Özder and Sağlam (2003) for *Adalia bipunctata* and *Coccinella septempunctata*, Michels and Behle (1991) for *Hippodamia sinuata* (in which the prey species effect on development rate disappeared at temperatures exceeding 20°C) and Wiebe and Obrycki (2002) for *Coleomegilla maculata* (and the lacewing *Chrysoperla carnea*). Sadeghi and Gilbert (1999), Mendes *et al.* (2002) and Petersen and Hunter (2002) studied larval performance in the hover-fly *Episyrphus balteatus*, in the anthocorid bug *Orius insidiosus* and in lacewings respectively, in relation to prey species.

Albuquerque *et al.* (1997) investigated and compared growth and development (and also reproduction) in two lacewings, one a specialist, the other a generalist, examining what alterations in these variables occurred when the predator species were given each other's prey species (see their paper for details).

Interference and Exploitation Competition and Other Interference Effects

Ecologists distinguish between competition through interference and competition through exploitation. In **interference competition** individuals respond to one another directly rather than to the level to which they have depleted the resource. In **exploitation competition** individuals respond, not directly to each other's presence, but to the level of resource depletion that each produces. With exploitation competition the intensity of competition is closely linked to the level of the resource that the competitors require, but with interference it is often only loosely linked (Begon *et al.*, 1996).

Larval predators show interference in the form of behavioural interactions. For example, larval dragonflies may interfere with one another's feeding through distraction (e.g. 'staring encounters' between dragonflies) and/or overt aggression (Baker, 1981; McPeek and Crowley, 1987; Crowley and Martin, 1989) (Figure 2.46). Such interactions are likely to result in reduced feeding or increased metabolic

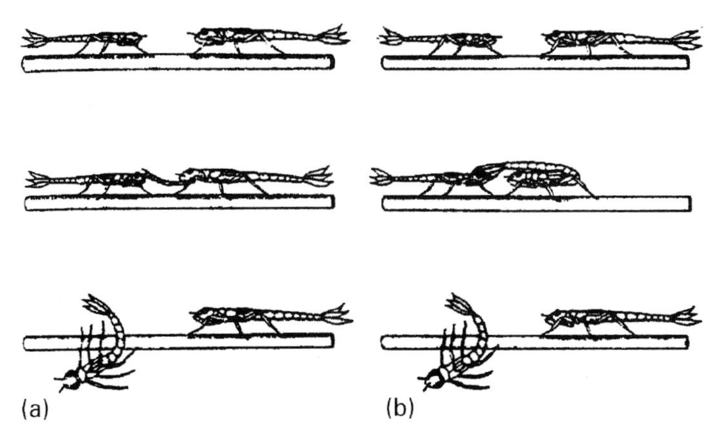

Figure 2.46 Aggressive interactions between damselfly larvae: (a) labial striking; (b) slashing with the gills. Larger larvae usually displace smaller ones, which may retreat by swimming off the perch. Source: Williams and Feltmate (1992). Reproduced by permission of CAB Publishing.

costs and therefore ultimately will cause reduced growth, development and survival.

Van Buskirk (1987) conducted an experiment to test whether density-dependent, interference-mediated reductions in growth, development and survival occurred in larvae of the dragonfly *Pachydiplax longipennis*. First instar larvae were raised in pools at initial densities of 38, 152 and 608 larvae/m^2, under two levels of prey availability (extra prey added daily to those already in pool; extra prey not added, i.e. food depletion likely to occur, pools in both cases containing the same initial density of prey), in a 3×2 factorial design. Van Buskirk found that with increasing predator density there was a decrease in growth and development rates, but he did not detect any statistically significant interactions between prey addition and predator density, suggesting that some form of interference, rather than prey exploitation, was important. Within the prey-added treatment, the *per capita* amount of prey available was greatest at low predator densities (since identical amounts of prey were available at all predator densities). If larvae were competing by exploitation alone, prey availability would have had a greater positive effect at low predator densities than at high predator densities, but this did not show up in the statistical analyses. Instead, prey availability increased survival by a similar amount at all predator densities. The positive effect of prey availability on survival suggests that food stress in the prey-absent larvae led to their becoming cannibalistic, the assumption being that larval dragonflies can survive long periods of time without prey, and thus mortality could not be attributed to starvation (Lawton *et al.*, 1980) (section 2.10). Direct evidence of cannibalism was not, however, obtained.

Baker (1989), using the 'condition' (an index of the relative mass per unit head width of larvae) of larval dragonflies, related larval growth to larval density in a series of field sites. He found that for most of the year there was little evidence of food limitation. His results are in contrast to those obtained in the study by van Buskirk (1987) and in studies by Pierce *et al.* (1985), Johnson *et al.* (1984) and Banks and Thompson (1987b), in which the data indicate aggressive interactions to be important in limiting food intake of larval Odonata in the field. Among the reasons for this discrepancy given by Baker are that in his study larval densities were not high enough for either exploitation competition or interference to occur. Baker also points to differences in methodology and interpretation between his study and those of other workers (see discussion in his paper).

Anholt (1990) points to 'asymmetries in the burden of refutation' in several studies of competition in larval Odonata and other animals. Authors, when they have been unable to find evidence of prey depletion, have concluded by default that interference is the primary cause of density-dependent growth, development and survival. That is, they have made the assumption that if it is not competition through exploitation, then it must be competition through interference. Anholt's study represents a significant departure from previous work on Odonata in that he attempted to disentangle the effects of interference and exploitation by manipulating the rates of the two processes. Anholt manipulated the frequency of interactions between larval *Enallagma boreale* by altering perch availability at a fixed density of predators. Anholt argued that increasing the abundance of perches (i.e. increasing habitat complexity) will reduce the frequency of larva-larva encounters and thereby reduce the intensity of interference competition without affecting the supply of planktonic prey, i.e. without depletion occurring. Anholt's experimental design was a fixed-effects analysis of variance: (a) with three factors (food availability, larval density, perch availability) completely crossed; (b) with two factors (larval density and food availability) crossed; and (c) with two factors (perch availability and starting instar) crossed. In Anholt's experiments, damselflies became more evenly distributed among available perches as the predator density per perch increased, demonstrating that there were behavioural responses to the manipulation of habitat complexity (a prediction made by Crowley *et al.*, 1987). Food supply and predator density strongly affected survival, but the proportion of the variance in survival attributable to the habitat complexity manipulation, i.e. interference, was very small. Furthermore, whilst there were significant density-dependent alterations in growth or development, they were not attributable to food-related interference competition. Thus, despite the overt nature of the interactions between individuals, their costs appear to be minimal. Anholt suggested that the density-dependent reduction in larval growth and development observed in his experiments could have been due to both resource depletion, i.e. exploitation, and resource depression. **Resource depression** is a term used to describe local reductions in prey availability that result from the prey minimising the risk of predation by becoming less active and/or altering their use of habitat space.

Gribbin and Thompson (1990) conducted laboratory experiments in which individuals of two instars (ones which commonly occur together in the field) of *Ischnura elegans* were maintained in small containers (transparent plastic cups) with a superabundance of prey (to avoid prey limitation) either: (a) in isolation, (b) with three larvae of the same instar, or (c) three larvae of different instars. Either one perch or a set of four perches was provided to larvae in each treatment, and the experiment was treated as a two-way analysis of variance with perch availability as one factor and larval combination as the second factor potentially influencing development and growth. Small larvae showed increased development times and decreased growth (measured as percentage increase in head width) when kept with large larvae, but similar effects were not evident when the small larvae were kept with other small larvae. Development time and size increases of large larvae were not significantly affected by the presence of small larvae, i.e. competition was asymmetric. Regardless of the instar combination used, reductions in growth and development (which were taken to be due to interference, since prey – approximately 200 *Daphnia magna* – was superabundant in all treatments) were lessened when there were more perches available, although only in a few cases was the lessening significant. Gribbin and Thompson found that in containers with only one perch, large larvae often occupied the perch, whilst the single, small larva positioned itself on the side of the cup where feeding efficiency was likely to have been reduced.

Hopper *et al.* (1996) investigated the consequences of cannibalism for growth and survival (see 2.10.2) of survivors in the dragonfly *Epitheca cynosura*. The eventual size of survivors from a high larval density, asynchronous treatment (asynchronous in hatching terms – asynchrony increases the likelihood of cannibalism, i.e. by older larvae) was greater than that of survivors from a low larval density, asynchronous treatment, while there was no difference in size between survivors from high and low larval density synchronous treatments.

For a study of interference and exploitation competition in a species of carabid beetle, see Griffith and Poulson (1993). Interference competition has been shown by Griffiths (1992) to occur between larvae of the ant-lion *Macroleon quinquemaculatus*. Note that *facilitation*, not interference, may occur between larval conspecifics in some predator species, e.g. nymphs of the pentatomid *Perillus bioculatus* (Cloutier, 1997).

Exploitation competition between two species of hover-fly was studied by Hågvar (1972, 1973).

The deleterious effects of competition on larval growth (and fecundity) can be expressed by plotting *k*-values (defined in subsection 7.3.4) against $\log_{10}$ predator density. When describing such effects, the terms 'scramble' and 'contest' competition are less appropriate than the terms 'exact compensation', 'over compensation' and 'under compensation' (Begon *et al.*, 1996; section 7.3).

Larvae may also show a reduction in feeding rates in the presence of higher-level predators (Murdoch and Sih, 1978; Sih, 1982; Heads, 1986). Such interference may reduce the rate of consumption of prey, even when the insects do not need to move in order to feed (Heads, 1986), with the potential result that growth, development and even survival may be adversely affected (Heads, 1986; Sih, 1982). See, however, Brodin and Johansson (2002). McPeek *et al.* (2001) showed that although the larvae of *Ischnura* and *Enallagma* ingest less food in the presence of a fish predator, interspecific differences in growth rate were primarily due to differences in the conversion efficiency of the species, i.e. the two genera differ in their physiological stress response to the presence of predators. Stoks (2001) concluded from his study of *Lestes sponsa* that predator-induced stress effects upon growth and development were due to lowered assimilation efficiency and/or a higher metabolic rate.

The early instar larvae of the waterboatman *Notonecta hoffmani* can suffer significant mortality due to predation from adult conspecifics (Murdoch and Sih, 1978; Sih, 1982), and the adult avoidance behaviour of larvae constitutes a form of interference. Sih (1982), in laboratory and field experiments, compared the behaviour of larvae when the adults were experimentally removed, with their behaviour in controls where adults were present. Early instar larvae avoided adults by altering their use of habitat space (spending less of the total time available in the central region of the pond or tub, where prey and adults occur at the highest densities), and some of the early instars also became less active. As a result of this behaviour, larvae of the first two instars experienced severely reduced feeding rates.

Host Size

Idiobionts

The concept of an individual host as a fixed 'parcel' of resource for a developing idiobiont parasitoid was introduced in Chapter 1 (subsection 1.5.6). For many idiobiont species host size determines the size (and/or mass) of the resultant parasitoid adult(s), as shown by data both on solitary and on gregarious species (Salt, 1940, 1941; Arthur and Wylie, 1959; Heaversedge, 1967; Charnov *et al.*, 1981; Greenblatt *et al.* 1982; Waage and Ng, 1984; van Bergeijk *et al.*, 1989; Corrigan and Lashomb, 1990). Development rate, however, is not necessarily positively correlated with the size of host oviposited in. For example, in *Trichogramma evanescens* development rate is highest in medium-sized eggs and lowest in small and large eggs (Salt, 1940), in *Elachertus cacoeciae* it is highest on fifth instar hosts and lower in fourth and sixth instars (Fidgen *et al.*, 2000), and in *Habrobracon hebetor* development time is unaffected by host larval size (Taylor, 1988a). The reasons for the lack of a clear relationship are complex, and the reader is referred to Mackauer and Sequeira (1993).

To investigate the influence of host size on growth and development in an idiobiont parasitoid species, present females (inseminated and, if necessary, uninseminated, to obtain data on both sexes) with hosts of different sizes and record the weight of the resultant adult progeny and the time taken from oviposition to adult eclosion (since adult eclosion is often influenced by light: dark cycles [Mackauer and Henkelman, 1975], observations should be carried out at the same time each day or under continous light conditions; video recording equipment can be used both to improve accuracy and to save time [Sequeira and Mackauer, 1992a]). If the parasitoid is a gregarious species, clutch size will have to be kept constant (section 1.6. describes clutch size manipulation techniques). One needs to bear in mind the possibly complicating effects of sex differences in food acquisition (and therefore growth and development) in broods of gregarious species. This problem can be partly circumvented by using uninseminated parent females, which will produce all-male egg clutches, but obtaining all-female clutches could prove very difficult (section 1.6).

For idiobiont parasitoids the age of the host may be a confounding factor. For example, some parasitoids that develop in host pupae may be able to utilise both very recently formed pupae and pupae within which the adult host is about to be formed. These different types of host pupa are likely to have the same external dimensions and similar mass but are likely to represent very different amounts of resource.

To determine whether host stage i.e. instar and not host size *per se* mainly accounts for any variation in growth or rate of development, parasitoids, e.g. idiobionts attacking larval Lepidoptera, can be presented with a range of host sizes within each host stage that overlaps with host sizes within the previous or subsequent stage.

Working over four trophic levels, Otto and Mackauer (1998) compared development of the idiobiont hyperparasitoid *Dendrocerus carpenteri* in its primary host *Aphidius ervi* which itself was reared in two aphid host species (*Acyrthosiphon pisum* and *Sitobium avenae*) of differing quality and growth potential. Within each aphid species, the authors found that terminal host size affected the size of *A. ervi*, which had a concomitant effect on the size of *D. carpenteri*. However, the development time of the hyperparasitoid was determined by the age of the *A. ervi* individual when it was attacked, and was longer in older hosts, which presumably was attributable to their reduced digestibility.

Koinobionts

During the initial phases of parasitism, hosts of koinobionts remain active and may continue feeding, growing and defending themselves (Mackauer and Sequiera, 1993). Put another way, for a koinobiont the host represents a dynamic resource. One might therefore not expect the same relationship between progeny size and host size at oviposition as exists for idiobionts (Mackauer, 1986; Harvey, 2000).

Sequeira and Mackauer (1992a,b) and Harvey *et al.* (1994, 1999) have shown, for different koinobionts, that adult parasitoid size (mass) is not a linear function of host size (mass) at oviposition across the full range of available host sizes (Figure 2.47a). In the solitary parasitoids *Aphidius ervi* and *Venturia canescens*, there is a linear increase in wasp size with increasing instar up to the penultimate instar, whereas in the final instar wasp size does not increase. By contrast, in *Cotesia rubecula*, which is also solitary, adult wasp size more generally decreases with instar parasitized. These variations in koinobiont development are linked to differences in host usage strategy (discussed by Harvey *et al.*, 2000). Whereas the larvae of most koinobionts obligatorily consume most (or all) host tissues before pupation, several endoparasitoid

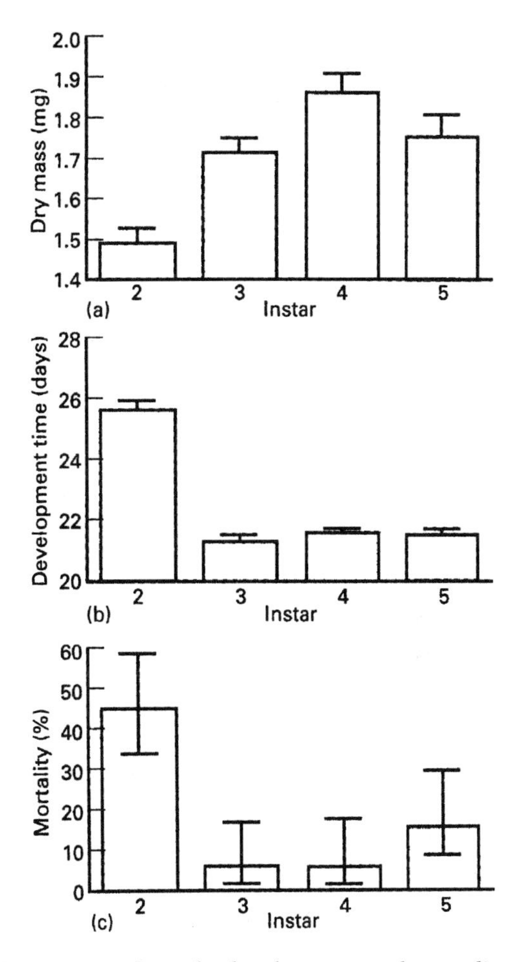

Figure 2.47 Growth, development and mortality of *Venturia canescens* (Ichneumonidae) reared in four instars of the moth *Plodia interpunctella*. (a) adult dry mass; (b) development time from oviposition to adult eclosion; (c) mortality. Source: Harvey *et al.* (1994). Reproduced by permission of The Ecological Society of America.

clades contain taxa whose larvae primarily consume host haemolymph before emerging through the side of a still-living host, to pupate externally. Since in the latter group only a fraction of the available host resources is consumed, the relationship between host size and parasitoid size may be more complex than for tissue-feeders. For example, parasitoid development may be more constrained by the availability of certain nutrients in the host haemolymph, rather than by host size *per se*.

The relationship between development rate and the size of the host when parasitised varies, being either linear throughout the whole range of available host sizes and highest in larger hosts (Fox *et al.*, 1967; Smilowitz and Iwantsch, 1973; Harvey *et al.*, 2000) or non-linear (Jones and Lewis, 1971; Avilla and Copland, 1987; de Jong and van Alphen, 1989; Harvey *et al.*, 1994, 2000) (Figure 2.47b).

Valuable insights into the effects of host stage on growth and development can be obtained by plotting the **growth trajectories** of both the host and the parasitoid (Sequeira and Mackauer, 1992b; Harvey *et al.*, 1994, 1999; Harvey and Strand, 2002) (Figure 2.48). Growth trajectories are studied by taking each host stage, dissecting parasitised hosts at various points in time after oviposition, separating the parasitoid larva from the host and measuring the dry weight of each. A trajectory is also plotted for unparasitised hosts. Using growth trajectories, Sequeira and Mackauer (1992b) showed that *A. ervi* responds to host-related constraints upon larval growth, and arrests host growth at a fairly fixed time

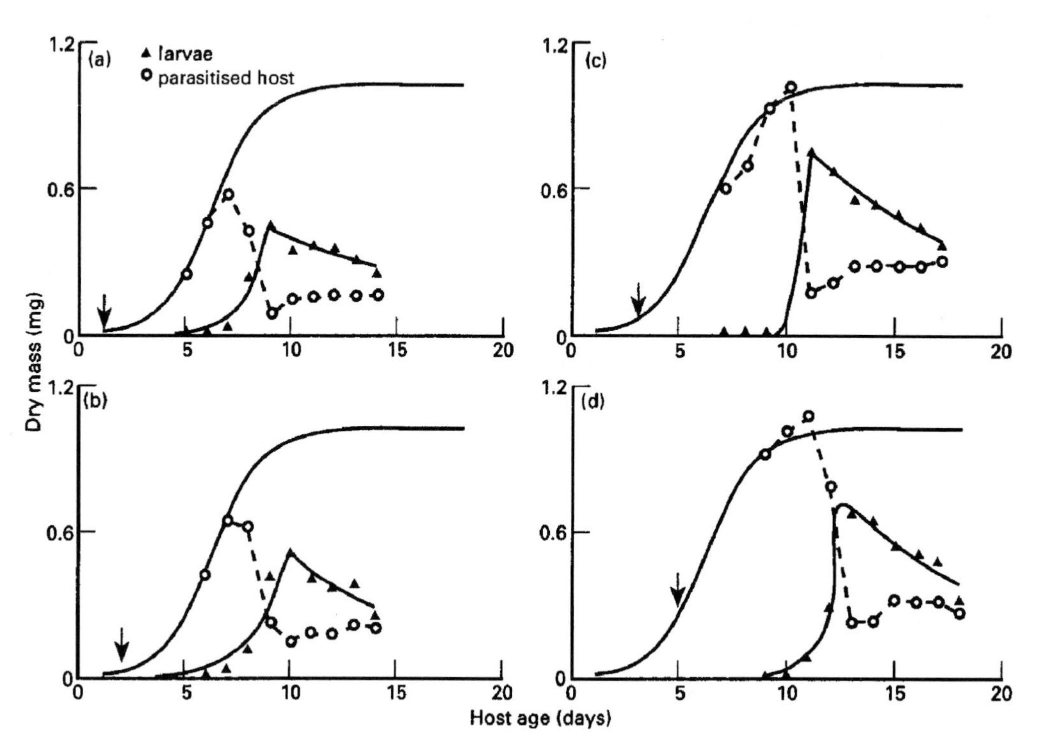

Figure 2.48 Growth trajectories of *Aphidius ervi* (▲) and of parasitised pea aphids (○) at different ages: (a) host nymphal instar one (24 h); (b) host nymphal instar two (48 h); (c) host nymphal instar three (72 h); (d) host nymphal instar four (120 h). The solid curve shows the corresponding trajectory of unparasitised aphids, samples of which were taken at various ages from birth to maturity. Arrows indicate the age of the host at oviposition. The turn-over point of the parasitoid growth trajectory corresponds to parasitoid age of 8 days. The trajectory of parasitoid larval growth provides a direct measure of host 'quality', reflecting the nutritional relationship between the two insects during the course of parasitism, and its shape will be characteristic of the parasitoid species. All curves will, however, be 'J'-shaped, there being two functionally distinct phases in the development of holometabolous insects: first there is an exponential growth phase as the parasitoid larva feeds and converts host tissues into its own body mass, then there is a negative exponential decay phase between pupation and adult eclosion, when feeding has stopped and there is differential mass reduction due to respiration, water loss and voiding of the meconium (Harvey *et al.*, 1994). Source: Sequeira and Mackauer (1992b). Reproduced by permission of The Ecological Society of America.

approximately 8 days after oviposition, at which point in time aphids parasitised as early instars have not reached their maximum size. In *A. ervi*, development time and adult mass covary positively (i.e there is a trade-off between development rate and growth) with an increase in host size from first to third instar, but they vary independently in parasitoids developing in fourth instar hosts. In the latter, adult mass does not increase but development rate does. Overall parasitoid development time is therefore approximately constant, whereas the largest wasps emerge from third and fourth instar aphids. The growth trajectories shown in Figure 2.48 indicate that in early instar hosts parasitoid growth and development rate are limited by the small size and growth potential of the host (compare, in Figure 2.48, the average mass attained by parasitised aphids with that attained by unparasitised aphids of equivalent age). By contrast, in fourth instar hosts excess resources are constantly available, thus allowing for an increase in development rate without an increase in adult mass.

As pointed out by Harvey *et al.* (1994), *A. ervi* may represent one end of a continuum of strategies among parasitoids, the other extreme being to delay parasitoid growth until the host reaches its maximum size (in which case we would expect parasitoid size to be unaffected by instar at oviposition but development rate to be highly variable). The latter pattern is exhibited by *Apanteles carpatus*, which attacks a wide range of sizes (representing all larval instars) of its host, the clothes moth *Tineola bisselliella*. Irrespective of host size at oviposition, the size of emerging wasps is fairly uniform, whereas development time increases exponentially with a decrease in host size, some wasps taking three months to complete their development in very small hosts (Harvey *et al.*, 2000). The strategy of *V. canescens* appears to lie somewhere along the continuum between the aforementioned two extremes (Harvey *et al.*, 1994; Harvey and Vet, 1997). See Harvey and Strand (2002) for a review of parasitoid developmental strategies.

As these studies have shown, by comparing the development of koinobionts in very small or otherwise nutritionally suboptimal hosts, it should be possible to elucidate the nature of trade-offs between life-history variables. The experimental protocol for studying the effects of host stage at oviposition upon growth and development is slightly more complex for koinobionts than for idiobionts inasmuch as the hosts need to be reared. Care must be taken to control for the effects of variations in host diet; Harvey *et al.* (1994) for example reared hosts with an excess of food. However, as pointed out by Mackauer and Sequeira (1993), there is an urgent need to examine the dynamics of parasitoid development under different constraints. These might include superparasitism, particularly in the case of gregarious species, where crowding intensifies competition with conspecifics for access to limited host resources (Wajnberg, *et al.*, 1990; Harvey, 2000). There is a need for more studies of the nutritional integration between host and parasitoid when hosts are reared on various food plants containing different concentrations of constitutively expressed or induced defensive chemical compounds (see below).

Host Species

Given that hosts of different species are likely to constitute different resources in both a qualitative and quantitative sense, we would expect parasitoid growth and development to vary in relation to the host species parasitised. This is indeed the case, as studies on idiobionts and a few koinobionts have shown (although few workers have measured growth together with development) (Salt, 1940; New, 1969; Dransfield, 1979; Moratorio, 1987; Bigler *et al.*, 1987; Taylor, 1988a; Ruberson *et al.*, 1989; Corrigan and Lashomb, 1990; Harvey and Thompson, 1995; Harvey and Vet, 1997; Harvey and Gols, 1998; Harvey *et al.*, 1999; Nicol and Mackauer, 1999; McNeill *et al.*, 1999; Seal *et al.*, 2002; Bazzocchi *et al.*, 2003).

Salt (1940), for example, showed how the size of adult progeny of *Trichogramma evanescens* varied with the species of moth within which larval development occurred (Figure 2.49).

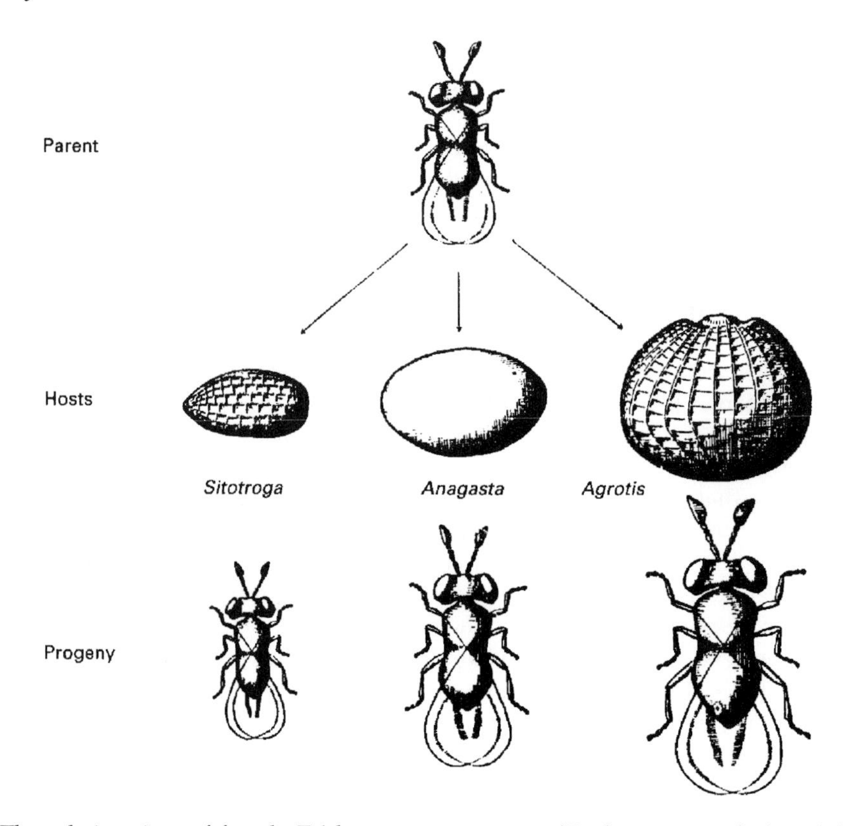

Figure 2.49 The relative sizes of female *Trichogramma evanescens* (Trichogrammatidae) and female progeny reared from different host species. The reader should note that the confounding effects of progeny clutch size were not controlled for in Salt's experiments (development was solitary in *Sitotroga* and *Anagasta* (now *Ephestia*) but was either solitary or gregarious in *Agrotis*). The female that emerged from the egg of *Agrotis* developed solitarily; nevertheless, females that developed gregariously in that host species were on average markedly larger than those that developed in either of the other two host species. Source: Salt (1940). Reproduced by permission of Cambridge University Press.

Moratorio (1987), working with *Anagrus mutans* and *A. silwoodensis*, showed that female progeny were larger when development occurred in the (large) eggs of *Cicadella viridis*, but were smaller when development occurred in the (small) eggs of *Dicrantropis hamata*. However, whereas *A. silwoodensis* develops fastest in *C. viridis*, *A. mutans* develops fastest in *D. hamata*, i.e. development rate and growth *counter*vary in relation to host species in *A. mutans*. Development rate and growth also countervary in relation to host species in *Telenomus lobatus* (Scelionidae): wasps develop more rapidly in eggs of *Chrysoperla* species than in eggs of *Chrysopa* species, but the adults attain a larger size in eggs of the latter genus, the eggs being larger than those of *Chrysoperla* (Ruberson *et al.*, 1989).

Similar findings have been reported in some koinobionts. In *V. canescens*, adult wasp size is positively correlated with the growth potential of the particular host species, although development time is extended in larger hosts (Harvey and Thompson, 1995; Harvey and Vet, 1997). By contrast, in *C. rubecula*, emerging wasps are larger and develop more rapidly in a smaller, habitual host (*Pieris rapae*), than in corresponding stages of a larger, factitious host (*P. brassicae*, see Harvey *et al.*, 1999). Harvey *et al.* (1999) found that *C. rubecula* arrested the development of *P. brassicae* larvae at an earlier stage (and smaller size) than that of larvae of *P. rapae*. This effect could be related to the fact that *P. rapae* is generally a much more suitable host for *C. rubecula*, which is rarely recovered in the field from other host species, including *P. brassicae*.

Further investigations should focus on the potential influence of host species on parasitoid development among host species of equivalent size (mass). If differences in performance are recorded, then this would suggest that the quality, rather than the quantity, of host resource affects growth and development.

Multitrophic Interactions and the Performance of Natural Enemies

It is well established that plants play an important role by mediating a suite of physiological interactions amongst the herbivores feeding on them and the natural enemies of the herbivores. Plants contain a bewildering array of toxic secondary compounds (Karban and Baldwin, 1997; Schoonhoven *et al.*, 1998), some of which negatively affect the herbivore's growth, development and survival (Giamoustaris and Mithen, 1995; van Dam *et al.*, 2000). These toxins are also frequently sequestered in the haemolymph or body tissues of resistant herbivores, thus providing them with the potential for some degree of protection against their natural enemy complex (Tullberg and Hunter, 1996; Wink *et al.*, 2000; Omacini *et al.*, 2001).

Several studies have reported that allelochemicals in the host's or the prey's diet negatively affect the growth, development, survival or morphology of their predators and parasitoids (Barbosa *et al.*, 1986; Duffey *et al.*, 1986; Gunasena *et al.*, 1990; Paradise and Stamp, 1993; Karban and English-Loeb, 1997; Havill and Raffa, 2000). In some cases one of the aforementioned life-history variables is negatively affected, whereas another is not (Karban and English-Loeb, 1997), and allelochemicals may reduce development rate and growth rate only when the prey are scarce (Weisser and Stamp, 1998). The effects of interspecific variation in plant quality may even work their way up to organisms in the fourth trophic level, such as primary parasitoids of insect predators or obligate hyperparasitoids (Orr and Boethel, 1986; Harvey *et al.*, 2003).

Experiments can be conducted in which growth and development of predators and parasitoids are measured when the carnivores are reared on separate cohorts of hosts or prey that have been fed on resistant and non-resistant strains of a cultivated plant, or on related species of wild plants. Of particular interest is the degree of adaptation shown by adapted specialist herbivores and their parasitoids, which in some cases perform better on more toxic plant species or genotypes (Harvey *et al.*, 2003). The effects of plant secondary compounds can also be investigated by incorporating the chemicals into the artificial diet of the herbivore (e.g. Campbell and Duffey, 1979; Reitz and Trumble, 1997; Weisser and Stamp, 1998; Williams, *et al.* 1998). Small amounts of a compound added to such a diet may even improve parasitoid larval performance (Williams *et al.*, 1988).

It should also be noted that fungal endophytes produce toxins that may affect larval parasitoid growth and/or development (e.g. see Barker and Addison, 1996; Bultman *et al.*, 1997).

Superparasitism

Introduction

In Chapter 1 (section 1.8) superparasitism was defined as the laying of an egg (by a solitary parasitoid) or a number of eggs (by a gregarious parasitoid) in an already parasitised host. In the case of a solitary parasitoid species, only one larva survives in each superparasitised host. In a gregarious species the number of survivors per host will depend on: (a) the total number of eggs the host contains or bears, and (b) the size of the superparasitised host (Beckage and Riddiford, 1978; le Masurier, 1991). This section is concerned with ways of studying the fitness consequences for *surviving* larvae, and asks how larval growth and development rate might be affected by superparasitism.

Solitary Parasitoids

Models of superparasitism as an adaptive strategy in solitary species (Visser *et al.*, 1990; van der Hoeven and Hemerik, 1990) have been based on the assumption that superparasitism has no fitness consequences for the surviving larva, i.e. it does not increase larval development time or reduce adult size. This would seem to be a reasonable assumption, since in solitary parasitoids **supernumerary larvae** (larvae in excess of the number that can ultimately survive, i.e. can complete development) are usually eliminated before they can utilise an appreciable amount of host resource. For example, Visser *et al.* (1992c) found no convincing evidence that *Leptopilina heterotoma* adults emerging from singly-parasitised hosts were larger than adults emerging from superparasitised hosts (see also Ueno, 1997). However, as pointed out by Bai and Mackauer (1992) and Harvey *et al.* (1993), superparasitism may have fitness consequences for the larvae of some parasitoid species. Simmonds (1943) and Wylie (1983), for example, reported that in *Venturia canescens* (Ichneumonidae) and *Microctonus vittatae* (Braconidae) larvae take longer to develop in superparasitised hosts than in singly-parasitised hosts, although neither author recorded the number of eggs contained per host. Similarly, Vinson and Sroka (1978), subjected hosts of *Cardiochiles nigriceps* (Braconidae) to varying numbers of ovipositions, recorded the time taken from oviposition to larval emergence from the host, and showed that as the degree of superparasitism increased, mean development time of the surviving larva increased (Table 2.1).

The fitness cost to *koinobionts* may be partly determined by the ability of the surviving larva to compensate for possibly reduced growth during embryonic and early larval development (when it may compete with the rival larva for host resources) by increasing growth later in development (Bai and Mackauer, 1992), and the same might apply to development rate. Bai and Mackauer (1992) carried out a simple experiment in which they subjected aphids to either one oviposition (singly-parasitised) or several ovipositions (superparasitised) by *Aphidius ervi*. They used unmated females, in order to control for the possible bias resulting from differential development (and survival) between male and female larvae. They then compared the total development time and adult weights in the different treatments. Interestingly, they found that *Aphidius ervi* gained 14% *more* dry mass in superparasitised hosts, i.e. growth was enhanced through superparasitism, and took no longer to develop, i.e. development rate was unaffected. The most likely explanation for this effect is that the superparasitised hosts ingested more food. As Bai and Mackauer point out, the fitness benefit, i.e. increased adult size gained by surviving larvae in superparasitised hosts, needs to be weighed against any costs in the form of reduced larval survival (subsection 2.10.2).

Table 2.1 Percentage of hosts yielding a larva, and the time taken from oviposition to larval emergence from the host, in the solitary parasitoid *Cardiochiles nigriceps* (Braconidae) parasitising *Heliothis virescens*. Source: Vinson and Sroka (1978).

Number of ovipositions per host	% of hosts yielding a parasitoid	Mean time (days) to emergence
1	92	12.3 ± 1.6
2	58	12.2 ± 1.9
3	63	14.7 ± 2.7
4	29	15.6 ± 2.5
5	27	15.9 ± 3.0
>5	21	16.9 + 3.4

As we noted above, adult size in *Leptopilina heterotoma* is not affected by superparasitism. In this parasitoid either compensation is complete or there is no initial reduction in growth as a result of superparasitism. Studying the trajectory of parasitoid larval growth (see above) would shed light on this.

Larval development rate may also be influenced by **heterospecific superparasitism (= multiparasitism)** (McBrien and Mackauer, 1990).

Experiments aimed at investigating the effects of superparasitism on larval growth (as measured by adult size) and development would involve exposing a recently parasitised host to a standardised female, and allowing the same or a different (conspecific or heterospecific) standardised female to deposit a specified number of eggs. The number of eggs laid in each case can be more easily monitored and controlled if the parasitoid is one of those species in which the female performs a characteristic movement during oviposition (subsection 1.5.6 and Harvey *et al.*, 1993). The time taken from oviposition to adult eclosion and the size or weight of emerging adults will need to be measured and the different treatments compared both with one another and with controls. The experiment could be expanded to take into account the possible effects of host size or host instar, as was done by Harvey *et al.* (1993). They showed that superparasitism in *Venturia canescens* reduced development rate in parasitoids reared from both third instar and fifth instar larval hosts (the moth *Plodia interpunctella)*, but that the reduction was greater in parasitoids reared from the later instar (Figure 2.50). The size of wasps reared from third instar hosts was unaffected by egg number (Figure 2.51a), but adult wasps from both of the superparasitised fifth instar treatments (two eggs, four eggs) were significantly smaller than those reared from singly-parasitised hosts (Figure 2.51b). Harvey *et al.* suggest that the reason superparasitism affected parasitoids from fifth instar hosts more than those from third instar hosts is that the fifth instar larvae were post-feeding, wandering larvae, i.e. their growth potential is zero. Parasitism of such hosts would be more like idiobiosis than koinobiosis, and the surviving larva would be less able to compensate for any negative effects of superparasitism.

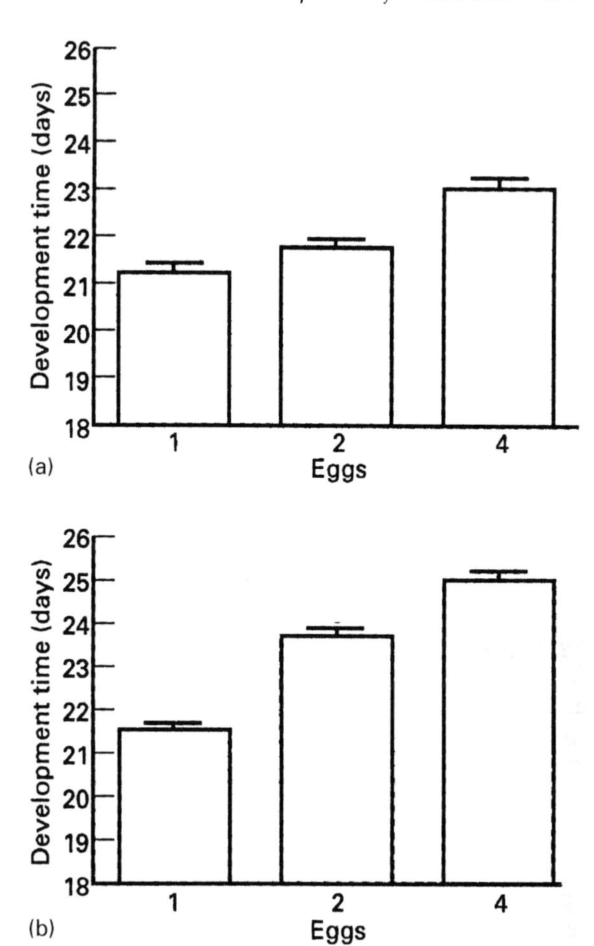

Figure 2.50 Effects of superparasitism on development in the solitary ichneumonid parasitoid *Venturia canescens*. Development time (number of days taken from oviposition to adult eclosion) of wasps reared from: (a) third instar; (b) fifth instar larvae of *Plodia interpunctella* containing one, two or four parasitoid eggs. Source: Harvey *et al.*, (1993). Reproduced by permission of Blackwell Publishing.

Gregarious Parasitoids

The fitness consequences of superparasitism already been touched upon, from both theoretical and experimental standpoints, in Chapter 1 (section 1.8). Leaving aside Allee effects (defined

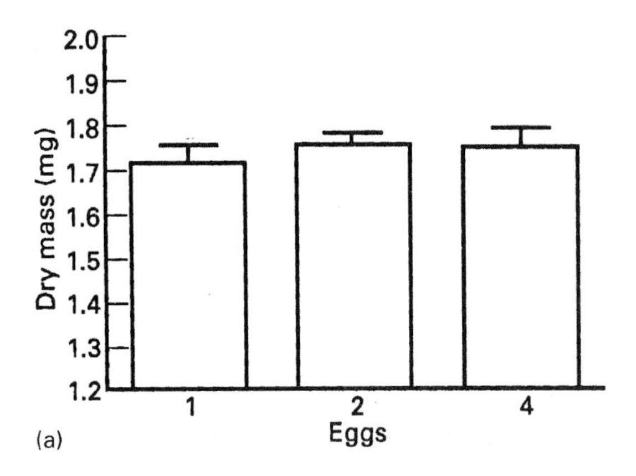

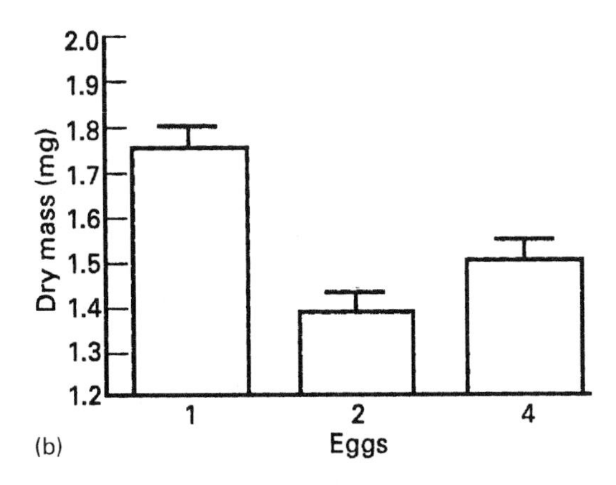

Figure 2.51 The effects of superparasitism on growth (as measured by adult dry mass) in the solitary ichneumonid parasitoid *Venturia canescens*. The dry weight of wasps reared from: (a) third instar; (b) fifth instar larvae of *Plodia interpunctella*, containing one, two or four eggs. Source: Harvey *et al.*, (1993). Reproduced by permission of Blackwell Publishing.

in section 1.6), superparasitism will intensify competition among larvae for host resources, with the result that the *per capita* growth and development rate of the parasitoid immatures will be reduced. This is at least what one would expect, although Nealis *et al.* (1984) found that increased larval density per host slowed development of *Cotesia glomerata* only slightly (le Masurier, 1991, found no significant effect of clutch size on development time in this species) and tended to *increase* the rate of development in *Pteromalus puparum*. Le Masurier (1991) also found no significant decrease in body size with increasing clutch size in a population of *C. glomerata* parasitising *Pieris brassicae*, although he did find such an effect in another population parasitising *Pieris rapae*.

Experiments aimed at investigating the effects of superparasitism on larval growth (as measured by adult size) and development in gregarious parasitoids would involve:

1. In the case of intraspecific superparasitism: exposing a recently parasitised host to a standardised female and allowing the same or a different conspecific female to oviposit a further egg or clutch of eggs;
2. In the case of multiparasitism, exposing a host recently parasitised by a female of one species to a female of another species.

In both cases, the time taken from oviposition to adult eclosion and the size or weight of emerging adults need to be measured and the different treatments (i.e. initial and second clutches of different sizes) compared with one another and with controls. With ectoparasitoids, eggs can be artificially added to existing clutches of various sizes (section 1.6, and Strand and Godfray, 1989). Assuming competitive equivalence of clutches produced by different females, the effects upon parasitoid growth and development of simultaneous oviposition by two conspecific females would be analogous to the effects of increasing the primary clutch, as in Figure 1.10. That is, an increase in the number of eggs laid per host would have a negative effect irrespective of whether the eggs are laid by one or by different females, provided all the eggs are laid at the same time. However, the competitive disadvantage of a second clutch may be underestimated from a fitness function curve that is based solely on initial clutches, if there is a significant time interval between the laying of initial and subsequent clutches (Strand and Godfray, 1989). Measurement of any such disadvantage, in terms of growth and development, to a second clutch requires the progeny from the two clutches to be distinguishable by the investigator. This is possible in those species in which there are mutant strains, e.g. the eye/

body colour mutant 'cantelope-honey' in *Habrobracon hebetor*. To ensure, when using mutants, that competitive asymmetries do not bias the results of experiments, reciprocal experiments should be carried out for each clutch size and time interval combination (Strand and Godfray, 1989). Molecular markers can also be used (see subsection 3.2.2).

The possibly complicating effects of sex differences in larval food acquisition also need to be borne in mind in experiments on gregarious parasitoids: compared with the adding of a female egg, the adding of a male egg to an existing clutch could have less of an effect upon fitness of the progeny in the initial clutch.

2.9.3 EFFECTS OF PHYSICAL FACTORS ON GROWTH AND DEVELOPMENT

Temperature

Development rate

Figure 2.52 shows the typical relationship between an insect's rate of development and temperature. There is a threshold temperature below which there is no (measurable) development; this threshold is sometimes referred to as the **developmental zero**. There is also an upper threshold above which further increases in temperature result in only small increases in development rate. The overall relationship is

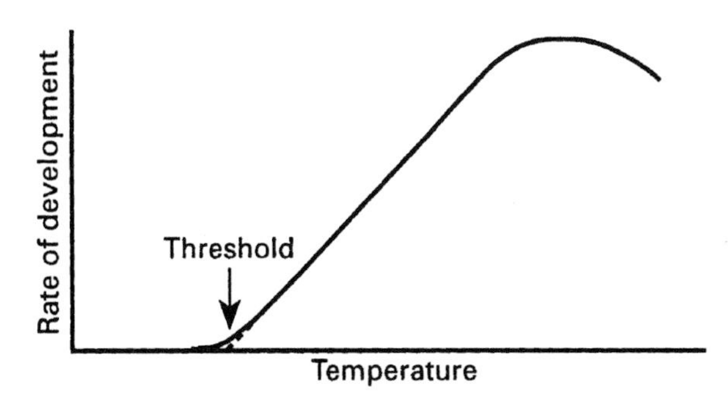

Figure 2.52 The rate of insect development as a function of temperature. Source: Gilbert *et al.* (1976). Reproduced by permission of W.H. Freeman & Co.

non-linear (Mills, 1981), but over the intermediate range of temperatures normally experienced by an insect species in the field, it is linear. As noted by Gilbert *et al.* (1976) why this should be so is a mystery, since rates of enzyme action (which are presumably basic to development) usually increase exponentially, not linearly, with increasing temperature.

The deleterious effects of a high temperature extreme depends on how long the insect is exposed to it. As pointed out by Campbell *et al.* (1974) with reference to the development rate-temperature relationship shown in Figure 2.52, temperatures within the high range (i.e. the part of the relationship where the curve decelerates) have a deleterious effect upon development only if the temperature is either held constant within the range or fluctuates about an average value within the range. If the temperature fluctuates about a daily average within the medium range (i.e. the linear part of the relationship) and the daily maximum reaches the high range, no deleterious temperature effect is observed.

Given the fact that the development rate versus temperature relationship is linear over the greater range of temperatures, the total amount of development that takes place during any given time period will be proportional to the length of time multiplied by the temperature above the threshold. With this physiological time-scale of **day-degrees** development proceeds at a constant rate, whatever the actual temperature. This concept is elaborated upon below.

To study the dependency of overall development rate on temperature in a parasitoid, expose hosts to female parasitoids at different constant temperatures (the range being chosen on the basis of field temperature records), and measure the time taken from oviposition to adult eclosion. To demonstrate the effect of temperature on overall larval development in a predator, provide different cohorts of larval predators with a fixed daily ration of prey, at different temperatures, from egg hatch to adult eclosion. With both parasitoids and predators the thermal requirements for development can be determined for particular stages, i.e. the egg (Frazer and McGregor, 1992, on coccinellids), each larval instar and the pupa.

The data obtained from the above experiments can be described by a linear regression equation of the form:

$$y = a + bT \quad (2.16)$$

where y is the rate of development at temperature T, and a and b are constants. If the regression line were to be extrapolated back, it would meet the abscissa at the **developmental zero, *t***, which may be calculated from $t = -a/b$. The total quantity of thermal energy required to complete development, the **thermal constant (*K*)** can be calculated from the reciprocal of the slope of the regression line, $1/b$.

Once t and K have been calculated from data obtained at *constant* temperatures, the rate of development under any *fluctuating* temperature regime can be determined by **thermal summation** procedures. Unit time-degrees (day-degrees or hour-degrees) above t are accumulated until the value of K is reached where development is complete. This can be done either by accumulating the mean daily temperature minus the lower threshold or by accumulating the averages of the maximum and minimum daily temperature minus the threshold (i.e. $\sum [\{(T_{max} - T_{min})/2\} - \text{threshold}]$). However, both of these methods will result in great inaccuracies if a temperature contributing to the mean lies outside of the linear portion of the relationship. Means, by themselves, give no indication of the duration of a temperature extreme: an apparent tolerable mean temperature may actually comprise a cyclical regime of two extremes at which no development is possible. A much more accurate method is to use *hourly* mean temperatures (Tingle and Copland, 1988).

Summation has been used by many workers, including Apple (1952), Morris and Fulton (1970), Campbell and Mackauer (1975), Lee *et al.* (1976), Hughes and Sands (1979), Butts and McEwen (1981), Osborne (1982), Goodenough *et al.* (1983), Nealis *et al.* (1984), Cave and Gaylor (1988), Rodriguez-Saona and Miller (1999) and Bazzocchi *et al.* (2003). However, the method has been much criticised as it has two inherent faults. First, the assumed linear relationship is known to hold as an approximation for the median temperature range only (Figure 2.52) (e.g. Campbell *et al.*, 1974, on aphid parasitoids; Syrett and Penman, 1981, on lacewings). Second, the lower threshold upon which summation is based is a purely theoretical point determined by extrapolation of the linear portion of the relationship into a region where the relationship is unlikely to be linear. The linear model is likely to underestimate development rates when average daily temperatures remain close to the threshold for long periods, although this can easily be corrected for (Nealis *et al.*, 1984). In an attempt to improve upon the thermal summation method, an algorithm was developed using a sigmoid function with the relationship inverted when the temperature is above the optimum (Stinner *et al.*, 1974). The assumed symmetry about the optimum is unrealistic, but Stinner *et al.* (1974) argue that the resultant errors are negligible. This algorithm has also been used in simulations for *Encarsia perniciosi* (McClain *et al.*, 1990a), fly parasitoids (Ables *et al.*, 1976) and other insects (Berry *et al.*, 1976; Whalon and Smilowitz, 1979; Allsopp, 1981). Ryoo *et al.* (1991) used a combination model involving upper thresholds to describe the development of the ectoparasitoid *Lariophagus distinguendus* (Pteromalidae).

In some cases the improvement in accuracy of simulations over the thermal summation method has been small or negligible, and it is questionable whether the use of complex models is necessary in relation to normal field conditions (Kitching, 1977; Whalon and Smilowitz, 1979; Allsopp, 1981). The method of matched asymptotic expansions was used to develop an analytical model describing a sigmoidal curve that lacks the symmetry about the optimum found in the algorithm of Stinner *et al.* (1974). Again, the authors concerned claimed excellent results (Logan *et al.*, 1976). However, comparisons of linear and non-linear methods to validate field data for *Encarsia perniciosi* showed no great differences (McClain *et al.*, 1990a).

Other non-linear descriptions of the development rate-temperature relationship have also been developed. These include the logistic curve (Davidson, 1944) and polynomial regression analysis (Fletcher and Kapatos, 1983). Polynomial regression analysis can be used to select the best-fitting curve to a given set of data. Successively higher order polynomials can be fitted until no significant improvement in *F*-value results. This approach was found useful in describing data for *Diglyphus intermedius* (Patel and Schuster, 1983) and mealybug parasitoids (Tingle and Copland, 1988; Herrera *et al.*, 1989). Higher order polynomials may produce unlikely relationships between data points and fluctuate widely outside them. Before selecting a particular fit, it should be examined over the entire range of the data. It may be better to choose one that has a comparatively poor fit but is biologically more realistic (Tingle and Copland, 1988).

Several authors have reported acceleration or retardation of development, when comparisons are made between development periods at cycling temperatures and at a constant temperature equivalent to the average of the cycling regime. The question of whether these effects are an artefact or are a real biological phenomenon is discussed by Tingle and Copland (1988).

Until recently, data on the development times of insects were almost always expressed in the form of means and standard deviations. This is not strictly justifiable as the distribution of eclosion times is not normal but shows a distinct skew towards longer developmental periods (Howe, 1967; Eubank *et al.*, 1973; Sharpe *et al.*, 1977). Several models have been developed which include a function to account for the asymmetrical distribution of development times (Stinner *et al.*, 1975; Sharpe *et al.*, 1977; Wagner *et al.*, 1984). Such models can be incorporated into population models (Barfield *et al.* (1977b), on *Habrobracon mellitor*). However, the poikilotherm model of Sharpe *et al.* (1977) did not give any great improvement in accuracy over day-degree models when predicting development of *Trichogramma pretiosum* (Goodenough *et al.*, 1983).

Biological control workers can use laboratory-obtained information on the effects of temperature on development in deciding which of several candidate species, 'strains' or 'biotypes' of parasitoids and predators to either introduce into an area or use in the glasshouse environment. In classical biological control programmes, the usual practice is to introduce natural enemies from areas having a climate as similar as possible to that in the proposed release area (Messenger, 1970; Messenger and van den Bosch, 1971; van Lenteren, 1986) (section 7.4). If there are several species, strains or biotypes to choose from, the one found to have a temperature optimum for development that is nearest to conditions in the introduction area should be favoured, all other things being equal.

A classic example of a biological control failure resulting from the agent being poorly adapted to the climate of the introduction area is the introduction of a French strain of *Trioxys pallidus* into California to control the walnut aphid. This parasitoid was poorly adapted to conditions in northern and especially central California where it never became permanently established. The French strain was unable to reproduce and survive to a sufficient extent in areas of extreme summer heat and low humidity. A strain from Iran was subsequently introduced and proved far more effective (DeBach and Rosen, 1991).

As will be explained in Chapter 7 (7.3.8), data on development rate-temperature relationships are used in population models to investigate dynamics and phenologies in a biological control context. Morales and Hower (1981) showed that they could predict the emergence in the field of 50% of the first and second generations of the weevil parasitoid *Microctonus aethiopoides* (Braconidae) by using the day-degree method. Goldson *et al.* (1998) applied a phenological model retrospectively to the phenology *Microctonus hyperodae* and its weevil host. McClain *et al.* (1990a) used the linear day-degree model and the sigmoid function model of Stinner *et al.* (1974) to predict the peaks of activity of parasitoids in orchards. The linear model predicted 8 of 13 peaks within ± 7 days, while the non-linear model was accurate for 7 of 13 peaks. Horne and Horne (1991) showed that simple day-degree models could account for the synchronisation of emergence of the encyrtid parasitoid *Copidosoma koehleri* and its lepidopteran host.

Growth Rate

Most studies on temperature relationships have dealt with development but have ignored growth. The relationship between growth rate and temperature in insects has been shown by direct measurement to increase linearly with temperature within the range of temperatures normally experienced by the insect in the field, in accordance with the following model:

$$\frac{1}{w}\frac{dw}{dt} = a(T - \theta) \qquad (2.17)$$

where T is the temperature, θ is the the threshold temperature below which no growth occurs, w is the larva's weight at time t, and a is a constant (Gilbert, 1984). Gilbert used this model to predict pupal weight, which determines fecundity, in the butterfly *Pieris rapae*. Tokeshi (1985) describes another method for estimating minimum threshold temperature and day-degrees required to complete growth, suitable for use with aquatic or terrestrial insects in either the laboratory or the field.

In some predator species the tendency is for successive larval instars to achieve a growth rate maximum at a higher temperature, e.g. in *Adalia bipunctata* the maxima recorded were 20°, 22.5°, 22.5° and 25°C for the first, second, third and fourth instars respectively (Mills, 1981). Mills (1981) suggested these differing optima could reflect the increasing temperatures experienced by the coccinellid larvae as they progress through the life cycle in the field. However, in other predators, there is no such tendency, e.g. in the damselflies *Lestes sponsa, Coenagrion puella* and *Ischnura elegans* maximum development rates were recorded at the same temperature for the last five instars (Pickup and Thompson, 1990).

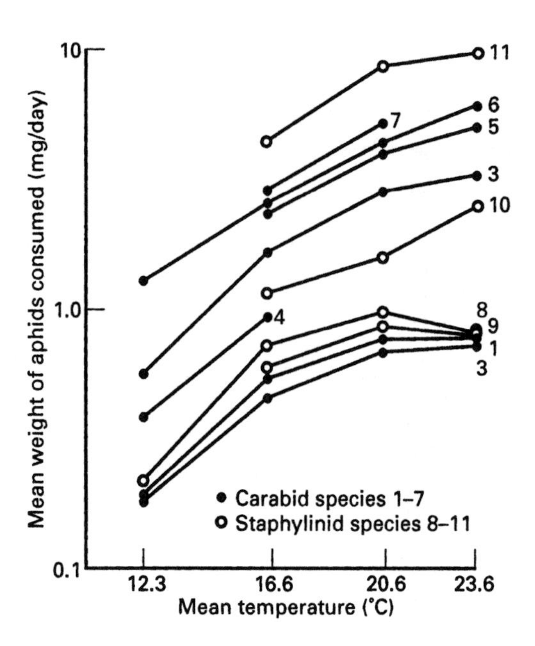

Figure 2.53 The effect of temperature on the mean weight of aphids (*Sitobion avenae*) consumed per day by eleven species of carabid and staphylinid beetles. In the experiments, individual beetles were given an excess of prey (first and second instar, in approximately equal proportions). Source: Sopp and Wratten (1986).

Interaction Between Temperature and Consumption Rate

Whilst temperature will affect growth and development rates of predators directly, one has to be aware that it can also exert an influence by changing the prey consumption rate (Mills, 1981; Gresens *et al*., 1982; Sopp and Wratten, 1986; Pickup and Thompson, 1990) (Figure 2.53). The rate at which food passes through the gut will be positively temperature-dependent, and this will affect consumption rate by affecting hunger (insect hunger is directly related to the degree of emptiness of the gut (Johnson *et al*., 1975)). B in equations 2.10–2.15 (representing in part basal metabolic costs) will also be temperature-dependent (Pickup and Thompson, 1990), and consumption rate will increase to counteract an increase in B.

To take any confounding effects of varying consumption rate into account when assessing

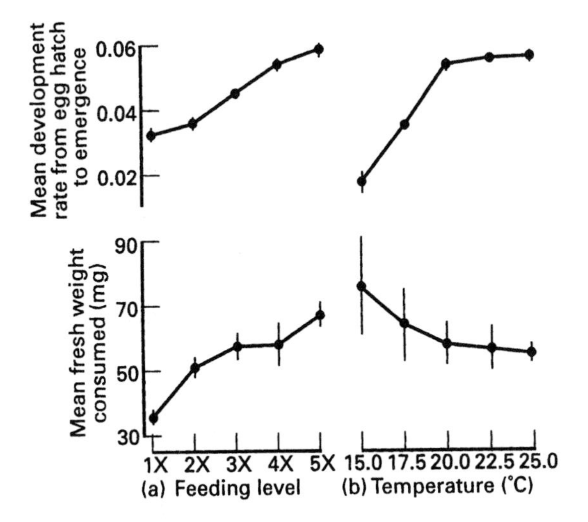

Figure 2.54 The mean rates of prey consumption and development of the immature stages of *Adalia bipunctata* (Coccinellidae) in relation to prey availability at: (a) a constant temperature (20°C) and various 'feeding levels' (the weight of prey corresponding to 1 times to 5 times the average teneral weight of the instar); (b) a range of temperatures, using one (4 times) feeding level. Source: Mills (1981). Reproduced by permission of Blackwell Publishing.

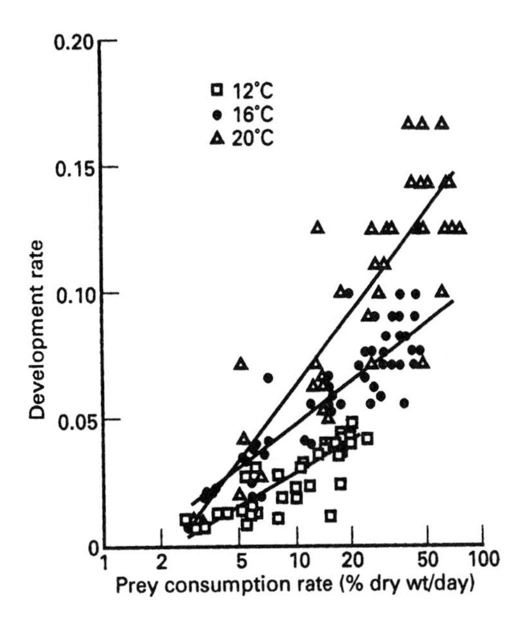

Figure 2.55 Development rate in relation to consumption rate (note log scale) in *Coenagrion puella* (Odonata: Zygoptera). Temperature affects development directly and indirectly by increasing the prey consumption rate. There is a clear interaction effect between temperature and consumption rate. Source: Pickup and Thompson (1990). Reproduced by permission of Blackwell Publishing.

the influence of temperature on growth and development rates in *Adalia bipunctata*, Mills (1981) compared: (a) the mean growth and development rates recorded at the experimental range of temperatures (i.e. fixed daily ration of prey) with (b) those predicted from Figure 2.41 (i.e. constant temperature regime) for the appropriate rates of consumption (Figure 2.54). With this analysis, Mills recorded significant deviations from the predicted growth and development rates, indicating that temperature does have a direct influence on growth and development.

A more straightforward approach to determining how consumption rate interacts with temperature to affect development rate involves plotting development rate against consumption rate, constructing regression lines for each temperature regime and then comparing the slopes of the lines. As can be seen from plots for the damselfly *Coenagrion puella* (Figure 2.55), higher consumption rates produce stronger developmental responses to increases in temperature.

Other Physical Factors

Diurnal predator larvae may, like the adults (subsection 2.7.4), show a reduction in daily consumption rate with decreasing photoperiod, and this will be reflected in a reduction in growth and development rates. Bear in mind, when varying photoperiod in experiments, that you may also be inadvertently varying the absorption of radiant energy by insects, thus altering their body temperature.

Larvae of terrestrial predators may, like the adults, increase their rate of prey consumption with decreasing humidity, which will cause them to grow larger and more rapidly. Predator larvae may develop faster in an incubator than in a large environment chamber, even at the same temperature, because of the lower humidity in the former (Heidari, 1989).

For a recent study of the effects of photoperiod on parasitoid development, see Urbaneja *et al.* (2001a).

2.10 SURVIVAL OF IMMATURES

2.10.1 INTRODUCTION

Below we discuss some factors that affect the survival of predator and parasitoid immatures. Parasitism and predation by heterospecifics are not considered (see Chapter 7 for practical approaches), whereas predation by conspecifics, i.e. cannibalism (Sabelis, 1992) is. Mortality of parasitoid juveniles is strongly dependent on that suffered by the hosts that support them. Hosts may be killed through predation, starvation and exposure to unfavourable weather conditions, and any parasitoids that are attached to or contained within the hosts will die. Price (1975) illustrated this relationship by reference to the host survival curves, which in insects are of either Type II or Type III (Figure 2.35), i.e. substantial mortality of hosts (very substantial in the latter case), and therefore of any parasitoid progeny they support, occurs by the mid-larval stage (see also Cornell *et al.*, 1998).

When investigating larval mortality, the possibility ought to be considered that some factors may cause higher mortality in one sex than in another. Some parasitoids allocate male eggs to small host individuals, e.g. pupae, and female eggs to large individuals (subsection 1.9.1). If small hosts suffer a higher degree of mortality from a predator than larger ones, then the survival rates of male and female parasitoids will differ.

2.10.2 EFFECTS OF BIOTIC FACTORS ON SURVIVAL OF IMMATURES

Food Consumption by Predators

By recording deaths of individuals within each instar in the food consumption experiment outlined earlier, the relationship between food consumption and survival can be studied. A model relating larval survival to prey availability was developed by Beddington *et al.* (1976b). If we assume that we are not dealing with a population of genetically identical individuals, then we would expect mortality through food shortage to take place at some characteristic mean ingestion rate μ_I with the population as a whole displaying variation about this mean value. Assuming that the proportion of the population experiencing 'food stress' is normally distributed about the mean, with standard deviation σ, then the proportion (S) of the larval population surviving to complete development within any particular instar of duration d, at an ingestion rate I, will be given by (Beddington *et al.*, 1976b):

$$S = \frac{1}{\sqrt{2\pi}} \int_{-\infty}^{z} \exp\left(-\frac{z^2}{2}\right) dz \qquad (2.18)$$

where $z = \dfrac{I - \mu_i}{\sigma_i}$.

Using equations 2.2 and 2.18, S may be expressed in terms of either consumption rate or prey density (Figure 2.56). The relationship shown in Figure 2.56c is shown by predators in the laboratory (Figure 2.57). As pointed out by Beddington *et al.*, whether a survival curve rises extremely rapidly or slowly depends on the range of prey densities over which experiments are carried out and the graphical scales chosen for plotting the data.

Mortality due to nutritional stress apparently occurs at feeding rates very much higher than the minimum rate necessary for growth and development, so that individuals that are growing normally (albeit slowly) at low feeding rates are apparently highly likely to suffer high mortality at the moult (Beddington *et al.*, 1976b). There may be no relationship between the overall survival rate between entering and leaving an instar (S) and the duration of each instar (d), as in *Blepharidopterus angulatus*, or S may decline in a variety of ways with increasing d (see examples in Beddington *et al.*, 1976b).

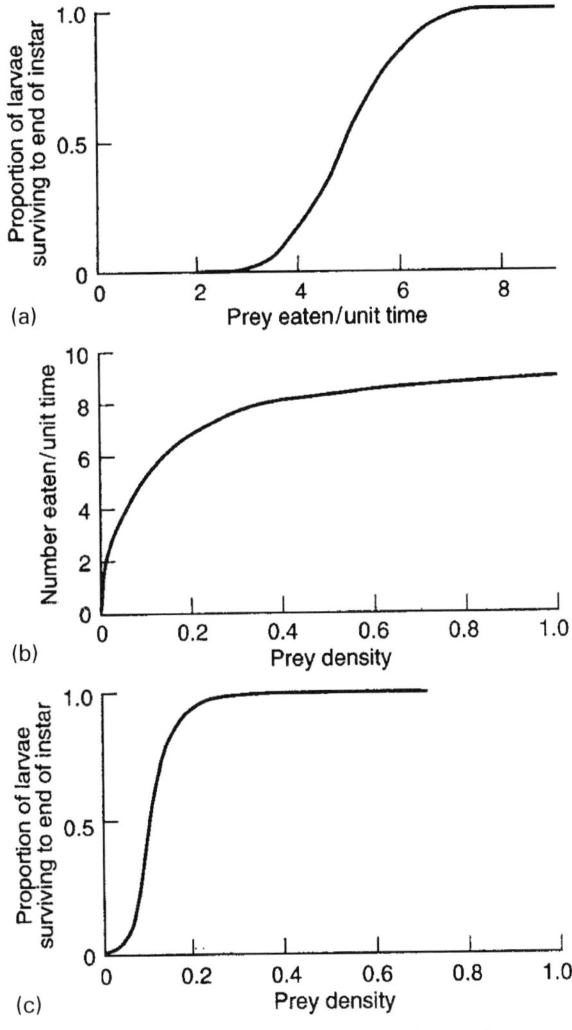

Figure 2.56 Hypothetical relationships between (a) the proportion of individual predators surviving to the end of an instar and their mean feeding rate during that instar; (b) predation rate and prey density; (c) the relationship obtained by combining (a) with (b). Source: Beddington *et al.* (1976b). Reproduced by permission of Blackwell Publishing.

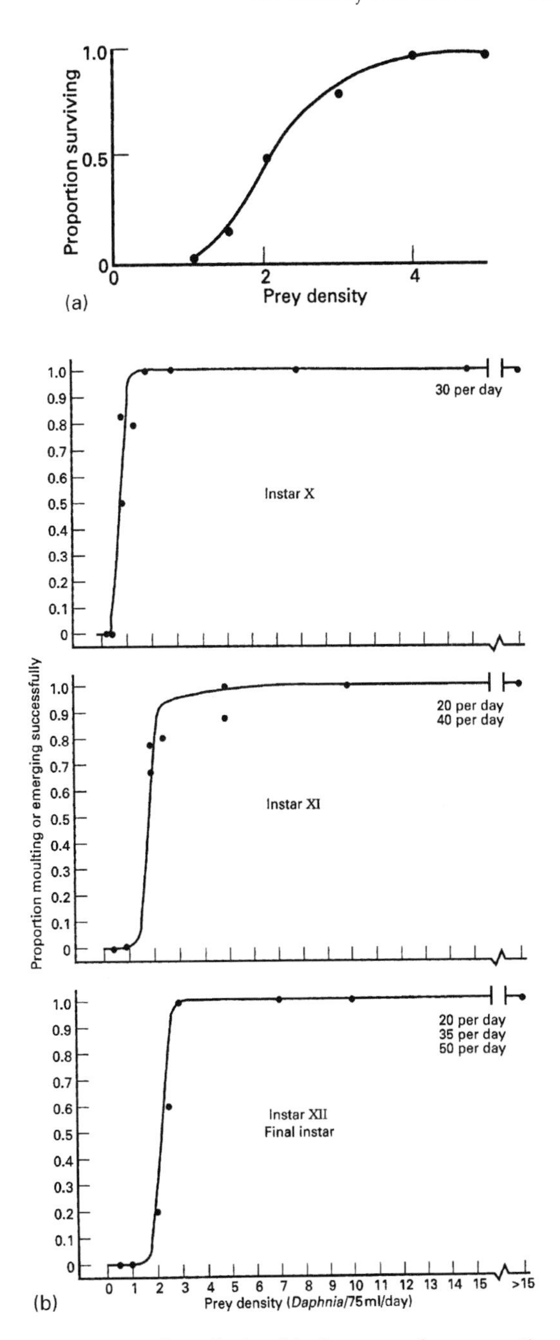

Figure 2.57 The relationship between the proportion of predators surviving to the end of an instar, and the mean density of prey available during that instar. (a) first instars of the coccinellid beetle *Adalia bipunctata* (data from Wratten, 1973); (b) tenth, eleventh and twelfth (final) instars of the damselfly *Ischnura elegans* (source: Lawton *et. al.*, 1980). Reproduced by permission of Blackwell Publishing.

Survival rates vary between successive instars at comparable prey densities. Figure 2.58 summarises the relationship between instar and the feeding rate at which 50% of a larval cohort survive, in four predatory insects and a spider. The plots indicate a constant increase in feeding rate between instars to maintain survival rates at 50%. In the case of *Ischnura elegans*, feeding

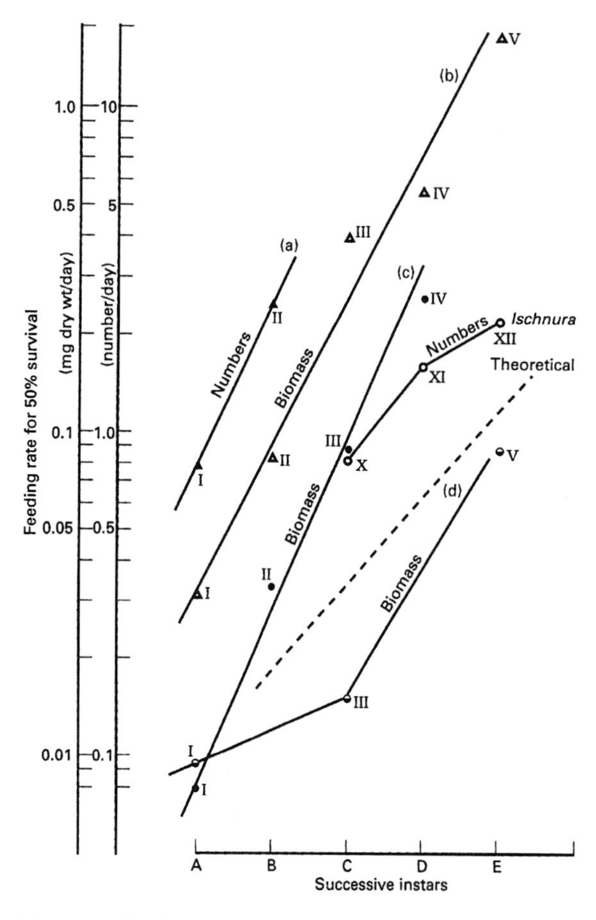

Figure 2.58 The relationship between instar number (in Roman numerals) and the consumption rate (feeding rate) at which 50% of a cohort of larvae in a particular instar successfully complete their development (LD_{50} for food stress): (a) *Adalia decempunctata* (Coccinellidae) (data from Dixon, 1959a); (b) *Notonecta glauca* (Notonectidae) (data from McArdle, 1977); (c) *Linyphia triangularis* (Arachnida: Linyphiidae) (data from Turnbull, 1962); (d) *Blepharidopterus angulatus* (Miridae) (data from Glen, 1973). The theoretical line is for larvae that double their body weight between instars, and where food-energy requirements increase by a 0.87 power of body weight. Source: Lawton *et al.* (1980). Reproduced by permission of Blackwell Publishing.

rates necessary to ensure 50% survival approximately doubled between instars ten and eleven, and increased by a factor of 1.4 between instars eleven and twelve (Lawton *et al.*, 1980). The theoretical line in Figure 2.58 is for larvae that double their body weight between instars and in which minimum requirements increase by a 0.87 power of body weight (which they do for larvae of a damselfly closely related to *Ischnura*. The steeper slopes shown by *Adalia, Notonecta* and the spider perhaps reflect the higher exponents in their metabolic rate:weight relationships and/or larger increases in average body weights between instars (Lawton *et al.*, 1980).

Prey Species

The relationship between larval survival and prey species has been most thoroughly investigated in coccinellid beetles (Hodek, 1973; Hemptinne *et al.*, 2000; Wiebe and Obrycki, 2002; Özder and Sağlam, 2003). Coccinellid larvae have been presented with *acceptable* prey of various species, and larval mortality measured (Hemptinne *et al.*, 2000, studied intraguild predation).

Consumption rate in relation to each prey species can be measured, and prey-related differences in survival correlated with differences in consumption rate. However, if there is a reduced rate of consumption on a prey species and survival is also low on that species, one cannot necessarily conclude that poor survival is a direct result of reduced consumption rate. Survival on the 'better' prey species may still be higher than on the 'poorer' one at equivalent consumption rates (Hodek, 1973). If it is, then the prey-related difference in survival may be due to factors such as differences in the size or qualitative attributes of the prey species.

Interference and Exploitation Competition, Cannibalism

Interference from conspecifics and predators can, through its effects on feeding rates, potentially reduce survival (Heads, 1986; Sih, 1982) (subsection 2.9.2). Exploitation is an obvious potential cause of mortality among larval conspecifics, as prey may be depleted to a level at which larvae experience nutritional stress.

An additional cause of mortality in the immature stages of dragonflies, damselflies, waterboatmen, coccinellid beetles, anthocorid bugs and antlions (Crowley *et al.*, 1987; Sih, 1987; Mills, 1982b; Agarwala and Dixon, 1992; Naseer and Abdurahman, 1990; Griffiths, 1992; Hopper *et al.*, 1996; Gagné *et al.*, 2002; Michaud, 2003) is cannibalistic behaviour. The adults and larvae of several Coccinellidae are known to be cannibalistic on eggs (Gagné *et al.*, 2002) [The neonate larvae of the coccinellid *Coleomegilla maculata lengi* prefer the eggs of conspecifics to aphid prey (Gagné *et al.*, 2002).] In dragonflies, cannibalism may result in the death not only of one of the interacting pair (same or smaller instar larva) but also both participants, since it could attract the attention of predators (Crowley *et al.*, 1987) (this also applies to non-cannibalistic interference). Hopper *et al.* (1996) showed that in the dragonfly *Epitheca cynosura* cannibalism among larvae was more important than exploitation competition in determining survival; they also found that when juveniles hatch asynchronously in close proximity, cannibalism is density-dependent (so can therefore contribute to population regulation), and they concluded that it can also increase population synchrony by exerting size-specific mortality on smaller individuals throughout development.

The effects of competition or cannibalism on survival in immature stages can be expressed as either percentage mortality plotted against predator density or as k-values for the mortality plotted against $\log_{10}$ predator density (Varley *et al.*, 1973) (subsection 7.3.4). If density-dependent mortality occurs, it will be shown within the upper range of densities only, i.e. there will be a threshold density of predators below which k is zero (Mills, 1982b) (or its value is slighty above zero, in which case one has to question whether the mortality recorded at low predator densities is entirely attributable to interference, exploitation or cannibalism). The manner in which k varies with $\log_{10}$ predator density indicates the nature of the density-dependence, i.e. exact, over- or under-compensation (subsection 7.3.4) and whether competition is of the scramble-type or contest-type (for explanations of these terms, see Varley *et al.*, 1973; Begon *et al.*, 1996).

Bear in mind that for the perpetrator, survival may be improved by cannibalism: larvae of the coccinellid *Cycloneda sanguinea* had a higher survival rate when fed on conspecific eggs, than when fed moth eggs (Michaud, 2003).

Destructive host feeding by parasitoids will very rapidly kill any parasitoid immatures contained within the host (Jervis and Kidd, 1986; Kidd and Jervis, 1991). Non-destructive host-feeding is unlikely to kill parasitoid immatures in the short-term, but could nevertheless reduce their life expectancy (e.g. see Heimpel and Collier, 1996; Ueno, 1997).

Host Size

As well as measuring growth and development of parasitoids, larval survival can also be recorded in relation to host size at oviposition. One might reasonably assume, for nutritional reasons, that generally for solitary idiobionts survival is highest in large hosts, although there could be cases where hosts above a certain size represent a resource in excess of the amount required by the larva to complete its development (in such cases larval survival may not be improved in the largest hosts, and it may even be reduced, e.g. due to putrefaction of the remaining host tissues) (see 2.9.2).

The relationship for koinobionts is likely to be more complex. For the solitary koinobionts *Lixophaga diatraeae* and *Encarsia formosa*, survival is highest in individuals that complete their development in medium-sized hosts (Miles and King, 1975; Nechols and Tauber, 1977). For the solitary koinobiont *Leptomastix dactylopii*, no significant differences were found between survival in different-sized hosts (de Jong and van Alphen, 1989). In *Venturia canescens* survival is highest in medium-sized hosts and lowest when the second instar host is oviposited in (Figure 2.47c). The probable reason for the lower survival in second instar hosts is injury to the host through insertion and removal of the ovipositor (this does not occur when later instars are attacked) (Harvey *et al.*, 1994).

A possible complicating factor in experiments is mortality from encapsulation, which may be higher in some stages than in others. Therefore, samples of hosts need to be taken and dissected during larval development to provide data on the frequency of encapsulation, so that this potential confounding factor can be controlled for in data analyses.

In those gregarious parasitoids in which progeny survival is 100% at the smallest clutch size, 100% survival might also occur at larger clutches if larger-sized hosts are utilised. The slope of the relationship might also be less steep in the case of larger hosts.

Host Age

Host age, rather than host size, could influence parasitoid survival. The effects of the two variables may, however, be difficult to disentangle. Survival in some egg parasitoids depends the age at which the host egg is attacked (Ruberson and Kring, 1993).

Host Species

Given that different host species are likely to constitute different resources, both qualitatively and quantitatively, parasitoid larval survival may vary with host species, the larvae (or eggs) dying through malnutrition, encapsulation (see **Host Physiological Defence Reactions**, below) or poisoning (e.g. if the host has sequestered toxins from its food plant, see **Multitrophic Interactions and The Performance of Immatures**, above).

In *Telenomus lobatus* percentage eclosion, i.e. survival of progeny, was higher from the eggs of *Chrysoperla* species than from eggs of *Chrysopa* species (Ruberson *et al.*, 1989). In the gregarious idiobiont *Habrobracon hebetor*, survival within clutches was density-dependent both on a small moth species, *Plodia interpunctella*, and on a large moth species, *Anagasta kuehniella*, but the density-dependence in the latter case applied only to very high (artificially manipulated) clutch sizes (Taylor, 1988a).

To investigate the effect of host species on larval survival, present females with hosts of different species and, if possible, of equivalent size and age. Maintain the hosts until the parasitoids pupate, and maintain the parasitoid pupae until the adults have ceased emerging. Any hosts that have received eggs but have not given rise to parasitoids should be examined (dissected in the case of endoparasitoids) for the remains of parasitoid eggs or larvae. Any parasitoid pupae that fail to produce adults should also be recorded. Sex differences in survival should also be sought (Ruberson *et al.*, 1989).

Hosts' Food Plant

See the subsection **Multitrophic Interactions and The Performance of Immatures** (above).

Superparasitism

Solitary Parasitoids

In solitary endoparasitoids, supernumerary larvae are eliminated (**contest competition**) either through physiological suppression or (more usually) through combat (Clausen, 1940; Fisher, 1971; Vinson and Iwantsch, 1980b; Quicke, 1997). This applies not only to self- and conspecific superparasitism but also to heterospecific superparasitism (= multiparasitism).

The first instar larvae of almost all solitary parasitoid wasp species are equipped with robust, often sickle-shaped mandibles (Fisher, 1961; Salt, 1961) (Figure 2.59). Fighting often takes place between larvae that are of approximately the same age, although in some species first instar larvae will attack and kill later instars that either have reduced mandibles or lack mandibles altogether (Chow and Mackauer, 1984, 1986). Note that the possession by first instar larvae of large mandibles does not necessarily mean that fighting is the sole mechanism employed in the elimination of rivals (Strand, 1986; Mackauer, 1990). Also, bear in mind that the first instar larvae of some facultatively gregarious species possess sharp mandibles, but the larvae do not practise siblicide (e.g. *Aphaereta pallipes*, Mayhew and van Alphen, 1999).

The mechanisms employed in the elimination of larval competitors in three solitary braconid parasitoids are summarised in Figure 2.60. As with other parasitoids, in cases of *intra*specific

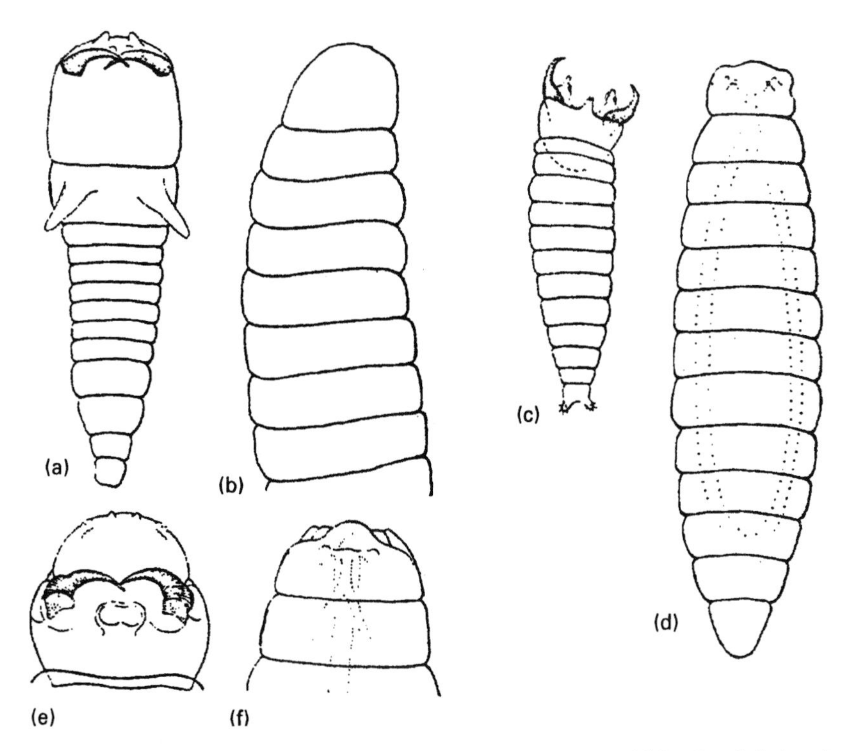

Figure 2.59 Larvae of parasitoid wasps that in the first instar have mandibles for fighting (a,c,e) but do not have such mandibles in the second instar (b,d,f). (a,b) *Biosteres fletcheri* (Braconidae); (c,d) *Psilus silvestri* (Diapriidae). (e,f) *Diplazon fissorius* (Ichneumonidae). Source: Salt (1961), reproduced by permission of The Company of Biologists Ltd.

larval competition the oldest larva generally survives and the younger larva dies, although this may not apply where the larval age difference is either very small or very large (in the latter case the older larva may have developed to the second, i.e. non-mandibulate instar by the time the second egg is either laid or hatches; Bakker *et al.*, 1985). See also Marris and Casperd (1996).

The 'oldest larva advantage' applies to some cases of *interspecific* larval competition among parasitoids but not to others (Tillman and Powell, 1992; Mackauer, 1990) (see below). Relative age differences can influence the outcome of an interaction, but the factors that appear to be more important in determining who survives are the particular competitive mechanism(s) and the developmental stage at which each comes into play. Bear in mind that: (a) the eggs of two species may be laid at the same time, but hatch at different times, and/or (b) the development rate of the larva may be greater in one species than in another, and these factors may determine the 'window of interaction'. For example, the braconids *Aphidius smithi* and *Praon pequodorum* require approximately the same amount of time to develop from oviposition to the second instar, but the embryonic period is much shorter in *Aphidius* than in *Praon*. This difference enables *Praon* to compete as a mandibulate first-instar larva with an older *Aphidius* larva. *Aphidius* larvae usually survive only if they have reached the end of the fourth (final) instar while *Praon* is still in the embryonic stage and thus unable to attack an older competitor (Chow and Mackauer, 1984, 1985).

A parasitoid species that wins under most conditions is described as **intrinsically superior** (Zwölfer, 1971, 1979). Ectoparasitoids tend to be intrinsically superior to endoparasitoids (Sullivan, 1971, gives one exception). The superiority

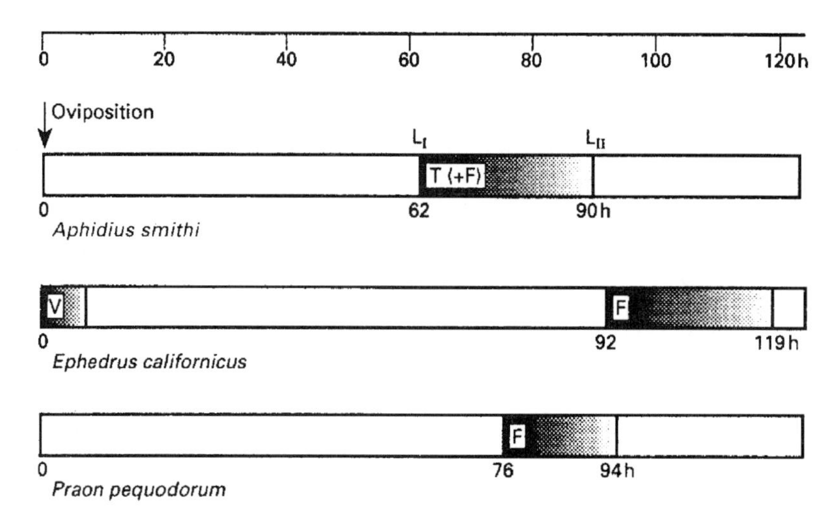

Figure 2.60 The mechanisms used in the elimination of competitors by the solitary braconid parasitoids *Aphidius smithi*, *Ephedrus californicus* and *Praon pequodorum* in pea aphids. F = fighting among first instar larvae (L_I); T = toxin released at eclosion of L_I; V = venom injected by female at oviposition. Median times taken from eclosion to L_I and L_{II} refer to parasitoid larvae developing in second-instar pea aphids at 20°C. Source: Mackauer (1990). Reproduced by permission of Intercept Ltd.

of ectoparasitoids is a result of envenomation and/or more rapid destruction of the host, rather than a result of the endoparasitoid being attacked directly (Askew, 1971; Vinson and Iwantsch, 1980b).

Collier *et al.* (2002) tested the hypothesis that relative egg size can be used to predict the outcome of 'intrinsic competition' between closely related parasitoid species (*Encarsia* spp.): a species with relatively large eggs should be superior to one with small eggs. The hypothesis was not supported by the experimental evidence: the species with the smaller eggs (*E. formosa*) prevailed in competition, irrespective of the order of exposure (however, *E. formosa* females killed the progeny of its superior larval competitor by host-feeding).

An experiment designed to investigate the relative competitive superiority of solitary endoparasitoid larvae in instances of superparasitism would involve varying the time interval between ovipositions (from a few seconds to many hours), either by the same parasitoid species or females of different species. If heterospecific superparasitism is being studied, then the sequence of species ovipositions can be reversed. Whatever the type of interaction being investigated, by taking regular samples of the superparasitised hosts and singly-parasitised hosts at successive points in time from the second oviposition and dissecting them, the following can be recorded:

1. The stage of development (embryonic or larval) already reached by the older parasitoid at the time of the second oviposition (determine this from dissection of singly-parasitised hosts);
2. The stage of development subsequently reached;
3. The stage of development of the younger parasitoid;
4. Which, if any, of the eggs or larvae are dead or alive (exceptionally, both may be dead, as suggested by the data in Table 2.1);
5. Any behavioural evidence of physical combat;
6. Whether either of the parasitoid immatures bear wounds (the latter may show signs of melanisation [Salt, 1961]).

Threshold time intervals for the different outcomes of competition (if there can be more than one outcome) can then be found. Note that for some interactions, the period of time between oviposition and the development of *the host* to a certain stage, indirectly determines which parasitoid species is the survivor. For example, in the case of *Trieces tricarinatus* and *Triclistus yponomeutae*, this interval determines the extent of development of the parasitoids after host pupation and the extent of development at the time of combat (irrespective of whether the host is singly- or multiparasitised, development of larvae beyond the first instar can only take place after host pupation) (Dijkerman and Koenders, 1988).

Instead of dissecting superparasitised hosts, the outcome of competition can be studied by rearing the parasitoids to the adult stage. However, in studies of intraspecific superparasitism, this method requires that distinguishable (preferably morphologically distinguishable) strains be used. This method would also prove useful for studying intraspecific superparasitism when the interval between ovipositions is so short that it is not possible, through dissection, to distinguish between the progeny of the first female and the progeny of the second female. For example, Visser *et al.* (1992c) measured the **pay-off from superparasitism** in the solitary parasitoid *Leptopilina heterotoma*. They used two strains of this species: a wild type with black eyes and a mutant with yellow eyes. Hosts parasitised by females of one strain were exposed to females of the other strain and the interval between ovipositions was varied. The sequence of ovipositions was reversed to take account of any competitive asymmetry between strains. The probability of a second female realising an offspring from superparasitism, i.e. the pay-off, was then calculated for each strain.

Harvey *et al.* (1993) examined whether parasitoid mortality from superparasitism varies with host instar in cases of near-concurrent oviposition by two conspecific females (*Venturia canescens*). Parasitoids were reared from third and fifth instar hosts (the moth *Plodia interpunctella*) containing either one, two or four parasitoid eggs. Parasitoid mortality was found to be significantly higher in fifth instar hosts than in third instar hosts, but within instars did not vary with egg number (Figure 2.61). The likely reason for the higher mortality in fifth instar hosts is that there is some physiological incompatibility between the parasitoid and fifth instar hosts associated with pupation (Harvey *et al.*, 1993).

Gregarious Parasitoids

In gregarious species where survival declines monotonically with increasing clutch size, the

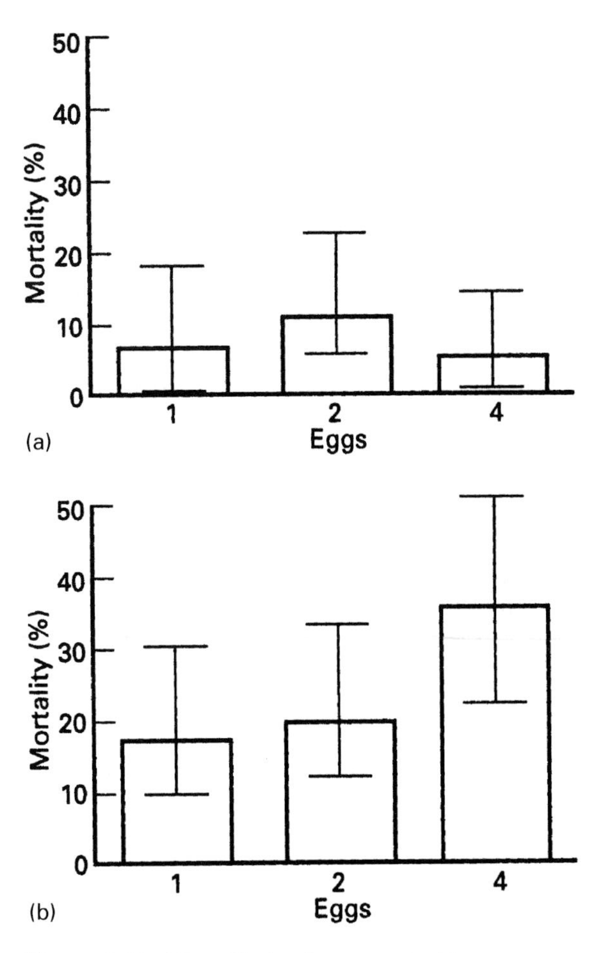

Figure 2.61 The effects of superparasitism on survival in the solitary ichneumonid parasitoid *Venturia canescens:* Mortality of parasitoids reared from (a) third instar; (b) fifth instar hosts, containing one, two or four parasitoid eggs. Encapsulation was not a complicating factor in the experiments. Source: Harvey *et al.* (1993). Reproduced by permission of Blackwell Publishing.

addition of an egg or clutch of eggs will (further) reduce percentage survival per host. The reduction will normally result from increased resource competition, since larvae of gregarious species do not engage in physical combat. In those species in which there is an Allee effect (1.6), there will be a threshold number of progeny per host below which all parasitoids die, so superparasitism of a host containing a clutch of eggs that is a number short of this threshold number is likely to *raise* the survival chances of the parasitoid immatures.

Assuming competitive equivalence of first and second clutches laid in or on a host, the effect on survival of simultaneous oviposition by two females would be analogous to the effects of increasing the initial clutch size (Strand and Godfray, 1989). However, in gregarious species mortality may vary not only with the number of eggs initially present but also with the time interval between ovipositions, i.e. it will depend on how soon superparasitism occurs after the laying of the initial clutch (Strand, 1986). Strand and Godfray (1989) demonstrated this for *Habrobracon hebetor*. In this species progeny survival within a second egg clutch, equal in size to the first, was approximately 42% (each clutch comprising 20 eggs), 78% (each clutch comprising 10 eggs) and 83% (each clutch comprising four) when the first and second clutches were 'laid' simultaneously (they were placed on hosts by the experimenter, see below). However, when the time between 'ovipositions' was 12 hours or more, progeny survival within the second clutch was reduced to less than 10% for clutches comprising either 10 or 20 eggs (Figure 2.62).

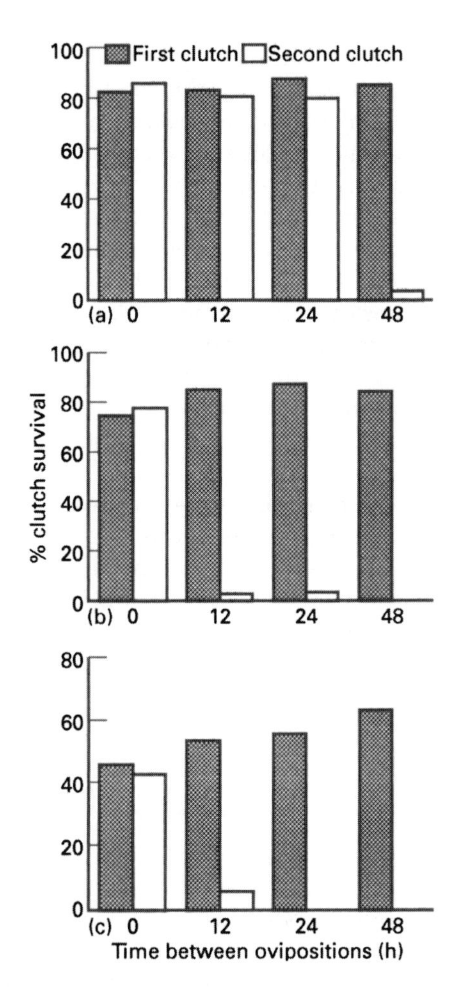

Figure 2.62 The relationship between progeny survival within first and second clutches of eggs, and the time between 'ovipositions' for starting clutches of (a) 4; (b) 10; (c) 20 eggs, in the gregarious parasitoid wasp, *Habrobracon hebetor* (Braconidae). First and second clutches were equal in size for each experiment. Source: Strand and Godfray (1989). Reproduced by permission of Springer-Verlag.

Experiments aimed at investigating the mortality effects of superparasitism in a gregarious species can be conducted along the lines described in the section on superparasitism in relation to growth and development rates. Sex differences in survival may be examined in such experiments; several studies (Vinson and Iwantsch, 1980b) have shown that with increased larval crowding there is a tendency for preferential survival of males.

Superparasitism in egg parasitoids can be investigated using *in vitro* techniques (Strand and Vinson, 1985; Strand *et al.*, 1986; Marris and Casperd, 1996); parasitoid eggs and larvae can be added to a volume of culture medium equivalent in volume to a host egg.

As noted in Chapter 1 (section 1.8), the survival chances of parasitoid immatures can in some case be *improved* by superparasitism. For example, of eggs of *Asobara tabida* laid in larvae of *Drosophila melanogaster*, 1% survive in singly parasitised hosts whereas 7% survive in superparasitised hosts (van Alphen and Visser, 1990), encapsulation being the principal cause of mortality in both cases. Van Strien-van Liempt (1983) measured the survival of *Asobara tabida* and *Leptopilina heterotoma* in multiparasitised *Drosophila* hosts and compared it with survival in singly parasitised hosts. Percentage survival in instances of multiparasitism was not always lower than survival in singly parasitised hosts; in most cases, multiparasitism provided a mutual survival advantage. In cases such as these where parasitoid survival is increased through superparasitism, the mechanism is thought to be exhaustion of the host's supply of haemocytes (see **Host Physiological Defence Reactions,** for further discussion).

Compared with an Israeli strain, a Californian strain of the aphelinid *Comperiella bifasciata* was subject to a higher encapsulation rate in red scale and also superparasitised more hosts. Blumberg and Luck (1990) suggested that since the risk of encapsulation is reduced in superparasitised hosts (see also Saggara *et al.*, 2000a), the higher degree of superparasitism shown by the Californian strain is a strategy to avoid encapsulation.

For a study of intra- and interspecific larval interactions among a subweb of dipteran (specialist and generalist tachinid) and hymenopteran parasitoids, and their consequences for parasitoid survival, see Iwao and Ohsaki (1996).

The mortality effects of superparasitism can be expressed as *k*-values (subsection 7.3.4).

Host Physiological Defence Reactions

Introduction

Endoparasitoid larvae and eggs may die owing to a reaction of the host's immune system. The term **immune system** is used in the loose sense that the hosts are capable of mounting a defensive response against foreign bodies. The response does not involve either a specific 'memory', with accelerated rejection of the second of two sets of an introduced foreign tissue, or a marked increase in the concentration of some specific humoral component, as has been shown for vertebrates. Thus, the probability of a parasitoid eliciting an immune response in an insect is independent of previous challenges (Bouletreau, 1986).

Host defence reactions are of several kinds (Ratcliffe, 1982; Ratcliffe and Rowley, 1987; Strand and Pech, 1995; Carton and Nappi, 1997; Quicke, 1997; Fellowes and Hutcheson, 2001, for reviews), but the most commonly encountered type of reaction is **encapsulation**. Usually in encapsulation the foreign invader becomes surrounded by a multicellular sheath composed of the host's haemocytes (Figure 2.63). Successive layers of cells can often be discerned, and on the outer surface of the parasitoid egg or larva there often develops a necrotic layer of melanised cells, representing the remnants of the blood cells that initiated the encapsulation reaction. The melanin deposits on the surfaces of encapsulated parasitoid eggs and larvae often provide the first clue to the occurrence of encapsulation (Figure 2.63a-c). Parasitoid immatures die probably from asphyxiation, although starvation may be the principal cause of death in some cases. Phagocytosis of parasitoid tissues gradually occurs, at least during the initial stages of encapsulation.

Parasitoids can resist, i.e. evade and/or supress the immune responses of their hosts (Askew, 1971; Quicke, 1997; Kraaijeveld *et al.*, 1998; Fellowes and Hutcheson, 2001). One means of evasion is the laying of eggs in refuges from encapsulation. Some parasitoids oviposit into specific host organs such as the nerve ganglia and salivary glands, where an egg cannot be reached by the host's haemocytes (Strickland, 1923; Salt, 1970; Rotheray, 1979; Dijkerman, 1988). Many early-instar parasitoid larvae, which are also potentially exposed to a host's immune defences, migrate to specific regions of the host after they hatch from the egg (the first-instar larvae of many ichneumonids use their caudal appendage for this purpose) (Salt, 1968). In other parasitoids, the immature stages

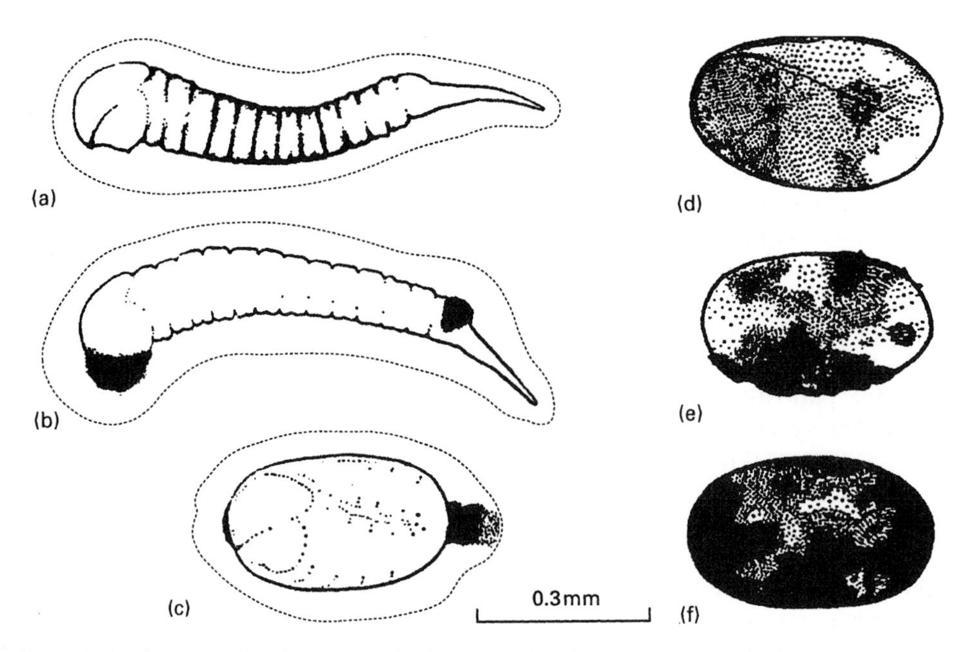

Figure 2.63 Encapsulation and melanisation: (a-c) encapsulated larvae and egg of *Venturia canescens* implanted in a non-host insect (deposits of melanin can be seen); (d-f) deposits of melanin on eggs 24, 32 and 48 hours after implantation in a non-host insect. Source: Salt (1970). Reproduced by permission of Cambridge University Press.

have surface properties that prevent encapsulation (Salt, 1968; Rotheram, 1967). The risk of encapsulation in a particular host species can be drastically increased by washing the surface of the parasitoid eggs using either solvents or water before they are artificially injected. Eggs of *Venturia canescens* removed from the ovarioles are encapsulated if they are artificially injected into the haemocoel of *Anagasta*, whereas eggs removed from the calyx region of the lateral oviduct do not become encapsulated. It is becoming increasingly clear that in some parasitoids the ovipositing female or her offspring are able to manipulate or disrupt the immune system (see reviews by Strand and Pech, 1995; Lavine and Beckage, 1996; Beckage, 1998a,b). Population genetic and dynamic aspects of encapsulation are discussed by Bouletreau (1986), Godfray and Hassell (1991), Kraaijeveld *et al.*, (1998) and Fellowes and Godfray (2000); see also Chapter 3 (subsection 3.5.3).

Encapsulation is usually studied *in vivo* in either laboratory-cultured or field-collected hosts. However, some workers have successfully used *in vitro* techniques (e.g. Ratner and Vinson, 1983; Benson, 1989; Lovallo *et al.* 2002).

Host Populations and Species

The ability of a host to encapsulate parasitoids is genetically determined (subsection 3.5.3) and there may be considerable variability between populations of a host species in encapsulation rate (Bouletreau, 1986; Maund and Hsiao, 1991; Kraaijeveld and van Alphen, 1994, 1995a; Hufbauer, 2001) (see also Dijkerman, 1990). For example, in *Drosophila melanogaster* there are clear differences between fly populations from different parts of the World with respect to the frequency with which *Leptopilina boulardi* is encapsulated (Bouletreau, 1986). Such effects are an important consideration when one is planning to release biological control agents (Maund and Hsiao, 1991).

The risk of encapsulation also varies between host species. The ability of a parasitoid species to avoid encapsulation may determine at least partly: (a) the range of host species that it parasitises in nature, and (b) different levels of successful parasitism recorded among these hosts (e.g. see Heimpel *et al.*, 2003). The relevance of this to classical biological control introductions is discussed by Alleyne and Wiedenmann (2001). Differential mortality in different host species due to encapsulation may have played an important role in the evolution of host specificity, including preferences, of many endoparasitoids. Dijkerman (1990) observed that the abundance of *Diadegma armillata*, a solitary endoparasitic ichneumonid, in the parasitoid complexes associated with *Yponomeuta* moths, varies among host species, being high in the complex associated with *Y. evonymellus* and very low in that associated with *Y. cagnagellus*. To determine whether this variation corresponds with the ability of each host species to encapsulate the parasitoid. Dijkerman used the following methods:

A. Parasitism experiments: Larvae of the different moth species were exposed to female *D. armillata*. Several days later, a sample of the hosts was taken and the insects dissected. The remaining hosts were maintained until the parasitoids emerged. By *dissecting* the hosts, the presence of parasitoid eggs or larvae was recorded, and the following noted:

1. The rate of infestation, i.e. the number of host larvae containing at least one egg of *D. armillata* as a percentage of the total number of larvae dissected;
2. Percentage encapsulation: [the number of encapsulated progeny divided by the total number of eggs found at dissection] × 100 (this measure of encapsulation efficiency might be less useful in cases where there is a high and variable degree of superparasitism among hosts, which was not the case in this study, see below).

By *rearing* hosts, the following were measured:

3. The rate of successful parasitism: [the number of host individuals yielding *D. armillata* adults divided by the total number of *Yponomeuta* yielding moths or parasitoids] × 100;
4. The percentage mortality of larvae: [the number of larvae dying during their development divided by the initial number of parasitoid larvae] × 100. [Note that if the mean number of parasitoid eggs per parasitised host significantly exceeds 1.0, a correction factor will need to be applied to the data to allow for the effects of parasitoid mortality through superparasitism.]

Simultaneously, under the same conditions, host larvae that were not exposed to parasitoids were reared to moth emergence. This was done to establish whether the results could be biased by a higher mortality of parasitised hosts, compared with unparasitised hosts, in rearings.

B. Dissections of field-collected, late instar, hosts: The following were recorded:

5. Percentage encapsulation (see above); percentage of successful attacks (successful at the time of dissection, not withstanding encapsulation later on), calculated as: [the number of parasitoid eggs or larvae recorded at dissection, divided by the total number of hosts dissected] × 100. To exclude the potentially confounding effects of time and place, comparisons were made only for samples collected at the same locality and same time of day.

Except for one species, *Y. evonymellus*, infestation rates and successful parasitism rates recorded in the laboratory were markedly different. In *Y. mahalebellus* and *Y. plumbellus* no wasps were reared despite infestation rates of 30% and 95%, whereas in *Y. evonymellus* almost all infested larvae yielded adult parasitoids. Since: (a) mortality of parasitised hosts was not different from that of control larvae, and (b) the mean number of parasitoid eggs per parasitised host was little more than 1.0, the differences between infestation and successful parasitism could be explained in part by encapsulation. The field dissections revealed that *Y. cagnagellus* suffers fewer successful attacks than *Y. evonymellus*, despite being the more abundant species at some localities.

The low successful parasitism in *Y. cagnagellus* corresponds with the very low probability of survival of *D. armillata* in that species.

An interesting footnote to Dijkerman's findings is the observation that all of the *Yponomeuta* species in which there was a high rate of encapsulation of *D. armillata* are considered to have diverged early in the evolution of the genus, whereas the more recently evolved moth species show either an intermediate rate of encapsulation or do not encapsulate eggs at all (Dijkerman, 1990).

Herard and Prevost (1997) undertook another study of *D. armillata*, in which encapsulation ability was examined in two *Yponomeuta* species; see their paper for details.

An alternative approach was taken by Benson (1989), who used an *in vitro* technique. He tested the eggs of three aphidophagous ichneumonid species (Diplazontinae) against the haemolymph of a range of hover-fly species. The host ranges and preferences (including behavioural preferences) of each species were already well known, and this enabled rank orders of reaction to be predicted. Haemolymph from a host species was mixed with insect tissue culture fluid and an egg of a diplazontine was added. When 24 hours had elapsed, the fluid was examined for changes in colour, the extent of the change, and the formation of a capsule. The predictions for each parasitoid species in different hosts and for each host species with different parasitoids were confirmed, strongly suggesting that differential host suitability has played a significant role in determining host specificity in diplazontine ichneumonids.

Heimpel *et al.* (2003) make a distinction between **'suitable' hosts**, in which most or all parasitoid progeny can complete development, and **'marginal' hosts**, in which a substantial fraction of host individuals is able to debilitate the immature parasitoids and survive, and point out that marginal hosts may act as a 'sink' for parasitoid eggs. The ecological significance of this effect was explored through modelling by Heimpel *et al.* (2003) (see their paper for details). Note, however, that 'suitability' was used by Heimpel *et al.* in a narrow sense for the purposes of their study; 'suitability' *sensu lato* encompasses contraints upon larval growth and survival, as well as upon survival.

Host Plant

The rate of encapsulation of a parasitoid in a particular host species may vary with the species of plant that the insect feeds on (Ben-Dov, 1972; Blumberg, 1991; Soussi and Le Ru, 1998, but see Blumberg *et al.*, 1995). For example, the scale insect *Protopulvinaria pyriformis* encapsulates a larger percentage of eggs *of Metaphycus stanleyi* when grown on *Hedera helix* or *Schefflera arboricola* than when grown on avocado plants (Blumberg, 1991). Similarly, the mealybug *Pseudococcus affinis* encapsulates a higher proportion of the eggs of the encyrtid *Anagyrus pseudococci* when reared on *Aeschynanthus ellipticus* than when reared on *Streptocarpus hybridus* (Perera, 1990). Blumberg *et al.* (1995) did not find a host plant effect for *Anagyrus pseudococci* in their study.

Host Stage and Age

With many endoparasitoids the probability of encapsulation occurring increases with host stage or host age (Berberet, 1982; van Alphen and Vet, 1986; Slansky, 1986; Van Driesche, 1988; Dijkerman, 1990; Strand and Pech, 1995; Sagarra *et al.*, 2000a). An explanation given by Salt (1968) for such a relationship is that earlier stages have fewer haemocytes available than later ones. Host stage does not affect the probability of encapsulation of *Habrolepis rouxi* (Encyrtidae) in its red scale hosts (Blumberg and DeBach, 1979).

Note that insect eggs lack a cellular defence response to foreign bodies (Salt, 1968, 1970; Askew, 1971; Strand, 1986; Quicke, 1997).

Superparasitism

The reduction in encapsulation ability of a host with superparasitism has already been discussed. Askew (1968) drew attention to this phenomenon. Explanations given in the literature are that the host is 'weakened' or that its supply of haemocytes becomes exhausted as a result of the increased parasitoid load.

Temperature

In some parasitoid species, the temperature at which the host is reared does not affect the frequency at which encapsulation occurs (e.g. *Habrolepis rouxi*: Blumberg and DeBach, 1979; *Aprostocetus ceroplastae*: Ben-Dov, 1972; *Anagyrus pseudococci*: Blumberg *et al.*, 1995; *Aphidius* spp.: Stacey and Fellowes, 2002), whereas in others it does (e.g. *Apoanagyrus diversicornis:* Van Driesche *et al.*, 1986; *Metaphycus stanleyi*: Blumberg, 1991) (see also Blumberg and Van Driesche, 2002). In *A. diversicornis* the rate of encapsulation is highest at the lower of two temperatures, whereas in *M. stanleyi* it is highest under high temperature regimes (Figure 2.64). It follows that for some species, there may be seasonal or geographical variations in encapsulation rate.

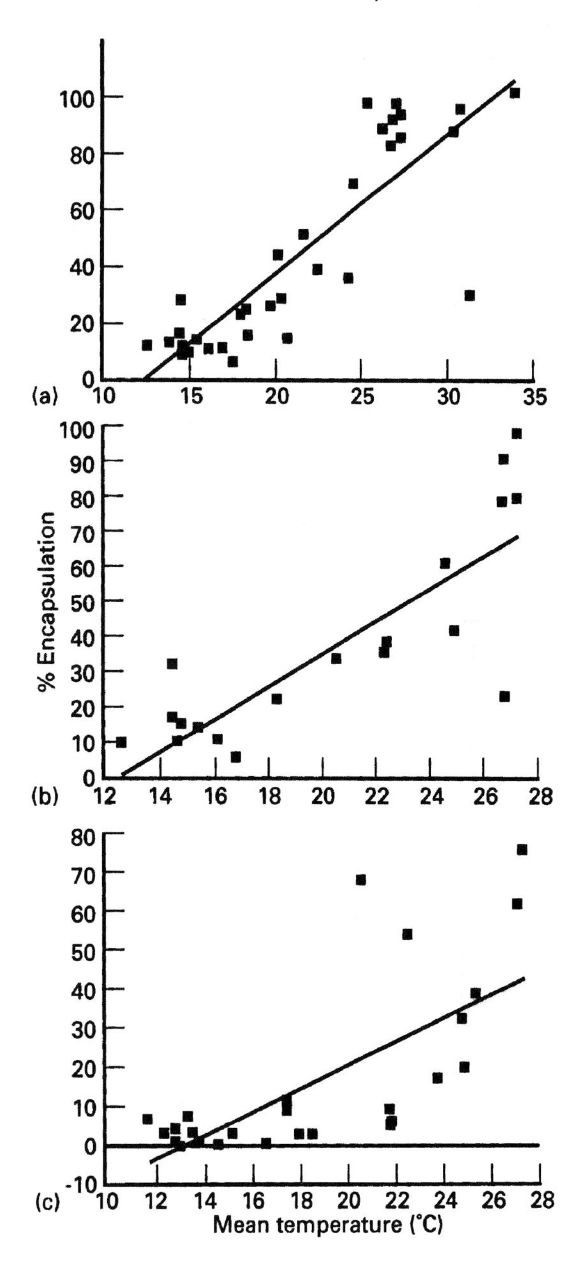

Figure 2.64 The relationship between the rate of encapsulation of eggs and mean temperature in *Metaphycus stanleyi* (Encyrtidae) parasitising the pyriform scale on: (a) *Hedera helix*, (b) *Schefflera arboricola*, under glasshouse conditions; (c) avocado in an orchard. Source: Blumberg (1991).

The Costs of Counterdefences to Host Resistance

While significant insights have been gained into the costs, to the host, of physiological resistance to parasitoid immatures (see Fellowes and Hutcheson, 2001, for a review), little is known about the costs of counterdefence in parasitoids. Kraaijeveld *et al.* (2001) sought evidence for the costs of counterdefence by *Asobara tabida* against *Drosophila* (see their paper for a protocol which involved artificially selecting populations); the only cost they could detect was the delay in hatching of the eggs (which results from them being embedded in host tissue – a defence against host haemocytes); this delay will, Kraaijeveld *et al.* conclude, reduce the chances of parasitoid survival if another parasitoid egg is laid in the same host. No cost was recorded in terms of either mean adult size, fat content or egg load of *A. tabida*.

2.10.3 EFFECTS OF PHYSICAL FACTORS ON SURVIVAL OF IMMATURES

Temperature

Parasitoids may be more hot/cold hardy than their hosts, in which case the lethal range of temperatures for the host will determine parasitoid survival. Prolonged exposure to extreme temperatures will kill the host first, and the parasitoid will then die as a result of starvation, anoxia or host putrefaction. Prior to death, a parasitoid's growth and development may be increased or decreased by the extreme temperature (Tingle and Copland, 1988).

On the other hand, parasitoids may be less hot/cold hardy than their hosts, such that they cannot tolerate the extremes of temperature that the host can tolerate, and so die as result of thermal stress. Parasitised hosts may theoretically even seek out warmer than optimal sites, raising their body temperature with the potential result that the parasitoid is killed (**'behavioural fever'**) (Karban, 1998; see Elliott *et al.*, 2002, and Ouedraogo *et al.*, 2003 on behavioural fever employed by locusts to suppress fungal pathogen infection).

Within the range of temperatures that are not *immediately* lethal to predator larvae, the lower the temperature, the longer totally starved larvae will be able to survive, and, in the case of larvae that have prey available, the less food larvae will require to stay alive (Lawton *et al.*, 1980).

Kfir and van Hamburg (1988) have shown that the outcome of heterospecific superparasitism can be influenced by temperature.

The influence of temperature on the host's ability to encapsulate parasitoids is discussed in the previous section.

Humidity

Low humidity can cause death of ectoparasitoid and predator larvae directly through desiccation (as with adults, small-bodied insects will be more prone to desiccation, all else being equal, due to their higher surface area to volume ratio), whereas high humidity can cause death indirectly by encouraging the growth of fungal pathogens.

Photoperiod

Photoperiod, because of its influence upon diurnal activity and therefore consumption rate, could affect survival of larval predators. Urbaneja *et al.* (2001a) found no evidence for an effect of photoperiod on survival in the parasitoid *Cirrospilus* sp. near *lyncus* (Eulophidae).

2.11 INTRINSIC RATE OF NATURAL INCREASE

2.11.1 INTRODUCTION

The parameter known as the intrinsic rate of natural increase describes the growth potential of a population under a given set of environmental conditions. It is often used, both by ecologists (Gaston, 1988) and by biological control workers (Messenger, 1964b), as a comparative statistic. In a biological control programme, practitioners may be faced with a choice of candidate parasitoid species; in the absence of other criteria they would select, for obvious reasons, the species with the greatest value for the intrinsic rate of natural increase (see subsection 7.4.3 for further discussion).

This population growth parameter is calculated, as described below, from age-specific survival and fecundity schedules. To understand first what it represents, we need to consider the most general of all population growth models, the exponential equation:

$$\frac{dN}{dt} = rN \qquad (2.19)$$

where N is the number of individuals in the population at any given time t, and r is the intrinsic rate of natural increase or the instantaneous *per capita* change in population size. Under conditions of an unlimited environment and with a stable age distribution, r is a constant.

For a given species, r can take a number of values. In theory at least, the species has an optimal natural environment in which its r will attain the maximum possible value, r_m, with a stable age distribution.

2.11.2 CALCULATING r_m FOR A PARASITOID WASP SPECIES

r_m is calculated by iteratively solving the following equation:

$$\sum_{x=0}^{n} e^{-r_m x} l_x m_x = 1 \qquad (2.20)$$

where x is the mid-point of age intervals in days, l_x is the fraction of the females surviving to the pivotal age x (or, put another way, the probability of a female surviving to age x), m_x is the mean number of female 'births' during age interval x per female aged x, and e is the base of natural logarithms. Trial r_m values are substituted into the above expression until the left hand side is (arbitrarily) close to 1.

l_x and m_x are calculated by tabulating (Table 2.2) age-specific fecundity and age-specific survival data obtained from cohort fecundity and survival experiments (subsections 2.7.2 and 2.8.1 discuss the experiments; a graphical display of such data is given in Figure 2.65). If we find from examination of the **life-table** that only 50% of wasps survive to the age of 5 days, then $l_5 = 0.5$. If we find that the average number of female offspring produced per individual alive during the age interval x is 25, then $m_{25} = 25$ (see caption to Table 2.2, for calculations based on another data set). The mean time taken from oviposition to adult eclosion, which can be measured in a separate experiment, is added to the pivotal age of each female. For example, this time period was 12.5 days for *Aphidius smithi* at 20.5°C (Mackauer, 1983). Parasitoid mortality during the immature stages also needs to be measured. In *A. smithi* this mortality was negligible, so the probability of being alive at pivotal age 12.5 days +1 day was set equal to 1.0 for all females (Mackauer, 1983). In *Aphidius sonchi* the time from oviposition to adult eclosion was 11.3 days and mortality of immatures was 8.0%, so the probability of being alive at pivotal age 11.3 days +1 day was set equal to 0.92 (Liu, 1985b).

Once the values for l_x and m_x are calculated, then the following population statistics can also be calculated (Messenger, 1964b):

1. The **gross reproductive rate**, $\text{GRR} = \Sigma m_x$ (the mean total number of eggs produced by females over their lifetimes, measured in females/female/generation);
2. The **net reproductive rate**, or 'basic reproductive rate' (the number of times a population will multiply per generation) $R_o = \Sigma l_x m_x$ (measured in females/female/generation);
3. The **finite capacity for increase**, $\lambda = e^{r_m}$ (the number of times the population will multiply itself per unit of time; measured in females/female/day);
4. The **mean generation time**, $(T = (\log_e R_o)/r_m$ (measured in days);
5. The **doubling time** $(\text{DT} = \log_e 2/r_m$ (the time, measured in days, required for a given population to double its numbers).

Using the data in Table 2.2, $r_c = 0.289$, $r_m = 0.296$, $\text{GRR} = 108$, $R_o = 71.2$, $\lambda = 1.344$, $T_c = 14.74$ (see below for explanation of this parameter), $T = 14.41$, $\text{DT} = 2.24$. Statistical and computational aspects of the estimation of r_m are discussed by Maia *et al.*, (2000). These authors also provide an SAS program that uses the jacknife technique.

Table 2.2 Hypothetical life-table for an experimental cohort of female parasitoids. x is the mid-point of age intervals (pivotal age) in days, l_x is the fraction of the females surviving to age x (in this example we assume no deaths occurred during development, so the proportion of females surviving to commence ovipositing is 1.0), and m_x is the mean number of female 'births' during age interval x per female aged x.

x	l_x	m_x	$l_x m_x$
12.5	1.0	12	12
13.5	0.9	14	12.6
14.5	0.8	18	14.4
15.5	0.7	22	15.4
16.5	0.5	25	12.5
17.5	0.3	13	3.9
18.5	0.1	4	0.4
			$\Sigma l_x m_x = R_o = 71.2$

r_m can be measured (in female/female/day) for each of a range of host densities. It increases with increasing host density (Mackauer, 1983; Liu, 1985b). In *Aphidius smithi* this increase is also reflected in λ and also DT, which was less than half as long at the highest than at the lowest host density (Mackauer, 1983). Because in both *A. smithi and A. sonchi* the ovipositional pattern and the pattern of survival were similar to one another at the different densities (Figure 2.65), host density showed no significant effect on *T*.

To obtain a true measure of the influence of host density on the parasitoid's population statistics, some authors have based the m_x values on the number of hosts *actually parasitised* ('effective

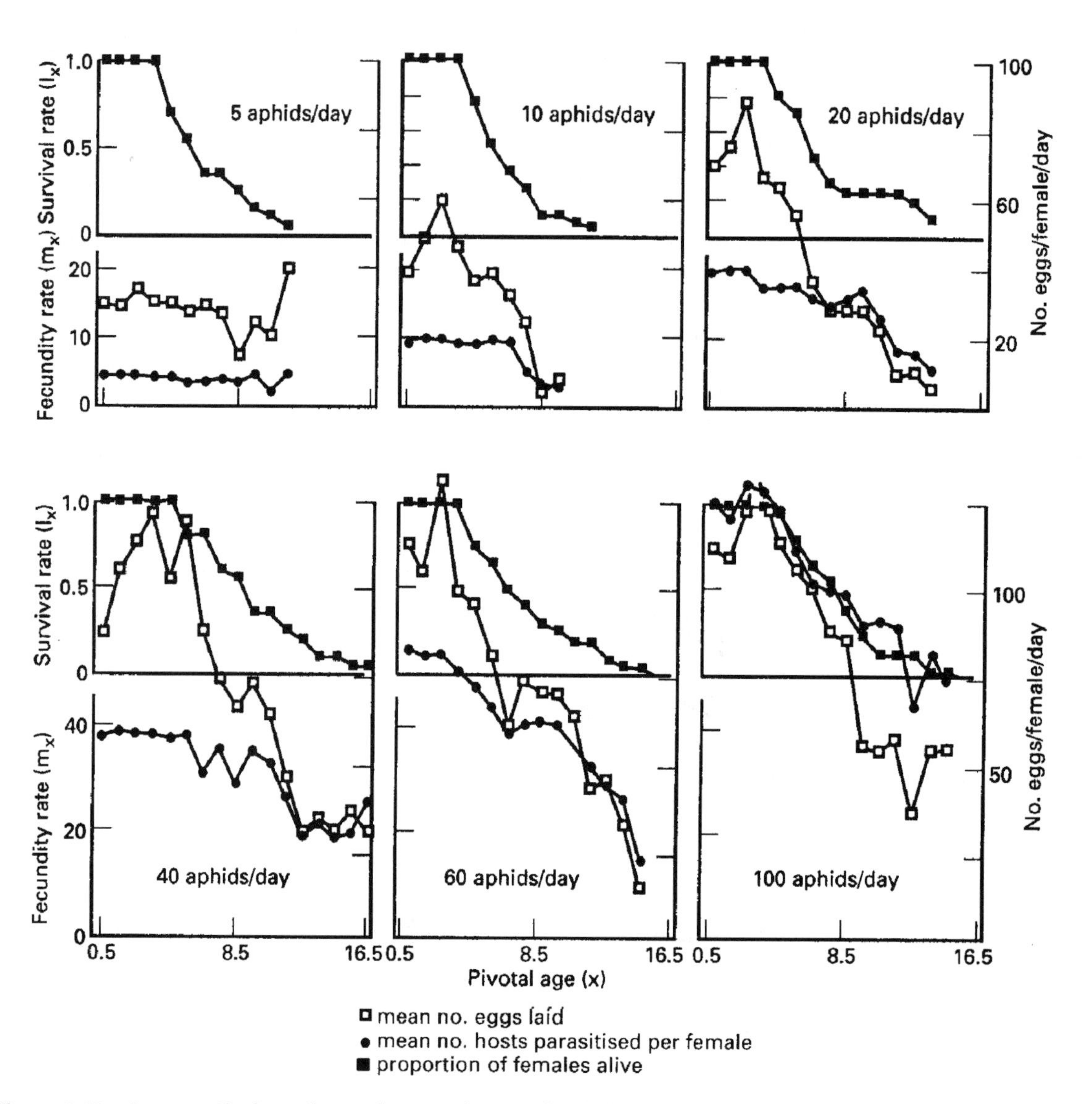

Figure 2.65 Age-specific fecundity and survival rates of *Aphidius smithi* provided with different densities of its host *Acyrthosiphon pisum*. Source: Mackauer (1983). Reproduced by permission of The Entomological Society of Canada.

eggs' of Messenger, 1964b). This takes account of superparasitism; thus the number of hosts parasitised can be assumed to equal the number of progeny eventually produced (ignoring cases where no parasitoid progeny succeed in developing in a parasitised host).

Another factor that needs to be taken into account is the sex ratio of the progeny. This can be achieved by multiplying all m_x values in the life-table by the **overall population sex ratio**, *P*, which is the proportion of females in all offspring produced. Regression of r_m on the natural logarithm of host density for different values of the sex ratio gives a series of parallel lines (Mackauer, 1983; Liu, 1985b; Tripathi and Singh, 1991) (Figure 2.66 shows regressions obtained for *Aphidius smithi)*. The variation in r_m as a function both of the parasitoid's sex ratio and of host density can be shown as a response surface (Figure 2.67 shows the response surface for *Aphidius sonchi)* (Mackauer, 1983, gives details of the statistical procedure involved in obtaining the response surface). As can be seen from Figure 2.69, r_m increases as either host density or sex ratio increases, and at a given value of *P*

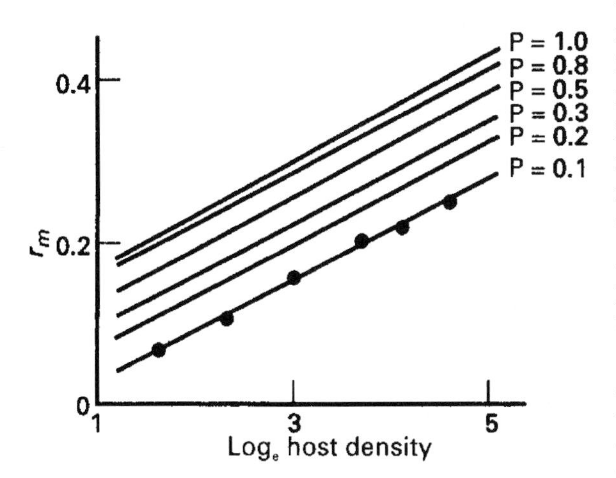

Figure 2.66 The relationship between the intrinsic rate of natural increase (r_m) *of Aphidius smithi* (Braconidae) and natural logarithm of host density, for different overall sex ratios. Source: Mackauer (1983). Reproduced by permission of The Entomological Society of Canada.

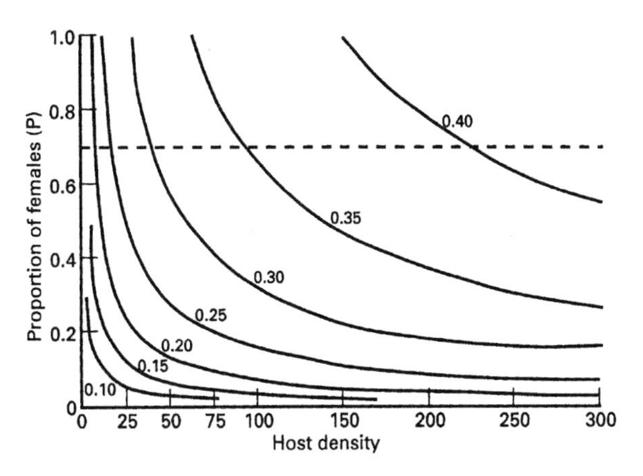

Figure 2.67 Response surface showing lines of equal r_m for *Aphidius sonchi* (Braconidae) for different host densities and parasitoid sex ratios (*P*, proportion of females). The broken line indicates a sex ratio of $P = 0.7$ observed in the laboratory. Source: Liu (1985b).

the rate of increase in r_m slows at higher host densities. In *A. sonchi* the deceleration in r_m at high densities is such that the percentage increase in host density required to obtain a given percentage increase in r_m is constant. For example, at $P = 0.70$, a 20% increase in r_m from 0.25 to 0.30 requires an increase in host density from 15 to 39 per day, i.e. 24 hosts, while a 20% increase in r_m from 0.30 to 0.36 requires an increase in host density from 39 to 101 per day, i.e. 62 hosts. This rule applies over the whole range of $0 \leq P \leq 1.0$, although the required increment in host density increases in absolute terms as the value of *P* declines. When $P = 0.40$, an increase of 48 hosts, from 30 to 78 per day, is required to obtain a 20% increase in r_m from 0.25 to 0.30.

The r_m of *Hyperomyzus lactucae*, the host of *A. sonchi*, is 0.3375. For a *P* value of 0.7, which is the sex ratio for *A. sonchi* in laboratory cultures, the parasitoid will achieve an r_m of 0.3378 at a host density of 74/day (preferably, the field sex ratio should be used in this computation, Mackauer, 1983). If the host density is increased to 200 per day, a sex ratio as low as 0.3 will yield an r_m of 0.3367, which is again close to that of the host.

Assuming an absence of superparasitism (which is typically higher at low densities), the parasitoid's realised m_x will be equal to its oviposition rate, so yielding values of r_m higher than those computed. The minimum host density required to eliminate egg wastage through superparasitism can be determined. Theoretically, at that density the parasitoid's r_m will reach a maximum value that can be computed by setting m_x equal to the daily totals of eggs laid at the highest oviposition rate (Mackauer, 1983, gives details of the statistical procedure involved).

Knowing how r_m varies in relation to factors such as host density (see above) and temperature (see below) can help biological control practitioners in deciding on the timing of introduction, for example in an inoculative release programme.

Equation 2.20 is not very 'transparent', that is, it is not particularly useful for any broad consideration of the relation between r_m and 'synoptic' life-history parameters such as generation times (Laughlin, 1965; May, 1976). A more useful statistic is r_c, **the capacity for increase**, which is an approximation for r_m. It is calculated as follows:

$$r_c = \frac{\log_e R_o}{T_c} \tag{2.21}$$

where T_c is the **cohort generation time**, defined as the mean age of maternal parents in the cohort at birth of female offspring (Laughlin, 1965; May, 1976) (for a discussion of the relationship between T and T_c, see May's paper):

$$(T_c = \sum_x l_x m_x / R_o)$$

Equation 2.21 is based on the assumption that the reproductive period is brief relative to the total life cycle, which results in a small error in the estimation of generation time. r_c is a good approximation for r_m when R_o and thus population size remains approximately constant, or when there is little variation in generation length, or for some combination of these two factors (May, 1976).

A relatively simple method for calculating values for r_c was developed by Livdahl and Sugihara (1984). It dispenses with the need to construct detailed survivorship and fecundity schedules, and uses indirect estimates of R_o and T_c. It assumes the organisms being studied to have a Type III survivorship curve for the whole life cycle, with high larval mortality and negligible adult mortality through the reproductive period; this assumption is only partly satisfied in the case of parasitoids, since in the laboratory there is likely to be low larval mortality while in the field there is likely, in many species, to be high mortality of females during the reproductive period. To use Livdahl and Sugihara's method, one only needs to observe cohorts during the maturation period in order to obtain measurements of the number of newly emerged adult females and their average size.

2.11.3 EFFECTS OF HOST/PREY SPECIES AND STAGE

Host stage and species, through their effects on body size in parasitoids, influence life-history variables such as fecundity and longevity, so they would be expected to affect r_m and r_c. This is indeed the case: see Cloutier *et al.* (2000) on r_m in *Aphidius nigripes*, and Yu *et al.* (1990) on r_c in *Encarsia perniciosi*. In both species, the intrinsic rate of natural increase/capacity for increase was higher in larger hosts.

Prey species can also be expected to influence the intrinsic rate of natural increase of predators, as has been confirmed, for example, by Venzon, (2002) for the bug *Orius laevigatus*.

2.11.4 EFFECTS OF TEMPERATURE

Since larval development rate, female survival and female fecundity vary with temperature (subsections 2.9.3, 2.8.4 and 2.7.3) we would expect r_m to vary also, which is the case. Figure 2.68 shows how r_m varies with temperature in three species of parasitoid and their aphid host (Force and Messenger, 1964).

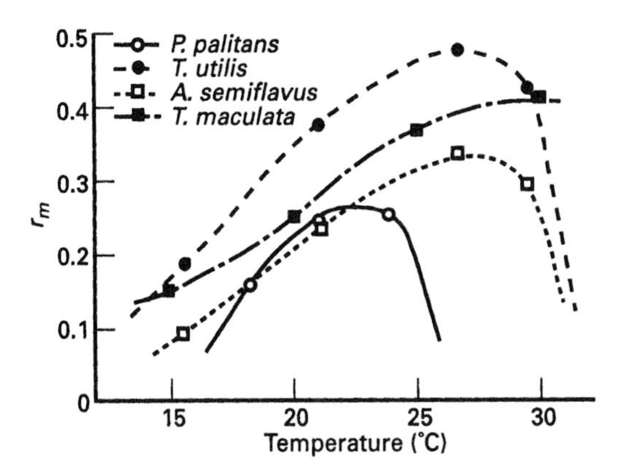

Figure 2.68 Comparison of intrinsic rate of natural increase (r_m) of the aphid parasitoids *Praon palitans*, *Trioxys utilis* (Braconidae), *Aphelinus semiflavus* (Aphelinidae) and their aphid host, *Therioaphis maculata*, over a range of constant temperatures. Source: Force and Messenger (1964). Reproduced by permission of The Ecological Society of America.

For examples of other studies, see Geusen-Pfister (1987) (*Episyrphus balteatus*), Cave and Gaylor (1989) (*Telenomus reynoldsi*), Lohr *et al.* (1989) (*Apoanagyrus lopezi*), Smith and Rutz (1987) (*Urolepis rufipes*), Mendel *et al.* (1987) (*Anastatus semiflavidis*); Miura (1990) (*Gonatocerus cinticipitis*) and Cocuzza *et al.* (1999b) (*Orius* spp.), Urbaneja *et al.* (2001b) (*Cirrospilus* sp.), Ren *et al.* (2002) (*Nephaspis oculatus*), Seal *et al.* (2002) (*Catolaccus hunteri*), and Roy *et al.* (2003) (*Stethorus punctillum*).

Siddiqui *et al.* (1973) provide a model to describe the relationship of $1/r_m$ to temperature. Using data for *Aphidius matricariae*, one of us (M.J.W. Copland) found the model to provide a good fit to the data over part of the temperature range only. A simple polynomial model could express the relationship much more accurately.

2.12 DORMANCY

2.12.1 INTRODUCTION

Life-cycles in most insects are characterised by profound season-related changes in growth, developmental and reproductive characteristics. Different species possess unique sets of ecophysiological responses that regulate seasonal cycles, facilitating temporal synchrony with seasonal variations in the availability and state of biotic and abiotic factors in their habitat (Tauber *et al.*, 1986; Tauber *et al.*, 1994). Understanding seasonal changes in the growth, development and reproduction of insect natural enemies is also an important tool in applied ecology. In particular, it is necessary to investigate the degree of synchrony between generations of pests and beneficial insects in order to determine the best strategies for successfully mass-rearing, storing and releasing biological control agents (van Lenteren, 1986; Chang *et al.*, 1996; Ringel *et al.*, 1998) (subsection 7.4.3). This is particularly true for parasitoids, most of which have very limited host ranges (Askew, 1971; Godfray, 1994; Quicke, 1997) and are therefore closely synchronised with successive generations of their hosts.

Particularly in temperate environments, insects enter a dormant state during unfavourable periods e.g. winter. Dormancy in insects occurs in a number of ways that differ both physiologically and ecologically. Types of dormancy are generally classified according to whether they are obligate and/or seasonally recurring (diapause and aestivation) or facultative in nature, occurring in direct response to certain stimuli (quiescence). Dormancy has been frequently reported among predators and parasitoids, where examples occur in all stages of development. Investigations of dormancy involving parasitoids are potentially much more complicated than in predatory insects, because host-parasitoid interactions occur over three, rather than two trophic levels. Furthermore, parasitoids are generally much more specific in their choice of hosts than predators are with their prey (Chapters 1 and 6). Factors which stimulate the onset of diapause, aestivation or quiescence in parasitoids may be perceived directly by the natural enemy or indirectly in response to physiological cues released by the host.

The dynamic effects and evolution of diapause in coupled parasitoid-host systems have been explored by Ringel *et al.* (1998) using theoretical modelling.

'Obligate' or 'Predictive' Dormancy: Diapause and Aestivation

Predictive dormancy is initiated in advance of adverse conditions, and most commonly occurs in response to predictable changes in seasonal environments (Müller, 1970). Two types of predictive dormancy have been described: **diapause** (during winter) and **aestivation, or summer diapause** (during summer). Tauber (1986) defines diapause as, a neurohormonally mediated dynamic state of low metabolic activity associated with reduced morphogenesis, increased resistance to environmental extremes, and altered or reduced behavioural activity. Diapause occurs during a genetically determined state of metamorphosis and generally in response to token environmental cues that precede the unfavourable condition.

The important point to bear in mind is that diapause-inducing stimuli are 'registered' commonly *before* the diapausing stage is reached. Diapause occurs in response to changes in, and interactions between, various biotic and abiotic factors including photoperiod, temperature, humidity and prey/host availability. Diapause termination also requires specific environmental conditions (Tauber *et al.*, 1993).

'Facultative' or 'Consequential' Dormancy: Quiescence

Quiescence is a reversible state of suppressed metabolic activity that occurs in response to environmental stimuli but does not involve preparatory hormonal or physiological changes in anticipation of environmental conditions. In many cases, timing and duration of quiescence are not fixed seasonally, but are highly variable and may last for many months or even years. The breaking of quiescence may require some kind of stimulation, signifying that the environment is favorable for development or activity.

Identifying, for a particular natural enemy, the nature of its dormancy, and establishing which biotic and physical factors play a role in its initiation, maintenance and termination, determining how these factors interact, and establishing which of the insect's life-stages are sensitive to predictive dormancy-inducing factors, can be very difficult, involving in some cases complex multifactorial experimental designs. Often, knowledge of the dormancy characteristics of related species can be helpful in simplifying experiments; for example it can help in narrowing down the list of candidate abiotic factors. We do not provide detailed advice on protocols here (for such information see the sources cited; the book, *The Ecology of Insect Overwintering* [Leather *et al.*, 1993], also should be consulted); instead, we provide a brief overview of diapause-inducing factors, supplemented with a few snippets of practical information.

2.12.2 EFFECTS OF BIOTIC FACTORS ON DORMANCY

Prey Availability and Quality (Predators)

Seasonal variation in the availability of prey has been reported to have a marked influence on the incidence of dormancy in predators. Aestivation in *Coccinella septempunctata* is stimulated by availability of suitable prey, as well as by a other factors (Kawauchi, 1985; Zaslavsky and Vagina, 1996). Polymorphic seasonal cycles in the lacewing, *Chrysoperla carnea*, are similarly influenced by the abundance of prey (Tauber *et al.*, 1986). In some predator species diapause incidence appears to depend on the type or the quality of the diet fed upon (Horton *et al.*, 1998). Since prey availability is, in many predators, likely to be linked to various biotic factors, it is important to try and devise an experimental design that enables the effects of the various factors to be disentangled (and interaction effects tested for), although this may be difficult or even impossible in many cases.

Host Physiology (Parasitoids)

Many parasitoids oviposit in nutritionally suboptimal early host stages, and their larvae exhibit **developmental arrest**, completing their development only after the host has moulted to the penultimate or even final instar (Vinson and Iwantsch, 1980a; Harvey *et al.*, 1994, 1999). Developmental delays are adaptive in a number of respects. Firstly, they reduce the selection pressure for a fixed maternal response at oviposition by allowing female parasitoids to attack a wide range of host stages rather than a single one (Cloutier *et al.*, 1991). Second, they ensure that the host reaches a critical size and physiological condition in which the parasitoid can complete its development (Hemerik and Harvey, 1999). Finally, they synchronise parasitoid and host generations intra- and interseasonally. There remains debate as to whether developmental delays are a form of diapause or of quiescence (Lees, 1955; Mellini, 1972; Godfray, 1994). Tauber *et al.* (1983) argue that host-mediated developmental arrest in the first instar parasitoid larva is a form of obligate diapause because it shares many characteristics associated with the diapause syndrome (see their paper, and also Doutt *et al.*, 1976).

Parasitoid larvae may also enter diapause in response to dormancy-related physiological changes in the host. For example, Polgar *et al.* (1991) and Christiansen-Weniger and Hardie (1997, 1999) examined factors influencing diapause induction in several braconid endoparasitoids attacking different morphotypes of their common aphid hosts (see also Polgar and Hardie, 2000). Parasitoids tended to enter diapause more in sexual hosts (oviparae) which occur in late summer, than in asexual hosts (virginoparae) which occur earlier in the season. Diapause appeared to be initiated mostly by hormonal differences between different aphid morphs. Diapause in idiobiont parasitoids has been reported to be influenced by the diapause status of their host in some associations (McNeil and Rabb, 1973; Strand, 1986), but not in others (Mackay and Kring, 1998).

The incidence of diapause among parasitoid progeny can vary with host species (Kraaijeveld and van Alphen, 1995b).

2.12.3 EFFECT OF PHYSICAL FACTORS ON DORMANCY

Photoperiod

Insect natural enemies, like other insects, are very sensitive to the duration and intensity of light exposure. In temperate regions, photoperiod is a major factor controlling diapause initiation, maintenance and termination in insects (Tauber *et al.*, 1983, 1986). Danilevskii (1965) defined the **'critical photoperiod'** as that which elicits a >50% response amongst individuals in a population.

Many heteropteran bugs overwinter in a state of reproductive diapause as adults, and typically diapause is induced by the photoperiod during nymphal development, although the adult stage may also be sensitive (see Yeargan and Barney, 1996; Ruberson *et al.*, 2000). The multivoltine coccinnellid *Coccinella septempunctata*, which is widely distributed over much of the Palearctic, undergoes aestivation as first generation adults, from April to August, in response to increasing day length (Sakurai *et al.*, 1986; Katsoyannos *et al.*, 1997) (this is immediately followed by a variable period of quiescence during winter). Photoperiod is also reported to be an important diapause-inducing stimulus for odonates (Norling, 1971; Pritchard, 1989).

In parasitoids, many studies have reported a key role for photoperiodic induction of dormancy (reviewed by Askew, 1971; Tauber *et al.*, 1983; Godfray, 1994; Quicke, 1997). Field sampling of hosts is a useful starting point for gathering tentative evidence of the role of photoperiod in diapause induction in bivoltine endoparasitoids (Jervis, 1980).

Temperature

Temperature is another important diapause-inducing stimulus for predators. In coccinellids, diapause may be stimulated by seasonal exposure to low temperatures (Kawauchi, 1985) or be due to an interaction of temperature and photoperiod (Ongagna and Iperti, 1994). The coccinellid *Rhyzobius forestieri* does not enter diapause, but the application of a cold shock at 8°C induces quiescence which can persist for several months if this condition is maintained (Katsoyannos, 1984).

In parasitoids, most studies have shown that temperature interacts with photoperiod in stimulating diapause induction (Brodeur and

McNeil, 1989; Pivnick, 1993; Polgar *et al.*, 1995), although some parasitoids may enter diapause in response to temperature alone (Wang and Laing, 1989). Temperature is also an important determinant in the breaking of diapause: for some species it may need to be low (amateur entomologists are well acquainted with the technique of 'chilling' insect pupae, in a refrigerator, for several weeks during the winter, before exposing them to warm indoor temperatures, to achieve a pre- or early-spring emergence of adults), whereas in others it may need to be high (e.g. see Hodek and Hodková, 1988; van den Meiracker, 1994; Ishii *et al.*, 2000).

To create more natural conditions in dormancy experiments, insects can be reared under gradually increasing temperatures (to stimulate the onset of summer aestivation) or gradually decreasing temperatures (to stimulate the onset of winter diapause).

The threshold temperature and the thermal constant (see subsection 2.9.3) for postdiapause development can be estimated for a parasitoid or predator (e.g. see Trimble *et al.*, 1990).

Moisture

Among the physical factors influencing dormancy in insects, the effects of moisture and/or humidity are the most poorly understood and least-studied. This is principally because the vast majority of phenological studies have been performed in the temperate zones, where photoperiod and temperature are considered, *a priori*, to play major roles. Evidence is accruing that moisture plays a vital role in the maintenance of dormancy in a range of predatory insects. For example, soil moisture, acting independently or in combination with photoperiod and temperature has been shown to influence rates of development or activity (Jayanth and Bali, 1993; Bell, 1994; Bethke and Redak, 1996; Sanon *et al.*, 1998; Nahrung and Merritt, 1999; see also review by Tauber *et al.*, 1998).

2.12.4 THE FITNESS COSTS OF DORMANCY

This is a little-explored area of insect natural enemy biology. Chang *et al.* (1996) revealed that post-diapause adults of the lacewing, *Chrysoperla carnea* experienced higher reproductive success than individuals which had overwintered in a state of quiescence. Moreover, first generation offspring of parents that had overwintered in diapause developed more rapidly and survived better than individuals whose parents had experienced quiescence. Ellers and van Alphen (2002) showed that in *Asobara tabida* an increase in diapause length led to higher mortality among diapausing pupae, together with decreases in egg load, fat reserves and dry weight of emerging adult females. See also Anderson (1962), on *Anthocoris nemorum*, and Leather *et al.* (1993) for a discussion of the costs of overwintering among insects generally.

2.13 INVESTIGATING PHYSIOLOGICAL RESOURCE ALLOCATION AND DYNAMICS

2.13.1 INTRODUCTION

This section is concerned with techniques used to study both: (a) the optimal strategy for the allocation, within the adult stage of parasitoids or predators, of carried-over physiological resources, i.e. those derived from the immature phase of the life-cycle; and (b) quantitative changes in these resources during adult life, in relation to variation in environmental factors such as food availability and quality, and host/prey abundance.

2.13.2 PATTERNS IN RESOURCE ALLOCATION

Intra- and interspecific differences in the pattern of resource allocation are of considerable interest, as they help ecologists and evolutionary biologists to understand why individuals and species differ in terms of key life-history traits. Negative correlations between the amounts of resources serving different life-history functions such as egg production and survival are particularly intriguing, as they imply the existence of trade-offs, and as such are evidence that life-histories are compromises. An associated goal of ecologists is to understand the integration of suites of life-history traits, and as is becoming apparent from the literature, studying patterns of resource allocation is the way forward in this quest.

Testable hypotheses relating to resource allocation include the following:

1. All else being equal, species whose females are longer-lived and which have higher resource intake prospects should invest more in building a 'sturdy body' or 'soma' (musculature and exoskeleton) at the expense of 'abdominal reserves' (principally reproductive organs and their contents i.e. eggs, together with fat body) (Boggs, 1981). Empirical support for Boggs' hypothesis comes from her study of three species of heliconiine butterflies (Boggs 1981) (see also Karlsson and Wickman, 1989; Wickman and Karlsson, 1989);
2. Among abdominal 'reserves' there will be a trade-off between those resources allocated to initial egg production and those allocated to survival (fat body and other reserves). This is predicted by general life-history, on the basis of between-function competition for limited resources (see Bell and Koufopanou, 1986; Van Noordwijk and De Jong, 1986; Smith, 1991). Empirical support for this hypothesis comes from the known differential allocation of carried-over larval resources to fat body storage and initial egg load in the parasitoid wasp *Asobara tabida* (Ellers and van Alphen, 1997).
3. As body size increases in parasitoid wasps, the total amount of 'abdominal reserves' increases, and allocation to both initial eggs (initial egg load) and stored reserves increases, but the increase in allocation to initial eggs is proportionately smaller than the increase in allocation to initial reserves. For an explanation of the adaptive significance of these relationships, and how they relate to ovigeny index, see Ellers and Jervis (2003).
4. Smaller parasitoid wasp individuals suffer disproportionately, in terms of survival, the costs of not feeding, because they emerge with smaller initial reserves. This is supported by Rivero and West's (2002) study of *Nasonia vitripennis* (see subsection 2.8.3).
5. Solitary parasitoid species should allocate relatively more resources to survival (as fat reserves) than gregarious species (Pexton and Mayhew, 2002; the hypothesis is based on optimal allocation theory, e.g. see Roff, 2002). This was supported by Pexton and Mayhew's study of two *Aphaereta* species.
6. Mothers should reduce egg provisioning with age (Begon and Parker, 1986; Roff, 2002). This is supported by Giron and Casas's (2003b) study of *Eupelmus vuilletti*.

2.13.3 RESOURCE DYNAMICS

Quantitative changes in resources will occur during adult life, in relation to environmental factors such as extrinsic nutrient availability and quality, and host/prey availability. Behavioural ecologists in particular are interested in these changes because they know foraging decisions to be physiologically state-dependent (Chapter 1), and they appreciate that foraging, mating behaviour and other activities (including dispersal) are constrained by nutrient (intrinsic and extrinsic) supply. Hypotheses relating to resource dynamics in insect natural enemies have been tested by Ellers *et al.* (1998, 2001), Olson *et al.* (2001), Rivero and West (2002), Ellers and van Alphen (2002), Giron and Casas (2003a) and Casas *et al.* (2003) (parasitoids), and Otronen (1995) (the predatory fly, *Scathophaga stercoraria*) (see also Legaspi *et al.*, 1996, on the predatory bug *Podisus maculiventris*).

2.13.4 THE TECHNIQUES

The techniques

Measuring Allocation, Among the Total Carried-over Resources, to 'Soma' and 'Abdominal Reserves'

This can be done by measuring the dry weight, the total nitrogen content, and the total carbon content of: (a) the head + thorax + legs + wings (collectively "soma", *sensu* Boggs 1981); and (b) the abdomen ("abdominal reserves" resource pool, *sensu* Boggs 1981).

The total amount of nitrogen in each body region can be measured using Kjeldahl digestion and subsequent Nesslerization (Minari and Zilversmit 1963), while the total amount of carbon can be measured using bomb calorimetry. Better still is elemental analysis using a CHN analyser.

Measuring Allocation, Among 'Abdominal Reserves', to Initial Egg Production and Survival, and Studying Resource Dynamics

Measuring resource allocation to initial egg production, and also the subsequent qualitative and quantitative changes that occur in reproductive tissues, can be determined using modifications of well-proven techniques (Van Handel, 1984, 1985a,b; Van Handel and Day, 1988; Olson *et al.*, 2000). Except in the case of small-bodied species (in which case separate individuals would have to be used), ovary protein content can be determined for one ovary, and both lipid and glycogen content determined for the other ovary. The Bradford dye-binding colorimetric micro-assay (Bradford, 1976) can be used for protein measurement, and lipid and glycogen measurement can be done using modifications of colorimetric techniques (vanillin reaction and chemical precipitation followed by hot anthrone reaction, Van Handel, 1985a,b; Van Handel and Day, 1988).

Measuring allocation to energy reserves, and also measuring alterations in the amounts of these resources, would involve measuring the quantities of lipid, glycogen, and stored sugars in the ovary-less abdomen (this would include haemolymph). Lipid and glycogen content can be measured as for the ovaries (see above); an alternative method of lipid measurement is ether extraction (Ellers, 1996; Eijs *et al.*, 1998; Ellers and van Alphen, 1997, 2002). Stored sugar content can be measured using the hot anthrone reaction (Olson *et al.*, 2001; Fadamiro and Heimpel, 2001).

The strategy of allocation, both from among the pool of 'abdominal reserves', could be influenced by: (a) nutrient intake prospects (see Chapter 8); (b) egg resorption capability (subsection 2.3.4); (c) thoracic musculature resorption capability (see Kaitala, 1988, and Kaitala and Huldén, 1990, for an example of flight muscle resorption in water-striders, and see Kobayashi and Ishikawa, 1993, for histological methodology); or (d) combinations of these (see Jervis and Kidd, 1986). Unless it is already one of the variables under consideration, body mass will need to be included as a covariable in data analyses. Phylogeny-based statistical methods (subsection 1.2.3) should be employed in the case of interspecific comparisons.

If one is interested in knowing the total level of energy reserves within an insect, these can be calculated by adding the energy content of carbohydrate to that of lipids, assuming 16.74 Joules per milligram of carbohydrate and 37.65 Joules per milligram of lipid (Casas *et al.*, 2003).

By studying carbohydrate and lipid dynamics in both field and laboratory experimental populations (freshly emerged, starved to death, fed *ad libitum*, partially starved), Casas *et al.* (2003) were able to show that *Venturia canescens* females are able to maintain a nearly constant level of energy over an extended foraging period, that they take sugars in the field, and also that lipid reserves may be limiting as lipogenesis does not occur in adults even under conditions of high sugar availability (all parasitoid wasps studied so far are unable to synthesise lipids from sugars in meaningful quantities, see Giron and Casas, 2003a).

2.14 TRACKING RESOURCES

Radiotracer studies, which been applied to other insects (e.g. see Boggs, 1997b, on Lepidoptera), are now being used to study the utilisation of extrinsic nutrients by parasitoid wasps (Rivero and

Casas, 1999; Giron *et al.*, 2002; Giron and Casas, 2003a). Rivero and Casas (1999) fed females of *Dinarmus basalis* on an artificial diet comprising a sugar + radiolabelled (^{3}H) amino acid solution. The liquid food was supplied in a capillary tube, and the weight of females was compared before and after feeding, so allowing the amount of radioactivity both in the insects themselves and in the eggs they laid to be related to the amount of food ingested. It was found that the maximum incorporation, into eggs, of labelled nutrients obtained via a discrete feeding event, occurred with a short period of time. However, it was also found that a large proportion of the nutrient input is stored and used gradually throughout the life of the parasitoid.

2.15 ACKNOWLEDGEMENTS

We wish to thank Jacques van Alphen, Francis Gilbert, Neil Kidd, Manfred Mackauer, Mark Shaw and David Thompson for commenting on parts of the original chapter, Mike Majerus for kindly answering some queries, John Morgan and Kevin Munn for providing technical assistance, and Michael Benson for kindly allowing us to quote from his Masters thesis.

GENETICS 3

L.W. Beukeboom and B.J. Zwaan

3.1 INTRODUCTION

Genetical research on insect natural enemies is rare and basically limited to parasitoid wasps and some coccinellid beetles. To our knowledge, no genetic studies have been done on other well-known groups such as parasitoid flies, robber-flies, scorpion flies or mantids. Even among the well-studied parasitic Hymenoptera, genetics has largely been neglected. For example, sex allocation strategies (see 1.9) have been intensively studied in this group, but hardly anything is known about the underlying genetic mechanism of sex determination. In many organisms there is geographic variation in the expression of biological traits, and in parasitic Hymenoptera such variation has been reported for sex ratio, host preference, oviposition strategies, superparasitism, diapause induction, virulence, developmental time, and other traits (see subsection 6.5), yet despite this broad suggestive evidence for the existence of genetic variation for many traits, relatively little is known about the underlying genetics.

A number of different reproductive modes are known from insects which have important consequences for the way in which traits are inherited. Whereas sexual reproduction leads to genetically heterogeneous offspring, certain forms of parthenogenesis lead to completely homozygous progeny within a single generation. Therefore, genetic knowledge of the reproductive modes of species is necessary for interpreting the genetic basis of variation in their life-history traits.

Many insect natural enemies are economically important as biological control agents, and knowledge of the genetic basis of traits is indispensable to biocontrol practitioners. Many traits may deteriorate rapidly through inbreeding when animals are taken into culture. Therefore, optimised culturing techniques and selection experiments may improve natural enemy efficiency. For many traits including morphology (e.g. wing size), physiology (e.g. the ability to escape the host's defence mechanism), biochemistry (e.g. insecticide resistance) and behaviour (e.g. activity patterns and mating preferences), a single gene (or a small number of major genes) has been shown to be responsible. These results are typically obtained from controlled genetic crosses and simple Mendelian genetics. Although still elaborate, molecular genetics nowadays makes it possible to isolate and characterise such genes. However, many other traits, such as life-history traits, appear to be coded by many genes each having a small effect (polygenic control). The study of such traits is the field of quantitative genetics, an area that is currently progressing rapidly through development of new molecular and computational techniques. In this chapter we present available genetic data of insect natural enemies with emphasis on new research approaches and techniques.

3.2 GENETIC MARKERS AND MAPS

3.2.1 INTRODUCTION

Genetic markers are alleles of different phenotype that can be used to monitor the inheritance of genes in genetic crosses. They are indispensable in the study of population genetics. Genetic structures of populations have traditionally been studied using enzyme electrophoresis. However, this technique is limited in the level of variation that it can detect. Owing to the recent progress in molecular biology, in particular the development of the Polymerase Chain Reaction (PCR), more sensitive DNA techniques have become available, such as Random Amplified Polymorphic DNA (RAPD) and Amplified Fragment Length Polymorphism (AFLP). Here, we will mention only briefly the main techniques, and refer to other works for practical details of the methods and data analysis. Applications of the techniques will be discussed throughout the chapter if appropriate and some are also discussed in chapters 5 and 6. Practical guidelines for allozyme electrophoresis are given by Steiner (1988), May (1992) and Avise (1994). A more extensive overview of molecular genetic techniques and applications in insects is given by Hoy (1994). Practical guides for DNA techniques include Sambrooke *et al.* (1989) and Diefenbach and Dveksler (1995). Books concerned with methods for analysing molecular and population data are those of Weir (1990), Hoelzel (1992) and Hartl and Clark (1989). Principles of genetic mapping are dealt with by Primrose (1995).

3.2.2 GENETIC MARKERS

Protein Markers

Allozymes

Enzyme electrophoresis is based on the separation of charged protein molecules in an electric field. The migration rate of a protein is not only related to its charge, but also to its size and shape. If the protein products of two alleles have different charges, their rates of migration will be different. This will be visible on a gel as separate bands after staining. Such allelic forms of the same locus are called **allozymes**. Most proteins examined are enzymes that catalyse specific biochemical reactions. They can consist of one single polypeptide unit (= **monomer**), two units (= **dimer**) or multiple units (= **polymers**). In addition, there can be more than one locus coding for the enzyme. Figure 3.1 shows the possible banding patterns of one-locus mono- and dimeric enzymes in homozygous or heterozygous state. More complicated patterns are explained in May (1992).

The process of enzyme electrophoresis can be divided into five steps (May, 1992): (1) *Extraction* of proteins from tissue samples which should be frozen at −70°C since most enzymes degrade easily; (2) *Separation* of allozymes in an electric field, which is applied in a starch or polyacrylamide gel and appropriate buffer; (3) *Staining* of the enzymes with specific substrates and dyes; (4) *Interpretation* of the observed banding patterns (Figure 3.1); (5) *Application* of the data to the specific research question.

A number of parameters can be used to quantify the observed variation. The **degree of polymorphism (*P*)** is the proportion of loci on which multiple bands are found. The **degree of heterozygosity (*H*)** is the proportion of loci that are heterozygous. The **genetic distance (*D*)** is a

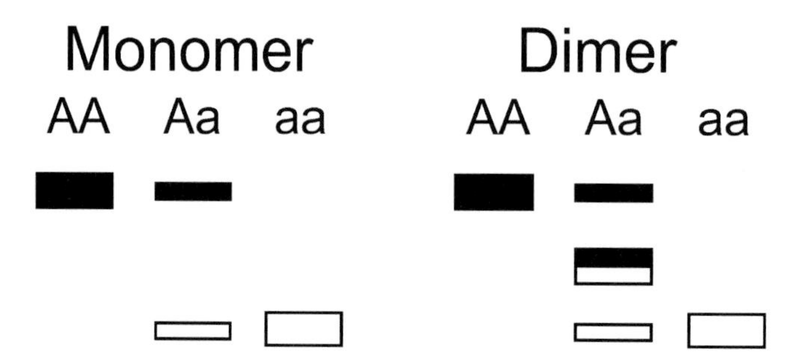

Figure 3.1 Possible allozyme banding patterns for one locus with two alleles (A, a). Both homozygotes and the heterozygote for a monomeric and dimeric enzyme are shown.

measure of gene diversity between populations and is estimated using genotype frequencies (see for details Hoelzel and Bancroft, 1992).

Previously, allozymes were used extensively in the study of insect natural enemies, but DNA-based techniques are more in vogue. Unruh *et al.* (1986) summarised measures of genetic variation for 66 species of Hymenoptera, and found that they maintain only about one-third of the level of heterozygosity of diploid insects (H = 0.037 *vs* 0.120). Parasitoid wasps were not found to be different from other Hymenoptera. Enzyme electrophoresis remains a useful technique for detecting endoparasitoids, measuring degrees of population differentiation and for distinguishing between strains or species.

Molecular Markers

Restriction Fragment Length Polymorphism (RFLP)

Restriction fragment length polymorphisms are obtained by digestion of DNA with restriction enzymes that cut DNA at specific recognition sequences (usually 4-6 base pairs in length). Fragments are visualised through standard DNA electrophoresis, blotting and probing. The technique requires large amounts of DNA, although it can also be used together with PCR amplification. It is currently used mostly to detect variants among known DNA segments, such as the mitochondrial genome (e.g. Vanlerberghe-Masutti, 1994).

Random Amplified Polymorphic DNA (RAPD)

The Random Amplified Polymorphic DNA technique (Williams *et al.* 1990) is a DNA polymorphism assay based on the PCR amplification of random DNA segments with single primers of arbitrary nucleotide sequence. Primers are short oligonucleotides that can attach to a single stranded DNA molecule to initiate DNA replication. The RAPD technique relies on the statistical chance that two complementary primer sites occur in the genome as inverted repeats enclosing a relatively short stretch of DNA (up to a few thousand base pairs). **Primers** are generally short (decamers) and their sequences arbitrarily chosen. This technique detects DNA polymorphisms in the absence of specific nucleotide sequence information. RAPD bands are inherited as dominant traits, so heterozygotes cannot be identified readily (Figure 3.2). Specific

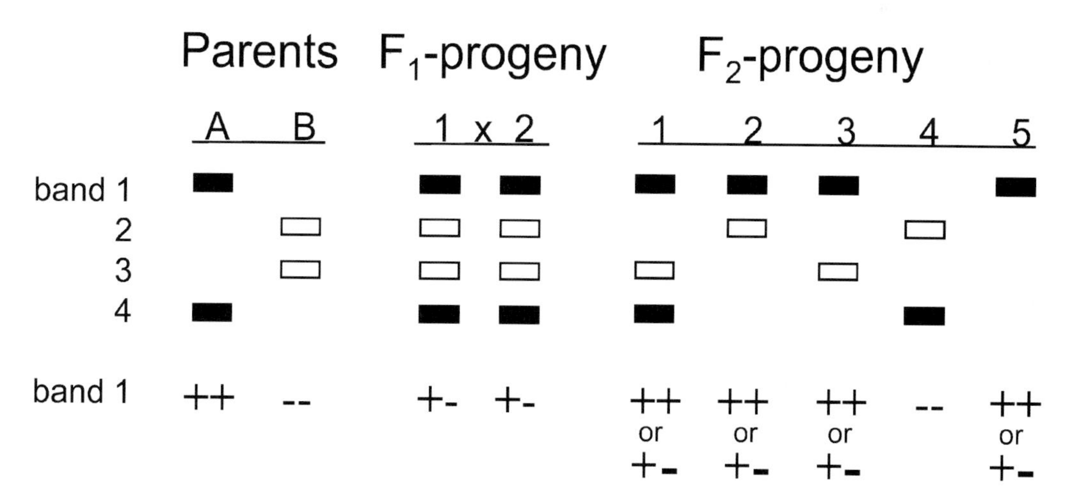

Figure 3.2 Inheritance of RAPD markers. Four bands are shown in a cross between two homozygous strains (A,B). All F_1-offspring are heterozygous for all bands and crossing of two such offspring yield F_2 progeny in which each band segregates in a 1:2:1 ratio. Five possible banding patterns of F_2 offspring are shown. Note, however, that because of dominance heterozygotes cannot be distinguished from homozygotes as shown for band 1 (+ = presence and − = absence of band).

advantages of the RAPD technique are its simplicity and rapidity: (1) only small amounts of DNA are required; (2) a universal set of primers can be used for many species; (3) no preliminary sequence information is needed; and (4) the number of primers is virtually unlimited, making detection of genetic polymorphism in even highly monomorphic populations possible. Its main drawback is its low reliability. Virtually all reaction steps and components can affect its reproducibility, including DNA extraction method, DNA and primer concentration, type of Taq polymerase and brand of thermocycler. Therefore, careful standardisation of conditions is necessary. The RAPD technique is especially useful for relatively simple tasks, such as species and strain identification, detection of parasitism and paternity analysis.

Amplified Fragment Length Polymorphism (AFLP)

The Amplified Fragment Length Polymorphism technique (Vos *et al.*, 1995) detects variation in the presence of genomic restriction fragments. It is based on restriction digests of DNA and PCR amplification of a subset of restriction fragments, and can as such be considered as a hybrid between the RFLP and RAPD techniques. It also combines the advantages of both techniques. It requires only small amounts of DNA, and no previous information about DNA sequences. Nevertheless, it is much more reliable than the RAPD technique because it is based on DNA restriction digests. DNA is digested with two enzymes, typically *Eco*RI and *Mse*I, which generate small DNA fragments (< 1kb). Fragments are ligated (by formation of a phosphodiester bond using the enzyme ligase) to common adapter sequences (short synthetic oligonucleotides) which serve as primer binding sites. Genomic fragments are amplified in two consecutive PCR reactions, a pre-amplification round using primers with one, and a selective amplification using primers with two or three selective nucleotides. The *Eco*RI selective primer is labelled and the PCR products visualised after electrophoresis. The AFLP technique is very powerful because it can detect many different bands within a single reaction. Moreover, varying the number and nature of selective bases can control the number of bands. Although the AFLP technique has been developed for plants, it appears equally useful for animals (e.g. Schneider *et al.*, 2002). It seems particularly promising for studies that require many polymorphic bands, such as studies of the genetic structure of populations and also paternity analysis. It is the preferred technique for establishing genetic linkage maps.

Minisatellites

Minisatellites consist of arrays of 15-60 bp which differ in copy number and are scattered throughout the chromosomes of many organisms. They are used for multi-locus DNA fingerprinting. They can be visualised using sequence specific probes in Southern blot assays after restriction digesting total genomic DNA (Bruford *et al.*, 1992). Many tandem repeat units contain a common core sequence in different organisms which makes it possible to use similar probes in different insect species. DNA fingerprinting based on minisatellites can be used to identify individual insects or their progeny, as well as for parentage analysis. The absence of knowledge about which bands belong to which loci makes the technique less convenient for estimating population genetic parameters such as genetic relatedness.

Microsatellites

Microsatellites are short DNA sequences of 2-8 base pairs that occur as arrays of several hundred copies in the genome of many organisms. They are among the most variable DNA sequences and are used extensively for DNA fingerprinting (Bruford *et al.*, 1992). They can be visualised with standard PCR and electrophoresis techniques using primers that are specific for the sequences of the flanking regions. Their main drawback is the large effort involved in obtaining such sequence information, because it requires construction of a genomic library, screening of clones for repeat sequences and subsequent sequencing of clones. Microsatellite-based fingerprinting is the most reliable method for identification of progeny and paternity analysis.

DNA Sequencing

DNA sequencing facilities are rapidly becoming broadly available. Sequencing DNA can yield information about variation at the nucleotide level. Because sequencing is both expensive and time-consuming, the fraction of the genome that can be analysed is limited. Which part of the genome will be sequenced is strongly dependent on the research question posed. DNA sequencing is typically applied in studies of variation at a higher level than the individual. Ribosomal DNA and the mitochondrial cytochrome genes are frequently used in phylogenetic analyses intended to separate strains or species (Field *et al.*, 1988; Gimeno *et al.*, 1997).

Phenotypic Markers

Phenotypic markers (or visible mutations) have traditionally been used in genetic studies of insects. The most abundant ones are eye-colour, body-colour and other morphological mutations. Eye-colour and body-colour mutations are known from many parasitoid wasps (e.g. Whiting, 1932, 1934, 1954; Baldwin *et al.*, 1964; Saul *et al.*, 1965; McInnis *et al.*, 1986). Examples of morphological mutants include a wingless mutant in the two-spot ladybird *Adalia bipunctata*, which lacks all or part of the elytra and flight wings (Marples *et al.*, 1993) and several "vestigial", "shrivelled" and "stubby" wing and "club" antennal mutants in parasitoid wasps (Baldwin *et al*, 1964; Whiting, 1935). Saul *et al.* (1965) described mutants for each of five linkage groups in the chalcidoid wasp *Nasonia vitripennis*. Most of the stocks of these mutants are now maintained in Rochester (USA) and Groningen (The Netherlands).

Mutants are generated by X-rays or chemical mutagenesis, such as ethylmethanesulfonate (EMS) administered during larval or adult stages. Protocols developed for *Drosophila* (Grigliatti, 1986) can readily be used for other organisms. Detailed information on dose-response curves of X-rays in *Nasonia* are given by Kayhart (1956) and Whiting (1967). They obtained eye-colour mutants with a frequency of up to 1%. Mutations are typically recessive to wildtype genes. Some are female sterile and less viable in homozygous state which requires specific crosses and propagation through heterozygous females. In haplodiploids, they can easily be detected in haploid males (see below).

3.2.3 BASIC GENETICS

The genetic constitution of an organism is called its **genotype**. It is based on the total amount of genetic material in a cell which is organised as a collection of genes on separate chromosomes that together make up the genome. An organism's **phenotype** is the physical manifestation of its genetic traits that result from a specific genotype and its interaction with the environment. When reproducing sexually, new combinations of genes and chromosomes can be generated through a number of processes: (1) *recombination between genes* through crossing-over between chromosomes; (2) *random assortment of chromosomes* in gametes during meiosis; and (3) *fusion of two different haploid genomes* contained in gametes of two parents (egg and sperm). If, in a diploid cell, two identical copies of a gene are present, the condition is termed **homozygous**, whereas the occurrence of two different copies is termed **heterozygous**. In a heterozygous cell, a dominant form of a gene (called allele) is expressed but the recessive form is not. Recessive alleles are only expressed in homozygous state. Dominant and recessive states are the extremes of a continuous scale and many genes are at least partly co-dominant (see section 3.5.2).

One-point Cross

The simplest genetic organisation of a trait is that it is coded by a single gene. Examples are most eye-colour mutants which are caused by mutations in single recessive genes. A first approach to reveal the genetic basis of a trait is to cross two individuals that carry different forms of the trait. Figure 3.3 shows an example of the inheritance of a single-gene trait, vestigial wings, in a diploid and a haplodiploid organism. The wild-type form is dominant and therefore F_1 heterozygotes have wildtype eyes. In diploids, crossing of two heterozygous F_1 individuals yield homozygous wildtype, heterozygous wildtype and homozygous vestigial winged individuals at frequencies of 1:2:1, which in this case translates into phenotypic ratios of 3 wildtype : 1 vestigial. In haplodiploids (males are haploid and females are diploid), F_1 females can reproduce unmated, resulting in haploid males with wildtype and vestigial wings in a ratio of 1:1. In haploids, there is no dominance/recessiveness, and genotypic and phenotypic ratios are equal.

An illustrative example of how to detect a recessive trait controlled by a single gene is the wingless morph in the diploid two-spot ladybird *Adalia bipunctata*, studied by Marples *et al.* (1993). These authors found a male with the elytra completely absent and the flight wings reduced to small buds. They mated it to a normal winged female. The 12 resulting F_1 offspring were all fully winged. This observation is consistent with a trait controlled by a single gene, but cannot yet exclude multiple genes or that the wingless individual had simply been damaged. Therefore, the authors mated F_1 individuals among themselves and obtained 14 wingless and 34 winged F_2 offspring, which is consistent with the 1:3 ratio expected for a single recessive allele.

Two-point Cross

Figure 3.4 shows an example of the inheritance of two single-gene traits, vestigial wings and ebony eye-colour with independent assortment. Both mutant alleles are recessive as in the one-point cross above. The test cross, which is typically used in gene-mapping studies (see below), involves mating the F_1 offspring to one of the parental lines in the diploid case and allowing virgin reproduction in the haplodiploid case. Four types of offspring are produced in equal numbers under both reproductive modes, i.e. each combination of the two traits.

Many quantitative traits are coded for not by a single gene but by several genes. Figure 3.5 shows the inheritance of a wing size trait that is based on two genes. The assortment of genes is similar as in Figure 3.4, but the phenotypes are different. This is caused by the fact that alleles on both loci are additive, i.e. their effects add up (see also Figure 3.19). Homozygous wildtype individuals have long wings whereas individuals homozygous for the "s" form have short wings. The heterozygous F_1 offspring are of intermediate wing length. In the diploid case, crossing two heterozygous F_1 individuals yields three types of intermediate forms (75%, 50% and 25% short) in addition to the parental short and long phenotypes (Figure 3.5a). Note that when genes are additive, these F_2 ratios are informative about the number of genes that underlie a certain trait (see Roff, 1997). In the haplodiploid case, one can breed a single unmated female. This results in males with three possible phenotypes (short, intermediate and long) in a ratio of 1:2:1 (Figure 3.5b).

Figure 3.6 shows the wing size distributions (corrected for body size) of parents and F_2 males of a cross between two haplodiploid *Nasonia* sibling species, the short winged *N. vitripennis* and the long-winged *N. giraulti* (Weston *et al.*, 1999). The F_2 distribution is intermediate to the parental ones, but still clearly bimodal. The 44% wing size variation in these species can be explained by a single major gene, but there are also several genes with minor effect (see section 3.5.5 below).

Three-point Cross

Thus far, we have considered genes to inherit independently from each other. However, genes that are located on the same chromosome may be linked and frequently segregate together. Only through the formation of cross-overs between genes on a single chromosome may such genes be uncoupled. Therefore, the frequency

	a			b		
	Diploid			**Haplodiploid**		
Parental strains	male	x	female	male	x	female
Genotype	*vg / vg*	x	+ / +	*vg*		+ / +
Phenotype	*vestigial*		wildtype	*vestigial*		wildtype
Gametes	*vg*		+	*vg*		+
F1 progeny						
Genotype	*vg* / +			*vg* / +		
Phenotype	wildtype			wildtype		
Test cross	F_1	x	F_1	virgin female		
Genotype	*vg* / +		*vg* / +	*vg* / +		
Gametes	+		+	+		
	vg		*vg*	*vg*		
F2-progeny	males + females			males only		
Genotype	+ / +	+ / *vg*	*vg / vg*	+		*vg*
Phenotype	wildtype	wildtype	*vestigial*	wildtype		*vestigial*
Expected ratio	1	2	1	1		1

Figure 3.3 Example of a one-point cross for a diploid (a) and haplodiploid (b) organism. A cross is performed between a male with mutant vestigial wings and a female with normal wildtype wings. The mutant *vestigial* allele is recessive, hence all F_1 offspring are genotypically heterozygous but phenotypically wildtype. In the diploid (a), the F_1 are crossed among themselves. The F_2 offspring consist of three genotypic classes in proportions 1:2:1 (i.e. homozygous +, heterozygous + *vg* and homozygous *vg* respectively), but only two phenotypic classes in proportions 3:1 (i.e. wildtype and *vestigial* respectively). In the haplodiploid (b), the F_1 virgin female produces haploid wildtype and *vestigial* sons in equal ratio.

	a		b	
	Diploid		**Haplodiploid**	
Parental strains	male A x	female B	male x	female
Genotype	*vg e / vg e* x	+ + / + +	*vg e*	+ + / + +
Phenotype	*vestigial ebony*	wildtype	*vestigial ebony*	wildtype
Gametes	*vg e*	+ +	*vg e*	+ +
F1 progeny				
Genotype	*vg e* / + +		*vg e* / + +	
Phenotype	wildtype		wildtype	
Test cross	F_1 x	A virgin female		
Genotype	*vg e* / + +	*vg e / vg e*	*vg e* / + +	
Gametes	+ +	vg e	+ +	
	vg +		*vg* +	
	+ *e*		+ *e*	
	vg e		*vg e*	
F2-progeny	males + females		males only	
Genotype	+ + / *vg e*	*vg* + / *vg e*	+ +	*vg* +
Phenotype	wildtype	*vestigial*	wildtype	*vestigial*
Expected ratio	1	1	1	1
Genotype	+ *e* / *vg e*	*vg e / vg e*	+ *e*	*vg e*
Phenotype	*ebony*	*vestigial ebony*	*ebony*	*vestigial ebony*
Expected ratio	1	1	1	1

Figure 3.4 Example of a two-point cross for a diploid (a) and haplodiploid (b) organism. A cross between a male of strain A with mutant *vestigial* (*vg*) wings and *ebony* (*e*) body colour, and a female with normal wildtype wings and body colour. The mutant *vestigial* and *ebony* alleles are recessive, hence all F_1 offspring are genotypically heterozygous and phenotypically wildtype. In the diploid (a), the F_1 are backcrossed to strain A. The genes must be located on different chromosomes because the four possible F_2 genotypes and phenotypes occur in a 1:1:1:1 ratio. In the haplodiploid (b) the F_1 females are bred as virgins and the F_2 males have four different genotypes and phenotypes in proportions 1:1:1:1.

at which two linked genes inherit independently is used as a measure for recombination. Recombination frequencies, in turn, are informative about the physical distance between genes (so called **map-distances**). Figure 3.7 shows an example of how to estimate recombination frequencies and establish map distances in a cross with three linked genes in *Drosophila* (see below).

Iso-female Lines

A powerful way of demonstrating the presence of genetic variation for a trait is through iso-female lines (Parsons, 1980). An **iso-female line** is a laboratory strain that is bred from a single mated female. In haplodiploid species, such as parasitoid wasps, a strain can be started from a single female if she can be kept alive long enough to mate with one of her sons. Typically, an array of lines is established from field-collected animals, and they are scored for a trait under identical laboratory conditions. The between-line variance component is a measure for the amount of genetic variation for the trait. One drawback of this method is that it results in inbreeding which can have a number of negative effects on fitness, particularly in diploid species. This is especially a problem in life-history research because inbreeding tends to cause spurious positive correlations between sets of traits (the best iso-female line is best in most traits). A second drawback is the fact that the estimate of genetic variation obtained from iso-female line comparisons does not distinguish between additive and dominance genetic variation (see section 3.5).

Artificial Selection

Another valuable genetic tool is artificial selection. Artificial selection can take place in populations in which individual variation has at least partly a genetic basis and by selectively breeding those individuals with the most extreme phenotype for the trait. For example, in a population of differently sized individuals, the largest can be chosen each time as parents to found the next generation. As generations pass, the selected line will become enriched for alleles that lead to larger size, and the average size of individuals will increase. Simultaneously, selection can take place for smaller size by breeding the smallest individuals each generation. If selection is carried on for a long enough period of time, two sub-populations may result that have non-overlapping body sizes. The responses to selection can be used to estimate variance components and heritabilities for the selected and correlated traits (section 3.5.3). Both the iso-female line and artificial selection techniques are further discussed in the context of genetic variation for sex ratio (section 3.4.2) and host-parasitoid interaction (section 3.5.3).

3.2.4 GENETIC MAPS

A genetic map outlines the total genome in terms of number of chromosomes and the order of loci along these chromosomes. The loci serve as landmarks on a geographical map: they indicate the particular chromosome as well as the position on the chromosome. Constructing a linkage map requires considerable effort, but it is well worth conducting. Apart from being valuable in providing a physical description of genome size and number of chromosomes for a species, genetic maps are essential for the mapping of (new) genes onto the genome. Such genes could be underpinning single locus traits (e.g. an eye-colour mutant) as well as polygenic traits (e.g. body size). Most quantitative traits are polygenic, and the technique of localising polygenes is known as **Quantitative Trait Loci (QTL) Mapping** (see below). In addition, genetic maps can be used for marker-based introgression experiments and for studying gene flow between natural or laboratory populations. Finally, an exciting prospect is the use of genetic maps to determine on which part of the genome natural selection is operating under field conditions.

Historically, detailed maps were only available in a few well-studied species with short generation times, such as *Drosophila melanogaster*. The genetic markers were mostly morphological mutants. For instance for

a

	Diploid		
Parental strains	male A	x	female B
Genotype	$s_1 s_2 / s_1 s_2$	x	$l_1 l_2 / l_1 l_2$
Phenotype	*short*		*long*
Gametes	$s_1 s_2$		$l_1 l_2$
F1 progeny			
Genotype	$s_1 s_2 / l_1 l_2$		
Phenotype	*intermediate*		
Test cross	F_1	x	F_1
Genotype	$s_1 s_2 / l_1 l_2$		$s_1 s_2 / l_1 l_2$
Gametes	$s_1 s_2$		$s_1 s_2$
	$s_1 l_2$		$s_1 l_2$
	$l_1 s_2$		$l_1 s_2$
	$l_1 l_2$		$l_1 l_2$

F2-progeny

		males + females			
	Gametes	$s_1 s_2$	$s_1 l_2$	$l_1 s_2$	$l_1 l_2$
Genotype	$s_1 s_2$	$s_1 s_2 / s_1 s_2$	$s_1 s_2 / s_1 l_2$	$s_1 s_2 / l_1 s_2$	$s_1 s_2 / l_1 l_2$
Phenotype		*short*	75% *short*	75% *short*	*intermediate*
Expected ratio		1	1	1	1
Genotype	$s_1 l_2$	$s_1 l_2 / s_1 s_2$	$s_1 l_2 / s_1 l_2$	$s_1 l_2 / l_1 s_2$	$s_1 l_2 / l_1 l_2$
Phenotype		75% *short*	*intermediate*	*intermediate*	25% *short*
Expected ratio		1	1	1	1
Genotype	$l_1 s_2$	$l_1 s_2 / s_1 s_2$	$l_1 s_2 / s_1 l_2$	$l_1 s_2 / l_1 s_2$	$l_1 s_2 / l_1 l_2$
Phenotype		75% *short*	*intermediate*	*intermediate*	25% *short*
Expected ratio		1	1	1	1
Genotype	$l_1 l_2$	$l_1 l_2 / s_1 s_2$	$l_1 l_2 / s_1 l_2$	$l_1 l_2 / l_1 s_2$	$l_1 l_2 / l_1 l_2$
Phenotype		*intermediate*	25% *short*	25% *short*	*long*
Expected ratio		1	1	1	1
Overall ratio	*short*	75% *short*	*intermediate*	25% *short*	*long*
	1	4	6	4	1

b

	Haplodiploid		
Parental strains	male x	female	
Genotype	$s_1\ s_2$	$l_1\ l_2\ /\ l_1\ l_2$	
Phenotype	*short*	*long*	
Gametes	$s_1\ s_2$	$l_1\ l_2$	
F1 progeny			
Genotype	$s_1\ s_2\ /\ l_1\ l_2$		
Phenotype	*intermediate*		
Test cross	virgin female		
Genotype	$s_1\ s_2\ /\ l_1\ l_2$		
Gametes	$s_1\ s_2$		
	$s_1\ l_2$		
	$l_1\ s_2$		
	$l_1\ l_2$		
F2-progeny	males only		
Genotype	$s_1\ s_2$	$s_1\ l_2$	
Phenotype	*short*	*intermediate*	
Expected ratio	1	1	
Genotype	$l_1\ s_2$	$l_1\ l_2$	
Phenotype	*intermediate*	*long*	
Expected ratio	1	1	
Overall ratio	short	*intermediate*	long
	1	2	1

Figure 3.5 Example of the inheritance of a trait that is coded for by two unlinked genes in a diploid (a) and haplodiploid (b) organism. A cross between a male of strain A with *short* (s_1 and s_2) wings and a female with *long* (l_1 and l_2) wings. The *short* and *long* alleles are co-dominant, hence all F_1 offspring are genotypically heterozygous and have intermediate wings. In the diploid (a), the F_1 crossed among themselves. There are sixteen possible genotypes and five phenotypes in proportions 1:4:6:4:1. In the haplodiploid (b) the F_1 females are bred as virgins and the F_2 males have four different genotypes and three phenotypes in proportions 1:2:1.

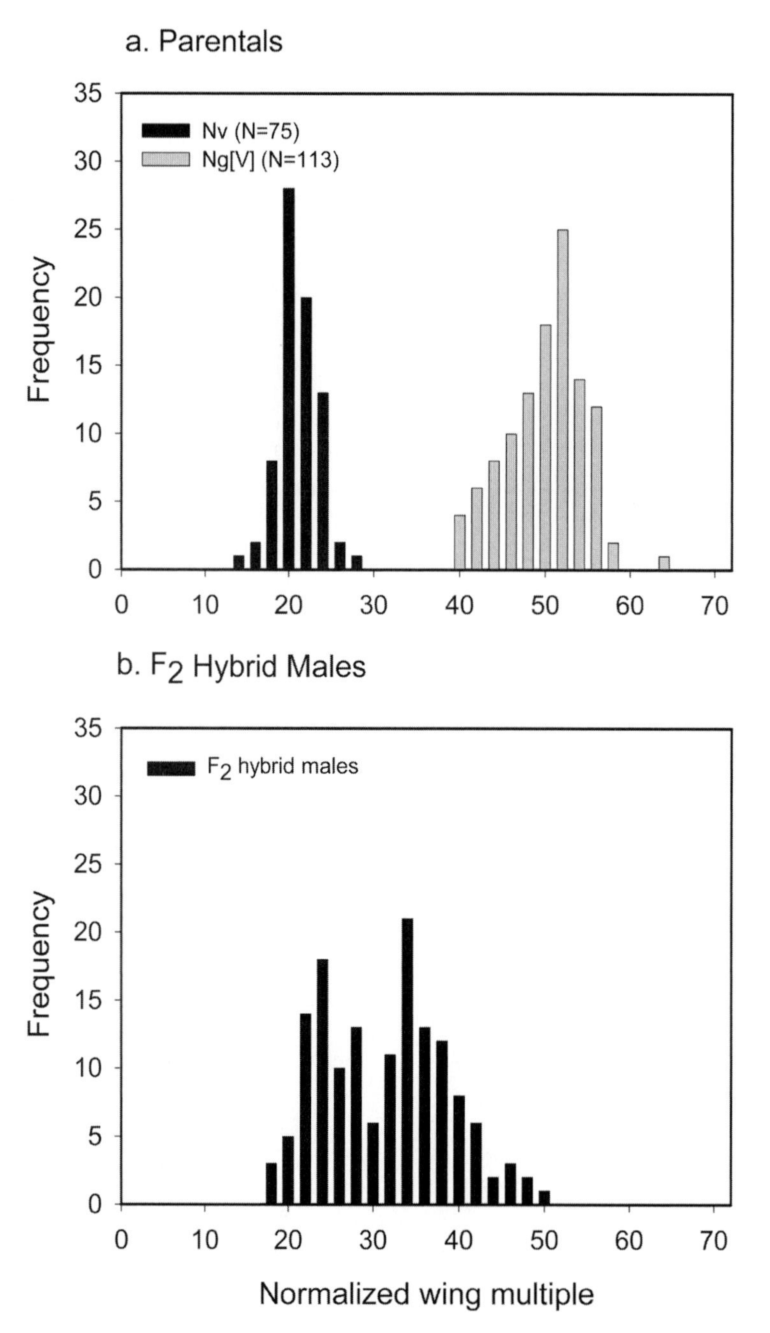

Figure 3.6 The distribution of fore wing sizes (corrected for body size) of parents and F_2 males of a cross between two haplodiploid *Nasonia* sibling species, the short winged *N. vitripennis* (NV) and the long-winged *N. giraulti* (Ng[V]). The F_2 distribution is intermediate to the parental ones and bimodal, indicating a single gene with major effect and one or more genes with small effect. Source Weston *et al.* (1999). Reproduced by permission of Blackwell Publishing.

Drosophila several hundred such loci have been described (Lindsley and Zimm, 1992), but fewer for the parasitoid wasps *Habrobracon hebetor* (Whiting, 1961) and *Nasonia vitripennis* (Saul *et al.*, 1965). Although later cryptic protein variation such as allozymes became detectable, neither these nor the major mutants were sufficiently abundant or polymorphic. However, with the advent of DNA-based molecular markers the construction of genetic linkage maps became feasible in virtually all species.

Basic Principle of Mapping

In constructing a genetic map, both the ordering as well as the distance between loci on chromosomes is determined. The order of the loci and the relative distance between them can be established through the analysis of marker allele segregation in test crosses. The elementary procedure is outlined below.

In order to maximise the efficiency of the map construction, the parental lines in the test crosses have to carry different alleles on each of the marker loci used, because then all markers will be informative. This is commonly the case for inbred laboratory stocks, or for populations from different geographical regions. Divergent selection lines can also be used, but depending on the number of generations of selection from a common ancestral stock, the number of informative markers may be limited (B. Zwaan, D. Conway and L. Partridge, pers. comm.). Once the choice of parental stocks is made, they are crossed and the resulting F_1 individuals are either back-crossed to both or one of the parents (i.e. the recurrent parent) or inter-crossed to produce the F_2 generation (Figure 3.7). For diploid organisms, the backcross design is most commonly used for dominant markers (e.g. presence or absence of RAPD markers) and the F_2 design for co-dominant markers (e.g. microsatellites). Since a large number of insect natural enemies are parasitoid wasps, it is noteworthy that, for linkage analysis in haplodiploid species, all meioses in the F_1 female are informative because recessive marker alleles can be scored (Antolin *et al.*, 1996). For instance, in constructing a linkage map for *hebetor* the sexually derived virgin F_1 females were given abundant hosts (moth larvae) to produce all *Habrobracon*-male broods. These male individuals were then genotyped for the markers (Antolin *et al.*, 1996).

Another way to ensure a high number of informative markers, is to use single pair crosses and score the genotype (and when used in combination with gene mapping approaches, the phenotype) of all the resulting progeny. In this case, even though particular markers are not fixed in the two parental lines, they may still be informative in a single pair cross. For instance, if the allele frequency of a marker gene is 0.9 for allele A_1 in parental line 1 and 0.1 in parental line 2, the chance of a A_1A_1 x A_2A_2 cross is, $(0.9)^2.(0.9)^2 = 0.656$. This makes even unfixed markers useful providing that enough markers are available. This approach has recently been successfully used in a genetic association study (Beldade *et al.*, 2002).

Once the progeny from the crosses have been genotyped, the recombination frequency for pairs of markers can be calculated. To do this, knowing the parental genotypes for each of the markers is convenient. However, this is not essential because the two most frequent genotypes for any pair of markers can be considered the parental types. When calculating the recombination frequencies it will become apparent which markers form one linkage group and which lie outside this linkage group (i.e. respectively, $c < 50$ and $c = 50$ Morgan; Figure 3.7). In principle, a linkage group represents one chromosome, and this can be confirmed with cytological data. Given that enough progeny have been scored so that the c-values are accurately calculated, within each linkage group the order of the markers can be established unambiguously.

To construct a map based on physical distance (expressed in centiMorgans, cM) the c-values have to be converted using a genetic mapping function. This is so because the c-values are not additive. Consider three linked loci A, B and C with recombination frequencies c_{AB}, c_{AC}, c_{BC}. The recombination frequency c_{AC} will be

			Diploid		
Parental strains			male A	x	female B
	Genotypes		+ + + / Y		*vg e r / vg e r*
	Phenotypes		wildtype		*vestigial ebony red*
	Gametes		+ + +		*vg e r*
			Y		
F1 progeny					
	Genotype		+ + + / *vg e r*		Y / *vg e r*
	Phenotype		wildtype female		*vestigial ebony red* male
Test cross			F_1 female	x	B male (or F_1 brothers)
	Genotype		+ + + / *vg e r*		Y / *vg e r*
	Gametes	parental	+ + +		*vg e r*
			vg e r		Y
		single cross-over I	*vg* + +		
			+ *e r*		
		single cross over II	*vg e* +		
			+ + *r*		
		double cross over	vg + r		
			+ *e* +		

F2-progeny

Genotype		Phenotype Number	Class	Class total	Frequency	Recombination frequencies
males	females					
+ + + / Y	+ + + / *vg e r*	wildtype	250	parental		
vg e r / Y	*vg e r* / *vg e r*	*vestigial ebony red*	260	parental	510	
vg + + / Y	vg + + / *vg e r*	*vestigial*	38	single I		
+ *e r* / Y	+ *e r* / *vg e r*	*ebony red*	42	single I	80	$a = 80 / 1000 = 0.080$
vg e + / Y	*vg e* + / *vg e r*	*vestigial ebony*	205	single II		
+ + *r* / Y	+ + *r* / *vg e r*	*red*	195	single II	400	$b = 400 / 1000 = 0.400$
vg + *r* / Y	*vg* + *r* / *vg e r*	*vestigial red*	4	double		
+ *e* + / Y	+ *e* + / *vg e r*	*ebony*	6	double	10	$c = 10 / 1000 = 0.010$
			Grant total		1000	

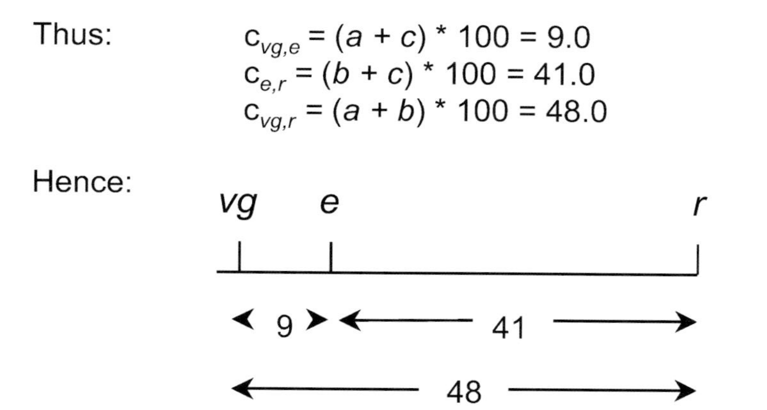

Figure 3.7 Example of a three-point cross with sex chromosome linkage for a diploid organism. A cross between a male of strain A that is wildtype for three mutations that are X-linked and a triple mutant female with *vestigial* (*vg*) wings, *ebony* (*e*) body colour and *red* (*r*) eyes. The mutant *vestigial, ebony* and *red* alleles are recessive, hence all F_1 offspring are genotypically heterozygous and phenotypically wildtype. The F_1 is either backcrossed to strain B or the F_1 females to their brothers. The F_2 is scored for their phenotype and it appears that the three genes are not segregating independently, but are linked. Eight different phenotypes can be distinguished that belong to four recombination classes. The class with the lowest numbers refers to double cross overs and from this the order of the markers can be seen immediately as *vg e r* (i.e. to separate *e* from *vg* and *r* you would need two recombination events). The frequencies of the recombination classes give information on what the parental genotypes were. The recombination frequencies can be calculated and measure the relative distance between the markers expressed in centiMorgans. Note that the sum of $c_{vg,e}$ and $c_{e,r}$ is larger than $c_{vg,r}$. This happens as a result of double cross overs (see text). For haplodiploids the results are similar to X-linkage in diploids, except that the Y chromosome is absent.

underestimated because recombination between B and C can mask the recombination between A and C in terms of the actual occurring genotype. This is the case when there is no interference, i.e. the presence of a crossover between chromatids does not influence the likelihood of another crossover occurring in the (close) vicinity. These aspects can be formulated as follows (Lynch and Walsh, 1998):

$$c_{AC} = c_{AB} + c_{BC} - 2(1 - \delta)\, c_{AB}c_{BC} \qquad (3.1)$$

With the interference parameter, δ, ranging between 0 (no interference) and 1 (complete interference). From (3.1) it follows that recombination values can only be considered additive when there is strong interference, or when the markers are closely linked (i.e. the fraction $2(1 - \delta)\, c_{AB}\, c_{BC}$ can be ignored). There are several mapping functions that predict physical distances from recombination frequencies, and we will mention the two most widely used.

Haldane's mapping function (Haldane, 1919) assumes random cross-over formation and no interference (Lynch and Walsh, 1998):

$$m = -\ln(1 - 2c_{AB})/2 \qquad (3.2)$$

with m being measured in morgans or centimorgans.

Kosambi's mapping function (Kosambi, 1944) permits some interference:

$$m = [\ln(1 + 2c_{AB})/(1 - 2c_{AB})]/4 \qquad (3.3)$$

For small recombination values ($c < 0.15$) the differences between Haldane's, Kosambi's or the $m = c$ mapping functions are negligible (Figure 3.8) (Kearsey and Pooni, 1996; Lynch and Walsh, 1998).

Construction of Linkage Maps

It is obvious that constructing linkage maps by hand is cumbersome and near to impossible when the number of markers involved is high. Therefore, the use of computers is indispensable and several packages have been developed (Stam, 1993), of which **Mapmaker** (Lander *et al.*, 1987) and **JoinMap** (Stam, 1993) are the most widely used (visit http://linkage.rockefeller.edu/soft/list.html for more examples). The programs construct the most likely linkage map, both in terms of ordering of markers as well as the distance between the markers. It can use least square procedures, maximum likelihood procedures, or both types.

For instance, JoinMap first calculates recombination frequencies and LOD scores for pairs of markers. **LOD** stands for the **logarithm of odds**, that is, the logarithm of the ratio of the probability that the two markers are linked over the probability that they are unlinked (Stam, 1993). Thus, the higher the LOD score the closer two markers are linked. Secondly, linkage groups are established using a user-set LOD threshold value, below which markers are considered to be unlinked. Starting with pairs of markers with the highest LOD scores, at any given time a marker is judged on its LOD score to be significantly linked to any of the markers already added to the map. If the marker is not linked to any of the existing linkage groups, a new group is created. This process will stop when all markers have been added to the map. Depending on the threshold LOD value, a unique set of grouped markers will result. Generally, a threshold LOD of 3 works well to ensure that the number of linkage groups corresponds with the actual number of chromosomes. Varying the LOD threshold value will show stability of grouping, and it will indicate sets of tightly linked markers. Thirdly, map distances are estimated. The c-values are converted into distances ('observed' distances) and the sum of the squares of the differences between the 'expected' and 'observed' distance is calculated. By trial and error the 'expected' distance can be changed until the sum of squares is minimal. These 'expected' distances, then, are the most likely map distances for this data set. For more details, see Stam (1993).

It is important to note that there are essential differences between the available computer packages. Mapmaker, for instance, is an interactive program, while only JoinMap has the potential of making an integrative map based on different marker data (e.g. combining

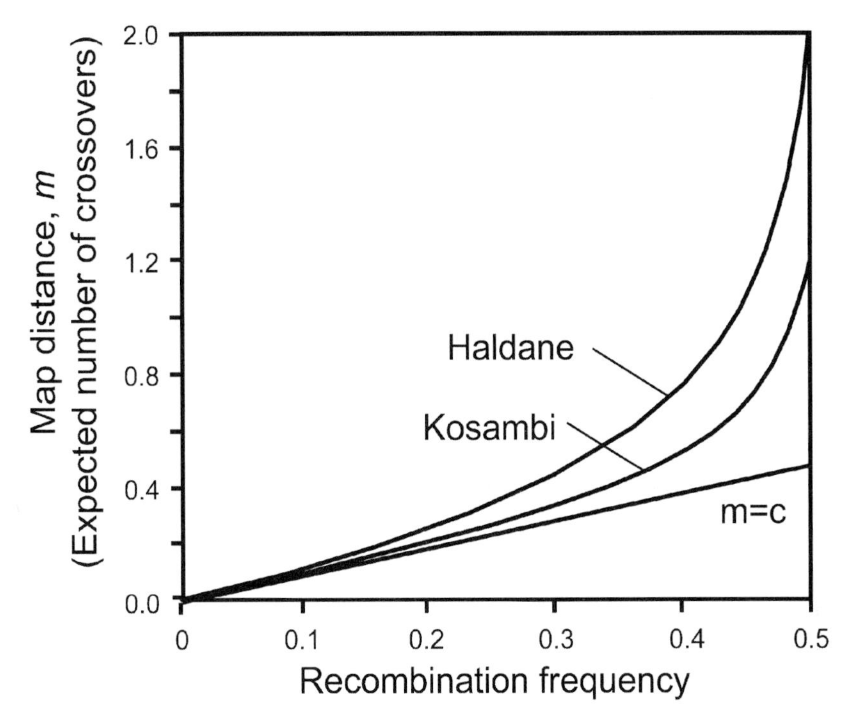

Figure 3.8 The relationship between map distance (m) and recombination value (c) for three mapping functions, Haldane's (Haldane, 1919), Kosambi's (Kosambi, 1944) and $m = c$.

morphological markers with microsatellite markers). However, the quality of the linkage map, first and foremost, depends on the quality of the data used: the map is only as good as the data. Large data sets and/or multiple estimates of the c-value for each pair of markers will greatly enhance the quality of the map.

Available Genetic Maps

To our knowledge, three linkage maps have been constructed for insect natural enemies, all of them for parasitoid wasps. The first linkage map was for *Habrobracon hebetor* with 79 RAPD-SSCP markers, a total length of 1156 cM and an average spacing of one marker per 17 cM (Antolin *et al.*, 1996; Figure 3.9). Recently, a RAPD marker-based linkage map was published for *Trichogramma brassicae*, with very similar features, 84 markers having an average spacing of 17.7 cM for a total length of 1330 cM (Laurent *et al.*, 1998). For *Nasonia*, a RAPD marker-based linkage map has been developed using the hybrid progeny of a cross between *N. vitripennis* and *N. giraulti* (Gadau *et al.*, 1999). Ninety-one markers covered a total genome length of 764.5 cM with an average space between the markers of 8.4 cM. Recently, Pannebakker *et al.* (submitted) described the linkage map of *Leptopilina clavipes*. Using AFLP markers, 5 linkage groups were found spanning a total distance of 219.5 cM.

Nature and Number of Markers

The linkage maps mentioned above all use RAPD techniques, probably because the markers are quickly generated. For the linkage maps to be useful for a variety of purposes, there needs to be additional incorporation of other types of marker. In general, such markers should preferably be highly polymorphic and co-dominant to allow the scoring of heterozygotes, and neutral with respect to fitness. The latter requirement

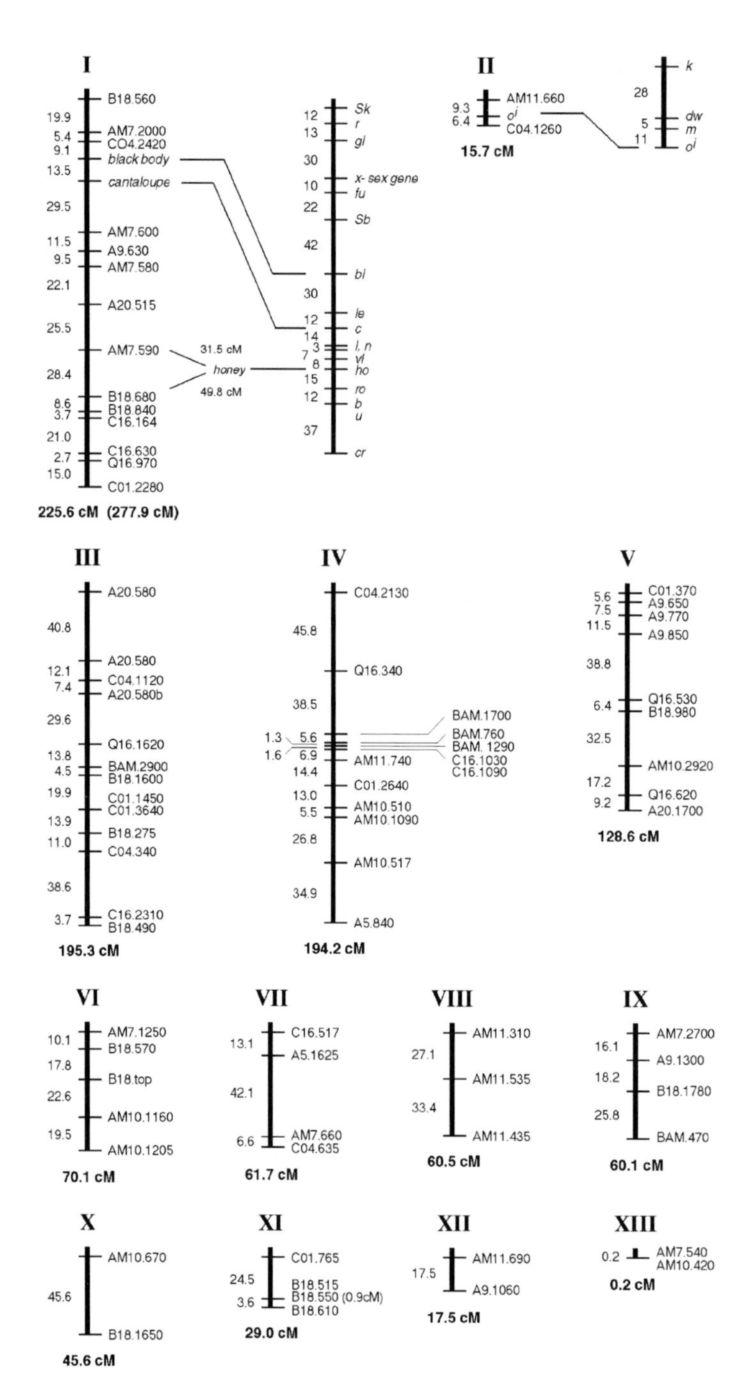

Figure 3.9 Linkage map of *Habrobracon hebetor*. There are 13 linkage groups (chromosomes). Groups I and II show the relationship between the maps based on molecular markers (see Antolin *et al.*, 1996) and morphological markers (see Whiting, 1961). Markers are shown on the right of each group and designated with the primer name and fragment size in bp of DNA. Distances between markers (cM) are shown at the left of each group, and total linkage group length appears at the bottom of each group. Source: Antolin *et al.* (1996). Reproduced by permission of the Genetical Society of America.

is not often mentioned, but the estimation of c-values will be biased if carriers of certain allele combinations are less/more likely to survive in the crosses. Molecular markers are therefore highly suitable for use in linkage maps, especially microsatellites that have many alleles per locus. As a drawback, the latter markers are much more difficult to develop. In contrast, most morphological mutants suffer from negative effects on general and reproductive fitness.

To obtain a general idea of how many markers (n) are needed to span a certain size of genome (L, cM) the following formula can be used (Lange and Boehnke, 1982):

$$n = \ln(1 - p)/\ln(1 - 2m/L) \qquad (3.4)$$

with p the fraction of loci within m map unit of some marker. This equation assumes a circular genome and corrections can be made for discrete chromosomes (Bishop *et al.*, 1983), thus incorporating chromosome ends. Ignoring chromosome ends tends to underestimate n (Lynch and Walsh, 1998). If, for instance, you want to know how many (additional) random markers you would need to get the spacing for the *Trichogramma* map down to 10 cM, you could use (3.4) as an approximation. For $p = 0.9$ (i.e. the probability that at least one marker is within 10 cM of a randomly chosen gene is 90%), the total number of markers would be 152. Therefore, 68 additional markers will have to be developed.

3.3 GENETICS OF REPRODUCTIVE MODE

3.3.1 INTRODUCTION

Most insects reproduce sexually, i.e. the haploid male gamete fuses with the haploid female gamete. There are two sexes, males and females that produce sperm and eggs through meiosis. **Hermaphroditism**, i.e. the male and female function combined in a single individual, is rare. Another common mode of reproduction in insects is **parthenogenesis**. Such asexual reproduction involves development of females from unfertilised eggs. Parthenogenetic reproduction has been found among insect natural enemy groups such as thrips (Thysanoptera), several non-parasitoid flies (Diptera), beetles (Coleoptera) and many Hymenoptera. In flies and beetles, parthenogenesis is frequently associated with polyploidy (Suomalainen *et al.*, 1987). This section discusses some of the genetic consequences of sexual and parthenogenetic reproduction found among insect natural enemies. It focuses mainly on parasitoid wasps because they are the only well-studied group of insect natural enemies.

3.3.2 ARRHENOTOKY

Haplodiploidy, in which males are haploid and females diploid, occurs in all ants, bees and wasps (Hymenoptera) and thrips (Thysanoptera), as well as in a few other insect species. The most common mode of haplodiploid reproduction is **arrhenotoky**, in which male progeny develop parthenogenetically from unfertilised (haploid) eggs and female progeny from fertilised (diploid) eggs. As a result, sons receive genetic material from their mother only, and are therefore 100% related to their mother and unrelated to their father. On the other hand, daughters receive one haploid copy of their genome from each of their parents and are therefore 50% related to their mother and 50% to their father (Figure 3.10). Unmated females can lay only haploid eggs and so produce progeny consisting solely of males. Mated females store sperm in a spermatheca and can control the sex of their offspring when ovipositing, by selectively releasing sperm to an egg as the latter passes down the common oviduct (Godfray, 1994)

3.3.3 THELYTOKY

A minority of species reproduce entirely parthenogenetically by **thelytoky**. With thelytoky there are no males, and unfertilised eggs give rise to diploid females, resulting in 100% relatedness of daughters to their mother (Figure 3.10). Thelytokous reproduction occurs sporadically in almost all major parasitoid families (Crozier, 1975; for overviews see Luck *et al.*, 1993; Quicke, 1997). Thelytokous reproduction can be recognised by initiating iso-female lines from field-collected individuals and scoring the

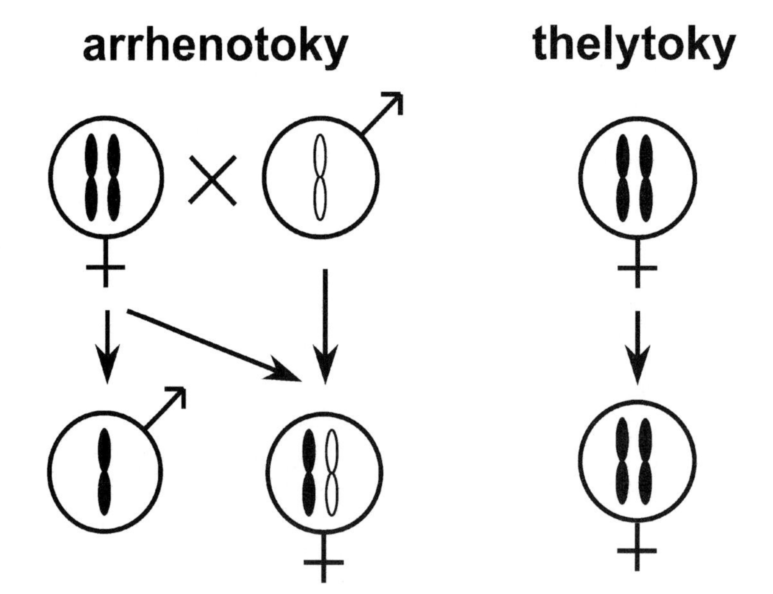

Figure 3.10 Modes of haplodiploid reproduction. In arrhenotoky males develop parthenogenetically from haploid unfertilised eggs and females from diploid fertilised eggs. In thelytoky females develop parthenogenetically from unfertilised diploid eggs.

sex of their progeny. Arrhenotokous females will produce sons and daughters when mated or exclusively sons when unmated, whereas thelytokous females will produce only daughters. It is highly advisable to screen a large number of offspring, because sex ratios in a particular species may be heavily female-biased for reasons other than thelytoky (section 1.9) and males may be missed (see below).

There are a number of cytological mechanisms by which thelytoky occurs and these have different effects on the genetic variation within and among offspring (Figure 3.11). Knowledge of these mechanisms is important for understanding to what extent genetic variation can be maintained in asexual species. A first indication of the actual mechanism may be obtained from parent-offspring analyses with molecular markers. Establishment of the exact mechanism will require careful cytological investigations. Depending on the timing of the process, i.e. pre-, peri- or post-meiotic, developing oöcytes in ovaries, or eggs that have recently been oviposited, need to be examined. Eggs are typically arrested in meiosis stage I during ovigenesis and resume development after oviposition (Went, 1982). The following are general descriptions of the most prevalent modes of thelytokous reproduction, but note that several modifications and aberrations of these processes have been described.

Apomixis

Apomictic parthenogenesis involves mitotic oögenesis, i.e. eggs are produced by a single mitotic division and there is no reduction division (Figure 3.11). This mode of reproduction is relatively rare in insects. It has been reported from several insect orders, including Orthoptera, Homoptera, Diptera and Hymenoptera, but not from parasitoids or predators (see Suomalainen *et al.*, 1987). Apomictically produced offspring are identical to their mother.

Automixis

Parthenogenesis by automixis is more common than by apomixis and occurs through meiotic oögenesis, i.e. a reduction division in combination with a process of chromosome doubling

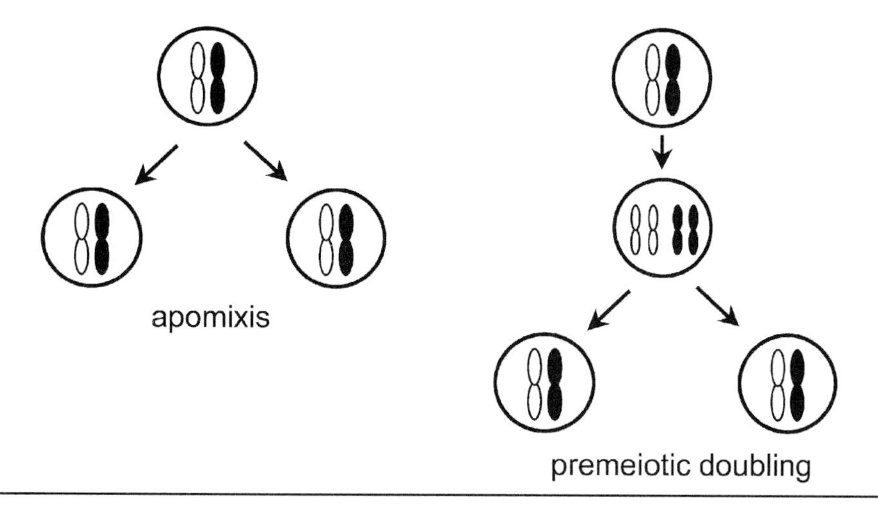

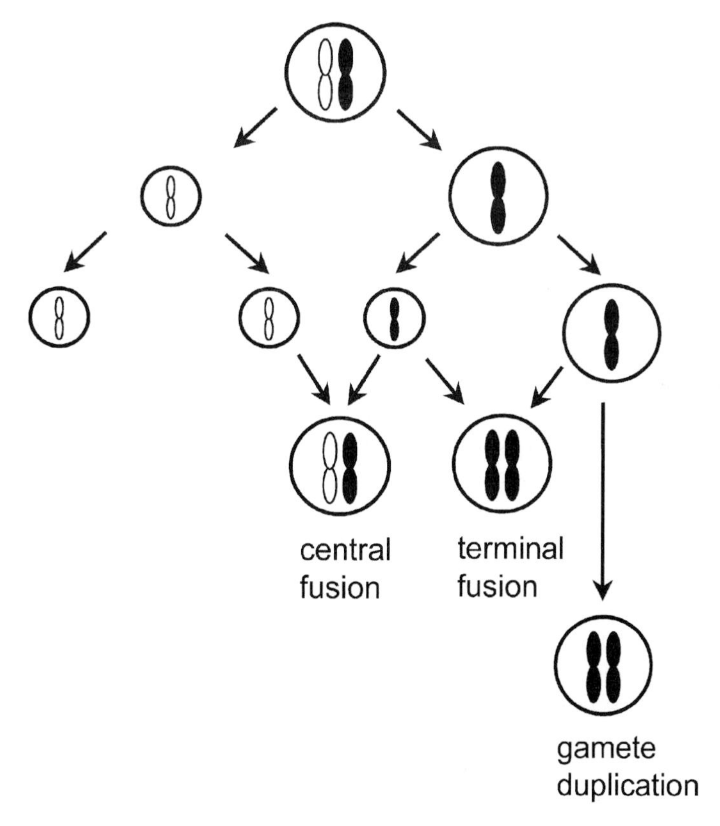

Figure 3.11 Several cytological modes of parthenogenesis. Apomixis is mitotic parthenogenesis. Premeiotic doubling consists of a duplication of all chromosomes before meiosis followed by a reduction division. Central fusion refers to the fusion of two second polar nuclei and central fusion to fusion of the egg nucleus with the second polar body. Gamete duplication is the doubling of chromosomes without cell division after meiosis. Apomixis and premeiotic doubling result in offspring identical to the mother. Gamete duplication lead to complete homozygosity in one generation. Under terminal and central fusion heterozygosity may be maintained but will decrease through crossing over.

known as **diploidy restoration**. Diploidy restoration can occur in a number of ways which have different consequences for the genetic variation among offspring.

Premeiotic Doubling

Premeiotic doubling is an automictic mechanism in which chromosome doubling occurs before meiosis and the first meiotic division restores the diploid state (Figure 3.11). If crossing over occurs, it has no effect because identical sister chromosomes pair. Consequently, offspring are identical to their mother and all heterozygosity is maintained as in apomixis. There are no examples from insects that are natural enemies of other insects (Suomalainen *et al.*, 1987).

Terminal Fusion

In meiosis there are two divisions of the oögonium (Figure 3.11). The first (meiosis I) is reductional and results in two haploid daughter cells, the egg cell and the first polar body. Both cells divide once more mitotically (meiosis II) resulting in two haploid daughter cells each. The daughter cells of the first polar body degenerate. The egg cell yields the egg and the second polar body. Whereas in the normal case this second polar body also degenerates, under terminal fusion it fuses with the egg cell yielding a diploid egg. Because these two cells arose from a single haploid egg cell in meiosis stage I in which crossing-over may have taken place, terminal fusion leads to increasing homozygosity of loci proximal of cross-overs (i.e. between centromere and cross-over). This type of thelytoky has been reported from the chalcid *Aphytis mytilaspidis* (Rössler and DeBach, 1973).

Central Fusion

With central fusion, one of the two daughter cells of the first polar body does not degenerate but fuses with the second polar body giving rise to the diploid egg. The egg cell and the other second polar body degenerate. Although this process can retain heterozygosity, it leads to an increase of homozygosity each generation. Whenever a cross-over occurs between both homologues in the oögonium, it can result in identical copies in the egg, depending on the assortment. Loci close to the centromere have a lower cross-over rate and therefore a higher chance to be retained in a heterozygous state. Central fusion is known from the ichneumonid wasp *Venturia canescens* (Speicher *et al.*, 1965; Beukeboom and Pijnacker, 2000).

Gamete Duplication

Gamete duplication refers to chromosome doubling during the first cleavage division of the mature egg. It basically involves **endomitosis**, i.e. chromosome replication without a following cell division. All cases of *Wolbachia*-induced thelytoky (see **Causes of Thelytoky** below) involve gamete duplication. It has been found in several parasitoid wasps including *Trichogramma*, *Muscidifurax* and *Encarsia* (reviewed in Stouthamer, 1997) and the *Drosophila* parasitoid *Leptopilina clavipes*. (Pannebakker *et al.* 2004). Because all chromosomes are simply doubled, this mechanism results in complete homozygosity within one generation.

Causes of Thelytoky

Although thelytoky is generally considered to be derived from arrhenotoky, the genetic factors that induce thelytoky are almost completely unknown. One cause of thelytoky is interspecific hybridisation (Nagarkatti, 1970; Pintureau and Babault, 1981; Legner, 1987). Nagarkatti and Fazaluddin (1973) crossed a female of *Trichogramma perkinsii* with a male of *T. californicum*, and the offspring comprised one thelytokous female, seven arrhenotokous females and ten males. Tardieux and Rabasse (1988) observed that females of *Aphidius colemani* produced daughter offspring thelytokously when mated to males of a closely related species, whereas only males were produced if courtship was interrupted and copulation prevented. Arrhenotokous reproduction was excluded because all daughters had the electrophoretic esterase pattern of their mother.

Another cause of thelytoky is the presence of microbes (Stouthamer *et al.*, 1990a). In several parasitoid wasp species thelytoky is caused by intracellular *Wolbachia* bacteria (see also subsection 6.5). Stouthamer showed that if such species are treated with antibiotics or exposed to high temperatures, which will kill the bacteria, they will produce both sons and daughters. Pijls *et al.* (1996) showed, by treating the thelytokous wasps with antibiotics, that the arrhenotokous population of *Apoanagyrus* (= *Epidinocarsis*) *diversicornis* attacking the cassava mealybug, *Phenacoccus manihoti* in Central South America is conspecific with morphologically identical but thelytokous populations in northern South America that attack *P. herreni.* Males were obtained that interbred successfully with females of the arrhenotokous population. However, this reversion to arrhenotokous reproduction is not always successful because males are frequently non-functional (see below).

Both arrhenotokous and thelytokous forms are known in several parasitoid wasp species and these forms occur either allopatrically or sympatrically (Stouthamer, 1993). In several such species, *Wolbachia* bacteria are absent and thelytoky must be caused by another mechanism. Although other parthenogenesis-inducing microorganisms might exist, it is likely that parthenogenetic egg production in such species has a genetic basis. In addition, genetic variation for parthenogenesis may be present in arrhenotokous populations, but this remains to be investigated. In *Spalangia endius*, Bandara and Walter (1993) found that virgin arrhenotokous females occasionally produced daughters from unfertilised eggs, but these daughters in turn reproduced arrhenotokously. This phenomenon was originally reported from *Habrobracon hebetor* by Speicher (1934). Similarly, Beukeboom *et al.* (1999) found daughter production from unfertilised eggs of arrhenotokous *Venturia canescens* females at a frequency of approximately 0.2%, but this ability was not transferred to the next generation. Thus, thelytokous reproduction seems to occur sporadically in arrhenotokous species, but whether this can lead to stable thelytokous lineages remains to be established.

The extent to which thelytokous species can be considered as purely clonal is questionable. In general, individuals are either sexual or obligatory asexual. However, in *Trichogramma* and *Aphytis*, thelytokous females were found to mate with sexual conspecific males and use their sperm to fertilise their eggs (Rössler and DeBach, 1973; Stouthamer and Kazmer, 1994). These examples concern *Wolbachia*-induced thelytoky and post-meiotic gamete duplication, which suggests that such forms of thelytoky are genetically isolated to a lesser degree than pre- and perimeiotic forms. Besides direct observation, evidence for gene flow between sexuals and asexuals may be obtained from genetic marker studies. Mitochondrial DNA, which is only maternally inherited, may be informative about the frequency at which asexuals arise from sexuals.

3.3.4 DEUTEROTOKY

Deuterotoky refers to female production with rare males. It differs from arrhenotoky in that both sexes develop from unfertilised eggs (Table 3.1). However, as pointed out by Luck *et al.* (1993), the distinction between thelytoky and deuterotoky is ambiguous, the reason being that some parasitoid wasp species originally designated as thelytokous have been found to produce small numbers of males. These males

Table 3.1 Reproductive modes in haplodiploid insects

Reproductive mode	Males from	Females from	Type
arrhenotoky	unfertilised eggs	fertilised eggs	sexual
thelytoky	absent	unfertilised eggs	asexual
deuterotoky	unfertilised eggs	unfertilised eggs	asexual

are produced when the maternal females have been exposed to high temperatures. Whilst they have been considered to be non-functional, there is evidence that in some cases they are not only capable of mating but also able to pass on their genes to progeny, which reproduce thelytokously. Some groups such as the gall-causing herbivorous Cynipidae alternate between two reproductive modes, i.e. deuterotoky and arrhenotoky or thelytoky. As already mentioned, the underlying genetics of these different modes of reproduction remains a challenging, as yet undiscovered, field.

3.4 GENETICS OF SEX DETERMINATION AND SEX RATIO

3.4.1 GENETICS OF SEX DETERMINATION

Sex Chromosomes

A variety of sex determining mechanisms is known from insects and includes male and female heterogamety, haplodiploidy, paternal genome loss and systems with X chromosome elimination (Metz, 1938; Hughes-Schrader, 1948; Crozier, 1971; Bull, 1983; Nur, 1989). The most widespread mechanism is **heterogamety**, in which one sex has two identical sex chromosomes (= homogametic sex, e.g. XX) and the other two different ones (= heterogametic sex, e.g. XY). The heterogametic sex is more often the male (XY) than the female (ZW). Female heterogamety is indicated with ZZ/ZW for distinction. Loss of the Y chromosome has resulted in an XO system in several insect groups such as for example praying mantids (White, 1954).

Heteromorphic sex chromosomes can easily be detected cytologically. Sex chromosomes are often morphologically different from the autosomes, i.e. they contain more heterochromatin, condense out of phase or are of different size. However, there are cases in which the sex determining genes are present on the autosomes and it is impossible to distinguish between sex chromosomes cytologically. For example, in the non-parasitoid *Megaselia scalaris* (Phoridae, some members of which are parasitoids), a single gene determines maleness and varies in its position in the genome (Mainx, 1964). Traut (1994) used phenotypic and molecular markers to map this gene to various chromosomes in different stocks.

Establishing Chromosome Numbers

A general introduction to animal cytogenetics is given by MacGregor (1993). The most suitable tissues for establishing chromosome numbers (**karyotypes**) are male and female gonads, but brain tissue of larvae can also be used (e.g. Traut and Willhoeft, 1990). It should be noted that other somatic tissues frequently undergo endopolyploidization, making chromosome counts both very difficult and unreliable. The number of **metaphase plates** (a phase of maximal contraction of chromosomes during cell division) may be increased by feeding the animals colchicine (0.15%) for a few hours (longer periods can lead to polyploidization of cells). One can use freshly eclosed adult males, but some authors have found the gonads of male larvae or pupae to yield more division figures. Females are typically allowed to lay eggs for a while before dissection of ovaries in order to ensure that they are in an active egg-laying condition.

Animals can be dissected in either water or Ringer's solution and the gonads immediately transferred to Carnoy's fixative (3 parts methanol or ethanol : 1 part glacial acetic acid, optionally one can add 2 parts chloroform). Gonads are then transferred to a drop of 45-70% acetic acid on an object glass, torn into fine pieces and then squashed. Several stains can be applied including: (a) 2.5% lacmoid (2.5% lacmoid in 1 part water : 1 part lactic acid : 1 part acetic acid), (b) 2% Giemsa (in Sörensen buffer: 55ml 1/15M Na_2HPO_4 plus 1/15M KH_2PO_4, pH 6,9), (c) Feulgen (=Schiff's) reagent: dissolve 1 g pararosaniline in 30 ml 1N HCl and 1 g $K_2S_2O_5$ in 170 ml demiwater, combine both solutions and destain for 24 h. in the refrigerator, shake solution with 600 mg Norit for 2 min and filter through Whatman paper, store in a tight bottle in refrigerator, solution must always smell of SO_2, if not add tiny amount of $K_2S_2O_5$), (d) carmine (add 4 g carmine and 1 ml concentrated HCl to 15 ml distilled water, boil and stir for 10 min, cool and add 95 ml of 85% alcohol, and filter; Snow, 1963) or (e) fluorescent 4′,6-diamidine-2-phenylindole-dihydrochloride (DAPI). Slides can be temporarily prevented from drying out by sealing them with nail polish. A more permanent method is to freeze them on dry ice, snap of the cover slip with a razor blade, air dry and seal them with DePeX (Sumner, 1972).

Van Dijken (1991) described a cytological technique for counting chromosomes of freshly laid eggs of the haplodiploid wasp *Apoanagyrus lopezi*. Eggs are dissected from hosts and placed in a droplet of 2% lacto acetic orcein (a chromosome stain made by dissolving 1.0 g of natural orceine in 10 ml of 85% lactic acid, 25 ml of glacial acetic acid and 15 ml of distilled water; this mixture is gently boiled for 1 h then cooled and filtered) and a cover slip gently placed over them. The eggs are then squashed to a monolayer after 25 min have elapsed and the cover slip sealed with nail varnish. After staining for 24 h at room temperature the chromosomes are examined and counted. Exact counts are unnecessary, as female have two sets of chromosomes which is easily distinguished from one set in males (Figure 3.12). The optimum time to fix and squash the eggs may vary between species (e.g. 18–24 h in *Apoanagyrus lopezi*), so it is advisable to make a series of egg squashes at different times following oviposition. In haplodiploids, unfertilised eggs of virgin females can be used as control to determine the haploid number of chromosomes. See also Ueno and Tanaka (1997).

Haplodiploidy

All Hymenoptera are haplodiploid; there are no heteromorphic sex chromosomes. Sex is determined by the number of chromosome sets present in the embryo.

Single-locus Complementary Sex Determination (slCSD)

Under single-locus Complementary Sex Determination (slCSD), sex is determined by multiple alleles at a single locus. Heterozygosity leads to female development, but hemizygotes and homozygotes develop into males. This mode of sex determination was first described by Whiting (1939, 1940, 1943) from *Habrobracon hebetor*. It has been demonstrated in all major suborders of the Hymenoptera (reviewed in Luck *et al.*, 1993; Cook, 1993a; Periquet *et al.*, 1993; Butcher *et al.*, 2000), but appears to be absent from the superfamily Chalcidoidea. The estimated number of alleles at the sex locus typically range from 10-20 (Cook and Crozier, 1995), but can be as high as 86 in fire ants (Ross *et al.*, 1993; see also Adams *et al.*, 1977, for methods of estimating the number of alleles).

Single locus CSD can easily be detected with inbreeding studies. The most straightforward method is **mother-son matings**. This is possible if females first reproduce as virgins and can be kept alive long enough to be mated with one of their sons. The resulting progeny will only carry two sex alleles and half of the fertilised (diploid) eggs will be homozygous for the sex allele and develop into **diploid males** (Figure 3.13a). Note that there will also be males developing from unfertilised (haploid) eggs. If mother-son crosses are not possible, **brother-sister matings** can be used. Such crosses can be matched (two-allelic) or unmatched (three-allelic) depending on whether the son inherited the same or a different allele as the daughter (Figure 3.13b). If both carry the same allele, the diploid offspring will be 50% homozygous and male (similar to mother-son crosses), but if they carry different alleles 100% of diploid eggs will be heterozygous and female. Because 50% of brother-sister matings will, by chance, be matched and 50% will be unmatched, on average 25% of fertilised eggs are expected to become diploid males in brother-sister crosses. This will be manifested in an increase in the sex ratio of inbred crosses. However, caution should be exercised because diploid males may be inviable. For example, in *Habrobracon hebetor*, diploid male embryos frequently die, but hatchability of such eggs can be restored under mineral oil (Petters and Mettus, 1980). It is therefore important to also compare brood sizes and the number of non-emerged hosts in inbred

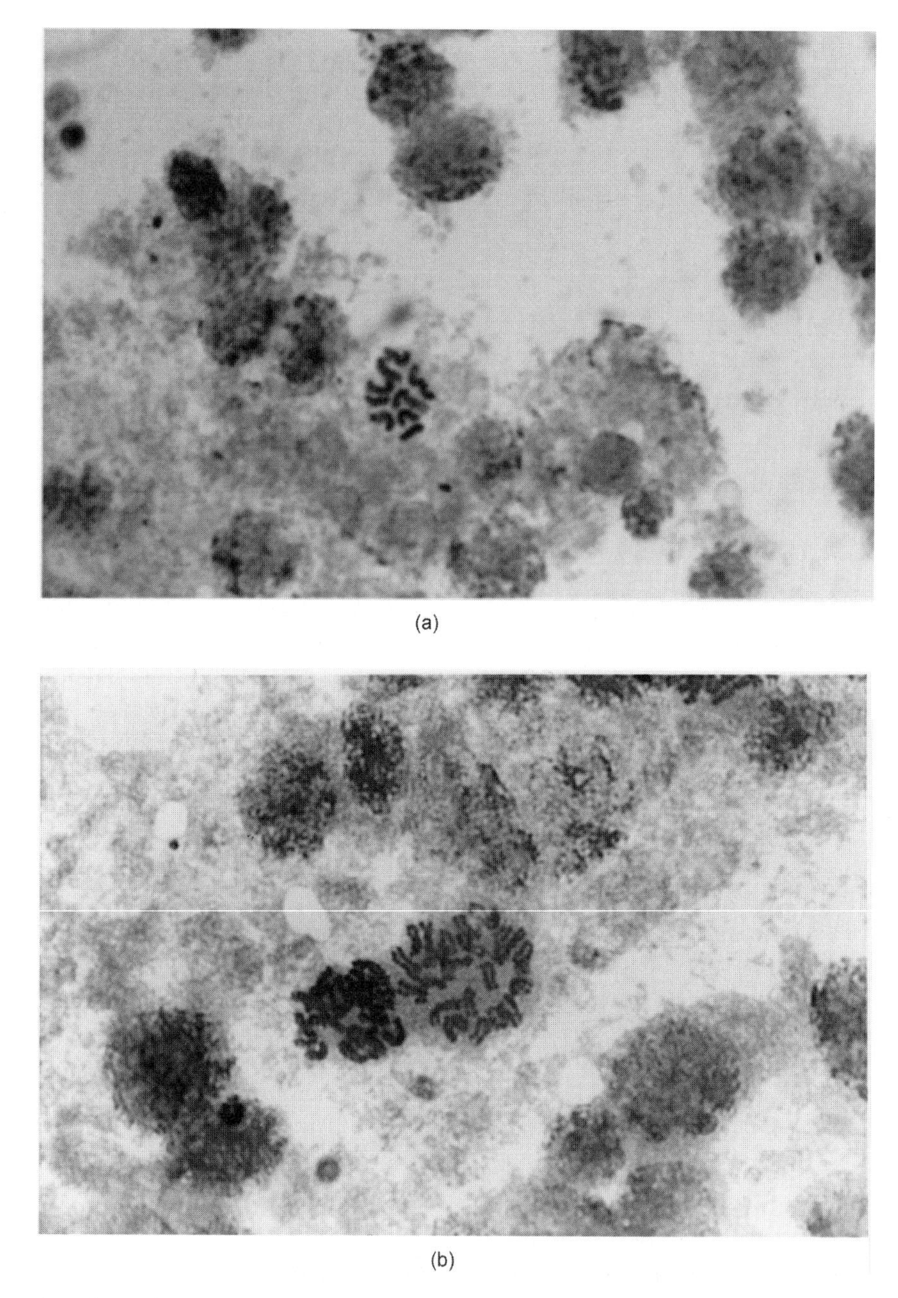

Figure 3.12 Metaphase chromosomes in eggs of *Apoanagyrus lopezi*, (a) haploid male and (b) diploid female. See text for methodology. Reproduced by kind permission of M.J. van Dijken.

and control (non-inbred) crosses (Beukeboom, 2001).

Diploid males can be recognised in a number of ways. Genetic identification is the most reliable method, i.e. through the use of genetic markers or cytology. Morphological identification is also possible but not always conclusive. Diploid males are typically larger than haploid

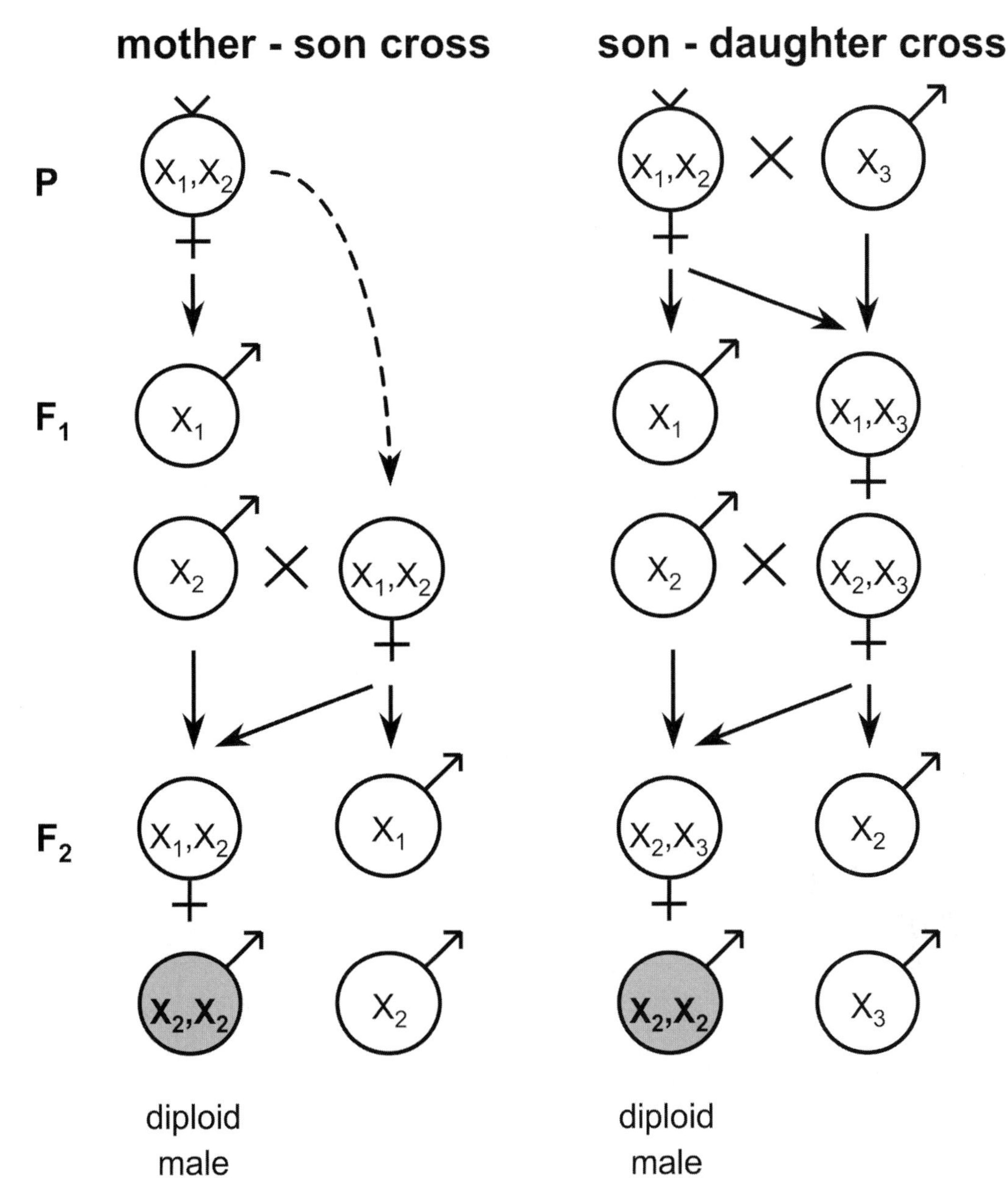

Figure 3.13 Inbreeding crosses to detect single locus Complementary Sex Determination. (A) Mother-son crosses. Diploid heterozygous females produce two types of hemizygous sons. In backcrosses with their mother both types yield 50% diploid homozygous sons among fertilized eggs. (B) Brother-sister crosses. Diploid heterozygous females are mated with their hemizygous brothers. Half of these crosses are matched (two-allelic) and result in 50% diploid homozygous sons among fertilized eggs, whereas the other half are unmatched (three-allelic, not shown) and do not result in diploid males. On average, 25% of fertilized eggs are expected to become diploid males in brother-sister crosses.

males, and because they have larger cells, their bristle spacing on the wings is larger (e.g. see Grosch, 1945). Although diploid males have been found in approximately 50 species of Hymenoptera, they are not always the result of homozygosity at the sex locus but, instead, may have arisen by mutation.

Species with a slCSD mode of sex determination are difficult to maintain in culture and are often lost due to a diminishing number of sex alleles and a concomitant increase in (sterile) diploid males (Stouthamer *et al.*, 1992b). Problems with culturing may, therefore, provide a first indication for this mode of sex determination. It also means that one should be cautious if using parasitoids that have been in culture for some time, when testing for slCSD. Cook (1993b) provided a quick test based on brood survival and secondary sex ratio to determine whether one is dealing with an inbred population that suffers from mortality of diploid males due to a single locus two allele system of sex determination (Figure 3.14). If 50% of fertilised eggs are male and fertilisation proportion is 100%, the primary sex ratio will be 50% (diploid) males in the absence of diploid male mortality, indicated by the right side of the graph. The left side of the graph corresponds to a situation in which all diploid males would die if fertilisation were 100%, resulting in a sex ratio of 0, and 50% of eggs developing into (surviving) females. Thus, if values fall outside the shaded area, there is insufficient mortality to generate the degree of female bias observed and slCSD is not possible. There are, however, a number of limitations to the test: (1) it only applies to species with female-biased sex ratios and measurable mortality; (2) it cannot establish the presence of diploid males; and (3) there may be reasons for reduced brood survival other than diploid male production, e.g. suboptimal hosts or host immune reactions. If this test is not possible or proves inconclusive, other methods need to be used such as those described above.

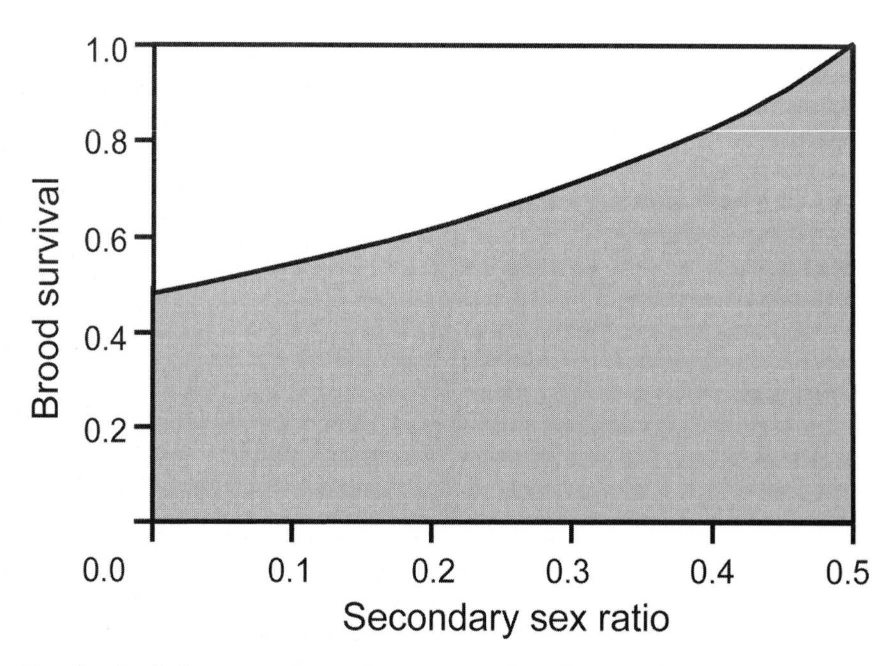

Figure 3.14 Test for single locus sex determination using brood survival and secondary sex ratio. Strains falling outside the shaded area cannot have single locus CSD. Points falling in the shaded area indicate that the experimental strains may have CSD. Source: Cook (1993b). Reproduced by permission of Blackwell Publishing.

Alternative models for sex determination

Single locus CSD is clearly absent in a number of wasp species. Prolonged inbreeding experiments over many generations with the pteromalid *Nasonia vitripennis* (Skinner and Werren, 1980) and the bethylid *Goniozus nephanditis* (Cook, 1993b) did not yield diploid males. Several alternative models have been proposed to explain sex determination in those species, including multi-locus CSD (mlCSD), maternal effect and genomic imprinting (reviewed in Cook, 1993a; Beukeboom, 1995). Dobson and Tanouye (1998) presented some recent evidence for a role of genomic imprinting in sex determination in *N. vitripennis*, but there are alternative interpretations of their results (Beukeboom *et al.*, 2000).

A slCSD or mlCSD mechanism of sex determination is likely to be absent in some species because it conflicts with their population biology or reproductive mechanism. Selection against CSD is expected with a population structure that is characterised by high inbreeding. This would impose a strong genetic load on a population because diploid males are typically sterile and less viable (Crozier, 1977; Cook and Crozier, 1995). Many parasitoid wasp species, including several chalcidoids, exhibit strong natural inbreeding (Hardy, 1994). In addition, certain forms of thelytokous reproduction lead to complete homozygosity (see above). This is inconsistent with CSD because all such offspring should develop into diploid males rather than females (reviewed by Cook, 1993a).

Gynandromorphs and Intersexes

A **gynandromorph** or **sex mosaic** is an individual which has both male and female characteristics. Such individuals have been reported from many insect orders. Some examples are shown in Figure 3.15. Gynandromorphs can arise in a number of ways. In diploid organisms, gynandromorphs usually result from loss of one sex chromosome in some cell lineages. In haplodiploids, gynandromorphs are often mosaic for haploid and diploid tissue. They can arise from mitotic loss of one set of chromosomes in certain cell lineages, or from fertilisation of one nucleus in a binucleate egg (post-cleavage fertilisation). Cold treatment of newly laid eggs can induce gynandromorphism in *Habrobracon* (Greb, 1933; Petters and Grosch, 1976). The behaviour of gynandromorphs is determined by the sex of the brain (Whiting, 1961; Clark and Egen, 1975).

Intersexes are individuals intermediate between a normal male and female (Goldschmidt, 1915). They are genetically uniform individuals and differ from gynandromorphs that have mixtures of male and female parts. They arise from a disturbed balance of the expression of female-and male-determining genes.

3.4.2 GENETICS OF SEX RATIO

Quantitative Genetics of Sex Ratio

Sex ratio is usually expressed as the proportion of males among progeny. Most diploid organisms have equal proportions of males and females (sex ratio = proportion males = 0.5). In contrast, haplodiploid organisms frequently have female-biased sex ratios (sex ratio < 0.5). The **primary sex ratio** refers to the initial proportion of male eggs immediately after oviposition. It can be determined using cytological techniques (van Dijken, 1991, see above). Due to differential mortality of the sexes during egg, larval or pupal development, the **secondary sex ratio** may be very different from the primary one.

Although there have been numerous theoretical and empirical studies of sex allocation decisions in the Hymenoptera (e.g. Charnov, 1982; Hardy, 1992, 1994a; Antolin, 1993; Wrensch and Ebbert, 1993), the genetic basis of sex ratios is much less studied. There are several approaches for studying the genetic basis of sex ratios: (1) breeding of iso-female lines; (2) selection experiments; and (3) quantitative genetic analyses. A first indication for the existence of genetic variation can be obtained from scoring progenies of field collected iso-female lines under similar laboratory conditions. Such geographical variations in sex ratio have been

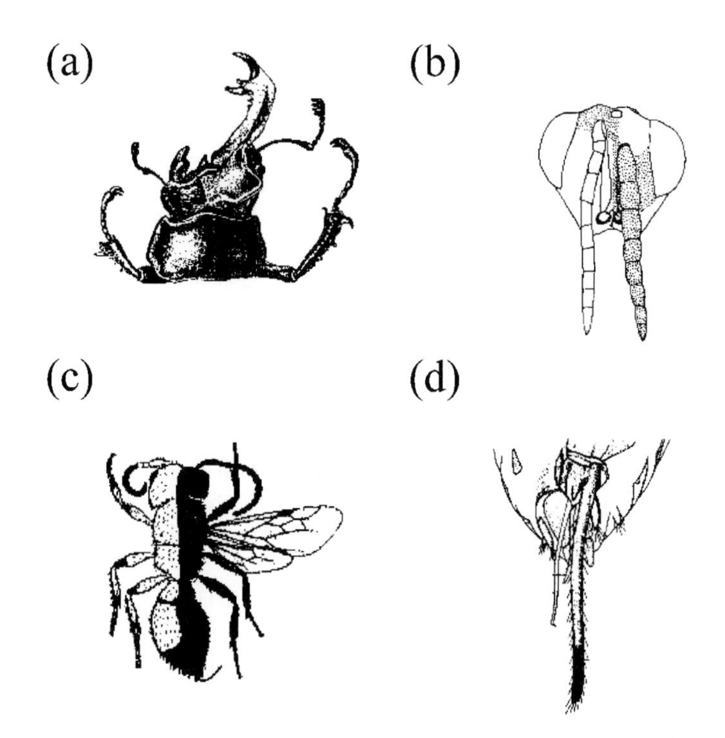

Figure 3.15 Examples of gynandromorphic animals. (a) Head and thorax of the stag beetle *Lucanus cervus*, right = male (head with sculptured ridges, large mandible, long antenna); left = female (head without ridges, small mandible, short antenna). Source: Stern (1968) after Dudich (1923). Reproduced by permission of Gustav Fisher Verlag GmbH & Co.; (b) Frontal view of the head of the chalcid *Hockeria rubra*, stippled = male (black colour) and non-stippled is female (orange colour). Source: Halstead (1988). Reproduced by permission of the Entomological Society of Washington; (c) Habitus of solitary wasp *Pseudomethoca canadensis*, right = male (black colour, 13 antennal segments, 2 ocelli, wings, male-type legs) and left = female (bright red colour, 12 antennal segments, no ocelli, no wings, female-type legs). Source: Stern (1968) after Wheeler (1910). Reproduced by permission of the Cambridge Entomological Club; (d) Ventral view of genitalia of *Habrobracon hebetor*, a complete set of male genitalia is present as well as a female ovipositor. Source: Stern (1968) after Whiting (1940). Reproduced by permission of The Biological Bulletin.

reported for a number of species (e.g. *Muscidifurax raptor* (Antolin, 1992b) and *Nasonia vitripennis* (Orzack and Parker, 1986, 1990; Orzack *et al.*, 1991), suggesting the existence of autosomal genetic variation for sex ratio.

The second approach is to attempt to select for sex ratio. Orzack and Parker (1986) were able to produce high (45% male) from low (10% male) sex ratio lines in *N. vitripennis* by selectively breeding over 15 generations those females which produced the highest proportion of males each generation (Figure 3.16). Similarly, Wilkes (1964) was able to select the sex ratio from the normal 8% to 95% males in *Dahlbominus fuliginosus*. In fact, in such experiments one selects for the proportion of eggs that are fertilised which may have very different underlying causes. Selection in *N. vitripennis* apparently involved major sex ratio genes as well as genes for behaviour of females (Parker and Orzack, 1985). In contrast, the results in *D. fuliginosus* were due to reduced functionality of sperm and had a physiological basis (Wilkes and Lee, 1965).

The third approach to the study of genetics of sex ratio is to use quantitative genetic techniques such as **parent-offspring regressions** to determine narrow and broad sense heritabilities. This is discussed below (see section 3.5). There are two phenomena that need to be considered

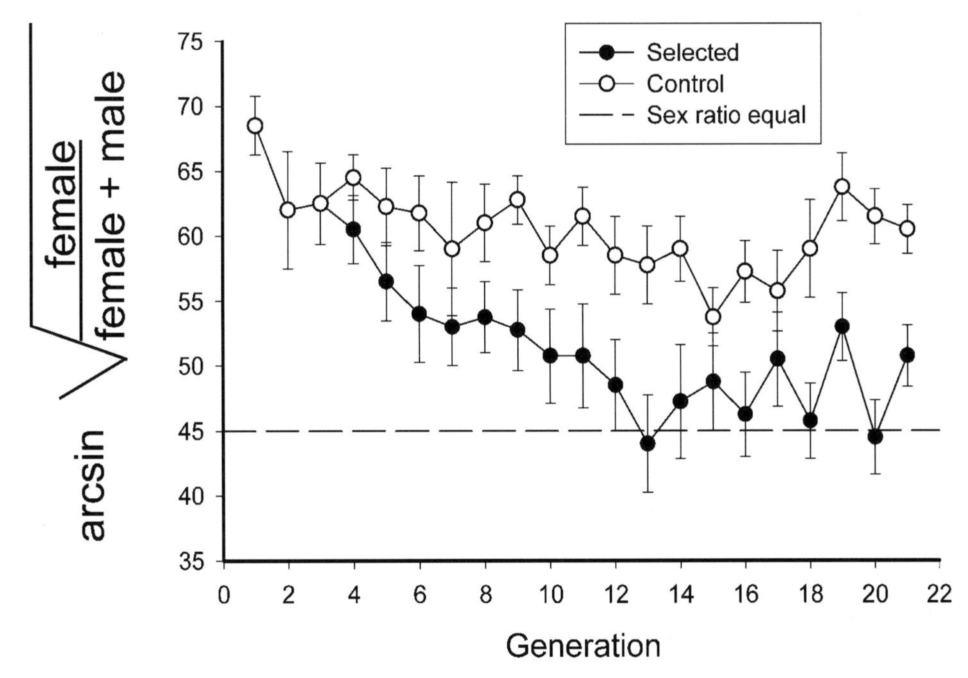

Figure 3.16 Response to selection on sex ratio in five lineages of *Nasonia vitripennis*. Data are expressed as the arcsin of the square root of the proportion females. The vertical bars indicate 95% confidence limits for the broods in each line and generation; selected lines (open circles) and control lines (solid circles). The horizontal dotted line represents equal (one to one) sex ratio. Individual selection for an increased proportion of males was performed on two replicate lines established from a heterozygous stock of wasps produced by two generations of crosses among wasps from five localities. From each generation, the ten broods in each line with the lowest proportion of females were chosen to contribute the parents for the next generation. In the control line, selected lines were chosen at random with respect to sex ratio (see for further details Parker and Orzack, 1985). Redrawn from Fig. 2B of Parker and Orzack (1985). Reproduced by permission of The Genetical Society of America.

in studies of the genetic basis of sex ratio, because they may lead to false conclusions. First, as Stouthamer *et al.* (1992b) and Luck *et al.* (1993) point out, one should be cautious with selection experiments for sex ratio in species with a CSD mechanism of sex determination. In such species, the number of sex alleles may rapidly decrease leading to a highly male-biased sex ratio. Clearly, in such cases, an increase in sex ratio is the result of the underlying sex determining mechanism rather than a response to selection on genes involved in the process of egg fertilisation. The second phenomenon is the widespread existence of nuclear and cytoplasmic **sex ratio distorters** that may overrule the plasticity in sex ratio response of individual females (see below).

More research on the genetic basis of sex ratios is needed. Little is known about genetic correlations between sex ratios and other life-history traits (e.g. diapause, Orzack and Parker, 1990). Moreover, many sex ratio models assume that sex ratio is controlled by many genes with small effect (Antolin, 1993), but this assumption largely remains to be tested. With the development of genomic map technology major progress in this field seems attainable.

Sex Ratio Distorters

Severe forms of sex ratio distortion other than thelytoky and diploid male production are known. The responsible agents vary from extra chromosomes residing in the nucleus to cytoplasmically transmitted organisms including bacteria, viruses and protozoans (see Hurst [1993] and Hurst *et al.* [1996] for an overview of cytoplasmic sex ratio distorters). As a rule they alter the sex ratio towards the sex through which they are transmitted, i.e. most increase the proportion of females because they are inherited through the egg cytoplasm (but see PSR, below). They have therefore been referred to as "selfish genetic elements" leading to "intragenomic conflict" between the autosomes (which are selected to inherit in a Mendelian fashion) and the sex ratio distorting element (Werren, 1987; Hurst, 1992). This theory predicts that suppressor genes to sex ratio distortion will evolve on the autosomes. Sex ratio distorters may be much more common than previously thought. There is some recent evidence from *Drosophila* that sex ratio distorters are frequently kept "in check" by locally adapted suppressor genes, but are expressed when crossed into a different genetic background (Atlan *et al*, 1997; Mercot *et al.*, 1995; Capillon and Atlan, 1999). Much more empirical work on the detection and dynamics of sex ratio distorting and suppressor genes is needed to understand fully their evolution and possible application in biological control.

Paternal Sex Ratio (PSR)

Perhaps the best known sex ratio distorter is the Paternal Sex Ratio (PSR) element (Werren *et al.*, 1981, 1987; Werren, 1991). It is a supernumerary chromosome in the parasitoid wasp *Nasonia vitripennis* that is present in some males. Males carrying the element cause females they mate with to produce all-male broods, even though their sperm fertilise the female's eggs. The PSR element destroys the chromosomes which are derived from the sperm nucleus after fertilisation of the egg. The maternal chromosomes are unaffected and because PSR survives itself, the resulting embryo develops into a haploid PSR-bearing male. Thus, PSR converts diploid (female) eggs into haploid PSR (male) eggs. It is considered a "selfish" genetic element because it completely eliminates its host's genes each generation (Nur *et al.*, 1988).

PSR was first discovered while attempting to select for variability in sex ratio control among field collected strains (Werren *et al.*, 1981). Werren and van den Assem (1986) showed that the trait was strongly correlated with egg fertilisation (Figure 3.17). Subsequent cytogenetic analyses showed that sperm from PSR-carrying males indeed entered the egg but that one set of chromosomes condensed into a chromatin mass during the first division of the egg (Figure 3.18) (Werren *et al.*, 1987). Using genetic markers, it was shown that the paternal chromosomes were always destroyed. Cytological investigation of testes from PSR males revealed a small supernumerary (or B) chromosome (Nur *et al.*, 1988) which is absent in control males (Figure 3.18). PSR males sometimes produced daughters among their offspring and this was shown to be due to occasional failure of the PSR chromosome to be included into sperm of carrier males (Beukeboom and Werren, 1993). Similar elements have been discovered in the parasitoids *Encarsia formosa* (Hunter *et al.*, 1993) and *Trichogramma kaykai* (Werren and Stouthamer, 2003).

Maternal Sex Ratio (MSR)

Another sex ratio distorter that has been recorded in *Nasonia vitripennis* is the Maternal Sex Ratio (MSR) factor (Skinner, 1982). MSR females produce broods that consist of daughters only, or rarely contain one or a few males. The nature of the MSR element is unknown, but it is cytoplasmically inherited and probably involves a mitochondrial variant that somehow affects the female's control over the spermatheca.

Male-killing Microbes

Maternally inherited microbes that kill male but not female hosts during embryogenesis are known from a number of insects, in particular coccinellid beetles and parasitoid wasps (Hurst,

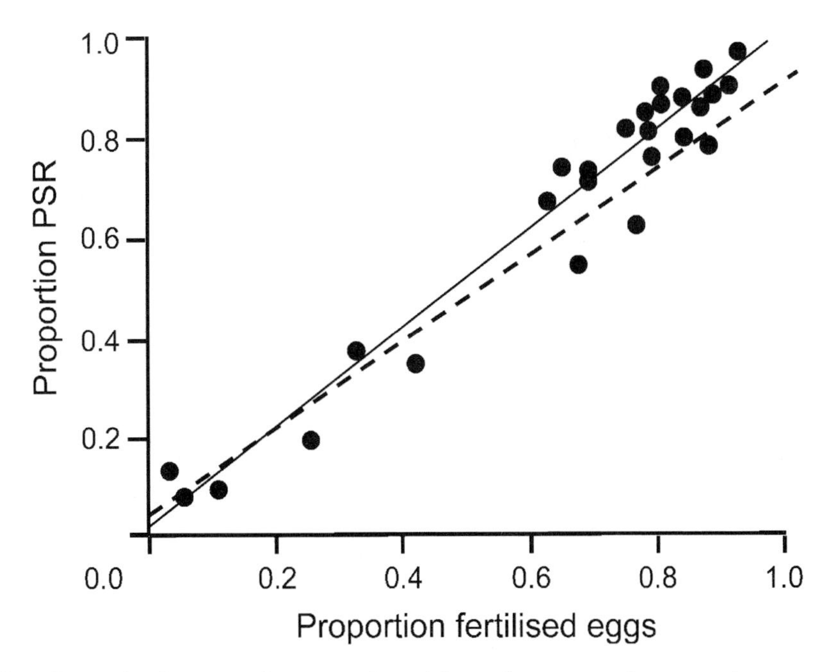

Figure 3.17 The relationship between the proportion of the male progeny that carry the paternal sex ratio factor (PSR) and the proportion of eggs fertilised by mated females of *Nasonia vitripennis*. The latter proportion was determined from control crosses in which male and female wasps were of various PSR-negative genetic strains (the percentage of females in the progeny being the percentage of fertilised eggs). The former proportion was determined for test crosses involving male wasps taken from a known PSR strain and females from the PSR-negative strains. The PSR trait was assayed as follows: females were mated with presumptive PSR males and the sex ratio of the resulting progeny scored. If a greater than 90% male brood resulted, the male parent was taken to be a PSR carrier. This phenotypic test involves a small bias since approximately 5% of control crosses also result in more than 90% males broods, as a result of inadequate mating. Where the assay proved ambiguous, at least five males from the F_1-brood were also tested, and if two or more produced all-male broods the grandparent was deemed to be a PSR carrier. The data presented are not corrected for the small bias. There is a strong linear relationship between PSR transmission and egg fertilisation, strongly suggesting that PSR is a factor transmitted from sperm to eggs upon fertilisation. In fact, PSR is an example of a parasitic B-chromosome (B-chromosomes are extra to the normal chromosome complement). PSR destroys the other paternal chromosomes in the early fertilised egg. It disrupts the normal haplo-diploid sex determination system of *Nasonia* by converting diploid (female) eggs into haploid eggs that develop into PSR-bearing males (Beukeboom and Werren, 1993). Source: Werren and Van den Assem (1986). Reproduced by permission of The Genetical Society of America.

1991). Several widely divergent microbial taxa are involved including *Rickettsia*, *Wolbachia*, spiroplasms and Flavobacteria (Hurst *et al.*, 1996). The presence of microorganisms can be readily determined, and their identity established, with molecular methods such as PCR amplification and the appropriate primers (see Chapter 6, subsection 6.5).

In the coccinellid *Adalia bipunctata*, it was noticed that some females produced female-biased sex ratios and that the success of hatching among the eggs laid by such females were low. This was shown to be caused by a *Rickettsia*-like bacterium that kills male embryos (Hurst *et al.*, 1992, 1993b). Recent evidence suggests that male-killing bacteria are common in aphidophagous coccinellids; Majerus and Hurst (1997) review the known examples and present predictions about which species groups are likely to carry male-killers.

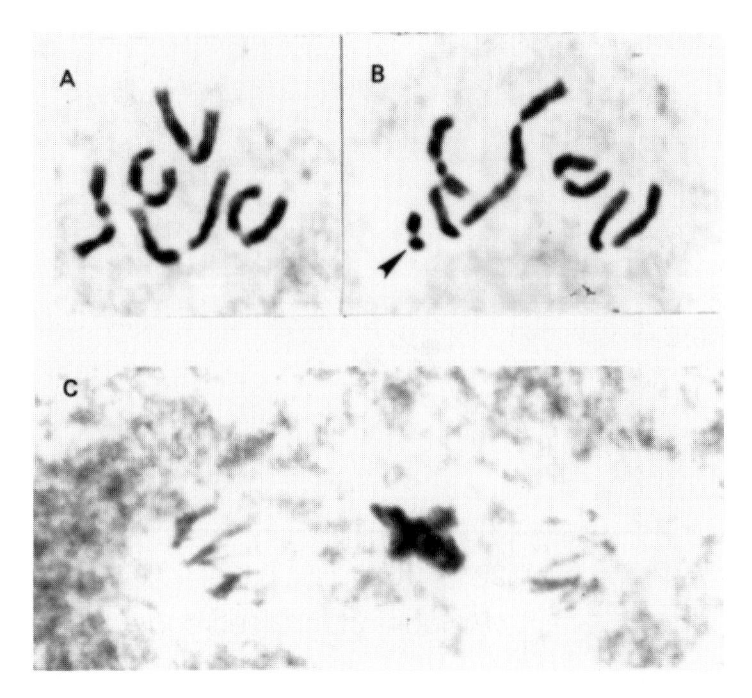

Figure 3.18 Appearance in spermatogonia, and action in fertilised eggs, of the PSR chromosome in *Nasonia vitripennis*; (a) Karyotype of a male with five chromosomes lacking PSR; (b) Karyotype of a male with five chromosomes carrying PSR (arrow). Source: Nur *et al.* (1988). Reproduced by permission of the AAAS; (c) The appearance of a dense chromatin mass of paternal chromosomes at the second mitotic division in a PSR fertilised egg. Haploid sets of maternal chromosomes are seen in anaphase. The paternal chromosomes do not participate in the mitotic division. Source: Werren *et al.* (1987). Reproduced by permission of Macmillan Journals Ltd.

Another bacterium, *Arsenophonus nasoniae*, is the cause of "son-killer" in *N. vitripennis* (Huger *et al.*, 1985; Skinner, 1985b; Werren *et al.*, 1986; Gherna *et al.*, 1991). It occurs at low frequencies in natural populations and causes all-female broods by killing male eggs only. The bacterium is present in the female's ovaries and transmitted via the eggs to offspring. It is also injected into the fly pupal host and transmitted horizontally by re-infecting other female larvae in case of superparasitism. Strictly speaking, son-killer affects *secondary* sex ratio.

Sex ratio distorting micro-organisms in insects can be detected by exposing them to a high temperature, e.g. 30°C (Luck *et al.*, 1993), but a better method is to feed infected females with antibiotics such as tetracycline or rifampicin, and then examining whether the egg hatch rate and the proportion of male progeny increases (Stouthamer *et al.*, 1990; Hurst *et al.*, 1992). Alternatively, one can stain eggs or haemolymph cells with the DNA stains DAPI or H33258 and examine them under the microscope for the presence of bacteria using cured strains as controls. Typically, hundreds of bacteria occur in a single egg (Breeuwer and Werren, 1990; Stouthamer and Werren, 1993; Hurst *et al.*, 1996).

Cytoplasmic Incompatibility

Another cause of sex ratio distortion is cytoplasmic incompatibility caused by *Wolbachia* bacteria. In several parasitoids (Luck *et al.*, 1993; Werren & O'Neill, 1997), strains are found that harbour *Wolbachia* bacteria (section 3.3.3). These bacteria are present in the egg cytoplasm. Successful fertilisation of eggs free of the bacterium can only be achieved by sperm from uninfected males, whereas eggs containing the bacterium can be fertilised by sperm from either infected or uninfected males (Breeuwer and Werren, 1990). In addition to this **unidirectional cytoplasmic incompatibility, bi-directional incompatibility** has been observed. This refers to the situation where two strains harbour different types of *Wolbachia* and both reciprocal crosses between strains are incompatible. The sperm chromosomes are destroyed in the fertilised egg in incompatible crosses (Breeuwer and Werren, 1990). Whereas this leads to no progeny at all in diploid organisms (because haploid eggs are inviable), it results in all-male broods in haplodiploids.

Wolbachia-mediated Thelytoky

Thelytoky in some parasitoid wasps is caused by *Wolbachia* bacteria. Stouthamer *et al.* (1990) demonstrated that thelytokous *Trichogramma* females began to produce sons after feeding on antibiotics. Similar reversion of thelytoky to arrhenotoky has been reported for a number of other species (reviewed in Stouthamer, 1997).

Other cases of thelytoky in parasitoid wasps may also be induced by micro-organisms. Several studies have found an increase in male production of thelytokous strains after exposing them to high temperatures (e.g. Wilson and Woolcock, 1960; Legner, 1985). They may revert to thelytoky when reared at lower temperatures. This is consistent with the hypothesis that the induction of thelytoky is dosage-dependent, i.e. high temperatures kill the bacteria and reduce their density to below a critical value, but their numbers may recover (Breeuwer and Werren, 1993; Hurst, 1993).

Feminisation

Wolbachia bacteria have also been reported to cause feminisation, i.e. change genotypically males into females. The best-studied case is that of the isopod *Armadillidium vulgare* (Rigaud, 1997), but it has also been reported from an insect, the Asian corn borer, *Ostrinia furnacalis* (Kageyama *et al.*, 1998). Although not yet reported from insect natural enemies, researchers working with such insects should be aware of the possibility.

3.5 QUANTITATIVE GENETICS

3.5.1 INTRODUCTION

Quantitative genetics studies the inheritance of traits that are of *degree* rather than of *kind* (Falconer and Mackay, 1996). **Quantitative traits (QTs)**, e.g. body size, usually show a continuous distribution of values that is approximately Gaussian. Understanding the genetics of QTs is essential for the study of any biological system because the majority of morphological, physiological, behavioural and life-history traits have quantitative characteristics.

The genetics underlying quantitative traits is often assumed to be polygenic: differences at many loci, each with a small effect, contribute to the genetic variation for the trait in the population under study. It is important to realise, however, that as few as three loci can produce normal distributions very much like the distributions shown by QTs (Thoday and Thompson, 1976). The implication of the underlying genetics, whether oligogenic or polygenic, and the continuous distribution of QTs, is that the individual genes cannot be identified by their segregation, hence Mendelian analysis does not apply. Moreover, QTs usually depend strongly on the environment in which the trait is measured or in which the organism has developed, or both. Below we will discuss how the genetic analysis of QTs has taken these aspects into account by using variance analysis and parent-offspring resemblance. Apart from classical quantitative genetics, we will also describe the possibility of using linkage maps to locate the actual genes underlying QTs, a process that is known as **quantitative trait loci (QTL) mapping**.

3.5.2 CLASSICAL QUANTITATIVE GENETICS

We will provide a brief outline of the principles of quantitative genetics, but for a more thorough introduction the reader can refer to a variety of textbooks (Falconer and Mackay, 1996; Kearsey and Pooni, 1996; Roff, 1997; Lynch and Walsh, 1998).

Quantitative traits are expressed in values. The **phenotypic value** (P) is the trait value measured on an individual. It is the sum of the **genotypic value** (G) and the **environmental effect** (E) (Falconer and Mackay, 1996):

$$P = G + E \tag{3.5}$$

The phenotypic value of an individual can be measured relative to the population mean. The population mean itself is expressed in terms of allele frequencies, genotypic values and the degree of dominance (population mean equals $a(p - q) + 2dpq$). Consider a locus with two alleles, A_1 with frequency p and A_2 with frequency q. We can assign values to the three possible genotypes A_1A_1, A_1A_2 and A_2A_2. Let this genotypic value for A_1A_1 be a and for A_2A_2 be $-a$. In the case that the alleles act additively, the genotypic value of A_1A_2 would be $((a + (-a))/2)$ equals zero. When there is dominance, the genotypic value would be different from zero. The genotypic value of the heterozygote A_1A_2 is usually called d (Figure 3.19). Therefore, d depends on the level of dominance and this is often expressed as the ratio d/a (Falconer and Mackay, 1996). Note that the genotypic value can be measured, but this is only practically possible when dealing with a single locus situation in which all genotypes can be phenotypically distinguished.

An important concept in quantitative genetics is the **breeding value** of an individual, which can be measured on the mean phenotypic value of its progeny (Falconer and Mackay, 1996). Therefore, the breeding value is the sum of the average effects of the parents genes on their progeny and this is what most researchers are interested in. Breeding values are a property of the gene and the population and depend on the degree of dominance and the allele frequencies. Only in the absence of dominance are breeding values and genotypic values the same.

Variances

As mentioned above, QTs show a continuous distribution of phenotypic values with a more or less normal distribution. In principle, the amount of variation around the population mean reflects both the effects of genes for the trait as well as the effect of the environment on the trait. Partitioning of variance components will show the relative importance of genetics and hence the potential for (natural) selection and/or evolution of the trait in that particular population. The total **phenotypic variance** (V_P) is:

$$V_P = V_G + V_E \tag{3.6}$$

where V_G and V_E are the **genetic variance** and the **environmental variance** respectively. The genetic variance is partitioned into the **additive** (V_A) and **interaction components** (within gene: the **dominance component** V_D; between genes: the **epistatic component**, V_I, that measures the variance due to interaction between genes). Hence:

$$V_P = V_A + V_D + V_I + V_E \tag{3.7}$$

By definition, additive effects are independent of the genetic background in which the alleles in question are placed. In contrast, the effect of alleles that show within and/or between gene interaction depends on the genotype in which they occur. Since parents pass on their genes and not their genotypes, the additive genetic component is the most significant if one wants

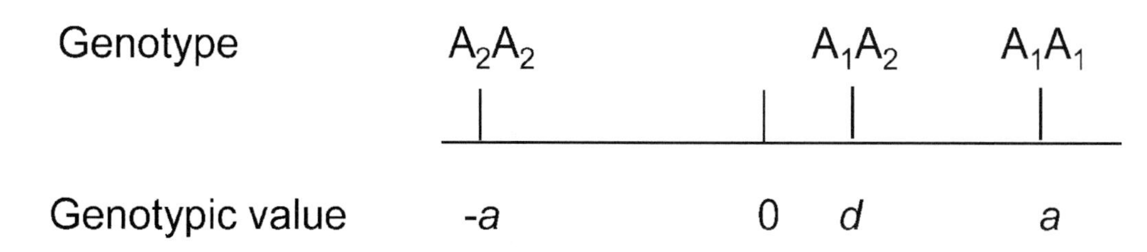

Figure 3.19 Schematic depiction of genotypic values for a one locus-two allele situation. a measures the additive effect and d is a measure of the dominance effect. When $d = 0$ the alleles are completely additive, when $d = a$, A_1 is completely dominant over A_2 and when $d = -a$, A_2 is completely dominant over A_1.

to predict the outcome of short-term selection on the trait and/or evolution of the trait. That is not to say the other components should be ignored. Dominance and epistatic variation can be converted into additive genetic variance due to bottlenecks or drift (Cheverud and Routman, 1996; Roff, 1997). In addition, studying the ratio between V_D and V_A is important, for example, for the study of the mechanisms behind the maintenance of genetic variation (Charlesworth, 1987).

Heritability

The **heritability (h^2)** is the ratio between the genetic variance and the total, phenotypic, variance. Hence it measures the degree to which a QT is genetically determined in a population. Two h^2 measures are distinguished, the **broad-sense h^2**:

$$h^2 = V_G/V_P = (V_A + V_D + V_I)/V_P \qquad (3.8)$$

and the **narrow-sense h^2**,

$$h^2 = V_A/V_P \qquad (3.9)$$

Following the reasoning above, the QT's narrow-sense h^2 is the most informative about the potential for selective responses in the QT.

The relationship between more QTs is measured as the correlation between two traits. The **genetic correlation** uses the covariance between two traits and can be established using several experimental designs (see below and Roff [1997] for details). Genetic correlations are an important issue in evolution in general and for the evolution of life-histories in particular. Negative genetic correlations between traits, or so called **trade-offs**, are especially important because they can impose (short-term) constraints on adaptive evolution in populations: a trait value can not increase or decrease without a concomitant decrease or increase in the other, negatively correlated, trait.

The above shows that the genetics of QTs can be satisfactorily described in terms of statistical variances. However, a major caveat associated with QTs is their strong environmental subjectivity. This extends to both the general effect of the environment on the individual's phenotypic trait value, as well as the differential response of genotypes to environmental change, a phenomenon known as **gene-environment interaction**. The careful reader will have noticed that the G x E interaction should be added to the variance components; however, usually, this component can be ignored (for more details see Falconer and Mackay [1996]; Lynch and Walsh [1998]). The general effect of the environment will seriously bias estimates of genetic variances and heritabilities. When, for instance, the rearing conditions are poor, at least several morphological, physiological and life-history traits will be negatively affected. The phenotypic variance will be increased, hence h^2 estimates will be deflated and any genetic component to V_P will be underestimated. Therefore, estimates of variance components and h^2 apply only to the environment in which they are measured and any extrapolation outside this environment should be done with caution. Having said all this, in evolutionary ecology heritability estimates should be done in the field. Such estimates have been reported to be lower in the field as compared to controlled laboratory conditions as a result of e.g. increased V_P (Hoffmann and Schiffer, 1998) or a decrease in V_A and increase in V_P (Simons and Roff, 1994). Surprisingly though, an extensive literature review showed that the h^2 estimates of morphological and life-history traits were higher in the field than in the laboratory, but that the reverse is true for behavioural traits (Weigenberg and Roff, 1996; Roff, 1997). Moreover, in these studies, a strong correlation existed between field and laboratory estimates of h^2. This general trend implies that the combination of effects on V_A and V_P should be measured and it also suggests that gene-environment interactions may affect V_G and V_A specifically. Note that V_G does not measure which gene is contributing to the genetic variance; consequently, although genetic variance estimates can be similar, very different genes may contribute to this variance. Studying G x E interaction is an interesting field in itself and has many implications for theories of evolution, such as the maintenance of genetic variation (Gillespie and Turelli, 1989).

Experimental Designs

Several experimental designs can be used to estimate the quantitative genetic parameters mentioned above. The concepts above and below are all based on the genetic relationship between relatives, or the likelihood that they share the same alleles. The degree to which a trait is associated between two individuals (the covariance) is generally given by (ignoring epistatic interactions):

$$cov = rV_A + uV_D \quad (3.10)$$

with r the probability of sharing the same allele, and u the probability of having the same genotype for a locus (Margolies and Cox, 1993; Falconer and Mackay, 1996). Obviously, the covariance depends on the genetic system that determines the relatedness between individuals, and is therefore very different for diploid and haplodiploid species (Margolies and Cox, 1993). Many insect natural enemies are parasitoid wasps, which are haplodiploid, hence the experimental designs will be discussed for both genetic systems.

Full-sib Design

This experimental design is very often used because it can be implemented in all biological systems. Single crosses are set up between a female and a male and the resulting families are raised separately. For the diploid situation, the covariance will be:

$$cov_{FSD} = 1/2V_A + 1/4V_D \quad (3.11)$$

This can be reasoned as follows. For a cross between A_1A_2 and A_3A_4, the four resulting progeny genotypes are, A_1A_3, A_1A_4, A_2A_3 and A_2A_4. Comparing any one genotype with all the possible genotypes (full-sib comparisons) shows that on average in half the cases an allele is shared and in one fourth of the cases a genotype is shared.

For the haplodiploid situation, we have to discriminate between males and females. Males are haploid, therefore by definition the within-gene interaction (V_D) is absent. The males will on average share half their alleles with each other, $r = 0.5$, hence (using equation [3.10]):

$$cov_{FSHM} = 1/2V_A \quad (3.12)$$

For females, the cross can be given as between A_1A_2 and A_3. The females in the progeny will be either A_1A_3 or A_2A_3. Therefore, following the reasoning above, $r = 3/4$ and $u = 1/2$, hence (using equation [3.10]),

$$cov_{FSHF} = 3/4V_A + 1/2V_D \quad (3.13)$$

An important point to notice in the full-sib design is that only for the haplodiploid males can the additive genetic variance be estimated. for all the other cases, estimates of genetic variance include the dominance deviation, hence heritability estimates will all be broad-sense. A full-sib design has successfully indicated genetic variation for oviposition behaviour in *Trichogramma maidis* (Chassain and Bouletreau, 1987), but the full significance of the variation for selective changes in nature remains obscure without knowing the individual components of this variation.

A technique that is often applied in parasitoid wasps is the use of iso-female lines (Parker and Orzack, 1985; Antolin, 1989; Prevost and Lewis, 1990; Cronin and Strong, 1995; Kraaijeveld *et al.*, 1998) (see 3.2.3). The main reason for this is, perhaps, that using the asexual production of males and then back-crossing it to the mother (Kraaijeveld *et al.*, 1998) is an easy way of obtaining near-isogenic lines. The genetic variance is then estimated from the between-lines mean squares in an analysis of variance. Again, this procedure estimates the V_G and broad-sense heritabilities. Cronin and Strong (1996) discuss the possibilities of natural and artificial selection using data from isogenic lines; however, their conclusions may be confounded by the lack of knowledge of the individual genetic variance components. Another potential problem with using iso-female lines is the interpretation of correlations between traits. Using iso-female lines will bias towards positive genetic correlations, because the best line (genotype) is best in everything and linkage between genes is mistaken for pleiotropic action of single genes.

A full-sib design has also been used to study phenotypic plasticity in adult abdominal colour pattern in the hover-fly *Eristalis arbustorum* (Ottenheim *et al.*, 1996). Offspring of single females were reared at six different temperatures, ranging from 8 to 26°C, and significant variation between females were found for the slope of colour pattern on temperature. This indicated genetic variation for plasticity.

Half-sib Design

The problem of the broad-sense genetic variance component can be circumvented using a half-sib design. For biological reasons, usually a male (**sire**) is mated to several females (**dams**) and this for a large number of males. Half-sib designs are more difficult to perform because it requires a biological system were multiple mated males are easily obtainable.

For the diploid situation, consider a male, A_1A_2 that has been mated with two females, A_3A_4 and A_5A_6. The resulting progeny of these families is A_1A_3, A_1A_4, A_2A_3 or A_2A_4, and A_1A_5, A_1A_6, A_2A_5 or A_2A_6 respectively. Comparing any genotype from one family with all the genotypes from the other family (half-sib comparison), shows that on average in one-fourth of the comparisons an allele is shared and in no case is a genotype shared ($r = 1/4$, $u = 0$). Hence (using equation [3.10]):

$$cov_{HSD} = 1/4V_A \qquad (3.14)$$

The haplodiploid cross can be exemplified as, A_1 (male) crossed with A_2A_3 and A_4A_5 (females). The resulting genotypes of the progeny are A_2 or A_3 (males) and A_1A_2 or A_1A_3 (females), and A_4 or A_5 (males) and A_1A_4 and A_1A_5 (females) respectively.

Clearly, half-sib males do not share any allele and can therefore not be used to estimate genetic variance components. For females, half-sibs share half of their alleles, but never the genotype ($r = 1/2$, $u = 0$). Hence (using equation [3.10]):

$$cov_{HSHF} = 1/2V_A \qquad (3.15)$$

Note that in all cases the additive genetic variance component can be directly estimated. Table 3.2 summarises the results and gives a practical guide to the interpretation of the nested analysis of variance (as performed by most statistical computer packages) and to the

Table 3.2 Nested analysis of variance for a half-sib design with n males mated with m females and each family containing r progeny. (A) ANOVA table with source of variation, degrees of freedom (df), mean squares (MS) and expected mean squares components (after Kearsey, 1996). (B) The relation between causal and mean squares components (after Margolies, 1993).

A			
Source	df	MS	Components
Between male half-sib family groups	n − 1	MS_M	$\sigma^2_W + r\sigma^2_{F(M)} + rm\sigma^2_M$
Between full-sib families within male half-sib family group	n(m − 1)	$MS_{F(M)}$	$\sigma^2_W + r\sigma^2_{F(M)}$
Within full-sib families	nm(r − 1)	MS_W	σ^2_W
Total	nmr − 1		

B		Causal components	
Components	Covariance estimate	Diploid	Haplodiploid
σ^2_M	cov_{HS}	$\frac{1}{4}V_A$	$\frac{1}{2}V_A$
$\sigma^2_{F(M)}$	cov_{FS}-cov_{HS}	$\frac{1}{4}V_A + \frac{1}{4}V_D$	$\frac{1}{4}V_A + \frac{1}{2}V_D$
σ^2_W	V_P-cov_{FS}	$\frac{1}{2}V_A + \frac{3}{4}V_D$	$\frac{1}{4}V_A + \frac{1}{2}V_D$
$\sigma^2_W + \sigma^2_{F(M)}$	cov_{FS}	$\frac{1}{2}V_A + \frac{1}{4}V_D$	$\frac{3}{4}V_A + \frac{1}{2}V_D$
$\sigma^2_W + \sigma^2_{F(M)} + \sigma^2_M$	V_P	$V_A + V_D$	$V_A + V_D$

calculations of the variance components. It is clear that in a half-sib design V_D can also be estimated. However, without replication, there is no direct test of its significance (Kearsey and Pooni, 1996).

A half-sib design was conducted to study the cost of chemical defence in the two-spot ladybird, *Adalia bipunctata* (Holloway *et al.*, 1993). All characters measured showed high levels of additive genetic variances. No clear negative covariances were found between the life history characters body weight and growth rate. This was thought to be due to sex dependent gene expression.

Another point of interest to raise is the environmental variance. This variance can be partitioned into a general environmental effect (V_{Ew}) and an environmental effect arising from siblings being raised in a common environment (V_{Ec}) (Falconer and Mackay, 1996). For the analysis of variance (the resemblance between relatives), only the latter component is important because it can cause differences between unrelated groups that are not the result of genetic differences between the groups, but are due to the common environment that individuals within each group share. Careful experimental design can take care of this, by rearing members from each family in more than one group and/or randomise members over the rearing environment (Falconer and Mackay, 1996).

Test Crosses

The above described half-sib design is known as the **North Carolina Experiment I** (NCI) (Kearsey and Pooni, 1996). Several varieties exist, such as the NCII, in which each male is crossed to each female. Naturally, this can only be done if the lines used in the crosses are (near)-isogenic. In this design the interaction variance between female and male half-sib families can be used to directly test for the significance of V_D (Kearsey and Pooni, 1996). Using **Triple Test Crosses (TTCs)** the between-gene epistatic interaction component (and in turn its components, like dominance by dominance interaction) can also be estimated. It can be used if one has highly inbred or isogenic lines and involves back-crossing of F_1 and F_2 progeny from two parental strains, to both the parental strains and their F_1 (for F_2 progeny). Discussing these complex designs (note for instance that four possible F_2 progeny can be made) in full is beyond the scope of this chapter; we recommend Kearsey and Pooni (1996) for an excellent introduction to this subject.

Parent-offspring Regression and Realised Heritability

Heritabilities can be calculated whenever broad or narrow-sense genetic variation estimates are available (equations [3.8], [3.9]). There are two additional ways to estimate h^2, using parent-offspring regression and realised heritabilities.

Parent-offspring regression is relatively straightforward. The phenotypic value is measured on preferably both parents and on their offspring. The values of the progeny are subsequently regressed on the values of the parents (using many families as individual data points). The slope (b) of the regression line is an estimate for h^2 (Falconer and Mackay, 1996).

The regression can be done using the mid-parent values and the average of the progeny. In that case, for diploids:

$$h^2 = b \tag{3.16}$$

Alternatively, values of the sons can be regressed on the fathers, and values of the daughters on the mothers. Since, in the diploid case, fathers and sons, and mother and daughters on average only share half their genes:

$$h^2 = 2b \tag{3.17}$$

If differences between son-father and daughter-mother regressions are apparent, it is an indication of maternal effects or genomic imprinting influencing the trait under consideration. "son" and "father" regression is not possible for haplodiploid species, because they do not share any genes (see above). In haplodiploids, the daughter-mother regression is identical to diploids. Parent-offspring regression for sex ratio traits in *Nasonia vitripennis* showed a heritability of around 0.10 (Orzack and Gladstone, 1994).

The selection response (R) in an experiment depends on the selection differential (S) between the mean value of the population and the mean value of the selected group, and the heritability of the trait:

$$R = h^2 S \tag{3.18}$$

Hence, knowing the selection response and the selection differential, the heritability can be estimated. This can be done in a one-generation experiment, but the so-called **'realised heritability'**, is usually calculated over several generations. The selection response is regressed on the cumulative selection differential over the generations, thus given the slope of the regression line as a direct estimate of h^2. The response should be measured against a certain control line, to take away the between generation general environmental effects. This estimation procedure assumes that the additive genetic variance remains unchanged over the experiment. When selection is successful this is obviously not the case, therefore, realised heritabilities should be estimated over the first part of the experiment where this assumptions is less likely to be violated. In haplodiploid species this h^2 estimate should be multiplied by a factor 3/2, because fewer genomes contribute to the selection response (Bulmer and Bull, 1982). The adjusted realised heritability for sex ratio in *Nasonia vitripennis* was on average 0.12 (Parker and Orzack, 1985). No selection response was found, however, in a similar experiment for sex ratio in *Trichogramma fasciatum* (Ram and Sharma, 1977).

General Considerations

Estimation of genetic variance components and heritabilities is notoriously difficult. In general, large sample sizes are needed, and depending on the questions one wants to answer, (Houle, 1992), the number of sires, dams, or progeny per dam should be optimised (Roff, 1996; 1997). The haplodiploid system requires fewer individuals to be measured than the diploid system to achieve the same level of efficiency, because r (the probability that individuals share the same allele) is greater in the former (see above). For the half-sib design, using a haplodiploid species requires half as many female per male as compared to using a diploid species because $r = 1/2$ (equation 3.15) and $r = 1/4$ (equation 3.14), respectively (Margolies and Cox, 1993).

According to population genetic theory, little additive genetic variation for total fitness exists in populations, above the amount generated by mutation. This is generally corroborated by the observation that life-history traits have lower heritabilities than morphological traits, with heritabilities of behavioural and physiological traits somewhere in between (Mousseau and Roff, 1987; Roff, 1997). Trade-offs between traits and $G \times E$ interactions for traits can, among other factors, significantly promote the maintenance of genetic variation (Roff, 1992; Stearns, 1992; Lynch and Walsh, 1998). As recently reviewed by Kraaijeveld *et al.* (1998) and discussed below, trade-offs in host-parasite systems play an important role in balancing the costs and benefits of resistance to the parasitoid, thus maintaining the genetic variation in resistance and virulence as observed between (geographical) distinct populations and in selection experiments. The potential genetic complexity of QTs closely related to fitness and the genetic correlations between the QTs are important aspects to study in order to further our understanding of host-parasite systems and are essential for implementing successful biocontrol programmes (see also Antolin, 1989).

3.5.3 GENETICS OF HOST-PARASITE INTERACTIONS

Introduction

The potential for co-evolution is particularly high in host-parasitoid systems. Parasitism directly affects the survival of the host's offspring (eggs, larvae or pupae) which is greatly reduced, mostly to zero. As a consequence, the reproductive success of the host is greatly reduced. Therefore, selection is strong for genetic variants that increase the survival of offspring either through avoidance or through resistance against the parasitoid. In turn, the reproductive success of the parasitoid directly depends on the survival of its eggs in the host. Hence, mutants with offspring that (more) efficiently circumvent the host's defence systems, i.e. are more virulent, will have a high(er) reproductive success. In summary, the reciprocal selection pressures on the parasitoid and the host are high and may result in a 'tit for tat' accumulation of adaptations and counter-adaptations.

A prerequisite for natural selection and co-evolution is the existence of genetic variation for both virulence and resistance. A first indication of the presence of such genetic variation comes from virulence and resistance differences between strains from different geographical regions. Especially if such variation shows longitudinal or latitudinal clinal patterns, it can be taken as evidence that different selection pressures are operating in the different populations. Moreover, reciprocal (genetic) interactions between hosts from one region and parasites from the other will indicate co-evolutionary selection and local adaptation. Geographical variation has been found for the virulence of the parasitoid *Asobara tabida* (Figure 3.20) and the resistance of its host *Drosophila melanogaster* (Kraaijeveld and van Alphen, 1995a). In this system the single most important factors determining this variation were considered to be the presence or absence of alternative hosts, and competition by other parasitoid species. Furthermore, only one reference strain of *D. melanogaster* was used which prevents any definite conclusions about local adaptation and co-evolution (Kraaijeveld *et al.*, 1998). As shown below, local adaptation has been found in the *Leptopilina boulardi – Drosophila melanogaster* system (Carton and Nappi, 1991).

The Traits

Host: Avoidance and Resistance

Hosts can reduce the cost of parasitism both through avoidance and resistance. Avoidance can be achieved by changes in the host's behaviour. For instance, larval activity could be reduced to avoid parasitoids that rely on vibrotaxis, or larvae can feed deeper into the substrate to avoid the parasitoid's ovipositor (subsection 5.5).

Resistance can involve both humoral and cellular responses. Some instances of humoral responses have been studied (see references in Kraaijeveld *et al.*, 1998) but the cellular response is much better investigated both at the physiological and molecular level. Immunology underlies the cellular response, such that certain cells (haemocytes) in the host's haemocoel attach to the foreign parasitoid egg and the continuing aggregation of such cells to the egg results ultimately in the death of the egg. This mechanism of encapsulation is a general defence mechanism seen in all arthropods (see subsection 2.10.2).

Parasite: Avoidance and Resistance

The strategy of the parasitoid for counteracting the host's defence mechanisms has also be divided into avoidance and resistance traits, although in some cases this distinction is not very transparent. **Avoidance** can be considered a passive strategy and refers to eggs that are not in contact with the haemolymph, such as eggs laid in tissue that is devoid of haemocytes or that cannot be reached by haemocytes. Ectoparasitoids also practice avoidance (Askew 1971).

Resistance is considered an active strategy for eluding the host's defences. Eggs are in contact with the hemolymph. At oviposition, the parasitoid female can inject substances that paralyse the immuno-responses. For example, injection of teratocytes by the female serves to deplete haemocytes by deflecting encapsulation from the egg to the teratocytes. Another resistance strategy is to lay more than one egg per host to quickly deplete the number of haemocytes. It has been shown for *Asobara tabida* that an egg laid in a *Drosophila* larva parasitised by *Leptopilina boulardi* has a higher chance of survival than one laid in an unparasitised host (Kraaijeveld, 1999). Other examples of resistance in the braconid *Asobara tabida* are fast-growing eggs that outcompete haemocytes that are not

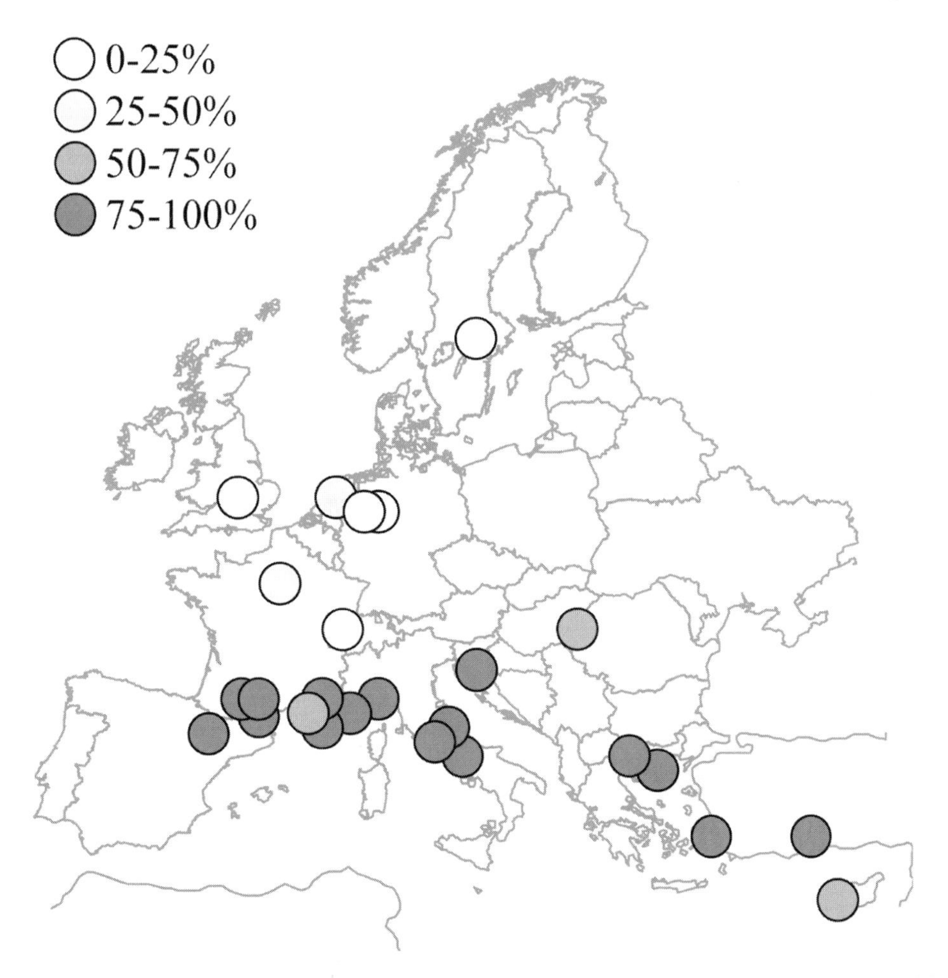

Figure 3.20 Geographical variation within western Europe in avoidance of encapsulation of the parasitoid *Asobara tabida* by a reference strain of its host *Drosophila melanogaster*. Source: Kraaijeveld and van Alphen (1994). Reproduced by permission of Cambridge University Press.

present in sufficient numbers to encapsulate the growing egg/embryo (Mollema, 1988) and sticky eggs that attach to host tissue which protects them from complete coverage by haemocytes (Kraaijeveld and van Alphen, 1994; Monconduit and Prevost, 1994). Finally, **molecular mimicry** has also been described in which the parasitoid egg is recognised as 'own' instead of 'foreign' (Feddersen *et al.*, 1986).

Intriguingly, other parasitoids use symbiotic association with viruses to attack the immune system of the host. The viruses are either securely integrated in the parasite's genome or end up in the parasite because of feeding in the host. The integrated viruses only replicate in the female parasitoid prior to oviposition. Other parasitoids use **virus-like particles** (**VLPs**, e.g. *Leptopilina boulardi*) to counteract the host immune reaction. VLPs have structures that superficially resemble viruses but that lack nucleic acids (Quicke, 1997). VLPs are used in several ways by the parasitoid. They can form a shield around the parasitoid egg in order to make it invisible for the host immune system (e.g. *Venturia canescens*). This is a form of molecular mimicry. The VLPs are coded for by the parasitoid genome and for *Venturia* the gene has been cloned (Theopold *et al.*, 1994). VLPs can also be used to directly attack the haemocytes. For instance, in *Leptopilina heterotoma* VLPs enter the haemocytes named lamellocytes and change the morphology and surface characteristics of these cells such that encapsulation is blocked (Rizki and Rizki, 1990).

Practical Aspects

Measuring Avoidance and Resistance

The traits involved in host and parasite avoidance and resistance are most accurately measured under controlled conditions. This usually entails rearing both the host and the parasite in laboratory situations. Most studies that measure the abilities of the host and the parasite use a single reference strain (Kraaijeveld *et al.*, 1998). For instance, if one wishes to assess the ability of parasitoids from different populations to successfully find the host and/ or circumvent a host's defences, the parasitoids are supplied with hosts from a single strain that has been reared under controlled conditions (e.g. Kraaijeveld and van Alphen, 1994). Such **reference strains** are usually genetically (near-)isogenic to avoid any possible variation in parasitoid performance caused by genetic variation in the host.

However, there is a serious problem with using a single, genetically uniform, reference strain: gene-environment interactions. This problem arises if the individuals that are being compared are reared in an environment to which they are not equally suited (Stearns, 1992). In this case, any differences found between the parasitoids in the above example may well be spurious. This is particularly a problem when one intends to measure negative genetic correlations, i.e. trade-offs (Partridge, 1989). Moreover, (near-)isogenic, hence inbred, lines may have reduced performance with respect to the trait(s) of interest. Although, this may not affect the relative ability scores of the parasitoids, it will certainly affect the absolute score values. As mentioned above, experimental conditions should be controlled. This is not to say, however, that no varying conditions should be applied, but that these should be applied in a controlled manner. Hence, to avoid the problems associated with the use of a single reference strain, more host strains should be used in order to investigate host source and parasitoid source interactions. This could be combined with research on local adaptation. For instance, Carton and Nappi (1991) reported that a central African population of *Leptopilina boulardi* was not successful in evading the host encapsulation defence. The parasitoid tested against the sympatric host population showed that 42 percent of the eggs were encapsulated. However, when tested against four allopatric host populations, encapsulation percentages ranged from 49 to 78. This is a clear example of local adaptation of the parasitoid to its host. This would have been missed had a single host strain been used.

Genetic Aspects

In principle, all standard genetic tools can be employed to study the genetics of host and parasite traits. To date, mostly analysis of iso-female lines, selection experiments and line crosses and segregation analysis are used to estimate genetic variation. In addition, Henter (1995) used a half-sib design to further partition the genetic variances for the virulence of *Aphidius ervi*. It was shown that 26 percent of the phenotypic variation was due to additive genetic factors. In general, the estimates of genetic variance and heritability (see section 3.5.2) were moderately high (0.2-0.7; for review see Kraaijeveld *et al.*, 1998). This shows that there is scope for selection on both the parasite and the host traits.

The genetic basis of virulence and resistance traits is still relatively unexplored. As mentioned, quantitative genetic tools have been employed because traits such as encapsulation responses can be considered to be quantitative. However, a quantitative trait does not necessarily implicate that it is polygenically determined with many loci with small effect. In fact, the *ability* to encapsulate might well be determined by a single gene.

Most early studies tend to conclude that encapsulation has a polygenic basis with the evidence resulting from line crosses (reviewed by Kraaijeveld *et al.*, 1998). Analysis of line crosses is subject to substantial errors e.g. when sample sizes are low (Roff, 1997; Lynch and Walsh, 1998). To analyse fully these instances, line crosses should be used in combination with gene mapping, a subject of the next section (3.5.4). In contrast, recently, evidence is mounting that the traits involved in several host-parasite systems are determined by major genetic factors.

One striking example of a single gene is the **sitter/Rover polymorphism** that affects the moving behaviour of the host *Drosophila melanogaster* (Sokolowski, 1980). The Rover allele is dominant over the sitter allele and Rover larvae actively move through the medium, while the sitter larvae are more sessile. As recently reviewed by Kraaijeveld *et al.* (1998) this genetic polymorphism has major consequences for the likelihood of being parasitised e.g. by *Asobara tabida*.

In the *Leptopilina boulardi-Drosophila melanogaster* system heritabilities were measured using the iso-female line procedure (Carton and Nappi, 1991). The heritability for encapsulation was estimated at 0.43, while the heritability for evading the encapsulation was estimated between 0.3 and 0.5. On the basis of these data one would be inclined to conclude that both traits have a polygenic basis. However, analysis of crosses between susceptible and resistant isogenic *Drosophila* lines (two parental line crosses, two reciprocal F_1 hybrids, eight reciprocal back-crosses and four reciprocal F_2 hybrids) strongly indicated that encapsulation is a monogenic trait with resistant being dominant over susceptible (Carton and Nappi, 1991). Recently, this has been confirmed in a detailed genetic study (Poirié *et al.*, 1999). The resistance *Leptopilina boulardi* gene was named *Rlb* and work is now concentrating on cloning the gene. To this end, detailed mapping experiments (see section 3.5.4) were performed using *Drosophila melanogaster* deletion strains (Lindsley and Zimm, 1992) and the avirulent strain of *Leptopilina boulardi* described above (Carton and Nappi, 1991). A detailed physical map has been obtained in the region 55C;55F3 (especially 55E2-E6;F3; Carton and Nappi, 1997; Poirié et al., 1999). Interestingly, in the encapsulation response of *Drosophila melanogaster* against *Asobara tabida* parasitism appeared also to be determined by a major gene (*Rat*, resistant to *A. tabida*, resistant allele is dominant; Poirié *et al.*, 1999). *Rat* was located 35 cM away from *Rlb*, but also on the second chromosome. It would be useful to study the polymorphism on *Rat* in relation to the clinal variation found in the encapsulation response of *Drosophila melanogaster* as discussed above (Kraaijeveld and van Alphen, 1995a). The apparent monogenic nature of the *Drosophila melanogaster* encapsulation response fits well with the interpretation of selection experiments for increased resistance in this species (discussed in Kraaijeveld *et al.*, 1998).

The genetics of resistance in *Leptopilina boulardi* has also been studied in more detail. Analyses of crosses between the African avirulent strain and a virulent strain indicated that the immune suppressive ability had a monogenic basis with incomplete dominance of the resistant allele (Dupas *et al.*, 1998). There was also evidence for the presence of minor genes involved in the resistant phenotype. These minor genes may well underpin the local adaptation of the African *Leptopilina* population to its sympatric host (see above). It would be interesting to explore the possibility that the major gene maps to the same position as the VLP gene (as yet uncloned). In this light it is interesting to note that in the African population, VLPs have been documented (Carton and Nappi, 1997) apparently without causing resistance to encapsulation. It is conceivable that the resistance gene found in Dupas *et al.* (1998) interacts with the VLPs to form the resistant phenotype.

As a final note, there may be a causal link between the genetic methods used and the genetic basis found in the experiments discussed above. Single iso-female or isogenic lines as used by Carton and Nappi (1991) are biased towards single gene effects (monogenic); they do not reflect the actual amount of genetic variation in the source population. Kraaijeveld and Godfray (1997) used (family) selection as a method to uncover the genetic basis of resistance. Their approach is more likely to pick up major and minor genes (polygenic) because it samples from the total amount of genetic variation available in the population. Moreover, the resistance and virulence phenotypes are a combination of several traits, which all have different genetic bases.

Maintenance of Genetic Variation: Costs and Trade-offs

The quantitative genetic parameters such as genetic variances and heritabilities for resistance and avoidance traits in hosts and parasitoids indicate high levels of genetic variation. Independent of whether the genetic basis is from monogenic to polygenic, this implies that in nature polymorphisms exist for the traits. This suggests that balancing factors are at play. Resistance to either parasitoid attacks or host defences is under strong selection because of the direct fitness benefits. The counteracting force could be the cost of resistance and/or avoidance. The costs and benefit trade-off and the balance between these depend on many biotic and abiotic factors such as season length, number of host species, number of parasitoid species *et cetera*.

The cost to the parasitoid may be absent or low when viruses are recruited to disarm the host defences. On the other hand, the parasitoid has to make sure that the viruses retain their symbiotic nature, which may involve mechanisms that are costly in terms of resources or energy. It is tempting to speculate that the VLPs are remnants of viruses resulting from natural selection for maintaining the benefits of the virus ability to interfere with the host immuno-response, but against the potentially harmful effects of the virus for the parasitoid. Information on the cost of resistance in parasitoids is essentially lacking at the moment. However, if we wish, for instance, to understand geographical patterns as described for *Asobara tabida*, it is essential that future research should focus on cost in parasitoids.

More research has been done on the cost of resistance in the host, mainly *Drosophila melanogaster*. As pointed out by Kraaijeveld *et al.* (1998), the critical issue is whether there are negative effects of having an encapsulation response when the chance of parasitism is low. Using selected *Drosophila melanogaster* lines Kraaijeveld and Godfray (1997) were the first to show a trade-off between being resistant and another life-history trait. Under limited food conditions the resistant genotypes were worse competitors that susceptible genotypes. No other life-history trait differences were found between the selected lines. Finding trade-offs is difficult in any system, hence the paucity of data on costs may not reflect the absence of costs, but rather the absence of carefully designed experiments. Again, experiments on costs and benefits of resistance under different ecological conditions are badly needed to further our understanding of the genetics of host-parasite interactions.

3.5.4 QTL-MAPPING

Principles

Variance analysis as discussed above, describes QTs in statistical terms. We do not know which genes or which alleles contribute to the genetic variance component and these may be different even when the values of variances are very similar for different populations of the same species. However, the development of molecular markers and linkage maps of these markers have ushered in the exciting field of molecular quantitative genetics. It is now potentially possible to elucidate quantitative polygenic traits in terms of the complex genetics of co-evolving genes and specific mechanisms. Genes that underly QTs are called **Quantitative Trait Loci (QTL)**.

The principles underlying **QTL-mapping** are extensions of standard mapping techniques of major genes (Falconer and Mackay, 1996; Lynch and Walsh, 1998; Figure 3.7). It uses the association between marker genes from a linkage map and the putative gene affecting the trait under consideration. However, as explained above, QTs are not described in discrete, qualitative classes, but in terms of phenotypic values, distributions and variances. Because several to many genes may underlie QTs, the association between the marker and any of these genes may be difficult to detect, depending on the heritability of the trait, the genotypic values of the alleles and the environmental variance. The key feature therefore is **linkage disequilibrium** (**LD**) between the marker and the QTL, with different marker genotypes having different expected phenotypic values because QTL are linked to the markers (Lynch and Walsh, 1998). Although in principle a straightforward analysis, practically, QTL-analysis involves sophisticated statistical techniques (see later).

Requirements

The two basic requirements are: (1) a linkage map of markers fulfilling the criteria as discussed under subsection 3.2.3 and (2) genetic variation for the QT for (closely-related) species, strains, populations or within populations. Of course, to acquire maximum LD in the progeny resulting from crosses between units that vary for the QT, the units will have to be fixed for both alternate alleles at the QTL and at the marker loci. Furthermore, ideally, alleles that increase the QT should be in one unit, and alleles that decrease the QT should be in the other. This is generally true for crosses between closely related species, and such species have been successfully used for QTL analysis (True *et al.*, 1997). It is an exciting prospect to use crossable *Drosophila* species to map genes for parasitoid resistance. However, only a handful of such pairs of species is likely to exist that produce both viable *and* fertile offspring. Therefore, divergent selection lines are often a good second bet to ensure optimal success for QTL mapping.

The parental strains in the crosses can be either still outbreeding or highly inbred so that they can be considered isogenic. For the reasons given above, the latter situation is preferred. Two possible crossing designs are most often used. For both, the parental lines are crossed to produce F_1 progeny. These then can be crossed either back to one or both of the parents, or crosses *inter se* to produce F_2. Only in the double backcross or F_2 design can additive effects of QTL be estimated unbiased when dominance effects are present (Falconer and Mackay, 1996). An F_2 design is more powerful than a backcross design, because in the latter only the heterozygous effect of a additive allele is detected, which is half that of the homozygous effect in the F_2 design (Falconer and Mackay, 1996). However, backcrossing designs applied to both parental lines take away most of this problem and have distinct advantages of their own. Recall that the choice for any particular design is also dependent on the genetic (dominant or co-dominant) properties of the markers used. In addition, notice that both the linkage map and the QTL analysis can be performed in one and the same experiment.

Subsequently, the resulting progeny are scored for their marker genotype on all of the marker loci and the individuals are scored for their phenotypic value. If a difference is found between marker genotype classes it can be inferred that a QTL is linked to the marker(s). This analysis was originally done using single marker analysis (Tanksley, 1993). This has one major drawback, namely the genotypic effect of the QTL is underestimated as a result of recombination between the marker and the QTL (underestimation by ($1-2c$); Falconer and Mackay, 1996). This problem was circumvented in the method of **interval-mapping**, in which the association between marker and phenotypic effect is considered using pairs of markers (Lander and Botstein, 1989). The choice of analytical procedure depends on the marker spacing (Tanksley, 1993). When the marker spacing is < 15 cM single marker and interval mapping give nearly identical results. For markers spaced > 20 cM but < 35 cM interval mapping provides the maximal benefit. For marker spacing > 35 cM even interval mapping is inefficient and gives poor results. We will now use some detail to explain interval mapping.

Interval Mapping

Consider two markers, M and N, with two alleles each, and a QTL, A, with two alleles in between the markers (after Falconer and Mackay, 1996). The recombination frequencies are c_{MN} between M and N, c_{MA} between M and A and c_{AN} between A and N. We assume complete interference, hence (following (1)), $c_{MN} = c_{MA} + c_{AN}$. The genotypes of the two parental lines are $M_1A_1N_1/M_1A_1N_1$ (with genotypic value *a*) and $M_2A_2N_2/M_2A_2N_2$ (with genotypic value –*a*) respectively. Hence, the F_1 progeny, $M_1A_1N_1/M_2A_2N_2$ (with genotypic value *d*). In this example, we backcross the F_1 with $M_1A_1N_1/M_1A_1N_1$. Table 3.3 shows all possible gametes and their frequencies based on the recombination values. Contrasting the four possible marker classes (double recombinations can be safely ignored if markers are closely spaced) provides an estimate of the QTL effect and its relative position between the two markers:

$$M_1N_1/M_1N_1 - M_1N_1/M_2N_2 = a - d \quad (3.19)$$

$$M_1N_1/M_1N_2 - M_1N_1/M_2N_1 = (ac_{AN} + dc_{MA})/c_{MN} - (ac_{MA} + dc_{AN})/c_{MN} \quad (3.20)$$

Equation 3.20 can be rearranged giving:

$$M_1N_1M_1N_2 - M_1N_1/M_2N_1 = (a - d)(c_{AN} - c_{MA})/c_{MN} \quad (3.20a)$$

Tests and Computer Packages

The marker class means will have associated variances, therefore statistical tests are required to decide whether the differences found between the markers classes are significant. There are broadly two approaches to significance testing: minimising the residual sum of squares in multiple regression, or maximising LOD scores.

Multiple Regression

In the multiple regression technique, the phenotypic value of each marker class is regressed on the unknown QTL parameters (*a*, *d*, and *c* values) (Haley and Knott, 1992). Starting with a marker interval (of known position and spacing), the recombination values between QTL and flanking markers can be iterated (in small steps) until *c* values are found that results in a significant regression and maximises the sum of squares of the fit (or minimises the residual error). The coefficients of the regression function are then the estimates of *a* and *d*. If the first interval that was considered does not yield a significant result, the next interval is considered and so on. We refer to Haley and Knott (1992) for a full description of the expression of the regression coefficients as a function of the QTL parameters in a F_2 design and to Kearsey and Pooni (1996) for practical guidance and examples.

Likelihood Analysis

Alternatively, a likelihood function is specified in terms of the observed data and the

Table 3.3 Expected gamete genotypes in a backcross design, their frequency and contribution to backcross progeny mean. Since we are interested in the differences between the marker classes, and not in to what extent each marker class contributes to the overall backcross progeny mean, actual marker class means are also given (after Falconer, 1996).

F_1 gamete type	Frequency	Genotypic value	Contribution to backcross progeny mean	Actual marker class mean
$M_1A_1N_1$	$(1 - c_{MN})/2$	*a*	$M_1N_1/M_1N_1 = a(1 - c_{MN})/2$	*a*
$M_1A_1N_2$	$c_{AN}/2$	*a*	$M_1N_1/M_1N_2 = (ac_{AN} + dc_{MA})/2$	$(ac_{AN} + dc_{MA})/c_{MN}$
$M_1A_2N_2$	$c_{MA}/2$	*d*		
$M_2A_1N_1$	$c_{MA}/2$	*a*	$M_1N_1/M_2N_1 = (ac_{MA} + dc_{AN})/2$	$(ac_{MA} + dc_{AN})/c_{MN}$
$M_2A_2N_1$	$c_{AN}/2$	*d*		
$M_2A_2N_2$	$(1 - c_{MN})/2$	*d*	$M_1N_1/M_2N_2 = d(1 - c_{MN})/2$	*d*

parameters to be estimated. Combinations of the parameters are then tried to find the maximum likelihood solution (usually in c steps of 0.01 to 0.05). This solution is then tested with the LOD ratio (see also under subsection 3.2.3) between the observed likelihood function and the likelihood function of no QTL segregating (Lander & Botstein, 1989). This log odds ratio is χ^2-distributed. In statistics, the level of significance is usually taken as $\alpha = 0.05$. However, QTL analysis will be done for n markers spread over the whole genome, hence the usual $\alpha = 0.05$ per test is clearly inappropriate. Corrections have to be made for multiple testing. The appropriate LOD threshold above which a QTL on a certain map position is judged significant depends on the size of the genome and the density of the markers (i.e. the number of tests performed). This threshold can be found by addressing the question: "what is the chance of finding a QTL (i.e. LOD scores passing the threshold) when no QTL is segregating in the population?" (Lander and Botstein, 1989). The threshold value differs according to whether the maps are dense or sparse maps, in other words whether LOD scores can be considered independent events or not. For a sparse map, the significance level per test is approximately α/n (Lander and Botstein, 1989; Lynch and Walsh, 1998). For dense maps, thresholds should be calculated depending on the map properties and a general formula is given in Lander and Botstein (1989). Threshold values can also be calculated using permutation tests (Doerge and Churchill, 1996). The original data set is reshuffled many times using the observed phenotypic values but these values are now randomly assigned to the genotypes. The QTL test statistic is calculated for each of these synthetic data sets (while no association exists between marker and trait) to give a frequency distribution of test statistics. An empirical threshold value can be found by finding the value of the test statistics belonging to the (1-α) percentile. For more general discussions of significance thresholds for QTL interval mapping, see Doerge and Rebaï (1996) and Lynch and Walsh (1998). Figure 3.21 shows a schematic plot of the LOD score against chromosome position as often found in many QTL papers.

Composite Interval Mapping

If more than one QTL is segregating in the population, the marker class variance includes a genetic component due to these QTLs unlinked to the interval. For instance, a QTL with a very large phenotypic effect may mask other QTLs in the populations (Kearsey and Farquhar, 1998). By using additional markers as co-factors, these effects can be removed. This is known as **composite interval mapping** that greatly increases the likelihood of detecting a QTL and also the precision of the estimated map position (Jansen and Stam, 1994; Zeng, 1994). In addition, there is a large body of literature on the statistical properties of testing for QTLs, such as the analysis for multiple traits (Jiang and Zeng, 1995; Korol *et al.*, 1995). An extensive review is that of Lynch and Walsh (1998). Moreover, an analysis of the genetic architecture, including epistatic interactions between QTL, is possible using multiple interval mapping (for details see e.g. Zeng *et al.*, 1999).

The multiple regression technique was developed because the maximum likelihood approach is computationally demanding and cannot be implemented in normal statistical packages (Haley and Knott, 1992). The tests are related, however (regression is maximum likelihood when errors are independent and follow a normal distribution), and yield nearly identical results (Lynch and Walsh, 1998). So, multiple regression can be easily performed. In addition, several computer packages are available (visit http://dendrome.ucdavis.edu/qtl/software.html for an extensive list), such as Map Manager (http://mcbio.med.buffalo.edu/mapmgr.html), and QTL cartographer (http://statgen.ncsu.edu/qtlcart/cartographer.html) (Basten *et al.*, 1997). The latter programme can be recommended because it uses a sequential procedure, starting with single marker analysis, followed by stepwise regression and finally interval mapping,

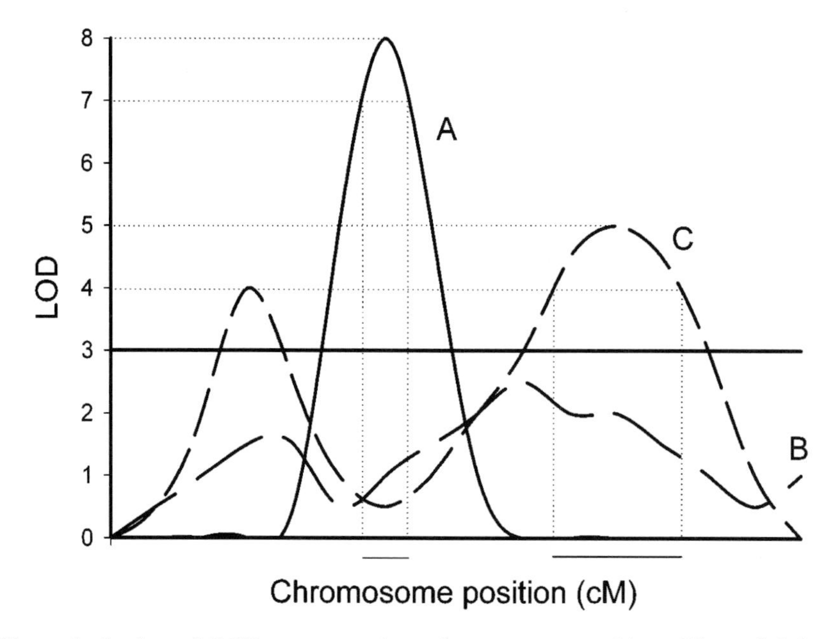

Figure 3.21 Theoretical plot of LOD score against chromosome position. The solid line indicates the significance threshold value (see text for details). The three lines indicate three traits. For trait A, a QTL with strong effect is found; for trait B no indication is found for a QTL; for trait C, there is statistical support for two QTLs. The precision of the map position of a QTL can be calculated as the one LOD support interval, which approximates 95% confidence intervals (Lynch, 1998). This interval can be found by subtracting one LOD from the peak of the LOD plot and subsequently projecting onto the X-axis the points where the LOD − 1 lines crosses the observed line (intervals indicated as horizontal lines on X-axis).

composite interval mapping, and multiple interval mapping.

General Considerations and Problems

Mapping QTLs for traits with high heritabilities is easier because they are less sensitive to environmental perturbations. However, even for traits with high heritabilities, no QTL can be found because detection depends on the heritability of single QTL. Hence, with many genes the individual locus heritability can be very low. Interestingly, at first sight the presence of QTL seems to falsify the polygenic nature of the QTs paradigm (see above). However there is a strong bias towards QTL with a large effect and such QTLs are usually mapped with reasonable precision (Kearsey and Farquhar, 1998). Probably due to these properties, the number of QTLs found generally equals the number of chromosomes (Kearsey and Farquhar, 1998). For more general observations on QTL we refer to elsewhere (Tanksley, 1993; Kearsey and Farquhar, 1998).

The above illustrates the largest problem associated with QTL mapping: detection of QTL is possible but the confidence intervals around the map position are large (the one LOD support interval is large, Figure 3.21 trait C). In connection with this, it is important to realise that a QTL does not necessarily mean a single locus: more than one gene may involve the peak found in the test statistics. Although developments in the statistical analysis have greatly helped to abate the consequences of this problem, increasing the precision of phenotypic value estimates is necessary. The power of the test is dependent on the within-marker-class standard deviation (Falconer and Mackay, 1996). This standard deviation can be reduced by measuring more individuals (or rather genotypes; note that the number of genotypes is restricted by the number of crossovers) or by creating **recombinant isogenic**

lines (RILs). In the latter, marker genotypes can be replicated many times and measured on several occasions or in different environments to study QTL-environment interactions. In *Drosophila melanogaster*, constructing RILs is straightforward using balancer chromosomes (Lindsley and Zimm, 1992). In addition, the haplodiploid system offers the unique possibility of creating (near) RILs (see above), but surprisingly they have not yet been exploited. Epistatic interactions are important in evolutionary theory (see above), but are difficult to detect in standard backcross and F_2 designs due to lack of statistical power. It is not surprising, however, that epistatic interactions are documented in experiments using RILs because of the potential of repeated measurements on RILs (e.g. Long *et al.*, 1995). The development of the multiple interval mapping techniques will facilitate the investigation of complex epistatic patterns.

To conclude this section, we will briefly touch upon the use of QTL mapping in outbred populations. Mapping in such populations is considerably more difficult than in inbred populations, because the power is much lower. For instance, not all parents are informative, i.e. some are fixed for the marker, but not for the QTL or fixed for the QTL but not for the marker. In addition, in one family a marker allele may be associated with one allele of the QTL, but in another family the same marker allele is associated with another allele at the QTL. How to deal with these problems is discussed in chapter 16 of Lynch and Walsh (1998).

Examples

The concept of QTL mapping was initiated from the field of plant and animal breeding, and it is therefore not surprising that most examples can be found there (e.g. Tanksley, 1993 for review). It was realised that knowing the position of genes influencing economically important traits (such as milk yield in cows) offered the potential of crossing such genes into breeding populations (introgression). In such crosses the phenotype of the individuals would not necessarily have to be measured, but the markers closely associated with the gene of interest could be used as a flag. Molecular and genetic identification of the identified QTL also offered the potential for unravelling the genetical, developmental and physiological pathways in which the QTL is involved.

Recently, the genetic basis of wing size differences between two *Nasonia* species has been investigated (Weston *et al.*, 1999; Gadau *et al.*, 2002). Males of *N. vitripennis* have short, rudimentary wings and are incapable of flying, whilst *N. giraulti* male wings are 2.4 times longer and the males can fly (although they have a low flight tendency). By curing the parental strains of their *Wolbachia* infections, the hybrid cross between the species resulted in viable and fertile F_1 females. *N. vitripennis* males of five eye colour mutant strains were crossed to *N. giraulti* females. The mutants belonged to either of one of the five linkage groups (chromosomes) of *N. vitripennis*. The resulting hybrid females were subsequently allowed to lay eggs on hosts to produce male offspring. As was mentioned above, all markers in haplodiploid males are informative and the males were scored for their wing size and eye-colour. The average size of the wing was compared between mutant colour eyes and wild-type eyes males within one mutant line cross. When a significant difference was found between these groups, it was concluded that a gene influencing wing size was located on the same chromosome on which the eye colour mutant was located. The frequency of eye colour mutants with large wings and wild-type wasps with small wings was used to calculate how close the wing size gene was placed next to the eye mutant (as in Figure 3.7). The findings from these crosses were corroborated in an introgression experiment. The major finding of this study is that apparently a gene (or several tightly linked genes) on chromosome IV explains 44% of the wing size difference between the two species and is located 1.4 map units from the eye colour mutant (Gadau *et al.*, 2002). Some caution should be exercised here, because no attempt has been made to locate the genes within one linkage group, and it should be remembered that *Nasonia* has a large genome.

Further studies will concentrate on the adaptive significance of the size differences and its evolution, together with the genetic analysis of differences in other traits, such as, courtship, male aggression and hosts preferences. It will be particularly interesting to see whether these traits are determined by at least one gene with large effect as well, or many genes with small effect.

Successful mapping of QTL has been reported for foraging behaviour in the honey bee, *Apis mellifera* (Hunt *et al.*, 1995). Selection was applied on the amount of pollen stored in combs of honey bee colonies. The selected strains were then crossed, and QTL mapping for the selected trait was performed using a backcross design and a RAPD marker-based map. Two QTL were identified located on two different linkage groups. Interestingly, in a separate cross the QTL effects were demonstrated by the co-segregation of marker alleles (associated with the QTL) and individual worker behaviour. Moreover, the alleles of the two marker loci were found in different frequencies in the different foraging task groups and the QTL appeared to affect both pollen load and nectar load.

Both the above examples illustrate that a QT does not necessarily have to be determined by many genes having a small effect. However, bear in mind the inherent bias towards genes with large effect in QTL analysis.

Concluding Remarks on QTL Mapping

Ensuring that most of the aforementioned problems are accommodated in the best possible way, QTL mapping is a potentially powerful and valuable tool for uncovering the genetic basis of QTs. Especially in the case where candidate genes are available, a better understanding can be developed for genetic variation of a trait in nature and why possible differences between populations or strains exist. Alternatively, in the absence of potential candidate genes in the interval where a QTL is mapped, the complete toolbox of molecular biology can be exploited to find the gene(s) underlying a QTL (Tanksley, 1993; Falconer and Mackay, 1996; Lynch and Walsh, 1998). To be frank, however, this is still a long way off for all but a few species.

MATING BEHAVIOUR 4

Ian C.W. Hardy, Paul J. Ode and Mike T. Siva-Jothy

4.1 INTRODUCTION

In sexually reproducing animals, reproduction entails **insemination** (the transfer of the male's sperm to the female) and **fertilisation** (the fusion of a sperm and an egg to create a diploid zygote). In many insect species, including predators and parasitoids, insemination and fertilisation are temporally separated, from minutes or hours to years (as in many social insects). The term "**mating behaviour**" refers to the behavioural events surrounding insemination, which ensure successful sperm transfer by the male and uptake by the female as well as, in many species, post-copulatory male behaviours that have evolved in response to sperm competition (Parker, 1970a). Alexander *et al.* (1997) divided mating behaviour into: pair formation, courtship, copulation, insemination and the events immediately following insemination, including temporary pair-maintenance, and we discuss mating behaviour mainly according to these divisions. Our discussion also includes components of mate competition and fertilisation. A typical sequence of mating behaviour, from the search for mates up to insemination and mate-guarding is shown in Figure 4.1. Figure 4.2 represents the behaviour of the parasitoid wasp *Cotesia rubecula* (Braconidae) as a specific example.

Here we take an approach that focuses on *mechanisms* or *causal relationships*. [This chapter is based largely on an earlier version by Hans van den Assem (van den Assem, 1996), who has retired]. The next chapter, on *Mating Systems* (Chapter 5), employs a complementary 'functional', or 'adaptive', approach. The taxonomic and biological diversity of insect natural enemies is enormous. Consequently, our choice of examples is selective, concentrating mainly on van den Assem's experiences and approaches in studying three species of chalcidoids: *Nasonia vitripennis* (Pteromalidae) and *Melittobia acasta* (Eulophidae), parasitoids of cyclorrhaphous fly pupae, and *Lariophagus distinguendus* (Pteromalidae), a parasitoid of the larvae of grain weevils. However, to mate successfully, most parasitoids and predators need to solve similar problems; these are: where to find a partner, how to recognise it, how then to behave so that insemination will result and, in the case of males, what measures (if any) to take so as to uphold or enhance a sperm monopoly in fertilisation (i.e. how to counteract sperm competition).

4.1.1 THE HOWS AND WHYS OF INVESTIGATING NATURAL ENEMY MATING BEHAVIOUR

Mating behaviour can be studied from several points of view, including an analysis of:

1. Its proximate **causation** and **organisation**, i.e. identifying which mechanisms underlie the production of behaviour and determining how they operate, which factors (internal and external) release behaviour, how incoming information is processed, and how motor activities become co-ordinated activities which can be observed as behavioural acts. This is the main approach taken in this chapter.
2. Its adaptive **function**, i.e. measuring the fitness consequences of behaviour, particularly in terms of reproductive success (discussed in Chapter 5).

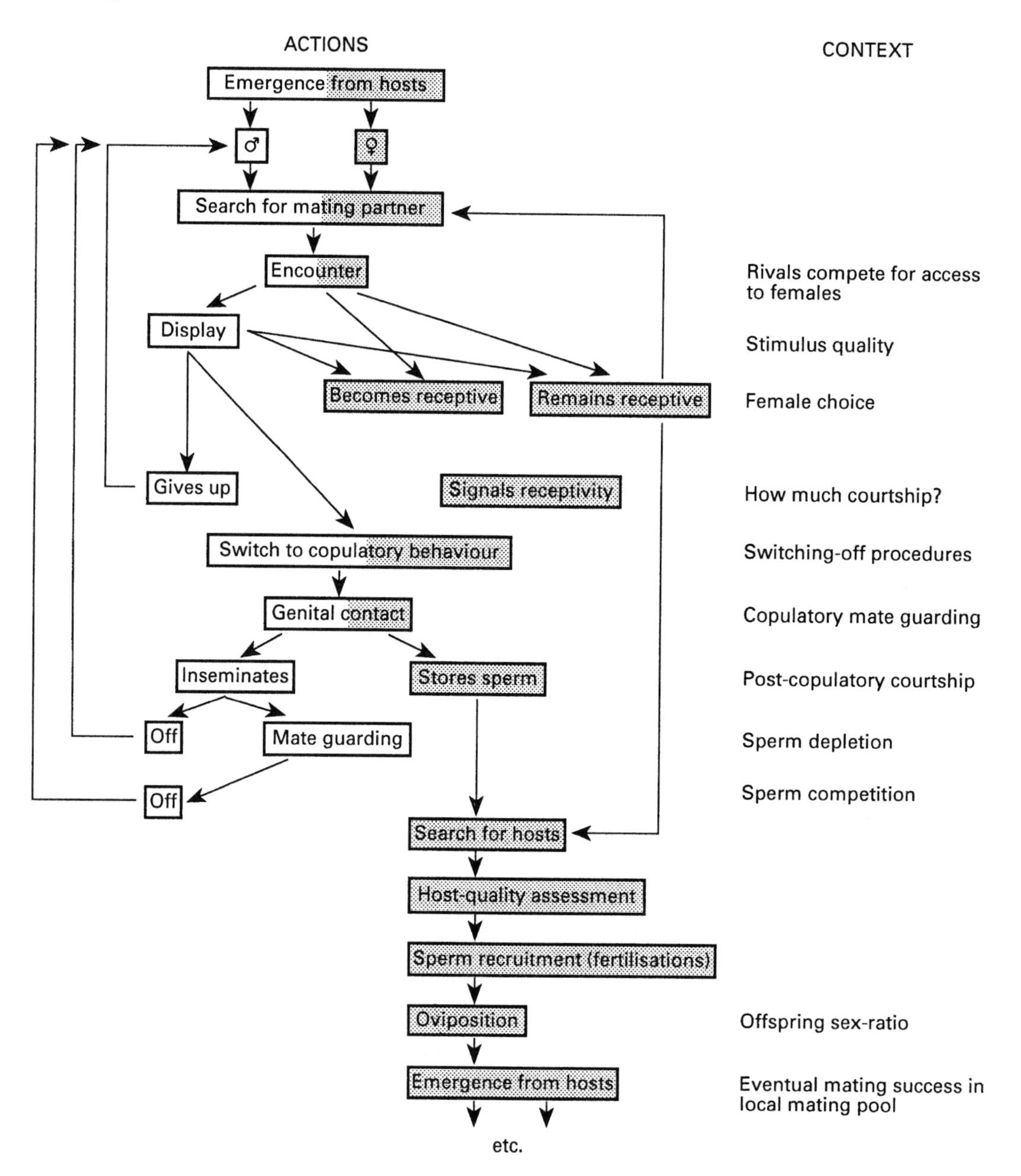

Figure 4.1 Schematic representation of the course of events in a typical sequence of mating behaviour. Shaded and unshaded parts of boxes denote contribution by female and male respectively.

3. Its **evolution**, i.e. the steps by which behaviour patterns have changed over evolutionary time (the '**evolutionary transformation series**') and have assumed their present form, and the directions any such developments have taken. This is discussed briefly at the end of this chapter.

There are several practical reasons for studying the mating behaviour of natural enemies:

1. Studies of mating behaviour can help to identify the attributes that make a given species an effective biological control agent. Such investigations are likely to be

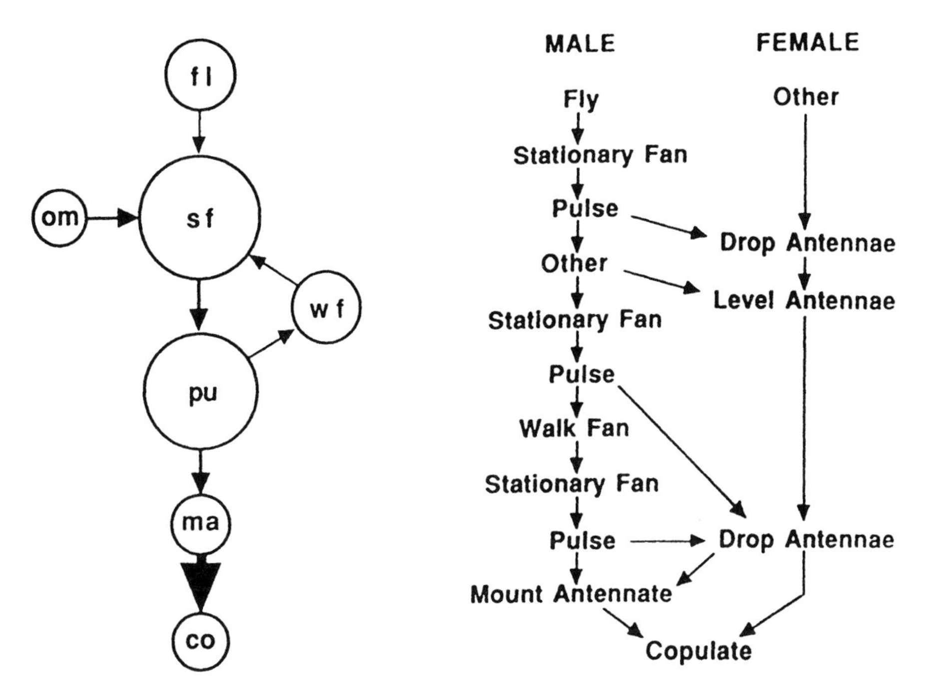

Figure 4.2 A specific example: mating behaviour in *Cotesia rubecula* (Hymenoptera: Braconidae), a solitary endoparasitoid with a polygynous mating system and a typical sequence of behaviour during courtship and mating. However, mating behaviour is more complex when males compete directly for mating opportunities (Field and Keller, 1993a, subsections 4.3.6.3 and 4.4.1, Figures 4.25 and 4.27). The left-hand flow diagram shows the sequence of male behaviour during successful courtship (using abbreviations defined in Table 4.1), the sizes of circles indicate the relative total durations of each behavioural component. The right-hand flow diagram shows interactions between male and female in typical courtship. Females emit a sex pheromone, and lower their antennae (when receptive). Males produce low-frequency acoustic emissions associated with wing movement (stationary fan) and 'pulses' of vibrational signals transmitted through the substrate, subsection 4.3.4). 'Other' behaviour by males occurs randomly, and is followed by stationary fan. There are usually three repetitions of the stationary fan-pulse behavioural act in a successful sequence. From Field and Keller (1993b), with permission from Plenum Publishing Corporation.

invaluable additions to the predictions made by population models of host-parasitoid dynamics (Luck, 1990). Studying mating behaviour may be critical in predicting the success of biological control introductions and establishments (e.g. Hopper and Roush, 1993) as well as in developing efficient techniques for the mass-rearing of insects that are required in large numbers at particular times, e.g. for field or greenhouse releases (Waage *et al.*, 1985; Hall, 1993; Heinz, 1998). Understanding responses of males to female sex pheromones may be useful in monitoring parasitoid populations (Suckling *et al.*, 2002).

2. Mating behaviour studies may also prove to be useful in the identification of a species to be introduced as a biological control agent (Gordh and DeBach, 1978). Given the high degree of host specificity of many species, it is essential that biocontrol practitioners are certain about the taxonomic identity of the insects they intend to use. Sibling species (morphologically identical or near-identical species that are difficult to distinguish morphologically) abound in the parasitic Hymenoptera, and an analysis of mating behaviour may be the easiest means of distinguishing between them because, in general, mating displays tend to be

species-characteristic (van den Assem and Povel, 1973; Matthews, 1975, see also Hunter *et al.*, 1996; Kazmer *et al.*, 1996; Heimpel *et al.*, 1997).

3. Studies of mating behaviour in insect natural enemies provide some of the clearest tests of several behavioural theories including mate choice and sexual selection (Godfray, 1994; Eberhard, 1996). Furthermore, parasitoid wasps are outstanding organisms with which to demonstrate the basic tenets of insect mating behaviour in the classroom (e.g. Barrass, 1976).

4.2 GENERAL METHODOLOGY

4.2.1 FIELD VERSUS LABORATORY STUDIES

Most detailed investigations are carried out in the laboratory because many parasitoids are very small, and observing them requires specialised equipment. Also, environmental conditions and insect material can be standardised. Important variables such as egg load (subsections 2.34–6) or time between successive matings are difficult to control in field experiments. In the absence of field observations, the extent to which laboratory mating behaviour constitutes natural behaviour can never be fully known, and conclusions regarding the function of behaviour will be tentative, even more so than those regarding causation (Ewing, 1983). Field investigations such as G.A. Parker's work on dung-fly mate searching (1970a,b,c,d, 1971, 1978) are valuable as they are less prone to experimental artifacts than laboratory studies. Parasitoid mating can in some cases be observed in detail under field conditions (Tagawa and Kitano, 1981; Takahashi and Sugai, 1982; Antolin and Strand, 1992; Field and Keller, 1993a), but often it is difficult to distinguish between individuals, which prevents assessment of an individual's success within a group. Relatively large, and thus easily visible, insects such as digger wasps (Sphecidae, Pompilidae, Tiphiidae) or dragonflies can be marked either with paint or tags glued to the dorsum. Such marks can even be applied to some quite small parasitoids (Driessen and Hemerik, 1992; Field and Keller, 1993a). In all cases, care must be taken to ensure that marks do not interfere with behaviour. Phenotypic markers such as eye- or body-colour mutants, and also resistance to insecticide, have also proven to be useful ways to investigate the behaviour of individuals (e.g. Ode *et al.*, 1995; Baker *et al.* 1998).

4.2.2 CULTURING INSECTS

Long-term culturing may influence insect behaviour. In *Nasonia vitripennis* for example, female responsiveness to certain male display stimuli altered after several generations of culturing, although there were no changes in stimulus production by males (van den Assem and Jachmann, 1982). Similar changes can occur in female receptivity thresholds and in the 'switching-off' mechanism (subsection 4.3.6), and are unlikely to be restricted to *Nasonia vitripennis*. However, Gordh and DeBach (1976) found no difference in sperm production and insemination potential between *Aphytis lingnanensis* (Aphelinidae) males from 25-year old cultures and males from field-collections. Similarly, Kazmer *et al.* (1996) argue that long-term culturing was not the cause of observed reproductive incompatibilities between strains of *Aphelinus asychis* (Aphelinidae). One way of circumventing the effects of long-term culturing is to keep separate mass cultures that are periodically mixed. Another way is to keep part of the stock in diapause for extended periods. Perhaps a better way is to periodically augment the culture with field-collected material.

4.2.3 STANDARDISING INSECT MATERIAL

As is true of most behavioural experiments, studies of mating behaviour should use individuals that have been standardised with respect to factors such as age, prior mating experience and body size that affect the willingness or ability to mate. If standardisation is not possible, these variables should be controlled for when interpreting observations.

Age differences are an important source of variation in courtship behaviour (Barrass, 1960a; Tagawa *et al.*, 1985). In many species of parasitoids, newly emerged females are unreceptive to mating attempts by males (Godfray, 1994). In *Habrobracon* (= *Bracon*) *hebetor* (Braconidae) both males and females are unreceptive to mating immediately after emergence, a behaviour that is thought to reduce inbreeding levels (Ode *et al.*, 1995). Furthermore, unmated adult females become unreceptive after a certain period in some species (Starý, 1970; Godfray, 1994 and references contained therein). Clearly, it is important to know the age of individuals, and this can be determined by monitoring pupae for adult emergence and subsequently holding the adult until any non-receptive period has passed.

Prior mating experience may also influence receptivity towards mating attempts. Females in some species are known to mate multiply but in other species females mate only once (e.g. Wilkes, 1966; Kitano, 1976; Hirose *et al.*, 1988; Ridley, 1993; Allen *et al.*, 1994; Ode *et al.*, 1995, Baker *et al.* 1998; Ruther *et al.* 2000; Chevrier and Bressac, 2002). Given that post-mating non-receptivity is common, it is critical to know the mating history of individual males and females that are to be used in mating studies. Virgin females can be obtained by isolating pupae in separate vials and holding each adult until it is to be used in an experiment.

There is also some evidence that body size may affect male and female mating behaviour and success. Van den Assem *et al.* (1989) demonstrated that small male and female *L. distinguendus* were at a disadvantage compared to larger conspecifics. However, small size is less of a disadvantage to males than females. Hunter *et al.* (1996) found that body size incompatibility in *Eretmocerus eremicus* (Aphelinidae) could prevent males from successfully copulating with receptive females. Antolin *et al.* (1995) showed that smaller *H. hebetor* males are less able to successfully copulate with females possibly because they are unable to elicit the proper mating receptivity response from the female. However the effects of body size on mating success were small, and another study on this species detected no effect of size on mating success (Cook *et al.*, 1994).

4.2.4 HANDLING INSECTS

Handling prior to experiments should be kept to a minimum or done in such a way that its effects are minimised or standardised. For example, wasps can easily be coaxed from one vial to another by using their positively phototactic responses. To transfer a wasp from a vial into an observation cell (see below), simply tap the vial base of the tube once or twice; some species of wasps will retract their legs and so fall out. In some species, males seem able to grip a surface more strongly than females (probably an adaptation to prevent being pushed away by rivals or by unwilling females). Another method is to capture a wasp between the soft hairs of a camel-hair brush, and introduce it into the observation cell by gently manipulating the brush.

4.2.5 EQUIPMENT FOR OBSERVING BEHAVIOUR

If only easily observable events are to be recorded (e.g. occurrence of copulation or duration of display), insects can be observed in vials or other small containers, but care should be taken because different ways of bringing wasps together (introducing a male into a vial containing a female, introducing a female into a vial containing a male, or introducing male and female into a fresh vial) are likely to have different effects on mating behaviour. In some parasitoids (e.g. *Nasonia vitripennis*), males apply substances (functioning as chemical signals, section 4.3.4) to the walls of vials.

Observing the mating behaviour of large parasitoids (e.g. Ichneumonidae) and many predators is possible with the naked eye, but most parasitoids are only a few millimetres long and some kind of magnification (usually between 6.5 and 40×, but 10× is usually sufficient) is needed. A hand-held lens, a dissection microscope or a video camera fitted with a suitable lens are three possibilities. At high magnification, reduction both in the field of vision and the depth of focus severely constrains the ability to observe many behavioural details.

When using microscopes or video cameras, insects must usually be kept within certain spatial limits and thus observation cells are often used, although large containers may be needed for many species including swarming insects. Cells can be made, for example, by cutting 7 mm-thick slices from transparent Perspex (plexiglass) tubing (internal diameter 25 mm). This diameter fills the visual field dissection microscopes at 10 × magnification. Cells can be capped with cover-slips (45 × 45 mm) and can be placed on a strip of paper to allow slight repositioning of the entire cell during observations (illustrated in van den Assem, 1996). Fauvergue *et al.* (1998a) overcame the limited field of view provided by a video camera by automating the camera's horizontal plane of movement to keep track of mate-searching parasitoids.

Computer-based software and hardware systems, such as *EthoVision* (Noldus *et al.*, 1995; reviewed by Johnston, 1998; and see http://www.noldus.com for updates) are also available for automated behavioural observations, although these are more effective for tracking the spatial positions of individuals than for observing fine details of behaviour. However, *EthoVision* also incorporates an event recorder (subsection 4.2.9), which allows such details to be added by a human observer, and can also perform basic statistical analysis or export data to other statistical packages.

4.2.6 NEGATIVE EFFECTS OF CONFINEMENT

Confining insects in small spaces, such as vials and observation cells, could have undesirable consequences. For example, chemical substances could accumulate and affect the further mating behaviour. In *Nasonia vitripennis*, courtship behaviour within a host puparium is inhibited, probably by chemical stimuli. This can be simulated by placing a large number of virgin females and a male in a vial or observation cell (van den Assem *et al.*, 1980a). The male begins courting immediately and copulates as soon as a female is ready; it then copulates with further females. At first, females may spontaneously exhibit receptivity (opening the genital orifice, section 4.3.6) while the male is engaged elsewhere, they may approach the male before exhibiting, or they may become immobile and adopt the posture on the spot. The male eventually stops walking and ceases displaying. However, when all wasps are transferred to a fresh vial, the male resumes his activities, only to become inactive some time later (Figure 4.3). This suggests that a chemical stimulus is involved. The practical implications are clear (at least for this species): mating behaviour is negatively affected by holding more than a few individuals in small closed containers for too long a period. In contrast, several gregarious species (e.g. *Melittobia* spp.) mate within host puparia and may occur in very large numbers. Inhibition does not occur within crowded vials. In such species, volatile olfactory signals (as in *Nasonia vitripennis*, section 4.3.4) would not operate efficiently and are likely to be absent.

4.2.7 CONTROLLING AMBIENT TEMPERATURE

Maintenance of proper ambient temperatures is important if representational behavioural acts are to be observed. Insect metabolism and activity are strongly temperature-dependent (e.g. see below), and ambient temperature is known to influence mating success in a number of insects (e.g. Larsson and Kustvall, 1990).

Artificial illumination will usually be required for microscope observations that, unfortunately, may influence behaviour. Fibre-optic illumination provides good lighting without also heating the observation area. Further measures can be taken to maintain constant temperature, such as fitting brass observation cells with Peltier units (illustrated in van den Assem, 1996), which transport heat through a semiconductor by a reversible current. Temperature is set and

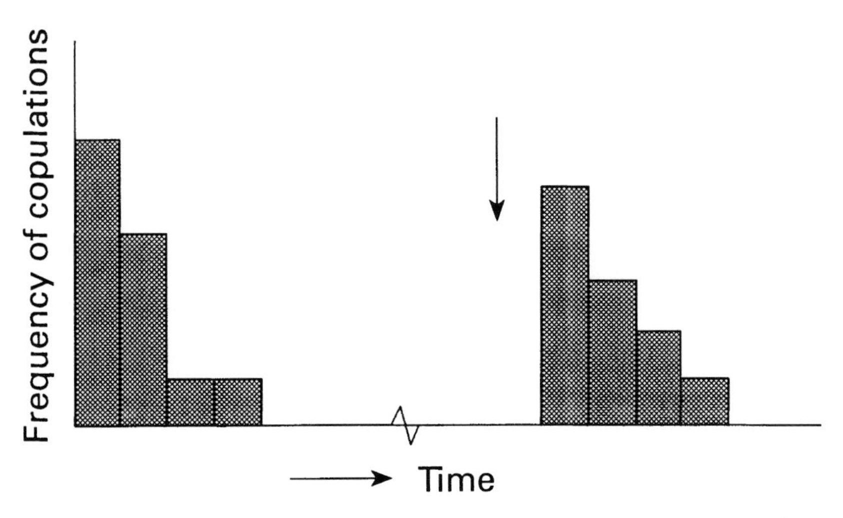

Figure 4.3 Typical temporal changes in frequency of copulations in a group consisting of one male and many female *Nasonia vitripennis* when kept in a small vial. Mating is resumed soon after the group is introduced into a fresh vial (vertical arrow).

monitored by a control unit, and is accurate to 0.2°C. Temperatures down to around −10°C can be obtained, but extra measures are required to dissipate heat. Extreme low temperatures can be used to immobilise wasps and so allow precise experimental manipulations (e.g. sealing of mouthparts, section 4.3.4). With less extreme low temperatures it is possible to slow down behaviour, so enabling complex movements to be analysed more accurately. For example, rates of display in *Nasonia vitripennis* can be halved if the temperature is set at 13°C, below which displays do not occur (Jachmann and van den Assem, 1996).

4.2.8 DESCRIBING BEHAVIOUR

To describe a behavioural sequence, it is useful to partition it into separate events or 'acts', with names assigned that are informative, unambiguous and not interpretative or descriptive of perceived ultimate function. The components of courtship behaviour of the parasitoid wasp *Cotesia rubecula*, distinguished and defined by Field and Keller (1993b), are given in Table 4.1 as an example (see also Figure 4.2). The records produced should reflect the dynamics of behaviour as accurately as possible. However, there are usually so many aspects of even a simple behavioural sequence that it is impossible to represent them all in one record. Usually a number of body components (e.g. limbs, antennae, head, abdomen) take part in a posture or a behavioural sequence such as a display. One approach is to record the sequence of events for each part of the body separately. One can subsequently examine how the movements of the sets of components are co-ordinated and determine whether any temporal relationships are recognisable. While motor patterns of movements may be relatively invariable, the intensity and orientation of these movements as well as the timing of successive events may vary greatly. Furthermore, events may occur in rapid succession or concurrently and these problems are compounded when studying the behaviour of interacting pairs of animals, in which case two observers may be required (e.g. Field and Keller, 1993b).

4.2.9 EQUIPMENT FOR RECORDING BEHAVIOUR

Making records of the most simple behavioural sequences require only a notepad and a pencil, plus a stopwatch if the timing of separate events,

Table 4.1 Components of *Cotesia rubecula* (Hymenoptera: Braconidae) courtship behaviour identified during observations prior to the main experimental observations and used to configure event recording software (section 4.2.9), the abbreviations in parentheses are used in Figure 4.2. From Field and Keller (1993b), with permission from Plenum Publishing Corporation.

Behaviour	Description
Female	
Drop-antennae	Head tilted forward, antennae held together and pressed down on substratum in front of body
Level-antennae	Antennae held parallel to substratum, intermediate between raised and drop-antennae
Walk	Moving forward, usually alternately touching the antennae to the substratum
Groom	Any cleaning of the body, including stroking the face and antennae with legs, or cleaning the body with mouthparts
Fly	Any airborne activity
Other	Any other behaviour (predominantly 'point', in which the wasp faced upwind with anterior portion of body raised and antennae raised and spread, which occurred at commencement of observation)
Male	
Fly (fl)	Any airborne activity
Stationary-fan (sf)	Standing with antennae raised and spread, wings raised and fluttering
Walk-fan (wf)	Walking with wings raised and fluttering
Pulse (pu)	Stationary (occasionally walking) wing-fanning interspersed with rapid pulsing movements of the abdomen and flexing of legs. Wings pushed down towards horizontal with each individual pulsing movement
Mount-antennae (ma)	Male placing fore and mid tarsi on dorsal surface of female gaster, rapidly antennating female head, and curling gaster underneath body to make genital contact
Groom	As for female
Walk-antennae	Palpating the substrate with both antennae while walking
Stationary-antennae	Palpating the substrate with both antennae while stationary
Wave-antennae	Stationary and waving antennae just above the female's body without contacting her
Other (om)	Any other behaviour, including pointing
Female and Male	
Copulate (co)	Male establishing genital contact and leaning upper part of body backward with fore legs only on female gaster, wings raised and antennae spread and vertical

or intervals between them, is of interest. However, many acts are very complex and rapid, and sequences of acts have to be recorded in some way if they are to be successfully described. Such records can be made by speaking into an audio tape recorder, but video recording is likely to be more useful because behavioural sequences can be repeatedly examined 'frame by frame' until all details have been noted, and also because there now exists the technology for downloading images onto a PC. Such records are indispensable for making precise descriptions of movements, and particularly for measuring temporal relationships between behavioural components. Once a movement has been seen in detail, it is much easier to recognise at normal speed and subsequently to notice any differences in performance. Filmed or videotaped records are very useful when comparing related species (section 4.6). There are certain disadvantages to the use of video recording; for example, it is usually not possible to view the entire field of action continuously, which becomes a problem when interactions between individuals are studied.

The alternative to the use of hand-written or videotaped records is machine encoding. Until relatively recently, the standard **event recorder** was a mechanical device comprising a collection of keys with corresponding pens, ink and paper (e.g. the 20-pen *Esterline Angus* recorder, described in van den Assem, 1996, and used by him for many years). Electronic event recorders have now replaced mechanical devices, and have become extremely sophisticated and well integrated with other research equipment. An early version, *The Soliprot* (illustrated in van den Assem, 1996) makes electronic records that can be read into a PC for analysis. A further advance is *CAMERA* (Haccou and Meelis, 1992, pages 1–2 and 374–375), a hardware and software system, including video, for recording behaviour and performing basic data analysis. Probably the most advanced event recorder is *The Observer* (and the *Observer Video-Pro*), a software package compatible with many hardware components such as standard computers, hand-held computers and video systems (Noldus, 1991; Noldus *et al.* 2000; see http://www.noldus.com for updates). Basic data analysis is possible, and data can readily be transferred to more powerful and general statistical packages. The event–recording capabilities of *EthoVision* (section 4.2.5) are compatible with those of *The Observer*. *The Observer* is reviewed by Davis (1993), Visser (1993) and Tourtellot (1992) and has been used in at least two studies of parasitoid mating behaviour (Field and Keller, 1993b; Allen *et al.*, 1994, Table 4.1). Other currently available event recorder systems are *The Spectator* and *Spectator GO!* (www.biobserve.com) and the *FIT-system* (flexible interface technique: Held and Krueger, 1999; Held *et al.* 1999; www.SmileDesign.ch). The FIT-system differs from the other event recorders in that the key pad is replaced with a touch-sensitive screen on which the researcher can draw a personal representation of the events to be recorded.

The males of many parasitoid wasp species produce acoustic stimuli during courtship (subsection 4.3.4). Such acoustic emissions can be recorded via the substratum using devices that act as transducers; these include electrodynamic microphones (van den Assem and Putters, 1980) and gramophone styli (Ichikawa, 1976). Experiments can be carried out in which recorded male signals are played back into the substrate (e.g. a plant), and the female's responses observed. Sivinski and Webb (1989) give many technical details on how to record acoustic signals of the braconid wasp *Diachasmimorpha longicaudata*. Webb *et al.* (1983) describe an analysis of the acoustic aspects of behaviour of the Mediterranean fruit-fly *Ceratitis capitata* (Tephritidae). See also Claridge *et al.* (1984) for aspects of recording the substrate-borne signals of insects such as leafhoppers and planthoppers.

4.3 PAIR FORMATION AND COURTSHIP

4.3.1 SEARCHING FOR MATES

Male and female parasitoids developing on the same host, or patch of hosts, with high levels of inbreeding (section 5.4.2) may not need to search extensively for mates. In *Nasonia vitripennis*, as in many gregarious or **quasi-gregarious** (solitary development on clumped hosts) species, newly emerged males remain on or near the host, and begin courting as soon as a female emerges. Males are able to locate a female within a host puparium (almost certainly using chemical cues) and characteristically position themselves on the puparium surface (King *et al.* 1969a). By contrast, in other gregarious species (e.g. *Melittobia* spp.; van den Assem, 1976a; Gonzalez *et al.*, 1985), females approach the males. Displays in *Melittobia* last relatively long (in some species up to about 30 minutes) before the male will copulate. Virgin females often attempt to position themselves between a courting pair. In *Melittobia*, males eclose before females and remain inside the host's puparium. Often, the first emerging male obtains most or all matings, after committing **fratricide**: the other males (usually sibs) are killed as pupae or eclosing adults (e.g. Abe *et al.* 2003). If two emerged males encounter each other, a fierce fight ensues. Mated females are no longer attracted to males and become positively phototactic and move away from the host. Because of the female-biased sex ratio (initially equal to 0.02–0.05 [proportion] males, but becomes even more biased due to fratricide), surviving males have plenty of mating opportunities (van den Assem *et al.*, 1982a).

In solitary species whose hosts do not occur in masses, mate searching by one or both sexes is necessary. Male parasitoids tend to search for females, rather than *vice versa*. Two reasons (both functional) for this are that **OSR operational sex ratios** (the ratio of males and females which are ready to mate at a given time; Emlen and Oring, 1977) is male-biased and that, due to haplodiploidy females do not have to mate in order to produce male offspring, and virgin reproduction may not be a disadvantage (section 5.3.3). Searching is likely to be concentrated in areas where the probability of mate finding is highest, i.e. sites where females emerge, feed or oviposit (Figure 4.4, section 5.3.4). Males may attempt to monopolise these sites, i.e. show territorial behaviour, provided that the sites are in short supply and are sufficiently clumped. Where this is not the case chemical, acoustic, tactile and visual stimuli may be involved in mate-finding (e.g. Cole, 1970; Vinson, 1978; Eller *et al.*, 1984; Kamano *et al.*, 1989; Field and Keller, 1993a, 1993b, 1994; Fauvergue *et al.*, 1995, 1998a, 1999; McNiel and Brodeur, 1995; Pompanon *et al.* 1997).

Chemical signals are usually volatile and provide long-distance information about the signaller's general, but not precise, location. However, virgin females of some species lay down trails of sex pheromones: *Aphelinus asychis* (Aphelinidae), a solitary parasitoid of aphids (Fauvergue *et al.*, 1995, 1998; Kazmer *et al.*, 1996), *Trichogramma brassicae* (Trichogrammatidae), a gregarious parasitoid of pyralid moth eggs (Pompanon *et al.*, 1997; Fauvergue *et al.*, 1998a), and *Ascogaster reticulatus* (Braconidae) an egg-larval parasitoid of tortricid moths (Kamano *et al.*, 1989). In *A. asychis*, males respond to an encounter with the pheromone by intensively searching in or near the marked area. Because the pheromone declines in activity within less than 24 hours it provides a reasonably precise cue to the location of virgin females. Fauvergue *et al.* (1998a) compared the response to trails and patches of pheromones deposited on the substrate by males of *A. asychis* and *T. brassicae.*

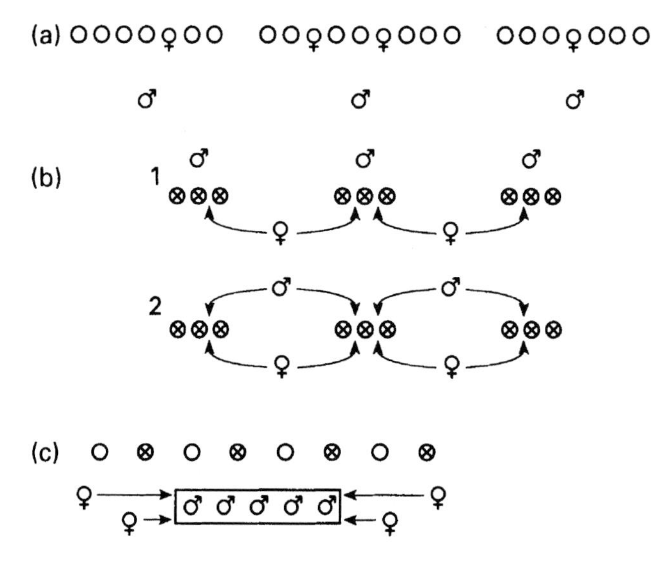

Figure 4.4 Schematic representation of different mate-searching strategies: (a) males monopolise sites where females emerge; (b1) males monopolise sites where females oviposit or search for food; (b2) males move continually between such sites; (c) sites cannot be monopolised, males aggregate and may signal in concert.

They predicted that male *A. asychis*, in which both sexes usually emerge in different localities, would follow pheromone trails to find females and that male *T. brassicae*, in which both sexes emerge from the same locality, would remain on pheromone patches in order to encounter females when they emerge. Contrary to expectation, male *A. asychis* do not follow trails any more than do male *T. brassicae* and male *T. brassicae* do not stay longer in pheromone patches than male *A. asychis*. In fact, the response to pheromone is quite similar in these species. One possible explanation is that there is a higher than expected degree of non-local mating in *T. brassicae* and that males of this species, like male *A. asychis*, commonly mate with females that have emerged in other localities, as observed in other species of *Trichogramma* (Kazmer and Luck, 1991; section 5.4.4).

Acoustic signals (produced by, or associated with, wing vibration) can give precise information on the signaller's locality, and small insects need to produce very high frequency sounds to communicate through air over anything but very short distances (Michelsen *et al.*, 1982; Michelsen, 1983). In structurally diverse environments such as vegetation, high frequency sounds are unsuitable for long-range communication due to scatter, reflection and interference, and so substrate-borne vibrations tend to be used. Some parasitoids produce such signals (e.g. Field and Keller, 1993b), but evidence that these function in the same way as long-range communication signals is lacking.

Passive visual signals are probably effective over longer distances only during daylight hours and where there is little cover (e.g. at the surface of bodies of water or in open spaces, many dragonflies exhibit brilliant sexually dimorphic colours or conspicuous markings, which function in mutual recognition (e.g. Grether, 1996). However, bioluminescent visual signals are particularly effective at night. Fireflies emit species-characteristic, coded flashes (Lloyd, 1971) and have specially adapted euconic compound eyes to perceive them (Chapman, 1972). All signalling incurs predation risks, for example the females of some firefly species attract heterospecific males by mimicking the female of that species, but then feed on, rather than mate with, the hapless male (Lloyd, 1975). Perception of visual cues over greater distances requires large eyes of sufficient acuity: in many swarming insects where being the first male to spot a female can mean the difference between reproductive success or failure, there is considerable sexual size dimorphism in eye size (e.g. honeybees, empidid flies, e.g. see Downes, 1970). Parasitoid wasps such as chalcidoids, proctotrupoids and cynipoids are too small to perceive long-range visual signals: only short-range (about 1 cm) visual communication is possible.

4.3.2 PROXIMATE FACTORS AND DISPLAY RELEASE

In general, a female parasitoid's presence is first detected by olfaction. Males increase their rate of antennal movement, vibrate their wings and commence walking rapidly. Similar reactions are obtained when males are presented either with homogenates of females on filter paper (Obara and Kitano, 1974; Yoshida, 1978) or with dismembered parts of females (Vinson, 1978; Field and Keller, 1994). *Anagrus* (Mymaridae) males will even attempt to copulate with a fine brush that has been wiped on virgin females (see Waloff and Jervis 1987). If females are killed in liquid nitrogen, immersed and agitated in a volatile solvent (e.g. acetone, diethyl ether, pentane, hexane, methylene chloride) and the solvent applied to filter paper and allowed to evaporate, the residue contains substances removed from the females. Males presented with this filter paper often perform the initial stages of display or even, as observed in *Pteromalus puparum*, produce a complete display and copulate *in vacuo* (van den Assem, 1996). Intact females taken from the solvent and dried, fail to produce a reaction, whereas dead females immersed in water and subsequently dried do produce a reaction. Clearly, males need not 'recognise' the object in its entirety in order to behave towards it as though it were a female: they react to chemical substances that release mating behaviour (van den Assem, 1996). Similar responses are found in insects that live in environments where visual communication is impaired. Males of the flour beetle, *Tenebrio molitor*, will readily court and copulate with a dried glass rod that has previously been dipped in ethanol in which receptive females have been immersed (e.g. Tschinkel *et al.*, 1967) suggesting that, as in *Pteromalus puparum*, visual cues play little role in releasing reproductive behaviour in males.

Attempts have been made to locate the source of the courtship-initiating pheromone in a number of parasitoids. The pheromone appears to be emanate from within widely different parts of the body. In several *Cotesia* species, a pair of glands in the female's reproductive system produce a sex pheromone (Tagawa, 1977, 1983; Weseloh, 1980; see also Field and Keller, 1994). In *C. glomerata* however, the pheromone emanates from a gastral segment (Obara and Kitano, 1974). In the braconid *Diaeretiella rapae* female gasters initiate courtship whereas other body parts do not (Askari and Alishah, 1979), while in the eulophid *Aprostocetus hagenowii* courtship is stimulated by extracts from the female thorax, but not from either the head or the gaster (Takahashi and Sugai, 1982). Using female pupae of different ages instead of adults, Yoshida (1978) found that pupae of the pteromalid *Anisopteromalus calandrae* secrete the courtship-eliciting pheromone, with a peak value around the red-eye stage (with older pupae there was a sharp decline in the effect). Full effects reappeared in adult females following emergence (a similar pattern has been reported for several Lepidoptera and Coleoptera). A simple experiment for distinguishing between chemical and visual stimuli in male courtship is illustrated in Figure 4.5. Visual stimuli may work over very short distances: *Nasonia vitripennis* males will follow a small black object that is moved to and from behind glass a short distance away (Barrass, 1960b).

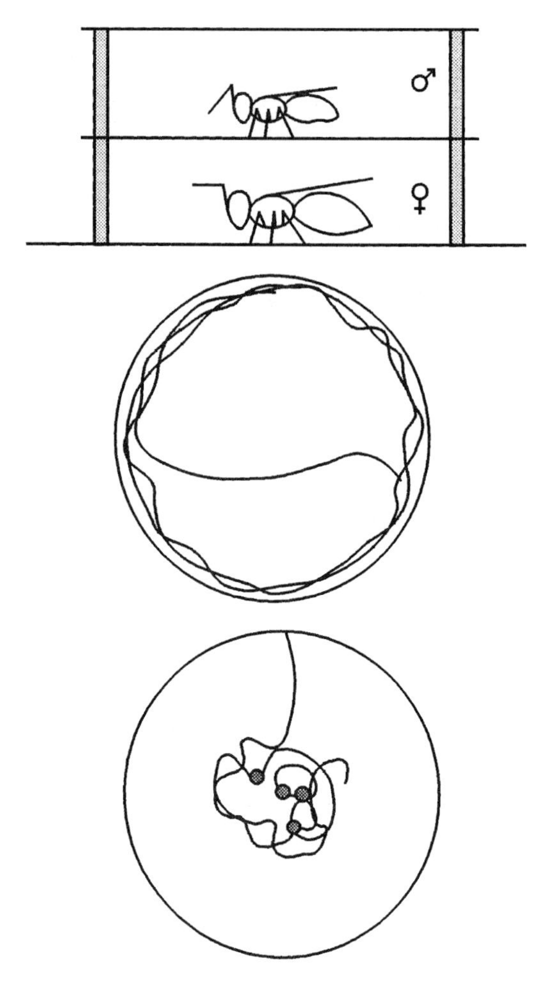

Figure 4.5 Experimental set-up for demonstrating the effect of chemical cues in the parasitoid *Anisopteromalus calandrae*. Two glass rings (20 mm diameter, 4 mm high) are separated either by a thin glass cover or a millepore filter. With the glass there is no obvious reaction on the part of the male, which is walking along the wall of the cell, with the filter he restricts searching to the area with pores, together with antennations and wing vibrations (dots) (source: Yoshida, 1978).

4.3.3 MALE ORIENTATION DURING COURTSHIP

Courting chalcidoid males either mount the female or remain alongside or facing the female (van den Assem, 1976b) (Figures 4.6 and 4.7).

Confrontation with a real female is often unnecessary for male courtship and mating behaviour to be elicited. Dummy females can be used to investigate which cues males respond to in courtship. A crude dummy made of cork, plywood or similar material may suffice, provided it carries the correct chemical cues (obtained from females that have been

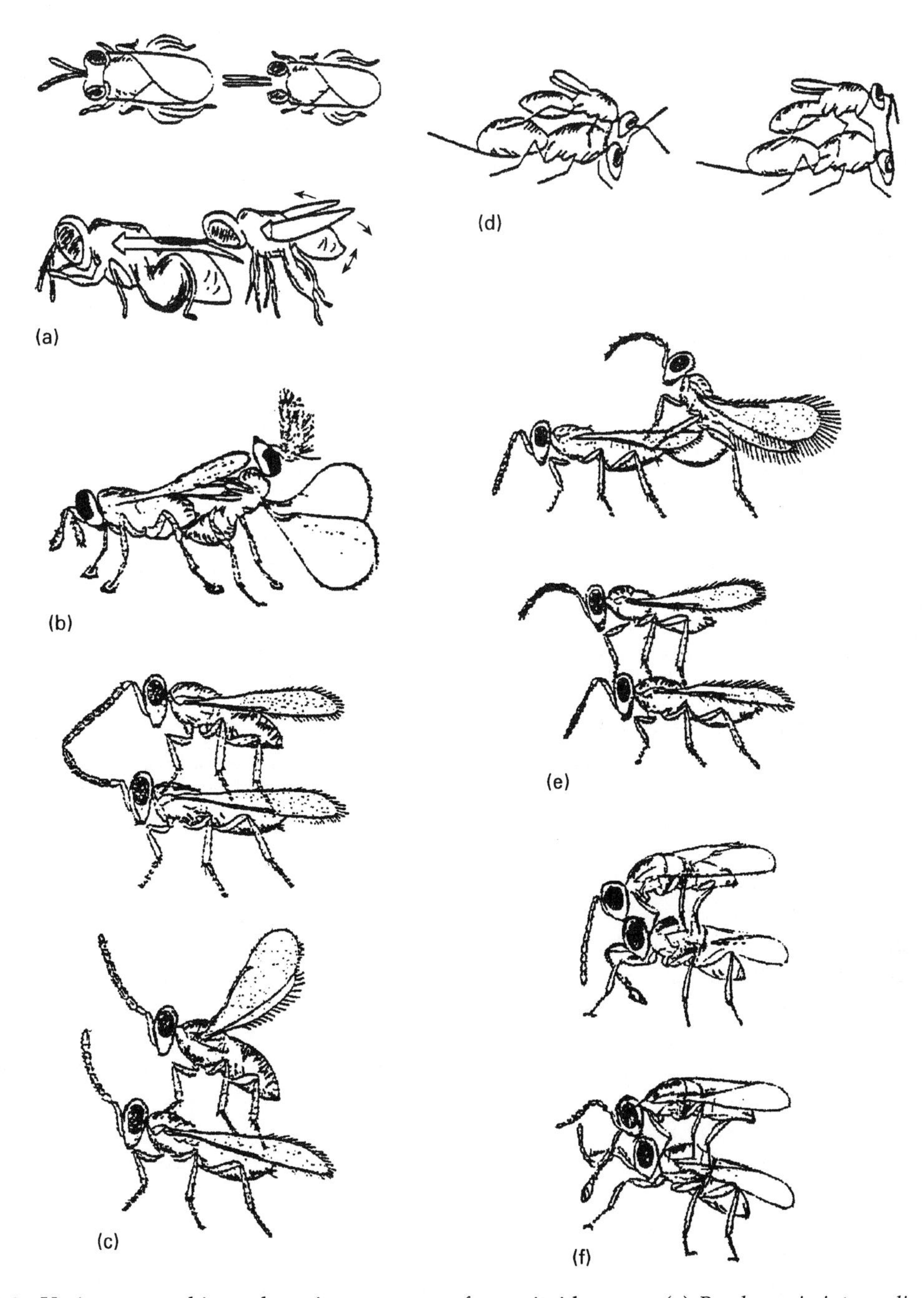

Figure 4.6 Various courtship and mating postures of parasitoid wasps: (a) *Brachymeria intermedia* (Chalcididae), male following female, and male performing courtship movements (wing pressing, and buzzing) (source: Leonard and Ringo 1978); (b) *T. evanescens* (Trichogrammatidae) (source: Hase 1925); (c) *Encarsia partenopea* (Aphelinidae), phases of post-copulatory courtship (source: Viggiani and Battaglia, 1983); (d) *Monodontomerus obscurus* (Torymidae), courting couple, male in low and in elevated position (source: Goodpasture 1975); (e) *Archenomus longiclava* (Aphelinidae), mating, and post-copulatory courtship (source: Viggiani and Battaglia, 1983); (f) *Amitus vesuvianus* (Platygasteridae), phases of pre-copulatory display (source: Viggiani and Battaglia, 1983).

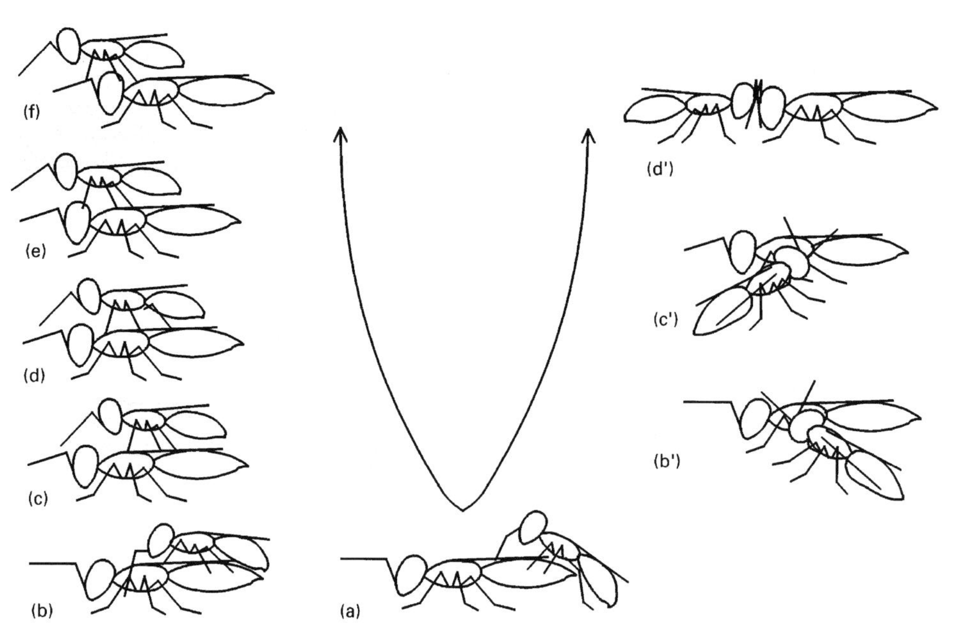

Figure 4.7 Phylogenetic transformation series for the male's courtship position in Chalcidoidea using arbitrarily chosen examples (where no references are given the information derives from J. van den Assem's observations). The plesiomorphic courtship position is similar to the mating position in: (a) *T. evanescens* (Trichogrammatidae) and *Choetospila elegans* (Pteromalidae). In the left hand branch the male mounts on the female, and his position moves gradually to the front; (b) *Spalangia cameroni* (Pteromalidae); (c) *Asaphes vulgaris* (Pteromalidae), *Sympiesis sericeicornis* (Eulophidae), *Aceratoneuromyia granularis* (Eulophidae); (d) *Pachycrepoideus vindemmiae* (Pteromalidae), *Vrestovia fidenas* (Pteromalidae); (e) *Nasonia vitripennis* (Pteromalidae), *Hobbya stenonota* (Pteromalidae), *Eupelmus spongipartus* (Eupelmidae), *Tetrastichus sesamiae* (Eulophidae), *Systole albipennis* (Eurytomidae); (f) *Anagyrus pseudococci* (Encyrtidae). In the right hand branch, the male courts on the substrate; (b′) *Achrysocharoides* species (Eulophidae) (Bryan, 1980); (c′) *Pediobius* species (Eulophidae), *Tachinaephagus zelandicus* (Encyrtidae); (d′) *Microterys ferrugineus* (Encyrtidae) (Parker and Thompson, 1925). There is probably a third direction of development: males remaining at the rear but having peculiarly elongated antennae that reach out to the front (van den Assem, 1986).

immersed and agitated in a volatile solvent, section 4.3.2). Within limits, its actual size is unimportant. However, males presented with a very crude dummy will rarely proceed beyond moving in an agitated manner while vibrating their wings, and will soon dismount (Yoshida, 1978; Yoshida and Hidaka, 1979; Tagawa and Hidaka, 1982). Yoshida and Hidaka (1979) tested whether males orient themselves in response to gravity. Dummies were attached to styrofoam at various angles: courtship position was determined by the female's posture relative to the substratum and not by gravity.

Male display position is mainly stereotyped, but may differ between taxa. *Nasonia vitripennis* males mount the female and place the fore-tarsi on the female's head (placement of the tarsi of the other legs depends on the relative size of the female). A stereotyped position requires specific stimuli for a correct orientation, and these can be identified using dummies on which different real body parts can be glued, in various positions (Figure 4.8). It is possible to determine what cues males use to arrive at and remain in the correct position, and release display behaviour. In *Nasonia vitripennis*, the position of the wings provides directional cues (to front or rear), and an object (antennae, ovipositor or an unnatural object) protruding at the end of a dummy seems necessary for display to be

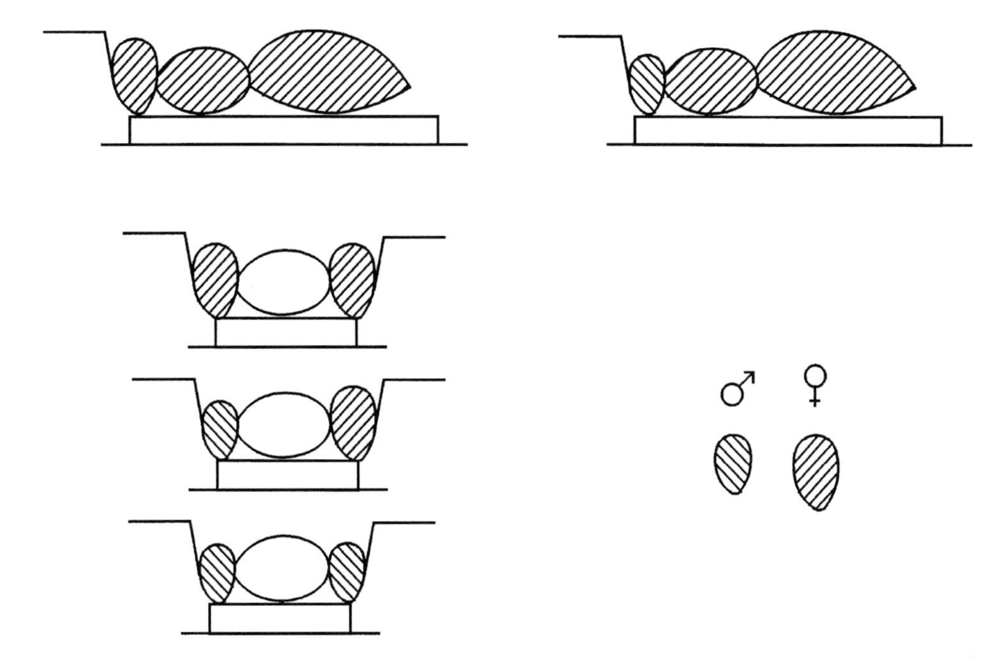

Figure 4.8 Examples of dummies that comprise male and female parts.

initiated (Figure 4.9). Chemical stimuli are also involved: a display directed at an ovipositor lasts for only a short period, after which the male turns around and moves to the front of the dummy. A male that courts a dummy comprising a female's body and a male's head will display for a shorter period of time than on a dummy that is comprised entirely of female body parts (J. van den Assem, pers. comm.). Similar dummies have been used to investigate the behaviour of other parasitoid species (*Cotesia glomerata* and *Anisopteromalus calandrae*; Yoshida and Hidaka, 1979; Kitano, 1975).

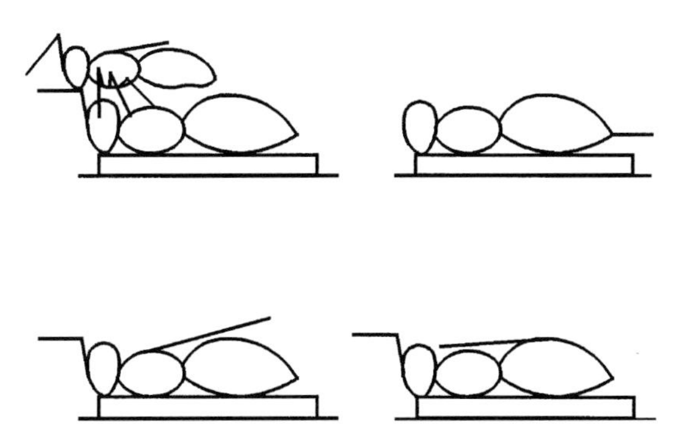

Figure 4.9 Examples of dummies used to determine what cues males use to orient themselves on females, remain in position and elicit display behaviour.

4.3.4 STIMULUS PRODUCTION BY MALES

Introduction

Once a male parasitoid has assumed the correct orientation relative to the female, and the female has become immobile (presumably in response to the male's behaviour), the male will commence courtship. In some parasitoids, courtship displays are visually inconspicuous whereas in others they are striking. According to Burk (1981), differences are likely to be correlated with the predominant mating system (section 5.4). Displays are likely to be relatively simple in species where the mating system is [illegible]d on male-controlled resources; M[illegible] do not display but attempt any nearby female. Signallir in species where mate enc or mating decisions are Signalling is most extr mating takes place wi males. However, these generalisations do not hold for parasitic wasps. Courtship in swarming species is not conspicuously complex (Nadel, 1987; Sivinski and Webb, 1989) whereas courtship in *Melittobia* seems to be very complex indeed (males are confined to the site of female emergence, there is a very high encounter rate, and the decision to mate is at least partly left to the female; van den Assem *et al.*, 1982a). Categorising behaviour as either simple or complex will be fraught with difficulties unless the relevant signals themselves have been identified.

Acoustic Stimuli

The males of many parasitoid wasp species vibrate their wings, both upon approaching the female and during courtship *sensu stricto*. *Nasonia vitripennis* males produce acoustic pulses of a constant quality throughout a display and the pulses coincide with wing vibrations (Figure 4.10). However, the wings themselves are not the source of the acoustic emissions because immobilising them, altering their surfaces, or removing them does not result in noticeable signal alteration. Similarly, males of the chalcidid wasp *Brachymeria intermedia* can be completely 'muted' by applying a tiny drop of superglue to the thoracic dorsum between the wing bases preventing the thorax from acting as a resonator (Leonard and Ringo, 1978).

The pattern of acoustic emissions produced during courtship is species characteristic (Figure 4.10), suggesting a display function. Such a function can be investigated by comparing the **courtship efficiency** (the amount of display necessary to induce receptivity in virgin females) and **mating success** (the proportion of mated females producing progeny of mixed sex) of normal males, mute-males and males with wings removed (van den Assem and Putters, 1980). In *Nasonia vitripennis* there is no difference between these three treatments, i.e. wing vibrations appear to play no role in courtship. Young mute males are able to induce receptivity within the same period as normal males and have similar mating success. However, females mounted by a mute male give a 'startle response', as if they were mounted without 'advance warning'. Older males are apparently more successful at courting females if they produce sound (van den Assem and

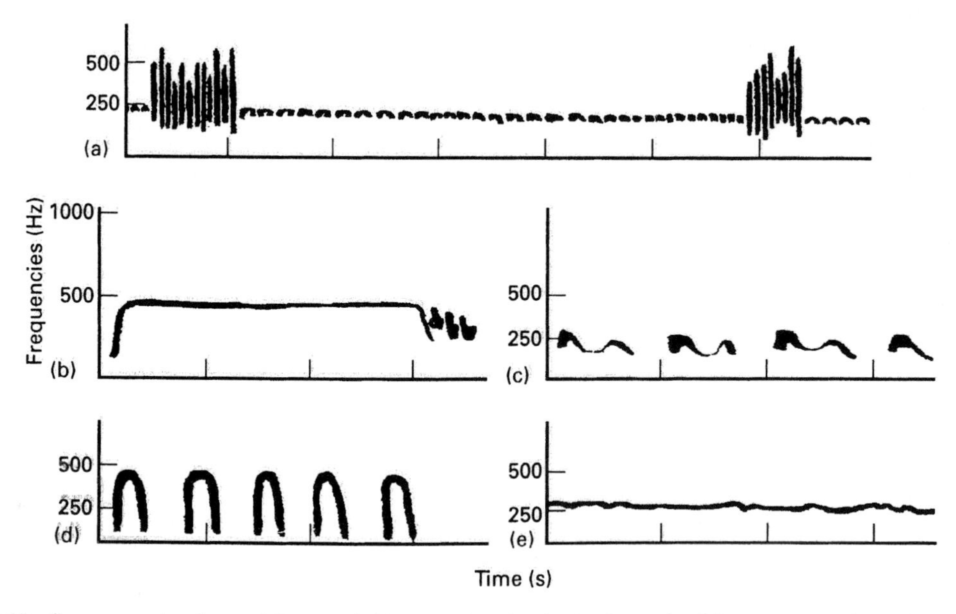

Figure 4.10 Sonagram tracings of the courtship sounds of: (a) *Nesolynx ales* (b) *Baryscapus daira*, (c) *Tetrastichus sesamiae* (all Eulophidae), (d) *Nasonia vitripennis*, (e) *Anisopteromalus calandrae* (both Pteromalidae).

Putters, 1980). Mute, old males were presented daily with a sequence of virgin females. *Nasonia vitripennis* recordings and 'white noise' were alternately broadcast at a barely audible level. Displays were more successful when accompanied by courtship sounds than by white noise, indicating that stimuli are normally airborne (i.e. sounds) and are not substrate-born. According to Barrass (1960a), old *Nasonia vitripennis* males vibrate their wings less than young males and are less successful in courtship due to the deteriorating production of courtship 'sounds'. However, a deterioration in pheromone production, and/or the "wafting" of that pheromone towards the female by the wing beating could also account for this observation. Adding the acoustic emissions of *Mesopolobus mediterraneus* to the display of old, mute *Nasonia vitripennis* males increased male success (van den Assem and Putters, 1980), suggesting that, in this case, emissions do not function in species recognition. However, the pitch of the *M. mediterraneus* emissions is similar to that of the *Nasonia vitripennis* emissions (although the temporal pattern is different), so pitch may be an important cue.

In *Cotesia rubecula*, wing-fanning by males produces both low-frequency airborne sounds and substrate-borne vibrations. Field and Keller (1993a) observed males courting females that had settled on cabbage leaves. When males were on leaves adjacent to those bearing females, female receptivity was never induced, but the same males were normally successful in inducing receptivity when placed onto the same leaves as the females; thus, vibrational courtship communication in this wasp species is very likely to be based on substrate-borne vibrations.

Chemical Stimuli

Chemical stimuli appear to play an important role in short-distance communication between males and females (section 4.3.2). It is usually easy to establish whether chemical stimuli are involved in display and to determine their origin. Simple experiments with *Nasonia vitripennis* demonstrated their importance in inducing female receptivity (van den Assem *et al.*, 1980a). The onset of receptivity coincides with head-nodding, a conspicuous display component, by the male (section 4.3.6). As the head moves up the mouthparts are extruded, and they are retracted when the head moves down. Males were manipulated so that nodding and mouthpart extrusion were prevented, both separately and in combination (Figure 4.11), without affecting other display components. The timing of courtship cycles (e.g. antennal sweeps were performed on time) was unchanged by gluing. Males with immobilised heads courted for the same amount of time as normal males to induce female receptivity. Males with sealed mouthparts were totally unsuccessful, despite normal courtship. However, if air from a vial containing courting, normal, untreated males and females was released over the females, they immediately became receptive. Air taken from empty vials or vials containing only males or only females was ineffective. Sealing a male's mouthparts appears to prevent the release of a pheromone that coincides with mouthpart extrusion.

Mouthpart extrusion probably serves a similar function in related pteromalids. Based on evidence from other Hymenoptera (e.g. honeybees and ants) it seems likely that the pheromones originate in the males' mandibular glands. This might be investigated by dissecting out the glands, agitating them in a solvent such as hexane and carrying out a behavioural assay, although the wasps' small size may be prohibitive. A solvent extract might also be used to study, by electroantennography, the receptors involved (the latter are probably located on the females' antennae), but again the small size of most parasitoid wasps is likely to generate practical problems. Contact pheromones, present on the cuticle, are probably very widespread. Because of their species-characteristic properties they play a role in 'recognition' of mating partners. Once a display is underway, a constant input of stimuli originating from contact pheromones is necessary to keep the display going.

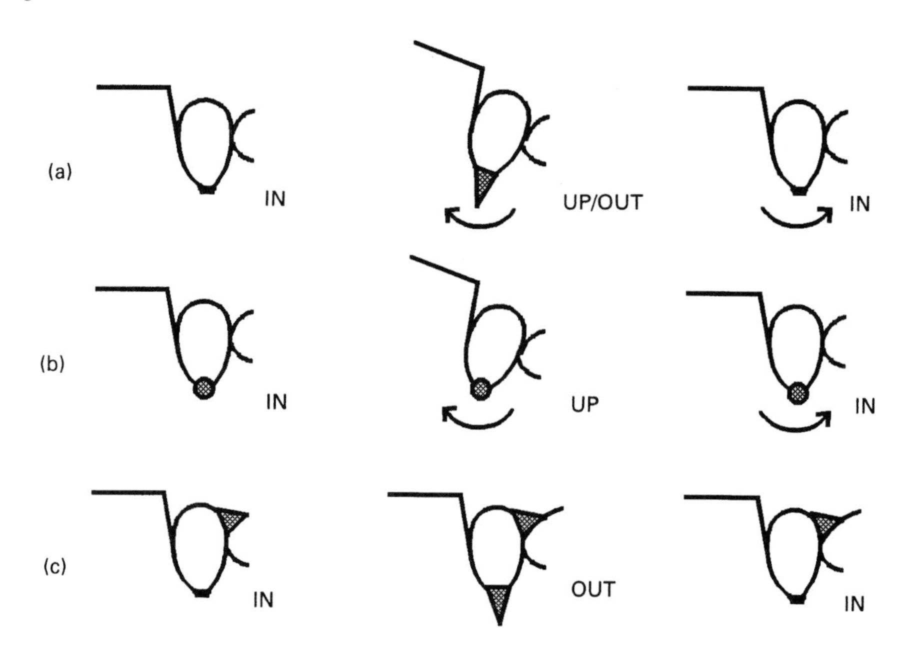

Figure 4.11 Head-nodding and mouthpart extrusion in *Nasonia vitripennis* males: (a) normal; (b) mouthparts sealed with superglue; (c) head movement prevented using superglue. Glue was applied to males immobilised by low (4°C) temperature and hardened by adding a small drop of water.

A recently developed, and presently virtually unexploited, method of simultaneously studying pheromone release and associated behaviour is to use time-course profiling of volatile release alongside observations or video recordings of behaviour. In this technique, originally developed by flavour chemists, insects are placed in an observation chamber which is connected to a mass spectrometer. Volatile chemicals emitted are analysed directly in the gas phase (Taylor *et al.*, 1995; Linforth and Taylor, 1998; Linforth *et al.*, 1996, Harvey *et al.*, 2000). The method also provides an (inexact) measure of the amount of volatile(s) released. To date, this method has been used to assess alarm pheromone release by bethylid wasps during female-female contests (I.C.W. Hardy, T.P. Batchelor, L. Evans and R.S.T. Linforth, unpublished), but not during mating behaviour.

Visual and Tactile Stimuli

The displays of many species of wasp involve leg movements, consisting of drumming, tapping, kicking and swinging. All may be used in tactile stimulation of the females. Males of some species have conspicuously decorated legs, for example those of *Mesopolobus* (Pteromalidae) bear coloured fringes or tufts which are moved in front of the female's compound eyes during courtship, suggesting a visual function (van den Assem, 1974). In the eulophid *Nesolynx albiclavus*, the male flexes his middle legs during a display. The tibiae are inclined steeply inwards and the tarsi steeply outwards (an unusual posture), and the male drums on the female's thorax (van den Assem *et al.*, 1982b). The tibial spur of the male's middle legs bears a basal triangular expansion from which long setae radiate, and the basitarsus bears remarkably long setae (Graham, 1989; Figure 4.12). The posture of the legs is such that these structures will make contact with the female during drumming. To discover whether they have any effect on female behaviour, one could remove either the setae or the legs, and compare the courtship success of altered males with that of normal ones.

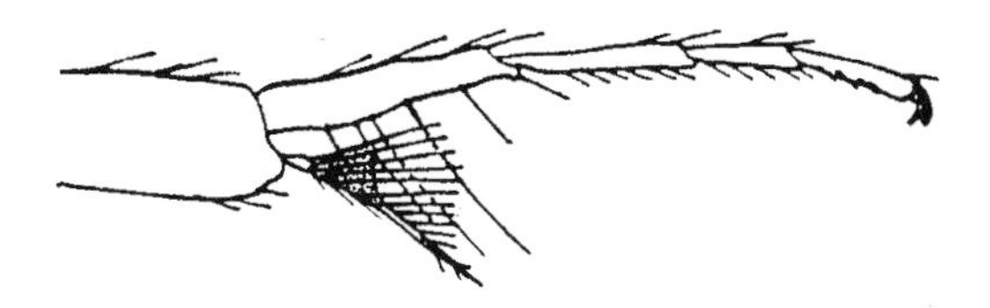

Figure 4.12 Distal part of tibia and tarsi of the middle leg of a male *Nesolynx glossinae*, showing a structure that probably provides specific stimuli during courtship From Graham (1989) reproduced by permission of Gem Publishing Co.

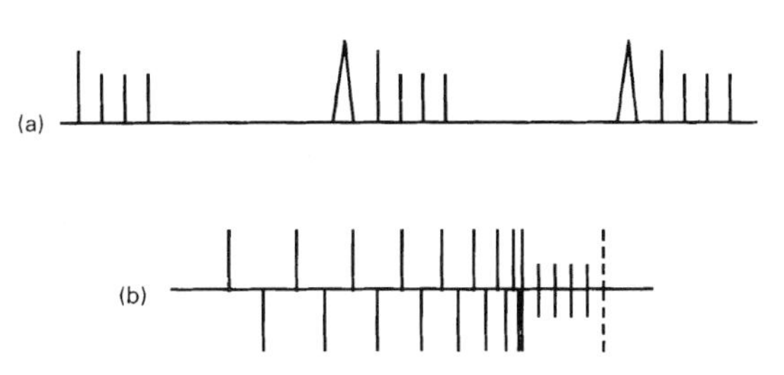

Figure 4.13 Schematic representation of: (a) 'cyclical', (b) 'finite' display. Time runs from left to right. The pteromalid *Nasonia vitripennis* produces a cyclical display. Triangles denote antennal sweeps, vertical lines denote head-nods. Single nods are separated by short pauses and are temporally clustered. First nods are denoted by longer lines than the following nods because they are more elaborate, last longer and coincide with the pheromone release. All series except the first are preceded by an antennal sweep. Series are separated by intervals. The period from the first nod of a series to the first nod of the subsequent series is termed a courtship cycle. A cyclical display consists of a repetition of identical or nearly identical cycles. The eulophid *Melittobia acasta* produces a finite display. Long vertical bars above the horizontal denote antennal movements (flagellar vibrations concluded by grasping of the female's antennae), those below the horizontal denote series of swinging movements with the hind legs in combination with a loss of antennal contact between the male and female. Alternations accelerate until they coincide. At this point (denoted by a series of shorter vertical bars above and below the horizontal) the antennae are stretched downwards and the hind legs moved up and down, rubbing against the female's thorax, instead of swinging to and fro. After a few seconds, these movements are followed by a brushing of the middle legs against the female's eyes (denoted by the dashed vertical line at the far right). A finite display does not consist of distinct, more or less identical cycles, but changes markedly over time, leading to a succession of 'phases' such as 'introduction' and 'finale'.

4.3.5 MALE INVESTMENT IN COURTSHIP

On meeting a conspecific female, males are usually ready to court immediately and thus to make an investment of time, energy and materials, but how much should a male invest? A thorough quantification of a male's time, and perhaps energy, investment would require separation of the courtship display into its components (subsection 4.2.8). These can be quantified in two ways: quick events that can be tallied to provide a total number of occurrences, and longer events whose duration can be measured. For a general account of how to quantify behaviour, see Martin and Bateson (1996).

In terms of overall temporal patterns, displays can be classified as either cyclical or finite (van den Assem, 1975). **Cyclical displays** involve a repetition of units of movement that may include various body components (Figure 4.13a). Successive units, or cycles, may be identical, but more often they are slightly different, e.g. in *Nasonia vitripennis* the number of head-nods per cycle changes according to a fixed pattern (Barrass, 1961). In **finite displays** quality changes during the performance because movements that were included earlier are omitted, to be replaced by new components. Finite displays have a predictable end, i.e. a finale, irrespective of whether female receptivity does or does not occur at that moment. Finite displays occur in *Melittobia* species (Figure 4.13b) and probably in four species of *Monodontomerus* (Torymidae) (Goodpasture, 1975), although no mention is made of the timing of female receptivity (section 4.3.6).

The cyclical display of *Nasonia vitripennis* males provides an excellent opportunity for a quantitative investigation of male courtship. Except for the first series, consecutive series of nods are preceded by an antennal sweep and are separated by intervals. Separate nods and series of nods can easily be counted and recorded, but between-series intervals require an automatic time-marker (section 4.2.9).

Additionally, males drum on the female's compound eyes with their front tarsi, and vibrate their wings (Barrass, 1960b). Drumming movements are much too rapid and too irregular to be counted separately, but can be quantified per bout, as present or absent. Wing vibrations can be quantified precisely using acoustic recording (section 4.2.9). The number of nods per series varies, as does the duration of intervals between series (general trends are illustrated in Figures 4.13, 4.14 and 4.15, see also Barrass, 1960b; Jachmann and van den Assem, 1993, 1996). Males exhibiting more nods per series exhibit fewer series up to the moment of dismounting. In measuring a *Nasonia vitripennis* male's display production, the first step would be to make external conditions constant by using an unreceptive or dummy female. Despite standardisation, males differ in the duration of displays (the period between mounting and giving-up) they perform, and there may also be variation within individuals observed on more than one occasion. Thus, performance is not exclusively controlled by external factors.

Some experiments require more rigorous standardisation. Assume that when males cease courting and walk away, the complex of internal factors (readiness to court) has reached a minimum and that this is similar for all males. Dismounting is followed by a refractory period until the male is ready to mount again. If the male and female remain together, the male terminates the pause by mounting the female

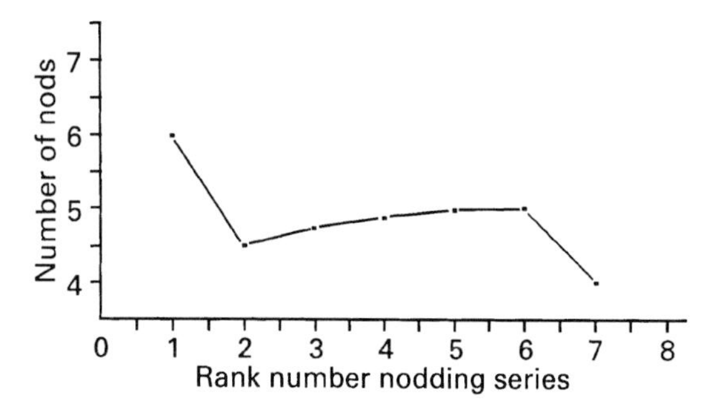

Figure 4.14 The number of nods in successive nodding series of courting *Nasonia vitripennis* males.

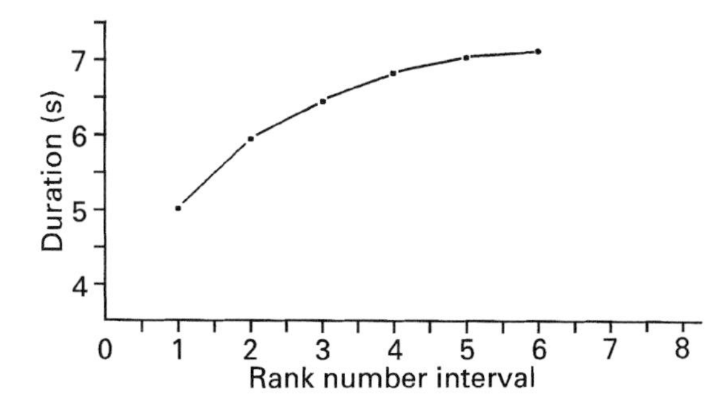

Figure 4.15 The duration of successive intervals between head-nod series in the courtship display of mate *Nasonia vitripennis*.

again. Alternatively, the experimenter can separate male and female, preventing courtship for a period, and then either reunite the partners or present the male with a different female. The second display is recorded in the same way. With pauses lasting less than 24 hours, the second display is always shorter than the first but a larger difference is found with a shorter pause (Figure 4.16). Time is not the only influential factor: males that have produced longer first displays (i.e. with more nodding series) produce longer second displays (Figure 4.17). These results suggest that something is 'used up' during displays, and that 'recovery' takes place during the pause. With longer pauses there is more recovery, but the recovery rate declines with time (van den Assem *et al.*, 1984).

Further conclusions are that males will court before they are able to display maximally, and that current display production is partly influenced by earlier displays. Thus, mating displays can be quite variable, even when performed by the same male. Clearly, for an understanding of display dynamics, records of motor patterns are insufficient – measurements of between-display intervals are also required (Figure 4.16). However, the performance of finite displays by *Melittobia* males appears to support the notion of an automaton: males produce a sequence of displays (directed towards receptive and unreceptive females) of roughly similar duration (Figure 4.18), independent of the duration of intervening pauses (van den Assem *et al.*, 1982a).

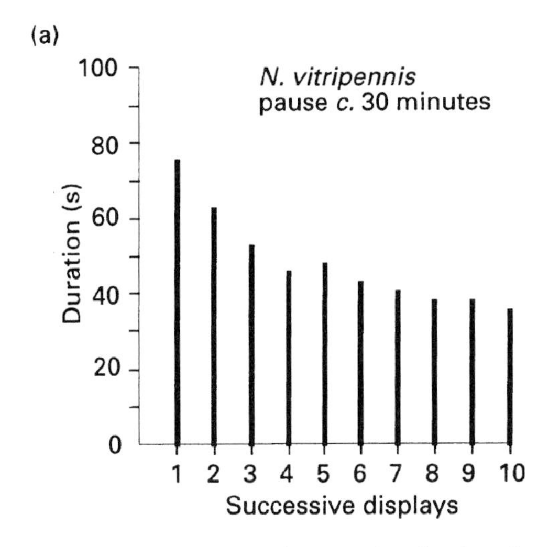

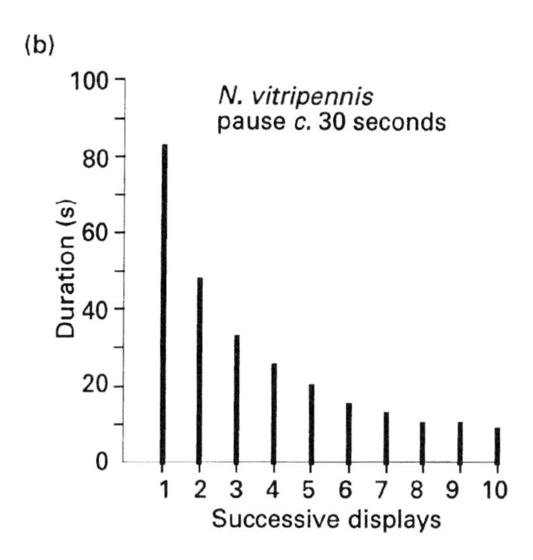

Figure 4.16 Durations of successive displays in male *Nasonia vitripennis* inter-display intervals (a) ca. 30 min., (b) ca. 30 s. Males cannot be described as simple courtship 'automata', as they do not produce unitary displays at any time.

Displays may come to an end when the male responds to external factors such as the onset of female receptivity (section 4.3.6). However, males that court dummies or unreceptive females (section 4.3.3) also give up for reasons other than being physically exhausted. The cessation of display coincides with (or occurs immediately after) head-nodding. It is possible that the readiness to court is incrementally reduced with the performance of each nod. A stiff hair struck against the courting male's dorsum (Figure 4.19) caused cessation of display and both the timing and duration of the interruption strongly influenced the numbers of extra series and nods produced. Hence, the display behaviour of *Nasonia vitripennis* (and presumably other species) follows simple rules (further discussed in Jachmann and van den Assem, 1993, 1996).

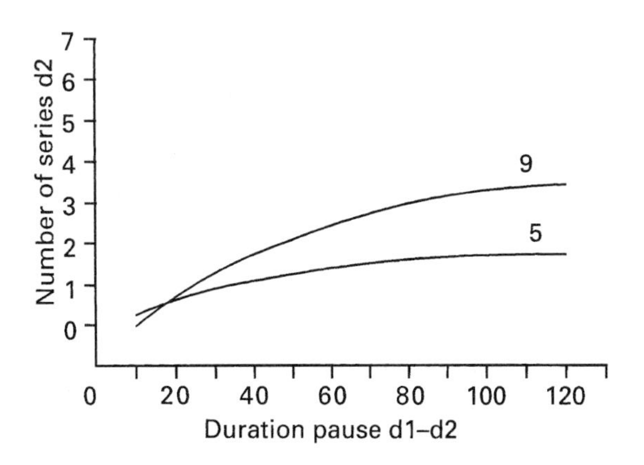

Figure 4.17 With pauses of equal duration, *Nasonia vitripennis* males that produce more (9) nodding series in the first display 'recover' more rapidly than those that produce fewer (5) series.

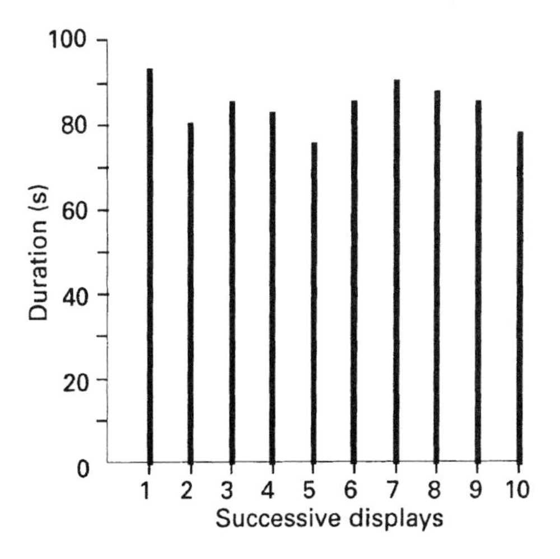

Figure 4.18 Courtship automata: duration of successive displays in male *Melittobia acasta*.

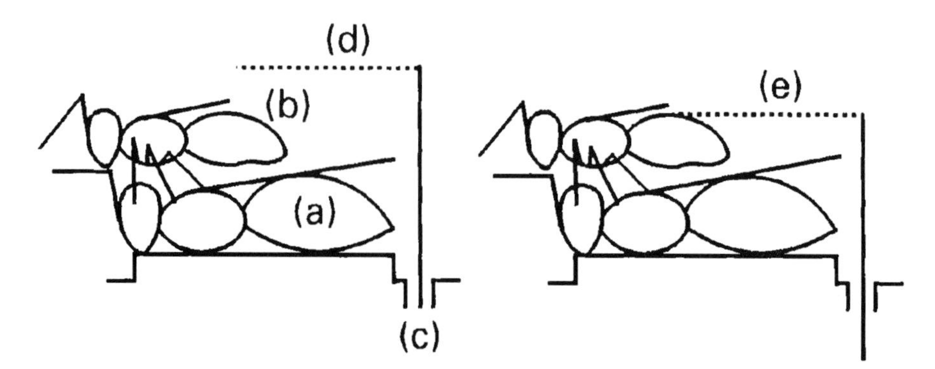

Figure 4.19 Apparatus for interrupting male display without causing dismounting: (a) dummy female, (b) courting male, (c) opening in bottom of observation cell and vertical rod, (d) hair, (e) hair in 'strike' position.

Van den Assem *et al.* (1984) found that *Nasonia* males that have courted several unreceptive females in succession produce only a short display with subsequent females, its duration depending on the length of time since previous dismounting. Males that had courted five females were either kept at room temperature (20°C) or placed in a freezer at −30°C (males were in styrofoam containers and thus not exposed to such low temperatures all the time) for 30 minutes, and were then presented with a sixth female at room temperature. Although the 'freezer' males were completely immobile and appeared to be dead, they became active surprisingly rapidly, but only then were presented with a female. 'Room temperature' males produced short displays as predicted, but 'freezer' males produced displays of the same length as those of inexperienced males, as if the effect of earlier performances had been lost. This rules out the possibility that something 'used up' during displays has to be either replaced or processed during pauses. If 'freezer' males were kept alone at room temperature for around 5 minutes before exposure to the sixth female, the effect of freezing was no longer evident. Thus, genuinely inhibitory processes, and not 'consumption' processes, play a role in terminating displays, but the physiological mechanisms are unknown. Unorthodox procedures may thus yield results that provide new insights into the causal processes underlying mating behaviour.

4.3.6 FEMALE RECEPTIVITY

Introduction

The mere presence of a male never results in overt female receptivity. Stimuli needed for the transition from latent to overt female receptivity have been discussed (section 4.3.4). Here, we consider more specific questions:

1. Are stimuli produced continuously or at intervals?
2. If at intervals, what is the temporal pattern?
3. Is the onset of receptivity random during a display, does it relate to a particular display movement, or is it induced by accumulated stimulation reaching a threshold?
4. Are courtship sequences 'chain reactions', with both participants reacting to one another step-by-step until copulation?

Simple chain reactions can be ruled out by observations of males courting dummies (section 4.3.3). The timing of successive display cycles appears to be a matter of internal control, but external stimuli are also of importance: courting *Nasonia vitripennis* males continuously monitor the position of the females' antennae (demonstrated with dummy females, subsection 4.3.6).

Direct observations are required to establish whether the onset of female receptivity coincides with the performance of a particular display movement by the male. However, this may not be obvious if, in cyclical displays, males repeat identical movements at high rates. In some ichneumonids, males vibrate their wings vigorously and repeatedly attempt to grasp the female's genitalia with their claspers. This 'wriggling' develops into copulation. A similar sequence is seen in *Trichogramma evanescens* (Hase, 1925; J. van den Assem, pers. comm.). Coincidences between display components and the onset of receptivity are easier to discern in cyclical displays with more differentiated motor patterns. Experiments involving mouthpart-sealed *Nasonia vitripennis* males demonstrate a chemical stimulus emanating from the extruded mouthparts (subsection 4.3.4). The onset of receptivity coincides with the first nod of a series (Figure 4.20a) and (almost) never with second or subsequent nods. Thus, a periodic production of essential stimuli induces receptivity. There is additional evidence: first nods differ in movement from subsequent nods, and females are able to perceive the stimuli continuously. Females may signal receptivity at any time once a threshold concentration of pheromone is reached. With finite displays, the production of releaser stimuli seems to occur once per display, which makes the timing of the switch from latent to overt receptivity predictable (Figure 4.20b). This is even stronger evidence of periodic production of stimuli.

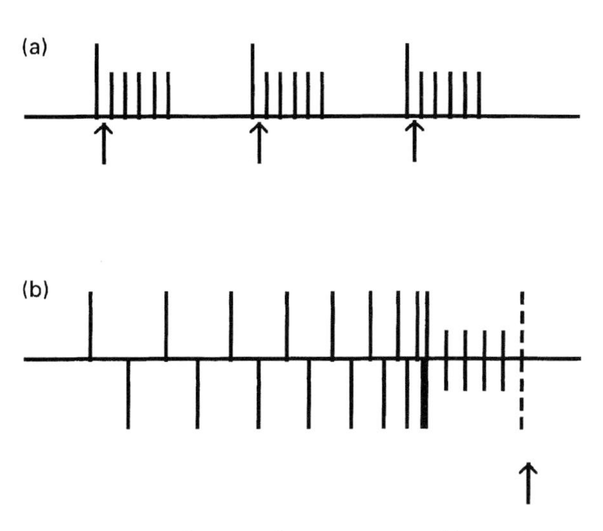

Figure 4.20 Onset of receptivity (denoted by upward-pointing arrows, for other symbols see Figure 4.13) according to performance of a particular event: (a) in the cyclical display of *Nasonia vitripennis* receptivity may occur at several points in a sequence, but always immediately following the performance of a first nod; (b) in the finite display of *Melittobia acasta* receptivity occurs at a unique point in time, at the end of a finale.

Receptive Posture

The switch from latent to overt receptivity is generally indicated by the female raising her gaster and opening the genital orifice. The posture varies between taxa. In eulophids, females 'sag in the middle' while raising the abdomen and direct the head and antennae upwards. In *Nasonia vitripennis* and related Pteromalinae, the head is lowered and the antennal flagellae are drawn towards the front of the head (Figure 4.21). In *Cotesia rubecula* the head is tilted forward and the antennae are lowered (Field and Keller, 1993b) while in *Eretmocerus eremicus* females become receptive without any clear

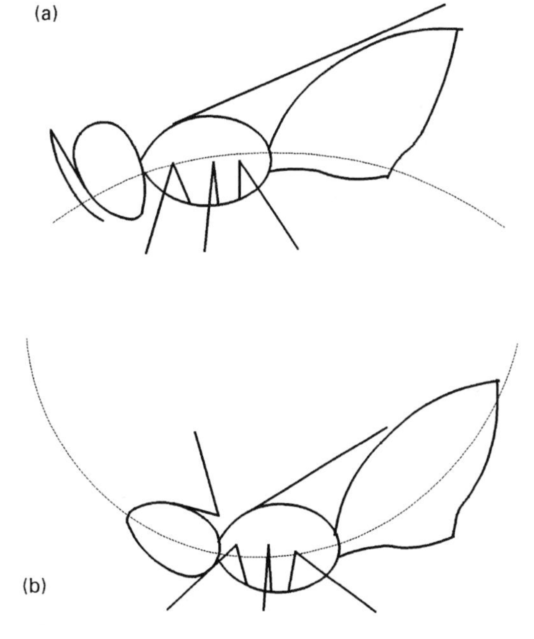

Figure 4.21 Female receptivity posture of (a) pteromalids and (b) eulophids.

Figure 4.22 Position of the courting male in different species of Pteromalidae and the possibility of the male perceiving directly the raising of the female's gaster.

changes in antennal position, while the abdomen is lifted to expose the genital orifice (Hunter *et al.*, 1996).

As soon as a female is receptive, the male ceases courting and switches to copulatory behaviour. The raising of the female's abdomen may provide a tactile stimulus to a mounted courting male in many species, but this is not the case in *Nasonia vitripennis* and related Pteromalinae (Figure 4.22), probably because of the anterior position a male adopts on a female. However, antennal folding would be easily perceived by the male. Van den Assem and Jachmann (1982) used a dummy with movable antennae, constructed from a freshly killed female, with antennae removed, and a rod bearing two pieces of wire representing antennae, which could be moved up and down (Figures 4.23 and 4.24). Males courted the inactive dummy as they did a living female and gave up after a similar period. When the 'pseudoantennae' were moved downwards, males ceased courting and attempted to copulate. This shows that antennal movements signal receptivity. Similar movements in other pteromalids probably have the same function.

Intraspecific Variability in Onset

Virgin females will usually become receptive when courted, sometimes even after very limited male display. In species that mate more than once, induction of further receptivity usually requires a longer display. Although there is no evidence, individual males may differ in their ability to stimulate a female. Limited evidence suggests that individual females differ in their sensitivity to stimuli: virgin female *Nasonia vitripennis* were allowed to become receptive and copulate with a male, but the male was removed before he regained the anterior position for post-copulatory courtship (during which females usually again signal receptivity, subsection 4.3.6). Females were then classified as those that had signalled initial receptivity with the male's first head-nod series, and those that did so after three or more series. All females that were presented with another male after 30 minutes signalled receptivity, although a longer display was required to release the response. The initial inter-female differences remained: females of the first category required less stimulation than those of the second category. On the other hand, males that had induced receptivity with the first nod of the first series

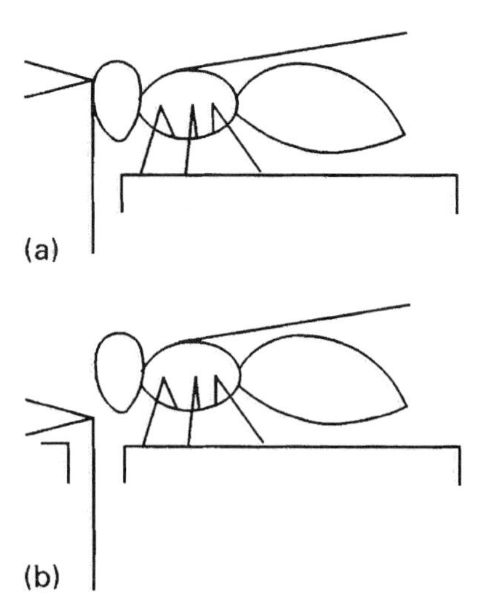

Figure 4.23 A dummy female *Nasonia vitripennis* with movable antennae, antennae in (a) courtship-eliciting position and (b) receptive position.

Figure 4.24 (a) A male *Nasonia vitripennis* performing courtship behaviour on a dummy with movable antennae. (b) The dummy's head in contact with the movable antennae. Because frequent renewal was necessary, dummies were fastened into the observation cell with a human hair strung between two clips. Further details are given in van den Assem and Jachmann (1982).

were not consistently more efficient compared with males that induced receptivity after several series, when presented with a second series of females half an hour later (van den Assem, 1986).

Switching-off of Receptivity

Sexual receptivity of female wasps can usually be induced only a limited number of times: at some point further matings will be refused. Females of many parasitoid wasp species (Gordh and DeBach, 1978; Ridley, 1993) but few predator species (e.g. Fincke *et al.*, 1987; Arnqvist, 1997) mate only once. Both internal and external factors are probably involved in the switching-off process that makes females permanently or temporarily unresponsive to courtship stimuli. Those related to insemination *sensu stricto* are obvious candidates; stored sperm can provide direct, 'external' stimuli (e.g. as a measure of the fullness of the spermatheca) (e.g. Thibout, 1975). However, an ejaculate contains materials other than sperm that might indirectly affect a female's receptivity (e.g. Leopold, 1976). The effect of spermathecal fullness would seem rather easy, and the effect of ejaculate substances relatively more difficult, to verify.

Ode *et al.* (1997) conducted an experiment to examine remating by females of the gregarious parasitoid *Habrobracon hebetor* (Braconidae). Virgin females, mutant for eye-colour (ivory), were presented with either an inexperienced mutant male or a wild-type male (black eyes) and were then presented with the opportunity to mate with a male of the opposite eye-colour on all successive days. Males and females were placed together for 15 minutes during which all copulatory events were noted. After each mating period, females were kept individually and given four hosts per day until death. All progeny were reared and the eye-colour of all daughters was noted to detect the occurrence of a second mating. Daughters possessed the eye-colour of their fathers. Only 0.5% (3/64) of females remated despite over 60% of them running out of sperm from the first mating. Several modifications of this technique could be performed for other species where remating patterns are less clear-cut. These might include interrupting copulations at various times to obtain 'filled' and 'partially filled' females or testing females for receptivity more than once a day.

Insecticide resistance can be used as a phenotypic marker to study multiple mating (**polyandry** in females, **polygyny** in males) in a species, but this involves performing insecticide bioassays, which can be time-consuming both to devise and carry out, and it obviously requires that members of the population possesses an insecticide resistance gene (see Baker *et al.* 1998 on *Anisopteromalus calandrae*, some individuals of which show resistance to malathion).

Allen *et al.* (1994) found that post-copulatory courtship and mate-guarding in *Aphytis melinus* (Aphelinidae) reduced the chance that females would mate with other males, but did not completely switch-off female receptivity to further mating. Field and Keller (1993a) found that female *Cotesia rubecula* remain receptive for a brief period after the first mating and that post-copulatory mimicking behaviour by the first male often distracts rival males during this period (Figure 4.25).

In many other groups of insects, the reappearance of receptivity in females seems to be under the direct control of hormones that are involved either in egg maturation or oviposition *per se*. Females may copulate either immediately after or before laying eggs whether or not there are sperm present in the spermatheca (e.g. dungflies [Parker, 1970a] and dragonflies [Waage, 1973]). In the cricket *Acheta domestica*, males transfer prostaglandin synthetase which induces faster rates of oviposition which in turn induces females to remate (references in Eberhard, 1997), and studies of *Drosophila melanogaster* have revealed similar male-transferred substances which increase female reproductive output (Chen *et al.*, 1988). Curiously, these substances have a negative effect on female fitness, but have presumably developed because any particular male is only interested in the female's reproductive value in the current batch of eggs.

In several groups of parasitic wasps the induction of female receptivity may itself initiate a process that inhibits the early reappearance of receptivity. For example, in *Leptopilina heterotoma* (Cynipidae) and *Lariophagus distinguendus* (van den Assem, 1969, 1970), *Nasonia vitripennis* (van den Assem, 1986) and *Habrobracon hebetor* (Ode *et al.*, 1997), females that had become receptive but were uninseminated (e.g. if mating was terminated before sperm transfer) were subsequently unreceptive.

4.4 COPULATION AND INSEMINATION

4.4.1 MALE READINESS TO COPULATE

Males can generally copulate without first courting. If a male encounters a female in the copulation posture (a living female or a dummy obtained by killing a copulating female in liquid nitrogen), he will proceed to copulate immediately. For example in *Nasonia vitripennis* '**sneaky males**' are able to copulate with females actively

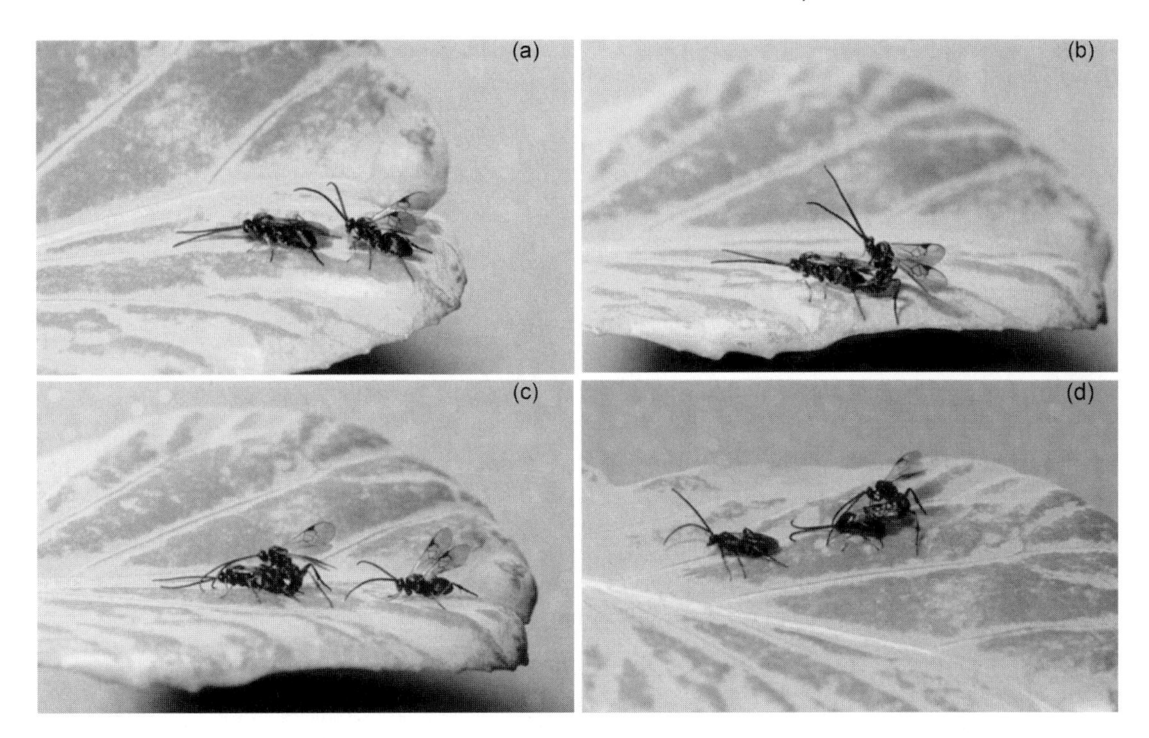

Figure 4.25 Post-copulatory female-mimicking behaviour in *Cotesia rubecula* (Braconidae): (a) female (left) signals receptivity to courting male (right); (b) female and male in typical copulatory position; (c) copulating male (centre) mimics female receptive position by lowering antennae in response to courtship by second male (far right); (d) following copulation, female (left) moves away, whilst the second male (right) is deceived by female-mimicking behaviour of the first male (centre). Reproduced from *Animal Behaviour*, 46 (6), S.A. Field and M.A. Keller, 'Alternative mating tactics and female mimicry as post-copulatory mate-guarding behaviour in the parasitic wasp *Cotesia rubecula*' pp. 1183–1189, 1993), by permission of the publishers, Elsevier Science, W.B. Saunders Company Limited and Churchill Livingstone.

courted by another male (Figure 4.26). Similarly, males may perform some part of the courtship sequence, but 'short-circuit' this by exploiting the courtship display of another male (Figure 4.27).

In *Nasonia vitripennis*, males re-assume the anterior position immediately following copulation

Figure 4.26 A courting *Nasonia vitripennis* couple, with a 'sneaky' male clasping the female's gaster.

and proceed to court afresh, and females usually signal receptivity for a second time. However, males never respond to a second signal but remain at the front and continue displaying for a short period before dismounting. Hence males react differently to identical external signals, depending on internal state which is influenced by previous behaviour. The copulatory act causes changes: males removed prior to or following the first receptivity signal, or just prior to genital contact, immediately attempt copulation when again confronted with the signal. The post-copulation refractory period is short. A male that is prevented from performing a post-copulatory display (by being brushed off the female at the termination of copulation) and which quickly mounts another female,

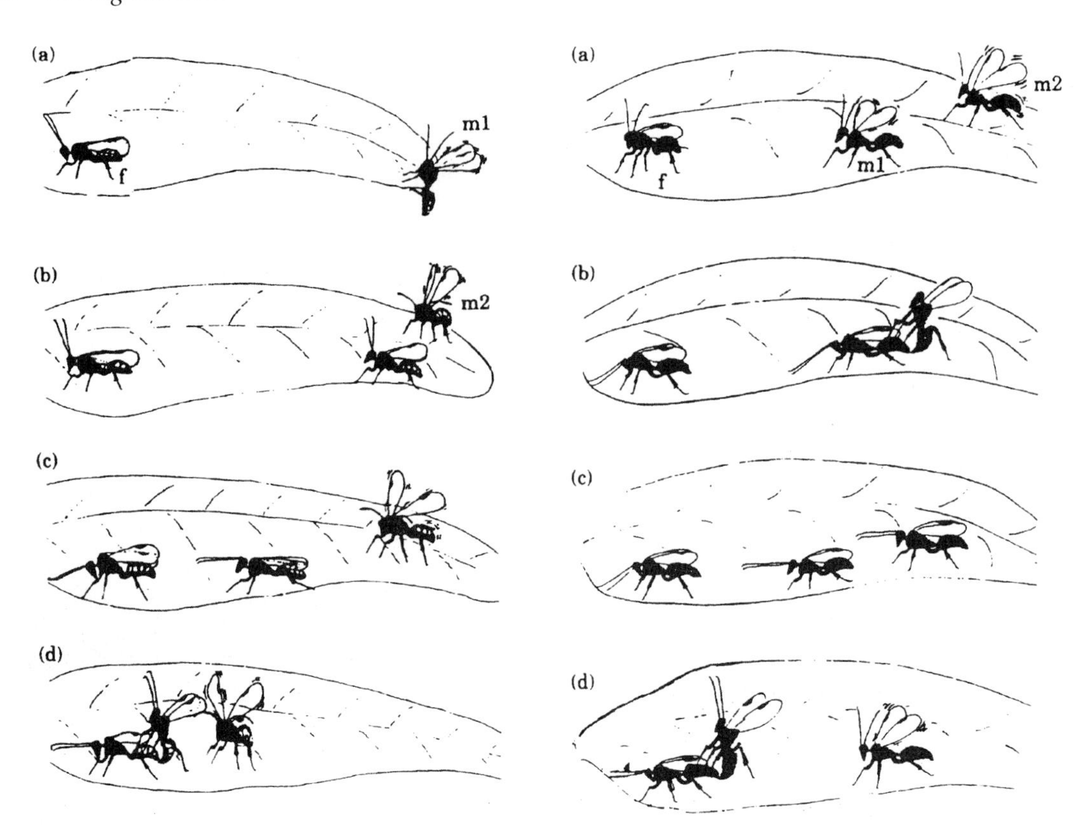

Figure 4.27 Two mating tactics in the braconid parasitoid wasp *Cotesia rubecula*, alternative to performing full courtship (Figure 4.2). In the **left hand sequence** a male exploits the courtship display of a rival male ('short circuit behaviour'): (a) Male 1 (m1) lands on a leaf bearing a female (*f*) and begins courting by wing–fanning. (b) After Male 2 (m2) lands and starts fanning, Male 1 stops fanning and waits, (c) Male 2 emits a 'pulse' of substrate-borne vibrations and Male 1 darts forward as the female becomes receptive (indicated by forward tilting of her head and lowered antennae), (d) Male 2 arrives at the copulating pair after performing the full courtship sequence. Short-circuiting occurs frequently in this species, and males employing such behaviour have a higher copulation success rate than rival males performing full-courtship. In the **right hand sequence** a male distracts a rival male by mimicking a female: (a) Two rival males (m1 and m2) court the same female (*f*), Male 2 has just 'pulsed'; (b) Male 1 adopts the female receptive position in response to the pulse. Male 2 is deceived into attempting to copulate with Male 1; (c) Male 1 darts away towards the female; (d) Male 1 and the female copulate while Male 2 resumes courtship. Female-mimicry occurs infrequently in this species and its success relative to alternatives is unknown. Reproduced from *Animal Behaviour*, 46 (6), S.A. Field and M.A. Keller, 'Alternative mating tactics and female mimicry as post-copulatory mate-guarding behaviour in the parasitic wasp *Cotesia rubecula*' pp. 1183-1189, 1993), by permission of the publishers, Elsevier Science, W.B. Saunders Company Limited and Churchill Livingstone.

will not back up if the second female is quick to signal (at the first nod of the first head-nod series). If a longer display is necessary to induce receptivity, the male will back up and copulate. The refractory state is thus not a female-specific, but a time-dependent, phenomenon.

After mating, the reproductive success of a male depends entirely on that of his mate. One way in which males could exert direct influence on the prospects of his sperm would be by choosing larger females in preference to smaller ones, since the former tend to have a higher fecundity (see section 2.7.3). We have not found any evidence of mate size preferences by male parasitoids, but the possibility has not been investigated systematically. Antolin *et al.* (1995) found some evidence for size-assortative mating in *Habrobracon hebetor* where smaller males were unable to mate effectively with large females.

4.4.2 DURATION OF COPULATION

Within pteromalid and eulophid species, there is no correlation between male or female body size and duration of copulation, and for individual males durations of successive copulations are similar. Furthermore, duration is not related to sperm depletion (subsection 4.4.4). There is, however, considerable interspecific variation (e.g. *Nasonia vitripennis* 15–20s, *Lariophagus* 40–80s, *Melittobia* 5-10s [all at 22°C, note that durations are temperature-dependent]). The duration of copulation seems to be pre-programmed in females. In several species, receptive females remain in the copulation posture for similar periods irrespective of whether a copulation has occurred or not, as demonstrated by observations of unmanipulated behaviour and by experiments in which males are brushed off receptive females (van den Assem, 1969, 1970, 1986). Obviously, a minimum period is required for delivery of the ejaculate. It seems unlikely that longer copulations involve transmission of greater volumes of sperm, or a slower flow; probably only part of the time is needed for sperm transmission. To investigate this, one could interrupt copulations at various stages and count the numbers of daughters produced or the amount of sperm transferred.

4.4.3 EJACULATION

Prior to ejaculation, mature sperm pass from the testes into two pairs of chambered vesicles. The proximate pair are smaller and thicker-walled than the distal pair and open into the ejaculatory duct. There is a sphincter muscle between the chambers. The proximate chambers become empty immediately after mating (Wilkes, 1965), and their combined volumes can be taken to correspond to a 'full ejaculate'. The minimum intervals between inseminations suggest that the proximate chambers can be refilled rapidly. A male is expected to invest his entire sperm production in one ejaculate only when mating occurs once per lifetime. Honeybee drones, for example, are lethally damaged by ejaculation and die shortly afterwards. A similar phenomenon would be expected in predator species where males are certain to be consumed when copulating (subsection 4.4.5). Male ejaculate in parasitoids comprises sperm and secretions from accessory glands (Leopold, 1976).

In *Nasonia vitripennis* the length of the male's protruded penis (aedeagus), in relation to the distance from the external opening of the vagina to the opening of the spermathecal duct, suggests that sperm are deposited at the duct's opening (Sanger and King, 1971). Sperm are found in the female's spermatheca within one minute of the termination of genital contact. In a number of parasitoid species, a mass of sperm swimming within the spermatheca can be observed under a light microscope at 100–1000x magnification (Wilkes, 1966; Nadel and Luck, 1985; Hardy and Godfray, 1990; Ode *et al.*, 1995; Kazmer *et al.*, 1996; Heimpel *et al.*, 1997; Figure 4.28). Sperm may move from the spermatheca's duct opening to the interior either by peristalsis or in response to a chemical gradient. There is movement in the opposite direction once sperm are used during fertilisation (King, 1962; King and Ratcliffe, 1969).

4.4.4 SPERM DEPLETION IN MALES

In general, males are able to inseminate more than one female. However, males that copulate with a number of females in rapid succession may temporarily deplete their supply of sperm (Wilkes, 1963; Laing and Caltagirone, 1969; Gordh and DeBach, 1976; Nadel and Luck, 1985; Ode *et al.*, 1997; Figure 4.28). In some parasitoid wasp species the sperm supply is partially replenished after a period of no mating, but males generally never regain their initial insemination potential, and full depletion reappears

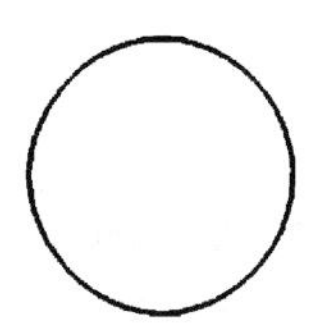 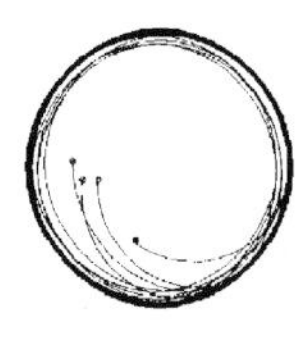 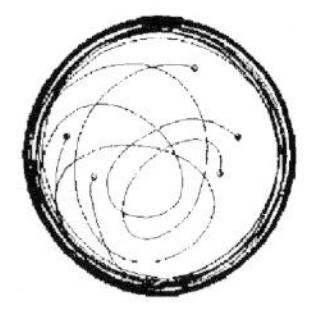  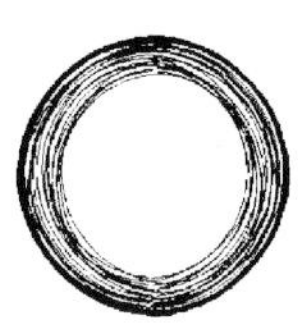

Figure 4.28 Band width of the sperm layer inside the spermatheca of *Pachycrepoideus vindemmiae* (Pteromalidae) after insemination by males at various stages of depletion. From Nadel and Luck (1985).

rapidly (Laing and Caltagirone, 1969; Ode *et al.*, 1996). In at least two species, *Nasonia vitripennis* (van den Assem, 1996) and *Habrobracon hebetor* (Ode *et al.*, 1996), larger-sized males have a larger insemination capacity. Males that copulate infrequently are depleted of sperm less rapidly than males that copulate frequently. Male *Pachycrepoideus vindemmiae* that copulate at 30 minute intervals reach a state of equilibrium where production equals ejaculation (Nadel and Luck, 1985).

The extent to which sperm depletion occurs in the field is largely unknown. The one exception we are aware of comes from a field study of *Habrobracon hebetor*: approximately 20% of the females were found to have no sperm in their spermathecae (Ode *et al.*, 1997). Many of the females brought into the laboratory, where they were given hosts daily until death, quickly ran out of sperm (as evidenced by the production of only sons), suggesting that sperm depletion is a biologically relevant phenomenon in this species.

Investigating lifetime sperm production in individual males is time–consuming; consequently, it has rarely been examined. In one study, of *Habrobacron hebetor* (Ode *et al.*, 1996), newly emerged males were allowed to copulate with ten virgin females in immediate succession each day until death. Each mated female was then given four hosts per day until death. Offspring from each of these hosts was allowed to complete development and the number of daughters was counted to determine the total numbers of daughters sired by the male. Larger males sired more daughters both per day and per lifetime than did smaller males. Nadel and Luck (1985) offer an alternative approach: dissection of spermathecae in Ringer's solution following insemination, and measuring the band-width of the sperm layer (Figure 4.28). In themselves, estimates of the quantity of sperm used by females are of limited interest, but are indispensable for an assessment of reproductive success (subsection 4.1.1). Sperm depletion of males is further discussed in subsection 5.4.7.

4.4.5 SEXUAL CANNIBALISM

A problem for males in some predatory insects is that they may be treated by females as prey and be killed before copulation. Three strategies may be adopted by males in order to maximise their inclusive fitness in such circumstances:

First, males can attempt to appease females with food, some other object or special signals (see Thornhill and Alcock, 1982). In empidid flies, there is a gradation of behaviour: (a) prey is devoured independently of courtship, i.e. no appeasement (e.g. *Platypalpus, Hybos, Empis trigramma, E. punctata*), (b) prey offered by the male is fed upon by the female during copulation (e.g. many *Empis* species), (c) prey or objects offered by the male is not fed upon by the female but acts as a releaser stimulus for copulation. In *Hilara*, an inanimate object is enclosed in a silken web (spun by the male, from glands in his fore tarsi; references in Cumming, 1994; see also Preston-Mafham, 1999), the web may be empty (i.e. it is merely a 'balloon', Kessell, 1955, 1959; also references in Cumming, 1994) or it contains prey that has been fed on by females during prior copulations and has little remaining nutritional value (references in Cumming, 1994).

Second, males can evolve special structures that restrain or immobilise the female prior to copulation. This is believed to be the evolutionary basis for the unusual arrangement of genitalia and claspers in male odonates (dragonflies and damselflies): by holding onto the female's head or prothorax, the male keeps her mouthparts at a safe distance. Secondary genitalia presumably evolved because the male's gonopore is relatively inaccessible during copulation.

Third, males can risk establishing genital contact without appeasement. This can be selected for if matings that result in the male being eaten lead to higher paternity than matings that are not fatal. This strategy is exemplified by praying mantids, the mating behaviour of which has long been known to involve sexual cannibalism (*Stagomantis carolina:* Howard, 1886; *Mantis religiosa* Fabre, 1907). In *M. religiosa*, the male approaches the female from behind by making almost imperceptible steps in her direction, unlike normal locomotion, then jumps suddenly on to the female's back clasping her with his raptorial fore legs. Once out of reach of the female's own raptorial legs, the male positions his genitalia close to the female's. Females must raise their genitalia for copulation. Mating proper lasts 4–5 hours, after which dismounts, out of the females' reach and may copulate again. However, whilst on a female's back the males may lose his grip and the female may strike, capture and consume the male (Roeder, 1935). Being eaten at this stage does not exclude establishing a genital contact and insemination (e.g. Maxwell, 1998), but the wisdom that decapitation is a necessary prerequisite for ejaculation is incorrect, as intact males perform just as well as decapitated males.

Liske and Davis (1984, 1987) suggested that sexual cannibalism in mantids is a laboratory artifact resulting from either confinement or feeding conditions, but Maxwell (1998) has reported sexual canabalism occurring in the field. Birkhead *et al.* (1988) showed that females may gain significantly in fitness by eating males. In *Hierodula membranacea*, the probability of the male being eaten correlates strongly with female diet: females maintained on 0.1 g (dry weight) of cricket prey per day (a non-starvation diet) ate males in 12/14 cases, whereas those maintained on 0.42 g consumed 1/5 males. The fecundities (measured by oöthecal weight) of females that had been maintained on poor diet and consumed their mate were significantly higher than those of poor-diet females that did not consume mates (but see Maxwell, 1998). Also, it was observed that males approach females and attempt copulation with extreme caution, mounting the female by leaping on to her back from well outside the range of her grasp, as in *M. religiosa* (see also Maxwell 1998).

4.5 POST-MATING EVENTS

4.5.1 SPERM USE BY FEMALES

In several species, females cannot use sperm immediately following insemination (Mackauer, 1976; van den Assem, 1977) and the period during which females are unable to fertilise eggs can vary considerably between individual female *Nasonia vitripennis* (van den Assem and Feuth de Bruijn, 1977). Newly emerged females may have no mature eggs in their ovaries (i.e. are extremely synovigenic, see subsection 2.3.4), so temporary post-insemination constraints may have little effect on their subsequent reproduction. In *Nasonia vitripennis*, the anatomy of the spermathecal duct is such that only one sperm is likely to descend at one time (Wilkes, 1965). Sperm depletion in mated females is discussed in subsection 5.4.7.

4.5.2 SPERM COMPETITION, DISPLACEMENT AND PRECEDENCE

Introduction

Sperm competition is the selective force that arises whenever the functional sperm of two or more males overlap in time (and usually in space) in a single fertile female (Parker, 1970a). Insects are predisposed to particularly high levels of sperm competition because female insects generally store sperm in the spermathecae, and they are usually polyandrous. Once the preconditions for sperm competition are met, the stage is set for the evolution of two antagonistic suites of male fitness traits (Parker, 1970a). Both of these sets of traits function to reduce sperm competition, by reducing the effects of and/or preventing ejaculate overlap. Examples of the ways male insects avoid sperm competition via the evolution of one, or both sets of these fitness traits are extremely well documented and are the subject of a number of comprehensive reviews (see Thornhill and Alcock, 1982; Smith, 1984; Simmons and Siva-Jothy, 1998).

The first set of traits functions to reduce the effects of sperm competition by enhancing the competitive ability of a male's ejaculate. Perhaps the most spectacular of traits in this category are the spinose appendages on the penis of certain damselflies (Figure 4.29) and dragonflies that are used to physically remove the sperm of rivals (stored in the female) before insemination (e.g. Waage, 1979a; Miller, 1984; Siva-Jothy, 1984). By removing rival sperm these males ensure they fertilise most, if not all, the eggs their mate lays in the subsequent oviposition bout. There are several lines of evidence that suggest variation in penis morphology in damselflies is related to variation in reproductive function which, in turn, is related to variation in the details of the mating system (see Waage, 1984: Siva-Jothy, 1984; Robinson and Novak, 1997).

The second set of traits favoured by sperm competition are those that prevent the female from remating and thus reduce the probablity that a particular male's sperm will have to compete in the future. There are various traits in this category, including behavioural mate-guarding (e.g. Waage, 1979b), the use of genital plugs to prevent male access to the female's genital tract (e.g. Lum, 1961; Labine, 1966) and the transfer of compounds along with the ejaculate that reduce

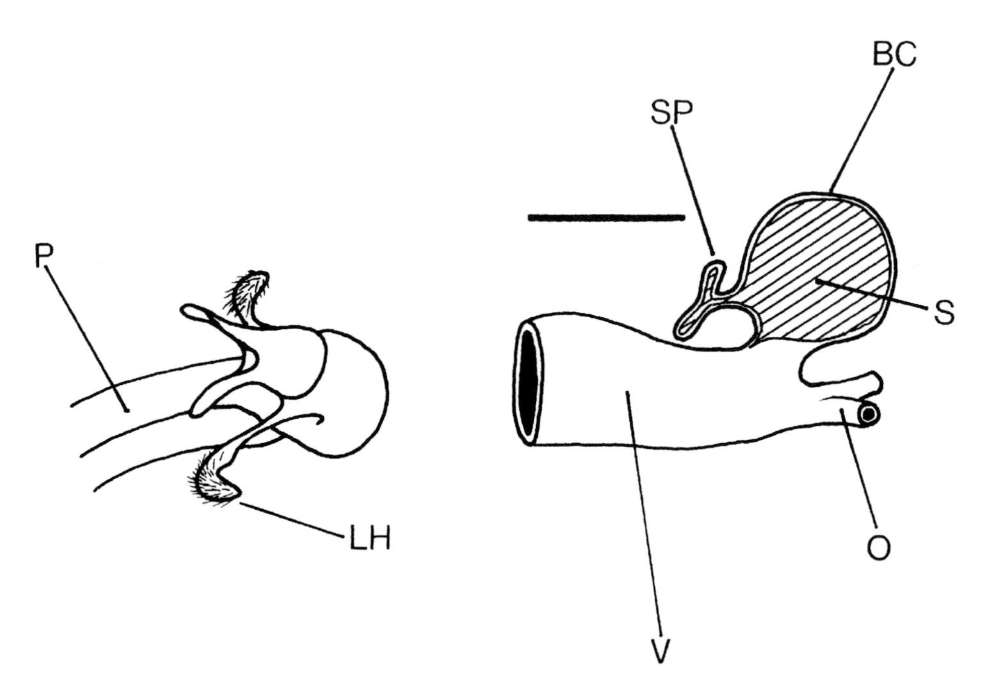

Figure 4.29 Genitalia of the damselfly *Mnais pruinosa pruinosa*. **Left panel:** distal part of the male intromittent organ. P = penis shaft; LH = lateral horns. The lateral horns bear recurved spines which the male uses to remove the sperm of rivals stored in the female's bursa copulatrix. Left is anterior and up is ventral. **Right panel:** A diagramatic representation of the female internal genitalia. V = vagina; O = oviducts, SP = spermathecae; BC = bursa copulatrix; S = stored sperm (hatched). Left is posterior and up is dorsal. Scale bar = 1 mm. (after Siva-Jothy and Tsubaki, 1989, by permisson of Springer-Verlag).

the female's receptivity to subsequent matings (e.g. Reimann *et al.*, 1967; Thornhill, 1976).

The proximate outcome of sperm competition is the non-random usage of the sperm a female receives from multiple males: in general the last male to mate with a female fertilises most of the eggs she lays in the ensuing oviposition episode (for a hymenopteran exception, see Elagoze *et al.* 1995). Sperm precedence is usually denoted by the term "$\mathbf{P_2}$" which is the proportion of eggs fertilised by the last male to mate. Estimates of P_2 based on the sterile male technique or genetic markers (both techniques described below) require the use of reciprocal mating treatments to account for the difference in competitive ability between sperm marked in different ways (Boorman and Parker, 1976). One interesting feature of the P_2 values measured for insects to date is that they show considerable intraspecific and interspecific variation (Lewis and Austad, 1994; Simmons and Siva-Jothy, 1998). Careful comparisons (controlling for phylogeny, subsections 1.2.3 and 5.3.4) of this variation across insect species revealed that it is negatively associated with the species-specific mean of P_2: in other words, species with strong last male sperm precedence (P_2 close to 1.0) have low variance (Simmons and Siva-Jothy, 1998). In general, species with high P_2 values (such as damselflies, e.g. Hooper and Siva-Jothy, 1996) have sperm competition mechanisms that completely negate rival sperm (e.g. Waage, 1979a). The scope for variance in P_2 values is therefore greatly reduced. In species with intermediate P_2 values the sperm from both males must be present in the female, so the scope for variation is greater, either through passive effects such as sperm mixing, or through other undocumented effects such as cryptic female choice or inter-ejaculate competition. Finally the males of species with low P_2 values tend to use mating plugs to protect their genetic investment. If the plug remains intact subsequent males cannot mate properly, so P_2 remains low. However, if the plug is breached (by traits also selected for under sperm competition), P_2 increases dramatically, resulting in the largest variance (Simmons and Siva-Jothy, 1998). The current interest in variance in P_2 values stems from theory proposing that females might exercise choice over sperm use to bias paternity in favour of particular mates (Eberhard, 1991; see below).

Another way of understanding why P_2 values vary is to clarify the nature of the mechanism underpinning sperm precedence. This problem has been approached through experimental and/or comparative studies of reproductive anatomy and physiology (see Siva-Jothy, 1984; Miller, 1991; Siva-Jothy and Tsubaki, 1989; Robinson and Novak, 1997). Such studies usually combine anatomical studies of male and female genitalia and their disposition during copulation, with manipulations, or observations, aimed at identifying sexually selected function during copulation. Because of the immense diversity of insect genitalia (almost certainly reflecting a similarly immense range of copulatory and sperm competition mechanisms) there is no single methodological approach for examining the mechanistic basis of sperm competition. However, examining the structure, dimensions, and articulations of the male genitalia usually provides valuable insights into the operational constraints and abilities during copulation and so may exclude, or suggest, certain mechanistic options. Moreover, male genitalia are often well-documented and described because of their importance in taxonomic studies. The dissection of *in copula* pairs that have been snap-frozen in liquid nitrogen at different times *in during copulation* then provides some idea of how the genitalia of males and females interact, further narrowing down mechanistic possibilities, and usually giving considerable insight into the function of otherwise abstract structures (e.g. Crudgington and Siva-Jothy, 2001). Because the female genitalia concerned with sperm competition are internal and consist mainly of endocuticle, their structure is better documented with careful dissection as well as serial sectioning of wax-embedded samples. Samples for sectioning should have sclerotised cuticle removed, be fixed in Bouin's fixative, and be subsequently embedded and sectioned (see Barbosa 1974; Bancroft and Stevens, 1991). Staining the sections with Haematoxylin and

Eosin provides good histological resolution in most insect tissues and, when combined with the spatial information inherent in the serial sections, will provide information to reconstruct the form, and to some extent histological function, of the female genital tract ('visualisation' software is now available, allowing serial section data to be converted into an integral whole). When all three approaches are combined the result is a fairly good understanding of the disposition, orientation, and interactions between the male and female genitalia during copulation. This information then provides the basis for more detailed hypothesis construction and testing.

Sperm Competition in the Odonata

The Odonata provide the paradigm for sperm competition mechanisms, largely because of several unique features of their mating system and reproductive anatomy that lend them to studies of this sort. Moreover, the mechanism they use to achieve sperm precedence is relatively easy to quantify. Male odonates displace the sperm of rivals within (e.g. Siva-Jothy, 1987; Miller, 1991) or outside (e.g. Waage, 1979a), the female's sperm storage organ using specialised penis morphology prior to insemination. For example, examination of the intromittent organ of male calopterygid damselflies (Figure 4.29) revealed a spine-covered tip that is of the correct length and dimensions for entering the female's sperm storage organs (Figure 4.29) during copulation. The male's spines are often covered with sperm after copulation, and Waage (1979) hypothesised that they are used to physically remove the stored sperm of rival males from the female's sperm storage organ prior to insemination. Waage predicted that, if copulating pairs were interrupted at different times during copulation, a diminishing sperm store in the female would be observed, followed by an abrupt increase (ejaculation). This is exactly what was found: at one point females *in copula* had no sperm in their sperm storage organs at all. The only logical conclusion that could be drawn was that males were removing rival sperm from storage within the female before they inseminated her with their own sperm. Subsequent sperm precedence and molecular genetical analysis (e.g. Siva-Jothy and Hooper, 1996; Hooper and Siva-Jothy, 1996) confirmed this inference. By displacing rival sperm from the place where it will be used to fertilise eggs during oviposition, and then placing their own sperm in that position, males ensure a high level of P_2 in the eggs the female subsequently lays. However, complete sperm removal does not appear to be the rule – several odonates exhibit partial sperm removal (McVey and Smittle, 1984; Siva-Jothy, 1987; Siva-Jothy and Tsubaki, 1989; Michiels, 1992; Siva-Jothy and Hooper, 1996): sperm precedence is still high in the eggs the female lays immediately after copulation, but declines with subsequent egg batches (eggs are laid in discrete clutches at approximately 3–5 day intervals) if the female can avoid remating (e.g. Siva-Jothy and Tsubaki, 1989) (Figure 4.30). Because the mating system of all the dragonflies studied in the context of sperm competition tend to centre upon oviposition resources which males defend, the opportunity for females to gain access to oviposition sites without remating seems slight. Interestingly, female oviposition without remating is apparent only in species where males cannot defend all the available resources, and it appears that some females actively seek out undefended resources, or they avoid remating. Recent studies have revealed that females actively choose where to oviposit (Hooper and Siva-Jothy, 1997), actively avoid remating prior to oviposition in some instances, and use sperm from matings other than their last one when they avoid remating (Siva-Jothy and Hooper, 1996). Unfortunately for the comparative biologist, the unique reproductive biology of odonates probably means that this mechanism of sperm competition is largely restricted to this order. There are examples of other mechanisms identified through empirical study (see Simmons and Siva-Jothy, 1998), but the vast majority of the remaining mechanistic literature is based around simulations of sperm mixing (e.g. Parker *et al.*, 1990).

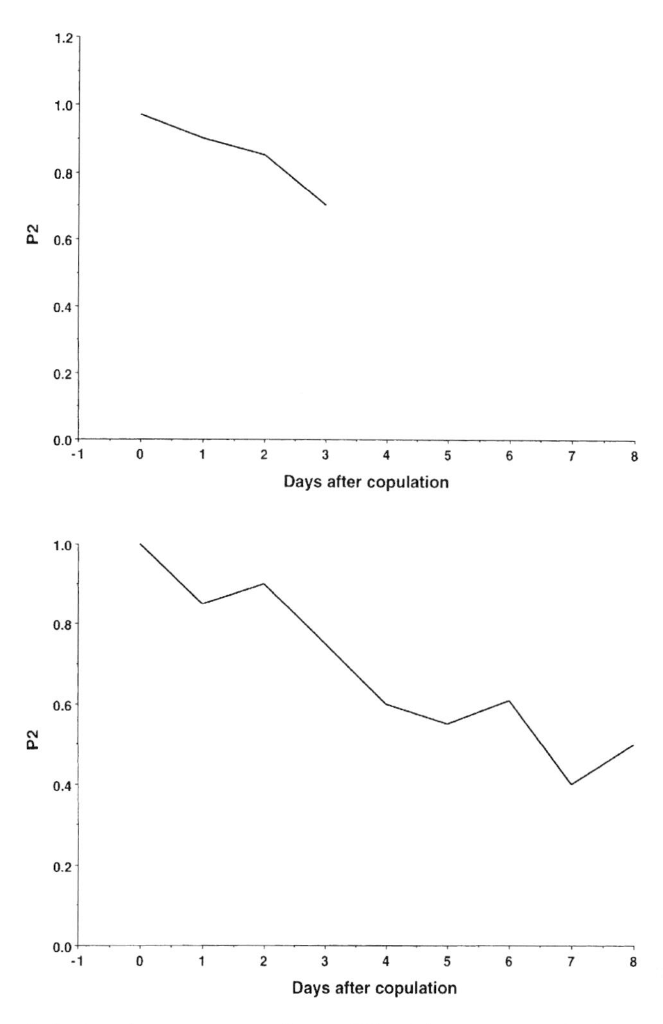

Figure 4.30 The decline in last male sperm precedence (P2) with time since the last copulation. Upper panel: The libellulid dragonfly *Nanophya pygmaea*. Males add their sperm to that of other males already in storage in the female's sperm storage organ. (after Siva-Jothy and Tsubaki, 1994). Lower panel: The calopterygid damselfly *Mnais pruinosa pruinosa*. Males partially remove the sperm of rivals that are storted in the female's sperm storage organs. (after Siva-Jothy and Tsubaki, 1989, by permisson of Springer-Verlag).

To date, the majority of studies have focused on male-based traits under selection from sperm competition, although recent years have seen a "paradigm riffle" (Gowaty, 1994) towards understanding female perspectives in sperm competition. Such studies concentrate on the genetic benefits females may receive from biasing reproductive decisions in their favour. Protocols have yet to be established, and are unlikely to be conserved given the dynamic nature of sexual selection, but some practical constraints have been proposed (Birkhead, 1998; Siva-Jothy and Hadrys, 1998). If female perspectives are evolutionarily important we would expect to: (a) see variation in paternity patterns (including sperm precedence); (b) be able to attribute some of that variance to female effects (that have a heritable basis); (c) be able to show a correlation between any variance attributable to female effects with somatic and/or gametic male trait(s); (d), be able to show that female–attributable shifts away from the male–determined patterns of paternity result in higher female fitness than if females are constrained to fertilise their eggs in a manner determined by males. These criteria are the best practical starting point we have at present.

The impetus for this new viewpoint has largely been generated by the work of Eberhard (1985, 1991) who argues that females are responsible for generating much of the variation in P_2 values through a process he dubs "cryptic female choice". The choice is "cryptic" because it cannot be recognised using the classical Darwinian criteria for identifying pre–copulatory female choice (Eberhard, 1996). Eberhard (1985, 1991, 1996) proposes that complex male genitalia, and complex and prolonged copulations, evolved as courtship traits upon which females base decisions about subsequent sperm usage. This hypothesis has some very strong supporting arguments. Once a male has inseminated a female, the female is largely in control of sperm storage, upkeep and usage: females have ample opportunity to manipulate ejaculates to their own ends, as demonstrated by the impressive control of sperm-use in female Hymenoptera. The main difficulty of this hypothesis is that it is difficult to test (but see Arnqvist, 1998, for an elegant supporting study) because many of the predictions generated are difficult to distinguish empirically from the effects of sperm competition (reviewed by Siva-Jothy and Hadrys, 1998). The benefits females receive from cryptic choice are purely genetic and there are major theoretical problems with the relative value of such benefits (see Andersson, 1994) in the face of other, more direct consequences of female mating decisions (see Waage, 1996, for a review).

Measuring Sperm Precedence

Sperm precedence values are measured using virgin females that have been double-mated to two males whose offspring can be easily identified. **Paternity assignment** involves the use of the sterile male technique (e.g. Boorman and Parker, 1976), genetic markers (e.g. van den Assem and Feuth de Bruijn, 1977; Eady, 1994), or, more recently, molecular genetical techniques (e.g. Hadrys *et al.*, 1992). All marker techniques have the inherent problem that sperm types arising from males bearing different markers often have differential fertilising ability. To control for this, mean sperm precedence values are usually obtained by conducting reciprocal crosses between pairs of males (each bearing one marker) with a single virgin female. By calculating the mean value of sperm precedence from two females receiving reciprocal mating order copulations from marked males, it is possible to control, to some extent, the effect of differential fertilising ability. In sterile mating treatments this value must be corrected by taking into account natural levels of infertility in normal males and the level of fertility in sterilised males (see Boorman and Parker, 1976).

The Sterile Male Technique

The sterile male technique involves exposing males to a dose of gamma- or X-irradiation sufficient to induce complete, or near-complete sterility, but insufficient to affect the males' behaviour. The **minimum effective sterilising dose** (MESD) can be determined by subjecting groups of males to increasing doses of radiation, mating them to virgin females and then scoring the percentage hatch among the resulting eggs. The relationship between percentage hatch and dose takes the form of a decelerating curve: the MESD is the inflection point were maximum sterilisation meets the minimum dose to achieve sterilisation. Once the MESD has been determined, virgin males can be assigned to either the sterile (S) or normal (N) groups. Sterile males are exposed to the MESD. Four types of treatment are then conducted: virgin females are double-mated to pairs of virgin males who are drawn from groups to result in the following pairwise treatments: SS, NN, SN or NS. Each male in a treatment copulates once in the specified mating order. The re-mating interval should reflect natural re–mating intervals, or the specific remating interval of interest: re–mating intervals can have a strong influence on the pattern of sperm precedence achieved (Simmons and Siva-Jothy, 1998), so laboratory studies need to control, and justify, this variable. Eggs are collected and scored as either sterile or normal at a point in embryological development when this distinction is clear (usually when eye-spots are developing in the embryo). Males exposed to the MESD produce sperm that bear dominant lethal mutations, so the zygotes they produce never develop past the earliest embryological stages and rarely develop eye-spots. Once the eggs have been scored as sterile or normal, P_2 can be calculated from Boorman and Parker's (1976) equation: $P_R = (1-x/p) + z/p \times [1-(x/p)]/[1-(z/p)]$, where P_R = proportion of eggs fertilised by the sterile male, p = fertility after NN mating, z = fertility after SS mating, and x = the number of viable eggs after an NS or SN mating. A remaining methodological problem is that the fertility of supposedly sterile males may increase over time following irradiation; this should thus be controlled for since it can lead to error in estimating P_2 (overestimation if the irradiated male was the first to mate, and underestimation if the irradiated male mated second) (Rugman-Jones and Eady, 2001).

Genetic and Molecular Genetic Markers

The use of genetic markers for sperm competition studies requires that the phenotypic markers are discontinuous Mendelian traits. Once inheritance characteristics have been determined, the expressed phenotypes can be used as markers. As long as the pattern of inheritance allows unambiguous assignation of offspring from controlled parental matings, then the marker can be used. An ideal situation occurs when the markers are the dominant and recessive allele at a single Mendelian locus. In such a system, virgin females expressing the homozygous recessive phenotype are mated with homozygous recessive and homozygous dominant males. Offspring can be easily assigned to the appropriate father, and the relative effect of mating order assessed.

A number of molecular genetical techniques has emerged in the last decade which enable the assignment of paternity. These range from the use of Random Amplified Polymorphic DNA (RAPD) to microsattelite markers. Whilst RAPDs is a potentially fast way of getting information out of an unknown genome (e.g. Hadrys *et al.*, 1992), it is beset with several important practical problems, not least of which is questionable inter-laboratory repeatability. Consequently, it has slowly fallen into disuse, despite early success. By far the most reliable and repeatable molecular method is to identify polymorphic microsattelite loci for the species under study. This can be a laborious process, requiring considerable expertise, if the PCR primers have not already been identified for the species in question. The process of identification involves finding repeat sequences in the target genome that are straddled by unique flanking sequences; these flanking regions are sequenced and PCR primers are constructed. Individuals can be identified on the basis of variation in the length of the repeat sequences that are identified by the primers in question. When a number of such loci are identifiable it is relatively easy to identify parentage accurately, even in relatively uncontrolled field situations: this technique has had considerable success in vertebrate studies, and is increasingly being used by entomologists. Its advantage is its powerful resolving ability, its drawback is the time and cost investment in identifying a panel of polymorphic loci. In most studies of sperm competition, this technique would be a sledgehammer to crack a nut (for example, the sterile male technique is quicker, cheaper and just as good at identifying P_2). Microsatellites are probably better suited to field studies of insects that are difficult to culture, or to studies of the dynamics between the sperm of several males in a single female over long periods of time.

Summary

There is a vast theoretical and empirical literature concerned with insect sperm competition, its evolutionary consequences, the resulting mechanisms and their consequences and the role that females may play in determining the parameters for the system. The phenomenon of sperm competition exists because of female reproductive behaviour (multiple mating) and female reproductive anatomy, and has a profound effect on male mating behaviour prior to, during and after copulation. The adaptive value of many reproductive behaviours are often only clear once they are viewed in the context of sperm competition and the ecology that determines the mating system.

4.5.3 MATING FREQUENCY AND LONGEVITY

Frequent mating may reduce an individual's longevity, indicating that there is a cost, in terms of survival, to reproduction (for practical advice on measurement of longevity see section 2.8). For both female and male fruit-flies (*Drosophila melanogaster*) significant, negative correlations between mating frequency and life-span have been recorded (Kummer, 1960; Partridge and Farquhar, 1981; Chapman *et al.*, 1995). In parasitoid wasps, a similar relationship may exist. Males of several parasitoid species continue mating throughout their lives (e.g. *Melittobia acasta:* J. van den Assem, pers. comm.; *Tetrastichomyia clisiocampae:* Domenichini, 1967). Males of a standard size can be presented with a number of virgin females, every day until death. Inseminated females are then kept alive and supplied daily with a surplus of hosts. The time spent by males in courting and mating, and the frequency of copulations, are controlled by the experimenter, while the number of sperm delivered to the females is assessed by counting the number of female progeny. Controls should include males that court as much (on average) but are prevented from copulating, and other males that do not engage in any reproductive activity. Gülel (1988) reported a significant decrease in longevity of *Dibrachys boarmiae* (Pteromalidae) males that copulated with five females per day and eventually became sperm-depleted, compared with males that mated only once per day. However, a control treatment ensuring equal levels of general motor activity by wasps was absent. In contrast, male longevity does not appear to be affected by mating frequency in *Habrobracon hebetor* (Ode *et al.*, 1996).

4.5.4 MATING AND EGG PRODUCTION

For diploid insects (in which all eggs need to be fertilised to produce progeny) insemination is likely to have a marked effect upon the behaviour of the female, releasing host-finding behaviour. However, insemination in haplodiploid insects such as parasitoid wasps is not necessarily expected to alter female behaviour significantly (but see Nishimura, 1997). No differences in oviposition behaviour are apparent between virgin and mated females of *Habrobracon hebetor* (Ode *et al.*, 1997) and *Lariophagus distinguendus* (J. van den Assem, personal communication). Female *Melittobia acasta*, once mated, can oviposit immediately and oviposition reaches a maximum in one or two days, but unmated females, when presented with a host, will sting it and host–feed but not oviposit (van den Assem, 1976b; van den Assem *et al.*, 1982a). Despite the presence of mature eggs in the ovaries, very few eggs are laid even after one or two days. Such females will mate as soon as a son has emerged, and only then commence ovipositing. Thus, being inseminated affects the oviposition behaviour of females. *Cotesia glomerata* (Braconidae) virgin females lay consistently smaller clutches of eggs than mated females, both in the field and the laboratory. Virgin females that were inseminated mid-way during an experiment began laying larger clutches (Tagawa, 1987). The adaptive significance of the virgin female/mated female clutch size difference is unclear. According to Tagawa (1987), the difference is not a reflection of female egg load, since females were dissected and their egg loads determined. Egg–staining methods, however, were not used by Tagawa to distinguish between fully mature and nearly mature eggs (see section 2.3.1). In *Nasonia vitripennis* locomotor activity is higher in mated than in virgin females, regardless of whether the female remained with her mate prior to testing. Despite being more active, mated females did not parasitise more hosts than virgin females (King *et al.* 2000).

4.6 COMPARATIVE STUDIES OF MATING BEHAVIOUR

4.6.1 MATING BEHAVIOUR AS A SOURCE OF TAXONOMIC CHARACTERS

The immense diversity of parasitoid wasps poses a major challenge to taxonomists. Sibling species abound, as do various degrees of incompatibility between field populations or laboratory strains, and in many cases it is difficult to establish the species status of populations. Molecular techniques are valuable in untangling problems of relatedness (e.g. Molbo *et al.*, 2002), but investigation of courtship behaviour characteristics may provide an alternative, or complementary, approach (e.g. Kazmer *et al.*, 1996). Many courtship displays exhibit combinations of features that can be used for identification purposes (Gordh and DeBach, 1978) and sophisticated and expensive equipment are not required for new detection.

When looking for species-diagnostic characters, one can compare entire courtship displays (van den Assem *et al.*, 1982a,b; van den Assem and Gijswijt, 1989; Hunter *et al.*, 1996). Records made with video equipment are useful as they allow repeated viewing and interspecific similarities and differences can be detected that might otherwise remain unnoticed. Besides the motor patterns of the appendages, it is the **temporal organisation of displays** (how they differ with respect to the order of appearance of components and to the number and lengths of intervals between components) that are particularly species-diagnostic.

For example, *Muscidifurax* was long considered to be a monotypic genus, with *M. raptor*, a cosmopolitan parasitoid of the house-fly (*Musca domestica*), its only representative. However, Legner (1969) reported reproductive isolation of several of many strains in field-derived laboratory cultures, and five species were later recognised using morphological characters. These are, however, very variable and difficult to observe (Kogan and Legner, 1970). When males and females of the same and different strains were introduced, interspecific differences in display movements (patterns of antennal movements (Figure 4.31) and in the duration of the intervals between successive display cycles) were readily observed, and the species were characterised behaviourally (van den Assem and Povel, 1973). There were also interspecific differences in males' treatment of either conspecific females or conspecific-like dummies, compared with their treatment of heterospecifics.

The case of *Nasonia vitripennis*, another pteromalid, is very similar and illustrates the state of knowledge of the group as a whole. *N. vitripennis* is probably the most intensively studied parasitoid species (Cousin, 1933; Whiting, 1967; Holmes, 1976; also references in Godfray, 1994). The genus was until recently considered to be monotypic, its sole representative being a cosmopolitan parasitoid of the pupae of cyclorrhaphan flies in a variety of habitats including manure, decaying carcasses and birds' nests (Werren, 1983). Through extensive collecting of fly puparia from birds' nests over a large area of the USA, three reproductively isolated strains were obtained (Darling and Werren, 1990). Members of different strains were sometimes collected from the same nest, or even from the same puparium. A morphological investigation led to the description of three species (Darling and Werren, 1990) but simple diagnostic differences (except for wing length in males) could not be provided. As with *Muscidifurax*, studies of courtship behaviour revealed reliable diagnostic characters, namely differences both in the overall temporal organisation of displays and in the details of motor patterns (van den Assem and Werren, 1994).

Mating behaviour was used, in conjunction with molecular techniques, to investigate reproductive incompatibility between strains of *Aphelinus asychis* (Aphelinidae), a parasitoid of aphids attacking wheat (Kazmer *et al.*, 1996). Cultures were set up from wasps collected in the field in five Mediterranean and two Central Asian localities. *Aphelinus asychis* is released as a biological control agent against aphids, and

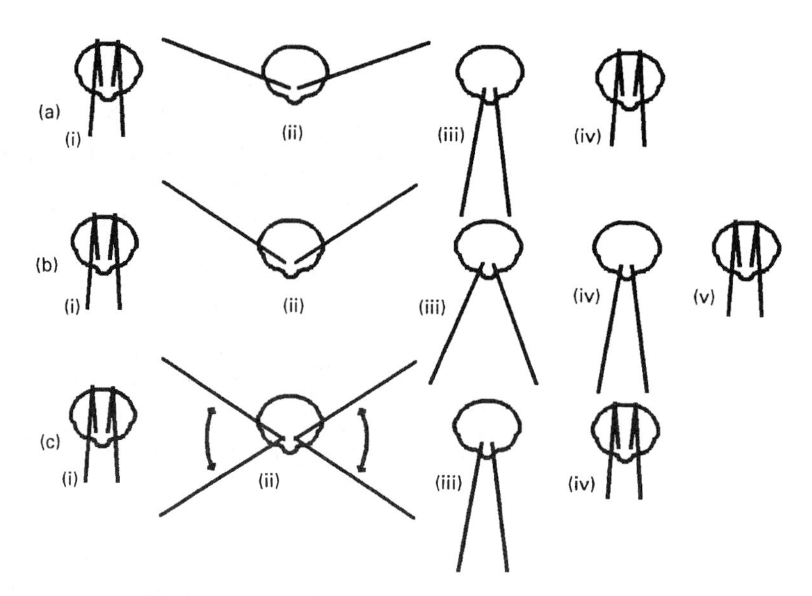

Figure 4.31 Species-characteristic display movements with the antennae in three sibling species of *Muscidifurax* (Pteromalidae): (a) *M. raptor*, (i) start position, (ii) spreading the antennae, (iii) extreme low position, (iv) end position, (b) *M. zaraptor*, (i) start position, (ii) spreading the antennae, (iii) intermediate position during the slow downward movements, (iv) end position; (c) *M. raptorellus*, (i) start position, (ii) waving episode, (iii) extreme low position, (iv) end position.

the success of control programmes is likely to be influenced by the presence of reproductive incompatibility between released wasps. Initial investigations revealed that there were three reproductively incompatible groups among these cultures, and more detailed investigations of response to sex pheromone, courtship behaviour and sperm transfer were carried out to determine the basis of the incompatibility. Males responded normally to the sex pheromones (section 4.3) of incompatible females and courted them actively. However, courtship by incompatible males did not usually induce receptivity by females, which did not stop moving or elevate their abdomens for copulation, and in most cases there was no genital contact or sperm transfer. Reproductive incompatibility is apparently the result either of the failure of females to recognise males as potential mating partners, or of active rejection of males. Molecular investigations showed that the genetic distances between the three reproductively compatible groups were larger than between compatible groups.

Hunter *et al.* (1996) also used a combination of behavioural observations and molecular (allozyme) techniques to investigate whether or not three North American populations of *Eretmocerus eremicus* (Aphelinidae), parasitoids of whitefly (Hemiptera: Aleyrodidae), belong to a single species. There were no interpopulation differences in the kinds of courtship behaviour performed, the sequence of behavioural acts, the duration or the frequency of components of courtship behaviour. There were, however, differences in behaviour when males and females from different populations were exposed to one another. Individuals from the Texas population did not copulate with individuals from Arizona or California, while California and Arizona individuals copulated (however, individuals from Arizona were relatively small, and in some cases wasps of differing sizes experienced difficulty in copulating) and there appear to be no post-copulatory reproductive barriers between these populations. Allozyme analysis (section 5.3.3) supported the results from mating behaviour experiments: the Texas population was genetically distinct from the California and Arizona populations, and so Hunter *et al.* (1996) concluded that the wasps studied belonged to two species.

Heimpel *et al.* (1997) used mating experiments in combination with life-history studies and mtDNA sequence analysis to distinguish between two populations of parasitoid identified as *Bracon hebetor* (now re-named *Habrobracon hebetor*). Parasitoids morphologically indistinguishable from *B. hebetor* were released in Barbados from 1970–1975 to control *Helicoverpa* (= *Heliothis*) spp. (Lepidoptera: Noctuidae). In the USA, *H. hebetor* is known as a parasitoid of phycitine moths *H.* (Lepidoptera: Pyralidae). The differences in host use suggested that the Barbados parasitoids and the American parasitoids might belong to a different species. Pairs of males and females from different 'strains' were placed together for 48 hours to allow mating, and a host was provided for oviposition on dissection, the spermathecae of the females were found to be empty, and all progeny reared from the hosts were male (arising from unfertilised eggs). The two 'strains' are thus reproductively isolated, and subsequent behavioural observations showed that no 'inter-strain' mating occurred: in one case a male produced a courtship display, but copulation was never observed. Immediately after wasps from different 'strains' had been observed, each individual of the pair was observed with a member of the same 'strain': in the 30/34 cases both individuals mated with a member of their own 'strain'. Genetic analysis showed that divergence between the 'strains' is consistent with interspecific differences. Heimpel *et al.* (1997) therefore concluded that the parasitoids from Barbados belong to a species reproductively distinct from *H. hebetor*.

4.6.2 MATING BEHAVIOUR AND PHYLOGENETICS

The aims of comparative studies of the mating behaviour of parasitoids and predators go beyond rendering assistance in problems of taxonomic identification. Such studies, if they use the appropriate analytical techniques, can also enable us to establish how behaviour evolved, i.e. changed during phylogeny. Parasitoids and predators, because of their high biological diversity, provide excellent material for comparative work. Morphological characters considered to plesiomorphic or apomorphic often correlate with behavioural traits that can be similarly labelled, even where the particular morphological characters (e.g. venation of wings, number of antennal segments etc.) do not seem to bear directly upon behavioural traits (e.g. orientation of the male during courtship, timing of release of receptivity-inducing stimuli etc.). Goodpasture (1975) arrived at the same conclusion when considering the phylogenetic relationships of *Monodontomerus* species (Torymidae), using character sets from mating behaviour, karyology and external morphology. The 'comparative method' can be used to test the significance of the aforementioned correlations between data sets (see Brooks and McLennan, 1991; Harvey and Pagel 1991; Mayhew and Pen, 2002; subsections 1.2.3 and 5.3.4).

Differences among parasitoid wasp taxa in the courting male's orientation with respect to the female can be viewed as indicators of a phylogenetic transformation series (Hölldobler and Wilson, 1983). In some groups, the male adopts the same position in courtship as in copulation, i.e. to the rear (*Trichogramma evanescens:* Hase, 1925; *Brachymeria intermedia:* Leonard and Ringo, 1978; *Spalangia nigra:* Parker and Thompson, 1925; *Spalangia endius:* van den Assem, 1986). In the majority of parasitoid wasps, however, positions during courtship and copulation are distinctly different: males court away from the mating position, and mostly perform either on top of the female or near to her on the substratum (Gordh and DeBach, 1978; Bryan, 1980; Grissell and Goodpasture, 1981; Orr and Borden, 1983). An hypothesised phylogenetic sequence of positional changes is summarised in Figure 4.7. The hypothesis remains to be tested, despite it appearing obvious which direction evolution has taken. For example, the tendency towards a more anterior courtship position is apparent in several families and subfamilies and may represent a response to a general selection pressure, perhaps more efficient communication at the front due to the presence of more diverse or more dense sense organs at the anterior ends of both sexes (van den Assem and Jachmann, 1982).

In a comparative study of mating behaviour, it is important to possess a phylogeny of the taxa involved; one may already exist in the literature, or willing taxonomists can provide a provisional tree. Mating behaviour traits need to constitute an independent data set, so the phylogeny needs to have been constructed using other characters. By 'mapping' mating behavior traits onto the tree, inferences can be drawn as to whether a particular trait has arisen independently in the evolution of the insect group, or is phylogenetically constrained. If there is evidence that a trait is phylogenetically constrained, then when making statistical comparisons it will be necessary to apply tests that take account of the potentially confounding effects of phylogenetic relatedness (see Harvey and Pagel 1991). Comparative studies of parasitoid mating are further discussed in subsection 5.3.4.

4.7 CONCLUSION

The study of insect mating behaviour continues to provide very valuable insights into evolutionary processes and relationships. Despite the increasingly reductionist nature of the biological sciences, it is clear that the study of insect behaviour, within a sound evolutionary framework, has major interpretative and heuristic value.

4.8 ACKNOWLEDGEMENTS

We dedicate this chapter to Hans van den Assem, and thank him for writing the original. We thank Hans, George Heimpel, Mark Jervis and Paul Rugman-Jones for suggestions and comments on our revision. Ian Hardy was funded by European Commission grants ERBFMBICT 961025 and 983172, Paul Ode by NSF DGE-9633975 and Mike Siva-Jothy partly by NERC grants GR9/03134 and GR3/12121.

MATING SYSTEMS 5

Ian C.W. Hardy, Paul J. Ode and Mike T. Siva-Jothy

5.1 INTRODUCTION

The term '**mating system**' is used to describe how males and females obtain mates in a population (Emlen and Oring, 1977; Thornhill and Alcock, 1982; Davies, 1991; Brown *et al.*, 1997). A particular mating system may be characterised by the events surrounding pair formation, courtship, copulation and the post-copulatory events (Brown *et al.*, 1997). Individual males and females engage in reproductive behaviours that maximise their own fitness, frequently to the detriment of their mates. Evolutionary biologists have come to regard events surrounding mating as a set of intra-sexual and inter-sexual 'battles' which reflect the sometimes common, sometimes differing, reproductive interests of males and females (e.g. Davies, 1991; Brown, *et al.* 1997; Choe and Crespi, 1997; Alonzo and Warner, 2000).

The mating systems of insect natural enemies are diverse and, despite the large number of studies carried out to date, there is still ample scope for further fruitful investigation. There are two main reasons why insect natural enemy mating systems have been investigated: (1) to understand the evolution of mating systems themselves, and (2) to understand the evolution of sex allocation decisions (which is likely to be greatly influenced by details of the mating system). An additional reason, which has received relatively little empirical attention, is an improved understanding of the conditions required for successful establishment of a natural enemy population (e.g. Hopper and Roush, 1993; Meunier and Bernstein, 2002).

Although these reasons are interrelated, something of a taxonomic dichotomy has developed in the study of mating systems. Workers on parasitoids (mostly Hymenoptera with a high degree of control over sex allocation due to haplodiploidy, see Chapter 3, section 3.3) have tended to investigate mating systems as a backdrop to sex ratio research. Workers on predatory insects (many of which may be constrained in their control of sex allocation) have tended to ask questions about mating systems *per se*, including considerations of sexual conflict. Of course, there are welcome exceptions to this 'dichotomy' but it will nevertheless be reflected in the structure of this chapter.

We begin by illustrating the diversity of mating systems among insect natural enemies (section 5.2). We next discuss general approaches to their study (section 5.3), and then review investigations of parasitoid biology that have examined one or more components of these mating systems (section 5.4). Section 5.4 is organised by 'issue', with methodology highlighted within each subsection. In section 5.5 we consider the mating systems of predatory, rather than parasitic, natural enemies.

In this chapter we take a more 'functional' approach than that taken in the companion chapter on *Mating Behaviour* (Chapter 4). We are more concerned with the *causes* and *consequences* of the mating patterns within a population, rather than with the fine details of the behavioural mechanisms involved. However, it will become clear that there is great overlap and productive interchange between these research areas.

5.2 MATING SYSTEMS THEORY

The majority of work aimed at understanding the evolution of insect mating systems takes

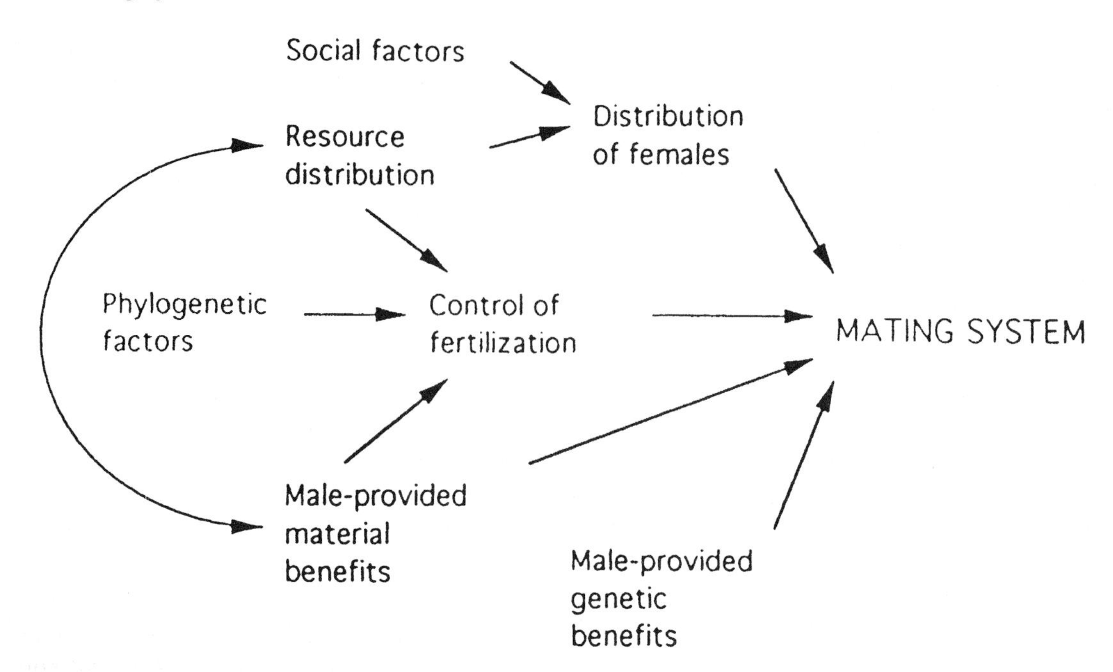

Figure 5.1 The main factors influencing animal mating systems. 'Phylogenetic factors' refers to the historical presence or absence of adaptations related to the control of fertilization in various lineages. (Source: Brown *et al.*, 1997). Reproduced by permission of Cambridge University Press.

Bateman's principle as its central tenet (Bateman, 1948). In short, this posits that: *females are the limiting sex since they cannot increase their fitness via multiple mating in the same manner as can males*. Consequently males tend to compete for access to females, and females tend to be choosy about their mating decisions. The combination of this basic evolutionary principle with the fact that ecological variables and constraints determine the spatial and temporal availability of females results in the evolutionary conditions that produce the mating system of an organism (see Emlen and Oring, 1977; Bradbury and Vehrencamp, 1977; Andersson, 1994). This ecologically-based view of mating systems has been particularly fruitful in understanding the evolution of diverse insect reproductive traits (e.g. Thornhill and Alcock, 1982; Choe and Crespi, 1997).

As previous authors have noted, there are various ways of classifying animal mating systems (e.g. Emlen and Oring, 1977; Thornhill and Alcock, 1982; Davies, 1991; Reynolds, 1996; Brown *et al.*, 1997; Godfray and Cook, 1997). In this chapter, we follow the mating classification scheme described by Brown *et al.* (1997; Figure 5.1) which stresses the role of sexual conflict. **Intra-sexual conflict** typically occurs before copulation. Males compete with each other for the opportunity to mate with females. Male-male competition is most intense when males can either defend one or more females or defend the resources which increase female reproductive success and to which females aggregate (Emlen and Oring, 1977). Females may compete for either genetic or material benefits such as nuptial gifts or resources such as favourable sites for offspring development. **Inter-sexual** conflict tends to occur over the level of parental investment provided by each sex and, consequently, is most intense after copulation. Which sex cares for the offspring depends on who is spatio-temporally close to the eggs at fertilisation and oviposition as well as how many parents are required to provide care for the successful development of offspring (Baylis, 1981). The sex with the highest rate of reproduction (due to faster rates of gametogenesis and/or lower investment in the form of parental care) will be the one most likely to compete for members of the opposite sex (Baylis, 1978; Reynolds, 1996). Depending on which sex is the primary care-giver, the mating system can be monogamous, polyandrous, or polygynous (Davies, 1991) (see below).

As noted above, studies of mating systems for parasitic Hymenoptera and insect predators have focused on different issues and this is also reflected in the classification schemes for mating systems in these two groups of organisms. Classifications of parasitoid mating systems are for the most part much simpler than those developed for predators. Parasitoid mating systems are primarily concerned with the issue of whether mating is restricted and local (mating opportunities are primarily in the natal area) or random and panmictic (mating opportunities extend throughout the population). Here we follow the mating classification adopted by Godfray and Cook (1997) which focuses on the site where mating occurs (see also Table 5.1). With few exceptions (e.g. Hamerski and Hall, 1989; Hardy and Blackburn, 1991; Field and Keller, 1993; Allen *et al.*, 1994) post-copulatory events (e.g. parental care or mate guarding) are absent in the parasitic Hymenoptera.

Given the diversity of life-history and feeding strategies in the remaining insect taxa (i.e. most organic diversity on the planet!) it is not surprising that the ecological constraints on mating systems are many and varied and, consequently, there is a broad range of mating system types (see Thornhill and Alcock, 1982, for review). The starting-point for classifying mating systems is the propensity of each sex to mate during its reproductive lifetime. Consequently, mating

Table 5.1 Summary of predicted sex ratio optima under different mating systems. Mating system is represented by two components: (a) the degree of local mating on the patch (Full = all mating is local, Partial = some mating is local and some non-local, None = panmixis), and (b) where there is at least some local mating, the mating structure within the patch itself. Sex ratio predicted under each mating system is briefly summarised relative to Hamilton's (1967) initial model of sex ratio under strict local mate competition with random mating within each patch. Modelling studies considering further influences on sex ratio optima under local mate competition are summarized in Godfray (1994, p167), Hardy (1994a) and Hardy *et al.* (1998), see also Kathuria *et al.* (1999). References: (1) Fisher, 1930, (2) Hamilton, 1967, (3) Werren, 1984, (4) Taylor and Bulmer, 1980, (5) Frank, 1986, (6) Nunney and Luck, 1988, (7) Werren and Simbolotti, 1989, (8) Stubblefield and Seger, 1990, (9) Ikawa *et al.*, 1993. (10) Greeff, 1996 (see also Denver and Taylor, 1995, Greeff and Taylor, 1997). Modified from Hardy (1994a) who also considers the influence of host quality

Mating Structure		
(a) Local mating	(b) Within patch	Sex Ratio
Full	Sib-mating biased	Female bias greater than Hamilton's prediction (7).
	Random	Hamilton's prediction: Sex ratio sensitive to foundress number, more female biased at lower foundress numbers (2,3,4,5,6,7,9).
	Sib-mating avoided	Less female bias than under random mating (8).
Partial	Sib-mating biased	Female bias decreased relative to equivalent conditions under fully local mating, and increased relative to partial local mating with random on-patch mating (6). (see also 10)
	Random	Female bias decreased compared to Hamilton's prediction (4,5,6,7,9). Different sex ratio optima for sibmated and non-sibmated mothers (10).
	Sib-mating avoided	Not formally studied (but see 10): we would expect little or no female bias.
None	None	Sex ratio equality (1).

systems are monogamous, polygynous, polyandrous or polygamous (see Krebs and Davies, 1993).

5.3 METHODS OF STUDYING MATING SYSTEMS

In this section we provide an overview of general approaches to studying the mating systems of insect natural enemies. Examples of applications of these methods in the study of particular mating systems are given in the following sections. Empirical approaches can be classified as 'behavioural', 'genetic' and 'comparative'. Theoretical models can also be employed to develop a framework of predictions that form a conceptual context for empirical work. The application of two or more of these approaches in concert is likely to prove synergistic.

5.3.1 THEORY AND MODELLING APPROACHES

The various modelling approaches used by evolutionary biologists can be classified as 'genetic' and 'non-genetic'. **Genetic models** are those that explicitly model a given behaviour, examining the conditions that favour the spread of a rare allele for a trait through the population. Antolin (1993) provides a summary of genetic models of sex ratio in subdivided populations. Non-genetic or **phenotypic** models examine a behavioural trait in terms of its contribution to an individual's fitness. Much behavioural ecology is based on non-genetic or phenotypic models, despite the recognition that population genetics underlies behavioural ecology (Grafen, 1991). Grafen (1991) provides a detailed discussion of phenotypic modelling and concludes that, although it may occasionally be incorrect, it is generally a well justified approach for studying the selective forces that influence character evolution. Phenotypic models of mating systems may focus on either the population or the individual. Examples of population-based phenotypic models of mating systems include the group of related sex allocation models along the continuum from local mate competition to random panmictic mating (Table 5.1). Hardy (1994a), Godfray (1994) and the sections below provide discussion and examples of mating systems that are intermediate *between* these two extremes.

Phenotypic models that describe mating systems in organisms practising parental care tend to focus on individuals. An important consideration when modelling mating systems is that individuals of one sex commonly compete amongst themselves for mating opportunities with members of the opposite sex. For a particular individual, the best solution to such problems is likely to depend on the behaviour of its competitor(s), but competitor behaviour will also be influenced by the behaviour of the individual. Game theoretic (or ESS) models (section 1.2.2) are frequently used to explore such situations. Within a population, a variety of different tactics to obtain copulations may coexist. At least three possibilities exist: (1) **alternative strategies** where the difference in tactics used by individuals has a genetic basis; (2) **mixed strategy**, where each individual exhibits a probabilistic mix of tactics; and (3) **conditional strategy**, where the tactic employed by a given individual is dependent on the status of that individual (Gross, 1996). While there are limited numbers of examples of populations with a mixed strategy, examples of a conditional strategy abound (Davies, 1991; Gross, 1996; Alonzo and Warner, 2000). In several mating systems such alternatives frequently exist, e.g. '**mate guarding**' (attempting to exclude competitors: Figure 5.2) or '**mate sneaking**' (obtaining copulations with females guarded by other males). Which alternative is adopted may simply reflect the limited capabilities of some individuals. For example, if successful mate guarding generates high fitness returns but males vary in mate guarding ability (e.g. dependent on body size), inferior males attempting to mate-guard fare very poorly indeed but gain intermediate fitness by adopting the tactic of sneaking mating opportunities with females guarded by superior males. Thus, inferior males may 'make the best of a bad job'. It may, however, be possible that alternative tactics generate equal fitness returns when occurring at particular relative frequencies, with

Figure 5.2 Mate guarding in the damselfly *Ceriagrion tenellum*. After copulation the male remains attached to the female's prothorax, with his abdominal cerci, and accompanies her during oviposition. By remaining attached in this way he ensures that other males cannot mate with her. The pair separates once the female has laid her current clutch of eggs and she therefore no longer provides an immediate opportunity for male reproductive success. Photo: M. Siva-Jothy.

the frequency of the alternatives determined and maintained by frequency dependent selection (e.g. Hamilton, 1979; Dominey, 1984; Cook *et al.*, 1997). For example, **'sneaky males'** may reap the same fitness returns as mate-guarding males when sneaking is relatively rare, but if sneaky males become more common in the population they become less successful. Alonzo and Warner (2000) provide a detailed discussion, and examples, of using models of dynamic state variable games to explore determinants of mating systems.

5.3.2 BEHAVIOURAL STUDIES

Relatively simple observations of behaviour, whether in an experimental or natural setting, can provide an enormous amount of information about components of mating systems. Behavioural studies have been the backbone of mating system studies in the past, and are likely to continue to play an important role. The methodology involved in behavioural studies does not need to be complex, but there have been recent advances in methods of recording and analysing behavioural data (e.g. Fauvergue *et al.*, 1999; Haccou and Meelis, 1992; section 4.2). A general introduction to methodology of behavioural studies is given by Martin and Bateson (1996). In general, mating system studies start with the careful observation of marked individuals throughout a reproductive period/ season and aim to document the nature and frequency of mate-encounters, the determinants of male copulatory success, the variance in male mating success and the female behaviours that determine the **operational sex ratio** (**OSR**, the ratio of sexually active males to sexually active females; for a recent discussion of OSR see Kvarnemo and Ahnestjö, 2002). Usually such studies are accompanied by investigation aimed at understanding the pattern of sperm precedence (section 4.5.2) which, when combined with an understanding of the pattern of egg-laying, enables the conversion of measures of mating-success to reproductive success. Measurements of female behaviour and physiology tend to concentrate on factors that determine female reproductive value and the female role in determining OSRs. The combination of these approaches provides, and has provided, the basis for describing and understanding an insect's mating system. The ultimate determinants of the mating system (inevitably ecological variables) can then be narrowed down and identified experimentally.

This mainly behavioural approach has been used recently in powerful combination with modern empirical approaches, such as genetic techniques for assessing levels of inbreeding and outbreeding and identifying paternity (section 5.3.3) and with data on further species, making use of recent advances in comparative methods (section 5.3.4). As noted above, there has long been a productive interchange between behavioural studies and theory-driven hypotheses.

5.3.3 GENETIC STUDIES

Introduction

Insect natural enemies are generally small, and the reproductive behaviours of individuals are usually difficult to observe, especially if mating takes place in localities other than the natal site. Great advances in the methodology of genetic studies in recent years offer a solution because important elements of mating systems can be estimated without observing behaviour. Despite these developments, and their applications to mating systems of animals such as birds, mammals and social insects (e.g. Honeycut *et al.*, 1991; Davies, 1992; Fjerdingstad and Boomsma, 1998), there have been relatively few genetic studies of insect natural enemy mating systems (e.g. Allen *et al.*, 1994; Molbo and Parker, 1996; Antolin, 1999). We expect the application of genetic techniques to be a major growth area (see Hewitt and Butlin, 1997, for a general account of the interface between genetics and behaviour).

Some studies of insect natural enemies have used genetic variation to characterise mating systems (population structure, paternity), implicitly assuming that mating systems have evolved for reasons unconnected with the measured variation itself (e.g. Kazmer and Luck, 1991; Allen *et al.*, 1994; Molbo and Parker, 1996). Other studies have focused on genetics as a force driving the evolution of mating systems (e.g. some form of inbreeding depression) (e.g. Antolin and Strand, 1992; Cook and Crozier, 1995; Godfray and Cook, 1997).

The Influence of Mating Systems on Genetic Variation

Introduction

The basic procedure is to obtain estimates of genetic variability at polymorphic loci and compare these to expectations from the Hardy-Weinberg theorem under population-wide, random mating (panmixis). The **Hardy-Weinberg ratio** is the ratio of genotype frequencies that evolve in infinite populations (>1000 individuals) when mating is random and neither natural selection nor genetic drift are operating (see e.g. Ridley, 1993a). Deviations from Hardy-Weinberg expectation imply that mating is non-random throughout the population, that the population is finite and/or that natural selection is operating. A major reason for non-random mating is that populations consist of isolated or partially isolated sub-populations (or demes).

Obtaining Estimates of Genetic Variation

Various techniques can be used to assess variation in DNA or its products (see section 3.2; an entomological introduction is provided by Cook, 1996, and see Kazmer, 1990; Hoy, 1994; Scott and Williams, 1994; Hadrys and Siva-Jothy, 1994). An established, economical and powerful method involves screening the electrophoretic mobility of enzymes (DNA products) (see Chapter 3, subsection 3.2.2, and Chapter 6, subsection 6.3.11).

Although enzyme markers can be used to study population structure and also the parentage or relatedness of individuals, they are often relatively invariant compared to DNA and, since they can be expressed at different points in the life-history of an insect, they can be unreliable if used to compare adults and early instars, especially in holometabolous insects.

There is an increasing range of techniques for the *direct* assessment of DNA variation: Random Amplified Polymorphic DNA [RAPDs], DNA fingerprints, DNA microsatellites and Amplified Fragment Length Polymorphisms [AFLPs] (section 3.2; and further reviews in Queller *et al.*, 1993; Avise, 1994; Cook, 1996; Mueller and LaReesa Wolfenberger, 1999). **DNA microsatellites** (repetitive dinucleotide sequences in which the number of repeats at a locus is variable) are a recent innovation in the use of DNA markers, and the technique is now sufficiently developed for routine use. Microsatellites have advantages over other genetic markers in that they are highly polymorphic single loci, thereby providing much information (examples of the application of microsatellites to mating system research are provided by Estoup *et al.*, 1994; Molbo *et al.* 2002). The technique is PCR-based, and so one can use minute quantities of DNA. Multiple variable loci occur but each locus can be studied independently, which aids interpretation and allows the use of existing computer programs for data analysis (e.g. Goodnight and Queller, 1999; see also http://gsoft.smu.edu/GSoft.html). The technique has one time-consuming step: the development of primers for individual repetitive loci for the species concerned. However, more and more primers are becoming available and expertise in primer development is now common.

In contrast, the RAPD and AFLP techniques avoid the problems of identifying and generating specific PCR primers. Primers, and even kits, are available "off the shelf" and promise the possibility of generating polymorphic DNA fragments from anonymous genomes. In principle, these techniques enable rapid analysis of relatedness/genetical population structure in systems where no knowledge of the organism's molecular genetics exists. Despite some impressive successes resulting from the application of RAPD to insect mating systems (e.g. Hadrys *et al.* 1992, 1993), the technique has largely fallen from grace. It is based on the statistical probability of finding DNA fragments of scorable length occuring between inverted repeat sequences of between 6 and 10 base-pairs long. The target genome is probed with a library of specially designed PCR primers which target the inverted repeat sequences with the aim of identifying primers that generate polymorphic banding patterns. The main problems are: (a) high sensitivity to contamination from other genomes (e.g. bacteria) and (b) difficulty in duplication of banding patterns between laboratories. In fact, obtaining repeatable results in the same laboratory (by careful adjustment of the reaction conditions) can take a considerable time investment. In contrast, AFLPs promise to be a more reliable way of generating polymorphic DNA fragments from anonymous genomes (Mueller and LaReesa Wolfenberger, 1999). This technique works by first cleaving the target DNA with a restriction enzyme and then adding a primer-target sequence to the cleaved ends of the fragmented DNA. These targets allow the restriction fragments to be amplified by PCR using the appropriate primer(s). AFLPs have been particularly well-utilised by researchers seeking to identify genetical population structure in plants (e.g. Travis *et al.* 1996), and consequently most of the commercially available kits are designed for use with plant genomes (the technique is sensitive to the size of the target genome). As a technique it is still relatively new, but it has been well received by molecular ecologists and has potential for studies of the genetical basis of population structure and insect mating systems.

The Influence of Genetics on Mating Systems

Introduction

There are several ways in which genetic systems possessed by insect natural enemies can influence the evolution of their mating systems. Here we discuss the ability of females to reproduce without mating and the genetic disadvantages of mating between close relatives. Individuals of one sex may also adopt an alternative mating tactic, which may have a genetic basis (e.g. Gross, 1996; section 5.3.1).

Virgin Reproduction

Many insect species have a chromosomal, or heterogametic, sex determination mechanism (both sexes are diploid) (e.g. Cook 2002). In such species, the sex chromosomes inherited at the time of fertilisation dictate whether an individual develops as a male or as a female; one sex is **heterogametic** and the other **homogametic**. Aside from severely constraining the production of biased sex ratios (Hamilton, 1979; Bull and Charnov, 1988, but see West and Sheldon, 2002; Hardy 2002), heterogamety requires that eggs must be fertilised in order to develop and thus females must mate (or self-fertilise) to reproduce. In contrast, other insect species, including many parasitoid wasps, are arrhenotokous (e.g. Cook 2002). Under arrhenotoky, fertilised (diploid) eggs generally develop into female offspring and unfertilised (haploid) eggs develop into males (see Chapter 3, section 3.3). Once mated, females store sperm (in the spermatheca) and potentially control the sex of offspring by regulating the fertilisation of eggs (hence the enormous interest in parasitoid sex ratios). Virgin females are also able to reproduce but are constrained to produce male offspring only. Empirical observations indicate that, with some exceptions, virgin parasitoids make ready use of this ability (e.g. Godfray and Hardy, 1993; Ode *et al.*, 1997).

Virginity has been studied in relation to mating systems using an ESS model (Godfray, 1990; subsection 5.3.1) and in relation to clutch size and developmental mortality under strict local mating (subsection 5.4.2) using a static optimality model (Heimpel, 1994; section 5.4.7). Tests of model predictions require estimates of the prevalence of virginity and there are several methods available (Godfray, 1988; Godfray, 1990; Hardy and Godfray, 1990; Godfray and Hardy, 1993; Hardy and Cook, 1995; Ode *et al.*, 1997; Hardy *et al.*, 1998a; West *et al.*, 1998). A direct method is to dissect females that are foraging for oviposition opportunities, or those that have dispersed from the natal site in species with strict local mating, and examine the contents of their spermathecae: females without sperm are either virgin or have been mated but become sperm-depleted. Alternatively, females can be provided with hosts suitable for daughter production, and the sex of their progeny examined. Indirect methods can be used for gregarious species in which sex ratio optima are likely to be female-biased or unbiased. Broods containing only males are likely to have been produced by virgin mothers and the proportion of all-male broods serves as an estimate of the proportion of virginity. Similarly, in species with strict local mating, females that develop in all-female broods (e.g. due to developmental mortality of males) will remain virgin, and the proportion of such females observed can be used as an estimator of the proportion of virginity in the population.

Recent empirical work by King (2002) found that female *Spalangia endius*, a parasitoid with a local mating population structure, produced a greater proportion of sons among their offspring in response to the presence of another mated female conspecific, but not when the conspecific was a virgin–a finding consistent with predictions of local mate competition theory (subsection 5.4.2).

Some parasitoid species are thelytokous: males are not produced and females arise from unfertilized eggs that become diploid by fusion of meiotic products (e.g. Luck *et al.*, 1993; Cook, 1993a; see subsection 3.3.2 for more detail). Obviously, thelytokous species do not have mating systems as mating does not occur. However, some species have thelytokous and sexual members. Thelytoky is sometimes caused by microbial (*Wolbachia*) infection in some parasitoids and can be cured by antibiotic or high temperature treatment (Stouthamer *et al.*, 1992a; 2002). In *Trichogramma* spp. (egg parasitoids) most infected populations consist of infected and uninfected individuals, but in all other documented cases infections have gone to fixation in the population (Stouthamer, 1997). In some parasitoid species, the males produced by cured females have often lost the full capacity to perform mating behaviour (Stouthamer, 1997). Further, thelytoky may be more likely to arise in species with high levels of inbreeding as it circumvents problems associated with producing highly female-biased and precise sex ratios (Cornell, 1988; Hardy, 1992; Cook, 1993a; Hardy *et al.*, 1998; section 5.4.5); this constitutes a further potential influence of mating systems on genetics (section 5.3.3).

Inbreeding Depression and Sex Determination in the Hymenoptera

Unlike diplodiploid species, sex in the Hymenoptera is not determined by sex chromosomes or the ratio of sex chromosomes to autosomes. Instead, Hymenoptera are haplodiploid, with males developing from unfertilised eggs and females developing from fertilized eggs (e.g. Cook 2002). The discovery of diploid males in *Habrobracon hebetor* 80 years ago (Whiting and Whiting 1925) prompted four models of sex determination in the Hymenoptera: single-locus **complementary sex determination (CSD)**, **multiple-locus CSD**, **genic-balance**, and **genomic imprinting**. Of these, single-locus CSD is thought to have the biggest impact on mating systems because its effects are exacerbated during inbreeding.

To date, diploid males have been found in 36 hymenopteran species (Cook, 1993a; Duchateau *et al.*, 1994; Carvalho *et al.*, 1995; Holloway *et al.*, 1999). Twenty-three of these cases are in the Apoidea (bees) and Vespoidea (wasps and ants). The remainder are found in the Tenthredinoidea (sawflies; 3 species), the Ichneumonoidea (8 species), the Chalcidoidea (1 species) and the Cynipoidea (1 species). Various techniques have been used to detect diploid males including the use of karyotyping, visible genetic markers, allozymes and molecular markers, and morphological characters (Cook, 1993a). All of these techniques are dependent on diploid males developing to the point at which they can be sexed (typically as pupae or adults). In the majority of species in which diploid species are known, viability of diploid males is similar to that of haploid males and diploid females. Yet, diploid male production may be much more prevalent than the number of known cases suggests; if diploid males are all inviable, diploid male production can go unnoticed. In this case, even if 'ploidy' can be established for eggs or larvae it may be very difficult, if not impossible, to assign sex to a diploid individual. One possible exception would be the identification of a sex-specific protein or imaginal disk tissue present early in development.

Many of the species known to produce diploid males appear to follow the single-locus CSD model (reviewed in Cook, 1993a), where sex is determined at a single locus (Whiting, 1939; 1943). Under single-locus CSD, individuals that are heterozygous (diploid) at the sex locus develop into females whereas individuals that are either hemizygous (haploid) or homozygous (diploid) at the sex locus develop into males. An important prediction of the single-locus CSD model is that diploid male production should increase with inbreeding (inbreeding increases levels of homozygosity at all loci including the sex locus). In addition, half of the fertilised offspring of a singly-mated mother should develop into diploid males (given that diploid males are produced). The impact that single-locus CSD has on a population mating system depends on the interaction between the amount of inbreeding, the number of sex alleles in the population, the number of times a female will mate, and whether diploid males are viable and can transfer sperm to females (Cook and Crozier, 1995). Species possessing single-locus CSD may avoid high rates of diploid male production by outbreeding; such appears to be the case with *Habrobracon* (= *Bracon*) *hebetor* (Antolin and Strand, 1992; Ode *et al.*, 1995). Outbreeding coupled with high sex allele diversity within populations can reduce the genetic load associated with single-locus CSD. Allelic diversity is expected to be maintained by strong frequency-dependent selection. The number of sex alleles in a population has been estimated using matched matings (e.g. Whiting, 1943;

Heimpel *et al.* 1999) as well as maximum likelihood approaches (e.g. Owen and Packer, 1994). Estimates in natural populations appear to be high enough to substantially reduce the production of diploid males. Finally, there may be selection for avoidance of mating with close relatives (e.g. Ode *et al.*, 1995) or for polyandry (e.g. Page, 1980). In species where females mate only once, females that mate with viable diploid males are constrained to produce only sons (e.g. Heimpel *et al.*, 1999); multiple mating increases the chance that females can produce daughters. Such considerations may be important when mass-rearing parasitoids for biological control programmes. Initial sampling procedures should involve obtaining naturally high allelic diversity in the founder culture of species with CSD, and measures such as maintaining large cultures or parallel smaller subcultures are recommended to prevent loss of allelic diversity during culturing (Stouthamer *et al.*, 1992b; Cook, 1993b).

The phylogenetic distribution of single-locus CSD suggests it is ancestral: it is found in the Ichneumonoidea, Apoidea, Vespoidea and Tenthredinoidea (Cook and Crozier, 1995). It is tempting to suggest that the mating systems of these groups should be confined to outbreeding. However, the prevalence of single-locus CSD as well as the type of mating systems found within each of these superfamilies is largely unknown. In at least some species of Ichneumonoidea (e.g. *Cotesia glomerata*; Kitano, 1976; Tagawa and Kitano, 1981) there is evidence suggesting a limited amount of local mating. Whether this species represents an exception or whether local mate competition (LMC, subsection 1.9.1) exists without inbreeding is unknown.

However, many parasitoid species inbreed with no apparent production of diploid males. In these species, sex appears to be determined by a mechanism other than single-locus CSD (Cook, 1993a). Single-locus CSD has been ruled out for 9 species of Chalcidoidea, Cynipoidea, and Chrysidoidea (see references in Cook, 1993a). It is possible that the presumed absence of single-locus CSD has allowed members of these superfamilies to evolve mating systems incorporating inbreeding and LMC. Again, it is important to note that differences in the mating systems between the Ichneumonoidea and the Chalcidoidea (for instance) are largely unknown.

Multiple-locus CSD may explain sex determination in some of these cases although to date there is no strong evidence for this mechanism. Multiple-locus CSD is similar to single-locus CSD except that more than one locus is involved in determining sex. Individuals homozygous at all of the sex loci are diploid males. Individuals heterozygous even at only some of the loci develop into diploid females. Inbreeding under multiple-locus CSD leads to increased diploid male production, but at a much slower rate than single-locus CSD. As a consequence, detection of multiple-locus CSD requires several generations of inbreeding. Multiple-locus CSD has been tested (and not detected) in only two species: *Nasonia vitripennis* (Skinner and Werren, 1980) and *Goniozus nephantidis* (Cook, 1993c).

The **genic balance theory** was first proposed by Cunha and Kerr (1957). It argues that sex is determined by a set of dosage-compensated maleness genes and a set of dosage-dependent femaleness genes. The effects of male genes are the same in haploid and diploid individuals. The effects of female genes in diploid individuals are twice that in haploid individuals. In haploid individuals, the effect of male genes is greater than the effect of female genes resulting in the development of males. Females develop from diploid eggs. This mechanism does not explain the production of diploid males nor has it received any empirical support.

Finally, a recent developed model of sex determination in the Hymenoptera is based on **genomic imprinting** (Poirié *et al.* 1993, cited in Beukeboom 1995). Under this model, one or more loci are imprinted paternally or maternally. Maternally imprinted loci are unable to bind with an unknown product that is inactive in the unfertilised egg. Males do not imprint the loci which are free to bind with this product, thereby activating it in fertilised eggs. Bound product is thought to activate a series of genes resulting in female development. This model has recently received support with empirical studies of *Nasonia vitripennis* and the sex ratio distorter PSR (Dobson and Tanouye 1998 subsection 3.4.2). Genomic imprinting, at least in this species, appears to explain the production of diploid males in species where single-locus CSD does not apply.

Although sex is determined by allelic diversity in probably the minority of haplodiploid species, it is thought to be the ancestral sex determination mechanism in the Hymenoptera and is better understood, both in terms of theory and in empirical work, than its alternatives: nevertheless, much remains to be discovered.

5.3.4 COMPARATIVE STUDIES

While some studies look for relationships between variables within species, others look for relationships across species: these are '**comparative studies**' (introduced in Chapter 1, subsection 1.2.3). Comparative studies may make use of the results of many behavioural and/or genetic studies and thus offer a means of assessing generality or testing otherwise untestable predictions, but often they sacrifice experimental rigour. In order to obtain data from sufficient species for meaningful comparisons, it will usually be necessary to use data obtained 'indirectly' from third parties (e.g. the literature). The data of interest from the literature are often collected for different reasons unrelated to purposes of a comparative study. As a result, the quality of data is often variable, sometimes seriously weakening the validity of the conclusions that can be drawn from comparative studies. Furthermore, information on within-species variation will be ignored when species averages are used.

Although the application of phylogeny-based comparative methods has become the new dogma, there is still much debate about exactly how phylogenetic information should be employed (Martins, 1996; Hansen, 1997; Price, 1997a), and there is continuing dissent over whether using phylogenetic information is indeed necessary (references in Harvey, 1996; see also Björklund, 1997; Price, 1997b). Price (1997b) argues that phylogenetic analyses remove cross-species correlations but replace them with 'evolutionary correlations' (the extent to which changes in one trait are accompanied by changes in the other) and that both measures are correlational, not causal, and can be influenced by confounding traits. He concludes that it will be valuable to compare species and contrast correlations. This is probably an already tacitly accepted practice in studies of insect mating (e.g. Cook *et al.*, 1997; West *et al.*, 1997; Boomsma and Sundström, 1998; Hardy and Mayhew, 1998a; West and Herre, 1998a; Fellowes *et al.*, 1999; Jervis *et al.*, 2001, 2003). Exactly how species data should be 'transformed' using phylogenetic information is still the subject of debate, but comparative studies will continue to be valuable. In view of the controversy surrounding the application of comparative statistical techniques, it is advisable to supplement phylogeny-based analyses with traditional ones, and to include the results from both types of analysis in any report of a comparative study (indeed, editors and reviewers may demand that you do so).

5.4 UNRAVELLING PARASITOID MATING SYSTEMS

5.4.1 INTRODUCTION

Natural enemies are generally small and, although desirable, it may be very difficult to study a mating system in its entirety. An overall idea of how common local or non-local mating is may be obtained by fitting a predictive model for partial local mating to sex ratio data (Debout *et al.* 2002, see also Read *et al.* 1992, 2002 for related discussions of the selfing rates of 'Protozoa'). Mating systems can be divided into a number of component parts and investigation of one of these can prove useful. Investigation of more than one component is, of course, even more useful. Good examples of studies studying several component parts of mating systems and employing a variety of techniques are Myint and Walter (1990) and Fauvergue *et al.* (1999). Ware and Compton (1994a, 1994b) used sticky traps to investigate dispersal of female fig wasps in the field: similar techniques could be employed to gain insight into parasitoid mating systems.

5.4.2 EMERGENCE AND MATING AT THE NATAL SITE

Introduction

Temporal and spatial patterns in the emergence of adult natural enemies are important aspects of mating systems. Offspring that develop in a group, whether sharing the same host (**gregarious**) or on separate hosts within a batch (**quasi-gregarious**), may have the opportunity of mating with other group members on maturity. Offspring developing singly in isolated hosts do not.

Emergence

The sexes may emerge simultaneously or males (**protandry**) or females (**protogyny**) may emerge first within an offspring group or population. We are unaware of any reports of protogyny in parasitoid wasps, but protandry is widespread. Males generally develop more rapidly than female conspecifics and are also generally smaller (Hurlbutt, 1987; Hirose *et al.*, 1988; Ramadan, *et al.* 1991; Godfray, 1994; Hardy and Mayhew, 1998a, subsection 2.9.2). Small male size may reflect selection for protandry, or protandry may be an incidental consequence of maternal sex allocation strategies that lead to males developing in smaller hosts (King, 1993).

Protandry in gregarious and quasi-gregarious species is often, but not always, associated with a high degree of mating at the emergence site, as males generally remain and mate with females as these emerge. Protandry can be studied by simple field or laboratory observations of developing broods or 'populations' (Kitano, 1976; Tagawa and Kitano, 1981; Nadel and Luck, 1985, 1992; Hirose *et al.*, 1988; van Dijken *et al.*, 1989; Kajita, 1989; Myint and Walter, 1990; Ramadan *et al.*, 1991; Pompanon *et al.*, 1995; Ode *et al.*, 1996; Fauvergue *et al.*, 1999; Hardy *et al.*, 1999, 2000; Loch and Walter, 1999, 2002). However, in some parasitoids mating may take place within the confines of the host or the host's covering (e.g. puparium), which may obscure some details of mating behaviour (Suzuki and Hiehata, 1985; Tepedino, 1988; Drapeau and Werren 1999). This complication does not occur with either quasi-gregarious endoparasitoids (e.g. Nadel and Luck, 1985, 1992; Myint and Walter, 1990) or ectoparasitoids, although ectoparasitoids mating within pupal cocoons may be difficult to observe (Tagawa and Kitano, 1981; Dijkstra, 1986; Hardy *et al.*, 1999). Protandrous male bethylids (ectoparasitoids of many Lepidoptera and Coleoptera of economic importance) often inseminate unemerged females by chewing an entrance hole in the female's cocoon (Griffiths and Godfray, 1988; Hardy and Mayhew, 1998a) and the behaviour, and the resulting holes, can be readily observed before female emergence. However, female bethylids may also mate post-eclosion (in *Goniozus triangulifer*, mating appears to be exclusively post-eclosion, Legaspi *et al.*, 1987).

Male and female emergence patterns have been studied in *Habrobracon hebetor*, a species noted above to exhibit severe inbreeding depression. In a laboratory study, Antolin and Strand (1992) found that males and females emerged and dispersed from a brood throughout the day; no difference was found in the emergence patterns of males and females. In some instances all females left the natal area before any males emerged, where in others all males left before all of the females had emerged. Most females left the natal area unmated.

Fauvergue *et al.* (1999) assessed emergence patterns of *Leptopilina heterotoma* and *L. boulardi* (Hymenoptera: Eucoilidae), parasitoids of frugivorous *Drosophila* larvae, by rearing out wasps from bananas exposed for two weeks in fruit orchards and then taken to the laboratory. Across all patches (bananas) males and females emerged simultaneously over about a ten-day period. Although in each species both sexes emerged from virtually every patch during this period, individual parasitoids frequently emerged without any conspecifics of the opposite sex emerging from the same patch on the same day. These observations were supplemented by a laboratory experiment in which offspring laid within a 24-hour period were reared out and individually observed until emergence using an automated image analysis system. Adults emerged over a five-day period and both species were highly protandrous. Similar studies on other parasitoids of flies have documented protandry as well as prolonged female emergence over several days (Legner, 1968; Nadel and Luck, 1985).

Local Mating

Some models developed to predict sex allocation optima assume that when the offspring of several females (foundresses) develop in spatial aggregation, mating occurs exclusively between group members (strict local mating) and also randomly, with respect to genetic relatedness (e.g. Hamilton, 1967; and see Table 5.1). There have been several tests of this relationship using individual parasitoid species (e.g. Werren, 1983; Waage and Lane, 1984; Strand, 1988); in general sex ratios are positively related to foundress number, as predicted (further discussed by Orzack, 1993; Godfray, 1994). Several comparative studies have examined cross-species relationships between foundress number and offspring sex ratio in species with strict local mating. The best-known of these have employed natural variation in foundress number in fig-pollinating wasps (Agaonidae), which are not insect natural enemies, but share much biology with parasitoid wasps (e.g. Herre, 1985, 1987; Herre *et al.*, 1997; West and Herre, 1998b). In these species (winged) females enter a fig, lay eggs and die (their corpses can be counted to estimate foundress number) while males are wingless and unable to disperse from the natal fig (providing what has been thought to be good evidence for strict local mating, however wingless males in some non-pollinating fig wasps do disperse; Greeff and Ferguson 1999; Cook *et al.*, 1999; Bean and Cook, 2001; and see Greeff *et al.* 2003 for evidence of male-dispersal in pollinating fig wasps). Their sex ratios support the predicted relationship across species (and also within species, Frank, 1985a; Herre, 1985; but see Kathuria *et al.*, 1999). Despite the enormous interest in parasitoid sex ratios, there have been few comparative studies and the evidence for strict local mating is less clear than for fig-pollinator wasps. Griffiths and Godfray (1988) examined a predicted relationship between sex ratio and clutch size in bethylid wasps, assuming both strict local mating and that offspring groups are produced by single foundresses. We discuss this more fully below (section 5.4.5).

Random Local Mating

The assumption of random within-group mating will be violated if mating with non-relatives is avoided and/or if non-siblings are less likely to encounter one another (due to asynchronous maturity resulting from asynchronous offspring production by mothers). Sex ratios more female-biased than under random mating are predicted (Nunney and Luck, 1988; Stubblefield and Seger, 1990; Table 5.1, Figure 5.3; see also Kinoshita *et al.* 2002). Conversely, if individuals prefer non-siblings as mates, sex ratios are predicted to be less female-biased (Stubblefield and Seger, 1990; Table 5.1). Of course, these predictions only apply to offspring groups produced by more than one foundress.

There have been few empirical investigations of within-group mating patterns. One of these examined mating patterns in fig-pollinating wasps (section 5.4.2). Mating is difficult to observe directly as it occurs within the fig. Frank (1985b) cut figs into halves, marked one male from each half and then joined two halves from different figs together. Males tended to remain in the half-fig of their origin, suggesting sibling-biased mating within figs and a possible explanation for observations of fig wasp sex ratios more biased than predicted by models assuming random within-group mating (Frank, 1985b).

Molbo and Parker (1996) used genetic techniques (allozyme polymorphism, subsection 5.3.3) to estimate foundress numbers in patches of hosts in natural populations of *Nasonia vitripennis*, a pteromalid wasp parasitising fly pupae. Males are protandrous and mate females on

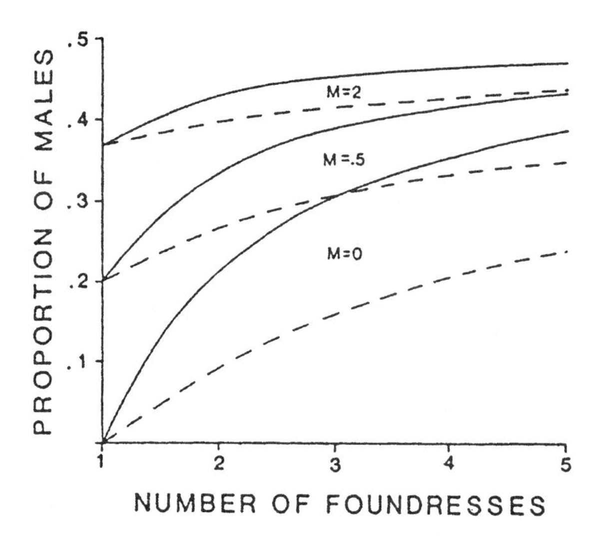

Figure 5.3 Predicted sex ratio optima in response to foundress number, maturation asynchrony and non-local mating. The model assumes haplodiploid genetics and that, after all females in the natal patch are mated, males disperse to seek virgin females in other patches. The degree of non-local mating is represented by 'M', the number of patches found by a dispersing male. Solid lines show the prediction assuming offspring maturation is synchronous and dashed lines are for asynchrony. Increasing foundress numbers, increasing non-local mating and increasing synchrony each lead to less female-biased sex ratios. (Source: Nunney and Luck, 1988). Reproduced by permission of Elsevier Science.

emergence (e.g. King *et al.*, 1969). Males also have reduced wings and are probably unable to disperse significantly from the emergence site (references in Hardy, 1994a). Within patches, adults emerged over one to eight-day periods (although most wasps emerged within four days), possibly because foundresses did not contribute offspring synchronously. As predicted by theory (Figure 5.3), sex ratios in patches with less synchronous emergence were more biased than in patches with shorter emergence periods, although there were too few data for formal testing (Molbo and Parker, 1996; Figure 5.4).

5.4.3 DISPERSAL FROM THE NATAL SITE

While properties such as protandry and mating at the natal site (local mating) can be demonstrated within the confines of a glass vial, such observations deny individuals the opportunity to disperse prior to mating, and thus may overestimate the importance of local mating. Modelling studies predict that dispersal and non-local mating can strongly influence parasitoid sex ratios (Table 5.1, Figure 5.3), but empirical investigations are relatively scarce (reviewed by Hardy, 1994a). Field observations are possible in some species (e.g. Tagawa and Kitano, 1981; Kajita, 1989; Loch and Walter, 1999). In the laboratory, the propensity for, and timing of, dispersal can be investigated by simply allowing emerging individuals to leave the vial, or similar methodology (Dijkstra, 1986; Myint and Walter, 1990; Forsse *et al.*, 1992; Fauvergue *et al.*, 1999; Hardy *et al.*, 1999, 2000; Loch and Walter, 2002). Dijkstra (1986), studying *Colpoclypeus florus* (Hymenoptera: Eulophidae), a gregarious parasitoid of tortricid leaf-roller larvae, Hardy *et al.*, (1999) studying *Goniozus nephantidis* (Hymenoptera: Bethylidae), a gregarious larval parasitoid of lepidopteran coconut pests, and Hardy *et al.* (2000) studying *Goniozus legneri*, a parasitoid of lepidopteran almond pests, all found that virtually all males and females dispersed from the natal site (Figures 5.5 and 5.6). Females collected post-dispersal can be provided with hosts (e.g. Hardy *et al.*, 1999, 2000) or the contents of their spermathecae can be examined following dissection (Nadel and Luck, 1985, 1992; Dijkstra, 1986; Ode *et al.*, 1998) to assess whether or not they have mated. Populations of parasitoids may have several levels of organisation ranging from individual broods to sub-populations to the entire population. Populations of *Habrobracon hebetor* are frequently subdivided and approximately half of the females disperse from subpopulations before they have mated (Ode *et al.*, 1998).

5.4.4 POST-DISPERSAL MATING AND IMMIGRATION: MATING AT THE NATAL SITE

Introduction

Although the above methods (section 5.4.3) can demonstrate dispersal and the prevalence of

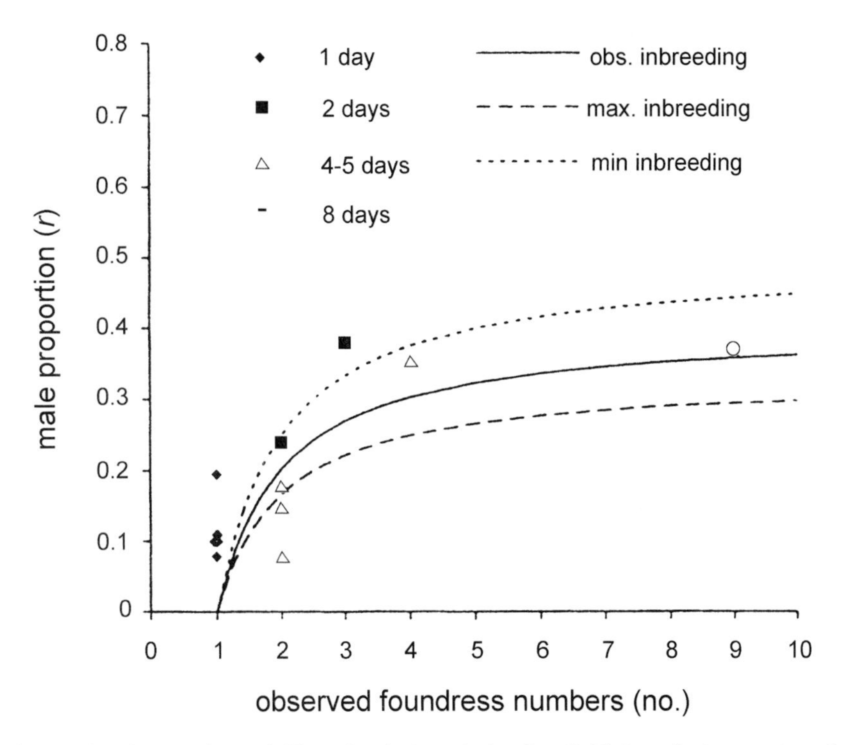

Figure 5.4 Sex ratios in patches of *Nasonia vitripennis* in the field in relation to foundress number and synchrony of emergence within each patch (1–8 days). Curves show the predicted optimal sex ratio with inbreeding coefficients of 0 and 1 (dashed lines) and observed inbreeding (F = 0.31) (solid line). (Source: Molbo and Parker, 1996). Reproduced by permission of The Royal Society of London.

local mating, they do not assess post-dispersal mating, i.e. it is unknown whether dispersing males obtain matings elsewhere or, similarly, whether females mate or re-mate after dispersal. They also do not allow the possibility of mating by immigrants at the natal patch. Several studies have been carried out in which a fuller range of behaviours is possible, ranging from laboratory cages, through much larger enclosed spaces to semi-natural and natural field populations. Comparative methods can also be used to explore relationships between the expected degree of local mating and the sex ratio (section 5.4.6).

Laboratory Cages

Vos *et al.* (1993) observed groups of *Leptomastix dactylopii*, a parasitoid wasp that develops quasi-gregariously in isolated patches of hosts (citrus mealybug) emerging in a laboratory cage. Almost all parasitoids left the natal patch before mating and at least some subsequent (off-patch) mating was observed. Both males and females were attracted to small traps containing members of the opposite sex. Fauvergue *et al.* (1999) observed emergence and dispersal of *Leptopilina heterotoma* and *Leptopilina boulardi* from patches within laboratory cages. Patches with unemerged conspecifics were also present and the numbers of individuals remaining on the natal patch, immigrating to other patches and elsewhere in the cage could be monitored. Similarly, pilot studies of *Goniozus nephantidis* in table-top sized cages indicate that males developing in all-male broods disperse and immigrate into broods of females. Non-sibling mating readily occurs, both in and around the females' natal area (S. Barnard and I.C.W. Hardy, unpublished).

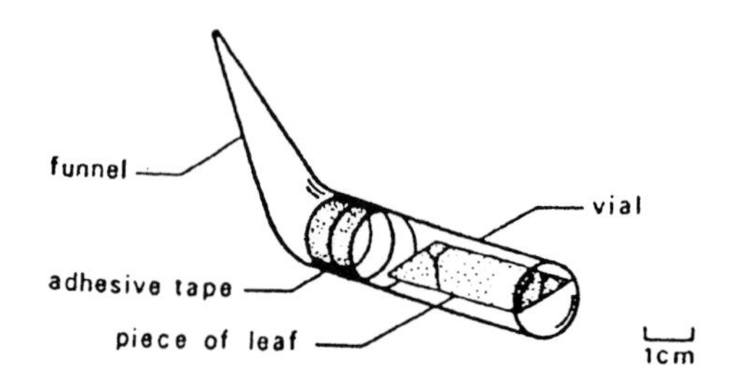
funnel
vial
adhesive tape
piece of leaf
1cm

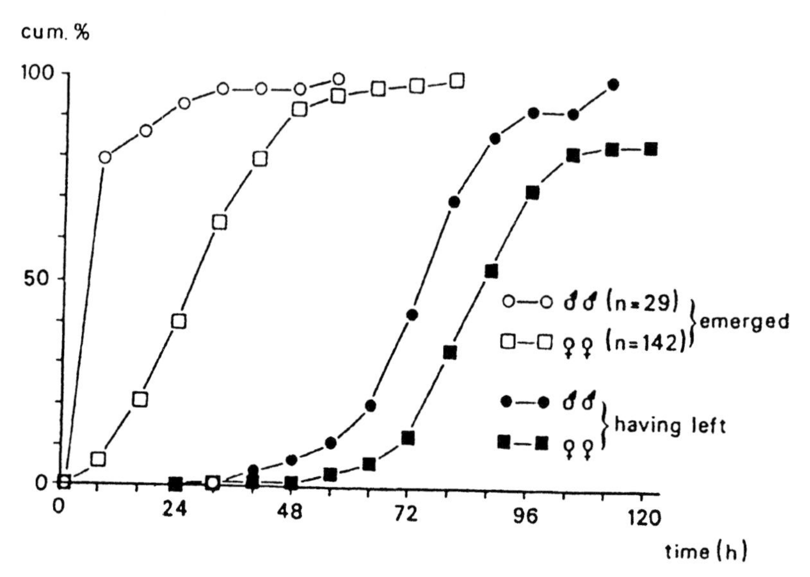
cum. %
100
50
0
0
24
48
72
96
120
time (h)
○—○ ♂♂ (n=29)
□—□ ♀♀ (n=142)
emerged
●—● ♂♂
■—■ ♀♀
having left

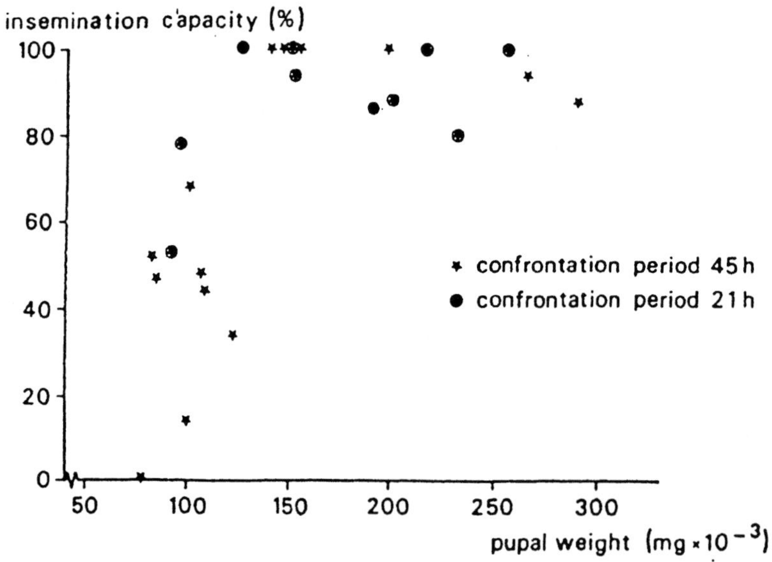
insemination capacity (%)
100
80
60
40
20
0
50
100
150
200
250
300
pupal weight (mg × 10⁻³)
confrontation period 45 h
confrontation period 21 h

Greenhouse Populations

Using greenhouse populations, Nadel and Luck (1992) investigated dispersal, immigration and mating on the natal patch and elsewhere in *Pachycrepoideus vindemmiae* (a quasi-gregarious pteromalid which parasitises *Drosophila* pupae) (Figure 5.7). Males were protandrous and often waited for females to emerge and mate (female *P. vindemmiae* mate only once, Nadel and Luck, 1985). Males also emigrated from broods and arrived at non-natal broods, before and after female dispersal in both cases. The majority of females had the opportunity of pre-dispersal mating with same-patch or immigrant males, although some virgin females were collected at oviposition sites. Males also migrated to oviposition sites, where some virgin females mated. Overall, there were 26-37% extra-brood matings overall. Although these observations demonstrate a complex mating system, it is difficult to assign relative importance to the observed mating system components with confidence, because of unnatural elements of the experiment (further discussed by Nadel and Luck, 1992; Hardy, 1994a).

Semi-natural and Natural Populations

Several studies of parasitoid mating systems have been carried out in 'agricultural' environments. It might be argued that such unnatural settings could influence aspects of the mating system under investigation, because: (a) population densities are likely to be unnaturally high; (b) parasitoids are relatively poorly co-evolved with their hosts; (c) host distributions within an agricultural setting may be distorted relative to 'natural' environments, However, some parasitoids have inhabited agricultural settings for thousands of years (e.g. *Habrobracon hebetor* has been associated with pests of stored grain for at least 3,500 years, Chaddick and Leek, 1972).

Myint and Walter (1990) examined aspects of the mating system of *Spalangia cameroni* (Hymenoptera: Pteromalidae), a quasi-gregarious parasitoid of housefly pupae, in 'field' investigations carried out in a poultry shed. How closely this represents natural conditions, and thus the quantitative importance of the results, remains unclear. Males generally emerged before females developing in the same batch of hosts, and most wasps left the natal site without mating (mating was observed when the emergence of males and females happened to coincide). Olfactometer (odour choice) experiments in the laboratory (section 1.5.2) showed that males were attracted to odours emanating from the host habitat. Males were also trapped in the field at patches of hosts suitable for oviposition by females (in both the presence and the absence of females) and at batches of parasitized hosts from which females were close to emerging. Sib-mating and local-mating are probably uncommon in natural populations.

Loch and Walter (1999) observed the eclosion and mating behaviour of *Trissolcus basalis* (Scelionidae), a quasi-gregarious parasitoid of homopteran eggs, emerging from host egg batches placed in the field. Males are protandrous and engage in aggressive contests for

Figure 5.5 Eclosion, local mating and dispersal from mixed sex broods in *Colpoclypeus florus*, a gregarious eulophid wasp in which broods develop within the web of the lepidopteran host (tortricidid leaf-roller larvae). **Upper panel:** Laboratory apparatus which allows dispersal. **Central panel:** Timing of eclosion and dispersal. Males eclose shortly before females and copulation takes place within the host's web. Both sexes disperse 2–8 days after eclosion. **Lower panel:** Insemination capacity of males. Male body weight (estimated by pupal weight) is positively correlated with the capacity to inseminate females (males were in contact with 15 females for a total of either 21 or 45 hours, although, to simulate natural emergence patterns, not all 15 females were presented simultaneously). Recently mated females do not re-mate, but it is not known whether they remain unreceptive. Although the mean sex ratio is female-biased, these investigations suggest that the mating system is unlikely to conform absolutely to 'strict local mating'. (Source: Dijkstra, 1986). Reproduced by permission of Brill Uitgeverij.

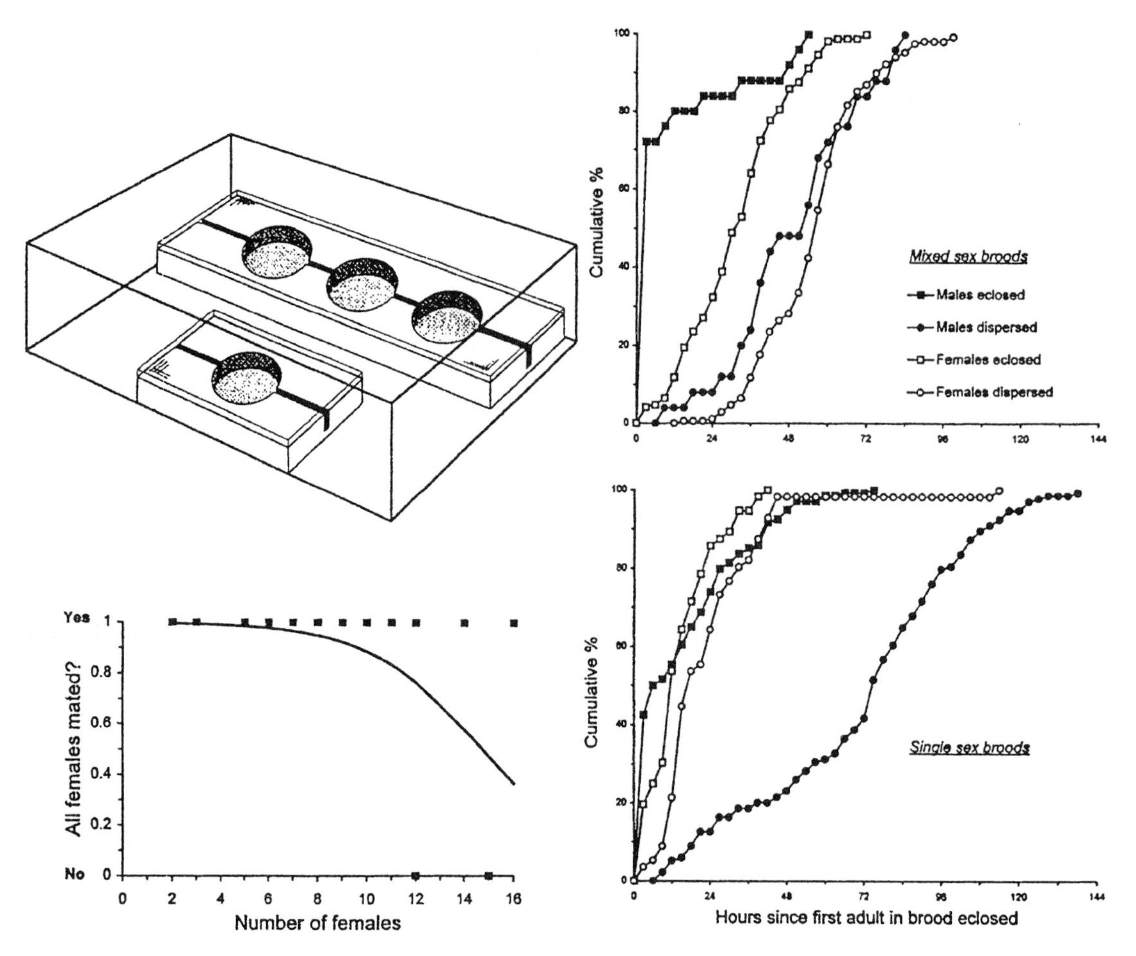

Figure 5.6 Eclosion, local mating and dispersal in *Goniozus nephantidis*, a gregarious bethylid wasp in which broods of siblings develop within narrow feeding galleries constructed by the lepidopteran host (oecophorid larvae which feed on coconut leaves). Broods may contain both sexes or, due to male mortality and subsequent female virginity, only one sex of offspring. Individual broods were allowed to mature in the central of three chambers in a 150mm long opaque plastic block decked with plexiglass (upper left panel). A 1 mm wide slot connected chambers and provided access to a larger transparent box from which dispersed individuals (those outside the block) were collected. A block with one chamber was also present to provide structure for dispersed adults. Observations were made every 3 hours. Dispersed females from mixed sex broods were allowed to produce offspring in order to assess their mating status. In mixed-sex broods, almost all females mated before dispersal (lower left panel). Males were generally protandrous, males dispersed from both mixed-sex and all-male broods, but males in all-male broods disperse more slowly (right hand panels). Virgin females dispersed from all-female broods (lower right panel). Although the mean sex ratio is female-biased, these investigations suggest that the mating system of *G. nephantidis* is unlikely to conform absolutely to 'strict local mating', although field observations and/or genetic studies are required. Lower left panel: the probability that all females are inseminated in relation to the number of females eclosing from 13 broods containing only one male. The data are binary (1 = all females inseminated, 0 = some females uninseminated) and the fitted curve is the probability of complete insemination, estimated using logistic regression. Although the two broods with incomplete insemination contained relatively large numbers of females the regression falls just short of significance, and in each of these only one female was uninseminated. Thus, a single male was sufficient to inseminate virtually all females, even when brood sizes were large. Similar results were obtained from *Goniozus legneri* using the same protocol (Hardy *et al.* 2000). (Source: Hardy *et al.*, 1999). Reproduced by permission of Blackwell Wissenschafts-Verlag.

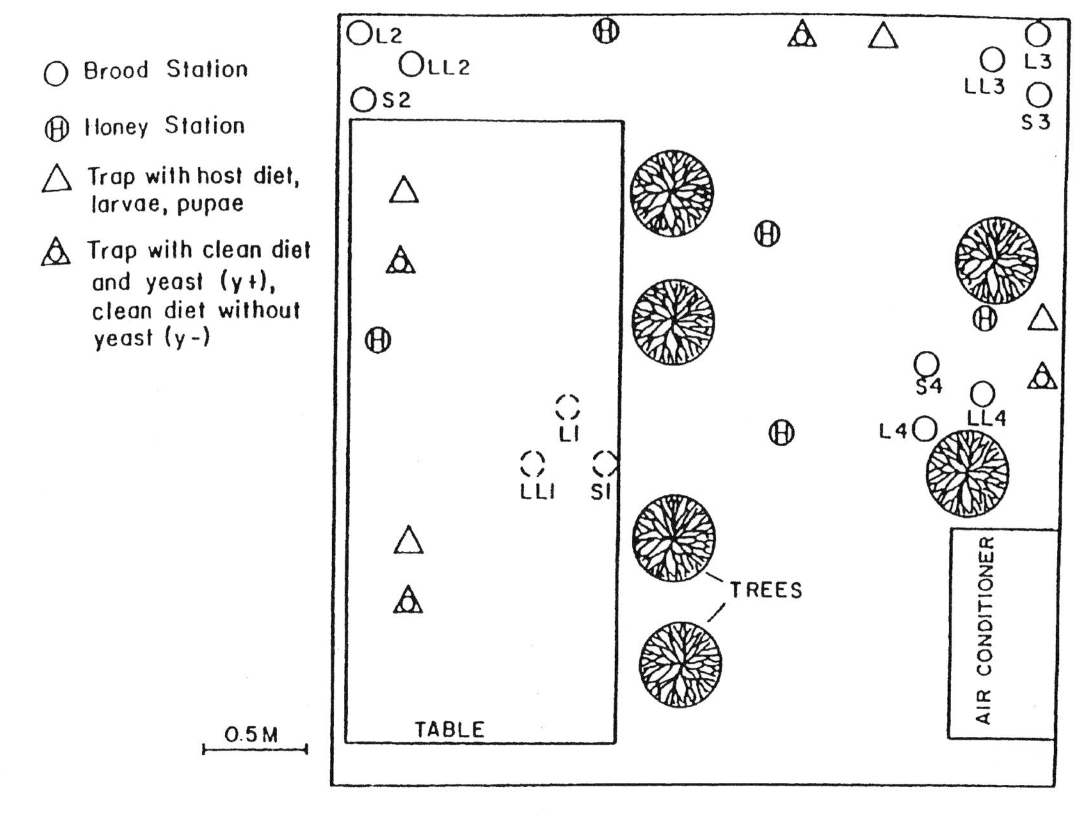

Figure 5.7 Plan of the experimental greenhouse used to observe mating and dispersal of *P. vindemmiae*. Four groups of three single-foundress broods were arranged on the floor on slightly raised 'brood stations (white plastic discs). Each group contained a small brood (S) of 5 hosts, a large brood (L) of 12 hosts and a large late brood (LL) of 12 hosts parasitised 2 days later than S and L. Wasps could only move between broods by flying (both sexes can fly). Wasps were not marked. Brood stations were checked every 2 hours during a 12 hour period for 6–7 days (wasps were found to be quiescent outside this period). Several oviposition site 'traps' were used to catch dispersed females, which were dissected to determine whether their spermathecae contained sperm. Honey was provided but wasps were not attracted to this. (Source: Nadel and Luck, 1992).

egg masses. Nearly 20% of females were not mated by the dominant male on emergence, 25% of females mated multiply, sometimes with multiple males, and both virgin and mated females moved slowly from the emergence site to its surroundings before dispersing. Males also dispersed from the natal site, both during and after the period of female emergence (Loch and Walter, 2002). Both local and non-local mating probably occur in nature, but the extent of outbreeding remains unknown.

Kitano (1976) and Tagawa and Kitano (1981) observed the eclosion and mating behaviour of *Cotesia* (as *Apanteles*) *glomerata*, a gregarious parasitoid of lepidopteran larvae. Observations were carried out in an agricultural system (cabbage fields). The sex ratios of most broods were female-biased, although there were some all-male broods. Males tended to remain at the natal site after emergence from both brood types. In mixed sex broods, males usually emerged first and began to search for females to court. Immigrant males arrived at both mixed sex and all-male broods. Recently-emerged females usually remained inactive near the natal site, mated within a few minutes and then dispersed. Local mating occurred but many females (40 + %) mated with immigrant males and some dispersed before mating and it is unknown whether these mated elsewhere, although other field studies (references in Ikawa *et al.*, 1993) have estimated that almost all females mate.

Van Dijken *et al.* (1989) observed emergence and mating of *Apoanagyrus lopezi*, a quasi-gregarious parasitoid of the cassava mealybug, in fields of cassava. Hosts occur in isolated batches. Males and females emerged from each batch over protracted periods and in random sequence. Experiments using cages of *A. lopezi* placed out in the field showed that males do not attract females or males, males are attracted to virgin females and less attracted to mated females. Males were also attracted to unparasitised hosts. *A. lopezi* probably has a panmictic mating system. Similarly, Powell and King (1984) placed cages containing adult *Microplitis croceipes* (Braconidae), a parasitoid of moth larvae infesting cotton, in cotton crops. Cages containing mated males and mated females were not attractive to free-living *M. croceipes*, but cages containing virgin females attracted males. Fauvergue *et al.* (1999) also used individuals in cages to study attraction between male and female *L. heterotoma* and *L. boulardi* in the field. Cages contained virgin females, mated females, males or no individuals (control). Virgin females were by far the most attractive to free-living conspecifics.

Kazmer and Luck (1991) (see also Antolin, 1999) used electrophoretic techniques (Kazmer, 1990; sections 3.2 and 5.3.3) to determine allozyme variation in populations of *Trichogramma pretiosum* and *Trichogramma* sp., parasitoids of lepidopteran eggs, and to assess the degree of non-sibling mating. *Trichogramma* sp. occurred as a natural population while *T. pretiosum* was collected from an agricultural population (tomato plots). Both species were sampled by rearing wasps from host eggs collected in the field. Although both species develop gregariously in isolated hosts and have female-biased sex ratios, only moderate frequencies (55–64%) of sib-mating were estimated. About one third of the total mating was non-local in *T.* sp. and 2–13% in *T. pretiosum*. Further evidence for male dispersal and non-local mating is that cages of virgin females placed out in the field attracted large numbers of males (cages containing mated females did not, Kazmer and Luck, 1991).

Eggleton (1990) observed the emergence of a female *Lytarnis maculipennis* (Ichneumonidae) (a parasitoid of wood inhabiting insects) in the field. Well before the female bored an exit to the surface of the wood, 4–6 patrolling males gathered to await her emergence. There were aggressive interactions between the males and one, which was larger than the others, was dominant and also appeared to mate with the female while the other males apparently did not.

5.4.5 POST-DISPERSAL MATING: NON-NATAL MATING LOCALITIES

Introduction

The aforementioned studies stress the importance of the natal site as a mating locality for both emergent and immigrant individuals. However, many species normally mate in other localities, such as oviposition sites, feeding sites and even arbitrary sites. It is important to obtain information on these in order to understand fully the mating system.

Mating at Oviposition Sites

Several studies have investigated the possibility that males seek mates at oviposition sites. Some of these are summarised above (van Dijken *et al.*, 1989; Myint and Walter, 1990; Nadel and Luck, 1992; section 5.4.4) and indicate that mating can occur at oviposition sites. However, Hardy and Godfray (1990) caught three species of parasitoids of *Drosophila* larvae at oviposition sites in the field. While in excess of a hundred females of each species were caught during the season, no males of two species and only one male of the third were caught. It is unlikely that oviposition sites are mating sites in these species. Males in at least one of these species do, however, disperse from the natal site and are attracted by virgin females (Fauvergue *et al.*, 1999).

Mating at Feeding Sites

Feeding during the adult stage is common among parasitoids. In many cases hosts suitable for oviposition may serve as food (e.g. Jervis and Kidd, 1986; Heimpel and Rosenheim, 1995; Heimpel and Collier, 1996) and thus feeding and oviposition sites are spatially coincident. However, they may not be spatially coincident and feeding sites may serve as a mating site. Many parasitoid wasp species have been observed to visit flowers, in many cases in order to feed from nectar. Jervis *et al.* (1993) observed 249 parasitoid species visiting and feeding at flowers: 18.5% of these were represented by both males and females and thus some may use flowers as mating sites, as has been observed in nectar-feeding species of *Agathis* (Braconidae) (Belokobylskij and Jervis, 1998). Males of the desert bee-fly *Lordotus miscellus* (Bombyliidae) defend rabbit bush plants (*Chrysothamnus nauseosus*) which the females of this parasitioid visit for feeding (Toft, 1984).

Mating at Arbitrary Sites: Swarming Behaviour ('lekking')

The males of several solitary parasitoid species swarm, e.g. *Blacus* species, *Diachasmimorpha longicaudata* (Braconidae) (Donisthorpe, 1936; Syrjämäki, 1976; van Achterberg, 1977; Sivinski and Webb, 1989), *Diplazon pectoratorius* (Ichneumonidae) (Rotheray 1981), *Aphelopus melaleucus* (Dryinidae) (Jervis, 1979) and some Encyrtidae, Pteromalidae and Eulophidae (Nadel, 1987; Graham, 1993). Pajunen (1990) suggests that swarming could have evolved from territorial behaviour when territorial systems broke down due to high population densities; however, this would not apply to all cases of swarming, e.g. *A. melaleucus* (M.A. Jervis, personal communication).

Nadel (1987) investigated mixed swarms of two species of encyrtid wasp (*Bothriothorax nigripes* and *Copidosoma bakerii*) and a species of pteromalid wasp (*Pachyneuron* sp.). Males were observed either to fly or to crawl over boulders. The swarms appeared at the same sites over a period of several weeks and contained several thousands of individuals of each species. In each species the vast majority were males. Swarms formed in the early morning and broke up during the night. Nadel established that females arrived for mating purposes: she dissected newly arriving females and found their spermathecae to be empty. Courtship and copulation occurred on boulders. Females were receptive upon arrival, were mated soon after, and became unreceptive.

Graham (1993) described the swarming behavior of eulophids (in particular *Chrysocharis gemma*) and torymids (mostly *Torymus* spp.). Swarms comprising many thousands of individuals, the overwhelming majority being female, were observed year after year upon and around the same clump of trees, and they were present for much of the reproductive season (such site fidelity is also reported by Svensson and Petersson, 2000, who studied predatory dance flies (Empididae)). The hosts of the observed species were not present locally. The ovaries of *Chrysocharis* females contained no mature eggs but it was not possible to ascertain whether or not their spermathecae contained sperm (which would indicate whether or not they were virgins) (van den Assem, 1996). The function, if any, of these swarms remains unknown: they may not be mating swarms at all. Syrjämäki (1976) observed swarms of *Blacus ruficornis* and found these never to contain females (>8000 individuals were examined). Similarly, Quimio and Walter (2000) found that swarms of the braconid wasp *Fopius arisanus* occurring in tree canopies were comprised of sexually immature adult males, while mature males were found in loose aggregations in lower-storey vegetation.

Males in a genuine mating swarm may release chemical signals in concert and thus amplify the attractive effects. In mixed species swarms, the signals of the different species may be very similar but interspecific communication is considered unlikely. Most probably, swarms are not formed in response to active long-distance signalling by the participants, rather, certain peculiarities of the site itself attract both sexes when the insects are receptive to mating. Nadel (1987) mentions that the sites she observed were on the only low, yet abrupt, peak within a 7.5 km radius of a mountain ridge. The sites were no more profitable in terms of food or oviposition than the surrounding areas, but they offered a distinct landmark and a certain degree of shelter from strong winds. **'Hilltopping'** behavior is known in many insects (Thornhill and Alcock, 1982). Likewise, many matings in *Habrobracon hebetor*, which attacks stored product pyralid moths, occur after dispersal from the natal site. Both males and females exhibit a prolonged period of unwillingness to mate during which time individuals disperse from the natal site (Ode *et al.*, 1995). Males aggregate on mounds on the surface of the grain and most matings appear to occur on these (Antolin and Strand, 1992). Sometimes, these aggregation sites are coincident with higher concentrations of hosts that will attract females.

As a final point regarding dispersal behaviour we would like to stress that, while several studies have documented the dispersal of males and females from a natal site or even a population, it is equally important to determine the fate of dispersing individuals. The one study that we are aware of concerns dispersal of the mymarid egg parasitoid, *Anagrus delicatus* (Antolin and Strong, 1987). By marking individuals, Antolin and Strong were able to document female-biased dispersal to small islands more than 1 km from the nearest potential source population. The number, sex, and mating status of immigrants founding a new population have all been modelled as important variables in determining whether a new population will become established or go extinct (Hopper and Roush, 1993) (see subsection 7.4.3 in relation to biological control introductions).

5.4.6 COMPARATIVE STUDIES OF LOCAL AND NON-LOCAL MATING

Growing numbers of authors have examined cross-species relationships between sex ratio and one important aspect of mating systems: the degree of **local mate competition** (mating at the natal site). Four studies have compared sex ratio and mating system across two congeneric species (Suzuki and Hiehata, 1985; Kazmer and Luck, 1991; King and Skinner, 1991; B. Stille and E.D. Parker, unpublished). Suzuki and Hiehata (1985) assessed mating system by direct observations, Kazmer and Luck, (1991) used genetic (allozyme) techniques (subsections 5.3.3 and 5.4.4), and King and Skinner used interspecific differences in wing morphology of *Nasonia vitripennis* and *N. giraulti* to infer male dispersal ability. Stille and Parker's study (on gall wasps; despite being herbivores, they share much biology with parasitoids) used both allozyme techniques and morphological differences. Differences in estimated levels of non-local mating and sex ratio were in qualitative agreement with theoretical predictions in three studies, but King and Skinner found differences in the opposite direction (the species in which males are flightless had less biased sex ratios than the species with winged males). The interpretation of all four studies suffers from complications (discussed in the original papers, and by Hardy, 1994a) and from the low statistical power obtained from comparing only two species (i.e. the high likelihood of interspecific correlations between sex ratio and mating structure occurring by chance). However, an advantage of the aforementioned studies is that they were specifically designed to be comparative studies of mating systems. Some larger scale comparative studies (see below) have been forced to use data from many sources, which, more often than not, were not gathered with respect to research on mating systems. Several recent studies have compared more than two species, thus improving statistical power, and have also employed phylogeny-based analyses (subsection 5.3.4).

Drapeau and Werren (1999, see also Drapeau 1999) compared mating behaviour and sex ratio in 17 strains of three species of *Nasonia*: *vitripennis, longicornis* and *giraulti*. They documented that within-host mating occurs in all three species, with the mean percentages of females mating within the host being 1.0%, 9.1% and 64.4% respectively. Higher levels of within-host mating are very likely to be correlated with higher levels of local mate competition, and thus more female biased sex ratios. Drapeau and Werren (1999) examined the sex ratios produced by single foundresses and found that *N. giraulti* had more biased sex ratios than *N. longicornis* and those of *N. vitripennis* were even less biased. More direct examination of the mating system thus accounted for the apparently anomalous result (King and Skinner, 1991, see above) that the species whose males had the longest wings (suggestive of a greater degree of male dispersal capability and thus less intense local mate competition) had the more female-biased sex ratio.

Hardy and Mayhew (1998a) extended Griffiths and Godfray's (1998) comparative study of the relationship between sex ratio and clutch size in bethylids to include more species and further consideration of bethylid mating systems. Based on the general biology of the Bethylidae, which indicates that most offspring develop in single foundress broods and that mating is predominantly local, Griffiths and Godfray (1988) made a working assumption of strict local mating. Using cross-species comparisons, they tested the prediction that average sex ratio will equal the reciprocal of average clutch size (i.e. one male per clutch). Sex ratio and clutch size data were obtained from the published literature. Griffiths and Godfray found qualitative support for the prediction but treated species data as statistically independent (with the justification that sex ratio is likely to be an evolutionarily labile character in bethylids; indeed there is much evidence for within-species adaptive variation). Although phylogenetic constraints upon sex allocation decisions are unlikely, confounding correlations with an unknown variable may have generated spurious significance (Ridley, 1989).

While local mating almost certainly predominates in the Bethylidae, males and unmated females in some species disperse and some non-local mating seems likely (Hardy and Mayhew, 1998a). Hardy and Mayhew reasoned that as bethylid offspring groups are usually produced by single foundresses, the influence of non-local mating on sex ratio should be marked (Nunney and Luck, 1988; Figure 5.3) and potentially detectable in addition to the relationship with clutch size (Figure 5.8). In the absence of direct evidence, sexual dimorphism in body size was used as an estimator of male dispersal ability relative to females (which must disperse from the natal site) as this potentially correlates with non-local mating. Dimorphism estimates were obtained by measuring specimens from field collections, cultures and, mainly, entomological museums. As predicted, sex ratios across species were less female-biased in species with larger males (large relative to expected male size across bethylids; males were usually smaller or similar in size to conspecific females). This result was due mostly to differences between two bethylid subfamilies: Epyrinae have larger males and higher sex ratios, and Bethylinae have smaller males and lower sex ratios. While the species values are consistent with the adaptive hypothesis, differences between the two subfamilies not considered may also account for the cross-species trend. To carry out a phylogeny-based analysis (subsection 5.3.4), an estimate of phylogeny was obtained from the taxonomic literature (Figure 5.8). This was poorly resolved within genera and consequently few independent contrasts were obtained (Figure 5.8). Analysis of contrasts found no significant relationship between sex ratio and dimorphism. Furthermore, there was also no significant relationship with clutch size in the most restrictive analysis, although it was confirmed in a phylogenetic analysis using data from further species. These results are probably attributable to the low statistical power obtained from a poorly resolved phylogeny. The relationships between sexual dimorphism and non-local mating and non-local mating and sex ratio are tantalisingly but inconclusively suggested.

A similar approach was taken in a field study testing the influences of foundress number and non-local mating on the sex ratios of non-pollinating fig wasps (West and Herre, 1998a; see also Hardy and Mayhew, 1998b; Fellowes *et al.*, 1999; Mayhew and Pen, 2002). Many non-pollinating species are parasitoids of pollinating fig-wasps. Unlike pollinator wasps (section 5.4.2), female non-pollinating fig-wasps lay eggs into a number of different figs from the outside. Consequently the number of foundresses contributing offspring to a particular fig is much

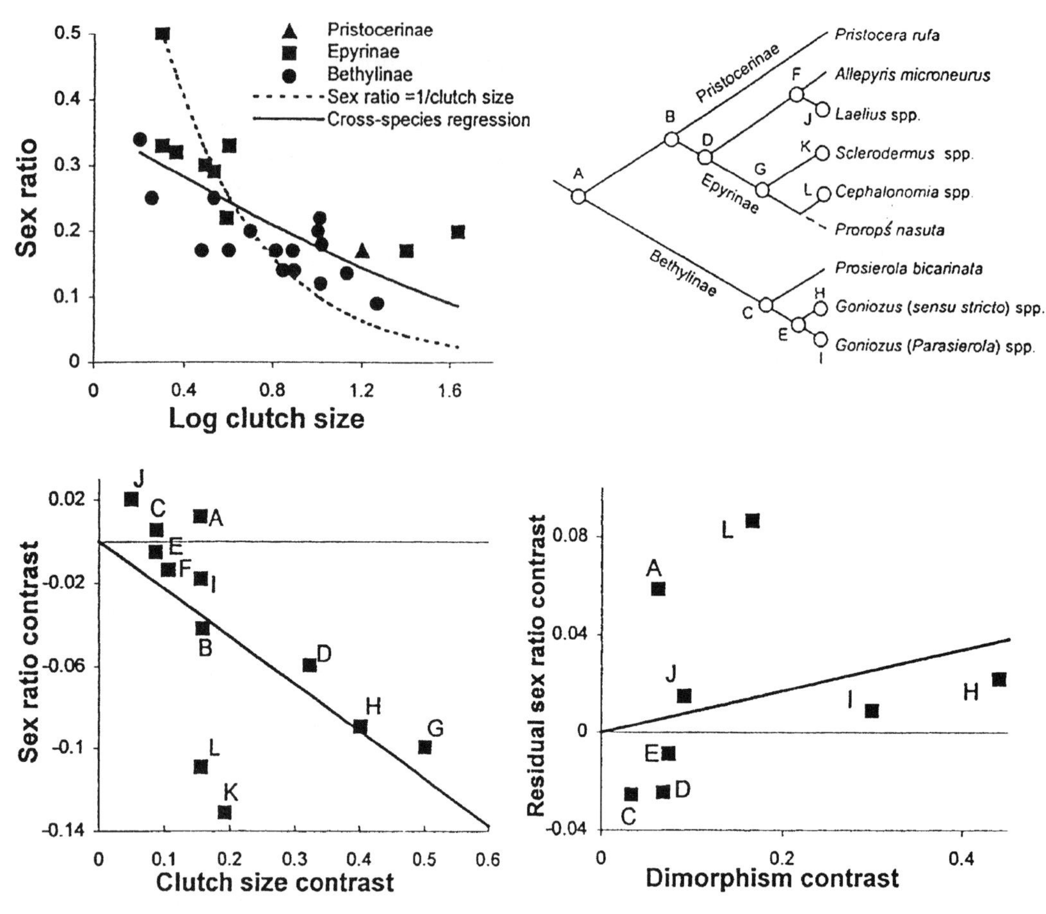

Figure 5.8 A comparative investigation of mating structure and sex ratio in bethylid wasps. Treating species as statistically independent samples (i.e. ignoring phylogeny constraints) found a negative relationship between sex ratio and clutch size (upper left panel) and that degree of sexual dimorphism (used as a proxy for male dispersal ability and thus an indirect estimate of mating structure) was also related to sex ratio, suggesting that non-local mating influences sex ratio optima. Since cross-species analyses may be flawed, the analysis was repeated using the phylogeny-based method of independent contrasts (section 5.4.4) generated using species data in conjunction with an estimate of bethylid phylogeny (upper right panel, the letters at the nodes refer to the contrasts plotted in the lower panels). Because the phylogeny estimate is poorly resolved, the number of independent contrasts for analysis is much reduced compared to the species values data. Phylogeny-based analysis confirmed the relationship between sex ratio and clutch size (lower left panel). The effect of sexual dimorphism was no longer significant, although sex ratio contrasts were generally higher when males were relatively larger (lower right panel). (Source: Hardy and Mayhew, 1998a (in which full analytical details are given). Reproduced by permission of Springer-Verlag.

less apparent than in pollinator species. West and Herre (1998a) predicted foundress number using a model considering three possible distributions (random, aggregated and even) of foundresses foraging for oviposition opportunities among figs. For all three scenarios, the proportion of figs in which a wasp species occurs is positively related to the average number of females laying eggs into each fig. They then used this proportion as an indirect estimate of foundress number in a test of the relationship between foundress number and sex ratio across 17 Panamanian species. Field samples found a positive relationship as predicted and as already

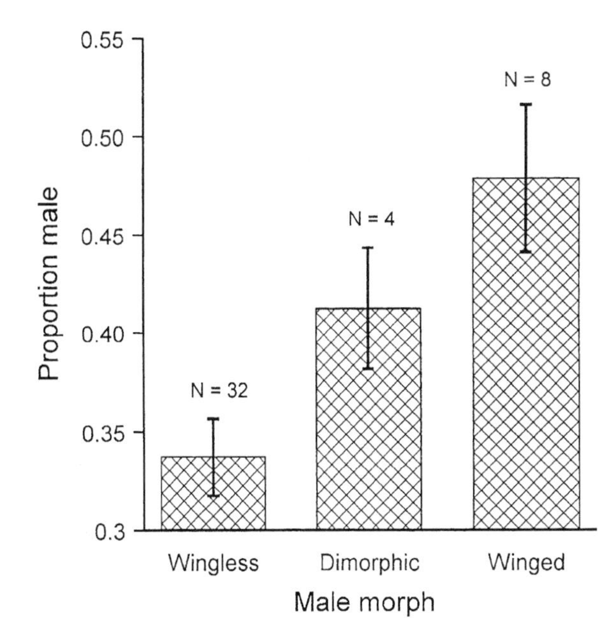

Figure 5.9 Mean sex ratios (± standard error) of 44 species of Old World non-pollinating fig wasps. In 32 species all males are wingless and unable to disperse from the natal fig, so mating is strictly local. In 8 species all males are winged and non-local mating is likely to be the norm. In 4 species both wingless and winged males are found, and a mixture of local and non-local mating is expected. In accordance with theoretical predictions, sex ratios are positively related to the expected degree of non-local mating. (Source: Fellowes *et al.*, 1999). Reproduced by permission of Springer-Verlag.

found among pollinator species (section 5.4.2). However, males in 10 species are wingless (as in pollinators) while in the other 7 males have wings. Both types mate within the fig, but winged males disperse concurrently with the females, presumably to forage for non-local mating opportunities (however, see Greeff and Ferguson (1999) and Bean and Cook, 2001, for recent evidence for dispersal of *wingless* male non-pollinating fig wasps, and Greeff *et al.* 2003 for pollinators). For a given estimated foundress number, the sex ratios of species with winged males were less biased than those of wingless-male species – a result consistent with predictions of theory considering non-local mating (Nunney and Luck, 1988; Figure 5.3).

West and Herre's cross-species analyses were supplemented with phylogeny-based methods (section 5.3.4). The phylogeny of the species is partially known from molecular studies (Herre *et al.*, 1996) but incomplete resolution resulted in only 9 contrasts. The significance of foundress number survived the loss in statistical power but no significant effect of male morphology was shown. As with the Bethylidae (Hardy and Mayhew, 1998a), this difference makes interpreting the data difficult because the cross-species result could be due to other differences between taxa that happen also to differ in male morphology, and reduced power from unresolved phylogenies emerges as a common problem.

In a closely related study, Fellowes *et al.* (1999) examined sex ratios of 44 species of Old World non-pollinating fig wasps, which included 8 species in which all males are winged, 32 in which all males are wingless and also 4 species in which males are dimorphic. The mean sex ratio of wingless-male species was lower than the mean sex ratio of male-dimorphic species, which was in turn lower than that of the winged-male species (Figure 5.9). The result that sex ratio is positively related to 'wingedness' was also found using phylogenetically based analysis despite an incompletely resolved estimate of phylogeny (Fellowes *et al.*, 1999, see also Hardy and Mayhew, 1999).

The aforementioned studies were aimed primarily at understanding relationships between sex ratio and mating system and employed sexual dimorphism as an indirect estimator of non-local mating. Related comparative investigations have focused more directly on sexual dimorphism itself and found relationships between wing dimorphism and the probable mating opportunities of winged and wingless males in fig wasps (Hamilton, 1979; Cook *et al.*, 1997). Female fig wasps have wings. In pollinator species all males are wingless. All males of a given non-pollinator species may have wings or be wingless (as in West and Herre's, 1998a, study, see above). Some non-pollinator species have both types of males (Hamilton, 1979; Cook *et al.*, 1987). With few known exceptions, winged males mate outside the fig and wingless males mate inside (see

Cook *et al.*, 1997; West and Herre, 1998a, for evidence of winged males mating within figs and for evidence of dispersal of wingless males see Greeff and Ferguson, 1999; Greeff *et al.* 2003). Since wings make movement inside the fig awkward and their manufacture requires resources, winglessness is thought to represent an adaptation to within-fig mating. In species with small broods (here 'brood size' describes the number of conspecifics developing in a fig, but these may be the progeny of a number of mothers), most females develop without any conspecific males sharing their natal fig and thus most mating must be non-local. Among these, species with only winged males predominate. Conversely, winged males are rare in species with larger broods: males are usually present in the natal fig and thus within-fig mating is more common. In male dimorphic (parasitic) species, winged males are more common in species with larger proportions of females developing in broods without males (Cook *et al.*, 1997).

Hamilton (1979) developed a model to explore more formally the observed relationship between male dimorphism and mating opportunities within each species. Assuming that each member of a brood is the progeny of a different mother and that females mate with either wingless males (before dispersal) or winged males (after dispersal), the model predicts that if the male morphs (alternative tactics, section 5.3.1) have equal fitness the proportion of winged males should be equal to the proportion of females which develop in figs without wingless males. As the assumption that each offspring in a fig is laid by a different mother is likely to be incorrect in the vast majority of cases, Greeff (1995, see also Greeff, 1998) extended the model to incorporate multiple-egg oviposition. When a female lays several eggs into a fig, her wingless sons may compete with each other for mating opportunities (Hamilton, 1967), while winged sons do not. Thus, individual wingless sons generate less fitness for mothers than do winged sons and Greeff (1995) consequently predicted that the proportion of winged males may exceed the proportion of females developing in broods without wingless males. In eight out of nine male dimorphic species examined by Cook *et al.*, (1997) the proportion of winged males equalled or exceeded the proportion of females developing in figs without wingless males. It is thus likely that the ratios of winged and wingless male morphs in these species have indeed been selected by the relative mating opportunities of the different morphs.

Finally, Eggleton (1991) used a morphometric analysis to predict the type of mating system found in the Rhyssini, a monophyletic tribe of ichneumonid wasps. He measured the shape of the male gaster of specimens from 9 species of Rhyssini that have been shown to exhibit one of three mating systems: males aggregate around emergence sites of females, males guard female emergence sites, and males gather around a recently emerged female. Males mate with females before female emergence in the first two systems. Eggleton found that the male gaster showed an allometric change in shape with an increase in body size in species exhibiting a mating system where males competed for mates before females emerged; no correlation was found with the other two mating systems. Eggleton argued that morphometric analyses can be used to predict the mating systems of species where no direct observations are currently available. How applicable this technique is to other taxa remains to be seen.

5.4.7 FEMALE RECEPTIVITY AND MALE ABILITY

Introduction

Females of haplodiploid species do not have to mate at all in order to reproduce, but unmated females can only produce male offspring and this may be a disadvantage. Selection may also favour females that mate multiply. Multiple mating will almost certainly be favoured in males, but males may be constrained by limited mating ability. In this section we examine these issues in relation to mating systems.

Constrained Oviposition

Virgin, as well as sperm-depleted, females are constrained to produce only sons. Whether constrained oviposition is advantageous, neutral or disadvantageous is predicted to depend on both the mating system and the sex ratios produced by other females (Godfray, 1990; see also Hardy and Godfray, 1990; Godfray and Hardy, 1993; Godfray, 1994; Heimpel, 1994; Godfray and Cook, 1997).

In panmictic populations with unbiased sex ratios, male and female offspring are of equal value and fitness costs to constrained reproduction are thus expected to be absent. Virgin reproduction might even be advantageous if there are energetic, time or mortality costs associated with mating, or with the production of female offspring (Nishimura, 1997; see also Chevrier and Bressac 2002). However, if constrained oviposition becomes common, the population sex ratio may become male-biased and male offspring will have less fitness than daughters due to frequency-dependent selection (Fisher, 1930). This may select for increased mating by eclosing females and/or for mated females to produce a higher proportion of daughters among their progeny (Godfray, 1990). Further, females might monitor the time between eclosion and mating and use this as an estimate of the level of virgin reproduction in the population. Assuming that mating status does not affect oviposition rate, an assumption that has been met in at least some species (e.g. Ode *et al.*, 1997), virgin reproduction becomes more common as the time between mating and eclosion increases and females, once mated, are predicted to produce more female-biased progeny sex ratios.

The most thorough examination of the role of constrained oviposition in parasitoid mating systems has been a series of laboratory and field studies of the stored products natural enemy, *Habrobracon hebetor* (Guertin *et al.*, 1996; Ode *et al.*, 1997). As noted earlier, individual *H. hebetor* females produce female-biased sex ratios, yet several mechanisms reduce the likelihood of inbreeding (Ode *et al.* 1995). Field studies, following two populations over three years, were conducted to examine whether the proportion of constrained (virgin or sperm depleted) females was consistently high enough to select for female-biased sex allocation and if mated (unconstrained) females were able to track fluctuations in the proportion of constrained females in the population (Ode *et al.*, 1997). Males and females were counted weekly at eight locations at each field site. Females were brought into the laboratory where they were given several hosts each day until they died. Examination of the progeny allowed determination of whether females had sperm at the time of collection as well as what sex allocation decisions they were making. At any given time, 20–30% of the females collected from the field were found to be constrained (either sperm depleted or virgin). Guertin *et al.* (1996) suggest that newly emerged females may experience a trade-off between remaining on the surface of the grain and finding a mate or going beneath the grain surface in search of hosts. This may explain the consistently high levels of virginity found in field populations. For each site and sample date, the observed proportion of constrained females was used to calculate a predicted sex ratio value (see Godfray, 1990) that was compared to the observed sex ratio decisions made by mated females from the collection date/site. While the proportion of constrained females appeared to explain much of the observed female-bias in individual sex allocation decisions, mated females did appear to be able to track fluctuations in the proportion of constrained females over the course of the season (Ode *et al.*, 1997). Likewise, mated females held overnight with either virgin or mated females produced similar sex ratios suggesting that females were not able to track levels of constrained oviposition and instead may have been responding to an evolutionary average.

The relationship between the time between eclosion and mating and the sex ratio has been tested in a laboratory study using *Aphelinus asychis* (Hymenoptera: Aphelinidae), a solitary parasitoid of aphids, in which mating is probably panmictic and the age of females at mating varies (Fauvergue *et al.*, 1995; 1998b). Virgin females were presented with mating opportunities at 1, 8 or 15 days after emergence, or were denied mating opportunities altogether (laboratory longevity is about 38 days). All females were provided with 100 hosts each day until death, and their progeny reared out and sexed. In accordance with the assumption that mating status does not affect oviposition rate, age at mating did not affect fecundity. As predicted, post-mating progeny sex ratio decreased overall, with increasing age at mating (Fauvergue *et al.*, 1998b; Figure 5.10). Similar patterns are reported in other parasitoids (Hoelscher and Vinson, 1971; Rotary and Gerling, 1973) but the interpretation of these results is hampered by uncontrolled variables (discussed in Fauvergue *et al.*, 1998b).

When populations experience high levels of local mate competition (LMC) (Hamilton, 1967, section 5.4.2) optimal progeny sex ratios are female-biased and virginity is thus expected to be extremely disadvantageous to individual females and mated females are predicted to make no, or very minor, adjustments to their progeny sex ratios in response to virgin reproduction in the population (Godfray, 1990). Godfray and Hardy (1993; see also Hardy and Godfray, 1990) explored the prevalence of virginity in relation to mating system across 24 parasitoid wasp species, using gregarious development as an estimator of local mating (gregarious species are generally expected to experience LMC while solitary species generally will not). Contrary to expectation, virginity was more prevalent among gregarious species (these comparisons were not phylogeny-based, subsection 5.3.4; and there are an increasing number of examples of gregarious species that do not experience LMC). However, it was implicitly assumed that developmental mortality was absent. Under LMC developmental mortality can lead to considerable levels of virginity in gregarious species, because if no males survive to maturity the resulting brood of females will be virgins. Extra 'insurance' males can be produced, but at the cost of daughters (Green *et al.*, 1982; Heimpel, 1994; Nagelkerke and Hardy, 1994; West *et al.*, 1997; Hardy *et al.*, 1998). Under optimal sex allocation, high levels of virginity are expected when clutch sizes are small and developmental mortality is common (Heimpel, 1994; West *et al.*, 1997). These predictions are generally supported by tests within gregarious parasitoids with a range of clutch sizes and developmental mortality rates (Morgan and Cook, 1994; Hardy and Cook, 1995; Hardy *et al.*, 1988, 1999), but comparative analyses of the correlation between mortality and virginity using the mean values from these species gives equivocal results (dependent upon whether cross-species or phylogenetically-based methods are used, section 5.3.4) (Hardy *et al.*, 1998). Virginity levels across 53 species of fig wasps are negatively correlated with brood size, using both cross-species and phylogeny-based comparative methods (West *et al.*, 1997).

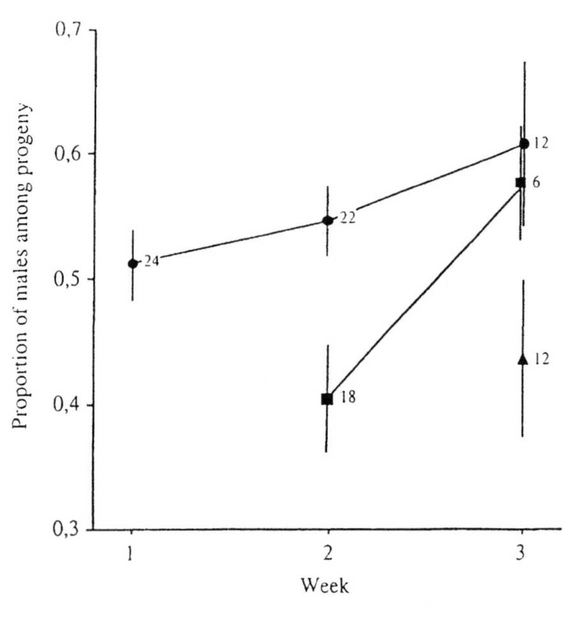

Figure 5.10 Progeny sex ratio (mean proportion males ± standard error, with sample sizes) as a function of time from emergence (age in weeks) of female *Aphelinus asychis* mated at one day old (filled circles), at 8 days (squares) and at 15 days (triangles). Generally, mated females produce generally lower sex ratios when mated at older age. (Source: Fauvergue *et al.*, 1998b). Reproduced by permission of Birkhäuser Verlag AG.

Female Sperm Depletion and Polyandry

Mated females that have used up sperm stored in their spermathecae are termed **'sperm-depleted'**.

Like virgin females, sperm-depleted females are unable to fertilise their eggs and are thus constrained to produce only male offspring. Depending on the mating system, this may or may not be disadvantageous (Godfray, 1990; section 5.4.5; see also Cook and Crozier, 1995, for the influence of sex determination mechanisms on selection for polyandry). When constrained oviposition is disadvantageous, sperm-depleted females are likely to re-mate (Leatemia *et al.*, 1995), and females with stored sperm may also mate again to reduce the probability of future sperm depletion (Chevrier and Bressac, 2002). However, there are exceptions to this pattern. In a study of field-collected females, nearly 75% of mated *Habrobracon hebetor* females became sperm-depleted at some point before the end of their reproductive lifespan as evidenced by the production of only sons (Ode *et al.*, 1997). Fewer than 5% of these females were willing to remate after sperm depletion. Therefore, once sperm depleted, females are likely to remain constrained to produce only sons for the rest of their lives.

A phylogeny-based comparative study, using data from 97 species of parasitoid wasps, found a significant association between polyandry (mating with multiple males) and gregarious development (Ridley, 1993b). Ridley suggested that females of gregarious species are polyandrous to increase within-brood genetic variation among their progeny which, he argued, would decrease resource competition between siblings feeding on the same host (e.g. by resource partitioning). However, genetic diversity is also expected to *promote* competition as brood members have divergent evolutionary interests (Hardy, 1994b). An alternative explanation for Ridley's result is that females are polyandrous because the extra sperm obtained reduce the risk of subsequent sperm depletion and that sperm depletion is a greater disadvantage for gregarious species than solitary species since gregarious species often inbreed while solitary mainly outbreed (Godfray, 1994; Hardy, 1994b). Possible sperm depletion appears to select for polyandry in at least one parasitoid species (Chevrier and Bressac, 2002; see Fjerdingstad and Boomsma, 1998, for an ant example). Furthermore, due to inbreeding and local mate competition, sex ratio optima of gregarious species will often be female-biased, and thus require more sperm (a greater proportion of eggs must be fertilised) than solitary species with similar fecundity. Males of gregarious species may mate many times in rapid succession and themselves become sperm-depleted (section 5.4.5), a further reason for gregarious polyandry if Godfray's explanation applies. Most parasitoid biologists would probably favour Godfray's explanation over Ridley's, since hosts are often entirely consumed by gregarious broods and significant resource partitioning seems unlikely. Further, polyandrous females would expected to seek genetically diverse mates if Ridley's explanation were correct, and this may influence the mating systems of gregarious species by encouraging outbreeding (further discussed by Hardy, 1994b).

One way of distinguishing between the alternative explanations would be to re-categorise the solitary species as 'truly solitary' (developing singly in hosts, hosts isolated) or 'quasi-gregarious' (developing singly in hosts, hosts in clumps). Interaction of developing siblings cannot occur in quasi-gregarious species (classified as solitary by Ridley) but in their mating systems they are likely to be similar to gregarious species: Godfray's hypothesis predicts polyandry while Ridley's hypothesis predicts monandry (Hardy, 1994b). Since only 2 of the 68 'solitary' species in Ridley's study are reported as polyandrous, if many of these 68 species were quasi-gregarious, the support would be for Ridley's hypothesis. Ridley's (1993b) study was hampered by the generally poor knowledge of parasitoid mating behaviour and any extensions of his work are likely to face similar problems. This again illustrates that, despite the vast knowledge of parasitoid biology in general, there is a great need for basic data on mating systems and mating behaviour.

Godfray's hypothesis could only explain the incidence of polyandry if female parasitoids do indeed become sperm-depleted. There is evidence that females can be very economical with their stored sperm, with usually probably one sperm being released per fertilised egg in *Habrobracon hebetor* (as *H. juglandis*) (Braconidae), a parasitoid of pyralid moths and *Dahl bominus fuscipennis* (Eulophidae), a parasitoid of sawflies (Speicher, 1936; Wilkes 1965). However, sperm depletion of females is reported from many laboratory studies (e.g. Gordh *et al.*, 1983; Hardy and Cook, 1995; Luft, 1996; Fauvergue *et al.*, 1998b; Pérez-Lachaud and Hardy, 1999). For example, Pérez-Lachaud and Hardy (1999) presented female *Cephalonomia hyalinipennis* (Bethylidae), a parasitoid of the Coffee Berry Borer, *Hypothenemus hampei* (Coleoptera: Scolytidae), with batches of hosts at regular intervals until the female died: 77% of females ran out of sperm before the end of their reproductive lives (females produced broods containing only adult males, and no further daughters). The probability of sperm depletion was positively correlated with the number of daughters produced (Figure 5.11). An estimate of the initial number of sperm stored can be obtained by assuming that only one sperm is used per fertilisation, and dividing the number of daughters produced prior to sperm depletion by the probability of egg to adult survival (Pérez-Lachaud and Hardy, 1999).

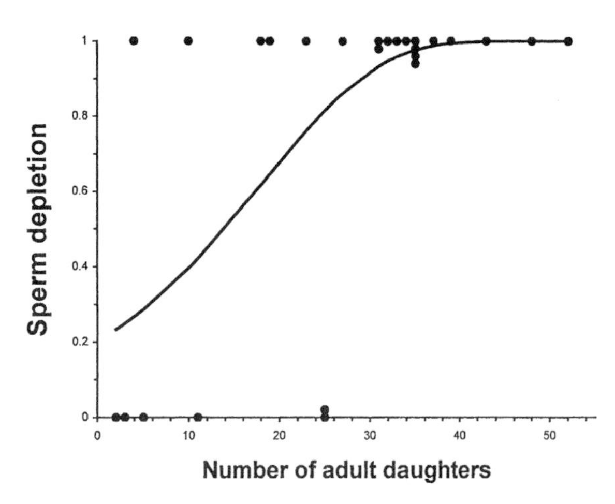

Figure 5.11 Sperm depletion of female *Cephalonomia hyalinipennis* in relation to the number of adult daughters produced (hosts were provided in batches of 10 hosts [4 prepupae plus 6 pupae] every 3–4 days). Data are binary (0 = female died with no evidence of sperm depletion, 1 = female was sperm depleted before death [only males emerged from a batch of hosts, and no further daughters were produced]). Where multiple data points overlap, some are displaced from their binary positions to show sample sizes. The curve illustrates the probability of becoming sperm-depleted, estimated using logistic regression. (Source: Pérez-Lachaud and Hardy, 1999). Reproduced by permission of Elsevier Science.

While laboratory evidence is useful, it must be borne in mind that under laboratory conditions females may be presented with an unnaturally large number of reproductive opportunities and sperm depletion may thus be overestimated and estimates may also be affected by the re-mating opportunities presented in the laboratory. Laboratory studies will generally be more useful for assessing the numbers of sperm that can be stored than for estimating the natural occurrence of sperm depleted females. The one example of estimations of sperm depletion in the field comes from *Habrobracon hebetor* (Ode *et al.*, 1997, see above for discussion).

Male Sperm Depletion and Polygyny

While natural selection may favour female virginity, monandry or polyandry, males will virtually always be selected to mate with many females (polygyny). A literature survey of parasitoid mating behaviour by Gordh and DeBach (1978) found that in all examined species (n = 48) males were polygynous (only 5/34 species of females were polyandrous). There have been several studies showing that body size has a positive influence on the number of females a male can successfully mate (e.g. Jones, 1982; King, 1987; van den Assem *et al.*, 1989; Godfray, 1994; Heinz, 1991; Ode and Strand, 1995). Patterns of spermatogenesis in male parasitoids are relatively poorly known: current evidence suggests that in some species spermatogenesis ceases before adult eclosion (e.g. Gerling and Legner, 1968), while in others it may continue during adult life (e.g. Wilkes, 1963; Laing and Caltagirone, 1969; Gerling and Rotary, 1974; Gordh and DeBach, 1976; Nadel and Luck, 1985) and, unusually, there may be a premating period during which adult males do not have sperm ready to ejaculate (Quimio and

Walter, 2000). [In social Hymenoptera, the testes normally degenerate upon male migration, the lifetime supply of sperm being stored in the seminal vesicles, but a few species have wingless males that continue spermatogenesis throughout life, and do not disperse and compete aggressively with other wingless males for within-nest mating opportunities, Heinze *et al.*, 1998.] Male parasitoids that mate many times, or many times in rapid succession may run out of sperm, either temporarily or permanently, and females receive no sperm or a reduced quantity (Schlinger and Hall, 1960; Wilkes, 1965; Laing and Caltagirone, 1969; Gordh and DeBach, 1976; Nadel and Luck, 1985; van den Assem, 1986; Dijkstra, 1986; Ramadan *et al.*, 1991; Ode *et al.*, 1996).

Nadel and Luck (1985) assessed the insemination capacity of male *Pachycrepoideus vindemmiae* (Hymenoptera: Pteromalidae), a quasi-gregarious pupal parasitoid of *Drosophila* and other flies. One-day old males were presented with a succession of ten 18-hour old virgin females (each female singly). Females were replaced with virgins immediately after mating. It took about one hour for a male to mate with all ten females. Mated females were either offered hosts and the sexes of their progeny recorded, or dissected and the contents of their spermathecae examined. The amount of sperm transferred on mating decreased as the number of females mated increased. In a similar experiment, males were presented with a rapid succession of mating opportunities, allowed to rest for 24 hours and then presented with further virgin females. A full complement of sperm was transferred to the latter female, showing that males replenish sperm supplies. The duration of the rest period was then varied: sperm replenishment following one mating was found to require about 30 minutes. Time between matings is thus an important influence on male insemination capacity. Nadel and Luck (1985) investigated the timing of adult emergence from batches of hosts (section 5.4.2). Males usually emerged one day before females. Females tended to emerge in the mornings and the mean interval between female emergence during daytime was 125 minutes. In 24% of cases, females emerged less than 30 minutes apart, but the majority of intervals were >30 minutes. Thus, in nature, males are probably able to fully inseminate the majority of females that emerge in their natal patch.

Similar methods have been used to examine sperm replenishment patterns in *Habrobracon lineatellae* (Laing and Caltagirone, 1969) and *H. hebetor* (Ode *et al.*, 1996). Ode *et al.* (1996) presented males either a high encounter rate of virgin females (until males refused to mate with an additional female; about twenty females per day) or a low encounter rate (one female per day). Body size of the male was measured (all females were of a standard size) and all mated females were given four hosts per day until death. By rearing the progeny, the timing of sperm depletion and replenishment patterns were determined as well as the number of daughters each male had sired. Furthermore, Ode *et al.* (1996) were able to show that male body size had an effect on sperm depletion rates. When mating opportunities were plentiful (10 matings per day), large males were able to deliver sperm to more females than small males both because they were able to mate with more females and were able to deliver more sperm to the females that they did mate. When mating opportunities were rare (1 per day), male size had no effect on the rate of sperm depletion.

Rather than presenting males with virgin females individually (Nadel and Luck, 1985; Ode *et al.*, 1996), females can be presented simultaneously, as was done by Dijkstra (1986) to assess the insemination capacity of *Colpoclypeus florus*. Individual males, of known weight, were presented with five females simultaneously for four hours, then 5 further virgin females were added and after a further four hours five more females were added. Males were in contact with females for 45 hours. The number, timing and duration, of female presentation was chosen on the basis of the lowest ratio of males to females normally observed and the timing of emergence and intersexual contact. In a further experiment, males were presented with females for 21 hours. Insemination was assessed by examining the contents of the females' spermathecae. The proportion of females inseminated did not depend on whether males were in contact with females for 21 or 45 hours, but was positively correlated with the weight of the males (Figure 5.5). Males were offered a further batch of five virgin females after the main experiment. The spermathecae of some of these females did not contain a full complement of sperm if the male was very small or had just mated many times. Dijkstra concluded that, under normal circumstances, a single male is sufficient to inseminate fifteen or more females. Limited male insemination capacity is unlikely to account for the observation that an increasing number of male eggs is laid in larger clutches; the high developmental mortality in *C. florus* is a more likely explanation (Hardy *et al.*, 1998).

Ramadan *et al.* (1991) used two methods to study male mating ability in four species of Opiinae (Braconidae), solitary parasitoids of tephritid fruit-fly larvae. In the first method, which was similar to Nadel and Luck's (1985) and Dijkstra's (1986), individual males were presented with ten virgin females for 24 hours and then the females were dissected for spermathecal examination. Replicates were carried out using males of different ages and sizes. Male ability peaked at around four days and in two species larger males were able to inseminate more females than smaller males. In the second method, 200 males and 200 females were placed together in cages (the studied species are solitary and may mate in swarms as well as immediately after eclosion). Every 24 hours, fifteen females from each cage were dissected (fifteen males were also removed and discarded to maintain an even sex ratio). The experiment was stopped when all fifteen females in a sample were inseminated. Most females mated during the first 24 hours (but some did not mate until more than 7 days old) and were observed to be polyandrous.

Some researchers have thus assessed male capacity at maximum rates of presentation of virgin females and then related this to the natural timing of female emergence (Nadel and Luck, 1985) while others first assessed emergence and then presented virgin females so as to mimic these patterns (Dijkstra, 1986). Note that these approaches are not applicable to species such as *Habrobracon hebetor* (Ode *et al.* 1996), where mating occurs at sites other than the site of emergence. An alternative, though closely related, approach was used by Hardy *et al.* (1999, 2000) and Loch and Walter (2002): eclosion and dispersal in broods of *Goniozus nephantidis* was monitored and dispersed females were provided with hosts to assess whether they were mated (section 5.4.2). Brood size varied naturally and many of the broods contained only one male. Virtually all females were inseminated (Figure 5.6). This method thus estimates the capacity of males to inseminate females as individuals mature and disperse naturally. No decisions about the timing and duration of presentation, or the number and age of individuals used, need be made by the investigators. Such decisions may influence the results if, for example, females have short periods of receptivity (section 4.3.6) or if male mating ability is age-dependent during the first few days of adult life (Hirose *et al.*, 1998; Ramadan *et al.*, 1991). However, the efficiency of this method is reduced by natural variation that may make some replicates less valuable (e.g. broods from which two males eclose) and important factors such as brood size (number of virgin females per male) are not under direct experimental control. A similar approach was taken by West *et al.* (1998) who examined insemination capacity in five species of fig wasp. Females that had emerged naturally in figs containing only one male were dissected. The maximum number of inseminated females in a fig was much greater than the mean number of females per fig. Males were also dissected to check whether their seminal vesicles contained sperm: none had empty vesicles. It is thus unlikely that males of these species become sperm-depleted.

5.5 PREDATOR MATING SYSTEMS

5.5.1 INTRODUCTION

The study of non-parasitoid insect mating systems has largely adopted the non-genetic, individual-based approach to defining and understanding an insect's mating system (see section 5.3.1). As outlined earlier, there are four basic types of mating system: monogamy, polygyny, polyandry and polygamy (see Alcock, 1998). Monogamy is a relatively rare mating system requiring special ecological constraints. One of the best examples comes from non-predators: there are termite species in which winged sexuals establish a new colony after their nuptial flight. The colony is founded by a single male and female, both of whom dig a new nest after removing their wings. The opportunity for further nuptial flights is therefore lost, and monogamy ensues in these primary reproductives: the male and female invest huge amounts of resource into their gonads and soon become immobile, reinforcing the reliance on their subterranean partner (see Wilson, 1974).

Polyandry is also an unusual phenomenon in insects, and again the best examples come from social insects (e.g. Wilson, 1971). Newly emerged honeybee queens restrict their nuptial flights to a few days after emergence when they rendezvous with the drones from other colonies in mating swarms that aggregate high above landmarks. Females need to mate several times in order to receive enough sperm to generate a long-lived, viable colony. The combination of queen rarity at mating swarms (i.e. the OSR is highly male-biased) and the need for multiple mating has led to the evolution of the peculiar copulatory mechanism in this species. Drones autotomise their genitalia once they have inserted their intromittent organ and released semen into the queen. This means that drones only get one opportunity at mating (they die soon afterwards) whilst females can mate with several males during their relatively short visits to mating swarms (Wilson, 1971). Since males only mate once, and females mate multiply, this is a polyandrous mating system.

Perhaps the best example of insect polygyny occurs in a predator, a protandrous eumenid wasp whose males defend clusters of female brood (Smith and Alcock, 1980). Males defend their cluster vigorously, and often violently, because it offers them the opportunity to mate with several females without incurring large search costs. This type of polygynous mating system (termed "female defence polygyny") is associated with female monogamy, the ability of females to mate soon after emergence and predictable clustering of receptive females.

Polygamy is, by and large, the rule in insect mating systems (see Thornhill and Alcock, 1982) with males and females mating multiply throughout their lives (and thereby providing the preconditions for sperm competition (section 4.5.2)). There is an ever-expanding list of variants on the polygamous theme. For example, when fertilisable females are plentiful and predictable in space and time, and are defendable, we usually observe a mating system where males defend a resource that is required by females and exchange access to it for copulations. A good example of this type of mating system comes from the predatory yellow dung-fly (*Scathophaga stercoraria*) where males defend spatially restricted oviposition sites (fresh dung) which females require in order to lay eggs. Males fight for areas of the dung pat that maximise their opportunity to encounter sexually receptive, gravid females as they arrive to oviposit (Figure 5.12). This polygamous mating system is termed **resource defence polygamy** and is relatively common (see below). Demonstrating resource defence polygamy first requires careful observations of the mating system, which should include measures of male size and mating success, as well as measures of those aspects of the putative resource that the observer feels is related to the mating success of the resource-holding male. In the majority of cases in insects-defended resources are usually a spatially and/or temporally restricted food and/or oviposition resource, so measures of size and age are usually relevant. Once correlations have been established between resource variables and mating success it is a relatively straightforward matter to manipulate the key resource variable and predict mating outcomes. This general approach has been used to identify quite obscure resource variables that play a central role in determining mating outcomes (e.g. Gibbons and Pain, 1992; Siva-Jothy *et al.*, 1995).

Figure 5.12 A pair of yellow dung-flies (*Scathophaga stercoraria*) copulate on a fresh dung pat. Males which are able to displace rivals from the areas around the fresh dungpat that maximise the probability of encountering a receptive, gravid female achieve relatively high reproductive success. By mounting, and hence guarding, females when they arrive, and copulating directly before these guarded females oviposit onto the dung, these males ensure they fertilise most of the eggs the female lays. The dung-fly mating system can be broadly described as "resource defence polygamy". Photo: M.Siva-Jothy.

The non-genetic, individual-based approach adopted by students of non-parasitic insect mating systems considers three major selective forces that underpin mating system evolution. First, and perhaps ultimately, ecological factors impose constraints upon the availability of receptive females. Second, the degree to which males compete for access to females as a consequence of female availability and third, the extent to which female behaviour (choosyness) influences male reproductive success. The first of these forces is a sufficiently intuitive, large, and well-reviewed area that we will not consider it further here (see Thornhill and Alcock, 1982, for review). We will, however, briefly consider the evolutionary forces that result in male and female traits (and the conflict that sometimes arises between males and females) that subsequently refine an insect species' mating system.

5.5.2 COMPETITION FOR FEMALES

The most overt manifestation of Bateman's principle (see section 5.2) is the enormous range of male sexual traits devoted to securing matings and ensuring fertilisation. In the first instance this phenomenal array of male traits is manifest as "attractiveness" traits such as acoustic (e.g. Bennet-Clarke, 1970), visual (e.g. Kimsey, 1980; Lloyd, 1966) and/or chemical signals (e.g. Breed *et al.*, 1980). However, when ecological factors dictate, males supplement, or forego, signalling by actively searching out females (e.g. Smith and Alcock, 1980). In more extreme instances, males are also selected to wait in the vicinity of resources that are themselves attractive to females. In general these resources tend to be feeding sites (e.g. Severinghaus *et al.*, 1981), oviposition sites (e.g. Siva-Jothy *et al.*, 1995), emergence sites (e.g. Gilbert, 1976) or conspicuous aspects of the habitat (landmarks) (e.g. Downes, 1970). As the OSR becomes more male-biased at these sites, selection will favour the evolution of agonistic behaviour in males directed at defending the resource that attracts females and/or the females themselves (Figure 5.12). Males that best defend the resource will be favoured since they will have a higher encounter rate with receptive females: the stage is now set for the evolution of additional morphological and behavioural traits that enhance a male's fighting ability. However, if the OSR is too male-biased the pay-offs of defence may be too low, or even negative (e.g. Alcock *et al.*, 1978). At such male-biased OSRs a mating system known as "**scramble competition**" generally evolves: males are selected to get to a receptive female as quickly as possible and not waste time or energy in fighting (e.g. Smith and Alcock, 1980). The ecological idiosyncrasies of particular species can also favour the evolution of alternative male mating behaviours within a mating system. Such mating systems support two (sometimes more) male tactics for securing mates (e.g. Johnson, 1982). These **alternative tactics** can be underpinned by different alleles; they can have a genetic basis which triggers a shift from one tactic to the other as conditions dictate; or they can be condition-dependent, in which case an individual is not obliged to show both tactics, but uses the one most appropriate for its condition (for a review see Cade, 1980). Variation within a mating system and the evolutionary conditions underpinning it, have been particularly attractive to game theory modelling (e.g. Dawkins, 1980).

Competition between males does not stop once a mate is secured. Most insect mating systems generate sperm competition (section 4.5.2), a selective force which favours the evolution of two suites of antagonistic male traits (Parker, 1970a): one set reduces the likelihood of the female remating (see Figure 5.2) whilst the other increases the competitiveness of a male's ejaculate. Both of these traits operate to reduce levels of sperm competition, and can have direct effects on the nature of mating systems. For example, if males are selected to reduce the chances of female remating to reduce the likelihood of future sperm competition, then the OSR will become more male-biased, and the foundation may be laid for an evolutionary arms race with females (since females may benefit from remating). Much of the variation in mating systems is generated by selection on males to enhance their ability to secure matings and fertilisations but it is clear from this last scenario that females also play a role in determining the characteristics mating systems. In recent years there has been a noticeable shift towards understanding female rôles in shaping mating systems, particularly in insects (Rowe *et al.*, 1994).

5.5.3 CHOOSY FEMALES

In any mating system where females gain a net benefit from being choosy (see Waage, 1996) about whom they mate with, males will be subjected to additional selection pressures. For example, females might be choosy because they benefit from gaining access to better nutritional (e.g. Thornhill, 1980a) or ovipositional (e.g. Siva-Jothy *et al.*, 1995) resources or, more controversially, because they obtain "good" male genes (see Andersson, 1994). These genes may be "good" because they have utility via parasite resistance (e.g. Zuk, 1988; Siva-Jothy and Skarstein, 1998) or developmental stability (manifested as "fluctuating asymmetry" e.g. Thornhill, 1992; Møller and Swaddle, 1997). Alternatively "good genes" may provide a benefit to the choosy female via the attractiveness they confer on her sons (e.g. Jones *et al.*, 1998). The challenge for evolutionary ecologists is to disentangle the relative importance of direct (i.e. resource-based) and indirect (i.e. gene-based) benefits to females. A good example of the pivotal role of female "choice" in determining the nature of a mating system comes from studies of scorpion flies (Mecoptera) (e.g. Thornhill, 1980a). Male fertilisation success is determined by the duration of copulation, which in turn is determined by the size of the prey item a male presents to a female (see Figure 5.13). Males that present larger nuptial gifts secure longer copulations, transfer more sperm, and also transfer receptivity-reducing compounds that delay female remating (Thornhill, 1980b). Female-based behaviour (nuptial feeding) results in selection favouring males that can secure and defend the largest nuptial gifts. Further to this fairly overt form of female-driven preference, Eberhard (1996) proposed that females may choose males cryptically by biasing fertilisation events (i.e. the outcome of sperm competition, section 5.2) in a particular male's favour. As well as influencing mating system parameters, cryptic female choice has been suggested as operating on the ability of a male to stimulate the female during courtship, and that this is a driving force for speciation via the rapid and divergent evolution of male genitalia (Eberhard, 1985).

The selective forces generated by sperm competition produce a number of male traits that

Figure 5.13 A male panorpid scorpion fly copulates with a female whilst she feeds on the cluster of dead insects he has captured, enveloped and suspended from a leaf. Males attract females by utilising pheromones and Thornhill (1980b) has shown that males with larger gifts are able to obtain higher levels of reproductive success because females prolong copulation in order to continue feeding. Photo: M. Siva-Jothy.

do not always function in the female's interest. A recent study has provided strong support for the idea that females may control some of the variation in sperm precedence patterns (Wilson *et al.*, 1997). By examining the variation in sperm precedence patterns in the bean weevil, *Callosobruchus maculatus*, using identical sets of paired matings between and within female sib-groups, Wilson *et al.* showed that some of the variance in sperm precedence (i.e. male fertilisation success) was explained by female genotype. Females therefore appear to be wresting control of fertility away from males. But male *C. maculatus* have their own agenda: work by Crudgington and Siva-Jothy (2000) has revealed that males deliberately damage the female's genital tract during mating. This male trait is believed to increase female reliance on her last mate's sperm by inducing reluctance to remate and increasing the remating interval. This is a particularly good example of sexual conflict since not only is the male trait easy to document and examine, but also the female shows a behavioural response (kicking), the experimental elimination of which results in increased costs for females. Examination of anatomical traits that may be involved in determining fitness consequences of mating have been dealt with in Chapter 4 (subsection 4.5.2). Another spectacular example of a reproductive conflict comes from studies of the ejaculate of *Drosophila melanogaster*. Male products transferred to the female in the ejaculate incapacitate rival sperm in storage in the female, and so have clear benefit to the copulating male (Harshman and Prout, 1994). However, they also reduce the longevity of the female (Chapman *et al.*, 1995). These substances are therefore the basis of a reproductive conflict of interest between male and female *D. melanogaster* that has led to an evolutionary arms race as intense as in any host-parasite system (Rice, 1996). Sexual conflict is a rapidly developing field of interest and, at present, there are few general rules about how to examine the phenomenon. The first step is always to identify a system that is likely to be under sexual conflict. Such a system will show male traits that benefit males but are costly to females. In some cases there may even be overt female behavioural responses (vigorous rejection or avoidance behaviours). It is then necessary to identify the male (inevitably) behavioural or anatomical traits that are likely to reduce female fitness. The next task is to document the cost to females of variation in those male traits. In a recent study of bed bugs Stutt and Siva-Jothy (2001) experimentally adjusted the male copulation rate to maintain maximum fertility with a minimum of copulations, and showed that male-controlled copulation rates resulted in a 25% reduction in female survivorship. This high cost will be the basis of a sexual conflict. At the time of writing there are only a handful of studies that unambiguously demonstrate sexual conflict, and each uses a very different approach because of the diverse nature of the adaptations in question. What all have in common is that they first demonstrate a reduction in female fitness (usually measured as longevity or fecundity) contingent on the operation of the male trait, and they subsequently demonstrate that variation in the male trait is causal *vis-a-vis* the cost to females.

5.6 CONCLUSIONS

With recent advances in evolutionary thinking, and the integrated use of currently available diverse investigational and analytical techniques, the study of natural enemy mating systems continues to provide us with: (a) insight into the evolutionary mechanisms that generate diversity in reproductive traits, (b) an understanding of the utility of those traits, and (c) an understanding of the determinants of reproductive success and optimal reproductive decisions in natural populations. Descriptions and studies of mating systems have largely been of interest to those seeking empirical support for evolutionary theory. Via effects on sex ratio and fecundity, mating systems affect influence natural enemy population dynamics (subsection 7.4.3); their potential for informing pest control strategies are probably underutilised.

5.7 ACKNOWLEDGEMENTS

We thank George Heimpel, Mark Jervis and Paul Rugman-Jones for comments and Xavier Fauvergue, Dave Parker, David Nash, James Cook, Mark Fellowes, Andrew Loch and Gimme Walter for unpublished material and help with Figures. Ian Hardy was funded by European Commission grants ERBFMBICT 961025 and 983172, Paul Ode by NSF DGE-9633975 and Mike Siva-Jothy partly by NERC grants GR9/03134 and GR3/12121.

POPULATIONS AND COMMUNITIES 6

K.D. Sunderland, W. Powell and W.O.C. Symondson

6.1 INTRODUCTION

In nature, any particular habitat contains animal and plant species which exist together in both time and space. Many of these species will interact with each other, for example when one species feeds on another or when two species compete for the same food or other resource. A group of species having a high degree of spatial and temporal concordance, and in which member species mutually interact to a greater or lesser extent, constitute a **community** (Askew and Shaw, 1986). The size and complexity of a community will depend upon how broadly that community is defined. For example, we could consider as a community the organisms which interact with each other within a particular area of woodland, the herbivore species which compete for a particular food plant or the complex of natural enemies associated with a particular prey or host species. Here, we are especially interested in communities of natural enemies, which are often surprisingly species-rich (Carroll and Risch, 1990; Hoffmeister and Vidal, 1994; Settle *et al.*, 1996; Sunderland *et al.*, 1997; Memmott *et al.*, 2000).

The animal species of a community obtain their food directly or indirectly from plants which are the primary producers of the community. Herbivores feed directly on plants whilst predators and parasitoids are either primary carnivores, feeding on herbivores, or secondary or tertiary carnivores, feeding on other predators or parasitoids. The successive positions in this feeding hierarchy are termed **trophic levels**. Thus, green plants occupy the first trophic level, herbivores the second level, carnivores which eat herbivores the third level, secondary carnivores the fourth level, and so on, although a species may occupy more than one level. For example, some insect parasitoids are facultative hyperparasitoids, thus having the potential to occupy two trophic levels. Similarly, some carabid beetles eat both insect prey and plant seeds, and because they are polyphagous predators, the insect prey consumed may consist of herbivores, detritivores or even other carnivores. In fact, since **omnivory** (feeding on more than one trophic level) is very common (Polis and Strong, 1996; Coll and Guershon, 2002) it is often more helpful to think in terms of **food webs (trophic webs)** (6.3.13) than trophic levels. Empirical studies show food webs to be much more complex than was previously thought (Williams and Martinez, 2000). Many generalist predators are known to kill and consume other predators (Rosenheim *et al.*, 1995) and parasitoids (Brodeur and Rosenheim, 2000). When the predators concerned also belong to the same **guild**, the interaction is termed **intraguild predation (IGP)** and the top predator benefits by obtaining a meal from the victim predator or parasitoid and by simultaneously reducing competition for the shared herbivore food resource (Polis *et al.*, 1989). There is a growing realisation that IGP amongst predators (Fagan *et al.*, 1998; Janssen *et al.*, 1998; Eubanks, 2001; Lang, 2003) and between predators and parasitoids (Sunderland *et al.*, 1997; Colfer and Rosenheim, 2001; Meyhöfer, 2001; Meyhöfer and Klug, 2002; Snyder and Ives, 2003) is very common in agricultural systems, but also occurs in diverse natural systems (Denno *et al.*, 2002; Woodward and Hildrew, 2002). Some parasitoid species can probably reduce the incidence of predation on their populations by avoiding habitat patches emanating olfactory cues associated with predators (Moran and Hurd, 1998; Taylor *et al.*, 1998; Raymond *et al.*, 2000). A given predator taxon can vary in its degree of impact on herbivore prey populations, and thereby on plant yield (i.e. vary in its propensity to initiate a significant **trophic cascade** – Polis, 1999), even within a year in one type of crop (Snyder and Wise, 2001). The incidence of trophic cascades is also quite variable in natural systems, as cascades depend on a number of interacting factors, including the relative colonisation rates of herbivores and predators in relation to the spatial distribution pattern of host plants (Thomas, 1989), the relative strengths of interactions in the upper trophic levels (Spiller and Schoener, 1990), and the productivity level of the system (Müller and Brodeur, 2002). Natural enemies in the higher trophic levels can exert a positive influence not only on plant productivity, but also on plant biodiversity (Moran and Scheidler, 2002; Schmitz, 2003). Trophic cascades are very common (Morin and Lawler, 1995; Halaj and Wise, 2001; Symondson *et al.*, 2002), and occurred in 45 out of the 60 independent cases (from 41 studies) analysed by Schmitz *et al.* (2000). Moran and Scheidler (2002) point out that trophic cascades operate in many systems ranging from simple (e.g. arctic) to very diverse (e.g. tropical rainforests), suggesting that diversity has little effect on the strength of top-down trophic interactions. Strong trophic interactions can exist in diverse systems that also encompass an abundance of weaker interactions. Available evidence suggests that intraguild predation and omnivory do not, in general, prevent generalist predators from initiating strong trophic cascades (Halaj and Wise, 2001). The probability of occurrence of any particular trophic cascade is currently very difficult to predict because of the complexity of the systems involved, and also because there is insufficient detailed knowledge of trophic interactions and mechanisms of community organisation. Amongst the many methods described in this chapter there are likely to be at least a few that are appropriate for studying trophic interactions in any given agricultural or natural system. In the realm of food webs, theory abounds but empirical evidence is scarce (Morin and Lawler, 1995); the methods described here will help to guide researchers in the collection of data vital for testing and developing theory. Techniques useful for discovering who eats whom are outlined in section 6.3. In addition, since symbiotic bacteria can also be transferred from one host species to another through the links in a food web (e.g. *Wolbachia* spp. ramifying through herbivore-parasitoid food webs), we describe, in section 6.5, methods for the detection and identification of *Wolbachia* and other symbionts in food webs. Phytophagy by natural enemies is discussed in Chapter 8.

Species within a community exist as **populations**. In its broadest sense the term population can be applied to any group of individuals of the same species occupying a particular space. This space may vary greatly in size, for example from a single tree to a wide geographical area, depending upon how the population is defined. It is important to define the spatial scale over which the population is to be studied at the start of an ecological investigation since the principal factors influencing the population dynamics of a species may vary depending upon spatial scale. For example, immigration and emigration may have a much greater influence on population persistence at small spatial scales than they do at larger ones (Dempster, 1989). The term 'population' is sometimes incorrectly used to refer to the combined numbers of a range of related species occupying a discrete area, for example the 'carabid population' of a field. Great care must be taken in the interpretation of changes in the abundance of such a 'population' since it will comprise a mixture of species with differing ecologies, and different species will be affected in different ways by the same environmental factors. The concept of a **metapopulation** as a collection of sub-populations of a species, each occupying a discrete habitat patch but with some level of genetic interchange, has received much attention in the study of species population dynamics (Gilpin and Hanski, 1991). This concept is discussed further in section 6.4.

In order to study a natural enemy population or community it is usual to select at random a representative group – a **sample** – of individuals on which to make the appropriate observations or measurements so that valid generalisations concerning the population or community as a whole can be made. Often, but not always, the sampled individuals need to be removed from their natural habitat using an appropriate collection technique. Most sampling methods are destructive, involving the physical removal of organisms from the study area. However, it is sometimes possible to sample by counting organisms *in situ* (6.2.6). It is essential that before starting a sampling programme, sampling techniques are chosen that are appropriate for both the type of ecological problem being investigated and the particular natural enemy species being investigated. Section 6.2 is devoted to a description of the most commonly used sampling techniques and their limitations, with reference to examples from the literature on natural enemies. Table 6.1 lists some applications of these techniques. Basset *et al.*, (1997) provide a key to assist ecologists in the choice of sampling methods suitable for one specific habitat – the tree canopy. Ausden (1996) gives a table which summarises the relative applicability of various sampling techniques for a range of invertebrate groups (but not exclusively natural enemies).

In summary, this chapter is concerned with the *sampling* and *monitoring*, in time and space, of natural enemy populations and communities, and describes a number of techniques that can be used for measuring or estimating the abundance of natural enemies, determining the structure and composition of communities and examining the spatial distribution of natural enemies in relation to their host or prey populations. Most of the techniques discussed can be used to obtain *qualitative* data relating to the predator or the parasitoid species present in a community or the prey/host species attacked by a particular natural enemy. Some of the techniques can also be used in obtaining *quantitative* estimates of natural enemy abundance or predation and parasitism, an aspect of natural enemy biology taken further in Chapter 7.

Estimates of animal numbers may be expressed in terms of either density per unit area or unit of habitat, and the unit of habitat can be an area of ground or a unit of vegetation such as a leaf or a whole plant. Estimates of this type are termed **absolute estimates of abundance** and must be distinguished from **relative estimates of abundance** which are not related to any defined habitat unit (Southwood and Henderson, 2000). Relative estimates are expressed in terms of trapping units or catch per unit effort and are influenced by other factors (e.g. climatic conditions) besides the abundance of the insect being sampled. When considering absolute estimates of insect abundance, the term population density may be applied to numbers per unit area of habitat and the term **population intensity** applied to numbers per leaf or shoot or host (Southwood and Henderson, 2000).

This chapter is concerned with practical techniques and we say little about the statistical analysis and interpretation of sampling data. Information on these topics can be found either elsewhere in this book or in the following publications: Cochran (1983); Eberhardt and Thomas (1991); McDonald *et al.* (1988); Perry (1989, 1998); Crawley (1993, 2002); Sutherland (1996); Krebs (1999); Southwood and Henderson (2000) and Dytham (2001).

6.2 FIELD SAMPLING TECHNIQUES

Our discussion of techniques is mainly confined to the sampling of insects from terrestrial habitats and from air. Sunderland *et al.* (1995a) discuss sampling options for determining the density of predators in agroecosystems, and Williams and Feltmate (1992), Ausden (1996) and Southwood and Henderson (2000) give fuller accounts of sampling methods for use in aquatic habitats.

6.2.1 PITFALL TRAPPING

A pitfall trap is a simple interception device consisting of a smooth-sided container which is

Table 6.1 Applications of different field sampling methods

Data Required	Sampling Technique	Comments
Absolute abundance	Pitfall traps	Do not provide data on absolute abundance
	Vacuum net	When used to sample a defined area or unit of habitat, calibration is necessary. Large, active insects may flee before sample taken. Multiple re-sampling or combination with ground search will increase accuracy
	Sweep net	Estimates of absolute abundance difficult to obtain
	Knock-down	For chemical knock-down, unit of habitat (e.g. whole plant) needs to be enclosed; calibration necessary; cannot be done when windy; ineffective for species enclosed within vegetation (e.g. galls, leaf-mines and silken retreats)
	Visual count	Labour-intensive; inects need to be conspicuous if census walk method used; efficiency varies with insect activity and observer; insects may be absent (e.g. underground) during daytime
	Mark-release-recapture	Important to satisfy a number of assumptions; choose appropriate calculation methods
Relative abundance	Pitfall traps	Factors affecting locomotor activity, catchability and escape rate by different species, need to be taken into account
	Vacuum net	Efficiency can change with height and density of vegetation
	Sweep net	A wide range of factors cause sampling variability; significant variability between operators
	Knock-down	Except for chemical knock-down, may not sample all species with the same efficiency
	Visual count	Labour-intensive; insects need to be conspicuous if census walk walk method used; efficiency varies with insect activity and observer
	Attraction	Difficult to define area of influence; insect responses may change with time
Dispersion pattern	Pitfall traps	Trap spacing important in minimising inter-trap interference; vegetation around individual traps can affect capture rates; some carabids aggregate in traps
	Vacuum net	Vegetation type can affect efficiency
	Sweep net	Disturbance can change dispersion pattern during sampling
	Knock-down	Chemical knock-down needs to be confined to defined sampling areas
	Visual count	Detectability needs to be constant over study area
	Examination of hosts for parasitism	Identification of immature stages may prove problematical
Phenology	Pitfall traps	Cannot detect immobile insects (e.g. during cold weather)
	Sweep net	Changes in vertical distribrution within the vegetation may result in non-detection
	Malaise trap	Provides useful information on flight periods
	Visual count	Changes in behaviour may affect ease of detection

Continued

Table 6.1 *(continued)*

Data Required	Sampling Technique	Comments
	Examination of hosts for parasitism	Rearing provides information on diapause characteristics
	Attraction	Responses to visual or chemical stimuli may be restricted to certain periods or behavioural states
	Sticky traps, Window traps	Only valid during active flight periods
Species composition	Pitfall traps	May not sample all species with the same efficiency; provide useful data on presence; lack of catch does not prove absence
	Vacuum net	Night samples need to be taken for nocturnal insects
	Sweep net	Only efficient for groups active in the vegetation canopy
	Knock-down	Very active fliers may escape, but, otherwise, the catch from chemical knockdown is not dependent on the activity of insects and is not influenced by "trap behaviour"
	Visual count	Most efficient for very conspicuous groups
	Attraction	Different species may not respond to the same visual or chemical stimuli
	Sticky/window traps	Species without active flying stage will not be recorded
Relative abundance of species	Pitfall traps, Vacuum net, Sweep net, Malaise trap, Knock-down, Attraction Visual count	Except for chemical knock-down, may not sample all species with the same efficiency
Locomotory activity	Pitfall traps	Linear pitfall traps provide useful information on population movements, especially direction of movements; best to combine with a marking technique
	Visual count	Can provide useful information, especially if combined with a marking technique
	Attraction	Attractant properties of trap can interfere with insect behaviour
	Mark-release-recapture	Provides useful information such as minimum distance travelled
	Sticky and window traps	Provide useful information on height of flight, as well as on direction

sunk into the ground so that its open top lies flush with the ground surface (Figure 6.1a). Invertebrates moving across the soil surface are caught when they fall into the container.

Pitfall traps (Figure 6.1b) are the most commonly used method of sampling ground-dwelling predators such as carabid and staphylinid beetles, spiders and predatory mites, but have occasionally been used to sample other groups of predators, such as dolichopodid flies (Pollet and Grootaert, 1987), pompilid wasps (Field, 1992), ants (Cherix and Bourne, 1980; Samways, 1983; Melbourne, 1999; Morrison, 2002), heteropteran bugs (Rácz,1983; Basedow, 1996) and harvestmen (Cherix and Bourne, 1980; Jennings *et al.*, 1984; Newton and Yeargan, 2002). Abundance data for fire ants (*Solenopsis* spp.) obtained by pitfall trapping may need to be interpreted with caution in cases where these ants are attracted to build nests under the pitfall traps (Penagos *et al.*, 2003). The containers that can be used as pitfall traps are many and varied, but round plastic pots

(a)

(b)

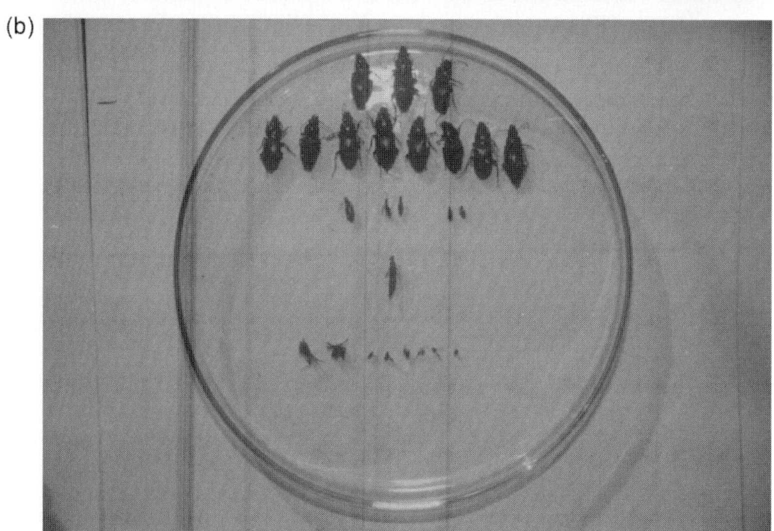

Figure 6.1 (a) A simple pitfall trap (with rain cover) for sampling predators moving across the soil surface, e.g. carabid and staphylinid beetles and spiders; (b) carabid beetles, staphylinid beetles and spiders caught in a single pitfall trap over a 7-day period in a cereal crop in southern England.

and glass jars with a diameter of 6-10 cm are the most popular. It is important to remember, however, that both trap size and trap material are known to influence trap catches, sometimes very strongly (Luff, 1975, Adis, 1979; Scheller, 1984).

A liquid preservative is often placed in the trap, both to kill and to preserve the catch, thereby reducing the risk of escape and preventing predation within the trap, particularly of small individuals by larger ones (conspecifics or heterospecifics). Amongst the preservatives that have been used are formaldehyde, alcohol, ethylene glycol, propylene glycol, acetic acid, picric acid, sodium benzoate and concentrated salt solutions, but there is evidence that some preservatives have an attractant or repellent effect on some insects and that such effects can differ between the two sexes of the same species (Luff, 1968; Skuhravý, 1970; Adis and Kramer, 1975; Adis, 1979; Scheller, 1984; Pekár,

2002). Lemieux and Lindgren (1999) showed that, although the efficiency of trapping carabid beetles did not differ significantly between concentrated brine (a super-saturated solution of sodium chloride) and antifreeze (ethylene glycol), the former was a poor preservative of predators. Addition of detergent can greatly increase the catch of small predators (such as linyphiid spiders), which may otherwise fail to break the surface film and thus be able to escape (Topping and Luff, 1995). Detergent may, however, reduce the catch of some species of carabid and staphylinid beetle (Pekár, 2002). Pitfall traps can be used to catch natural enemies whose DNA is to be analysed for determination of diet (subsection 6.3.12), or in a genetics study. In such cases it is recommended to test the effect of the trapping fluid/preservative on the DNA of the target natural enemy. Gurdebeke and Maelfait (2002) found that DNA of the spider *Coelotes terrestris* was adequately preserved when 96% ethanol was used as trapping fluid in a funnel pitfall trap (the funnel reduced the evaporation rate of ethanol). RAPD profiles could not be generated from spiders collected in formaldehyde. If pitfall trapping is used as a source of predators for stable isotope analysis (6.3.1) traps should be emptied at least every 48 hours (Ponsard and Arditi, 2000) because isotopic content changes during decomposition.

It is sometimes necessary, for example when carrying out mark-release-recapture studies (subsection 6.2.10), to keep specimens alive within the trap. In such cases it is advisable to place some kind of material in the bottom of the trap to provide a refuge for smaller individuals. Compost, small stones, leaf litter, moss or even polystyrene granules may be used for this purpose. Small pieces of meat may also be added to the traps to reduce cannibalism and interspecific predation (Markgraf and Basedow, 2000), but this may introduce the complication of differential species-specific olfactory attraction (subsection 6.2.8) which will make the results more difficult to interpret in some types of study. Live-trapping may yield different relative species compositions of predators such as spiders and carabid beetles compared with kill-trapping, and so the two methods are not necessarily comparable (Weeks and McIntyre, 1997).

At the end of a trapping period it is advisable to replace traps with fresh ones so that the catch can be taken *en masse* back to the laboratory. Plastic pots with snap-on lids are readily available commercially, and are convenient when large numbers of traps need to be transported. To prevent the sides of the hole from collapsing during trap replacement, a rigid liner is useful, and liners can be readily made from sections of plastic drainpipe of an appropriate diameter. In open habitats it is also advisable to place covers over traps in order both to prevent birds from preying on the catch, and to minimise flooding during wet weather. Covers for simple, round traps can be made from inverted plastic plant pot trays, supported by wire. Such covers may increase the catch of carabid and staphylinid beetles that are behaviourally adapted to finding refuge under stones and logs. Soil-coring tools, including bulb planters, can be used to make the initial holes when setting pitfall traps, but it is essential to ensure that the lip of the trap is flush with the soil surface. When traps are operated over a prolonged period of time, some maintenance work is often necessary. For example, in hot, dry weather some soils crack and shrink, creating gaps around the edge of traps, thus reducing their efficiency.

A number of variations on the conventional pitfall trap have been developed in attempts to improve their efficiency or to tailor them for particular habitats and target taxa. Lemieux and Lindgren (1999) tested a trap (modified from a design by Nordlander (1987)) with fifteen 1.3 cm diameter holes around the trap circumference at 2 cm below the upper rim. This trap incorporated a tight-fitting lid and was buried so that the lower 3 mm portion of the holes was below ground level, thus predators could enter the trap through the holes only. The overall catch of carabid beetles was not different from that of conventional pitfalls, but the Nordlander trap had the advantage of excluding vertebrates and reducing the dilution of trapping fluid caused by rainfall. Several workers have used linear traps made from lengths of plastic or metal guttering to increase catches and to obtain directional information on insect movements (6.2.11). Sigsgaard *et al.* (1999) separated two plastic cup pitfalls (positioned back to back) by a barrier of semi-transparent hard plastic sheet to obtain data on directional movement of spiders, carabids and ants entering and leaving a rice field, and a similar design was used by Hossain *et al.* (2002) to record spiders and carabid beetles moving from harvested to unharvested plots of alfalfa. Two cup pitfalls may be placed a short distance apart and joined by a solid barrier which diverts walking insects into the traps at either end (Wallin, 1985; Jensen *et al.*, 1989; French *et al.*, 2001), or the trap can be placed at the apex of V-shaped fences (Culin and Yeargan, 1983; Markgraf and Basedow, 2000). Winder *et al.* (2001) found that the use of five traps arranged in a cross formation, and connected by plastic guide barriers (0.5 m lengths of lawn edging sunk 5 cm into the ground), was a more efficient design than single traps for catching some species of carabid and staphylinid beetle, and lycosid spiders (but not linyphiid spiders). Mechanical devices have been used to allow automatic, time-based sorting of trap catches (Williams, 1958; Ayre and Trueman, 1974; Barndt, 1976; Alderweireldt and Desender, 1992; Chapman and Armstrong, 1997, see also subsection 6.2.11), whilst Heap (1988) suspended UV-emitting, fluorescent light tubes above his traps to increase the catch rate. A funnel trap can be used for sampling in litter, where litter is placed on a gauze mesh across the mouth of the funnel, and predators fall through the mesh and are directed by the funnel into a pitfall trap (Edwards, 1991) (Figure 6.2). Lund and Turpin (1977) used a funnel pitfall containing liquid nitrogen to freeze carabid beetles as soon as they entered the trap. Immediate freezing prevented predation within the trap and arrested decomposition and enzymatic breakdown of the gut contents of the beetles (which were to be tested serologically (subsection 6.3.9) for evidence of predation on cutworm larvae). One litre of liquid nitrogen per pitfall lasted about 24 hours. A similar funnel-pitfall trap (modified to kill entering predators with carbon tetrachloride vapour) was used by Sunderland and Sutton (1980) to collect predators of woodlice in dune grassland. The funnel pitfall of French *et al.* (2001) contained insecticidal cattle ear tags to kill trapped carabid beetles. Hengeveld (1980) used a funnel trap to direct carabids into a tube containing 4% formalin solution. Luff (1996) considered funnels to be valuable as a way of removing the lip of the pitfall. A funnel was found to prevent carabid beetles hanging onto the trap edge, and it also reduced the likelihood of trapped beetles escaping. Newton and Yeargan (2002) added a varnished hardboard apron surrounding the funnel, and this design was successful for monitoring harvestmen (Opiliones) in soybean, grass and alfalfa crops (see also the experiments of Epstein and Kulman, 1984). Funnel traps were found to be up to three times more efficient per centimetre of trap diameter, compared to standard cup pitfalls, for catching carabid and staphylinid beetles and lycosid spiders, but no significant improvement in efficiency was reported for ants and linyphiid spiders (Obrist and Duelli, 1996). "Hanging desk traps" are pitfalls for collecting from rock faces. They have a flat collar attached that are hung at a slight angle to the rock face (to prevent rain water

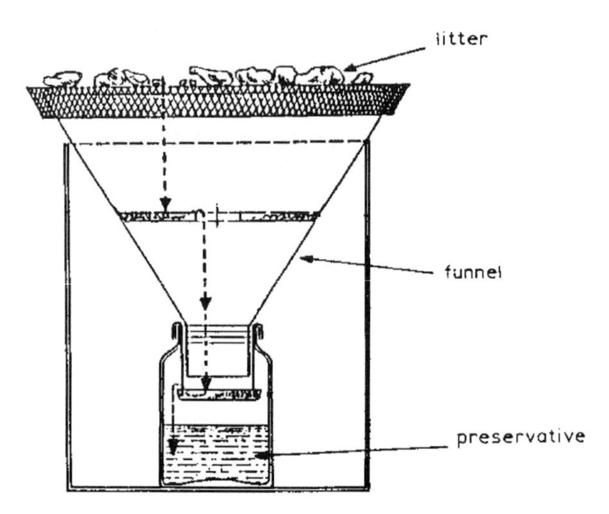

Figure 6.2 A litter-covered pitfall trap. Source: Edwards (1991). Reproduced by permission of Elsevier Science.

flowing in) with the inner edge of the collar taped to the rock and covered with small stones (Růžička and Antuš, 1997). So far, they have been used to assess spider diversity in mountainous areas (Růžička, 2000). Weseloh (1986a) sampled the ground beetle *Calosoma sycophanta* on tree trunks using traps that consisted of plastic tree bands that directed climbing beetles into waxed paper cups attached to the tree. A similar method, reported by Hanula and Franzreb (1998), utilised an inverted metal funnel attached to the tree trunk. Arthropods, including spiders and harvestmen (Phalangida), crawling up the tree trunk, pass through the funnel and enter a collection container attached to its upturned spout. A "stalk trap", employing the same principle but on a smaller scale, was used by Lövei and Szentkirályi (1984) to trap carabid beetles ascending and descending the stalks of maize plants. Bostanian *et al.* (1983) used a "ramp pitfall" that has two metal ramps leading to a rectangular pitfall, and a metal roof. Several of these traps were run in an untreated carrot field with traditional circular plastic pitfalls running concurrently. The latter caught more carabids plus a range of other invertebrates, but the ramp trap caught 99.5% carabids and was selective, catching the larger species (virtually none <5 mm). This type of trap saves sorting time for researchers interested only in large carabids. The mechanism of size selection is not known, but small species may turn back because of the steepness of the ramp, or be repelled by the darkness. Loreau (1987) devised a pitfall trap to sample adults and larvae of carabid beetles within the soil (Figure 6.3). A 7 centimetre diameter plastic pot is contained within a plastic cylinder which has four 1.5 cm windows at whatever depth it is desired to sample from. The top of the cylinder, which is enclosed with a plate, is set at the soil surface, or the surface of the litter. Predators enter the cylinder through the windows and fall into the pot which contains preservative. A wire is attached to the pot so that the trap can be emptied without removing the cylinder. Other minor variations in the design of hypogean pitfalls have been reported (Kuschel, 1991; Owen, 1995). Epstein and

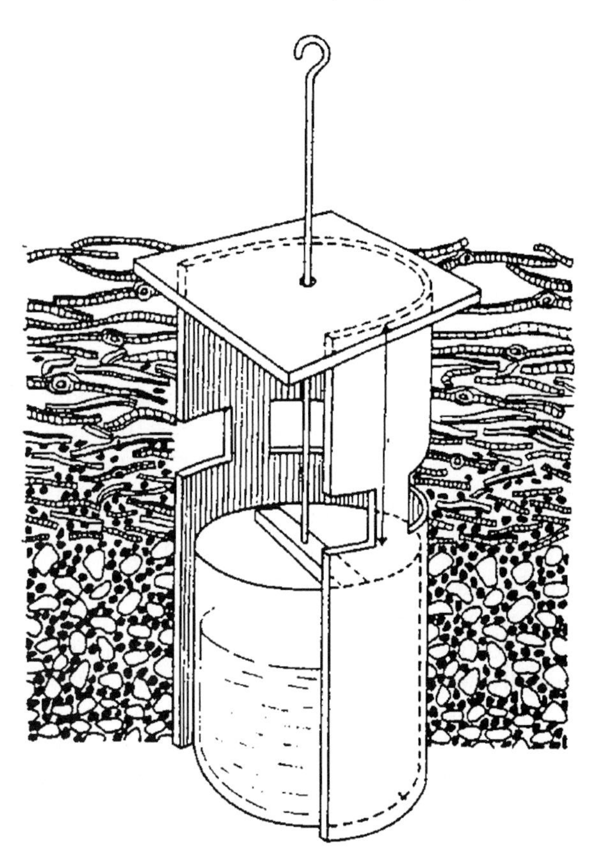

Figure 6.3 A pitfall trap to study the vertical distribution and activity of adults and larvae of carabid beetles within the soil. Source: Loreau (1987). Reproduced by permission of Urban & Fischer Verlag.

Kulman (1984) experimented with traps surrounded by plywood aprons, either resting on the rim of the pitfall trap or raised above it, allowing entry by carabids from above or below the apron. Although these trap designs caught fewer beetles than conventional traps, it was shown that trap design can significantly affect both the numbers and ratios of different species caught. Epstein and Kulman suggest that more than one trap design should be used in parallel where activity-density of a range of species is being measured. Dormann (2000) designed a trap that can catch predators in periodically flooded habitats, such as peat bogs and salt marshes. The trap is attached to a slider rod in a tube set vertically in the soil. When the water level rises (e.g. an incoming tide) a Styropor®

flotation device below the trap cup pushes the cup up against the roof of the trap, sealing it off from the water (aided by formation of a bell of trapped air). When the water recedes the trap sinks back to soil surface level and capture of walking predators can resume. Traps were not damaged even at times of high tide when there was strong wave action.

Several traps, which may be regarded as equivalent to pitfall traps, have been developed for sampling aquatic invertebrates (Southwood and Henderson, 2000); that designed by James and Redner (1965) is particularly useful for collecting predatory water beetles. Floating pitfall traps have been devised, and used successfully, to monitor the activity of large wolf spiders of the genus *Pirata* which are adapted for walking on the surface film of water (Renner, 1986).

A common practice in recent years has been the erection of **physical barriers** in the field, usually to enclose defined treatment areas within which pitfall trapping is carried out (Powell *et al.*, 1985; Holopäinen and Varis, 1986) (Figure 6.4a,b). Caution must be exercised in the use of such barriers in arable crop fields because some predators invade the fields from field boundaries during spring and early summer (Pollard, 1968; Coombes and Sotherton, 1986). The erection of full barriers too early in the year would exclude these species from the enclosed areas, resulting in erroneous data on predator communities. The use of physical barriers can reduce catches of carabid beetles by as much as 35–67% over a growing season (Edwards *et al.*, 1979; de Clercq and Pietraszko, 1983; Holopäinen and Varis, 1986). By contrast, some carabid beetle species emerge as adults from the soil within arable fields (Helenius, 1995), and if emerging populations are high many individuals will attempt to disperse away from overcrowded areas. Barriers may prevent this dispersal, resulting in artificially high predator densities within enclosed areas (Powell and Bardner, 1984). Pitfalls inside smaller, replicated fenced and mesh-covered areas ("**fenced pitfalls**" – Sunderland *et al.*, 1995a) can, however, be used to produce a more reliable estimate of predator abundance than would be the case for unfenced pitfalls. The catch over a period of several weeks (Ulber and Wolf-Schwerin, 1995) is assumed to represent a high proportion of those predators that were initially present in the fenced area (where this has been measured for carabid beetles it is usually more than 80%, Bonkowska and Ryszkowski, 1975; Dennison and Hodkinson, 1984; Desender *et al.*, 1985; Holland and Smith, 1997; 1999), but after this period of time the fenced pitfalls need to be moved a few metres (i.e. re-set) if the species composition and density of predators are to be continuously monitored (if not moved, they become, effectively, emergence traps, see subsection 6.2.7). Density estimates from this method compare well with other methods (Gist and Crossley, 1973; Baars, 1979a; Dennison and Hodkinson, 1984; Desender and Maelfait, 1986). Mommertz *et al.* (1996) compared a variety of sampling methods and concluded that fenced pitfalls should be used for predators in arable crops, but that unfenced pitfalls are a valuable additional method for sampling the larger predators. Holland and Smith (1999) showed a linear relationship between fenced and unfenced traps in the number of six species of carabid beetle caught, but no such relationship for linyphiid spiders. Staphylinid beetles were better represented in fenced than unfenced traps and carabid species composition varied between the two trap types, with smaller species being more evident in the fenced traps (Holland and Smith, 1999). Fences or isolators are usually constructed of wood (Desender and Maelfait, 1983; Dennison and Hodkinson, 1984; Holland and Smith, 1999), metal (Bonkowska and Ryszkowski, 1975; Grégoire-Wibo, 1983ab; Scheller, 1984; Helenius 1995) or plastic (Sunderland *et al.*, 1987b), and the mesh covering needs to be very tightly sealed because small climbing predators can be attracted to the traps and enter through small gaps, thus inflating the density estimate (Sunderland *et al.*, 1995a). Fenced pitfalls have been used to estimate the density of carabid beetles (Basedow 1973; Bonkowska and Ryszkowski, 1975; Baars, 1979a, Dennison and Hodkinson, 1984; Desender and Maelfait 1986; Helenius 1995; Ulber and Wolf-Schwerin,

(a)

(b)

Figure 6.4 (a) Polyethylene barriers surrounding experimental plots containing pitfall traps to restrict immigration by carabid beetles. (b) Close-up of polyethylene barriers.

1995), staphylinid beetles, and spiders (Basedow and Rzehak, 1988). Connecting the pitfalls with small barriers within the fenced area will increase capture efficiency (Durkis and Reeves, 1982; Ulber and Wolf-Schwerin, 1995; Figure 6.5). Further aspects of pitfall trap methodology are discussed in Luff (1975, 1987), Adis (1979) and Scheller (1984).

Pitfall trap catches are a function of predator abundance, activity and trappability (= probability of capture of an individual in the population, Melbourne, 1999), and so changes in any of these variables will affect the capture rate. Locomotor activity in natural enemies and methods for its investigation are discussed in section 6.2.11. Thomas *et al.* (1998) recorded four peaks

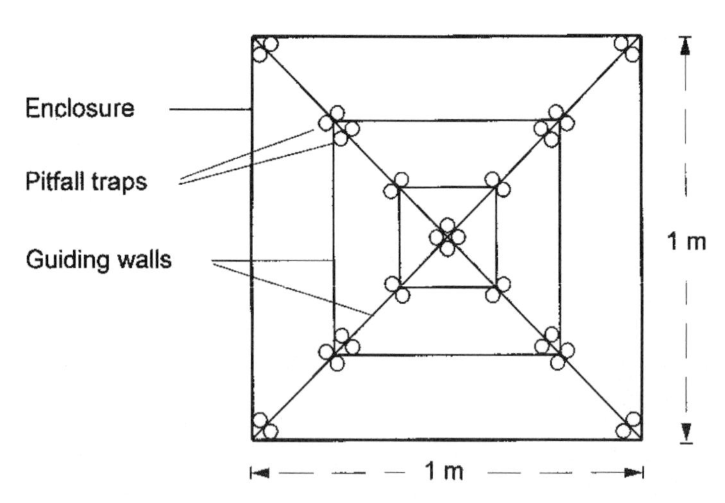

Figure 6.5 Fenced-pitfall trapping comprising an enclosure with 36 pitfall traps and guiding walls (Source: Ulber and Wolf-Schwerin, 1995) Reproduced by permission of Aarhus University Press.

in pitfall catch of the carabid *Pterostichus melanarius* during a period when density (estimated by mark-release-recapture, see section 6.2.10) showed a smooth increase to a plateau. The peaks in pitfall catch were correlated with periods of rainfall. The number of individuals present in a trap at the end of a trapping period is determined by both the capture rate (the rate at which individuals fall into the trap) and by the escape rate (the rate at which individuals manage to escape from the trap). Both capture rate and escape rate depend on the predator species (and its sex and hunger level) and they are also influenced by a variety of factors including the size, shape, pattern and spacing of trap, presence of a trap cover, type of preservative used, vegetation density, soil type, soil surface texture, food availability and weather conditions (Heydemann, 1962; Luff, 1975, Uetz and Unzicker, 1976, Thomas and Sleeper, 1977, Adis, 1979, Müller, 1984, Niemelä *et al.*, 1986, Benest 1989b; Abensperg-Traun and Steven, 1995; Work *et al.*, 2002). In a mowed weedy field in the Czech Republic the average increase in catch of carabids per 1°C increase in temperature was 6.3% (Honěk, 1997).

There is also evidence that some beetles aggregate in pitfall traps (Thomas and Sleeper, 1977; Benest, 1989a), probably in response to aggregation or sex pheromones or to defensive secretions, and this can result in considerable inter-trap variability in catches (Luff, 1986). Raworth and Choi (2001), however, found no significant attraction of the carabid *Pterostichus melanarius* to traps that already contained this species, but a maximum of three ''bait'' individuals were used, and responses to higher densities remain to be investigated. Penagos *et al.* (2003) considered that male carabids (*Calosoma calidum*) were attracted to traps containing a female and that trapping results were affected by this sexual attraction. Inter-trap variability is also affected by weeds, ants, carcasses of small mammals that have fallen into the trap, soil type and the pH of the topsoil (Törmälä, 1982; Honěk, 1988; Powell *et al.*, 1995). All of these influences should be considered when interpreting pitfall trap data.

Pitfall traps do not provide absolute measures of predator density because the number of individuals caught depends partly on their locomotor activity. Therefore, the abundance of a species as measured by pitfall trap catches has been termed its **activity density** or **activity abundance** (Heydemann, 1953; Tretzel, 1955; Thiele, 1977). The activity density of a species may provide, to some extent, a measure of the predator's role in an ecosystem (for example in capturing prey), since this role sometimes depends on its mobility as well as on its frequency (Thiele, 1977). Considerable caution is, however, needed here, because some predators have a "sit-and-wait" foraging strategy, and intense predator activity can be associated with the search for mates rather than for food, or even with the need to escape from an inimical environment. Exclusive reliance on pitfall trapping in ecotoxicological studies is inadvisable because pesticides can directly increase (Jepson *et al.*, 1987) or decrease (Everts *et al.*, 1991) activity, or indirectly affect activity by altering food availability (Chiverton, 1984). In some cases (e.g. carabid beetles) hungry predators have been shown to be more active than satiated predators (Mols, 1987), and the relationship between degree of hunger and activity level (and, thus probability of capture) can vary significantly according to habitat type (Fournier and Loreau, 2000). Locomotor activity, and therefore capture rate, often varies between the two sexes of a species, so that reliable estimates of sex ratio within predator populations are also difficult to obtain from pitfall data. For example, most of the spiders which are active on the ground in cereal crops are males, and so more males than females are caught in pitfall traps (Sunderland, 1987; Topping and Sunderland, 1992). As a confounding factor, there is also good evidence to suggest that sex ratios of some carabids caught in traps will vary according to the state of nutrition of the predators (Szyszko *et al.*, 1996). Females may be less active than males when well fed, because both sexes spend less time searching for food but the males continue to search for females. When hungry, females may be more active than males because they have a greater food requirement related to egg production.

Pitfall traps have been used to compare activity densities for the same species in different habitats and at different times of year but it must be remembered that the activity densities of different species are not necessarily comparable (Bombosch, 1962). It is difficult to separate the influences of activity and abundance on trap catches, but, when trapping is continued throughout the year, whole year catches may be linearly related to density for some individual species (Kowalski, 1976; Baars, 1979a). Using this "annual activity" approach, analyses are restricted to within-species and within-habitat comparisons, (not within-season) and are also not appropriate for all species. Loreau (1984a), for example, showed that the annual activity of the carabid *Pterostichus oblongopunctatus* was not linearly related to its population density. The "within-habitat" restriction was emphasised by a study of pitfall catches of ants in grassland, where an area of 80 cm radius centred on each trap was either unmodified, cleared of litter, or cleared of all vegetation (Melbourne, 1999). These habitat manipulations had a statistically significant effect on abundance, relative abundance, species richness and species composition of the catch. Trappability increased, overall, as the habitat became more open, but there was considerable interspecific variation in response to degree of clearing (Melbourne, 1999). If sufficient data are available for a particular species of natural enemy, it may be possible to estimate population density from pitfall catches. Raworth and Choi (2001) measured the rate and direction of movement of the carabid *Pterostichus melanarius* in relation to temperature in the laboratory. They used these data in a simulation model of beetle movement which was then calibrated and validated using mark-release-recapture during July in a grid of pitfall traps in a raspberry field (which had a flat surface and few weeds, and so was in this respect similar to laboratory arenas). The model showed that pitfall catch should increase linearly with beetle density, with the slope of the relationship being temperature-dependent. The authors stress that the equations and data are restricted to *P. melanarius* under the specified conditions, but this approach for converting pitfall catch to density estimates could be applied to other species and habitats, if the considerable amount of work entailed were justified in particular cases.

It is difficult to obtain from pitfall trap catches an accurate picture of the relative abundance of different species within a community, because different species are caught at different rates and also escape at different rates (Jarošík, 1992; Topping, 1993; Lang, 2000). Mommertz *et al.*

(1996) found that even carabid beetles of the same size and genus (*Poecilus*) exhibited large differences in trappability in an artificial wheat field in the laboratory. Video recording techniques in the laboratory (Halsall and Wratten, 1988a) and the field (N. Paling, personal communication) have also been used to show that predator species vary greatly in trappability and the capacity to escape from pitfalls. Jarošík (1992) attempted to use pitfall trap data to compare patterns of species abundance in communities of carabid beetles from different habitats, but concluded that pitfall trap data were inadequate for this purpose. Pitfall trapping in unfenced areas usually gives very different results to other methods (such as D-vac, photoeclectors, ground search in quadrats, soil flotation, fenced pitfalls) for the species composition, size distribution, sex ratio and relative abundance of polyphagous predators (Bonkowska and Ryszkowski, 1975; Lohse, 1981; Desender and Maelfait 1983,1986; Dennison and Hodkinson, 1984; Basedow and Rzehak, 1988; Dinter and Poehling, 1992, Topping and Sunderland, 1992; Andersen, 1995; Dinter, 1995; Ulber and Wolf-Schwerin, 1995; Lang 2000; Arneberg and Andersen, 2003). For example, Lang (2000) found that the abundances of carabid beetles and lycosid spiders were overestimated by pitfalls, but the abundances of staphylinid beetles and linyphiid spiders were underestimated, compared to catches made with photoeclectors (subsection 6.2.7). Pitfalls have sometimes been used in the study of ant communities. Lobry de Bruyn (1999) reviews the utility and limitations of pitfalls for estimating ant biodiversity. Pitfall trapping does provide useful presence/absence data, and could complement other collection methods in estimating the natural enemy biodiversity of a particular region. For example, Wesolowska and Russell-Smith (2000) caught 44 species of jumping spider (Salticidae) in pitfall traps in a variety of habitat types in Mkomazi Game Reserve (Tanzania). This compared with 34 species caught by hand and 15 species by sweep netting. However, approximately half of the species collected by each method were collected by that method alone, and each method collected a similar proportion of rare species. Species lists derived from extensive trapping programmes can be used, with the aid of modern ordination techniques (Pielou, 1984; Digby and Kempton, 1987), to classify different habitats based on their carabid communities, or to identify environmental factors which are influencing species distributions (Eyre and Luff, 1990; Eyre *et al.*, 1990). Large scale trapping programmes can entail the commitment of a large amount of resources and it becomes necessary to optimise the use of these resources. For spiders in an agricultural landscape, Riecken (1999) found that using fewer traps resulted in less of a reduction in the total number of species caught than did reducing the duration of sampling.

With the addition of appropriate analytical techniques, pitfall trapping programmes can be used to estimate the biodiversity of a group of predators in a particular habitat. Samu and Lövei (1995), for example, used pitfalls to collect ground-dwelling spiders in a Hungarian apple orchard. Simulation of increased sample size was made by computer sub-sampling of the original data set. An asymptotic function (taken from the theory of island biogeography) was used to describe the sampling curve and the value for the asymptote was considered to be a realistic estimate of the total number of species present. Brose (2002) showed that sampling effort can be reduced if carefully-selected non-parametric estimators are used to calculate the species richness that would be obtained from a more intensive pitfall sampling programme. It should be remembered, however, that it is likely that some species of the natural enemy group(s) of interest will not be amenable to capture in pitfall traps. Estimating species richness of invertebrate groups that are active both on the ground and in vegetation requires multiple sampling techniques, such as pitfall trapping combined with suction sampling or sweeping (Standen, 2000).

Information on the spatial dispersion patterns and density of predators living on the soil surface can be obtained using pitfall traps either spaced in a grid system (Ericson, 1978; Gordon and McKinlay, 1986; Niemelä, 1990; Bohan *et al.*, 2000), laid out as transects through heterogeneous habitats (Wallin, 1985), or as concentric circles at fixed distances from a central point (Parmenter and MacMahon, 1989; Buckland *et al.*, 1993). Traps within a regular grid system may interfere with each other, the central traps catching fewer individuals than the outer traps (Scheller, 1984). Trap spacing is therefore important, interference increasing as between-trap distances decrease. In Canadian conifer forests it was determined that traps needed to be at least 25 m apart to reduce depletion effects (Digweed *et al.*, 1995). Measurement of the spatial aggregation of invertebrates may be radically affected by between-trap distances within a sampling grid. Bohan *et al.* (2000) found that, at 4 m and 8 m scales, pitfall trap catches of the carabid *Pterostichus melanarius* were randomly distributed, but at a 16 m scale catches were spatially aggregated. The scales at which aggregation occurs are likely to be very different between species and habitats and will change over time. The catch may also be increased if the surrounding ground is disturbed during trap establishment ("digging-in effects") (Digweed *et al.*, 1995). Similarly, within heterogeneous habitats, differences in vegetation type and density around traps will affect capture rates, hindering the investigation of dispersion patterns. Consideration also needs to be given to aggregation that may occur in response to infochemicals (subsections 1.5.2 and 1.12.2). Such aggregation may be independent of trap positions and can vary with between-trapping periods (Luff, 1986).

Despite their limitations, pitfall traps remain a very useful sampling tool for obtaining both qualitative and quantitative data on those predators which are active on the soil surface, providing that the objectives of the sampling programme are clearly defined and that the many factors which can influence the catch rate are considered during data interpretation. Advice on the interpretation of pitfall data is given by Luff (1975, 1987); Adis (1979); Ericson (1979); Baars (1979a).

Some small-bodied natural enemies live on or in leaf-litter (e.g. some diapriid and eucoilid parasitoids). Pitfall trapping is unlikely to be useful in sampling such insects, and a leaf-litter sampling technique (subsection 6.2.7, and Southwood and Henderson, 2000) might be attempted.

6.2.2 VACUUM NETTING

Although several different types of vacuum net have been developed, nearly all operate on the same principle. They employ a fine mesh net enclosed in a rigid sampling head which is attached to a flexible tube. The tube is connected to a fan which is driven by an electric or petrol-fuelled motor. The fan draws air through the flexible tube via the net in the sampling head, sucking small arthropods on to the net from the vegetation enclosed by the sampling head. The sampling tube and its head are quite wide (sampling approximately $0.1\,m^2$) in some machines (Dietrick *et al.*, 1959; Dietrick, 1961; Thornhill, 1978; Duffey, 1980) (Figure 6.6a), but narrower (approximately $0.01\,m^2$) in others (Johnson *et al.*, 1957; Heikinheimo and Raatikainen, 1962; Southwood and Pleasance, 1962; Arnold *et al.*, 1973; Henderson and Whitaker,1976; Macleod *et al.*, 1995) (Figure 6.6b) Machines with wide tubes require a more powerful motor in order to attain the required air speed of at least 90 km/h through the collecting head. Many of the machines currently in use in the UK are of the wide-tube variety and are driven by two-stroke, lawn-mower, petrol-fuelled engines which are mounted on rucksack frames so that they can be carried on the back of the operator (Thornhill, 1978). Taubert and Hertl (1985) designed a small vacuum net driven by a battery-powered electric motor, presenting a lighter load for the operator. New designs of vacuum insect net continue to be devised and evaluated (Summers *et al.*, 1984; Holtkamp and Thompson, 1985; De Barro, 1991; Wright and Stewart, 1992; Macleod *et al.*, 1994,1995; Samu and Sárospataki, 1995a; Toft *et al.*, 1995). The modified Allen-vac has a collecting bag held sufficiently far inside the apparatus that it does not become snagged in dense woody scrub or desert thorn scrub (Osborne and Allen, 1999).

(a)

(b)

Figure 6.6 (a) Wide-nozzled ($0.1\,m^2$), petrol engine-driven vacuum sampler; (b) entomologists vacuum sampling rice insects in a paddy in Indonesia. (b) reproduced by permission of Anja Steenkiste.

The corn KISS (keep-it-simple-sampler) (Beerwinkle *et al.*, 1999), is a hand-held modified leaf blower that directs air across a maize plant and into a net, which is also attached to the blower (i.e. this is a blower rather than a vacuum device, and the plant is sandwiched between blower and net). To sample a plant the KISS is lifted from the base to the top of the plant in a sweeping motion. Beerwinkle *et al.* (1999) found that the KISS was as efficient as Berlese extraction (6.2.7) for sampling mobile predators (such as spiders, ladybird larvae, nymphal predatory bugs, and lacewing larvae) exposed on the plant surface. The aquatic equivalent of this device is the "air-lift sampler", which blows air at the substrate and forces dislodged insects into a mesh cup (Williams and Feltmate, 1992).

Vacuum nets can be used in a number of different ways to collect arthropods - either natural enemies or hosts and prey - from vegetation. A commonly used method involves pressing the sampling head to the ground over the vegetation, (providing this is not too tall or dense), and holding it in place for a defined period of time (e.g. 10 seconds), a process which may be repeated several times within a specified area to form a single sample. This method is appropriate for wide-nozzled machines, and by measuring the size of the sampling head the catch can be related to a finite area of vegetation, so giving an absolute measure of species densities.

An alternative method of using a wide-nozzled suction sampler is to hold the sampling head at an angle to the ground whilst pushing it through the vegetation over a defined distance. As it brushes through the vegetation the advancing sampling head dislodges arthropods which are then sucked into the net. This method is particularly useful when collecting insects from crops planted in discrete rows as it allows a specified length of row to be sampled. Graham *et al.* (1984), however, found that placing the sampling head vertically over plants was more efficient than sweeping it through the vegetation for sampling parasitoid wasps (Mymaridae) and predatory heteropteran bugs (Nabidae) in alfalfa. Motorised vacuum nets have also been used to sample natural enemies from the foliage of fruit trees (Suckling *et al.*, 1996; Green, 1999; Gurr *et al.*, 1999).

Another approach involves enclosing an area of vegetation with a bottomless box or cylinder and using a narrow-nozzled sampler to remove insects from the enclosed vegetation, the soil surface and the interior walls of the box (Johnson *et al.*, 1957; Southwood and Pleasance, 1962; Smith *et al.*, 1976; Henderson and Whitaker, 1976; Törmälä, 1982; Summers *et al.*, 1984; Heong *et al.*, 1991; Wright and Stewart, 1992; Toft *et al.*, 1995; Schoenly *et al.*, 1996a; De Kraker *et al.*, 1999). If the vegetation is tall, it can then be cut and removed before sampling a second time. Wright and Stewart (1992) adapted a commercial garden leaf-blower (Atco Blow-Vac) as a narrow-nozzled suction sampler and used it to extract insects from areas of grassland enclosed by an acetate sheet cylinder. Compared with a wide-nozzled suction sampler, the sampling head of which covered the same area of ground as the acetate cylinder, the narrow-nozzled apparatus extracted significantly more predatory beetles and spiders from the vegetation. Haas (1980) used repeated suction sampling within a gauze-covered isolator in grassland, and Marston *et al.* (1982) used a D-vac to remove invertebrates after knock-down by permethrin inside cages in a soybean crop. Highly active species may escape from the sampling area when the sampler approaches, and researchers have devised a number of ingenious modifications of technique in an attempt to circumvent this problem. These range from a cylindrical cage on the end of a pole that can be dropped over a sugar beet plant from a distance, before vacuum sampling the enclosed area (Hills, 1933), to various more sophisticated **"drop trap" methods** operating on the same basic principle (Turnbull and Nicholls, 1966; Gromadzka and Trojan, 1967; Mason and Blocker, 1973). Duelli *et al.* (1999) took suction samples from the inside of a cubic tent placed over the vegetation. They found that by quickly tilting the cube over the vegetation against the wind they could trap most predators inside the tent, including fast-flying adult flies. The problem of large, active predators fleeing from an area to be sampled due to the noise and vibration of the approaching researchers (Uetz and Unzicker, 1976; Samu and Kiss, 1997) is common to most of the field sampling techniques described in this chapter, and "drop trap" methods could be more widely considered. Other modifications to suction sampling techniques include adjuster bars to enable sampling at various vertical distances above ground (Kennedy *et al.*, 1986), extension tubes to reduce mortality of natural enemies in the collecting net (Topping and Sunderland, 1994), and a flange within the D-vac head to reduce escape rates of large predators (Yeargan and Cothran, 1974).

After each sample is taken with a vacuum net device, the collecting net is usually removed from the sampling head. The net can either be tied off (to secure the catch) and replaced with a fresh one, or the catch can be transferred to a polythene bag so that the net can be re-used. Moreby (1991) described a simple modification to the net and sampling head which speeds up the transfer of samples to polythene bags. In order to reduce the risk of losses due to predation within the net or bag, the catch should be killed whilst still in the field. This can be done by placing the tied-off nets into bags already containing a chemical killing agent or by adding wads of cloth or paper, impregnated with killing agent, to the catch once it has been placed in the polythene bag. For some purposes however, for example the rearing of parasitoids from hosts (subsection 6.3.6), the catch may need to be kept alive.

On returning to the laboratory, catches may be sorted by hand but it is often useful to pass the sample through a series of sieves before hand-sorting, especially if organisms of a fixed size are being counted. Hand-sorting can be extremely time-consuming, particularly if very small insects are being sampled (Figure 6.7). Consequently, some workers have used Berlese funnels or flotation techniques to speed up the process (Dietrick *et al.*, 1959; Marston, 1980). Further details on the construction of vacuum nets and on their use may be found in Johnson *et al.* (1957); Dietrick *et al.* (1959); Dietrick (1961); Southwood and Pleasance (1962); Weekman and Ball (1963); Arnold *et al.* (1973); Thornhill (1978); Kogan and Herzog (1980); Southwood and Henderson (2000).

The efficiency of vacuum nets varies considerably in relation to the biomass (Bayon *et al.*, 1983) and vertical stratification (Sunderland and Topping, 1995) of the natural enemies being sampled, and the height, density and type of vegetation being sampled (Henderson and Whitaker, 1976). Dense vegetation may affect efficiency by reducing airflow and by forming a mat under which natural enemies can cling and avoid capture. This problem can be reduced by taking a suction sample, then immediately cutting the vegetation to a low height and re-sampling (Dinter, 1995; Hossain *et al.*, 1999). When this protocol was applied to lucerne, recapture rates of marked ladybirds and predatory heteropterans were increased from approximately 0.7 to 0.9 (Hossain *et al.*, 1999). Fenced pitfall traps (6.2.1) can be a useful alternative to suction sampling for estimating the density of large carabid beetles, because these traps are appropriate for nocturnal as well as diurnal species and are less affected than vacuum nets by the density of vegetation in the sampled habitat (Holland and Smith, 1999). Generally, small, winged insects such as Diptera and adult Hymenoptera 'Parasitica' are sampled with the greatest efficiency by suction sampling, and Poehling (1987), working in cereal crops, concluded that it is a suitable method for sampling these types of insect. An investigation was made (W. Powell, unpublished data) into the efficiency of a wide-tubed vacuum net (area of sampling head: $0.1\,m^2$) for sampling adult parasitoids in flowering winter wheat. Known numbers of parasitoids were released into large field cages ($9\,m \times 9\,m \times 6\,m$) and allowed to settle for several hours before sampling. Comparisons of expected and actual catches indicated a sampling efficiency of over 90%. Henderson and Whitaker (1976) investigated the efficiency of a narrow-tubed vacuum net used in conjunction with a bottomless box which enclosed a $0.5\,m^2$ area of grassland. Again, efficiency was highest for Diptera (79-98%) and Hymenoptera (60-83%) and was lowest for Acarina (12-40%). Sampling efficiency tended to decline with increasing grass height for most of the groups sampled. Vacuum nets are also regarded as an efficient means of sampling adult parasitoids in soybean crops (Marston, 1980). Duffey (1974) found the efficiency of the D-vac to be only 14-58% for spiders in pasture. He also suspected that invertebrates were sucked into the nozzle from beyond the $0.09\,m^2$ nozzle area. This was confirmed by Samu *et al.* (1997), who caught three times more spiders in an alfalfa transect of 48 vacuum insect net subsamples compared with suction sampling an enclosed part of the crop of the same area as the transect. This lateral suction effect can be corrected by mathematical conversion (Pruess *et al.*, 1977) or by sampling within small enclosures (see above). The efficiency of a "Blower-vac" suction sam-

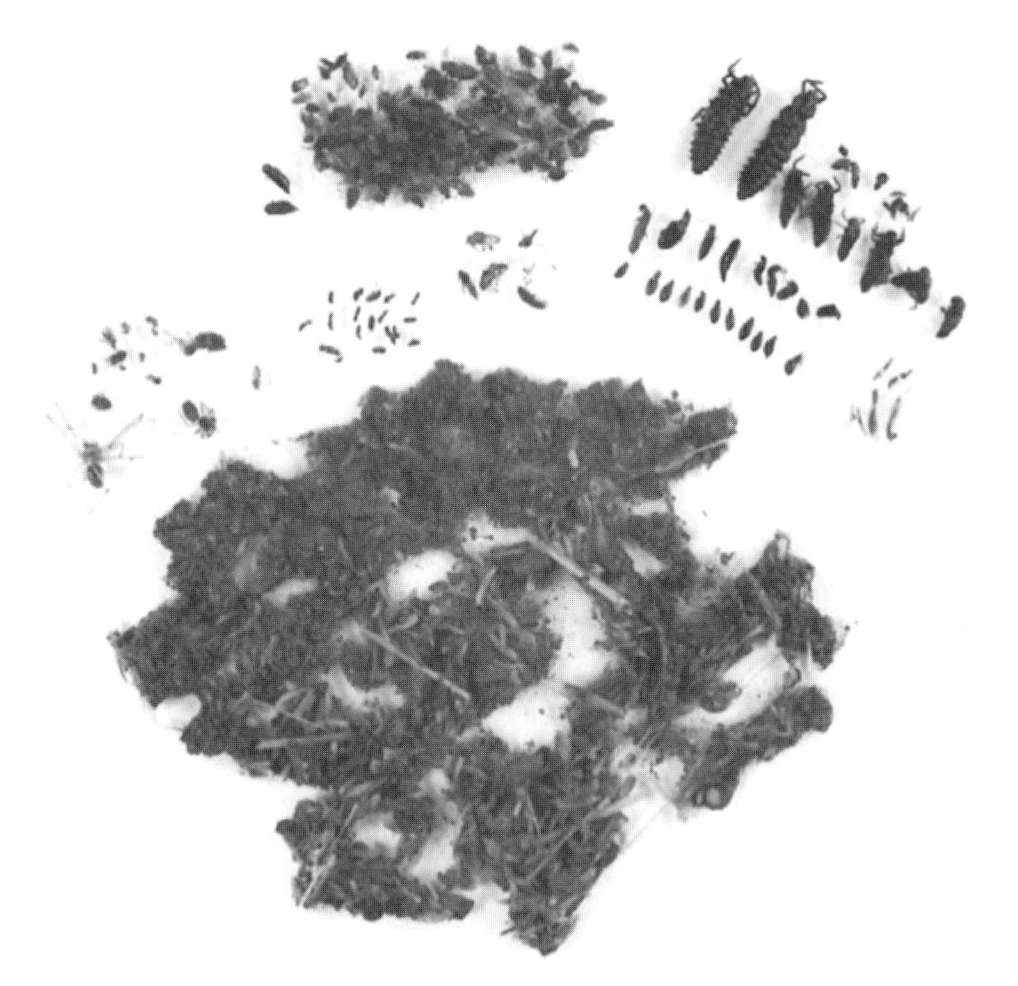

Figure 6.7 Contents of a vacuum sample taken in a cereal field in July in southern England. The catch shown is the result of five 10 s samples, each taken from an area of 0.1 m^2. The catch is preserved in 70% ethanol and is divided into Hemiptera (mainly aphids), Coleoptera (including larvae), Diptera, Hymenoptera 'Parasitica', spiders and soil and plant debris.

pler was 60-70% for pests and natural enemies in apple trees, when compared with destructive sampling. This was superior to the use of a beating tray which failed to detect half the species present, but *Orius* bugs were under-represented (Suckling *et al.*, 1996).

The time of day when samples are taken can also influence vacuum net catches, because the locomotor activity of most natural enemies varies during the day. Vickerman and Sunderland (1975) used a vacuum net and sweep nets to compare the diurnal and nocturnal activity of predators in cereal crops. More adults and larvae of staphylinid beetles (*Tachyporus* spp.) were caught during the hours of darkness than in the daytime, as were spiders, hover-fly larvae and earwigs (*Forficula auricularia*). Whitcomb (1980) used a vacuum net to sample spiders in soybean fields, but Sunderland (1987) advocated a combination of vacuum netting and ground-searching to estimate spider densities in crops, because some species are more accurately assessed by the latter technique than by the former, whereas the reverse is true for other species. More robust predators, such as adult beetles, and those which can rapidly move out of the way of the advancing operator, are less efficiently sampled by a vacuum net (Sunderland *et al.*, 1987b). Green (1999) compared spider diversity from diurnal and nocturnal vacuum net sampling of citrus trees with that from continuous pitfall sampling of the ground below the trees. Green concluded that a combination of sampling methods and timings were vital to avoid a biased interpretation of the composition of spider assemblages.

Because of the variability in sampling efficiency in relation to different natural enemy groups, vacuum netting is not always useful for comparing the relative abundances of different taxa in communities. When using a vacuum net in a sampling programme it is important, whenever possible, to calibrate the data collected by comparing them with data obtained using an absolute sampling method (Smith *et al.*, 1976; Pruess *et al.*, 1977; Whitcomb, 1980; Dewar *et al.*, 1982; Dinter and Poehling, 1992; Topping and Sunderland, 1994; Dinter 1995; Sunderland and Topping, 1995). Usually, once this is done, the vacuum net can be used to obtain absolute estimates of predator population densities by sampling discrete units of vegetation. A series of samples taken in a regular grid pattern can then be used to investigate the spatial dispersion pattern of individual natural enemy species.

A further limitation on the use of vacuum nets ought to be mentioned. Moisture on the vegetation being sampled severely reduces net efficiency, and so use either during wet weather or following a heavy dew is not advised, since large numbers of insects tend to adhere to the sides of the net and to the inner surfaces of the collecting head (Törmälä, 1982). Not only does this make their efficient removal from the net difficult, but delicate arthropods (e.g. aphids, leafhoppers, linyphiid spiders) can be damaged, making their identification to sex or species either difficult or impossible. De Barro (1991) claimed that wet vegetation becomes less of a problem if a Blower/Vac sampler with a high nozzle air velocity, that expels free water through the blower tube, is used. Schoenly *et al.* (1996a) sampled from irrigated rice fields using a suction device (vacuuming $0.8\,m^3$ air min^{-1}) that delivered arthropods and water through a hose into a plastic reservoir containing a nylon mesh strainer. Ethanol was then used to wash the collected arthropods (including parasitoids, spiders, and predatory bugs) from the strainer into a collecting tube.

A lightweight (7.8 kg), petrol-engined vacuum sampler without a net ("Vortis", manufactured by the Burkard Manufacturing Company Ltd, of Rickmansworth, UK, http://www.burkard.co.uk/homepage.htm) operates on a different principle from many other samplers. Air, instead of being drawn in through the sampling nozzle, enters above ground level. Static vanes above the air inlet create a vortex, lifting insects into an expansion chamber, from where they are deposited into a detachable, transparent collecting vessel outside the apparatus. As well as being lightweight and therefore very portable, the device has several advantages over vacuuum nets:

1. Suction pressure remains constant even after many hours of operation. Because of the way the insects are accumulated, a fall-off in suction pressure does not occur (the nets in other devices need to be repeatedly unclogged of accumulated insects and debris in order to avoid a reduction in suction pressure);
2. Because of the type of suction mechanism the sampler employs, the nozzle can be applied continuously to the ground during sampling, thus saving time (with vacuum nets the nozzle has to be repeatedly lifted from the ground during sampling, to allow the entry of air);

Vacuum sampling of natural enemies from air has also been carried out (see suction traps in 6.2.11).

6.2.3 SWEEP NETTING

A sweep net comprises a fine-meshed, cone-shaped net mounted on a rigid, circular or D-shaped frame which is attached to a short handle. It is commonly used for collecting arthropods from vegetation (especially herbaceous vegetation) because it is inexpensive, highly portable and easy to use, and also because it allows the rapid collection of a large number of insects. As the name suggests, sampling with the net involves sweeping it rapidly through the vegetation so that the rigid frame dislodges insects which are then caught in the moving net. It may be either swept backwards and forwards in a simple arc or made to follow a more sinuous track, for example a figure-of-eight. However, it is important to define a standard sweeping technique before commencing a sampling programme, because both the method and the pattern of sweeping can significantly affect capture rates (DeLong, 1932; Kogan and Pitre, 1980; Gauld and Bolton, 1988). For example, Gauld and Bolton (1988) point out that with Hymenoptera 'Parasitica' the particular sweeping technique used can account for as much as a twenty-fold difference in catch size and a corresponding difference in the diversity of wasps caught. It is advisable to practice one's chosen technique in order to achieve a reasonable level of consistency.

Each sampling will normally consist of a fixed number of sweeping movements of fixed speed and fixed duration over a predetermined path through the vegetation. After the sample has been taken, less active natural enemies may be selectively extracted from the net using an aspirator. Alternatively, if the species being collected are active fliers, the sample can be transferred to a polythene bag and extracted later in the laboratory. As with vacuum net sampling, it is advisable to kill the catch immediately after capture to prevent losses from predation, unless the insects are specifically required to be kept alive.

Sweeping is a method particularly suitable for the collection of small hymenopteran parasitoids, especially chalcidoids, proctotrupoids, cynipoids and braconids (Noyes, 1989), and a sweep net has been designed specifically for this purpose (see Noyes, 1982, for details).

Although they are frequently used for sampling arthropods in crop fields, sweep nets are subject to considerable sampling variability because their efficiency is affected by a range of factors. These include the distribution, density, activity and life-stage of the organism being sampled (Ellington *et al.*, 1984) as well as vegetation height and density, and climatic conditions. In addition, the proficiency of different operators often varies significantly. It is also very difficult to relate a sweep net sample to a finite unit or area or volume of vegetation, making absolute estimates of population densities almost impossible to obtain using this method alone. Nevertheless, Tonkyn (1980) developed a formula which he used to express sweep net data as the number of insects caught per unit volume of vegetation sampled, thereby facilitating comparisons with other sampling methods.

Sedivy and Kocourek (1988), studying variation in the species composition of herbivore, predator and parasitoid communities in alfalfa crops, compared the sampling efficiency of a sweep net with that of a vacuum net. They concluded that larval lacewings (Chrysopidae) were caught equally well by both methods, but that the sweep net was more efficient in capturing adult ladybirds (Coccinellidae) and adult hover-flies (Syrphidae), whereas the vacuum net was more efficient at capturing adult lacewings, nabid bugs and parasitoid wasps. The same two methods have also been compared in relation to the sampling of spiders in soybean fields, where 34% fewer spiders were collected by the sweep net than by the vacuum net (LeSar and Unzicker, 1978). In cereal crops, Poehling (1987) found the sweep net to be more suitable than the vacuum net for sampling ladybird and hover-fly larvae. As with vacuum nets, different natural enemy species are caught with different efficiencies by the sweep net, and the time of day when sampling is carried out affects sweep net catches in the same way as it affects vacuum net catches (Vickerman and Sunderland, 1975) (subsection 6.2.2). Wilson and Gutierrez (1980) assessed the efficiency of a sweep net in the sampling of predators in cotton, comparing this method with visual counting carried out on whole plants. The sweep net was only 12% efficient in estimating total predator numbers compared with the visual counts. When individual species were considered, the sweep net was the more efficient method only for detecting lacewing adults, because these insects were easily disturbed and tended to fly away during visual counting. In addition, the vertical distribution of predators on the cotton plants affected the efficiency of the sweep net, which was most efficient for catching those species with a distribution biased towards the top of the crop canopy. Fleischer *et al.* (1985) likewise concluded that the sweep net is an inefficient method for sampling predators in cotton. Sweeping was found to be inappropriate for sampling taxa (such as Gerridae and Lycosidae) that are normally found at the base of rice hills (Fowler, 1987a). Elliott and Michels (1997) developed regression models to convert sweep net data into density estimates for ladybirds in alfalfa. When plant height was included in the model, the coefficient of determination of the regression for estimating adult density was 0.93, and the authors concluded that sweep netting was an efficient method for estimating ladybird densities in alfalfa IPM programmes.

Mayse *et al.* (1978) compared sweep netting with direct observation and the use of a **"clam trap"** for sampling arthropods in soybean. The trap is a 30 cm × 90 cm hinged wooden frame supporting a cloth or organdy bag open on three sides (Figure 6.8). It is brought down rapidly over a row of soybean plants and clamped shut with latches. The plants are then cut through at ground level after which the organdy bag (containing plants and arthropods) can be transferred to a plastic bag for transport to the laboratory, where the arthropods are extracted by washing with soapy water, and are retained in a sieve. Direct observation and clam trapping yielded similar results for the number of species recorded in different fields, but the results from sweep netting were much more variable, probably due to changes in vertical distribution of the arthropods sampled.

In summary, sweep nets, while frequently used as a tool for sampling arthropods in field crops, (and even, occasionally, in tree canopies, see Basset *et al.*, 1997) are limited in their

Figure 6.8 (a) a clam trap used for sampling arthropods from a 30 cm segment of a row crop such as soybean; (b) use of clam trap in a soybean field (Source: Mayse *et al.*, 1978) Reproduced by permission of The Entomological Society of Canada.

usefulness for investigations of natural enemy populations, and so should be used with caution, particularly if quantitative information is being sought. Also, like vacuum nets, they perform poorly under wet conditions. Further descriptions of sweeping techniques and the factors which influence sweep net catches are given by DeLong (1932); Saugstad *et al.* (1967); Kogan and Pitre (1980); Southwood and Henderson (2000).

6.2.4 MALAISE TRAPPING

Malaise traps, first designed by René Malaise (Malaise, 1937), are tent-like interception devices that are particularly useful for obtaining large quantities of insect material for faunal surveys, studies of the relative abundance of species, phenological studies, studies of diurnal activity patterns, and taxonomic investigations (Steyskal, 1981; Gressitt and Gressitt, 1962; Townes, 1962; Butler, 1965; Ticehurst and Reardon, 1977; Matthews and Matthews, 1971, 1983; Sivasubramaniam *et al.*, 1997). They are especially recommended for the collection of the adults of entomophagous insects such as Asilidae, Dolichopodidae, Empididae, Pipunculidae, Syrphidae, Tachinidae, and certain parasitoid Hymenoptera (Benton, 1975; Owen, 1981; Owen *et al.*, 1981; Gilbert and Owen, 1990; Belshaw, 1993; Quinn *et al.*, 1993; Hågvar *et al.*, 1998). Spider catches are usually small, but Malaise traps are useful for monitoring species that are not active on the ground and therefore are not caught in pitfall traps. Koomen (1998) caught 61 species of spider in one trap over two years. Traps are nowadays constructed of fine mesh fabric netting, and incorporate a vertical panel (matt black in colour) that serves to direct insects upwards to the roof apex (a commercial supplier of this type of trap is Marris House Nets, of Bournemouth, UK) (Figure 6.9). Alternatively, traps can be constructed from inexpensive lightweight materials, such as folding wood tripods secured with a bolt and wingnut and covered with spunbonded polyester netting. Platt *et al.* (1999) describe a 1.8 kg trap constructed from these materials that is easy to install and folds compactly for transport. Insects entering Malaise traps

Figure 6.9 A Malaise trap of the type manufactured by Marris House Nets, UK.

accumulate at the highest point within the roof (i.e. the apex) and pass eventually into a collecting bottle or jar that contains preservative (usually 70-95% ethanol or isopropyl alcohol or a sodium benzoate solution with detergent). 200 ml of 70% ethanol is sufficient for weekly collections in shaded habitats, but 250 ml are needed in habitats exposed to full sun (Cresswell, 1995). If diel activity cycles are under study, it is possible to replace the collecting bottle with a time-sorting mechanism (based on a quartz clock motor powered by a 1.5 V battery) which directs the 24 hour catch into 12 perspex tubes at 2 hour intervals (Murchie *et al.*, 2001). Individual Malaise traps can be highly variable in their catch rates (Longino and Colwell, 1997). For most kinds of insect, siting and orientation of traps is likely to be crucial; boundaries between different vegetation types, e.g. the edges of forest clearings, should be used, to exploit the fact that insect flight paths tend to be concentrated in such areas, while the collecting chamber end of the trap ought to point towards the Sun's zenith to exploit the positively phototactic responses of the insects. Malaise traps are normally positioned on the ground, but traps with rigid frames can be hoisted into tree canopy habitats (Faulds and Crabtree, 1995). However, Stork and Hammond (1997) showed that Malaise trapping in the canopy of an oak tree recovered half as many species of beetles as fogging (6.2.5), and some of the most abundant species, known to be associated with oak, were absent. Unless diurnal activity patterns are being investigated, Malaise traps can be left for several days before emptying, although some workers report catches so large that daily collection is necessary (Gilbert and Owen, 1990). Malaise traps can be modified to record insects entering the trap from two opposite directions. Traps modified in this way were used by Hossain *et al.* (2002) to record parasitoids, ladybirds and hover-flies dispersing from harvested into unharvested plots of lucerne.

Malaise traps do not provide data on the absolute abundance of insects, but they can yield useful biodiversity data. Gaston and Gauld (1993) reported on a network of Malaise traps operated at seventeen sites in Costa Rica for more than a hundred Malaise trap-years. One hundred and fifty species of Pimplinae (Ichneumonidae) were caught, and this was considered to be an accurate estimate of species richness, because: (a) when the catches of the seventeen sites were successively added together in a random sequence the curve for cumulative number of species plotted against cumulative number of sites approached an asymptote; (b) few extra species were added by casual collection using other methods at other sites in Costa Rica. This assessment of biodiversity was achieved in spite of Pimplinae being scarce in Costa Rica (overall mean of only six individuals caught per trap per month). Maes and Pollet (1997) considered Malaise traps to be useful for large-scale inventories of Dolichopodidae, but less suitable for ecological studies of these flies. This is because most dolichopodids fly for only short distances and a large number of traps would be necessary to counteract the resulting variability of catch.

An interception device similar in operation to the Malaise trap was devised by Masner and Goulet (1981). It incorporates a vertical polyester net treated with a synthetic pyrethroid insecticide. Intercepted insects crawling on the net are killed by the insecticide and fall onto a plastic tray placed beneath. A clear polythene roof minimises rain damage and deflects positively phototactic insects back onto the net. In the field, this design provided a larger catch of small-bodied Hymenoptera than did a Malaise trap (Masner and Goulet, 1981). Masner (in Noyes, 1989), and Campos *et al.* (2000), later improved the efficiency of the trap by setting a yellow trough into the ground below the intercepting vertical net. Noyes (1989) used a Masner-Goulet trap without treating the vertical net with insecticide, and obtained poor catches of parasitoid wasps compared with a Malaise trap, in tropical rain forest. Basset (1988) devised an apparatus that combined a Malaise trap with a window trap (6.2.11) for sampling the canopy faunas of rainforest trees. Natural enemies either flew or crawled into the Malaise trap or hit the plexiglass sheet of the window trap and fell into the collection vessel (containing 20% ethylene glycol) below. Natural enemies reacted differently to the two trap components, which therefore varied between species in collection bias, and the results underscored the desirability of employing more than one sampling technique in community studies. In spite of this, spiders were under-represented in the composite trap compared with canopy fogging (6.2.5). Basset *et al.* (1997) describe other variants of composite flight-interception traps for use in tree canopies. When comparing Malaise traps and window traps (as separate pieces of apparatus) at a forest edge, Schneider and Duelli (1997) found that window traps were more effective for catching spiders, but Malaise traps were more efficient for Hymenoptera and Dermaptera.

Malaise traps need to be located away from ant's nests, as the ants can severely reduce catches.

Drift nets (the aquatic equivalent of the Malaise trap) can be anchored in streams at any required depth and used to sample aquatic insects moving downstream with the current. Daytime-only sampling could underestimate abundance, since much drift occurs at night (Williams and Feltmate, 1992).

6.2.5 KNOCK-DOWN

Introduction

Knock-down involves the dislodgement of insects from their substratum (usually vegetation), causing them to fall onto either a tray, a funnel, a sheet or a series of such devices situated beneath.

Mechanical Knock-down

A common method employed in sampling invertebrates from vegetation is to either shake plants or beat them with a stick, causing the invertebrates to fall on to either a white cloth or polythene sheet laid on the ground, or onto a beating tray (a device resembling an inverted white umbrella, Figure 6.10) (Jervis, 1979, 1980a). The fallen invertebrates are then either collected by hand or with an aspirator. Alternatively, they may be beaten into either a large plastic funnel fitted with a collecting jar containing a preservative (Basset *et al.*, 1997; Rieux *et al.*, 1999), a cloth funnel that directs them into a plastic zip-lock bag (Roltsch *et al.*, 1998), or a semi-rigid polythene funnel which can be folded (Morris *et al.*, 1999a).

This method is useful for estimating numbers of slow-moving arthropods which are easily dislodged from the plant, but it is advisable to calibrate the technique against other, absolute, sampling methods (Kogan and Pitre, 1980). The density of spiders knocked off soybean plants onto a ground cloth was not significantly different from the density estimated from plant fumigation (Culin and Rust, 1980). Majer *et al.* (1996) found that branchlet shaking was a viable alternative to chemical knockdown for sampling natural enemies such as spiders and small wasps (but not for tiny predatory mites). They reported that sixty 30 cm-long branchlets could be sampled per tree in 15 min and that four workers were able to sample ten trees adequately in a day. Fleischer *et al.*, (1985) found that the beating method gave estimates of predator density in cotton that were less than one-third the size of those obtained by removing bagged plants to the laboratory. Beating has also been used to sample spiders from apple trees (Dondale *et al.*, 1979) and to assess the numbers of adult anthocorid bugs, lacewings and ladybirds feeding on psyllids in pear trees (Hodgson and Mustafa, 1984). Beating and sweeping have been used to collect spiders from the forest canopy. This was facilitated by operating from a "**canopy raft**" (i.e. a 580 m^2 inflatable hexagonal platform transported to the canopy by an air-inflated dirigible) (Basset *et al.*, 1997). Beating is more commonly used than sweep netting to sample arthropods in trees, but Radwan and Lövei (1982) found that beating ladybirds from apple trees collected only 30% of the beetles observed by visual searching. Coddington *et al.* (1996) also found that beating was less productive than visual search for estimating spider species richness on hardwood trees. It is sobering that the samples were quite distinct taxono-

Figure 6.10 Use of a beating tray to collect insects from a tree canopy.

mically and that the two methods, although applied to the same vegetational strata, seemed to be accessing different components of the arachnofauna.

Pale-coloured insects may prove difficult to locate on white fabric such as that in a standard beating tray. This problem may be overcome by using a dark-coloured fabric.

When beating is carried out in warm, sunny conditions, winged insects that fall on to the collecting sheet or tray surface tend to fly off from the latter very readily. It is therefore advisable in such circumstances to have a small team of workers who can remove insects as they land. An alternative is to use a beating tray modified to reduce the problem of escaping insects. Morris and Campos (1996) used a transparent polythene cone attached to a plastic ring which could be moved as easily as a sweep net. A detachable container was fitted below the cone to trap insects that were beaten into the apparatus. Newton and Yeargan (2002) determined the density of nocturnal harvestmen (Opiliones) in soybean by beating the plants at night (illuminated by headlamps) within a defined area temporarily enclosed by wooden boards. Dislodged harvestmen, and individuals secreted in the litter, were then collected from the ground.

Belshaw (1993) used a beating technique, together with a box trap (Figure 6.11) to collect resting Tachinidae from ground vegetation.

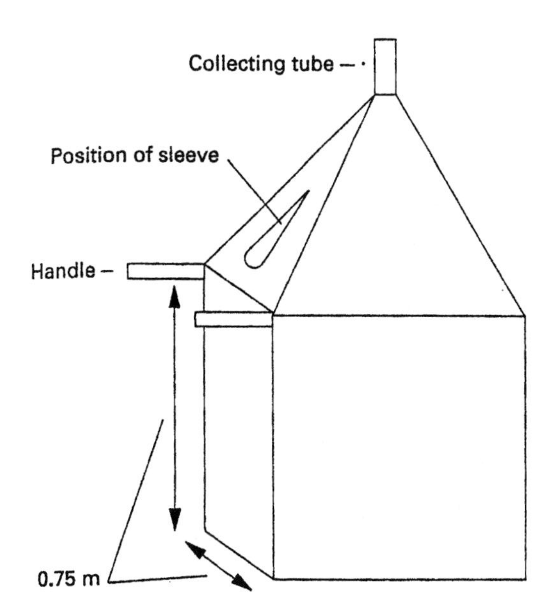

Figure 6.11 Box trap used by Belshaw (1993) to collect insects, disturbed by beating, from ground vegetation. Reproduced by permission of the Royal Entomological Society of London.

The box trap comprises a muslin-covered, open-bottomed metal frame. Using the handles, the device is placed quickly on the ground, and the vegetation thus enclosed is beaten with a stick. Winged insects (Tachinidae, in the case of Belshaw's study) disturbed in this way are collected either in the removable tube situated at the apex of the trap, or (via the sleeve, using an aspirator) from the insides of the cage. Because the box trap covers a known area of vegetation, Belshaw's technique can be used to obtain absolute measures of the density of insects. As the insects are not collected from beneath the plant, it could be argued that this is not, strictly speaking, a knock-down technique.

Comparison of beating sheets, sweeping and pitfalls in peanut fields showed that pests and natural enemies were represented to different degrees, on a species by species basis, using the three methods. It is recommended that at least two methods be used to assess the effect of a pest management practice on pests and beneficials (Kharboutli and Mack, 1993).

Chemical Knock-down

Natural enemies in tree canopies can be sampled directly and precisely from large expensive structures such as tower cranes (Parker *et al.*, 1992) and canopy rafts (Basset *et al.*, 1997), but, since most research budgets preclude the use of such equipment, chemical knock-down techniques provide a valuable alternative.

Insecticide knock-down, pioneered by Roberts (1973) and popularly known as **fogging**, is being used increasingly in faunistic and other investigations of parasitoids and predators, particularly in forest canopy habitats and hedgerows (Neuenschwander and Michelakis, 1980; Noyes, 1984; Askew, 1985; Noyes, 1989; Joyce *et al.*, 1997; Tobin, 1997; Watt *et al.*, 1997; Stork *et al.*, 2001). A pyrethroid insecticide (e.g. Resilin E, Resilin 5OE; Noyes, 1989) fog is sprayed from the air by plane or helicopter, or is released from a device that is either held by the operator (Figure 6.12a) or is hoisted into the tree canopy using a system of ropes and pulleys. Fitting a radio-controlled servo unit to the release valve of the fogger prevents wastage of insecticide and poor targetting of fog as the machine is hoisted, or when its exhaust pipe is pointing in the wrong direction (Stork and Hammond, 1997). Adis *et al.* (1998) list the commercial fogging machines currently available. Insects can be collected by placing, beneath the canopy, either a sheet on the ground (this method may, however, lead to problems with ants plundering the catch) or several funnel-shaped trays slung from ropes (Figure 6.12b) (Longino and Colwell, 1997; Stork and Hammond, 1997; Watt *et al.*, 1997; Stork *et al.*, 2001). Researchers have used a wide range of combinations of insecticides, delivery systems and arthropod collection methods, and Adis *et al.* (1998) make constructive recommendations for standardisation of protocols so that future data can be reliably compared. Catch size is not related to insecticide concentration in a simple manner and stronger concentrations do not necessarily increase the catch (Stork and Hammond, 1997).

Figure 6.12 Chemical knock-down of insects from the forest canopy, using the 'fogging' technique: (a) application of the fog; (b) devices used to collect insects as they drop from the canopy. Reproduced by permission of Nigel Stork.

After five minutes of fogging an Australian rainforest canopy with natural pyrethrins, Kitching *et al.* (1993) collected falling arthropods for 4 hours and noted that 63% of the total catch was obtained by the first hour, 78% by the second and 90% by the third. Similarly, Stork and Hammond (1997) reported that 80% of the catch from birch trees in England were obtained within 75 – 85 min. It is not sensible to re-sample the same tree until recolonisation has occurred (Basset *et al.*, 1997). Faunal recovery rates vary considerably due to factors that are incompletely understood, but may often exceed ten days (Stork and Hammond, 1997). Morning fogging collects fewer unwanted "tourist" species than does evening fogging, at least in UK deciduous woodland (Stork and Hammond, 1997). Fogging of a whole tree canopy does not provide information relating to particular strata within the canopy, cannot be used effectively under windy conditions (when ground windspeeds are a gentle 5 km h^{-1} canopy windspeeds above 7 m can be in excess of 15 km h^{-1}), and loses any material that does not fall vertically (Majer *et al.*, 1996; Basset *et al.*, 1997). An approximate estimation of the stratum from which the natural enemies originate can, however, be obtained by concentrating the fogging in specific strata or by suspending the collecting funnels at various heights (Kitching *et al.*, 1993). More precise distributional data can be obtained by "**restricted canopy fogging**", which involves trapping the foliage in a 60,000 cm^3 conical plastic enclosure, exposing it to carbon dioxide at a pressure of 1 bar for twenty minutes, and then shaking the enclosure vigorously for a further ten minutes. Narcotised natural enemies are transferred to alcohol via a collector attached to the bottom of the enclosure. This method was found to extract 80 – 90% of arthropods from the foliage (Basset, 1990). A comparison of canopy fogging with restricted canopy fogging suggested that the latter collected more spiders and small invertebrates but fewer active fliers such as Diptera. Restricted canopy fogging, however, cannot be applied to wet foliage, does not collect invertebrates from the trunks of trees (Basset, 1990), and the foliage is disturbed when the enclosure is positioned (Basset *et al.*, 1997). Stork and Hammond (1997) observed that some beetles and bugs will fly or drop off foliage as an escape reaction to even the minor disturbance of a sudden movement by the researcher, and these "nervous" species are, therefore, likely to be underestimated by restricted canopy fogging. Floren and Linsenmair (1998) carried out tree-specific sampling of Coleoptera by fogging trees that had 100 m^2 cotton roofs fitted over their crowns, to exclude beetles from other trees of the higher canopy.

Based on experience with preparing an inventory of ant species in rain forest, Longino and Colwell (1997) noted that fogging is efficient in generating a large number of specimens in a short period of time, compared with Malaise traps, which must be left in the field for a long time to achieve the same result.

Mist-blowing is similar to fogging, but less dependent on wind speed and direction to transport the insecticide. Mist-blowing uses ultra-low-volume techniques to generate a fine mist which is propelled into the canopy (Chey *et al.*, 1998). Use of a fluorescent marker showed that, with this technique, insecticide coverage of both leaf surfaces was similar and there was little variation between species of tree, but, when applied from the ground, the mist did not reach much higher than 7 metres (Chey *et al.*, 1998).

Chemical knock-down was used to sample parasitoids of Oriental fruit-fly in guava orchards in Hawaii. Large plastic sheets were put under trees, a pyrethrin mist was blown into trees, which were then shaken vigorously for four minutes, and the parasitoids subsequently removed from the sheets. These parasitoids are poor flyers and unable to avoid the pyrethrin, which gives very fast knock-down (Stark *et al.*, 1991).

Fogging can also be used for extracting insects from field crops; for example, Kogan and Pitre (1980) used a large, plastic fumigation cage in the collection of arthropods from soybeans.

6.2.6 VISUAL COUNTING

Although relative estimates of natural enemy densities are useful when investigating the effects of experimental treatments, it is often desirable to obtain absolute density assessments for natural enemy populations in the field (Chapters 1, 7 and 8). The most common way of doing this, especially for conspicuous life-stages, is to visually count individuals *in situ*, on either a defined-area or period-of-search basis. This can extend to the visual searching of entire small trees to record absolute numbers of selected natural enemy species (Wyss, 1995).

Coddington *et al.* (1996), in a study of spider species richness, compared visual search of the foliage of hardwood trees with beating, and visual search of the litter with litter extraction. Visual search yielded more species than other methods but there was a significant degree of variation in density and species richness attributable to the level of collecting experience of the four individuals conducting the survey. For the counting of less conspicuous stages or those in concealed locations, plants or other parts of the habitat can be removed to the laboratory for either extraction or examination (Southwood and Henderson, 2000) (6.2.7).Visual counting is also a good method for examining spatial dispersion patterns of natural enemy species, provided that detectability of insects remains constant across the study area.

Lapchin *et al.* (1987) compared three methods of estimating densities of hover-flies and ladybird beetles in winter wheat crops:

1. An observer walked at a steady pace through the crop and counted all adults and larvae seen within a period of 2 min (**census walk method**);
2. An observer made a detailed search of plants and soil surface in a defined area, and, immediately afterwards, a second observer searched the same area. Predator density was then calculated using De Lury's method (Laurent and Lamarque, 1974): $P = C_1^2/(C_1 - C_2)$, where P is predator population density, C_1 is the number of predators collected by the first observer and C_2 the number collected by the second. The method assumes equal capture efficiency for the two counts, and no emigration or immigration between counts.
3. Plants were cut and removed to the laboratory for detailed examination.

The visual searches carried out in the field (1. and 2.) proved to be unsuitable for estimating the densities of hover-fly larvae because of the insects' cryptic colouration, their relative immobility and their habit of resting in concealed positions on plants. The population density of hover-fly larvae estimated by detailed searches in the field was less than 1% of that estimated from plant removal and examination (6.2.7). However, the detailed searches using De Lury's method were adequate for assessing absolute numbers of ladybird adults and of final-instar larvae, although they were not reliable for earlier instars. The census walks provided less accurate estimates of absolute density but they were satisfactory for determining seasonal trends in predator numbers, especially numbers of adult ladybirds.

Chambers and Adams (1986) also used detailed visual searches to estimate numbers of hover-fly and of ladybird eggs and larvae in winter wheat. Plant shoots were searched *in situ* and additionally the shoots in $0.1\,m^2$ quadrats were cut and counted and the resulting stubble, together with the soil surface and any weeds present, was searched. Mowes *et al.* (1997) showed that counting predators on tillers of winter wheat was nearly as accurate as "absolute" density sampling (suction sampling in a cage and examination of cut vegetation in the laboratory) for Coccinellidae, Syrphidae and Chrysopidae, but not for the majority of groups of polyphagous predators. However, detailed visual searching is labour-intensive, and Poehling (1987) concluded that, although it was the most accurate method of estimating ladybird and hover-fly numbers in cereal crops, it was not appropriate when large numbers of experimental plots needed to be assessed in a short time. It is especially labour-intensive in habitats with dense vegetation. Duffey (1974) found that 60-90 man-minutes were needed to search $0.5\,m^2$ of calcareous grassland and remove the resident spiders. Mobile predators on exposed surfaces (such as tree trunks and branches) are more evident and counting is easier. Peng *et al.* (1999) estimated abundance of the green ant, *Oecophylla smaragdina*, on cashew by counting the number of ant trails on the main branches and by ranking trails according to the amount of ant activity present.

Frazer and Gilbert (1976) used a census walk method to assess adult ladybird densities in alfalfa fields. This entailed walking along either side of each crop row and counting all beetles that were visible. However, because these counts were influenced by the weather, beetles being more active and therefore more easily seen when temperatures were high, they also searched 30 cm lengths of crop row more thoroughly. Even so, they found that these counts never exceeded 25% of the beetles actually present, because the ladybirds spend most of their time in the stubble at the base of the crop. This finding highlights the influence of insect activity on the efficiency of visual searches and census walks in the field, active individuals being far easier to see than inactive ones. Both the degree of predator hunger and the time of day can affect the activity of ladybirds (Frazer and Gill, 1981; Davis and Kirkland, 1982). Time of day can also affect counts of hover-fly larvae; twice as many were found during detailed night-time searches compared with day-time searches of plants and the soil surface in an oat crop (Helenius, 1990). A further factor influencing visual field assessments is the observer's experience with the method. When assessing adult ladybirds in strawberry crops using census walks, Frazer and Raworth (1985) noted significant differences between the numbers recorded by different observers. Similarly, variation among observers accounted for a large proportion of the error in predator density estimates from visual searches in cotton crops (Fleischer *et al.*, 1985). Because of the effect of hunger on activity, Frazer (1988) advocates the use of sampling procedures which are specifically designed to estimate only the numbers of active, hungry individuals where predictions of future predation rates are required. Weseloh (1993) used a census walk method to determine the abundance of ants in different plots.

It is possible to obtain data on the densities of adult parasitoids using visual counting in the field, if the parasitoids are either large-bodied and conspicuous or slow-moving. Many parasitoids search the food plants of their hosts by walking, so facilitating visual observations in the field. The numbers of adult *Diadegma* ichneumonid wasps foraging over Brussels sprouts plants in an experimental plot were recorded by a pair of observers using binoculars (Waage, 1983). The observers checked all sides of each plant but did not approach closer than 5 m, to avoid disturbing the wasps. Hopper *et al.* (1991) counted adult *Microplitis croceipes* (Braconidae) along 60 m rows of cotton plants, catching each individual with a hand net in order to establish its sex.

Obviously, census walk methods are only appropriate for highly conspicuous insects, and the methodology that has been developed for monitoring butterfly populations in Britain (Pollard, 1977) could be applied to certain natural enemies, such as ladybirds, hover-flies, dragonflies or large-bodied parasitoid species. Because weather affects flight activity, butterfly census walks are not done when the temperature is below 13°C, and between 13°C and 17°C they are only done in sunny conditions (Pollard, 1977). It is also important to define the distance limits, on either side of the observer, within which the counts are made during a census walk. During the butterfly counts use is made of natural features such as footpaths or forest rides, but for census walks through crops, boundaries can be defined in terms of crop rows or by placing markers, such as bamboo canes, along the route. The butterfly census walk data are used to calculate indices of abundance for individual species at each census site. The mean count per census walk is calculated for each week and these are summed to give the index of abundance for the season. Population changes from year to year can then be assessed by comparing these indices.

In some cases, parasitised insects can be readily distinguished from healthy individuals, either in the later stages of parasitism, or immediately after death, and their numbers assessed by visual counting. An obvious example is the 'mummification' of parasitised aphids. Aphid 'mummies', resulting from parasitism by aphidiine braconids (Figure 6.13) or Aphelinidae, act as protective cells within which larval parasitoids complete their metamorphosis to the adult stage. Mummies are conspicuous within aphid colonies and are therefore readily counted *in situ*. Consequently, mummy counts

Figure 6.13 Aphid 'mummies': *Sitobion avenae* parasitised by *Aphidius rhopalosiphi* (Braconidae) on ears of wheat.

are frequently used to estimate aphid parasitoid abundance (Lowe, 1968; van den Bosch *et al.*, 1979; Carter and Dixon, 1981; Messing and Aliniazee. 1989; Feng and Nowierski, 1992; Hance, 1995; Lapchin *et al.*, 1997). However, it is advisable to combine this sampling technique with other methods of assessing parasitoid populations, such as the rearing or dissection of parasitoids from samples of live aphids (6.2.9, 6.3.5, 6.3.6), since mummy counts alone can give misleading results for several reasons:

1. Some parasitised aphids leave their colonies, and may even leave the food plant altogether, prior to mummification (Powell, 1980; Dean *et al.*, 1981; Brodeur and McNeil, 1992);
2. Many aphid parasitoids pass through diapause in the aphid mummy and the proportion doing so increases as the season progresses (Singh and Sinha, 1980), leading to the accumulation of mummies in the study site, and so increasing the likelihood that the same individuals are counted in successive samples;
3. Heavy rain or strong wind may dislodge mummies from plant leaves.

Visual searching of plants in the field or of cut plants in the laboratory has been used to assess predators on a variety of other field crops including soybeans (Kogan and Herzog, 1980), sorghum (Kirby and Ehler, 1977), brassicas (Smith, 1976; Horn, 1981; Landis and Van der Werf, 1997) and cotton (Fleischer *et al.*, 1985). Fleischer *et al.* (1985) placed cylindrical cloth bags over whole cotton plants, tying the bottom securely around the base of the main stem. The bags were left collapsed on the ground for a week and then two people rapidly pulled the bags up over the entire plants, tying off the top before removing bagged plants to the laboratory for subsequent visual examination.

Soil-surface-dwelling polyphagous predators are more difficult to assess visually in the field because many of them are active mainly at night, and spend much of the daytime concealed in the litter on the soil surface, under stones or in the soil itself. Brust *et al.* (1986a,b) estimated ground predator numbers in corn by placing metal quadrats (13 x 75 cm) over plant rows and visually searching the surface litter and the soil to a depth of 0.5-1.0 cm. Winder *et al.* (1994) searched within quadrats, then used a trowel to transfer weeds, stones and plant roots to a plastic tray for further examination.

Sunderland *et al.* (1986a, 1987a) estimated linyphiid spider densities in cereal fields by searching the crop and the top 3 cm of soil within 0.1 m^2 quadrats, and collecting the spiders with an aspirator. Nyffeler (1982) used 1 m^2 quadrats for counting large spiders on vegetation, but 0.04 – 0.16 m^2 quadrats for counting small spiders on the soil surface. Harwood *et al.* (2001a; 2003, 2004) used 0.008 m^2 mini-quadrats, which had sufficient spatial precision to enable predators and prey numbers to be quantified both within web-sites of linyphiid spiders and separately in areas between web-sites. At this scale it is possible to analyse invertebrate densities within confined microhabitats, such as within and between rows of a cereal crop. Use of quadrats may underestimate the density of large, active adult spiders (such as Lycosidae) which can flee from the area of disturbance. Nyffeler and Benz (1988a) used a different method to estimate numbers of linyphiid spiders in winter wheat fields and hay meadows; they counted the number of webs within randomly-placed quadrats, but conceded that their counts were probably underestimates, because some spiders occupied cracks in the soil. Toft *et al.* (1995) used potato starch powder to reveal webs, then measured the distances from a random point in the field to the ten nearest webs. "Distance methods" (Krebs, 1999; Southwood and Henderson, 2000) were then used to estimate the density of web-building spiders; e.g. $N_1 = n/\pi\Sigma(X_i^2)$, where N_1 = estimate of population density from point-to-web data, n = sample size, and X_i = distance from random point to nearest web. Key deficiencies of this method are that not all individual spiders produce webs, and some species produce webs on a daily basis, these may remain intact for several days or be destroyed immediately by heavy rain.

Sunderland *et al.* (1987b) attempted to achieve an accurate estimate of total predator density in cereal crops by combining a range of sampling techniques. Insects were first extracted using a vacuum net with a sampling head which covered 0.1 m^2. The sampled area was then isolated by means of a metal cylinder which was driven into the ground to a depth of 8-10 cm, and the plants within this area were cut and removed to the laboratory for close visual examination. The soil surface was searched visually and the predators were collected with an aspirator. Next, pitfall traps were set in the enclosed area which was further isolated by sealing the top of the metal cylinder with a fine mesh net. It was concluded that any single sampling method will underestimate predator density, and that a more accurate estimate is obtained by combining different methods. However, in any system the advantages of intensive sampling methods must be balanced against the disadvantages of excessive labour requirements. Increased effort at each sample site must be traded off against numbers of replicate sites that can be managed, potentially reducing the statistical power of an ecological study.

6.2.7 EXTRACTION OF NATURAL ENEMIES FROM LIVING PLANT TISSUES, LEAF-LITTER AND SOIL (INCLUDING EMERGENCE TRAPS AND SOIL-FLOODING)

Some natural enemies spend part of their life-cycle concealed, along with their prey or hosts, within plants. The prey- or host-infested parts of the plants can be removed for subsequent dissection. In some cases, the plants infested by the hosts or prey, e.g. gall-causing insects and leaf-miners, are relatively easy to distinguish from uninfested ones, and these alone need to be collected. However, plants containing the early stages of gall-causers, and those containing borers, can be difficult or impossible to recognise. The removal of whole plants or parts of plants for subsequent close examination is the most efficient way of counting life-stages which are not highly mobile. Harris (1973) assessed several methods for estimating numbers of aphidophagous cecidomyiid fly larvae attacking aphid colonies on a variety of plants. Visual searches in the field proved difficult when aphid colonies were very dense or occurred in protected situations such as curled leaves or galls. An alternative method was to place samples of aphid-infested plant material into polythene bags and keep these in the laboratory for 2-3 days. Subsequently, older predator larvae left the aphid colonies and could be counted as they crawled on the inner surface of the bag. However, the most efficient method was to shake samples with 70% ethyl alcohol in polythene bags and then wash them into a plastic dish where insects could be brushed from the plant material with a soft brush. After removal of larger individuals the samples were washed through a series of filters to retrieve the eggs and small larvae. Field searches and incubation in polythene bags were respectively only 11% and 32% as efficient as the alcohol-washing method.

Collection of plant parts can be extended even to sampling from large trees. Abbott *et al.* (1992) used a cherrypicker to cut branchlets from the crowns of jarrah (*Eucalyptus marginata*) at 14 m above ground. Each branchlet was allowed to fall into an open plastic bag held just below the branchlet, invertebrates were killed immediately by adding to the bag a wad of cotton wool soaked in ethyl acetate, and the contents of the bag were later sorted and identified in the laboratory. Pettersson (1996) sawed branches off spruce trees, transported them wrapped in plastic, and then collected spiders by hand as the branches dried out in the laboratory. Similarly, Halaj *et al.* (1998) harvested and bagged 1 m-long tree branches in a survey of spider abundance and diversity in forest canopies.

Predators or parasitoids concealed in soil or leaf-litter can be studied by taking a sample of the concealing medium and extracting the natural enemies either by hand (Dubrovskaya, 1970; Franke *et al.*, 1988; Thomas *et al.*, 1992), by extraction equipment (Duffey, 1962; Kempson *et al.*, 1963; Edwards and Fletcher, 1971; Bonkowska and Ryszkowski, 1975; Pollet and Desender, 1985; Hassall *et al.*, 1988; Helenius *et al.*, 1995; Pfiffner and Luka, 2000) such as a Tullgren funnel apparatus (Workman, 1978; Chiverton, 1989), or by soil washing and flotation (Desender *et al.*, 1981; Dennison and Hodkinson, 1984; Sotherton, 1984, 1985), or by various combinations of these methods (Alderweireldt, 1987; De Keer and Maelfait, 1987, 1988). If the aim is to quantify species composition, it should be noted that the efficiency of heat extraction techniques, such as Tullgren funnels, can vary greatly between taxa of invertebrates, and even between species within a family (Snider and Snider, 1997). The Winkler/Mozarski eclector is a suitable litter extractor for expeditions because it does not require a source of power and is lightweight (five eclectors can be transported easily in a rucksack). This extractor exploits the escape responses of disturbed invertebrates which pass through the nylon mesh of a litter bag into a collecting vessel placed below (Besuchet *et al.*, 1987). Edwards (1991) reviewed many of these extraction techniques and concluded that dry funnel techniques are most efficient for micro- and mesoarthropods, but flotation techniques are more efficient for macroarthropods. To sample ants in forest and grassland habitats, Morrison (2002) searched under rocks, logs and litter, then extracted ants from the leaf litter using Berlese funnels and the Winkler method. Sotherton (1984) used a spade to dig out soil cores (20 cm x 20 cm), to a depth of 35 cm or to bedrock, which were extracted by breaking them up into a large container of saturated salt solution. Organic matter, which floated, was removed using a fine-mesh sieve. Predators were then extracted by hand. Spence and Niemelä (1994) used a similar method to collect carabids from litter. Litter from 0.25 m^2 of forest floor was placed in a large container of water and carabids were collected as they came to the surface. This litter washing technique was 96% efficient (based on recovery of marked beetles added to "litter samples"), and estimates of carabid density for litter washing were twice those for Tullgren funnel extraction. The efficiency of extraction can be determined by adding known numbers of natural enemies to a unit of defaunated habitat, or, alternatively, by carefully searching the processed habitat sample for unextracted corpses. The first method is expected to understimate and the second to overestimate true efficiency (Sunderland *et al.*, 1995a) and so the correct efficiency value is likely to fall between the two estimates.

Removal of samples of habitat from aquatic systems can be achieved using a freeze-corer, Ekman grab, or multi-core sampler (see details in Williams and Feltmate, 1992). Over the years, various devices have been designed for extracting samples of vegetation from aquatic habitats (Hess, 1941; Gerking, 1957; McCauley, 1975; James and Nicholls, 1961; Williams and Feltmate, 1992). An extensive review of these and other techniques for sampling aquatic habitats can be found in Southwood and Henderson (2000). The substratum from areas of stream beds, delineated by quadrats, can be removed with the aid of a Surber sampler (see Surber, 1936, for details) and taken to the laboratory for extraction of invertebrates, including predators (Hildrew and Townsend. 1976, 1982; Woodward and Hildrew, 2002). Since the Surber sampler does not catch all the invertebrates disturbed from the stream bed, it cannot be used for absolute population estimates (Southwood and Henderson, 2000).

Adult insects emerging from soil (e.g. from pupae contained therein) can be collected using an **emergence trap** (Southwood and Henderson (2000), describe various designs). Such traps comprise a metal, plastic or opaque cloth box that covers a known area of the soil surface. Insects are collected in transparent vials situated at either the sides or the top of the box. Emergence traps can also be constructed from wooden or perspex frames (e.g. enclosing 0.5 - 1 m^2), covered with fine mesh, and containing two to four pitfall traps emptied weekly or fortnightly (Helenius *et al.*, 1995; Purvis and Fadl, 1996; Holland and Reynolds, 2003). When the emergence trap is in the form of a large pyramidal tent, with a collecting chamber at the apex, it is sometimes termed a **photoeclector**. Photoeclectors have been used extensively for monitoring and collecting natural enemies in crops (Funke, 1971; Törmälä, 1982; Bosch, 1990; Büchs, 1991, 1993; Wehling and Heimbach, 1991; Kleinhenz and Büchs, 1993). A pitfall trap can be used inside the photoeclector (Mühlenberg, 1993) to remove ground predators and prevent predation on emerging natural enemies, but, to be effective, it should be positioned near the wall of the photoeclector (Kromp *et al.*, 1995). If emergence traps are to be used in studying phenologies, it should be borne in mind that the construction of the trap will influence the microclimate above the enclosed area of soil. All traps tend to reduce daily temperature fluctuations (the deeper the trap, the smaller the fluctuations), and the development rate of pupae may be affected (Southwood and Henderson, 2000). Emergence traps have been used to study ground predators such as carabid beetles (Helenius and Tolonen, 1994; Helenius, 1995), spiders (Duffey, 1978), predatory Diptera (Delettre *et al.*, 1998) and parasitoid wasps (Jensen, 1997; Moore, 2001). If traps are emptied frequently at regular intervals, temporal emergence patterns can be recorded. There are few data on trap efficiency, but an exception is the study of Moore (2001) which showed that 80 of 100 adult *Bracon hylobii* (a braconid parasitoid of the large pine weevil, *Hylobius abietis*) released into an emergence trap were captured within 6 hours.

Arboreal emergence traps (Basset *et al.*, 1997) can be used to investigate predators living in or on the bark of trees. Reeve *et al.* (1996) used such traps to monitor *Thanasimus dubius*, a clerid beetle predator of the southern pine beetle (*Dendroctonus frontalis*). One-metre vertical sections of pine trunk were enclosed with polyethylene screening leading down to funnels constructed of galvanized metal screening, which terminated in collecting jars filled with ethylene glycol solution. This method was used for two years and demonstrated a number of distinct emergence periods for *T. dubius*. Time-sorting arboreal photoeclectors attached to the trunks of oak and pine trees demonstrated a range of diel activity patterns in the spider community of this habitat. The photoeclector was a funnel of black cloth attached to a plastic tube, with a sampling jar mounted above. An electric motor, powered by two car batteries, was used to drive a turntable which changed the sampling jar at 6 hour intervals (Simon *et al.*, 2001).

Aquatic insects, emerging as the adult flying stage, move upwards through the water from the substrate and can be caught in a Mundie pyramid trap or a Hamilton box trap. These are essentially cages placed on the substrate which direct emerging insects into collecting jars at the apex of the trap. Alternatively, simpler and cheaper floating emergence traps (e.g. cups suspended from styrofoam board) can be used to catch the insects when they reach the surface (Williams and Feltmate, 1992).

Soil flooding techniques can be used to extract predators from soil *in situ* (Desender and Segers, 1985; Brenøe, 1987; Basedow *et al.*, 1988; Kromp *et al.*, 1995). Isolators are sunk into the ground, the vegetation examined and removed, then 2 litres of water are added to each $0.1\,m^2$ isolator. Predators that emerge are removed and then another 2 litres of water are added and any late-emerging predators are also removed. The process is reasonably rapid (taking about 10 min per $0.1\,m^2$ isolator), but large volumes of water must be transported to the site and the method is ineffective under some soil conditions (Sunderland *et al.*, 1995a). Gradual flooding of soil samples taken back to the laboratory, a standard technique for measuring slug biomass in the field (Glen *et al.*, 1989), has also proved useful for extracting carabid larvae (Thomas, 2002).

Efficient extraction of tiny predators, such as phytoseiid mites, from plant substrates requires a degree of inventiveness. To remove phytoseiid mites from apple leaves the sample of leaves was soaked in water and detergent for 2hours then leaves were washed on both sides and the washings directed through a filtering apparatus to be caught eventually on a small circular net, which could be transferred directly to a stereomicroscope. Examination of processed leaves showed that all mites had been successfully removed by this procedure (Jedlicková, 1997). As an alternative to water and detergent, an aqueous solution containing ethyl alcohol, sodium hypochlorite and liquid detergent was found to be effective for removing phytoseiid mites from lime leaves (Childers and Abou-Setta, 1999). Overwintering phytoseiids were removed from sections of wood from Australian grapevines using a microwave oven. Sections of wood in plastic bags were given 6 min at 120W and this brought the mites out of their hiding places in fissures in the wood. The method is expected to be more efficient than Berlese funnel extraction and also dislodges Tenuipalpidae, Bdellidae, Anystidae, Parasitidae and Tydeidae (James *et al.*, 1992). Mites can be dislodged from leaves using a brushing machine (Chant and Muir, 1955). Leaves are passed between a pair of contra-rotating cylindrical brushes and dislodged mites fall onto a rotating sticky disc. This method is quick, efficient and allows accurate identification of species and developmental stage. If the samples are examined immediately after brushing, living mites can be distinguished from dead ones. A commercial mite-brushing machine is produced by Leedom Engineering, 1362 Casa Court, Santa Clara, California 95051, USA (M.G. Solomon, personal communication).

6.2.8 SAMPLING BY ATTRACTION

Introduction

It is possible to exploit the attraction responses shown by natural enemies towards certain stimuli (section 1.5) in order to develop sampling techniques which can provide information on phenology and relative abundance. However, data collected by attraction for comparative purposes need to be treated with caution because different species will vary in their level of response to the same stimulus. Obviously, absolute estimates of insect abundance cannot be obtained using these methods.

Visual Attraction

Many winged insects are attracted to certain colours during flight, most commonly to yellow. Consequently, shallow, coloured trays or bowls filled with water, in which the attracted insects drown, are frequently used as traps to sample flying insects (Southwood and Henderson, 2000) (Figure 6.14a,b). A preservative, such as 10% formalin or 1% formaldehyde, or sodium benzoate (Wratten *et al.*, 2003), can be used in the **water traps**, and it is advisable to add a few drops of detergent if the target natural enemies are small and otherwise able to walk on the surface film, as was found to be the case for small dolichopodid flies (Pollet and Grootaert, 1996). Principally designed to attract phytophagous insects, yellow-coloured water traps also catch some groups of natural enemies, with varying levels of efficiency. They have been used in cereal crops where adult hover-flies were the most abundant predators caught, although adult lacewings and adult parasitoids also occurred in the traps (Storck-Weyhermüller,

Figure 6.14 Yellow water traps: (a) trap used by entomologists at Rothamsted to sample insects such as adult hover-flies and parasitoids in a potato crop; (b) traps arranged in an experimental cereal field (arranged perpendicularly to the field edge), used by entomologists at the University of Southampton to measure the within-field distribution and abundance of adult hover-flies.

1988; Helenius, 1990). The relative abundance of adult hover-flies, anthocorid bugs and parasitoids has been assessed in Brussels sprouts fields using yellow water traps (Smith, 1976). Yellow (Delettre *et al.*, 1998) and white (Pollet and Grootaert, 1987;1996) water traps have been used to monitor predatory Empidoidea (i.e. dolichopodid and empidid flies) in natural habitats. Pollet and Grootaert (1993) reported that yellow water traps caught the largest number of individuals of Empidoidea, but white ones provided a more reliable estimate of species diversity in woodland. Such generalisations obscure the variation that can occur between species. For example, Pollet and Grootaert (1987) caught 60 species of Dolichopodidae in coloured water traps in a humid woodland, and, although the majority of species were caught in white or red traps, *Sciapus platypterus* showed a marked preference for blue traps. Blue-green traps raised 60 cm above the soil surface were found to be the best option for catching arboreal Empidoidea (Pollet and Grootaert, 1994).

The attractiveness of yellow water traps to individual insects will depend to some extent on the latter's physiological condition. Adult hover-flies for example, are caught more readily when they are newly-emerged and/or hungry, and when food sources are scarce (Schneider, 1969; Hickman *et al.*, 2001). Similarly, females of the parasitoid *Cotesia rubecula* are more likely to search out yellow targets when starved, than when fed on sugar solution (Wäckers, 1994).

Traps of various colours have been used to catch adult hover-flies (Dixon, 1959; Sol, 1961, 1966). Water traps of 32 different colours were tested in a UK cauliflower crop. The colour "fluorescent yellow" caught significantly more hover-flies than other colours (Finch 1992), and it was also demonstrated in the laboratory that the hover-fly *Episyrphus balteatus* is attracted to yellow more than white, green or blue (Sutherland *et al.*, 1999). Schneider (1969) suggested that colour preference may be influenced by the colour of the most abundant flowers in bloom at the time of sampling. Sol (1966) defined the colours used in his water trap studies on the basis of human interpretations of colour, but a better approach, adopted by other workers (such as Kirk, 1984; White *et al.*, 1995; Wratten *et al.*, 1995), is to define colours in terms of ultraviolet reflectance spectra. Even apparently colourless surfaces (e.g. translucent plastic) may have UV reflectance properties (M.A. Jervis, personal communication). Kirk (1984) recorded two species of predatory fly of the genus *Medetera* (Dolichopodidae) in water traps of seven different colours, and found most in white and blue traps, with very few in yellow traps. Although the rank order of the catches for each colour was similar in both species, the proportions caught by each colour of trap were different. Obviously, great care must be taken when estimating the relative abundances of different species from coloured trap catches. Wratten *et al.* (1995) compared the attractiveness of water traps of different colour to two species of hover-fly in New Zealand. *Melangyna novaezelandiae* preferred yellow (also later confirmed by Bowie *et al.* (1999) for *Melangyna viridiceps* in Australia), but *Melanostoma fasciatum* showed no significant preference between yellow, white and blue. Trap reflectance was measured with a UV-visible spectrocolorimeter. Daily trap emptying gave 15% larger catches than infrequent emptying. This may have been because the accumulation of dust, leaves and insects in the infrequently emptied traps reduced trap reflectance.

The attraction of hover-flies to coloured water traps is based on their visual attraction to flowers as pollen and nectar sources. Many adult parasitoids also feed on nectar and pollen (Chapter 8) and are therefore likely to respond to coloured traps when foraging for such resources, although it has been suggested by Helenius (1990) that some parasitoids, such as those attacking aphids, may lack a behavioural response to yellow traps, because they feed on host blood or honeydew (but see Jervis *et al.*, 1993). Visual cues, such as certain colours, may very well be used in host habitat location by adult parasitoids. When a glass prism was used to split a beam of light entering a clear plastic box which contained a group of adult aphid parasitoids (*Aphidius rhopalosiphi*), the parasitoids all moved to the region of the box illuminated by the visible yellow-green band of the spectrum (W. Budenberg and W. Powell, unpublished data). Goff and Nault (1984) tested the pea aphid parasitoid, *Aphidius ervi*, for its response to transmitted light and recorded the strongest response to green (wavelength 514 nm).

When used in field crops, water traps are usually positioned level with, or just above, the top of the crop canopy, but they have also been used successfully when placed near the ground between the trees in orchards, forest plantations or natural forest habitats (Noyes, 1989). Arthropod activity, and hence size of catch in water traps, can vary with height above ground. Vega *et al.* (1990) describe an adjustable water trap that can be positioned at any height up to 2.2 m. Traps of an unspecified colour were used to sample adult anthocorid bugs, which attack psyllids in pear orchards (Hodgson and Mustafa, 1984). For some of the species caught, there was a significant relationship between the numbers of anthocorids in the water traps and the numbers present on the trees, as estimated by beating. Yellow pan traps are particularly efficient in the sampling of parasitoid wasps, such as Ceraphronidae, Scelionidae, Platygastridae, Diapriidae, Mymaridae and Encyrtidae (Masner, 1976; Noyes, 1989). Stephens *et al.* (1998) found that more parasitoid wasps were caught in yellow pan traps than in white ones.

The attractiveness of certain colours to insects has also been employed in the sampling of predators and parasitoids with **sticky traps** (Wilkinson *et al.*, 1980; Weseloh, 1981, 1986b; Neuenschwander, 1982; Moreno *et al.*, 1984; Trimble and Brach, 1985; Ricci, 1986; Samways, 1986; Bruck and Lewis, 1998; Heng-Moss *et al.*, 1998; Hoelmer *et al.*, 1998). Trimble and Brach (1985) examined the effect of colour on sticky trap catches of *Pholetesor ornigis*, a braconid parasitoid of leaf-miners. Yellow (of various shades) and orange traps were significantly more attractive than red, white, blue or black traps, and spectral composition appeared to be more important than reflectance. Yellow was also more attractive than other colours to *Macrocentrus grandii*, a braconid parasitoid of the European corn-borer, *Ostrinia nubilalis* (Udayagiri *et al.*, 1997). In contrast, high reflectance, regardless of colour, seemed to be important in the attraction of tachinid parasitoids to sticky traps placed in forests (Weseloh, 1981). The aphelinid parasitoid, *Aphytis melinus*, was caught in greater numbers on green and yellow sticky cards than on white, blue, fluorescent yellow, black, red or clear cards (Moreno *et al.*, 1984), whilst eight times as many *Aphidius ervi* (Braconidae Aphidiinae) were attracted to green sticky cards than to gold, blue, white, red, black or orange cards (Goff and Nault, 1984). Coloured balls of various sizes (to simulate fruit) were coated with sticky material and hung from guava trees in Hawaii. Some species of parasitoid of the Oriental fruit-fly, *Bactrocera* (=*Dacus*) *dorsalis*, were attracted to balls of specific colours (Vargas *et al.*, 1991).The opiine parasitoids of this pest were more attracted to yellow than to darker colours, and capture rates on red spheres were very low (Cornelius *et al.*, 1999). Coloured papers were inserted into glass tubes coated with sticky material (Tanglefoot®) to investigate the colour preference of naturally-occurring *Trichogramma* in sorghum, as part of an investigation of how these parasitoids locate their host habitat (Romeis *et al.*, 1998). Coloured sticky traps can also be used to study diel variation in parasitoid flight activity. Idris and Grafius (1998) showed that the numbers of *Diadegma insulare* (a parasitoid of the diamondback moth, *Plutella xylostella*) caught at different times of day on white sticky traps was directly proportional to the numbers recorded by direct observation.

The ladybird beetles *Coccinella transversalis* and *Adalia bipunctata* were more attracted to yellow than to seven other colours, and preferred pure yellow to yellow-white hues (Mensah, 1997). Other ladybirds (*Coleomegilla maculata* and *Coccinella 7-punctata*) also preferred yellow, but the lacewing *Chrysoperla carnea* was trapped equally on red, green, white and yellow sticky cards (Udayagiri *et al.*, 1997). Ricci (1986) examined catches of ladybird beetles on yellow sticky traps which were being used to monitor pest populations in olive groves and in safflower and sunflower fields. Attached to some of these traps were infochemical (subsection 1.5) lures, used to increase further their attractiveness to target pest species, but, curiously, in the oilseed crops more ladybirds were caught on unbaited traps than on those with lures. When yellow sticky traps and yellow water traps were placed horizontally on the ground in a wheat field, similar numbers of adult hover-flies were caught on both, but more hover-fly larvae were caught in the water traps (Bowie *et al.*, 1999). Harwood *et al.* (2001a) placed 7.5 cm^2 sticky traps on the soil surface of a winter wheat crop to monitor the activity density of linyphiid spiders and their prey. In this case the traps were painted black to resemble in colour the ground surface and *not* be attractive, since the aim was to quantify the normal traffic of these arthropods over the surface of the soil. Although small, these traps did capture significant numbers of spiders (0.6 $trap^{-1}$ day^{-1} at web sites and approximately half this value at non-web sites), and so could be useful in quantitative studies of both intra- and inter-specific interactions between spiders. As estimators of population density, however, they suffer from similar deficiencies to those described for pitfall traps (subsection 6.2.1), and would be best-used in conjunction with other techniques, such as mini-quadrats. Sticky traps and quadrats were seen as complementary (Harwood *et al*, 2001a), in that the former measure cumulative activity-density over 24 hours (including, therefore, the night) and are efficient at catching flying insects, whereas the latter provide a measure of the absolute densities of invertebrates (including those that are inactive and hiding under soil and stones etc) at a single point in time, but miss many flying or fast-moving arthropods that escape during sampling.

Insects can be removed from sticky traps using white spirit (McEwen, 1997), kerosene (Mensah, 1997), toluene, heptane, hexane, xylene, ethyl acetate (Murphy, 1985), citrus oil, vegetable oil or various other solvents (Miller *et al.*, 1993). Coloured sticky traps can be placed in most habitats, need only be replaced occasionally, and, since they are inexpensive, can be used in large numbers if the aim is to assess biodiversity and capture rare species. Colunga-Garcia *et al.* (1997), for example, caught thirteen species of coccinellid on yellow sticky traps.

Light traps have also been investigated as a means of monitoring abundance of predators such as adult lacewings and ladybirds (Bowden, 1981; Perry and Bowden, 1983; Honěk and Kocourek, 1986), carabid (Kádár and Lövei, 1992; Yahiro, 1997; Kádár and Szentkirályi, 1998), staphylinid (Markgraf and Basedow, 2002) and cantharid beetles (Löbner and Hartwig, 1994). If cool white light (15W fluorescent bulbs emitting light with major peaks in the visible region around 440 and 590 nm, with some UV) is used, catches of non-predatory species (such as moths) are reduced, resulting in more efficient capture and handling of predators such as lacewings and bugs (Nabli *et al.*, 1999). It should be noted that light trap efficiency may also vary with illumination, so that changes in activity and abundance may be obscured if catches are not corrected for variations in the intensity of background illumination provided by moonlight (Bowden, 1981). Moon phases were not, however, found to affect light trap catches of ladybirds, lacewings, and nabid bugs (Nabli *et al.*, 1999). Information is limited to nocturnally active species and trap effectiveness varies between species (Bowden, 1982). Directional light traps have been used in forests to study arthropod stratification at different heights (Basset *et al.*, 1997). An air-inflated dirigible gliding slowly over the forest canopy, illuminated by 500-W lights and towing a large net, has also been used as an ambitious mobile light trap (Basset *et al.*, 1997). Subaquatic light traps are available for sampling positively phototropic aquatic insects from the water mass at night (Williams and Feltmate, 1992).

Nocturnally-active Hymenoptera 'Parasitica' are attracted to **blacklight traps** (emitting UV light of wavelength 320-280 nm) and these may be useful for determining relative abundance and seasonal distribution (Burbutis and Stewart, 1979). In a comparison of attraction to light of different wavelengths, ladybirds (species not given) were most attracted to blacklight traps (Nabli *et al.*, 1999). The pyrgotid (Diptera) parasitoids of scarabaeid beetles are nocturnal and difficult to collect, but they are attracted to light. Four UV light traps were used to determine species composition and phenology of these elusive parasitoids in Texas (Crocker *et al.*, 1996). Some species of predatory water beetle are attracted to UV and there is potential for monitoring them with blacklight traps. A solar electric module can be used to charge a 12V battery to power the blacklight trap, and this enables the trap to be run in remote areas with minimal maintenance (Gerber *et al.*, 1992).

Gauld and Bolton (1988) note that light trapping is a valuable collecting method for parasitoid wasps in tropical habitats, where a significant proportion of the fauna is nocturnal.

Olfactory Attraction

The McPhail trap (McPhail, 1939; Steyskal, 1977; McEwen, 1997) (Figure 6.15), a device commonly used to monitor the numbers of olive pests such as olive-fly (*Bactrocera oleae*), can also be used to collect the adults of certain lacewings (Neuenschwander and Michelakis, 1980; Neuenschwander *et al.*, 1981). The trap is suitable only for lacewing species whose adults are non-predatory (Neuenschwander *et al.*, 1981). The attractant used is either protein hydrolysate (to which borax is added as an insect preservative) or ammonium sulphate. Protein hydrolysate solutions, and perhaps even tryptophan solutions (van Emden and Hagen, 1976; Jervis *et al.*, 1992b; McEwen *et al.*, 1994),

Figure 6.15 A McPhail trap in an olive tree canopy (M.A. Jervis).

might in future also be employed as attractants, in studies of lacewings and other natural enemies, for use in conjunction with visually-based trapping devices, e.g. coloured sticky traps, to enhance trapping efficiency.

There is obvious scope for the use of **infochemicals** (section 1.5) to monitor natural enemy populations in the same way as they are currently used to monitor pests (Pickett, 1988). Delta traps baited with virgin females of aphid parasitoids (Braconidae, Aphidiinae) caught large numbers of conspecific males when placed in cereal crops (Decker, 1988; Decker *et al.*, 1993) (Figure 6.16a). The use of pheromone traps could provide information on seasonal population trends and on the dispersal behaviour of some species of natural enemy.

Some parasitoids are attracted by the sex pheromones of their hosts (Powell, 1999). Adult female aphid parasitoids of the genus *Praon* were caught in large numbers in water traps that were combined with a source of synthetic aphid sex pheromones (Hardie *et al.*, 1991, 1994; Powell *et al.*, 1993) (Figure 6.16b). Similarly, aphelinid parasitoids were collected on sticky traps baited with either synthetic sex pheromones or virgin females of the San Jose scale, *Quadraspidiotus perniciosus* (Rice and Jones, 1982; McClain *et al.*, 1990). Sticky traps baited with 'Multilure', an aggregation pheromone for the bark beetle *Scolytus multistriatus*, attracted a range of parasitoid wasps which attack both eggs and larvae of the beetle (Kennedy, 1979). Other possibilities for exploiting infochemicals include the use of aggregation pheromones for monitoring Coleoptera, particularly carabid beetles (Thiele, 1977; Pickett, 1988), and the use of plant-derived chemicals as attractants. Caryophyllene ($C_{15}H_{24}$), a volatile sesquiterpene given off by cotton plants, has been used in delta traps to monitor lacewing adults and the predacious beetle *Collops vittatus* (Malachiidae) (Flint *et al.*, 1979, 1981). Sex pheromones of predators may also prove useful. Legaspi *et al.* (1996) caught females of *Podisus maculiventris* (Heteroptera, Pentatomidae) in traps that utilised the male sex pheromone of this species. Synthetic attractant pheromones have also been used to catch *Podisus distinctus* in Brazil (Aldrich *et al.*, 1997). Predatory adult wasps, such as Sphecidae, also feed on plant nectar and are strongly attracted to commercial pest traps (e.g. white bucket traps

(a)

(b)

Figure 6.16 (a) Delta-shaped pheromone trap baited with virgin female aphidiine braconid parasitoids. Captured male parasitoids, that have been attracted by the female sex pheromone, can be seen in the water tray; (b) Petri dish pheromone trap with aphid sex phermone lure consisting of a small glass vial containing aphid sex pheromone. Female parasitoids are attracted by the pheromone and are caught in the dish which contains water with a small amount of detergent added.

for monitoring armyworm adults, *Spodoptera frugiperda*) baited with synthetic floral extracts (Meagher and Mitchell, 1999). Food volatiles (e.g. isobutanol and acetic acid) will attract wasps such as *Vespula germanica* and *Vespula pensylvanica* to traps that kill by drowning (Landolt, 1998), such as the Yellowjacket Trappit Dome trap (Agrisense, Fresno, California). This is a plastic trap similar in shape to the McPhail trap (Fig. 6.15) which has a transparent top half and an opaque yellow lower half. Wasps enter through the bottom and drown in a detergent solution to which various other compounds can be added. Vespine species vary in their responses to olfactory cues and combinations of food volatiles may be more effective attractants than single compounds (Day and Jeanne, 2001).

Opiine braconid parasitoids of tephritid fruit-flies were caught in orchards in Hawaii using 7 cm diameter yellow polypropylene balls (to simulate ripe fruit). Each ball was split open and slices of ripe guava inserted, then the ball was covered in sticky material. Although the effectiveness of this trap was reduced somewhat by attraction of non-target Diptera and Hymenoptera, this design of trap is considered to have potential for basic population surveys, monitoring of augmentative releases, dispersal studies and behavioural studies of host-habitat finding by these biocontrol agents (Messing and Wong, 1992). Imported fire ants, *Solenopsis invicta*, were monitored in fields by placing out containers smeared with commercial dog food, ground beef or sardines. The containers were then capped and frozen for subsequent counting of ants (Lee *et al.*, 1990; Hu and Frank, 1996; Rossi and Fowler, 2000). Morrison (2002) examined baits (a freeze-killed cricket plus sugar solution on an 8 cm diameter plastic lid) a few hours after placement, and estimated the number of individuals of each ant species by allocating them to nine abundance classes. Commercial cat food placed on 10 cm × 15 cm white cards, and left *in situ* for 30 min before examination, was used to monitor ant populations in deciduous forests (Weseloh, 1993). Baits consisting of cottonwool buds soaked in 20% sugar solution, and mashed canned tuna in oil, were used to monitor ant activity in tropical upland rice (Way *et al.*, 2002). Ants (especially *Pheidole* spp. and *Lepisiota* spp.) tend to respond more strongly to protein baits (e.g. powdered fish) than to carbohydrate baits (e.g. molasses) (Sekamatte *et al.*, 2001). Carabid beetles were attracted to pitfall traps baited with fish or pig skin (Deng and Li, 1981), and predators of horn-fly, *Haematobia irritans*, were attracted to pitfall traps baited with bovine manure (Hu and Frank, 1996).

Attraction to Sound

Sound traps, broadcasting calls of mole crickets (*Scapteriscus* species), were used to attract phonotactically orientating tachinid parasitoids (*Euphasiopteryx depleta*) of these hosts (Fowler, 1987b). Each sound trap was covered with a plastic bag coated with an adhesive (Tanglefoot®), which trapped the landing *E. depleta* females. The method yielded useful data on the abundance, mobility and phenology of these parasitoids. These sound traps also attracted adults of *Megacephala fulgida*, a specialist tiger beetle predator of *Scapteriscus* spp. (Guido and Fowler, 1988). Walker (1993) used an "artificial cricket" to attract and sample females of the tachinid *Ormia ochracea*, which is a parasitoid of field and mole crickets (*Gryllus*, *Orocharis* and *Scapteriscus*). The "artificial cricket" consisted of an electronic sound synthesiser, an amplifier and a speaker mounted in a metal box. When songs were broadcast, flies arrived within seconds. Speaker stations only 14 m apart in an apparently uniform pasture consistently produced very different fly catches (varying 2.5-fold), but the reasons for this remain obscure.

Using Hosts and Prey as 'Bait'

Hosts of parasitoids can be used as 'bait' to detect the presence and level of activity of adult parasitoids in the field. Although this method is considered under the heading of sampling by attraction, the natural enemies may not locate the hosts/prey by attraction *per se*.

For monitoring parasitoids of the eggs of the white stem-borer (*Tryporyza* (= *Scirpophaga*) *innotata*) in Javan rice, plants from an insectary infested with 2-6 egg masses were placed out in the field for 1-5 days, then retrieved and reared out (Baehaki 1995). Similar methods were used to study parasitoids attacking eggs of leaf-folder moths (Pyralidae) (De Kraker *et al.*, 1999) and the brown planthopper, *Nilaparvata lugens* (Fowler, 1987a) in rice fields. To investigate parasitoids of the eggs of linyphiid spiders, Van Baarlen *et al.* (1994) placed spider eggsacs in the field in lidless Petri dishes and retrieved them seven days later for rearing out parasitoids. The percentage of *Erigone* spp. eggsacs parasitised by *Gelis festinans* (Ichneumonidae) determined by this method was not significantly different from that for naturally occurring eggsacs, and it was concluded that the Petri dish introduction method is a simple and efficient means of assessing parasitism in the field (Van Baarlen *et al.*, 1994). Ellers *et al.* (1998) devised a trap baited with apple sauce, yeast and *Drosophila* larvae, to monitor the braconid wasp *Asobara tabida*. Potted cereal seedlings infested with cereal aphids were used by Vorley (1986) to detect winter activity of aphid parasitoids in pasture and winter cereal crops. The pots were left in the field for 14 days after which they were retrieved, and the surviving aphids reared in the laboratory until parasitised individuals mummified. Hyperparasitoids can be investigated in the same way by exposing secondary hosts that contain primary parasitoids.

Baits can also be used to monitor the activity of predators on vegetation, on the soil surface and under the soil. Bugg *et al.* (1987) stapled egg masses of armyworm (*Pseudaletia unipuncta*) to radish foliage to monitor predation pressure in weedy and weed-free plots. Other researchers have used artificial prey, such as dipteran puparia (e.g. *Drosophila, Sarcophaga, Musca*) stuck onto small pieces of cardboard (Speight and Lawton, 1976; Brown *et al.*, 1988; Burn, 1989; Carcamo and Spence, 1994; Chapman and Armstrong, 1996), freeze-killed aphids (Dennis, 1991) or tethered caterpillars, to monitor predator pressure on the ground. Brust *et al.* (1986a) secured larvae of pest Lepidoptera to the soil surface using a 25 cm nylon thread attached to a 25 cm long wooden stake. They claimed no effect of tethering on the behaviour of the larvae or the predators. Weseloh (1990), however, showed that tethering doubled the predation rate by ants on larvae of the Gypsy Moth (*Lymantria dispar*). If *in situ* observations of the baits are carried out, either by human presence (Brust *et al.*, 1986b), or by video surveillance (Lys, 1995), the species exerting the observed predation pressure can be determined (subsection 6.2.6). Weseloh (1990) used tethered larvae in small cages with a narrow slit that only ants could enter, compared with uncaged tethered larvae, to determine that a fifth of the predation experienced by Gypsy Moth larvae was caused by ants. Hu and Frank (1996) seeded artificial cowpats with eggs of the horn-fly, *Haematobia irritans*, and showed that most predation of the later stages of this pest was by the red imported fire ant, *Solenopsis invicta*. Similar methods can be used to monitor predation within the soil itself. Small plastic vials, sealed with a cap, but with small holes near the top, were buried in grassland soil. The holes were large enough to allow ants to come and go, but too small to allow escape of the "bait" prey organisms placed inside. Mealworms as bait were used to map the vertical distribution of four ant species, and a range of other prey species were used to determine their attractiveness and vulnerability to ants. Little is known about predator-prey relationships within the soil, and so this technique may also prove to be very useful in this context (Yamaguchi and Hasegawa, 1996) (subsection 6.3.2). Eggs of Western corn rootworm, *Diabrotica virgifera virgifera* (Coleoptera, Chrysomelidae) were placed in the soil in mesh packets of various mesh sizes to detect predation by mites, carabids and ants. Unfortunately, many eggs were also damaged in fine-mesh control packets in the field, and so the method needs further development (Stoewen and Ellis, 1991).

Attraction to Refuges

Natural enemies may be attracted to refuges (such as dark, damp, cool places where nocturnal predators can find stress-free conditions during the heat of a summer's day), or they may encounter them by random movement and then remain there. This behaviour can be exploited to trap natural enemies and thus obtain information about the presence of a species in a habitat. Since the method relies on behaviour of the natural enemy, which may vary with habitat, season, age of the natural enemy, and many other factors, it cannot be used to prove absence from a habitat, nor will it provide reliable data on relative abundance, unless carefully calibrated for the target species concerned. Mühlenberg (1993) describes "trap stones", made of concrete or Plaster of Paris, containing channels sealed by a cover plate, that can be placed on the ground to trap predators. Pekár (1999) and Bogya *et al.* (2000) used cardboard paper traps, 30 cm x 100 cm, wrapped around tree trunks to monitor overwintering spiders in orchards, and Roltsch *et al.* (1998) used similar traps to study spider overwintering in vineyards. Similarly, earwigs (*Forficula auricularia*) were monitored in an orchard by strapping 4 cm x 5 cm PVC tubes containing corrugated paper rolls to the trunks of pear trees (Sauphanor *et al.*, 1993). This trap design exploits the positive thigmotaxis and need for a diurnal refuge which is characteristic of earwigs (Dermaptera). Folded burlap bands (30 cm wide) tied onto tree trunks with string have been used to monitor predators of fruittree leaf-roller, *Archips argyrospila* (Braun *et al.*, 1990). Carabid densities in the field can be raised by the provision of corrugated iron sheets, which they use as refugia, leading to increased control of mollusc pests (Altieri *et al.*, 1982). A similar principle has been recommended for the control of slugs in horticulture, where boards of wood within plots containing vulnerable plants are used as refuges by both slugs and carabids (Symondson, 1992). The aim here is to increase encounter rates between predator and prey, to provide refuges for the predators within the crop, and to enable slugs to be both monitored and collected from beneath the boards. In aquatic systems, substrate-filled trays, leaf packs or multi-plate samplers can be used as refuges for sampling aquatic insects (Williams and Feltmate, 1992).

6.2.9 SAMPLING THE IMMATURE STAGES OF PARASITOIDS

Methods for detecting the presence of the immature stages of parasitoids (eggs, larvae and pupae) occurring upon or within hosts, are usually employed when estimating the impact of parasitism on host populations (Chapter 7), but they are also used in studies of foraging behaviour (Chapter 1), parasitoid life cycles and phenologies, parasitoid fecundity (section 2.7), spatial distribution of parasitism (subsection 7.3.10), parasitoid-host trophic relationships (subsection 6.3.5), and host physiological defence mechanisms (subsection 2.10.2).

Parasitised and unparasitised hosts can sometimes be easily distinguished visually. Insects parasitised by ectoparasitoids can often be recognised easily by the presence of eggs and larvae on their exterior. Hosts parasitised by endoparasitoids are generally less easily distinguishable from unparasitised ones, especially during the early stages of parasitism. During the later stages of parasitism, hosts may undergo alterations in integumental colour; for example some green leafhoppers become orange or yellow in colour (Waloff and Jervis, 1987). Aphid mummies (subsection 6.2.6) are normally a distinctly different colour from live aphids (Figure 6.13). If parasitised hosts cannot be distinguished at all from unparasitised ones by external examination, then **host dissection** can be used to detect the presence of parasitoid immature stages (section 2.6). For example, Fowler (1987a) dissected 1,500 eggs of brown planthopper, *Nilaparvata lugens*, from rice fields, and found high levels of parasitism by *Anagrus* spp. (Mymaridae) and lower levels by *Oligosita* sp. (Trichogrammatidae). Similarly, Murphy *et al.* (1998) examined eggs of the grape leafhopper (*Erythroneura elegantula*) under a dissecting microscope for presence of a developing parasitoid, *Anagrus epos* (Mymaridae). However, host dissection is a time-consuming activity, and, for species attacked by a complex of closely-related parasitoids, identification of larvae, and particularly eggs, may prove either difficult or impossible.

The immature stages of insect parasitoids have received little taxonomic study. Some morphological information is available concerning the larvae of the following groups: Hymenoptera (Finlayson and Hagen, 1977); ichneumonoids (Short, 1952, 1959, 1970, 1978); pimpline ichneumonids (Finlayson, 1967); ichneumonine ichneumonids (Gillespie and Finlayson, 1983); braconids (Ĉapek, 1970, 1973), eurytomids (Roskam, 1982; Henneicke *et al.*, 1992); aphidiine braconids (O'Donnell, 1982; O'Donnell and Mackauer, 1989; Finlayson, 1990); pipunculids (Benton, 1975: several genera; Albrecht, 1990: *Dorylomorpha*; Jervis, 1992: *Chalarus*); Rhinophoridae (Bedding, 1973). In a few groups, some species can be distinguished on the basis of structural differences in their eggs, e.g. *Eurytoma* spp. (Claridge and Askew, 1960) and Rhinophoridae (Bedding, 1973). If one takes a large taxonomic group, such as the family Tachinidae, published descriptions of the immature stages are available for several species. There is, however, a dearth of synthetic taxonomic studies that provide identification keys.

The colour and form of aphid mummies, the shape of the parasitoid exit-hole, and characteristics of the meconial pellets (waste products deposited by parasitoid larvae prior to pupation), can be used to identify the parasitoids involved, at least to genus (Johnson *et al.*, 1979; Powell, 1982). Certain parasitoids which attack hosts feeding in concealed locations within plant tissues often leave clues to their identity, such as exuviae or meconial pellets, within the hosts' feeding cells.

Host dissection is less reliable as a method for the detection of parasitoid immatures in cases where small parasitoid stages occur inside relatively large hosts, and accuracy can vary considerably between different people performing dissections (Wool *et al.*, 1978). Therefore, the most popular method for the detection of parasitoids in a sample of hosts is to maintain the latter alive in the laboratory until the adult parasitoids emerge (subsection 6.3.6). This approach avoids the problem of identifying the parasitoid immatures but, from the standpoint of obtaining estimates of either percentage parasitism or parasitoid adult density, the problem arises of a time delay being introduced between sampling and obtaining the population estimate - parasitised hosts are removed from the influence of other mortality factors, e.g. predation, fungal pathogens, multiparasitism, host physiological defence mechanisms (subsection 2.10.2), which would normally operate in the field after the sampling date. Therefore, the number of emerging adult parasitoids may not give an accurate estimate of the density of adults emerging in the field. The problem of estimating percentage parasitism is discussed further in Chapter 7.

The reliability of the rearing method as a means of obtaining information on the occurrence of parasitoid immatures in hosts depends upon the ease with which the hosts can be cultured. Any hosts which die during the rearing process, before parasitoid emergence is expected, should be dissected and examined for the presence of parasitoids. Dissection of larvae of alfalfa weevil (*Hypera postica*) and several species of Miridae for the presence of parasitism was compared with rearing out of parasitoids. Parasitism rate as measured by dissection was 12–44% higher than that measured by rearing. This was partly because hosts and parasitoids can be attacked by entomogenous fungi, and rearing of these doubly-attacked hosts produces no parasitoids. There are, however, advantages and disadvantages to using both methods (Table 6.2). Dissection may fail to detect parasitoid eggs, but these eventually register by rearing out (which also makes it easier to identify the species of parasitoid involved) (Day, 1994).

An X-ray machine was used to detect successfully the gregarious endoparasitoid *Pteromalus puparum* and the solitary endoparasitoid *Brachymeria ovata* in pupae of *Pieris rapae*. Hundreds of pupae could be screened simultaneously, but the early stages of parasitism were not detected (Biever and Boldt, 1970). Using magnetic resonance microimaging, it was possible to see a larva of the braconid parasitoid *Dinocampus coccinellae* within the body of a living ladybird, *Coccinella 7-punctata* (Geoghegan *et al.*, 2000). This method may, however, be better suited for studying parasitoid development and behaviour within living hosts than for mass-screening hosts to determine the incidence of parasitism. Near-infrared spectroscopy correctly identified (80–90% success rate) which house-fly puparia

Table 6.2 Comparison of dissection and rearing methods for determining the occurrence of parasitoid immatures in hosts. Source: Day (1994.), reproduced by permission of the Entomological Society of America.

Objectives	Best Method	
	Dissection	Rearing
Accurate mortality data		
1. Parasitism and disease incidence	+	
2. Detects combinations: superparasitism, multiple parasitism, and parasite-disease interactions	+	
3. Avoids additional or confounding mortality caused by stress, inadequate food, and crowding	+	
Identification of mortality factors		
1. Parasitoids (adults required)		+
2. Disease (e.g. spores required)		+
3. Detects kleptoparasitoids	+	
4. Detects hyperparasitoids	+[a]	+
5. Stinging mortality		+[b]
6. Host-feeding mortality		+[b]
Measures other factors		
1. Reduced feeding by parasitised hosts		+[b]
2. Mortality of host in different instars and stages		+[b]
3. Diapause of host and parasitoid		+
4. Sterilisation of host	+	
Convenience and logistics		
1. Food for host not needed	+	
2. Requires less expertise (usually)		+
3. Work load can be delayed or extended	+[c]	
4. Prompt results	+[d]	
5. Host survival is low when reared	+	
6. Detection of a newly-established parasite		+[e]

[a]Ectoparasitoids only, unless special methods are used. [b] Laboratory studies only, not practical for field-collected hosts [c] If hosts are frozen [d] Total time required is not listed because it will vary with the species. In general, dissections require a more concentrated effort, but rearing extends over a long period and is less efficient, so the time spent per parasitoid detected is likely to be similar for the two methods. [e] Because much larger sample sizes are usually possible, which increases the odds of finding scarce parasitoids, and immature parasitoids often cannot be identified to species

were parasitised by parasitoid wasps, even when parasitoids were in the early stages of development. Differences between the absorption spectra of parasitised and unparasitised pupae may have been due to differences in moisture, chitin or lipid compositions (Dowell *et al.*, 2000).

For researchers who have access to gas chromatography equipment, hydrocarbon profiles can provide a reliable method for identifying parasitoids, before or after leaving the host. Geden *et al.* (1998) used cuticular hydrocarbon profiles to identify three species of *Muscidifurax* (Pteromalidae, parasitoids of house-fly and stable-fly pests) which are difficult to separate morphologically. Furthermore, they showed that specimens of *M. raptor* from five countries and three continents could all be identified by the method. The method can be successful with adult parasitoids and with pupal exuviae (Carlson *et al.*, 1999). Techniques are described in Phillips *et al.* (1988) and other applications of the method are listed in Symondson and Hemingway (1997).

Recently, electrophoretic, immunological and DNA-based techniques have been developed for detecting parasitoid immature stages within their hosts (Powell and Walton, 1989; Stuart and Greenstone, 1996; Demichelis and Manino, 1998) (subsections 6.3.9, 6.3.11, 6.3.12). Tilmon *et al.* (2000) used a two-stage molecular approach for determining the rate at which three species of *Peristenus* (Hymenoptera: Braconidae) were parasitising *Lygus lineolaris* (Heteroptera: Miridae) in an alfalfa field. They used PCR (polymerase chain reaction) to target part of the mitochondrial cytochrome oxidase I (COI) gene. Using mitochondrial genes ensured that there was a high copy number per cell while the part of the COI gene chosen has low intraspecific variability. Primers were designed that amplified DNA from all three *Persistenus* species and these primers were used to detect parasitoids within *L. lineolaris* nymphs. Restriction digests of the PCR product were then carried out to determine which of the three parasitoid species had been present. This procedure took days (compared with months for the rearing-out method, see subsection 6.3.6) and also gave the researchers the flexibility to store samples until convenient for processing (Tilmon *et al.*, 2000). DNA-based methods are also being investigated for the rapid identification of taxonomically-difficult parasitoids, such as *Trichogramma* species (Menken and Raijmann, 1996; Van Kan *et al.*, 1996; Stouthamer *et al.*, 1999a; Chang *et al.*, 2001) and *Muscidifurax* species (Taylor and Szalanski, 1999), and, similarly, for predators such as phytoseiid mites (Navajas *et al.*, 1999). Silva *et al.* (1999) present a dichotomous key, based on PCR of ITS-2 rDNA, for the rapid differentiation of five *Trichogramma* species from Portugal. RAPD-PCR assays of the encyrtid parasitoid *Ageniaspis citricola* suggested that it may in reality be two species, one occurring in Thailand and the other in Taiwan (Hoy *et al.*, 2000). Information of this sort can be crucial to the success of biological control programmes.

6.2.10 MARK-RELEASE-RECAPTURE METHODS FOR ESTIMATING POPULATION DENSITY

Many of the sampling methods discussed above provide only relative estimates of predator and parasitoid population densities, and of those that provide absolute estimates, some only work well for conspicuous or less active species. Inconspicuous or very active natural enemies, such as carabid beetles and lycosid spiders, are more difficult to count and an alternative method of estimating population levels is to estimate densities using mark-release-recapture data. These data are obtained by live-trapping a sample of individuals from the population, marking the insects so that they can be distinguished from uncaptured individuals, releasing them back into the population, and then re-sampling the population. An absolute estimate of population density can be calculated from the proportion of re-captured, marked individuals in the second sample, provided that a number of assumptions are satisfied. The main ones are:

1. That marking neither hinders the movement of an individual nor makes it more susceptible to predation or any other mortality factor;
2. That marks are retained throughout the trapping period;
3. That marked and unmarked individuals have an equal chance of being captured;
4. That, following release after marking, marked individuals become completely and randomly mixed into the population before the next sample is taken.

The original method of estimating total population size from mark-release-recapture data was devised by Lincoln (1930) for the study of waterfowl populations. The standard Lincoln Index formula is:

$$N = \frac{an}{r} \tag{6.1}$$

where N is the population estimate, a is the number of marked individuals released, n is the total number of individuals captured in the sample, and r is the number of marked individuals captured.

The Lincoln Index relies upon the sample of marked individuals, which is released back into the population, becoming diluted in a random way, so that the proportion of marked individuals in a subsequent sample is the same as the proportion of marked individuals originally released within the total population. However, the original Lincoln Index method assumes, unrealistically, that the population is closed, with no losses or gains either from emigration and immigration or from deaths and births, during the period between the taking of consecutive samples. Since the Lincoln Index was devised, a number of modifications and alternative methods of calculating population density have been developed. The best known are those of Fisher and Ford (1947); Bailey (1951, 1952); Craig (1953); Seber (1965); Jolly (1965); Parr (1965); Manly and Parr (1968); Fletcher *et al.* (1981). These methods, together with the formulae used to estimate population sizes, are described and discussed in detail by Seber (1973); Begon (1979, 1983); Blower *et al.* (1981); Pollock *et al.* (1990); Sutherland (1996); Krebs (1999) and Southwood and Henderson (2000). Begon (1983) reviews the use of Jolly's method, based on 100 published studies, and discusses potential alternatives. The computer programme 'Capture' can be used to select the most appropriate model for a specific set of circumstances (Otis *et al.*, 1978; White *et al.*, 1982; Samu and Kiss, 1997; Kiss and Samu, 2000). A practical problem that is often encountered with carabids (Ericson, 1977; Kromp and Nitzlader, 1995; Samu and Sárospataki, 1995b) and lycosids (Samu and Kiss, 1997) is that, in spite of a large expenditure of effort devoted to marking individuals, the recapture rate of some species can be less than 10%. Begon (1979) provides tables of minimum sample sizes of marked and recaptured individuals for stated levels of accuracy of the population estimate.

Among natural enemies, carabid beetles are the group most often subjected to mark-release-recapture studies, but mainly for the purpose of investigating activity and dispersal (subsection 6.2.11). However, there have also been several attempts at estimating carabid (Ericson, 1977, 1978; den Boer, 1979; Brunsting *et al.*, 1986, Nelemans *et al.*, 1989; Hamon *et al.*, 1990; Thomas *et al.*, 1998), staphylinid (Frank, 1968) and lycosid (Greenstone, 1979a; Kiss and Samu, 2000) population densities using mark-release-recapture data. In a ten-year study of the carabid *Nebria brevicollis*, 11,521 beetles were individually marked using a branding technique (Nelemans *et al.*, 1989) (marking methods are discussed more fully in subsection 6.2.11). The beetle population was sampled continuously using pitfall traps, and two methods were used to calculate population sizes: Jolly's method, as modified by Seber (1973), and Craig's method (Craig, 1953). Of the two, Craig's method gave the better estimates, those obtained using Jolly's (1965) method being much too low. All the methods used to estimate population sizes from mark-release-recapture data assume constant catchability over the trapping period, but Nelemans *et al.* (1989) found significant between-year differences in recapture probabilities. Also, the frequency distribution of recaptures deviated significantly from that predicted by a Poisson distribution, more beetles than expected failing to be recaptured and more than expected being recaptured three or more times. Consequently, there is a danger of significant errors occurring in estimates of carabid numbers based on mark-release-recapture data from pitfall trapping, Jolly's method in particular tending to underestimate populations (Nelemans *et al.*, 1989; den Boer, 1979). The size of errors arising from variability in the chances of recapture will depend to some extent on the species being studied. Since the behaviour of individuals affects catchability in pitfall traps, it is sometimes advisable to treat the sexes separately (Ericson, 1977; Samu and Sárospataki, 1995b; Samu and Kiss, 1997). Furthermore, the handling procedure during marking and release will affect the activity of beetles following release. This can be mitigated, to some extent, by leaving an interval of a few days between successive trapping periods to allow marked beetles time to redistribute themselves within the population before being recaptured (Ericson, 1977).

Because carabid beetles are highly mobile, some marked individuals are likely to leave and re-enter the trapping area, thereby biasing population estimates (Ericson, 1977). To avoid this problem, enclosures have been used during some carabid mark-release-recapture studies (Loreau, 1984a; Brunsting *et al.*, 1986; Nelemans *et al.*, 1989) (subsection 6.2.1). In a similar way, a field cage was used to assess the accuracy of Jolly's method for estimating populations of the ladybird beetles *Coccinella californica* and *Coccinella trifasciata* in alfalfa and oat crops (Ives, 1981). Beetles were captured in experimental plots using a visual searching method; these were then marked at the site of capture with spots of enamel paint, and were then released immediately after marking. Estimates of population density were calculated using Jolly's method and, in the caged plots, these proved to be very accurate, although estimates of populations in open plots were likely to be less precise. When there was no limitation on flight, it was considered necessary that aphid densities in the study area should be high enough to provide adequate food for the ladybirds, thus preventing dispersal. A limitation noted in Ives' study was the amount of time spent marking captured beetles, which restricted the numbers that could be caught in a sampling period, thereby reducing the accuracy of the mark-release-recapture method. To alleviate this problem, visual counts were done whilst walking through the plots, in addition to the counts made whilst marking. A similar method was used to estimate wolf spider population densities in an estuarine salt marsh (Greenstone, 1979a). As an alternative to enclosure, records of the distance moved by marked predators within a trapping grid (e.g. a grid of pitfalls) can be used to calculate the "area of influence" of traps or grid (Kuschka *et al.*, 1987; Franke *et al.*, 1988). Area of influence can also be estimated by defining concentric rectangles within the grid and noting at which size of rectangle the density is stabilised (Loreau and Nolf, 1993). Population size (from mark-release-recapture) divided by area of influence provides an estimate of population density.

Some mark-release-recapture studies have been done on hover-fly populations, but on phytophagous and saprophagous rather than predatory species (Neilsen, 1969; Conn, 1976). Adult narcissus bulb flies, *Merodon equestris*, were marked on the tibiae with cellulose paint applied with a fine blade of grass (Conn, 1976), a method that could also be used with aphidophagous syrphids. In this study, population size was estimated using Jolly's method and the Fisher-Ford method. Despite recapture rates of 30–50%, both methods gave large errors in daily estimates of population size because samples were small, and the total amount of variation in estimates using Jolly's method was usually two to three times higher than that obtained by the Fisher-Ford estimates. Multiple regression analyses revealed that 35–40% of the variation in the Fisher-Ford estimates and 35–45% of the variation in the Jolly estimates was attributable to the effect of variation in temperature.

Adult parasitoids are much more difficult to mark for mark-release-recapture studies because many of them are small bodied and difficult to handle. However, some success has recently been achieved by labelling braconids, mymarids and aphelinids with fluorescent dust or trace elements (Jackson *et al.*, 1988; Jackson and Debolt, 1990; Hopper and Woolson, 1991; Corbett *et al.*, 1996; Bellamy and Byrne, 2001). The trace elements were added to artificial diets on which the hosts of the parasitoids were reared. Fleischer *et al.* (1986) studied patterns of uptake and elimination of rubidium by the mirid bug *Lygus lineolaris* by spraying mustard plants with varying rates of Rb. A concentration of 200 ppm RbCl added to the diet of its *Lygus* spp. hosts provided Rb-labelled *Leiophron uniformis* which could be distinguished from wasps collected from field populations for 6–8 days (Jackson and Debolt, 1990). Similarly, *Microplitis croceipes* adults labelled with rubidium or strontium via the diet of their hosts, *Helicoverpa* spp., at a concentration of 1000–2000 ppm, were distinguishable from field-collected, unlabelled wasps for up to 20 days (Hopper and Woolson, 1991). However, there is a danger that high levels of these elements could affect the biology of the labelled insects. Jackson *et al.* (1988) noted that labelling *Anaphes ovijentatus* (Mymaridae) with high doses of rubidium (1000 ppm RbCl), via the eggs of its *Lygus* spp. hosts, tended to reduce longevity and fecundity slightly. Strontium was found to be superior to rubidium for labelling parasitic Hymenoptera because: (a) background levels are lower, (b elimination rates are less, (c) it does not affect parasitoid development and behaviour, and (d) it is less expensive (Gu *et al.*, 2001). The radioisotope ^{32}P was used to label *Trichogramma dendrolimi*, a parasitoid of lepidopteran eggs, in order to evaluate the impact of mass releases against a tortricid moth (Feng *et al.*, 1988). In this case, no adverse effects of the radioisotope labelling on longevity, reproduction or sex ratio of the parasitoid were detected.

By applying unique marks to each individual insect in a mark-release-recapture study, additional information on longevity and survival rates can be obtained. Conrad and Herman (1990) used data from a mark-release-recapture study of dragonflies to calculate daily survival estimates using the Manly-Parr method (Manly, 1971; Southwood and Henderson, 2000), which were then converted to daily expected life-span estimates using the formula of Cook *et al.* (1967): Expected life-span $= -1/\log_e$ (Survival). Conn (1976) calculated the longevity of phytophagous hover-flies from mark-release-recapture data.

6.2.11 METHODS USED IN INVESTIGATING NATURAL ENEMY MOVEMENTS

Predators need to locate their prey in order to feed, and parasitoids must find their hosts in order to lay eggs on/in them. Biocontrol practitioners need to determine how far parasitoids can move between host areas and areas of non-host foods to decide, for example, where to site plants for the supply of pollen and nectar (Chapter 8). Prey/host location generally requires spatial displacement on the part of the predator or parasitoid, and this may be either directed (e.g. movement towards the source of a stimulus) or random. Such locomotor activity is a major factor influencing the population dynamics of natural enemies because it affects the rate of encounters with prey and host patches and so influences the amount of predation or parasitism. It occurs as a result of behavioural responses to stimuli which may be internal, physiological stimuli, e.g. hunger (Mols, 1987), or external, environmental stimuli, e.g. infochemicals (section 1.5). In the main, host- and prey-location behaviour involving spatial dispacement constitutes **trivial movement**, i.e. it is restricted to the habitat normally occupied by the insects, whereas **migratory movements** take the insect away from its original habitat (Southwood, 1962), although there are notable exceptions such as some mymarid wasps. The techniques described here can be applied to either type of movement, although so far as trivial movements are concerned, foraging behaviour at a low level of host and prey patchiness is not dealt with (Chapter 1 discusses this aspect of behaviour).

The study of insect movement in the field is often difficult; in general, it is easier to measure the consequences of movement than it is to examine the process itself. For example, it is easier to record that an individual has shifted from one location to another during the period between two sampling occasions than it is to record the path taken by that individual, or the speed at which it travelled.

Some measure of the level of locomotor activity within a population can be obtained by intercepting moving individuals. For a given population, the higher the level of activity the more individuals will be intercepted within a given time. Pitfall traps, for example (subsection 6.2.1), can be used to study the activity of ground-dwelling predators, such as carabid beetles. In order to study the diurnal activity patterns displayed by carabid beetles and other surface-living predators, automatic, time-sorting pitfall traps have been developed and used in a variety of habitats (Williams, 1958; Ayre and

Trueman, 1974; Barndt, 1976; Luff, 1978; Desender *et al.*, 1984; De Keer and Maelfait, 1987; Kegel, 1990; Alderweireldt and Desender, 1990; Bayram 1996). Individuals falling into the trap are channelled into fresh collecting tubes or compartments after predetermined time intervals, which may be as short as two hours (Ayre and Trueman, 1974; Kegel, 1990). Funke *et al.* (1995) used time-sorting pitfall traps at the tops of two-metre-high artificial tree trunks to investigate the diel periodicity of flying predators in deciduous woodland. These authors also devised ring-shaped pitfall traps, and deployed them, with and without artificial tree trunks in the middle, to determine which species of predator (carabid and staphylinid beetles, spiders, harvestmen and centipedes) use trunk silhouettes for orientation when moving on the ground.

Population survival can depend on the rate of exchange of individuals between subpopulations. For non-flyers this exchange is by walking. To measure relative rates of dispersal by walking, five species of carabid were individually marked and released in the centre of 7 m radius fenced areas then caught in pitfalls at the edge. Circles were set up in areas of high and low density vegetation (Klazenga and De Vries, 1994). Similarly, a 10 m diameter circle of pitfalls was placed such that half the circle was in an oat field and half in adjacent grassland. Carabids were captured and marked with coloured paint (a different colour for each habitat) then released. 11% were recaptured and about 4% were recaptured in a different habitat from the one where they were originally caught (Kajak and Lukasiewicz, 1994). The same approach, using 10 m diameter enclosures, was employed to investigate the effects of insecticides on the movement of several carabid species (Kennedy *et al.*, 1996; Kennedy and Randall, 1997). Circles of 10 m–20 m radius were used by Ellers *et al.* (1998) who released marked braconid parasitoids (*Asobara tabida*) in the centre of the circle and investigated dispersal ability in relation to the body size of individual wasps. Chapman and Armstrong (1996) employed pitfalls, together with time-specific enclosure of some plots with plastic barriers, and deduced that some carabid beetles moved from intercropped to monocropped plots at night. This methodology showed that, at night, the carabid *Pterostichus melanarius* will move a short distance out from dense vegetation into adjacent open crops, which could justify the use of grass strips and weed strips for enhancing biocontrol within fields (Chapman *et al.*, 1999). **Linear pitfall traps** can also be used to gain information on direction of movement (Duelli *et al.*, 1990). If such traps are positioned along the boundary between two distinct habitats, they will detect major population movements across the boundary, and this could be related to changing conditions, such as levels of prey availability, within habitats. Linear traps placed at increasing distances from a habitat boundary will give some indication of rates of immigration into that habitat, e.g. rates of colonisation of arable fields from overwintering refuges (Pausch *et al.*, 1979). Petersen (1999) sited a vertical polyethylene barrier in an arable field parallel to the field edge; the barrier was set in a zig-zag pattern and pitfalls were sunk in corners of the barrier on the edge side to record spring emigration of the carabid, *Bembidion lampros*, from overwintering sites in a grass bank. Fagan (1997) studied the movement of two species of mantid out of experimental plots. The plots were surrounded by sticky plastic sheets which acted as absorbing boundaries. The pattern of emigration for *Mantis religiosa* was stepped and a "boundary-flux" diffusion model was inappropriate, but cumulative dispersal of *Tenodera sinensis* was sigmoidal and the diffusion model provided a good fit. The model assumes that all individuals are identical, exhibit Brownian random movement at a constant rate, and that the area is homogeneous (Fagan, 1997). Diffusion models have been used to predict the effectiveness of agroecosystem diversification for enhancing natural enemies. The spatial scale at which enhancement occurs is influenced by mobility (represented by the diffusion coefficient) of the natural enemy (Corbett and Plant, 1993). Coloured water traps placed, for example, at different distances from a field edge, may enable the distribution of flying predators and parasitoids to be mapped (Hradetzky and Kromp, 1997).

Interception traps, such as **window traps** and sticky traps, can be used to monitor the activity and movements, including the direction of movement and height above ground, of flying insects. A window trap consists of a sheet of transparent material, such as glass or plastic, supported in a vertical position, at an appropriate height above the ground, by means of a rigid frame (Figure 6.17a). Flying insects hit the window and fall into a water-filled tray fixed to its lower edge. Lengths of plastic guttering, fitted with end stops, make convenient trays. A drainage hole bored into the base of the tray, and closed by means of a rubber or plastic bung, allows it to be emptied at the end of each sampling period. Catches can then be sorted in the laboratory. The "vane-style" window trap (made from two intersecting pieces of clear PVC plastic arranged over a funnel and collecting vessel) can be set on the ground or hung in the canopy of a tree (Juvonen-Lettington and Pullen, 2001). Separate trays are normally fitted to each side of standard window traps to provide information on flight direction. Directional information can be increased by using two traps positioned at right angles to one another. Data from window traps have been used: (a) to relate the flight activity of carabid beetles to wind direction and to the reproductive state of females (van Huizen, 1990), (b) to monitor the immigration of carabids into newly-formed polders in the Netherlands, and thereby detect the establishment of new populations (Haeck, 1971), (c) to monitor levels of predator activity in cereal fields (Storck-Weyhermüller, 1988), and (d) to detect periods of aerial dispersal by spiders (De Keer and Maelfait, 1987), carabid beetles (Markgraf and Basedow, 2000) and staphylinid beetles (Markgraf and Basedow, 2002). Hance (1995) used directional window traps to show that more ladybirds, *Coccinella 7-punctata*, were entering a maize field from an adjacent orchard than from an adjacent fallow field. He also noted that some carabid species (such as *Demetrias atricapillus* and *Bradycellus verbasci*) were well represented in window traps but were very rarely caught in pitfall traps in the same area. Thus window traps are an important supplement to pitfalls for documenting the carabid fauna of a region. Window traps can also be hung from the trunks of trees to intercept strong-flying arthropods, such as wasps and predatory beetles, that are flying towards the tree (Hanula and Franzreb, 1998).

Another type of flight interception trap, used to monitor the movement of staphylinid beetles in a Canadian raspberry plantation, comprised 1.8 m^2 window screen mesh (1.5 mm) held over galvanised metal pans containing 2% formalin solution (Levesque and Levesque, 1996). Boiteau *et al.* (1999) constructed a 15 m-high sampling tower holding 40 interception traps (10 per side), placed at regular intervals above ground. Each trap was a plywood board, painted yellow, with an antifreeze-filled gutter below. The tower was sited in a pasture in an agricultural area and showed that, for thirteen species of ladybird, more than half the flights were at or below 3.8 m. The tower was also used to monitor aerial dispersal and vertical flight profiles of other predators, including Carabidae, Staphylinidae and Neuroptera (Boiteau *et al.*, 2000ab).

Sticky traps work on a similar principle to window traps but consist of a sticky surface to which the intercepted insects adhere upon impact (Figure 6.17b). Any suitable surface, either flat or curved, can be coated with a weatherproof adhesive, several types of which are commercially available. Sticky traps supported on poles at various heights were used to monitor the activity of predatory anthocorid bugs in pear orchards (Hodgson and Mustafa, 1984). Very few anthocorids were caught on traps placed within the orchard, but those placed at its edges successfully detected two peaks of flight activity during the year, and the peaks are believed to represent movement to and from hibernation sites. The airborne dispersal of the predatory mite *Phytoseiuliis persimilis* and its prey *Tetranychus urticae* was investigated using sticky traps made from microscope slides coated with silicon grease (Charles and White, 1988). These were clipped in a vertical position onto pieces of

(a)

(b)

Figure 6.17 Traps for interception of flying insects: (a) window trap; (b) sticky trap. The window trap consists of a sheet of transparent plastic; the insects are caught in sections of guttering containing water with a small amount of detergent.

wooden dowelling, which were arranged in the form of a horizontal cross at the top of a supporting pole. A network of 29 yellow sticky traps was used to study dispersal of coccinellid beetles on a 330 ha farm. The study showed that *Propylea 14-punctata* and *Adalia bipunctata* appeared first on traps placed near grass banks and hedgerows, but *Coccinella 7-punctata* was more evenly distributed (Zhou *et al.*, 1994). Greenstone *et al.* (1985) compared horizontal sticky wires with vertical sticky panels for monitoring ballooning spiders above a soybean field. Wire traps underestimated numbers of small money spiders (Linyphiidae). Traps subsequently used (Greenstone, 1990) comprised 16 sheets of 25 cm x 50 cm, 12.7 mm mesh, galvanised hardware cloth, coated with adhesive, and stapled to a wooden frame. They were hung by clamps from parallel horizontal pairs of clothes lines attached to vertical posts set in the ground. This design enabled rapid replacement of traps, which was done weekly. In some habitats sticky traps can also be useful for monitoring natural enemies that are walking or running, rather than flying. Docherty and Leather (1997) attached two polythene bands, covered with sticky Oecotak™, to the trunks of pine trees to estimate the numbers of spiders and harvestmen (Opiliones) ascending and descending the tree trunks. Van Laerhoven and Stephen (2002) suspended a series of sticky-coated wire mesh traps from a rope at 4 metre intervals to a height of 16 metres on pine tree trunks to catch foraging adult parasitoids of the Southern Pine Beetle *Dendroctonus frontalis*. They noted that, for maximum efficiency, traps should be in contact with the trunk because parasitoids disperse by walking as well as flying. Braun *et al.*, (1990) applied 10 cm bands of sticky Tanglefoot® directly to the trunks of trees to monitor movement of the natural enemies of fruittree leafroller, *Archips argyrospila*. Directional arboreal photoeclectors (subsection 6.2.7) can also be used to study upward or downward migration of arthropods on tree trunks (Basset *et al.*, 1997).

Sticky trap catches of parasitoids and predators can sometimes be increased by using colour as an attractant (Neuenschwander, 1982; Moreno *et al.*, 1984; Trimble and Brach, 1985; Samways, 1986; Antolin and Strong, 1987) (see subsection 6.2.8), but it is questionable whether meaningful information on natural patterns of movement can be obtained using coloured sticky traps, because of the likelihood of the traps' attractant properties interfering with the natural behaviour of insects. The material comprising interception traps, even colourless, translucent ones, should be tested for its degree of UV reflectance, if such traps are to be used for investigations of insect movements.

The direction and height of flight of green lacewings in alfalfa fields was recorded using wire-mesh sticky traps attached to poles, as were data on both diel and seasonal patterns of flight activity (Duelli, 1980). Similarly, wire-mesh traps, measuring 1 m^2 and placed at a range of heights, were used to monitor insect flight activity in cereal crops (Kokuba and Duelli, 1986). The main predators caught were adult ladybirds, hover-flies and lacewings, but there was evidence that different species preferred to fly at different heights, so that the choice of trap height can be very important. On a larger scale, sticky nets measuring 2.5 m high by 1.5 m wide were used to monitor the flight activity of two ladybird species around experimental plots of alfalfa and oats (Ives, 1981). Captures of marked beetles in these nets helped to demonstrate that reductions in ladybird numbers recorded in certain plots were due to the emigration of beetles, rather than to beetle mortality.

Sticky material (e.g. Tanglefoot®) can also be used in a different way to study small-scale movements of walking natural enemies. Rosenheim and Brodeur (2002) used Tanglefoot® to create funnel shapes on papaya petioles. Ladybird larvae (*Stethorus siphonulus*) detected and avoided the sticky walls of the funnel and were channelled into a holding area (delimited by a solid barrier of Tanglefoot®) on the petiole, because it was easy for them to enter by the wide end of the funnel but very difficult to exit by the narrow end. These funnel traps required only minor repair over a ten-day period, including light rainfall, and enabled the authors to determine the stage during larval development at which *S. siphonulus* exits from papaya leaves.

Nets are more efficient than solid interception traps for catching weak flyers, because they interfere much less with wind flow. It was estimated that the wire-mesh sticky traps used to catch green lacewings reduced wind speeds by only 10% (Duelli, 1980). Therefore, Malaise traps constructed of netting, and the Masner-Goulet trap (subsection 6.2.4), might be useful for monitoring the flight activity and movement patterns of insect parasitoids. The vertical distribution of aeronautic spiders and mites at heights up to 100 m was established by attaching nets to three tall radio masts (Freeman, 1946). A different approach to netting natural enemies is to employ mobile nets, which provides great versatility in the choice of horizontal and vertical sample zones. Greenstone *et al.* (1991) carried 0.62 m diameter nets attached to an automobile and a light aircraft (Figure 6.18). A slow-flying (72 km h^{-1}) fixed-wing aircraft did not destroy small or delicate natural enemies, and the pair of nets attached to such an aircraft sampled at a a rate of 38,910 m^3 h^{-1} (compared with automobile sampling at 21,420 m^3 h^{-1}). Natural enemies sampled by this method included spiders, predatory flies (Dolichopodidae), wasps, ants and parasitoid wasps, staphylinid and coccinellid beetles and anthocorid bugs. Many of these were still alive when nets were emptied (Greenstone *et al.*, 1991). Other options for flying aerial nets, for which methods have been devised, include helium-filled balloons, parafoil kites, radio-controlled model aircraft and helicopters (Reynolds *et al.*, 1997).

Water-filled tray traps (deposition traps) were used to monitor aerial dispersal of spiders (Topping and Sunderland, 1995; Weyman *et al.*, 1995; Thomas and Jepson, 1997,1999). Each trap was a 10 cm deep, 1 m^2, fibreglass tray, filled with water and ethylene glycol (20:1) plus 1% detergent, fitting inside a 1.6 m^2 metal tray containing the same fluid (Figure 6.19). The inner tray sloped towards a central hole to which a muslin tube was attached and tied. The outer tray acted as a barrier (moat) to prevent spiders walking from the crop to the inner tray, which therefore received only aerial immigrants (Topping and Sunderland, 1995). These deposition traps caught more spiders than sticky sheets of the same area laid flat on the ground (deposition:sticky was 1.9:1). Spiders in the vegetation may have snagged their ballooning lines on the edge of the deposition trap and then been able to run across the moat into the inner tray. Deposition traps have the advantage that they can be left unattended for long periods, but they are more labour-intensive to operate than rotor and suction traps (below) (Topping and Sunderland, 1995).

Suction traps (see Southwood and Henderson, 2000, for general design) have been used to determine the phenology and vertical location of ballooning spiders (Salmon and Horner, 1977; Dean and Sterling, 1990; Sunderland, 1991; Blandenier and Fürst, 1998; Topping and Sunderland, 1998; Thorbek *et al.*, 2002), carabid and staphylinid beetles (Sunderland, 1992) and the green lacewing, *Chrysoperla carnea* (Perry and Bowden, 1983). Traps were located variously in fields, at the edge of fields, and on the tops of buildings. Holzapfel and Perkins (1969) caught spiders, ants, parasitoid wasps, staphylinid beetles and predatory Heteroptera in an electric suction trap on board ship during nine cruises across the Pacific. The traps used in these studies were the Johnson-Taylor suction trap (Johnson and Taylor, 1955), the 46 cm-diameter Enclosed- cone Propeller Suction Trap (Taylor, 1955) and the Rothamsted Insect Survey 12.2 m suction trap (Taylor, 1962). These traps have motor-driven fans that draw air down a tube, and the arthropods are segregated from the air by a filter and then deposited in a container filled with preservative. Miniature fan traps have been used to monitor the movement of parasitoids. Bellamy and Byrne (2001) deployed 102 such fan traps to study dispersal of the aphelinid *Eretmocerus eremicus*. Each trap had a 12 V DC 0.12 A fan which drew air into a 7.6 cm- long PVC pipe containing a plastic collecting vial with an organdy base; wasps were drawn into

(a)

(b)

Figure 6.18 (a) Insect net deployed in flight showing details of harness, reducing pulley, and main line weak link; (b) frame for mounting an insect net atop an automobile (Source: Greenstone *et al.*, 1991.) Reproduced by permission of The Entomological Society of America.

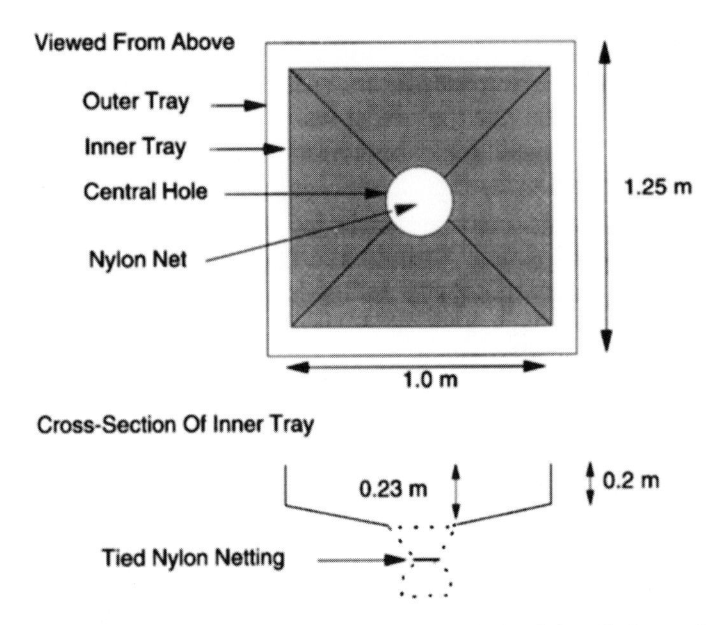

Figure 6.19 Plan of a deposition trap. Source: Topping and Sunderland (1995). Reproduced by permission of Aarhus University Press.

the trap and held against the base of the vial by suction. The cost of materials for such small battery-operated fan traps is about US$10, and they are ideally suited for parasitoid studies in that they capture insects intact and alive and are insufficiently powerful to catch larger insects (Hagler *et al.*, 2002), which results in more efficient processing of the catch. Collecting vials can be quickly and easily capped, removed and replaced.

Zhang *et al.* (1997) caught flying carabids (*Harpalus rufipes*) in **blacklight traps** (subsection 6.2.8) and found that 97% were sexually immature, suggesting a pre-reproductive dispersal phase in this species. It is likely that these beetles disperse by flight over great distances since flights of tethered beetles in the laboratory typically lasted for about two hours (Zhang *et al.*, 1997).

Vertical-looking radar (VLR), which can automatically monitor the horizontal speed, direction, orientation, body mass and shape of arthropods intercepting the beam (Smith and Riley, 1996) could, potentially, provide useful information on the aerial migration of natural enemies.

Techniques such as blacklight and Malaise traps rely mainly on the behaviour of the natural enemy. Suction traps and sticky traps tend to be sensitive to weather conditions and the size of the organism. To circumvent some of these problems, a large **rotary interception trap** was devised (Figure 6.20). It had a high efficiency and had the advantage that the sampling area was precisely defined. The rotary trap was more efficient than a co-located suction trap at catching Syrphidae, Staphylinidae and parasitic Hymenoptera (Topping *et al.*, 1992).

If the vertical aerial distribution of natural enemies is being investigated, then it is important to bear in mind that the number of insects trapped at a particular trap elevation will be a function of both the spatial density of the insects at that elevation and the downwind component of their ground speed (Duelli, 1980). The latter depends upon the wind velocity (V_W) and the flight velocity (air speed) (V_L) of the insects. If the flight course coincides with the wind direction, the two velocities sum, but if the insects fly on a course at an angle α to the wind direction, the downwind component of the

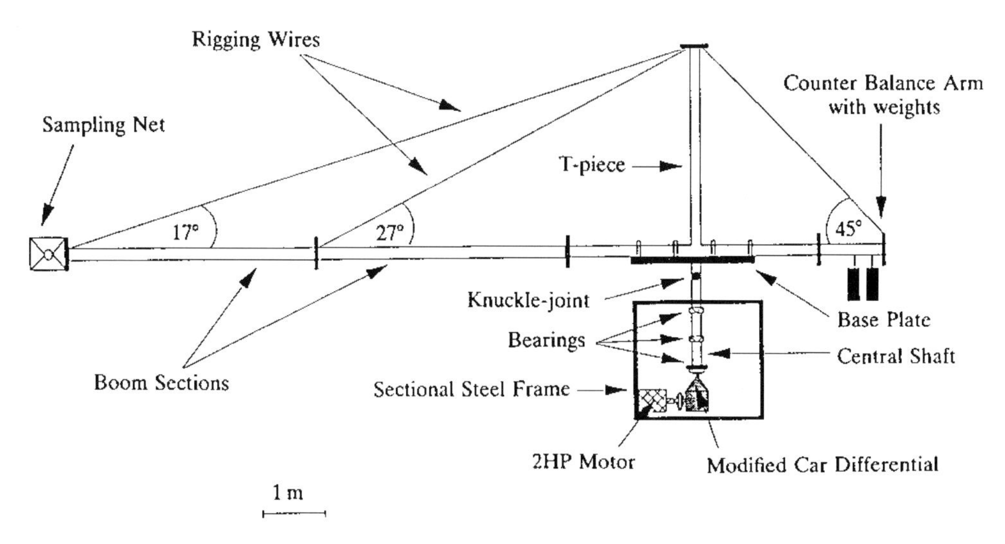

Figure 6.20 Diagrammatic representation of a rotary sampler. Source: Topping *et al.* (1992). Reproduced by permission of The Association of Applied Biologists.

ground speed is $V_W + V_L.\cos\alpha$ (see Duelli, 1980). If, as is likely, wind velocity increases with height from the ground (Figure 6.21a), the numbers of trapped insects (N_T) need to be corrected for each trap elevation, to obtain the relative densities (N_F) per unit volume of air, as follows (see Figure 6.21b):

$$N_F = N_T/(V_W + V_L.\cos\alpha) \qquad (6.2)$$

Interception traps can also be used to study the movement of aquatic predators which are carried along in flowing water (Elliott, 1970; Williams and Feltmate, 1992; Southwood and Henderson, 2000). These devices generally take the form of tapering nets, with or without a collecting vessel attached, which are either positioned on the stream bed (Waters, 1962), or are designed to float (Elliott, 1967).

Valuable information on natural enemy activity in the field can be gained by monitoring the movements of marked individuals (Scott, 1973). The commonest of the marking techniques used for this purpose is the application of spots of paint (usually enamel paint because some other types of paint may have toxic effects), which has been used to mark carabid beetle adults (Jones, 1979; Perfecto *et al.*, 1986; French *et al.*, 2001) and larvae (Nelemans, 1988), cantharid beetle larvae (Traugott, 2002), lycosid spiders (Hackman, 1957; Dondale *et al.*, 1970; Samu and Sárospataki, 1995b; Samu and Kiss, 1997), linyphiid spiders (Samu *et al.*, 1996), araneid spiders (Olive, 1982) and adult ladybirds (Ives, 1981). Tiny protonymphs of predatory phytoseiid mites have even been marked, using watercolours (Schausberger, 1999). Acrylic paint was used to mark *Episyrphus balteatus* (Syrphidae) for short-duration dispersal studies. Individuals did not appear to be adversely affected and were seen foraging normally for 90 min after marking (MacLeod, 1999). Alternative fluid markers are inks (Kromp and Nitzlader, 1995) and typewriter correction fluid (Holland and Smith, 1999; Raworth and Choi, 2001). However, these marks may be lost, particularly by carabid species which burrow in the soil or squeeze themselves into cracks or under stones. It is advisable, therefore, to test the durability of any marking system in preliminary trials. Raworth and Choi (2001) advocate painting identical marks on each elytron, since the probability of losing both marks (at least for short-duration studies of epigeal beetles) is very low. More robust marking systems have been used with (or are suitable for) carabid

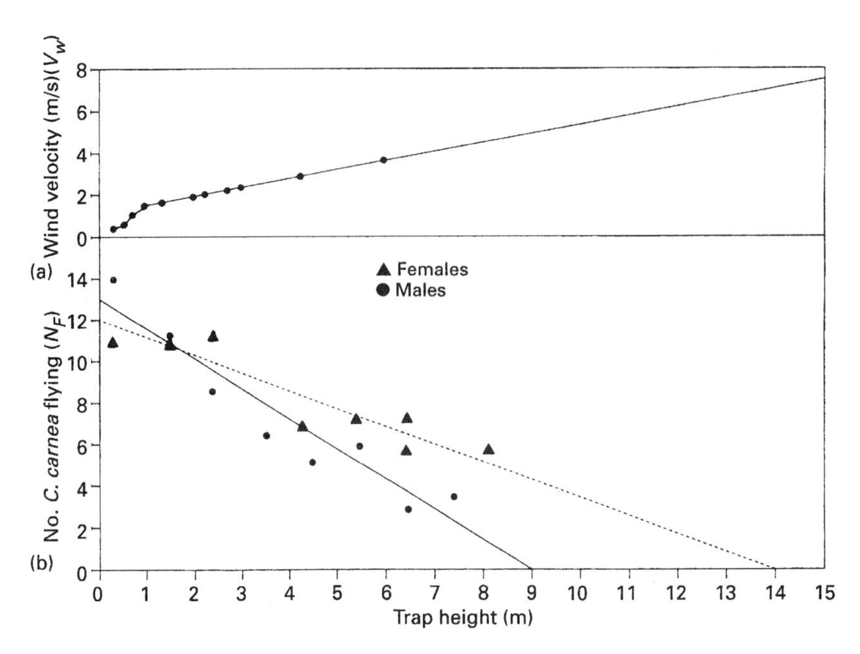

Figure 6.21 Correcting the results of sticky trapping (traps set at varying heights above the ground) for the effects of wind velocity, in the green lacewing, *Chrysoperla carnea*, in alfalfa fields. This figure shows: (a) wind velocity (V_w, see text) as a function of height in an alfalfa field. In a night with average wind speed, the 'boundary layer' (the layer within which the insect's air speed exceeds wind speed) is at 50 cm. Above 1 m the ground speed is mainly a function of the wind speed; (b) the relative numbers (N_F, see text) of insects flying at a particular elevation. The numbers can be estimated from the vertical distribution of trapped insects by correcting for wind speed, air speed of the lacewing, and the angle between wind direction and flight course (see text for details). The corrected vertical distribution of females in the oviposition phase and of males, are treated here as linear regressions. Source: Duelli (1980). Reproduced by permission of Springer-Verlag.

beetles, including branding with a surgical cautery needle (Figure 6.22a,b), or a fine-pointed electric soldering iron (Manga, 1972; Ericson, 1977; Nelemans *et al.*, 1989; Mikheev & Kreslavskii, 2000), etching the elytral surface with a high-speed drill (Best *et al.*, 1981; Wallin, 1986; Thomas *et al.*, 1998; Fournier and Loreau, 2002), cutting notches in the edges of the elytra and thorax with a small medical saw (Benest, 1989a; Loreau and Nolf, 1993), cutting the tips off the elytra (Wallin, 1986) and etching the dorsal surface of the elytra with an engineering laser (Griffiths *et al.*, 2001). It is important, especially when using a technique which involves physical mutilation, to ensure that marked individuals have the same survival rate as unmarked individuals and that their subsequent behaviour is unaffected.

A system of mark positions can be devised that will allow large numbers of a natural enemy species to be individually marked (Samu *et al.*, 1996; Thomas *et al.*, 1998; Holland and Smith, 1999; Southwood and Henderson, 2000). For example, Kiss and Samu (2000) marked the dorsum of the cephalothorax and abdomen of lycosid spiders with dots of enamel paint in such a way that the trap in which the spider was collected was coded by position of the dots, the date of trapping was coded by colour, and a supplementary dot was added if the spider was recaptured.

Information on minimum distances travelled and speed and direction of movement can be obtained from the recapture data of marked individuals (Evans *et al.*, 1973; Ericson, 1978; Greenstone, 1979a; Best *et al.*, 1981; Nelemans,

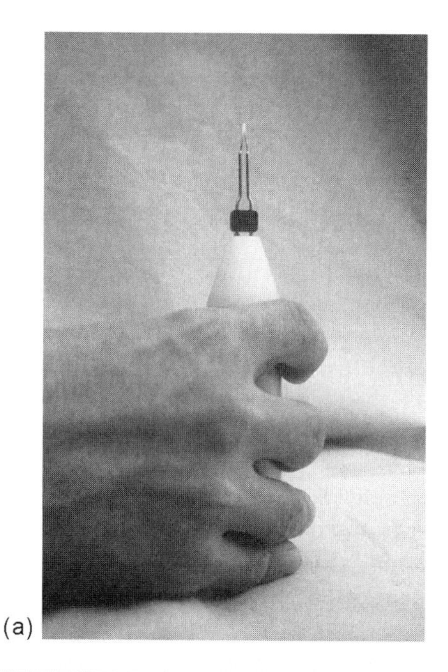

Figure 6.22 (a) Surgical cautery needle used for marking the elytra of carabid beetles; (b) carabid beetle (*Pterostichus melanarius*) marked using the needle.

1988; Mader *et al.*, 1990; Lys and Nentwig, 1991; Zhang *et al.*, 1997; Thomas *et al.*, 1998). Evans *et al.* (1973) studied the movements of adult mutillid wasps in an open sandy habitat. The area was divided into a grid using wooden stakes and patrolled daily by an observer. Adult wasps were caught, marked with spots of coloured paint (model aircraft dope) using the head of an insect pin, and then released. Individuals could be recognised by the colour and positions of paint spots. Similarly, valuable data about the dispersal behaviour of parasitoid wasps attacking a tephritid fly on thistles were obtained by paint-marking (Jones *et al.*, 1996). Jones *et al.* marked adult parasitoids with acrylic paint, and the location of individuals on thistles within a study site was recorded daily. Laboratory tests showed that marking reduced longevity of one species slightly, but the others were unaffected. The average recapture rate was 22%. The activity of wolf spiders in an estuarine salt marsh was investigated in a similar way; the spiders were caught using an aspirator and individually marked with enamel paint (Greenstone, 1979a). During an investigation of the population dynamics of the dragonfly *Calopteryx aequabilis*, Conrad and Herman (1990) studied movement patterns in relation to adult territoriality by marking insect wings with a drawing pen so that individuals could be recognised. The study was carried out along a 635 metre section of a stream, which was divided into 5 metre sectors, delineated with small numbered flags. Observers patrolling the stream noted the sex, age category and sector-location of all individuals sighted. Population movements were analysed using a modification of Scott's (1973) method, which allows separation of the velocity and distance components of movements.

Marking can be dispensed with in cases where the species to be studied is new to an area. For example, Coll and Bottrell (1996) studied the eulophid *Pediobius foveolatus*, which parasitises Mexican Bean Beetle (*Epilachna varivestis*, Coccinellidae), and does not overwinter in Maryland where the sudies were conducted. The authors released parasitoids at the edges of field plots, and, after various periods of time (depending on the experiment), they vacuumed all plants in the plots to record the movement of the parasitoids (Coll and Bottrell 1996). In an analogous study, *Lysiphlebus cardui* (Braconidae, Aphidiinae) adults were released in a mountain area of Germany before the natural populations were active. Potted thistles containing colonies of the host aphid (*Aphis fabae cirsiiacanthoidis*) were distributed at up to 470 metres from the release point, but female *L. cardui* were found to move a maximum of 20 metres to a new host patch (Weisser and Völkl, 1997). In the context of the release of new species of parasitoid, hosts as 'bait' (subsection 6.2.8) can be used to monitor the progress of dispersal. **'Sentinel' egg masses** of the spruce budworm (*Choristoneura fumiferana*) were placed out at designated locations in a Canadian forest and later retrieved to the laboratory to rear out parasitoids. These egg masses showed that the released parasitoids *Trichogramma minutum* dispersed 19 metres in 5 days (Smith, 1988). The braconid parasitoid *Cotesia glomerata* was introduced into North America to control the caterpillar pest *Pieris rapae* in brassica fields, but there were fears that it might disperse into woodland and reduce populations of the native butterfly *Pieris virginiensis*. Benson *et al.* (2003) put out **sentinel larvae** of *Pieris rapae* and *Pieris napi* in woodland sites, and, after subsequent dissection revealed no *Cotesia* eggs or larvae in the sentinel hosts, they concluded that *C. glomerata* (even though present in adjacent meadows) do not forage in forested habitats and so are not a threat to *P. virginiensis*. To study dispersal of the parasitoid *Venturia canescens* in relation to environmental heterogeneity, Desouhant *et al.* (2003) released wasps (marked on the thorax with a dot of acrylic paint) from a central point and monitored their speed and pattern of spread out to 97 traps (moth larvae in food medium on 1 metre-high poles set 10 metres apart). This design allowed the authors to relate dispersal to various factors such as sunlight intensity, amount and type of vegetation (Desouhant *et al.*, 2003). Monitoring dispersal of mass-reared natural enemies of the same species as occur naturally in the area of release would be facilitated if the reared individuals carry a genetic marker, such as an eye or body colour mutation (Akey, 1991), which is noticeable visually, or which can be detected unambiguously by biochemical means (Symondson and Liddell, 1996a).

Some attempts have been made to obtain information on the movement of individuals without interrupting that movement. Baars (1979b) labelled adult carabids with an iridium isotope, (^{192}Ir), by mixing it with quick-drying enamel paint. Using a portable scintillation detector, it was possible to track the movements of up to ten individuals over the same period of time in the field. However, the beetles frequently lost their paint marks and there was some evidence from laboratory tests that the radiation affected beetle mortality rates.

The chemical composition of natural enemies (**chemoprinting**) might, potentially, be used to identify their area of origin, and thus be used to determine, indirectly, the minimum distance dispersed. This might be successful if they were to disperse as adults after pre-adult development on hosts/prey living on soils and vegetation with a characteristic chemical fingerprint. Wavelength dispersive X-ray fluorescence spectrometry can be used to make quantitative measurements of the elemental composition of individuals (Bowden *et al.*, 1979). Dempster *et al.* (1986) used this method to measure the relative composition of nine chemical elements in adult Brimstone butterflies (*Gonepteryx rhamni*), and they found differences between sexes, sites, seasons and years. For this method to be of use, the chemical pattern obtained by pre-adult stages should be stable in the adult, and inter-site variation should be greater than intra-site variation. Unfortunately, in the Dempster *et al.* (1986) study, the differences gained in pre-adult stages were quickly masked as the adults aged and fed. Sherlock *et al.* (1986) also failed to demonstrate clear source-related chemoprints in the cereal aphid *Rhopalosiphum padi*.

The proportions and composition of **stable isotopes**, (e.g. deuterium), in the bodies of migrating animals may also reflect their area of origin (Hobson, 1999). Hobson *et al.* (1999) described the first application of the stable isotope technique to infer migration of an invertebrate. They showed that the stable hydrogen isotopic composition of wing keratin (metabolically-inert after eclosion) of monarch butterflies (*Danaus plexippus*) was highly correlated with that in their milkweed host plants, which, in turn, was related to stable geographical patterns of deuterium in rainfall. Using these techniques it was possible to determine, approximately, the natal latitude in eastern USA of monarchs collected after a period of migration (e.g. when overwintering in Mexico) (Hobson *et al.*, 1999). These techniques could prove useful for studying migration or long-distance aerial dispersal of natural enemies such as hover-flies and spiders. The chemical elements C, N, S, H and O all have more than one isotope and isotopic compositions can be measured with great precision using a mass spectrometer (Peterson and Fry, 1987). Stable C and N isotopes have been used in entomological studies. Carbon, for example, is fixed in C_3 plants (such as cotton) by the Calvin photosynthetic pathway and in C_4 plants (such as sorghum) by the Hatch-Slack pathway. C_3 and C_4 plants contain different ratios of isotopes ^{13}C and ^{12}C, and these ratios are maintained at higher trophic levels. Using a stable isotope mass spectrometer, Ostrom *et al.* (1997) showed that isotope values (isotopic ratios for C and N) were sensitive to change of laboratory diet by the ladybird *Harmonia variegata*, within a period of three weeks. These authors then documented diet shifts (alfalfa-, wheat- and maize-based feeding) in field-collected ladybirds (*Coleomegilla maculata*), which implied local dispersal of the beetles between habitats.

Instead of relying on natural labelling (e.g. chemical fingerprint or stable isotope ratio), natural enemies can be labelled artificially with one or more specific trace elements (e.g. rubidium and strontium). This technique of **elemental marking** is defined as the establishment of an elemental concentration in an individual, significantly higher than that carried by the indigenous population (Akey, 1991). This method can be applied without having to handle the insect (as is necessary with paints, dyes, powders etc.), and it does not entail environmental contamination (as for radioactive isotopes). An atomic absorption spectrophotometer and graphite furnace permits analysis of micrograms of sample. Parasitoids and predators labelled with rubidium can be distinguished from controls for 4 to 20 days, depending on species. The method is useful for a wide range of parasitoids and predators (Graham *et al.*,1978; Prasifka *et al.*, 2001). After plots of cotton and sorghum were treated with foliar solutions of rubidium chloride a range of ladybirds, spiders, predatory beetles and predatory bugs became labelled and were found to have moved between fields. Recapture rates were three to four times higher than for the use of fluorescent dust markers in a comparable system (Prasifka *et al.*, 1999; 2001). Elemental marking usually does not affect the biology of the labelled animal. One exception to this was a reduction in longevity of male parasitoids, *Anaphes iole* (Mymaridae) labelled with 1000 ppm rubidium. There were no adverse effects, however, on the parasitoid *Microplitis croceipes*, triple-labelled with caesium, rubidium and strontium at 1000 ppm (Jackson 1991). Corbett *et al.* (1996) treated prune trees with rubidium in the autumn. Adult prune leafhoppers, *Edwardsiana prunicola*, became labelled and their mymarid parasitoid, *Anagrus epos*, carried a label 3.9 times greater than the background level of rubidium. Emerging parasitoids retained the label throughout their lifetime. Such labelling could, potentially, be used to quantify dispersal of the parasitoid from prune to grapevines, but, unfortunately, there was overlap in rubidium level between treated and natural populations of the parasitoid, and probabilistic statistical techniques will be needed to quantify dispersal in this context. Detection of marked eggs from marked mothers is possible, and so this technique can provide information on survival, dispersal and oviposition (Akey, 1991).

Small natural enemies may not be ideal candidates for marking with paints and tags, and detection of trace elements requires specialised equipment. An alternative is to mark by immersing the insects in rabbit immunoglobulin (IgG) and then assay (after dispersal) by ELISA (enzyme-linked immunosorbent assay) (see subsection 6.3.9) with goat anti-rabbit IgG. This marker can remain detectable for at least eight days. This is, however, an external marker, which will be lost on moulting or during development to a different life stage (Hagler *et al.*, 1992a). Internal protein markers can be used for marking fluid-feeding natural enemies (Hagler and Jackson, 2001). Whitefly parasitoids (*Eretmocerus eremicus*) marked both internally (fed on honey solution containing rabbit IgG) and externally (exposed to a mist of rabbit IgG solution from a medical nebulizer), were released into a cotton field and 40% of recaptured individuals (during 32 h after release) were carrying the label. This study provided valuable information about gender-related dispersal, diel activity pattern, and distance moved by this parasitoid (Hagler *et al.*, 2002).

Radio telemetry, a standard method for studying vertebrate dispersal, can also be used for studies of large invertebrate predators. Riecken and Raths (1996), fixed 0.7 g transmitters (with an aerial length of 5 cm and a battery life of 28 days) to the elytra of *Carabus coriaceus* (the largest species of carabid beetle in West Germany) using silicone glue. Heavy rain sometimes caused transmitter malfunction, but, despite such problems, the authors were able to detect beetles at a maximum range of 400 m. This method, unlike harmonic radar (below), allows transmission at different frequencies, and, therefore, the individual tracking (without capture and disturbance) of more than one specimen in the same area (Riecken and Raths, 1996). This property of radio telemetry was exploited by Janowski-Bell and Horner (1999) to track individual tarantula spiders (*Aphonopelma hentzi*) in a Texas scrub habitat. Transmitters weighing 0.6–0.8 g were attached to adult male spiders weighing 2.5–7.5 g, and half of these spiders retained their transmitters for three or more days. The signal was received at a distance of several hundred metres and spiders could be detected in burrows and under rocks. Males were observed to move up to 1300 m during 18 days when searching for females. Although the possibility that the transmitter and aerial affected behaviour to some extent could not be excluded, tagged spiders were, nevertheless, observed to enter burrows and crawl past resident females, as normal (Janowski-Bell and Horner, 1999).

Remote sensing techniques, such as **harmonic radar**, can also be used to track the movement of natural enemies. The natural enemy target of harmonic radar tracking must carry a transponder (antenna plus diode), which detects radio waves transmitted from a hand-held device, and re-radiates some of the received energy at an harmonic frequency. This method does not require a battery to be attached to the target, which means that extremely lightweight tags can be developed (Kutsch, 1999). An harmonic radar system has been used to track the movements of large carabids over the ground (Mascanzoni and Wallin, 1986; Wallin and Ekbom, 1988; Wallin and Ekbom, 1994; Lövei *et al.*, 1997), and the foraging flights of bumblebees, *Bombus terrestris* (Osborne *et al.*, 1999). Improved tags are now available for reflecting harmonic radar. The 0.4 mg tags, developed recently, are an order of magnitude lighter than those used on bees, and two orders of magnitude lighter than those used on carabids. Detection is possible from a distance of 50 m. The tag increased the weight of a tachinid fly by 0.1–0.9% and, 48 hours after release, the fly had moved 100 m (Roland *et al.*, 1996). In general, tags weighing up to 1.5% of insect body weight should have no deleterious effects on flight performance (Riley *et al.*, 1996; Roland *et al.*, 1996; Boiteau and Colpitts, 2001). **Metal detection equipment**, employing a pulse field induction loop, is many times less expensive than harmonic radar. It is most suitable for detecting microhabitat preferences and local movement (e.g. to find pupation sites) of relatively sedentary cryptic invertebrates. Piper and Compton (2002) tagged beetle larvae with tiny steel tags (1 × 3 mm; 0.35 mg) and released them in leaf litter. Using a hand-held metal detector powered by a 12V battery, they were able to detect 90% of the released larvae after five months in the field. The detection range is only 3–7 cm, but tagged animals can be detected through litter and soil, and laboratory experiments showed no significant effect of tagging on movement of larvae through leaf litter (Piper and Compton, 2002).

Other marking techniques which have been used to monitor the movements of natural enemies include: (a) numbered plastic discs glued to the dorsum of adult or larval carabids (Nelemans, 1986, 1988; Lys and Nentwig, 1991), (b) coloured and numbered bee tags attached to adult *Tenodera aridifolia* mantids (Eisenberg *et al.*, 1992) and adult paper wasps (Schenk and Bacher, 2002) (c) spots of typewriter correction fluid applied to carabids (Wallin, 1986; Hamon *et al.*, 1990; Holland and Smith, 1997), (d) spots of acrylic paint applied to the pronotum of adult eucoilid and braconid parasitoids with a fine brush (Driessen and Hemerik, 1992; Ellers *et al.*, 1998), (e) numbers written on the wings of damselflies with a drawing pen (Conrad and Herman, 1990), (f) fluorescent powder applied to green lacewings (Duelli, 1980), and (g) stable isotopes fed to coccinellid adults in drinking water (Iperti and Buscarlet, 1986). Driessen and Hemerik (1992) attempted to mark adult *Leptopilina clavipes*, a eucoilid parasitoid of *Drosophila* flies, with micronised **fluorescent dust**. Although the dust worked well with the drosophilids, it was unsuccessful as a means of marking the parasitoids because it was rapidly lost from the smooth surface of the exoskeleton and elicited prolonged preening behaviour in the wasps. A potential problem with the use of dusts is that they may become accidentally transferred to unmarked individuals of the same species during the recapture process. Fluorescent dusts have also been used to detect visits by parasitoids to specific plants by applying the dusts to the leaves of the plant and then searching for traces of the dust on parasitoids under UV illumination (Ledieu, 1977). Corbett and Rosenheim (1996) used parasitoid mark-recapture data to estimate the **"diffusion" parameter** D, which is the random component of movement, independent of forces such as wind, attraction to resources and repulsion from conspecifics. Grape leaves containing leafhopper (*Erythroneura elegantula*) eggs, parasitised by the mymarid *Anagrus epos*, were put in a bag and gently rolled for 1 min in a mixture of 100 ml Day-Glo^R fluorescent powder and 1 kg walnut husks (as a carrier). As they emerged from the host egg, 85% of parasitoid adults marked themselves with a few minute particles of powder, which lasted at least 48 h, and permitted mark-recapture studies in vineyards (recaptures on yellow sticky traps). Marks were too small for detection in UV light, and a microscope was needed. Since only a few particles of marker adhered, minimal effects on survival and behaviour were expected (Corbett and Rosenheim,1996).

Self-marking with fluorescent powder was also used for opiine braconid parasitoids of tephritid fruit-flies in Hawaii in a study of dispersal in the field (Messing *et al.*, 1993), but, unfortunately, this marker was found to greatly increase adult parasitoid mortality, compared to controls. Garcia-Salazar and Landis (1997) studied the effects of wet and dry Day-Glo^R fluorescent marker, at three doses, on the survival and flight behaviour of *Trichogramma brassicae*. In the dry method, powder was added to glass beads in a small plastic cup and shaken to build up a deposit on the walls. The beads and excess marker were removed and wasps were added and allowed to self-mark through contact with the walls. In the wet method, the marker was suspended in ethyl alcohol and swirled to coat the walls. Marking did not affect wasp survival, but the flight response was reduced at the highest dose for the dry method. More individuals, however, retained the mark for 24 h in the dry method. Adults of *Spalangia cameroni*, a hymenopteran parasitoid of the house-fly *Musca domestica*, were allowed to self-mark in sawdust containing Day-Glo® fluorescent powder. Particles of fluorescent powder could be detected on the parasitoids with the naked eye for about four days, and marking reduced their survival only slightly, from eleven days to ten days (Skovgård, 2002).

Prasifka *et al.* (1999) marked predators with fluorescent dust in a sorghum field and made predator collections one day later in an adjacent cotton field. Examination of predators (mainly *Orius* bugs), under a dissecting microscope, revealed particles of dust, and demonstrated that predators had dispersed from the senescing sorghum crop into cotton. Six colours of fluorescent dust were used, with each colour being applied to a crop zone at a specific distance from the sorghum-cotton interface, so that the distance moved by predators could be determined. Dust was applied at 4.8 g per m of row at 3.5 kg cm^{-2} of pressure, through a compressed-air sandblast gun, with a spray nozzle internal diameter of 4.4 mm.

Thousands of individuals of the staphylinid beetle *Aleochara bilineata* were marked with DayGlo pigment and released weekly from 24 Canadian gardens. Three percent of marked beetles were recaptured with barrier pitfalls, water traps and interception traps in the gardens. Recaptured beetles had dye particles adhering to the intersegmental membranes of the abdomen, and at the bases of wings, head and coxae. Capture of marked beetles in control gardens (where no releases had been made) showed that they were capable of flying at least 5 km (Tomlin *et al.*, 1992). Judging by experience with marking grasshoppers (Narisu *et al.*, 1999), large natural enemies, such as carabid beetles, marked with fluorescent powder, should be detectable under UV light at night, without the need for recapture, which would enable dispersal of the subject to be monitored without disturbance. Narisu *et al.* (1999) reported that 64% of marked grasshoppers were relocated with UV light at night, compared with only 28% by visual searching during the daytime. Schmitz and Suttle (2001) marked individuals of a large lycosid spider, *Hogna rabida*, with fluorescent powder and recorded (from a distance of 2.5 m using binoculars) horizontal and vertical distances moved every five minutes until the spider noticed the presence of an observer (which could take up to 6 h).

Natural enemies with pubescent cuticles could be marked with **pollen** of a species that does not occur in the study area. Many pollens can be identified to species with a microscope, they are durable and some have evolved to adhere tenaciously to the insect exoskeleton. In some contexts ingested pollen can also yield valuable information about natural enemy dispersal. Wratten *et al.* (2003) dissected the guts of adult hover-flies, stained the contents with saffranin (White *et al.*, 1995), and were able to detect pollen of *Phacelia tanacetifolia* which the hover-flies had taken from strips of this flower planted at the edge of fields. This method was exploited to quantify the degree of inhibition of hover-fly dispersal attributable to various types of hedges and fences in the farmed landscape. *Phacelia* pollen was detectable in the hover-fly gut for less than 8 hours, which enabled calculation of a crude estimate of rate of dispersal (Wratten *et al.*, 2003).

Vital dyes could prove useful in dispersal studies. Braconidae were labelled with the vital dye acridine orange, through their food. The alimentary canal and eggs of these parasitoids became labelled within 24 h, and could be seen fluorescing under an epifluorescent microscope. Tissues continued to fluoresce for 6 weeks after death. Parasitoid longevity was unaffected, and no behavioural effects were observed. Oviposited eggs of some species were labelled, and the label persisted for about 2 days. Stained eggs hatched normally and eventually produced adult offspring (Strand *et al.*, 1990).

Topham and Beardsley (1975) labelled tachinid parasitoids of sugarcane weevils with a **radioactive marker**, in an attempt to monitor the distance travelled by the adult flies between oviposition sites within crops and nectar food sources at field margins (section 6.5). Adults of the eulophid wasp *Colpoclypeus florus*, a parasitoid of leaf-rollers, were fed on honey-water containing the radioisotope ^{65}Zn. Labelling did not affect fecundity, but it significantly reduced longevity and the viability of oviposited eggs (Soenjaro 1979). Emerging adults of the egg parasitoid *Trichogramma semifumatum* were fed on ^{32}P honey-water solution. Three-and a-half million wasps were released in California alfalfa and sampled with sweep net and vacuum insect net. Only a few hundred were recaptured, due to rapid dispersal and dilution of marked wasps in the area, but tagged wasps lived at least 15 days after release, and the study provided useful data on rates of dispersal (Stern *et al.*, 1965). *Trichogramma ostriniae* (an egg parasitoid of the Asian corn-borer, *Ostrinia furnacalis*) were dipped or sprayed with ^{32}P, because less than 25% became marked if allowed to feed on radioactive honey-water. The label did not affect longevity or reproduction, but 63–83% of the label was lost over five days in the field, probably due to self-grooming. The method has potential for short-term ecological studies in situations where radioactive markers are permitted (Chen and Chang, 1991). In some countries there are, nowadays, environmental concerns and regulations concerning the use of radioactive markers in field studies (Greenstone, 1999).

Southwood and Henderson (2000) provide further information on marking techniques and the use of codes, which allow the recognition of large numbers of individuals.

In the context of commercial rear-and-release biocontrol programmes, quantitative assessments of factors that affect searching efficiency (such as flight initiation and walking activity) of parasitoids is an important aspect of quality control (Bigler *et al.*, 1997). A simple laboratory test was devised and showed that strains of *Trichogramma brassicae* varied considerably in flight initiation (percentage flyers and non-flyers in a three day post-emergence period at 25°C). Results from field cage studies, using cages with clear plastic walls and sticky strips, confirmed the laboratory results (Dutton and Bigler, 1995). The laboratory flight-test apparatus used by Ilovai and van Lenteren (1997) to monitor quality of mass-reared parasitoids (*Aphidius colemani*) consisted of a glass cylinder with the internal wall coated with repellent, and the exterior wall shaded to exclude light. Mummies were placed at the bottom of the cylinder and emerging parasitoids flew up to the light at the top of the cylinder, where they were caught on a glass plate, coated with adhesive. Only 40–65% of emerging adults were found to be capable of flight.

The technology is available to study patterns of diel activity efficiently in the laboratory, with a combination of robotics and video image analysis, in real time. A video camera can be moved, automatically, to partition its time between a number of subjects. A system such as this was used to analyse diel cycles of locomotory activity by *Trichogramma brassicae*, measuring the percentage of time spent moving, linear speed, angular speed, and sinuosity of trajectory. Forty individuals were studied, with one 5 s recording every 5 min (Allemand *et al.*, 1994). For most parasitoids the last stage of host-finding is by walking, and parasitisation rate is heavily influenced by rate of walking and searching. With this in mind, Wajnberg and Colazza (1998) video-recorded the walking path of *T. brassicae*, and used an automatic tracking computer device to transform the path to X-Y coordinates (with an accuracy of 25 points s^{-1}). The track width was set at two reactive distances (this species can perceive host eggs at a mean reactive distance of 3.7 mm), and the area searched per unit time was computed with each surface unit counted once, even if the wasp crossed its own path. *T. brassicae* was found to search, on average, at $28\,mm^2\,s^{-1}$. The method can be used to guide genetic selection of mass-reared *Trichogramma* species. Using similar technology, Van Hezewijk *et al.* (2000) found a significant amount of variation between strains of *Trichogramma minutum*, and within strains over time.

DNA-based techniques (subsection 6.3.12) may enable dispersal of natural enemies released in biocontrol programmes to be distinguished from that of native natural enemies of the same species, and also permit tracking of the genetic marker through successive generations (which would not be possible with other marker systems). Gozlan *et al.* (1997) attempted to use RAPD-PCR (subsection 6.3.12) to distinguish between strains of three species of *Orius* (predatory heteropteran bugs) used in augmentative release programmes. Species were readily distinguished, but a high degree of polymorphism prevented discrimination between strains. This method was used successfully, however, by Edwards and Hoy (1995), to monitor insecticide-resistant biotypes of *Trioxys pallidus* (a parasitoid of the walnut aphid, *Chromaphis juglandicola*) released into orchards, and to distinguish them from wild populations. The resistant biotype disappeared from one orchard by a year after release, but it was detected for three years in two other orchards. It might be possible to use PCR techniques to monitor dispersal of virus-labelled natural enemies. De Moraes *et al.* (1998) used PCR to detect, successfully, nucleopolyhedroviruses in homogenates of hemipteran predators up to 48 days after application of these viruses in a soybean field. Although peak levels of virus detection occurred at 10–45 days after application, the virus was found in predator homogenates five days before it was found in the host larvae (velvetbean caterpillar, *Anticarsia gemmatalis*), suggesting that the foraging predators became externally contaminated with virus particles. Similarly, Smith *et al.* (2000) used PCR to detect genetically-engineered baculoviruses in predators (spiders, ladybirds and heteropterans) five days after a cotton field was sprayed with these viruses. If natural enemies, before release, were deliberately surface-contaminated with a unique (e.g. genetically-manipulated) virus, this could be useful for short-term dispersal studies (baculoviruses are usually inactivated by UV light after two or three days).

6.3 DETERMINING TROPHIC RELATIONSHIPS

6.3.1 INTRODUCTION

In nature, the trophic interrelationships of invertebrates within a community rarely, if ever, consist of simple food chains (section 6.1). They often comprise an extensive feeding web composed of several trophic levels. The comparative ease with which parasitoids of endophytic hosts can be reared from, and their remains (egg chorions, larval exuviae) located within, structures such as leaf mines, stem borings and galls, together with the consequent certainty with which parasitoid-host relationships can be discerned, has meant that the most detailed studies of food webs involving insects have been on gall- formers, leaf-miners and stem-borers (Askew and Shaw, 1986; Redfern and Askew, 1992; Claridge and Dawah, 1993; Memmott *et al.*, 1994; Dawah *et al.*, 1995 and Martinez *et al*, 1999). Food webs associated with endophytic insects are also attractive to researchers because of their greater complexity. Askew (1984) has documented more than fifty species of cynipid gall wasps, and their associated parasitoids, on oak and rose galls in Britain, and Redfern and Askew (1992), Claridge and Dawah (1993), Memmott *et al.* (1994), Dawah *et al.* (1995) and Lewis *et al.* (2002) give diagrammatic representations of a range of food webs based on gall-formers, stem-borers and leaf-miners. Complex webs are also now being constructed for non-endophytic pests such as aphids, their parasitoids and secondary parasitoids (Müller *et al.*, 1999) and well as for Lepidoptera (Henneman and Memmott, 2002). They have also been used to answer specific ecological questions, such as the degree to which parasitoid biocontrol agents are penetrating natural communities (Henneman and Memmott, 2002). Further considerations on food webs and their construction can be found in subsection 6.3.13.

Described below are several methods currently used (see Table 6.3 for a summary of their particular uses and relative merits) for elucidating the trophic relationships between invertebrates and their natural enemies from the standpoint of: (a) the host or prey, i.e. the species composition of its natural enemy complex and (b) the natural enemy, i.e. its host or prey range. Some of these methods may also be used when the trophic relationships are already known, either to confirm that a relationship exists in a particular locality, and/or to record the amount of predation and parasitism. Their use in quantitative studies is largely dealt with in Chapter 7 (subsections 7.2.11-14). In cases where a predator is likely to be feeding in more than one ecosystem during a relatively short period of time, analyses using **stable isotopes** of carbon and nitrogen (Fantle *et al.*, 1999; Post *et al.*, 2000; Post, 2002) might be applicable for determining the relative contribution of prey from each ecosystem to the overall diet of the predator (but without any precision concerning exactly which species have been consumed). Stable isotope analyses applied to body tissues can summarise some general aspects of the diet of an individual over a long period of time, as compared with the various methods of gut contents analysis (subsections 6.3.7–6.3.12) which describe foods consumed during one or

Table 6.3 Comparison of available methods for investigating qualitative aspects of trophic relationships in natural enemy complexes and communities. Advantages and disadvantages of each method are discussed in more detail in the text (section 6.3)

Method	Advantages	Disadvantages
Field observation	Immediate and usually unequivocal results obtained with minimal equipment	Time-consuming and labour-intensive; small, hidden, nocturnal, subterranean or fast-moving invertebrates difficult to observe; possibility of disturbance to invertebrates during observation; high cost of video equipment; unlikely to observe predation by highly-mobile predators at natural (low) densities
Collection of prey and remains from within and around burrows and webs	Prey items conveniently located in or around nest/burrow/web	Often involves disturbing predators; applicable to few natural enemies; invertebrates caught in webs may not be prey
Tests of prey/host acceptability	Results usually unequivocal. Defines prey/host that a predator/parasitoid is physically and behaviourally *capable* of attacking	Identification of parasitoid immatures to species level often not possible due to lack of descriptions and keys; false negatives a problem; may not apply to prey choices in the field (false positives)
Rearing parasitoids from hosts	Results usually unequivocal	False negatives a problem, if sampling is not sufficiently extensive and intensive; hosts of koinobionts need to be maintained until parasitoid larval development is complete
Gut dissection and faecal analysis	Simple and quick to perform with minimal equipment; samples can be stored either frozen or in preservative for long periods prior to testing; direct information on prey consumption in the field	Can only be used if predator or parasitoid ingests solid, identifiable parts of prey or hosts that possess solid parts; identification of prey remains not always possible; possibility of misleading results due to scavenging and secondary predation

Immunological analyses of gut contents	Accurate, highly-sensitive and reproducible techniques available; cell lines generating monoclonal antibodies can be retained for future use, giving reproducibility to assays over time; sample can be stored frozen for long periods prior to testing; large numbers of predators can be analysed rapidly	Time-consuming to develop and calibrate, requires specialist equipment and expertise; risk of false positives through cross-reactions, scavenging (and, possibly, secondary predation)
Prey labelling	Simpler to perform than either electrophoretic or immunological tests; dyes easily detected; samples can be stored frozen for long periods prior to testing	Expensive equipment required to detect some labels; radioactive labels a hazard to operator and environment; usually involves severe disturbance to the system under study; labels can be lost rapidly; label passes through trophic levels by a variety of routes (e.g. coprophagy)
Electrophoresis	Can be accurate and sensitive once standardised; large numbers of samples can be analysed rapidly	Expensive, time-consuming to calibrate, and requires specialist equipment and expertise; risk of false positives through scavenging and secondary predation; inappropriate for highly-polyphagous species (coincidence of diagnostic bands)
DNA-based techniques	Can potentially provide highly-accurate data identifying predation on different species, biotypes and populations	Requires molecular biology facilities; DNA is degraded rapidly in the gut of a predator; risk of false positives through scavenging, secondary predation and (where PCR is involved) cross-contamination; cannot yet distinguish between stages

a few meals. Stable isotope ratios (subsection 6.2.11) are useful natural tracers of nutrient flows in ecosystems (Lajtha and Michener, 1994; Ponsard and Arditi, 2000) and could be used to provide a broad summary of how organic matter flows through a food web (subsection 6.3.13), including cryptic food webs in soil (Eggers and Jones, 2000; Scheu and Falca, 2000). $^{15}N/^{14}N$ ratios, for example, tend to increase with each successive link in a food chain (e.g. plant – herbivore – predator) because deamination and transamination enzymes preferentially remove light (^{14}N) amine groups and the $^{15}N/^{14}N$ ratio in organisms becomes progressively heavier than that of atmospheric nitrogen. Excreted nitrogen is, therefore, lighter than body tissue nitrogen. Interpreting the isotopic signature of a consumer in terms of its trophic position needs to be done in relation to an appropriate isotopic baseline. Post (2002) discusses methods for generating an isotopic baseline and evaluates assumptions underlying estimation of trophic position. In the northern temperate lake ecosystems studied by Post (2002), snails were a suitable baseline for the littoral food web and mussels for the pelagic food web. Lycosid spiders, *Pardosa lugubris*, were shown to have higher $^{15}N/^{14}N$ ratios than their prey, suggesting that stable isotope analysis may be appropriate for assessing the trophic position of the organisms tested (Oelbermann and Scheu, 2002). ^{15}N content in predators, however, is affected by food quality and starvation (Oelbermann and Scheu, 2002). Such effects need to be taken into account in determining trophic position. In addition, scavenger species often have ^{15}N levels within the range of values found for predators (Ponsard and Arditi, 2000), which needs to be acknowledged when interpreting results. Stable isotope ratios have been used to elucidate features of rocky littoral food webs that were not detected by analysis of gut contents (Pinnegar and Polunin, 2000). Using stable isotope analysis of carbon and nitrogen, McNabb *et al.* (2001) found that linyphiid and lycosid spiders in experimental cucurbit gardens treated with straw and manure mulches appeared to be preying largely on detritivores (such as Collembola). They caution, however, that ^{13}C values of Collembola were not significantly different from those for spotted cucumber beetle (*Diabrotica undecimpunctata howardi*), which was a major herbivore in the system under study. Thus, in some cases, stable isotope analysis may lack the precision to discriminate between consumption of detritivores and herbivores.

Unbiased sampling of predators that are to be assayed by post-mortem methods (e.g. gut dissection, immunoassays, electrophoresis, DNA techniques) should, ideally, employ at least one sampling method relying on predator activity (e.g. sticky trap, Malaise trap, pitfall), run for at least 24 hours (to accommodate the predator's diel activity cycle), and one that will include inactive predators (e.g. ground search). The latter is not, however, always practicable, for example with large carabids that are frequently found at densities well below one per square metre, in which case data should be interpreted with caution. Violent methods (e.g. sweeping, beating, suction sampling) are not advisable in circumstances where damaged target prey could contaminate the predators and produce false positives (Crook and Sunderland, 1984). Dry pitfall traps should be avoided if there is a significant degree of entry of the traps by the target prey (i.e. if the density of prey in the traps is greater than that found in the same area of open ground), because artificial confinement of prey with hungry predators may produce an erroneous assessment of prey consumption in the field. This can sometimes be overcome by using dry traps with a mesh insert, so that larger predators are separated physically from smaller prey. There is evidence that this works well for negatively-phototactic prey, such as earthworms, which remain in the bottom of the trap, but less well for positively-phototactic prey, such as aphids, which can climb up the sides of the trap, and thus come within reach of the predators once more (W.O.C. Symondson, D.M. Glen, M.L. Erickson, J.E. Liddell and C. Langdon, unpublished data). An additional problem can be predation by one predator upon another within the trap, where the predator consumed has previously fed on the target prey. Hengeveld (1980), for example, found the incidence of spider remains in the guts of carabid beetles to be much greater when the beetles were collected in live-trapping pitfalls than in funnel traps where the beetles were killed instantly in 4% formalin solution. The inclusion of stones, soil or mesh can help to minimise this problem in dry traps. If the predator of interest is, for example, a large carabid, then holes in the bottom of the trap can allow small predators, and other target or non-target prey, to escape. Wet traps can also lead to biased results. Water can attract prey, such as slugs (and possibly many insects during dry weather), to the local area of the trap leading to increased consumption by predators in that area (Symondson *et al*, 1996a). It has also been shown that when slugs drown in a trap, carabid beetles drowning in that same water may ingest proteins released by the dying or dead prey, again leading to false positives (W.O.C. Symondson, unpublished data). Trapping reagents that cause protein denaturation (e.g. formalin, alcohol) should be avoided. Regular emptying of traps, and transport of predators on ice, will reduce the rate of microbial degradation of the gut contents of dead predators (Sunderland, 1988), and slow down digestion in live predators caught in dry traps (Symondson *et al.*, 1996a).

With most of the post-mortem methods described below, it is advisable to carry out some initial laboratory studies with all predator species that will be collected from the field for testing. Predators that have just eaten a large meal of the target prey should be tested. This initial screening will identify any species that have an extreme degree of extra-oral digestion (Cohen, 1995), and which will therefore often be negative in the test, even though they may be important consumers of the target prey in the field. The rates at which prey remains decay within predators are discussed in section 6.3.9.

False negatives are a problem common to all the methods; that is, due to undersampling, some parasitoid or predator species comprising an insect's natural enemy complex may be overlooked, as may certain insect species comprising a natural enemy's prey/host range. More serious, however, is the problem of false positives with several of the methods discussed here (e.g. tests of prey and host acceptability, immunoassays, prey/host labelling, electrophoresis). Positives in post-mortem tests indicate consumption of the target prey, but not necessarily predation, because the label can enter the predator by other routes, such as scavenging and secondary predation (Sunderland, 1996), or by consumption of autotomised parts of the prey (e.g. the slug *Deroceras reticulatum* sheds the posterior part of its foot and escapes when attacked by a large generalist carabid beetle, which then eats the autotomised organ – Pakarinen, 1994), or by trophallaxis (Morris *et al.*, 2002). Recent investigation of an aphid-spider-carabid system suggested that, at least in this system, secondary predation will cause little error in predation estimates; detection of secondary predation using a sensitive anti-aphid monoclonal antibody (subsection 6.3.9) only occurred where carabids were killed immediately after consuming at least two spiders, which were, in turn, eaten immediately after consuming aphids (Harwood *et al.*, 2001b). Under certain circumstances there may be no requirement to separate predation from scavenging. In the study by Symondson *et al.* (2000), monoclonal antibodies were used to assess the importance of earthworms simply as an alternative source of non-pest food that could help maintain populations of the carabid *Pterostichus melanarius* when pest density was low. In this scenario, dead earthworms were as useful as live ones.

6.3.2 FIELD OBSERVATIONS ON FORAGING NATURAL ENEMIES

This method is the simplest and most unequivocal for gathering evidence on the trophic relationships within both predator-prey and parasitoid-host systems. It falls, generally, into two areas. First, **direct observation** with the naked eye and binoculars in the field (Holmes, 1984), and second, **remote observation** using video recording apparatus. However, there are a number of practical drawbacks to these techniques: they are often very labour-intensive, data often accumulate very slowly, and data may be of limited value if the trophic relationship itself is disturbed during observations. Furthermore, many natural enemies are nocturnal, live concealed in the soil, or are very small, creating difficulties for observation. Sometimes predator-prey interactions under a closed crop canopy within arable ecosystems are the focus of study, making observation without disturbance particularly difficult. Such drawbacks are especially important when studying minute, but often very fast moving, parasitoids. An added complication when attempting to record oviposition by endoparasitoids is that it is sometimes difficult to assess whether or not an ovipositor insertion has in fact resulted in an egg being laid (subsection 1.5.6).

Despite the drawbacks outlined above, direct field observation has been the technique used in a number of studies of predation and parasitism. Workers have either used prey that has deliberately been placed out in a habitat, or they have used naturally-occurring populations of prey. Greenstone (1999) tabulates studies in which direct observation has been used to determine the spectrum of spider species attacking pests.

Predation of the eggs of cotton bollworms, *Helicoverpa zea*, was investigated by attaching batches of eggs to the upper surface of cotton leaves, using fresh egg white and a small paint brush (Whitcomb and Bell, 1964). The eggs were observed directly for 12 hour periods during daylight and all acts of predation were recorded. Predators were identified whenever possible, and the type of feeding damage inflicted was noted. Ants tended to remove eggs completely, explaining the disappearance of eggs from cotton plants recorded in earlier experiments. Using a similar approach, Buschman *et al.* (1977) allowed adults of the velvetbean caterpillar, *Anticarsia gemmatalis*, to lay eggs on soybean plants in laboratory cages, and then placed the egg-laden plants out in soybean fields. The plants were continuously observed through the daylight hours by a team of observers working in 2 hour shifts.

Weseloh (1989) identified potential ant predators of gypsy moth larvae by placing moth larvae on the forest floor in an area where a particular ant species was seen to be foraging, and then noting the ants' behavioural reaction to the larvae. Morris *et al.* (1998), in their study of predation by ants (*Tapionoma nigerriumus*) on the olive moth, *Prays oleae*, recorded the number and proportion of ants carrying this prey, as they approached the nest along one of their many trails, during 10 min intervals throughout the day. Interestingly, far fewer ants were seen carrying *P. oleae* than might have been suggested by analyses using ELISA (Morris *et al.*, 1999b; Morris *et al.*, 2002). This was explained by the fact that ants participate in trophallaxis, the exchange of food between individuals, and illustrates very well the need to complement biochemical approaches with direct observation and/or behavioural studies. Riddick and Mills (1994) used binoculars, and a red flashlight, to observe carabid predation of tethered codling moth (*Cydia pomonella*) larvae in an apple orchard. Binoculars enabled the researchers to monitor the larvae from a distance of several metres and allowed identification of the carabids, but without disturbing their behaviour. Similarly, Villemant and Ramzi (1995) used binoculars to observe predation on egg masses of *Lymantria dispar* on cork oak. A large wolf spider, *Hogna rabida*, was the focus of observation by Schmitz and Suttle (2001), but this species was found to retreat into the litter if researchers approached within 1.5 metres. To circumvent this problem, individual spiders were dusted with fluorescent powder and observed from 2.5 metres using binoculars.

Nyffeler *et al.* (1992) observed diurnal predation of the cotton fleahopper, *Pseudatomoscelis seriatus* (Heteroptera, Miridae), by walking through cotton crops (subsection 6.2.6). During 1 hour observation periods the number of predators without prey, the number of predators with fleahopper prey, and the number of predators with alternative prey, were all counted. Predators carrying prey were collected, and both the predator and its prey were identified later in the laboratory. Nyffeler (1999) collated data from 31 similar published observational studies of spiders attacking pests in the field. Lavigne (1992) observed the predatory behaviour of Australian robber-flies (Diptera, Asilidae) in pastures, using two main approaches: continuous observation of single flies over extended periods, and transect walks to record as much behaviour as possible. Feeding flies were collected, identified, and finally released, once the prey had been removed for identification. Similarly, Halaj and Cady (2000) made 30-60 minute walks from 08:00 h to 11:00 h and 21:30 h to 00:30 h (using headlamps), and recorded feeding activity by harvestmen (Opiliones) in a soybean field and an adjacent hedgerow. When feeding was observed, the harvestman and its prey were preserved in alcohol for later identification in the laboratory. Van den Berg *et al.* (1997) made day-time visual observations of predator feeding rate on soybean aphids during ten-minute periods. Their data supported calculation of functional responses for a ladybird, *Harmonia arcuata*, and a rove beetle, *Paederus fuscipes*, but the foraging of spiders, ants and crickets was found to be too cryptic for application of this method.

Holmes (1984) examined individual colonies of the cereal aphid *Sitobion avenae* on selected ears of wheat, identified by a plastic label, and re-examined them at regular intervals, to observe the searching behaviour of staphylinid beetles. The data not only confirmed that aphids were consumed, but also provided quantitative information on the rate of prey consumption by predators.

Griffiths *et al.* (1985) studied nocturnal carabids in cereals, using arenas placed out in the field at night, and illuminated with red light, to which the carabids do not respond. Red light illumination is used to minimise disturbance to the predator and prey but allow observation. It is, however, important to note that not all arthropods are insensitive to red light (Sunderland, 1988). Harvestmen (Opiliones) and springtails (Collembola) are known exceptions. Many such studies do not yield quantitative data, since the prey has to be removed to be identified, and the predator's normal behaviour is disturbed (subsection 7.2.11 discusses quantitative aspects of this technique).

Diel activity patterns of a squash bug egg parasitoid, the scelionid *Gryon pennsylvanicum*, were recorded at 3 hour intervals in a squash field. A range of behaviours, which included "probing-ovipositing", was recorded by direct observation. A red flashlight was used at night (Vogt and Nechols, 1991). Landis and Van der Werf (1997) observed sugar beet plants during the daytime for 200 minutes per day, and took care not to touch the plants, or cast shadows on them. Aphids were observed being eaten on six occasions, including by cantharid and anthicid beetles.

Rosenheim *et al.* (1993), using field enclosures, showed that predation by heteropteran bugs on lacewing larvae (intraguild predation) can release cotton aphids from effective biological control. Because of the possibility that enclosures might distort the normal behaviour of the predators in this system, Rosenheim *et al.* (1999) later made 450 hours of direct visual observations in cotton fields, and demonstrated that lacewing larvae (*Chrysoperla carnea*) were, indeed, killed frequently by predatory Heteroptera (various species of *Orius, Geocoris, Nabis* and *Zelus*).

With sufficient ingenuity, even predation occurring below the soil surface can be monitored directly. Brust (1991) observed underground predation of larvae of southern corn rootworm (*Diabrotica undecimpunctata*) on maize roots. He positioned a plexiglass plate close to the roots, and had a bag of soil that could be removed to allow root observation at 2 hour intervals for 24 hours. This method showed that carabid larvae (in the genera *Harpalus* and *Pterostichus*) were significant predators of first to third instar rootworm larvae. A similar method was used successfully to investigate predation of eggs of southern masked chafer (*Cyclocephala lurida*) and Japanese beetle (*Popillia japonica*) by ants in turfgrass (Zenger and Gibb, 2001). Such approaches are excellent for observation of predatory behaviour and confirming trophic links but provide limited quantitative information.

Time-lapse video recording techniques, which reduce the arduous and time-consuming nature of direct observation, are now available (Wratten, 1994). These techniques are very useful for observing attacks on relatively sedentary prey, but are not appropriate, in most cases, for recording attacks on small active prey, or for monitoring the foraging behaviour of highly mobile natural enemies in complex three-dimensional habitats, such as plant canopies. Recording is possible at very low light intensities, or under infra-red light, thus reducing disturbance (Howling and Port, 1989, Howling, 1991; Schenk and Bacher, 2002). Amalin *et al.* (2001), using a video time-lapse cassette recorder and red light illumination (5.0 Lux) in the laboratory, discovered that nocturnal clubionid spiders are able to detect leafminers through the leaf epidermis and then either puncture the mine and eat the larva *in situ*, or cut a slit in the mine and remove the prey. Laboratory studies of the foraging behaviour of some slug species have been carried out by several workers (Bailey, 1989; Howling and Port, 1989; Howling, 1991) and the methods could be adapted for studying the foraging behaviour of predators. Halsall and Wratten (1988b) used time-lapse video equipment to compare the responses of polyphagous predators to plots of high and low aphid density in a winter wheat field. Cameras were focussed on a 9 cm x 11 cm patch of moistened silver sand within each plot, and night illumination was provided by 100 W IR lights, powered from a portable generator. The nocturnal carabid, *Anchomenus dorsalis* (= *Agonum dorsale*), was found to make a greater number of entries into the high than the low aphid density plot. Similar methods were used in a field of winter wheat by Lys (1995), who focussed the camera on a 10 cm x 10 cm area of ground where *Drosophila* puparia stuck to cardboard had been placed. In addition to securing data on relative predation rates by diurnal and nocturnal predators, this study provided fascinating insights into the agonistic behaviour of carabids under field conditions (e.g. *Poecilus* (= *Pterostichus*) *cupreus*) attacked and repelled conspecifics as well as smaller predators). The simultaneous use of multiple video cameras in the field can enable a more efficient collection of quantitative data on predation and parasitism in some systems. Meyhöfer (2001) used a system of 16 cameras equipped with infrared diodes for night vision, a video multiplexer and a time-lapse video recorder to study predation of aphid mummies in a sugar beet field. Camera input routed via the multiplexer was stored on a single videotape (720 hours of observation on a 3 hour tape). Schenk and Bacher (2002) employed a similar range of equipment to study predation on sedentary late instar larvae of the shield beetle, *Cassida rubiginosa*, on creeping thistle. Equipment was checked several times daily, and, when necessary, cameras were re-focussed or missing larvae replaced. Examination of the video tapes showed that 99% of predation events were due to the paper wasp, *Polistes dominulus*. Technological advances now allow the input of up to 72 cameras to be stored directly to computer hard disc (Meyhöfer, 2001). Security and protection of video equipment is an important consideration in field studies. Practical advice concerning equipment, arenas and automatic analysis can be found in Varley *et al.*, (1993).

Because an individual act of predation is often completed in a relatively short period of time, and predators spend only part of their time foraging for food, the number of records of predatory acts occurring during an observation period can be very low (Sterling, 1989). Consequently, the likelihood of observing any acts of predation is often low for many predator species. It is sometimes possible to increase the rate of field data accumulation by artificially increasing prey densities. In an investigation of spider predation on the rice green leafhopper, *Nephotettix cincticeps*, Kiritani *et al.* (1972) placed, in paddies, rice plants that had been artificially infested with unnaturally high numbers of leafhopper eggs. Observers patrolled the rice plots after the eggs had begun to hatch, and recorded occurrences of spiders feeding on leafhoppers, noting the species of both the predator and the prey involved. Similarly, Way *et al.* (2002) planted potted rice plants infested with the brown planthopper, *Nilaparvata lugens*, into tropical upland rice fields and observed predation by ants both on the plants and on the ground below. Direct observation can be used to estimate predation rates (Greenstone, 1999), if data are collected, not only on the percentage of predators seen with prey, but also on the handling time (the period from attack to cessation of feeding), and on the time available for prey capture in the field (Nyffeler and Benz, 1988b).

6.3.3 COLLECTION OR EXAMINATION OF PREY AND PREY REMAINS

If the predator does not consume all of its prey, the remains of the cadaver may display distinctive marks that enable the perpetrator to be identified. For example, Neuroptera leave a pair of small holes, Heteroptera a single hole, and ants a peppering of small holes (Mills, 1997). Andow (1990) studied, in the laboratory, the nature of the damage caused to egg masses of *Ostrinia nubilalis* (Lepidoptera, Pyralidae) by various predators (Figure 6.23). Adults of the coccinellid *Coleomegilla maculata* consumed the entire egg mass, leaving only the basal and lateral parts of the egg chorion. The staphylinid beetle *Stenus flavicornis* chewed around the edges of an egg mass and killed many more eggs than were actually consumed. Larvae of the lacewing *Chrysopa* sp. left distinctive feeding holes in the egg chorion. The heteropteran bug *Orius insidiosus* did not completely consume the egg, and extensive melanisation occurred near the site of attack. The phytoseiid mite *Amblyseius* sp. also caused melanisation, but of more restricted extent. Andow (1992) used this information to interpret damage to sentinel egg masses in maize crops, and showed that *C. maculata* predation was greatest in plots established after ploughing, compared with *Chrysopa* predation, which was greatest in no-till plots. Hossain *et al.* (2002) used sentinel moth eggs of *Helicoverpa* spp. (live eggs attached to sections of stiff paper which were stapled to lucerne plants) to provide an index of egg predation intensity in plots of harvested and unharvested lucerne. To avoid egg hatch in the field, eggs less than 24 hours old were used, and were transported to the field at 9°C in a portable refigerator. Egg cards were then retrieved from the field after 24 hours exposure to predation (Hossain *et al.*, 2002). When prey are extremely small, scanning electron microscopy (SEM) can be used successfully to categorise characteristic predator-specific feeding traces in the laboratory, and then screen prey taken from the field to determine the incidence of predation by different species of predator. Using the SEM in this way, Kishimoto and Takagi (2001) showed that staphylinid beetles of the genus *Oligota* were the main predators of spider mite eggs on pear trees in Japan. Feeding traces have also been used to identify the carabid, staphylinid and mammalian predators of winter moth (*Operophtera brumata*) pupae in oak woodland (East, 1974). Caterpillars of fruittree leaf-roller, *Archips argyrospila*, have a characteristic blackened and shrunken appearance after being fed on by carabid beetles, and this was used to estimate predation rates after examination of infested foliage (Braun *et al.*, 1990). Similarly, gypsy moth (*Lymantria dispar*) pupae with characteristic large, jagged wounds, were known to have been attacked by the carabid beetle *Calosoma sycophanta* (Fuester and Taylor, 1996). Examination of damaged egg masses of *Lymantria dispar* on cork oak showed cork dust and faecal pellets characteristic of oöphagous beetle larvae, *Tenebroides maroccanus* (Trogossitidae) (Villemant and Ramzi, 1995). Mummies of *Aphis gossypii* containing large ragged-edged holes were considered to have been attacked by coccinellid beetles (Colfer and Rosenheim, 2001). Predation marks on collected mines were used to identify which predators had attacked citrus leaf-miner, *Phyllocnistis citrella* (Amalin *et al.*, 2002). Ants, for example, slit open the mine and pull out the prey but lacewing larvae pierce the mine and suck out the fluid contents of the prey. For a few species, visual inspection reveals a clear difference between parasitised and unparasitised hosts that

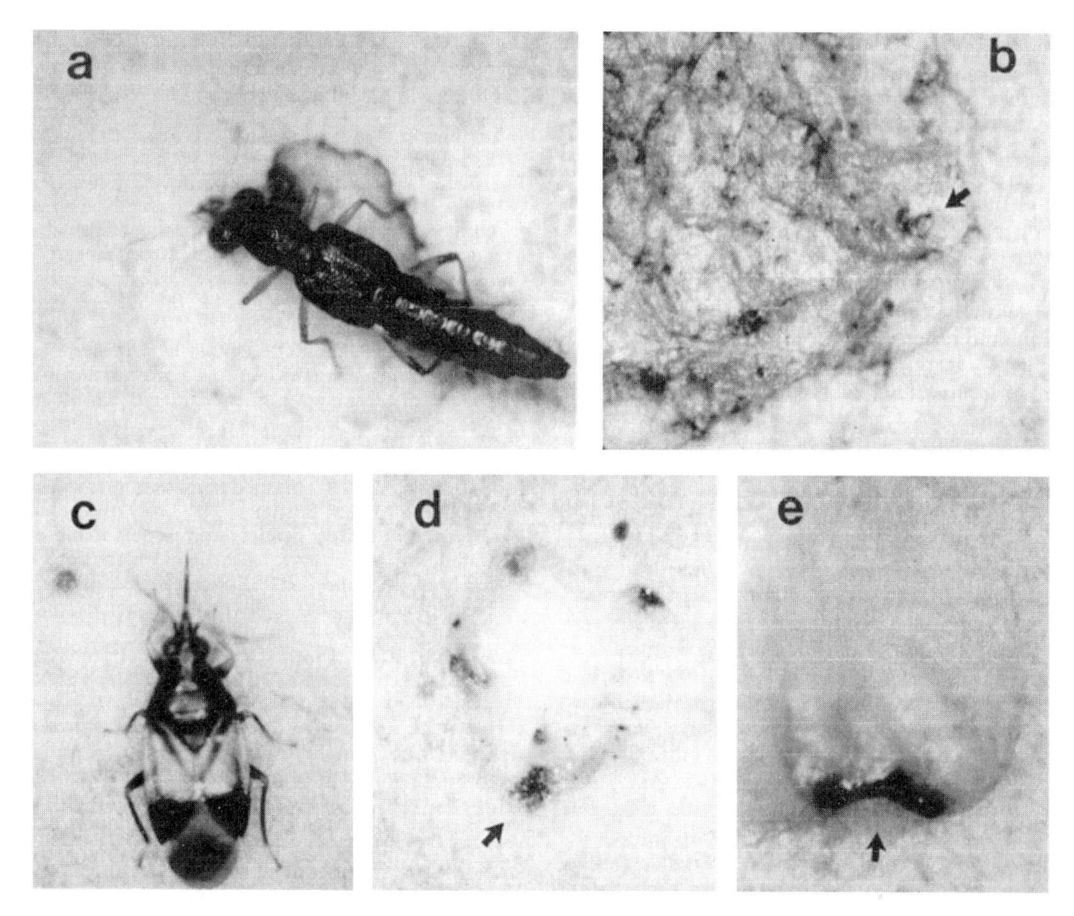

Figure 6.23 Damage caused to egg masses of *Ostrinia nubilalis* (Lepidoptera: Pyralidae) by various predators; (a) predation by the staphylinid beetle *Stenus flavicornis*, (b) predation by a lacewing larva (*Chrysopa* sp.), arrow marks a feeding hole, (c) predation by the heteropteran bug *Orius insidiosus*, (d) damage from predation by *O. insidiosus*, arrow marks fungal growth at feeding hole, (e) damage from mite predation, arrow marks damage Source: Andow (1990). Reproduced by permission of The Entomological Society of America.

can be assigned to a particular species of parasitoid. For example, healthy pupae of the small white buttefly, *Pieris rapae*, are green, but those attacked by *Pteromalus puparum* are brown (Ashby and Pottinger, 1974). Pupae of the glasshouse whitefly, *Trialeurodes vaporariorum*, when attacked by the parasitoid *Encarsia formosa*, turn from white to black after several weeks when the parasitoid pupates (Sunderland *et al.*, 1992).

With the workers of some predatory ant species it is possible to gently remove food items with forceps (Ibarra-Núñez *et al.*, 2001) from the jaws of the ants as they return to the nest (Stradling, 1987). Rosengren *et al.* (1979), in a detailed study of wood ants (*Formica rufa* group) in Finland, used this method to investigate the diet of *Formica polyctena* (Table 6.4). Vogt *et al.* (2001) exposed the underground foraging trails of fire ants (*Solenopsis invicta*) to collect the ants together with the material they were carrying. Unfortunately, the manual method of food item collection disturbs the ants, and some prey items may be missed. Skinner (1980) overcame these problems when studying the feeding habits of the wood-ant *Formica rufa* (Formicidae), by using a semi-automatic sampling device which collected the ant 'booty' as it was carried back

Table 6.4 Pooled data from four nests showing number of prey items (n) expressed as a percentage of total items (%) carried to the nest of *Formica polyctena* during two periods in 1978 (Source: Rosengren *et al.*, 1979.) Reproduced by permission of the International Organisation for Biological and Integrated Control.

	30th June – 3rd July		27th July – 31st July	
	n	%	n	%
Homoptera (mostly aphids)	135	20.4	15	2.4
Adult Diptera (mostly Nematocera)	122	18.4	147	23.4
Sawfly larvae (Tenthredinidae)	73	11.0	50	7.9
Lepidoptera larvae	74	11.2	43	6.8
Unidentified carrion	53	8.0	78	12.4
Adult Coleoptera (mostly Cantharidae)	51	7.7	34	5.4
Ant workers	33	5.0	52	8.3
Ant reproductives	2	0.3	38	6.0
Lumbricidae (number of pieces)	23	3.5	21	3.3
Homoptera Auchenorrhyncha	19	2.9	22	3.5
All other groups	76	11.5	129	20.5
Total number of items	661	99.9	629	99.9

to the nest. With this method, an ant nest is surrounded by a barrier which forces the ants to use the exit and entry ramps provided. The incoming ants are then treated in either of two ways:

1. They are periodically directed through a wooden box with an exit-hole sufficiently large to allow the ant, but not its booty, to pass through. Prey left in the box may then be retrieved and identified;
2. They are allowed to fall off the end of the entry ramp into a solution of 70% alcohol.

Naturally, the latter treatment, because it kills the ants, can be used for short periods only.

The remains of the prey of larval ant-lions (Neuroptera: Myrmeleontidae) can be located, either buried in the predator's pit, or on the sand surface close to the pit, and can be identified, as shown by Matsura (1986). The diet of first instar larvae of *Myrmeleon bore* was far less varied than that of second and third instar larvae.

Food items may be either observed *in situ* on, or manually collected from, the webs of spiders. This method was used by Middleton (1984) in a study of the feeding ecology of web-spinning spiders in the canopy of Scots Pine (*Pinus sylvestris*), and by Sunderland *et al.* (1986b) when measuring predation rates of aphids in cereals by money spiders (Linyphiidae). Alderweireldt (1994) collected linyphiid spiders that were carrying prey in their chelicerae, and prey remains were collected from their webs, in Belgian maize and ryegrass fields. Jmhasly and Nentwig (1995) made regular daytime surveys of spider webs in a field of winter wheat. Each web was visited at 45 minute intervals in an attempt to record prey cadavers in the webs before spiders had time to discard the remains. The timing of web observations can be optimised in relation to environmental conditions and spider biology. Miyashita (1999), for example, visited webs of *Cyclosa* spp. in hedgerows between 11:00 h and 13:00 h, because about half of the webs were destroyed by 16:00 h (due to the action of wind and damage by prey) and webs were rebuilt only at night. Pekár (2000) collected entire webs of the theridiid spider *Theridion impressum*, and captured prey were picked out and identified later in the laboratory. In all these studies, examination of webs was confined to the daylight hours, because small prey items in delicate webs on or near the ground cannot be recorded reliably at night. The prey spectrum at night could be different from that which has been recorded during the daytime, and so techniques need to be developed to investigate nocturnal predation.

Some ant species make refuse piles which could be examined to determine their diet. Caution is needed here, however, since Vogt *et al.* (2001) found that the refuse piles of *Solenopsis invicta*, which were dominated by remains of Coleoptera, were entirely different from material seen to be carried into the nest, which was dominated by larvae of Lepidoptera. Prey and hosts of fossorial wasps, such as Sphecidae and Pompilidae, may be analysed by examining the prey remains within the nest. Larval provisioning by the wasp *Ectemnius cavifrons*, a predator of hover-flies, was investigated by Pickard (1975), who removed prey remains from the terminal cell of a number of burrows in a nest. Other fossorial wasps may be investigated using artificial nests: i.e. already-hollow plant stems, or pieces of dowelling or bamboo that have been hollowed out by drilling (Cooper, 1945; Danks, 1971; Fye, 1965; Parker and Bohart, 1966; Krombein, 1967; Southwood and Henderson, 2000). The stem or dowelling is split, and then bound together again, with, for example, string or elastic bands. The binding may subsequently be removed to allow examination of the nest contents. Details of one of the aforementioned methods, together with photographs of artificial nests and individual cells, are contained in Krombein (1967). Thiede (1981) describes a trap-nesting method involving the use of transparent acrylic tubes. An entrance trap that separates yellowjacket wasps (*Vespula* spp.) entering and leaving the nest has been devised. Those entering can be diverted into a gassing chamber and anaesthetised with carbon dioxide to enable their prey load to be removed and identified. The wasps treated in this way eventually recover, and 30–60 wasps can be sampled per hour (Harris, 1989). The large tube of an entrance trap can be fitted to the nest entrance at night, using black polythene and soil. Then, an inner collecting container (with integral funnel) can be inserted into the large tube during daytime to trap returning foragers (Harris and Oliver, 1993).

6.3.4 TESTS OF PREY AND HOST ACCEPTABILITY

Information on host or prey range in a natural enemy species can be obtained by presenting the predators or parasitoids, in the laboratory, with potential prey or hosts, and observing whether the latter are accepted (e.g. Goldson *et al.*, 1992; Schaupp and Kulman, 1992, for studies on parasitoids). For visually-hunting predators, it may be possible to use video-imagery techniques that enable the prey stimulus to be controlled with precision, and standardised. Clark and Uetz (1990) showed that the jumping spider *Maevia inclemens* did not discriminate between live prey and a simultaneous video image of prey displayed on Sony Watchman micro-television units. These spiders stalked and attacked the televised prey. Arachnids have flicker fusion frequency values in a similar range to that of humans, and so images are percieved as moving objects, rather than as a series of static frames. Some insects, however, have critical flicker fusion frequency values greatly in excess of 60 Hz (the human range is 16-55 Hz and arachnids span the range 10-37 Hz) and would not be good subjects for this methodology.

Simple laboratory assessments of prey and host acceptability should always be treated with caution, because of the artificial conditions under which they are conducted (Greenstone, 1999). They can only indicate *potential* trophic relationships, and other factors need to be taken into consideration when extrapolating to the field situation. For example, it is necessary to establish that the natural enemy being investigated actually comes into contact with the potential prey or host species under natural conditions. The predator or parasitoid may only be active within the habitat for a limited period during the year, and this may not coincide with the presence of vulnerable stages of the prey, or host, in that habitat. In order to avoid testing inappropriate predator/prey or parasitoid/ host combinations it is thus advisable to establish whether there is both spatial and temporal synchrony before carrying out laboratory acceptability tests (Arnoldi *et al.*, 1991). The placing of insects in arenas, such as Petri dishes and small cages, also raises serious questions regarding the value of the method. Confining potential prey or hosts in such arenas alters prey and host dispersion patterns, and, in particular, it is likely to reduce the opportunities for prey to escape from the natural enemy, so increasing the risk of false positives being recorded. Some insect species may rarely, if ever, be attacked in the field by a particular natural enemy species, because they are too active to allow capture by the latter. For some types of predator, such as web-building spiders, prey acceptability tests can be carried out in the field. The same strictures about testing appropriate prey types still apply to field testing, but at least the behaviour of the predator is natural and unconstrained. Henaut *et al.* (2001), for example, used an aspirator to gently blow live potential prey organisms into the orb webs of spiders in a coffee plantation. They used prey types that were abundant in the plantation and recorded the extent to which the web retained the prey, as well as capture and consumption rates by the spiders.

Laboratory studies are particularly useful in determining prey size choice, helping to determine an upper limit to the size of prey that the predator is capable of tackling (Greene, 1975; Loreau, 1984b; Digweed, 1993; McKemey *et al.*, 2001), and, sometimes, a lower limit below which single, unaggregated prey are ignored (Greene, 1975). Finch (1996) showed that carabids (Figure 6.24) with mandibles above a certain size were incapable of manipulating, and therefore of consuming, the eggs of the cabbage root-fly, *Delia radicum*. These limits on prey size can be particularly relevant in

Figure 6.24 Size range of carabid beetles in relation to predation of the eggs of cabbage root fly, *Delia radicum*
Source: Finch (1996). Reproduced by permission of The Royal Entomological Society.

cases where size determines the species range or life-stage that will be attacked by a polyphagous predator. McKemey *et al.* (2001) found that the carabid *Pterostichus melanarius* had a preference for smaller slugs within laboratory arenas, but this result was not supported by choice experiments within a crop. Within the structurally complex habitat of a crop of wheat no slug size preferences were found (McKemey *et al.*, 2003), apparently because the smallest slugs had greater access to refugia from the beetles within cut wheat stems and clods of soil. Again, therefore, laboratory studies can provide misleading data; they can show which sizes of prey the predators are capable of killing, but not necessarily the preferences the predators have in the field.

Predators are often also scavengers. A predator that is known to consume a particular prey type may be (with respect to that prey type) a scavenger, rather than a true predator. This is an important distinction if the impact of predation on the prey species is being inferred from the post-mortem examination of field-collected predators. Laboratory studies are appropriate for determining whether a predator has the physical and behavioural capacity to kill a particular potential prey species (Greenstone, 1999; Sunderland, 2002). Sunderland and Sutton (1980) showed that eight species of predator that regularly consumed woodlice (Isopoda) in the field (as determined from serological investigations), were, in fact, unable to kill even tiny woodlice in laboratory tests, and were therefore unlikely to have been predators of woodlice in the field. Halaj and Cady (2000) frequently observed harvestmen (Opiliones) eating earthworms (Lumbricidae) in the field, but laboratory tests showed that even the largest species of harvestman was incapable of killing earthworms of a size that were regularly consumed in the field. Therefore, harvestmen in the field were probably scavenging. Some species of ant that were shown (by ELISA) to contain remains of the olive moth, *Prays oleae*, were thought to have obtained these remains primarily by scavenging (Morris *et al.*, 2002).

Even when a parasitoid stabs the host with its ovipositor, the host may still be rejected for oviposition (McNeill *et al.*, 2000) (subsection 1.5.7). It is therefore important that egg release is confirmed during host acceptability tests involving parasitoids. In a few cases this can be done by observing the behaviour of the parasitoid. Otherwise, it can be done either by dissecting hosts shortly after exposure to parasitoids, to locate eggs within the host's body (see section 2.6), or by rearing hosts until parasitism becomes detectable. Furthermore, just because an insect species is accepted for oviposition, does not mean that the species is suitable for successful parasitoid development. Under laboratory conditions many parasitoids will, if given no alternative, oviposit in unsuitable "hosts", but subsequent monitoring of the progeny of the parasitoids will show that they have been killed by the host's physiological defences (subsection 2.10.2).

When the introduction of an exotic natural enemy species into a new geographical area is being contemplated (in a classical biological control programme), it is important to determine whether the natural enemy is potentially capable of parasitising or preying on members of the indigenous insect fauna (section 7.14). This can only be done safely through laboratory acceptability tests. The braconid parasitoid *Microctonus hyperodae* was collected from South America and screened for introduction into New Zealand as a biological control agent of the weevil *Listronotus bonariensis*, a pest of pasture (Goldson *et al.*, 1992). Whilst in quarantine, the parasitoid was exposed to as many indigenous weevil species as possible, giving priority to those of a similar size to the target host. In the tests, 25 to 30 weevils of each species were exposed to a single parasitoid for 48 h in small cages, after which the weevils were held in a larger cage to await the emergence of parasitoids. In a second series of trials, the parasitoids were given a choice of target hosts (*L. bonariensis*) and test weevils in the same cages. It is important to carry out such a choice test, since in non-choice laboratory trials natural enemies may attack insect species which they would normally ignore in field situations, thereby giving a false impression of the natural enemy's potential behaviour following field release. See also Chapter 7, subsection 7.4.4.

6.3.5 EXAMINATION OF HOSTS FOR PARASITOID IMMATURES

The principle behind this technique is that by examining (or dissecting, in the case of endoparasitoids) field-collected hosts, parasitoid immatures, or their remains, located upon or within hosts, they can be identified to species, thus providing information on the host range of parasitoids, and the species composition of parasitoid complexes. Unfortunately, the requirement that the parasitoid taxa involved be identified can rarely be satisfied. Few genera, and even fewer higher taxa, have a published taxonomy (i.e. keys and descriptions to the species developed for their immature stages, subsection 6.2.9), and, where such taxonomies are available, they are rarely complete. Nevertheless, informal taxonomies can be developed in conjunction with the rearing method described in the next subsection (6.3.6), by associating immature stages, or their remains, with reared adult parasitoids.

6.3.6 REARING PARASITOIDS FROM HOSTS

One of the most obvious ways of establishing host-parasitoid trophic associations is to rear the parasitoids from field-collected hosts. Smith (1974), Gauld and Bolton (1988), and Shaw (1997) give general advice, and Jervis (1978) and Starý (1970) describe methods of rearing parasitoids of leafhoppers and aphids, respectively. The ease with which this can be done varies with parasitoid life-history strategy. It is usually far easier to rear idiobionts (subsection 1.5.7) from eggs, pupae or paralysed larvae, than it is to rear koinobionts (subsection 1.5.7), since the hosts of koinobionts usually require feeding. When supplying plant material to phytophagous hosts of koinobionts during rearing, it is very important to ensure that the material does not contain individuals of other insect species from which parasitoids could also emerge, as this can lead to erroneous host-parasitoid records. The risk of associating parasitoids with the wrong hosts is particularly acute when parasitoids are reared from hosts in fruits, seed heads and galls, as these may contain more than one herbivore, inquiline or the host species. Such material should be dissected after the emergence of any parasitoids, so that the host remains can be located and identified, and the parasitoid-host association determined correctly.

Whenever possible, hosts should be reared individually in containers, so that emerging adult parasitoids can be associated with particular host individuals. This allows the recording of accurate data on any parasitoid preferences for particular host developmental stages or sexes, and it also avoids potential problems arising from the failure of the entomologist to distinguish between closely-related host species. The choice of suitable containers will depend upon the host involved, but very often simple boxes, tubes or plastic bags will suffice (Smith, 1974; Jervis, 1978; Starý, 1970; Grissell and Schauff, 1990; Shaw, 1997; Godfray *et al.*, 1995). The addition of materials such as vermiculite, Plaster of Paris, or wads of absorbent paper, to rearing containers is recommended to avoid problems such as growth of moulds associated with the accumulation of excessive moisture.

Some parasitoids, including several Braconidae (Shaw and Huddleston, 1991), cause changes in host behaviour, that often result in the movement of parasitised individuals from their normal feeding sites prior to death, as was recorded, for example, for the aphid parasitoid, *Toxares deltiger* (Powell, 1980). A significant proportion of the mummies formed by several aphid parasitoid species occur some distance away from the aphid colony, and even away from the food plant. They tend to be missed during collection, leading to inaccuracies in measurements of parasitism, and underestimation of the size of the parasitoid complex (Powell, 1980). As an insurance against the possible complicating effects of parasitoid-induced changes in host behaviour, it is advisable to collect and rear a random sample of apparently-healthy hosts, in addition to obviously-parasitised individuals. With aphids, it is also advisable to rear parasitoids from both live aphids and mummies, because some hyperparasitoids oviposit into the mummy stage, whereas others oviposit into the larval parasitoid prior to host mummification (Dean *et al.*, 1981).

Parasitoids may also be collected from the field after they have killed, and emerged from, the host. The larvae of some species leave the body of their host and pupate, either upon the host remains, or on surrounding vegetation. The host remains (for subsequent identification) and the parasitoid pupae (which may be in cocoons or the exuvium of the last larval instar) may be collected, and parasitoids reared from the latter in individual containers. Where gregarious species are concerned, the pupae from a particular host individual should be kept together.

The importance of keeping detailed records of all parasitoid rearings cannot be stressed enough. At the time of collection from the field, the identity and developmental stage of the host should be recorded, along with habitat and food plant data, location and date (Gauld and Bolton, 1988). Where possible, the date of host death, together with that of parasitoid emergence, should be recorded. When adult parasitoids are retained as mounted specimens, the remains of the host and parasitoid pupal cases and cocoons should also be retained, to aid identification. Such remains are best kept in a gelatin capsule, which can be impaled on the pin of the mounted parasitoid. If a previously-unrecorded host-parasitoid association is recorded during rearing, it is very important to provide 'voucher' specimens for deposition in a museum collection.

In order to ensure, as far as possible, that all the species in the parasitoid complex of a host are reared, samples of parasitised hosts need to be taken from a wide area, from both high- and low-density host populations, from a variety of host habitats, and over the full time-span of the host life-cycle, or at least the time-span of the host stage that one is interested in (for example, the pupal stage, if only pupal parasitoids are the subject of interest). A parasitoid species may be present in some local host populations, but not in others. Its absence may be due to climatic unsuitability, or because of an inability of the parasitoid to locate some populations. Ecologists refer to the **constancy** of a parasitoid species (Zwölfer, 1971): the probability with which a species may be expected to occur in a host sample. Constancy is expressed as the percentage, or the proportion, of samples taken that include the species, e.g. see Völkl and Starý (1988). Host species with a wide geographical distribution require more extensive sampling to reveal the total diversity of their parasitoid complex than do host species with a restricted distribution (Hawkins, 1994). We can reasonably assume that the relationship between the extent of the host's distribution and parasitoid species richness is linear (B.A. Hawkins, personal communication).

Major temporal changes may occur in the species composition of a parasitoid complex. Askew and Shaw (1986) refer to the example of the lepidopteran *Xestia xanthographa*. Samples of larvae of this host taken 10 weeks apart yielded very different species of parasitoid (Table 6.5). Another factor to consider is sample size. The probability of a parasitoid species being reared from a host will depend on the percentage parasitism inflicted, and also on the size of the host sample taken. The larger the sample of hosts from a locality, the greater the probability that all the parasitoid species in the local complex will be reared. This relationship is probably asymptotic, and sample sizes of 1000 hosts, or larger, provide good estimates of parasitoid species richness that are not so strongly dependent on sample size (B.A. Hawkins, personal communication). Martinez *et al.* (1999) discuss the amount of sampling effort required to reveal the structural properties of parasitoid food webs.

As well as providing information on host range, parasitoid complex, and community structure, rearing of parasitoids can also yield valuable data on parasitoid phenologies. For

Table 6.5 Parasitism of *Xestia xanthographa* collected near Reading, UK, on two dates during spring 1979. The figures take no account of larval-pupal parasitoids (Source: Askew and Shaw, 1986.) Reproduced by permission of Academic Press Ltd.

		Sampling Dates	
		2nd March	12th May
Numbers of unparasitised larvae		188	50
Numbers of larvae parasitised by			
Tachinidae	*Periscepsia spathulata* (Fallén)	24	0
	Pales pavida (Meigen)	1	0
Braconidae	*Glyptapanteles fulvipes* (Haliday)	8	0
	Cotesia hyphantriae Riley	1	0
	Meteorus gyrator (Thunberg)	1	0
	Aleiodes sp. A	46	0
	Aleiodes sp. B	10	0
Ichneumonidae	*Hyposoter* sp.	1	0
	Ophion scutellaris Thomson	0	6
Total larvae sampled		280	56
Percentage parasitism		33	11

example, by noting the times of emergence of larval parasitoids from the hosts, and the time of adult emergence, Waloff (1975) and Jervis (1980b) obtained valuable information on the timing and duration of adult flight period, and on the incidence of diapause, in several species of Dryinidae and Pipunculidae.

Rearing can also be a valuable source of material for taxonomic study. If a larval taxonomy of the parasitoids is to be developed (subsection 6.2.9), association of parasitoid adults with larvae can usually only be achieved reliably through rearing. By associating reared adults with their puparial remains, Benton (1975), and Jervis (1992), developed a larval taxonomy for a number of Pipunculidae. Some parasitoids are more easily identified through examination of their larval/puparial remains than they are through examination of the adult insects (Jervis, 1992).

6.3.7 FAECAL ANALYSIS

For predators that have faeces which contain identifiable prey remains, faecal analysis can be used to estimate dietary range. This method has been applied to aquatic insects (Lawton, 1970; Thompson, 1978; Cloarec, 1977; Folsom and Collins, 1984). Folsom and Collins (1984) collected larvae of the dragonfly *Anax junius* and kept them until their faecal pellets were egested. The pellets were subsequently placed in a drop of glycerin-water solution on a microscope slide and teased apart. Prey remains were identified by reference to faecal pellets obtained through feeding *Anax* larvae on single species of prey, by examination of whole prey items and by examination of previously published illustrations of prey parts. The data obtained were used to compare the proportion of certain prey types in the diet, with the proportions found in the aquatic environment. Carabid beetles often consume solid fragments of their prey (subsection 6.3.8), and may retain these in the gut for days or weeks before defaecation of the meal is completed (Young and Hamm, 1986; Sunderland *et al.*, 1987a; Pollet and Desender, 1990). This makes field-collected carabids suitable candidates for faecal analysis in the laboratory. However, unless there is some reason for the beetles to be retained alive (e.g. they may be of an endangered species), gut dissection (subsection 6.3.8) will be more convenient than faecal analysis as a means of assessing carabid diet. Other types of predator, however, may be very difficult to dissect, in which case faecal analysis can be very useful. Phillipson (1960) examined 1367 faecal pellets from the harvestman *Mitopus morio* (Opiliones), and was able to determine the relative contribution of plant and animal material to its diet. Burgess and Hinks (1987) checked the faeces of several hundred field-collected crickets (*Gryllus pennsylvanicus*) and found that 0–28% (depending on collection site and season) had been feeding on adult flea beetles (*Phyllotreta cruciferae*), judging by characteristic remains of elytra, antennae and metathoracic legs. These authors had earlier determined, from laboratory experiments (subsection 6.3.4), that *G. pennsylvanicus* is a voracious predator of *P. cruciferae*. Rate of production of faeces has also been used in an attempt to estimate predation rates by ladybirds (Honěk, 1986), and to estimate the time of the most recent meal in the field by carabids (Young and Hamm, 1986).

6.3.8 GUT DISSECTION

If a predator is of the type that ingests the hard, indigestible, parts of its prey, simple dissection of the gut might easily disclose the prey's identity, as is indeed the case for a number of predators. The main advantages of this method for investigating trophic relationships are the simplicity of equipment required and the immediacy of results.

Dissection is usually performed using entomological micropins or fine watchmaker's forceps, which are used to remove and transfer the digestive tract (the crop, proventriculus, mid gut and hind gut, Chapter 2, subsection 2.2.2) to a microscope slide where it can be teased apart and the prey fragments identified. As with faecal analysis (subsection 6.3.7), the prey remains found in field-collected predators can be compared with those on reference slides prepared by feeding a predator with a single, known type of prey. Predators need to be killed as soon as possible after collection, otherwise the gut contents may be lost due to the predator defecating or regurgitating. *Sialis fuliginosa* loses a proportion of its gut contents by regurgitation when placed in preservative (Hildrew and Townsend, 1982) and many carabids will regurgitate if handled roughly. Not all invertebrates lose gut contents in this manner, but this possibility must be borne in mind when carrying out gut dissection. Some aquatic predators collected in nets continue feeding in the net, so need to be narcotised immediately upon capture, e.g. *Chaoborus* larvae (Pastorok, 1981). Similar precautions are needed for samples taken from terrestrial pitfall traps.

Gut dissection has been used to study the diet of a range of predators: (a) *aquatic insects* (Fedorenko, 1975; Bay, 1974; Hildrew and Townsend, 1976, 1982; Pastorok, 1981; Woodward and Hildrew, 2001, 2002), (b) *carabid beetles* (Sunderland and Vickerman, 1980; Hengeveld, 1980; Dennison and Hodkinson, 1983; Chiverton, 1984; Pollet and Desender, 1986; Luff, 1987; Sunderland *et al.*, 1987a; Dixon and McKinlay, 1992; Holopäinen and Helenius, 1992; Holland *et al.*, 1996), (c) *staphylinid beetles* (Kennedy *et al.*, 1986; Chiverton, 1987), (d) *coccinellid beetles* (Eastop and Pope, 1969; Ricci, 1986; Triltsch, 1997), (e) *earwigs* (Dermaptera) (Crumb *et al.*, 1941; Skuhravý, 1960), (f) *harvestmen* (Opiliones) (Dixon and McKinlay, 1989), and (g) *centipedes* (Chilopoda) (Poser, 1988). When dissecting carabid beetles, it can be worthwhile to examine not only the crop, but also the proventriculus and rectum, because characteristic prey fragments (such as the S-shaped chaetae of lumbricid earthworms) can lodge here for longer periods than in the crop (Pollet and Desender, 1990). Sunderland (1975) examined the crop, proventriculus and rectum of a variety of predatory beetles (Carabidae, Staphylinidae). In a similar study, the crop and the hindgut of two New Zealand carabid species were placed separately on microscope slides and examined in 50% glycerol at up to 400 × magnification. Food remains were identified, as far as possible, using microscopical preparations from reference collections (grass and weed seeds, other plant material, invertebrates from a range of sampling methods) for the areas where the carabids were caught (Sunderland *et al.*, 1995b). These methods were also used to investigate diet of the carabid *Pterostichus versicolor* from heathland. Beetles were caught in pitfall traps and dissected. Slides were prepared of the fore- and the hindgut. Reference slides were made by feeding beetles with known prey, and also by macerating known arthropods (Bruinink, 1990). Gut contents of freshwater aquatic predatory invertebrates can also be identified from reference slides (Woodward and Hildrew, 2002).

Although, in studies such as the above, many fragments found in the gut could be placed in general prey categories, identification of prey remains to species has often only been possible if distinctive pieces of prey cuticle (e.g. aphid siphunculi) remained intact. Recognisable fragments generally found in the guts of carabid and staphylinid beetles include: chaetae and skin of earthworms; cephalothoraces of spiders; claws, heads and/or antennae from Collembola; aphid siphunculi and claws; sclerotised cuticle, mandibles and legs from beetles and the head and tarsal claws of some Diptera (Figure 6.25). Hengeveld (1980) and Chiverton (1984), in their studies of Carabidae, also had to group most dietary components, assigning only a few prey

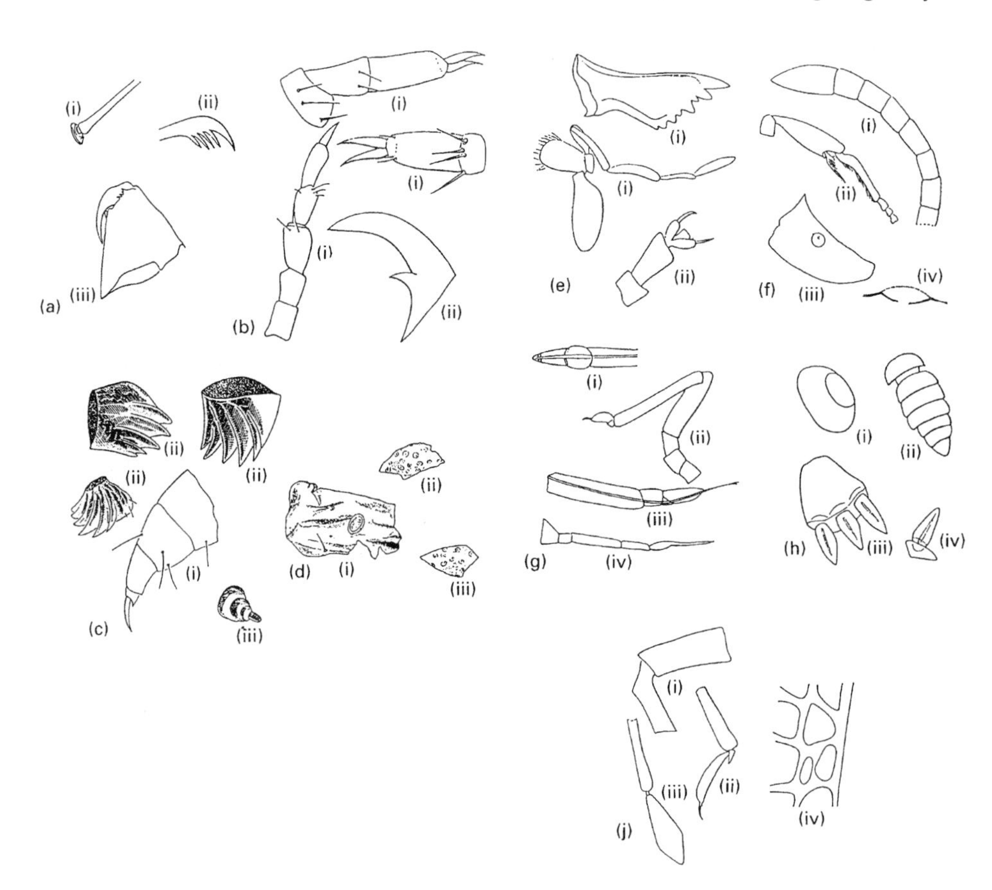

Figure 6.25 Examples of prey fragments found in dissections of carabid beetle guts: (a) lycosid spider, (i) bristle, (ii) claw, (iii) chelicera; (b) carabid or staphylinid beetle larva, (i) legs, (ii) mandible; (c) lepidopterous (?) larva, (i) leg, (ii) mandibles, (iii) unidentified part; (d) fragment of exoskeleton of (i) lepidopterous (?) larva, (ii) heteropteran bug; (e) components of ant's mouthparts (i) and tarsus (ii); (f) ant's (i) antenna, (ii) leg, (iii) exoskeleton (fragment), (iv) ocellus, (g) components of aphid's (i, iii) mouthparts, (ii) leg, (iv) antenna; (h) bibionid fly, (i) antennal segment, (ii) antenna, (iii, iv) parts of leg; (j) *Acalypta parvula* (Hemiptera: Tingidae), (i, ii) parts of leg, (iii) part of antenna, (iv) fragment of forewing. Source: Hengeveld (1980). Reproduced by permission of E.J. Brill (Publishers) Ltd.

remains to species. Pollet and Desender (1988) were able to identify, to species, the remains of some Collembola in the guts of carabid beetles. The efficiency of detecting large Collembola was 67% (based on recovery of Collembola mandibles), and a semi-quantitative analysis of consumption was therefore possible. Similarly, Dixon and McKinlay (1989) identified, to species, some of the aphid remains from the guts of harvestmen, and were able to estimate that some individual *Phalangium opilio* contained the remains of at least five aphids.

An obvious difficulty with the gut dissection method is the digestion of prey prior to dissection. The abilities of the different investigators to recognise the prey remains will influence comparative studies, as will the ability to distinguish between fragments from the same or different individuals within a prey species or group.

Although the gut contents of predator species probably derive mainly from predation, carrion-feeding (scavenging) is also known to take place in some species. Therefore, a knowledge of carrion availability may also be useful in predicting the diet of a species. While scavenging is clearly problematical in a study of the effects of predation on pest population dynamics, it is not a problem for studies that set out to assess the role of alternative animal foods in sustaining predator populations (Symondson *et al.*, 2000). Alternative foods may help to sustain generalist predators in a crop when pest numbers are low, allowing the predators to survive, ready to feed on the pest when it becomes available (Murdoch *et al.*, 1985).

The acquisition by a predator of prey materials through **secondary predation** (i.e. feeding upon another predator species which itself contains prey items) can also produce misleading results. Although countable remains of certain ingested structures (e.g. the mandibles of beetles or the shells of slugs) may provide some quantitative information, these parts of the prey may well be avoided by the predators. Whether they are eaten, even by a predator that is not a fluid-feeder and which consumes more solid remains, may depend upon the size of the indigestible parts (biasing the data towards small prey) or the state of hunger of the predator. An awareness of these problems is also required with the faecal analysis method described earlier.

The majority of predators are obligate fluid-feeders, and other predators are facultative fluid-feeders, with an unknown proportion of their ingestion confined to prey fluids. The following techniques (antibody, electrophoretic and DNA-based analyses) are suitable for investigating food consumption by fluid-feeders (Symondson, 2002b).

6.3.9 ANTIBODY TECHNIQUES

Most predators are solely or partly fluid-feeders, and so their diet cannot be investigated using the aforementioned faecal analysis and gut dissection methods. Ecologists have, for over three decades (Loughton *et al.*, 1963), used a variety of antibody methods to study predation (see reviews by Boreham and Ohiagu, 1978; Boreham, 1979; Frank, 1979; Sunderland, 1988; Sopp *et al.*, 1992; Greenstone, 1996; Symondson and Hemingway, 1997; Symondson, 2002a). Applications of antibody methods dealt with here include:

1. Determination of the composition of the predator complex of a single (or small number of) prey species
2. Quantification of predation on a target pest by one or more known predators.
3. Detection and identification of parasitoid immatures within or upon hosts.

[A fourth category, the determination of the dietary range of a predator, has rarely been attempted because of the costs and the amount of labour involved with raising and characterising separate antisera (or monoclonal antibodies) against each potential prey species. Indeed, this would probably no longer be ethically acceptable because, inevitably, large numbers of mammals (see below) would have to be used. Cross-reactivity between antisera and non-target prey species can be high, and the procedures required to overcome this problem become more complicated the greater the number of prey species involved. Dennison and Hodkinson (1983), however, attempted to do this, and raised eleven separate antisera against Enchytraeidae, Nematoda, Lumbricidae, Diplopoda, Isopoda, Araneae, Acarina, Collembola, Diptera, mouse carrion and fungal mycelia. Similarly, a range of antisera have been used frequently to identify the sources of blood meals in haematophagous insects (e.g. Ngumbi *et al.*, 1992; Blackwell *et al.*, 1994). Development of DNA-based techniques may allow us, in future, to study dietary ranges (see subsection 6.3.12)].

Serological methods are based on the principle that antibodies raised in a mammal (usually a mouse or a rabbit), against antigens of an invertebrate prey species, can be used to detect antigens from that species in the gut contents of the predator. Material (tissues, haemolymph) is taken from an invertebrate species and injected into a mammal, triggering an immune response (see below). The antibodies generated are then combined, in various ways, with prey materials present in the guts of field-caught predators, to test for the presence of molecules specific to particular invertebrate taxa.

Antibody techniques can help to establish which of an assemblage of predator species preys on a crop pest (1, above). A survey of the predator community within the prey's habitat is carried out, so that possible natural predators can be identified. Materials (antigens) are taken from the target prey species and antibodies raised against them. These antibodies are then used to test the guts of field-caught predators, to establish whether those predators feed on the pest species in the field, to determine what proportion of predator individuals contain prey remains and, where possible, to measure the quantity of such remains (e.g. Symondson *et al.*, 1996a, 2000) (2, above). [Subsection 7.2.13 deals with the application of antibody techniques in studies of predator-prey population dynamics].

Applying antibody techniques to the detection and identification of parasitoids (3, above) is relatively straightforward. Antibodies are raised to the immature stage(s) of the parasitoid, and field-collected hosts are screened for the presence of antigens (e.g. Stuart and Greenstone, 1996, 1997).

The major attractiveness of studies using antibodies is their potential for extreme specificity due to biological recognition at the molecular level. Cross-reactions with antibodies produced to molecular configurations common to 'target' and 'non-target' species - an important drawback to some serological methods – may, in the case of polyclonal antisera (which contain hundreds or thousands of different antibodies), be overcome by **absorption** (Sunderland and Sutton, 1980; Symondson and Liddell, 1993a), a process whereby unwanted antibodies are precipitated out of the antiserum by incubation with extracts of the unwanted, cross-reacting, species. Methods that seek to increase the strength of the ELISA signal may also amplify cross-reactivity (see description of ELISA protocols, below). Schmaedick *et al.* (2001) raised antibodies against *Pieris rapae* (Lepidoptera) which they conjugated to biotin. They then used a commercially-available streptavidin-alkaline phosphatase conjugate to bind the enzyme (alkaline phosphatase) to the biotin and hence to the antibodies. Following addition of an enzyme substrate the signal was strong, but there was a high level of cross-reaction with another lepidopteran pest, *Trichoplusia ni*. This was removed by preabsorbing the anti-*P. rapae* antibodies with *T. ni* proteins. Absorption can also be achieved by affinity purification, i.e. by passing the antiserum through a column to which cross-reacting proteins have been bound. Cross-reacting antibodies become attached to the column, but purified target antibodies pass through (Schoof *et al.*, 1986). Alternatively, since absorption techniques reduce the titre of the antiserum, other methods, such as electrophoresis (subsection 6.3.11), may be used to identify and separate out from a 'target' prey or parasitoid species extract, any protein component specific to that species. This is then used to raise a more specific antiserum (Tijssen, 1985; Buchholz *et al.*, 1994; Schultz and Clarke, 1995). Fast protein liquid chromatography (FPLC) can also be used to purify protein antigens (Arnold *et al.*, 1996). Such methods cannot guarantee specificity, because even species-specific proteins may well possess on their surface **epitopes** (antibody binding sites) that are present on very different proteins in other species, and, therefore, absorption or affinity purification may still be necessary. Laboratory procedures for many immunochemical techniques are described by Rose and Friedman (1980), Wilson and Goulding (1986), Hudson and Hay (1989), Hillis and Moritz (1990), Roitt *et al.* (2001) and Hay and Westwood (2002).

Monoclonal antibodies (MAbs), which are now rapidly replacing the use of polyclonal antisera in predation studies, do not need to be raised against a purified antigen. As the antibodies selected will be specific for a single epitope configuration, other epitopes on the same or other proteins, that might be common to many species, will not affect specificity. The chances that a monoclonal antibody will cross-react with something else are clearly much lower than one of the hundreds or thousands of different antibodies in a polyclonal antiserum doing so. If, however, a cross-reaction is subsequently found, absorption is not an option and a new MAb must be generated.

Production of Antibodies

Raising Polyclonal Antisera and Monoclonal Antibodies

A general protocol for the production of a polyclonal antiserum is given by Symondson and Hemingway (1997). Polyclonal antibodies are usually raised by repeated subcutaneous injections of antigens into a rabbit, followed by the taking of a small quantity of blood from the immunised animal after a suitable period. The blood is allowed to clot (at 37°C), and the serum separated by centrifugation. Inoculation using a single injection of antigen in this manner stimulates an immune response, with the production of primarily IgM antibodies. A secondary response is induced by further injections, using the same antigen, at a later date. This secondary response is faster and stronger than the first and yields higher levels of the more specific and useful IgG antibody. When using weakly antigenic compounds, the levels of antibodies produced may be increased by co-immunisation with an adjuvant, by increasing the duration of the antigenic challenge, or by increasing antigenicity of the compound by conjugation with larger proteins. Practical details concerning antibody production techniques can be found in Liddell and Cryer (1991) and Delves (1997).

Antibodies are most readily raised to large proteins and polysaccharides, but can be generated against almost any biological materials. High molecular weight proteins, with many epitopes, will generate a multitude of different antibodies, hence the term **polyclonal antisera**. The relative amounts of each antibody clone in the serum are determined by a variety of factors, thus no two polyclonal antisera, even to the same antigen, are the same. This makes comparison of field results obtained by using different antisera to the same target (over time or between laboratories) difficult, and is a primary reason for the move to using monoclonal antibodies (see below). Polyclonal antisera cannot generally be used to quantify predation upon a single pest species within a habitat shared by related species and genera. This is because the antibodies will bind to different degrees with material from different species, the strength of the reaction being generally related to the taxonomic distance between the species involved. To overcome this problem, isolation of a single antibody-producing clone (monoclone) is required. Production of monoclonal antibodies (MAbs) requires access to dedicated cell culturing facilities, a laminar flow cabinet, and an incubator with CO_2 control (Symondson and Hemingway, 1997). However, once these have been established, monoclonal antibody techniques offer several advantages over those employing polyclonal antisera. One of these is the reproducibility of assays over time. In contrast to polyclonal sera, monoclonal cell lines can produce almost limitless supplies of uniform antibodies, and the monoclonal cell lines (hybridomas) producing those antibodies may be stored under liquid nitrogen and used to generate further supplies of the specific antibody at a later date (Liddell and Cryer, 1991).

An example of the value of this "standardisation of reagent" effect, is that a single MAb was used, by different research teams, to detect bollworm egg predation by spiders in both North America and India (Sigsgaard, 1996; Ruberson and Greenstone, 1998). Similarly, a monoclonal antibody raised to detect predation by carabid beetles on cereal aphids (Symondson *et al.*, 1999b) is now being used by other groups in the UK, Switzerland and the USA to measure predation on aphids by both carabids and spiders. Another major advantage is that monoclonal antibodies can be used to detect prey at a range of taxonomic levels. As well as being raised against broad groups, such as gastropods, aphids or earthworms (Symondson *et al.*, 1995, 1999b, 2000), antibodies have been made that detect prey at the level of genus (e.g. Symondson and Liddell, 1993c; Bacher *et al.*, 1999), species (Hagler *et al.*, 1993; Symondson and Liddell, 1996b; Symondson *et al.*, 1997, 1999c) species aggregate (Symondson *et al.*, 1999a), stage (Hagler *et al.*, 1991; Greenstone and Trowell, 1994; Mendis *et al.*, 1996; Goodman *et al.*, 1997) and even instar (Greenstone and Morgan, 1989). A MAb was produced that was highly specific to eggs and adult females of the pink bollworm, *Pectinophora gossypiella* (Hagler *et al.*, 1994), while another was specific to egg antigen of the whitefly *Bemisia tabaci*. Its reactivity appeared to be confined to the egg and adult female stages (which carry eggs or vitellogenins) (Hagler *et al.*, 1993). A similar, but more species-specific antibody, against *B. tabaci* (Symondson *et al.*, 1999c) also reacted with female, but not male, whiteflies. A *Lygus hesperus* (Miridae) species- and egg-specific MAb was prepared, and some predators were shown to feed on eggs (or gravid females) of this pest in the field. Young eggs, containing a higher proportion of vitellin, were more responsive to the MAb than older eggs with less vitellin (Hagler *et al.*, 1992b). A range of different species- or genus-specific antibodies were raised against the eggs of slugs (Mendis *et al.*, 1996), and were shown to react with mature, egg-producing adult, but not immature, slugs. These antibodies were used to detect predation on eggs by small species of carabid beetle that are thought to be incapable of attacking mature slugs (they are incapable of overcoming the mucus produced by the slugs) (Mendis, 1997). Fifteen MAbs were produced using egg, larval and adult vine weevil, *Otiorhynchus sulcatus*, and some were found to be specific to egg or adult antigens, and others to two or three life-stages. Carabids from soft fruit plantations did not cross-react with these MAbs (Crook, 1996). Occasionally, MAbs which appear to be highly specific can show surprising cross reactions. A MAb against the slug *Deroceras reticulatum* gave no positive reaction with its congener *D. caruanae*, but cross-reacted strongly with the New Zealand flatworm *Artioposthia triangulata*, and with the millipede *Polymicrodon polydesmoides* (Symondson and Liddell, 1996b). This emphasises the need for extensive specificity data, even of MAbs, before natural enemies from the field are routinely tested. MAbs in the past have had very short detection periods, but one which detected three genera of pest slugs in the UK was found to be capable of detecting a meal in the carabid *Pterostichus melanarius* for 2.5 days at 16°C. When *P. melanarius* was allowed to feed on earthworms after consuming a slug (*Deroceras reticulatum*), the decay rate of slug antigens was significantly reduced (and hence the detection period extended) compared to beetles with no access to alternative prey (Symondson and Liddell, 1995). Another MAb, specific to earthworms, was also detectable in *P. melanarius* for 2.5 days at 16°C (Symondson *et al.*, 2000). *Polistes metricus* (Vespidae) adults were allowed to prey on *Helicoverpa zea* fifth instars, the wasps were assayed with a MAb, and there was found to be an exponential decay of prey antigen in the wasps. In this situation detection half-life was considered to be a more valid measure of detection than the absolute detection period (Greenstone and Hunt, 1993). Similar **decay curves** were found in many other studies (e.g. Symondson and Liddell, 1996c; Symondson *et al.*, 1997, 1999a), and need to be calibrated, separately, for each predator species, before such information can be incorporated in any quantitative models of predation. For example, Naranjo and Hagler (2001) found that the detection half-life for *Orius insidiosus* fed on eggs of *Pectinophora gossypiella*, and tested with an anti-*P. gossypiella* MAb, was more than twice as long as for another predatory heteropteran, *Geocoris punctipes*. In some instances there can be an initial increase in detectability with time, as more epitopes are revealed during digestion, followed by the usual exponential decay curve (Symondson *et al.*, 1999b). It has been shown that calibration of decay curves to determine detection periods must be done separately for each sex (Symondson *et al.*, 1999a). Meal size can affect the detection period and Agustí *et al.* (1999a) found that, although an exponential decay model was appropriate for predatory heteropteran bugs (*Dicyphus tamaninii*) that had eaten one *Helicoverpa armigera* egg, a linear model gave a better fit for individuals that had consumed ten eggs. Further discussion of the factors affecting detection periods can be found in Symondson and Hemingway (1997).

Standard methodologies for **production and use of monoclonal antibody** are described in Blann (1984), Goding (1986), Liddell and Cryer (1991) and Liddell and Weeks (1995), and are illustrated in Figure 6.26. The spleen from a mouse, immunised with an antigen (such as an insect protein), is used as a source of sensitised *B* lymphocytes. These antibody-producing cells, which are capable of producing the enzyme hypoxanthine guanine phosphoribosyl transferase ($HGPRT^+$), are fused with myeloma cells, cancerous cell lines capable of indefinite growth *in vitro*, but lacking the enzyme ($HGPRT^-$). Fusion is usually effected with polyethylene glycol, and the mixture of fused and unfused cells seeded into culture plates containing HAT culture medium (hypoxanthine, aminopterin and thymidine). Cells can normally synthesise DNA either by the main biosynthetic pathway (*de novo*) or by the 'salvage pathway', using preformed bases in the presence of HGPRT. Aminopterin, in the HAT medium, is a metabolic poison which completely inhibits the *de novo* pathway. Thus, only hybrid cells can survive in HAT medium. Myeloma cells will die, because they lack genes for the production of the HGPRT, and cannot use the salvage pathway. Lymphocytes will die because they cannot be cultured *in vitro*. Hybridomas will survive because they contain the enzyme genes (i.e. they are $HGPRT^+$), allowing DNA synthesis, and inherit the ability to grow indefinitely *in vitro* from their myeloma parent. Therefore, after 10-14 days, hybridomas are the only cells surviving. An alternative to HAT is Azaserine-hypoxanthine (O-diazoaceteyl-L-serine-hypoxanthine), which operates in a similar manner, but which is less toxic to hybridoma cells, increasing the potential number of surviving clones. Over the next 7–14 days individual cultures are tested for the production of the required specific antibody. Non-productive cultures are rejected, and aliquots from productive ones frozen down at each stage, to allow selection of other useful clones or in case of later contamination by fungi or bacteria in the cultures.

Finally, single clones are obtained, either by diluting down the cultures and distributing them so that each well of a culture plate contains only one hybridoma cell (cloning by limiting dilution), or else diluting in a weak nutrient agar, so that individual clones may be observed growing and can then be picked out individually using a Pasteur pipette. At such low cell numbers during cloning, growth must be supported either by introducing thymocyte feeder cells, or, preferably, using a hybridoma cloning supplement, such as BM Condimed H1 (Roche Diagnostics, Sussex, UK), which is prepared from the supernatant of a mouse lymphoma cell line, and contains the necessary growth factors and cytokines. Larger quantities of monoclonal antibody are then produced *in vitro*, either in tissue culture flasks or in bioreactors.

A potential area for the further development of conventional antibody procedures lies in the use of **recombinant phage antibodies** and **antibody engineering** (Liddell and Symondson, 1996; Symondson *et al.*, 1996b; Symondson and Hemingway, 1997). These techniques may allow a more rapid, and less labour-intensive

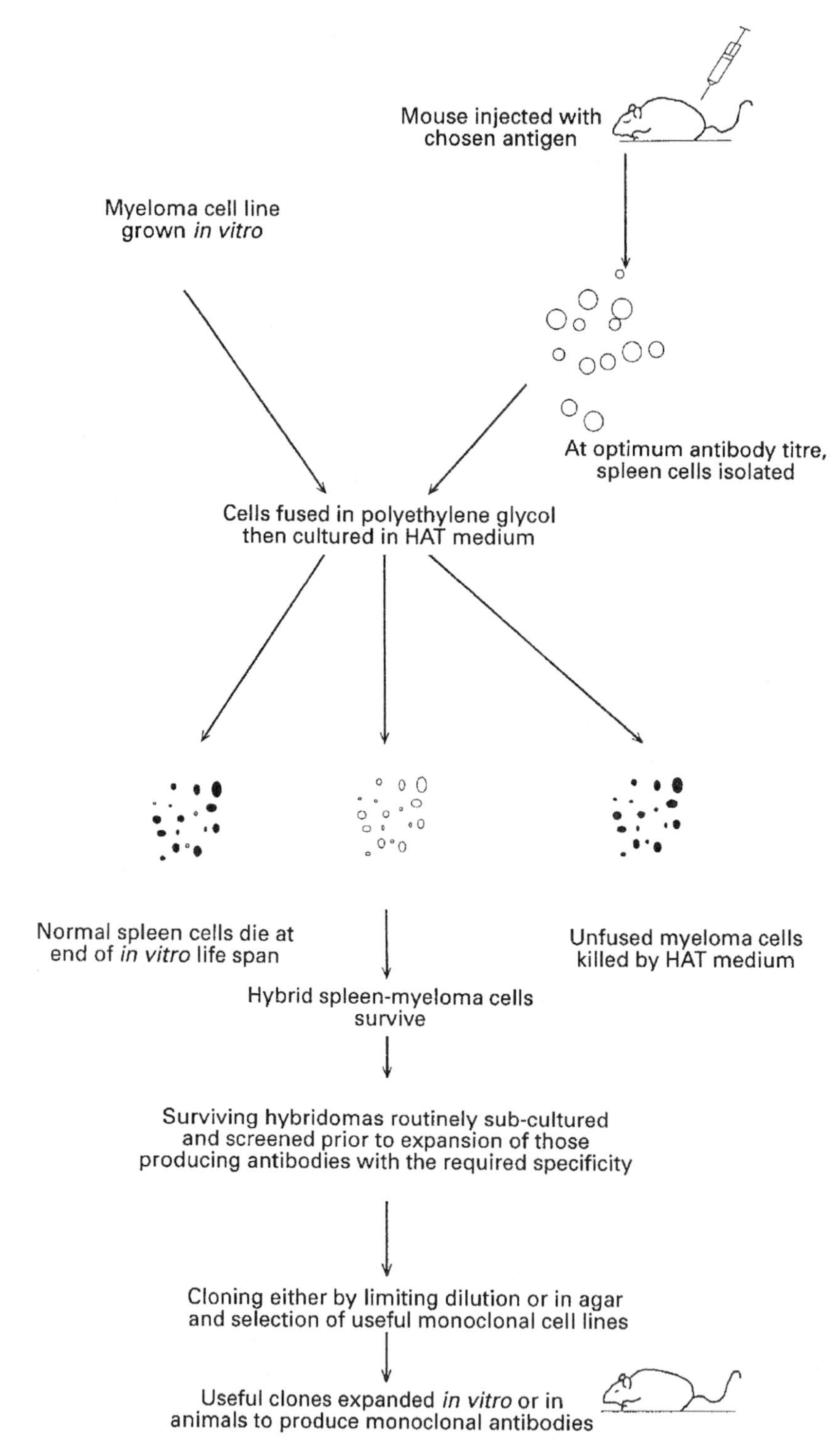

Figure 6.26 General principles of monoclonal antibody production.

production of monoclonal antibodies, and provide the opportunity to 'engineer' (genetically manipulate) antibodies for specific purposes, such as improving prey species specificity (Symondson and Hemingway, 1997). This technology has not made the leap from laboratory to field application in entomological research and will probably never do so now that less complex molecular techniques are becoming available (subsection 6.3.12).

Once antibodies have been raised to a specific prey antigen, immunoassays fall into four categories: precipitation, agglutination, complement fixation (all of which were used in predation studies in the past, but none of which appear in the literature any longer), and labelled antibody tests. The latter exploit the signal amplification properties of a label to increase sensitivity. Precipitation and agglutination tests were used in a variety of natural enemy studies (Table 6.6), whilst complement fixation tests have been largely confined to the analysis of blood meals in haematophagous insects. However, all three have, to a large extent, been replaced by labelled antibody assays, which offer greater sensitivity. Both polyclonal and monoclonal antibodies are now widely used in enzyme-linked immunosorbent assays (ELISA). However, it must be recognised that monoclonal antibody production is still comparatively costly, and initially time-consuming, so the need for specificity in any particular study must be balanced against the extra costs involved with using monoclonal antibodies. Reviews of assay systems suitable for use in entomological studies can be found in Boreham and Ohiagu (1978), Washino and Tempelis (1983), Sunderland (1988), Greenstone (1996), Symondson and Hemingway (1997) and Symondson (2002a).

Precipitation, Agglutination and Complement Fixation Methods

Precipitation tests rely on the fact that antigens may be reacted with specific antibodies to form complexes sufficiently large to be precipitated out of solution. Such precipitation may be performed either in liquids (Nakamura and Nakamura, 1977; Gardener *et al.*, 1981; Dennison and Hodkinson, 1983), or gels, such as agarose or polyacrylamide (Tod, 1973; Lund and Turpin, 1977; Whalon and Parker, 1978; Kuperstein, 1979; Lesiewicz *et al.*, 1982). The component antigens can be separated electrophoretically before reaction with antibodies (Pettersson, 1972; Hance and Rossignol, 1983), or an electric field can be used to increase the migration rate of the reactants (Sergeeva, 1975; Leslie and Boreham, 1981; Allen and Hagley, 1982; Doane *et al.*, 1985; Calver *et al.*, 1986). Techniques are described in Crowle (1980), and some early examples of their use in studies of insect predators are given by Frank (1979). Though simple, these techniques lack sensitivity (Greenstone, 1979b).

Agglutination occurs when antigens, and specific antibodies to those antigens, are present. Antibodies may be attached to particles, e.g. latex (Ohiagu and Boreham, 1978), or erythrocytes, both of which have been used in predation studies. Techniques are described by Nichols and Nakamura (1980), and Greenstone (1977, 1996).

Complement fixation techniques can be highly sensitive, and are described by Palmer (1980) and Staak *et al.* (1981). However, the latter authors also point out that the specificity of the antiserum used cannot be increased by absorption, because complement-fixing immune complexes, resulting from absorption, interfere with the test.

Labelled Antibody Immunoassays

Introduction

The increased sensitivity conferred by the signal amplification of antibody labelling has meant that all the methods described above have been superseded, to a large degree, by labelled antibody immunoassays, such as **enzyme-linked immunosorbent assay (ELISA)**. It is, however, important to appreciate that the extreme sensitivity of ELISA does not necessarily result in increased specificity. Increased specificity is best achieved by using monoclonal, rather than polyclonal, antibodies.

Table 6.6 Examples of the application of precipitation, agglutination and complement fixation, in studies of predacious and other insects

Method	Test	Predator/Prey System	References
Precipitation	Ring Test (Inter-face Test)	Ground-living beetles (predominantly Carabidae and Staphylinidae)	Dennison and Hodkinson (1983)
		Predators of *Aedes cantans*	Service (1973)
	Single Radial Immunodiffusion	Lycosid spider *Pardosa sternalis*	McIver (1981)
		Cereal aphid *Rhopalosiphum padi*	Pettersson (1972)
	Double diffusion (Ouchterlony technique)	Predators of *Lygus lineolaris*	Whalon and Parker (1978)
		Carabid predation on *Eurygaster integriceps*	Kupersterin (1979)
	Immunoelectrophoresis	Cereal aphid *Rhopalosiphum padi*	Pettersson (1972)
	Rocket Immunoelectrophoresis	Predation on beetle *Dendroctonus frontalis*	Miller *et al.* (1979)
	Crossover Immunoelectrophoresis	Predation on *Dendroctonus frontalis,*	Miller *et al.* (1979)
		Eldana saccharina,	Leslie and Boreham (1981)
		Inopus rubriceps	Doane *et al.* (1985)
Agglutination	Latex Agglutination	Predation on aphid *Acyrthosiphon pisum* by ladybird beetle *Coccinella septempunctata*	Ohiagu and Boreham (1978)
	Passive Haemagglutination Inhibition Assay	Lycosid spider *Pardosa ramulosa*	Greenstone (1977, 1983)
Complement Fixation	Complement Fixation	Blood meals of tsetse-flies	Staak *et al.* (1981)

In ELISA, prey antigen is detected when it binds (directly or indirectly) with labelled antibodies, hence detection of the label is indicative of the presence of the antigen. Because the amount of label is directly proportional to the amount of antigen in the sample, the test may be regarded as quantitative, a major advantage over the tests described previously. Various labels are available, including radioisotopes, fluorescent compounds, and enzymes which catalyse specific reactions to give a coloured product. Rose and Friedman (1980) describe the use of radio-immunoassays in medicine, but, to date, this method has not been widely applied in entomology, due to expense and safety considerations.

Preparation of Natural Enemies for Immunoassays (ELISA)

The method of preparing insect material for analysis depends on the size of the predator being studied. In the case of smaller predators, a whole-body homogenate is prepared, whereas with larger predators, such as carabid beetles, the gut (or even just the foregut) is required. The contents of the foregut are likely to be most useful for analysis, since this part of the gut is used mainly for storage, and so less digestion of prey materials will have taken place. The excised gut is diluted with a suitable buffer solution (e.g. PBS, Phosphate Buffered Saline), agitated, and then centrifuged. The resulting supernatant is then used as a stock solution, which may be stored frozen and re-used for several years (especially if stored at $< -70°C$). Laboratory-starved predators are used to calibrate the assay. It is also essential to include positive controls (i.e. samples of the prey or predators that have just consumed the target prey). In the preparation of positive controls for sucking predators, it is necessary to be aware that if the predator is removed from the prey prematurely, much of the antigenic material may still be within the prey, even though the predator has nearly finished feeding. This is because fluid-feeding predators, such as spiders, can have a feeding cycle with a sucking phase alternating with a relaxing phase (Cohen, 1995). During the latter the extracted food is released back into the prey (Pollard, 1990).

For protocols involved in preparing insect material for studies of parasitism, see Stuart and Burkholder (1991), and Stuart and Greenstone (1996).

Fluorescence Immunoassay

Antibodies can be coupled with a fluorescent compound, whilst the test antigen is held within a matrix, such as cellulose acetate discs. Following incubation in the conjugated antiserum, the discs may be read under UV light. More commonly, however, the antigen is bound to a solid phase, such as a polystyrene microtitration plate. This assay is both very sensitive and quantitative, although problems of contrast were encountered by Hance and Grégoire-Wibo (1983) in their studies of aphid predation by carabid beetles.

Enzyme-linked Immunosorbent assay

The basis of ELISA is that some enzymes, commonly horseradish peroxidase or alkaline phosphatase, may be coupled with antibody, in such a way as to produce conjugates which retain both immunological and enzymatic activity. Such conjugates are stable for a period of many months, and can be reacted with prey antigen which has been previously fixed to a solid phase (directly or by antibody) to form a complex where a specific antibody-antigen reaction occurs. Antigen binding may be either direct or indirect (see below). Whichever assay is used, the non-reacting components are flushed away between each step. The fixed complex is then incubated with an enzyme substrate under standardised conditions, and the enzyme catalyses the conversion of the substrate to form a coloured product, the intensity of which can be measured in a plate spectrometer (Figure 6.27). ELISA has been widely modified for differing studies. Many of the variations are reported and described by Voller *et al.* (1979), and references dealing with their entomological applications are cited in Sunderland (1988), Greenstone (1996), Symondson and Hemingway (1997), and Symondson (2002a). These authors discuss ELISA formats commonly used in studies of predation. For purely qualitative studies, positive and negative controls may be sufficient, but for quantitative studies (subsection 7.2.13) an antigen dilution series should be included on each ELISA plate, both for quantitative assessment (through regression analysis), and to take account of any inter-plate variation in results. The message amplification effect of the enzyme component makes ELISA a much more sensitive test than those tests based upon precipitation (Miller, 1981). Examples of the use of ELISA for analysis of field-collected invertebrates are given by Sunderland (1988), Greenstone (1996), Symondson and Hemingway (1997) and Symondson (2002a). These authors emphasise the suitability of ELISA for routine testing of large numbers of individuals. When optimised, the amount of antigen required per test is small enough to allow an individual predator meal to be tested against a wide range of antisera (Symondson and Hemingway, 1997).

In a **Direct ELISA**, (Figure 6.28a) antigen is bound directly to a solid phase ((i) in Figure 6.28a), a well on a polystyrene microtitration

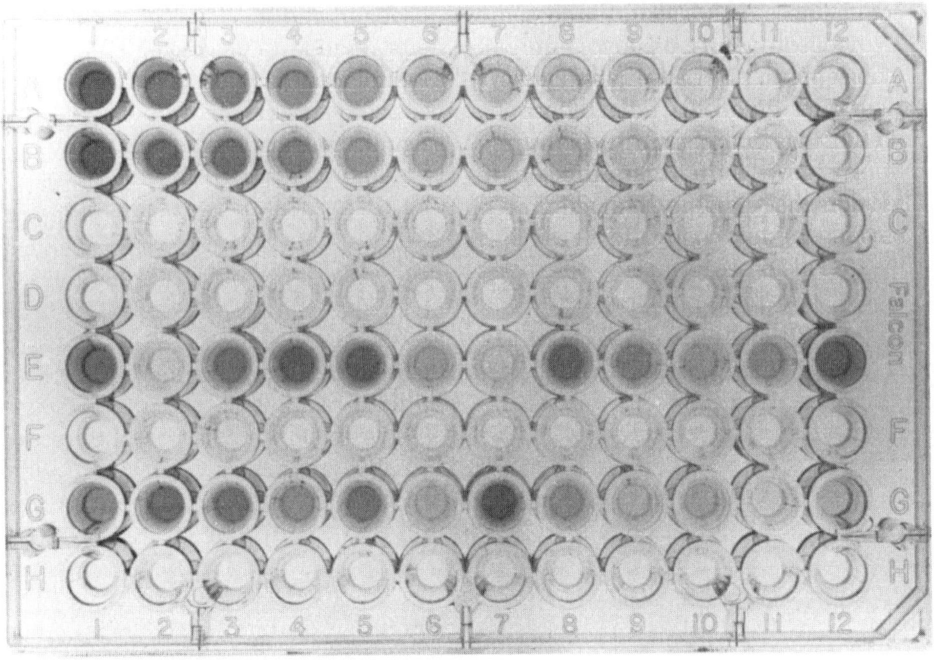

Figure 6.27 An ELISA plate used by Symondson and Liddell (1993b). Contained in rows A–D: a 1.5 x dilution series of antigen (slug haemolymph), with the highest concentrations towards the left; E–G (replicated in rows F and H): diluted samples extracted from the crops of carabid beetles. Rows A, B, E and G were treated with an anti-haemolymph antiserum, and rows C, D, F and H with non-immune serum. Background readings, and heterologous reactions between non-immune serum and the antigen, can be subtracted from the readings obtained with the specific antiserum. Regression of $\log_e$ haemolymph concentration versus $\log_e$ absorbance readings produces an equation by which the concentration of haemolymph in the unknown samples can be calculated.

plate, where it is reacted with enzyme-labelled antibody (ii). The intensity of colour produced following the addition of enzyme substrate is then measured and is proportional to the amount of antigen present. Although the assay is apparently simpler than those discussed below, it does involve the extra step of conjugating a specific antibody with an enzyme label before starting. Possibly, for this reason, it has been little used in predation studies.

In the **Double Antibody Sandwich ELISA** (Figure 6.28b), non-conjugated antibody, specific for a target prey, is attached to the solid phase (i) and the unknown antigen solution (the diluted gut contents of a predator, which may or may not have eaten the target prey) added. The captured antigen (ii) then reacts with enzyme-labelled antibody (again specific for the target prey) (iii) and, on addition of the enzyme substrate, the intensity of the coloured product formed is again proportional to the amount of unknown antigen. This method is more specific than Direct ELISA, since it relies on the presence of two epitopes on each antigen molecule, rather than one. Specificity may be increased still further if the epitopes are different and the antigen has to match two different binding sites. This method has been employed in a number of ecological studies (e.g. Chiverton, 1987; Sunderland *et al.*, 1987a; DuDevoir and Reeves, 1990; Sopp *et al.*, 1992).

With the **Indirect ELISA** (Figure 6.28c), prey-specific antibody (ii) is reacted with antigen that is bound directly to the ELISA plate (i). A commercial anti-rabbit immunoglobulin conjugated with enzyme is then added (iii), and followed by the appropriate enzyme substrate. This method is essentially similar to the direct

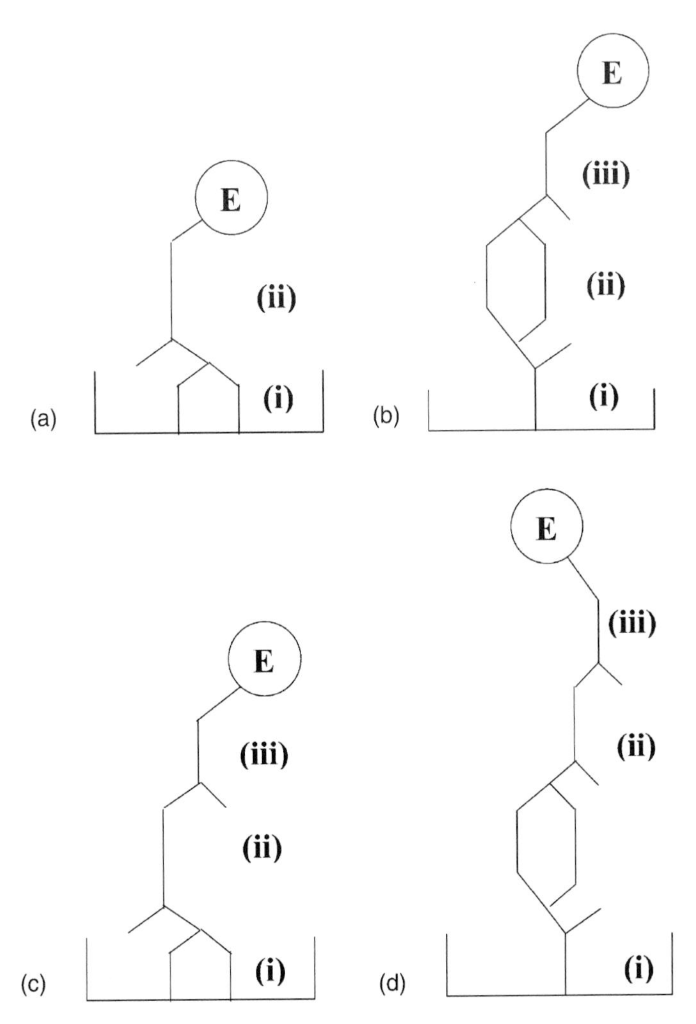

Figure 6.28 (a) Direct ELISA; (b) Double Antibody Sandwich ELISA; (c) Indirect ELISA; (d) Indirect Double Antibody Sandwich ELISA

method, but utilises commercially available anti-species conjugate, thus saving on specific antiserum. The colour change mediated by the enzyme is again proportional to the quantity of antigen present. In the indirect ELISA, absorbance can be considerably affected by pH, because pH-sensitive inhibitors in the antigen preparation can prevent the antigen from binding to microtitre plates under some pH conditions (Arnold *et al.*, 1996). The pH of prey antigen in predators can be variable. Cohen (1993), for example, found that the pH of prey of the heteropteran bug *Zelus renardii* increased as it was being pre-orally digested. In practice, phosphate buffered saline (PBS, pH 7.2) has proved to be an effective medium for binding prey remains of insects and molluscs retrieved from the guts of insects and spiders (e.g. Symondson and Liddell, 1993a). The pH of the gut contents has little bearing on binding properties when the antigen is diluted, for example, 1:20,000 (w/v) in PBS (Symondson and Hemingway, 1997). The indirect ELISA has been used to good effect in many field studies (e.g. Hance and Rossignol, 1983; Hagler and Naranjo 1994a,b; Symondson and Liddell,

1993b; Sigsgaard, 1996; Symondson *et al.*, 1996a; Lim and Lee, 1999; Symondson *et al.*, 2000).

The qualitative aspects of all the above methods have been tested and compared by Crook and Payne (1980), who found the indirect method to be the most sensitive, and the double antibody sandwich method to be the most specific. The sensitivity of the indirect method allows samples from a single predator to be divided among several microplates (Naranjo and Hagler, 1998).

A fourth method of ELISA is the **Indirect Double Antibody Sandwich ELISA** (Figure 6.28d). As with the double antibody sandwich, non-conjugated antibody (e.g. raised in a rabbit) is first bound to the substrate (i), then antigen is added. Thereafter, the captured antigen is reacted with prey-specific antibody from another species (e.g. raised in a mouse) (ii), which, in turn, reacts with a commercial anti-mouse immunoglobulin conjugated with enzyme (iii). The coating antibody (i) must not be the same species as the second antibody (ii), otherwise the conjugate (iii) will bind to (i) as well as (ii). The intensity of the colour of the resulting product is again used to measure the amount of antigen present. Once again, specificity is increased by use of a two-site assay. The use of the commercial conjugate, both here and in the indirect ELISA, amplifies the signal, because several anti-mouse antibodies (carrying enzyme) will, in practice, attach to each mouse antibody.

The major advantage offered by ELISA lies in the fact that, when calibrated, it can be used quantitatively. A disadvantage for field entomologists is that there is sometimes detectable colour development, in the control, so that a quantitative colour reading must be taken, using a spectrometer or plate reader. Statistical procedures are then applied to these readings and used to set confidence limits for determination of positive or negative results (Fenlon and Sopp, 1991; Symondson *et al.*, 2000). It is possible to eliminate such background by reducing the concentration of reagents, hence permitting visual positive/negative scoring of results, but only by reducing the sensitivity of the assay.

Stuart and Greenstone (1990, 1996) applied an **immunodot assay**, based on ELISA, to their studies of predator gut contents. A simplified version of the immunodot assay, employing nitrocellulose membranes, onto which predator homogenates are applied, was found to be fast, sensitive and economical (Greenstone and Trowell, 1994). This technique dispenses with the need for expensive apparatus, and tests can be scored by eye as positive or negative. Alternatively, a densitometer (Stott, 1989), or Chroma Meter, can be used to quantify dot blot assays. For the latter, quantitative results are almost identical to ELISA, but the apparatus is small, portable and battery-operated and so is useful in situations where an ELISA reader is not available (Hagler *et al.*, 1995). Tissue print immunoblotting was found to be faster and more sensitive than a double antibody sandwich ELISA for detecting cereal aphid feeding in German winter wheat by carabids and staphylinids, using a polyclonal antiserum (Löbner and Fuchs, 1994). However, the sensitivity of different techniques depends critically upon the precise conditions used, and the sensitivity achieved even by different workers using the same ELISA format, varies enormously. Agustí *et al.* (1999a) tested a squash blot assay (not previously used in predation studies), with a MAb specific to heliothine vitellin, in an attempt to detect heteropteran predation of *Helicoverpa armigera* eggs. In this assay, heteropteran individuals were placed on a dry nitrocellulose membrane, squashed with a glass plate, and the residues removed with a brush. Subsequent procedures were as for the dot blot assay. Unfortunately, although individual eggs could be detected, the high background noise from the squashed predators prevented detection of ingested eggs (i.e. non-target proteins saturated the competitive binding sites).

Because it combines speed, specificity and sensitivity, ELISA has been widely adopted for arthropod predation studies. Examples of applications in predation studies can be found in Fichter and Stephen (1979), Ragsdale *et al.* (1981), Crook and Sunderland (1984), Lövei *et al.*, (1985), Service *et al.* (1986), Sunderland *et al.* (1987a), Greenstone and Morgan (1989),

DuDevoir and Reeves (1990), Cameron and Reeves (1990), Hagler *et al.*, (1992b), Hagler and Naranjo, (1994a), Symondson *et al.* (1996a), Symondson *et al.*, 2000) and Bohan *et al.* (2000).

ELISA has also been applied to the investigation of parasitoids. Assessing parasitism by host dissection is labour-intensive, and accuracy varies with the operator and size of both host and parasitoid. Allen *et al.* (1992) compared host dissection with ELISA. Cocoons of the braconid *Pholetesor ornigis* (which attacks the spotted tentiform leaf-miner, *Phyllonorycter blancardella*, a lepidopteran pest of apple) were used to prepare an antiserum. The parasitoid antigens were chemically extracted, without trituration, directly into the ELISA plate using sodium dodecyl sulphate mercaptoethanol and EDTA buffered with TRIS. In many cases, the host larvae were cleared by this extraction buffer sufficiently to enable parasitoid larvae to be seen with transmitted light. Dissection, followed by ELISA, showed that false positives were not encountered. Keen *et al.* (2001), using ELISA to detect the pteromalid wasp *Muscidifurax raptorellus* in house-fly pupae, were able to detect parasitism for 7 to 21 days after the pupa had been stung. Stuart and Burkholder (1991), using indirect ELISA, developed monoclonal antibodies specific to two parasitoids of stored products pests. The antibodies reacted with all life-stages and both sexes of the parasitoids. Stuart and Greenstone (1996), used an immunodot assay with a monoclonal antibody for the detection of second instar *Microplitis croceipes* in *Helicoverpa zea*. They also developed a MAb capable of detecting larval and adult stages of several species of parasitoid wasp in the caterpillars *H. zea* and *Heliothis virescens*, but which did not cross-react with noctuid pests. This technique was considered useful in screening for parasitism in general, in relation to assessing the need for sprays in IPM programmes (Stuart and Greenstone, 1997).

Sunderland *et al.* (1987a) compared ELISA with gut dissection for the detection of feeding by polyphagous predators on cereal aphids. They found that, when the same individuals were tested using both methods, agreement between the two tests was poor for those individuals that had eaten aphids. This suggests that the rate of voiding solids from the gut is not directly proportional to the rate of disappearance of prey antigens during digestion. Furthermore, the relative efficiencies varied according to the species of predator tested, and secondary predation was found to be a potential problem with both techniques. In general, secondary predation and scavenging have the potential to cause erroneous interpretations of results from post-mortem methods (such as gut dissection, antibody methods, electrophoresis and DNA-based techniques) that are used to determine predation, rather than just consumption (Sunderland, 1996). However, an experiment by Harwood *et al.* (2001) suggests that the importance of secondary predation as a source of error may have been exaggerated, and that by the time prey proteins have been ingested successively by two predators they will usually have been denatured beyond recognition. Dennison and Hodkinson (1983) made a comparison between precipitin tests and gut dissection, and found many more trophic links using antibodies.

Hagler (1998) compared the efficiency of five different immunoassay techniques. *Hippodamia convergens* (Coccinellidae) were allowed to eat one or five eggs of the pink bollworm, and were then tested. The double antibody sandwich was the only method that detected a single egg in all individuals immediately after feeding. Disadvantages of the dot blot method are: (a) that it is more difficult to quantify than ELISA (e.g. scanning densitometer or Chroma-Meter are expensive and the method labour-intensive), (b) requires more immunoreagents. The indirect ELISA was affected by the protein concentration of the predator samples. Those that contained $>125\,\mu g$ of protein gave negative reactions. However, such problems can be easily overcome by preparing gut samples on a weight-to-volume basis and stabilising protein concentrations in the controls (e.g. Symondson *et al.* 2000). The Western Blot assay (SDS-PAGE (subsection 6.3.11), followed by dot blot) was very time-consuming and costly, but was considered to have potential for testing (by pooling MAbs in the assay) whether individual predators have consumed a range of prey. Agustí *et al.* (1999a) found that the dot blot assay (when used with a heliothine vitellin-specific MAb to detect predation of *Helicoverpa armigera* eggs by heteropteran predators) was more rapid, sensitive and economical, than an indirect ELISA. They considered that dot blot could be more sensitive because of the higher capacity of nitrocellulose membranes (compared to polystyrene plates) to fix proteins.

6.3.10 LABELLING OF PREY AND HOSTS

Potential prey can be labelled with a chemical which remains detectable in the predator or parasitoid (Southwood and Henderson, 2000). By screening different predators within the prey's habitat for the presence of the label, the species composition of the predator complex can be determined.

Various labels are available for studying predation and parasitism. These include radioactive isotopes, stable isotopes, rare elements and dyes (usually fluorescent), which are introduced into the food chain, where their progress is monitored. The label is injected directly into the prey, or it is put into the prey's food source. Parasitoid eggs can be labelled by adding a marker to the food of female parasitoids. Appropriate field sampling can then reveal whether suspected predators have eaten labelled prey, or whether suspected hosts have been oviposited in by a particular parasitoid species.

Collembola have been labelled with radioactive ^{32}P through their food. This did not affect Collembola survival rate, or the probability of predation by carabids. In a laboratory experiment, predation rates estimated by **radiotracers** were similar to those estimated from a comparison of Collembola survival in containers with and without predators (Ernsting and Joosse, 1974). *Helicoverpa armigera* and *H. punctigera* eggs were radiolabelled by injecting adult females with ^{32}P, and the fate of these eggs was then monitored in the food chain (Room, 1977). Both adults and larvae of the bean weevil, *Sitona lineatus*, were labelled with ^{32}P by allowing them to feed on broad bean plants which had their roots immersed in distilled water containing the radiolabel (Hamon *et al.*, 1990). The labelled weevils were then exposed to predation by carabid beetles within muslin field cages, and the levels of radioactivity shown by the carabids subsequently measured using a scintillation counter. Greenstone (1999) lists six other studies in which radionuclides were used to detect consumption of pests by spiders. Breene *et al.* (1988) labelled mosquito larvae with radioactive ^{32}P and released them into simulated ponds where spiders were present After 48 h spiders were removed to assess levels of radioactivity, and 30-77% had eaten the labelled prey. Spiders were observed preying on the mosquito larvae by grasping them from beneath the surface of the water, pulling their bodies through the surface tension and consuming them.

Labelling may also be used to investigate intraspecific trophic relationships, such as maternal care. Adult female lycosid spiders (*Pardosa hortensis*) were labelled for a month with radioactive tritium and leucine through their food (*Drosophila* flies). When they produced cocoons (which are carried attached to the spinnerets), the cocoons were removed and replaced with unlabelled cocoons, which the mothers adopted readily. Twenty days later, the spiderlings inside the cocoon were found to be labelled with radioactive tritium, and parts of the cocoon were labelled with radioactive leucine (an amino acid found in cocoon silk). This finding suggests that the mother periodically opened the cocoon, regurgitated fluid into it for the spiderlings, and then repaired the opening with fresh silk (Vannini *et al.*, 1993). Monitoring of radioactive labels may be done using a Geiger counter or a scintillation counter, to measure α- and/or β- emissions (Hagstrum and Smittle, 1977, 1978), or by autoradiography (McCarty *et al*, 1980). In autoradiography, the sample containing the potentially labelled individuals is brought into contact with X-ray sensitive film, and dark spots on the developed film indicate the position of labelled individuals. However, although simpler to perform than either serological or electrophoretic methods, hazards to both the environment, and the operator, posed by radiolabelling, mean that the method needs to be confined to laboratory studies or is used in the field such a way that the labelled insects can be safely and reliably recovered at the end of the experiment.

The cereal aphid *Sitobion avenae* was marked through its food with the **stable isotope** ^{15}N, samples being analysed with an elemental analyser coupled to a mass spectrometer (Nienstedt and Poehling, 2000) (see also subsection 6.2.11). The label was detected successfully in a linyphiid spider (*Erigone atra*) and a carabid beetle, *Anchomenus dorsalis* (= *Agonum dorsale*), in laboratory tests. The isotope had no effect on the environment, or on the behaviour of the labelled animals. Predators that had eaten two aphids had a significantly higher level of marker than those that had eaten one aphid, and the marker was detectable in predators for at least eleven days (Nienstedt and Poehling, 1995). ^{15}N can also be used to mark parasitoids (by rearing them on hosts that have been raised on a diet incorporating ^{15}N). Parasitoid adults are expected to retain the label throughout their life because nitrogen is a major constituent of fibrin and chitin (Steffan *et al.*, 2001). Utilisation of stable isotopes of carbon and nitrogen as natural tracers of nutrient flows in ecosystems is discussed in subsection 6.3.1.

Rare elements such as rubidium, or strontium, may be used as markers, in a similar way to radioactive elements (Shepard and Waddill, 1976; Graham *et al.*, 1978). For example, Gypsy moth larvae were successfully labelled with rubidium, and their phenology and survival were not affected. The label was retained for five days in these larvae, and adult carabids (*Carabus nemoralis*) eating larvae in laboratory trials acquired the tag. Rubidium concentration in beetles was positively correlated with number of larvae eaten, and negatively with the number of days since feeding (Johnson and Reeves, 1995). In such studies the path of the rare element through the food chain is monitored with the aid of an atomic absorption spectrophotometer. Unfortunately, while such markers may be retained for life, and self-marking is possible using labelled food sources, the equipment necessary for detection is expensive, both to purchase and to run, as well as requiring trained operators.

Prey can be labelled with **rabbit immunoglobulin** (IgG), then predators that feed on the labelled prey can be detected with a goat anti-rabbit IgG. The method is only appropriate for predators with chewing mouthparts (that ingest the exoskeleton of their prey), as only 30% of sucking predators were positive 1 hour after feeding. (Hagler and Durand, 1994).

Fluorescent dyes offer another means of marking prey. Hawkes (1972) marked lepidopteran eggs with such dyes during studies of predation by the European earwig, *Forficula auricularia*. An alcoholic suspension of dye was sprayed onto eggs, which were then eaten by the earwigs. Thereafter, the earwig guts were dissected out and examined under UV light. Although such dyes are useful, being both simple and inexpensive to use, they are relatively short-lived, and possible repellant effects, due to either the dye or the carrier, must be taken into account. The dye is usually voided from the predator gut within a few days. Hawkes (1972) found no evidence of dye retention in any of the internal structures of earwigs dissected. In a few specific cases, naturally-coloured prey can be seen inside the gut of a living predator. For example, remains of citrus red mite (*Panonychus citri*) imparted a red colouration to the gut of the phytoseiid predator *Euseius tularensis*, that did not fade until more than 8 hours after ingestion (Jones and Morse, 1995). These authors devised a ten-point gut colouration rating scale to score the rate of decline of this natural label, after a meal, at a range of temperatures. Pigments from the mites, *Panonychus ulmi* and *Bryobia arborea*, can be seen in the guts of predators, such as lacewing larvae (Chrysopidae), that accumulate digestive wastes from large numbers of prey (Putman, 1965). Some of these ingested mite pigments fluoresce brilliantly in UV light. Pigments from very small numbers of mites can be detected by chromatography in small predators, such as anthocorid bugs, predatory thrips, and theridiid spiders, but, unfortunately, consumption of other prey can obscure the pigments (Putman, 1965).

A potential alternative approach might be to **live-stain** host or prey organisms. Lessard *et al.* (1996) marked protists with a fluorescent DNA-specific stain (DAPI – 2,4-diamadino-6-phenylindole) (see subsection 3.4.1) which had no ill effects on their swimming behaviour or growth rate, and which was visualised easily in guts of their fish predators (larval walleye pollock, *Theragra chalcogramma*) using epifluorescence microscopy. It is not yet known whether this method could be used easily to study predation or parasitism (or, indeed, dispersal, see subsection 6.2.11) by arthropod natural enemies.

Major potential problems with the use of labels are secondary predation (Putman, 1965; Sunderland *et al.*, 1997), scavenging and trophallaxis. The label may be recorded in several predatory and scavenger species within the prey's habitat, but only some of these may be directly responsible for prey mortality. Other food chains may also become contaminated (e.g. in the Johnson and Reeves (1995) study, rubidium 'leaked' from both larvae and beetles in their frass), further adding to the confusion. An additional disadvantage is that, in most cases, prey must be handled, to some extent, to achieve the labelling, and this may disturb the system under investigation. Released prey may not have time to re-establish themselves, either in terms of spatial separation, or in their favoured microhabitats, possibly making them more vulnerable to predation and other biotic mortality factors (such as parasitism and disease).

6.3.11 ELECTROPHORETIC METHODS

Introduction

Since the first edition of this book there has been a rapid increase in the use of molecular and antibody techniques for determining trophic relationships, and a parallel decline in the use of electrophoresis. Although these alternative approaches have many advantages (in terms of sensitivity and specificity) (subsection 6.3.9, 6.3.12), the necessary facilities and skills to undertake, for example, PCR, or to raise monoclonal antibodies, are not always available, and/or may not be required, to answer particular questions. Additionally, electrophoretic techniques are less expensive than DNA methods (Loxdale and Lushai, 1998). Here, therefore, we have retained a shortened section on electrophoresis with numerous references to further reading.

As with serology, both the dietary range of a predator species, and the composition of a prey's predator complex, can be determined using electrophoresis. An advantage of electrophoresis over immunological techniques is that the latter require enough prey materials to make the antiserum. This can be difficult with, for example, tiny mites as the prey (Murray *et al.*, 1989; Solomon *et al.*, 1996). In practice, electrophoresis is now being replaced rapidly by monoclonal antibody-based immunology and DNA techniques. Electrophoresis is defined as the migration of colloids (usually proteins) under the influence of an electric field. The rate of ion migration is proportional to the electric field strength, net charge on the ion, size and shape of the ion, and viscosity of the medium through which it passes (Sargent and George, 1975). Therefore, application of an electric field to a protein mixture in solution, e.g. a predator's gut contents, will result in differential migration towards one or other of two (positive + and negative −) electrodes. The direction of migration is, of course, dependent on charge, and, since proteins can exist as zwitterions (ions whose net charge depends upon the pH of the medium), conventional electrophoresis is performed under buffered conditions (i.e. constant pH). Zone electrophoresis is a modification of this idea, whereby the mixture of molecules to be separated is sited as a narrow zone, or band, at a suitable distance from the electrodes, so that, during electrophoresis, proteins of different mobilities travel as discrete zones, which then gradually resolve as electrophoresis proceeds.

Development of this basic concept has led to a wide range of electrophoretic techniques (vertical and horizontal slab, thin layer, two-dimensional, gradient pore and isoelectric focusing), utilising a variety of supporting media (cellulose acetate, alumina, agarose, starch and polyacrylamide). Isoelectric focusing, the second most important method after conventional electrophoresis, separates proteins which differ mainly in their charge. Proteins tested migrate in a pH gradient, and separate at their respective isoelectric points (Sargent and George, 1975; Hames and Rickwood, 1981). Thus, given that prey species will differ in the proteins they contain, the pattern produced by electrophoresing the gut contents of a polyphagous predator will, with either of these methods, provide information on the composition of the predator's diet.

Here, electrophoretic methodology is described briefly, followed by examples of how the techniques outlined have been used in some studies of natural enemies.

Electrophoretic Methodology

Many texts are available which deal with the general methods and techniques used in electrophoresis, e.g. Chrambach and Rodbard (1971), Gordon (1975), Sargent and George (1975), Hames and Rickwood (1981), Richardson *et al.* (1986), Pasteur *et al.* (1988), Hillis and Moritz (1990), and Quicke (1993). The reader is referred to these texts for detailed descriptions and experimental protocols, and to Menken and Ulenberg (1987), Loxdale and den Hollander (1989), Loxdale (1994), Symondson and Liddell (1996a), and Symondson and Hemingway (1997) for reviews of the application of electrophoretic methods in agricultural entomology.

A discussion of the relative merits of different types of gel may be found in Richardson *et al.* (1986) (see also Menken and Raijmann, 1996). Methodologies for polyacrylamide gel electrophoresis (PAGE) will be described below, since this method has been used by a variety of workers on a range of different insect groups. The recipes and procedures are based upon those used by Loxdale *et al.* (1983) in their studies of cereal aphid population genetics.

Most enzyme groups studied using electrophoresis are involved in the fundamental functions of all living organisms. Therefore, failure to detect a particular enzyme is usually a technical problem related to sample preparation, storage or electrophoretic conditions, rather than absence of the enzyme in the organism. Also, it is important that all samples are standardised and treated in exactly the same way.

Insect samples for electrophoretic analysis should be stored at –70°C, or in liquid nitrogen, until required. When dealing with larger insects, it is advisable to dissect out specific tissue, rather than use the whole organism for analysis. The advantage of concentrating on a specific tissue is that it minimises the introduction of different allozymes, as different tissues vary qualitatively in their allozyme content. In gut content analysis, such studies might involve determining which parts of the prey the predator actually ingests.

For the purpose of detecting prey remains in the guts of large predators, the gut, or gut region, should be excised and macerated in buffer to produce an homogenate (preferably the fore gut should be used; Giller (1984) investigated rates of digestion and found that prey material could be readily distinguished in the fore gut (or crop) of *Notonecta glauca*, but not after it has reached the mid gut, where most digestion takes place). In the case of the large fluid-feeding carabid beetle, *Carabus violaceus*, Paill (2000) obtained crop contents in the field by squeezing the abdomen of the beetle until it regurgitated a sample of crop fluid, which was then collected with a micropipette. Milked beetles were released alive. With small predators, a whole-body homogenate is prepared. To minimise enzyme degradation within a sample it is advisable to homogenise samples in a tray of ice. Because of the difficulty of homogenising large numbers of small insects, such as aphids and whiteflies, the insects can, as an alternative to homogenisation, be squashed individually with a plastic rod, in the middle of a narrow strip of filter paper, and then placed in a pocket of the gel (Wool *et al.*, 1984). Alternatively, microhomogenisers can be constructed from capillary tube and stainless steel wire (Murray *et al.*, 1989). Sample preparation is discussed in detail in Richardson *et al.* (1986).

When determining the dietary range of a predator species, the suspected prey species, and also starved samples of the predators, need to be characterised individually by electrophoresis. It is necessary to establish if there is any variation in banding pattern within a given prey species. To aid the interpretation of banding patterns, preliminary feeding experiments are desirable, to identify potential problems such as: (a) changes in the mobility of the marker enzyme, due to its modification in the predator gut, (b) predator and prey having a common food source, (c) parasitism of prey, and (d) selective feeding (e.g. predatory bugs may extract only haemolymph from their prey). For quantitative studies, data are also needed on detectability periods in relation to temperature (Murray *et al.*, 1989).

A study carried out by Walrant and Loreau (1995) illustrates the limitations of electrophoresis for certain types of investigation. These authors compared electrophoresis with gut dissection for the same individuals of the carabid beetle *Abax parallelepipedus* (=*ater*). Overall, both methods detected, equally well, that a meal was present, but microscopy was better able to provide an identification of what had been eaten. Where dissection showed that an individual had more than one prey species in its gut, the electrophoretic gel often could not be interpreted. For highly polyphagous species such as this, it is difficult to build an adequate set of reference gels, and there is likely to be confusion in the interpretation of banding patterns. For example, Traugott (2003) was unable to distinguish between larvae of Diptera and Lepidoptera in the diet of soldier beetle (*Cantharis* species) larvae.

When using vertical electrophoresis, the sample density is increased by the addition of either glycerol, or sucrose, to the homogenising solution. This ensures that the sample will sink into, and remain at the bottom of, the wells. A few crystals of bromophenol blue may also be used as a tracking dye. A detergent such as sodium dodecyl sulphate (SDS), or Triton X-100, will solubilise membrane-bound proteins and eliminate aggregates, and should be added to both the homogenising solution and the gel buffer. Triton X-100, unlike ionic surfactants (e.g. SDS), does not cause denaturing of proteins which results in a concomitant loss of activity in the case of enzymes. Nicotinamide-adenine dinucleotide phosphate (NADP), and P-mercaptoethanol, may also be added to the homogenising solution, in some cases, to stabilise NADP-dependent dehydrogenases, and to reduce oxidative changes, respectively. Finally, it may sometimes be necessary to buffer the homogenising solution with Tris buffer (Tris (hydroxymethyl) aminomethane).

After homogenisation, the sample is centrifuged to remove any solid matter prior to electrophoresis. The supernatant is transferred to sample wells in the gel medium using a microsyringe.

An electrophoresis apparatus consists of a DC power supply connected to electrodes in buffer-filled tanks. The samples to be separated are applied to the gel, and power is supplied from units which provide constant voltage, but limit current, and thus protect gels from excessive power levels. Since high voltages are used in the running of gels, extreme care must be exercised, to avoid electrical hazards.

Details of buffers and gel preparation can be found in Loxdale *et al.* (1983), or in many of the other texts mentioned above.

The application of samples to prepared gels, prior to an electrical potential difference being applied, is referred to as loading. In the absence of a stacking gel, the initial position of the samples is termed the 'origin' (position of proteins at time zero), whereas, when a stacking gel is employed, the origin is at the interface between the stacking and running gels. It is important not to overload the wells, as excess material in one track may cause distortion in band running, or create enzyme band shadows (possibly due to lateral diffusion), in adjacent tracks. Such shadows may be recorded as false positives. It is advisable to perform test runs with differing volumes and concentrations of sample, to determine the best values for optimum sharpness and clarity in any particular case. Also, it may be necessary to run a standard (e.g. a parthenogenetic aphid, Loxdale *et al.* (1983), or marker proteins, Murray *et al.*, 1989) of known mobility on each gel, so that a mobility ratio for each band can be calculated with respect to this standard.

Electrophoretic running conditions for a selection of enzyme systems are given in Loxdale *et al.* (1983). When assaying general proteins, gels are usually run at room temperature, but gels separating isoenzymes are run under refrigerated conditions (5–10°C). For most purposes, the voltage is kept constant (usually automatically by the power supply), and the current is allowed to fall as electrophoresis proceeds.

Following electrophoresis, gels are removed from the cassettes and placed in staining solution. Recipes for a wide range of stains may be found in Shaw and Prasad (1970), Harris and Hopkinson (1977), Loxdale *et al.* (1983), Richardson *et al.* (1986), and Pasteur *et al.* (1988). However, it must be noted that, in many laboratories, these basic recipes have been extensively modified to suit individual requirements. Gels are destained for five minutes in freshly prepared 7% acetic acid. Samples are usually recorded by scoring diagram, gel preservation, or photography (Figure 6.29). Light-sensitive gels should be scored immediately, and then discarded, whereas all other types of gel can be scored and then stored refrigerated in sealed, evacuated polythene bags. In scoring gels, the number and mobility of bands produced is recorded. This involves measuring the migration distance of individual bands (in mm) from the gel origin. A mobility ratio (MR) can then be calculated by comparison with a band of standard mobility.

Applications of Electrophoresis

Determination of Predation and Dietary Range

Murray and Solomon (1978) and Corey *et al.* (1998) used PAGE gradient gels to determine dietary range in several mite and insect predators, including the bugs *Anthocoris nemoralis*,

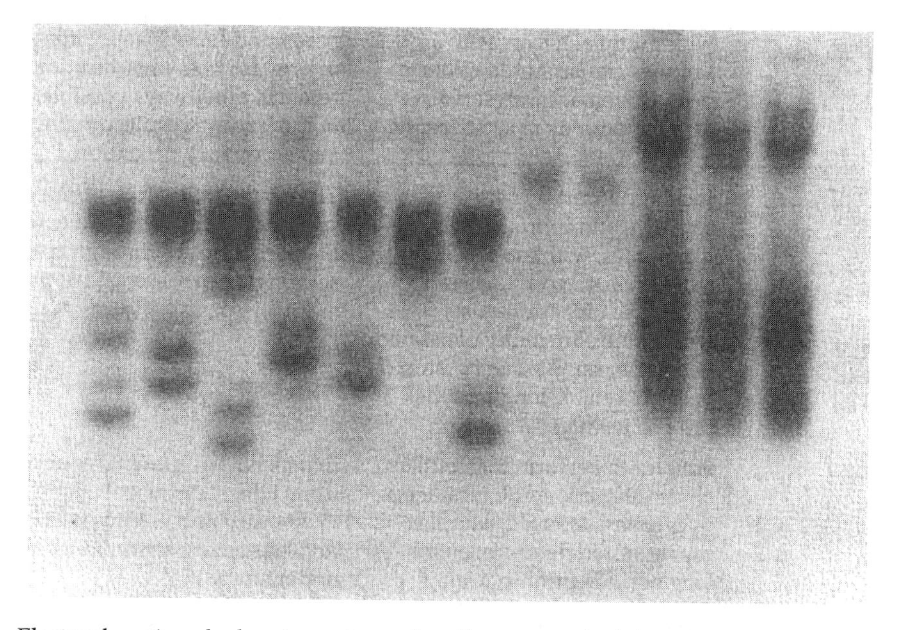

Figure 6.29 Electrophoretic gel, showing esterase banding patterns of three parasitoid species attacking the grain aphid, *Sitobion avenae*. Left to right: 1–7 *Aphidius picipes*; 8–9 *Ephedrus plagiator*; 10–12 *Praon volucre*. (Source: Powell and Walton, 1989.) Reproduced by permission of Oxford University Press.

Orius minutus, O. insidiosus and *Blepharidopterus angulatus.* Fitzgerald *et al.* (1986) developed methods to quantify predation on mites using gradient gels. Quantities of target prey remains in predators can also be determined by transmission densitometry (Lister *et al.*, 1987). The scanning laser densitometer can not only measure absorbance, but also provide an accurate quantification of migration distance for each band (Schulz and Yurista, 1995). Giller (1982, 1984, 1986) used disc gel electrophoresis to examine the dietary range of the water boatman, *Notonecta glauca*. Giller concluded that, provided 'diagnostic' bands were available for prey, the method was a rapid means of determining dietary range. Van der Geest and Overmeer (1985) evaluated the technique for predaceous mite-prey studies. They found that different prey spider mite species could be detected in predators, but that there were problems with the detection of eriophyid and tydeid mites. With the eriophyids, 30 to 40 mites have to be present in a predator's gut for detection to be successful, whereas with the tydeids the esterase patterns were not very distinct, and the patterns sometimes overlapped with those of the predator. Electrophoresis showed that a larger predator, the anthocorid bug *Orius vicinus*, had consumed many eriophid pests in an apple orchard (Heitmans *et al.*, 1986). Dicke and De Jong (1988) used electrophoresis to determine the prey range and measure the preference of the phytoseiid mite *Typhlodromus pyri* for different mite prey. Electrophoresis was shown to be, in principle, a means of determining dietary range in carabid beetles (Lövei, 1986), using polyacrylamide gradient slab gel and staining for esterase activity. A similar protocol was used by Schelvis and Siepel (1988), who studied feeding by larvae of *Pterostichus oblongopunctatus* and *P. rhaeticus* on a range of different prey, and later by Walrant and Loreau (1995). The latter authors prepared 57 reference gels representing the esterase banding patterns for a broad range of prey that might have been consumed by the predator under study, *Abax parallelepipedus* (=*ater*). Mini-polyacrylamide gels and isoelectric focusing were used to show that 50% of adults and larvae of a large fluid-feeding carabid beetle, *Carabus violaceus*, had fed on the pest slug *Arion lusitanicus* in grassland (Paill, 2000). Using gradient gels and phosphoglucose isomerase and esterase staining, Traugott (2003) determined that field-collected soldier beetle larvae (*Cantharis* spp.) had eaten mainly earthworms. Electrophoresis also has potential as a means of detecting host- and pollen-feeding by adult parasitoids (Jervis *et al.*, 1992a), and can be applied to aquatic (Schulz and Yurista, 1995) as well as terrestrial systems.

Detection and Identification of Parasitoids in Hosts

Electrophoresis is useful for the detection and identification of parasitoids in hosts (Menken, 1982; Walton *et al.*, 1990ab; Demichelis and Manino, 1998). Wool *et al.* (1978) used PAGE to detect the presence of *Aphidius matricariae* (Braconidae: Aphidiinae) in a glasshouse population of the aphid *Myzus persicae*. These authors obtained the most promising results when staining for the enzymes esterase (EST), malate dehydrogenase (MDH) and malic enzyme (ME), since all these enzymes gave diagnostic banding patterns for parasitised aphids. MDH was later used on starch gels to assess levels of parasitism in field populations of the rose aphid *Macrosiphum rosae* (Tomiuk and Wöhrmann, 1980). A more comprehensive study of parasitism of cereal aphids was carried out by Castañera *et al.* (1983), who examined the enzyme banding patterns produced by the parasitoid *Aphidius uzbekistanicus* in laboratory populations of the grain aphid *Sitobion avenae*. They tested 14 enzyme systems and agreed with Wool *et al.* (1978) that the most useful systems were EST, MDH and ME, but also obtained reasonable results with glutamate oxaloacetate transaminase (GOT). This work was followed up by Walton *et al.* (1990b), who used EST to monitor parasitism in field populations of *S. avenae*. EST, aconitase (ACON), leucine amino peptidase (LAP) and 6-phosphoglucose dehydrogenase (6PGDH) proved useful for detecting the endoparasitoid *Glypta fumiferanae* (Ichneumonidae) in larvae of the western spruce budworm *Choristoneura occi-*

dentalis (Castrovillo and Stock, 1981). In this case, ACON and 6PGDH produced monomorphic bands, whereas EST and LAP were polymorphic. For the routine screening of host populations in order to assess percentage parasitism, enzyme systems giving single host bands are quicker and easier to interpret (Castrovillo and Stock, 1981). Leafhoppers from grapevine and potato in the Italian Alps were assayed by PAGE, and α-glycerophosphate dehydrogenase was found to be the best enzyme system for detection of dryinid parasitoids in these hosts (Demichelis and Manino, 1995, 1998).

Wool *et al.* (1984) examined six enzyme systems for detecting the endoparasitoids *Encarsia lutea* and *Eretmocerus mundus* (Aphelinidae) in the whitefly *Bemisia tabaci*. EST again produced the best results, although weakly-staining bands were also produced using MDH and aldolase (A0).

Höller and Braune (1988) used isoelectric focusing to assess percentage parasitism of *Sitobion avenae* by *Aphidius uzbeckistanicus*. Both host and parasitoid showed specific bands for MDH, permitting clear detection of parasitism.

In order to interpret electrophoretic estimates of percentage parasitism, it is necessary to know which developmental stages of the parasitoid can be detected reliably by the technique. When establishing which enzyme systems were best for detecting parasitism of *S. avenae*, Castañera *et al.* (1983) deliberately used aphids containing final-instar larvae, but they stressed the importance of determining the earliest detectable stage of the parasitoid for the purposes of field monitoring. In laboratory studies, Walton *et al.* (1990a), using three parasitoid species, *Aphidius rhopalosiphi*, *Praon volucre* and *Ephedrus plagiator*, failed to detect the egg stage, and they concluded that estimates of percentage parasitism should be corrected to take this into account.

In most cases, comparison of results using electrophoresis with those from mummy-counting and host rearing have shown electrophoretic estimates to be very similar to estimates obtained from host rearing. This is to be expected, since both methods sample the same parasitoid developmental stages, and both involve the removal of live aphids from the field, and hence from the risk of further parasitism. However, Höller and Braune (1988) compared estimates of percentage parasitism by *A. uzbeckistanicus* obtained from isolectric focusing, with measurements obtained from the rearing of field-collected hosts, and found poor agreement between the two methods. The possible reasons for this discrepancy are discussed in their paper.

As a sampling tool for the detection and measurement of parasitism in insect pest populations, electrophoresis has a number of obvious advantages over rearing and host dissection. Parasitoids can be detected at an early stage in their development, and the species involved can usually be identified without the need to rear parasitoids through to the adult stage. This enables accurate measurements of parasitism to be obtained very rapidly, so enhancing their value in forecasting and decision-making in pest management. Walton (1986), in screening cereal aphid populations, was able to test around 100 aphids per day, when staining for esterase activity alone. In their evaluation of electrophoresis, Castrovillo and Stock (1981) established that a single technician could process 150 budworm larvae per day.

In studies concerned with the detection and measurement of parasitism, it is necessary to identify those enzyme systems which produce distinctive banding patterns for host and parasitoid. In some cases, diagnostic parasitoid bands may have similar mobilities to bands produced by the host. The existence of enzyme systems with clearly-defined parasitoid bands allows the rapid detection of parasitoid immature stages within a sample of hosts. In many cases there is the added advantage of parasitoid species identification based on the banding patterns.

Discrimination Between Parasitoids

Electrophoresis can be used to discriminate between several closely related species that may be morphologically very similar, or indistinguishable, particularly as immature stages within their hosts. Castañera *et al.* (1983) examined the adults of five aphidiid wasp species

(*Aphidius rhopalosiphi*, *A. picipes*, *Ephedrus plagiator*, *Praon volucre* and *Toxares deltiger*) associated with the cereal aphid *S. avenae*, and established that they could be identified on the basis of their EST banding patterns. Distinctive MDH patterns were also obtained by Tomiuk and Wöhrmann (1980) for two aphidiid parasitoids (*Aphidius rosae* and *Ephedrus* sp.) of the rose aphid, *Macrosiphum rosae*.

Höller *et al.* (1991), using isoelectric focusing with EST, found that they could discriminate *Aphidius rhopalosiphi* and an unidentified, but closely-related, species.

Polypeptides with different amino acid sequences may possess the same net charge and, consequently, display the same electrophoretic mobility (Berlocher, 1980, 1984). This means that enzyme bands effectively represent only phenotypes and not genotypes. Hence the term **electromorph** (a polypeptide characterised by its electrophoretic mobility) is often preferred to **allozyme** (defined as any of one or more variants of an enzyme coded by different alleles at the same gene locus), especially when the genetic basis of the bands seen on the gel is complex, and not easily elucidated by formal crossing experiments.

Isoenzymes are alternative forms of an enzyme produced by different loci, and may exist in a number of different allozyme states. At any given locus with segregating alleles producing monomeric enzymes, homozygotes display either slow or fast (SS, FF) bands, whereas heterozygotes (SF) have both bands. Where the coded enzyme is polymeric, (i.e. di-, tri-, tetra-), very complex heterozygous banding patterns may be produced, involving multiple bands (five in the case of a tetramer) (Harris and Hopkinson, 1977; Richardson *et al.*, 1986). Any, or all, of such banding patterns may assist discrimination between species (Berlocher, 1980; Ayala, 1983).

6.3.12 DNA-BASED TECHNIQUES

Introduction

Molecular technology is rapidly permeating many important areas of biology, and insect ecology is no exception. DNA techniques have largely replaced electrophoresis, and other protein-based methodologies, in insect taxonomy for distinguishing between morphologically similar species, for constructing phylogenetic relationships, and for examining the stucture of populations. This is mainly because DNA can provide, reliably, far more information, at all levels, than electrophoresis or any other previous approach (Symondson and Hemingway, 1997, Symondson 2002b). DNA can be extracted from minute samples of insects (Hemingway *et al.*, 1996), and the techniques are sufficiently fast for use in relation to pest management programmes (Loxdale *et al.*, 1996). Useful manuals providing details of basic techniques can readily be found (e.g. Berger and Kimmel, 1987; Sambrook *et al.*, 1989). Much is now known about how different regions of the genome evolve at different rates (Avise, 1994), and, hence, which target genes are most appropriate for examining differences at the level of the species, population or individual (e.g. Moritz *et al.*, 1987; Liu and Beckenbach, 1992; Brower and DeSalle, 1994; Simon *et al.*, 1994; Lunt *et al.*, 1996; Caterino *et al.*, 2000). A range of techniques has been developed, that are appropriate for use in entomological studies (Loxdale *et al.*, 1996; Loxdale and Lushai, 1998). However, application of DNA techniques to the study of trophic interactions is still relatively new, and few of the potential approaches available have been exploited to date. Here, we concentrate on techniques that have proved successful in studies of predation and parasitism, confining ourselves to studies which seek to detect and quantify interactions between organisms, rather than on the genetics of the predators and prey or parasites and hosts. There have been so few studies in this area, to date, that it would be inappropriate to recommend any single protocol. In general, methods for extracting DNA to detect parasitism and predation are the same as those in the many papers dealing with the molecular taxonomy of insects. For other molecular procedures, refer to the specific papers quoted.

Detection of Parasitoids Within Hosts

The rapid detection, identification and quantification of parasitoid early-stages within their hosts is a major concern for practitioners of IPM, who wish to avoid pesticide intervention wherever possible, and who also wish to monitor the spread of newly-introduced biological control agents. Simple dissection is slow, labour-intensive and innacurate, and hence a range of biochemical (immunological and electrophoretic – see above) and molecular techniques have been developed.

One of the simplest, fastest and least expensive molecular approaches is to use **RAPD-PCR (Random Amplified Polymorphic DNA – Polymerase Chain Reaction)** (Micheli *et al.*, 1994; Rollinson and Stothard, 1994). The great advantage of this approach is that short random primers are selected initially, without the need for prior knowledge of target sequences. Although PCR conditions must be optimised, primers can be screened rapidly to find those that are capable of revealing reproducible, species-specific banding patterns on a gel for parasitoids within their hosts.

Black *et al.* (1992) detected parasitoids in aphids (*Diaeretiella rapae* in *Diuraphis noxia* and *Lysiphlebus testaceipes* in *Schizaphis graminum*), and the parasitoid patterns were species-specific. Parasitoid DNA was detectable by six days after parasitism. PCR and arbitrary primers were used to distinguish between eight strains of *D. rapae*, three strains of *Aphidius matricariae*, and three other parasitoid species. Most primers give distinctly different patterns with different species, and this should allow determination of the species involved in parasitism in field collections.

A more precise approach is to use primers for a target sequence within a specific gene. Much attention has been directed at genomic ribosomal gene clusters (rDNA), of which there are hundreds of tandemly-repeated copies per cell. The level of conservation of these regions is such that universal primers are now available for amplifying sequences from a broad range of insects. Within these clusters, internal transcriber spacer regions (ITS1, ITS2) have been shown to reveal useful differences in length and sequence between closely-related species (e.g. Black *et al.*, 1989; Proft *et al.*, 1999; Chen *et al.*, 1999; Taylor and Szalanski, 1999). Amornsak *et al.* (1998) used PCR to amplify the ITS2 regions of the moth pests *Helicoverpa armigera* and *H. punctigera*, and a parasitoid of their eggs, *Trichogramma australicum*. The DNA sequence data obtained from the wasp were then used to create primers that permitted the specific amplification of *T. australicum* DNA from parasitised moth eggs. It was possible to detect parasitism only 12 hours after the eggs were parasitised, and the authors considered that the method is likely to prove useful for evaluating the potential of *T. australicum* for biocontrol of *Helicoverpa* species in Australia. Zhu and Greenstone (1999), and Zhu *et al.* (2000), also used ITS2 regions to distinguish between *Aphelinus albipodus*, *A. varipes*, *A. hordei*, *A. asychis* and *Aphidius colemani*. DNA of *A. asychis*, *A. hordei* and *Aphidius colemani* was detectable in the Russian wheat aphid, *Diuraphis noxia*, as soon as one day after parasitism. This compares well with electrophoresis, which could not detect the parasitoid until more than three days after parasitisation (Walton *et al.*, 1990a).

Other potential targets for molecular detection of parasitism are the 12S and 16S ribosomal RNA genes (rRNA) within mitochondria. With many hundreds of copies per cell, mitochondrial genes, again, potentially provide a large target for PCR. The rRNA genes are known to expose relatively recent evolutionary events (Hillis and Dixon, 1991), and have been shown to be effective at revealing differences between closely-related species of Hymenoptera (Whitfield and Cameron, 1998).

Although a PCR stage increases the chance of detection by amplifying the target sequence, other approaches have been used successfully. Clearly, if PCR can be omitted, this could potentially increase the rate at which hosts might be screened. Greenstone and Edwards (1998) developed a species-specific hybridisation probe to a multiple-copy genomic DNA sequence from the braconid wasp *Microplitis croceipes*, and used it to detect parasitism of *Helicoverpa zea*. Despite the lack of a PCR stage, it could detect first and second instar larvae as efficiently as host dissection, but was not effective for detecting parasitoid eggs. The probe reacted with the congener *M. demolitor*, but not with a representative growth stage of another parasitoid of this host (*Cotesia marginiventris*), or with *H. zea* itself. Developing a hybridisation probe is complex and expensive, but performing the assay is simpler and less expensive than use of ELISA. The authors reported that the whole procedure from host collection to evaluation of results is possible in around 24 hours.

Detection of Prey Remains Within Predators

DNA methods are also starting to be used in predation studies. There is an important role for such methods, which are, potentially, an efficient means of determining predator diet both qualitatively and quantitatively.

The sources of blood meals by haematophagous insects, such as mosquitoes, ticks and crab lice, have been detected using such techniques (Coulson *et al.*, 1990; Tobolewski *et al.*, 1992; Gokool *et al.*, 1993; Lord *et al.*, 1998). Prey DNA in the guts of predators and host-feeding parasitoids is likely to be in a highly degraded state, and so markers based on PCR techniques offer the most promise.

Zaidi *et al.* (1999) demonstrated that multiple-copy genomic DNA sequences, that confer insecticide resistence on mosquitoes, *Culex quinquefasciatus*, could be detected for at least 28 hours after the mosquitoes were fed to carabid beetles, *Poecilus* (=*Pterostichus*) *cupreus*. Detection periods were shown to depend upon the length of DNA sequence, with successful amplifications of 146 bp and 263 bp sequences. Longer sequences could not be amplified reliably from gut samples, suggesting that short sequences survive intact for a longer period during the digestion process. The prey were equally detectable whether the beetles had eaten one mosquito or six, digested for zero or 28 hours. DNA was extracted from the whole predator in this experiment, yet PCR was able to amplify the target sequences from the much smaller prey within the digestive systems of the carabid. Having demonstrated that the 40–50 fold replication of genomic multiple-copy genes allowed clear detection of semi-digested prey DNA within predators, Zaidi *et al.* speculated that the far greater copy number of mitochondrial genes per cell would make these attractive targets. This was confirmed by Chen *et al.* (2000), who used PCR primers to amplify mitochondrial COII sequences (77 to 386 bp) from six species of cereal aphid, and were able to detect aphid remains in the guts of predators. After feeding on a single *Rhopalosiphum maidis* aphid, then on five aphids of the related *R. padi*, a 198 bp fragment, specific for *R. maidis*, could be amplified from 50% of coccinellid predators after 4 hour, and 50% of chrysopid predators after 9 hour. The authors claim this technique to be superior to the use of monoclonal antibodies, in being more certain of success, faster, and less expensive to develop, yet of similar sensitivity and specificity. The overall costs of developing probes and assaying predators from the field appear to be comparable with those for ELISA.

Agustí *et al.* (1999b) also found that detection periods were strongly affected by sequence length. Primers were designed to amplify **sequence-characterised-amplified-regions (SCARs)** derived from a randomly-amplified polymorphic DNA (RAPD) band. Immediately after feeding ten eggs of the target prey, *Helicoverpa armigera*, to the predator, *Dicyphus tamaninii* (Heteroptera: Miridae), SCAR primers could amplify successfully 600 and 254 bp fragments, but not a larger 1100 bp sequence using a third set of primers. After four hours digestion in *D. tamaninii*, only the 254 bp sequence could be detected in 45% of fed predators. In specificity tests, the primers failed to amplify a 254 band from any of the other species tested (five lepidopterans, two whiteflies and two predators), but, in two cases, did amplify sequences of different sizes. SCAR primers were also prepared from DNA of the glasshouse whitefly (*Trialeurodes vaporariorum*), and a short fragment (310 bp) was used to detect whitefly DNA in the gut of *Dicyphus* *tamaninii* (Agustí *et al.*, 2000). At 25°C, 60% of predators were positive four hours after eating ten *T. vaporariorum* nymphs, and the primer detected eggs and adults of this species, but also gave faint bands with other species of whitefly and aphid (Agustí *et al.*, 2000). Hoogendoorn and Heimpel (2001) prepared four primer pairs, of different length, that were specific to the European corn-borer, *Ostrinia nubilalis*. Using the shortest primer, a meal of corn-borer eggs was detectable in the ladybird *Coleomegilla maculata* for up to 12 hours, and longer primers gave shorter detection periods. The authors suggested that such a set of primers could be used to determine the range of times since feeding by predators in the field.

Attempts by Johanowicz and Hoy (1999) to detect 16S mitochondrial genes from *Wolbachia*, eaten by the mite predator *Amblyseius reductus* together with their host, the prey mite *Tetranychus urticae*, failed, probably because the chosen primers were designed to amplify a relatively long, 900 bp sequence. Greatorex (1996) prepared primers to detect segments of DNA of the prey mite *Orthotydeus caudatus* in its predator (the phytoseiid mite *Typhlodromus pyri*) using RAPD-PCR. Some primers amplified a range of mite species, but one pair were specific to *O. caudatus*, which it could detect in samples containing a 100,000th of a mite. The rate of digestion of prey DNA by the predator was not determined.

Future approaches could, potentially, be borrowed from studies of predation in vertebrate systems. PCR amplification of microsatellite markers (i.e. tandemly-repeated short nucleotide sequences) have been used successfully to detect predation in vertebrate systems. Scribner and Bowman (1998) used this method to identify species of waterfowl preyed upon by glaucous gulls. Some remnants of DNA from consumed goslings were estimated to be detectable 8–16 hour after ingestion. Asahida *et al.* (1997) used PCR and restriction analysis of mitochondrial DNA to identify stone flounder (*Kareius bicoloratus*) DNA in the stomach of the sand shrimp (*Crangon affinis*), but for only up to five hours after ingestion. Their method can only detect the presence or absence of target DNA, but it has the potential for use in the simultaneous detection of multiple prey species. Other examples of detection of vertebrate prey remains in predator guts or faeces have been reviewed in Symondson (2002b). DNA-based techniques have been used to detect predation by arthropods on natural populations of prey in the field (Agustí *et al.*, 2003).

6.3.13 FOOD WEB CONSTRUCTION AND ANALYSIS

The methods described above can help both to determine and to quantify predator-prey and parasitoid-host relationships. However, to appreciate fully the significance of these relationships for the organisation and functioning of a community, techniques for integrating and analysing these data are required. Food webs (trophic webs) can be used to map all the trophic interrelationships of interest in a particular system. This gives a more or less static representation of a community (Hall and Raffaelli, 1993), and manipulation experiments are preferable for investigating the dynamics of community organisation (Memmott *et al.*, 1994). Unfortunately, for many communities, it is not feasible to manipulate component species on the scale required, and, in these cases, the detection of patterns in food webs is the next best means of gaining insights into community structure and function. Methods still need to be developed to describe efficiently the more complex and dynamic aspects of food webs (Polis *et al.*, 1996), including the many categories of potential indirect effects (Abrams *et al.*, 1996). Progress is likely to come, initially, from parasitoid webs, because host-parasitoid trophic relationships are more easily quantified than is the case for predators and their prey (Godfray and Müller, 1999; Lewis *et al.*, 2002). Dietary information for predators is usually incomplete, and often reflects the amount of observational effort expended (e.g. the number of prey species of the scorpion *Paruroctonus mesaensis* reached 100 on the 181st survey night, and never reached an asymptote after 2000 person-hours of observation over five years, Polis, 1991). The application to food web research of post-mortem techniques (subsections 6.3.8–6.3.12) for assessing predation is only just beginning (Memmott *et al.*, 2000).

Memmott and Godfray (1994) list a number of different types of food web;

1. "**connectance web**" – only the existence of interactions between host/prey and natural enemy species is recorded (i.e. a topological food web – Yodzis and Winemiller, 1999; Woodward and Hildrew, 2001) (Figure 6.30)
2. "**semi-quantitative web**" – the relative numbers of (or % parasitism by) each natural

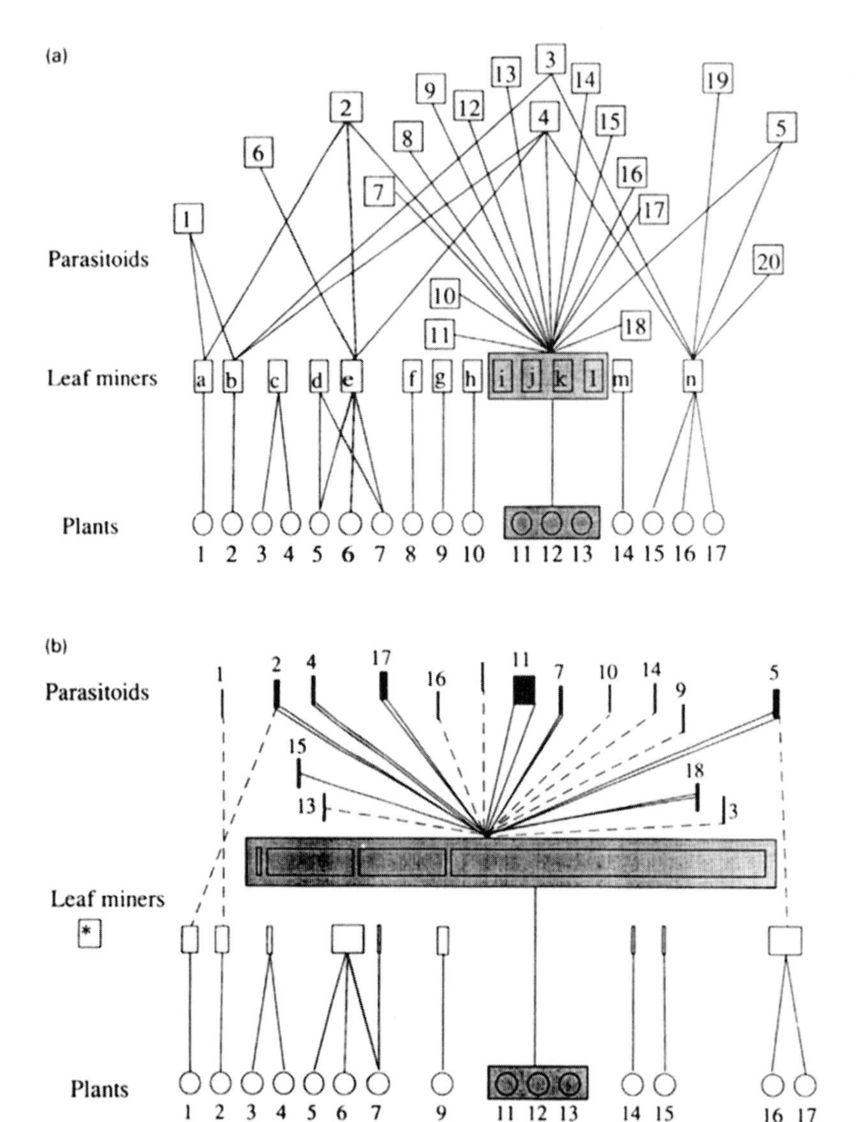

Figure 6.30 Comparison of a connectance web and a quantified web for a plant-hispine-parasitoid food web. (a) connectance web displaying all interactions, (b) quantified web showing the absolute abundance of leaf miners, parasitoids and interaction strength. The first rectangle (*) in the leaf miner level represents 30 miners-parasitoids reared per 1000 m^2. Numbers and letters each refer to different species of plant, leaf miner and parasitoid. Source: Memmott and Godfray (1993). Reproduced by permission of CAB Publishing.

enemy species per host/prey species are recorded (e.g. Dawah *et al.*, 1995)

1. "**quantitative web**" – the densities of host/prey, natural enemies and the strengths (i.e. % parasitism) of host/prey-natural enemy interactions are recorded (e.g. Müller *et al.*, 1999; Schönrogge and Crawley, 2000; Lewis *et al.*, 2002) (Figure 6.30)
2. "**source web**" – is centred on a plant or herbivore and extends to all higher trophic levels (e.g. Hövemeyer, 1995; Memmott *et al.*, 2000).
3. "**sink web**" – is centred on a natural enemy and includes all relevant lower trophic levels (e.g. Schoenly *et al.*, 1991).
4. "**community web**" - is non-centred and includes all species (e.g. Schoenly *et al.*, 1991).

Webs can also be site-specific, time-specific (e.g. for a single crop growth stage), or cumulative (gathered over many occasions within a discrete area). The IRRI cumulative rice web, for example, contains 687 taxa and more than 10,000 trophic links (Schoenly *et al.*, 1996a). Some published webs combine data from localities, seasons or years, ("**composite webs**") (e.g. Dawah *et al.*, 1995), but there is a danger that this may obscure subtle patterns in web structure. Web studies vary in the degree of their trophic resolution (e.g. in some studies facultative hyperparasitoids are not distinguished from other parasitioids, Lewis *et al.*, 2002). Webs may be constructed from either taxonomic species or **trophic species** (i.e. functional groups that contain organisms that eat and are eaten by the same species within a food web, Cohen and Briand, 1984), and less sampling effort is required to characterise webs accurately in the latter case (Martinez *et al.*, 1999). Trophic-species (= **trophospecies**) are currently delimited subjectively, and it would be better to use objective quantitative methods, but the identification of optimal methods to do so is proving to be elusive (Yodzis and Winemiller, 1999).

Web data are useful in a variety of contexts (Memmott and Godfray, 1994), such as: (a) comparison between hosts in different feeding locations, (b) comparison between early- and late-successional habitats, (c) comparison between tropical and temperate regions, (d) study of the determinants of parasitoid host range, (e) as a guide to the probability of apparent competition between hosts (Rott and Godfray, 2000; Lewis *et al.*, 2002) (subsection 7.3.10), (f) to probe the intricacies of community structure and function (e.g. degree of compartmentation, influence of keystone species, role of the frequency distribution of different body sizes, Woodward and Hildrew, 2002), (g) to assess the effect of alien hosts on native parasitoids and native hosts (Schönrogge and Crawley, 2000) and (h) to quantify prenetration of food webs in native habitats by introduced alien parasitoids (Henneman and Memmott, 2002). Comparison of collections of food webs may uncover patterns in the structure of natural communities (Rott and Godfray, 2000). Study of food webs is also likely to guide the development of models into profitable and realistic directions (Cohen *et al.*, 1994; Memmott and Godfray, 1994). Food web data can also be valuable in underpinning other studies. West *et al.* (1998), for example, used knowledge of leaf-miner-parasitoid and aphid-parasitoid-hyperparasitoid food webs in an investigation of the potential mechanisms of horizontal transfer of various strains of *Wolbachia* bacteria, which can cause parthenogenesis in parasitoids (section 6.5). Valladares and Salvo (1999) compared plant – leaf-miner (Diptera: Agromyzidae) – parasitoid webs in a natural area (woodland and grassland), and an agricultural area (containing various crops, including potatoes, brassicas and beans). They considered that exercises of this sort could be useful in identifying parasitoids with potential for the biological control of pests.

Memmott and Godfray (1993) list some useful summary statistics for food webs, which facilitate comparisons (e.g. Schmid-Araya *et al.*, 2002) between webs: (i) number of species and trophic levels, (ii) number of trophic interactions divided by the number of possible interactions (**connectance**) (Warren, 1994, reviews various definitions of connectance), (iii) average **interaction strength** between species (Laska and

Wootton, 1998, identify four different theoretical concepts of interaction strength), (iv) ratios of number of species at different trophic levels, (v) degree of **compartmentation** (i.e. where species interactions are arranged in blocks and within-block interactions are strong but between-block interactions are weaker, see Raffaelli and Hall 1992), (vi) number of species feeding at more than one trophic level (**omnivory**), and (vii) incidence of **trophic loops** (e.g. A attacks B attacks A, Polis *et al.*, 1989). Some of these summary statistics are sensitive to the degree of taxonomic and trophic resolution of the web, but the majority are stable over a wide resolution range (Sugihara *et al.*, 1997). Various types of **robustness analysis** can be used to quantify the extent to which sampling effort may bias these web summary statistics. Lewis *et al.* (2002), for example, investigated the effects of omitting the least frequent linkages from the dataset (84 parasitoid species attacking 93 species of leaf-mining insect in tropical forest in Belize). They concluded that their sampling programme had uncovered the majority of host and parasitoid species in the community under study, but not all the interactions among species. Quantitative webs can give strong indications about which species are likely to be **keystone species** (e.g. species whose interactions dominate the system numerically), and may also be useful for testing the current hypotheses of food web theory (Müller *et al.*, 1999). Not all web types are suitable for estimating the general properties of community webs. Hawkins *et al.* (1997), for example, showed that webs based on single, or few, sources underestimated the number of links per species, compared with full webs, and produced inconsistent estimates of the ratios of number of species at the different trophic levels.

Practical problems to be aware of when constructing a web include: (a) the difficulty of making large collections of uncommon hosts for rearing (Schönrogge and Crawley, 2000); this is a limitation for a connectance web, but less serious for a quantitative web (where interactions between rare prey/hosts and their natural enemies do not have a strong numerical influence), (b) sample bias in relation to parasitized versus unparasitized hosts, and between parasitoid species that take very different times to develop in the host, or bias against collecting parasitoids that attack very young or very small hosts, (c) bias due to differential ease of rearing-out (Rott and Godfray, 2000; Lewis *et al.*, 2002), and (d) failure to sample unexpected host locations (e.g. leaf-miners pupating on the ground) (Memmott and Godfray, 1994). Some of these problems can be overcome by following the suggestions given in subsections 6.3.5 and 6.3.6. Müller *et al.* (1999) stress the importance of obtaining quantitative information specifically for food web construction, rather than adapting data that was previously collected for some other purpose. This directed approach also means that data collection can employ common methodologies, shared between different communities, which will then facilitate more meaningful comparisons between these communities.

It should be pointed out that detection of some of the rarer interactions in the web will only become apparent after very intensive and extensive surveying. For example, Dawah *et al.* (1995) demonstrated the existence of tertiary parasitism in grassland galler-parasitoid communities in the UK, but parasitoids of hyper-parasitoids were recorded for only 63 out of 64,781 observations. Woodward and Hildrew (2001) showed that, in food web studies involving predation, a large number of gut contents samples are needed to achieve an acceptable degree of accuracy. They constructed yield-effort curves for the dominant predators in their study system (an acid stream flowing through a lowland heath) to examine how the cumulative percentage of feeding links detected increased with cumulative number of guts examined. Only species that had reached an asymptote for their feeding links were included in the calculation of food web statistics, and, in most cases, this required several hundred gut contents samples per species.

Space does not permit discussion of all the methodological variants that have been employed in the construction of different types of food web, for various natural enemies and communities. The study of Müller *et al.* (1999), however, will be described as an example of some procedures for sampling and analysis that can be very useful. These authors constructed monthly quantitative parasitoid webs for aphid-parasitoid relationships (based on 71 species of aphid and parasitoid) in an abandoned damp field in England over a period of two years (Godfray and Müller, 1999). Random plant units (usually a whole plant or a grass ramet) in each square of a 20×20 m grid were taken twice a month, and the numbers of aphids and mummies on these were counted. Aphid and mummy densities were calculated after measuring the number of plant units per m^2. Mummies were placed in gelatin capsules until the parasitoids emerged. Monthly webs were visualised in diagrams (Figure 6.31) showing the relative abundance (indicated by bar width) of different aphid species in the centre of the diagram, with the relative abundance of primary parasitoids below, and the relative abundance of secondary parasitoids above that of the aphids, and with wedges indicating the relationships between species. These illustrations were produced with low-level graphics routines in the computer package *Mathematica*. The Dirichlet distribution (Goodhardt *et al.*, 1984) was found to be an effective means for summarising species compositions in the different webs, and showed that there were many rare, and few common, species of both aphids and parasitoids. Parasitoid-overlap graphs (Figure 6.32) were constructed to summarise the patterns of shared parasitism. In these graphs, vertices, representing different

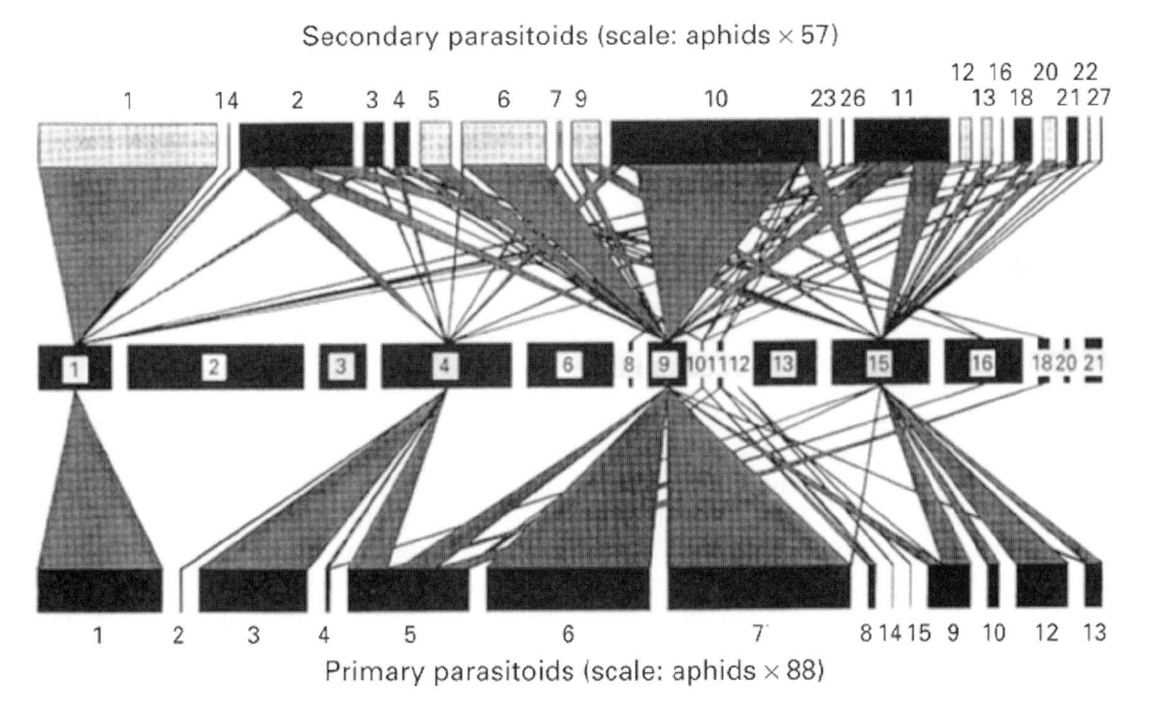

Figure 6.31 Example of an aphid-parasitoid web; July 1994 in an abandoned field in southern England. Relative aphid abundances are shown in the centre with primary parasitoids below and secondary parasitoids above (hyperparasitoids in grey and mummy parasitoids in black). Each number refers to a different species. Species densities are shown to scale. Source: Müller et al. (1999). Reproduced by permission of Blackwell Publishing.

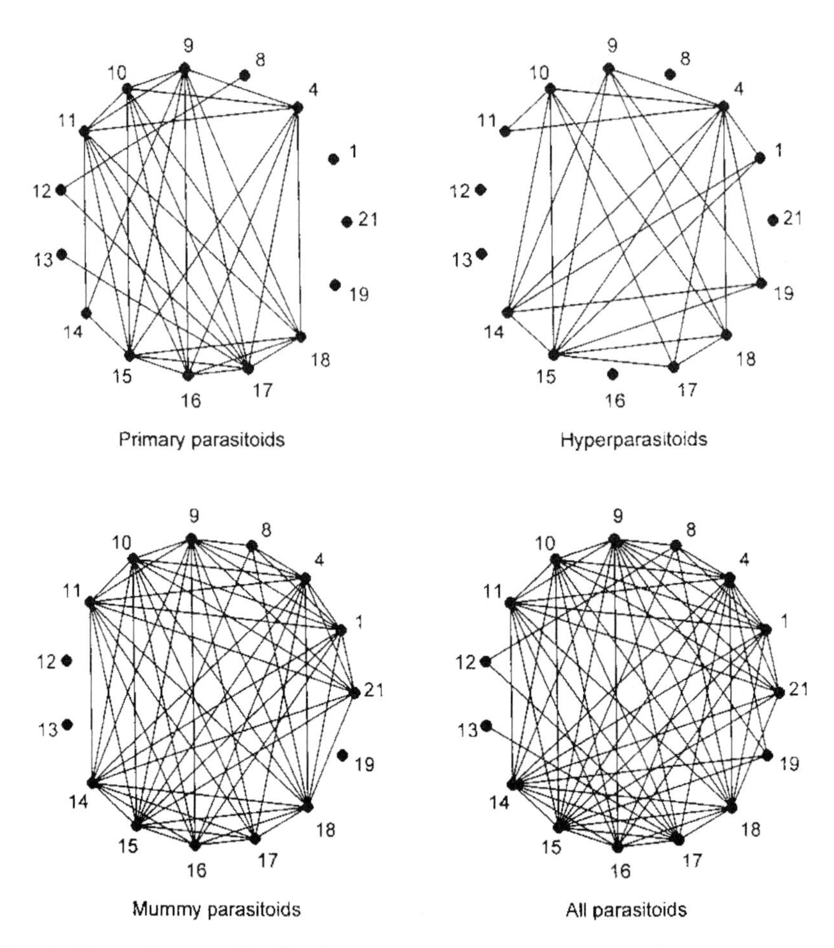

Figure 6.32 Parasitoid-overlap graphs for an aphid-parasitoid web in an abandoned field in southern England. The vertices represent the aphids from which at least one species of parasitoid were obtained. Vertices are joined by edges when the aphids share at least one species of parasitoid. Each number refers to a different species. Source: Müller *et al.* (1999). Reproduced by permission of Blackwell Publishing.

aphid species, are linked by lines if both are attacked by a species of the same category of parasitoid (i.e. primary parasitoids, hyperparasitoids that attack the primary parasitoid within the living aphid, and mummy parasitoids that attack the mummy irrespective of whether it contains a primary or a secondary parasitoid). These graphs illustrate the potential for aphid species on different host plants to interact indirectly through shared parasitoids (e.g. **apparent competition** – Holt and Lawton, 1994) (subsection 7.3.10), and, so, they provide a simple visual indication of the potential cohesiveness of the community. Polyphagous parasitoids may exert more influence on a given host species if alternative host species are present in the habitat (Lawton, 1986; Settle and Wilson 1990), and parasitoid-overlap graphs summarise the extent to which diverse host reservoirs are available to this category of parasitoid. Müller *et al.* (1999) developed a new type of graph (the **quantitative parasitoid-overlap diagram**) to show the relative importance of any one species of aphid as a source of parasitoids to attack other species of aphid (Figure 6.33). In these diagrams, which are modifications of the traditional parasitoid-overlap graph, the size of each vertex circle (again representing an aphid species) is pro-

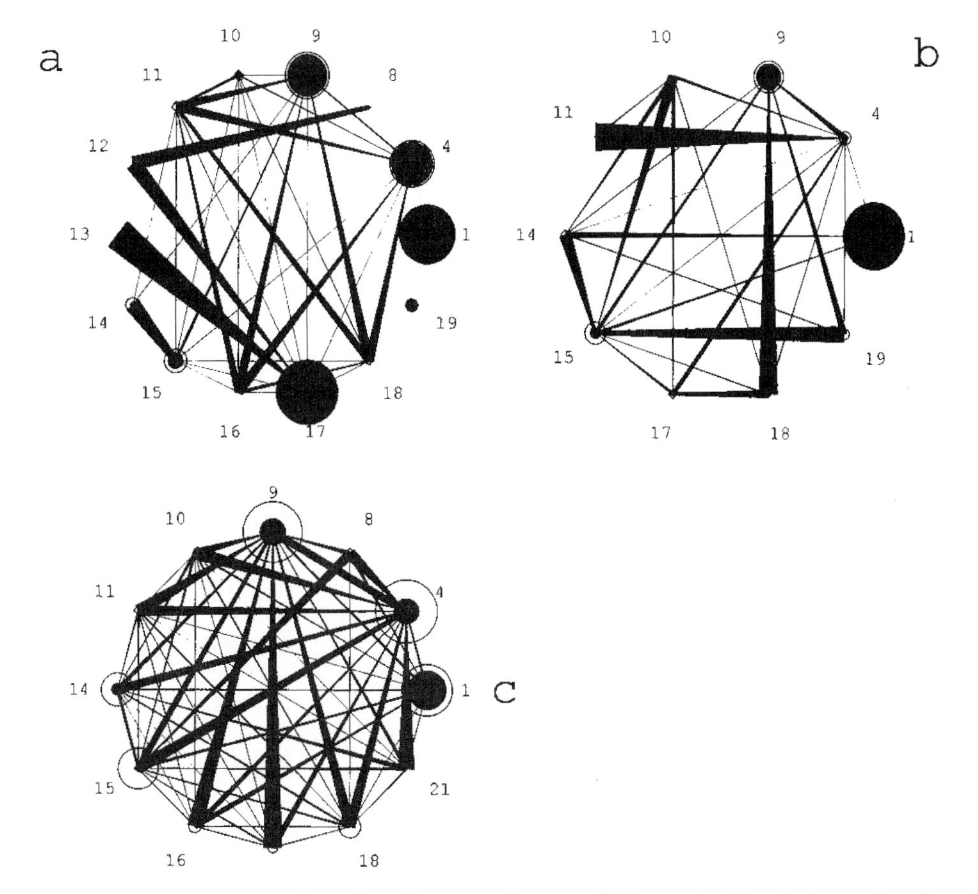

Figure 6.33 Quantitative parasitoid-overlap diagrams for an aphid-parasitoid web in an abandoned field in southern England. The vertices represent aphids and they are linked if the two species share a parasitoid. The size of each vertex is proportional to aphid abundance. The extent to which the vertices are coloured black, and the strength and symmetry of the links, measure the importance of each aphid species as a source of parasitoids attacking other species of aphids and itself. Each number refers to a different species. (a) primary parasitoids, (b) hyperparasitoids, and (c) mummy parasitoids. Source: Müller *et al.* (1999). Reproduced by permission of Blackwell Publishing.

portional to aphid abundance, and the degree of vertex shading and the width of the links are related to the degree to which each aphid species is a source of parasitoids attacking itself or other aphid species. These diagrams were useful in the Müller *et al.* (1999) study, in showing that aphid-primary parasitoid dynamics of the common aphid species were not strongly linked, whereas mummy parasitoids attacking the common aphid species were not dominated by individuals that had developed in those common species (i.e. few aphid species acted as the main source of their own mummy parasitoids). Mummy parasitoids were thus shown to be important agents of linkage in the community between distinct aphid-parasitoid-hyperparasitoid sets. Quantitative parasitoid-overlap diagrams have also been used to display the extent to which leaf-miner hosts (twelve species of *Phyllonorycter* attacking alder, willow, oak and beech) are a source of their own parasitoids, or a source of parasitoids attacking congeners (Rott and Godfray, 2000). Graphs of predator overlap (links joining predators that share prey) and prey overlap (links joining prey that share predators) can also be constructed. In a study of changes in a stream food web after invasion by a new top predator (the dragonfly *Cordulegaster boltonii*), Woodward and Hildrew (2001) found that pre- and post-invasion predator overlap graphs could be represented in one dimension (termed "interval graphs"), but that the prey overlap graphs were non-interval because several species could not be placed in one dimension with other species in the graph. Interval graphs denote a high degree of interconnectedness (in this case a high degree of generalist feeding), which non-linear modelling suggests may stabilise food webs (McCann *et al.*, 1998). In the acid fish-free stream studied by Woodward and Hildrew (2002) all six invertebrate predators ate nearly every invertebrate taxon smaller than themselves. To summarise this system, and its dynamic changes, they presented a series of monthly intraguild food webs and niche overlap webs (the latter based on Pianka's Niche Overlap Index, Pianka, 1973). In these web diagrams circles (representing predator species) varied in area according to predator density, and vertical location of circles was used to indicate the relative body-size of each predator species. Intraguild predation in this system was shown to be strongly asymmetric (with larger predators dominant) and niche overlap decreased as discrepancy in body-size increased.

In spite of important advances in the analysis of parasitoid webs, there remains a need to integrate information about, not only parasitoids, but also predators and pathogens that attack the same species of host (Godfray and Müller, 1999). Interactions between these natural enemy groups (Sunderland *et al.* 1997) could also influence community dynamics. In general, over-simplification of food web descriptions in the past may have misled theorists into making erroneous assumptions and generalisations (Polis, 1991; Hall and Raffaelli, 1993).

Quantifying the degree of compartmentation of food webs can be a useful step towards understanding their dynamics. A keystone predator, for example, can maintain a diversity of prey species by preventing a single dominant prey species from out-competing a range of other prey species, which form the base of a whole compartment of prey, predator and hyperpredator species. Thus, knowledge of compartmentation enables prediction of cascade effects arising from removal or inhibition of keystone predator species (Raffaelli and Hall, 1992). Raffaelli and Hall calculated the trophic similarity between all species in a web using the Jaccard coefficient, $S_{ij} = a/a+b+c$ (where a is the number of species that interact with i and j, b is the number of species interacting with i only, and c is the number of species interacting with j only), and then graphed the frequency distribution of these coefficients. Webs with no compartments exhibit more or less unimodal distributions of trophic similarity, whereas, for compartmented webs, the distributions will be U-shaped, bi- or polymodal. They then carried out an ordination analysis (Hill, 1973) on the similarity matrix, to identify which species comprised the compartment.

Food web methodology can be applied, profitably, in ecotoxicological investigations, as a powerful summary of system changes following pesticide applications. Schoenly *et al.* (1996b) constructed plot-specific food webs for a plot of irrigated rice in the Philippines, sprayed with the synthetic pyrethroid deltamethrin, compared with an unsprayed control plot. Species composition and abundance were determined from vacuum netting (subsection 6.2.2), and trophic relationships were taken from the larger "Philippines cumulative rice web": i.e. 546 taxa and 9319 consumer-resource links at 23 sites (Cohen *et al.*, 1994). Schoenly *et al.* calculated the magnitude and direction of differences in percentage herbivores, and natural enemies of herbivores, on fourteen sampling dates, and the mean food chain length (mean number of links of maximal food chains from basal species to top predator), of sprayed and unsprayed webs for each date. The insecticide was found to cause a dramatic increase in percentage herbivores, and a reduction in mean food chain length. These changes were associated with pest outbreaks.

6.4 GENETIC VARIABILITY IN FIELD POPULATIONS OF NATURAL ENEMIES

Individuals within a population usually vary genetically, and this variation can be expressed both in the insects' morphology, and in a range of biological attributes, such as behaviour (Roush, 1989). Genetic variation may, thus, have considerable influence on the killing efficiency of predators and parasitoids. Genetic variability appears to be the rule, rather than the exception, within animal populations (Ayala, 1983). For example, studies of proteins,using electrophoretic techniques (subsection 6.3.11) have shown that very high levels of allelic genetic variation exist in many populations (McDonald, 1976). Relative assessments of genetic variability within populations can be obtained using molecular techniques, including electrophoretic (protein) markers, and DNA-based markers, especially microsatellites (subsection 6.3.12). Electrophoretic methods still have the advantage of being cheaper, and simpler to use, than DNA-based methods, and the results are often easier to analyse and interpret (Loxdale and Lushai, 1998). Allozyme frequencies and levels of heterozygosity, obtained by electrophoresis, have been used to compare different populations, e.g. of aphid parasitoids (Unruh *et al.*, 1989; Atanassova *et al.*, 1998). However, the use of protein markers requires fresh material, or samples that have been kept deep frozen, whereas DNA markers can be used with old, stored, or dry samples, or samples preserved in alcohol (Loxdale and Lushai, 1998). Microsatellites have been increasing in popularity, but it has been suggested that, because they are assumed to be predominantly selectively neutral, they are not the most suitable technique for population studies aimed at investigating historical patterns of genetic variation and gene flow (Queller *et al.*, 1993; Jarne and Lagoda, 1996). Powell and Walton (1989) review the use of electrophoresis in the study of intraspecific variability in hymenopteran parasitoids, and Loxdale and Lushai (1998) review the molecular marker techniques currently available.

Most populations are patchily distributed, due to discontinuities in the distribution of food, or other essential resources. When investigating the population dynamics of a predator or parasitoid, it is important to determine whether one is dealing with a local population or a metapopulation. In a **local population**, although its distribution may be discontinuous, individuals move regularly between patches, and the population is largely redistributed every generation (Taylor, 1991). In other words, there is considerable interaction between individuals throughout the population. A **metapopulation** comprises a series of local populations between which movements of individuals are infrequent, and random, so that the local populations have largely independent dynamics (Taylor, 1991). Hence, there is a high rate of genetic exchange between groups of individuals occupying separate habitat patches within a local population, but a low rate of genetic exchange between local populations. Analysis of molecular marker frequencies within, and amongst, samples of individuals collected from discrete habitat patches or geographic areas, or from different host species (in the case of parasitoids), could provide indirect evidence of the levels of genetic exchange between populations. Identical, or very similar, frequency scores would suggest high rates of genetic exchange and *vice versa*.

However, some caution is necessary, because differences in allozyme or other marker frequencies may not always be indicative of low rates of movement between populations. To illustrate this, let us consider a hypothetical parasitoid attacking three different host species, which occupy different local habitats within the same area (e.g. an aphid parasitoid attacking pest aphids on two different crops, and a non-pest aphid on plants in neighbouring hedgerows). When examined electrophoretically, allozyme frequencies are consistently different between samples of parasitoids collected from different hosts, some allozymes only appearing in samples from one of the hosts. This might suggest that discrete parasitoid populations exist on each host species, with very little movement of individuals between hosts. However, the same result would occur if only individuals with certain genotypes moved regularly between host species. Allozyme frequency differences would still be maintained, despite high rates of movement between populations. Powell and Wright (1992) have suggested that this situation exists for some oligophagous aphid parasitoids, their populations comprising a mixture of specialist and generalist individuals. The genotype of a generalist allows it to recognise and attack all the host species, whereas that of a specialist allows it to recognise and attack only part of the host range, possibly a single species. This hypothesis is consistent with the concept of **population diversity centres**, proposed for aphid parasitoids by Němec and Starý (1984, 1986), who advocate that the 'main', or original, host of a parasitoid, is the one on which it shows the greatest degree of genetic diversity. On this basis, oligophagous, and evolutionarily 'young', parasitoid species, will show the greatest genetic variability, whilst monophagous, and evolutionarily 'ancient' species, will show the least (Němec and Starý, 1983). Bossart and Prowell (1998) discuss the limitations of current molecular methodologies for estimating population structure, especially advocating caution in the interpretation of data provided by these indirect methods when estimating gene flow. They also recommend, strongly, that indirect molecular methods should be more closely linked with direct assessments of insect movements between populations. Neigel (1997) reviews strategies for estimating gene flow from genetic markers.

There have been very few studies of the population structure or gene flow among local populations of natural enemies (Hopper *et al.*, 1993; Roderick, 1996). Between 1978 and 1986, numerous releases were made, in the USA, of the ladybird beetle *Harmonia axyridis*, which was collected originally from several different locations in Asia (Krafsur *et al.*, 1997). These introductions appeared to have failed, since post-release surveys did not produce any recoveries. However, the beetle has since reappeared at a range of locations, many of which are far removed from the original release sites. Krafsur *et al.* (1997) used gel electrophoresis to estimate genetic variability and gene flow amongst eight widely-separated populations of *H. axyridis*. They found a high level of gene diversity in each sample, which they considered to be indicative of there having been no substantial population bottlenecks during the beetles' establishment, whilst genetic similarities among the widely-distributed populations argued for a single source, and high levels of gene flow. Theoretically, a high dispersal rate, especially in heterogeneous environments, is a requisite for a useful biological control agent (Murdoch, 1990; Roderick, 1992 see subsection 7.4.3), implying that successful natural enemies should remain panmictic. Vaughn and Antolin (1998) used heritable Random Amplified Polymorphic DNA, (RAPD)-PCR, in conjunction with Single Strand Conformation Polymorphism (SSCP) analysis, to investigate the local population structure of the aphid parasitoid *Diaeretiella rapae* attacking hosts on cereal and brassica crops. In contrast to the ladybird example above, the results indicated that the parasitoid population was subdivided, with most variation occurring between fields in an area separated by less than 1 kilometre, rather than between areas separated by as much as 23 kilometres. Estimates of effective migration rates between populations were low, suggesting that there was little gene flow between populations in the different fields sampled. Further studies are needed, especially to relate dispersal rates and gene flow to host population densities. It is likely that parasitoids remain in restricted local areas when hosts are abundant, but that females, in particular, disperse more readily at low host densities.

Understanding the genetic structure of predator and parasitoid populations, and the effects of ecological factors, such as habitat fragmentation, on gene flow and variability, is important, because the behavioural traits that determine the effectiveness of biological control agents are often genetically variable. Roush (1989, 1990) and Hopper *et al.* (1993) discuss genetic considerations in the use of entomophagous insects as biological control agents. The genetic variability of a range of parasitoid behavioural traits has been demonstrated in a number of laboratory studies. Traits that have been shown to be variable, genetically, include sex allocation for *Nasonia vitripennis* (Parker and Orzack, 1985; Orzack and Parker, 1990), the ability to evade encapsulation for *Leptopilina boulardi* (Carton *et al.*, 1989), odour-conditioned ovipositor probing behaviour for *L. boulardi* (Pérez-Maluf *et al.*, 1998), the circadian rhythm of locomotory activity for *Leptopilina heterotoma* (Fleury *et al.*, 1995), the area searched per unit time for *Trichogramma brassicae* (Wajnberg and Colazza, 1998), and patch-time allocation for *Telenomus busseolae* (Wajnberg *et al.*, 1999) (see Chapter 3). Results of the study using *T. busseolae* indicated that there was genetic variability in the response of female parasitoids to information acquired during intra-patch foraging, which affected patch time allocation.

Many of the studies of genetic variability in biological traits in insects use iso-female lines (subsection 3.2.3). Statistical analysis of the variation observed within, and between, lines provides an indication of heritability, and the existence of significant variation in the trait. Each iso-female line is established from the progeny of a single mating pair, and they are especially useful for studies on species having a high reproductive rate and short generation time (e.g. many parasitoids), because this allows the rapid expansion of 'families' into large populations (Hoffmann and Parsons, 1988). The use of iso-female lines can provide an heritability estimate, but it is advised that lines are tested within five generations after they are set up, and they should be maintained at reasonably large (>50) population sizes (Hoffmann and Parsons, 1989).

Although the existence of intraspecific genetic variation has long been recognised, considerable confusion still exists over infraspecific terminology, with 'subspecies', 'variety', 'race', 'strain' and 'biotype' all being widely used, but rarely defined clearly (Berlocher, 1984). Gonzalez *et al.* (1979) proposed that the term **biotype**, alone, should be used to designate genetic variants in parasitoids, but Claridge and den Hollander (1983) argued that the term has no biological basis, and is, therefore, redundant. Because of this confusion in terminology, it is obviously important to describe, and quantify, any intraspecific variability with precision, when investigating population structure.

6.5 *WOLBACHIA* BACTERIA (AND OTHER INTRACELLULAR SYMBIONTS) AND PARASITOIDS AND PREDATORS (by R.D.J. Butcher and M.A. Jervis)

6.5.1 INTRODUCTION

In addition to a wealth of gut-associated microbes, arthropods, including insect parasitoids and predators, are infected by a diverse array of intracellular symbionts that broadly encompass the continuum between pathogenic parasite through to obligate mutualist. Whilst some of these symbionts may be restricted to host-derived specific tissues, (e.g. the primary symbionts of aphids [*Buchnera*] and other phloem-feeders within mycetocytes [Buchner, 1965]), others are more diversely spread across arthropod hosts. In contrast to the 'mycetocyte symbionts' like *Buchnera*, these 'guest symbionts' (Ebbert, 1993) are able to undergo horizontal transmission and colonise a more diverse array of insects in different food webs. *Wolbachia* bacteria are possibly the most widespread of these 'guest symbionts' with repeated estimates suggesting that around 20–30% of all insect species are at least partially infected (Werren *et al.*, 1995a,b; Cook & Butcher, 1999; Werren & Windsor, 2000), representing 36–40% of all insects when weighted for insect order/family diversity. However, if the *wsp* gene (see below) is truly *Wolbachia*-specific, unweighted infection estimates may be as high as 76% of insects (Jeyaprakash and Hoy, 2000). *Wolbachia* are not, however, restricted to insects and infect a diverse but patchy range of arthropods and nematodes, including insect parasitoids and their hosts (see reviews by Werren, 1997; Cook and Butcher, 1999; Stouthamer *et al.*, 1999b; Werren, 2000).

Coupled with the diverse array of phenotypes *Wolbachia* seem to display within their arthropod hosts ranging from pathogen through reproductive parasite to mutualists, they have rapidly assumed the position of being a general arthropod symbiont model (e.g. O'Neill *et al.*, 1997; Werren, 1997; Stouthamer *et al.*, 1999b; Werren, 2000; Bourtzis and Miller, 2003).

Wolbachia are a monophyletic clade and proposed genus (type species: *Wolbachia pipientis* Hartig and Wolbach, from germ line tissue of *Culex pipiens*), comprising obligatory intracellular, cytoplasmically-inherited Gram-negative bacteria (Rickettsiae in the α-Proteobacteria).

The transmission mode of any symbiont will greatly affect its biology and dynamics. For *Wolbachia*, whilst excluded from sperm and being maternally transmitted through the oöplasm to the progeny (**vertical transmission, VT**), several lines of evidence suggest that significant stable **horizontal transmission (HT)** (i.e. transmission between different species of host) of *Wolbachia* has occurred (reviewed in Werren, 1997; Cook and Butcher, 1999; Stouthamer *et al.*, 1999b). The presence of closely related *Wolbachia* isolates within phylogenetically diverse host arthropods (but not sibling species) suggests that HT was (and may still be) long-term evolutionarily important, if not essential (*Wolbachia* infections are maintained for shorter periods than the host speciation rate), with pure VT an evolutionary dead end. However, such assumptions are based upon the absence of either significant recombination relative to gene divergence rates, or host-dependent selection upon the gene sequences used. That aside, the current and future optimal evolutionary transmission routes are unknown but are probably dependent upon the balance between, on the one hand, the degree of recent host coevolution (and genome diminution/genetic drift) and, on the other hand the probability of future recombination (which requires **horizontal transfer** – i.e. the initial stage of HT involving transfer into somatic tissue and/or haemocoel – of different isolates into the same host).

Although *Wolbachia* always occur in their host's reproductive tissues, as expected and required in an obligatory VT symbiont, at least some somatic tissues are infected in a wide range of insects (e.g. Dobson *et al.*, 1999; Kondo *et al.*, 1999; Butcher *et al.*, 2004a), as are the egg cells of some spiders (Oh *et al.*, 2000). Therefore, in contrast to the type species' host, *Culex pipiens*, somatic tissue infections by *Wolbachia* may be widespread if not general and will likely play no role in VT, at least in adult somatic tissues. However, somatic tissue-derived *Wolbachia* are equally capable of experimental transfection of new hosts as germ line-derived *Wolbachia* (Butcher *et al.*, 2004a), and thus are potential sources for horizontal transfer vectors. Additionally, *Wolbachia* can survive for prolonged periods outside the host intracellular environment as long as it is protected from prolonged exposure to elevated hyperosmotic (>500 mOsM) or temperature (>24°C) stress (Butcher *et al.*, 2004b), allowing a potentially broad range of vectors.

Whether this is the general case for other vertically transmitted intracellular arthropod symbionts is currently unknown, but in general, for horizontal transfer, and thus HT, to occur between species an ecological interaction between the different species is required. Therefore, organisms within a feeding community are likely candidates for HT of symbionts such as *Wolbachia* due to their frequent and close associations with one another and with potential vectors. Moreover, the close developmental associations between insect parasitoids and their hosts make parasitoids particularly attractive candidates for HT vectors (Werren *et al.*, 1995a,b). For example, in contrast to other natural enemies such as spiders, mantids, scorpion flies or robber-flies, in the case of endoparasitoids the parasitoid embryo or larva can die within the host allowing host survival and development and thus transfer of symbiont from parasitoid to host as well as *vice versa* (see subsection 6.5.3, below).

Currently, little is known about the evolutionary and population biology of *Wolbachia*, such as how and what limits the range and rate of HT (see below for stages), to what extent host nuclear and/or cytoplasmic components limit these actions of *Wolbachia*, how this has (co)evolved, and the frequency and importance of recombination between isolates. However, with genome diminution and coevolution with the host a likely occurrence, it is clear that HT of *Wolbachia* will be important in its biology and that, in conjunction with other work, studies of *Wolbachia* within food webs and sub-webs will be of considerable value in elucidating some of these facets of *Wolbachia* biology including its relative stable HT rates, potential vectors and consequently recombination rates.

This section is mainly concerned with the detection and identification (isolate 'typing') of *Wolbachia* in herbivore-parasitoid food webs, and how the data obtained can shed light on whether stable transmission of the bacteria among host insects (including the hosts of the parasitoids, and the parasitoids themselves) has been largely vertical, or horizontal, over evolutionary time. However, *Wolbachia* are not only representative of just one clade (genus) of a diverse array of arthropod intracellular symbionts, but they also may interact (cooperatively or competitively) with other symbionts within the same host or population. Given that host-parasitoid interactions tend to vary geographically and temporally, the distribution and population dynamics of *Wolbachia* infections may consequently depend in part upon the infection histories (including HT) of other symbionts, at least geographically or temporally. Therefore, in this section we also urge studies to be broadened to concurrently include the infection frequencies and HT of other arthropod intracellular symbionts, as well as *Wolbachia* as a model system, in food webs and to include several geographically distinct sample sites of each food web studied. Although from hereon we refer principally to *Wolbachia*, the same principles apply to other VT symbionts.

For any largely VT symbiont, significant coevolution between the symbiont and its arthropod host is expected to occur, leading to a high degree of congruency between symbiont and host phylogenies. In contrast, if largely HT (i.e. transmitted by the parasitoids themselves or *via* other vectors), then a low degree of congruence between symbiont and insect (herbivore and associated parasitoids) phylogenies would be expected (Cook and Butcher, 1999). Correspondingly, a high degree of congruence between the symbiont phylogenies of the hosts and the parasitoids would be expected if parasitoids were the principal HT vectors, but this would decrease as the role of other vectors became more significant. Thus comparisons of symbiont and host molecular phylogenies should shed some light on the principal transmission mode and may detect potential HT vectors.

Note that by using comparative molecular phylogenies this approach principally evaluates stable transmission methods, essentially ignoring transient infections and transfers, so other processes may compound it. For example, recombination between transient multiple infections may lead to the apparent phylogenetic appearance of *de novo* and/or multiple infections on the one hand, or allow efficient HT to new arthropod hosts on the other hand. However, it is important to distinguish between horizontal transfer, the initial stage of HT that laboratory experiments have revealed parasitoids to be vectors for (Heath *et al.*, 1999; Huigens *et al.*, 2000; Butcher *et al.*, 2004c,d) and HT, which can be detected phylogenetically. Simplistically, HT comprises several discrete stages, broadly classed as:

1. Initial horizontal transfer by the vector to a new host, presumably usually into somatic tissue or the haemocoel.
2. Infiltration (infection) of host gametogenic tissue, which, for example, requires crossing cellular membranes and, perhaps before then, avoidance of the new host immune system in the haemocoel *en route* to germ line tissue (although prior haemocyte infection will limit this).
3. Acquisition of efficient VT, including the ability both to correctly proliferate in germ line tissue and to segregate into the gametes so as to pass the infection on to each new generation.
4. (Assuming HT is rare) induction of altered host reproductive state or inclusive fitness

to a sufficient level to enable invasion of (spread within) the new essentially uninfected host population. Even so, this may only be possible in localised or substructured populations, such as by Bartonian waves (e.g. Turelli and Hoffmann, 1999) in order for infections to spread deterministically (Caspari and Watson, 1959).
5. Successful competition or cooperation with sympatric intracellular symbionts within the new host to avoid exclusion. In the case where the pre-existing symbiont(s) is (are) the same genus, this may entail recombination to form new isolates.

Failure at any one or more of these stages, which may well be host and/or *Wolbachia* isolate-dependent, will prevent the *Wolbachia* infection spreading into the new host population and thus prevent stable HT and its subsequent detection. Consequently, although the magnitude of any host-dependent, or host-*Wolbachia* co-dependent, influences upon each of these stages is currently unknown, HT is likely to represent only a small but variable subset of actual horizontal transfer events (Butcher *et al.*, 2004a,d).

Although this section is restricted to studies of *Wolbachia* in food webs to elucidate the principal transmission mode and potential vectors for HT, including insect parasitoids, there is, of course, a wide array of other valuable *Wolbachia* (and other symbionts in general) -based research topics of particular relevance to readers of this book which, along with references to practical protocols, include:

(1) **Parthenogenesis induction** (**PI**) in the parasitoid host (Stouthamer *et al.*, 1990, 1993; Pijls *et al.*, 1996; but importantly see also Zchori-Fein *et al.*, 2001);

(2) Interactions between *Wolbachia* and the parasitoid's or its victim's sex ratio including **cytoplasmic incompatibility** (**CI**), male-killing and feminisation of genetic males (Bordenstein and Werren, 1998; Vavre *et al.*, 2000, 2001; Jiggins *et al.*, 2000, 2001a; Hiroki *et al.*, 2002; Kageyama *et al.*, 2002; Sasaki *et al.*, 2002; Jiggins, 2002a);

(3) Mating strategies and parasitoid sex determination systems. For example, PI-*Wolbachia* may be excluded from species with complementary sex determination (Cook and Butcher, 1999), or select against multiple matings, in contrast to the possible selection for multiple matings under CI-*Wolbachia* (Stouthamer *et al.*, 2001). Additionally, potential female lekking in extreme female-biased population sex ratios due to male-killing *Wolbachia* has been reported (Jiggins, 2002a);

(4) Altered gene flow between sympatric and allopatric populations potentially leading to biotypes/ecotypes, cytoplasmic sweeps (incongruence between mitochondrial and nuclear DNA sequence based phylogenies), including its possible role in speciation (Sinkins *et al.*, 1997; Werren, 1998; Turelli and Hoffmann, 1999; Ballard, 2000; Bordenstein *et al.*, 2001);

(5) Interactions with other "selfish" genetic elements within the host population (e.g. Cook and Butcher, 1999; Rokas, 2000);

(6) The effects of *Wolbachia* infection on host fitness parameters (including sex ratio): (Hsiao, 1996; Bordenstein and Werren, 2000; Fleury *et al.*, 2000; Dedeine *et al.*, 2001; Tagami *et al.*, 2001; Grenier *et al.*, 2002; Stouthamer and Mak, 2002; Wenseleers *et al.*, 2002; Abe *et al.*, 2003).

Additional protocols or potential interest to readers include: (a) enrichment and experimental transfection by microinjection of *Wolbachia* (Braig *et al.*, 1994; Chang and Wade, 1996; Grenier *et al.*, 1998; van Meer *et al.*, 1999b; Pintureau *et al.*, 2000; Butcher *et al.*, 2004a,b); (b) establishing *Wolbachia* in insect tissue cultures (Dobson *et al.*, 2002; Noda *et al.*, 2002); and (c) spatio-temporal dynamics of within-insect infections using quantitative PCR (Ijichi *et al.*, 2002).

6.5.2 DETECTION AND IDENTIFICATION OF *WOLBACHIA* IN PARASITOIDS

There are three main methods for detecting *Wolbachia* and other symbionts in herbivores and parasitoids, although only the latter two used together are useful for more than a preliminary investigation.

Experimental Manipulation

The presence of bacteria including, but not diagnostic for, *Wolbachia* in a parasitoid wasp species can be inferred when, after treatment with *Wolbachia*-sensitive antibiotics (e.g. tetracycline, rifampicin or sulphamethoxazole) but not *Wolbachia*-insensitive ones (e.g. kanamycin, gentamycin, erythromycin and penicillin / ampicillin), significant alteration to the progeny sex ratio occurs, either directly, or following the four possible crosses between infected and uninfected parents. For example, thelytokous wasps produce sexual females and males and are therefore 'cured' (not necessarily permanently) of their thelytoky (e.g. see Stouthamer *et al.*, 1990, 1993; Pijls *et al.*, 1996), or treated females produce a reduced egg hatch rate and/ or female-biased sex ratio when mated to untreated males compared to the other three possible crosses (Vavre *et al.*, 2000, 2001). However, this type of approach alone is equivocal, as it: (a) does not exclude other bacteria that may be the true causative agent(s) of the phenotype (e.g. Zchori-Fein *et al.*, 2001) or are co-required for stable HT; (b) does not distinguish single from **multiple infections**; (c) is restricted to host species that can be cultured; and (d) will overlook *Wolbachia*-host pairings that have less extreme effects (e.g. Bourtzis *et al.*, 1998; Hariri *et al.*, 1998). Consequently, it is at best restricted to preliminary screening of thelytokous species alongside diagnostic detection (see below).

Microscopy

Thin sections of host tissues (principally ovaries from recently eclosed females) are fixed and stained with either: (a) bacterial stains such as extensive Giemsa, lacmoid (followed by bleach treatment) or carbol-fuschin/malachite green (e.g. Welburn *et al.*, 1987; Bourtzis *et al.*, 1996), or (b) the DNA stains DAPI and or Hoescht-33258 (O'Neill and Karr, 1990) following fixation and removal of RNA (Butcher *et al.*, 2004b). However, it is important to note that, used alone, this is not diagnostic for *Wolbachia*. Whilst, for example, the presence of only small cocci or rod-shaped microbes would exclude morphologically unrelated symbionts such as protists (microsporidia) and Mollicutes (spiroplasma), these morphologies could also belong to a diverse array of other intracellular bacteria, including for example other Rickettsiae, the parthenogenesis-inducing CFB (Zchori-Fein *et al.*, 2001) and the "son-killer" *Arsenophonus* (Huger *et al.*, 1985). Furthermore, it also does not discriminate between different *Wolbachia* isolates, and thus single from multiple infections. Indeed, whilst multiple *Wolbachia* infections within a host may be fairly common (e.g. Jeyaprakash and Hoy, 2000; Malloch *et al.*, 2000; Reuter and Keller, 2003), the presence of multiple infections of different symbionts within the same host insect also appears to be polymorphic but fairly common within populations (R.D.J. Butcher, unpublished). Diagnostic specificity is thus still absolutely required and, besides PCR-based DNA amplification and sequencing (see below), may be attained by incorporating immunocytochemistry with anti-*Wolbachia* antibodies (e.g. anti-*wsp*; Kose and Karr, 1995; Dobson *et al.*, 1999) or by *in situ* hybridisation with fluorescent or radiolabelled RNA probes derived from *Wolbachia*-specific DNA. Indeed, the use of differently labelled isolate-specific (e.g. *wsp* gene) DNA probes may allow quantification and localisation of different *Wolbachia* isolates within multiply infected tissues. However, in contrast to real-time PCR (see below), when coupled with z-plane confocal microscopy these methods both: (a) allow the spatial quantification of *Wolbachia* loads within cells or embryos (e.g. O'Neill and Karr, 1990), which in some cases may be more relevant than the average cell or tissue *Wolbachia* densities (e.g. Clark *et al.*, 2003); and (b) prevent the registering of false positives due to the presence of PCR amplifiable (degrading) bacterial DNA in the absence of an actual infection.

Note, however, that for arthropods with multiple symbiont infections the *Wolbachia* isolate(s) must be segregated from each other and the other symbionts in order to ascertain, via Koch's postulates, a definitive role for *Wolbachia* in the observed phenotype as opposed to the other symbionts. In the laboratory, segregation of multiple infections may be aided by induced diapause (Perrot-Minnot and Werren, 1999), exposure to differential antibiotic doses, elevated temperatures, or limiting titre transfections (Butcher *et al.*, 2004b). However, for samples that cannot easily be adapted to laboratory culture, this is restricted to the existence of naturally occurring infection polymorphisms at a frequency that will allow the isolation of sufficient single infected individuals. The draw back here is that if single infections occur in field populations at low frequencies, the sample sizes required to attain sufficient samples of each infected class may become both time- and resource-limiting.

Polymerase Chain Reaction (PCR) Amplification and DNA Sequencing

With the current inability to make axenic cultures (outside an intracellular environment) of most intracellular microbes including *Wolbachia*, DNA-based PCR amplification by *Wolbachia*-specific primers is the most widely used method for detecting *Wolbachia* isolate(s) in parasitoids and other insects. Coupled with sequencing of the cloned PCR amplicon(s) it is the only method that leads to accurate identification of each *Wolbachia* isolate. Classically, sequence divergence of the 16S rDNA gene is used to define clades and, especially, genus status of uncultureable microbes (e.g. Weisberg *et al.*, 1991; Roux and Raoult, 1995), and this is also the case for *Wolbachia*. However, beyond genus/clade identification, the 16S and 23S rDNA genes (O'Neill *et al.*, 1992; Rousset *et al.*, 1992) are too conserved, and (in the same operon) the 5S rDNA gene and associated SR2 region (van Meer *et al.*, 1999a) are too small, for phylogenetic discrimination between different *Wolbachia* isolates. Consequently, these have fallen out of favour, whilst the paucity of information, due to few studies, on the diversity of the *gro*-E (Masui *et al.*, 1997), *dna*-A and flanking gene regions (Bourtzis *et al.*, 1994; Sun *et al.*, 1999) has, unfortunately, tended to exclude these gene fragments from initial selection. Therefore, albeit if somewhat circuitously, the *fts*-Z (Holden *et al.*, 1993; Werren *et al.*, 1995a), and the highly polymorphic *wsp* (Braig *et al.*, 1998; Zhou *et al.*, 1998) gene fragments are currently the principal or sole sequences used. Note that some investigators have employed both (e.g. see Rokas *et al.*, 2002; Kittayapong *et al.*, 2003), although it is possible that this situation may change after the sequencing of the complete genomes of several different *Wolbachia* isolates, but it seems likely that *wsp* will remain as the favoured gene fragment due to its highly polymorphic nature (H.R. Braig, personal communication). However, the nature of the selection pressure on *wsp* is unknown and may be partially host-dependent (Braig *et al.*, 1998; Jiggins *et al.*, 2001b; Jiggins, 2002b; Jiggins *et al.*, 2002) or lie within a recombination hot-spot, whilst its specificity for the *Wolbachia* clade has been questioned (R.D.J. Butcher and J.M. Cook, unpublished). Therefore, the validity of *wsp* for comparative phylogenetic analysis, especially for ascertainment of HT, is unclear (indeed, it may be compromised) and should always be used in conjunction with other gene fragments.

That aside, PCR amplification, amplicon cloning and sequencing also allow the detection and resolution of multiple *Wolbachia* infections within a single host (e.g. Vavre *et al.* 1999; Malloch *et al.*, 2000; Jamnongluk *et al.*, 2002; Reuter and Keller, 2003; Van Borm *et al.*, 2003), and have been used to estimate *Wolbachia* infection levels by semi-quantitative methods (see Sinkins *et al.*, 1995). However, the technique has been largely superseded by real-time quantitative fluorescent PCR using molecular beacon probes (or dsDNA intercalating fluorochromes such as Cyber Gold) to obtain estimates of average cell or tissue *Wolbachia* densities. Although sensitive PCR methods for amplification of *Wolbachia* are being developed (e.g. Jeyaprakash and Hoy, 2000; Kittayapong *et al.*, 2003; Butcher *et al.*, 2004b), care must be taken with the interpretation of both apparently uninfected and multiply infected tissues. Apparent absence of infection may be due to the existence of infection levels below the threshold of the PCR (which hence must be evaluated and stated), a not insignificant occurrence in field samples from populations exposed to elevated temperatures, diapause, natural antibiotics (e.g. Pijls *et al.*, 1996; Van Opijnen and Breeuwer, 1999; Perrot-Minnot and Werren 1999; Giorgini, 2001; Hurst *et al.*, 2001) or competing symbionts (e.g. microsporidia; R.D.J. Butcher, unpublished). With apparent multiple infections it is necessary to consider first the possibility of artefacts being introduced by either differential amplification efficiencies between *Wolbachia* isolates, or by PCR-mediated recombination (Malloch *et al.*, 2000), and then additionally the role of actual recombination leading to the co-presence of recombinants and thus higher levels of apparent multiple infections (Malloch *et al.*, 2000; Reuter and Keller, 2003) (see **Phylogenetic Tree Construction** in subsection 6.5.3, below).

Note also that residual *Wolbachia* DNA (e.g. derived from degraded xenogenic tissue, such as host tissue in an enclosing parasitoid or predator) in the absence of bacteria (i.e. the absence of a *Wolbachia* infection) may still be sufficient to create a PCR amplicon and thus a false positive result. Samples determined to be positive by PCR amplification should therefore be validated either by microscopic examination (see above) or, for laboratory cultures, by confirming that the infection is inheritable (detectable over another generation).

With the potentially large numbers of samples from a food web to screen, cloning and sequencing all *Wolbachia* positive amplicons is likely to be heavily demanding upon resources and time. However, screening for the presence of *Wolbachia* and primary sub-typing to determine whether multiple or different *Wolbachia* isolates are present may be combined using sensitive PCR amplification (e.g. Jeyaprakash and Hoy, 2000; Kittayapong *et al.*, 2003; Butcher *et al.*, 2004b) followed by restriction enzyme digestion and resolution (RFLP) by Single Strand Conformation Polymorphism (SSCP)-, or Denaturing Gradient Gel Electrophoresis (DGGE) -PAGE. Sequencing is thus reduced to only novel isolate(s), and these then become additional reference standards. The choice of DGGE or SSCP resolution of the PCR-RFLP sub-typing (e.g. Hein *et al.*, 2003) will probably depend upon the symbiont(s) and gene fragment examined. The resolution of PCR-RFLP fragments by DGGE (and its sister methods TGGE and CDGE) requires prior knowledge of the gene sequence to design optimal GC clamps, specific long GC primers, and considerable development to fully optimise resolution (as well as, ideally, specialised DGGE-PAGE tanks). However, for *Wolbachia* using the *wsp* gene reliable discrimination between amplicons at the 1% sequence divergence level using the whole (~580bp) *wsp* amplicon has been attained (R.E. Ellwood-Thompson, personal communication), and this may well be improved by prior appropriate RFLP analysis. Thus PCR-RFLP-DGGE using *wsp* may offer a real alternative to cloning and sequencing all positive amplicons for *Wolbachia*.

However, if other symbionts or gene fragments are included, the time and resources required to optimise DGGE resolution for multiple symbionts and genes are likely to be highly restrictive and offer little advantage to direct sequencing (or T-RFLP). In these cases, or until RFLP-DGGE conditions are established for all arthropod symbionts and phylogenetically useful gene fragments, the SSCP approach to resolving PCR-RFLP fragments may be the current choice for other symbionts since it does not require either prior knowledge of the gene sequence or specialised primers, and can detect single base pair changes across all of the gene sequence independent of the relative GC content. Note, however, that different single base pair sequence changes within the same restriction fragment cannot always be resolved from each other in larger fragments by SSCP and thus, in contrast to DGGE (which can resolve well up to 200–400 bp), gene fragments must be reduced to less than 200–250 bp, requiring a greater number of restriction enzymes. This point becomes more relevant with the use of less divergent genes. For example resolution of amplicons at the 0.5% sequence divergence level will discriminate between most, but not all, *Wolbachia* isolates/strains using the highly divergent *wsp* gene fragment, but is unlikely to do so using the more conserved *fts*-Z or *gro*-el gene fragments. Note that the definition of a true *Wolbachia* strain, i.e. the distinction between intra- and inter- strain polymorphism of these gene fragments, is currently unresolved.

6.5.3 IS THERE HORIZONTAL TRANSMISSION OF *WOLBACHIA* IN FOOD WEBS?

With respect to parasitoids as vectors, cross-species (host-host, parasitoid-host and parasitoid-parasitoid) HT of intracellular symbionts such as *Wolbachia* within a food web can, theoretically, occur in at least four main ways, but note that the same principles apply to other potential vectors as well as symbionts:

1. Host-host transfer occurs *via* a vector (from hereon, a parasitoid) that remains uninfected. An uninfected parasitoid, probing an infected individual host species with its ovipositor (to investigate, sting and/or lay an egg), acquires *Wolbachia* bacteria or *Wolbachia*-containing host cells that either adhere to the outer surface of its ovipositor or passively enter the latter, and then passes these on to the next uninfected host(s) that it probes, or oviposits into.

2. Host-parasitoid transfer occurs when an uninfected parasitoid acquires *Wolbachia* from the infected host tissue that it is developing in (endoparasitoids), or feeding upon (ectoparasitoids). Perhaps an injury has allowed access to the parasitoid's haemocoel or, in the case of endoparasitoids, the *Wolbachia* enters exposed imaginal discs and developing gametic tissues. Alternatively, an uninfected parasitoid may acquire *Wolbachia*, or *Wolbachia*-infected host tissue, whilst probing an infected host (as in 1 above), and then deposit her egg(s), together with the *Wolbachia*, into the next host, where her progeny acquire the infection, as outlined above.

3. Parasitoid-to-host transfer is the reciprocal of (2) above, with the uninfected host acquiring *Wolbachia* from an infected parasitoid by suppressing, physiologically, the development of the *Wolbachia*-infected parasitoid (or parasitoid mortality occurs following oviposition for any other reason, i.e. unsuccessful parasitism), and the host acquires the infection. Alternatively, *Wolbachia* (free bacteria or infected cells) could be passed from the ovipositor of an infected parasitoid, during probing of the host, without oviposition.

4. Parasitoid-to-parasitoid transfer may occur as a result of progeny of an infected parasitoid species losing in a case of heterospecific superparasitism (multiparasitism), and the *Wolbachia* in the losing parasitoid's tissues eventually being acquired by the uninfected victor. Alternatively, for gregarious species, in the absence of larval fighting following multiparasitism, damage induced by any other method to both infected and uninfected parasitoid larvae may allow acquisition of the infection *via* transfer of *Wolbachia* or *Wolbachia*-infected cells to the surviving uninfected parasitoid.

Note that: (a) the above four routes would also apply to the acquisition of a new *Wolbachia* isolate by any insect already supporting one or more different *Wolbachia* isolates resulting in multiple infections; (b) in most of these cases the recipient host's germ line tissue would not be directly probed by the *Wolbachia*-carrying vector, and thus HT also entails successful penetration (infection) of gametic tissue by the newly transferred *Wolbachia* (see above).

Parasitoids have been implicated as HT vectors for *Wolbachia*: (a) in general, based upon broadly similar phylogenetic relationships between hosts and parasitoids (Werren *et al.*, 1995a,b), and (b) specifically in a few cases where parasitoid-host pairs share *Wolbachia* isolate(s) with very similar (Vavre *et al.*, 1999; R.D.J. Butcher and J.M. Cook, unpublished data) or identical (Noda *et al.*, 2001; R.D.J. Butcher, unpublished data) sequences. However, the latter observations of a few cases of a close association between *Wolbachia* isolates in the parasitoid and associated host are potentially equivocal as they could be expected by chance, given that most closely related *Wolbachia* isolates are found in phylogenetically diverse hosts. That aside, evidence of a role for parasitoids as vectors for the horizontal transfer of *Wolbachia* has been obtained experimentally (Heath *et al.*, 1999;

Huigens *et al.*, 2000; Butcher *et al.*, 2004c,d). Since in several cases the *de novo* infections were stable over at least 20 generations, this supports the hypothesis that parasitoids may act as HT vectors (Butcher *et al.*, 2004c). The corollary, that most of the *de novo* infections did not lead to stable infections, illustrates that most parasitoid-mediated interspecies horizontal transfers are likely to fail to lead to stable HT (Heath *et al.*, 1999; Butcher *et al.*, 2004c,d). However, even accounting for this low predicted HT rate, evidence for HT of *Wolbachia* by parasitoids still lacks convincing support from the field study of arthropod guilds where congruency between *Wolbachia* isolates in hosts and parasitoids is conspicuously lacking. Indeed, the few existing preliminary studies of parasitoid-host sub-webs and webs are either inconclusive or unsupporting of parasitoids as HT vectors for *Wolbachia* (Schilthuizen and Stouthamer, 1998; West *et al.*, 1998; Schoemaker *et al.*, 2002; Kittayapong *et al.*, 2003).

However, these food web studies used, in the main, remarkably low sample sizes (typically 1–5 individuals per species), with poor or undefined PCR detection thresholds (resolution), robbing them of valid interpretation of the negative results. Furthermore, given the likely relative rareness of cross-species HT (in contrast to horizontal transfer rates) (Butcher *et al.*, 2004c,d), food webs may be noisy due to the compounding presence of both other vectors for HT and due to recombination between transient multiple infections (see "Phylogeny construction" in subsection 6.5.3 below).

For example, horizontal transfer of *Wolbachia* with long-term stable infections, and thus potential HT, has been obtained; (a) following oral ingestion during cannibalism/predation of infected insects (Butcher *et al.*, 2004c), especially in conjunction with non-lethal midgut-epithelial damaging agents such as germinating microsporidia spores and *Bacillus thuringensis* endotoxins (R.D.J. Butcher and D.J. Wright, in preparation); and (b) by haemolymph-feeding ectoparasitic mites (Mesostigmata) upon *Apis mellifera mellifera* (*Varroa destructor* and *Acarapis woodi*) and *Bombus* sp. (*Locustacurus buchneri*), (R.D.J. Butcher, M.E. Murnion and W.G.F. Whitfield, in preparation). The latter cases are interesting, as in several populations the proposed ectoparasitic mite vectors appeared to never become infected themselves but rather vectored the *Wolbachia* between their insect hosts (method 1, above), presumably by transfer of infected haemolymph or haemocytes. They thus illustrate the need to look beyond shared *Wolbachia* infections between proposed vector and host in phylogenetic studies to identify potential vectors.

In addition to other pitfalls in obtaining evidence of HT that are discussed by Cook and Butcher (1999), careful analysis is required in teasing out potential HT from food web phylogenies. Intraspecies horizontal transfer of *Wolbachia*, experimentally shown to occur for several parasitoid species following conspecific superparasitism (Huigens *et al.*, 2000; Butcher *et al.*, 2004c,d), can be difficult to detect. Firstly, for insect species infected with only a single *Wolbachia* isolate, where mitochondrial DNA sequence diversity exists (mitochondrial haplotypes or mitotypes) within the population, recent intraspecies HT will serve to reduce or prevent associations between mitochondrial haplotypes and *Wolbachia* infection, assuming no selective advantage between any mitotype and *Wolbachia* exists, subject to the rate of HT exceeding drift. However, to be conclusive, this requires knowledge of both the mitochondrial and *Wolbachia* gene evolution rates to allow distinction between the continuum bounded by: (i) recent HT of *Wolbachia* amongst existing mitotypes of a host species, and (ii) an ancestral infection of a single mitotype host with subsequent mitochondrial and *Wolbachia* divergence over time. For *Wolbachia* genes in general, and especially *wsp*, divergence rates with time are neither known, nor known to be constant. Of course, recent intra species HT where no mitotype differences within the population exist will not be detected, and will serve only to aid stochastic spread of the infection. Secondly, where more than one *Wolbachia* isolate infects the host population, intraspecies horizontal transfer may allow for the acquisition (even if only briefly) of multiple *Wolbachia* infections within an individual host allowing competition and/or recombination between them to occur. Subsequent non-random loss of infections will give the appearance of an unrelated infection pattern, or, if not lost, the appearance of higher levels of multiple infections. Thus, unless recombinants can be identified and their lineage traced, the resulting phylogenetic signals may be contradictory. The latter case, the existence of separate *Wolbachia* isolates in different individuals within a food web, may occur following either a recent cross-species transfer of a different *Wolbachia* isolate into the (sympatric) insect study population, or following invasion into the study population of infected insects from a relatively ancient allopatric cross-species HT event, and this may vary within the same food webs but from different geographical localities.

Studies of the same food webs but from different localities, will also help to answer other questions. For example, in laboratory experiments, between-species *Wolbachia* horizontal transfer rates were found to be dependent upon the parasitoid-host life-history (highest under superparasitism of solitary endoparasitoids, declining through multiparasitism and gregarious ectoparasitism, lowest in solitary ectoparasitoids)(Butcher *et al.*, 2004c; and see Huigens *et al.*, 2000), whilst stable transfections, and thus assumed probable HT rates, were partially dependent upon the phylogenetic distance between donor and recipient species (Butcher *et al.*, 2004d). Although somewhat intuitive, these proposed life-history-dependent horizontal transfer, and host-dependent HT, rates still remain to be validated in the field, and comparative analysis not only within, but also between geographically distinct food web and sub-web studies may be a worthwhile option to check these hypotheses despite the potential noise from, and reduced resolution due to, other HT vectors in addition to parasitoids.

Likewise, the laboratory-based observations of *Wolbachia* HT following oral ingestion (Butcher *et al.*, 2004c), especially in the co-presence of infective microsporidia spores (R.D.J, Butcher, in preparation) still await field confirmation. The screening of other predators in food webs, such as scorpion flies or robber-flies, mantids and spiders may thus be a worthwhile addition especially if polyphagous (infecting a broad host range) microsporidia are present, although the low HT efficiency may be of some concern. Especial care is required in screening predators to ensure no cross-contamination with prey remains (e.g. gut contents) occurs (i.e. false PCR amplicons).

Phylogenetic Tree Construction

Having screened members of a well-characterised natural food web for *Wolbachia*, the next stage is the construction of both *Wolbachia* and host insect phylogenies for further analysis. Single gene-based phylogenies are somewhat equivocal due to their potential to track the gene tree and not the species tree. For diploid-diploid sexual (eukaryote) organisms, single mitochondrial based loci have been used because they are 1/4 the effective population size relative to nuclear genes and thus are more likely to track the species tree (Moore, 1995; Palumbi *et al.*, 2001). However, this assumption is less applicable for arrhenotokous (such as hymenopteran) species with, simplistically, an effective mitochondrial population size of 1/3, 5/12 and 1/2 for sexual, cyclical parthenogenes and asexuals, respectively; and is clearly inapplicable for prokaryotic symbionts such as *Wolbachia*, and even amitochondrial protists such as microsporidia. Additionally, recent *Wolbachia*-mediated cytoplasmic sweeps may drastically reduce mitochondrial genetic diversity in infected populations (relative to uninfected) requiring the use of both nuclear and mitochondrial loci for phylogenetic analysis (e.g. Ballard, 2000). Thus, ideally, both arthropod and *Wolbachia* analysis should be based upon multiple gene phylogenies, the former including nuclear and mitochondrial loci.

The choice of gene fragments to construct the arthropod phylogeny requires consideration both of the likely genetic divergence at each locus and of the phylogenetic distances between food-web members. Likewise, the number of gene fragments required per species will require consideration of the extra resolution afforded in relation to the time and resources available. In the absence of such information a starting point would be to use the nuclear D1-D3 region of the 28S rDNA (e.g. Belshaw *et al.*, 2001) and long wave opsin (Mardulyn and Cameron, 1999), and the mitochondrial cytochrome oxidase (subunits I and / or II) and cytochrome b (Simon *et al.*, 1994). These could then be supplemented where required with nuclear ITS-2 and microsatellites or AFLP markers, and other mitochondrial haplotypes derived from NADH genes for increasingly finer resolution of sibling species or ecotypes/biotypes. Initial screening for mitotype differences could be performed by PCR-RFLP-SSCP analysis to reduce the sequencing load.

Likewise, for the multiple gene based *Wolbachia* phylogenies, ideally, gene fragments that are equidistant across the genome should be used to minimise bias due to recombination but, currently, this is practically restricted to *wsp* with initially one out of *gro*-E, *dna*-A and *fts*-Z, although inclusion of another *Wolbachia* gene may subsequently be required for support of critical or interesting nodes. For finer scale resolution, it may be possible to use phage insertion sites (ISW1) by inverse PCR (Masui *et al.*, 1999) to discriminate recent *Wolbachia* isolates and recombination, in conjunction with a comparative phylogenetic analysis of the WO prophage itself (Masui *et al.*, 2000). However, the phylogenetic complexity of this approach, for example to take account of the possibility of phage HT between arthropod hosts without *Wolbachia*, currently remains unresolved and thus equivocal, and makes this potentially a research topic in its own right.

Recombination rates within *Wolbachia* are unknown, but for the *wsp* gene recombination has been established to occur (Reuter and Keller, 2003), following earlier equivocal (Werren and Bartos, 2001) and potentially invalid (Jiggins *et al.*, 2001b) claims to its existence. As already mentioned, the result of PCR-mediated or actual recombination can lead to ambiguity over *Wolbachia* lineages and thus HT studies and so needs to be clarified. PCR-mediated recombination can only occur in multiply infected hosts (mixed template PCR) and can easily be checked for (Malloch *et al.*, 2000; Reuter and Keller, 2003), and excluded by reamplification with specific primers. However, sequences additionally need to be checked for actual recombination events and lineages traced not only within an individual or host species, but within the food web in general, using any of several programs (e.g. LARD; see Reuter and Keller, 2003). Additionally, since older recombinations with subsequent HT will probably not be recovered due to divergence and drift, *Wolbachia* phylogenies, as already stated, should be based upon multiple gene fragments, with further analysis performed with the exclusion, as required, of sequences that are highly incongruent with others. Although recombination in *Wolbachia* genes other than *wsp* has not been recorded to date, no systematic study has been performed, and with a lower divergence rate in other genes recombination events would be less easily detected. Thus it is not clear if *wsp* resides in a recombination or research hot-spot, but certainly fine-scale phylogenies generated with *wsp* should be viewed with extra caution since, in addition to the above reasons, it may display host-specific evolution traits (Braig *et al.*, 1998; Jiggins, 2002b; Jiggins *et al.*, 2002).

Translocation of *Wolbachia* genes into the host nuclear genome is known to occur, but its frequency is unknown (Kondo *et al.*, 2002), raising the potential phylogenetic problem of using nuclear pseudogene sequences. Thus *Wolbachia*, like host mitochondrial gene fragments, must be checked for pseudogene characters (stop codons and indels) and excluded. Ideally, FISH, or *in situ* hybridisation-based cytological analysis, should additionally be used to confirm restriction of the *Wolbachia*/mitochondrial gene fragment sequences used in phylogeny construction to cytoplasmic commensuals/mitochondria and not the host nucleus/chromosomes; and, if cultures of the insect are available, the loss of symbiont PCR amplicons following curing by antibiotic treatment should be checked for. However, especially with large food webs, and therefore large sample sizes, the latter two methods may not always be practicable, but should at least be considered for critical or outlying data.

Having thus obtained gene fragment sequences purged of recombinants or nuclear pseudogenes, the sequences can then be used for phylogeny construction. There is no currently accepted optimal methodology for phylogeny construction from gene sequences. The use of maximum likelihood, maximum parsimony (PAUP 4.10; Swofford, 1998) and the Metropolis-coupled variant of Markov chain Monte Carlo Bayesian analysis (Larget and Simon, 1999) (Mr Bayes v2.0; Huelsenbeck and Ronquist, 2001), as a standard approach, although slightly computer time-demanding, should allow production of different phylogenies robust enough for evaluation of the data, but the key point is to use more than a single methodology and check for congruency between the different phylograms obtained for each gene fragment. If a high degree of congruency and bootstrap/jacknife support exists between the different gene phylograms, then phylogenies based upon all gene sequences, with equal weighting should be constructed, and then one should proceed as described below. Incongruence between gene phylograms may be investigated by altering weightings of each gene fragment in the combined phylogeny to obtain the most parsimonious trees (phylograms).

Food Web Analysis by Comparative Phylogenetics

Once suitable phylogenies are obtained, analysis can proceed as follows:

1. Focus on sub-webs:

(a) **On individual host species and their respective parasitoid complexes (i.e. 'source' sub webs)**. If none of the parasitoid species in a complex contains the same *Wolbachia* as their shared host species as defined by sequence analysis of all *Wolbachia* gene fragments analysed (Figure 6.34a), then, simplistically, we may speculate that they have not recently acquired their stable *Wolbachia* isolate(s) from that host. Of course, comparative phylogenetics cannot address unstable transfer with stochastic loss, and it is likely to overlook detection of most cases of non-recent recombination (depending upon the number and size of gene fragments assayed):

(b) **On individual parasitoid species and the host species within their host range (i.e. 'sink' sub-webs)**. If a parasitoid species is found not to contain the same *Wolbachia* strain as any of its hosts (Figure 6.34b), this suggests that it has not recently acquired *Wolbachia* from these hosts; however, it may have acted as a *Wolbachia* vector.

2. Construct phylogenies for the *Wolbachia* in the parasitoids and in their associated hosts. Take each trophic level in turn (i.e. herbivores or parasitoids). If VT has been the predominant mode of transmission (i.e. bacteria and insects have cospeciated to a significant extent), then *Wolbachia* strains from closely related insect species ought to cluster closely together in the tree (i.e. they will tend to form a monophyletic group). Considering both herbivores and parasitoids, if herbivore and parasitoid *Wolbachia* strains are found to cluster closely together, this suggests that host-parasitoid HT has taken place, involving acquisition of infection by the parasitoids, as opposed to mechanical transmission. However, once again, care is needed, as outlined in (1) above, especially in considering other differential sources of HT.

3. Construct both a *Wolbachia* and a host insect phylogeny and assess the degree of congruence between the two (Mitter *et al.*, 1991; Page, 1993). If there is a strong degree of congruence, then we conclude that HT has taken place among the insect species. A caveat applies here: as Mitter *et al.* (1991) have reasoned, phylogenies can show significant concordance without coevolution being the cause. It is thus necessary to know the times at which the various lineages of both bacteria and insects split. If coevolution has been largely responsible for the observed distribution of *Wolbachia* among hosts, then,

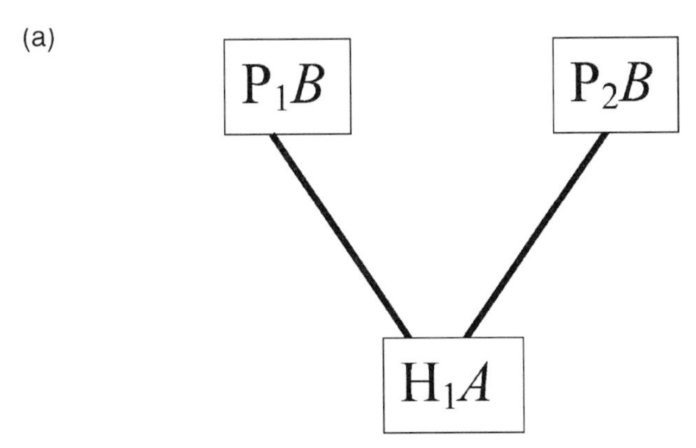

Parasitoid species 1 and 2 have not acquired *Wolbachia* from their shared host species H_1, but may have transferred the bacterium between themselves through multiparasitism

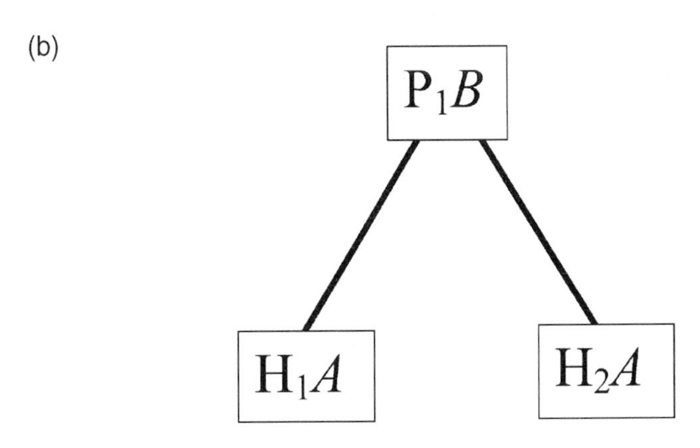

Parasitoid species 1 has neither: (i) acquired *Wolbachia* from either of its two host species; nor (ii) donated the bacterium, but may have been a *Wolbachia* vector from H_1 to H_2 (*or vice-versa*) without infecting itself

Figure 6.34 Inferring acquisition and transfer of *Wolbachia* in subwebs (considering only singly-infected insects).

not only should the branching patterns of the two cladograms be concordant, but also the dates of the lineage-splits should correspond closely.

4. Where food webs have been sampled from different geographical locations (or time-points), and especially if other intracellular symbionts have been analysed, the above analyses on each food web can then be compared between the different geographical or temporally isolated webs, both to test the robustness of correlates and to reveal potential specific interactions.

Caveats: Other Symbionts

There are five caveats relating to the aforementioned approach:

1. Where the *Wolbachia* in the parasitoids are different, it is assumed that there has not been post-transfer 'speciation' of the bacterium. [The actual divergence rates of the currently used *Wolbachia* gene fragments within and between different arthropod hosts are unknown, and thus the distinction between a polymorphism within a given isolate versus a different isolate is far from clear.]
2. Real sub-webs will not necessarily conform to either of the aforementioned patterns, making interpretation difficult.
3. It is assumed that the trophic connections within the food web are both stable, and, as a whole, well known, and that transfers involving hitherto undocumented parasitoid and host species can be ruled out. With the analysis limited by the validity of the constructed food web, it is thus important to thoroughly characterise the food web first.
4. It assumes parasitoids are the only, or main, vector of HT, and it excludes other sources of HT that might differ in their web species usage (e.g. ectoparasitic mites, microsporidia etc.; but see (4) under **Foodweb Analysis by Comparative Phylogenetics**, in subsection 6.5.3 above).
5. It is assumed that there is no preferential and differential loss of *Wolbachia* between species, following establishment of multiple infections by horizontal transfer, that would lead to parasitoids and hosts having different final *Wolbachia* isolates (including recombinants), despite having undergone HT.

In this light, it is of particular note that little is known about the population biology of intracellular symbionts, especially concerning cooperation or competition within or between different *Wolbachia* isolates and other symbionts that share the same host. Preliminary surveys in three host-plant associated insect guilds (associated with *Brassica* sp., *Urtica* sp. and thistles (principally *Cirsium* sp.) reveal non-random infection patterns within insect populations, suggesting that both strong competitive (rarely if ever co-occur) and cooperative (almost exclusively co-occur) interactions between symbionts within the same insect host may be relatively common (R.D.J. Butcher, unpublished data), but this and its effects upon *Wolbachia* HT rates, if any, remain to be clarified. Therefore, it may prove fruitful to perform a preliminarily screening for other intracellular microbes; if these are found to be relatively common, then they should be included, along with *Wolbachia*, within the food web analysis, and, a check should be made for non-random distribution patterns.

The detection, identification and quantification of other intracellular symbionts are principally evaluated by PCR amplification and DNA sequencing, in conjunction with microscopy (subsection 6.5.2 above). Note, however, the need for careful aseptic dissections to exclude exoskeleton- and gut-associated microbes, and stringently clean PCR for eubacterial amplifications.

Microsporidia can be detected by cytological techniques with valuable taxonomic information obtained from sporont and meront karyotype morphology (Vavra and Larsson, 1999; Becnel and Andreadis, 1999), as well as from tissue distribution. However, unequivocal species diagnosis and isolate identification currently requires PCR amplification and sequencing of the small (16S rDNA) sub-unit rDNA for species identification (Weiss and Vossbrink, 1998), with the large (28S rDNA) sub-unit and juxtaposed ITS region providing additional isolate identification. Although 'conserved' primers are available for the small and 5′ part of the large rRNA genes (Weiss and Vossbrink, 1998; but note that there are sequence errors in their Table 2), these are not optimal for all microsporidia species, and amplification of the complete 28 S rDNA currently requires inverse PCR or direct sequencing in the absence of contaminating host DNA. Additionally, care is required to optimise the PCR to avoid cross-amplification of the host insect 18 S rDNA without loss of amplification of the microsporidia DNA. Usually, host-derived cross-primed amplicons, if present, are of a different size, but confirmation of the infection can only be attained by cloning and sequencing of the amplicon. Ideally, sufficient spores should first be obtained for purification, DNA extraction, PCR amplification and sequencing, to develop more specific primers for each microsporidia isolate in the guild.

Eubacteria may be amplified, with stringently clean PCR techniques, following careful aseptic tissue dissections to avoid gut and extra-skeletal microbes, using general eubacterial 16S rDNA primers (Unterman *et al.*, 1989; Weisberg *et al.*, 1991) and identified by sequence analysis, although slightly more specific primers for some groups, such as rickettsiae and spiroplasma, also exist (e.g. see within Balayeva *et al.*, 1995; Jiggins *et al.*, 2000; Tsuchida *et al.*, 2002).

6.5.4. EXISTING DATA AND SUGGESTED FUTURE STUDIES (CONSIDERATION OF FOOD WEB TYPE, LOCATION, AND MINIMAL SAMPLE SIZES)

To date, only a few parasitoid-host food web studies have been performed with respect to *Wolbachia* infection (Schilthuizen and Stouthamer, 1998; West *et al.*, 1998; Schoemaker *et al.*, 2002; Kittayapong *et al.*, 2003), and none for other arthropod symbionts. Moreover, these studies have either been on incomplete food webs, or involved very low sample sizes for most of the constituent arthropod species, despite the serious lack of evidence (or even the existence of evidence to the contrary) that *Wolbachia* either: (a) attains fixation in all field populations of all hosts, or (b) only exists as a single isolate within a single sympatric species. Thus, the field is wide open for thorough, in-depth studies.

For horizontal transfer, and thus HT, to occur between species an ecological interaction between the different species is required. Therefore, organisms within a feeding community are likely candidates for HT of symbionts such as *Wolbachia*, due to their frequent and close associations with each other and potential vectors. Despite these clear reasons for food web studies for intracellular symbiont HT, including *Wolbachia*, such studies are likely to be both time- and resource-costly when the food web is already well characterised, itself a long-term undertaking. Since the limited studies conducted to date have failed to provide clear support for the parasitoid HT vector hypothesis for *Wolbachia*, the value of intensive food web studies for *Wolbachia* HT studies may, therefore, be seriously questioned, yet, in contrast, screenings of parasitoid-host and predator-prey sub-webs have provided some tentative evidence (Vavre *et al.* 1999; Noda *et al.*, 2001; Kittayapong *et al.*, 2003; R.D.J. Butcher and J.M. Cook, unpublished data), dangling a carrot of potential support and the prospect of a better understanding. We suggest the situation may be improved by: (a) avoiding food webs largely centred upon host insects not known to support *Wolbachia* infections – such as aphids (although these may of course be ideal for studies into the HT of aphid secondary symbionts); (b) preliminary screening, to ascertain whether infection frequencies within a food web are high enough for reasonable detection, (c) focusing on taxonomically homogeneous food webs (see below); and (d) examining other symbionts at the same time where they co-occur at high enough frequencies to merit inclusion.

Negative results, even after PCR quantification, require care in interpretation. Cook and Butcher (1999) argue that food web studies cannot provide evidence that completely rules out HT between hosts and parasitoids (or shared hosts and shared parasitoids), because horizontal transfers that satisfy all five stages outlined in the introduction in subsection 6.5.1 above (i.e. are both stable over enough generations to be detected phylogenetically, and sufficiently spread within a population that they are detected with a relatively small sample size, see below) are likely to be rare. Certainly experimental between-host *Wolbachia* transfer under "ideal" laboratory culture conditions frequently leads to infection of the host with a novel *Wolbachia* isolate which is lost over successive generations (Butcher *et al.*, 2004d). Such transfections are often more sensitive (in terms of reduced

VT efficiency) to the effects of biotic and abiotic factors representative of field, as opposed to ideal laboratory, conditions (R.D.J. Butcher, unpublished). Thus, horizontal transfer may be common but with, for example, poor subsequent VT efficiencies, HT is consequently rare due to failure of one or more of the steps required to establish an infection following horizontal transfer (see subsection 6.5.1 above). In addition, competition between, and possible recombination amongst, different *Wolbachia* isolates, following multiple infections, may not be the same in different host species, resulting in the establishment of separate and distinct stable infections between parasitoids and hosts, despite horizontal transfer and all other stages of HT having occurred. The use of several gene sequences, and preferably a MLST (multi-locus sequence typing) approach, will allow detection of recombination between *Wolbachia* isolates within a host or host guild.

However, taxonomically homogeneous parasitoid-host food webs, mainly comprised of members of one family (e.g. Eurytomidae, see Dawah *et al.*, 1995), are likely to be less subject to some of these drawbacks, and so deserve special attention, as the taxonomic barriers to VT and HT ought to be low compared to those existing in a taxonomically highly diverse web, such as one that includes host insects belonging to several major taxa.

Symbiont (including *Wolbachia*) infections within a host species may vary geographically (e.g. *Drosophila simulans* and *D. melanogaster*; reviewed in Hoffmann and Turelli, 1997), or both temporally and geographically (e.g. *Plutella xylostella*, *Spodoptera litura* and *Anthophila fabriciana* and associated parasitoids [R.D.J. Butcher, unpublished]). Therefore, studies within the same guilds but including different geographical locations may yield differing but informative results, especially if they encompass intracellular symbionts in general rather than *Wolbachia* only (see [4] in "Food web analysis by comparative phylogenetics" in subsection 6.5.3).

Finally, it is important to note that not all individuals of an insect species may carry *Wolbachia* at levels above the threshold of detection, and these will produce false negatives in screening. Non-infection may result from **'curing'** by exposure to high temperatures (e.g. Pijls *et al.*, 1996; Van Opijnen and Breeuwer, 1999; Giorgini, 2001; Hurst *et al.*, 2001), low temperature or diapause (Perrot-Minnot and Werren, 1999), or natural antibiotics, which is likely to result in localised patches of uninfected insects, or from imperfect maternal transmission of *Wolbachia* due to competing intracellular parasites (R.D.J. Butcher, in preparation). Indeed, whilst VT efficiency is typically estimated at 90-100% in lab cultures, depending on both host species and *Wolbachia* isolates, field estimates are sparse but, where available, are lower (e.g. Turelli and Hoffmann, 1995; Hoffmann *et al.*, 1998). Additionally, as mentioned above, different (allopatric) populations of the same species may harbour different *Wolbachia* infections. For the aforementioned reasons, it is imperative to screen at least 20 (50–100 may be required) individuals of each species per locality (analysis of infection profile frequencies will enable determination of the minimal sample sizes needed), to sample from several disparate localities, and to record and use the frequency(ies) of infection(s) in the analysis. All samples should be randomised and not clumped to help to ensure they are from as many matrilines as possible.

6.6 ACKNOWLEDGEMENTS

We wish to thank: Rob Butcher and Mark Jervis for writing the *Wolbachia* section, Charles Godfray for commenting on the latter, Mark Walton for contributions to other sections, Rhianedd Ellwood-Thompson and Andy Weightman for advice on *Wolbachia*, Brad Hawkins for sharing his thoughts concerning parasitoid communities, Mike Solomon for some mite techniques references, Hugh Loxdale, Nigel Stork and Anja Steenkiste for providing photographic material, Jan Cawley and Vyv Williams for photographic assistance, and the HRI Library staff for tracing publications. KDS and WP gratefully acknowledge funding by the UK Department for Environment, Food and Rural Affairs.

POPULATION DYNAMICS 7

N.A.C. Kidd and M.A. Jervis

7.1 INTRODUCTION

The reasons for studying the population dynamics of insect natural enemies are basically twofold. Firstly, predators and parasitoids are an important component of terrestrial communities (LaSalle and Gauld, 1994), so therefore are of central interest to the ecologist who attempts to unravel the complexity of factors driving the dynamics of species interactions. Secondly, the knowledge gained from studies of predator and parasitoid populations may be of immense practical value in insect pest management (Hassell, 1978, 2000b; DeBach and Rosen, 1991; Van Driesche and Bellows, 1996).

In this chapter, we aim to demonstrate how ecologists and biological control researchers can assess the role of natural enemies in insect population dynamics, and how the information obtained (together with that gained from other biological studies) can be put to use in terms of biological control practice. We begin by reviewing methods for demonstrating and quantifying predation and parasitism (section 7.2). We then examine the different techniques for determining the effects of natural enemies on insect population dynamics (section 7.3). Finally, we examine ways in which this and other information can be used in choosing appropriate biological control agents for introduction (section 7.4). The reader should note that we make no attempt at providing a comprehensive review of insect natural enemy population dynamics – that would require a book in itself (a very large one at that!). Thus, there are several topics to which we make only brief reference, or do not mention at all. In most cases, the reasons for omission are simply either that the topic has received adequate coverage in other texts, or that it is not specifically related to insect natural enemy population dynamics. Readers mainly concerned with the theoretical background to parasitoid and predator population dynamics should also consult the books by Hassell (1978, 2000), Hawkins and Cornell (1999) and Hochberg and Ives (2000).

7.2 DEMONSTRATING AND QUANTIFYING PREDATION AND PARASITISM

7.2.1 INTRODUCTION

In this section we are concerned with a variety of techniques that can be applied to field, and in some cases laboratory, populations of natural enemies and their prey for the following purposes:

1. To demonstrate that natural enemies have a significant impact upon host and prey populations (subsections 7.2.3 to 7.2.9);
2. To measure rates of predation and parasitism in the field and/or the laboratory, and to provide indices of predation and parasitism (subsections 7.2.10 to 7.2.15).

The techniques used for 1. and 2. provide respectively: (a) a preliminary assessment of the impact of parasitoids and predators upon host and prey populations, and (b) quantitative information which can be used to further our understanding of the role of natural enemies in insect population dynamics. In the latter case, additional methodologies are required to achieve the objective; these are discussed in detail in section 7.3.

We begin by discussing the introduction of natural enemies in classical biological control. Strictly speaking, this is not a technique *per se* for demonstrating that natural enemies have a significant impact on host/prey populations. However, we include it because: (a) it can provide dramatic evidence of the impact of natural enemies upon insect populations, and (b) it can be simulated in the laboratory.

7.2.2 NATURAL ENEMY INTRODUCTIONS IN CLASSICAL BIOLOGICAL CONTROL

Some of the best demonstrations of the effectiveness of natural enemies are provided by cases of so-called **'classical biological control'**. Classical biological control may be defined as the importation of a natural enemy into a geographic region to control an insect species. Typically, the latter is an exotic species that has become established without its adapted natural enemy complex, and because the local natural enemies are ineffective, it has become a pest. For a classical biological control programme to be completely successful, the natural enemy has to reduce the pest populations to a level where the latter no longer inflict economic damage. In ecological terms, the natural enemy is required both to depress the pest population below a certain level and to prevent it from again reaching that level by promoting stability (Figure 7.1 and Figure 7.2b) (for a dissenting view regarding the necessity for stability in natural enemy-pest population interactions, see Murdoch *et al.*, 1985; Kidd and Jervis, 1997; and section 7.4). Successful biological control resulting from parasitoid introduction can be simulated in the laboratory, as shown in Figure 7.2b (also Figure 7.3).

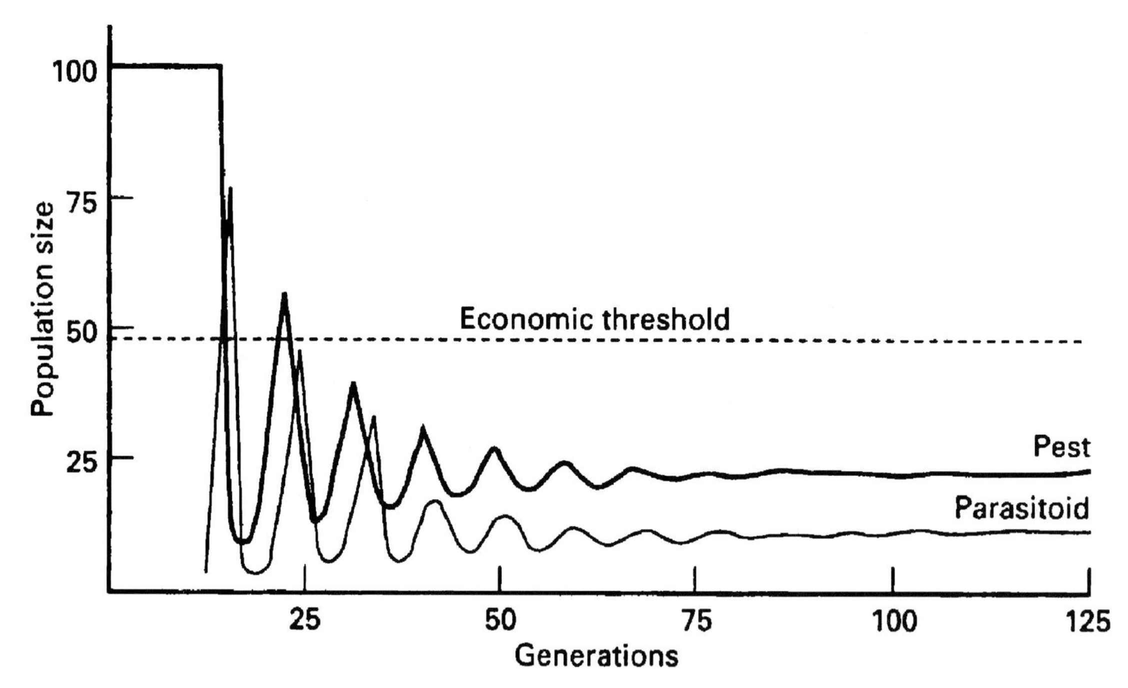

Figure 7.1 Hypothetical example of successful control of a pest population by an introduced parasitoid or predator. In this case, the pest population in the first generation is at an outbreak level (100), well above the economic threshold. The parasitoid or predator is introduced after 10 generations. As its numbers increase, pest numbers decrease. Initial oscillations in both populations decrease with time, and a stable equilibrium is attained at which the pest population remains suppressed to well below the economic threshold. Source: Greathead and Waage (1983), reproduced by permission of The World Bank.

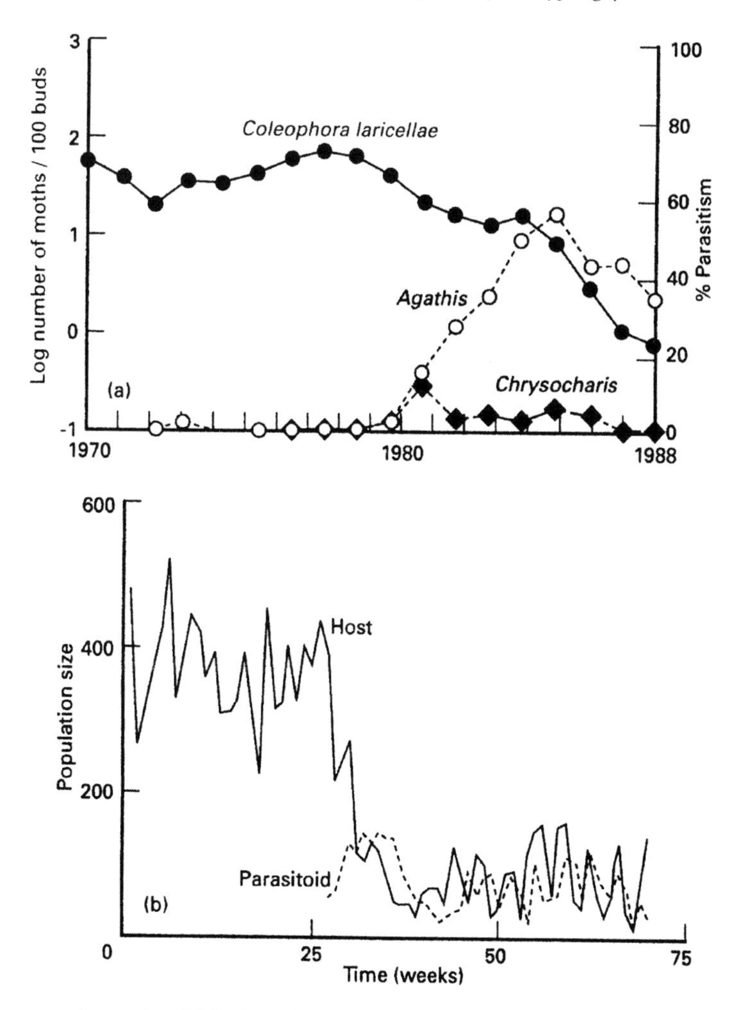

Figure 7.2 Examples, one from the field, the other from the laboratory, of the successful biological control of an insect pest following introduction of parasitoids: (a) Two parasitoid species, *Agathis pumila* (Braconidae) and *Chrysocharis laricinella* (Eulophidae), were introduced into Oregon, U.S.A., against the larch case-bearer, *Coleophora laricellae* (Lepidoptera). The figure shows the combined data for 13 plots over 18 years (source: Ryan, 1990). Reproduced by permission of the Entomological Society of America. (b) The pteromalid *Anisopteromalus calandrae* was introduced into a laboratory culture of the bruchid beetle *Callosobruchus chinesis* 26 weeks after the beetle culture was started (source: May and Hassell, 1988). Reproduced by permission of the Royal Society.

Classical biological control has served as a paradigm for the role of predators and parasitoids in insect herbivore population dynamics. Natural enemy introductions have been viewed as ecological experiments on a grand scale, allowing comparison, under field conditions, of the population dynamics of insect species in the presence and absence of natural enemies (Strong *et al.*, 1984). Many biocontrol practitioners (e.g. DeBach and Rosen, 1991; Waage and Hassell, 1982) consider there to be no fundamental difference between the successes achieved using exotic natural enemies in classical biological control and the action of indigenous species ('natural control' *sensu* Solomon, 1949, and DeBach, 1964). Classical biological control is thus seen as simply isolating a process that is taking place around us all the time. Using

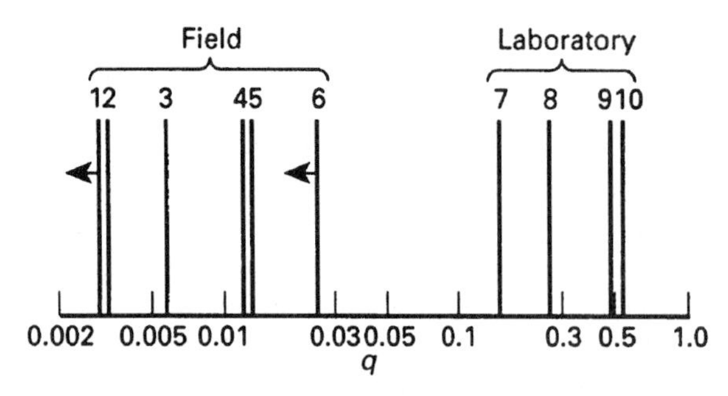

Figure 7.3 The degree to which a pest population may be suppressed by an introduced parasitoid, in six field and four laboratory parasitoid-host systems. The degree of depression in each case is expressed as a q-value, q being defined as the average abundance of the host in the presence of the parasitoid (i.e. post-introduction: N^*) divided by the host's average abundance in the absence of the parasitoid (i.e. pre-introduction: K). Arrows imply minimum estimates of the degree of depression. Note that doubts have been cast upon the role of *Cyzenis albicans* (5) in the direct control of the pest (the winter moth, in Canada) (subsection 7.3.4). Source: Beddington *et al.* (1978) (see that paper for the identity of all the species involved), reproduced by permission of Macmillan Magazines Ltd.

a database analysis of the results of key factor analyses (see subsection 7.3.4) and of historical data from natural enemy introductions, Hawkins *et al.* (1999) showed that classical biological control is, in fact, not strictly a 'natural' phenomenon, because it: (a) overestimates the extent to which parasitoids exert top-down control (subsection 7.3.10) on insect populations, and (b) results most often from the formation of a single, strong link in simplified food webs, in contrast to the 'natural control' that results from multiple links in more complex webs. With these caveats in mind, studies of classical biological control introductions can nevertheless shed considerable light on 'natural control' by predators and parasitoids.

The degree to which a host population may be reduced in abundance by an introduced parasitoid was examined by Beddington *et al.* (1978), who used a simple measure, $q = N^*/K$, where N^* is the average abundance of the host in the presence of the parasitoid (i.e. post-introduction) and K is the average abundance of the host prior to introduction of the parasitoid. Beddington *et al.* calculated q-values for six different field parasitoid-host systems (cases of successful biological control) and four laboratory systems. Figure 7.3 shows the calculated q-values to be of the order of 0.01; that is, the host populations were depressed to about one hundredth of their former abundance. [With the arrowhead scale (not included in Beddington *et al*,'s analysis), abundance was reduced by one hundredth but then rose again, settling at under one sixtieth. (Itioka *et al.*, 1997).] Note that the degree of depression of pest abundance required for successful biological control will vary from case to case, because economic thresholds are determined not only by pest density, but also by pest impact, crop value and other factors (Figure 7.4) (Waage

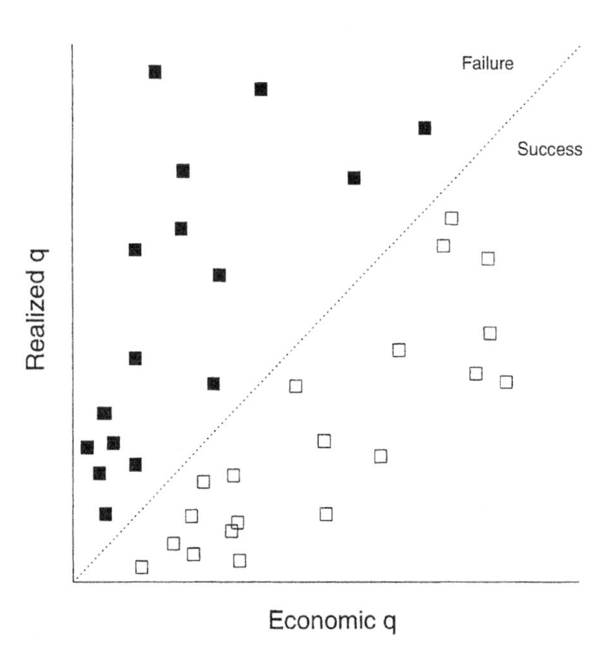

Figure 7.4 Hypothetical plot of outcomes of biological control programmes in terms of realised level of pest suppression and the threshold level required for the programme to be an economic success. q refers to the Beddington *et al.* (1975) index of equilibrium pest density after a classical biological control introduction, divided by equlibrium pest density prior to introduction (see text and Figure 7.3). From Hochberg and Holt (1999). Reproduced by permission of Cambridge University Press.

and Mills, 1992; Hochberg and Holt, 1999, see also section 7.4).

In the past, the value of detailed and precise quantitative data on the effects of natural enemy introductions on pest populations was not fully appreciated (Waage and Greathead, 1988; May and Hassell, 1988), with the result that often only anecdotal evidence on the effects of introductions has been available for some programmes (section 7.4). Notable exceptions are the introduction of the tachinid parasitoid *Cyzenis albicans*, introduced to control the winter moth in Canada (Embree, 1966), the release of parasitoid wasps against the larch casebearer in the USA (Ryan, 1990, 1997) (Figure 7.2a), the release of *Encarsia partenopea* against the whitefly *Siphoninus phillyreae* in the USA (Gould *et al.*, 1992a,b) (Figure 7.5), and the release of two parasitoid wasps against the arrowhead scale (*Unaspis yanonensis*) in Japan (Itioka *et al.*, 1997).

Another criticism that can be aimed at some programmes is that depression of the pest population cannot necessarily be attributed to the introduced natural enemy. That is, introduction and depression may be merely coincidental. For example, in programmes involving whitefly pests, reductions in pest density following parasitoid release were reported but the workers concerned failed to provide proper controls to demonstrate that the introduced natural enemies were indeed responsible for the depression (Gould *et al.*, 1992a). Even where experimental controls are employed in biological control programmes, it is likely that, due to the rapid spread of the natural enemy, comparisons of test and control plots are possible for a brief period only. This problem arose with the monitoring of releases of *Encarsia partenopea* against the whitefly *Siphoninus phillyreae* in California (Figure 7.5). Parasitoids were released in May, and by midsummer had appeared at all control sites (4–11 km away from the nearest release sites) (Gould *et al.*, 1992a). A problem of this type could be difficult to overcome, since too wide a separation of control and test sites reduces the validity of comparisons.

Nowadays there is an increasing awareness of the need for detailed and precise quantitative data on the effects of introductions, and classical biological control programmes are tending to be much more carefully documented through the routine collection of population data. However, this usually applies only to biological control programmes with good funding and well-trained staff. Another constraint upon the gathering of pre- and post-release population data is the great rapidity with which many pest problems arise. Examples of insects having very rapidly become serious pests are the mealybug *Rastrococcus invadens* in west Africa, and the psyllid *Heteropsylla cubana* in the Pacific region, Asia and elsewhere.

Traditionally, most quantitative studies measure population densities over a number of seasons before and after release (section 7.3). Nowadays, however, there is an awareness of the usefulness of undertaking detailed studies of within-season changes in population density (Gould *et al.*, 1992b; Itioka *et al.*, 1997) and also of recording changes in pest age-structure immediately following parasitoid introduction (Gould *et al.*, 1992a, see Figure 7.5).

At present, there is no standard protocol (at least not a sufficiently detailed one; see Neuenschwander and Gutierrez, 1989; Waterhouse, 1991; Van Driesche and Bellows, 1996) for the quantification of the impact of natural enemies in classical biological control programmes. Such a protocol, if developed, would probably be restrictive, given the diversity in ecology that exists among insect pests.

7.2.3 EXCLUSION OF NATURAL ENEMIES

Exclusion methods have been widely employed in assessing the impact of insect natural enemies on host and prey populations under field conditions. The principle behind their use is that prey populations in plots (any habitat unit, from part of a plant to a whole plant or a group of plants) from which natural enemies have been eradicated and subsequently excluded will, compared with populations in plots to which natural enemies are allowed access: (a) suffer lower predator-induced mortality or parasitism, and (b) if the experiment is continued for a long

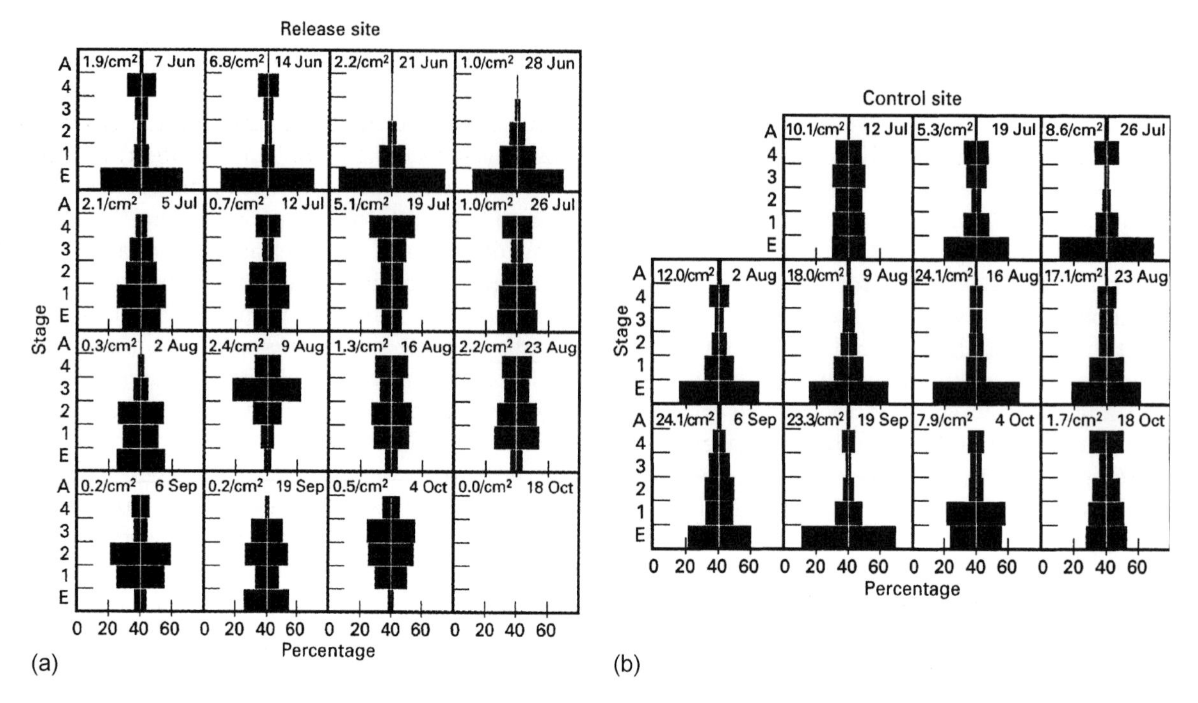

Figure 7.5 An example of classical biological control where post-introduction changes in pest age-structure (histograms) were monitored. The parasitoid wasp *Encarsia partenopea* (Aphelinidae) was introduced into California as an attempt to control the whitefly *Siphoninus phillyreae*. The whitefly, a pest of ornamental shrubs and fruit trees, was first recorded in the USA in 1988, and since then spread rapidly within California and into neighbouring states . Gould *et al.* (1992a) used several study sites, which were divided into release (test) and non-release (control) sites. In the release sites, parasitoids were released in large numbers over a period of several weeks, commencing 10th May. In all sites, densities of the pest's immature stages (eggs, nymphs) and adult stages were monitored, while in the release sites parasitism by *E. partenopea* was also recorded. Densities of the pest (numbers/cm^2, in the top left-hand corner of each graph) remained at low levels at the release sites, whereas at the control sites they were increasing by the beginning of summer. Shown here are changes in age-structure for (a) one release site and (b) one control site. After the parasitoid became abundant at a site (the parasitoid eventually dispersed to, and became established in, control sites), the pest population contained a decreasing proportion of young stages, as result of the parasitoid killing (through parasitism) fourth instars, so reducing recruitment of eggs to the whitefly population. Observe that in (a) the decline in the proportion of immature stages was more marked, and occurred much earlier than in (b). The initial increase in density of whiteflies in (a) is attributable to oviposition by female whiteflies that were already present at the time of parasitoid release. Source: Gould *et al.* (1992a). Reproduced by permission of Blackwell Publishing.

enough period, increase more rapidly and reach higher levels. The results of some exclusion experiments are shown in Figure 7.6.

Usually, the starting densities of prey are made equal in both the test and the control plots (for consistency's sake, we refer here to the exclusion plots as 'test' plots, and the non-exclusion plots as 'control' plots, since what is being tested is the effect of excluding natural enemies, not of including them; note that not all authors use the same nomenclature). Exclusion experiments may be conducted for periods of several days to several weeks. A long experimental period will be required if test-control differences in prey equilibrium densities (section 7.3) are to be compared.

Various exclusion devices have been employed; they include mesh cages placed over individual plants or groups of plants (Figure 7.7), mesh cages in the form of sleeves

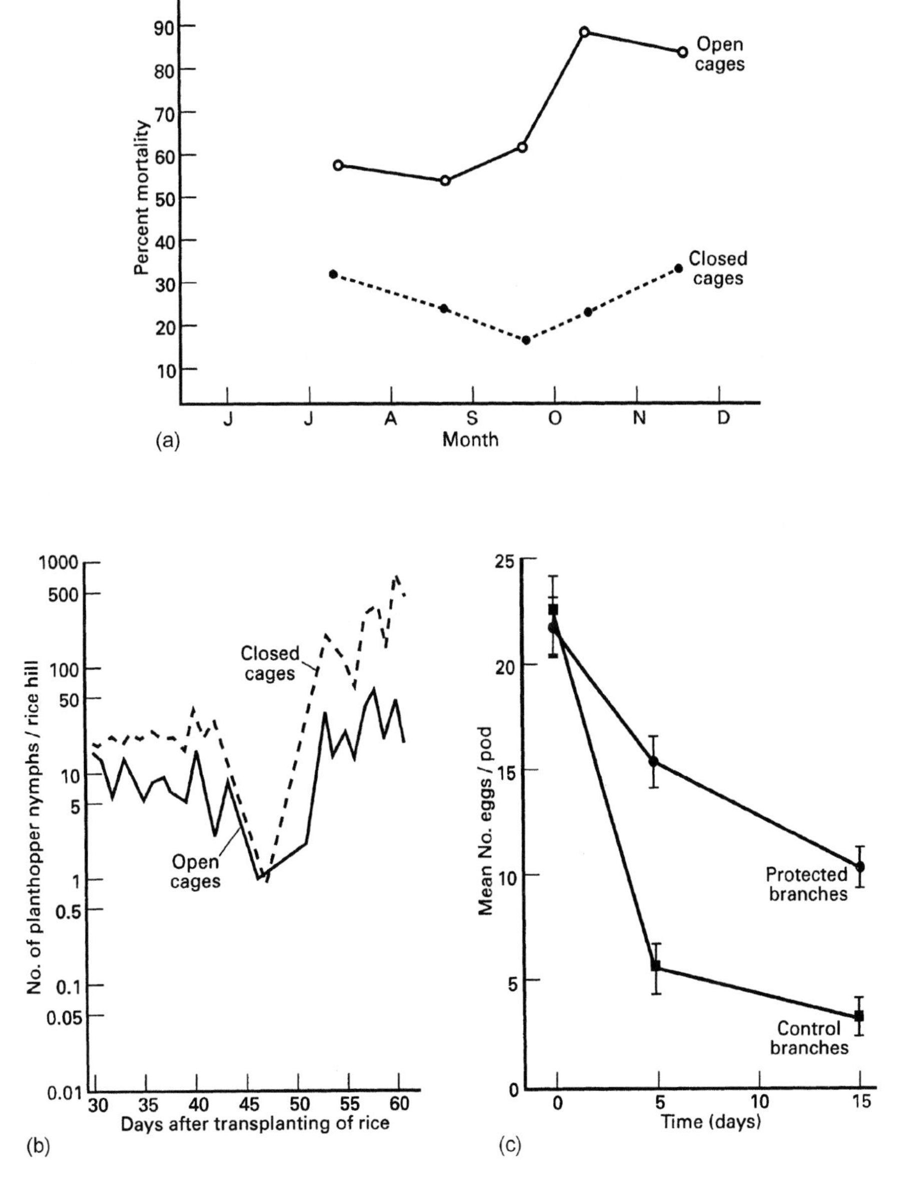

Figure 7.6 Effects of excluding predators from prey: (a) percentage mortality of California red scale (*Aonidiella aurantii*) on *Hedera helix*, in open clip cages that permitted the entry of parasitoids, and in closed cages that excluded the parasitoids (source: DeBach and Huffaker, 1971). (b) Population changes in the rice brown planthopper (*Nilaparvata lugens*) in closed and open field cages after all the arthropods had been removed and replaced with 25 first-instar planthopper nymphs per rice plant. Exclusion was not perfect: large numbers of predators were recorded in the 'closed' cages towards the end of the experiment, so the experiment ought to be considered as an 'interference' experiment (source: Kenmore *et al.*, 1985). (c) The mean number of bruchid beetle eggs per pod of *Acacia farnesiana* on protected and unprotected branches at days 0, 5 and 15 of the experiment. Protection of the branches was achieved by wrapping a 10 cm-wide band of tape around the base, and applying a sticky resin to the tape (source: Traveset, 1990). (a) reproduced by permission of Plenum Publishing Corporation, (b) by permission of The Malaysia Plant Protection Society, (c) by permission of Blackwell Publishing.

Figure 7.7 Exclusion cages in use in a rice paddy in Indonesia. Reproduced by kind permission of Anja Steenkiste.

placed over branches or leaves, clip cages attached to leaves, greased plastic bands tied around tree branches and trunks, and vertical barriers (walls constructed of polythene, wood or hardboard) around plants. For details of construction, consult the references cited below. The precise type of device used will depend upon the natural enemies being investigated, and whether the aim is either to exclude all natural enemies (so-called **total exclusion**) or to exclude particular species or groups of species (so-called **partial exclusion**). For example, a terylene mesh/gauze cage placed over a plant ought, if the mesh size is sufficiently small, to exclude all aerial and surface-dwelling insect natural enemies, from the largest to the smallest. By increasing the size slightly, small parasitoid wasps may be allowed in, while increasing the mesh size further will allow larger types natural enemy to enter also, and so on. By first examining the ability of tiny *Anagrus* wasps (Mymaridae) to pass through terylene meshes of different mesh sizes, colleagues at Cardiff were able to decide on the appropriate size of mesh for excluding all natural enemies of rice brown planthopper other than the egg parasitoids (Mymaridae and Trichogrammatidae) whose impact was being investigated (Claridge *et al.*, 2002). By having a cage with its sides raised slightly above the ground, predators such as carabid beetles and ants may be allowed access to insect prey such as aphids on cereals, whereas adult hover-flies and many types of parasitoid will be denied access. Conversely, a trench or a wall may prevent access to prey by ground-dwelling predators but allow access by aerial predators and parasitoids.

The exclusion devices can be placed around or over already existing populations of prey, in which case the density of the prey at the start of the experiment will need to be recorded. Preferably, prey-free individual plants, plant parts or plots of several plants (any prey already present are cleared by hand removal or by using low-persistence insecticides) can be loaded with set numbers of prey. The latter approach has the advantage that equivalent starting densities of prey/hosts in test and control plots can be more easily ensured, and also any parasitoid immature stages present within hosts can be eliminated from within the test plots. It may also be necessary to employ a systemic insecticide when eradicating prey such as leafhoppers or planthoppers from a plot, in order that any prey eggs present within plant tissues are killed; of course, loading with prey cannot take place until one can be sure that the plant is free of the insecticide.

In some cases, simply comparing prey densities on caged plants with those on uncaged plants may produce misleading results, because:

1. Prey within the cages may be protected to some extent from the mortality or other deleterious effects of weather factors such as as rainfall or wind;
2. In the two treatments microclimatic conditions may be very different. Cages, even ones constructed largely of nylon or terylene mesh, may alter the microenvironment (light intensity, humidity, wind speed, temperature) surrounding the plant (Hand and Keaster, 1967) to such a degree as to influence the impact of natural enemies, either: (a) directly by affecting the physiology, the behaviour and consequently the searching efficiency of the predators or parasitoids, or (b) indirectly by affecting the behaviour, e.g. spatial distribution, and physiology, e.g. rate of development, longevity, fecundity, of the prey. Changes in prey behaviour and physiology can be brought about by microclimate-induced alterations in plant physiology.

In order to determine whether microclimatic effects on prey are likely to confound the results of an exclusion experiment, the effects of caging on prey population parameters such as fecundity and survival should be investigated. Frazer *et al.* (1981b), for example, investigated whether the observed increase in densities of pea aphid (*Acyrthosiphon pisum*) in exclusion cages (to as much as five times the levels recorded in uncaged plots) was due to an effect of caging upon aphid fecundity. No significant differences in the fecundity of aphids were found between caged and uncaged insects. Furthermore, a simulation model (subsection 7.3.8) showed that for a change in fecundity alone to be responsible for the difference in densities of prey between test and control plots, fecundity would have to have been three times the maximum rate ever observed.

In order to separate the effects of microclimate and natural enemy exclusion upon prey populations, exclusion devices that are either: (a) as similar as possible in construction, or (b) very different in construction but which nevertheless provide similar microclimatic conditions in their interiors, may have to be employed in both test and control treatments, with the obvious proviso that predators need to be allowed adequate access to prey in the control treatment. For example, in assessing the impact of the egg parasitoids *Anagrus* spp. (Mymaridae) and *Oligosita* (Trichogrammatidae) upon planthopper populations, exclusion cages can be constructed that have a very small mesh size to prevent such tiny parasitoids from entering (see above), while almost identical cages with a slightly larger mesh size can be constructed to allow the parasitoids to enter but prevent entry of larger types of natural enemy (Fowler, 1988; Claridge *et al.*, 2002) (see above). In assessing the impact of parasitoids on insect herbivores on trees, gauze sleeve cages can be used on tree branches, the test cages being tied at both ends to exclude parasitoids, and the control cages being left open at both ends to allow parasitoids to enter (DeBach and Huffaker, 1971). However, insects such as hover-flies are deterred from ovipositing on branches in open-ended sleeves (Way and Banks, 1968). Way and Banks used rather dissimilar test and control cages in controlling for the effects of microclimate. The test cages had walls of terylene mesh, whereas the control cages had walls of wooden slats. Despite the major difference in construction, microclimate was similar in the two cage types.

One solution to the problem of achieving a closely similar cage design in the different treatments is not to bother providing natural enemies with access routes to the interior of the control cages, but to carry out an **exclusion/inclusion experiment**. Such an experiment would involve the use of identical cages in the two treatments and the caging of a known number of predators and parasitoids with prey/hosts in the 'control' treatment (Lingren *et al.*, 1968). This type of experiment has the added advantage that the densities of natural enemies will be more precisely known. *Per capita* predation and parasitism rates can be calculated (Dennis and Wratten, 1991) and, provided the densities used reflect those normally recorded in the field (this includes taking account of aggregative responses; Dennis and Wratten, 1991), useful estimates of *per population* rates of predation and parasitism can be obtained. A major disadvantage of exclusion/inclusion experiments is that in the control cages the dispersal of natural enemies is likely to be severely restricted or prevented. Long-distance approach behaviour of foraging predators and parasitoids to prey and hosts e.g. in response to kairomones (subsection 1.5.1), may also be interfered with;

3. If the prey are mobile, both immigration and emigration of prey/hosts may be different between the test and the control treatments (restricted or prevented altogether in the test treatment, normal in the control). In order to rule out the possibility that aphid population numbers in fully-caged cereal plots were augmented as a result of emigrant alatae re-infesting the plots, Chambers *et al.* (1983) removed all alate (winged) aphids that settled on the insides of some of the test cages whilst allowing the aphids to remain in another. Removal of alatae was found not to alter the pattern of population change in the cages. Therefore, re-infestation of shoots inside cages was unlikely to have been a cause of the cage/open plot differences in population numbers observed by Chambers *et al.* in their study (Figure 7.8a).

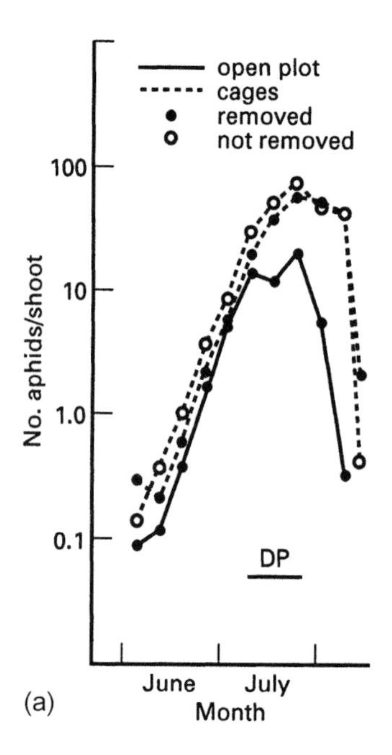

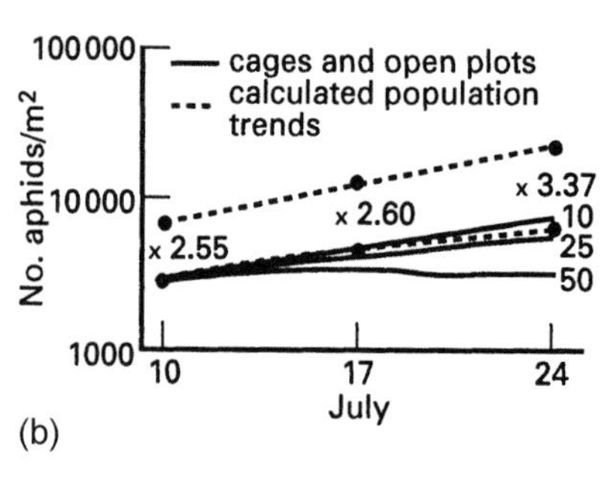

Figure 7.8 (a) Total numbers of cereal aphids in cages where alate adult aphids were removed from the inner walls and roof, in cages where they were not removed, and in the adjacent open plot. DP denotes the period of divergence between treatments. (b) Aphid populations in cages and open plots, and calculated population trends for different *per capita* predation rates. Also shown is the difference, expressed as a multiple, between populations in cages (containing aphids but not predators) and in open plots during the DP. Note log scales. Source: Chambers *et al.* (1983). Reproduced by permission of Blackwell Publishing.

Exclusion methods have a number of other important potential limitations:

4. Even where the microclimate is the same in the different treatments, it may be so different from ambient conditions that prey/host populations are severely affected, and any results obtained bear little relation to natural processes. The effects of caging upon microclimate can be assessed using instrumentation of various kinds (see Unwin and

Corbet, 1991, for a review). If caging is found to affect microclimate significantly, then it may be possible to provide some means of ventilation, e.g. an electrical fan, to maintain ambient temperature and humidity. Effects on light intensity may be minimised by choosing the appropriate type of screening material;

5. Whilst it is often possible to establish whether a particular **guild** (a group of species attacking the same host or prey stage/stages) of natural enemies has a significant impact upon prey populations, it may not be possible to determine which particular species of that guild is mainly responsible for the effect. Unless direct observations shed light on which species is responsible, information will be required on the relative abundance of different natural enemy species within a locality. Where the immature stages of parasitoids can be identified to genera or species, dissections of hosts (subsection 6.2.9) in the control plots may allow determination of the parasitoid species that usually contributes most to parasitism and help show whether an increase in host numbers in the test plots is due to the exclusion of that species. The problem of attribution of predatory or parasitic impact is a minor one where the natural enemy complex is known to comprise only one or two species in a locality;
6. If, in the test plots, prey numbers (e.g. of aphids) increase, they may do so to such an extent that predator species (e.g. coccinellids, hover-flies) other than the ones that are excluded (e.g. carabid beetles) are attracted preferentially into the test plots, i.e. through an aggregative response (subsection 1.14.7) by the predator or parasitoid. The impact of the excluded natural enemy species may thus be underestimated. This limitation also applies to the use of barriers and trenches, where the enclosed plants are exposed to invasion by a variety of aerial predators;
7. Whilst exclusion methods can reveal that natural enemies have a significant impact upon prey populations, other methods generally need to be applied before the predator-prey interaction can be properly quantified. The results need to be related to the density of predators present in the habitat, if realistic estimates of predation rates are to be obtained. Exclusion experiments provide minimal information, if any, on the *dynamics* of the predator-prey or parasitoid-host interaction, a limitation that applies also to several of the methods described below. This problem can be at least partly overcome by the construction of paired life-tables for the insects in test and control plots (Van Driesche and Bellows, 1996; Itioka *et al.*, 1997).
8. One hundred per cent exclusion of natural enemies is sometimes difficult to achieve, with the result that zero predation or parasitism in test plots is not recorded (e.g. see Kenmore *et al.*, 1985). Either during or at the end of an exclusion experiment, it is important to check for the presence of natural enemies in the test plots (see caption to Figure 7.6b), and to count the numbers of any such insects that have succeeded in gaining entry to the latter. Exclusion methods employing devices that are far from 100% effective in excluding natural enemies are, strictly speaking, interference methods (see below).

Other serious problems that may be encountered by experimenters include: (a) plants outgrowing their cages; expanding cages can be devised to counter this problem (Nicholls and Bérubé, 1965); (b) plants in test cages deteriorating very rapidly due to the abnormally high prey densities reached; little can be done to remedy this problem, which can severely limit the duration of the experiment.

Exclusion methods have been used to assess the impact of predators and parasitoids on populations of a wide variety of prey and host insects including: (a) *aphids* (Way and Banks, 1968; Campbell, 1978; Edwards *et al.*, 1979; Aveling, 1981; Frazer *et al.*, 1981b; Chambers *et al.*, 1983; Carroll and Hoyt, 1984; de Clercq, 1985; Kring *et al.*, 1985; Hance, 1986; Dennis and Wratten, 1991; Hopper *et al.*, 1995;

Bishop and Bristow, 2001); (b) *mealybugs* (Neuenschwander and Herren, 1988; Boavida *et al.*, 1995), (c) *armoured scale insects* (DeBach and Huffaker, 1971; Itioka *et al.*, 1997); (d) *soft scale insects* (Smith and DeBach, 1942; Bishop and Bristow, 2001); (e) *planthoppers* (Kenmore *et al.*, 1985; Rubia and Shepard, 1987; Fowler, 1988; Rubia *et al.*, 1990; Claridge *et al.*, 2002); (f) *pond-skaters* (water-striders) (Spence, 1986); (g) *aquatic stoneflies and chironomids* (Lancaster *et al.*, 1991); (h) *beetles* (Sotherton, 1982; Sotherton *et al.*, 1985; Traveset, 1990); (i) *flies* (Burn, 1982); (j) *moths* (Sparks *et al.*, 1966; Lingren *et al.*, 1968; van den Bosch *et al.*, 1969; Irwin *et al.*, 1974; Rubia and Shepard, 1987; Steward *et al.*, 1988; Rubia *et al.*, 1990); (k) *butterflies* (Ashby, 1974).

The results of exclusion experiments can be quite dramatic. For example, the numbers of brown planthopper nymphs on rice plants in test cages reached twelve times the level attained in control cages, even though exclusion of predators proved to be imperfect (Kenmore *et al.*, 1985; Figure 7.6b). In Campbell's (1978) study of the hop aphid *(Phorodon humuli)*, aphid numbers reached around $1000/0.1\,m^2$ in test cages, whereas in uncaged control plots they declined virtually to zero. In exclusion/inclusion experiments carried out by Lingren *et al.* (1968), adult bollworm moths were introduced into test and control cages, and in the control cages different types of predator were subsequently introduced. The number of moth eggs in the test cages reached a level ten times higher than that recorded in the control cages containing the predators *Geocoris punctipes* (Lygaeidae) and *Chrysoperla* spp. (Chrysopidae).

Even where a marked difference in prey numbers is observed between test and control treatments, and the possible confounding effects of factors other than predation can be discounted, it is important to establish whether the predators in question really do have the potential to produce the test versus control plot difference observed. This requirement was appreciated by Chambers *et al.* (1983). As well as testing for the effects of aphid emigration, parasitism, fungal disease and cage microclimate, they sought to determine whether the *per capita* daily predation rates of aphid-specific predators were sufficiently high to have accounted for the differences in aphid numbers they recorded between fully-caged and open plots (Figure 7.8a and above). Using information on: (a) aphid rate of increase in the absence of predators (i.e. data were obtained from aphids in the cages); (b) predator densities in the open (control) plots; and (c) *per capita* daily predation rates of predators (published values), Chambers *et al.* were able to calculate population trends for aphid populations exposed to predation (Figure 7.8b, see Chambers *et al.* for method of calculation). They established that the predation rate that would be required to bring about the observed cage/open plot difference lay within published values.

7.2.4 INSECTICIDAL INTERFERENCE

The phenomenon of pest resurgence brought about by the application of insecticides, and inadvertent elimination of a pest's natural enemies reveals dramatically the impact the latter normally have (DeBach and Rosen, 1991; Shepard and Ooi, 1991). With this effect in mind, insecticides have been used as a method of assessing the effectiveness of natural enemies.

With the insecticidal interference method, the test plots are treated with an insecticide, so as to eliminate the natural enemies, and the control plots are untreated. The insecticide used is either a selective one, or a broad-spectrum one that is applied in such a way as to be selective, affecting only the natural enemies. Depending on the duration of the experiment, repeated applications of the insecticide may be required, to prevent immigrating natural enemies from exerting an impact upon prey in the test plots. Drift of insecticides onto control plots also needs to be carefully avoided. The results of some insecticidal interference experiments are shown in Figure 7.9.

Some limitations of the method are that:

1. In the test plots not only the natural enemies but also the prey may be affected by the insecticides, so confounding the results of the experiment. The numbers of prey may

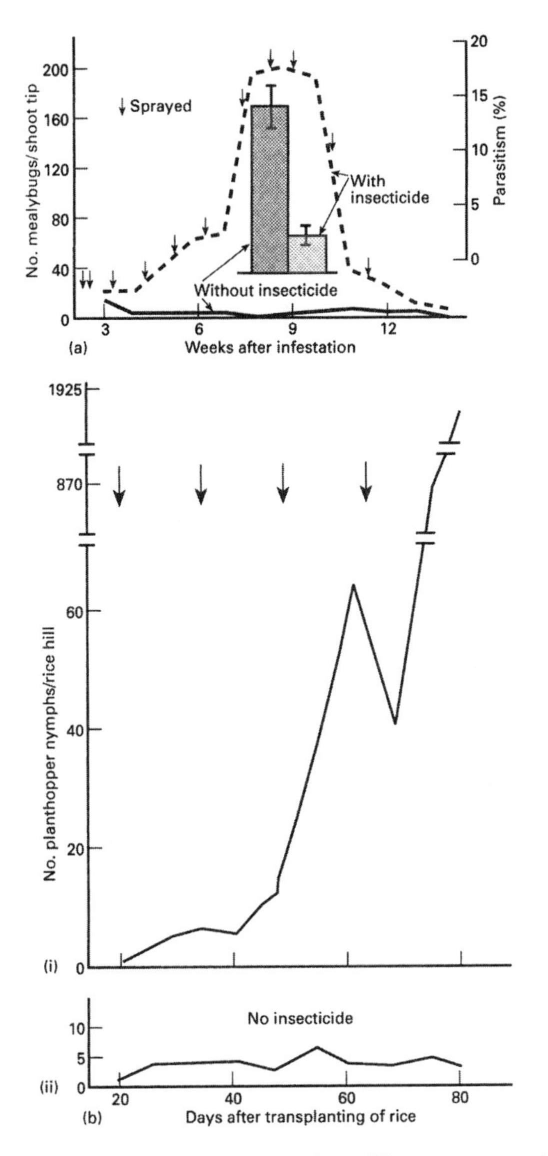

Figure 7.9 (a) Cassava mealybug (*Phenacoccus manihoti*) population development in insecticide-treated and untreated plots, together with mean levels of parasitism (histograms) (source: Neuenschwander and Herren, 1988). (b) Population changes in rice brown planthopper (*Nilaparvata lugens*) in: (i) a plot treated with four sprays (arrowed) of the insecticide decamethrin, and (ii) an untreated plot. Planthoppers were sampled from 40 rice hills per plot, using a vacuum net (subsection 6.2.2) (source: Heinrichs *et al.*, 1982). (a) reproduced by permission of The Royal Society, (b) by permission of The Entomological Society of America.

be inadvertently reduced due to the toxic effects of the insecticide (i.e. either the insecticide turns out not to be selective in action, or drift of a broad-spectrum insecticide has occurred) or they may be inadvertently increased due to some stimulatory, sublethal, effect of the insecticide upon prey reproduction (e.g. prey fecundity may be increased). Insecticides can be tested in the laboratory for their possible sublethal effects upon prey reproduction (Meuke *et al.*, 1978; Kenmore *et al.*, 1985).

2. In the test plots, 100% elimination of natural enemies is often not achieved (e.g. see Kfir, 2002), and so the full potential of natural enemies to reduce prey numbers is underestimated;
3. Limited information is provided on the dynamics of the predator-prey interaction, even where densities of natural enemies are known (see **Exclusion Methods**).

The main advantages of the method are that the possibly confounding effects of microclimate can be ruled out, and very large experimental plots can be used.

As an alternative to blanket spraying of test plots, an **insecticide trap method** can be used. Ropes of plaited straw treated with insecticide, trenches dug in the soil and containing formalin solution or insecticide-soaked straw, or some other insecticide-impregnated barrier, can severely reduce the numbers of natural enemies entering test plots. One treatment used by Wright *et al.* (1960) and Coaker (1965) in studying beetle predators of the cabbage root-fly *Delia radicum*, involved the placing of insecticide-soaked straw ropes along the perimeters of test plots. Whilst it was not 100% efficient, the latter treatment had a dramatic effect upon predator numbers, and also significantly affected prey numbers in test plots.

The insecticide interference method has been used to assess the impact of parasitoids and predators upon populations of: (a) *aphids* (Bartlett, 1968); (b) *armoured scale insects* (DeBach, 1946; 1955; Huffaker *et al.*, 1962; Huffaker and Kennett, 1966); (c) *leafhoppers and planthoppers*

(Kenmore *et al.*, 1985; Ooi, 1986); (d) *flies* (Wright *et al.*, 1960; Coaker, 1968); *moths* (Ehler *et al.*, 1973; Eveleens *et al.*, 1973; Kfir, 2002); (e) *thrips* (Nagai, 1990) and (f) *spider mites* (Plaut, 1965; Readshaw, 1973; Braun *et al.*, 1989).

7.2.5 PHYSICAL REMOVAL OF NATURAL ENEMIES

As its name suggests, physical removal involves just that: predators are removed either by hand or with a hand-operated device, each day, from test plots. The method is a variant of exclusion, described above. Large, relatively slow-moving predators can simply be picked off plants by hand, while small, very active predators and parasitoids can be removed using an aspirator. This method has advantages in that microclimatic confounding effects can be ruled out (since cages are not used), and the contribution of particular natural enemy species to parasitism and predation can be relatively easily assessed. However, the method also has disadvantages in that:

1. Removal of natural enemies is very labour-intensive; for the method to provide more than just a crude measure of natural enemy effectiveness, a 24 hour per day watch needs to be kept on plants, and several observers need to be involved in removing insects;
2. Removal of natural enemies may involve disturbance to prey and thereby increase prey emigration;
3. Predators and parasitoids, before they are detected and removed, may have the opportunity to kill or parasitise hosts;
4. Like exclusion, the method provides limited information on the dynamics of the predator-prey interaction, even where densities of natural enemies are known (see **Exclusion Methods**).

Hand removal has been used to evaluate the effectiveness of aphid predators (Way and Banks, 1968 (Figure 7.10); Pollard, 1971). Luck *et al.* (1988) suggest it can be used as a calibration method for interference and exclusion methods.

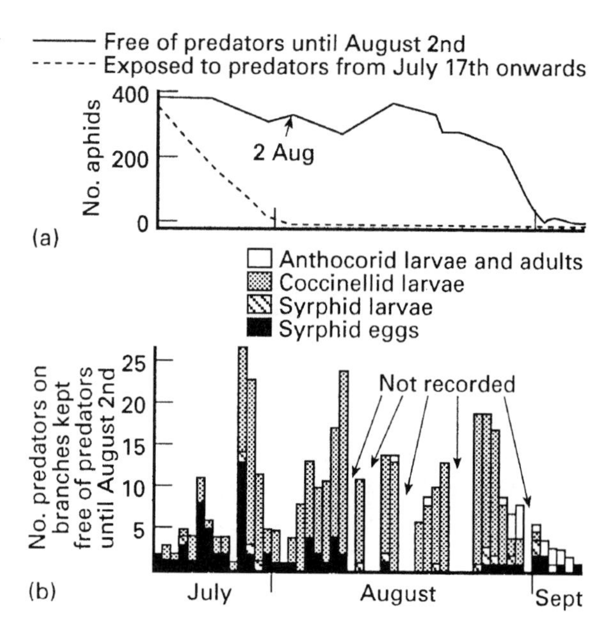

Figure 7.10 Effect of hand-removal of predators from colonies of *Aphis fabae*: (a) numbers of aphids on six branches of *Euonymus europaeus* kept free of predators until August 2nd, compared with numbers of aphids on branches exposed to predators from July 17th onwards; (b) numbers of different predators found on, and subsequently removed from, branches up to 1 2nd August, and numbers that were found on, and allowed to remain, after 2nd August. In this experiment, crawling predators were excluded from the branches by a grease band. Source: Way and Banks (1968), reproduced by permission of The Association of Applied Biologists.

7.2.6 BIOLOGICAL 'CHECK' METHOD

This interference method exploits the fact that honeydew-feeding ant species, when foraging for honeydew sources and tending homopteran prey, interfere with non-ant predators and parasitoids, either causing them to disperse or killing them. In one set of plots, ants are allowed to forage over plants, whereas they are excluded from the other set. Natural enemies have access to both types of plot, but they are subject to interference by ants in the former. The method can be used with prey that do not produce honeydew, provided either natural or artificial honeydew is made available to the ants. This method has several of the disadvantages of other interference methods and exclusion methods.

7.2.7 EGRESS BOUNDARIES

Egress boundaries are simple devices which allow predators to move out of, but not into, plots and thereby reduce predator numbers (see caption to Figure 7.11). These devices were used by Wratten and Pearson (1982) in assessing the effectiveness of predators of sugar beet aphids. Predator numbers within the plots were monitored using pitfall traps (subsection 6.2.1). A many-fold difference in aphid numbers was eventually recorded between test and control plots (40 aphids per plant and no more than 0.3 aphids per plant, respectively).

With egress boundaries (and ingress boundaries, see subsection 7.2.8 below), it is difficult to attribute differences in prey mortality between test and control plots to particular densities of predators, as the densities vary continuously over time. Therefore, it is difficult to calculate predation rates.

7.2.8 PREDATOR ENRICHMENT

With this method, the numbers of predators in the test plots are artificially boosted whereas the numbers in the control plots are not. Predator numbers in the test plots are enhanced by means of **ingress boundaries**, devices which allow predators to move into but not out of the plots. This method was employed by Wright *et al.* (1960) and Coaker (1965) in assessing the effectiveness of predatory beetles attacking cabbage root-fly (Figure 7.11), and by Wratten and Pearson (1982) in assessing the effectiveness of various predators of sugar beet aphids. Wratten and Pearson found that, by using their ingress boundaries, total numbers of harvestmen (Opiliones) captured (by pitfall-trapping) in test plots were 45% higher than in control plots, whereas the numbers of staphylinids, coccinellids and lycosids were increased by a

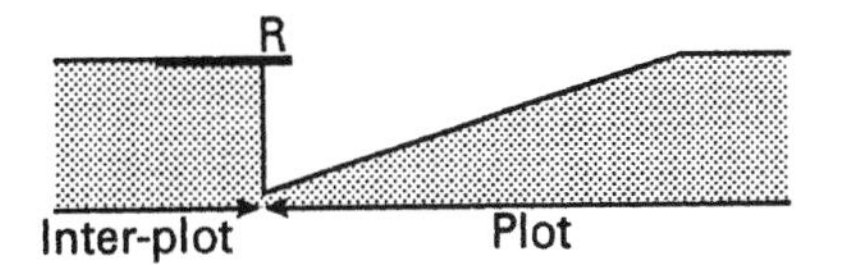

Figure 7.11 Cross-section through boundary used to allow predators to move into, but not out of, experimental plots (i.e. ***ingress* boundary**), in a study of predation of cabbage root-fly (*Delia radicum*). R = roofing felt. The device can be easily converted into an ***egress* boundary** if the roofing felt is suspended from the plot margin, and the sloping part of the trench is on the margin of the interplot. Source: Wright *et al.* (1960), by permission of The Association of Applied Biologists.

maximum of 14%. However, the numbers of aphids eventually recorded in the test and control plots did not differ greatly.

7.2.9 PREDATION AND PARASITISM OF PLACED-OUT PREY

With this method, known densities of prey are placed out in the field for a set period of time, and the numbers of dead or parasitised individuals recorded. The main conditions applying to the method are that prey ought to be placed out in as natural a fashion as possible, using natural densities, locations and spatial arrangements, so that they are neither more nor less susceptible to predation than usual. Also, an alteration in the overall density of prey in the field habitat (and therefore a perturbation to the system) ought to be avoided by having the artificially placed prey replace an equivalent number of prey simultaneously removed from the habitat. To enable the prey to be identified at the end of the experimental period, they may need to be either marked in some way or, if they are mobile, tethered. The marking or tethering technique ought not to affect (increase or decrease) the acceptability of prey to predators. Burn (1982) placed out eggs (stained with Bengal Rose) of the carrot fly (*Psila rosae*) to measure predation by beetles, and Weseloh (1990) placed out larvae (tethered with long thread) of the gypsy moth (*Lymantria dispar*) to measure predation by a complex of predators. Burn

(1982) determined beforehand whether staining of eggs affected the readiness of predators to eat treated eggs. Weseloh (1990), using a type of cage that allowed ants to enter but prevented moth larvae from escaping, compared the degree of predation recorded for tethered larvae with that recorded for untethered larvae. He found the tethered larvae to be more susceptible to predation by ants than untethered ones, and so he used a correction factor to apply to the mortality rates he recorded for tethered larvae placed out in open sites.

Ôtake (1967, 1970) devised a method, involving the use of artificially infested plants containing eggs of known age, to measure field parasitism of planthopper eggs by *Anagrus* (Mymaridae) wasps. The plants were exposed in the field for a set time period, and were then returned to the laboratory and dissected to determine the numbers of parasitised and unparasitised eggs. This **'trap plant' method** was used by Fowler *et al.* (1991) and Claridge *et al.* (1999) to investigate various aspects of egg parasitism of rice-associated planthoppers and leafhoppers.

Provided the conditions set out above are satisfied, or some correction for bias in results can be applied, this method can provide useful data on the effectiveness of natural enemies. Weseloh (1990) concluded that the estimates of daily *per population* predation rates that he obtained by placing out tethered larvae were, if suitably corrected for bias, comparable with estimates obtained by other methods.

The main usefulness of the method, however, lies in providing comparative data, especially indices of predation and parasitism. For example, it can shed light on the relative effectiveness of different natural enemy species within a habitat, or on the effectiveness of a particular natural enemy species in different parts of a habitat (Fowler *et al.*, 1991; Speight and Lawton, 1976). Speight and Lawton (1976) used the method to examine the influence of weed cover on predation by carabid beetles within a habitat. Their study is also interesting in that artificial prey, *Drosophila* pupae killed by deep-freezing, were used.

The term **'prey enrichment'** has been used to describe experiments involving the placing out of prey without the removal of existing prey.

Paired life-tables can be constructed for test (hosts or prey added) and control plots (Van Dreische and Bellows, 1996).

7.2.10 LABELLING OF PREY

With this method, prey are labelled with a dye, a radioactive isotope or a rare element (subsection 6.3.10 describes labelling methods) and released into the field to expose them to natural predation. After an appropriate period of time has elapsed, predators are collected from the field, screened for the label, and the amount of label present quantified. The *per capita* consumption rates of predators are calculated by measuring the label 'burdens' of the insects, and if the field density of predators is known, *per population* estimates of predation can also be estimated. For details, see McDaniel and Sterling (1979).

The technique has little to recommend it, in view of the following:

1. Radioactive labels can be hazardous to health;
2. It is difficult to ensure that all prey carry the same amount of label – there is normally considerable variation;
3. The same level of radioactivity can result from consumption of different numbers of prey;
4. The rate of excretion of the label from an individual predator appears to depend upon the quantity of food subsequently eaten;
5. Labelling can affect the susceptibility of prey to predation. Earwigs (*Forficula auricularia*), for example, prefer to feed on undyed as opposed to dyed eggs of the cinnabar moth (*Tyria jacobaeae*) (Hawkes, 1972);
6. The label can easily and rapidly spread through the insect community by various routes, including excretion, honeydew production, trophallaxis (i.e. by ants), moulting, scavenging on dead prey, and secondary predation;

7. The protocol can be very labour-intensive and, where rare elements and radioisotopes are used, specialised equipment is required;
8. Field populations of prey are disturbed.

Prey labelling has been used to quantify predation by natural enemies of aphids (Pendleton and Grundmann (1954), moths (Buschman *et al.*, 1977; McDaniel and Sterling, 1979; Gravena and Sterling, 1983) and isopods (Paris and Sikora, 1967).

7.2.11 FIELD OBSERVATIONS

With this method, predation is quantified by making field observations, either directly or using video recording techniques, of predators *in situ* (subsections 6.2.6 and 6.3.2 describe methods). Kiritani *et al.* (1972) estimated the number (n) of rice leafhoppers killed by spiders per rice hill per day as follows:

$$n = FC/P \qquad (7.1)$$

where F is the number of predators seen feeding per rice hill during the observation period, C is the total amount of feeding activity in 24 hours expressed in terms of the specified period of observation, and P is the probability of observing predation (the average amount of time, in hours, taken to eat a prey individual, divided by 24 hours). A series of values of n were plotted against time and the area under the curve taken as the total number of prey killed. As noted by Southwood (1978), this method relies upon a high degree of accuracy in observing all instances of predation at a given moment and on values of C and the time taken to eat prey being fairly constant.

Edgar (1970) measured predation by wolf spiders (Lycosidae) in a similar manner to Kiritani *et al.* (1972), while Sunderland *et al.* (1986b) quantified predation of web-spinning money spiders (Linyphiidae) differently, as follows:

$$n = prk \qquad (7.2)$$

where n is the number of aphids killed/m^2/day, p is the proportion of ground covered by webs, r is the rate of aphid falling/m^2, and k is the proportion of aphids entering webs that are killed or die (determined from field observations and laboratory experiments). Using this approach, it was shown that aphid populations could be reduced by spider predation by up to approx. 40%.

Waage's (1983) work on foraging by ichneumonid parasitoids (subsection 6.2.6) shows direct observation to have considerable potential as a method for measuring rates of parasitism, at least for some medium– to large-bodied parasitoid wasp species.

As noted in subsection 6.3.3 the prey 'booty' collected by ants can be taken from the insects upon their return to the nest. A mechanical or photoelectric counter, as suggested by Sunderland (1988), or video recording equipment can enable predation rates to be calculated. The particular prey population being exploited by ants can easily be located by following the insects' trails, so that the prey's population density can be measured.

Video recording of predation and parasitism is likely to prove most fruitful if either the prey are relatively sedentary (e.g. some predators of aphids) or the predators are sedentary (e.g. ant-lion larvae, tiger-beetle larvae).

7.2.12 GUT DISSECTION

Gut dissection (subsection 6.3.8), is one of the simplest techniques for measuring ingestion and predation rates. Also, being a **'post-mortem method'** like serology and electrophoresis discussed below, it has the advantage over methods involving experimental manipulations of the predator-prey system that the results apply directly to an undisturbed, natural system (Sunderland, 1988).

The proportion of dissected predators containing remains of a particular prey in their guts can provide a crude index of *per population* predation rates. More meaningful measures can be obtained by counting the number prey remains, corresponding to prey individuals, present within the guts of predators (e.g. number of prey head capsules), and recording also the throughput time of prey in the gut.

Sunderland and Vickerman (1980) used gut dissection in evaluating the relative effectiveness of different predators of cereal aphids, by multiplying the proportion of such insects that contained aphid remains during the aphid population increase phase by the mean density of predators (ground examination samples) at this time. The species with the highest indices were considered to be the most valuable in constraining the build-up of aphids in cereal fields.

For other studies employing the technique, see Andow (1992) and Cook *et al.* (1994).

7.2.13 SEROLOGY

Field predation rates

The serological methods discussed in subsection 6.3.9 in relation to the determination of dietary range in field-collected predators have mainly been aimed at quantifying field predation. Various models are available (Table 7.1); these are based, to varying extents, upon the following variables: predator density, the proportion of predators testing positive for the prey, the detection period of the prey remains in the predator's gut (= the 'prey antigen half-life'), the prey biomass recovered, the mean proportion of the meal remaining, the *per capita* predation rate as measured in the laboratory or outdoor insectary, and the *per capita* predation rate measured as a function of prey density. For reviews of the models, see Sopp *et al.* (1992) and Naranjo and Hagler (2001).

Although shown by Sopp *et al.* (1992) to provide more accurate estimates of predation rate than its predecessors, their model (5 in Table 7.1) still involves several assumptions (some of which are common to other models) which, when violated, will reduce the accuracy of the predation rate estimates:

1. The detection periods measured in the laboratory are realistic estimates of those in the field;
2. The term ft_{DP} relates to the mean time since ingestion of prey materials for the population under study;
3. The predator takes discrete meals which are digested and voided before another meal is taken. If the meal comprises several prey individuals or, in the case of those predators with long detection periods, several meals are taken in rapid succession, they are regarded as a single meal, and so predation rate will be underestimated;
4. There is not a variable degree of partial prey consumption;

Table 7.1 Predation rate models employed with serological methods.

Model	Source
1. pd/t_{DP}	Dempster (1960, 1967)
2. pr_id	Rothschild (1966)
3. pr_id/t_{DP}	Kuperstein (1974, 1979)
4. $[\log_e(1-p)]d/t_{DP}$	Nakamura and Nakamura (1977)
5. Q_od/ft_{DP}	Sopp *et al.* (1992)
6. $pdr_i(N)/t_{DP}(\theta)$	Naranjo and Hagler (2001)

r = *per population* field ingestion or predation rate (biomass or numbers of prey; to convert the former to the latter, the mean weight of individual prey in the field needs to be known).
r_i = *per capita* ingestion or predation rate measured in the laboratory or an outdoor insectary.
p = the proportion of field-collected predators found to contain prey remains.
d = predator population density.
t_{DP} = detection period for prey in the predator's gut (a function of temperature, see below)
Q_o = the quantity of prey recorded in the predator's gut (note that the immunodot assay technique (subsection 6.3.9) cannot be used to record this, see Greenstone, 1995).
f = the proportion of food remaining in the predator's gut.
N = prey density
θ = temperature

5. The predator density in the field is accurately known (this is rarely the case);
6. Both the predator sample and the amount of prey biomass present in predator guts are representative of the predator population as a whole. This is related to sample size and the sampling regime adopted;
7. There is no cross-reactivity between the prey species and non-target prey species (if cross-reactivity is a problem, it can be overcome by the use of monoclonal antibodies, subsection 6.3.9)
8. The presence of prey remains is the result of predation and not of scavenging, secondary predation or feeding on alternative prey.

The Naranjo and Hagler (2001) model (6 in Table 7.1) incorporates more biological realism by using the predators' functional response. Comparisons were made with the Dempster (1960) and the Nakamura and Nakamura (1977) models; these were found to either overestimate (model 1) or to underestimate (model 4), predation rates. Comparisons were not made with the Rothschild (1966), Kuperstein (1979) and Sopp *et al.* (1992) models. However, Naranjo and Hagler argue that the first two of these (models 2 and 3) would have greatly overestimated prey attack rates in their study system, and that the last (5, i.e. Sopp *et al.*'s) has limited applicability, due mainly to the problems inherent in using Q_0 (it remains to be seen whether Naranjo and Hagler are correct on this point).

It is important to know how the detection period (t_{DP}) can vary with: (a) meal size, (b) temperature, and (c) the presence of non-target prey items in the gut. Sopp and Sunderland (1989) demonstrated the effects of (a) and (b) on the detection period and antigen decay rate (the rate of disappearance of detectable food) in the beetles *Bembidion lampros* (Carabidae) and *Tachyporus hypnorum* (Staphylinidae) and the spider *Erigone atra* (Linyphiidae). Previously starved predators were fed freshly-killed aphids and were then kept at one of a range of temperatures for varying periods. The proportion of prey remaining in the gut at intervals after feeding was measured and plotted (Figure 7.12a). Curves were fitted to the (transformed) data (Figure 7.12b) and the detection period estimated (this is just one method of detection period measurement; Symondson and Liddell, 1993e, provide a review).

Sopp and Sunderland (1989) concluded the following from their study and other studies:

1. Usually, within a predator species, the detection period declines with increasing temperature; larger species tend to have longer detection periods, possibly because of the larger meal sizes, but within a species meal size appears to have little effect upon the detection period. Spiders have very long detection periods, even at high temperatures, perhaps because of their ability to store partially digested food in gut diverticula;
2. In most predators the antigen decay rate follows a negative exponential form; the majority of the detectable antigens disappear within one-third of the detection period. Agustí *et al.* (1999a) found that, whereas an exponential decay model was appropriate for predatory heteropteran bugs (*Dicyphus tamaninii*) that had eaten one *Helicoverpa armigera* egg, a linear model gave a better fit for individuals that had consumed ten eggs.

Symondson and Liddell (1993e) point out that the delectability of invertebrate remains in the crops of predators such as carabid beetles is influenced not only by the antigen decay rate but also the residence time of a meal in the crop and gizzard (i.e. the fore gut). If the rate of through-put of prey material happens to be less than the antigen decay rate, then there will be a discrepancy between the true proportion of prey material present and the amount estimated from an ELISA. Quantification of this discrepancy would provide a means of estimating original meal size, when the time since feeding can be estimated; such quantification requires crop weight loss and antigen decay rate to be measured as separate variables (for protocol, see Symondson and Liddell's paper).

Harwood *et al.* (2001b) have shown that it is possible to test whether secondary predation is

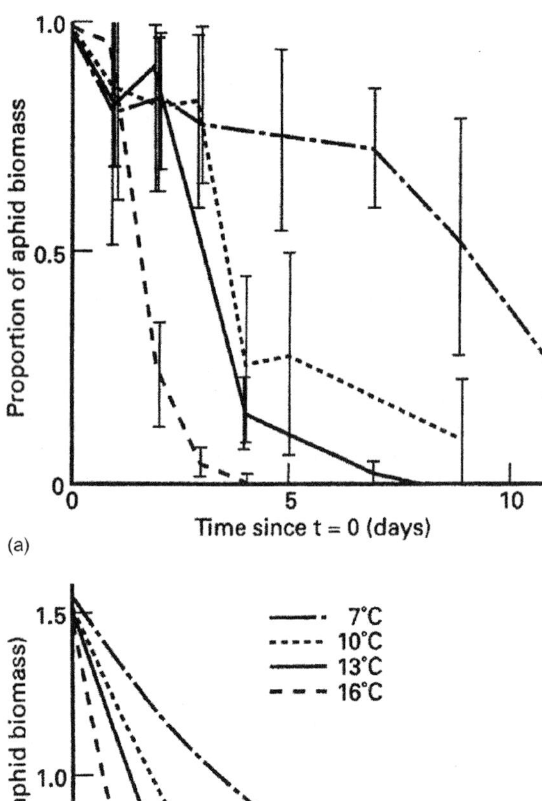

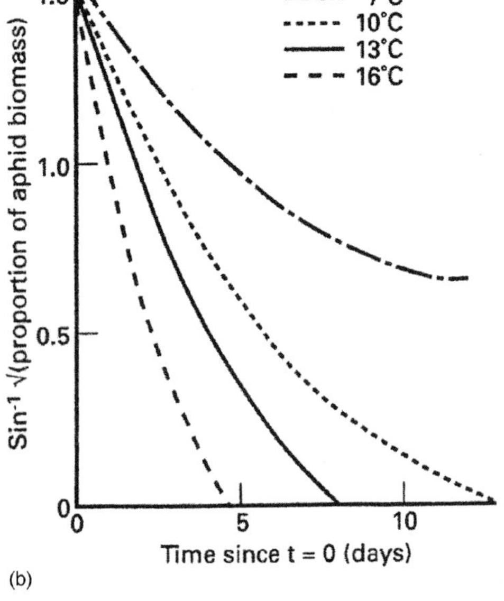

Figure 7.12 (a) The proportion of aphid biomass present in the gut, immediately after feeding, that is subsequently detected at various time intervals in the carabid beetle *Bembidion lampros*. (b) Antigen decay rate curve. Symondson and Liddell (1993e) expressed the antigen decay rate differently, and also took account of the loss in weight of the predator's crop in estimating meal size (see their paper for details). Source: Sopp and Sunderland (1989).

likely to be a significant confounding factor in a particular study, if a serological technique is applied in such a way as to maximise the possibility of such detection.

Predation Indices

Sunderland *et al.* (1987a) compared different polyphagous predators in terms of their probable value as cereal aphid predators, by calculating, for each predator species, the following index:

$$P_g d / D_{max} \tag{7.3}$$

where P_g is the percentage of predators testing positive using ELISA, D_{max} is the maximum period over which prey antigens can be detected in any individual of a given species, and d is the mean predator density. Spiders tended to have the highest indices.

7.2.14 ELECTROPHORESIS

Electrophoresis, like serology, has been used in quantifying predation by fluid-feeding arthropod predators, albeit less commonly (for a review, see Solomon *et al.*, 1996). As with serology and gut dissection, the proportion of predators testing positive for prey contents can easily be determined (subsection 6.3.11), but to obtain meaningful quantitative information on predation, the quantity and detection period of prey materials ingested also need to be known. We have little further to say about electrophoresis, as it has been superseded by ELISA. The latter is not only a far more sensitive method for determining the quantity of prey proteins in the guts of predators, but also it requires less time to test gut contents.

With the aforementioned indirect and post-mortem methods (electrophoresis, serology, gut dissection, labelling of prey) converting ingestion rate to predation rate can generate serious errors. For example, if scavenging occurs, the true predation rate will be overestimated. Sunderland (1996), in discussing sources of potential error in estimating predation rates, points out that the latter (predation being defined loosely) can also be *underestimated*, because predators may kill or wound prey without ingestion occurring (e.g. 'wasteful killing' by satiated predators, see Johnson *et al.*, 1975). For a discussion of the multiplicity of factors that can lead to inaccurate estimates of predation rates, see Sunderland's (1996) review.

7.3 THE ROLE OF NATURAL ENEMIES IN INSECT POPULATION DYNAMICS

7.3.1 INTRODUCTION

Having reviewed some of the methods by which insect mortality due to natural enemies can be quantified, we now turn our attention to the more difficult task of assessing its dynamic significance. Mortality factors acting on an insect population can cause three possible dynamic changes. They can:

1. Affect the average population density;
2. Induce fluctuations in numbers;
3. Contribute to the regulation of population numbers.

Of the three, it is undoubtedly the contribution that natural enemies make to population regulation which has most occupied the minds of ecologists over the years.

Factors which regulate population numbers can act either by:

1. Returning populations towards a notional **equilibrium** number after some perturbation (i.e. **stabilising** population numbers);
2. Restricting population numbers within certain limits, but allowing fluctuations in numbers (e.g. cycles) within those limits (Murdoch and Walde, 1989).

For a factor such as parasitism or predation to regulate, the strength of its action must be dependent on the density of the population affected. That is, it needs to be **density-dependent**, its *proportional* effect being greater at high population densities and smaller at low densities (Figure 7.13; cf. **density-independent** factors). Density-dependence operates through **negative feedback** on population numbers, which may involve changes in the rates of reproduction, dispersal and immigration as well as changes in mortality. If the proportion of hosts parasitised varies with changing host density, either temporally or spatially (subsection 7.3.10), this can profoundly affect the dynamics of the interaction. As we shall see (subsections 7.3.4, 7.3.7), density-dependent factors can also affect average population levels and can, under certain conditions, induce perturbations too (subsection 7.3.4).

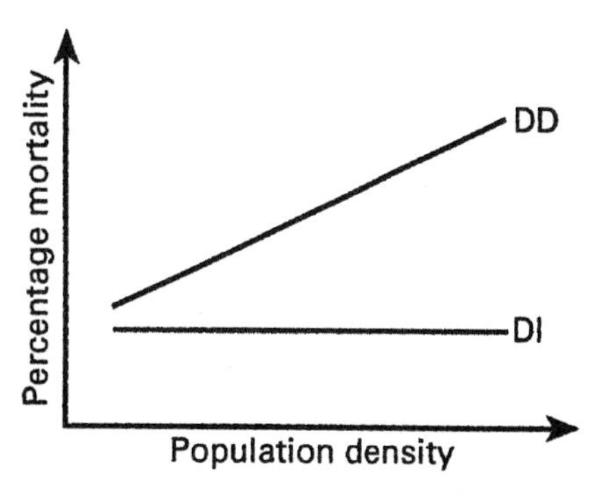

Figure 7.13 The negative feedback effect of a density-dependent mortality factor (DD) in which proportional mortality increases with population density (*cf.* density-independent factors (DI) in which proportional mortality is unrelated to population density).

We should make it clear at this stage that our discussion of population dynamics and population regulation is specifically aimed at issues and techniques relating to the actions of natural enemies. In recent years a number of issues which do not focus *directly* on natural enemies have received a great deal of attention in the literature, for example, the detection of density-dependence from time-series data (Godfray and Hassell, 1992; Holyoak, 1994; Rothery *et al.*, 1997; Hunter and Price, 1998; Turchin and Berryman, 2000; Berryman and Turchin, 2001), and the nature and significance of deterministic chaos (May, 1974a; Gleick, 1987; Berryman, 1991; Logan and Allen, 1992; Godfray and Grenfell, 1993; Hastings *et al.*, 1993; Cavalieri and Kocak, 1995; Desharnais *et al.*, 2001). We make no attempt to cover these important topics, but simply refer the reader to the references given above.

We begin by addressing the problems associated with using percentage parasitism estimates to assess the impact of parasitoids on host populations (subsection 7.3.2). We then discuss what is perhaps the simplest (but least insightful) technique of assessing the impact of natural enemies, that of comparing their numbers with those of the prey or host populations (subsection 7.3.3). We then review the more conventional methods of life-table analysis (subsection 7.3.4) and show how simple population models can be derived from the information obtained. The limitations of the life-table approach are discussed, showing the need for supplementary field experiments (e.g. convergence and factorial experiments) (subsection 7.3.5). Next, we discuss how the important methodology of experimental component analysis can be applied using both analytical and simulation models (subsections 7.3.6, 7.3.7, 7.3.8, 7.3.9), and go on to examine some of the more contentious issues which have developed out of this approach (subsection 7.3.10).

7.3.2 THE PROBLEM OF 'PERCENT PARASITISM'

A point which is perhaps worth stressing at this stage is that the importance of natural enemies in host or prey population dynamics may have little to do with the degree of mortality which they cause *per se*, a fact which is often misunderstood by researchers in pest management. Many publications, for example, have reported high 'percent parasitism' in insect pest populations, the clear implication being that this mortality is likely to contribute, in a major way, to reducing average population levels and/or to regulating populations. Unfortunately, such inferences may not be justified, for reasons which will become apparent later on in this chapter.

'Percent parasitism' may also be a poor measure of the impact of parasitoids on host population dynamics for a number of other reasons. First, as Van Driesche (1983) pointed out, the number and timing of samples taken are usually inadequate for the task. To assess a parasitoid's contribution to host population mortality, it is the percentage attacked for the *generation* which must be determined and this may best be done within the context of a complete life-table study of the host population (subsection 7.3.4). Furthermore, percent parasitism does not take account of other forms of parasitoid-induced mortality, such as host-feeding (Jervis and Kidd, 1986; Jervis *et al.*, 1992a), which may sometimes outweigh parasitism in their contribution to host mortality. The degree of temporal synchrony of parasitoid and host population may also be an important factor in determining how well sampling estimates generational levels of parasitism. Using a series of simple theoretical models Van Driesche (1983) was able to establish that:

1. Where susceptible hosts are all present before parasitoids begin ovipositing, and the parasitoid oviposition period does not overlap with the start of parasitoid emergence (Figure 7.14a), then the peak percent parasitism sampled can give a good estimate of generational percent parasitism;
2. Where the situation in (1.) prevails, but hosts begin to develop to the next (unsusceptible) stage before all parasitoids have emerged, this will cause the peak percent parasitism to overestimate generational parasitism (Figure 7.14b);
3. Where the situation in (1.) prevails, but parasitoids begin to emerge before all parasitoid oviposition is complete, then peak percent parasitism will underestimate generational parasitism (Figure 7.14c);
4. If hosts enter the susceptible stage gradually and concurrently with parasitoid oviposition, and if host entry and exit do not overlap appreciably and parasitoid oviposition and emergence do not overlap appreciably (point X in Figure 7.14d), then a sample of percent parasitism at this point can accurately estimate generational parasitism;
5. If hosts enter the susceptible stage gradually and concurrently with parasitoid oviposition, but if both hosts and parasitoids enter and leave the system at rates other than in (4.), then samples of percent parasitism will bear little relation to generational percentage parasitism.

All of the above conclusions are based on the assumption that host mortality is caused solely

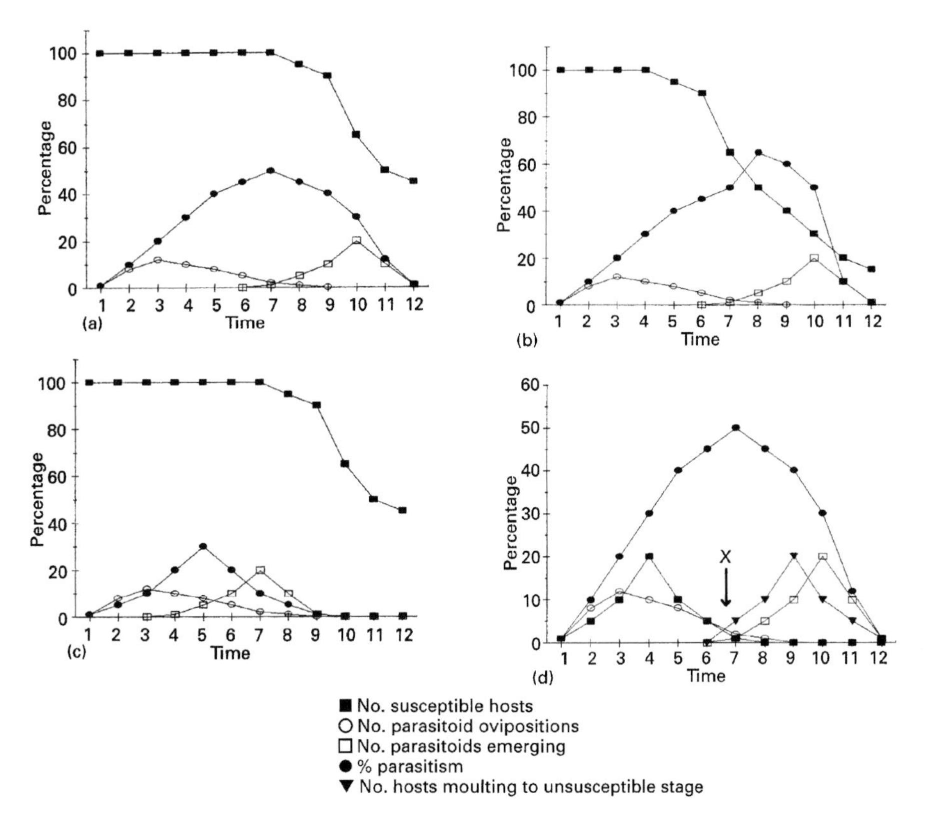

Figure 7.14 Synchrony of parasitoid and host populations may affect the accuracy of estimates of generational percent parasitism (see text for an explanation). Adapted from Van Driesche, 1983)

by parasitism. Where this restriction does not apply (possibly most cases!), correspondence between samples and generational parasitism levels will be even harder to determine. Also, if the sampling method is in any way selective towards either parasitised or unparasitised hosts (subsection 6.2.9), this will introduce a further error into the estimate (Van Driesche, 1983). Van Driesche *et al.* (1991) suggested some ways of circumventing the above problems. One is to measure recruitment to both the host and the parasitoid (parasitised hosts) populations continuously, total recruitment to both populations being found by summing the recruitment values for all intervals. The ratio of total parasitoid recruitment to total host recruitment provides an unbiased estimate of total losses to parasitism. Another method uses death rate measurements from field samples. If individuals are collected at frequent intervals, reared under field temperatures, and the proportion dying from each cause recorded from one sample to the next, then the original percentage of the sample that was parasitised can be estimated. Gould *et al.* (1990) and Buonaccorsi and Elkinton (1990) provide equations for the calculations. The method requires that all hosts have entered the susceptible stage before the first sample and that no host recruitment occurs during the sampling period. Details and examples of these and other techniques can be found in Van Driesche and Bellows (1988), Bellows *et al.*, (1989a) Van Driesche *et al.* (1991) and Ruiz-Narvaez and Castro-Webb (2003). Ruiz-Narvaez and Castro-Webb (2003) devised a statistical method for estimating percentage parasitism when host and parasitoid phenologies are unknown.

7.3.3 CORRELATION METHODS

In field populations a useful preliminary indication of the impact of natural enemies can often be obtained by statistically correlating their numbers against those of their prey or hosts. Significant positive or negative correlations may imply some causative association, which can then be tested by further investigation. Correlation alone, of course, should not be taken as proof of causation. A high positive correlation may indicate a degree of prey specificity on the part of the predator (Kuno and Dyck, 1985), which might be expected to show a rapid numerical response to variations in prey density (subsection 7.3.7 gives a definition). Heong *et al.* (1991), for example, found that the numbers of heteropteran bugs and spiders, which are major predators of Homoptera Auchenorrhyncha in rice, correlated positively with the numbers of Delphacidae and Cicadellidae. A positive correlation would also be accentuated by a low predator attack rate and/or a prey species with a relatively slow rate of population growth (Figure 7.15a).

Negative correlations, on the other hand, may indicate a slow or delayed numerical response by the predator to changing prey density. These responses are commonly shown by highly polyphagous predators which may 'switch' to feeding on a prey type only after it has increased in relative abundance in the environment (section 1.11). Negative correlations are also more likely to be associated with prey species which tend to show rapid changes in abundance, or with predators having a high attack rate (Figure 7.15b). For example, negative correlations between aphids and coccinellid beetles are frequently found on lime trees during the summer, and can be explained by the rapid rate of increase in the aphid population in the spring, coupled with the slow rate of response by the coccinellids (Dixon and Barlow, 1979). Later in the season predator numbers increase, forcing the already declining aphid population to crash (Figure 7.16a). Syrphid predators, on the other hand, can show a rapid numerical response to increasing cereal aphid populations, producing a positive within-season correlation (Chambers and Adams, 1986) (Figure 7.16b).

The tentative conclusions afforded by correlation techniques should only be drawn with extreme caution, and then only with a detailed appreciation of the biologies of the species involved. In particular, it must be remembered that correlations can be created just as easily by predator populations *tracking* changes in prey numbers, as by *bringing about* those changes. Also, absence of any correlation should not be taken to imply that predators do not have any

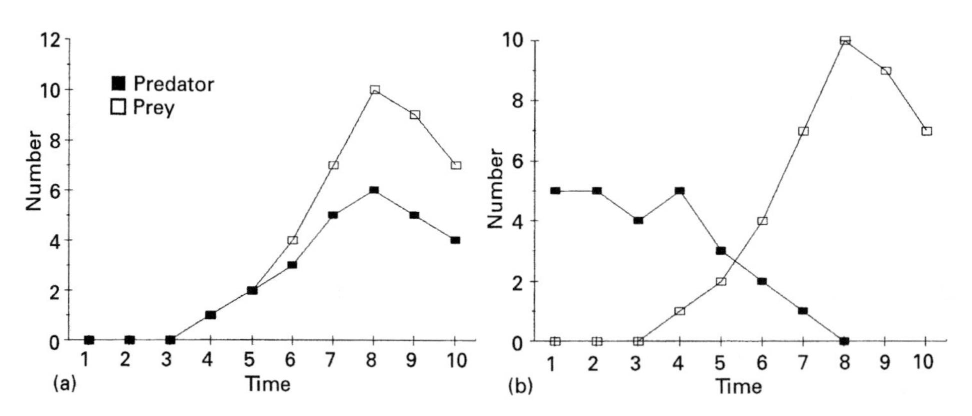

Figure 7.15 Relationships between predator and prey population numbers which produce either positive or negative correlations: (a) a positive correlation between predator and prey numbers produced by a slow rate of prey increase coupled with a relatively low predator attack rate, such that prey numbers are not reduced, while predator numbers are still rising; (b) a negative correlation between predator and prey numbers caused by predators depressing prey numbers, which only increase after predator numbers have declined.

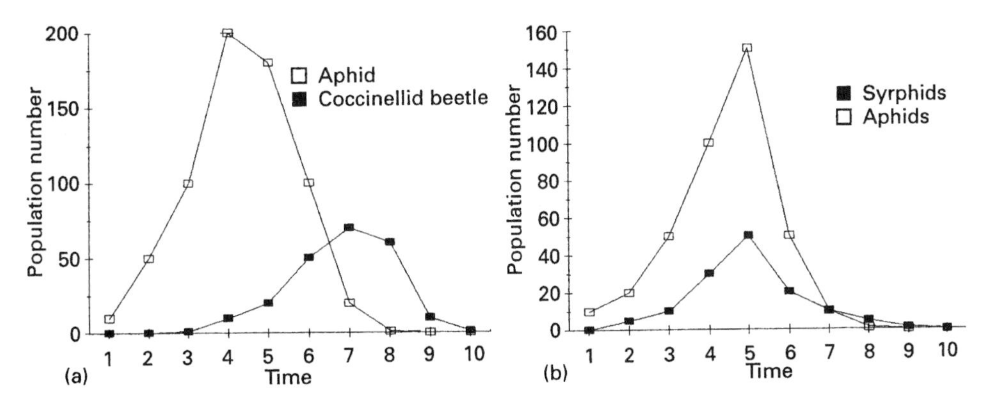

Figure 7.16 (a) A negative relationship between aphid and coccinellid beetle numbers on lime trees. Predator numbers increase slowly in response to aphid numbers and only reach their highest densities after aphid numbers have already declined; (Schematic representation based on information given in Barlow and Dixon, 1979). (b) Syrphids show a rapid numerical response to increasing cereal aphid numbers, declining as aphid numbers decline. This produces a positive correlation between predator and prey numbers. Schematic representation based on Chambers and Adams (1986).

impact. We might expect a lack of any correlation in cases where predators have features intermediate to the aforementioned extremes.

7.3.4 LIFE-TABLE ANALYSIS

Introduction

The concept of the life-table has already been introduced in section 2.11.2, in relation to the calculation of intrinsic rates of increase. Here, we are concerned with using life-tables of a somewhat different nature to determine how specific mortality factors (e.g. a particular natural enemy species) affect prey or host population dynamics. For example, is the mortality density-dependent or density-independent? Is there evidence for delayed or over-compensating density-dependence? In short, does mortality from this source tend to regulate numbers at, or disturb numbers from, a certain level? To answer these and related questions, we need to take life-tables apart and analyse the specific mortalities separately. Because some insect populations (e.g. aphids) tend to have generations which overlap in time, while others do not, two quite different approaches have been developed for each category, respectively the time-specific life-table and the age-specific life-table.

Age-specific Life-Tables

The life-table approach pioneered by Pearl and Parker (1921), Pearl and Miner (1935) and Deevey (1947) was extended to insects with discrete generations by the single-factor analysis of Morris (1959) and the **key factor analysis** of Varley and Gradwell (1960) (the latter sometimes incorrectly referred to as *k*-factor analysis). Of the two methods, the latter has been most widely used in population ecology (Podoler and Rogers, 1975) and will be the one concentrated upon here. For those readers interested in the Morris method, details are provided by Southwood and Henderson (2000). Varley and Gradwell's method is given a very detailed treatment suitable for the beginner in Varley *et al.* (1973). As the latter book is now, alas, out of print, we feel it is worthwhile discussing the procedures in detail, especially since there have been subsequent developments.

The usefulness of the Varley and Gradwell approach depends on the availability of sequential life tables for a number of generations of a univoltine population. In temperate regions, for example, it is commonly the case that insect populations overwinter as eggs and develop through a number of discrete stages in the spring and summer (Figure 7.17). The adults then mature in the autumn to lay a new generation of overwintering eggs before dying. In this situation, generations remain completely separate. By obtaining population density estimates for the numbers entering each stage in the life cycle, it is then possible to construct a **composite life-table**, consisting of a sequence of independent life-tables for each generation (Table 7.2). The numbers entering each stage can be estimated in two different ways: (a) by direct assessment of recruitment (for example, by measuring fecundity or fertility, section 2.7), or (b) by indirect calculation from counts of stage densities. Several techniques are available which provide an estimate by the second route, and these are reviewed by Southwood and Henderson (2000). The graphical method of Southwood and Jepson (1962), for example, involves plotting the density of a stage against time and dividing the area under the plot by the average duration of the stage (mean development time). This yields an unbiased estimate of the number entering the stage if there is either no mortality, or the mortality occurs only at the end of the stage. Any mortality during the stage will result in underestimation. Bellows *et al.* (1989b) provide an extension to this method which can be used for interacting host and parasitoid populations. A number of other methods are discussed by Manly (1990).

It should be noted from Table 7.2 that the actual density estimates of numbers entering each stage are retained in the life-table, rather than corrected to a common starting number (*cf.* Table 2.2). The reason for this will become clear. Where stage mortalities can be partitioned into a number of definable causes, these are quantified separately in the table. In this way it may be possible to build similar life-tables for particular natural enemies. Varley *et al.* (1973) provide a number of rules to follow in the construction of the table. These are:

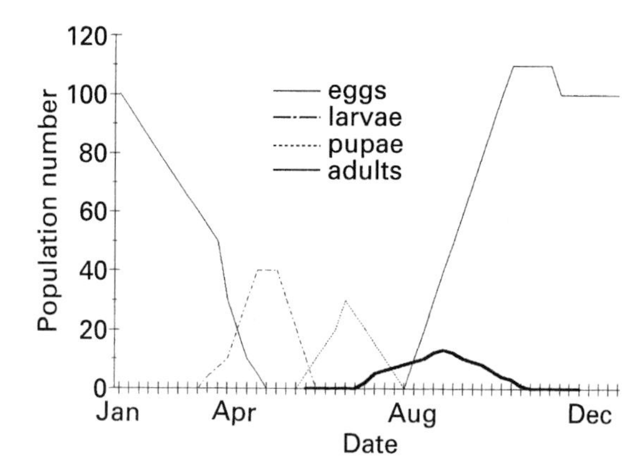

Figure 7.17 Schematic life-cycle of a typical temperate-zone univoltine insect population.

1. Where mortalities are reasonably well separated in time, they are treated as if they are entirely separated with no overlap;
2. Where events overlap significantly in time, they can be considered as if they are exactly contemporaneous;
3. All insects must be considered either as live and healthy or, alternatively, as dead or cer-

Table 7.2 Composite life-tables for six generations of a hypothetical insect population with discrete generations. Each k-value is calculated as $k = \log_{10}$ before mortality $-\log_{10}$ after mortality. $K = k_1 + k_2 + k_3$ (Note: whilst such life-tables have traditionally been presented in columns, putting them in rows (as is done here) makes spreadsheet regression calculations easier.)

Year	Eggs	k_1	Larvae	k_2	Pupae	k_3	Adults	K
1	1000	0.824	150	0.398	60	1.080	5	2.302
2	800	0.426	300	0.685	62	1.190	4	2.301
3	1200	0.681	250	0.455	50	0.824	12	1.960
4	700	0.942	50	0.204	50	0.699	10	1.845
5	500	0.553	140	0.301	70	0.766	12	1.620
6	1200	1.000	120	0.150	85	1.230	5	2.380

tain to die from some cause. For example, parasitised larvae are scored as certain to die, with the parasitoid recorded as the cause of death;

4. No insect can be killed more than once. Where hosts are attacked by two parasitoid species, death of the host is credited to the first parasitoid. If the second parasitoid emerges as the victor, it is taken as the cause of death of the first parasitoid. The second attack is thus entered in the life-table of the first parasitoid but not in that of the host.

Although somewhat arbitrary, rules such as these are necessary to balance the budget. However, as we explain below, conclusions from the analysis may unfortunately be sensitive to the rules adopted.

By converting the data in Table 7.2 to logarithms ($\log_{10}$), we can calculate for each successive mortality, in any generation:

$k = \log_{10}$ number before mortality - $\log_{10}$ number after mortality

where k is a measure of the proportion dying from the action of the mortality factor. In practice these calculations are easily carried out using a spreadsheet programme (Table 7.2 caption), which can also be used for the regression analyses (see below). Within each generation, we can thus determine a sequence of k-values, $k_1, k_2, k_3, \ldots k_n$, corresponding to each successively acting defined mortality up to the adult stage (Table 7.2). Strictly speaking, this should be up to the stage before reproduction begins, any pre-reproductive mortality being counted as separate k-factors. Mortality during the adult stage can be counted as one or more k-factors acting on the adults, or alternatively as a k-mortality acting on the next generation of eggs (Varley *et al.*, 1973). The final post-reproductive mortality to act on a generation, i.e. that which brings generation numbers to zero, contributes nothing to between-generation variation in numbers and is not included in the analysis. To do so would cause two problems. First, we are dealing here with the $\log_{10}$, of numbers, so how would we treat zero values? Second, the final reduction in adult numbers to zero, is by its nature density-dependent. In a sense, the ultimate extreme of regulation is to return a population to an equilibrium of zero! We illustrate the point by including this spurious density-dependence in our analysis (Figure 7.20). The sum of all the k-values up to, but not including this last mortality, provides us with a measure of total generation mortality K, i.e.

$$k_1 + k_2 + k_3 \ldots\ldots\ldots k_{n-1} = k \qquad (7.8)$$

The advantages of using k-values instead of percentage mortalities lie in the ease of calculation and the fact that k-values can be added to give a measure for total generation mortality (K) (adding percentages would have no meaning).

Two basic questions can be answered from an analysis of the table at this stage:

1. Which factor or factors contribute most to variations in mortality from generation to generation, i.e. the so-called **key factor(s)** causing population change?
2. Which factors contribute to regulation of population numbers?

Key Factors

The answer to the first question can often be obtained from a graphical representation of the data. Plotting the k-values against generation may be enough to reveal the key factor(s) causing population change (Figure 7.18). Here, variations in k_3 between generations most closely follow variations in overall mortality (K), indicating that k_3, is the key factor. Note that the key factor is not necessarily the factor causing greatest mortality (k_1 in this case).

Sometimes, a simple graphical inspection may not be enough to reveal the key factor, in which case the statistical method of Podoler and Rogers (1975) can be employed. This involves regressing each k-value against total generation mortality (K), the mortality with the greatest slope (b) being the key factor. In our example k_3 is confirmed as the only significant key factor (Figure 7.19). Where more than one factor is found to contribute, a hierarchy of significance can be constructed.

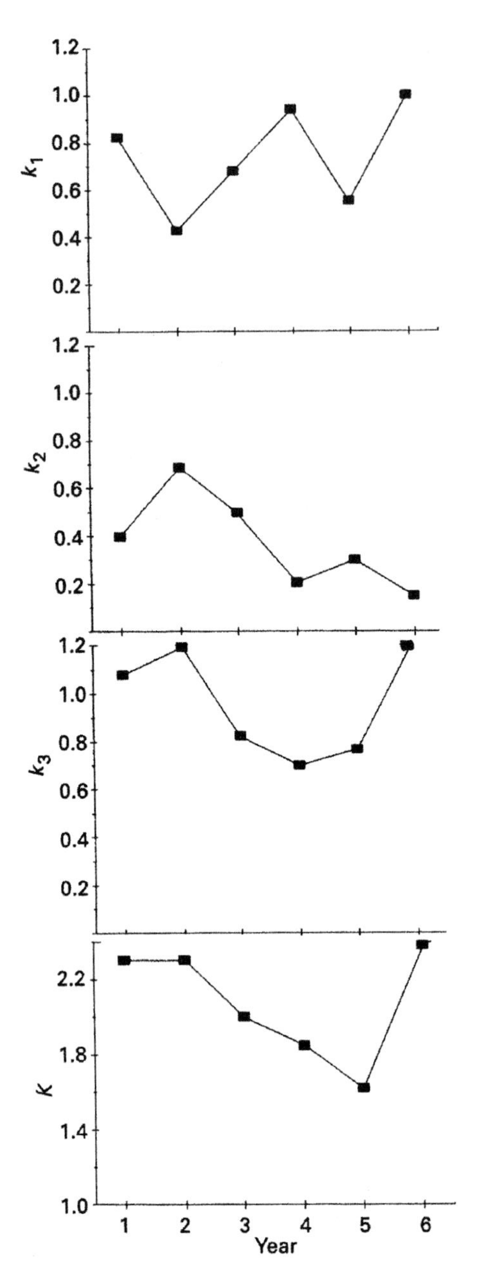

Figure 7.18 Key factor analysis of the mortalities acting on a hypothetical insect population (see Table 7.2 for data).

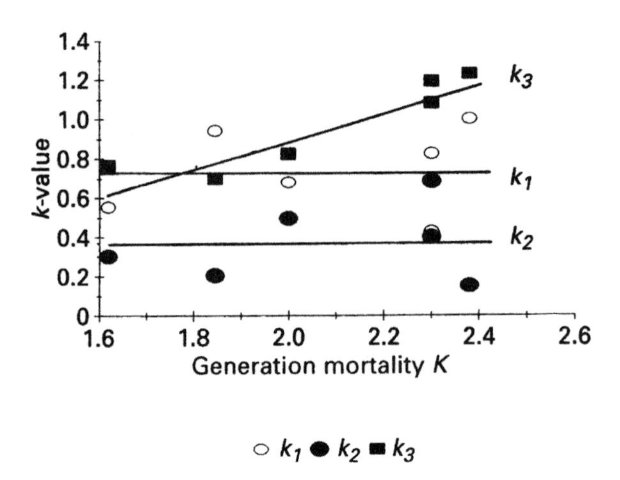

Figure 7.19 Podoler and Rogers' method for identifying key factors. The factor with the greatest slope (k_3 in this case) is the key factor causing population change ($k_3 = 0.68K - 0.46$; $R^2 = 0.84$).

Strictly speaking, the Podoler and Rogers' procedure for identifying key factors is not statistically valid, in that it contravenes the basic rules of regression. These are that the axes should be independent of each other and the independent variable should be error-free. Clearly, K consists of the k-values against which it is being regressed, and it is also subject to sampling error. Where the regression relationship of the putative key factor is not clear cut, a simpler expedient may be to use the correlation coefficients, which are not subject to the same restrictions. In this case, the key factor would be the one with the highest correlation between k and K, (maximum $r=1$). Manly (1977) devised an alternative method based on multiple regression analysis, whilst the problems of sampling error have also been considered by Kuno (1971). As we shall now see, a similar problem with regression is confronted in the detection of density-dependence from life-table data.

Detecting Density-dependence

Assessing which factors contribute to regulation of the population again involves plotting each k-value, this time against the $\log_{10}$ density on which it acts (i.e. *before* the mortality). In our example (Figure 7.20) the plot of k_1 against log density of eggs contains six data points, corresponding to each generation. Similarly, k_2 is plotted against $\log_{10}$ density of new larvae, again with six data points, and so

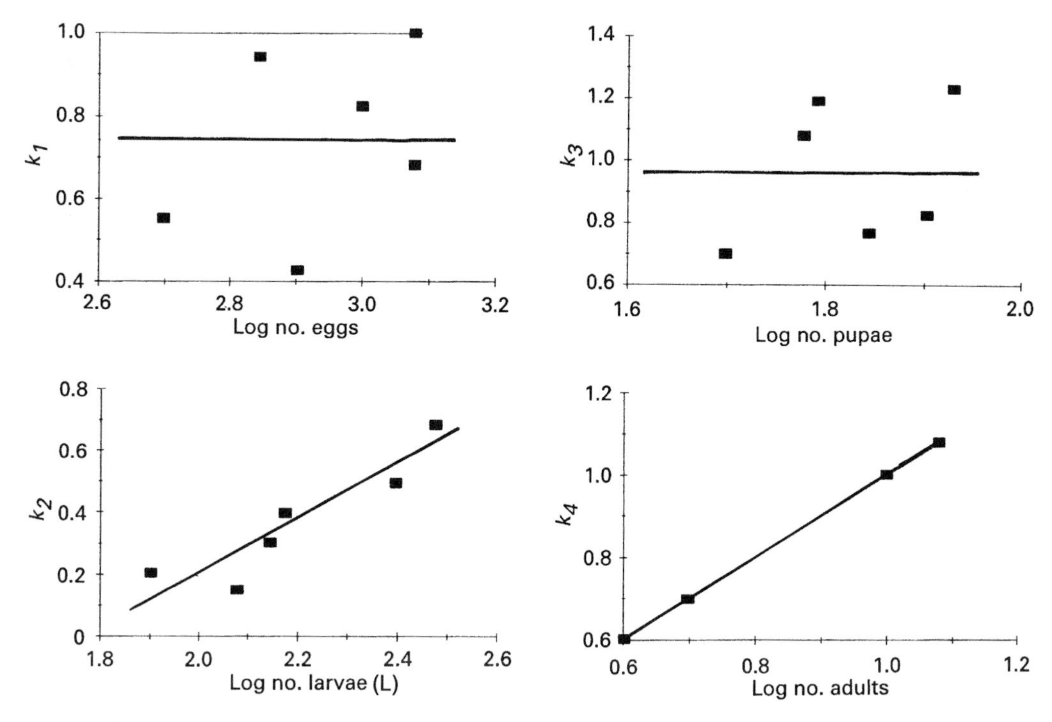

Figure 7.20 The identification of density-dependent factors from life table data. k-values for the different mortalities are plotted against the population densities on which they acted. In this case, only k_2 is significantly density-dependent. (k_2 = 0.86L–1.52; R^2 = 0.84; k_1 = 0.74; k_3, = 0.96.) k_4 is the last mortality to act, bringing numbers down to 0 (or in this case 1, which was used to make the log calculations workable). This mortality is, by its nature, always density-dependent (see text), but is not included in the analysis, as it contributes nothing to population variation or regulation.

on. Remembering that each k-value is a measure of *proportional* mortality; positive relationships for any of these plots would indicate that mortality is acting in a density-dependent fashion. A horizontal slope would indicate density-independence, while a negative slope would indicate inverse density-dependence. Regression analysis is generally employed to calculate the significance of the slopes. Here, the only significant density-dependence is found in k_2. However, the problem of statistical validity (mentioned above in relation to Podoler and Roger's method) again arises. As k-values are calculated in the first place from $\log_{10}$ densities, the two axes are not independent. Moreover, the independent variable ($\log_{10}$ density), estimated from population samples, is not error-free. To overcome the problem, Varley and Gradwell (1968) suggest a 'two-way regression' test, which involves both the regression of $\log_{10} N_t$ (initial density) on $\log_{10} N_{t+1}$ (final density) and $\log_{10} N_{t+1}$ on $\log_{10} N_t$. If both regressions yield slopes significantly different from $b = 1$ and are on the same side of the line, then the density-dependence can be taken as real. This method may be unnecessarily stringent (Hassell *et al.*, 1987; Southwood *et al.*, 1989), requiring that density-dependence remains apparent when all sampling errors are assumed to lie firstly in the estimates of N_t, then in N_{t+1}. Bartlett (1949) provided an alternative regression method in which sampling errors are distributed between both axes.

If density-dependence is accepted, then the regression coefficients can be taken as a measure of the *strength* of the density-dependence. The closer b is to 1, the greater the stabilising effect of the mortality. A slope of $b=1$ will compensate perfectly for any changes in density at this stage (**exact compensation**), while a slope of $b<1$ will be unable to compensate completely for any changes (**undercompensation**). Slopes of $b>1$ imply **overcompensation**, the significance of which will become clear later.

A further insight into the nature of density-dependence can also be obtained by again plotting each k-value against the $\log_{10}$ density on which it acts, but in a **time sequence** (Varley and Gradwell, 1965; Figure 7.21). Different factors trace a different pattern depending on their mode of action; density-independent factors show an irregular, zigzag pattern (Figure 7.21a), while direct density-dependent factors show a more discernible straight-line pattern of points clustered within a narrow band (Figure 7.21b). A spiral pattern (Figure 7.21c) indicates delayed density-dependence, in which the action of the k-mortality is not felt until one or two generations hence. Insect parasitoids frequently act in this way for reasons which will be explained in subsection 7.3.7. Manly (1988) provided a statistical test for spiral patterns based on a comparison of the internal angles of the spiral.

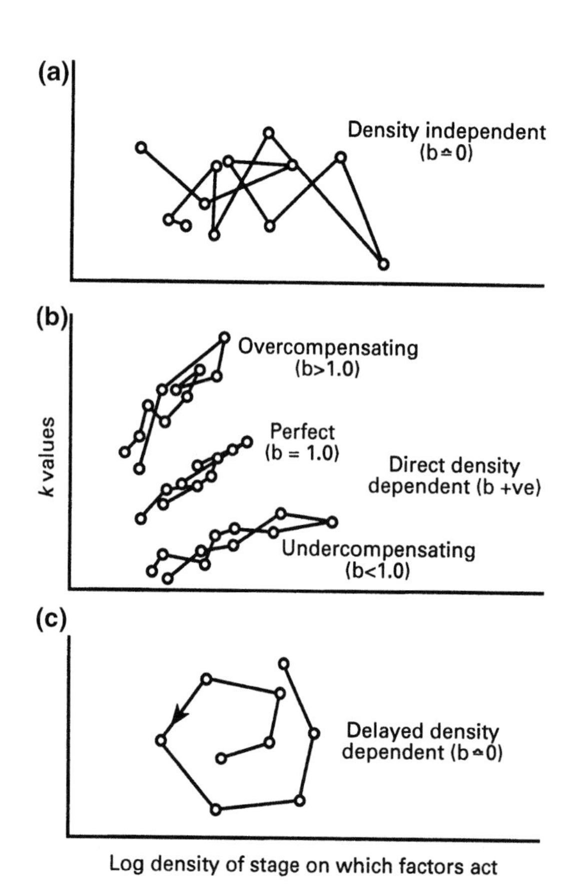

Figure 7.21 Time sequence plots showing how density relationships can be identified from the patterns produced. Source: Southwood (1978).

A Simple Inductive Model

At this point the 'formal' methodology associated with key factor analysis has been fully described, but further insights into how k-mortalities affect population dynamics can be derived from a simple **inductive model** constructed using the information obtained above (inductive models are those based on particular case studies, which yield general insights into population dynamics; philosophically, induction is the process of arguing from the particular case to the general case (*cf.* deduction, deductive models, subsection 7.3.7).

We begin by linking the numbers in each life stage to the next, through the mortalities expressed by $k_1, \ldots k_{n-1}$ as follows:

$$k_1 = m_1 - E_t + c_1 \quad (7.9)$$

$$L_t = E_t - k_1 \quad (7.10)$$

$$k_2 = m_2\, L_t + c_2 \quad (7.11)$$

$$P_t = L_t - k_2 \quad (7.12)$$

$$k_3 = m_3 P_t + c_3 \quad (7.13)$$

$$A_t = P_t - k_3 \quad (7.14)$$

where E_t, L_t, P_t and A_t are the $\log_{10}$ numbers of eggs, larvae, pupae and adults respectively at time t (m values are the regression constants for each equation, and c values are constants). Assuming a 50:50 sex ratio, we can find the $\log_{10}$ number of females (F) from:

$$10^{F_t} = 10^{A_t}/2 \quad (7.15)$$

or

$$F_t = A_t - 0.30103 \quad (7.16)$$

In our example, $k_1 = 0.74$, $k_2 = 0.86L_t - 1.52$ and $k_3 = 0.96$. The number of eggs laid by adults can be estimated from either: (a) cohort fecundity experiments (performed in the laboratory [subsection 2.72] and/or in the field), (b) dissection of females and estimating potential fecundity (subsection 2.72), or (c) regression of eggs in year $t+1$ against the estimated number of females ($A/2$) in year t. Assuming the following relationship between female numbers and eggs deposited (Figure 7.22):

$$E_{t+1} = 0.86F_t + 2.1 \quad (7.17)$$

We now have a series of equations which can be used sequentially to simulate dynamic changes from one generation to the next, over as many years as we require. Note that the **model** as it stands is completely **deterministic** in that it takes no account of the potential variation in the relationships, i.e. particular values for variables on the right hand side of the equations produce only one possible value for the variable on the left hand side. **Stochastic models**, on the other hand, *do* take account of the variability in the relationships, by including mathematical terms to describe chance events which may affect one or more of the relationships in the model. In this case, particular values for variables on the right hand side of the equations may produce a number of possible values for the variables on the left hand side. The methodology of stochastic modelling is discussed further in subsection 7.3.8, and a good introductory treatment can also be found in Shannon (1975).

Simulations of the model with different starting densities of eggs show that numbers approach an equilibrium within 2–3 generations, i.e. are strongly regulated (Figure 7.23a). Proof that regulation is provided by k_2 can be obtained by altering equation (7.12) such that k_2 becomes density independent ($k_2 = c_2$). Here, numbers either increase indefinitely or decrease to zero, depending on the other parameter values (Figure 7.23d), i.e. regulation is removed. Alternatively, increasing the strength of density-dependence by increasing the slope of the regression relationship between k_2 and L_t (e.g. $b = 1.2$ in equation (7.12), can produce oscillations of decreasing amplitude which eventually return to equilibrium (Figure 7.23b). Increasing the b-value even further (e.g. $b = 2.4$), however, can result in oscillations of increasing amplitude leading to the extinction of the population (Figure 7.23c). Thus, density-dependence is confirmed to be potentially either stabilising or destabilising in its effect, depending on its strength. It is also apparent that the weaker the density-dependence, the higher the equilibrium value becomes.

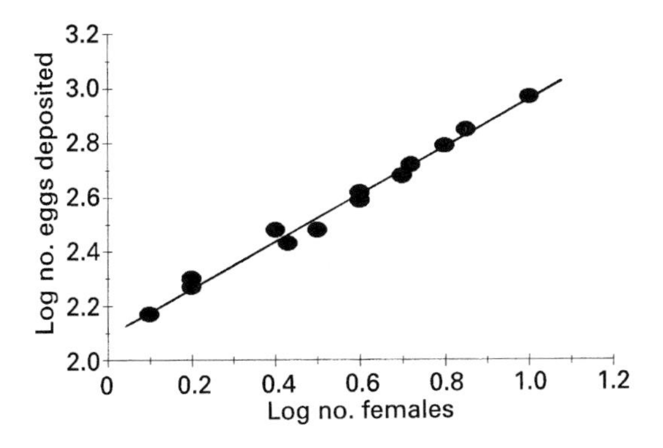

Figure 7.22 The relationship between female numbers and reproduction used in model 5.3.4. ($E = 0.86F + 2.1$; $R^2 = 0.99$).

A Case Study: The Winter Moth

To appreciate the considerable number of studies on which key factor analysis has been performed, the reader is referred to Podoler and Rogers (1975), Dempster (1983), Price (1987), Stiling (1987, 1988) and Hawkins *et al.* (1999). There is no doubt, however, that it is Varley and Gradwell's own study (1968, 1970) of the winter moth *(Operophtera brumata)*, together with the various follow-up studies in England and Canada, which have made this perhaps the best understood and most widely-quoted example. It is worth reviewing briefly some of the features of this study, as it serves to illustrate some of the potential problems in

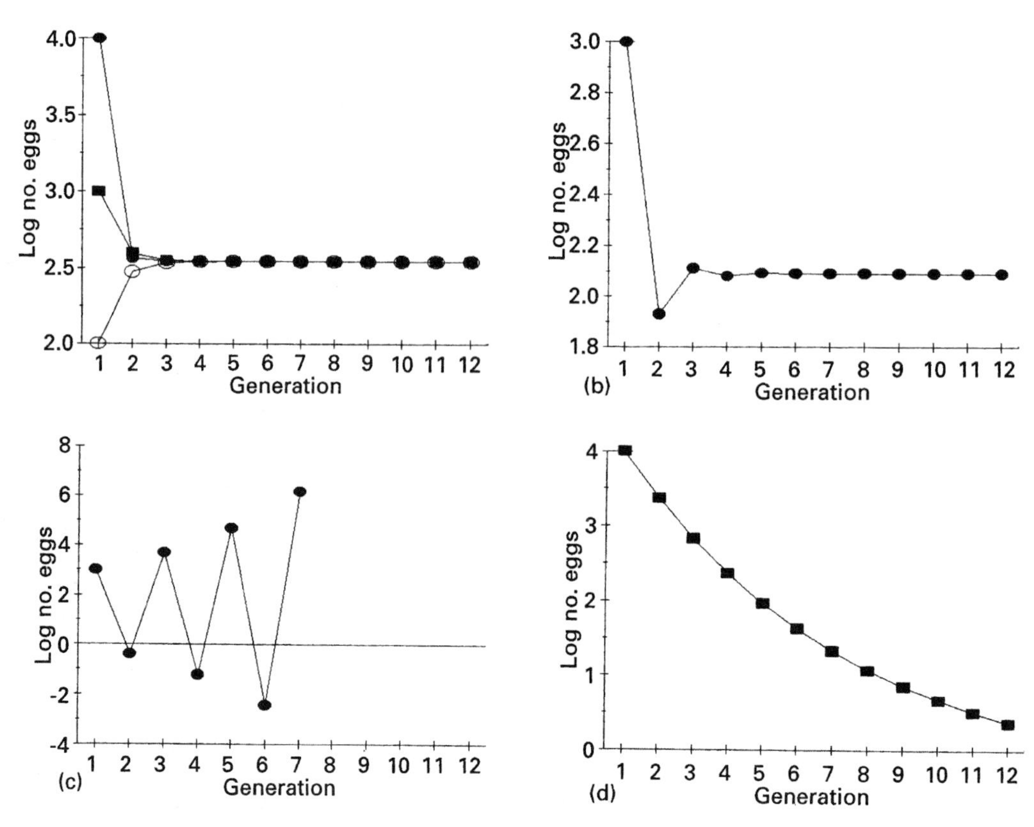

Figure 7.23 Predicted egg numbers over 12 generations: (a) with different starting densities of eggs; (b) with density dependence of larval mortality increased from $b = 0.86$ to $b = 1.2$; (c) with density dependence of larval mortality increased from $b = 0.86$ to $b = 2.4$, and (d) with density dependence of larval mortality removed ($b = 0$).

using key factor analysis, which we shall discuss shortly.

The winter moth feeds on a wide range of mainly deciduous trees, and occasionally defoliates oaks. The life-cycle at Wytham Wood, near Oxford, UK, where Varley and Gradwell's study was carried out, is as follows: eggs are laid in early winter in the tree canopy and hatch in spring to coincide with bud-burst; the caterpillars feed on the foliage until fully grown, whereupon they descend to the forest floor on lines of silk and pupate in the soil; adults emerge in November and December, the females ascending the trees to mate, the females than ovipositing in crevices on the bark. There is therefore one generation each year.

Data collected between 1950 and 1962 reveal that 'winter disappearance' (k_1), during the period between the egg stage and that of the fully grown larvae, is the key factor inducing population variation between years. Parasitism, disease, and predation (k_2–k_6) are relatively insignificant in this respect (Figure 7.24). The only significant regulating factor to be detected, however, was predation on pupae (k_5, Figure 7.25), subsequently shown to be caused mainly by shrews and ground beetles (Frank, 1967a,b; East, 1974; Kowalski, 1977). Parasitism showed no sign of being density-dependent, either at the larval stage (k_2) or at the pupal stage (k_6), leading the authors to suggest that the wide variations in densities from year to year, caused by the key factor 'winter disappearance', may be obscuring a possible delayed density-dependent relationship. The lack of any detectable regulating potential by the larval parasitoid

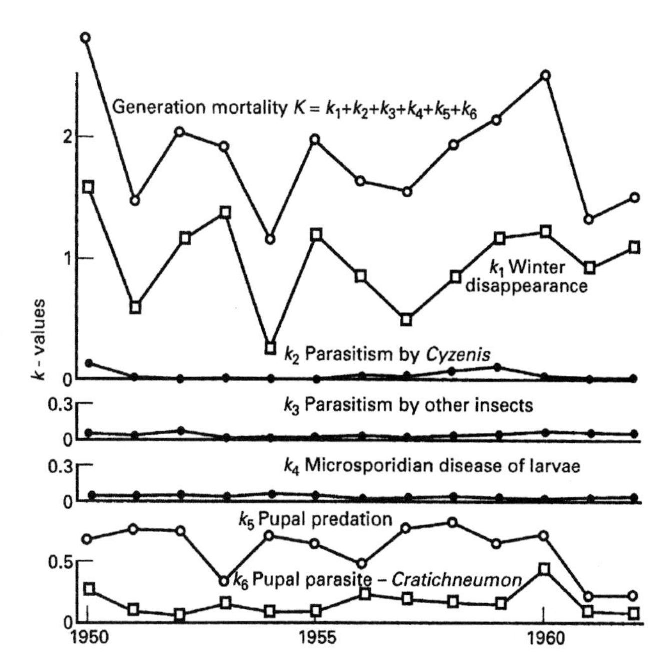

Figure 7.24 Key factor analysis of the mortalities acting on the winter moth. Source: Varley *et al.*, (1973). Reproduced by permission of Blackwell Publishing.

Cyzenis albicans (Diptera: Tachinidae) (k_2) was particularly surprising as this tachinid fly had previously been introduced in 1955 as a very effective biological control agent against winter moth in Nova Scotia, Canada (Embree 1966, 1971). This difference could perhaps be explained by higher levels of *Cyzenis* mortality in the UK. The parasitoid, although attacking the moth in the larval stage, continues to develop within the moth pupae throughout the summer and early winter and is therefore exposed to the same mortality factors as the moth pupae. Varley and Gradwell recorded as much as 98% mortality of *Cyzenis* puparia. This is higher than that for winter moth pupae, but understandable as *Cyzenis* spends 4–5 months longer in the soil, emerging in the spring.

A population model for the winter moth and its main parasitoids, *Cyzenis* (k_2) and the ichneumonid wasp *Cratichneumon culex* (k_6,), was developed by Varley *et al.* (1973), using basically the same approach which we elaborated above, but with two important differences. First, the variations in k_1 could not be predicted, so the observed values were used instead. Second, parasitism (k_2 and k_6) were modelled using the 'area of discovery' concept (subsection 7.3.7) rather than the simple regression relationships shown in Figure 7.25. There was good agreement between the model output and estimated field densities of the winter moth and its two parasitoids (Figure 7.26), although it has to be pointed out that testing the accuracy of a population model against the same data from which it is constructed, is not considered to be good modelling practice (subsection 7.3.8). However, collection of independent field data for acceptable validation of such life-table models is likely in many cases to prove impracticable, possibly involving years of extra work. This is one of a number of drawbacks associated with the Varley and Gradwell approach, which we shall now consider in detail.

Disadvantages of the Approach

The difficulty of obtaining additional field data for model validation highlights the single

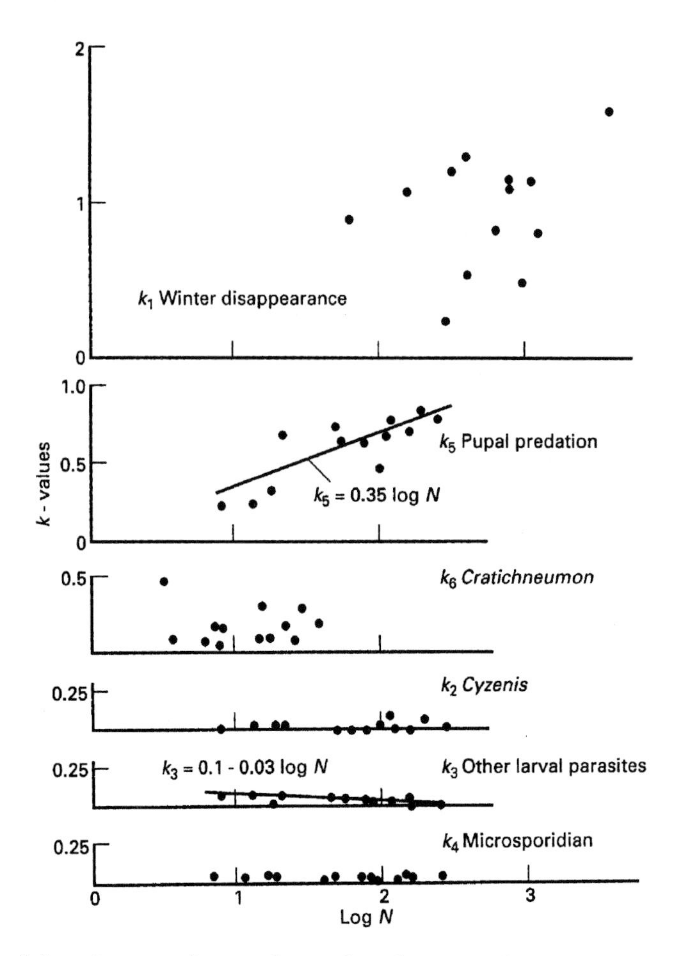

Figure 7.25 k-values of the winter moth mortalities plotted against the population densities on which they acted. k_1, k_2, k_4 and k_6 are density-independent; k_3 is weakly inversely density-dependent; k_5 is strongly density-dependent. Source: Varley *et al.*, (1973). Reproduced by permission of Blackwell Publishing.

biggest problem of the whole approach, namely that of securing a long enough sequence of data to perform the analysis with a reasonable likelihood of detecting statistically significant relationships (Hassell *et al.*, 1987). For insect populations having one generation a year, we may be contemplating the commitment of 15–20 years to a study, with no guarantees of success. The population processes affecting the main species may also change over the period of the study, with the result that key factors or density-dependent factors may alter or become obscured. Moreover, the method depends heavily on knowing all of the important factors to include in the study at the outset. There is not much scope for incorporation of new components at a later stage. There are a number of other problems as outlined in (A) to (F) below:

A: Contemporaneous and sequential mortalities Difficulties can arise when several agents act contemporaneously on a stage or when the precise sequence in which they act is unclear. Clearly, changes in the proportion killed by one agent will affect the number available to be attacked by other agents. Whether they are assumed to act concurrently or sequentially will have an important bearing on the results of the analysis. Buonaccorsi and Elkinton (1990) provide methods for estimating contemporaneous mortality factors using **marginal attack rates**.

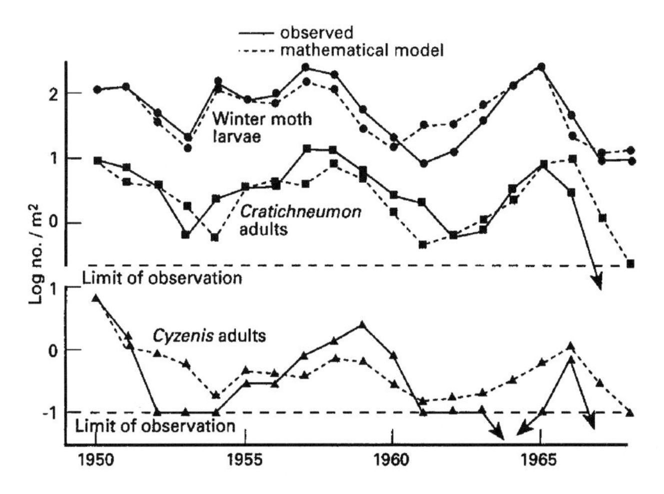

Figure 7.26 Observed changes in density of the winter moth and its two main parasitoids, and the densities predicted by the mathematical model. Source: adapted from Varley *et al.*, (1973). Reproduced by permission of Blackwell Publishing.

The marginal attack rate of a mortality factor is equivalent to the proportion of the population which would be killed by the factor acting alone, instead of in combination with other factors (Bellows *et al.*, 1992; Elkinton *et al.*, 1992). The methodology can also be extended to give estimates of *k*-values (see Bellows *et al.*, 1992 for a review). Assumptions about the sequence in which the mortalities act can strongly affect conclusions drawn, a point made forcefully by Putman and Wratten (1984) who audaciously illustrated their argument with a re-analysis of Dempster's (1975) study of the cinnabar moth (*Tyria jacobaeae*). In the original study, which assumed starvation of larvae to precede predation, Dempster's analysis showed starvation to be the key factor. Putman and Wratten reversed the sequence of these mortalities and found that predation became the key factor instead. We may question the justification for Putman and Wratten's re-ordering of the sequence of mortalities in this study, but we cannot ignore the point of the demonstration!

B: Composite mortalities Some of the mortality categories in the life-table may contain or mask a number of others which could be important key or regulating factors. This is particularly likely to be the case with poorly understood, wide categories, such as 'winter disappearance' in the winter moth example. Varley *et al.* (1973) accounted for this variable mortality as being mainly due to asynchrony between egg hatch and tree bud burst. Late opening of buds deprives young larvae of leaves to feed on, leading to death or emigration. However, other unstudied processes may also have had a part to play, for example, variations in adult fecundity, egg mortality from a number of possible sources, predation of early instars, etc..

C: Proving causation As Price (1987) has pointed out, the methods of life-table analysis, based as they are on correlation, do not always provide an unambiguous picture of cause and effect relationships. We can distinguish between **Type A density-dependence**, which is *causally* related to changes in population density and **Type B density-dependence** which is only *statistically related* (Royama, 1977), and to prove the former we need ideally to obtain life-table data both in the presence and absence of the suspected agent. Biological control introductions offer one potentially productive source of information on the causative role of natural enemies, but few 'before and after' life-table studies have in fact been carried out (subsection 7.2.2). Ryan (1997), however, has described a good example in the larch casebearer, *Coleophora laricella* (Lepidoptera), in North America, against which two parasitoid species were introduced from Europe. Life-tables were carried out both before and after the introductions, and from those it became clear that one of the parasitoids, *Agathis pumila* (Braconidae) was the key factor inducing the decline of the moth. The parasitoid also appeared to be acting in a delayed density-dependent fashion, stabilising moth numbers at low densities. Another successful example, which we have already alluded to, is the study by Embree (1966,1971) of the introduction of two winter moth parasitoids, *Cyzenis albicans* and *Agrypon flaveolatum* (Ichneumonidae) into Nova Scotia. An opportunity to carry out a similar study was more recently afforded by the introduction of the winter moth to western Canada (Roland, 1990; see also Roland, 1994). The same two parasitoid species were introduced to British Columbia from 1979 to 1981, host populations subsequently declining to one-tenth of their peak density. Parasitism, mostly from *C. albicans*, rose from zero before introduction to around 80% in 1984 and declined thereafter to 47% in 1989. Mortality of pupae, interestingly, rose during the same period to a level higher (>90%) than that caused by parasitism, suggesting a strong interaction between parasitism and subsequent mortality of unparasitised pupae. This effect was subsequently found to be present also in the Nova Scotia data. Roland suggested three possible explanations; these were:

1. Pupae parasitised by *Cyzenis* are present in the soil for twice as long as unparasitised pupae, so the greater availability of pupae after parasitoid introduction may be attracting higher numbers of pupal predators;
2. Predation and parasitism do not act independently of each other, predation rising in the presence of parasitism;
3. Pupal mortality factors in the soil have a minor effect at high population density and only exert a major effect after populations have declined.

To unravel the factors responsible, Roland carried out experiments with placed-out moth pupae (subsection 7.2.8), pitfall traps (subsection 6.2.1) to measure predator activity, and exclusion cages of different mesh size (subsection 7.2.3) to determine which predator sizes, if any, account for pupal mortality. Staphylinid beetles were found to be the most likely contenders, being also important predators of winter moth pupae in Britain. The results of Roland's experiments suggest that explanations (2) and (3) apply. Both have a part to play, predators showing a preference for unparasitised pupae (loading survival in favour of parasitoids and against the moth), predation becoming a major factor only after the parasitoid-induced decline (for an update on the winter moth analysis, see Roland, 1994).

D: Interaction effects Roland's analysis highlights the difficulties created when interaction effects occur between mortality factors. This is a problem which conventional life-table analysis is not equipped to deal with, assuming as it does that factors operate independently of each other. The only effective solution is to carry out factorial exclusion experiments (subsection 7.3.5) both in the presence and in the absence of the suspected interacting agents, under a range of relevant conditions, e.g. population density.

E: Compensatory effects An alternative method for evaluating the role of natural enemies from life-table data, discussed by Price (1987), might be to develop survivorship curves for cohorts of insects and to subtract from these the effects of specific natural enemies. A comparison could then made to assess the important contribution of natural enemies to mortality (see Figure 7.27). As Price himself points out, however, this type of analysis is likely to lead to very misleading conclusions, as it fails to recognise the possibility of compensation in the system. For example, removal of a high mortality due to natural enemies may be compensated for by a relatively higher mortality from other factors such as starvation or adverse weather conditions. An understanding of these potential compensatory mechanisms is crucial and again can only be gained adequately by experimentation.

F: Difficulties in detecting density-dependence It is possible for strongly regulated populations to show little variation from equilibrium, and this may make statistical detection of the

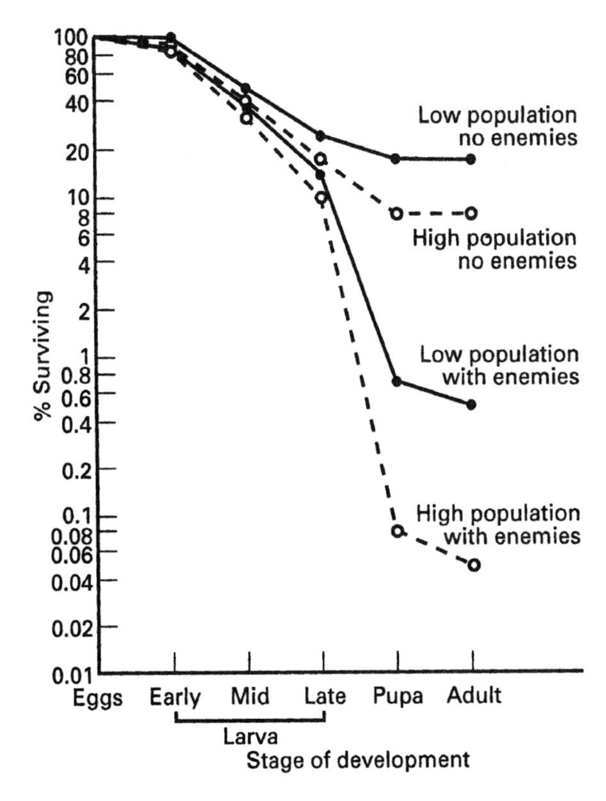

Figure 7.27 Survivorship curves for low and high populations of spruce budworm, *Choristoneura fumiferana*, together with those in which the effects of natural enemies have been removed. Source: Price (1987). Reproduced by permission of Elsevier Science.

processes of regulation difficult using traditional life-table methods (Gould *et al.*, 1990). Equally, stochastic variation may also obscure underlying density-dependent processes. Dempster (1983), for example, analysed 24 sets of data on Lepidoptera and could find in only three cases evidence of density-dependent mortality from natural enemies. He concluded that most insect populations are unlikely to be regulated by predators or parasitoids. However, Hassell (1985), using a simple model, showed that density-dependence can be present but remain undetected because of natural stochastic variation obscuring the relationships (see, however, Mountford, 1988). Hanski (1990) provides a useful review of the various problems inherent in detecting density-dependence from life-table and time-series data, together with a number of the statistical methods which have been proposed (see also Pollard *et al.*, 1987).

Being aware of the aforementioned pitfalls above is crucial before embarking on any population study based on age-specific life-tables, but it is in the nature of such long-term studies that unforeseen problems are likely to arise and may be difficult to correct after starting. For further, more detailed, treatments of age-specific or stage-structured life-table analysis, the reader is referred to Manly (1990).

Time-Specific Life-Tables

Time-specific (or vertical) life-tables are more suitable for use with populations in which the generations overlap, due to a short development time of the immature stages relative to the reproductive period of the adults (Kidd, 1979). Such populations (humans and aphids being examples) tend, after a period of time, to achieve a **stable age distribution** (Lotka, 1922) in which the proportion of the population in each age group or stage remains constant. In this situation, all the ecological processes affecting the population are, at least in theory, operating concurrently. This means that the relative numbers in each age group at any instant in time provide an indication of the proportional mortality from one age group to the next. However, we cannot deduce from this what mortality factors are operating, or whether any regulation is occurring, so the value of a time-specific life-table is limited in this respect.

Estimating mortality from parasitism may be easy to do with discrete generations (Varley, *et al.*, 1973; Van Driesche and Taub, 1983; but see subsection 7.3.2), but is more difficult when generations overlap. Van Driesche and Bellows (1988) provide an analytical method for doing this. Hughes (1962, 1963, 1972) developed a technique based on the time-specific life-table approach, which could be used for analysing aphid populations with a stable age (i.e. instar) distribution. Using a graphical method to compare population profiles at successive physiological time intervals (**Hughes' method**), Hughes was able to partition the mortalities acting on the different instars, for example, parasitism, fungal disease and 'emigration'. As Hughes (1972) pointed out, however, there is no easy way of estimating errors in the construction of these life-table diagrams. In fact, the whole technique is critically dependent on the assumption of a stable age distribution. Although Hughes provided a simple statistical (χ^2) method to test the validity of the assumption, Carter *et al.* (1978) showed it to be insensitive to significant changes in the age distribution. Applying a more stringent test to Hughes' own field data for the cabbage aphid, *Brevicoryne brassicae* upon which his technique was developed (Hughes, 1962, 1963), Carter *et al.* (1978) found that these populations never achieved a stable instar distribution. Although Hughes' method has been widely used and was recommended for the International Biological Programme's study of the aphid *Myzus persicae* (Mackauer and Way, 1976), it should now only be used with extreme caution. Readers interested in the detailed methodology should consult Hughes (1972) and Carter *et al.* (1978).

Whilst Hughes' method is now considered to be of limited applicability, his work did lead directly to the development of the earliest simulation models for analysing insect populations with relatively complex population processes. For field populations with overlapping or partially overlapping generations, the use of such models is now the only sensible way forward. These techniques are discussed in detail below (subsection 7.3.8).

Variable Life-Tables

The term 'variable life-table' (or 'time-varying life-table') has been used to describe a particular class of computer-based, age-structured population model, in which the birth and survival rates experienced by each age-class change in a realistic way (Gilbert *et al.*, 1976). The population life-table is in fact computer-generated from reproduction and survival relationships obtained in the field or laboratory, and as such becomes the output of the exercise rather than forming the basis of the analysis. The technique has therefore more in common with the methodology of simulation modelling than with that of life-table analysis, and will be discussed further in subsection 7.3.8.

7.3.5 MANIPULATION EXPERIMENTS

Convergence Experiments

The problems of detecting density-dependence from life-table data have already been discussed (subsection 7.3.4). One way of testing directly whether density-dependent mechanisms are operating is to carry out a 'convergence experiment' (Nicholson, 1957) in which densities of comparable subpopulations are manipulated to achieve artificially high or low levels and are then monitored through time. Convergence to a common density is then taken as evidence for density-dependent regulation. What constitutes an artificially high or low population density in the context of this type of experiment will vary according to the species under study, and can only be adequately assessed from some historical knowledge of past densities. Practical difficulties in manipulating densities of some species may also limit the usefulness of this technique. Amongst successful studies, Brunsting and Heessen (1984) manipulated densities of the carabid predator *Pterostichus oblongopunctatus* within enclosures in the field and found evidence for convergence within two years. Criticisms can be levelled at this technique in that enclosures may prevent emigration or immigration of beetles, leading to spurious mortality from 'artificial' sources. In this particular study, however, care was taken to note that beetle motility was naturally low and remained low even at the enhanced densities, there being no evidence for density-induced emigration. Gould *et al.* (1990) manipulated densities of gypsy moth by artificially loading eight forest areas with different densities of egg masses to achieve a wide range of infestation levels. This method revealed previously undetected density-dependent mortality in the larval stage, primarily due to two parasitoid species. Orr *et al.* (1990) also provide a good example of a convergence experiment carried out in the laboratory using the freshwater predatory bug, *Notonecta*.

Factorial Experiments

Factorial experiments are used to determine whether factors potentially capable of limiting population numbers combine in a simple additive way, or show more complex patterns of interaction (Hilborn and Stearns, 1982; Arthur and Farrow, 1987; Mitchell *et al.*, 1992). There are three criteria for successfully carrying out factorial experiments:

1. At least two factors need to be manipulated to at least two levels each;
2. A sufficiently long time-series of data must be available to assess equilibrium population levels around which numbers fluctuate;
3. There must be replication (Mitchell *et al.*, 1992).

Mitchell *et al.* (1992) examined the interaction between resource levels (food and food/water ratios) and three population levels (zero, low, high) of the parasitoid *Leptopilina heterotoma* on laboratory populations of *Drosophila melanogaster*. This provided 12 different experimental combinations of the three potentially interacting factors. Both food and wasps showed significant effects on equilibrium levels, but without any significant interaction. With this type of experiment involving census data collected over time, the problem of serial autocorrelation is encountered (Arthur and Farrow, 1987), which makes the use of analysis of variance inappropriate. This can be circumvented using GLIM (see Mitchell *et al.*, 1992; Crawley, 1993, 2002).

7.3.6 EXPERIMENTAL COMPONENT ANALYSIS

As explained in subsection 7.2.15, this approach is based on the assumption that the complexities of ecological interactions, such as those involving predation and parasitism, can be quantified in terms of a relatively small number of dynamic processes (Southwood and Henderson, 2000). Each process, reduced to its component parts, can be investigated experimentally and described by a series of equations. The equations describing all the component processes can then be incorporated into a system or population model, the accuracy of which can then be assessed by comparing its behaviour with real observations.

The so-called **components of predation** can be investigated experimentally using the important distinction between functional and numerical responses (subsection 7.2.15). To assess the significance of these responses, particularly to predator-prey and parasitoid-host population equilibrium levels and stability, two different modelling approaches can be adopted; one based on simple analytical models, the other involving the construction of more elaborate simulation models.

7.3.7 PUTTING IT TOGETHER: ANALYTICAL MODELS

Incorporating the Components of Predation

To assess the impact of parasitism or predation on an insect population, the information on functional and numerical responses needs to be incorporated into population models. Analytical models are usually based on systems of relatively simple equations which can be 'solved', usually by rearrangement, to provide straightforward answers. Some population models, however, have systems of equations which are too complex for solution, and so the only way of obtaining useful insights is to perform simulations with the model under differing conditions, for example by changing parameter values. Of course, there is no reason why models capable of analytical solution cannot also be used for simulation. Analytical solutions tend to be more tractable when a simple **deductive modelling** approach is adopted (deductive models are those based on very general, often intuitive, concepts, which can be useful in providing insights which might apply to particular case studies; philosophically, **deduction** is the process of arguing from the general case to the particular case (*cf.* induction, inductive models, subsection 7.3.4).

Whilst populations with overlapping generations and stable age distributions can be modelled analytically in continuous time using differential equations, this method is less suitable for the bulk of insect populations, at least in temperate regions, which have **discrete generations**, i.e. separated in time. A more appropriate modelling format is provided by difference equations which model population change in discrete time-steps, i.e. $N_{t+1}=f(N_t)$. The discrete time model which has been most widely used in insect population ecology is the host-parasitoid model of Nicholson and Bailey (1935), hereafter referred to as the **Nicholson-Bailey model**. Originally developed to explore the dynamic implications of parasitoid searching behaviour, the model has been extensively elaborated in recent years to examine other features of parasitoid, (and predator) biology (Hassell, 1978, 2000b; Waage and Hassell, 1982; Hassell and Waage, 1984; Godfray and Hassell, 1988; May and Hassell 1988; Hassell and Godfray, 1992; Mills, 2001). No doubt part of the appeal of this model lies in the simplicity with which it purports to capture the essence of the parasitoid-host interaction. Using a time-step of one generation, the model takes the following form:

$$N_{t+1} = F.\exp(-aP_t) \quad (7.18a)$$

$$P_{t+1} = N_t[1 - \exp(-aP_t)] \quad (7.18b)$$

where N_t and P_t are the numbers of hosts and parasitoids respectively at time t, F is the host net rate of increase in the absence of parasitism and a is the parasitoid's **area of discovery**, which is essentially the proportion of the habitat which is searched in the parasitoid's lifetime. A number of assumptions about the parasitoid and its host are implicit in these equations:

1. Generations of both populations are completely discrete and fully synchronised; the time-step (t) is therefore one generation;
2. One encountered host leads to one new parasitoid in the next generation;
3. The parasitoid is never egg-limited;
4. The area of discovery (= searching efficiency) is constant;
5. Each parasitoid searches the habitat at random.

This latter assumption is catered for in the model by using the Poisson distribution to distribute attacks at random between hosts. The zero term of the distribution (e^{-x}, where x is the mean number of attacks per host) defines the proportion of the host population escaping attack, in this case e^{-aP}, or $\exp(-aP)$. The proportion attacked is therefore $1-\exp(-aP)$. A more detailed description of the derivation of these equations is not provided here as it has already been covered in a number of texts (e.g. Varley *et al.*, 1973; Hassell, 1978).

Following Hassell (1978), the equilibrium populations N^* and P^* can now be found by setting $N_{t+1}=N_t=N^*$ and $P_{t+1}=P_t=P^*$ giving the analytical solution:

$$N^* = F \quad (7.19a)$$

$$P^* = \frac{\log_e}{a} \quad (7.19b)$$

The equilibrium levels of both populations thus change with respect to both F and a.

However, the Nicholson-Bailey model *is inherently unstable*, a fact that can be confirmed either by simulation (Figure 7.28a) or by **stability analysis** (Hassell and May, 1973; Hassell, 1978, 2000b) (stability analysis is a technique which has been widely used in the analysis of deductive models, but the mathematics beyond the scope of this book; we refer readers to the appendices in Hassell and May, 1973, and Hassell, 1978). When perturbed from equilibrium, the model produces oscillations of increasing amplitude, which in the real world would result in the extinction of one or both populations (Figure 7.28a). Stability in the model could easily be produced, however, by the incorporation of a density-dependence component into F, to simulate, for example, competition between hosts for food resources. Progressively increasing the degree of density-dependence produces in the first instance **stable limit cycles** (Figure 7.28b) followed by **damping oscillations** (Figure 7.28c). Whilst density-dependence in F

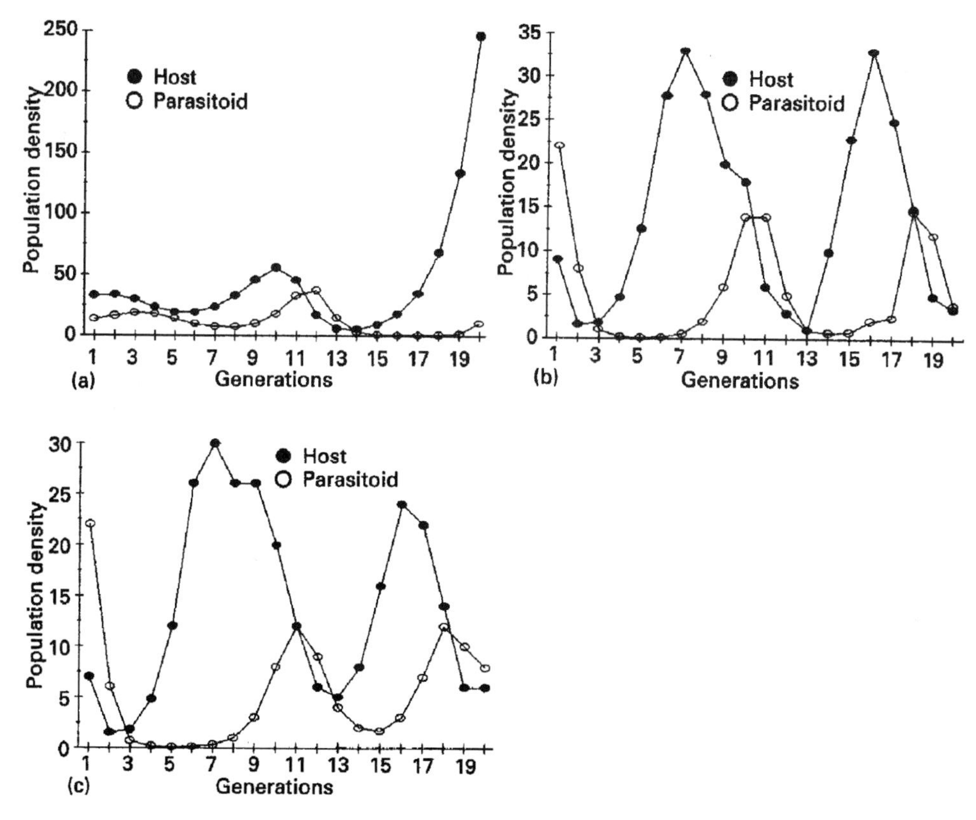

Figure 7.28 (a) Typical numerical changes predicted by the Nicholson-Bailey model. The incorporation of (increasing) density-dependence into the model results (b) in cyclical oscillations within an upper and lower boundary (limit cycles), followed by (c) damping oscillations which approach an equilibrium.

would be a reasonable component to incorporate, it does not advance our understanding of how features of parasitoid biology such as searching behaviour influence dynamics. To achieve that, more detailed descriptions of, for example, functional and numerical responses (see below) need to be incorporated into the model.

To facilitate the incorporation of the components of predation into the Nicholson-Bailey model, it is useful to begin with a more generalised form of the model (Hassell, 1978, 2000):

$$N_{t+1} = FN_t f(N_t P_t) \quad (7.20a)$$

$$P_{t+1} = N_t[1 - f(N_t P_t)] \quad (7.20b)$$

in which survival of hosts is a function of both host and parasitoid numbers. This survival function can now be explored in relation to the following components:

A: Functional responses: The assumption of a constant searching efficiency described by one component, a or a', is of course an oversimplification, given what we know of the way in which parasitoid and predator attack rates change with prey density (section 1.10). The way in which handling time affects this relationship have already been discussed (section 1.10) and can be described by **Holling's 'disc' equation:**

$$\frac{N_e}{P_t} = \frac{a'TN_t}{1 + a'T_h N_t} \quad (7.21)$$

where N_e is the number of hosts encountered. This Type 2 functional response (section 1.10) can now be incorporated into the Nicholson-

Bailey model using the so-called **random parasite equation** first described by Rogers (1972):

$$N_a = N_t\left[1 - \exp\left\{-\frac{a'TP_t}{1 + a'T_hN_t}\right\}\right] \quad (7.22)$$

Predator versions of equation 7.22 were also developed by Royama (1971) and Rogers (1972) to take account of gradual prey depletion during each interval t. Prey eaten by predators do not remain exposed to further encounters, in contrast to hosts which may be re-encountered and thus incur additional T_h costs to the parasitoid. Reproducing the Royama (1971) equation:

$$N_a = N_t[1 - \exp\{-a'P_t(T - T_h(N_a/P_t))\}] \quad (7.23)$$

Note that here N_a is present on both sides of the equation. The simplest way of finding N_a for given values of the other parameters is by **iteration** (i.e. by repeatedly substituting different values of N_a until both sides of the equation balance). Detailed mathematical derivations for both equations (7.22) and (7.23) are given by Hassell (1978). As the Type 2 functional response can be seen to be inversely density-dependent when percentage parasitism is plotted against prey density (Figure 1.15b), it is perhaps not surprising that when incorporated into the Nicholson-Bailey model as:

$$f(N_tP_t) = \exp[-(-a'TP_t)/(1 + a'T_hN_t)] \quad (7.24)$$

for parasitoids, its effect is to further destabilise the model. In general, the greater the ratio T_h/T the greater the destabilising effect, while the original Nicholson-Bailey model is re-established when $T_h = 0$ (Hassell and May, 1973). Although T_h thus determines both the degree of destabilisation and the maximum attack rate (= the plateau), both could equally well be influenced by egg-limitation (Hassell and Waage, 1984; section 1.10).

To explain the Type 3 (i.e. sigmoid) functional response, Hassell (1978) suggested a model which assumes that only a' varies with prey density, such that $a' = bN_t\,(1 + cN_t\,)$, with b and c constants. This gives a sigmoid analogue to the disc equation:

$$N_e/P_t = \frac{bN_t^2T}{1 + cN_t + bT_hN_t^2} \quad (7.25)$$

where N_e is the number of prey encountered. For parasitoids, where hosts remain to be re-encountered:

$$N_a = N_t\left[1 - \exp\left(-\frac{bTN_tP_t}{1 + cN_t + bT_hN_t^2}\right)\right] \quad (7.26)$$

which can easily be incorporated into the Nicholson-Bailey model. Hassell (1978) also provides an alternative equation for predators, where prey are gradually depleted with time (see also Hassell *et al.*, 1977):

$$N_a = N_t\left[1 - \exp\left\{-\frac{bP_t}{c}\left(T - \frac{T_hN_a}{P_t} - \frac{N_a}{bN_tP_t(N_t - N_a)}\right)\right\}\right] \quad (7.27)$$

N_a can again be found by iteration (see equation 7.23). Both equations (7.26) and (7.27) produce similar sigmoid relationships, but (7.27) is easier to use. Intuitively, such sigmoid responses might be expected to have a stabilising influence on population interactions, where the equilibrium falls within the density-dependent part of the response (Figure 1.15c). This was demonstrated by Murdoch and Oaten (1975) using continuous time differential equations (see below) with no time-delays. However, the time-delay of one generation inherent in the Nicholson-Bailey model is sufficient to prevent any sigmoid functional response, of the form of equation above, from stabilising an interaction (Hassell and Comins, 1978).

The above conclusion is, of course, restricted to those predators or parasitoids that are are prey- or host-specific, i.e. there is a **coupled interaction** between parasitoid host. Generalist predators and parasitoids, because they attack several prey species, are involved in a looser interaction, so the situation for these insects is somewhat different (Hassell, 1986). Here, predators may display switching behaviour (section 1.11). Hence, neither the predator numerical response nor its population density is likely to be dependent on the abundance of any one prey type. The interaction between a generalist 'predator' and a *single* prey species was modelled by Hassell and Comins (1978) using equation (7.26) for a parasitoid population at equilibrium:

$$P_{t+1} = P_t = P^* \qquad (7.28)$$

This model was found to have two equilibria (Figure 7.29), the lower (S) being locally stable, while the upper (R) was unstable, such that prey exceeding R escaped parasitoid control and increased indefinitely. The important point in this model is that the total response of the parasitoid shows a sigmoid relationship with prey density. This can be achieved by a sigmoid functional response and a constant parasitoid density, as in the model, or by a rising numerical response to prey density, coupled with either a Type 2 or a Type 3 functional response (Figure 7.30).

At first sight the incorporation of functional responses into simple models seems to be fairly

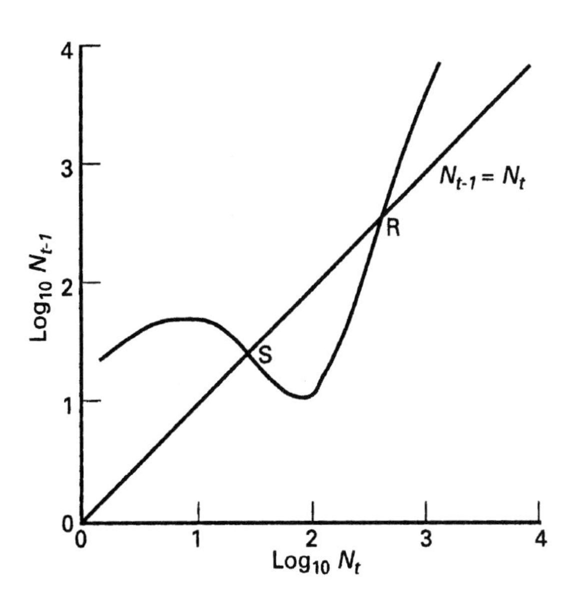

Figure 7.29 A population growth curve for equation 7.26. The intersections with the 45° line are the lower potentially stable equilibrium (S) and the upper unstable equilibrium (R). Source: Hassell and Comins (1978). Published by permission of Elsevier Science.

straightforward, but there are a number of complications which the reader needs to be aware of:

1. For a particular predator, the functional response is likely to vary with age or size of both predator and prey, and the model may have to be modified to incorporate the effects of age-structure (see **Other Analytical Modelling Approaches**, below);
2. Simple laboratory experiments to assess functional responses over a short time period (e.g. 24 hours) should be used in *generation-based* models with caution, as they may give a misleading impression of the predator's *lifetime* functional response. This, in the context of the Nicholson-Bailey model, is the relevant component if we are interested in understanding the effects of functional responses on population dynamics (Waage and Hassell, 1982; Kidd and Jervis, 1989). The problem may be avoided, however, when we consider the functional response in relation to predator aggregation in patchy environments (see below).

B: Aggregative responses: Although the Nicholson-Bailey model assumes random search by parasitoids (i.e. each host has the same probability of being parasitised), in reality natural enemies tend to show an aggregative response (defined in subsection 1.14.2). Examples have been widely reported and reviewed by a number of authors (Hassell *et al.*, 1976; Hassell, 1978, 2000b; Krebs and Davies, 1978; Lessells, 1985; Walde and Murdoch, 1988). This behaviour has already been discussed in relation to foraging behaviour (section 1.14), but here we are concerned with its implications for population dynamics.

Hassell and May (1973) modelled the effects of parasitoid aggregation in a simple way by first distributing hosts and parasitoids between n patches, then considering each patch as a sub-model of the Nicholson-Bailey model (Table 7.3). Thus, in each patch i, there is a proportion α_i of hosts and β_i of parasitoids.

As can be seen in Table 7.3, a greater proportion of parasitoids is placed in the high

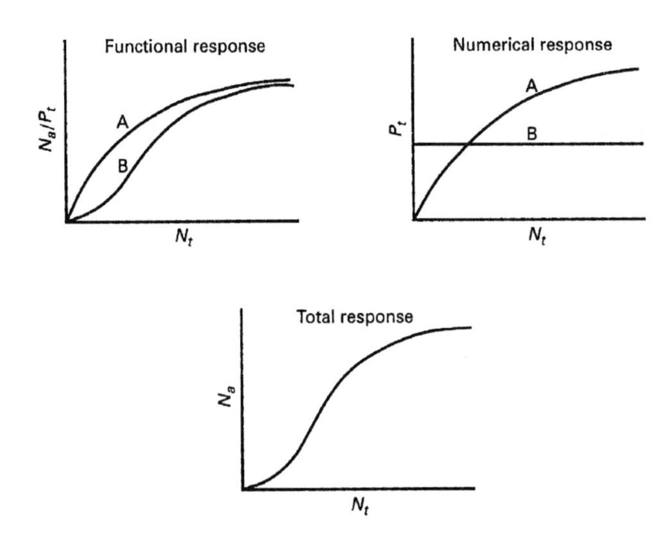

Figure 7.30 Alternative ways of achieving a sigmoid total functional response between prey eaten (N_a) by P_t predators and prey density N_t. The total response may be achieved by combining either response A or B with numerical response A or from functional response B with no numerical response. Source: Hassell, (1978). Reproduced by permission of Princeton University Press.

density patches of prey than in the low density ones. The parasitism function now becomes:

$$f(N_t P_t) = \sum_{i=1}^{n} \left[\alpha_i \exp(-a\beta_i P_t)\right] \qquad (7.29)$$

where a is the searching efficiency per patch. Equation (7.29) redistributes hosts and parasitoids as in the above scheme at the beginning of each generation. In the case where parasitoids are distributed evenly over all patches, we regain the property of random oviposition as in the original model. The stability analysis of Hassell and May (1973) shows that the model may now become stable with a sufficiently uneven prey distribution and enough parasitoid aggregation in high density host patches. To allow easier analysis of the properties of the model, Hassell and May (1973) used a single high host density patch (α) and distributed the rest of the host population evenly amongst the other patches $[(1-\alpha)/(n-1)]$. The parasitoid distribution was defined by a single **'aggregation index'** μ where:

$$\beta_i = c\alpha_i^{\mu} \qquad (7.30)$$

c being a normalisation constant. The degree of parasitoid aggregation is now governed by μ, $\mu=0$ corresponding to random search and $\mu=\infty$ to the situation where all parasitoids are in the high host density patch. Stability is now affected by precise values of μ, F, α and $n-1$ (Figure 7.31).

Table 7.3 The proportional distribution of hosts (α) and parasitoids (β) between n patches to incorporate the aggregative response into the Nicholson-Bailey model

	Patch				
	1	2	3	4	5
α	0.5	0.2	0.1	0.1	0.1
β	0.8	0.1	0.05	0.03	0.02

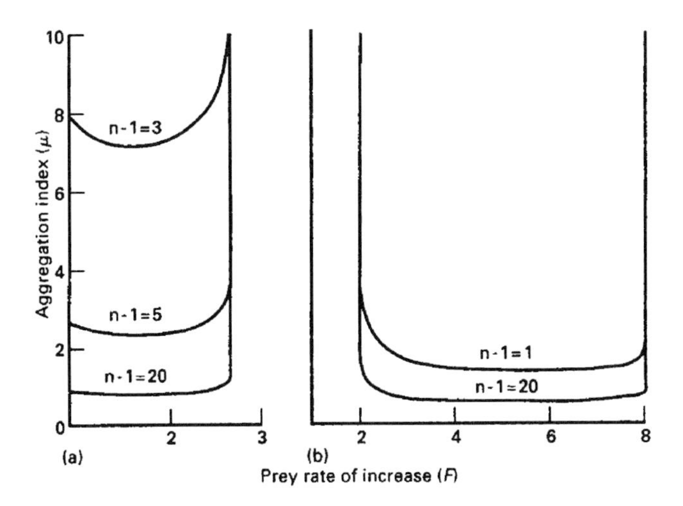

Figure 7.31 Stability boundaries between the aggregation index μ and the prey rate of increase F, for different values of $n-1$: (a) $\alpha = 0.3$; (b) $\alpha = 0.7$. Source: Hassell and May (1973). Reproduced by permission of Blackwell Publishing.

A more general ('phenomenological') model has also been developed by May (1978), to capture the essential features of parasitoid aggregation without the detail. This model uses the negative binomial to distribute parasitoid encounters between hosts. Thus:

$$N_{t+1} = FN_t[1 + (aP_t/k)]^{-k} \qquad (7.31a)$$

$$P_{t+1} = N_t\left[1 - (1 + (aP_t/k))^{-k}\right] \qquad (7.31b)$$

Here the parameter k (the exponent of the negative binomial) describes parasitoid aggregation, being strongest when $k \to 0$ and weakest when $k \to \infty$ (random). May's model, and variants thereof, has been used by a number of authors to include an aggregative response component into population models (Beddington *et al.*, 1975, 1976a, 1978; Hassell, 1980b). Hassell (1978, 2000b) provides a good account of the development and application of this model.

The stabilising potential of aggregation by predators and parasitoids must therefore temper our previous conclusions regarding the significance of functional responses, as measured in single-patch experiments (section 1.10). We can envisage the Type 2 and Type 3 response curves as essentially a 'within patch' phenomenon, with searching between patches defined by the aggregative response (Hassell, 1980a,b) (see also section **What is Searching Efficiency?**, below).

C: Mutual interference: The incorporation of mutual interference (subsection 1.14.3 gives a detailed discussion) into the Nicholson Bailey model was first carried out by Hassell and Varley (1969), using the inverse relationship between parasitoid searching efficiency and the density of searching parasitoids shown in equation (1.6). Removing the logarithms, this becomes:

$$a = QP_t^{-m} \qquad (7.32)$$

where Q and m are constants. When this equation is substituted into the Nicholson-Bailey model, it gives the equations:

$$N_{t+1} = FN_t \exp(-QP_t^{1-m}) \qquad (7.33a)$$

$$P_{t+1} = N_t[1 - \exp(-QP_t^{1-m})] \qquad (7.33b)$$

This modification has the effect of producing a stable equilibrium given suitable values of m and F (Figure 7.32). The higher the mutual interference constant m, and the lower F, the more likely stability becomes. Q has no effect on stability, but does affect the equilibrium level.

More elaborate, but behaviourally more meaningful, mathematical descriptions of

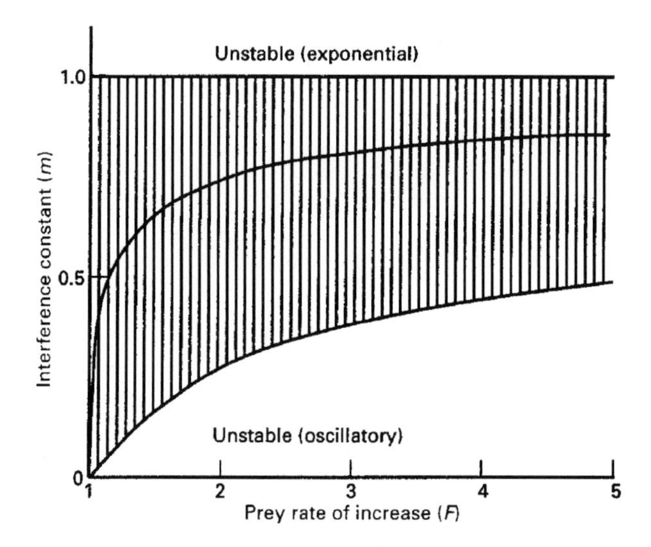

Figure 7.32 Stability boundaries for equation (7.33) in terms of the interference constant m and the prey rate of increase F. The hatched area denotes the conditions for stability, approached through exponential damping above the central curve and oscillatory damping below the curve. Source: Hassell and May, (1973). Reproduced by permission of Blackwell Scientific Publishing.

mutual interference have also been developed (Rogers and Hassell, 1974; Beddington, 1975) and explored in deductive models. While the precise stability conditions in these models may differ from the earlier version, the essential conclusion, that mutual interference can be a powerful stabilising force, remains intact.

D: Ratio-dependent functional responses: So far, searching efficiency (= attack efficiency) has been discussed in terms of separate attributes relating to either host (functional response) or parasitoid density (mutual interference). Some authors have instead proposed combining these effects in terms of the *ratio* of parasitoids (or predators) to hosts (or prey), to form the **ratio-dependent functional response** (Getz, 1984; Ginzburg, 1986; Arditi and Ginzburg, 1989). With respect to the Nicholson-Bailey model, this would replace the function $f(N_tP_t)$ with $f(N_t/P_t)$. Ratio-dependent functional responses have been developed by Beddington (1975), DeAngelis *et al.* (1975) and Getz (1984), sharing the general form:

$$N_a/P_t = N_t/P_t\{1 - \exp[-abP_t/(c + aN_t + bP_t)]\} \quad (7.34)$$

where a is the search rate, b the maximum number of hosts attacked per parasitoid and c is a constant. How far the functional response can be generalised in this way has been strongly debated in recent years (Murdoch and Briggs, 1996; Abrams, 1997; Abrams and Ginzburg, 2000; Hassell, 2000b), not least because of the way published functional response data have been re-analysed to fit the ratio-dependent format (Arditi and Akcakaya, 1990; Hassell, 2000b). What would help to settle at least some of the issues are some new studies specifically designed to detect ratio-dependence. Mills and Lacan (2004) provide a multifactorial protocol for parasitoid-host interactions.

Whatever the rights and wrongs of the ratio-dependence argument, the debate serves to emphasise the need to consider both the effects of hosts *and* those of parasitoid density on the overall functional response, the latter having perhaps been relatively neglected in the past (Mills and Lacan, 2004).

E: Numerical responses: The aggregative response could be said to be a form of numerical response, but here we use the term numerical response to refer specifically to changes in predator numbers from one generation to the next. In the Nicholson-Bailey model this is simply achieved by making the proportion of hosts killed by parasitoids in each generation, $[1-\exp(-aP_t)]$, equivalent to the number of parasitoids in the next generation. Each host killed therefore produces one living parasitoid. Gregarious larval development can be easily catered for by incorporating an additional component c into equation 7.18b, such that:

$$P_{t+1} = cN_t[1 - \exp(-aP_t)] \qquad (7.35)$$

Here, c is the average number of adult parasitoids to emerge from each parasitised host ($c=1$ for solitary parasitoids and $c>1$ for gregarious parasitoids, Waage and Hassell, 1982). As incorporated above, c has no effect on stability, but raising its value depresses the equilibrium (Waage and Hassell, 1982). However, c can also be further elaborated to cater for certain factors which may influence the number of parasitoids emerging per host. For conspecific superparasitism in solitary parasitoids, where only one larval parasitoid can survive, the situation will usually reduce to equation (7.18b) with $c=1$ (exceptionally, superparasitism results in the death of both rivals in a host, in which case, $c<1$). If there is mortality from factors such as multiparasitism or encapsulation, c will be <1. Where clutch size affects larval survival adversely in gregarious species, this effect can be incorporated by replacing c in equation (7.35) with $c(1-\delta F)$, for values ≥ 0, where F is clutch size and δ is a constant defining the strength of density-dependence. This expression assumes a negative linear relationship between clutch size and progeny survival (although a negative exponential relationship may be more realistic), and will have a regulating effect both on the parasitoid population and the parasitoid-host interaction. To obtain realistic values for parameter c would require the construction of detailed life-tables for the parasitoid (see Hassell, 1969; Escalante and Rabinovich, 1979).

An additional parameter s can also be incorporated into equation (7.35) to take account of variation in the sex ratio of the parasitoid progeny. Thus:

$$P_{t+1} = scN_t[1 - \exp(-aP_t)] \qquad (7.36)$$

where s is the proportion of parasitoid progeny that are female (Hassell and Waage, 1984). Again, changes in s will have no effect on stability, but smaller values (i.e. male-bias in progeny) will raise the equilibrium. Density-dependence in s (subsection 1.9.2) (to incorporate density-dependence, s has to be altered to another form, i.e. $s=f$ [parasitoid density]), has a stabilising influence on the parasitoid-host interaction (Hassell *et al.*, 1983; Hassell and Waage, 1984; Comins and Wellings, 1985; Mills and Getz, 1996).

For predators, the above models are inappropriate, as there is no simple relationship between the prey death rate and the predator rate of increase. The rate of increase of a predator population will depend on (Lawton *et al.*, 1975; Beddington *et al.*, 1976b; Hassell, 1978):

1. The development rate of the immature stages;
2. The survival rate of each instar;
3. The realised fecundity of the adults.

The biotic and abiotic factors affecting each of these components are considered in detail in sections 2.9, 2.10 and 2.7 respectively (Chapter 2). To build a general model of the predator rate of increase we would need to incorporate the effects of prey consumption on development and survival of the different instars, and on adult fecundity. This task would be beyond the scope of analytical modelling (Hassell, 1978), being more suited to simulation (subsection 7.3.8). Beddington *et al.* (1976b), however, took a simpler approach whilst retaining some of the features of predator reproduction. Adult fecundity, F, was related to the number of prey eaten during the predator's life by the equation:

$$F = c[(N_a/P_t) - \beta] \qquad (7.37)$$

where N_a is the number of prey attacked, c is the efficiency with which consumed prey are converted to new predators and β is the threshold prey consumption needed for reproduction to start (see also equation [2.3] and related discussion). The model therefore takes account, in a simple way, of the predator's need to allocate some of the prey biomass assimilated to growth and maintenance (2.8.3). Incorporating this equation into the Nicholson-Bailey model yields the equations:

$$N_{t+1} = N_t \exp[r(1 - N_t/K) - aP_t] \quad (7.38a)$$

$$P_{t+1} = c\big[\{N_t[1 - \exp(-aP_t)]\} - \beta P_t\big] \quad (7.38b)$$

with the rate of increase of the prey population, in the absence of predation, defined by $(1-N_t/K)$, which includes a density-dependent feedback component (K being the carrying capacity). Of course, handling time, predator aggregation and mutual interference are not included. Nevertheless, the model can be used to show the stability differences between predator-prey models where $\beta > 0$ and those of parasitoid-host models where $\beta = 0$. The important effect of increasing β is to reduce the range of stable parameter space in the model. Furthermore, where $\beta = 0$, the model is **globally stable**, i.e. it returns to equilibrium irrespective of the degree of perturbation. With $\beta > 0$, only **local stability** is apparent, i.e. equilibrium is re-attained only when perturbation is within certain limits. Thus, as predators need to eat more prey before reproducing, the chances of a stable interaction diminish.

Kindlmann and Dixon (2001) give examples of predator-prey systems where numerical and functional responses may be irrelevant to the system dynamics: predator reproduction should perhaps be correlated with the age of the prey patch rather than the number of prey present.

F: Other components: The number of relevant components of predator-prey and parasitoid-host interactions which could be examined in simple analytical (Nicholson-Bailey-type) models is potentially very large. We have attempted to summarise the approach with reference to some of the more widely discussed examples. Others which have been examined include: (a) differential susceptibility of hosts to parasitism (i.e. variability in host escape responses or physiological defences, temporal asynchrony between parasitoid and host) (Kidd and Mayer, 1983; Hassell and Anderson, 1984; Godfray *et al.*, 1994); (b) parasitoid host-feeding (e.g. Yamamura and Yano, 1988; Kidd and Jervis, 1989; Murdoch *et al.*, 1992b; Briggs *et al.*, 1995; Jervis and Kidd, 1999); (c) competing parasitoids (e.g. Hassell and Varley, 1969; May and Hassell, 1981; Taylor, 1988b); (d) hyperparasitism (e.g. Beddington and Hammond, 1977); (e) multiple prey systems (e.g. Comins and Hassell, 1976); (f) host-generalist-specialist interactions (Hassell, 1986; Hassell and May, 1986); (g) combinations of parasitoids, hosts and pathogens (e.g. May and Hassell, 1988; Hochberg *et al.*, 1990; Begon *et al.*, 1999); (h) parasitoid egg-limitation (Mills and Getz, 1996; Mills, 2001); (i) dispersal between breeding sites (Weisser and Hassell, 1996) and local mate competition (Meunier and Bernstein, 2002); (j) incidence of diapause (Ringel *et al.*, 1998).

Some of the multi-species interactions are discussed further in section 7.3.10. Some of the aforementioned authors have taken an **individual-based approach**. This recognises that properties such as parasitoid life-history, physiology and behaviour vary among individuals. It has involved (a) **state-structuring** (taking account of the physiological basis of foraging decisions e.g. size of egg load influences the decision whether to feed or oviposit, and (b) **stage-structuring** (taking account of size- and stage- variation among hosts, and the selective behaviour shown by parasitoids). Murdoch *et al.* (1992b), Brigg's *et al.* (1995, 1999) and Shea *et al.* (1996) in particular should be consulted.

Recent reviews of parasitoid-host and predator-prey models are provided by Mills and Getz (1996), Barlow and Wratten (1996), Barlow (1999), Berryman (1999), Hochberg and Holt, (1999), Hochberg and Ives (2000), Mills (2000, 2001), and Hassell (2000b).

What is Searching Efficiency?

Having reviewed the essential behavioural components involved in searching by natural enemies (functional and aggregative responses, mutual interference), we are perhaps in a better position to answer the question of what is meant by the term 'searching efficiency'. This term has most often been used synonymously with 'attack rate', i.e. a more efficient predator kills more prey per unit time than a less efficient one. However, particularly in the context of population models, the detailed usage of the term has varied considerably. In the original Nicholson-Bailey model the area of discovery a defines a lifetime searching efficiency, whereas the 'attack' coefficient a' in equation (7.21), defines an 'instantaneous' searching efficiency in terms of numbers of prey attacked per unit time, T. In a patchy environment, however, searching efficiency is sensitive to two factors: (a) the patch-specific searching ability of the predators, and (b) the extent to which the distribution of the predators is non-random (Hassell, 1982a). To take account of this, Hassell (1978, 1982b) proposed a model for overall searching efficiency of predators (or parasitoids) where prey are gradually depleted:

$$a' = \frac{1}{n}\sum_{i=1}^{n}\left[\frac{1}{P_t T_i}\log_e\left(\frac{N_i}{N_i - N_{ai}}\right)\right] \qquad (7.39)$$

where n is the number of patches, N_a is the number of prey attacked, and N_i, N_{ai} and T_i are the number of hosts available, number of hosts parasitised, and the time spent searching, respectively on the ith patch. This equation represents an important step forward, as it views the functional response as essentially a within-patch phenomenon, i.e. occurring on a small spatial and temporal scale. Laboratory experiments to measure functional responses have, of course, been carried out on exactly this scale (section 1.10), so the within-patch interpretation affords a more realistic correspondence between experiment and modelling. It also circumvents the need for an average *lifetime* functional response measure to use in single-patch models, such as the original Nicholson-Bailey model (see above). Potentially, the patch model defined by equation (7.39) could be further expanded to include variation in the functional response with prey size, predator age and a range of other components, but the complexity involved would make this procedure more appropriate for the simulation approach discussed below (subsection 7.3.8).

An experimental measure of overall searching efficiency of a predator can be obtained by:

1. Estimating the average amount of time spent per predator in each patch (this can be calculated from the average number of predators found in each patch during a predetermined time period or over the course of the experiment (e.g. 24 hours);
2. Recording the average number of prey killed in each patch (providing a range of prey densities between patches).

These estimated parameter values can then be substituted in equation (7.39) to find a'. A suitable patch scale and experimental arena will need to be chosen, and this will depend on the prey species involved. Wei (1986), for example, used five clusters of rice plants, each set into a glass tube of water and interconnected by slender wooden strips to facilitate searching by the mirid bug *Cyrtorhinus lividipennis* for its prey, the brown planthopper *Nilaparvata lugens*. Ideally, a lifetime measure of searching efficiency could be obtained by repeating the procedure for each day of the predator's immature and adult life, and then either totalling or averaging the values of a' obtained. In practice, this may be very difficult due to the time involved and the large number of prey needed during the experiment.

Hassell and Moran (1976) proposed for parasitoids a measure of **'overall performance'**, A, that takes account of larval survival.

$$A = \frac{1}{P_t}\log_e[N/(N - P_{t+1})] \qquad (7.40)$$

where P_t and P_{t+1} are the densities of searching parasitoids in successive generations, and N is the number of available hosts. Clearly, a major constraint on the effectiveness of a parasitoid is likely to be mortality during the immature stages (subsection 2.10.2). 'Overall performance' may thus provide a more useful measure of the relative usefulness of different parasitoid species in biological control (Hassell, 1982b) (subsection 7.4).

Other Analytical Modelling Approaches

So far, we have concentrated our attention on ways of expanding the Nicholson-Bailey model to include more realistic components of predation. Although space does not permit a detailed discussion, it should be mentioned that other analytical modelling frameworks are available and have been used successfully to gain insights into the dynamics of predator-prey interactions. We now briefly discuss two of these to provide the reader with a lead into the literature.

Models structured in terms of differential equations, to encompass continuous-time changes, have a long pedigree, beginning with the predator-prey interactions of Lotka and Volterra (Lotka, 1925; Volterra, 1931) (Berryman, 1999, provides a review). Host-parasitoid versions are also available (Ives, 1992; Hassell, 2000b). It has been shown that, where parasitoids aggregate in patches of high host density, such models are usually unstable (Murdoch and Stewart-Oaten, 1989), in contrast to their discrete-time counterparts. This serves to highlight the fact that there are important differences between the two model types, and also emphasises the point that the behaviour of analytical models can be highly sensitive to minor variations in their construction (Ives, 1992; Berryman, 1992). Perhaps more importantly, however, it points to a number of limitations of the discrete-time format, where artificial time-jumps of one generation are not only imposed on growth and mortality, but also on the behavioural attributes of the natural enemy (e.g. spatial redistribution within the habitat). It seems likely that future modelling work will increasingly place greater emphasis on the more flexible (but mathematically more complex) continuous-time format incorporating, for example, stage-structure (= age-structure) and developmental delays in the predator attack rate (Mills and Getz, 1996).

Continuous-time models incorporating age-structure have been developed by Nisbet and Gurney (1983), Gurney and Nisbet (1985) and extended to cover host-parasitoid systems (Murdoch *et al.*, 1987; Godfray and Hassell, 1989; Gordon *et al.*, 1991). Murdoch *et al.* (1987) showed that the incorporation of an age class which is invulnerable to parasitism into such a model with overlapping generations can, under certain circumstances, promote stability (subsection 7.3.2).

Discrete-time models incorporating age-structure have also been developed (Bellows and Hassell, 1988; Godfray and Hassell, 1987, 1989), using an elaboration of equation 7.31. Godfray and Hassell (1987, 1989) used this model form to demonstrate that parasitism can act to separate the generations of a host population, when otherwise they would tend to overlap. Whether host generations were separated in time depended on the relative lengths of host and parasitoid life-cycles.

An alternative modelling format which also allows for the incorporation of age-structure in populations is provided by matrix algebra (Leslie, 1945,1948). Matrices can also be used to incorporate quite complex age-specific variations in fecundity, survival, development and longevity, thus encompassing populations which show either discrete or overlapping generations. This flexibility also makes the matrix approach extremely suitable for simulation modelling (see below). The matrix methodology is also easy to use and has the advantage over computer-based models of having an easily communicated, mathematical notation. Williamson (1972) provides a good introduction to the use of the technique in population dynamics (see also Buckland *et al.* 2004).

7.3.8 PUTTING IT TOGETHER: SIMULATION MODELS

Introduction

The main value of analytical models has been to provide insights into the general possibilities of population dynamics and how they might alter with changing conditions. Simulation models, on the other hand, attempt to mimic the detailed dynamics of particular systems and involve a somewhat different methodology. Simulation models of population dynamics can be constructed at different levels of complexity, from a relatively simple expansion of the analytical approach described in the previous section (Godfray and Hassell, 1989), to extremely elaborate systems of interlocking equations involving large numbers of components. However, they all share a common methodology for testing their accuracy **(validation)** and for assessing their behaviour **(experimentation and sensitivity analysis)** which will be discussed in detail. Whilst they can be formulated in conventional mathematical notation, simulation models are often constructed in practice as computer programs, which facilitate the complex calculations involved and also present the output in a readily accessible format. The models can be constructed in continuous or discrete time, again using systems of differential or difference equations respectively. The discrete time format has tended to be favoured by modellers interested in simulating the most complex insect population systems, involving age- or stage-specific fecundities and mortalities (Stone and Gutierrez, 1986, and Crowley *et al.*, 1987 give continuous-time examples). The aim of such inductive models is to encapsulate the detail of the particular system in question, with the emphasis on realism and accuracy. If successful, the model can often be used in decision-making, for example, in integrated pest management programmes.

To illustrate the way in which age-structure can be incorporated into a relatively simple simulation model, the model of Kidd (1984),which can be used to simulate populations with either discrete or overlapping generations, is discussed. The model considers a hypothetical population reproducing asexually and viviparously. Each individual is immature for the first three days, becomes adult at the beginning of the fourth day, reproduces on the fifth and sixth days, and then dies (Figure 7.33). The population is divided, therefore, into six one-day age groups, each adult producing, say, two offspring per day. To simulate population change from day to day, the computer:

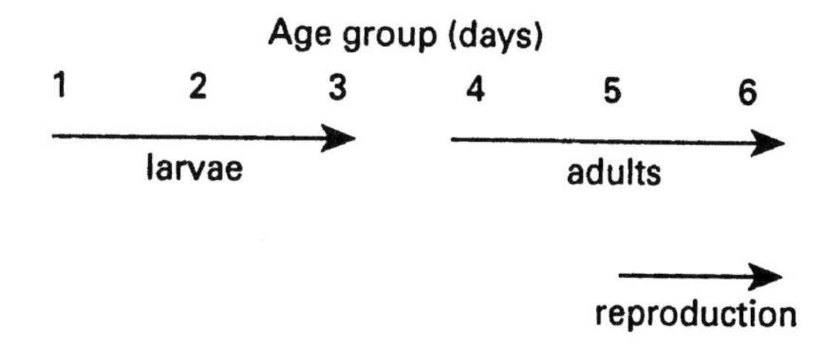

Figure 7.33 Life-history characteristics of an hypothetical population with discrete generations. Source: Kidd (1979). Reproduced from *Journal of Biological Education* by permission of The Institute of Biology.

1. Dimensions an array with six elements;
2. Places the initial number in each element of the array (initial age structure);
3. Calculates the total reproduction [REPMAX = (2* number of adults in age group 4) + (2* number of adults in age group 5) + (2* number of adults in age group 6)];
4. Ages the population by one day (this is done by moving the number in each age group N into age group N + 1);
5. Puts the number reproduced (REPMAX) into the one-day-old age group.

The model operates with a time-step of one day, and steps 3. to 5. can be repeated over as many days as required to simulate population growth. This very simple model can then be further elaborated to include additional components, such as mortality acting on each age group (e.g. either a constant proportional mortality, a uniformly-distributed random mortality or a density-dependent function) (Figure 7.34). By changing the length of immature life relative to reproductive life the model can be used to explore the behaviour of populations with either discrete, partially overlapping, or fully overlapping generations (see Kidd, 1979, 1984 for details and for a BASIC program).

This very simple deterministic simulation model provides the basic format for a number of modelling approaches, including the variable life-table models of Hughes and Gilbert (1968)

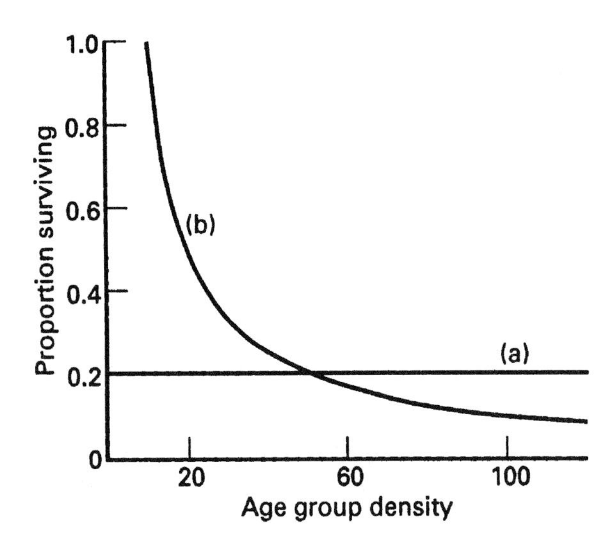

Figure 7.34 The effects of: (a) a density-independent factor: (b) a density-dependent factor used in the simulation model. Source: Kidd (1979). Reproduced from *Journal of Biological Education by* permission of The Institute of Biology.

and Gilbert *et al.* (1976) (see above). Variable life-table models have generally been used to determine how the relationships governing birth and death processes affect population dynamics in the field, and they rely on intensive laboratory and field observations and experiments during the model development and validation stages (Gilbert *et al.*, 1976; Gutierrez *et al.*, 1990; see Getz and Gutierrez, 1982 for an historical review and examples). The sequence of steps involves:

1. Estimating intrinsic relationships, such as development rates and growth rates;
2. Estimating extrinsic biotic relationships, including density-dependent effects, effects of natural enemies etc.;
3. Estimating abiotic effects such as weather factors.

Ideally, the accuracy of the model needs to be tested at each stage, before the modeller can confidently proceed with the incorporation of more complex components. In this way, the model increases progressively in realism and complexity, without sacrificing accuracy.

To illustrate the process, including the important techniques of validation and sensitivity analysis, the population study of Kidd (1990a,b) on the pine aphid, *Cinara pinea is* used as an example. This species infests the shoots of pine trees (*Pinus sylvestris*) and can be cultured in the laboratory as well as studied in the field. The first task was to construct a relatively simple model to simulate the changing pattern of aphid numbers on small trees in the laboratory. This model incorporated a number of relationships obtained from observation and experiment, including: (a) an increase in the production of winged migratory adults with increasing population density; (b) a decrease in growth rates (adult size) and development rates with crowding and poor nutrition; (c) a decline in fecundity with smaller adult size. The effect of variable temperature on development was included by accumulating day-degrees above a thermal development threshold of 0°C (subsection 2.9.3).

Model Validation

The output from this prototype model (Figure 7.35) was compared with the population changes on four small saplings using a 'least-squares' goodness-of-fit test. Testing the accuracy of model output against real data is the process of **model validation** and should involve some statistical procedures (Naylor, 1971), although many population modellers have in the past relied on subjective assessment of similarity (e.g. Dempster and Lakhani, 1979). One frequently used statistical method is to compare model output with means and their confidence limits for replicated population data (Holt *et al.*, 1987). If the model predictions fall within the confidence range, then it can be assumed that the model provides an acceptable description of the data. Frequently, however, the replicates of population data show divergent behaviour, and for the model to be useful it needs to take this variability into account. This was the case with the pine aphid: the four populations on the small trees behaved differently, and simply to average the data for each sampling occasion would have,

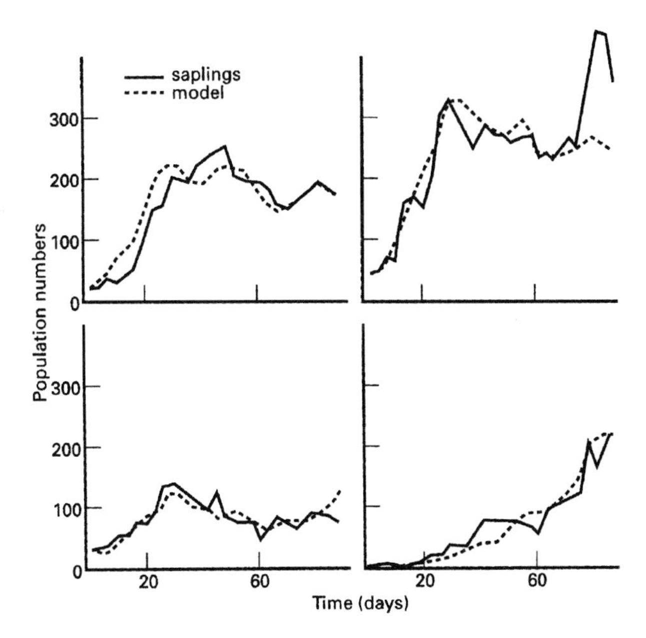

Figure 7.35 Population dynamics of pine aphids on four laboratory saplings and population numbers predicted by the simulation model. Source: Kidd (1990a). Reproduced by permission of the Society for Population Biology.

at best, lost valuable information and, at worst, been statistically meaningless. The 'least-squares' method used (see Kidd 1990a for details) took into account the variability both within and between the trees and, in fact, the model was able to explain 83% of the variation in the data. The model thus seemed to provide an acceptable description of aphid population behaviour in the laboratory, so was it adequately validated at this stage? The answer is *no* for another important reason: *for acceptable validation, models need to be compared with population data which have been independently collected and not used to provide data for the construction of the model.* The four sapling populations, in this case, had yielded data which had been used in the model. For acceptable validation an independent set of populations on four trees was used and here the model explained 78% of the variability within and between trees (Figure 7.36).

Experimentation and Sensitivity Analysis

At this stage it was possible to use the model to assess the relative contribution of each component to the aphid's population dynamics in the laboratory. This involved manipulating or removing particular components and observing the behaviour of the system. It was also possible to reveal those components to which model behaviour was particularly sensitive and which might repay closer investigation. For example, changing the reproductive capabilities of the adults, not surprisingly, affected the rate of population increase, but this effect was extremely sensitive to the nature of density-dependent nymphal mortality. Population growth rates and the periodicity of fluctuations were also found to be sensitive to changing development rates, mediated through changes in plant quality.

Having achieved a sufficient degree of accuracy in simulating laboratory populations, it was possible to then incorporate the complexities associated with the field environment. In the first instance, this meant: (a) revising some components of the aphid/ plant interaction to make them more appropriate for field trees; (b) including the more extreme variations in temperature associated with the field; (c) incorporat-

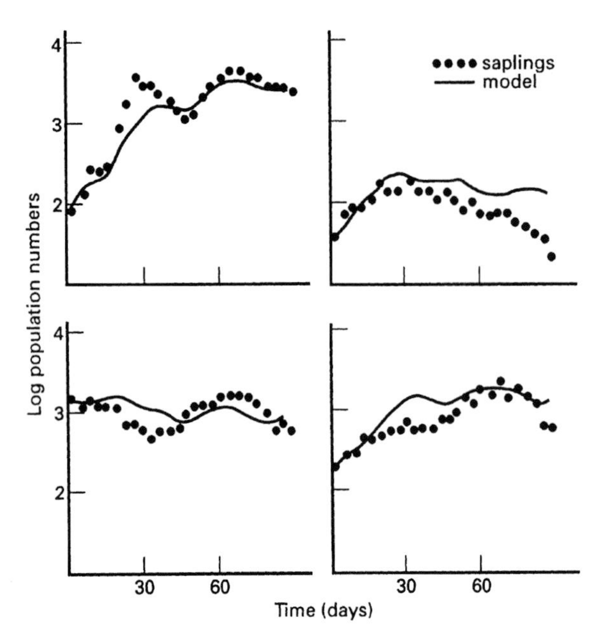

Figure 7.36 Population dynamics of pine aphids on four independent laboratory saplings and population numbers predicted by the simulation model. Source: Kidd (1990a). Reproduced by permission of the Society for Population Biology.

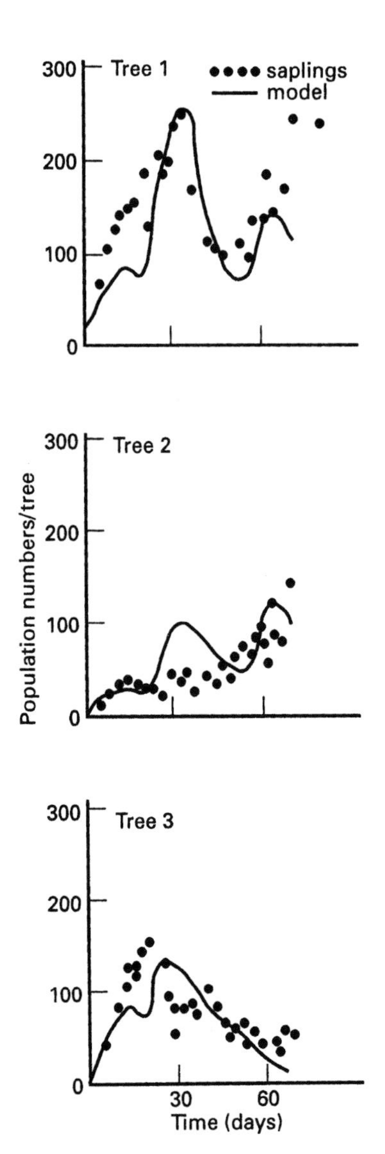

Figure 7.37 Pine aphid population dynamics on three field saplings from which predators were excluded and population numbers predicted by the simulation model. Source: Kidd (1990b). Reproduced by permission of The Society for Population Biology.

ing weather effects. At this stage mortalities due to natural enemies were excluded. Output from the revised model was then compared with populations on three field saplings covered by cages designed to exclude predators (Kidd, 1990b). At this stage the model was able to account for 52% of the numerical variation within and between trees (Figure 7.37); this is acceptable given the greater innate variability of field data. The model also predicted a pattern of numbers which was very close indeed to that of aphid populations on mature field pine trees, at least in the early season (Figure 7.38). Where predictions diverged from reality later in the season, this could probably be taken to reflect the impact of natural enemies which only become apparent after June.

While the model, as it stands, is purely deterministic in its construction, a stochastic element could have been included in any of the components by defining one or more parameters, not as constants, but in terms of their mean values and standard deviations. A random number generator could then have been used to produce a normally-distributed random number each time a parameter was used in the model. In this way, biologically meaningful variation could be reproduced. A number of standard computer programs are available for generating random

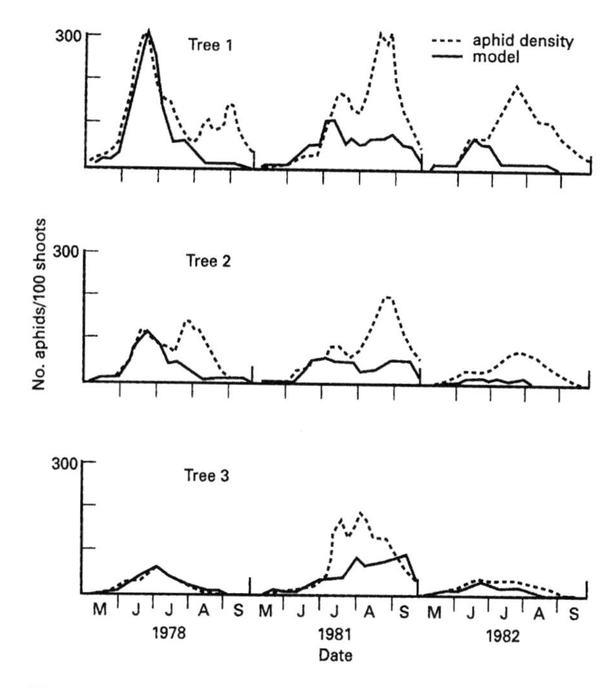

Figure 7.38 Pine aphid population densities on three mature field trees during three years, together with densities predicted by the model. (Source: Kidd, 1990b.) Reproduced by permission of the Society for Population Biology.

numbers for a range of possible distributions. Sometimes it is desirable in a simulation model to reproduce an entire or partial distribution of data, rather than a single random value. This technique, known as **Monte Carlo Simulation**, can easily be carried out using the random number generator in an iterative fashion.

We have dwelt at some length on this example in order to show the general procedures involved in a simulation study. Similar examples concerning aphid populations are provided by Hughes and Gilbert (1968), Gilbert *et al.* (1976), Barlow and Dixon (1980), and Carter *et al.* (1982) and for other insect groups by Gutierrez *et al.* (1988a,b) (cassava mealybug and cassava green mite), Holt *et al.* (1987) (brown planthopper), and Sasaba and Kiritani (1975) (green rice leafhopper). Clearly, there may be considerable variations in model construction, depending on the nature of the problem to be solved. For example, Stone and Gutierrez (1986) developed a model to simulate the interaction between pink bollworm and its host plant (cotton), which also included a detailed plant growth submodel (see also Gutierrez *et al.*, 1988a,b; Gutierrez *et al.*, 1994). Incorporating the effects of variable temperatures on development into simulation models has also been achieved in a number of different ways Stinner *et al.*, 1975; Frazer and Gilbert, 1976; Pruess, 1983; Wagner *et al.*, 1984, 1985; Nealis, 1988; Comins and Fletcher, 1988; Weseloh, 1989a; Kramer *et al.* 1991). With models based on a physiological time scale, for example, it is clearly impractical to use a time-step as short as one day-degree (one day may be the equivalent of about 20 day-degrees). In this situation, a physiological time-step corresponding to a convenient developmental period may be used. For *Masonaphis maxima*, Gilbert *et al.* (1976) used the **quarter-instar period or 'quip'** of 13.5 day-degrees Farenheit.

Incorporating the Components of Predation

Having constructed a basic simulation model, how can we best incorporate the components of predation or parasitism? This can be done in each of three ways:

1. By applying simple field-derived mortality estimates for predation (or parasitism) (subsection 7.3.2) in the prey (host) model; this tells us nothing of the dynamic interaction between predator and prey, however;
2. By constructing submodels for predators and parasitoids, involving age-structure if necessary, and including components of searching behaviour (e.g. functional and aggregative responses and numerical responses as already described in subsections 2.7.3, 2.9.2, 5.3.6 and 7.3.7); by constructing natural enemy submodels, but in this case using empirically derived estimates of predation and its effects.

These methods are now discussed in turn:

Method 1

Vorley and Wratten (1985) used a variable life-table model of the cereal aphid, *Sitobion avenae* to predict population growth in the absence of natural enemies. The difference between predicted and actual numbers in the field was attributed to 'total mortality', i.e. parasitism plus 'residual mortality'. By discounting the effects of parasitism (estimated from dissections), residual mortality could then be calculated. By running the model with only residual mortality acting, the effects of parasitism on the dynamics of the aphid population could finally be estimated. Clearly, this technique, in common with others in this category, is capable only of assessing the 'killing power' of a mortality source during the period for which data have been collected. It has no reliable predictive power (i.e. it is an **interpolative** rather than an **extrapolative method**) and yields no information on the dynamic interaction between parasitoid and host. A similar approach was adopted by Carter *et al.* (1982) for parasitoid and fungal mortality of cereal aphids.

Method 2

A. **Age-specific:** Age-specific submodels have frequently been developed for parasitoids and predators for use in both theoretical models (see Godfray and Hassell, 1989, and Kidd and Jervis, 1989; for parasitoids) and in field or crop-based simulation studies (Yano, 1989a,b, for parasitoids and Gilbert *et al.*, 1976 for parasitoids and hover-flies). These submodels are generally constructed in the same format as the main population model on which they act. For example, in the *Masonaphis maxima* model of Gilbert *et al.* (1976), the parasitoid *Aphidius rubifolii* was found to have an egg development time of six quips. In the theoretical model of Kidd and Jervis (1989), time and age groups were measured in days, the life-history features of the host and parasitoid having the structure shown in Figure 7.39.

B. **Searching behaviour:** As functional response relationships have usually been derived from short duration experiments (e.g. 24 hours; section 1.10), they are likely to be more meaningful when incorporated into simulation models with a short time-step of, for example, one day, rather than the one-generation time-step of the Nicholson-Bailey model (subsection 7.3.7). In the model of Kidd and Jervis (1989), parasitoids searched for hosts sequentially, either at random between patches, or in selected high host density patches. Type 2 functional responses were incorporated for both feeding and oviposition in a biologically realistic way, by allowing each parasitoid a maximum available search time per day (10 hours), the 'efficiency of search' being constrained by feeding handling time, oviposition handling time and egg-limitation. A full BASIC program listing for this model is provided by Kidd and Jervis (1989), to which we refer readers interested in developing the individual-based queueing techniques described.

Few attempts have been made to measure the searching efficiency of natural enemies in the field (e.g. Young, 1980; Hopper and King. 1986; Jones and Hassell, 1988; Weisser *et al.*, 1999, Schenk and Bacher, 2002), probably due in large part to the obvious technical difficulties of confining particular densities of predators/parasitoids and prey/hosts within localised patches. Where the Holling disc equation (equation (7.21)) has been used to model a parasitoid-host interaction, parameter values have sometimes been estimated from field data. This can be done by iteration, i.e. by using a range of alternative parameter values in repeated simulations to find which fit the data best (Ravlin and Haynes, 1987). The difficulty here is that values of a' and T_h which produce accurate simulations may bear no resemblance to laboratory estimates of these parameters, casting some doubt on the realism of at least some components in the model. An alternative approach has been used by some workers (Godfray and Waage, 1991; Barlow and Goldson, 1993) to capture the essence of parasitoid search in a way that can potentially be used to describe their dynamic interactions

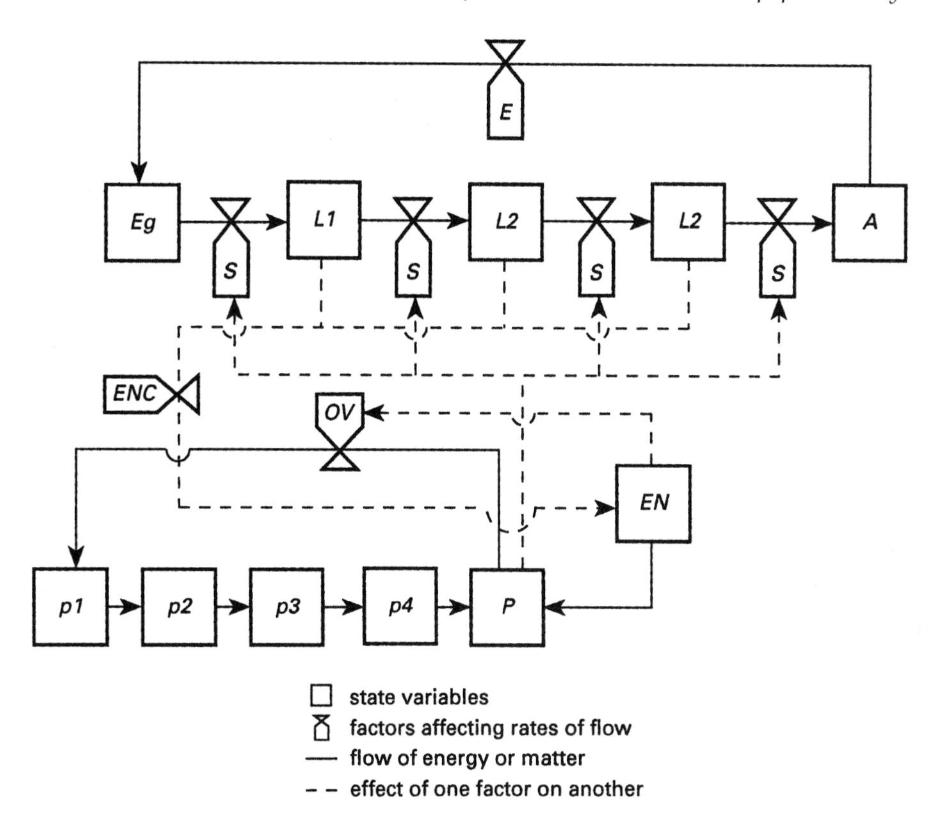

Figure 7.39 Simplified relational diagram of the parasitoid-host system: *Eg*, host egg age group; *L1-L3*, host larval age group; *A*, host adults; *S*, survival rate; *E*, host reproductive rate; *ENC*, encounter rate,. *p1-p4*, immature parasitoid age groups; *P*, adult parasitoids; *OV*, rate of parasitoid oviposition; *EN*, individual parasitoid energy stocks. Source: Kidd and Jervis (1989). Reproduced by permission of The Society for Population Biology.

with host populations in the field. This is done using the simple Nicholsonian component $[1-\exp(-aP^m)]$ to describe the proportion of hosts attacked during a defined time period, where P is the number of parasitoids, a is searching efficiency and m defines the strength of density-dependence in parasitoid attack (see equation (7.33). Barlow and Goldson (1993) used this term to model the interaction between the weevil pest of legumes, *Sitona discoideus* and an introduced braconid parasitoid, *Microctonus aethiopoides*. Thus:

$$q = 1 - \exp(-aP^m) \qquad (7.41)$$

where q is the estimated proportion of hosts parasitised. Leaving aside life-cycle complications, parameter values for a and m were simply estimated from the relationship between parasitoid densities (estimated from percentage parasitism) with percentage parasitism in the following generation (Figure 7.40). To do this, equation (7.41) was first linearised by rearrangement using the following steps:

$$1 - q = \exp(-aP^m) \qquad (7.42a)$$

$$\log_e(1 - q) = -aP^m \qquad (7.42b)$$

$$-\log_e(1 - q) = aP^m \qquad (7.42c)$$

$$\log_{10}[-\log_e(1 - q)] = \log_{10} a + m \log_{10} P \qquad (7.42d)$$

Thus, $\log_{10}[-\log_e(1-q)]$ can be regressed against $\log_{10}P$ with slope m and intercept $\log_{10}a$ to find values of m and a. The model derived by Barlow and Goldson (1993) gave acceptable

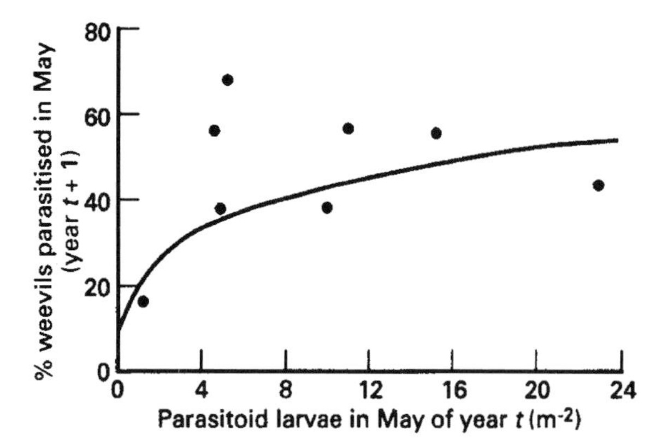

Figure 7.40 The relationship between the density of parasitised legume weevils in year t and the percentage parasitised in year $t+1$. Source: Barlow and Goldson (1993). Reproduced by permission of Blackwell Scientific Publishing.

predictions of parasitoid and host numbers over a ten-year period. This is an example of a **model of intermediate complexity** (Godfray and Waage, 1991; see also subsection 7.4.3).

A. **Numerical responses:** As indicated in subsection 7.3.7, the rate of increase of a predator population depends on a number of components, namely, the development rate of the immature stages, the survival rate of each instar, and the fecundity of the adults. To build a comprehensive submodel of the predator rate of increase we would need, therefore, to incorporate the various factors which might affect prey consumption (e.g. prey density, handling time, aggregative response, mutual interference), together with the effects of prey consumption on development and survival of the different instars, and on adult fecundity. The influence of prey consumption on predator growth and development, survival and fecundity has already been discussed at length, together with experimental methodologies and descriptive equations (subsections 2.9.2, 2.10.2, 2.7.3). Although the modelling of these processes is technically feasible with computer simulation, to obtain sufficiently detailed information on all of the components involved would be a potentially very time-consuming operation. To our knowledge, the task has not yet been fully carried out for any single predator species in the formal detail prescribed by Beddington *et al.* (1976b). However, Crowley *et al.* (1987) describe a detailed *sub*-model applicable to damselfly predators, but without an equally detailed model for the *Daphnia* prey. Nevertheless, this clearly remains an important challenge for those interested in modelling predator-prey processes, as a computer submodel, based on general theory and incorporating all of the components of predation in this way, would have immense practical as well as theoretical value.

Method 3

For simulation models specifically designed to mimic field population dynamics, the components of predation and parasitism have usually been incorporated into submodels in a more pragmatic way relevant to the particular system and problem in question. As a wide variety of approaches has been adopted, with no single unified methodology, we can only present a number of examples to illustrate the possibilities. For parasitism acting in their *Masonaphis maxima* model, for example, Gilbert *et al.* (1976) first established the duration (in quips, see above) of the four developmental stages of *Aphidius rubifolii* (egg, larva, pupa and adult). Adult females were assumed to search at random, ovipositing a constant (average) number of eggs per quip, in susceptible third and fourth instar and young adult aphids. No account was taken of how parasitoid oviposition might vary with aphid density, so it is perhaps not surprising that the model, at least in its early versions, provided poor predictions of aphid parasitism.

For syrphid predation, on the other hand, Gilbert *et al.* (1976) assigned predator larvae to four developmental periods, each of 10 quips duration and consuming 1, 2, 6 and 15 average-sized aphids per quip. This provided a consumption rate of 240 aphids for each syrphid to complete development, a figure broadly agreeing with observations. In the model, syrphid eggs were simply laid in proportion to aphid densities, thus generating the number of larvae attacking the aphids. Aphids were assumed to be attacked at random, and were subtracted from the model during each time-step in proportion to the voracity and number of each syrphid age group present. No account was taken of predator starvation or survival, but, even so, Gilbert *et al.* found this simple model gave reasonably accurate predictions of predation in the field.

Some variations on this **'maximum consumption' method** have been carried out on other systems. Wratten (1973) and Glen (1975) respectively examined the effectiveness of the coccinellid, *Adalia bipunctata* and the mirid bug, *Blepharidopterus angulatus* as predators of the lime aphid, *Eucallipterus tiliae* using the methodology developed by Dixon (1958, 1970). The sequence of tasks involved in this approach are:

1. To monitor the age distribution and population densities of prey and predators at regular intervals (of, say, one week) over a period of months or years;
2. To measure the efficiency of capture (% encounters resulting in capture) of each predator instar in relation to the different prey instars in the laboratory;
3. To estimate the amount of time spent searching per day (for *A. bipunctata* this meant time exceeding a minimum temperature threshold for activity, corresponding approximately to the 16 hours of daylight);
4. To estimate the percentage of searching time spent on areas with prey (44% for *A. bipunctata*);
5. To calculate the area traversed/day (= distance travelled in 1 hour at mean temperature *16*0.44*R [R being the width of perception of the predator]);
6. To find the % of time spent on areas already searched;
7. To calculate the 'area covered' from 5. and 6., taking into account time wasted by covering areas already searched;
8. To estimate the time spent feeding as a proportion of the total time available.

Components 3.to 5. are all determined empirically from observations in the laboratory. It can be seen that some of the above procedures follow Nicholson and Bailey's (1935) methodology for calculating the 'area of discovery' (subsection 7.3.7) and effectively take into account handling time and aggregation of predation, but not mutual interference. Negative effects on the predators are again simply estimated from experiments to determine the minimum density of aphids required for predator survival at different stages (see reference to Gilbert *et al.*, 1976, above).

For *A. bipunctata* (Wratten, 1973), the number of aphids of a particular instar consumed by each coccinellid instar per week (N_a) could be found from the equation:

$$N_a = 0.01A^*D^*0.01E^*C_o^*n^*C_f \qquad (7.43)$$

where A is the area covered/larva/week, D is the density of aphids, E is the capture efficiency, C_o is a correction factor for aphid distribution, n is the density of larvae, and C_f is the correction factor for the proportion of available time spent feeding up to satiation. This calculation was made for every combination of coccinellid and aphid instar in the populations on each sampling date, so as to arrive at an overall number of aphids in each instar removed per week. Assuming that these aphids would have remained in the population in the absence of predation, Wratten (1973) was able to estimate the potential aphid population size and structure in the absence of predation from information on the development times and reproductive patterns of the aphids, obtained by confining them in leaf cages. The difference between the observed aphid population and the predicted population in the absence of predation thus demonstrated the effect of coccinellid mortality. A later study incorporating these components of coccinellid predation, together with similar ones for the mirid predator (Glen, 1975), into a simulation model of the lime aphid (Barlow and Dixon, 1980), showed that both species can have a destabilising effect on aphid numbers.

Clearly, where prediction of the prey death rate over a limited period (e.g., one season) is the sole aim of the model, fairly crude representations of predation may suffice. However, where a longer time-frame is being simulated, especially one involving 'coupled', i.e. monophagous predator-prey interactions, a more accurate submodel incorporating a predator numerical response will be needed. A number of approaches which could be used to achieve this end are described in detail for aphid-coccinellid interactions by Frazer *et al.* (1981a), Gutierrez *et al.* (1981), Mack and Smilowitz (1982) and Frazer and Raworth (1985). In practice, none of these studies attempted to simulate the interaction beyond one field season. To take just one example in detail, Frazer *et al.* (1981a) derived empirical relationships for coccinellid reproduction and survival by confining adult beetles with aphids in screen-walled cages in alfalfa fields. Ives (in Frazer *et al.*, 1981a) had found in the laboratory that female coccinellids required 1.3 mg live weight of aphids/quip for maintenance, additional prey being converted to eggs at a rate of 0.7 eggs/mg of aphid. A direct relationship could then be established between the predation rate and the reproductive rate of adult females. Overall survival from egg to adult was estimated by comparing the expected numbers of eggs laid in the cages with the number of beetles recovered at emergence, and was subsequently found to show a sigmoid relationship with aphid density (using total aphid density during the first larval instar of the beetle, although a running average could also have been used).

7.3.9 COMBINING TESTABILITY AND GENERALITY

We have emphasised above the distinction between analytical and simulation models, and also the difference between the deductive and inductive approaches. The latter, essentially philosophical, distinction is equivalent to that proposed by May (1974b), who referred instead to **strategic** and **tactical models**. Strategic models attempt to describe and abstract the general features of population dynamics while ignoring the detail. Tactical models, on the other hand, are developed to explain the complex dynamics of particular systems, being particularly useful in population management programmes. Tactical models, however, may not readily lead to general conclusions about population dynamics. Although they can readily be tested against real system behaviour (subsections 7.3.4 and 7.3.8), the testing of strategic models is more problematical, as more than one model can often be invoked as a plausible explanation of a particular phenomenon. Murdoch *et al.* (1992a) suggest a way of making models both testable and general, by first building tactically-orientated models, then progressively stripping out the detail, whilst testing the new models at each stage to determine the loss in predictive capacity. Murdoch *et al.* (1992a) provide an example of how the methodology might be used by referring to the work of Murdoch and McCauley (1985) and McCauley and Murdoch (1987 and related papers) on *Daphnia*-algae interactions. A similar approach has also been advocated by Berryman (1990) and Kidd (1990c).

7.3.10 MULTISPECIES INTERACTIONS

Introduction

So far in this chapter our consideration of how to model the role of natural enemies in population dynamics has been largely confined to two-species interactions, whether they be predator-prey or parasitoid-host in nature. We need to be aware, of course, that these simplified models ignore a whole range of potential influences from the wider multispecies environment. In recent years, some of these effects have begun to be explored in more depth, using either a modelling or an experimental approach.

Apparent Competition

It has long been recognised that two species sharing a resource that is in short supply are likely to enter into a state of competition (resource competition) which may result in one species being disadvantaged and possibly even excluded by the other. What has now become clear is that a similar outcome, known as **apparent competition**, may occur if the two species are *not* in resource competition, but share a natural enemy. Here, the increase in abundance of an alternative prey leads to a numerical increase in the predator, with increased levels of attack on both prey species. Over time, one prey species may be eliminated. While this phenomenon has been extensively explored through analytical modelling (see Bonsall and Hassell 1999, for a review), empirical evidence has tended to be circumstantial and anecdotal (but see below). Bonsall and Hassell (1999) also provide a critical review of some of these studies and their experimental designs. Experimental conditions may be difficult to maintain in practice, particularly where the species involved are especially mobile (Morris *et al.*, 2001). A good field demonstration of apparent competition (with appropriate experimental design) involving two aphid species and their shared coccinellid predator is provided by Müller and Godfray (1997). See also van Nouhuys and Hanski (2000) and Morris *et al.* (2001) for examples involving two primary parasitoid species with a shared secondary parasitoid.

Apparent Mutualism

Apparent competition is not the only 'indirect' interaction mediated by shared natural enemies. Higher densities of one prey species may, through switching, satiation or egg-depletion, cause predation pressure on the other prey species to be relaxed. The second prey species can thus benefit from the presence of the first, an interaction known as **apparent mutualism** (Holt, 1977; Holt and Kotler, 1987; Abrams and Matsuda, 1996). Hoogendoorn and Heimpel (2002) reported on one such interaction between two coccinellids with a shared parasitoid. One of the coccinellids was a relatively poor host for the parasitoid, acting as a mortality 'sink' for its eggs. This effect acted to reduce parasitism on the other coccinellid and raised its equilibrium levels, as demonstrated by an analytical model of the interaction.

Competing Predators

The reciprocal three-species interaction of two predators sharing a single prey species has been analysed extensively using the simplified parasitoid-host format of the Nicholson-Bailey model (see Hassell, 2000b, for a review). Here, a host species is attacked by two parasitoid species, each of which aggregates its attacks independently of the host and the other parasitoid. Using the model of May (1978) (described in 7.3.7 above), with the parameter k describing the degree of aggregation of attacks, the conditions for stability were shown to depend on: (a) the sequence of parasitoid attack, (b) the outcome of multiparasitism and (c) the stability of the individual host-parasitoid links. The possibility of the three species coexisting is greatly reduced if only one parasitoid contributes strongly to stability, while coexistence becomes impossible if the two parasitoids attack randomly (unless some other density-dependence is introduced). May and Hassell (1981) also consider the effects on equilibrium levels (as opposed to stability), while Hassell (2000b) discusses these outcomes in relation to the practical issue of multiple biological control introductions (see also section 7.4).

Interactions Across Three Trophic Levels

'Bottom-up' and 'Top-down' Effects

The relative extent to which insect herbivore populations are constrained by natural enemies or by plant resources, i.e. whether the herbivores are mainly 'top-down' (natural enemy) or 'bottom-up' (plant resource) constrained, has recently been subject to much discussion. The debate has centred upon the argument put forward in the classic paper by Hairston *et al.* (1960) that herbivore (*sensu lato*) populations tend not to be food-limited. This has often been interpreted as saying that the world is green because herbivores are maintained at low abundances by top-down, rather than bottom-up forces. The observation that significant intra- and interspecific competition occur rather infrequently in insect herbivores, has lent weight to the top-down argument (but see Denno *et. al.*, 1995), as have the dramatically successful cases of classical biological control (Strong *et al.*, 1984). Most food web and population dynamics theory applied to insect herbivores has assumed that bottom-up forces have, in general, minimal effects on abundance (Hawkins, 1992).

However, it is increasingly being argued that the importance of bottom-up effects ought not to be underestimated; while plants may appear to be a superabundant resource for insect herbivores, in reality they are not, because food quality varies significantly and may thus act as a major constraint, both upon herbivore abundance (e.g. Hunter and Price, 1992; Ohgushi, 1992; Wratten, 1992) and upon herbivore quality (from the standpoint of the growth, development, fecundity and survival of the natural enemies consuming their tissues, see Thaler, 2002, and subsection Chapter 2). Bottom-up constraints are now receiving more equitable treatment in the study of herbivore population dynamics and it is increasingly being accepted that, instead of it being a case of either top-down or bottom-up, an interaction between the forces is the more realistic view to take (e.g. Kato, 1994; Stiling and Rossi, 1997; Kidd and Jervis, 1997). However, certain barriers to progress continue to impede our understanding of the relative roles of top-down and bottom-up forces in tri-trophic systems.

Foremost of these barriers is the semantic confusion existing over the dynamic nature of the 'forces'. Some authors refer to these in the sense of 'regulation', where density-dependent effects act to increase stability and/or persistence. Others continue to use the term 'control', which could mean either 'regulation' or simple density-independent population *suppression*. When trying to disentangle the relative roles of top-down and bottom-up constraints, it is clearly important to understand which processes are being analysed (Hassell *et al.*, 1998). Hunter *et al.* (1997) used two-way analysis of variance on 16 years of time-series data for the winter moth, in an attempt to partition out the relative contributions of top-down and bottom-up forces to variations in numbers. 'Tree' and 'year' were taken as main effects in the analysis, with Hunter assuming that a measurable proportion of tree variation could be attributed to bottom-up effects (budburst phenology = *density-independent*), while a proportion of year-to-year variation could be assigned (using correlation analysis) to delayed *density-dependent* top-down effects. Thus, by confusing different processes, Hunter failed to separate out the relative contributions of bottom-up and top-down processes to *either* regulation or population suppression. A number of statistical problems with Hunter's study have also been detected and these are discussed in detail by Hassell *et al.* (1998).

Despite its problems, Hunter's analysis has been of some value in pointing to the need for adequate population data, rigorous analysis and clarity of definitions, before significant progress in this area of the bottom-up/top-down debate can be made. In a more recent paper, Hunter himself makes these arguments (Hunter, 2001), and outlines some possible approaches which may circumvent some of the difficulties. These may involve time-series analysis, life-table analysis (taking into account female oviposition preferences) and experimental methods. The latter are likely to involve detailed factorial experiments in which plant resources and predation pressure are manipulated. Stiling and Rossi (1997), for example, manipulated small island populations of *Asphondylia* flies, their four species of parasitoid, and the seaside plant, *Borrichia frutescens*, on which the flies produce galls. Using two-way repeated-measures analysis of variance, the effects of parasitism and plant quality (as manipulated by nitrogen fertiliser) on the number of fly galls could be determined. A significant interaction between parasitism and plant nitrogen was detected, with parasitism being important only on the high-nitrogen plants where galls were abundant. Of course, this study was not designed to separate out regulating from non-regulating influences, but it does serve as a reminder of the complex interactions between natural-enemy and plant-mediated effects which may occur.

As we have argued previously (Kidd and Jervis, 1997), the issue of 'regulation' may not actually be of major importance in the bottom-up/top-down debate. What is more at issue is whether particular populations are routinely constrained by plant resources alone, or whether the impact of natural enemies acts to keep numbers below the resource ceiling. In an analysis of 32 forest pest species, we found that 12 were reported to have population densities largely determined by resource constraints, while in 20 cases natural enemies were considered to have a major role (Kidd and Jervis, 1997). Even in the 12 resource-limited cases, however, we argued that natural enemies may have a more important role to play than simple mortality or percentage parasitism estimates might suggest. Using a simple simulation model, it was possible to demonstrate some profound and unexpected dynamical effects from the interaction between resource limitation and relatively low predator-induced mortalities (see Kidd and Jervis, 1997 for details). This emphasises the point that the interactions between bottom-up and top-down effects are likely to be complex in many cases, and may continue to defy our simplistic attempts to assess their relative contributions. Walker and Jones (2001) provide a recent synthesis of the subject.

Bottom-up effects will not not necessarily be exerted by the host's or prey's food plant *per se*, but by fungal endophytes, which will influence herbivore and parasitoid/predator performance (subsection 2.9.2) and ultimately affect herbivore and natural enemy abundance and dispersion patterns. Therefore, they could influence parasitoid-host population dynamics. See Omacini *et al.* (2001).

Modelling Bottom-up/Top-down Interactions

Despite the wealth of modelling techniques available (see sections 7.7 and 7.8), surprisingly few attempts have been made to model *natural* tritrophic systems of the plant-herbivore-predator variety (see, however, Barlow and Dixon, 1980; Kidd, 1990c; Larsson *et al.*, 1993; Mills and Gutierrez, 1999). We suspect that this may be partly due to a lack of information on the plant-herbivore side of the interaction. The same, however, is not true of agriculture, where both analytical and simulation models have been constructed to help integrate biological control programmes into crop-pest systems (e.g. Gutierrez *et al.*, 1984, 1988a, 1994; Holt *et al.*, 1987; Mills and Gutierrez, 1999), or to understand the effects of plant resistance on the pest-natural enemy interaction (Thomas and Waage, 1996). In the latter case, it has been shown by both modelling and experimentation that the presence of natural enemies (usually parasitoids) can enhance (additively or synergistically) the effects of plant resistance. As parasitism tends to be density-independent, its proportional effect is greater on partially resistant crop varieties. More complex tritrophic outcomes were explored by Hare (1992) using a series of graphical models, but a more rigorous theoretical treatment (and review) is provided by Thomas and Waage (1996).

Higher-level Tritrophic Interactions

Moving up one trophic level, in our survey of tritrophic interactions, brings us to a consideration of the interaction between a prey species, its predator (*A*) and a further predator species (*B*), attacking predator *A*. Its simpler host-parasitoid-hyperparasitoid equivalent has been considered, at least theoretically. Nicholson and Bailey (1935) extended their basic host-parasitoid model to include a randomly attacking hyperparasitoid, so that equation 7.18 now becomes:

$$N_t + 1 = F\exp(-aP_t) \quad (7.44a)$$

$$P_{t+1} = N_t[1 - \exp(-aP_t)][\exp(-aQ_t)] \quad (7.44b)$$

$$Q_{t+1} = [1 - \exp(-aP_t)][1 - \exp(-aQ_t)] \quad (7.44c)$$

where Q is the hyperparasitoid population.

The effect of including Q is to raise the host equilibrium, although the interaction remains unstable. However, including a density-dependent component to the host rate of increase F stabilises the interaction and allows a more meaningful analysis of the effects of Q. As Beddington and Hammond (1977) have shown, Q always raises the host equilibrium and the range of parameter space allowing stability is smaller than in the absence of Q. However, stability is more likely if the hyperparasitoid has a higher searching efficiency than the primary parasitoid, P.

May and Hassell (1981) have also explored the above interaction, where both parasitoids aggregate their searching behaviour. In this situation, stability can occur in the absence of host density-dependence. The interaction is always stable if both parasitoid species strongly aggregate their searching behaviour and if Q has a higher searching efficiency than P. In all cases the effect of the hyperparasitoid is to raise the host equilibrium, a conclusion which has serious implications for biological control (see section 7.4). See also Ives and Jansen (1998) for a stochastic approach to modelling host-parasitoid-hyperparasitoid interactions.

Intraguild Predation

When a species feeds on more than one trophic level it is showing **omnivory**, a feature which appears to be relatively common in natural communities (Polis *et al.*, 1989), but is of debatable significance in stabilizing community structure. A particular category of omnivory is shown by **intraguild predation**, where two predator species potentially compete for the same prey, but one of them also feeds on its competitor (Polis *et al.*, 1989). The overall result is thought to be a reduction in predator pressure and a decrease in top-down control of the shared prey (Rosenheim *et al.*, 1995; Snyder and Ives, 2001). In a more detailed analysis using analytical models, Holt and Polis (1997) were able to demonstrate that for the three-species interaction to be stable, the **intermediate predator** (the one acting as both predator and prey) has to be the better competitor for the shared prey. If the latter becomes scarce, the **top predator** (the one acting solely as a predator) may become extinct, as a result of its inferior competitive ability. As the prey becomes more abundant, the risk of extinction of the *intermediate* predator increases as a result of apparent competition with the shared prey. Empirical evidence to support these predictions is scarce, possibly due to the same difficulty in carrying out suitably designed experiments as were discussed in relation to apparent competition (see above). Demonstrating that it is not impossible, however, Finke and Denno (2003) carried out replicated 2×2 factorial (randomised block) experiments to test for intraguild predation between a wolf spider and a mirid bug predator, feeding on a shared planthopper prey. In support of theory, the presence of both predators reduced planthopper numbers *less* than did each predator in isolation. Mirid numbers were suppressed by the presence of spiders, but spider numbers were unaffected by the mirids, a result again predicted by theory, for higher prey densities (see also section 7.4 for a discussion of this study in relation to biological control).

Rosenheim *et al.* (1995) provide a comprehensive review of intraguild predation in insects, and review also empirically-based simulation models and general analytical models of intraguild predation in relation to biological control (see also section 7.4; also see Müller and Brodeur, 2002).

7.3.11 SPATIAL HETEROGENEITY AND PREDATOR-PREY MODELS

One of the most important conceptual advances has been the realisation of the importance of spatial heterogeneity in the dynamics of predator-prey and parasitoid-host interactions. It is now well established from deductive modelling that direct density-dependent relationships from patch to patch, resulting from the aggregative responses of natural enemies, can be a powerful stabilising influence (Hassell, 1978, 1980b, 2000a,b). Whilst optimality theory (section 1.12) predicts that such patterns of **direct**

spatial density-dependence ought to be common (Comins and Hassell 1979; Lessells, 1985), surveys of published studies suggest that they are in fact in the minority. Examples of the opposite, i.e. **inverse spatial density-dependent** relationships, where predation or parasitism are concentrated in the lowest density patches, are just as common, as are examples with no relationship at all, i.e. **density-independent** spatial relationships (Morrison and Strong, 1980, 1981; Lessells, 1985; Stiling, 1987; Walde and Murdoch, 1988). An inverse spatial density-dependent relationship may be found if hosts or prey in high density patches are less likely to be located than those in low density patches, perhaps due to greater host or prey concealment. Price (1988) showed this for parasitism by *Pteromalus* of the stem-galling sawfly *Euura lasiolepis* on willow (*Salix lasiolepis*). Even without this concealment effect, it is still theoretically possible for parasitism among patches to be density-independent or inversely density-dependent (Lessells, 1985). A mechanistic explanation (subsection 1.2.1) for this lies in the balance between two counteracting processes (Hassell *et al.*, 1985): (a) the spatial allocation of searching time by parasitoids in relation to host density per patch, and (b) the degree to which exploitation is constrained by a relatively low maximum attack rate per parasitoid within a patch. Inverse density-dependent parasitism can theoretically result from insufficient aggregation of searching time by female parasitoids in high density patches to compensate for any within-patch constraints on host exploitation imposed per parasitoid by time-limitation, egg-limitation or imperfect information on patch quality. Density-independent relationships, on the other hand, can result if processes (a) and (b) are in balance.

These observations would appear to undermine the importance of spatial density-dependence as a regulating factor in natural predator-prey and parasitoid-host systems, but Hassell (1984) was able to demonstrate, using the approach encapsulated in equations (7.31) and (7.32) above, that patterns of inverse spatial density-dependence can be just as stabilising as patterns of direct spatial density-dependence; whether direct or inverse relationships have the greater effect depends upon the characteristics of the host's spatial distribution (Hassell, 1985; also Chesson and Murdoch, 1986). In fact, it is now apparent that even where parasitism between patches is density-independent, the spatial distribution of parasitism, if sufficiently variable, can also promote stability (Chesson and Murdoch, 1986; Hassell and May, 1988). The biological interpretation seems to be that as long as some patches of hosts or prey are protected in **refuges** from natural enemy attack, whether in high density or low density patches, stability remains possible. This may be so, but we should exercise caution in making instant biological inferences from the behaviour of such models – other interpretations may be equally plausible or preferable (McNair, 1986; Murdoch and Reeve, 1987).

The conclusion derived from deductive models that different spatial patterns of parasitism have a powerful stabilising potential opened up a whole new area of study and debate in population ecology. Consequently new methods for studying spatial patterns of parasitism in the laboratory and field began appearing. These are now reviewed in turn before we address some of the more contentious issues surrounding the subject.

Detecting Spatial Density-Dependence

Laboratory Methods

The general methodology for studying spatial variation in parasitism or predation in the laboratory is described in subsection 1.14.2. Hassell *et al.* (1985) provide a good example of how two species of parasitoid attacking bruchid beetles (*Callosobruchus chinensis*) in experiments can show inverse spatial density-dependence. The beetle itself is a pest of legumes and commonly breeds in stored dried pulses. Black-eyed beans (*Vigna unguiculata*) were used in these experiments, which were carried out in clear perspex arenas (460 mm × 460 mm × 100 mm). Twenty-five patches of equal size were marked out on white paper sheets in an hexagonal grid, with 75 mm spacing between the centres. Different densities of beans containing 13-day-old hosts were allocated at random to patches and the number of beans in each patch made up to 32 with the addition of the required number of uninfested beans. Twenty-five parasitoids of one species were introduced into each arena and, after 24 hours, infested beans transferred to vials to await parasitoid emergence. The variable pattern of density-dependence found might be explained by both: (a) the allocation of parasitoid searching time in patches of different host density, and (b) the maximum attack rate per parasitoid constraining the extent of exploitation within patches. Parasitoids showed no tendency to aggregate in some patches over others, while their maximum attack rates per patch were limited by handling time constraints (see above) (Hassell *et al.*, 1989b give further details; see both papers for methods of culturing the beetle and its parasitoids).

Field Methods

Traditional techniques for analysing field population data for density-dependence seldom take account of spatial variation in patterns of mortality. The Varley and Gradwell method (subsection 7.3.4), for example, explores variations in *k*-mortalities over several generations, but takes no account of spatial variation amongst subunits of the population within a generation (Hassell,1987; Hassell *et al.*, 1987). Stiling (1988) surveyed 63 life-table studies on insects of which about 50% failed to detect any density-dependence acting on the populations. Hassell *et al.* (1989a), however, argued that density-dependence may still have been present but undetected due to: (a) the inadequate length of time over which some studies were conducted, and (b) the inability of the analyses to take account of spatial variations in patterns of mortality amongst subpopulations.

If traditional methods are inappropriate, how should we go about detecting spatial density-dependence in the field? The first problem is to decide on a suitable spatial scale (e.g. leaf, twig, tree) for the collection of data. The most appropriate scale is that at which natural enemies recognise and respond to variations in host density (Heads and Lawton, 1983; Waage, 1983, also sections 1.5 and 1.12). As this may be difficult to determine initially, samples are best taken in a hierarchical manner, so that analyses can be carried out at a number of different scales afterwards (Ruberson *et al.* 1991). Within each level of patchiness, patch density can then be related by regression analysis to percentage mortality or *k*-value, although the statistical validity of regression in this context is questionable (subsection 7.3.4). Hails and Crawley (1992) proposed an alternative logistic regression analysis based on generalised linear interactive modelling. Using the cynipid wasp, *Andricus quercuscalicis*, which forms galls on Turkey oak, as a test system, the method was able to detect spatial density-dependence in 15% of cases, with 66% of those being inversely density-dependent. Hails and Crawley manipulated patch densities by controlling the oviposition of adults on the buds of the trees. A similar study was carried out by Cappuccino (1992), who manipulated densities of another gall-making insect, the tephritid fly *Eurosta solidaginis*. Spatial variation in predation of the fly by a beetle, *Mordellistena* (Mordellidae), was noted at three scales, together with parasitism of the beetle. Interestingly, spatial variation in beetle parasitism depended, not on beetle patch density, but on the density of the fly.

Pacala *et al.* (1990) and Hassell *et al.* (1991) showed, again using the simple deductive models of host-parasitoid systems, that the contribution of spatial heterogeneity in parasitism to stability can be assessed using a simple rule. This states that the **coefficient of variation squared (CV^2)** ($=$ [variance/mean]^2) of the density of searching parasitoids close to each host must exceed approximately unity for the heterogeneity in parasitism to stabilise the interaction, i.e. $CV^2 > 1$. Moreover, CV^2 can be partitioned into the component of heterogeneity that is independent of host density (C_i) and the component that is dependent on host density (C_D), such that $CV^2 = C_i C_D - 1$. To estimate CV^2 directly the local density of searching parasitoids needs to be known. In most field systems, however, this is impracticable and consequently little information on this parameter is available (Waage, 1983). However, a considerable body of information is already available on percentage parasitism as a function of local host density and a procedure was provided by Hassell and Pacala (1990) to estimate the relevant parameters required from these data. The reader is warned, however, that the calculations require some mathematical facility, and the procedure is only applicable to restricted types of host-parasitoid interaction, as Hassell and Pacala were careful to point out. Readers interested in applying the technique should consult Pacala *et al.* (1990), Pacala and Hassell (1991), Hassell *et al.* (1991) and Hassell (2000b).

While Hassell and Pacala (1990) provide a detailed account of how to derive CV^2 from field data using 65 examples, it needs to be pointed out that most studies on spatial distribution of parasitoids have only been conducted over a very short time-span (e.g. one generation) (reviews by Lessells, 1985; Stiling, 1987; Walde and Murdoch, 1988; Hassell, 2000b). Observed spatial distribution patterns may not therefore be typical of the interaction (Redfern *et al.*, 1992), and data collected over a number of generations or years is more likely to give a representative picture. Redfern *et al.* (1992) studied patterns of spatial density-dependence amongst parasitoids of two tephritid fly species over a period of seven years. CV^2 values were calculated for total parasitism and each parasitoid species separately. The CV^2 values (together with their statistical significance) were found to be highly variable from year to year, making it difficult to draw conclusions. What drives fluctuations in CV^2 from generation to generation is unclear, but will need to be fully elucidated, if techniques such as the CV^2 rule are to have wider applicability. A further problem with the method is that it only applies to interactions where parasitism is of overriding importance and other regulating effects upon the host are negligible. How the latter might affect the stability conditions of the interaction are unclear and may be a major limitation of the technique. A recent review and assessment of the debate is given by Hassell (2000b).

Spatial Density-Dependence Versus Temporal Density-Dependence in Regulation

As spatial (i.e. within-generation) density-dependence or heterogeneity in the pattern of natural enemy attack appears to be a potentially powerful stabilising mechanism, we need to examine its relationship with temporal (i.e between-generation) density-dependence, upon which conclusions about regulation have traditionally been based. We have already seen that conventional methods for the analysis of life-table data are unsuitable for the detection of regulation resulting from spatial density-dependence (Hassell, 1985), so that many previous conclusions regarding the failure of natural enemies to regulate in particular systems may eventually need to be revised. Hassell (1985) postulated that much of the regulation in natural populations is likely to arise from within-generation variation in parasitism and predation, with only weak dependence on between-generation variation.

Not all authors have agreed with this view, however. Dempster and Pollard (1986) have argued that regulation must ultimately depend on temporal density-dependence and should therefore be detectable, at least in principle, by conventional analyses. They doubted that spatial density-dependence leads necessarily to temporal density-dependence. This raises interesting questions about the relationship between spatial and temporal density-dependence and the way the former operates. For regulation to occur, some temporal feedback is required. Using the discrete generation model of DeJong (1979), Hassell (1987) was able to show that, in the absence of stochastic variation, spatial density-dependence acting within generations translates directly into temporal density-dependence acting between generations. With variability added to the parameters governing spatial distribution and survival, however, temporal density-dependence becomes obscured and is less likely to be detected by the conventional method of plotting mortality against population density from generation to generation. Murdoch and Reeve (1987) also point out that we should not necessarily expect to detect spatial density-dependence from life-table data on prey or host populations, as spatial density-dependence acts on the *natural enemy* population, not the host or prey. The authors demonstrated their point by a closer analysis of the terms of the Nicholson-Bailey model which indicate that stability results from a decline in parasitoid efficiency as parasitoid density increases. Unless life-tables also contain data for the relevant natural enemies, then this information would be overlooked.

Hassell *et al.* (1987) analysed 16 generations of life-table data for the viburnum whitefly *(Aleurotrachelus jelinekii)* and found no evidence for temporal density-dependence. Spatial density-dependence of egg mortality between leaves was apparent in eight generations. Again using a variant of the DeJong model to simulate the whitefly population, Hassell *et al.* (1987) claimed to show that this spatial density-dependence could regulate the population in the absence of temporal density-dependence. Stewart-Oaten and Murdoch (1990), however, disputed this conclusion, arguing that the DeJong model has implicit temporal density-dependence which Hassell *et al.* (1987) had overlooked. With this temporal density-dependence removed, Stewart-Oaten and Murdoch were able to show that the spatial density-dependence in the resulting model can indeed lead to stability, but that destabilisation is also a strong feature at low population levels and when whitefly clumping increased. The Stewart-Oaten and Murdoch model thus demonstrates the important point that stability in population models is often sensitive to both changes in parameter values and to any subtle variations incorporated (Murdoch and Stewart-Oaten, 1989; Reeve *et al.*, 1989).

We have tried to provide a brief overview of the subject of spatial heterogeneity and aggregation in predator-prey systems Much of the progress made in the study of their roles is directly attributable to the use of deductive models, but as we have seen, the conclusions which can be derived from these models are, to some extent, dependent both on how the models are constructed and on the parameter values used. Paradoxically, while progress using this approach can be rapid, with new ideas and hypotheses being generated, often as many questions are raised as are answered. Thus, much of the subject remains speculative at his stage, and firm conclusions regarding the precise role of spatial heterogeneity in population dynamics remain elusive.

Metapopulation dynamics

Any consideration of spatial heterogeneity in population dynamics inevitably impinges on the concept of the metapopulation, which was introduced and defined in section 6.1. To recap briefly, **local populations** can be defined as units within which local population processes (e.g. reproduction, predation etc.) occur, and within which movements of most individuals are confined. **Regional populations** (or **metapopulations)** are collections of local populations linked by dispersal. Between-patch variations in parasitism and predation, as discussed above, are deemed to influence dynamics at the local population level, but regional metapopulation effects may also be important, although the distinction between such 'within-population' and 'between-population' processes may be somewhat artificial in many systems (Taylor, 1990).

Dispersal between local populations has frequently been proposed to account for the persistence of regional populations despite unstable fluctuations or extinctions at the local level (DeAngelis and Waterhouse, 1987, and Taylor, 1990, give reviews). For predator-prey systems, two different approaches have been used to model this situation (see Taylor, 1990, for details):

1. Those in which extinctions and recolonisations of local 'cells' (= populations) occur frequently – the so-called **cell occupancy models**;
2. Those in which within-cell dynamics are described explicitly by standard predator-prey models.

These models consistently show that persistence at the regional level can be enhanced by dispersal between local populations, provided that: (a) populations fluctuate asynchronously between cells, (b) predator rates of colonisation are not too rapid relative to those of the prey, and (c) some local density-dependence is present. While the degree of density-dependence may be quite low, resulting in frequent local extinctions, the metapopulation may persist for a long time (Hanski *et al.*, 1996; Kean and Barlow, 2000). These conclusions are consistent with the results of a number of laboratory studies exploring the effects of spatial structure and dispersal on persistence of predator-prey systems (Huffaker, 1958; Pimentel *et al.*, 1963; Holyoak and Lawler, 1996; Holyoak, 2000; Bonsall *et al.*, 2002). Pimentel *et al.* (1963), for example, examined the interaction between a parasitoid wasp and its fly host in artificial environments consisting of small boxes connected by tubes. The interaction persisted longer with more boxes and with reduced parasitoid dispersal. Bonsall *et al.* (2002a) developed a similar system of interconnecting boxes (Figure 7.41) to study a bruchid beetle-parasitoid metapopulation interaction, with comparable results. While agreement with theory may be encouraging, the small scale on which these experiments, by necessity, have to be carried out, is unlikely to reflect processes at the regional metapopulation level. The results may be equally well explained by local population, between-patch, spatial dynamics. Similarly, Murdoch *et al.* (1985) invoked metapopulation processes to explain persistence of a number of field predator-prey and parasitoid-host systems, despite apparent local extinctions. However, as Taylor (1990) points out, extinction in these examples was either not proven or occurred at a scale more consistent with local population processes.

Evidence for the importance of metapopulation processes in field predator-prey systems is still relatively scarce, despite a number of recent attempts at detection in the field (see Walde, 1994; Harrison and Taylor, 1997; Davies and Margules, 1998). One of the best-studied examples involving an arthropod predator-prey system is provided by the Glanville Fritillary butterfly, *Melitaea cinxia*, and its specialist braconid parasitoid, *Cotesia melitaearum*, occupying the Åland Islands (Finland) (Lei and Hanski, 1997, 1998). In the study area, around 1700 suitable patches of dry meadow were available for colonisation, of which several hundred were

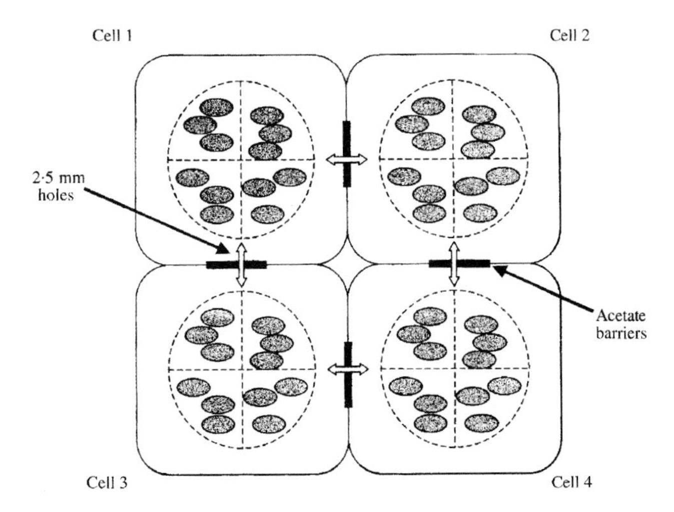

Figure 7.41 Schematic diagram of the experimental four-cell metapopulation studied by Bonsall *et al.* (2002a), Cells were linked and dispersal was restricted using acetate gates. Reproduced by permission of Blackwell Publishing.

occupied by the butterfly and about 15% by the parasitoid. As predicted by theory, the dynamics of the butterfly were greatly influenced not only by the area and isolation of patches, but also by the parasitoid, which, provided that hyperparasitism was not too high, increased the risk of local extinction in the butterfly population. Clearly, more detailed empirical studies of this nature are required to provide something of a 'reality check' to a burgeoning theoretical literature. Despite the current imbalance between theoretical and practical studies, there is no doubt that the metapopulation will remain a critical consideration for the future of population dynamics. For reviews, see Hanski (1999) and Hassell (2000a,b).

Metapopulations and Multispecies Interactions

The metapopulation concept has been extended to include multispecies interactions, including some of those described in section 7.3.10. So far, these efforts have been largely confined to relatively simple host-parasitoid systems and explored using analytical models (see Hassell *et al.*, 1994; Comins and Hassell, 1996; Wilson and Hassell, 1997). As a general rule, it has been shown that adding a third species to a two-species interaction (an extra host, a competing parasitoid or a hyperparasitoid) can result in stable coexistence, although under stricter conditions than for a two-species system and always dependent on some form of 'fugitive coexistence'. Hassell (2000b) provides a detailed review.

7.4 SELECTION CRITERIA IN CLASSICAL BIOLOGICAL CONTROL

7.4.1 INTRODUCTION

In this section, we deal mainly with the criteria used in the selection of natural enemy species for introduction in classical biological control programmes (defined in subsection 7.2.2). Some of these criteria will, however, apply to agent selection in other types of biological control programme (see section 8.6, and below). We realise that the majority of readers may never actively practise biological control, but we hope that what follows will provide a framework for research such that they can nevertheless make an indirect contribution to the subject.

Several steps are involved in a classical biological control programme (Beirne, 1980; DeBach and Rosen, 1991; Waage and Mills, 1992; Van Driesche and Bellows, 1996; Neuenschwander, 2001):

1. **Evaluation** of the pest problem in the region targeted by the programme. Information should ideally be gathered on: (a) the precise identity of the target pest species and its area of origin; (b) the distribution and abundance of the pest; (c) crop yield loss and economic thresholds of pest numbers; (d) the identity and ecology of indigenous natural enemies that may have become associated with the pest in its exotic range; (e) the climatic features of the target region.
2. **Exploration** in the pest's region of origin. It should ideally involve: (a) determination, through quantitative surveys, of the composition of the natural enemy complex associated with the pest in its natural habitat(s) (see section 6.3 for methods of recording parasitism). Samples of parasitoids and predators should be taken from areas of low, as well as high, pest abundance; (b) quantification of the impact of these natural enemies on populations of the pest (see sections 7.2 and 7.3 for methods); (c) determination of their degree of specificity, i.e. their host/prey ranges (see section 6.3 for methods). Exploration may also involve determining which natural enemies are associated with taxonomically related pests (the search for 'new-associations', see below). Of the range of species available, one or more are chosen on the basis of their potential to control pest populations; in a typical classical biological control programme, only a small fraction of the natural enemy complex of a pest is chosen for future study (i.e. the biological control practitioners apply selection criteria even at this early stage in the programme – **first stage selection**). The constraints upon selecting a larger fraction include: (a) the limited

resources (finances, time) available to practitioners; (b) the unsuitability of certain species due to their polyphagous habits (which may result in non-target effects, see 7.4.4, below) or their potential to interfere with other enemy species (facultative hyperparasitoids, intraguild predators); (c) the difficulty sometimes encountered, in multiple parasitoid introductions, of achieving establishment of a species, following establishment of another (Ehler and Hall, 1982; Waage, 1990; Waage and Mills, 1992; see below). If 'new associations' are to be employed, then one or more members of the natural enemy complex of a species taxonomically related to the pest, but not of the complex associated with the pest itself, will be selected in a programme.

3. **Quarantine**. This involves maintaining the chosen agents in an escape-proof facility in the target (i.e. non-native) region. One of the purposes of quarantine is to eliminate any associated hyperparasitoids, parasitoids of predators or insect and plant pathogens that have inadvertently been introduced into cultures along with the agents. Further selection of agents may occur during this phase, based on ease of handling and culturing and on other criteria (see below) (**second stage selection**).
4. **Release** of suitable agents occurs after clearing through quarantine. A multidisciplinary approach is required here, involving skills in population ecology, population genetics, meteorology, engineering and sometimes even aeronautics (agents may be released by aircraft).
5. **Monitoring** then takes place (ideally, monitoring of the host population should commence before any agents are introduced; see sections 7.2 and 7.3 for discussion of monitoring methods). This requires knowledge of pest and crop life-histories and phenologies, as well as expertise in population dynamics. Evidence should be sought of the successful establishment of agents (e.g. see Brewer *et al.*, 2001): crops and associated vegetation needs to be scrutinised, and signs of predation and parasitism looked for. A pest sampling programme should be carried out, to determine what (if any) effects the introduced agents have had/are having on pest population dynamics and demography. The use of control plots is essential, as correlation does not mean causation (see subsection 7.2.2).
6. **Evaluation** involves assessing the degree of attainment of the programme aims and objectives, asking the following questions: (a) has the introduction proved successful in controlling the pest? (b) has control been partial, substantial or complete (*sensu* DeBach, 1971)? (c) has the programme been cost-effective (economic analysis, e.g. see Bokonon-Ganta *et al.*, 2002)? (d) has the local (human) community benefited? (e) have there been any non-target effects (e.g., have the population dynamics of other organisms been affected (see section 7.4.4)?

A more detailed discussion of the steps involved in classical biological control programmes is provided by Van Driesche and Bellows (1996). Ways of improving the outcome of classical biological control programmes are discussed by Kareiva (1990), Waage and Barlow (1993), Barlow (1999), Freckleton (2000), Mills (2000) and Shea and Possingham (2000) (see also Fagan *et al.*, 2002). Shea and Possingham (2000) applied stochastic dynamic programming, linked to a metapopulation approach, in identifying optimal release strategies (number and size of releases), and they derived useful rules of thumb that can enable biological control workers to choose between management options. Van Lenteren *et al.* (2003) discuss the data requirements for assessing establishment by imported natural enemies.

7.4.2 WHAT WORKED (OR DID NOT WORK) PREVIOUSLY – THE USE OF HISTORICAL DATA

The history of biological control shows that it has been largely an 'art', aided by the knowledge of what worked successfully in the past either against the same pest species or a taxonomically related species (van Lenteren, 1980; Waage and Hassell, 1982). Biological control is becoming much more of a science, and historical databases, although deficient in many respects (see below), can still prove useful to practitioners in helping them both to make generalisations and to erect hypotheses, informed by ecological theory (Greathead, 1986; Waage and Greathead, 1988). The principal database of classical biological control introductions is BIOCAT, developed at the CAB International Institute of Biological Control (now part of CABI Bioscience). It is the most comprehensive database on biological control developed to date (Greathead and Greathead, 1992). Unfortunately, like other databases, it is composed of records whose reliability is in some cases questionable. As Waage and Mills (1992) point out, "the record of classical biological control is troubled by erroneous identifications of pests and natural enemies, occasional errors in dates and places and the arbitrary (and sometimes entirely incorrect) interpretation of success".

Nevertheless, analyses carried out with the BIOCAT database are providing useful insights into the habitat-, pest- and natural enemy-related factors that influence the success of natural enemy introductions (e.g. see Greathead, 1986; Mills, 1994; Jervis *et al.*, 1996a; Hawkins and Cornell, 1994; Kidd and Jervis, 1997; Hawkins *et al.*, 1993, 1999). For example, the results of analyses tell us that introductions are particularly successful against Homoptera and Lepidoptera (Greathead, 1986; Mills, 2000), and that success rates are substantially greater in exotic, simplified, managed habitats than in more natural habitats, particularly when involving parasitoids (Hawkins *et al.*, 1999). The database has also been widely used to test ecological hypotheses concerning enemy-victim interactions (e.g. see Hawkins, 1994; Kidd and Jervis, 1997; Mills, 2001), including the hypothesis that the action of natural enemies in classical biological control is not fundamentally different from that of indigenous natural enemies (Hawkins *et al.*, 1999) (see subsection 7.2.2).

For details of how analyses of the BIOCAT database are conducted (including ways of minimising bias), see Stiling (1990), Hawkins (1994), Mills (1994), Jervis *et al.* (1996a), Kidd and Jervis (1997) and Hawkins *et al.* (1999). Comparative statistical techniques (subsection 1.2.3) should be employed to overcome pseudoreplication.

7.4.3 NATURAL ENEMY ECOLOGY AND BEHAVIOUR

Introduction

During the exploration phase of a biological control programme, a decision will need to be made as to whether the natural enemies are to be collected from the pest species or from other, taxonomically closely related, species. The **theory of new associations** (Pimentel, 1963; Hokkanen and Pimentel, 1984) states that natural enemy-pest interactions will tend to evolve towards a state of reduced natural enemy effectiveness, and that natural enemies not naturally associated with the pest (i.e. species presumed to be less coevolved with the target pest) either because they do not come from the native area of the pest or because they come from a related pest species, may prove more successful in biological control. Hokkanen and Pimentel (1984) analysed 286 successful introductions of biological control agents (insects and pathogens) against insect pests and weeds, using data from 95 programmes, and concluded that new associations were 75% more successful than old associations. However, the validity of this conclusion was called into question by more refined analyses (Waage and Greathead 1988; Waage, 1990). The latter showed that new associations can be as effective as old ones once the natural enemy is established, but that establishment of a natural enemy on a new host is very difficult. Therefore, in practice, studies on the target pest in its region of origin remain the most promising approach to finding an effective biological control agent rapidly, but the potential usefulness of new associations should also be considered (Waage and Mills, 1992).

If only a fraction of the natural enemy complex of a pest can be used in classical biological control, it is essential that the potentially most important and least 'risky' species (i.e. in terms of their potential impact on pest populations and their threat to non-target organisms, respectively) are used from among the candidates available (Waage and Mills, 1992). Decisions on the relative merits of natural enemies can be based on **reductionist criteria** and **holistic criteria** (Waage, 1990). As we shall see, many of these criteria are based mainly or entirely on theory. Gutierrez *et al.* (1994) question the reliance of biological control practice on theory, arguing that the latter has contributed little either to increasing the rate of success of introductions or to an understanding of the reasons for failures (see also Kareiva, 1990, and Barlow, 1999). As we have already mentioned, ecological theory has provided useful insights into the performance of past biological control programmes, identifying some factors that influence success and failure, and it has provided guidance for future programmes. A decade on, Gutierrez *et al.*'s point concerning the contribution of ecological theory to biological control practice remains a valid one, but this is not to say theory should be discarded. Instead, ways should be sought of applying theory more effectively, so that the record of success in classical biological control programmes can be improved. Mills (1994) estimated that of 1,450 unique pest-introduced parasitoid combinations, only 38% resulted in establishment (i.e. successful colonisation) of the agent, and of 551 parasitoid introductions that have resulted in establishment and have provided some degree of control, only 17% have been successful – clearly, there is considerable room for improvement in classical biological control. Constraints upon the contribution of ecological theory to biological control, and some ways of overcoming them, are considered below.

Reductionist Criteria

Introduction

The reductionist approach involves selecting agents on the basis of particular biological attributes. Reductionist criteria are mostly derived from the parameters of the analytical parasitoid-host or predator-prey population models discussed in 7.3, in particular those parameters which are important to the lowering of host or prey population equilibria and/or which promote population stability. Stability of the natural enemy-pest interaction is seen as desirable: (a) because it reduces the risk that the pest population will be driven to (local) extinction by the control agents, which themselves would otherwise become locally extinct; and (b) it reduces the likelihood of the pest population exceeding the economic threshold.

Despite their frequent use in theoretical studies, reductionist criteria have rarely been used in practice, one reason being the difficulty of estimating parameter values (Godfray and Waage, 1991). As Waage and Mills (1992) point out, laboratory measures of parameters are unlikely to reflect the values of parameters in the field, where the environment is more complex (although, as shown in Chapter 1 of this book, with a little thought, experiments can in some cases be designed which take environmental complexity more into account).

A further criticism that can be aimed at the reductionist approach concerns the validity of separating a parasitoid's or predator's biology into components, and of assuming that there are natural enemies that have 'ideal combinations' of such components (see below). The whole organism, not its component parts, is what forms the basis for prediction of success in biological control (Waage and Mills, 1992). However, models have been developed (see subsections 7.3.7, 7.3.8) in which biological attributes are assembled in a more realistic fashion (Hassell, 1980a; Murdoch *et al.*, 1987; Gutierrez *et al.*, 1988a, 1994; Godfray and Waage, 1991). Even so, a major difficulty with many models is that a burdensome number of parameters has to be estimated in order for the model to be operated. A notable exception to this is the Godfray and Waage (1991) model of the mango mealybug-parasitoid system, which incorporated a few, easily measurable parameters such as the stage of the host attacked by the different parasitoids (*Gyranusoidea tebygi* and *Anagyrus* sp.), age-specific development rates for host and parasitoids, age-specific survivorship of hosts in the field, adult longevities and daily oviposition rates. Some parameters such as searching efficiency were more difficult to measure, and so a range of realistic values was tested in each case using sensitivity analysis. Godfray and Waage (1991) categorise their model as one of intermediate complexity, in that it is more complex than the analytical models of theoretical ecology e.g. the Nicholson-Bailey model (subsection 7.3.7) but less complex than many detail-rich simulation models, e.g. that of Gilbert *et al.* (1976) (subsection 7.3.8).

Godfray and Waage's (1991) was a **prospective model** (i.e. constructed prior to introduction) as opposed to a **retrospective model** (i.e. constructed following introduction). Prospective models are a potentially very useful tool in decision-making in biological control (Godfray and Hassell, 1991), but an important constraint upon their development and use in biological control programmes is the need for practitioners to achieve control of the pest as rapidly as possible (Waage, 1990). Mills and Gutierrez (1996) devised a prospective model for the dynamics of a heteronomous hyperparasitoid in a cotton-whitefly parasitoid-system.

Thus, selection of agents based on comparisons of species' attributes will probably never be a top priority; indeed so far it has been very rare for candidate species available in culture and cleared through quarantine, *not* to be introduced (Waage, 1990). Nevertheless, pre-introduction studies on candidate species, aided by modelling, still have a significant contribution to make to classical biological control programmes; at the very least they could bring about a re-ordering of the sequence of introduction of species destined for release, i.e. help to move the more effective agents to the front of the queue. Bearing in mind that programmes usually end before all candidate agents have been released, prioritising agents on the basis of their likely efficacy would ensure that the 'best' species are released (Waage, 1990).

Before we go on to discuss which attributes of candidates are considered to be particularly desirable, it is important to remind the reader that most biological control models are based on equilibrium population dynamics, and that in using such models, one seeks to determine to what extent a low, stable pest equilibrium can be achieved. The assumption that low, stable host equilibrium populations result in good pest control was questioned by Murdoch *et al.* (1985) who argued that, instead of considering biological control in terms of local population dynamics, theoreticians and practitioners should view it terms of the more biologically realistic scenario of metapopulation dynamics (subsection 7.3.10). By doing so, local instability – and thus a high risk of extinction of local populations – is not necessarily a problem for biological control practitioners, as persistence may be possible in the metapopulation. Note that whereas successfully controlled pests often appear to fluctuate because of local extinctions (Luck, 1990; *Aonidiella aurantii* being a notable exception, see Murdoch *et al.*, 1985), metapopulation persistence is perceived to be generally the case following successful natural enemy establishment. Furthermore, there is no evidence that any failure in biological control has been the result of a lack of the persistence of a successfully established agent (Waage, 1990). However, even if metapopulation dynamics ensure persistence, local instability can remain a pest management problem: high temporal variation in pest densities increases the risk of the pest population exceeding the economic threshold (see Murdoch, 1990). The level of this risk depends on the degree to which the host population is suppressed by the biocontrol agent: if the degree suppression is such that pest numbers, despite fluctuating greatly, never exceed the economic threshold, then high temporal variation in pest numbers can be acceptable; indeed, successful pest control is theoretically possible through suppression alone (Kidd and Jervis, 1997).

Another point concerning the equilibrium population dynamics basis of most biological control models is that the models will be more appropriate for some pests than for others (Godfray and Waage, 1991). In some agroecosystems, e.g. arable crops and glasshouse systems, cultural practices and/or seasonal factors will prevent the population ever coming close to equilibrium, i.e. **transient dynamics** will per-tain. Thus, some of the selection criteria discussed below have little or no relevance to certain crop systems. Even in classical biological control, it is questionable whether biological control introductions ever attain equilibrium, at least at the local population level (Murdoch *et al.*, 1985), and thus a measure of the **transient impact** of an introduced parasitoid may be more appropriate for predicting control potential (Hochberg and Holt, 1999; Mills, 2000, 2001).

Biological control models have mainly been of the deterministic analytical type. Murdoch *et al.* (1985) suggested the use of **stochastic boundedness models** as a preferable alternative. With the latter, emphasis is placed on estimating the probabilities that either host or parasitoid (or prey and predators) may become extinct, or that the pest exceeds an economic threshold, rather than on stability analysis (Chesson, 1978b; 1982).

Listed below are several attributes (some of which are common model parameters) of natural enemies considered to be among the most desirable for biological control, based on theoretical modelling, practical considerations and past experience. Implicit to the reductionist approach is the notion that any combination of life-history parameters is possible (Waage, 1990). However, it is becoming increasingly apparent, through studies on natural enemies, that there are trade-offs between one attribute and another, for example, that between adult reproductive capacity and larval competitive ability in parasitoids (see **Holistic Criteria**). It would therefore be better to concentrate on 'suites' of often counterbalancing attributes. Waage argues that for reductionist criteria to be useful, they need to be derived at a higher level where the traits are integrated in the patterns or strategies that we observe in nature, and he cites one approach as being the computation of intrinsic rates of natural increase (r_m) (see section 2.11 for methods).

High Searching Efficiency

The density responsiveness of candidate biological control agents has been compared through short-term (24 hour) functional response experiments (see section 1.10 for design). For example, if one is comparing the potential effectiveness of two parasitoid or predator species, the species with the higher maximum attack rate (which is set by handling time and/ or egg-limitation) may be selected, as it will, all else being equal, depress the pest equilibrium to a greater extent. Sigmoid functional responses, because they result in density-dependent parasitism or predation, are potentially stabilising at low pest densities. However, in a coupled interaction responses have to be very pronounced, and the destabilising time-delays in the population interaction small, for the stabilisation to be marked (Hassell and Comins, 1978; subsection 5.3.7). Using the BIOCAT database, Fernández-Arhex *et al.* (2003) tested for, but were unable to detect, a relationship between the form of the functional response and success in classical biological control (Type 2 versus Type 3).

As pointed out by Waage and Greathead (1988), the natural enemy functional response offers a good conceptual framework for understanding the action of agents in inundative releases. Ratio-dependent functional responses (subsection 7.3.7) are now being considered as more relevant (Mills and Lacan, 2004).

Spatial Heterogeneity in Natural Enemy Attack

There are two issues here: (a) the degree to which a refuge from parasitism contributes to host suppression, and (b) the degree to which a refuge contributes to population stability. The following discussion refers to proportional refuges (which are generally thought to be more realistic approximations of variation in risk to parasitism than constant number refuges, see Hassell, 1978, 2000b, and Holt and Hassell, 1993). Note that the refuge can be an attribute of the host (or its food plant), rather than of the parasitoid (see references in Mills and Getz, 1996), but that even if the former applies, other parasitoid characteristics can determine the population dynamic effects of the refuge.

Concerning suppression, whatever the nature of the refuge, if it is very large, most of the hosts can escape from parasitism, and so the impact of the parasitoid population on the host population will be small no matter what other attributes the biological control agent may possess (Hochberg and Holt, 1999; Mills, 2000, 2001). Hawkins *et al.* (1993), through a BIOCAT database analysis, provided evidence to support the view that the larger the refuge from parasitism, the lower the probability of success in classical biological control. In this case, the maximum level of (percentage) parasitism achieved in the target region was used as the measure of refuge size (for a critique, see Myers *et al.*, 1994).

Mills (2001) showed that the ability of a parasitoid to suppress a host population depends on the size of the host refuge from parasitism, the host net rate of increase (F in the Nicholson-Bailey model) and on whether the parasitoid is egg-limited (limited in its attacks by the number of eggs it has available for laying) or 'host-limited' (i.e. limited in its attacks by the female's ability to find hosts) (for a model, see Mills' paper). Even in the absence of a refuge, an egg-limited parasitoid will be unable to suppress the abundance of a host if the latter has a high F. However, such a parasitoid can suppress host abundance substantially if the host has a sufficiently low F and a minimal refuge from parasitism. [A sufficiently low F could, theoretically, be achieved by the use of partially resistant crop cultivars]. With a host-limited parasitoid, substantial host suppression can occur when the refuge is sufficiently small in relation to F (Hochberg and Holt, 1999).

Concerning the effects of refuges on population stability, spatial heterogeneity in parasitism, through its host refuge effect, is recognised as being one of the major stabilising factors that can lead to the persistence of parasitoid-host and predator-prey populations in models of parasitoid-host interactions (subsection 7.3.10) (but see Murdoch and Stewart-Oaten, 1989). This applies both to aggregation in patches of high or low host or prey density and to aggregation in patches independent of host or prey density. Heterogeneity in parasitism, whether the result of aggregation or other (i.e. host- and/or habitat-related) factors, can have a stabilising effect at even the lowest population levels, unlike host resource limitation or mutual interference (Hassell and Waage, 1984; May and Hassell, 1988) (see subsection 7.3.7).

Because of its stabilising potential, parasitoid aggregation has been promoted as a primary selection criterion for parasitoids (Murdoch *et al.*, 1985 and Waage, 1990), but it is debatable whether it ought to be promoted as an independent criterion. Firstly, there is the question of whether the operation of a local stability-promoting factor, of whatever kind, is necessary for successful biological control (see above). Secondly, refuges have generally been shown in models to raise equilibrium population levels (Murdoch, 1990). Not only in the case of refuges but also with regulating factors such as density-dependent sex ratio, there is a strong trade-off between the degree of stability and the degree of host suppression. This phenomenon, has been termed the **paradox of biological control** (Luck, 1990; Arditi and Berryman, 1991). However, Arditi and Berryman (1991) showed that the paradox can be resolved with the Lotka-Volterra model if a ratio-dependent functional response is assumed (see subsection 7.3.7). While ratio-dependent functional responses have been disputed to occur in nature (see Abrams, 1994; Murdoch, 1994) there is evidence that they can occur in some parasitoids (including species attacking hosts whose dynamics can be described by Lotka-Volterra type rather than Nicholson-Bailey type models) (Hoddle *et al.*, 1998; Jones *et al.*, 1999; Faria *et al.*, 2000; Hoffmann *et al.*, 2002; Mills and Lacan, 2004), although they may be a highly constrained trait, phylogenetically speaking (Mills and Lacan, 2004). The question remains of how the paradox of biological control can be resolved through the Nicholson-Bailey modelling framework.

The implications of natural enemy aggregation for biological control have generally been considered at the local population level. Ives and Settle (1997), however, employing a metapopulation model, found that as predator aggregation increases in fields in which pest abundance is high, pest equilibrium densities increase. This is because high-pest density fields are late in the pest population trajectory, so predators have less of an effect on the maximum pest density achieved, and these fields retain predators that could be more effective, in pest control terms, by moving to more recently infested fields.

Despite the attention that spatial pattern of attack has received from researchers, there is very little empirical information relating to the spatial patterns of attack by natural enemies used in biological control (Mills, 2000). Exceptions include parasitoids attacking California Red Scale (*Aonidiella aurantii*). Murdoch and co-workers (Murdoch, 1994; Murdoch *et al.*, 1996) carried out experimental manipulations of both the distribution and the abundance of the scale insect on individual citrus trees, and concluded that the spatial heterogeneity in parasitoid attack that characterises this agent-pest system did not account for either local stability or the success of reduction in scale abundance.

Temporal Heterogeneity in Natural Enemy Attack

For parasitoids and predators, hosts and prey are a temporally heterogeneous resource, altering in both vulnerability and quality as they pass through the various stages in their life-cycle. Parasitoids respond to this, practising differential allocation of progeny (sex ratio, clutch size) (Murdoch *et al.*, 1987, 1992, 1997; Godfray, 1994; Briggs *et al.*, 1995; subsection 1.5.7). Murdoch *et al.* (1997) argue, from modelling of stage-dependent parasitoid oviposition behaviour, that these parasitoid responses have a common outcome: stabilising delayed density dependence in the *per capita* recruitment rate of the parasitoid population. Size-selective destructive host feeding (Kidd and Jervis, 1991) reinforces the delayed density dependence in *per capita* recruitment rate that is induced by size-selective clutch size allocation (for evidence of the latter, see Murdoch *et al.*, 1997; but also see Murdoch *et al.*, 1992b)

Murdoch *et al.*'s (1987) stage-structured parasitoid-host model (in which either the adults or the juveniles of the pest can be made invulnerable to attack by the parasitoid) incorporates a developmental delay in both the host and the parasitoid (Figure 7.42). The stability of this model depends on the length of the parasitoid time lag, T_2, relative to the duration of the invulnerable stage, T_A. The parasitoid's time lag is destabilising; the longer the developmental period of the parasitoid is relative to that of the host, (i.e. high T_2/T_A), the more difficult it is to obtain stability. A longer development time also leads to exponential increases in the pest equilibrium. Therefore, Murdoch (1990) considered a short parasitoid development time to be a desirable

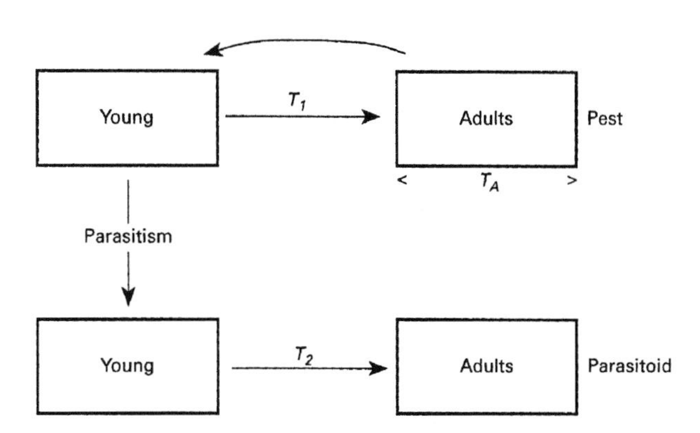

Figure 7.42 Diagram of a parasitoid-pest model in which both species have an immature stage lasting for T_1 and T_2 days respectively. The adult pest is invulnerable to attack by the parasitoid, and lives for an average of T_A days. Source: Murdoch *et al.* (1990). Reproduced by permission of Intercept Ltd.

attribute of a parasitoid species for biological control. For methods of measuring development times of parasitoids, see subsection 2.9.2.

A High Maximum Level of Parasitism in the Host's Native Range

Hawkins and Cornell (1994) using the BIOCAT database, obtained a positive, albeit 'noisy', correlation between maximum level of parasitism in the pest's country of origin, and the degree of success in classical biological control. In Hawkins and Cornell's data set, a cut-off in control outcomes exists at maximum parasitism rates of approximately 35%; below this level biological control very rarely achieves economic success (Hochberg and Holt, 1999). Although conditions in the native environment and those in the exotic one will be similar to some extent, there will be enough differences to make strict extrapolation problematical. Nevertheless, all else being equal, if a parasitoid species achieves a maximum level of parasitism many times higher than that of another, then based on Hawkins and Cornell's results, there would be a strong case for employing the former.

Kean and Barlow (2000) confirmed the relationship between parasitism level and biological control in parasitoid-host models, but they also found that if the host's r_m is low (a rare condition for many pests), substantial host reduction does not necessarily require high levels of parasitism.

A High Parasitoid Fecundity

There is no clear empirical evidence that a high parasitoid fecundity increases the probability of agent establishment, although a future analysis taking phylogenetic confounding effects into account might reveal a different result (Lane *et al.*, 1999). Establishment rate declines from egg through to prepupa in parasitoids attacking Lepidoptera, and this may well be the result of the decline in fecundity observed for species attacking successively later stages in the host life cycle (Price, 1975; Mills, 1994). Establishment rate increases at the pupal stage, possibly due to the trend towards polyphagy among pupal parasitoids of Lepidoptera (they can utilise non-target host species).

Lane *et al.* (1999) and Mills (2001), through modelling, identified parasitoid fecundity (a key determinant of **attack capacity** – the maximum number of hosts that a parasitoid can attack in its lifetime, see Mills, 2001) as a major factor in pest suppression, irrespective of whether transient or equilibrium dynamics best represent the real dynamics of agent-pest interactions. This view is supported by Lane *et al.*'s (1999) literature survey, which revealed a positive correlation between attack capacity and success of classical biological control introductions against Lepidoptera (there was no correlation for parasitoids of Homoptera). Lane *et al.* also showed that, in their model (which incorporated parasitoid fecundity limitation, a refuge from parasitism, and a density-dependent host population growth function) a high fecundity should provide stable control of a host population over a wider range of parameter space. Lane *et al.* stress, however, that low fecundity should not always be considered to be a disadvantage in a biological control agent (see their paper for details).

Gregariousness/Number of Female Parasitoids Produced Per Clutch

Parasitoid gregariousness, or the number of female parasitoids produced per clutch, has been identified as a major factor in pest suppression (Mills, 2001). Even small increases in gregariousness lead to dramatic reductions in host abundance. Support for this conclusion came from an analysis of the BIOCAT database (Mills 2001). Gregarious species were found to represent a significantly greater proportion of parasitoid introductions that have led to success than of those that have led to failure.

A High Degree of Seasonal Synchrony With the Host

Populations of hosts and parasitoids with discrete generations frequently show imperfect phenological synchrony, with the result that some host individuals experience a reduced or even zero risk of parasitism, i.e. there is a temporal refuge effect. Compared with perfect synchrony, imperfect synchrony will result in a raising of the host equilibrium level. Models developed by Münster-Swendsen and Nachman (1978) and Godfray *et al.* (1994) have shown that it will also stabilise the parasitoid-host population interaction.

The degree of 'phenological matching' between parasitoid and pest can be investigated in the laboratory, by subjecting the insects to a range of temperatures and/or photoperiods (Goldson and McNeill, 1992). Parasitoid populations taken from different localities from within the host's native range may show different diapause characteristics and therefore will show different degrees of synchrony with pest populations that originate from only one locality.

Seasonal synchrony with the host is not an important selection criterion in biological control programmes involving inoculative release and inundative release, since synchrony can be achieved by the grower through the release of parasitoids when most hosts are in the susceptible stage for parasitism (van Lenteren, 1986).

Parasitoid Guild (Host Stages Attacked/Killed)

Mills (1994) found, through a BIOCAT database analysis of parasitoid introductions against Lepidoptera, that: (a) it is easier to achieve establishment with parasitoids that attack earlier stages of host development, although pupal parasitoids also perform well; (b) the overall success of egg parasitoids is poor (they have the highest probability of establishment – probably due to the egg parasitoid category of his database being dominated by *Trichogramma* spp., which tend to be polyphagous and therefore can utilise non-target hosts (c) the earlier the parasitoid completes development in (i.e. kills) later host stages, the higher the success rate.

A High Intrinsic Rate of Natural Increase (r_m)

As noted above, the agent's intrinsic rate of population increase has been proposed as a primary selection criterion, in that it comprises an integrated suite of natural enemy traits and is therefore biologically more realistic (but note that Neuenschwander, [2001, in reviewing the biological control programme against cassava mealybug, concluded that r_m was, in retrospect, a poor predictor of agent effectiveness in that particular case).

Modelling by Hochberg and Holt (1999) has shown that r_m (which they estimated from a partial derivative of their host refuge model) is enhanced by a greater searching efficiency, a greater attack capacity (maximum number of hosts attacked over the parasitoid's lifetime) and a greater mean number of parasitoids emerging from a parasitised host. It was also shown that that in highly productive environments (high K, the host's carrying capacity of the environment), it is parasitoid fecundity alone that determines the conversion of hosts to parasitoids and therefore the transient impact of parasitism on the host population. If the clutch sex ratio is biased towards females (as is often the case for gregarious parasitoids), then a gregarious species will have a higher r_m than a solitary species with the same fecundity (see Mills, 2001).

As pointed out by Huffaker *et al.* (1977), it is a common error to conclude that a natural enemy having a lower r_m than that of its host or prey would be a poor biological control agent. The parasitoid (or predator) need only possess an r_m high enough to offset that part of the host's r_m that is not negated by predation or parasitism (and that part not negated by other mortality the natural enemy may inflict, e.g. through host-feeding and host-mutilation, see Jervis *et al.*, 1992a).

Rapid Numerical Response/Short Generation Time Ratio

The calculation of r_m values yields no insights into the dynamic interaction between natural enemy and the pest, whereas the numerical response does (subsection 7.3.7). Whilst some biological control workers have assumed a rapid numerical response to be a desirable feature of a natural enemy, this conclusion needs some qualification. While it is likely that a slow numerical response to a pest with a high population growth rate will lead to delayed density-dependence and limit cycles (subsection 7.3.7), a rapid numerical response to slow pest population growth may result in overcompensating fluctuations and decreased stability. Therefore the potential value of a numerical response needs to be assessed in relation to the population characteristics of the target pest.

Modelling by Godfray and Hassell (1987) in relation to parasitoids, and Kindlmann and Dixon (1999) in relation to predators, has pointed to the role of the relative lengths of natural enemy and pest generation times in determining equilibrium levels. Godfray and Hassell's (1987) simulations indicate a slight raising of pest equilibrium density when the **generation time ratio (GTR**, the ratio of the natural enemy's generation time to that of its host/prey**)** is greater than one, while Kindlmann and Dixon's (1999) simulations reveal that the suppressive effect of a predator is inversely related to the natural enemy's development time (see Kindlmann and Dixon's, 1999, 2001, papers for a functional explanation). Predators are considered to be less effective than parasitoids: q-values (subsection 7.2.2) for introduced predators are an order of magnitude larger than those for parasitoids (Table 1, in Kindlmann and Dixon, 1999) – a difference considered to be due in large part to the difference in GTR. Note that predators are, on the whole, better at controlling long-lived pests (smaller GTR) such as scale insects than shorter-lived ones such as aphids (longer GTR) (Kindlmann and Dixon, 2001).

Mode of Reproduction

Modes of reproduction in parasitoids are discussed in section Chapter 3 (section 3.3) Stouthamer (1993) considered the merits of arrhenotoky and thelytoky in parasitoid wasps from the standpoint of classical biological control; some of his conclusions were that:

1. Arrhenotokous forms (species or 'strains') will be able to adapt more rapidly to changed circumstances. If environmental conditions in the area of introduction are different from those in the form's native range, arrhenotokous wasps may have the advantage;
2. Assuming that a thelytokous form and an arrhenotokous form produce the same number of progeny, the thelytokous form will (all else being equal): (a) have a higher rate of population increase, and (b) depress pest populations to a lower level (see parameter s in the modified Nicholson-Bailey model (subsection 7.3.7);
3. Arrhenotokous forms must mate to produce female offspring; therefore, in situations where the wasp population density is very low, males and females may have problems encountering one another (an Allee effect). Thelytokous forms are therefore better colonisers.

A reaction-diffusion model comparing arrhenotokous parasitoids with sexually reproducing diploid ones predicted that haplodiploidy permits successful establishment in parasitoid populations that are 30% smaller – diploid populations suffer more from an Allee effect (Hopper and Roush, 1993). Mills (2000) outlines a post-introduction protocol for assessing the influence of mating on parasitoid establishment (see also Hopper, 1996). It includes releasing cohorts of increasing size (released as mature pupae) in spatially replicated locations, then dissecting the resulting female parasitoids at host patches, to assess whether they have been inseminated or not (see subsection 4.4.3).

Destructive Host-Feeding Behaviour

In the literature on biological control, the occurrence of non-concurrent destructive host-feeding behaviour (in which different host individuals are used for feeding and oviposition, see section 1.7) has been given as a reason for assuming a particular parasitoid species to be a potentially effective biological control agent. This is not an unreasonable assumption to make, in view of the fact that: (a) this type of host feeding behaviour is an additional source of mortality to parasitism, and (b) several parasitoid species have been shown to kill far more hosts by host-feeding than by parasitism. However, theory informs us that destructive host feeding is an undesirable attribute in a biological control agent (Jervis *et al.*, 1996a):

1. With regard to *establishment rate*, one model predicts destructive host-feeders to be no better than other types of parasitoids, while another model predicts them to be worse (Jervis *et al.*, 1996a; Jervis and Kidd, 1999);
2. With regard to *success rate*, destructive host feeders are predicted to be inferior, compared to other parasitoids, in suppressing host abundance, due to their lower numerical response (for a review of models used, see Jervis *et al.*, 1996a, and Jervis and Kidd, 1999).

Analyses of the BIOCAT database reveal destructive host-feeders to be either just as likely or more likely to become established than other parasitoids, and to be more successful in controlling the host compared to other parasitoids. The conclusion drawn is that destructive host-feeders are as good as, and probably no worse than, other parasitoids (Jervis *et al.*, 1996a).

Collier and Hunter (2001) suggest that destructive host-feeding behaviour might influence multiple agent interactions and pest suppression in biological control, where parasitoids feed on hosts that have been parasitised by a potential parasitoid competitor.

High Dispersal Capability

Techniques for studying dispersal by natural enemies are discussed in subsection 6.2.11. If a natural enemy has a high ability to disperse (either as an adult or as an immature stage within the host), then it can be expected to spread rapidly from the initial release point. Thus, fewer resources (time, money) may need to be invested in large numbers of point releases over a region to ensure that the natural enemy becomes established over a wide area. Another reason for favouring high dispersal capability in classical biological control agents is that it can minimise a time-delay in re-invasion of areas where the enemy has, for reasons of local instability, become extinct; a significant delay can allow the pest population to reach undesirable levels.

Wilson and Hassell (1997) have shown, through modelling, that demographic stochasticity increases the probability of extinction of small local populations, and that because of this, higher dispersal rates are required to ensure persistence of the metapopulation.

Fagan *et al.* (2002) suggest there may be a life-history trade-off between spatial spread rate and suppressive ability. They found evidence, albeit weak, for their hypothesis, but they do not explain how the trade-off would operate at the organismal level.

In biological control programmes that involve inundative releases of parasitoids, the parasitoids are used as a kind of 'biopesticide', so it is important that the insects do *not* disperse rapidly away from the crop. Parasitoids can be encouraged to remain within the crop either by 'pre-treating' females with host kairomones so as to stimulate search following release (Gross *et al.*, 1975), by applying kairomones directly to the crop to act as arrestants (defined in section 1.5) (Waage and Hassell, 1982) or by applying non-host foods (section 8. 1) to the crop.

High Degree of Host Specificity/Absence of Intraguild Predatory Behaviour

Practical approaches to studying host specificity in parasitoids and predators are discussed in subsections 1.5.7 and 2.10.2, and section 6.3.

One explanation for the poor performance, overall, of predators compared with parasitoids in classical biological control in perennial crop systems is their tendency to be more polyphagous. Among introductions of coccinellids, success rates are higher for monophagous species than for polyphagous ones (Dixon, 2000). It is argued that, because of parasitoid or predator polyphagy, a pest cannot be maintained at low equilibrium populations, as the natural enemy will concentrate on the more abundant alternative host or prey species (see switching behaviour, section 1.11). However, as Murdoch *et al.* (1985) point out, a polyphagous natural enemy can survive in the absence of the pest in the event of the latter's local extinction, and it can therefore be ready to attack the pest when it re-invades. For this reason, polyphagy may not be as undesirable an attribute in classical biological control as it is commonly assumed to be. However, polyphagous natural enemies pose risks to non-target organisms (see subsection 7.4.4).

An argument has been made, in relation to phytoseiid mite predators, for using those generalist predators that have a high degree of plant-specificity (McMurtry, 1992).

A lack of host specificity is not a problem in programmes aimed at pests of protected crops i.e. in glasshouses, as the environment is a simple one, usually containing unrelated pest species at each of which different indigenous parasitoid species are targeted (van Lenteren, 1986).

The general perception is that restraint should be exercised in using either: (a) facultative hyperparasitoids – these are dynamically equivalent to 'intrinsically superior' primary parasitoids (but see Rosenheim *et al.*, 1995), although the predicted dynamic consequences of introducing them vary with the particular type of model used (see Table 7.4); or (b) predators (including host-feeding parasitoids) that not only attack the pest but also attack other natural enemy guild members (intraguild predation), as they can seriously interfere with suppression (see Polis and Holt, 1992; Rosenheim *et al.*, 1995; Murdoch *et al.*, 1998; Rosenheim, 2001), also subsection 7.3.10). Snyder and Ives (2001), in a series of manipulation experiments involving an indigenous parasitoid-generalist predator-aphid system, showed how the generalist predator, which acted primarily as an intraguild predator, can disrupt 'natural' control. In a similar study (Snyder and Ives, 2003) the same researchers showed that the impact of a specialist and several generalists was additive rather than disruptive, although a simulation model fitted to their data suggested that longer-term experiments would have revealed non-additive effects. Even so, except in cases where predators strongly attacked aphid mummies, combined control by both types of natural enemy was predicted to be more effective than with either acting separately. Population cage experiments by Hunter *et al.* (2002), involving an autoparasitoid attacking a whitefly host and also a heterospecific parasitoid and conspecifics, showed that there was no disruption of biological control. Finke and Denno (2002) showed how the structural characteristics of the herbivore's habitat can mediate the effects, upon planthoppers, of intraguild predation by wolf spiders upon mirid bugs. In contrast to structurally simple laboratory 'habitats', more complex habitats increased the combined effectiveness of the predators in suppressing planthopper populations. This effect was attributed to the existence of a refuge for mirids within structurally more complex vegetation (thatch-rich as opposed to thatch-free): it was found that in complex salt-marsh habitats the predatory mirids were relatively more abundant than in simple ones. Finke and Denno's findings suggest both that in classical biological control the dynamic significance of intraguild predation will vary according to the type of agroecosystem involved and/or the type of habitat management practised, and that habitat effects can be tested for through simple experiments.

High Degree of Climatic Adaptation

The optimum range of temperatures or humidities for development, reproduction and survival of a candidate biological control agent (subsections 2.5.2, 2.9.2, 2.7.3 and 2.10.2) may be different from that of the pest, and the parasitoid may either fail to become established or prove ineffective owing to the direct or indirect effects of climate in the region of introduction. The conventional wisdom is that a parasitoid species should be selected for which climatic conditions in the region of introduction are optimal (DeBach and Rosen, 1991) (subsection 2.9.3). This view is supported by the database analysis

Table 7.4 Implications of competition theory for classical biological control (adapted from Murdoch *et al.* 1998).

Theory	Outcome	Recommendation for Release(s)
Simple models (exploitation competition)	Best competitor gives most control	All species; best competitor wins
Enemies interfere (interference competition)	Coexistence can lessen suppressive effect	Best agent, not winner
Enemies interfere and are also self-limiting		
*Intra*specific > *inter*specific limitation	Coexistence reduces pest density	Many species
*Intra*specific < *inter*specific limitation	Added species may lessen suppressive effect	Best agent, not best competitor
Stage-structured pest population	Winner may lessen suppressive effect on key pest stage and even lessen the effect on the total host population	Best agent, not best competitor

of Stiling (1993) which showed that the climatological origin of parasitoids has a large influence on establishment rate. However, the climatic adaptation criterion should not be rigidly applied: *Apoanagyrus lopezi*, which successfully controlled cassava mealybug in West Africa, originated from Paraguay, where the climate is very different (Gutierrez *et al.*, 1994; Neuenschwander, 2001).

Having determined the thermal requirements of an agent and/or knowing the climatic conditions in areas where it has already successfully invaded, a **climate diagram** can be used to predict where the insect is likely to become established. Samways (1989) describes such an approach for a coccinellid predator of scale insects. However, **climatic matching programmes** such as CLIMEX (Sutherst and Maywald, 1985) offer a more rapid method, not only for predicting the establishment prospects of species with known thermal requirements, but also for identifying sites for exploration (see Worner *et al.*, 1989).

Mills (2000) recommends investigating, post-importation, the role of climatic matching as follows: either (a) release fixed numbers of parasitoids from a single climatically characterised founder population along a climatic gradient in the target region, or (b), using either unique genetic markers (subsection 3.2.2) or morphometric markers (Phillips and Baird, 1996) for different geographic strains of a single parasitoid species, release, in combination, equal numbers of several strains at a series of climatically different locations in the target environment. The latter method can allow the success of local establishment to be related to the degree of climatic match between original and target localities for each strain.

Ease of Handling and Culturing

Greathead (1986) concluded, from an analysis of the BIOCAT database, that the most important factors in choice of natural enemy in classical biological control programmes have, perhaps, been ease of handling and availability of a technique for culturing the insects. The case of biological control of the mango mealybug, *Rastrococcus invadens*, is an illustration of how ease of rearing can influence selection. Two encyrtid parasitoids, *Gyranusoidea tebygi* and *Anagyrus* sp., were being considered for introduction into West Africa. Despite the latter species being the dominant parasitoid in rearings from field-collected mealybugs in India, the former species was selected as the first candidate for introduction, owing to the ease with which it could be cultured (see Waage and Mills, 1992 for a discussion). A reason given by Waage (1990) for the more extensive use of Ichneumonidae compared with Tachinidae in programmes aimed at controlling exotic Lepidoptera is the greater difficulty encountered in culturing the latter parasitoids (other dipteran parasitoids such as Pipunculidae are notoriously difficult to culture). It is noteworthy that the ranking of culturable agents for introduction usually follows the sequence in which they are established in culture (Waage, 1990).

Practical approaches to rearing and culturing parasitoids and predators are discussed by Waage *et al.* (1985).

Holistic Criteria

Introduction

The holistic approach to the selection of agents considers less the properties of the agent and emphasises instead the interactions between candidate species and between agents and mortalities acting on the pest in its area of introduction. One important consideration in this approach is that the relationships between natural enemies in biological control releases need to be viewed as dynamic, not static. Examples of this approach are:

Collecting Parasitoids From Non-outbreak Areas in the Native Range of the Pest

Selection of agents can begin during the exploration phase of a programme by confining exploration to low density populations of the host. The species composition of parasitoid complexes varies with the density of the host population (subsection 6.3.6). It has been argued that parasitoids collected from host population outbreak areas may not necessarily be those best-suited to maintain the pest at low densities in its exotic range (Pschorn-Walcher, 1977; Fuester *et al.*, 1983; Waage, 1990; Waage and Mills, 1992). To increase the likelihood of obtaining the 'better' species, Waage (1990) and Waage and Mills (1992) recommend the use of experimental host cohorts placed out in the field (subsections 6.2.8 and 7.2.9).

Selection of Agents that Follow, Rather than Precede, Major Density-dependent Mortalities in the Pest Life-cycle

Additional density-dependent mortalities acting later in the pest's life cycle can influence the effect of mortality caused by a natural enemy that attacks the pest earlier on (May *et al.*, 1981; May and Hassell, 1988). Indeed, if the density-dependence is over-compensating, too high a level of mortality caused by an early-acting parasitoid can lead to an increase in the host population above the parasitoid-free level! A density-dependent mortality acting upon a host – whether due to intraspecific competition (van Hamburg and Hassell, 1984) or the action of natural enemies (Hill, 1988) – can be described by the following model (Hassell, 1975):

$$S = N(1 + aN)^{-b} \tag{7.45}$$

in which S is the number of survivors, N is the initial prey density, a is a constant broadly indicating the densities at which survival begins to fall rapidly, and b is a constant that governs the strength of the density-dependence ($b = 1$ is perfect compensation, $b < 1$ is under-compensation, and $b > 1$ is over-compensation, subsection 7.3.4). Figure 7.43 shows, for the stem-borer *Chilo partellus* (Lepidoptera), a hypothetical example where S is plotted against N, for three density-dependent functions with different values of b (N in this case refers to larval densities). When $b = 1$ (curve A), the density-dependence tends to compensate for any early-acting (egg) parasitism, as long as the initial larval density is not reduced to lie on the steeply rising part of the curve. When $b < 1$ (curve B), however, there is only partial compensation, and egg parasitism will always reduce the numbers of larvae ultimately surviving. When

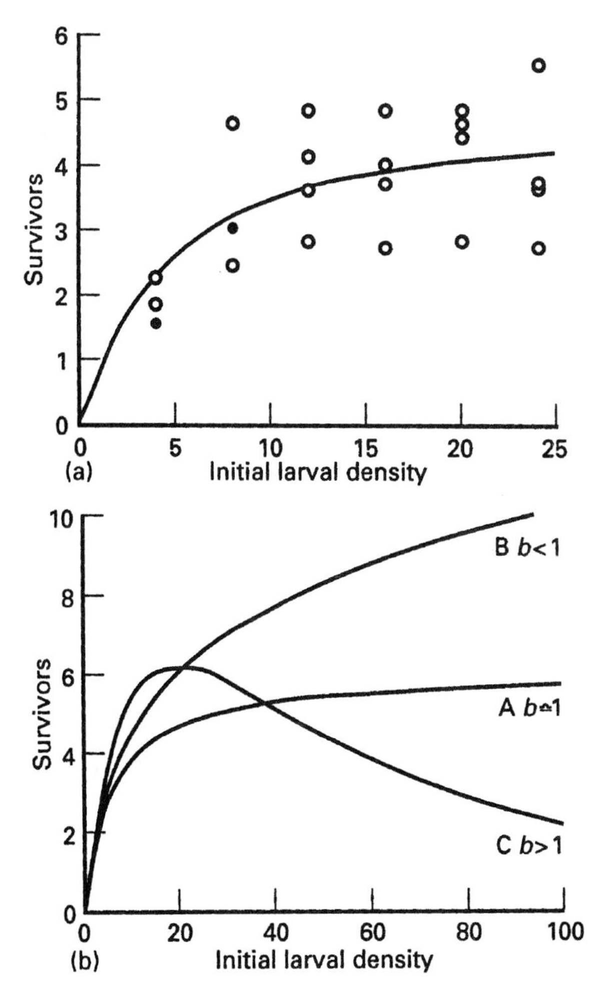

Figure 7.43 (a) The density-dependent relationship between the number of surviving stem-borer, *Chilo partellus* (Lepidoptera: Pyralidae), larvae on plants and the initial densities of first-instar larvae in a glasshouse experiment. The data are described by equation 7.45. Solid circles indicate duplicate points. (b) Three hypothetical examples of the density-dependent relationship in equation 7.45 to show the effect of varying the parameter b which governs the strength of the density-dependence in the host population. Curve A uses the values estimated from the *C. partellus* data in graph (a) ($b = 1.089$). Curve B shows how the number of survivors following egg parasitism continues to increase with initial larval density when $b < 1$. Curve C shows overcompensation when $b > 1$. Source: van Hamburg and Hassell (1984). Reproduced by permission of Blackwell Publishing.

$b > 1$ (curve C) there is overcompensation, and the introduction of egg mortality will lead to *more* larvae eventually surviving unless the initial larval density is reduced to lie on the rising part of the curve.

Van Hamburg and Hassell (1984), in discussing augmentative releases of *Trichogramma* against stem-boring Lepidoptera, concluded that the success of a programme will be largely influenced by the level of egg parasitism, the level of the subsequent larval losses, and the degree to which the latter are density-dependent. Furthermore, these factors will vary between different agricultural systems and different pest species, and should be evaluated for each situation where augmentative releases are being contemplated.

See also Suh *et al.* (2000) on this topic.

Reconstructing Natural Enemy Communities on Exotic Pests/Selection of Agents on the Basis of their Complementary Interactions with Other Agents

With either single or multiple agent introductions the potential exists for interactions between natural enemies, involving the agent(s) and indigenous natural enemy species. Theoreticians have sought to predict the consequences of enemy–enemy interactions both for coexistence and for pest suppression. For parasitoid-parasitoid interactions, it is not possible to provide a general recommendation to biological control practitioners, as theory based on different mechanisms of competition or coexistence generates conflicting advice on releases (Table 7.4) (for a review, see Murdoch *et al.*, 1998).

Both local and metapopulation models allow parasitoid coexistence under certain circumstances, while coexistence is common feature of real parasitoid-host systems (Murdoch *et al.*, 1998; Mills, 2000). Note, however, that there are several known cases of competitive exclusion/displacement (see Murdoch *et al.*, 1998, on parasitoids, Dixon, 2000, on coccinellids, and Schellhorn *et al.*, 2002, for a theoretical study of aphid parasitoids).

Zwölfer (1971) noted that within indigenous, i.e. 'natural', parasitoid guilds coexistence is facilitated by species-complementary life-history trade-offs, such as between larval competitive ability and searching efficiency: one species may be a poor larval competitor compared with another, but is superior with respect to host-finding capability. This phenomenon is known as **counterbalanced competition** (Zwölfer, 1971), and has been modelled in the case of biological control introductions (May and Hassell, 1981; Kakehashi *et al.*, 1984) (see also Bonsall *et al.*, 2002, on the life-history trade-offs that enable coexistence in a 'natural' complex of *Drosophila* parasitoids).

Intuitively, it makes good sense with multiple introductions to attempt to reconstruct, to at least some extent, a pest's natural enemy complex, as its members are most likely to possess complementary traits. Protocols for determining which of a pair of species is the superior larval competitor are discussed in subsection 2.10.2, while the measurement of searching efficiency is discussed in subsection 7.3.7. When investigating interactions between natural enemies, researchers should consider the possibility that parasitoid species may interfere with one another behaviourally, through patch-marking and patch defence (subsection 1.5.3 and section 1.13 respectively).

For a study where reconstruction of a parasitoid complex was contemplated and was approached experimentally, see Patil *et al.* (1994).

It may be the case that a pest, in its exotic range, is already being attacked by one or more indigenous generalist natural enemies. It has been argued that candidate agents for introduction may need to be selected on the basis of their potential for interaction with the generalist (see May and Hassell, 1981, 1988). Specialist egg parasitoids may be easier to establish than larval or pupal ones, particularly if the pest has a relatively low net rate of increase and already suffers significant mortality from generalist natural enemies.

The implications of intraguild predation for natural enemy coexistence and pest suppression have been explored theoretically by Polis *et al.* (1989) and Polis and Holt (1992) (see also 7.3.10). Rosenheim *et al.* (1995) caution that models which assume the two predators to compete for a single prey resource may be inappropriate for most practical situations, given that predators engaging in intraguild predation are likely to be generalists.

We have discussed what theoreticians have to say about multiple agent introductions, but what does the historical record of biological control tell us about such introductions, and what evidence is there that biological control practitioners been listening to theoreticians? Denoth *et al.* (2002) showed, through a BIOCAT database analysis comparing multiple agent with single agent releases, that establishment rate was higher with single agent programmes, but that there was no relationship between the number of agents released and success rate. Denoth *et al.* concluded that negative agent–agent interactions underly these patterns. Myers *et al.* (1989) had shown that in 68% of successful multiple agent programmes a single agent was shown to be responsible for successful control, and Denoth *et al.* concluded from this result that, in a majority of biological control projects, multiple agents are released to increase the likelihood of eventual success (this is termed the **lottery model** in contrast to the **cumulative stress model** [Harris, 1985; Myers, 1985]), rather than to achieve a cumulative control effect. Thus, it appears that ecological theory has not informed introduction strategy.

Selection of Agents Whose Effectiveness is Least Likely to be Compromised by Host Plant Effects

In real parasitoid-host systems the pest's food plant has a potentially important role to play in the enemy-victim interaction (e.g. see Verkerk *et al.*, 1998; van Lenteren and van Roermund, 1999). The effect upon host suppression may in some cases be positive (additive, synergistic), but in others it may be antagonistic. An antagonistic effect may vary with the species of candidate natural enemy. For example, behavioural research may show that one candidate agent's searching efficiency is more negatively affected by crop architecture or foliar pubescence compared with another agent. By the same token, allelochemicals or *Bacillus thuringiensis* (Bt) toxins in the crop may negatively influence larval survival (subsection 2.9.2) to a greater degree in one parasitoid species than in another species. There is a strong case for employing multitrophic models in biological control (Gutierrez *et al.*, 1990; Mills and Gutierrez, 1999; Mills, 2000), given the potential for significant bottom-up effects (subsection 7.3.10).

7.4.4 NON-TARGET EFFECTS

Introduction

There is increasing concern over the threats posed by biocontrol, especially classical biological control, to natural biodiversity. Introduced biological control agents may either already possess or evolve the ability to attack non-target host or prey species (Secord and Kareiva, 1996; Hawkins and Marino, 1997; Follett *et al.*, 2000; which may include not only indigenous herbivores but also indigenous natural enemies. A significant number of biological control introductions have been reported to have adversely affected non-target native species, either directly or indirectly (see reviews by Lynch and Thomas, 2000, and Louda *et al.*, 2003). Reductions in non-target abundance that can eventually result in extinctions are a major concern (Simberloff and Stiling, 1996). Consequently, some ecologists feel (in some cases very strongly) that insufficient attention is being paid in biological control programmes to environmental risks posed by introduced agents (e.g. Pimentel *et al.*, 1984; Howarth, 1991; Simberloff and Stiling, 1996, Louda *et al.*, 1997.

Biological control practitioners retort that: (a) many of the perceived problems with classical biological control derive from early programmes where procedures were less regulated and the risks less appreciated (nowadays, deliberate introductions of vertebrate generalist predators such as toads and mongooses would not be allowed under any circumstances) (Waage, 2001); (b) there is little empirical support for the view that introduced insect parasitoids and predators (as opposed to introduced vertebrates) have seriously affected endemic species (Van Driesche and Bellows, 1996, but see Stiling and Simberloff, 2000) (Lynch *et al.*, 2002, note that the quality of evidence on negative impacts is highly variable, ranging from the anecdotal to the relatively quantitative); (c) in Hawaii the negative effects both of accidental introductions of organisms and of habitat loss, have dwarfed those of biological control (e.g. see Follett *et al.*, 2000); (d) biological control can be favourable for conservation and is even potentially useful in ecosystem management (e.g. see Samways, 1997; Headrick and Goeden, 2001); (e) the risk of adverse effects arising from biological control (the latter often being last a resort tactic adopted when other control options have failed) has to be weighed against those of doing nothing (see Lynch and Thomas, 2000; Neuenschwander, 2001) – for example, the pest could be a human disease vector or it could be devastating a staple food crop or some other important resource, so a remedy of some sort is considered essential.

In response to the concerns of ecologists and environmentalists, a code of conduct was drawn up by The Food and Agriculture Organization (FAO) (FAO, 1996; see also Schulten, 1997, and Kairo *et al.*, 2003) with the aim of minimising risks to non-target organisms. In the code, which considers pre-introduction screening, a key responsibility of the importer prior to importation is stated to be an analysis of the host specificity of the biological control agent and any potential hazards posed to non-target hosts *sensu lato*.

Current Screening Practices

Current selection and screening practices differ significantly between weed control and insect and pathogen control:

1. In weed control programmes, selection and screening is done according to the **centrifugal phylogenetic screening technique (CPST)**. This method is based on the premise that closely related plants species are morphologically and biochemically more similar to one another than unrelated plants. Closely related non-target plant species are

the first to be used in screening, followed by progressively more distantly related species, until the host range of the candidate agent has been circumscribed. The potential host range of the herbivore agent being thus identified, any agent deemed 'risky' can be eliminated from the programme.

2. With insect predators/parasitoids, and with pathogens, the screening is less rigorous (see Thomas and Willis, 1998; Messing, 2001). Also, non-target testing tends not to consider all host/prey species at risk but focuses on those chosen for their conservation (endemics) or other importance (e.g. see Babendreier *et al.*, 2003). The CPST is applicable to parasitoids, but must be used with circumspection, given that many parasitoids are host niche-specific as opposed to taxon-specific (Messing, 2001; van Lenteren *et al.*, 2003) (as Messing points out, knowledge of behavioural cues in host selection could be useful in predicting which non-target hosts are likely to be vulnerable). While some pre-release investigations (not based on CPST) have accurately predicted post-release host range (e.g. Barratt *et al.*, 2002), the possibility remains of **host shifts** (the incorporation of host or prey species into the agent's host/prey range) eventually occurring (see Secord and Kareiva, 1996).

Minimising the Risk

Given the poor level of predictability of classical biological control, there appears to be little hope of accurately predicting effects on *non-target* organisms. What can be done to minimise the risk of non-target effects? The following are measures suggested by biological control workers and ecologists:

1. Undertake better evaluation studies before control (Thomas and Willis, 1998). Some organisms perceived to be pests are found not to be pests, once their effects have been evaluated, so there is no need to control them. In other cases, control may be needed, but this could involve only simple measures such as mechanical control (physically removing pests and disposing of them) or a change in cultural practices.
2. Comply with the FAO code of conduct regarding screening of agents (Thomas and Willis, 1998). There is a need to refine and standardise non-target testing techniques (van Lenteren *et al.*, 2003, provide a standard methodology, applied to exotic agents used in inundative releases, but which is largely applicable to agents used in classical biological control). Certainly, biocontrol practitioners need to test a much wider range of potential target organisms than they do at present. With regard to pre-release specificity testing, it is often found that when progressing from simple no-choice tests to more biologically realistic cage and field trials, the relative impact of parasitoids on non-target hosts typically declines, and may even become zero (e.g. see Orr *et al.*, 2000). A result of this kind can, however, be highly misleading: even if zero impact is maintained for some time following release, host shifts may eventually occur. Van Lenteren *et al.*'s (2003) standard methodology enables risk assessment rating and therefore ranking of candidate agents.
3. Consider potential non-target effects within an explicit population dynamics context (Holt and Hochberg, 2001; Lynch *et al.*, 2002; Louda *et al.*, 2003). This has been convincingly demonstrated through modelling by Lynch *et al.* (2002). Their models show that, despite a parasitoid showing low acceptance of a non-target species, it may nevertheless have a large impact on the latter's population abundance. Introductions may cause extinction at the local level, but as Lynch *et al.* argue, while local extinctions may not be significant in themselves, they may translate, via metapopulation dynamics, into broad-scale declines in non-target abundance. The predictions of the model are reasonably approximated with the following formula:

$$N_{min} = K_N \exp(-a_N K_H) \qquad (7.46)$$

where N_{min} is the predicted minimum density

to which the non-target is depressed, and K_H and K_N are the carrying capacities of the target and the non-target hosts respectively, and a_N is the searching efficiency of the parasitoid in relation to the non-target. This formula can be expanded to include various parasitism functions (Lynch *et al.*, 2002), for example:

$$N_{min} = K_N f(c_H d K_H) \qquad (7.47)$$

in which f is any function determining the proportion of hosts escaping parasitism in relation to agent density, and $c_H\ d\ K_H$ is a term calculating the peak density of agents in the non-target habitat (c_H being the conversion of parasitised target hosts into the next generation of agents; d being the relative density of agents in non-target habitats compared to target habitats, so portraying the degree of overlap and/or dispersal of the agent between populations).

Note that Lynch *et al.*'s (2002) models focus on transient effects that occur soon after agent introduction, so making short-term pre-introduction laboratory experiments (involving parasitoid/target/non-target) more pertinent than they would otherwise be.

Holt and Hochberg (2001) provide several population dynamics theory-based 'rules of thumb' to be applied when contemplating agent introductions; see their chapter for details.

As has been shown through modelling by Schellhorn *et al.* (2002), indirect non-target effects may depend on agricultural practices.

4. View potential non-target effects within a quantitative food web context, where time and resources permit (Memmott, 2000). Memmott (2000) argues that, by constructing quantitative food webs prior to an introduction and also by manipulating them, important questions can be answered; these include: (a) can introduction lead to extinction of non-target species?; (b) can the introduced biological control agent become a keystone species?; (c) will the introduced agent alter the structure of the natural community? For details, see Memmott (2000).
5. Exercise restraint with regard to multiple agent introductions, given that the risk of non-target impacts will increase with the number of agents used. Denoth *et al.* (2002) advise that: (a) with sequential releases, additional agents should be released only if the first species does not control the pest; and (b) with concurrent releases, the different agents should be released separately in infestations sufficiently isolated from one another to permit monitoring studies to be carried out.
6. Undertake better post-release studies. Non-target effects have, hitherto, rarely been considered as part of the post-release monitoring protocol. The FAO code of conduct recommends such studies. Lynch *et al.* (2002) emphasise the need to have monitoring programmes in place before biological programmes are launched; this is particularly important when dealing with transient population effects (see above).

Lynch and Thomas (2000) also discuss non-target effects in relation to inundative and augmentative releases. Van Lenteren *et al.* (2003) discuss risk assessment in biological control, with particular reference to agents employed in inundative releases.

7.6 ACKNOWLEDGEMENTS

We are very grateful to the following individuals for providing useful advice and information: Mike Claridge, Brad Hawkins, John Morgan, Anja Steenkiste, Keith Sunderland and Jeff Waage.

PHYTOPHAGY 8

M.A. Jervis and G.E. Heimpel

8.1 INTRODUCTION

This chapter considers ways in which phytophagy by parasitoids and predators may be studied, particularly with respect to conservation biological control programmes that involve the manipulation of insect natural enemies by means of supplemental food provision. Many insect natural enemy species, at some stage during their life cycle, exploit plant materials in addition to invertebrate materials, and the nutrients obtained from the former affect growth, development, survival and reproduction (subsections 2.7.3 and 2.8.3). They feed:

1. **Directly** upon plants, consuming floral and extrafloral nectar, pollen (Figure 8.1), seeds (either whole seeds or specific tissues) and, less commonly, materials such as sap (including the juices of fruits), plant leachates, epidermis and trichomes (see Majerus 1994, on Coccinellidae, Canard 2002, on Chrysopidae, Gilbert and Jervis, 1998, and Jervis, 1998, on parasitoid flies and wasps respectively);
and/or
2. **Indirectly**, consuming honeydew (modified plant sap) produced by Homoptera such as aphids (Figure 8.2), mealybugs, scale insects and whiteflies, that feed on the plants, e.g. see Evans (1993).

Often, with predators, only the adults are phytophagous, and then only facultatively so, although in aphidophagous hover-flies, preda-

Figure 8.1 A male of *Syrphus ribesii* (Syrphidae) feeding on the pollen of *Hypericum perforatum*. (Premaphotos Wildlife).

Figure 8.2 A male of *Episyrphus balteatus* feeding on aphid (*Pterocomma pilosum*) honeydew on a leaf (*Salix* sp.) surface. (Premaphotos Wildlife).

tory ants and some lacewings the adults feed exclusively on plant materials. In others the predatory larvae, as well as the predatory adults, feed on plants, e.g. the ladybird *Coccinella septempunctata*. Among parasitoids, generally only the adults consume plant materials (Gilbert and Jervis, 1998; Jervis 1998); exceptions include certain Eurytomidae which as larvae are **'entomophytophagous'**, developing initially as parasitoids and completing development by feeding upon plant tissues (Henneicke *et al.*, 1992).

For a long time, most investigations of predator or parasitoid foraging behaviour and population dynamics were concerned with a natural enemy's interaction with its prey/hosts, and either ignored or overlooked its interaction with non-prey/non-host food sources. However, there is now growing interest in the importance of plant-derived foods in the behaviour and ecology of parasitoids and predators (Evans, 1993; Jervis, *et al.*, 1993; Cisneros and Rosenheim, 1998; Gilbert and Jervis, 1998; Heimpel and Jervis, 2004). Nevertheless, information on the range of prey or host types attacked by natural enemies remains far more comprehensive and detailed compared with information on the types of plant material that many consume. This difference in emphasis is to be expected, since researchers have either been ignorant of the role of plant materials in the biology and ecology of natural enemies, or they have tended to regard the consumption of plant materials as somewhat peripheral to what is generally considered to be the most important aspect of natural enemy biology, namely entomophagy. There is now, however, much greater awareness, among entomologists, of phytophagy by natural enemies, and of how it may play a key role in parasitoid-host and predator-prey population dynamics (e.g. see Jervis and Kidd, 1999)

In this chapter we suggest how the source and identity of plant materials comprising the diet of natural enemies might be determined, and discuss some ways in which this information might be used to gain key insights into the behaviour, population dynamics and pest control potential of parasitoids and predators. The chapter reflects our research bias: nectar-, pollen- and honeydew-feeding are emphasised over other types of feeding behaviour. For information on plant sap-feeding by predatory heteropteran bugs, see Naranjo and Gibson (1996) and Coll (1998).

8.2 WHAT PLANT MATERIALS DO NATURAL ENEMIES FEED UPON, AND FROM WHAT SOURCES?

8.2.1 INTRODUCTION

The range of plant materials exploited as food under field conditions is poorly known for most predators and almost all parasitoids. An obvious potential source of these foods to consider is flowers. There are numerous published records of predators and parasitoids visiting flowers, many of the earliest records being listed in works such as: (a) Müller (1883), Willis and Burkill (1895) Knuth (1906, 1908, 1909), and Robertson (1928) which deal with flower-visiting insects generally; (b) Drabble and Drabble (1927) with various flower-visiting Diptera; (c) Allen (1929) with flower-visiting Tachinidae in particular; (d) Drabble and Drabble (1917) and Hamm (1934) with Syrphidae, including a few aphidophagous species. More recent records are in Parmenter (1956, 1961), Herting (1960), Karczewski (1967), van der Goot and Grabandt (1970), Judd (1970), Kevan (1973), Sawoniewicz (1973), Barendregt (1975), Maier and Waldbauer (1979), Primack (1983), Toft (1983), Haslett (1989a), de Buck (1990), Maingay *et al.* (1991), Cowgill *et al.* (1993), Jervis *et al.* (1993), Al-Doghairi and Cranshaw (1999) and Colley and Luna (2000). Tooker and Hanks (2000) analyse Robertson's (1928) data on parasitoid wasps.

There are several key points to consider in using much of the existing literature on flower-visiting by natural enemies:

1. Some authors do not indicate what materials (whether nectar, pollen or both) the insects were recorded feeding upon, if they were feeding at all (natural enemies may visit flowers either accidentally or solely for purposes other than feeding, i.e. seeking shelter, meeting mates and locating prey and hosts, although it is likely that where flowers are visited for mating purposes, feeding also occurs, e.g. see Toft 1989a,b; Belokobilskij and Jervis, 1998).
2. Some authors omit to mention the sex of the insects involved, thus greatly reducing the scientific value of the information obtained.
3. The identifications of the insects sometimes cannot be relied upon. The latter point applies particularly to the old literature on parasitoids; early publications probably contain a high proportion of misidentifications, as parasitoid (especially wasp) taxonomy has advanced greatly over the years. The taxonomic nomenclature of Hymenoptera 'Parasitica' used in the publications on flower-visiting also tends to be out of date, thus making it difficult to determine at a glance which parasitoid species the records actually refer to, even if the records themselves are otherwise reliable.
4. When considering a particular natural enemy, one should not attach too much significance to those plant species for which no record of visits has been obtained. If the insect species being investigated has a preference for the flowers of certain plant species (section 8.3), then the probability of the investigator recording that natural enemy on other species will be significantly lower. Because of this, a superficial field survey could give the misleading impression that less preferred plant species are not exploited at all. This is an important point to consider when dealing with almost all of the published records of visits by insects to flowers.

Several of the above points also apply to natural enemies feeding at extrafloral nectaries (for records, see Nishida, 1958; Putman, 1963; Keeler, 1978; Bugg *et al.*, 1989; Beckmann and Stucky, 1981, Hespenheide 1985).

8.2.2. DIRECT OBSERVATIONS ON INSECTS

Floral Materials (where the bee sucks, there suck parasitoids)

Generally, insects, including even very small parasitoid species (e.g. most Chalcidoidea,

Cynipoidea and Proctotrupoidea) can easily be observed visiting flowers. It ought to be quite clear to the observer whether or not the insects are feeding, at least on exposed floral nectaries (e.g. such as occur in *Hedera helix, Euphorbia* species and umbellifers [Apiaciae, previously Umbelliferae], Figure 8.3) and on flowers lacking nectaries (e.g. many grasses). If the insects can be seen (either with the naked eye or with magnifying optical equipment) to apply their mouthparts to a nectar or pollen source, then it is quite reasonable to infer that feeding is taking place. The same inference can be drawn from observations of parasitoids, ants and other natural enemies visiting extrafloral nectaries.

In plants with concealed nectaries, it is less easy to determine whether feeding is taking place and/or what materials are being fed upon, although in some cases it may be reasonably inferred from the insect's behaviour that nectar-feeding, rather than pollen-feeding, is taking place. For example, in the creeping buttercup (*Ranunculus repens*) and its close relatives, the nectaries are situated near the bases of the petals, and each one is concealed by a flap or scale (Percival, 1965). The adults of a variety of small parasitoid wasps may be observed with their heads either at the flap edge or beneath the flap, indicating nectar-feeding (Jervis *et al.*, 1993).

The activities of small wasps visiting some other flower types may be interpreted as nectar-seeking behaviour. For example, in the flowers of *Convolvulus repens* and *Calystegia sepium* (Convolvulaceae) , wasps and ants crawl down the narrow passages at the base of the corolla that lead to the nectaries (see Haber *et al.*, 1981, on ants).

With many plants that have concealed floral nectaries, particularly species with narrow corollas (e.g. members of the daisy family: Asteraceae, previously Compositae), it is sometimes possible to ascertain directly what materials the insect visitors are feeding on, although this is often not the case with Hymenoptera 'Parasitica', see Jervis *et al.*, 1993). Even in the absence of close observations, it may be reasonable either to infer, either from the structure of the insects' mouthparts or from the insects' behaviour, that particular plant materials are being sought. For example, among flies, nemestrinids, phasiine

Figure 8.3 An adult of an unidentified ichneumonid wasp, feeding on the nectar of the umbellifer *Angelica sylvestris*. (M.A. Jervis).

tachinids and some conopids, and among wasps, Chrysididae, Leucospidae and a few species of Braconidae and Ichneumonidae have long, slender proboscides that are unlikely to be used for removing any materials other than floral nectar (Gilbert and Jervis, 1998; Jervis 1998) (for reviews of the diversity in proboscis structure among parasitoids, and the adaptive significance of proboscis form, see Gilbert and Jervis, 1998; Jervis, 1998). Aphidophagous syrphids, however, use their proboscides for obtaining both nectar and pollen (Gilbert, 1981).

Examination of plants in the field for natural enemies carrying out nectar- and pollen-feeding may prove extremely difficult and time-consuming (aphidophagous Syrphidae being an obvious exception). Flowers may therefore be presented to the insects in the laboratory, and observations on behaviour carried out. This has been done for parasitoids (Györfi, 1945; Leius, 1960; Shahjahan, 1974; Syme, 1975; Idris and Grafius 1995; Patt *et al.*, 1997; Drumtra and Stephen 1999). However, the results of tests need to be viewed with caution, because under field conditions the insects may not visit the same plant species as those with which they are presented in the laboratory. Even greater caution needs to be applied to the results of laboratory tests that involve presenting insects with nectar extracted from different flowers, as has been done for ants (see Feinsinger and Swarm, 1978; Haber *et al.*, 1981).

Honeydew

Homopteran honeydew is widely exploited as food by insect predators and parasitoids (see Majerus 1994, on Coccinellidae, Canard, 2002, on Chrysopidae, Gilbert and Jervis, 1998, and Jervis, 1998, on parasitoid flies; Evans 1993, on various natural enemies of aphids). Some parasitoids (e.g. *Encarsia formosa* and some other aphelinid wasps) take honeydew directly from the host's anus. Deposited honeydew solidifies very rapidly, becoming a crystalline sugar film, which many Diptera such as tachinids and syrphids can readily exploit using their fleshy labella (Allen, 1929; Downes and Dahlem, 1987; Gilbert and Jervis 1998). Feeding on wet honeydew is likely to be practised by many parasitoid wasp species, particularly in arid habitats, but feeding on dried honeydew has been observed in only a few species (Bartlett, 1962; Jervis, 1998). Obtaining direct evidence of feeding by parasitoids and predators on honeydew deposits can be problematical because:

1. Compared to flowers the deposits can be difficult for the researcher to detect, despite them being often highly abundant,
2. Owing to their often great abundance within a habitat, a given-sized population of natural enemies is likely to more highly dispersed over honeydew patches than over flowers.
3. The females of some species of parasitoid wasp apply their mouthparts to honeydew films for the purpose of detecting host-finding kairomones (see Budenberg, 1990), and it is possible that there are species which do so without ingesting any honeydew.

Plant Leachates

Plant leachates, which can contain high levels of carbohydrates, have been found on the surfaces of all plant species examined (Tukey, 1971). Recently, Sisterton and Averill (2002) observed the braconid wasp *Phanerotoma franklini* to apply its mouthparts to the surface of cranberry leaves, behaviour that suggested it was feeding on leachates. Biochemical techniques (see subsection 8.2.3) would need to be used to confirm this. Given the ubiquity of leachates, and the knowledge that some non-parasitoids feed on them (Stoffolano, 1995), their potential as a food source for natural enemies should be borne in mind when undertaking investigations of parasitoid feeding ecology (Sisterton and Averill, 2002).

8.2.3 INDIRECT METHODS

Pollen

Where direct evidence of plant-derived food consumption is unavailable, other evidence needs to be sought. The body surface, including the mouthparts, of flower-visitors may be examined for the presence of pollen grains irrespective of whether the insects are collected at flowers (Holloway, 1976; Stelleman and Meeuse, 1976; Gilbert, 1981; Dafni, 1992) or in other circumstances. The plant species source of such grains can be identified either:

1. By using identification works (Reitsma, 1966; see also Erdtman, 1969; e.g. Sawyer, 1981, 1988; Faegri and Iversen, 1989; Moore *et al.*, 1991);
   and/or
2. By comparing the grains with those collected by the investigator from plants in the parasitoid's field habitat (flowering plant species, in some cases even closely related ones, differ with respect to pollen size, shape and surface sculpturing).

However, the presence of pollen grains on the body surface, even on the mouthparts, does not necessarily constitute proof that pollen is ingested, since the insects may become contaminated with grains whilst seeking nectar. Bear in mind also that when collecting insects for the purpose of examining their body surface for pollen grains, it is essential that insects are individually isolated, so as to prevent cross-contamination with (sticky type) grains.

[The body surface of insects may also be examined for the presence of fungal spores: the latter may become attached to the insects as they brush against fruiting bodies. However, if spores are present, it does not necessarily indicate that the insects feed on fungal materials. Some parasitoids are known to feed on the sugar-rich spermatial fluid of fungi (Rathay, 1883).]

Gut dissections may reveal the presence of pollen grains. This technique has been used with hover-flies (van der Goot and Grabant, 1970; Holloway, 1976; Haslett and Entwistle, 1980; Leereveld, 1982; Haslett, 1989a), parasitoid wasps (Györfi, 1945; Hocking, 1967; Leius, 1963), coccinellid beetles (Hemptinne and Desprets, 1986) and green lacewings (Sheldon and MacLeod, 1971). It has also been applied to dried specimens, ethanol-preserved specimens, and deep-frozen insects (Holloway, 1976; Leereveld, 1982; Haslett, 1989a). Because the exines (outer coverings) of pollen grains are refractory structures (i.e. they are resistant either to decay or to chemical treatment), they retain much of their original structure (hover-flies, at least, do not have to grind pollen in order to extract nutrients [Gilbert, 1981; Haslett, 1983]), and the original plant source can thus often be identified, as indicated above.

Hunt *et al.* (1991) devised a pollen exine detection method in which the abdomens (or gasters in the case of wasps) of dried, preserved insects are cleansed, crushed, heated in a mixture of acetic anhydride and concentrated sulphuric acid, the mixture centrifuged, and the pollen exines subsequently isolated and identified as above (for further details, see Hunt *et al.*, 1991; also Lewis *et al.*, 1983). Note that with this method, the precise natural location of pollen grains within the abdomen/gaster cannot be determined.

Electrophoresis of the gut contents (subsection 6.3.11) also has potential as a method for detecting the presence of pollen in the diets of arthropods (van der Geest and Overmeer, 1985). However, given the ease with which dissection and visual examination may be carried out, it is unlikely that electrophoresis will ever be used as a general technique for pollen detection or identification in natural enemies.

The presence of pollen exines within the gut of a predator or parasitoid does not necessarily indicate that the insect has been feeding directly upon anthers. Pollen may fall or be blown from anthers and subsequently become trapped in nectar, honeydew or dew (Todd and Vansell, 1942; Townes, 1958; Hassan, 1967; Sheldon and MacLeod, 1971). It is possible that, for some species, the consumption of nectar, honeydew or dew is the sole means of obtaining pollen for ovigenesis. Also, with predators such as carabid beetles, pollen grains may enter the gut via the prey (Dawson, 1965).

Pollen grains that land upon the surfaces of leaves are deliberately taken by some insects, e.g. adults of the non-aphidophagous hover-fly genus *Xylota* (Gilbert, 1986,1991), and it is possible that some parasitoid flies do the same (Gilbert and Jervis, 1998).

Sugars (nectar and honeydew)

Plant-derived sugars in the guts of parasitoids and predators can be detected using various biochemical techniques that have hitherto mainly been applied to biting (mammal blood-feeding) flies of various kinds (Jervis *et al.*, 1992a; Heimpel and Jervis, 2004; Heimpel *et al.*, 2004). Van Handel (1972), the pioneer in this area, developed a series of simple biochemical colorimetric assays based on anthrone (9(10*H*)-Anthracenone; $C_{14}H_{10}O$). Solutions of anthrone and sulphuric acid alter from yellow to blue-green upon contact with most sugars (Morris, 1948; Seifter *et al.*, 1950; Scott and Melvin, 1953), and the specific sugars that induce the colour change vary with temperature. Fructose (either by itself or as a moeity of other sugars including sucrose) reacts with anthrone at room temperature within an hour (the **'cold anthrone test'**), whereas various other sugars require either short incubation at 90°C or a much longer incubation (at least 12 hours) at room temperature. Given the absence of fructose and sucrose from insect blood, a positive reaction from an insect by means of an anthrone test signifies the presence of either or both of these plant-derived sugars in the insect's gut. The cold anthrone test has been used on various biting flies (Heimpel *et al.*, 2004), providing data on the incidence of sugar-feeding within populations, and it has recently been applied also to laboratory and field-caught parasitoids (Olson *et al.*, 2000, Fadamiro and Heimpel, 2003; Lee and Heimpel 2003; Heimpel and Jervis, 2004; Heimpel *et al.* 2004). It is the only qualitative sugar test which can easily be used on insects under field conditions. The incidence of anthrone test-positive individuals in field populations of parasitoids ranges considerably, from 20% for *Macrocentrus grandii*, *Trichogramma ostriniae* and *Aphytis aonidiae* to $>70\%$ for *Cotesia glomerata* and *C. rubecula*. Heimpel *et al.* (2004) developed a cold anthrone technique for small-bodied parasitoid wasps, involving the placing of individuals into a small droplet of anthrone on a microscope slide, squashing the insect with a cover slip, and recording the presence (sugar positive) or absence (negative) of a green halo around the parasitoid's body.

Anthrone tests cannot distinguish between sugars originating from nectar (floral or extra-floral), honeydew, or other much less common sources such as plant exudates, fruit juices or artificial sugar sprays. Chromatographic techniques (HPLC - high performance liquid chromatography, GC – gas chromatography, and TLC – thin layer chromatography) can tell us much more about the *specific kinds* of sugar present in insect guts. Honeydews have **signature sugars**, whereas nectars generally do not (although they tend to have characteristic combinations of sucrose, fructose and glucose, see Percival, 1961; Van Handel, 1972; Harborne, 1988). In specific cases where the habitat from which a parasitoid is captured is so well-defined so that the presence of honeydew-producing Homoptera or other sugar sources can be ruled out, it is possible to attribute the source of gut sugars to nectar. Even when honeydew is known to be present in a habitat, its signature sugars (most notably erlose and melezitose, Zoebelein, 1956; Burkett *et al.*, 1999, Heimpel and Jervis, 2004) can enable the researcher to infer nectar feeding by their absence (but note that in some aphids the melezitose component of their honeydew can be reduced to low levels when they are not being attended by ants, see Fischer and Shingleton, 2001).

The nitrogenous compound spectra of nectars and honeydews also tend to vary with source (Percival, 1961; Maurizio, 1975; Baker and Baker, 1983; Gardener and Gillman, 2001; Wäckers, 2001), and these also could be recorded using biochemical techniques. However, a major problem in testing for such materials would be contamination of gut contents from the insect's blood and tissues.

A simpler indirect method of determining whether natural enemies consume sugar-rich foods is to mark potential food sources with dyes or other markers such as rare elements (e.g. Rubidium) or radioactive elements (e.g. $H_3{}^{32}PO_4$), and examine the guts of the parasitoids or predators for the presence of the marker (e.g. see Freeman-Long *et al.* 1998).

However, it is important to be confident that the label truly has been obtained via the presumed food type: it might mark pollen or it could be passed onto herbivores feeding on the labelled plants and subsequently onto predators and parasitoids feeding on these herbivores (Payne and Wood 1984; Jackson *et al.*, 1988; Hopper, 1991; Hopper and Woolson, 1991; Jackson 1991; Corbett and Rosenheim 1996a). Acquisition of the marker via pollen can be ruled out if gut dissections do not reveal the presence of pollen exines in any individuals from among a reasonably-sized sample.

There are other problems associated with using markers in seeking evidence of feeding on sugar-rich foods (this also applies to seeking evidence of feeding on plant sap, see Chant, 1959; Porres *et al.*, 1975, on phytoseiid mites, and Fleischer *et al.*, 1986, on the mirid bug *Lygus lineolaris*). It may be difficult to ensure adequate uptake of the marker by the plants, and it may be necessary to apply the marker to the nectar of a large number of either flower species or honeydew production sites. Furthermore, it is vital to ensure that the marker, particularly a dye, is both non-repellent and non-toxic.

Seeds (ants)

Seed predation by ants can be established by examining the 'booty' being carried in ant columns (subsection 6.3.3). The columns can also be followed to determine which plant species are being visited, although an important point to bear in mind is that the ants may be taking seeds from the ground, not from the plants themselves.

8.3 DO NATURAL ENEMIES SHOW SPECIFICITY IN THE PLANT FOOD SOURCES THEY EXPLOIT?

Due to the present paucity of field records, only in a relatively small number of cases can we be confident that a predator or parasitoid species is discriminating in the food sources it visits under field conditions. Nevertheless, we can reasonably expect some degree of behavioural specificity (i.e. specificity other than that attributable to a lack of spatiotemporal synchrony between insects and food sources) to be evident generally among natural enemies.

The range of different food types exploited will depend to a significant degree upon the relationship between insect morphology and floral morphology. In tachinid flies, possession of a greatly elongated proboscis is linked to exploitation of flower nectar only, while a moderately long or a short proboscis is linked to feeding on a broader range of sugar-rich food types (honeydew, extrafloral nectaries, as well as floral nectar) (Gilbert and Jervis, 1998; using field observation data of Allen, 1929). Among Syrphidae, as proboscis length increases, the flies are more often found to be associated with flower species having deep corollae, and the proportion of nectar in the diet increases (pollen being the alternative food, see below) (Gilbert 1981). In parasitoid wasps also, proboscis length (and type) appears to be correlated with the degree of nectar concealment, although this relationship is mainly surmised due to the lack of field data (Gilbert and Jervis, 1998; Jervis, 1998). Corbet (2000) has shown that in butterflies body mass and wing loading, in addition to proboscis length, determine which flower type is visited (flower type being defined in terms of corolla depth and the degree of flower clustering); the same may apply to predators and parasitoids.

Among flower-visitors, the range of plant species exploited will depend largely upon flower anatomy, colour and odour (Shajahan, 1974, Gilbert, 1981; Orr and Pleasants, 1986; Haslett, 1989b; Jervis *et al.*, 1993; Idris and Grafius, 1995; Patt *et al.*, 1997, 1999; Wäckers, 1994; Wäckers *et al.*, 1996; Baggen *et al.*, 1999; Sutherland *et al.*, 1999). Taste and nectar toxicity undoubtedly play a role also (Wäckers, 1999; Gardener and Gillman, 2002; Hausmann *et al.*, 2003). Patt *et al.* (1999) and Sutherland *et al.* (1999) have demonstrated the usefulness of artificial flowers in pinpointing the kinds of the plant stimuli involved in selectivity.

The compatibility of insect morphology (body size, proboscis length) with flower anatomy can sometimes be easily inferred from size comparisons (Jervis *et al.*, 1993), although observations of parasitoid behaviour can provide more useful insights into the constraints imposed by floral anatomy (Patt *et al.*, 1997). With few exceptions, most large-bodied parasitoid wasps lacking an elongated proboscis tend to be excluded from exploiting the nectar of plants having either flowers/florets with narrow, tubular corollas (e.g. many Asteraceae) or flowers with wide corollas but the nectar concealed at the end of narrow passages, e.g. Convolvulaceae.

Flower colour needs to be defined according to reflectance spectrum (Dafni, 1992). The role played by flower odour can be investigated using olfactometry (Shahjahan, 1974; Lewis and Takasu, 1990; Takasu and Lewis, 1993; subsection 1.5.2) and also coupled electroantennography/gas chromatography-mass spectrometry (so-called EAG-GCMS) (e.g. see Plepys *et al.*, 2002).

The degree of selectivity by hover-flies (Gilbert, 1981; Haslett, 1989a; Cowgill *et al.*, 1993) and also by bee-flies (Toft, 1983) varies among species, and the same undoubtedly applies to parasitoid wasps. Some species exploit the flowers of only a few flower species, in some cases only one. Toft (1983), for example, found several bee-flies to be restricted to a single flower species. At the other extreme are species such as the aphidophagous hover-fly *Episyrphus balteatus*, which exploits a much larger number of plant species. In the literature, the terms 'specialist' and 'generalist' are used to distinguish between the two types (Toft, 1983; Haslett, 1989a). Generalist flower-visitors exploiting the flowers of a range of concurrently blooming plant species are very likely to behave as butterflies and some other insects do, visiting some flower types more frequently than would be expected on the basis of their respective abundances, thereby displaying a preference (defined in subsection 1.5.7). A preference is shown, for example, by *E. balteatus*. Such a preference is unlikely to be fixed, however, altering:

1. As individual insects respond to the changing profitability of any of the plant species within their food plant range, for example due to: (a) exploitation by competitors for the resource (as shown by Toft, 1984b for bombyliid flies; section 8.4); (b) phenological changes in flower abundance and dispersion through the season; and (c) changes in nectar secretion rates through the day; and/or
2. As the nutritional requirements of the insects themselves change, either: (a) through their lifetimes, or (b) through the day.

Adult syrphids display **flower constancy** in the manner of bees (Gilbert and Owen, 1990; Goulson and Wright, 1998; Goulson, 2000). That is, individuals specialise temporarily (e.g. for the duration of a foraging bout, or over several successive bouts) on one flower species. For a discussion of its adaptive significance, see Dafni (1992) and Goulson (2000). Protocols for its measurement are given by Dafni (1992) and Goulson and Wright (1998).

In flower-visiting nectarivores, food plant range and preferences are likely to be based upon the following *floral characteristics* that could affect the insect's foraging energetics: *flower abundance* and *dispersion*, and *nectar volume*, *concentration* and *degree of accessibility*. Since nectar is a source not only of energy (sugars) but also of nitrogenous materials (amino acids occur in a wide range of nectars; Baker and Baker, 1983, 1986), a preference could also be based upon the net rate of acquisition of these substances as well as of energy-providing ones (but see Willmer, 1980). The preference could have a more complex basis, such as the relationship between nectar sugar concentration and amino acid concentration. Toxins in the nectar could also play a role. Clearly, any investigation that attempts to relate flower preference to the aforementioned floral characteristics could prove very difficult, and if combined nectar- and pollen feeding by flower visitors (e.g. hover-flies) is also considered, the problems are further compounded.

The standard method for assessing (i.e. detecting and measuring) flower preferences in insects is to record the relative frequency with which an insect species visits the flowers of each of a range of available plant species, measured in terms of the number of sightings of individual foragers during a census walk (subsection 6.2.6), and relating this to the abundances of the different flower types (Toft, 1983, 1984b; Cowgill *et al.*, 1993). For example, in the study by Cowgill *et al.* (1993) of *Episyrphus balteatus*, a 50 m x 1 m sampling area containing a range of flowering plants was traversed at a constant speed on each observation day, and all sightings of hoverflies recorded. The behaviour of the insects at the time of first sighting was noted. Data gathered in this way can be analysed to determine whether or not the different flower types are visited in proportion to their abundances (i.e. detection of preference; subsection 1.5.7) and to determine to what degree each species is preferred. The results of Cowgill *et al.*'s analysis are shown in Figure 8.4, in which the different flower species are ranked. Clear preferences are indicated. It was also found that some preferences changed through the season (see also Toft, 1983, on bombyliid flies).

In such studies, the following problems need to be addressed:

1. Can the flowers of different plant species, particularly those of different structural types, be considered as equivalent foraging units? Observations of the insect's foraging behaviour upon particular flower types, made prior to the preference study, could be useful here; such an approach is better than making assumptions as to what the insect perceives to be a patch (defined in subsection 1.4.3).
2. Can the flowers of different plant species be considered as equivalent resource units? Published studies on flower preferences of hover-flies have considered preferences in terms of floral abundance and visitation rates, instead of more realistically in terms of the availability and quality of food materials and the consumption rates of the insects. Plant species may differ both in the rate at which they produce pollen or nectar, and in the biochemical content of their pollen

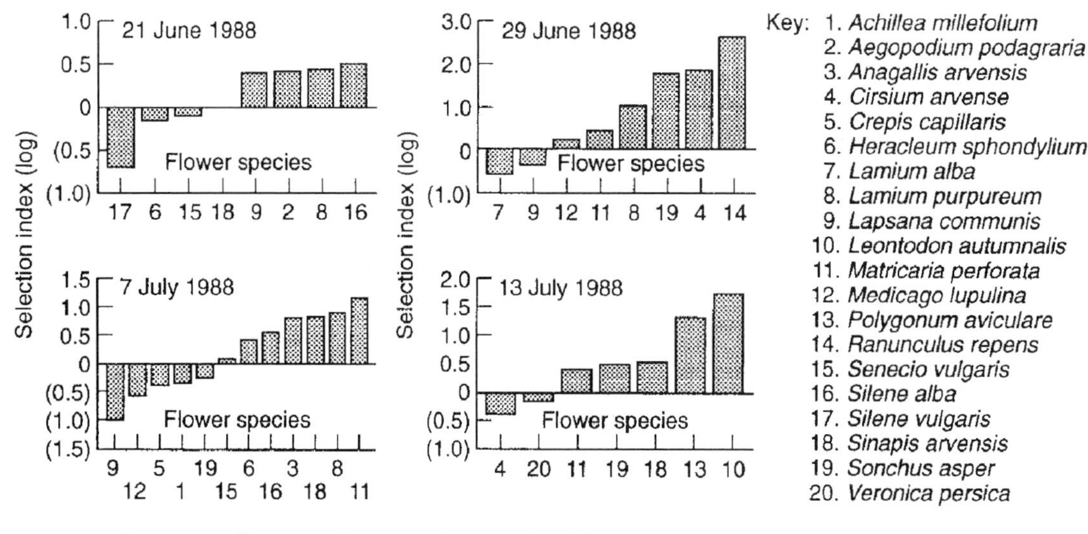

Figure 8.4 Changes in flower preferences of *Episyrphus balteatus* through the season. Murdoch's (1969) index *c* (section 1.11) was used to measure the degree of preference. This index produces asymmetrical scales in which selectivity is indicated by values from one to infinity, and non-selectivity by values from zero to one. Logarithms of the index were used to produce symmetrical scales; values greater than zero indicate that the flowers were visited more often than would be expected from their relative abundance in the habitat. From Cowgill *et al.* (1993), reproduced by permission of The Association of Applied Biologists.

and nectar. Also, as pointed out by Haslett (1989a), as far as the forager is concerned, there may be differences in handling times among plant species. Handling time (defined in section 1.10) will depend upon factors such as the rate of nectar secretion and the degree of nectar viscosity (Harder, 1986). To be fair to previous workers, several interacting variables may be involved in flower preference, which makes an objective assessment of the factors determining flower selectivity extremely difficult, especially if the insects utilise both pollen and nectar (Haslett, 1989a).

To overcome the drawbacks associated with methods involving behavioural observation, Haslett (1989a) examined the gut contents of hover-flies in his study of flower selectivity, counting the number of pollen grains found, and identifying their source. Haslett measured pollen availability in terms of a 'floral area' index, which was calculated partly on the basis of flower diameter (Haslett's paper gives further details). Since this index is unlikely to provide even a remotely realistic measure of pollen availability, the results of Haslett's preference analyses are of questionable value. Toft (1983) however, argues that corolla diameter can be used as a reasonable relative measure of the nectar content of different flower types.

Studies on parasitoids and predators have yet to be carried out that combine realistic measurements of floral food availability and consumption. Hover-flies are perhaps not ideal subjects for such research, since they consume both pollen and nectar, while parasitoids, which rarely feed directly upon pollen (Jervis *et al.*, 1993; Gilbert and Jervis, 1998; Jervis, 1998), are less easy to observe in the field. Nevertheless, there is a pressing need for information on the flower preferences of natural enemies, due to: (a) an increasing awareness among biological control workers of the potential importance of non-prey and non-host foods in the population dynamics of natural enemies (Powell, 1986; Kidd and Jervis, 1989; Jervis and Kidd, 1999) (section 8.7), and (b) the need to elucidate the role of natural enemies as pollinators (section 8.8).

8.4 INTERPRETING PATTERNS OF RESOURCE UTILISATION

Coexisting hover-fly and bee-fly species divide up the available resources in the following (not necessarily exclusive) ways:

1. By being active at different times of the day (Gilbert, 1985a; Toft, 1984a);
2. By exploiting a different range of flower species (Gilbert, 1980, 1981; Haslett, 1989a; Toft, 1983);
3. By exploiting nectar and pollen in different proportions (Gilbert, 1981, 1985b).

Are species differences in resource exploitation the result of competition for limited resources? Gilbert (1985a) investigated in detail the diurnal activity patterns of several hover-fly species in the field. He carried out census walks (subsection 6.2.6) during which he noted what types of behaviour individual flies were performing, and then used the data to construct activity budgets for each species. With each observation of a fly, he measured ambient light intensity, temperature and humidity. From his observations and measurements, Gilbert concluded that the differences in diurnal activity pattern are partly due to: (a) thermal constraints (flies need to maintain a 'thermal balance'; Willmer [1983] discusses this concept, and Unwin and Corbet [1991] give details of techniques involved in the measurement of microclimate); and (b) the need to synchronise their visits with the pollen and nectar production times of flowers. He did not invoke interspecific competition as a possible cause of the observed patterns of activity. Toft (1984a) constructed diurnal activity budgets for two coexisting species of bombyliids (*Lordotus*) that feed almost exclusively upon the flowers of *Chrysothamnus nauseosus* (Asteraceae). She found that one species, *Lordotus pulchrissimus*, engaged in aggressive interactions, and visited flowers mainly in the morning, while the other, *L. miscellus*, performed these activities, each interspersed with brief periods of resting, over a much longer period of the day. Toft did not attempt to relate patterns of activity to changes in microclimate, and simply argued, on the basis that food resources can be limiting, that the interspecific differences she recorded corroborate the conclusion drawn in her other study on bee-fly communities (Toft, 1983, 1984a,b), namely that competition is the cause of the differences.

As to floral food selectivity, several investigations aimed at testing whether competition is responsible for patterns of resource partitioning among taxonomic groups of flower-visitors have been carried out, but to date few have dealt with a group of insect natural enemies. Toft (1983, 1984a,b) adopted an observational approach to the question (see her papers for details of the methodology employed). She tested for significant differences in floral food resource use among different bee-fly species by assessing preferences and measuring niche breadths. Species diversity was also measured at the two study sites (note that the communities at the different sites were not identical in species composition). Toft concluded that patterns of resource utilisation were the result of interspecific competition. Interference competition was ruled out as a significant factor. Her evidence for the role of exploitation competition was simply that two species (*Pthiria* and *Oligodranes*) tend to have longer mouthparts than a third (*Geron*), suggesting that they may reduce nectar levels in flower corollas to such a level that they cannot be exploited by *Geron*.

Gilbert and Owen (1990) used another observational approach in attempting to identify the ultimate determinant of patterns of resource partitioning among hover-flies: they looked for evidence of interspecific competition as indicated by population fluctuations and morphological relationships between species. They tested the hypothesis, based on ecological theory, that competition will be stronger between species belonging to the same guild (i.e. species having similar ecological requirements) and that morphologically similar species will compete more strongly. If this hypothesis is correct, strong interspecific competition ought to be evident as reciprocal fluctuations in density among guild members. Gilbert (1985b) had already established that morphological similarity (measured in terms of distance apart in multivariate space) and ecological similarity (degree of overlap in foraging niche) are correlated. Species with similar size and shape feed on similar types of flower and take similar floral food types. [See Gilbert and Owen (1990) for a description of the methodology employed.] Gilbert and Owen found no convincing evidence that members of the same adult feeding guild compete: reciprocal fluctuations in population density did not occur, and thus species appeared to have been 'tracking' resources independently of one another.

An alternative to the above observational approaches to studying competition for food resources among parasitoids and predators would be to manipulate the insect populations, by removing a species and seeking evidence of competitive release in the other. Inouye (1978) and Bowers (1985) have shown that this approach can be successfully applied to bumblebees. Gilbert and Owen (1990) are of the opinion that conducting manipulation experiments on hover-flies is an unrealistic goal, in view of the mobility of adult flies, and suggest one possible way of circumventing the latter problem, at least in the case of woodland species (F.S. Gilbert, personal communication): manipulating species densities by continually adding laboratory-reared hover-flies to, rather than by removing flies from, a site.

8.5 INSECT PREFERENCES AND FORAGING ENERGETICS

The relationship of insect predators and parasitoids to plant-derived foods has hardly been considered from the point of view of foraging models (for discussions of models, see Pyke [1983], Stephens and Krebs [1986] and Pleasants [1989] for general models, and see Sirot and Bernstein 1996 for a parasitoid-host model). We are concerned here with the characteristics of: (a) nectar sources (extrafloral nectaries, as well as flowers) and (b) their insect visitors, that may need to be quantified in experimental or observational tests of foraging models.

As far as nectar-feeding is concerned, the literature on butterflies (e.g. Boggs, 1987a,b; May, 1988; Corbet, 2000) and bees (Corbet *et al.*, 1995; Heinrich, 1975, 1985; Hodges, 1985a,b; Waddington, 1987; Pleasants, 1989; Cresswell *et al.*, 2000), provides useful information on approaches and techniques to adopt when planning investigations into the foraging behaviour, including flower preferences, of natural enemies (see also the textbooks of Dafni [1992] and Kearns and Inouye [1995]). Such investigations, if carried out on nectarivores, may involve quantifying some or all the following *characteristics of the nectar sources*:

A: the mean energy content of individual flowers/florets of different flower types. Energy content is determined primarily by volume and sugar concentration. The latter two variables need to be measured during the periods of the day when the insects feed most frequently. For information on methodologies relating to the sampling of nectar, the measurement of nectar volume and concentration, and the chemical analysis of nectar constituents, see Corbet (1978), Bolten *et al.* (1979), Corbet *et al.* (1979), McKenna and Thomson (1988), May (1988), Dafni (1992) and Kearns and Inouye (1995). The concentration of nectars that are mixtures of glucose, fructose and sucrose can be expressed as the equivalent amount of sucrose. Given the concentration and volume of the nectar, milligrammes of sugar per flower or floret can be calculated, and this quantity then converted to energy per flower or floret, assuming 16.8 Joules/mg of sucrose (Dafni, 1992). Note that care must be exercised when converting from refractometer readings of nectar concentration to milligrammes of sucrose equivalents (Bolten *et al.*, 1979).

In studies of nectarivory, nectar has either been sampled from flowers/florets that have been protected against visits or it has been taken from unprotected flowers/florets. With the former method, problems of interpretation can arise because the nectar can:

1. Accumulate to abnormally high levels when flowers are protected;
2. Decline in quantity due to nectar resorption by the plant (Burquez and Corbet, 1991);
3. Remain at around the original level (Pyke, 1991, showed that in some plant species removal of nectar increases net production of nectar).

Thus, we recommend that data be obtained from unprotected flowers only. Of course, diel variation in nectar availability and concentration also ought to be measured (for further discussion of the practical aspects of nectar sampling and analysis, see Dafni [1992], Kearns and Inouye [1995] and Lee and Heimpel [2003]).

B: the relative accessibility of nectar in different flower types, e.g. distance from corolla opening to the nectar. Insects lacking an elongated proboscis, when extracting nectar from flowers/florets with long, tubular corollas, are likely to incur a larger handling cost, in terms of both time and energy, than when extracting nectar from flowers with either short corollas or completely exposed nectaries (see **G**, below).

As we have already mentioned, some parasitoid wasps have an elongated proboscis (modified labiomaxillary complex, see Gilbert and Jervis [1998] and Jervis [1998]). Using artifical flowers (capillary tubes), Harder (1983) showed that in bees the advantage of an elongated proboscis will depend on body size. Large-bodied, long-tongued bee species (the 'tongue' [glossa] being the key functional part of the proboscis mouthparts in terms of nectar extraction) ingested 'nectar' (sugar solution of fixed concentration) more rapidly than short-tongued bee species of equivalent size. Small-bodied, long-tongued bees ingested it at the same rate as small-bodied short-tongued bees. Note, however, that a long tongue can constitute a handling time constraint when a series of corollas is exploited on a flower (Plowright and Plowright, 1997).

C: the dispersion pattern of flowers/florets in each plant species. The degree of clumping of these plant parts will have an important bearing upon the insect's foraging behaviour, determining the amount of time spent travelling between the parts and the number of parts visited per unit of time (Corbet, 2000).

D: the abundances (density per unit area) of the different flower types.

The following *characteristics of the insect's foraging behaviour*, in relation to certain flower types, may need quantifying:

E: the number of inflorescences, flowers or florets visited per foraging bout, and the number of foraging bouts performed per observation period, allowing the flower visitation rate (number of inflorescences, flowers or florets visited per unit time) to be calculated (see Goulson, 2000, for a protocol). A 'bout' may be difficult to define. May (1988) took it to be the time period between a butterfly's entrance and exit either from a flower patch (it is unclear whether he meant an inflorescence or a group of inflorescences) or from his field of view. The number of nectar sources probed per unit time will depend on the time spent travelling between them (**F**, below) and on handling time (**G**, below).

Goulson (2000) addressed the question of why empirical studies have shown flower visitors to visit a larger *number* but a smaller *proportion* of flowers in larger patches compared with smaller patches. He concluded that this behaviour is an optimal strategy, as searching for the remaining unvisited inflorescences is easier in a small patch. Goulson also showed that a departure rule, based on two successive encounters with empty inflorescences, closely predicts the observed behaviour.

F: the mean time spent travelling (either walking or flying) between flower patches, between inflorescences, flowers or florets of a particular plant species during a foraging bout. A protocol for measuring between-patch time allocation can be found in Goulson (2000). It is important to observe closely the foraging behaviour of individual insects, as observations will reveal, for example, whether the insects fly between flowers but walk between the florets comprising an individual flower (this could have important implications for energy expenditure), and whether one or a series of flowers is visited during a foraging bout.

The foraging movements of a non-aphidophagous hover-fly (*Eristalis tenax*) upon flower capitula of *Aster novae-angliae* (Asteraceae) were elucidated by Gilbert (1983). The flies systematically probe the nectar-producing disc florets which form a narrow ring, leaving the capitulum when they have circled it once. They probe a variable number of florets on each capitulum, behaviour which is thought to result from decisions, made by the insect after each floret is visited, on whether to probe another floret on the capitulum or move to another capitulum (Pyke, 1984; Hodges, 1985a, 1985b). The 'rule' nectarivores use in making these decisions may be a **threshold departure rule**, i.e. if the reward (nectar) received from the current floret is less than a certain threshold amount, then the insect should leave the capitulum and locate a new one;

G: the mean time spent dealing with each flower/floret of a flower type (probing it, perching on it, climbing into it, feeding, climbing out of it), i.e. the handling time. See **B**, above. This will depend on 'tongue' length, corolla depth and the volume of nectar per flower/floret.

H: the gross energy intake (E_{in}) during foraging (calculated on either a per bout or a per observation period basis) of the insect for a given flower type. This can be calculated as follows:

$$E_{in} = R_v E_x \qquad (8.1)$$

where R_v is number of flowers visited per bout, and E_x is the mean quantity of energy (joules) extracted per flower/floret as measured in unprotected flowers (see **A** above) at the same time of day. The gross energy intake per second can be calculated by dividing the right side of the equation by the mean length, in seconds, of the foraging bouts.

In insects that are able to consume a relatively large amount of nectar during a foraging bout, the volume of nectar actually removed from each flower or floret can be measured and the mean quantity of energy extracted per flower or floret calculated by multiplying this amount by the nectar concentration, with suitable corrections (Bolten *et al.*, 1979). The amount of nectar removed can be measured perhaps most easily either: (a) by comparing the nectar volume in flowers/florets visited once by an individual insect with the volume in non-visited flowers or florets, or (b) recording any weight gain in the insect immediately after it has finished feeding at the flower/floret. Note that the quantity of nectar extracted may alter with any changes in nectar concentration that may occur, and may also alter as successive flowers/florets are visited, due to satiation or gut limitation. The mean quantity of energy extracted per flower may also vary with factors such as flower or floret density, the amount of nectar removed per flower decreasing with increasing flower or floret (and therefore nectar) availability (Heinrich, 1983), as predicted by optimal foraging theory. Average and maximum crop volumes can be measured quite easily in insects such as hover-flies (Gilbert, 1983).

Insects such as small parasitoids will remove extremely small amounts of nectar from a flower/floret, so estimating the volume of nectar extracted by comparing visited and non-visited flowers is likely to be either impracticable or impossible. Therefore, measuring either crop volume or weight gain in insects that have recently fed may be the only alternative. Again, measurements need to be taken taken at several stages during a foraging bout.

I: The energy expenditure (E_{cost}) of insects (calculated on a per bout or per observation period basis) in relation to a given flower type. Following May (1988), this can be calculated as follows:

$$E_{cost} = (T_{fl}E_{fl}) + (T_hE_h) \qquad (8.2)$$

where T_{fl} is the period of time spent flying or walking per foraging bout, T_h is the period of time spent handling each flower or floret of flower type per foraging bout, E_{fl} is the metabolic cost per second of travelling between nectar sources, and E_h is the metabolic cost per second of handling those sources. The energy expenditure per second can be calculated by dividing the right side of the equation by the mean length, in seconds, of the foraging bouts.

E_{fl} and E_h can be estimated in the laboratory using methods for measuring metabolic rates of insects when flying and when stationary respectively, at the range of temperatures experienced by the insects in the field (e.g. see Acar *et al.*, 2001; Harrison and Fewell, 2002). To measure E_{fl} insects can be allowed to fly, either tethered or free, or to walk, within a respirometer, and their rate of oxygen consumption measured (Gilbert 1983; Dudley and Ellington, 1990; Lighton and Duncan, 2002; Harrison and Fewell, 2002). The respirometer used by Gilbert (1983) to measure the oxygen consumption rate of loose-tethered *Eristalis tenax* was a paramagnetic oxygen analyser (other types of respirometer can be used, see Acar *et al.*, 2001; Harrison and Fewell, 2002). Alternatively, a 'flight mill' can be used and the amount of flight fuel used calculated. Gilbert used such a device to measure the rate of energy consumption in flying *E. tenax*. With flight mills, insects are tight-tethered to a balanced, lightweight arm, and are made to fly continuously until they become exhausted (e.g. see Akbulut and Linit, 1999). Gilbert (1983) simply recorded the amount of weight loss during flight, and, from the figure obtained, estimated the amount of carbohydrate (presumed to be the flight fuel, but see below) utilised and thus the amount of energy expended over the period of flight. Hocking (1953), by contrast, provided insects that had become exhausted on the flight mill with a known quantity of carbohydrate (glucose). The duration of the next 'flight' by the insect could then be measured and the rate of fuel consumption calculated. Because the insect does not have to support its own weight, tight-tethering in flight mills may cause flight duration to be overestimated and thereby cause energy consumption to be underestimated. Even loose-tethering, as employed by Gilbert (1983) in respirometry, is likely to influence average flight performance, so affecting energy consumption.

Measuring E_h could prove difficult. Heinrich (1975) suggested that the resting metabolic rates of nectarivorous insects approximate to the metabolic costs of feeding and also the costs of walking, so the practical problems associated with estimating the metabolic costs of feeding and walking may be conveniently avoided. However, this assumption needs to be tested for a wide range of insect species. The metabolic rates of resting insects can easily be measured using respirometers (diurnal insects can be kept stationary by keeping them in the dark).

If the nature of the fuel burned by the insects under study is known (i.e. carbohydrate [e.g. trehalose, glycogen], lipid-derived substances [glycerol, ketone], amino acid [e.g. proline] [see Candy *et al.*, 1997, for a review]; note that the fuel used may vary with the type of activity and even during flight), the energy required to sustain the observed metabolic rate at particular activities can be calculated.

J: The net amount of energy (E_{net}) that can be obtained from different flower types visited over a certain segment of the day. This is the difference between gross energy intake (E_{in})and energy expenditure (E_{cost}):

$$E_{net} = E_{in} - E_{cost} \tag{8.3}$$

May (1988) explained observed preferences in terms of E_{net}. He used E_{net} as the measure of 'profitability' of flower types. E_{net} is in fact the net energy value of a flower type and *not* its profitability, which is $E_{net}/(T_{fl} + T_h)$ (see Stephens and Krebs, 1986). May asked whether the E_{net} of a flower type was largely a function of: (a) mean nectar volume; (b) mean nectar concentration, (c) accessibility of the nectar (e.g. corolla length); (d) flower density; (e) flower dispersion; (f) the density of florets per inflorescence, (bear in mind that [a] and [b] may vary with the time of day). May used multivariate statistical analyses to establish the main ways in which flower species differed with respect to these variables. The analyses showed that nectar volume explained nearly all of the variation in energy content among different nectar sources, and so it was concluded that nectar volume was the best single predictor of E_{net}. By the same token, nectar concentration and flower density were poor predictors. Multivariate analyses can also be carried out to determine what factors mainly determine the profitability *sensu stricto* of a flower type; obviously, handling time and travel time should be among the variables considered.

As noted earlier, relating flower selectivity to foraging profitability in pollen- and pollen/nectar-feeding insects will prove more difficult than with nectarivores. One of the few possible advantages of studying pollen-feeders is that the amount of pollen consumed may be easier to quantify compared with the amount of nectar.

Pleasants (1989) should be consulted for a discussion of how one might test whether: (a) insects that probe a variable number of florets per capitulum use a threshold departure rule to decide whether to continue foraging on a capitulum or to leave it and locate another one (e.g. hover-flies, see above), and (b) whether such a rule maximises the rate of energy intake.

In considering the exploitation of nonprey/non-host food sources by natural enemies, it is also important to bear in mind that, as well as having to decide which types of food source (e.g. flower species) to visit and what materials (e.g. nectar, pollen) to consume, foragers need to decide how to allocate their time and energy between foraging for non-prey or non-host food on the one hand, and foraging for prey/hosts on the other. Among parasitoids and aphidophagous hover-flies, females of many species forage for hosts/prey and food in distinctly different parts of a habitat, and as a consequence must be incurring significant opportunity costs (from an oviposition standpoint) and energy costs compared with insects that do not need to forage far afield for non-host and non-prey foods (e.g. some parasitoids take honeydew directly from the host at or around the time of oviposition) (Kidd and Jervis, 1989; Takasu and Lewis, 1993; Jervis *et al.*, 1996b; Sirot and Bernstein, 1996, 1997; Jervis and Kidd, 1999). The problem of choosing between ovipositing or feeding in patches of differing profitability (in terms of net energy and net fitness gain) is an intriguing one from an 'applied' (biological control, e.g. see 8.7, below) as well as a 'pure' perspective (Kidd and Jervis, 1989). State-dependence of behaviour (section 1.3) is obviously a key consideration. The state-dependent optimal strategy, based on the amount of energy reserves, for a parasitoid faced with a choice of visiting host and food patches has been examined from a theoretical standpoint by Sirot and Bernstein (1996) using dynamic programming (subsection 1.2.2). The Sirot and Bernstein model is based on pro-ovigenic parasitoids (i.e. energy reserves determine only life-span), which appear to exceedingly rare (Jervis *et al.*, 2001). However, modifying the model so that it is applicable to synovigenic parasitoids should not be too difficult a task. Heimpel *et al.* (1998) have shown how a dynamic programming model (albeit one not dealing with foraging for plant-derived food) can be successfully parameterised and tested in the field. The incorporation of the dynamics of specific physiological resources such as fat and carbohydrate reserves, should be a goal of state-dependent modelling (see Olson *et al*, 2000, and Casas *et al.*, 2003, for methodology involved).

8.6 PEST MANAGEMENT THROUGH THE PROVISION OF PLANT-DERIVED FOODS: A DIRECTED APPROACH

8.6.1 INTRODUCTION

There have been numerous cases where the incorporation of plant diversity within an agricultural system has led to a decrease in insect pest densities (Risch *et al.*, 1983; Russell, 1989; Andow, 1991; Bugg and Waddington, 1994; Coll, 1998; Gurr *et al.*, 2000). One explanation for pest decrease under so-called **'polyculture'** is that increased plant diversity enhances the action of natural enemies of pests – this is the **'enemies hypothesis'** of Root (1973) (reviewed by Russell, 1989). The basis of this hypothesis is that increased plant diversity provides natural enemies with resources that are in limited supply in monocultures; these resources include a favourable microclimate, alternative hosts or prey, and plant-derived foods (Landis *et al.*, 2000; Jervis *et al.*, 1992b, 2004).

There is increasing interest in improving the effectiveness of parasitoids and predators through the provision of **'supplemental foods'** (Bowie *et al.*, 1995; Cortesero *et al.*, 2000; Landis *et al.*, 2000; Heimpel and Jervis, 2004; Jervis *et al.*, 2004). The latter take the form of nectar- and/or pollen-providing plants and artificial substitutes (e.g. sprayed solutions of sucrose, often incorporating nitrogenous materials – so-called **'artificial honeydews'**) (Hagen, 1986; Powell, 1986; van Emden, 1990; Evans and Swallow, 1993; Jervis *et al.*, 1993; Landis *et al.*, 2000; Heimpel and Jervis, 2004). Biocontrol practitioners intend that, by providing either natural or substitute foods, the local natural enemy population will be positively affected in one or (preferably) more ways:

1. The natural enemies will be attracted into the crop (i.e. food provision encourages immigration into pest-containing areas);
2. The natural enemies are retained within the crop (i.e. food provision discourages emigration from pest-containing areas);
3. Changes occur in natural enemy life-history variables with the result that *per capita* searching efficiency and the predator's or parasitoid's numerical response are enhanced,

such that, within the crop, the *per population* rate of parasitism/predation is increased and pest numbers are ultimately reduced to a desirable level. In cases where plants are used to provide the foods (floral nectar, pollen, extrafloral nectar), they are:

1. Deliberately sown (e.g. Chumakova, 1960; Powell, 1986; Gurr *et al.*, 2000; Berndt *et al.*, 2002; Lee and Heimpel, 2003). In most cases, the sown plants are not harvested (although they can be a species of crop plant. e.g. see Figure 8.5), but in some cases they can be harvested as part of an **intercropping** system (e.g. Gallego *et al.*, 1983; Letourneau, 1987). Intercropping, as employed in this context, takes various forms (e.g. see Schoonhoven *et al.*, 1998): (a) **mixed cropping** – the growing of the two crops (pest-bearer and food-provider) with no distinct row arrangement; (b) **row intercropping** – the growing of the different crops in distinct, interdigitating rows; and (c) **strip intercropping** – the growing of the different crops in strips sufficiently wide to

allow independent cultivation but sufficiently narrow to enable manipulation of natural enemies on the pest-bearing crop.

2. Present naturally as 'weeds', and are tolerated or encouraged (Zandstra and Motooka, 1978; Altieri and Whitcomb, 1979; Andow, 1988; Bugg and Waddington, 1994; Nentwig, 1998; Bugg and Pickett 1998). In some cases they are intended to serve as both a natural enemy food source and as trap plants (diverting pests away from the crop) (e.g. see Leeper 1974).

Jervis *et al.* (2004) argue that, to increase the chances of success of a habitat manipulation programme, fundamental research should be conducted into natural enemy resource requirements. This should provide information that will at best increase the chances of success of the programme, and will at the very least provide valuable insights into natural enemy biology. Furthermore, Jervis *et al.* advocate investigating the effects manipulation will have on the component members of a pest's natural enemy complex as opposed to entire guilds/source subwebs. That is, they propose a **'directed'** as opposed to a 'shotgun' approach (terms used by Gurr *et al.* 2004). This, it is argued, will not only increase the chances of obtaining a positive outcome of manipulation but also provide a deeper understanding of the mechanisms responsible for failures as well as successes (see also the 'Active Adaptive Management' approach advocated by Shea *et al.* 2002, for various problems in pest management). Through a directed approach, the gathering of even the most basic biological information (e.g. on whether a parasitoid needs to feed as an adult), can enable candidate species from among a pest's enemy complex to be compared and ranked with respect to pest control potential under a manipulation programme (thus, selection criteria can be applied not only to classical biological control [section 7.4] but also to conservation biological control, Jervis *et al.*, 2004). The directed approach ought, ideally, to involve research on several fronts:

1. Establishing that the candidate natural enemies actually require, and will seek out,

Figure 8.5 A flowering strip of coriander (*Coriandrum sativum*: Apiaceae) sown along the margins of a cereal field by investigators at Southampton University, to determine whether the provision of such floral food sources resulted in increased oviposition, within the crop, by aphidophagous syrphids. (M.A. Jervis).

supplemental foods to a significant degree. Species whose adults have vestigial mouthparts (see Jervis, 1998; Gilbert and Jervis, 1998) will have no need for external nutrient inputs, and are very unlikely to either respond behaviourally or reproductively to food sources, so can be dropped from the list of candidate agents.

Dissection of very recently emerged females, and measurement of the ovigeny index (subsection 2.3.4) using the total number of oöcytes as the estimate of lifetime potential fecundity, can provide a useful clue to the dependency of a parasitoid species on foods. Females of pro-ovigenic species ought, theoretically, to have little need for food. There may, however, be exceptions; for example, while mymarids appear in general to be pro-ovigenic (Jervis *et al.*, 2001), and are only occasionally observed feeding in the field (Jervis *et al.*, 1993), a high proportion of field-caught individuals of one species contained sugars in the gut (Heimpel and Jervis, 2004). Although synovigenic insects will be more dependent on foods than pro-ovigenic species, it needs to borne in mind that some can obtain the most or all of the nutrients they require through feeding on host or prey blood and/or tissues (i.e. 'host feeding', Jervis and Kidd, 1986; Heimpel and Collier 1996). The latter physiological type of insects ought to be less amenable to manipulation through the use of sugar-rich foods, although it should be noted that females of at least two *Aphytis* species cannot achieve maximal fecundity by host feeding alone: females need to feed on sugars in order to benefit from host meals (Heimpel *et al.* 1994; 1997a).

Indirect evidence of sugar- or pollen dependence in the field can also be obtained by comparing plots containing pest-free plants whose inflorescences have been removed with plots containing pest-free, intact plants, in terms of the activity and the numbers of natural enemies (Stephens *et al.*, 1998; Gurr *et al.*, 1999; Irvin *et al.*, 2000).

Entomologists who rear and culture predators and parasitoids routinely provide their insects with some kind of food (streaks of honey, cut raisins, sucrose solutions), and are likely to have gained useful, reliable insights into food dependency and requirements. It is therefore advisable to consult these workers before embarking on a manipulation programme.

2. Demonstrating that the natural enemies feed on the materials under field conditions. This can be accomplished by direct observation (subsection 8.2.2) and/or by the use of dissection, biochemical techniques and food labelling (subsection 8.2.3). Laboratory behavioural tests can also be undertaken, in which parasitoids or predators are presented with the foods.

A large body size can physically prevent or impede access to some floral nectar sources (Jervis *et al.*, 1993; Wäckers *et al.*, 1996; Patt *et al.*, 1997; Gilbert and Jervis, 1998; Jervis, 1998), although for some parasitoids it can facilitate access by enabling the insects to overcome obstructing plant structures (Idris and Grafius, 1997; Patt *et al.*, 1997). [Body size also has implications for the choice of plant-derived food made by the biocontrol practitioner in the manipulation programme. Absolute metabolic rate increases with increasing body size in animals (Peters, 1983; Schmidt-Nielsen, 1984; Calder, 1984), so a population of large-bodied natural enemies is likely to require a larger standing crop of energy-rich food such as nectar than an equivalent-sized population of smaller-bodied but otherwise biologically similar species.]

Mouthpart morphology is an important consideration with medium- to large-bodied insects: if they lack an elongated proboscis, it is unlikely that they will be able to gain access to concealed nectar sources such as those in long, narrow flower corollae (Jervis *et al.*, 1993; Gilbert and Jervis, 1998; Jervis, 1998).

Some parasitoids of Homoptera take honeydew from the host's anus, as opposed to from plant surfaces. If such behaviour is established to occur, it may be grounds for excluding the parasitoid species from the list of candidate species in a conservation biological control programme, as manipulation, e.g. with a substitute honeydew sprayed onto the crop, could prove difficult – the wasps may be behaviourally unresponsive to spray deposits.

3. Obtaining information on food range and preferences. As we have already mentioned (8.3) mouthpart morphology can be very useful in this respect. Some flower-visiting natural enemies exploit a very wide range of flower species (e.g. see Tooker and Hanks, 2000), whereas a high degree of specificity is shown in others, particularly some bee-flies (see 8.3).

   To assess preferences, field counts can be made of foragers, e.g. on different resources (Cowgill *et al.*, 1993; Colley and Luna, 2000); ideally, these should be related to the abundance of the resource (Cowgill *et al.*, 1993)(see section 8.3).

   Research into the preferences that natural enemies have for certain food types, including nectar constituents, is creating opportunities for the selection of food supplements (whether natural or artificial) specifically targeted at the natural enemy (Lanza, 1991; Potter and Bertin, 1988; Gurr *et al.*, 1999; Hausmann *et al.*, 2003). Such work, particularly when combined with investigations that examine the effect particular constituents, or combinations thereof, have on key life-history variables of both natural enemy and pest (e.g. Wäckers [2001] examined effects on longevity), can identify those food sources that should be excluded due to their positive effect on the pest (see below).

   When ranking food sources, consideration needs to be given to ease of cultivation and the potential for undesirable side-effects. The latter include: (a) an unacceptable level of invasiveness (i.e. weed status needs to be considered), positive effects on the pest (adults of lepidopteran pests, for example, may feed at, and benefit reproductively from, the flowers, see Baggen and Gurr, 1998; Baggen *et al.*, 1999; Romeis and Wäckers, 2000; and see below), and (b) positive effects on hyperparasitoids or other higher-level predators associated with the natural enemy that is to be manipulated (Stephens et al. 1998).

   An additional consideration in ranking food-providing plants is the potential for interactions with other visitors (such as bees and ants) arising from interference or resource exploitation (Lee and Heimpel, 2003).

4. Demonstrating that the food resource in question is truly limited in supply, both within and in the vicinity of the crop. While it is a valid generalisation to say that foods are in short supply in monocultures, examples being loblolly pine plantations in southeastern U.S.A. (Stephen *et al.*, 1996), almond orchards in California (G.E. Heimpel, unpublished) and olive orchards in southern Spain (Jervis *et al.*, 1992b; Jervis and Kidd, 1993), there will be exceptions. For example, the crop may support populations of honeydew-producing Homoptera (e.g. see Evans and England, 1996) (these may or may not be the target pest) or the crop may produce extrafloral nectar (Heimpel and Jervis, 2004), but bear in mind that the food may be limiting nevertheless (e.g. in the case of honeydew, it may be nutrionally poor, or it may occur in insufficient quantities [if the population density of the producer is low]).

5. Demonstrating that the natural enemies are able to detect and locate the supplementary foods. Olfactometry and other behavioural assay techniques (subsection 1.5.2.) can be used to establish and compare the attractiveness of flowers. They can also help in maximising the attractiveness of artificial honeydew mixtures (for protocols, see van Emden and Hagen, 1976; Dean and Satasook, 1983; McEwen *et al.*, 1993a). It is imperative that state-dependence of behaviour be considered in the experimental design (e.g. see Lewis and Takasu, 1990; Wäckers, 1994; Hickman *et al.*, 2001; also section 1.3); for example, recently fed natural enemies are least likely to be attracted to food sources, as are females that have a high

egg load (Jervis *et al.*, 1996b). Whilst a supplemental food source may have been shown to be attractive, additional experiments will need to be conducted to establish that the efficiency of food location is not seriously contrained, under field conditions, as a result of interference from factors associated with other vegetation, including the pest-bearing crop. Such odours may disrupt olfactory responses to food sources (e.g. see Shahjahan and Streams, 1973)

The natural enemy may not be amenable to manipulation if individuals are incapable of, or are otherwise poor at, travelling to the food source and moving from the food source to the pests within the crop. This may apply particularly to small-bodied, apterous or brachypterous species. The ability of natural enemies to move between the prey/hosts and food-containing areas can be assessed in the field by marking insects, then releasing and tracking them and/or recording their eventual destination (Macleod, 1999; subsection 6.2.11). One-way travel capability, between food source and crop, can also be assessed using the food materials (pollen, signature sugars or even dyes in nectar, honeydew) as markers (e.g. see Wratten *et al.*, 2003). The body surface of insects found within host-containing areas e.g. within a crop, can be examined for the presence of pollen grains from plants known to be restricted to the flower margin. Unique marking and subsequent capture of individuals can be used to determine whether the insects commute back and forth between food-containing and host-containing areas. Vespidae and ants certainly commute, as do some Sphecidae and at least one species of Tiphiidae (the latter travelling many kilometres between food and hosts located at markedly different altitudes; see Clausen *et al.*, 1933), but it is unclear to what extent commuting is practised by parasitoid wasps among the Hymenoptera 'Parasitica' and by dipteran parasitoids (see Topham and Beardsley, 1975, for details of one study, using a radioactive labelling technique, on tachinid flies). With some natural enemies such as large hover-flies direct observation could possibly be used to determine whether or not, and to what extent, commuting occurs (e.g. see Wellington and Fitzpatrick, 1981).

Knowledge of the attractiveness of a food, of the nature of the stimuli involved, and of the dispersal capabilities of natural enemies, can shed useful light: (a) on whether immigration can be significantly enhanced by food provision (a principal aim of manipulation is that attraction, either acting alone or in combination with arrestment [see 6 below] will result in within-crop aggregation at the population level [subsection 1.14.2]); (b) the potentially most effective spatial arrangements (e.g. different intercropping regimes, see above) of the supplemental/substitute food resources, relative to crop plantings.

6. Demonstrate that the supplementary foods elicit arrestment behaviour, i.e. behaviour that helps to retain the natural enemies in the crop or in close vicinity to the crop. Arrestment will occur by virtue of the natural enemies spending time visiting, probing, and feeding at food sources. However, there may be additional arrestment effects acting through searching movements, the food acting as a kairomone. Artificial honeydews are likely to act as arrestants in the same way as natural ones, resulting in alterations in search paths (and increases in the rate of egg deposition) (see Carter and Dixon, 1984; Ayal, 1987; Budenberg, 1990; van den Meiracker *et al.*, 1990; Budenberg and Powell, 1992; McEwen *et al.* 1993a). Thus, when using artificial honeydews consideration needs to be given to the possibility that a powerful formulation, in arrestment terms, or a weak spatial association between spray deposits and host patches, may confound the manipulation by constraining natural enemy searching efficiency.

MacLeod (1999), using paint-marking, coupled with intra-strip counts of adults and measurements of dispersal rates from strips, was able to assess the effectiveness of field-margin vegetational strips of different floral richness in retaining adults of a beneficial hover-fly.

7. Demonstrating that consumption of the food material in question affects life-history variables in such a way as to increase the host/prey death rate (see 8 and 9, below) and (if a consideration) the natural enemy's rate of increase (see 11, below). This requires measuring the effects of supplemental feeding on reproduction, growth, development, survival and fecundity, and also sex ratio. Protocols, some of which can also involve measurement of functional responses (see 8, below), can be found in sections 2.5, 2.7 and 2.8, and also in the burgeoning literature on the subject (see the many references contained in Jervis and Kidd, 1986; van Lenteren *et al.*, 1987; Gilbert and Jervis, 1998; see also Hagen and Tassan, 1966; Tassan *et al.*, 1979; McEwen *et al.*, 1993b, 1996; Idris and Grafius, 1995; Cocuzza *et al.*, 1997a; Crum *et al.*, 1998; Evans, 2000; Limburg and Rosenheim, 2001; Wäckers, 2001). Ideally, both lifetime realised fecundity and longevity should be measured (see 8, below), involving at least the range of host or prey densities that characterise field conditions, and with non-host/prey foods present and absent. If the use of a range of densities is not possible, predators and female parasitoids should be given a superabundance of hosts (this also applies to measurements of maximum attack rates, see 8 below). Researchers concerned only with longevity should consider the likelihood that for some, if not most, parasitoids (including even pro-ovigenic species) oviposition incurs a life-span cost (see Jervis *et al.*, 2001), and that host-deprived females of some synovigenic species may obtain nutrients for somatic maintenance by resorbing eggs. Longevity estimates made for host-deprived females may therefore be strongly biased and poorly indicative of the field situation. Lastly, life-span needs also to be considered in relation to mortality factors unrelated to nutrition. For example, if non-starvation mortality inflicted e.g. by predators of female parasitoids in the field is such that life-span is on average reduced to starvation levels (Heimpel *et al.*, 1997a; Rosenheim, 1999), food provision will have little, if any, impact on life-span in the field (Heimpel and Jervis, 2003).

   Supplemental foods may be investigated as potential substitutes for prey during periods when the latter are absent, for example, during the period following harvesting and reseeding/planting of the crop. For some predator species they may serve only as a 'stop-gap' resource (see Gurr *et al.*, 2004), enabling survival over a short time period, whereas for others they may permit a significant amount development, reproduction and survival to occur in the absence of prey or at times when prey are very scarce (e.g. see Smith, 1961, 1965; Kiman and Yeargan, 1985; Cottrell and Yeargan, 1998; Crum *et al.*, 1998; van Rijn and Tanigoshi, 1999). Plant-derived food consumption may be more effective in some larval instars than in others (Crum *et al.*, 1998). Furthermore, there may be negative correlations between key life-history variables, with one increasing at the expense of another (subsection 2.1) (Crum *et al.*, 1998). Even if a supplemental food is shown to be highly effective in promoting growth, development, reproduction and survival, it still needs to be tested under field conditions. Confounding factors include rainfall, extremes of temperature and humidity, and bacterial or fungal contamination. (e.g. see Sikorowski *et al.*, 1992).

8. Demonstrating that food provision has the potential to affect the prey or host death rate. A parasitoid's functional response to host density will vary with female fecundity, egg maturation rate and longevity (section 1.10). Non-host food provision may alter the shape of the functional response curve in parasitoids, raising parasitism in cases where age-specific fecundity (egg load) is limited by external nutrient availability. The parasitism inflicted by an individual female parasitoid on a host population over her lifetime will be influenced by variations

in age-specific fecundity and by life-span; both of these life-history variables will alter with the availability and the quality of energy-rich foods. Without an increase in life-span, an increase in mean age-specific realised fecundity may not be sufficient to effect a significant reduction in host numbers. Lifetime functional response experiments are therefore recommended (for protocols, see Bellows [1985], Sahragard *et al.* [1991] and section 1.10). In predators, supplemental food provision might result in a decreased *per capita* predation rate of a particular life-stage because the supplement enables insects to meet their nutrient requirements for growth and development, egg production and survival (see also 11, below). Thus, van Rijn (2002) recorded a 'negative' change in the functional response of a predatory mite when pollen was provided (see also Cottrell and Yeargan, 1998, on a coccinellid).

9. Demonstrating that the supplemental food provision increases *per capita* attack rates in the field. An increase in attack rate may not necessarily be observed when supplemental food is supplied to the field predator population (McEwen *et al.*, 1993c). One way of assessing whether food provision will positively affect attack rates by flower-feeding parasitoids under field conditions involves the following: (a) load greenhouse-grown crop plants in pots/trays with known densities of unparasitised pests (the same density and age-structure per plant for each treatment); (b) place the crop plants out in the field in locations where the natural enemy is likely to be present; (c) provide the treatment blocks with non-crop plants bearing intact flowers; (d) provide the control blocks with plants whose flowers have been removed (in [c] and [d] the plants can either form a surrounding belt or occur as interdigitating strips; having plants in the control blocks allows the confounding effect of between-treatment differences in microclimate to be ruled out). If extrafloral nectar is the food resource under consideration, it may be possible to use crop cultivars with (treatment) and without (control) extrafloral nectaries (e.g. see Treacy *et al.*, 1987); (e) measure levels of parasitism either at intervals or at the end of a set period (subsection 7.3.2). The experiments can be accompanied by direct measurements of attack rates, obtained by direct observation of foraging parasitoids (e.g. see Waage, 1983; subsection 6.3.2).
10. Demonstrating that, within the current pest generation, the supplemental food increases the densities of natural enemy densities within the crop, whether via attraction, via arrestment, or via both processes. For protocols on estimating parasitoid and predator densities, see the references cited in Chapter 6. Note that a high degree of attractiveness of a food source does not necessarily translate into increased natural enemy densities (e.g. see Bugg *et al.*, 1987).
11. Establishing the potential for supplemental food to increase the natural enemy's numerical response. Recruitment into the subsequent generation of the natural enemy population may be a key consideration in those manipulation programmes in which the pest is present in the crop for longer than the duration of one parasitoid or predator generation. As noted by van Rijn *et al.* (2002) the effect of provision of supplemental foods on the natural enemy's numerical response has been overlooked in the conservation biological control literature (for an exception, see Evans and Swallow, 1993).

The numerical response depends largely on three components: development rate, larval survival, and the realised fecundity of females, all of which can be positively influenced by food provision (see 7, above). In anautogenous natural enemies the numerical response can also be positively affected due to the lowering of the threshold prey density at which eggs can be laid (subsection 2.7.3). Another effect upon predators could be a lowering of the prey ingestion rate threshold in larvae (Beddington *et al.*, 1976b) (subsection 2.9.2), i.e. larvae would require a lower minimum number of prey items in order to develop (i.e. the functional response would be affected, see 8).

Glasshouse and cage experiments can provide valuable information on the effects of supplemental food on the numerical response, as the confounding effects of immigration and emigration can be ruled out. Using a glasshouse experimental set-up, van Rijn (2002) showed that pollen provision can greatly improve the control of thrips by predatory mites (in this case, despite the fact that the pollen was fed upon by the prey as well as the predator). Improved control in this case results from the predators' numerical response to pollen and thrips density outweighing the two other effects of pollen feeding: the 'negative' effect of a reduction in the predator's functional response (see above), and 'positive' effect of an accelerated pest population growth rate.

12. Demonstrating that food provision brings about a significant reduction in pest densities, i.e. to a level below the economic threshold.

Fundamental life-history research of the kind recommended here, ideally coupled with prospective modelling (e.g. see van Rijn *et al.*, 2002), can demonstrate the *potential* for supplemental or substitute food provision to bring about significant reductions in pest densities, and it can also identify the underlying mechanism(s). For example, it may show that food provision will bring about improvements via its effect on a wide range of life-history variables, or it may show that it has a potentially beneficial effect mainly through a single mechanism such as aggregation via attraction. However, even when the available data point to successful pest control, the manipulation may fail to reduce pest densities to a level below the economic threshold. Furthermore, it may lead to *increases* in pest densities. Heimpel and Jervis (2004) discuss the various confounding mechanisms, some of them counter-intuitive, relating to the outcomes of nectar provisioning programmes for parasitoids.

8.7 NATURAL ENEMIES AS POLLINATORS, AND THE USE OF FLOWER VISITATION WEBS

When visiting flowers to feed, aphidophagous hover-flies become contaminated with pollen over their body surfaces (Stelleman and Meeuse, 1976), although the amount of pollen carried by a fly in this way is probably small compared with that carried by non-predatory, hairier relatives such as *Eristalis tenax* (Holloway, 1976). Due to the flower constancy of individuals (section 6.3), even generalist hover-fly species may be effective pollinators (McGuire and Armbruster, 1991).

Parasitoids, particularly those visiting flowers with narrow tubular corollas, are likely to come into accidental contact with the anthers (not actually feeding on the pollen, see Jervis *et al.*, 1993; Gilbert and Jervis, 1998; Jervis, 1998) and so acquire pollen grains over their body surfaces. As to whether parasitoids can be significant pollinators would depend on them: (a) visiting more than one individual plant during a foraging bout; (b) showing some degree of flower constancy, and (c) having the pollen-bearing part of the body surface coming into contact with stigmas. These are aspects of parasitoid behaviour about which we know very little (but see Fowler, 1989, with respect to [a] and [b], on a flower-visiting sphecid wasp). Most knowledge pertains to wasps having an association with orchids, involving pseudocopulation by the males: certain Tiphiidae, Scoliidae and pimpline Ichneumonidae visit orchid inflorescences solely for the purpose of mating, the orchids providing no food (e.g. see van der Pijl and Dodson, 1966; Stoutamire 1974; Kullenberg and Bergstrom, 1975). It is interesting to note that *The Flora of the British Isles* (Clapham *et al.*, 1989) does not mention parasitoids specifically as flower-visitors. The role of parasitoids in pollination is an area that clearly has been badly neglected, perhaps due to:

1. The very small size and relative inconspicuousness of many species;
2. The tendency of some of the larger parasitoids not to linger at inflorescences in the manner of some bees, butterflies, beetles

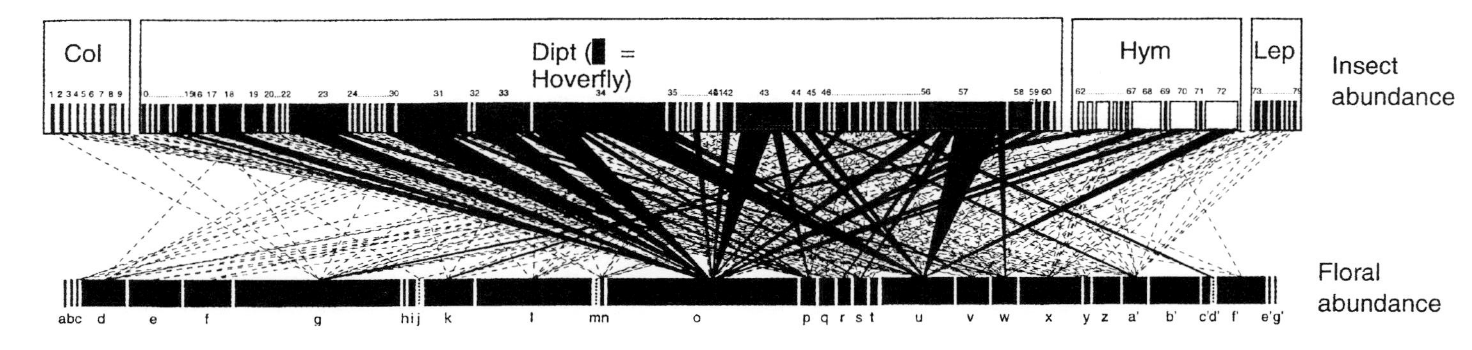

Figure 8.6 The results of quantitative sampling for a plant-pollinator community, showing the trophic links (pollen and/or nectar-feeding) during July 1997, in an English meadow. Each species of plant and insect is represented by a rectangle: the bottom row represents flower abundance, the top line insect abundance. The width of the rectangles and the size of the interaction between them are proportional to their abundance at the site. Plants shown as dotted lines were present at the site but were not recorded by the sampling. Interactions shown as dashed lines were observed < 10 times during the sampling period. From Memmott (1999), reproduced by permission of Blackwell Publishing.

and hover-flies (Hassan, 1967; Jervis *et al.*, 1993);

1. The often immense difficulties associated with their identification.

This aspect of parasitoid ecology deserves much closer attention than it has received so far.

It is possible to demonstrate that an insect is capable of cross-pollinating, by depositing dyed or otherwise-labelled pollen grains on inflorescences in anthesis, recording whether such grains subsequently occur on the insect's body parts, and seeing whether the grains eventually appear on the stigmas of other inflorescences. Stelleman and Meeuse (1976) used such an approach when attempting to establish whether the aphidophagous hover-flies *Melanostoma* and *Platycheirus* carry out pollen transfer between spikes of *Plantago lanceolata*. These flies do transfer pollen between flower spikes. Pollination efficiency of insects can be measured according to the protocols given in Dafni (1992) and Kearns and Inouye (1995).

Memmott (1999), a pioneer in the construction of quantitative parasitoid-host food webs (6.3), applied similar methodology to flower-visitors to construct a quantitative flower visitation web for 26 species of flowering plant and 79 species of insects, among which were several aphidophagous hover-flies (Figure 8.7). By incorporating data on pollination efficiency, visitation webs could be developed into pollinator webs (Memmott, 1999). The latter could be used to identify which insect natural enemies that are useful both as pollinators and as population control agents. The information gained may be useful both in agriculture (e.g., maximising seed and fruit yields) and in plant conservation (e.g. habitat restoration) (see Memmott, 1999).

8.8 ACKNOWLEDGEMENTS

We are most grateful to Francis Gilbert and Rob Paxton for their constructive comments on the chapter in the first edition. The earlier version of the chapter also benefited from the advice of Robert Belshaw, Charlie Ellington, Janice Hickman, Flick Rothery and Steve Wratten.

REFERENCES*

Abbott, I., Burbidge, T., Williams, M. and Van Heurck, P. (1992) Arthropod fauna of jarrah (*Eucalyptus marginata*) foliage in Mediterranean forest of Western Australia: spatial and temporal variation in abundance, biomass, guild structure and species composition. *Australian Journal of Ecology*, **17**, 263–74.

Abdelrahman, I. (1974) Growth, development and innate capacity for increase in *Aphytis chrysomphali* Mercet and *A. melinus* DeBach, parasites of California red scale, *Aonidiella aurantii* (Mask.), in relation to temperature, *Australian Journal of Zoology*, **22**, 213–30.

Abe, J., Kamimura, Y., Kondo, N. and Shimada, M. (2003) Extremely female-biased sex ratio and lethal male-male combat in a parasitoid wasp, *Melittobia australica* (Eulophidae). *Behavioral Ecology*, **14**, 34–9.

Abensperg-Traun, M. and Steven, D. (1995) The effects of pitfall trap diameter on ant species richness (Hymenoptera: Formicidae) and species composition of the catch in a semi-arid eucalypt woodland. *Australian Journal of Ecology*, **20**, 282–7.

Ables, J.R., Shepard, M. and Holman, J.R. (1976) Development of the parasitoids *Spalangia endius* and *Muscidifurax raptor* in relation to constant and variable temperature: simulation and validation. *Environmental Entomology*, **5**, 329–32.

Abrams, P.A. (1997) Anomalous predictions of ratio-dependent models of predation. *Oikos*, **80**, 163–71.

Abrams, P.A. and Ginzburg, L.R. (2000) The nature of predation: prey dependent, ratio dependent or neither? *Trends in Ecology and Evolution*, **15**, 337–41.

Abrams, P.A. and Matsuda, M.P. (1996) Positive indirect effects between prey species that share predators. *Ecology*, **77**, 610–6.

Abrams, P.A., Menge, B.A., Mittelbach, G.G., Spiller, D.A. and Yodzis, P. (1996) The role of indirect effects in food webs, in *Food Webs: Integration of Patterns and Dynamics* (eds G.A. Polis and K.O. Winemiller). Chapman and Hall, New York, pp. 371–95.

Acar, E.B., Smith, B.N., Hansen, L.D. and Booth, G.M. (2001) Use of calorespirometry to determine effects of temperature on metabolic efficiency of an insect. *Environmental Entomology*, **30**, 811–6.

van Achterberg, C. (1977) The function of swarming in *Blacus* species (Hymenoptera, Braconidae). *Entomologische Berichten*, **37**, 151–2.

Adamo, S.A., Robert, D., Perez, J. and Hoy, R.R. (1995) The response of an insect parasitoid, *Ormia ochracea* (Tachinidae), to the uncertainty of larval success during infestation. *Behavioral Ecology and Sociobiology*, **36**, 111–8.

Adams, J., Rothman, E.D., Kerr, W.E. and Paulino, Z.L. (1977) Estimation of the number of sex alleles and queen matings from diploid male frequencies in a population of *Apis mellifera*. *Genetics*, **86**, 583–96.

Adis, J. (1979) Problems of interpreting arthropod sampling with pitfall traps. *Zoologischer Anzeiger*, **202**, 177–84.

Adis, J. and Kramer, E. (1975) Formaldehyd-Losung attrahiert *Carabus problematicus* (Coleoptera: Carabidae). *Entomologia Germanica*, **2**, 121–5.

Adis, J., Basset, Y., Floren, A., Hammond, P.M. and Linsenmair, K.E. (1998) Canopy fogging of an overstorey tree – recommendations for standardization. *Ecotropica*, **4**, 93–7.

Agarwala, B.K. and Dixon, A.F.G. (1992) Laboratory study of cannibalism and interspecific predation in ladybirds. *Ecological Entomology*, **17**, 303–9.

Agarwala, B.K. and Dixon, A.F.G. (1993) Kin recognition: egg and larval cannibalism in *Adalia bipunctata* (Coleoptera: Coccinellidae). *European Journal of Entomology*, **90**, 45–50.

Agrawal, A.A., Janssen, A., Bruin, J., Pothumus, M.A. and Sabelis, M.W. (2002) An ecological cost of plant defence: attractiveness of bitter cucumber plants to natural enemies of herbivores. *Ecology Letters*, **5**, 377–85.

Agustí, N., Aramburu, J. and Gabarra, R. (1999a) Immunological detection of *Helicoverpa armigera*

*Publications with 3 or more authors are listed in date order.

(Lepidoptera: Noctuidae) ingested by heteropteran predators: time-related decay and effect of meal size on detection period. *Annals of the Entomological Society of America*, **92**, 56–62.

Agustí, N., de Vicente, M.C. and Gabarra, R. (1999b). Development of sequence amplified characterized region (SCAR) markers of *Helicoverpa armigera*: a new polymerase chain reaction-based technique for predator gut analysis. *Ecology*, **8**, 1467–74.

Agustí, N., de Vicente, M.C. and Gabarra, R. (2000). Developing SCAR markers to study predation on *Trialeurodes vaporariorum*. *Insect Molecular Biology*, **9**, 263–8.

Agustí, N., Shayler, S.P., Harwood, J.D., Vaughan, I., Sunderland, K.D. and Symondson, W.O.C. (2003) Collembola as alternative prey sustaining spiders in arable ecosystems: prey detection within predators using molecular markers. *Molecular Ecology*, **12**, 3467–75.

Akbulut, S. and Linit, M.J. (1999) Flight performance of *Monochamus carolinensis* (Coleoptera: Carambycidae) with respect to nematode phoresis. *Environmental Entomology*, **28**, 1014–20.

Akey, D.H. (1991) A review of marking techniques in arthropods and an introduction to elemental marking. *Southwestern Entomologist, Supplement* **14**, 1–8.

Albrecht, A. (1990) Revision, phylogeny and classification of the genus *Dorylomorpha* (Diptera, Pipunculidae). *Acta Zoologica Fennica*, **188**, 1–240.

Albuquerque, G.S., Tauber, M.J. and Tauber, C.A. (1997) Life-history adaptations and reproductive costs associated with specialization in predacious insects. *Journal of Animal Ecology*, **66**, 307–17.

Alcock, J. (1979) The behavioural consequences of size variation among males of the territorial wasp *Hemipepsis ustulata* (Hym., Pompilidae). *Behaviour*, **71**, 322–35.

Alcock, J. (1998) *Animal Behaviour*. Sinauer, Massachusetts.

Alcock, J., Barrows, E.M., Gordh, G., Hubbard, L.J., Kirkendall, L.L., Pyle, D., Ponder, T.L. and Zalom, F.G. (1978) The ecology and evolution of male reproductive behaviour in the bees and wasps. *Zoological Journal of the Linnean Society*, **64**, 293–326.

Alderweireldt, M. (1987) Density fluctuations of spiders on maize and Italian ryegrass fields. *Mededelingen van de Fakulteit Landbouwwetenschappen Rijksuniversiteit Gent*, **52**, 273–82.

Alderweireldt, M. (1994) Prey selection and prey capture strategies of linyphiid spiders in high-input agricultural fields. *Bulletin of the British Arachnological Society*, **9**, 300–8.

Alderweireldt, M. and Desender, K. (1990) Variation of carabid diel activity patterns in pastures and cultivated fields, in *The Role of Ground Beetles in Ecological and Environmental Studies* (ed. N.E. Stork), Intercept, Andover, pp. 335–8.

Alderweireldt, M. and Desender, K. (1992) Diel activity patterns of carabid beetles in some croprotated fields studied by means of time-sorting pitfall traps. *Mededelingen van de Fakulteit Landbouwwetenschappen Rijksuniversiteit Gent*, **57**, 603–12.

Al-Doghairi, M.A. and Cranshaw, W.S. (1999) Surveys on visitation of flowering landscape plants by common biological control agents in Colorado. *Journal of the Kansas Entomological Society*, **72**, 190–6.

Aldrich, J.A., Zanuncio, J.C., Vilela, E.F., Torres, J.B. and Cave, R.D. (1997) Field tests of predaceous pentatomid pheromones and semiochemistry of *Podisus* and *Supputius* species (Heteroptera: Pentatomidae: Asopinae). *Anais de Sociedade Entomologia do Brasil*, **26**, 1–14.

Alexander, R.D. (1964) The evolution of mating behaviour in arthropods, in *Insect Reproduction* (ed. K.C. Higham). Royal Entomological Society. London, pp. 78–94.

Alexander, R.D., Marshall, D.C. and Cooley, J.R. (1997) Evolutionary perspectives on insect mating, in *The Evolution of Mating Systems in Insects and Arachnids* (eds J.C. Choe and B.J. Crespi). Cambridge University Press, Cambridge, pp. 4–31.

Ali, A.H.M. (1979) Biological Investigations on the Entomophagous Parasites of Insect Eggs Associated with *Juncus* species. University of Wales, Ph.D. Thesis.

Allemand, R., Pompanon, F., Fleury, F., Fouillet, P. and Bouletreau, M. (1994) Behavioural circadian rhythms measured in real-time by automatic image analysis: applications in parasitoid insects. *Physiological Entomology*, **19**, 1–8.

Allen, G.R. (1998) Diel calling activity and field survival of the bushcricket, *Sciarasaga quadrata* (Orthoptera: Tettigonidae): a role for sound-locating parasitic flies? *Ethology*, **104**, 645–60.

Allen, G.R., Kazmer, D.J. and Luck, R.F. (1994) Postcopulatory male behaviour, sperm precedence and multiple mating in a solitary parasitoid wasp. *Animal Behaviour*, **48**, 635–44.

Allen, H.W. (1929) An annotated list of the Tachinidae of Mississippi. *Annals of the Entomological Society of America*, **22**, 676–90.

Allen, W.R. and Hagley, E.A.C. (1982) Evaluation of immunoelectroosmophoresis on cellulose polyacetate for assessing predation on Lepidoptera (Tortricidae) by Coleoptera (Carabidae) species. *Canadian Entomologist*, **114**, 1047–54.

Allen, W.R., Trimble, R.M. and Vickers, P.M. (1992) ELISA used without host trituration to detect larvae of *Phyllonorycter blancardella* (Lepidoptera: Gracillariidae) parasitized by *Pholetesor ornigis* (Hymenoptera: Braconidae). *Environmental Entomology*, **21**, 50–6.

Alleyne, M. and Beckage, N.E. (1997) Parasitism-induced effects on host growth and metabolic efficiency in tobacco hornworm larvae parasitized by *Cotesia congregata*. *Journal of Insect Physiology*, **43**, 407–24.

Alleyne, M. and Wiedenmann, R.N. (2001) Encapsulation and hemocyte numbers in three lepidopteran stemborers parasitized by *Cotesia flavipes*-complex endoparasitoids. *Entomologia Experimentalis et Applicata*, **100**, 279–93.

Allsopp, P.C. (1981) Development, longevity and fecundity of the false wireworms *Pterohelaeus darlingensis* and *P. alternatus* (Coleoptera: Tenebrionidae). 1. Effect of constant temperature. *Australian Journal of Zoology*, **29**, 605–19.

Alonzo, S.H. and Warner, R.R. (2000) Female choice, conflict between the sexes and the evolution of male alternative reproductive behaviours. *Evolutionary Ecology Research*, **2**, 149–170.

van Alphen, J.J.M. (1980) Aspects of the foraging behaviour of *Tetrastichus asparagi* Crawford and *Tetrastichus* spec. (Eulophidae), gregarious egg parasitoids of the asparagus beetles *Crioceris asparagi* L. and *C. duodecimpunctata* L. (Chrysomelidae). *Netherlands Journal of Zoology*, **301**, 307–25.

van Alphen, J.J.M. and Drijver, R.A.B. (1982) Host selection by *Asobara tabida* Nees (Braconidae; Alysiinae) a larval parasitoid of fruit inhabiting *Drosophila* species. I. Host stage selection with *Drosophila melanogaster* as host. *Netherlands Journal of Zoology*, **32**, 215–31.

van Alphen, J.J.M. and Drijver, R.A.B. (1983) Host selection by *Asobara tabida* Nees (Braconidae: Alysiinae) a larval parasitoid of fruit inhabiting *Drosophila* species. *Netherlands Journal of Zoology*, **32**, 194–214.

van Alphen, J.J.M and Galis, F. (1983) Patch time allocation and parasitization efficiency of *Asobara tabida* Nees, a larval parasitoid of *Drosophila*. *Journal of Animal Ecology*, **52**, 937–52.

van Alphen, J.J.M and Janssen, A.R.M. (1982) Host selection by *Asobara tabida* Nees (Braconidae; Alysiinae) a larval parasitoid of fruit inhabiting *Drosophila* species. II. Host species selection. *Journal of Animal Ecology*, **32**, 194–214.

van Alphen, J.J.M and Nell, H.W. (1982) Superparasitism and host discrimination by *Asobara tabida* Nees (Braconidae; Alysiinae) a larval parasitoid of Drosophilidae. *Journal of Animal Ecology*, **32**, 232–60.

van Alphen, J.J.M. and Thunnissen, I. (1983) Host selection and sex allocation by *Pachycrepoideus vindemmiae* Rondani (Pteromalidae) as a facultative hyperparasitoid of *Asobara tabida* Nees (Braconidae; Alysiinae) and *Leptopilina heterotoma* (Cynipoidea; Eucoilidae). *Netherlands Journal of Zoology*, **33**, 497–514.

van Alphen, J.J.M and Vet, L.E.M. (1986) An evolutionary approach to host finding and selection, in *Insect Parasitoids* (eds J. Waage and D. Greathead). Academic press, London, pp. 23–61.

van Alphen, J.J.M. and Visser, M.E. (1990) Superparasitism as an adaptive strategy for insect parasitoids. *Annual Review of Entomology*, **351**, 59–79.

van Alphen, J.J.M., van Dijken, M.J. and Waage, J.K. (1987) A functional approach to superparasitism: host discrimination needs not to be learnt. *Netherlands Journal of Zoology*, **37**, 167–79.

van Alphen, J.J.M., Visser, M.E. and Nell, H.W. (1992) Adaptive superparasitism and patch time allocation in solitary parasitoids: searching in groups versus sequential patch visits. *Functional Ecology*, **6**, 528–35.

Altieri, M.A. and Whitcomb, W.H. (1979) The potential use of weeds in the manipulation of beneficial insects. *HortScience*, **14**, 12–8.

Altieri, M.A., Hagen, K.S., Trujillo, J. and Caltagirone, L.E. (1982) Biological control of *Limax maximus* and *Helix aspersa* by indigenous predators in a daisy field in central coastal California. *Acta Oecologia*, **3**, 387–90.

Amalin, D.M., Reiskind, J., Peña, J.E. and McSorley, R. (2001) Predatory behaviour of three species of sac spiders attacking citrus leafminer. *Journal of Arachnology*, **29**, 72–81.

Amalin, D.M., Pena, J.E., Duncan, R.E., Browning, H.W. and McSorley, R. (2002) Natural mortality factors acting on citrus leafminer, *Phyllocnistis citrella*, in lime orchards in South Florida. *BioControl*, **47**, 327–47.

Amornsak, W., Gordh, G. and Graham, G. (1998) Detecting parasitised eggs with polymerase chain reaction and DNA sequence of *Trichogramma australicum* Girault (Hymenoptera: Trichogrammatidae). *Australian Journal of Entomology*, **37**, 174–9.

Andersen, J. (1995) A comparison of pitfall trapping and quadrat sampling of Carabidae (Coleoptera) on river banks. *Entomologica Fennica*, **6**, 65–77.

Anderson, M. (1994) *Sexual Selection*. Princeton University Press, Princeton.

Anderson, N.H. (1962) Studies on overwintering of *Anthocoris* (Hem., Anthocoridae). *Entomologist's Monthly Magazine*, **98**, 1–3.

Andow, D.A. (1988) Management of weeds for insect manipulation in agroecosystems, in *Weed*

Management in Agroecosystems: Ecological Approaches (eds M.A. Altieri and M. Liebman). CRC Press, Baton Rouge, pp. 265–302.

Andow, D.A. (1990) Characterization of predation on egg masses of *Ostrinia nubilalis* (Lepidoptera: Pyralidae). *Annals of the Entomological Society of America*, **83**, 482–6.

Andow, D.A. (1992) Fate of eggs of first-generation *Ostrinia nubilalis* (Lepidoptera: Pyralidae) in three conservation tillage systems. *Environmental Entomology*, **21**, 388–93.

Anholt, B.R. (1990) An experimental separation of interference and exploitative competition in a larval dragonfly. *Ecology*, **71**, 1483–93.

Anthony, C.D. (2003) Kinship influences cannibalism in the wolf spider, *Pardosa milvina*. *Journal of Insect Behavior*, **16**, 23–36.

Antolin, M.F. (1989) Genetic considerations in the study of attack behaviour of parasitoids, with reference to *Muscidifurax raptor* (Hymenoptera: Pteromalidae). *Florida Entomologist*, **72**, 15–32.

Antolin, M.F. (1992a) Sex-ratio variation in a parasitic wasp. Reaction norms. *Evolution*, **46**, 1496–510.

Antolin, M.F. (1992b) Sex ratio variation in a parasitic wasp. II. Diallele cross. *Evolution*, **46**, 1511–24.

Antolin, M.F. (1993) Genetics of biased sex ratios in subdivided populations: models, assumptions, and evidence. *Oxford Surveys in Evolutionary Biology*, **9**, 239–81.

Antolin, M.F. (1999) A genetic perspective on mating systems and sex ratios of parasitoid wasps. *Researches on Population Ecology*, **4**, 29–37.

Antolin, M.F. and Strand, M.R. (1992) The mating system of *Bracon hebetor* (Hymenoptera: Braconidae). *Ecological Entomology*, **17**, 1–7.

Antolin, M.F. and Strong, D.R. (1987) Long-distance dispersal by a parasitoid (*Anagrus delicatus*, Mymaridae) and its host. *Oecologia*, **73**, 288–92.

Antolin, M.F., Ode, P.J. and Strand, M.R. (1995) Variable sex ratios and ovicide in an outbreeding parasitic wasp. *Animal Behaviour*, **49**, 589–600.

Antolin, M.F., Bosio, C.F., Cotton, J., Sweeney, W., Strand, M.R. and Black, W.C. (1996) Intensive linkage mapping in a wasp (*Bracon hebetor*) and a mosquito (*Aedes aegypti*) with single-strand conformation polymorphism analysis of random amplified polymorphic DNA markers. *Genetics*, **143**, 1727–38.

Anunciada, L. and Voegele, J. (1982) The importance of nutrition in the biotic potential of *Trichogramma maidis* Pintureau and Voegele and *T. nagarkattii* Voegele et Pintureau and oösorption in the females. *Les Trichrogrammes, Les Colloques de L'INRA*, **9**, 79–84 (In French).

Apple, J.W. (1952) Corn borer development and control on canning corn in relation to temperature accumulation. *Journal of Economic Entomology*, **451**, 877–9.

Arditi, R. (1983) A unified model of the functional response of predators and parasitoids, *Journal of Animal Ecology*, 52, 293–303.

Arditi, R. and Akcakaya, H.R. (1990) Underestimation of mutual interference of predators. *Oecologia*, **83**, 358–61.

Arditi, R. and Berryman, A.A. (1991) The biological control paradox. *Trends in Ecology and Evolution*, **6**, 32.

Arditi, R. and Ginzburg, L.R. (1989) Coupling in predator-prey dynamics: ratio dependence. *Journal of Theoretical Biology*, **139**, 311–26.

Arneberg, P. and Andersen, J. (2003) The energetic equivalence rule rejected because of a potentially common sampling error: evidence from carabid beetles. *Oikos*, **101**, 367–75.

Arnold, A.J., Needham, P.H. and Stevenson, J.H. (1973) A self-powered portable insect suction sampler and its use to assess the effects of azinphos methyl and endosulfan in blossom beetle populations on oilseed rape. *Annals of Applied Biology*, **75**, 229–33.

Arnold, P.I., Serafy, J.E., Clarke, M.E. and Schultz, D.R. (1996) An immunological study of predation on hatchery-reared juvenile red drum (*Sciaenops ocellatus* Linnaeus): description of an ELISA and predator-prey studies in nature. *Journal of Experimental Marine Biology and Ecology*, **199**, 29–44.

Arnoldi, D., Stewart, R.K. and Boivin, G. (1991) Field survey and laboratory evaluation of the predator complex of *Lygus lineolaris* and *Lygocoris communis* (Hemiptera: Miridae) in apple orchards. *Journal of Economic Entomology*, **84**, 830–36.

Arnqvist, G. (1997) The evolution of water strider mating systems: causes and consequences of sexual conflicts, in *The Evolution of Mating Systems in Insects and Arachnids* (eds J.C. Choe and B.J. Crespi). Cambridge University Press, Cambridge, pp. 46–163.

Arnqvist, G. (1998) Comparative evidence for the evolution of genitalia by sexual selection. *Nature*, **393**, 784–86.

Arthur, A.P. and Wylie, H.G. (1959) Effects of host size on sex ratio, development time and size of *Pimpla turionellae* (L.) (Hymenoptera: Ichneumonidae). *Entomophaga*, **4**, 297–301.

Arthur, W. and Farrow, M. (1987) On detecting interactions between species in population dynamics. *Biological Journal of the Linnean Society*, **32**, 271–9.

Artiss, T. (2001) Structure and function of male genitalia in *Libellula*, *Ladona* and *Plathemis* (Anisoptera: Libellulidae). *Odonatologica*, **30**, 13–27.

Asahida, T., Yamashita, Y. and Kobayashi, T. (1997) Identification of consumed stone flounder, *Kareius bicoloratus* (Basilewsky), from the stomach contents of sand shrimp, *Crangon affinis* (De Haan) using mitochondrial DNA analysis. *Journal of Experimental Marine Biology and Ecology*, **217**, 153–63.

Ashby, J.W. (1974) A study of arthropod predation of *Pieris rapae* L. using serological and exclusion techniques. *Journal of Applied Ecology*, **11**, 419–25.

Ashby, J.W. and Pottinger, R.P. (1974) Natural regulation of *Pieris rapae* Linnaeus (Lepidoptera: Pieridae) in Canterbury, New Zealand. *New Zealand Journal of Agricultural Research*, **17**, 229–39.

Askari, A. and Alishah, A. (1979) Courtship behavior and evidence for a sex pheromone in *Diaeretiella rapae*, the cabbage aphid primary parasitoid. *Annals of the Entomological Society of America*, **72**, 749–50.

Askew, R.R. (1968) A survey of leaf-miners and their parasites on laburnum. *Transactions of the Royal Entomological Society of London*, **120**, 1–37.

Askew, R.R. (1971) *Parasitic Insects*. Heinemann, London.

Askew, R.R. (1984) The biology of gall wasps, in *Biology of Gall Insects* (ed. T.N. Ananthakrishnan). Edward Arnold, London, pp. 223–71.

Askew, R.R. (1985) A London fog. *Chalcid Forum*, **4**, 17.

Askew, R.R. and Shaw, M.R. (1986) Parasitoid communities: their size, structure and development, in *Insect Parasitoids* (eds J. Waage and D Greathead). Academic Press, London, pp. 225–64.

van den Assem, J. (1969) Reproductive behaviours of *Pseudeucoila bochei* (Hym., Cynipidae). *Netherlands Journal of Zoology*, **19**, 641–48.

van den Assem, J. (1970) Courtship and mating in *Lariophagus distinguendus* (Hym., Pteromalidae). *Netherlands Journal of Zoology*, **20**, 329–52.

van den Assem, J. (1971) Some experiments on sex ratio and sex regulation in the pteromalid *Lariophagus distinguendus*. *Netherlands Journal of Zoology*, **21**, 373–402.

van den Assem, J. (1974) Male courtship patterns and female receptivity signal of Pteromalinae, with a consideration of some evolutionary trends. *Netherlands Journal of Zoology*, **24**, 253–78.

van den Assem, J. (1975) The temporal pattern of courtship behaviour in some parasitic Hymenoptera, with special reference to *Melittobia acasta*. *Journal of Entomology*, **50**, 137–46.

van den Assem, J. (1976a) Queue here for mating: waarnemingen over het gedrag van ongepaarde *Melittobia* vrouwtjes ten opzichte van een mannelijke soortgenoot. *Entomologische Berichten*, **36**, 74–8.

van den Assem, J. (1976b) Male courtship behaviour, female receptivity signal, and the differences between the sexes in Pteromalinae, and comparative notes on other Chalcidoids. *Netherlands Journal of Zoology*, **26**, 535–48.

van den Assem, J. (1977) A note on the ability to fertilize following insemination (with females of *Nasonia vitripennis* van den Hym., Chalcidoidea). *Netherlands Journal of Zoology*, **27**, 230–5.

van den Assem, J. (1986) Mating behaviour in parasitic wasps, in *Insect Parasitoids* (eds J Waage and D Greathead). Academic Press, pp. 137–67.

van den Assem, J. (1996) Mating behaviour, in *Insect Natural Enemies: Practical Approaches to their Study and Evaluation* (eds M. Jervis and N. Kidd). Chapman and Hall, London, pp. 163–221.

van den Assem, J. and Feuth de Bruijn, E. (1977) Second matings and their effect on the sex ratio of the offspring in *Nasonia vitripennis* (Hym. Pteromalidae). *Entomologia Experimentalis et Applicata*, **21**, 23–8.

van den Assem, J. and Gijswijt, M.J. (1989) The taxonomic position of the *Pachyneurini* (Chalc. Pteromalidae) as judged by characteristics of courtship behaviour. *Tijdschrift voor Entomologie*, **132**, 149–54.

van den Assem, J. and Jachmann, F. (1982) The coevolution of receptivity signalling and body size in the Chalcidoidea. *Behaviour*, **80**, 96–105.

van den Assem, J. and Povel, G.D.E. (1973) Courtship behaviour of some *Muscidifurax* species (Hym., Pteromalidae): a possible example of a recently evolved ethological isolating mechanism. *Netherlands Journal of Zoology*, **23**, 465–87.

van den Assem, J. and Putters, F.A. (1980) Patterns of sound produced by courting chalcidoid males and its biological significance. *Entomologia Experimentalis et Applicata*, **27**, 293–302.

van den Assem, J. and Werren, J.H. (1994) A comparison of the courtship and mating behaviour of three species of *Nasonia* (Hym., Pteromalidae). *Journal of Insect Behaviour*, **7**, 53–66.

van den Assem, J., Gijswijt, M.J. and Nübel, B.K. (1980a) Observations on courtship and mating strategies in a few species of parasitic wasps (Chalcidoidea). *Netherlands Journal of Zoology*, **30**, 208–27.

van den Assem, J., Jachmann, F. and Simbolotti, P. (1980b) Courtship behaviour of *Nasonia vitripennis* (Hym. Pteromalidae): some qualitative evidence for the role of pheromones. *Behaviour*, **75**, 301–7.

van den Assem, J., in den Bosch, H.A.J. and Prooy, E. (1982a) *Melittobia* courtship behaviour, a comparative study of the evolution of a display. *Netherlands Journal of Zoology*, **32**, 427–71.

van den Assem, J., Gijswijt, M.J. and Nubel, B.K. (1982b) Characteristics of courtship and mating behaviour used as classificatory criteria in

Eulophidae-Tetrastichinae. *Tijdschrift voor Entomologie*, **125**, 205–20.

van den Assem, J., Putters, F.A. and Prins, T.C. (1984a) Host quality effects on sex ratio of the parasitic wasp *Anisopteromalus calandrae* (Chalcidoidea, Pteromalidae). *Netherlands Journal of Zoology*, **34**, 33–62.

van den Assem, J., Putters, F.A. and van der Voort Vinkestijn, M.J. (1984b) Effects of exposure to an extremely low temperature on recovery of courtship behaviour after waning in the parasitic wasp *Nasonia vitripennis*. *Journal of Comparative Physiology*, **155**, 233–7.

van den Assem, J., van Iersal, J.J.A. and los den Hartogh, R.L. (1989) Is being large more important for female than for male parasitic wasps? *Behaviour*, **108**, 160–95.

Atanassova, P., Brookes, C.P., Loxdale, H.D. and Powell, W. (1998) Electrophoretic study of five aphid parasitoid species of the genus *Aphidius* (Hymenoptera: Braconidae), including evidence for reproductively isolated sympatric populations and a cryptic species. *Bulletin of Entomological Research*, **88**, 3–13.

Atlan, A., Merçot, H., Landré, C. and Montchamp-Moreau, C. (1997) The sex-ratio trait in *Drosophila simulans*: geographical distribution of distortion and resistance. *Evolution*, **51**, 1886–95.

Ausden, M. (1996) Invertebrates, in *Ecological Census Techniques* (ed. W.J. Sutherland). Cambridge University Press. Cambridge, pp. 139–77.

Austin, A.D. (1983) Morphology and mechanics of the ovipositor system of *Ceratobaeus* Ashmead (Hymenoptera: Scelionidae) and related genera. *International Journal of Insect Morphology and Embryology*, **12**, 139–55.

Austin, A.D. and Browning, T.O. (1981) A mechanism for movement of eggs along insect ovipositors. *International Journal of Insect Morphology and Embryology*, **10**, 93–108.

Austin, A.D. and Field, S.A. (1997) The ovipositor system of scelionid and platygastrid wasps (Hymenoptera: Platygastroidea): comparative morphology and phylogenetic implications. *Invertebrate Taxonomy*, **11**, 1–87.

Aveling, C. (1981) The role of *Anthocoris* species (Hemiptera: Anthocoridae) in the integrated control of the damson-hop aphid. *Annals of Applied Biology*, **97**, 143–53.

Avilla, J. and Albajes, R. (1984) The influence of female age and host size on the sex ratio of the parasitoid *Opius concolor*. *Entomologia Experimentalis et Applicata*, **35**, 43–7.

Avilla, J. and Copland, M.J.W. (1987) Effects of host age on the development of the facultative autoparasitoid *Encarsia tricolor* (Hymenoptera: Aphelinidae). *Annals of Applied Biology*, **110**, 381–9.

Avise, J.C. (1994) *Molecular markers, Natural History and Evolution*. Chapman and Hall, New York.

Ayal, Y. (1987) The foraging strategy of *Diaeretiella rapae*. 1. The concept of the elementary unit of foraging. *Journal of Animal Ecology*, **56**, 1057–68.

Ayala, F.J. (1983) Enzymes as taxonomic characters, in *Protein Polymorphism: Adaptive Significance and Taxonomic Significance* (eds G.S. Oxford and E. Rollinson), Systematics Association Special Volume No. 24. Academic Press, London, pp. 3–26.

Ayre, G.L. and Trueman, D.K. (1974) A battery operated time-sort pitfall trap. *The Manitoba Entomologist*, **8**, 37–40.

van Baaren, J. and Boivin, G. (1998) Learning affects host discrimination behavior in a parasitoid wasp. *Behavioral Ecology and Sociobiology*, **42**, 9–16.

van Baaren, J. and Nenon, J.P. (1996) Host location and discrimination mediated through olfactory stimuli in two species of Encyrtidae. *Entomologia Experimentalis et Applicata*, **81**, 61–9.

van Baaren, J., Landry, B.L., and Boivin, G. (1999) Sex allocation and larval competition in a superparasitizing solitary egg parasitoid: competing strategies for an optimal sex ratio. *Functional Ecology*, **13**, 66–71.

Baars, M.A. (1979a) Catches in pitfall traps in relation to mean densities of carabid beetles. *Oecologia*, **41**, 25–46.

Baars, M.A. (1979b) Patterns of movement of radioactive carabid beetles. *Oecologia*, **44**, 125–40.

Babendreier, D. and Hoffmeister, T.S. (2002) Superparasitism in the solitary ectoparasitoid *Aptesis nigrocincta*: the influence of egg load and host encounter rate. *Entomologia Experimentalis et Applicata*, **105**, 63–9.

Babendreier, D., Kuske, S. and Bigler, F. (2003) Non-target host acceptance and parasitism by *Trichogramma brassicae* Bezdenko (Hymenoptera: Trichogrammatidae) in the laboratory. *Biological Control*, **26**, 128–38.

Bacher, S., Casas, J., Wackers, F. and Dorn, S. (1997) Substrate vibrations elicit defensive behaviour in leafminer pupae. *Journal of Insect Physiology*, **43**, 945–52.

Bacher, S., Schenk, D. and Imboden, H. (1999) A monoclonal antibody to the shield beetle *Cassida rubiginosa* (Coleoptera, Chrysomelidae): a tool for predator gut analysis. *Biological Control*, **16**, 299–309.

Baehaki, S.E. (1995) The use of egg masses for egg parasitoid monitoring of white rice stem borer (*Tryporyza* (*Scirpophaga*) *innotata*). *Indonesian Journal of Crop Science*, **10**, 1–10.

Baggen, L.R. and Gurr, G.M. (1998) The influence of food on *Copidosoma koehleri* (Hymenoptera: Encyrtidae), and the use of flowering plants as a habitat management tool to enhance biological control of potato moth, *Phthorimaea operculella* (Lepidoptera: Gelechiidae). *Biological Control*, **11**, 9–17.

Baggen, L.R., Gurr, G.M. and Meats, A. (1999) Flowers in tri-trophic systems: mechanisms allowing selective exploitation by insect natural enemies for conservation biological control. *Entomologia Experimentalis et Applicata*, **91**, 155–61.

Baggen, L.R., Gurr, G.M. and Meats, A. (2000) Field observations on selective food plants in habitat manipulation for biological control of potato moth by *Copidosoma koehleri* Blanchard (Hymenoptera: Encyrtidae), in *Hymenoptera: Evolution, Biodiversity and Biological Control* (eds A.D. Austin and M. Dowton). CSIRO, Melbourne, pp. 388–95.

Bai, B. and Mackauer, M. (1990) Oviposition and host-feeding patterns in *Aphelinus asychis* (Hymenoptera: Aphelinidae) at different aphid densities. *Ecological Entomology*, **15**, 9–16.

Bai, B. and Mackauer, M. (1992) Influence of superparasitism on development rate and adult size in a solitary parasitoid *Aphidius ervi*. *Functional Ecology*, **6**, 302–7.

Bai, B. and Smith, S.M. (1993) Effect of host availability on reproduction and survival of the parasitoid wasp *Trichogramma minutum*. *Ecological Entomology*, **18**, 279–86.

Bailey, N.T.J. (1951) On estimating the size of mobile populations from recapture data. *Biometrika*, **38**, 293–306.

Bailey, N.T.J. (1952) Improvements in the interpretation of recapture data. *Journal of Animal Ecology*, **21**, 120–27.

Bailey, P.C.E. (1986) The feeding behaviour of a sit-and-wait predator, *Ranatra dispar* (Heteroptera: Nepidae): optimal foraging and feeding dynamics. *Oecologia*, **68**, 291–7.

Bailey, S.E.R. (1989) Foraging behaviour of terrestrial gastropods: Integrating field and laboratory studies. *Journal of Molluscan Studies*, **55**, 263–72.

Baker, H.C. and Baker, I. (1983) A brief historical review of the chemistry of floral nectar, in *The Biology of Nectaries* (eds B. Bentley and T. Elias). Columbia University Press, New York, pp. 126–52.

Baker, J.E., Perez-Mendoza, J. and Beeman, R.W.E. (1998) Multiple mating potential in a pteromalid wasp determined by using an insecticide resistance marker. *Journal of Entomological Science*, **33**, 165–70.

Baker, R.L. (1981) Behavioural interactions and use of feeding areas by nymphs of *Coenagrion resolutum* (Coenagrionidae: Odonata). *Oecologia*, **49**, 353–8.

Baker, R.L. (1989) Condition and size of damselflies: a field study of food limitation. *Ecology*, **81**, 111–19.

Bakker, K., Eysackers, H.J.P., van Lenteren, J.C. and Meelis, E. (1972) Some models describing the distribution of eggs of the parasite *Pseudeucoila bochei* (Hym., Cynip.) over its hosts, larvae of *Drosophila melanogaster*. *Oecologia*, **10**, 29–57.

Bakker, K., van Alphen, J.J.M., van Batenberg, F.H.D., van der Hoeven, N., Nell, N.W., van Strien-van Liempt, W.T.F.H. and Turlings, T.C. (1985) The function of host discrimination and superparasitism in parasitoids. *Oecologia*, **67**, 572–6.

Bakker, K., Peulet, P. and Visser, M.E. (1990) The ability to distinguish between hosts containing different numbers of parasitoid eggs by the solitary parasitoid *Leptopilina heterotoma* (Thomson) (Hym. Cynip.). *Netherlands Journal of Zoology*, **40**, 514–20.

Balayeva, N.M., Eremeeva, M.E., Tissot-Dupont, H., Zakharov, I.A. and Raoult, D. (1995) Genotype characterisation of the bacterium expressing the male-killing trait in the ladybird beetle *Adalia bipunctata* with specific Rickettsial molecular tools. *Applied and Environmental Microbiology*, **61**, 1431–7.

Baldwin, W.F., Shaver, E. and Wilkes, A. (1964) Mutants of the parasitic wasp *Dahlbominus fuscipennis* (Zett.) (Hymenoptera: Eulophidae). *Canadian Journal of Genetics and Cytology*, **6**, 453–66.

Ballard, J.W.O. (2000) Comparative genomics of mitochondrial DNA in *Drosophila simulans*. *Journal of Molecular Evolution*, **51**, 64–75.

Bancroft, J.D. and Stevens, A. (1991) *Theory and Practice of Histological Techniques*. Churchill Livingstone, Edinburgh.

Bandara, H.M.J. and Walter, G.H. (1993) Virgin production of female offspring in a usually arrhenotokous wasp, *Spalangia endius* Walker (Hymenoptera: Pteromalidae). *Journal of the Australian Entomological Society*, **32**, 127–8.

Banks, M.J. and Thompson, D.J. (1987a) Lifetime reproductive success of females of the damselfly *Coenagrion puella*. *Journal of Animal Ecology*, **56**, 815–32.

Banks, M.J. and Thompson, D.J. (1987b) Regulation of damselfly populations: the effects of larval density on larval survival, development rate and size in the field. *Freshwater Biology*, **17**, 357–65.

Barbosa, P. (1974) *Manual of Basic Techniques in Insect Histology*. Autumn Publishers, Massachusetts.

Barbosa, P.J.A., Saunders, J.A., Kemper, R. Trumbule, J. Olechno, J. and Martinat, P. (1986) Plant allelochemicals and insect parasitoids: Effects of nicotine on *Cotesia congregata* and *Hyposoter annulipes*. *Journal of Chemical Ecology*, **12**, 1319–28.

Barendregt, A. (1975) Boemvoorkeur bij Zweefvliegen (Dipt., Syrphidae). *Entomologische Berichten*, **35**, 96–100.

Barfield, C.S., Bottrell, D.C. and Smith, J.W. Jr. (1977a) Influence of temperature on oviposition and adult longevity of *Bracon mellitor* reared on boll weevils. *Environmental, Entomology*, **6**, 133–7.

Barfield, C.S., Sharpe, P.J.H. and Bottrell, D.G. (1977b) A temperature driven development model for the parasite *Bracon mellitor* (Hymenoptera: Braconidae). *Canadian Entomologist*, **109**, 1503–14.

Bargen, H., Saudhof, K. and Poehling, H.M. (1998) Prey finding by larvae and adult females of *Episyrphus balteatus*. *Entomologia Expermentalis et Applicata*, **87**, 245–54.

Barker, G.M. and Addison, P.J. (1996) Influence of clavicipitaceous endophyte infection in ryegrass on development of the parasitoid *Microctonus hyperodae* Loan (Hymenoptera: Braconidae) in *Listronotus bonariensis* (Kuschel) (Coleoptera: Curculionidae). *Biological Control*, **7**, 281–7.

Barlow, C.A. (1961) On the biology and reproductive capacity of *Syrphus corollae* Fab. (Syrphidae) in the laboratory. *Entomologia Experimentalis et Applicata*, **4**, 91–100.

Barlow, N.D. (1999) Models in biological control, in *Theoretical Approaches to Biological Control* (eds B.A. Hawkins and H.V. Cornell). Cambridge University Press, Cambridge, pp. 43–68.

Barlow, N.D. and Dixon, A.F.G. (1980) *Simulation of Lime Aphid Population Dynamics*. Pudoc, Wageningen.

Barlow, N.D. and Goldson, S.L. (1993) A modelling analysis of the successful biological control of *Sitona discoideus* (Coleoptera: Curculionidae) by *Microctoniis aethiopoides* (Hymenoptera: Braconidae) in New Zealand. *Journal of Applied Ecology*, **30**, 165–8.

Barlow, N.D. and Wratten, S.D. (1996) Ecology of predator-prey and parasitoid-host systems: progress since Nicholson, in *Frontiers of Population Ecology* (eds R.B. Floyd, A.W. Sheppard, and P.J. De Barro). CSIRO, Melbourne, pp. 217–43.

Barndt, D. (1976) Untersuchungen der diurnalen und siasonalen Activitat von Kafern mit einer neu entwickelten Elektro-Bodenfalle. *Verhandlungen des Botanischen Vereins der Provinz Brandenberg*, **112**, 103–22.

Barrass, R. (1960a) The courtship behaviour of *Mormoniella vitripennis*. *Behaviour*, **15**, 185–209.

Barrass, R. (1960b) The effect of age on the performance of an innate behaviour pattern in *Mormoniella vitripennis*. *Behaviour*, **15**, 210–18.

Barrass, R. (1961) A quantitative study of the behaviour of the male *Mormoniella vitripennis* towards two constant stimulus situations. *Behaviour*, **18**, 288–312.

Barrass, R. (1976) Rearing jewel wasps *Mormoniella vitripennis* (Walker) and their use in teaching biology. *Journal of Biological Education*, **10**, 119–26.

Barratt, B.I.P., Ferguson, C.M., Goldson, S.L., Phillips, C.M. and Hannah, D.J. (2000) Predicting the risk from biological control agent introductions: a New Zealand approach, in *Nontarget Effects of Biological Control* (eds P.A. Follett and J.J. Duan). Kluwer Academic Publishers, Dordrecht, pp. 59–75.

Bartlett, B.R. (1962) The ingestion of dry sugars by adult entomophagous insects and the use of this feeding habit for measuring the moisture needs of parasites. *Journal of Economic Entomology*, **55**, 749–53.

Bartlett, B.R. (1964) Patterns in the host-feeding habit of adult Hymenoptera. *Annals of the Entomological Society of America*, **57**, 344–50.

Bartlett, B.R. (1968) Outbreaks of two-spotted spider mites and cotton aphids following pesticide treatment. 1. Pest stimulation vs. natural enemy destruction as the cause of outbreaks. *Journal of Economic Entomology*, **61**, 297–303.

Bartlett, M.S. (1949) Fitting a straight line when both variables are subject to error. *Biometrics, 5*, 207–12.

Basedow, T. (1973) Der Einfluss epigäischer Rauparthropoden auf die Abundanz phytophager Insekten in der Agrarlandschaft. *Pedobiologia*, **13**, 410–22.

Basedow, T. (1996) Phenology and population density of predatory bugs (*Nabis* spp.; Heteroptera: Nabidae) in different fields of winter wheat in Germany, 1993/94. *Bulletin SROP/WPRS*, **19(3)**, 70–6.

Basedow, T. and Rzehak, H. (1988) Abundanz und Aktivitätsdichte epigäischer Rauparthropoden auf Ackerflächen – ein Vergleich. *Zoologischer Jahrbucher, Abteilung für Systematik Okologie und Geographie der Tiere*, **115**, 495–508.

Basedow, T., Klinger, K., Froese, A. and Yanes, G. (1988) Aufschwemmung mit Wasser zur Scnellbestimmung der Abundanz epigäischer Raubarthropoden auf Äckern. *Pedobiologia*, **32**, 317–22.

Basset, Y. (1988) A composite interception trap for sampling arthropods in tree canopies. *Journal of the Australian Entomological Society*, **27**, 213–19.

Basset, Y. (1990) The arboreal fauna of the rainforest tree *Agyrodendron actinophyllum* as sampled with restricted canopy fogging: composition of the fauna. *Entomologist*, **109**, 173–83.

Basset, Y., Springate, N.D., Aberlenc, H.P. and Delvare, G. (1997) A review of methods for sampling arthropods in tree canopies, in *Canopy Arthropods* (eds N.E. Stork, J. Adis and R.K. Didham). Chapman and Hall, London, pp. 27–52.

Basten, C.J., Weir, B.S. and Zeng, Z.-B. (1997) *QTL cartographer: a reference manual and tutorial for QTL mapping*. Department of Statistics, North Carolina State University, Raleigh.

Bateman, A.J. (1948) Intra-sexual selection in *Drosophila*. *Heredity*, **2**, 349–68.

Battaglia, D., Poppy, G., Powell, W., Romano, A., Tranfaglia, A. and Pennacchio, F. (2000) Physical and chemical cues influencing the oviposition behaviour of *Aphidius ervi*. *Entomologia Experimentalis et Applicata*, **94**, 219–27.

Bay, E.C. (1974) Predator-prey relationships among aquatic insects. *Annual Review of Entomology*, **19**, 441–53.

Baylis, J.R. (1978) Paternal behaviour in fishes: a question of investment, timing or rate? *Nature* **276**, 738.

Baylis, J.R. (1981) The evolution of parental care in fishes, with reference to Darwin's rule of male sexual selection. *Environmental Biology of Fishes*, **6**, 223–51.

Bayon, F., Fougeroux, A., Reboulet, J.N. and Ayrault, J.P. (1983) Utilisation et intérêt de l'aspirateur "D-vac" pour la détection et le suivi des populations de ravageurs et d'auxiliaires sur blé au printemps. *La Défense des Végétaux*, **223**, 276–97.

Bayram, A. (1996) A study on the diel activity of *Pardosa* spiders (Araneae, Lycosidae) sampled by the time-sorting pitfall trap in different habitats. *Turkish Journal of Zoology*, **20**, 381–87.

Bazzocchi, G.G., Lanzoni, A., Burgio, G. and Fiacconi, M.R. (2003) Effects of temperature and host on the pre-imaginal development of the parasitoid *Diglyphus isaea* (Hymenoptera: Eulophidae). *Biological Control*, **26**, 74–82.

Beach, R.M. and Todd, J.W. (1986) Foliage consumption and larval development of parasitized and unparasitized soybean looper, *Pseudoplusia includens* (Lep.: Noctuidae), reared on a resistant soybean genotype and effects on an associated parasitoid, *Copidosoma truncatellum* (Hym: Encyrtidae). *Entomophaga*, **31**, 237–42.

Bean, D. and Cook, J.M. (2001) Male mating tactics and lethal combat in the nonpollinating fig wasp *Sycoscapter australis*. *Animal Behaviour*, **62**, 535–42.

Beckage, N.E. (1998a) Modulation of immune responses to parasitoids by polydnaviruses. *Parasitology*, **116**, S57–S64.

Beckage, N.E. (1998b) Parasitoids and polydnaviruses – an unusual mode of symbiosis in which a DNA virus causes host insect immunosuppression and allows the parasitoid to develop. *Bioscience*, **48**, 305–11.

Beckage, N.E. and Riddiford, L.M. (1978) Developmental interactions between the tobacco hornworm *Manduca sexta* and its braconid parasite *Apanteles congregatus*. *Entomologia Experimentalis et Applicata*, **23**, 139–51.

Beckage, N.E. and Riddiford, L.M. (1983) Growth and development of the endoparasitic wasp *Apanteles congregatus*: dependence on host nutritional status and parasite load. *Physiological Entomology*, **8**, 231–41.

Beckmann Jr, R.L. and Stucky, J.M. (1981) Extrafloral nectaries and plant guarding in *Ipomoea pandurata* (L.) G.F.W. Mey (Convolvulaceae). *American Journal of Botany*, **68**, 72–9.

Becnel, J.J. and Andreadis, T.G. (1999) Microsporidia in insects, in *The Microsporidia and Microsporidosis* (eds M. Wittner and L.M. Weiss). ASM Press, Washington, pp. 447–501.

Bedding, R.A. (1973) The immature stages of Rhinophorinae (Diptera: Calliphoridae) that parasitise British woodlice. *Transactions of the Royal Entomological Society of London*, **125**, 27–44

Beddington, J.R. (1975) Mutual interference between parasites or predators and its effect on searching efficiency. *Journal of Animal Ecology*, **441**, 331–40.

Beddington, J.R. and Hammond, P.S. (1977) On the dynamics of host-parasite-hyperparasite interactions. *Journal of Animal Ecology*, **46**, 811–21.

Beddington, J.R., Free, C.A. and Lawton, J.H. (1975) Dynamic complexity in predator-prey models framed in difference equations. *Nature*, **251**, 58–60.

Beddington, J.R., Free, C.A. and Lawton, J.H. (1976a) Concepts of stability and resilience in predator-prey models. *Journal of Animal Ecology*, **45**, 791–816.

Beddington, J.R., Hassell, M.P. and Lawton, J.H. (1976b) The components of arthropod predation. II. The predator rate of increase. *Journal of Animal Ecology*, **45**, 165–85.

Beddington, J.R., Free, C.A. and Lawton, J.H. (1978) Modelling biological control: on the characteristics of successful natural enemies. *Nature*, **273**, 513–19.

Beerwinkle, K.R., Coppedge, J.R. and Hoffman, C. (1999) A new mechanical method for sampling selected beneficial and pest insects on corn – the corn KISS. *Southwestern Entomologist*, **24**, 107–13.

Begon, M. (1979) *Investigating Animal Abundance – Capture-recapture for Biologists*. Edward Amold, London.

Begon, M. (1983) Abuses of mathematical techniques in ecology: applications of Jolly's capture-recapture method. *Oikos*, **40**, 155–8.

Begon, M. and Parker, G.A. (1986) Should egg size and clutch size decrease with age? *Oikos*, **47**, 293–302.

Begon, M., Harper, J.L. and Townsend, C.R. (1996) *Ecology: Individuals, Populations and Communities, 3rd edn*. Blackwell, Oxford.

Begon, M., Sait, S.M. and Thompson, D.J. (1999) Host-pathogen-parasitoid systems, in *Theoretical*

Approaches to Biological Control (eds B.A. Hawkins and H.V. Cornell). Cambridge University Press, Cambridge, pp. 327–48.

Beirne, B.P. (1980) Biological control: benefits and opportunities, in *Perspectives in World Agriculture*. Commonwealth Agricultural Bureaux, Slough, pp. 307–21.

Beldade, P., Brakefield, P. M. and Long, A. D. 2002. Contribution of Distal-less to quantitative variation in butterfly eyespots. *Nature*, **415**, 315–8.

Bell, C.H. (1994) A review of diapause in stored-product insects. *Journal of Stored Products Research*, **30**, 99–120.

Bell, G. and Koufopanou, V. (1986) The cost of reproduction. *Oxford Surveys in Evolutionary Biology*, **3**, 83–131.

Bell, H.A., Marris, G.C., Smethurst, F. and Edwards, J.P. (2003) The effect of host stage and temperature on selected developmental parameters of the solitary endoparasitoid *Meteorus gyrator* (Thun.) (Hym., Braconidae). *Journal of Applied Entomology*, **127**, 332–9.

Bellamy, D.E. and Byrne, D.N. (2001) Effects of gender and mating status on self-directed dispersal by the whitefly parasitoid *Eretmocerus eremicus*. *Ecological Entomology*, **26**, 571–7.

Bellows Jr, T.S. (1985) Effects of host and parasitoid age on search behaviour and oviposition rates in *Lariophagus distinguendus* Forster (Hymenoptera: Pteromalidae). *Researches on Population Ecology*, **27**, 65–76.

Bellows Jr, T.S. and Hassell, M.P. (1988) The dynamics of age-structured host-parasitoid interactions. *Journal of Animal Ecology*, **57**, 259–68.

Bellows Jr, T.S., Van Driesche, R.G. and Elkinton, J. (1989a) Life tables and parasitism: estimating parameters in joint host-parasitoid systems, in *Estimation and Analysis of Insect Populations* (eds L. McDonald, B. Manly, J. Lockwood and J. Logan). Springer-Verlag, Berlin, pp. 70–80.

Bellows Jr, T.S., Van Driesche, R.C. and Elkinton, J.S. (1989b) Extensions to Southwood and Jepson's graphical method of estimating numbers entering a stage for calculating mortality due to parasitism. *Researches on Population Ecology*, **31**, 169–84.

Bellows Jr, T.S., Van Driesche, R.C. and Elkinton, J.S. (1992) Life table construction and analysis in the evaluation of natural enemies. *Annual Review of Entomology*, **37**, 587–614.

Belokobylskij, S.A. and Jervis, M.A. (1998) Descriptions of two new species of *Agathis* (Hymenoptera, Braconidae, Agathidinae) from Spain, with a record of mating by one species on flowers. *Journal of Natural History*, **32**, 1217–25.

Belshaw, R. (1993) Malaise traps and Tachinidae (Diptera): a study of sampling efficiency. *The Entomologist*, **112**, 49–54.

Belshaw, R., Lopez-Vaamonde, C., Degerli, N. and Quicke, D.L.J. (2001) Paraphyletic taxa and taxonomic chaining: evaluating the classification of braconine wasps (Hymenoptera: Braconidae) using 28S D2–3 rDNA sequences and morphological characters. *Biological Journal of the Linnean Society*, **73**, 411–24.

Ben-Dov, Y. (1972) Life history of *Tetrastichus ceroplastae* (Girault) (Hymenoptera: Eulophidae), a parasite of the Florida wax scale, *Ceroplastes floridensis* Comstock (Homoptera: Coccidae), in Israel. *Journal of the Entomological Society of South Africa*, **35**, 17–34.

Benedet, F., Bigot, Y., Renault, S., Pouzat, J. and Thibout, E. (1999) Polypeptides of *Acrolepiopsis assectella* cocoon (Lepidoptera: Yponomeutoidea): an external host-acceptance kairomone for the parasitoid *Diadromus pulchellus* (Hymenoptera: Ichneumonidae). *Journal of Insect Physiology*, **45**, 375–84.

Benest, G. (1989a) The sampling of a carabid community. 1. The behaviour of a carabid when facing the trap. *Revue d'écologie et de Biologie du Sol*, **26**, 205–11.

Benest, G. (1989b) The sampling of a carabid community. II. Traps and trapping. *Revue d'écologie et de Biologie du Sol*, **26**, 505–14.

Bennet-Clarke, H.C. (1970) The mechanism and efficiency of sound production in mole crickets. *Journal of Experimental Biology*, **52**, 619–52.

Benson, J., Pasquale, A., Van Driesche, R. and Elkinton, J. (2003) Assessment of risk posed by introduced braconid wasps to *Pieris virginiensis*, a native woodland butterfly in New England. *Biological Control*, **26**, 83–93.

Benson, J.F. (1973) Intraspecific competition in the population dynamics of *Bracon hebetor* Say (Hymenoptera: Braconidae). *Journal of Animal Ecology*, **42**, 105–24.

Benson, M. (1989) The Biology and Specificity of the Host-parasitoid Relationship, With Reference to Aphidophagous Syrphid Larvae and their Associated Parasitoids. University of Nottingham, M.Phil. Thesis.

Benton, F. (1975) Larval Taxonomy and Bionomics of some British Pipunculidae. Imperial College, University of London, Ph.D. Thesis.

Berberet, R.C. (1982) Effects of host age on embryogenesis and encapsulation of the parasite *Bathyplectes curculionis* in the alfalfa weevil. *Journal of Invertebrate Pathology*, **40**, 359–66.

Berberet, R.C. (1986) Relationship of temperature to embryogenesis and encapsulation of eggs of *Bathyplectes curculionis* (Hymenoptera: Ichneumonidae) in larvae of *Hypera postica* (Coleoptera: Curculionidae).

Annals of the Entomological Society of America, **79**, 985–8.

Berberet, R.C., Bisges, A.D. and Zarrabi, A.A. (2002) Role of cold tolerance in the seasonal life history of *Bathyplectes curculionis* (Hymenoptera: Ichneumonidae) in the Southern Great Plains. *Environmental Entomology*, **31**, 739–45.

Bergelson, J. (1985) A mechanistic interpretation of prey selection by *Anax junius* larvae *(Odonata:* Aeschnidae). *Ecology*, **66**, 1699–705.

Berger, S.L. and Kimmel, A.R. (1987) *Methods in Enzymology. Volume 152. Guide to Molecular Cloning Techniques.* Academic Press, London.

Berjeijk, van K.E., Bigler, F., Kaashoek, N.K. and Pak, G.A. (1989) Changes in host acceptance and host suitability as an effect of rearing *Trichogramma maidis* on a factitious host. *Entomologia Experimentalis et Applicata*, **52**, 229–38.

Berlocher, S.H. (1980) An electrophoretic key for distinguishing species of the genus *Rhagoletis* (Diptera: Tephritidae) as larvae, pupae or adults. *Annals of the Entomological Society of America*, **73**, 131–7.

Berlocher, S. H. (1984) Insect molecular systematics. *Annual Review of Entomology*, **29**, 403–33.

Bernal, J.S., Luck, R.F. and Morse, J.G. (1998) Sex ratios in field populations of two parasitoids (Hymenoptera: Chalcidoidea) of *Coccus hesperidum* L. (Homoptera: Coccidae). *Oecologia*, **116**, 510–8.

Bernal, J.S., Luck, R.F. and Morse, J.G. (1999) Host influences of sex ratio, longevity, and egg load of two *Metaphycus* species parasitic on soft scales: implications for insectary rearing. *Entomologia Experimentalis et Applicata*, **92**, 191–204.

Berndt, L.A., Wratten, S.D. and Hassan, P.G. (2002) Effects of buckwheat flower on leafroller (Lepidoptera: Tortricidae) parasitoids in a New Zealand vineyard. *Agricultural and Forest Entomology*, **4**, 39–45.

Berrigan, D. (1991) The allometry of egg size and number in insects. *Oikos*, **60**, 313–21.

Berry, I.L., Foerster, K.W. and Ilken, E.H. (1976) Prediction model for development time of stable *flies. Transactions of the American Society of Agricultural Engineers*, **19**, 123–7.

Berryman, A.A. (1990) Modelling Douglas-fir tussock moth population dynamics: the case for simple theoretical models, in *Population Dynamics of Forest Insects* (eds A.D. Watt, S.R. Leather, M.D. Hunter and N.A.C. Kidd). Intercept, Andover, pp. 369–80.

Berryman, A.A. (1991) Chaos in ecology and resource management: What causes it and how to avoid it. in, *Chaos and Insect Ecology* (eds J.A. Logan and F.P. Hain) Virginia Experimental Station Information Series 91–3. Virginia Polytechnic Institute and State University, Blacksburg.

Berryman, A.A. (1992) The origins and evolution of predator-prey theory. *Ecology*, **73**, 1530–35.

Berryman, A.A. (1999) The theoretical foundations of biological control, in *Theoretical Approaches to Biological Control* (eds B.A. Hawkins and H.V. Cornell). Cambridge University Press, Cambridge, pp. 3–21.

Berryman, A.A. and Turchin, P. (2001) Identifying the density-dependent structure underlying ecological time series. *Oikos*, **92**, 265–70.

Bess, H.A. (1936) The biology of *Leschenaultia exul* Townsend, a tachinid parasite of *Malacosoma distria* Hubner. *Annals of the Entomological Society of America*, **29**, 593–613.

Bess, H.A., van den Bosch, R. and Haramoto, F.H. (1961) Fruit-fly parasites and their activities in *Hawaii. Proceedings of the Hawaii Entomological Society*, **17**, 367–78.

Best, R.L., Beegle, C.C., Owens, J.C. and Oritz, M. (1981) Population density, dispersion, and dispersal estimates for *Scarites substriatus, Pterostichus chalcites*, and *Harpalus pennsylvanicus* (Carabidae) in an Iowa cornfield. *Environmental Entomology*, **10**, 847–56.

Besuchet, C., Burckhardt, D.H. and Löbl, I. (1987) The "Winkler/Moczarski" eclector as an efficient extractor for fungus and litter Coleoptera. *The Coleopterist's Bulletin*, **41**, 392–4.

Bethke, J.A. and Redak, R.A. (1996) Temperature and moisture effects on the success of egg hatch in *Trirhabda geminata* (Coleoptera: Chrysomelidae). *Annals of the Entomological Society of America*, **89**, 661–6.

Beukeboom, L.W. (1995) Sex determination in Hymenoptera: a need for genetic and molecular studies. *Bioessays*, **17**, 813–17.

Beukeboom, L.W. (2001) Single-locus complementary sex determination in the ichneumonid *Venturia canescens* (Gravenhorst) (Hymenoptera). *Netherlands Journal of Zoology*, **51**, 1–15.

Beukeboom, L.W. and Pijnacker, L.P. (2000) Automictic parthenogenesis in the parasitoid wasp *Venturia canescens* revisited. *Genome*, **43**, 939–44.

Beukeboom, L.W. and Werren, J.H. (1993) Transmission and expression of the parasitic paternal sex ratio (PSR) chromosome, *Heredity*, **70**, 437–43.

Beukeboom, L.W., Driessen, G., Luckerhoff, L. Bernstein, C., Lapchin, L. and van Alphen, J.J.M. (1999) Distribution and relatedness of sexual and asexual *Venturia canescens* (Hymenoptera). *Proceedings Experimental and Applied Entomology of the Netherlands Entomological Society*, **10**, 23–28.

Beukeboom, L.W., Ellers, J. and van Alphen, J.J.M. (2000) Absence of single locus complementary sex determination in the braconid wasps *Asobara tabida* and *Alysia manducator*. *Heredity*, **84**, 29–36.

Biever, K.D. and Boldt, P.E. (1970) Utilization of soft X-rays for determining pupal parasitism of *Pieris rapae*. *Annals of the Entomological Society of America*, **63**, 1482–3.

Bigler, F., Meyer, A. and Bosshart, S. (1987) Quality assessment in *Trichogramma maidis* Pinteureau et Voegele reared from eggs of the factitious hosts *Ephestia kuehniella* Zell. and *Sitotroga cerealella* (Olivier). *Journal of Applied Entomology*, **104**, 340–53.

Bigler, F., Suverkropp, B.P. and Cerutti, F. (1997) Host searching by *Trichogramma* and its implications for quality control and release techniques, in *Ecological Interactions and Biological Control* (eds D.A. Andow, D.A. Ragsdale and R.F. Nyvall). WestView Press, Oxford, pp. 240–53.

Birkhead, T.R. (1998) Cryptic female choice: criteria for establishing female sperm choice. *Evolution*, **52**, 1212–18.

Birkhead, T.R., Lee, K.E. and Young, P. (1988) Sexual cannibalism in the praying mantis *Hierodula membranacea*. *Behaviour*, **106**, 112–18.

Bishop, D.B. and Bristow, C.M. (2001) Effect of Allegheny mound ant (Hymenoptera: Formicidae) presence on homopteran and predator populations in Michigan jack pine forests. *Annals of the Entomological Society of America*, **94**, 33–40.

Bishop, D.T., Cannings, C., Skolnick, M. and Williamson, J.A. (1983) The number of polymorphic DNA clones required to map the human genome (ed. B.S. Weir), Marcel Dekker, New York, pp. 181–200.

Björklund, M. (1997) Are 'comparative methods' always necessary? *Oikos*, **80**, 607–12.

Black, W.C., McLain, D.K. and Rai, K.S. (1989) Patterns of variation in the rDNA cistron within and among world populations of a mosquito *Aedes albopictus* (Skuse). *Genetics*, **121**, 539–50.

Black, W.C., Duteau, N.M., Puterka, G.J., Nechols, J.R. and Pettorini, J.M. (1992) Use of random amplified polymorphic DNA polymersae chain reaction (RAPD-PCR) to detect DNA polymorphisms in aphids. *Bulletin of Entomological Research*, **82**, 151–9.

Blackburn, T.M. (1991a) A comparative examination of lifespan and fecundity in parasitoid Hymenoptera. *Journal of Animal Ecology*, **60**, 151–64.

Blackburn, T.M. (1991b) Evidence for a 'fast-slow' continuum of life-history traits among parasitoid Hymenoptera. *Functional Ecology*, **5**, 65–74.

Blackman, R.L. (1967) The effect of different aphid foods on *Adalia bipunctata* L. and *Coccinella 7-punctata* L. *Annals of Applied Biology*, **59**, 207–19.

Blackwell, A., Luntz, A.J.M. and Mordue, W. (1994) Identification of bloodmeals of the Scottish biting midge, *Culicoides impunctatus*, by indirect enzyme-linked immunosorbent assay (ELISA). *Medical and Veterinary Entomology*, **8**, 20–4.

Blandenier, G. and Fürst, P.A. (1998) Ballooning spiders caught by a suction trap in an agricultural landscape in Switzerland. *Proceedings of the 17th European Colloquium of Arachnology, British Arachnological Society, Buckinghamshire, UK*, 177–86.

Blann, A.D. (1984) Cell fusion and monoclonal antibodies. *The Biologist*, **31**, 288–91.

Blower, J.G., Cook, L.M. and Bishop, J.A. (1981) *Estimating the Size of Animal Populations*. George Allen and Unwin, London.

Blumberg, D. (1991) Seasonal variations in the encapsulation of eggs of the encyrtid parasitoid *Metaphycus stanleyi* by the pyriform scale, *Protopulvinaria pyriformis*. *Entomologia Experimentalis et Applicata*, **58**, 231–7.

Blumberg, D. and DeBach P. (1979) Development of *Habrolepis rouxi* Compere (Hymenoptera: Encyrtidae) in two armoured scale hosts (Homoptera: Diaspididae) and parasite egg encapsulation by California red scale. *Ecological Entomology*, **4**, 299–306.

Blumberg, D. and Luck, R.F. (1990) Differences in the rates of superparasitism between two strains of *Comperiella bifasciata* (Howard) (Hymenoptera: Encyrtidae) parasitizing California red scale (Homoptera: Diaspididae): an adaptation to circumvent encapsulation? *Annals of the Entomological Society of America*, **83**, 591–7.

Blumberg, D. and Van Driesche, R.G. (2002) Encapsulation rates of three encyrtid parasitoids by three mealybug species (Homoptera: Pseudococcidae) found commonly as pests in commercial greenhouses. *Biological Control*, **22**, 191–9.

Blumberg, D., Klein, M. and Mendel, Z. (1995) Response by encapsulation of four mealybug species (Homoptera: Pseudococcidae) to parasitization by *Anagyrus pseudococci*. *Phytoparasitica*, **23**, 157–63.

Boavida, C., Neuenschwander, P. and Herren, H.R. (1995) Experimental asssessment of the impact of the introduced parasitoid *Gyranusoidea tebygi* Noyes on the Mango Mealybug *Rastrococcus invadens* Williams, by physical exclusion. *Biological Control*, **5**, 99–103.

den Boer, P.J. (1979) The individual behaviour and population dynamics of some carabid beetles of forests, in *On the Evolution of Behaviour in Carabid Beetles* (eds P.J. den Boer, H.U. Thiele and F. Weber), Miscellaneous Papers 18, University of Wageningen. H. Veenman and Zanen, Wageningen, pp. 151–66.

Boggs, C.L. (1981) Nutritional and life-history determinants of resource allocation in holometabolous insects. *American Naturalist*, **117**, 692–709.

Boggs, C.L. (1987) Ecology of nectar and pollen feeding in Lepidoptera, in *Nutritional Ecology of Insects, Mites, Spiders and Related Invertebrates* (eds F. Slansky Jr and J.G. Rodriguez). Wiley-Interscience, New York.

Boggs, C.L. (1992) Resource allocation: exploring connections between foraging and life history. *Functional Ecology*, **6**, 508–18.

Boggs, C.L. (1997a) Reproductive allocation from reserves and income in butterfly species with differing diets. *Ecology*, **78**, 181–91.

Boggs, C.L. (1997b) Dynamics of reproductive allocation from juvenile and adult feeding: radiotracer studies. *Ecology*, **78**, 192–202.

Bogya, S., Marko, V. and Szinetar, C. (2000) Effect of pest management systems on foliage- and grass-dwelling spider communities in an apple orchard in Hungary. *International Journal of Pest Management*, **46**, 241–50.

Bohan, D.A., Bohan, A.C., Glen, D.M., Symondson, W.O.C., Wiltshire, C.W. and Hughes, L. (2000). Spatial dynamics of predation by carabid beetles on slugs. *Journal of Animal Ecology*, **69**, 367–79.

Boiteau, G. and Colpitts, B. (2001) Electronic tags for tracking of insects in flight: effect of weight on flight performance of adult Colorado potato beetles. *Entomologia Experimentalis et Applicata*, **100**, 187–93.

Boiteau, G., Bousquet, Y. and Osborn, W.P.L. (1999) Vertical and temporal distribution of Coccinellidae (Coleoptera) in flight over an agricultural landscape. *Canadian Entomologist*, **131**, 269–77.

Boiteau, G., Bousquet, Y. and Osborn, W.P.L. (2000a) Vertical and temporal distribution of Carabidae and Elateridae in flight above an agricultural landscape. *Environmental Entomology*, **29**, 1157–63.

Boiteau, G., Osborn, W.P.L., Xiong, X. and Bousquet, Y. (2000b) The stability of vertical distribution profiles of insects in air layers near the ground. *Canadian Journal of Zoology*, **78**, 2167–173.

Bokonon-Ganta, A.H., van Alphen, J.J.M. and Neuenschwander, P. (1996) Competition between *Gyranusoidea tebygi* and *Anagyrus mangicola*, parasitoids of the mango mealybug, *Rastrococcus invadens*: interspecific host discrimination and larval competition. *Entomologia Experimentalis et Applicata*, **79**, 179–85.

Bolten, A.B., Feisinger, P., Baker, H.C. and Baker, I. (1979) On the calculation of sugar concentration in flower nectar. *Oecologia*, 41, 301–4.

Bombosch, S. (1962) Untersuchungen über die Auswertbarkeit von Fallenfangen. *Zeitschrift für Angewandte Zoologie*, **49**, 149–60.

Bommarco, R. (1998) Stage sensitivity to food limitation for a generalist arthropod predator, *Pterostichus cupreus* (Coleoptera: Carabidae). *Environmental Entomology*, **27**, 863–9.

Bonkowska, T. and Ryszkowski, L. (1975) Methods of density estimation of carabids (Carabidae, Coleoptera) in field under cultivation. *Polish Ecological Studies*, **1**, 155–71.

Bonsall, M.B. and Hassell, M.P. (1999) Parasitoid-mediated effects: apparent competition and the persistence of host-parasitoid assemblages. *Researches on Population Ecology*, **41**, 59–68.

Bonsall, M.B., French, D.R. and Hassell, M.P. (2002a) Metapopulation structures affect persistence of predator-prey interactions. *Journal of Animal Ecology*, **71**, 1075–84.

Bonsall, M.B., Hassell, M.P. and Aesfa, G. (2002b) Ecological trade-offs, resource partitioning, and coexistence in a host-parasitoid assemblage. *Ecology*, **83**, 925–34.

Boomsma, J.J. and Sundström, L. (1998) Patterns of paternity skew in *Formica* ants. *Behavioral Ecology and Sociobiology*, **42**, 85–92.

Boorman, E. and Parker, G.A. (1976) Sperm (ejaculate) competition in *Drosophila melanogaster*, and the reproductive value of females to males in relation to mating status. *Ecological Entomology*, **1**, 145–55.

Bordenstein, S.R. and Werren, J.H. (1998) Effects of A and B *Wolbachia* and host genotype on interspecies cytoplasmic incompatibility in *Nasonia*. *Genetics*, **148**, 1833–44.

Bordenstein, S.R. and Werren, J.H. (2000) Do *Wolbachia* influence fecundity in *Nasonia vitripennis*? *Heredity*, **84**, 54–62.

Bordenstein, S.R., O'Hara, F.P. and Werren, J.H. (2001) *Wolbachia*-induced incompatibility precedes other hybrid incompatibilities in *Nasonia*. *Nature*, **409**, 707–10.

Boreham, P.F.L. (1979) Recent developments in serological methods for predator-prey studies. *Entomological Society of America, Miscellaneous Publication*, **11**, 17–23.

Boreham, P.F.L. and Ohiagu, C.E. (1978) The use of serology in evaluating invertebrate prey-predator relationships: a review. *Bulletin of Entomological Research*, **68**, 171–94.

Bosch, J. (1990) Die Arthropodenproduktion des Ackerbodens. *Mitteilungen der Biologische Bundesanstalt für Land- und Forstwirtschaft*, **266**, 74.

van den Bosch, R., Leigh, T.F., Gonzalez, D. and Stinner, R.E. (1969) Cage studies on predators of the bollworm in cotton. *Journal of Economic Entomology*, **62**, 1486–9.

van den Bosch, R., Horn, R, Matteson, P., Frazer, B.D., Messenger, P.S. and Davis, C.S. (1979) Biological control of the walnut aphid in California: impact of the parasite *Trioxys pallidus*. *Hilgardia*, **47**, 1–13.

Bossart, J.L. and Prowell, P.D. (1998) Genetic estimates of population structure and gene flow: limitations, lessons and new directions. *Trends in Ecology and Evolution*, **13**, 202–6.

Bostanian, G., Boivin, G. and Goulet, H. (1983) Ramp pitfall trap. *Journal of Economic Entomology*, **76**, 1473–5.

Bouletreau, M. (1986) The genetic and coevolutionary interactions between parasitoids and their hosts, in *Insect Parasitoids* (eds J. Waage and D. Greathead). Academic Press, London, pp. 169–200.

Bourtzis, K. and Miller, T.A. (2003) *Insect Symbiosis*. CRC Press, Boca Raton.

Bourtzis, K., Nirgianaji, A., Onyango, P. and Savakis, C. (1994) A prokaryotic dna-A sequence in *Drosophila melanogaster*: *Wolbachia* infection and cytoplasmic incompatibility among laboratory strains. *Insect Molecular Biology*, **3**, 132–42.

Bourtzis, K., Nirgianaki, A., Markakis, G. and Savakis, C. (1996) *Wolbachia* infection and cytoplasmic incompatibility in *Drosophila* species. *Genetics*, **144**, 1063–73.

Bourtzis, K., Dobson, S.L., Braig, H.L. and O'Neill, S.L. (1998) Rescuing *Wolbachia* have been overlooked. *Nature*, **391**, 852–3.

Bowden, J. (1981) The relationship between light- and suction-trap catches of *Chrysoperla carnea*. (Stephens) (Neuroptera: Chrysopidae) and the adjustment of light-trap catches to allow for variation in moonlight. *Bulletin of Entomological Research*, **71**, 621–9.

Bowden, J. (1982) An analysis of factors affecting catches of insects in light traps. *Bulletin of Entomological Research*, **72**, 535–56.

Bowden, J., Brown, G. and Stride, T. (1979) The application of X-ray spectrometry to analysis of elemental composition (chemoprinting) in the study of migration of *Noctua pronuba* L. *Ecological Entomology*, **4**, 199–204.

Bowers, M.A. (1985) Experimental analyses of competition between two species of bumble bees (Hymenoptera: Apidae). *Oecologia*, **67**, 224–30.

Bowie, M.H., Wratten, S.D. and White, A.J. (1995) Agronomy and phenology of 'companion plants' of potential for enhancement of insect biological control. *New Zealand Journal of Crop and Horticultural Science*, **23**, 423–7.

Bowie, M.H., Gurr, G.M., Hossain, Z., Baggen, L.R. and Frampton, C.M. (1999) Effects of distance from field edge on aphidophagous insects in a wheat crop and observations on trap design and placement. *International Journal of Pest Management*, **45**, 69–73.

Bradburne, R.P. and Mithen, R. (2000) Glucosinolate genetics and the attraction of the aphid parasitoid *Diaeetiella rapae* to *Brassica*. *Proceedings of the Royal Society of London B*, **267**, 89–95.

Bradbury, J.W. and Vehrencamp, S.L. (1977) Social Organisation and Foraging in emballonurid bats III. Mating Systems. *Behavioral Ecology and Sociobiology*, **2**, 1–17.

Bradford, M.M. (1976) A rapid and sensitive method for the quantification of microgram quantities of protein utilizing the principle of protein-dye binding. *Analytical Biochemistry*, **72**, 248–54.

Braig, H.R., Guzman, H., Tesh, R.B. and O'Neill, S.L. (1994) Replacement of the natural *Wolbachia* symbiont of *Drosophila simulans* with a mosquito counterpart. *Nature*, **367**, 453–55.

Braig, H.R., Zhou, W.G., Dobson, S.L. and O'Neill, S.L. (1998) Cloning and characterization of a gene encoding the major surface protein of the bacterial endosymbiont *Wolbachia pipientis*. *Journal of Bacteriology*, **180**, 2373–8.

Brakefield, P. (1985) Polymorphic Müllerian mimicry and interactions with thermal melanism in ladybirds and a soldier beetle: a hypothesis. *Biological Journal of the Linnean Society*, **26**, 243–67.

Braman, S.K. and Yeargan, K.V. (1988) Comparison of developmental and reproductive rates of *Nabis americoferus*, *N. roseipennis* and *N. rufusciilus* (Hemiptera: Nabidae). *Journal of the Entomological Society of America*, **81**, 923–30.

Branquart, E. and Hemptinne, J.-L. (2000) Development of ovaries, allometry of reproductive traits and fecundity of *Episyrphus balteatus* (Diptera: Syrphidae). *European Journal of Entomology*, **97**, 165–70.

Braun, A.R., Bellioti, A.C., Guerrero, J.M. and Wilson, L.T. (1989) Effect of predator exclusion on cassava infested with tetranychid mites (Acari: Tetranychidae). *Environmental Entomology*, **18**, 711–4.

Braun, D.M., Goyer, R.A. and Lenhard, G.J. (1990) Biology and mortality agents of the fruittree leafroller (Lepidoptera: Tortricidae), on bald cypress in Louisiana. *Journal of Entomological Science*, **25**, 176–84.

Breed, M.D., Smith, S.K. and Gall, B.G. (1980) Systems of mate selection in a cockroach species with male dominance hierachies. *Animal Behaviour*, **28**, 130–4.

Breene, R.G., Sweet, M.H. and Olson, J.K. (1988) Spider predators of mosquito larvae. *Journal of Arachnology*, **16**, 275–7.

Breeuwer, J.A.J. and Werren, J.H. (1990) Microorganisms associated with chromosome destruction and reproductive isolation between two insect species. *Nature*, **346**, 558–560.

Breeuwer, J.A.J. and Werren, J.H. (1993) Cytoplasmic incompatibility and bacterial density in *Nasonia vitripennis*. *Genetics*, **135**, 565–74.

Brenøe, J. (1987) Wet extraction – a method for estimating populations of *Bembidion lampros* (Herbst) (Col., Carabidae). *Journal of Applied Entomology*, **103**, 124–7.

Brewer, M.J., Nelson, D.J., Ahern, R.G., Donahue, J.D. and Prokrym, D.R. (2001) Recovery and range expansion of parasitoids (Hymenoptera: Aphelinidae and Braconidae) released for biological control of *Diuraphis noxia* (Homptera: Aphididae) in Wyoming. *Environmental Entomology*, **30**, 578–88.

Briggs, C.J., Nisbet, R.M., Murdoch, W.W., Collier, T.R. and Metz, J.A.J. (1995) Dynamical effects of host-feeding in parasitoids. *Journal of Animal Ecology*, **64**, 403–16.

Briggs, C.J., Murdoch, W.W. and Nisbet, R.M. (1999) Recent developments in theory for biological control of insect pests by parasitoids, in *Theoretical Approaches to Biological Control* (eds B.A. Hawkins and H.V. Cornell). Cambridge University Press, Cambridge, pp. 22–42.

Broad, G.R. and Quicke, D.L.J. (2000) The adaptive significance of host location by vibrational sounding in parasitoid wasps. *Proceedings of the Royal Society of London B*, **267**, 2403–9.

Brodeur, J. and McNeil, J.N. (1989) Biotic and abiotic factors involved in diapause induction of the parasitoid, *Aphidius nigripes*. *Journal of Insect Physiology*, **35**, 969–74.

Brodeur, J. and McNeil, J.N. (1992) Host behaviour modification by the endoparasitoid *Aphidius nigripes:* a strategy to reduce hyperparasitisrn. *Ecological Entomology*, **17**, 97–104.

Brodeur, J. and Rosenheim, J.A. (2000) Intraguild interactions in aphid parasitoids. *Entomologia Experimentalis et Applicata*, **97**, 93–108.

Brodeur, J. and Vet, L.E.M. (1994). Usurption of host behaviour by a parasitic wasp. *Animal Behaviour*, **48**, 187–92.

Brodin, T. and Johansson, F. (2002) Effects of predator-induced thinning and activity changes on life history in a damselfly. *Oecologia*, **132**, 316–22.

Brookes, C.P. and Loxdale, H.D. (1985) A device for simultaneously homogenising numbers of individual small insects for electrophoresis. *Bulletin of Entomological Research*, **75**, 377–8.

Brooks, D.R. and McLennan, D.A. (1991) *Phylogeny, Ecology and Behavior: A Research Program in Comparative Biology*. University of Chicago Press, London.

Brose, U. (2002) Estimating species richness of pitfall catches by non-parametric estimators. *Pedobiologia*, **46**, 101–7.

Brower, A.V.Z. and DeSalle, R. (1994) Practical and theoretical considerations for choice of a DNA sequence region in insect molecular systematics, with a short review of published studies using nuclear gene regions. *Annals of the Entomological Society of America*, **87**, 702–16.

Brown, J.J., Kiuchi, M., Kainoh, Y. and Takeda, S. (1993) In vitro release of ecdysteroids by an endoparasitoid, *Ascogaster reticulatus* Watanabe. *Journal of Insect Physiology*, **39**, 229–34.

Brown, R.A., White, J.A. and Everett, C.J. (1988) How does an autumn applied pyrethroid affect the terrestrial arthropod community?, in *Field Methods for the Study of Environmental Effects of Pesticides* (eds M.P. Greaves, B.D. Smith and P.W. Grieg-Smith), BCPC Monograph No. 40. BCPC. Farnham, Surrey, pp. 137–46.

Brown, W.D., Crespi, B.J. and Choe, J.C. (1997) Sexual conflict and the evolution of mating systems, in *The Evolution of Mating Systems in Insects and Arachnids* (eds J.C. Choe and B.J. Crespi). Cambridge University Press, Cambridge, pp. 352–37.

Browning, H.W. and Oatman, E.R. (1981) Effects of different constant temperatures on adult longevity, development time, and progeny production of *Hyposoter exiguae* (Hymenoptera: lchneumonidae). *Annals of the Entomological Society of America*, **74**, 79–82.

Bruck, D.J. and Lewis, L.C. (1998) Influence of adjacent cornfield habitat, trap location, and trap height on capture numbers of predators and a parasitoid of the European corn borer (Lepidoptera: Pyralidae) in central Iowa. *Environmental Entomology*, **27**, 1557–62.

Bruford, M.W., Hanotte, O., Brookfield, J.F.Y. and Burke, T. (1992) Single-locus and multilocus DNA fingerprinting, in *Molecular Genetic Analysis of Populations* (ed. A.R. Hoelzel). IRL Press, Oxford, pp. 225–69.

Bruinink, P.J. (1990) Some notes on the diet of the ground beetle *Pterostichus versicolor* Sturm (Coleoptera, Carabidae). *Polskie Pismo Entomologiczne*, **60**, 153–66.

Brunsting, A.M.H. and Heessen, H.J.L. (1984) Density regulation in the carabid beetle *Pterostichus oblongopunctatus*. *Journal of Animal Ecology*, **53**, 751–60.

Brunsting, A.M.H., Siepel, H. and van Schaick Zillesen, P.G. (1986) The role of larvae in the population ecology of Carabidae, in *Carabid Beetles. Their Adaptations and Dynamics* (eds P.J. den Boer, M.L. Luff, D. Mossakowski and F. Weber). Gustav Fischer, Stuttgart, pp. 399–411.

Brust, G.E. (1991) A method for observing belowground pest-predator interactions in corn agroecosystems. *Journal of Entomological Science*, **26**, 1–8.

Brust, G.E., Stinner, B.R. and McCartney, D.A. (1986a) Predation by soil-inhabiting arthropods in intercropped and monoculture agroecosystems. *Agriculture, Ecosystems and Environment*, **18**, 145–54.

Brust, G.E., Stinner, B.R. and McCartney, D.A. (1986b) Predator activity and predation in corn agroecosystems. *Environmental Entomology*, **15**, 1017–21.

Bryan, G. (1980) Courtship behaviour, size differences between the sexes and oviposition in some *Achrysocharoides* species (Hym., Eulophidae), *Netherlands Journal of Zoology*, **30**, 611–21.

Buchholz, U., Schmidt, S. and Schruft, G. (1994) The use of an immunological technique in order to evaluate the predation on *Eupoecilia ambiguella* (Hbn.) (Lepidoptera: Cochylidae) in vineyards. *Biochemical Systematics and Ecology*, **22**, 671–7.

Buchner, P. (1965) *Endosymbiosis of Animals with Plant Microorganisms*. Interscience Publishers, New York.

Büchs, W. (1991) Einfluss verschiedener landwirtschaftlicher Produktionsintensitäten auf die Abundanz von Arthropoden in Zuckerrübenfeldern. *Verhandlungen der Gesellschaft für Ökologie*, **20**, 1–12.

Büchs, W. (1993) Auswirkungen unterschiedlicher Bewirtschaftlungsintensitäten auf die Arthropodenfauna in Winterweizenfeldern. *Verhandlungen der Gesellschaft für Ökologie*, **22**, 27–34.

de Buck, N. (1990) Bloembezoek en bestuivingsecologie van zweefvliegen in het bijzonder voor Belgie. *Studiedocumenten van het Koninklijk Belgisch Instituut voor Natuurwetenschappen*, 60, Koninklijk Belgisch Instituut voor Natuurwetenschappen, Brussels, pp. 134.

Buckland, S.T., Anderson, D.R., Burnham, K.P. and Laake, J.L. (1993) *Distance Sampling*. Chapman and Hall, London.

Budenberg, W.J. (1990) Honeydew as a contact kairomone for aphid parasitoids. *Entomologia Experimentalis et Applicata*, **55**, 139–48.

Bugg, R.L. and Pickett, C.H. (1998) Introduction: enhancing biological control – habitat management to promote natural enemies of agricultural pests, in *Enhancing Biological Control* (eds C.H. Pickett and R.L. Bugg). University of California Press, Berkeley, pp. 1–23.

Bugg, R.L. and Waddington, C. (1994) Using cover crops to manage arthropod pests of orchards: a review. *Agriculture, Ecosystems and Environment*, **50**, 11–28.

Bugg, R.L., Ehler, L.E. and Wilson, L.T. (1987) Effect of common knotweed (*Polygonum aviculare*) on abundance and efficiency of insect predators of crop pests. *Hilgardia*, **55**, 1–52.

Bugg, R.L., Ellis, R.T. and Carlson, R.W. (1989) Ichneumonidae (Hymenoptera) using extrafloral nectar of faba bean (*Vicia faba* L., Fabaceae) in Massachusetts. *Biological Agriculture and Horticulture*, **6**, 107–14.

Bull, J.J. (1983) *The Evolution of Sex Determining Mechanisms*. Benjamin/Cummings, Menlo Park, California.

Bull, J.J. and Charnov, E.L. (1988) How fundamental are Fisherian sex ratios? *Oxford Surveys in Evolutionary Biology*, **5**, 98–135.

Bulmer, M.G. and Bull, J.J. (1982) Models of polygenic sex determination and sex ratio control. *Evolution*, **36**, 13–26.

Bultman, T.L., Borowicz, K.L., Schneble, R.M., Coudron, T.A. and Bush, L.P. (1997) Effect of a fungal endophyte on the growth and survival of two *Euplectrus* parasitoids. *Oikos*, **78**, 170–6.

Buonaccorsi, J.P. and Elkinton, J.S. (1990) Estimation of contemporaneous mortality factors. *Researches on Population Ecology*, **32**, 1–21.

Burbutis, P.P. and Stewart, J.A. (1979) Blacklight trap collecting of parasitic Hymenoptera. *Entomology News*, **90**, 17–22.

Burgess, L. and Hinks, C.F. (1987) Predation on adults of the crucifer flea beetle, *Phyllotreta cruciferae* (Goeze), by the northern fall field cricket, *Gryllus pennsylvanicus* Burmeister (Orthoptera: Gryllidae). *Canadian Entomologist*, **119**, 495–6.

Burk, T. (1981) Signaling and sex in acalyptrate flies. *Florida Entomologist*, **64**, 30–43.

Burk, T. (1982) Evolutionary significance of predation on sexually signalling males. *Florida Entomologist*, **65**, 90–104.

Burkett, D.A., Kline, D.L. and Carlson, D.A. (1999) Sugar meal composition of five North Central Florida mosquito species (Diptera: Culicidae) as determined by gas chromatography. *Journal of Medical Entomology*, **36**, 462–7.

Burkhard, D.U., Ward, P.I. and Blanckenhorn, W.U. (2002) Using age grading by wing injuries to estimate size-dependent adult survivorship in the field: a case study of the yellow dung fly *Scathophaga stercoraria*. *Ecological Entomology*, **27**, 514–20.

Burkhart, B.D., Montgomery, E., Langley, C.H. and Voelker, R.A. (1984) Characterization of allozyme null and low activity alleles from two natural populations of *Drosophila melanogaster*. *Genetics*, **107**, 295–306.

Burn, A.J. (1982) The role of predator searching efficiency in carrot fly egg loss. *Annals of Applied Biology*, **101**, 154–15.

Burn, A.J. (1989) Long-term effects of pesticides on natural enemies of cereal crop pests, in *Pesticides and Non-target Invertebrates* (ed P.C. Jepson). Intercept, Dorset, pp. 177–93.

Burquez, A. and Corbet, S.A. (1991) Do flowers reabsorb nectar? *Functional Ecology*, **5**, 369–79.

Bursell, E. (1964) Environmental aspects: temperature, in *The Physiology of the Insecta* (ed. M. Rockstein). Academic Press, New York, pp. 283–321.

Burstone, M.S. (1957) Polyvinyl pyrrolidone. *American Journal of Clinical Pathology*, **28**, 429–30.

Buscarlet, G. and Iperti, L.A. (1986) Seasonal migration of the ladybird *Semiadalia undecimnotata*, in *Ecology of Aphidophaga* (ed. I. Hodek). Academia, Prague and Dr. W. Junk, Dordrecht, pp. 199–204.

Buschman, L.L., Whitcomb, W.H., Hemenway, R.C., Mays, D.L., Nguyen Roo, Leppla, N.C. and Smittle. B.J. (1977) Predators of velvetbean caterpillar eggs in Florida soybeans. *Environmental Entomology*, **6**, 403–7.

van Buskirk, J. (1987) Density-dependent population dynamics in larvae of the dragonfly *Pachydiplax longipennis:* a field experiment. *Oecologia*, **72**, 221–5.

Butcher, R.D.J., Whitfield, W.G.F. and Hubbard, S.F. (2000) Single-locus complementary sex determination in *Diadegma chrysostictos* (Gmelin) (Hymenoptera: Ichneumonidae). *Journal of Heredity*, **91**, 104–11.

Butcher, R.D.J., Meikle, L., Hubbard, S.F. and Whitfield, W.G.F. (2004a) The distribution of *Wolbachia* within, and transfection from the haemocoel into, somatic and germ line tissues. *Molecular Ecology*, (in press).

Butcher, R.D.J., Hubbard, S.F. and Whitfield, W.G.F. (2004b) Purification and survival of insect-associated *Wolbachia* outside the host cytoplasm: osmotic and temperature sensitivity. *Insect Biochemistry and Molecular Biology*, (in press).

Butcher, R.D.J., Hubbard, S.F. and Whitfield, W.G.F. (2004c) Life-style related parasitoid-mediated horizontal transfer of *Wolbachia* across insect taxa. *Heredity*, (in press)

Butcher, R.D.J., Hubbard, S.F. and Whitfield, W.G.F. (2004d) Parasitoid-mediated horizontal transmission of *Wolbachia* is restricted by the host co-dependent vertical transmission efficiency and altered host reproduction. *Heredity*, (in press).

Butler, G.D. (1965) A modified Malaise insect trap. *Pan-Pacific Entomology*, **41**, 51–3.

Butler, G.D., Hamilton, A.G. and Lopez Jr., J.D. (1983) *Cardiochiles nigriceps* (Hymenoptera: Braconidae): Development time and fecundity in relation to temperature. *Annals of the Entomological Society of America*, **76**, 536–38.

Butts, R.A. and McEwen, F.L. (1981) Seasonal populations of the diamondback moth in relation to day-degree accumulation. *Canadian Entomologist*, **113**, 127–31.

Cade, W. (1975) Acoustically orienting parasitoids: fly phonotaxis to cricket song. *Science*, **190**, 1312–13.

Cade, W. (1980) Alternative male reproductive behaviours. *Florida Entomologist*, **63**, 30–44.

Calatayud, P.A., Auger, J., Thibout, E., Rousset, S., Caicedo, A.M., Calatayud, S., Buschmann, H., Guillaud, J., Mandon, N. and Bellotti, A.C. (2001) Identification and synthesis of a kairomone mediating host location by two parasitoid species of the cassava mealybug *Phenacoccus herreni*. *Journal of Chemical Ecology*, **27**, 2203–17.

Calder, W.A. 1984. *Size, Function and Life History*. Harvard University Press. Cambridge, Massachusetts.

Calver, M.C., Matthiessen, J.N., Hall, G.P., Bradley, J.S. and Lillywhite, J.H. (1986) Immunological determination of predators of the bush fly, *Musca vetustissima* Walker (Diptera: Muscidae), in south-western Australia. *Bulletin of Entomological Research*, **76**, 133–9.

Camara, M.D. (1997) Predator responses to sequestered plant toxins in buckeye caterpillars: are tritrophic interactions locally variable? *Journal of Chemical Ecology*, **23**, 2093–106.

Cameron, E.A. and Reeves, R.M. (1990) Carabidae (Coleoptera) associated with Gypsy moth, *Lymantria dispar* (L.) (Lepidoptera: Lymantriidae), populations, subjected to *Bacillus thuringiensis* Berliner treatments in Pennsylvania. *Canadian Entomologist*, **122**, 123–9.

Campbell, A. and Mackauer, M. (1975) Thermal constants for development of the pea aphid (Homoptera: Aphididae) and some of its parasites. *Canadian Entomologist*, **107**, 419–23.

Campbell, A., Frazer, B.D., Gilbert, N., Gutierrez, A.P. and Mackauer, M. (1974) Temperature requirements of some aphids and their parasites. *Journal of Applied Ecology*, **11**, 431–8.

Campbell, B.C. and Duffey, S.S. (1979) Tomatine and parasitic wasps: potential incompatibility of plant antibiosis with biological control. *Science*, **205**, 700–2.

Campbell, C.A.M. (1978) Regulation of the Damson-hop aphid *(Phorodon humuli* (Schrankl) on hops *(Humulus lupulus* L.) by predators. *Journal of Horticultural Science*, **53**, 235–42.

Campos, W.G., Pereira, D.B.S. and Schoereder, J.H. (2000) Comparison of the efficiency of flight-interception trap models for sampling Hymenoptera and other insects. *Anais da Sociedade Entomologica do Brasil*, **29**, 381–9.

Canard, M. (1983) Photoperiodic sensitivity in the larvae of the chrysopid, *Nineta flava*. *Entomologia Experimentalis et Applicata*, **34**, 111–18.

Canard, M. (1986) Is the Iberian lacewing *Chrysopa regalis* a semivoltine species? *Ecological Entomology*, **11**, 27–30.

Canard, M. (2002) Natural food and feeding habits of lacewings, in *Lacewings in the Crop Environment* (eds P. McEwen, T.R. New and A.E. Whittington). Cambridge University Press, Cambridge, pp. 116–29.

Canard, M., Semeria, Y. and New, T.R. (1984) *Biology of the Chrysopidae*. W. Junk, The Hague.

Candy, D.J., Becker, A. and Wegener, G. (1997) Coordination and integration of metabolism in insect flight. *Comparative Biochemistry and Physiology B*, **117**, 497–512.

Čapek, M. (1970) A new classification of the Braconidae (Hymenoptera) based on the cephalic structures of the final instar larva and biological evidence. *Canadian Entomologist*, **102**, 846–75.

Čapek, M. (1973) Key to the final instar larvae of the Braconidae (Hymenoptera). *Acta Instituti Forstalis Zvolenensis*, **1973**, 259–68.

Capillon, C. and Atlan, A. (1999) Evolution of driving X chromosomes and resistance factors in experimental populations of *Drosophila simulans*. *Evolution*, **53**, 506–17.

Cappuccino, N. (1992) Adjacent trophic-level effects on spatial density dependence in a herbivore- predator-parasitoid system. *Ecological Entomology*, **17**, 105–8.

Carbone, S.S. and Rivera, C.C. (2003) Egg load and adaptive superparasitism in *Anaphes nitens*, an egg parasitoid of the *Eucalyptus* snout-beetle *Gonipterus scutellatus*. *Entomologia Experimentalis et Applicata*, **106**, 127–34.

Carcamo, H.A. and Spence, J.R. (1994) Crop type effects on the activity and distribution of ground beetles (Coleoptera, Carabidae). *Environmental Entomology*, **23**, 684–92.

Carlson, D.A., Geden, C.J. and Bernier, U.R. (1999) Identification of pupal exuviae of *Nasonia vitripennis* and *Muscidifurax raptorellus* parasitoids using cuticular hydrocarbons. *Biological Control*, **15**, 97–106.

Carroll, C.R. and Risch, S. (1990) An evaluation of ants as possible candidates for biological control in tropical agroecosystems, in *Agroecology: Researching the Ecological Basis for Sustainable Agriculture* (ed. S.R. Gliessman). Springer Verlag, New York, pp. 30–46.

Carroll, D.P and Hoyt, S.C. (1984) Natural enemies and their effects on apple aphid, *Aphis pomi* DeCeer (Homoptera: Aphididae), colonies on young apple trees in central Washington, *Environmental Entomology*, **13**, 469–81.

Carter, M.C. and Dixon, A.F.G. (1984) Honeydew, an arrestment stimulus for coccinellids. *Ecological Entomology*, **9**, 383–7.

Carter, N. and Dixon, A.F.G. (1981) The 'natural enemy ravine' in cereal aphid population dynamics: a consequence of predator activity or aphid biology? *Journal of Animal Ecology*, **50**, 605–11.

Carter, N., Aikman, D.P. and Dixon, A.F.G. (1978) An appraisal of Hughes' time-specific life table analysis for determining aphid reproductive and mortality rates. *Journal of Animal Ecology*, **47**, 677–89.

Carter, N., Dixon, A.F.C. and Rabbinge, R. (1982) *Cereal Aphid Populations: Biology, Simulation and Prediction*, Pudoc, Wageningen.

Carton, Y. (1978) Olfactory responses of *Cothonaspis* sp. (parasitic Hymenoptera, Cynipidae) to the food habit of its host (*Drosophila melanogaster*). *Drosophila Information Service*, **53**, 183–4.

Carton, Y. and Nappi, A. (1991) The *Drosophila* immune reaction and the parasitoid capacity to evade it: genetic and coevolutionary aspects. *Acta Oecologia*, **12**, 89–104.

Carton, Y. and Nappi, A.J. (1997) *Drosophila* cellular immunity against parasitoids. *Parasitology Today*, **13**, 218–27.

Carton, Y., Capy, P. and Nappi, A.J. (1989) Genetic variability of host-parasite relationship traits: utilization of isofemale lines in a *Drosophila simulans* parasitic wasp. *Genetics Selection Evolution*, **21**, 437–46.

Carvalho, G.A., Kerr, W.E. and Nascimento, V.A. (1995) Sex determination in bees. XXXIII. Decrease of xo heteroalleles in a finite population of *Melipona scutellaris* (Apidae, Meliponini). *Revista Brasileira de Genética*, **18**, 13–16.

Casas, J. (1990) Foraging behaviour of a leafminer parasitoid in the field. *Ecological Entomology*, **14**, 257–65.

Casas, J. and Hulliger, B. (1994) Statistical analysis of functional response experiments. *Biocontrol Science and Technology*, **4**, 133–45.

Casas, J., Gurney, W.S.C., Nisbet, R. and Roux, O. (1993) A probabilistic model for the functional response of a parasitoid at the behavioural timescale, *Journal of Animal Ecology*, **62**, 194–204.

Casas, J., Nisbet, R.M., Swarbrick, S. and Murdoch, W.W. (2000) Eggload dynamics and oviposition rate in a wild population of a parasitic wasp. *Journal of Animal Ecology*, **69**, 185–93.

Casas, J., Driessen, G., Mandon, N., Wileaard, S., Desouhant, E., van Alphen, J., Lapchin, L., Rivero, A,. Christides, J.P. and Bernstein, C. (2003) Energy dynamics in a parasitoid foraging in the wild. *Journal of Animal Ecology*, **72**, 691–7.

Caspari, E. and Watson, G.S. (1959) On the evolutionary importance of cytoplasmic sterility in mosquitoes. *Evolution*, **13**, 568–70.

Castañera, P., Loxdale, H. D. and Nowak, K. (1983) Electrophoretic study of enzymes from cereal aphid populations. II. Use of electrophoresis for identifying aphidiid parasitoids (Hymenoptera) of *Sitobion avenae* (F.) (Hemiptera: Aphididae). *Bulletin of Entomological Research*, **73**, 659–65.

Castelo, M.K., Corley, J.C. and Desouhant, E. (2003) Conspecific avoidance during foraging in *Venturia canescens* (Hymenoptera:Ichneumonidae): The roles of host presence and conspecific densities. *Journal of Insect Behavior*, **16**, 307–18.

Castrovillo, P. J. and Stock, M.W. (1981) Electrophoretic techniques for detection of *Glypta fumiferanae*, an endoparasitoid of western spruce budworm. *Entomologia Experimentalis et Applicata*, **30**, 176–80.

Caterino, M.S., Cho, S. and Sperling, F.A.H. (2000) The current state of insect molecular systematics: a thriving Tower of Babel. *Annual review of Entomology*, **45**, 1–54.

Cavalieri, L.F. and Kocak, H. (1995) Chaos: a potential problem in the biological control of insect pests. *Mathematical Biosciences*, **127**, 1–17.

Cave, R.D. and Gaylor, M.J. (1988) Influence of temperature and humidity on development and survival of *Telenomus reynoldsi* (Hymenoptera: Scelionidae) parasitising *Geocoris punctipes* (Heteroptera: Lygaeidae) eggs. *Annals of the Entomological Society of America*, **81**, 278–85.

Cave, R.D. and Gaylor, M.J. (1989) Longevity, fertility, and population growth statistics of *Telenomus reynoldsi* (Hymenoptera: Scelionidae). *Proceedings of the Entomological Society of Washington*, **91**, 588–93.

Cezilly, F. and Benhamou, S. (1996) Optimal foraging strategies: a review. *Revue d'écologie-la Terre et la Vie*, **51**, 43–86.

Chaddick, P.R. and Leek, F.F. (1972) Further specimens of stored products insects found in ancient Egyptian tombs. *Journal of Stored Products Research*, **8**, 83–86.

Chambers, R.J. and Adams, T.H.L. (1986) Quantification of the impact of hoverflies (Diptera: Syrphidae) on cereal aphids in winter wheat: an analysis of field populations. *Journal of Applied Ecology*, **23**, 895–904.

Chambers, R.J., Sunderland, K.D, Wyatt, I.J. and Vickerman, C.P. (1983) The effects of predator exclusion and caging on cereal aphids in winter wheat. *Journal of Applied Ecology*, **20**, 209–24.

Chan, M.S. and Godfray, H.C.J. (1993) Host-feeding strategies of parasitoid wasps. *Evolutionary Ecology*, **7**, 593–604.

Chang, N.W. and Wade, M.J. (1996) An improved microinjection protocol for the transfer of *Wolbachia pipientis* between infected and uninfected strains of the flour beetle *Tribolium confusum*. *Canadian Journal of Microbiology*, **42**, 711–14.

Chang, S.C., Hu, N.T., Hsin, C.Y. and Sun, C.N. (2001) Characterization of differences between two *Trichogramma* wasps by molecular markers. *Biological Control*, **21**, 75–8.

Chang, Y.F., Tauber, M.J. and Tauber, C.A. (1996) Reproduction and quality of F_1 offspring in *Chrysoperla carnea*: differential influence of quiescence, artificially-induced diapause, and natural diapause. *Journal of Insect Physiology*, **42**, 521–28.

Chant, D.A. (1959) Phytoseiid mites (Acarina: Phytoseiidae). I Bionomics of seven species in south-eastern England, II. A taxonomic review of the family Phytoseiidae, with descriptions of thirty eight new species. *Canadian Entomologist*, **91**, 12–49.

Chant, D.A. and Muir, R.C. (1955) A comparison of the imprint and brushing machine methods for estimating the numbers of fruit tree red spider mite, *Metatetranychus ulmi* (Koch), on apple leaves. *Report of the East Malling Research Station for 1954*, 141–5.

Chapman, P.A. and Armstrong, G. (1996) Daily dispersal of beneficial ground beetles between areas of contrasting vegetation density within agricultural habitats. *Proceedings of the Brighton Crop Protection Conference – Pests and Diseases 1996, BCPC, Farnham, Surrey UK*, 623–8.

Chapman, P.A. and Armstrong, G. (1997) Design and use of a time-sorting pitfall trap for predatory arthropods. *Agriculture, Ecosystems and Environment*, **65**, 15–21.

Chapman, P.A., Armstrong, G. and McKinlay, R.G. (1999) Daily movements of *Pterostichus melanarius* between areas of contrasting vegetation density within crops. *Entomologia Experimentalis et Applicata*, **91**, 477–80.

Chapman, R.F. (1998) *The Insects: Structure and Function*. Cambridge University Press, Cambridge.

Chapman, T., Liddle L.F., Kalb, J.M., Wolfner, M.F. and Partridge, L. (1995) Cost of mating in *Drosophila melanogaster* females is mediated by male accessory gland products. *Nature*, **373**, 241–4.

Charles, J.G. and White, V. (1988) Airborne dispersal of *Phytoseiulus persimilis* (Acarina: Phytoseiidae) from a raspberry garden in New Zealand. *Experimental and Applied Acarology*, **5**, 47–54.

Charlesworth, B. (1987) The heritability of fitness, in *Sexual Selection: Testing the Alternatives* (eds J.W. Bradbury and M.B. Andersson). John Wiley and Sons, New York, pp. 21–40.

Charnov, E.L. (1976) Optimal foraging: attack strategy of a mantid. *American Naturalist*, **110**, 141–51.

Charnov, E.L. (1979) The genetical evolution of patterns of sexuality: Darwinian fitness. *American Naturalist*, **113**, 465–80.

Charnov, E.L. (1982) *The Theory of Sex Allocation*. Princeton University Press, New Jersey.

Charnov, E.L. and Skinner, S.W. (1985) Complementary approaches to the understanding of parasitoid oviposition decisions. *Environmental Entomology*, **14**, 383–91.

Charnov, E.L. and Stephens, D.W. (1988) On the evolution of host selection in solitary parasitoids. *American Naturalist*, **132**, 707–22.

Charnov, E.L., Los-den Hartogh, R.L., Jones, W.T. and van den Assem, J. (1981) Sex ratio evolution in a variable environment. *Nature*, **289**, 27–33.

Chase, J.M., Leibold, M.A. and Simms, E. (2000) Plant tolerance and resistance in food webs: community-level predictions and evolutionary implications. *Evolutionary Ecology*, **14**, 289–314.

Chassain, C. and Bouletreau, M. (1987) Genetic variability in the egg-laying behaviour of *Trichogramma maidis*. *Entomophaga*, **32**, 149–57.

Chelliah, J. and Jones, D. (1990) Biochemical and immunological studies of proteins from polydnavirus *Chelonus* sp. near *curvimaculatus*. *Journal of General Virology*, **71**, 2353–9.

Chen, C.C. and Chang, K.P. (1991) Marking *Trichogramma ostriniae* Pang and Chen (Hymenoptera: Trichogrammatidae) with radioactive phosphorus. *Chinese Journal of Entomology*, **11**, 148–52.

Chen, P.S., Stumm-Zollinger, E., Aigaki, T., Balmer, J., Bienz, M. and Bohlen, P. (1988) A male accessory gland peptide that regulates reproductive behaviour of female *D. melanogaster*. *Cell*, **54**, 291–98.

Chen, X., Li, S., Li, C.B. and Zhao, S.Y. (1999) Phylogeny of genus *Glossina* (Diptera: Glossinidae) according to ITS2 sequences. *Science in China Series C – Life Sciences*, **42**, 249–58.

Chen, Y., Giles, K.L., Payton, M.E. and Greenstone, M.H. (2000) Identifying key cereal aphid predators by molecular gut analysis. *Molecular Ecology*, **9**, 1887–98.

Cherix, D. and Bourne, J.D. (1980) A field study on a super-colony of the Red Wood Ant *Formica lugubris* Zett. in relation to other predatory arthropods (spiders, harvestmen and ants). *Revue Suisse de Zoologie*, **87**, 955–73.

Chesson, J. (1978a) Measuring preference in selective predation. *Ecology*, **59**, 211–15.

Chesson, J. (1983) The estimation and analysis of preference and its relationship to foraging models. *Ecology*, **63**, 1297–304.

Chesson, P.L. (1978b) Predator-prey theory and variability. *Annual Review of Ecology and Systematics*, **9**, 323–47.

Chesson, P.L. (1982) The stabilizing effect of random environments. *Journal of Mathematical Biology*, **15**, 1–36.

Chesson, P.L. (1984) Variable predators and switching behaviour. *Theoretical Population Biology*, **26**, 1–26.

Chesson, P.L. and Murdoch, W.W. (1986) Aggregation of risk: relationships among host-parasitoid models. *American Naturalist*, **127**, 696–715.

Cheverud, J.M. and Routman, E.J. (1996) Epistasis as a source of increased additive genetic variance at population bottlenecks. *Evolution*, **50**, 1042–51.

Chevrier, C. and Bressac, C. (2002) Sperm storage and use after multiple mating in *Dinarmus basalis* (Hymenoptera: Pteromalidae). *Journal of Insect Behaviour*, **15**, 385–98.

Chey, V.K., Holloway, J.D., Hambler, C. and Speight, M.R. (1998) Canopy knockdown of arthropods in exotic plantations and natural forest in Sabah, north-east Borneo, using insecticidal mist-blowing. *Bulletin of Entomological Research*, **88**, 15–24.

Chiappini, E. and Mazzoni, E. (2000) Differing morphology and ultrastructure of the male copulatory apparatus in species-groups of *Anagrus* Haliday (Hymenoptera: Mymaridae). *Journal of Natural History*, **34**, 1661–76.

Childers, C.C. and Abou-Setta, M.M. (1999) Yield reduction in 'Tahiti' lime from *Panonychus citri* feeding injury following different pesticide treatment regimes and impact on the associated predacious mites. *Experimental and Applied Acarology*, **23**, 771–83.

Chiverton, P.A. (1984) Pitfall-trap catches of the carabid beetle *Pterostichus melanarius*, in relation to gut contents and prey densities, in treated and untreated spring barley. *Entomologia Experimentalis et Applicata*, **36**, 23–30.

Chiverton, P.A. (1987) Predation of *Rhopalosiphum padi* (Homoptera: Aphididae) by polyphagous predatory arthropods during the aphids' pre-peak period in spring barley. *Annals of Applied Biology*, **111**, 257–69.

Chiverton, P.A. (1989) The creation of within-field overwintering sites for natural enemies of cereal aphids. *Proceedings of the 1979 British Crop Protection Conference – Weeds, BCPC, Farnham, Surrey*, 1093–6.

Chloridis, A.S., Koveos, D.S. and Stamopoulus, D.C. (1997) Effect of photoperiod on the induction and maintenance of diapause and on development of the predatory bug *Podisus maculiventris* (Hem: Pentatomidae). *Entomophaga*, **42**, 427–34.

Choe, J.C. and Crespi, B.J. (eds) (1997) *The Evolution of Mating Systems in Insects and Arachnids*. Cambridge University Press, Cambridge.

Chow, A. and Mackauer, M. (1991) Patterns of host selection by four species of aphidiid (Hymenoptera) parasitoids: influence of host switching. *Ecological Entomology*, **16**, 403–10.

Chow, A. and Mackauer, M. (1999) Marking the package or its contents: host discrimination and acceptance in the ectoparasitoid *Dendrocerus carpenteri* (Hymenoptera: Megaspilidae). *Canadian Entomologist*, **131**, 495–505.

Chow, F.J. and Mackauer, M. (1984) Inter- and intraspecific larval competition in *Aphidius smithi* and *Praon pequodorum* (Hymenoptera: Aphidiidae). *Canadian Entomologist*, **116**, 1097–107.

Chow, F.J. and Mackauer, M. (1985) Multiple parasitism of the pea aphid: stage of development of parasite determines survival of *Aphidius smithi* and *Praon pequodorum* (Hymenoptera: Aphidiidae). *Canadian Entomologist*, **117**, 133–4.

Chow, F.J. and Mackauer, M. (1986) Host discrimination and larval competition in the aphid parasite *Ephedrus californicus*. *Entomologia Experimentalis et Applicata*, **41**, 243–54.

Chrambach, A. and Rodbard, D. (1971) Polyacrylamide gel electrophoresis. *Science*, **172**, 440–51.

Christiansen-Weniger, P. and Hardie, J. (1997) Development of the aphid parasitoid, *Aphidius ervi*, in asexual and sexual females of the pea aphid, *Acyrthosiphon pisum*, and the blackberry-cereal aphid, *Sitobion fragariae*. *Entomophaga*, **42**, 165–72.

Christiansen–Weniger, P. and Hardie, J. (1999) Environmental and physiological factors for diapause induction and termination in the aphid parasitoid, *Aphidius ervi* (Hymenoptera: Aphidiidae). *Journal of Insect Physiology*, **45**, 357–64.

Chumakova, B.M. (1960) Additional nourishment as a factor increasing activity of insect-pests parasites. *Proceedings of All Union Scientific Research Institute of Plant Protection*, **15**, 57–70. (In Russian).

Chyzik, R., Klein, M. and Bendov, Y. (1995). Reproduction and survival of the predatory bug *Orius albidipennis* on various arthropod prey. *Entomologia Experimentalis et Applicata*, **75**, 27–31.

Cisneros, J.J. and Rosenheim, J.A. (1998) Changes in foraging behaviour, within-plant vertical distribution, and microhabitat selection of a generalist insect predator: an age analysis. *Environmental Entomology*, **27**, 949–57.

Clapham, A.R., Tutin, T.G. and Moore, D.M. (1989) *Flora of the British Isles*. Cambridge University Press, Cambridge.

Claridge, M.F. and Askew, R.R. (1960) Sibling species in the *Eurytoma rosae* group (Hym., Eurytomidae). *Entomophaga*, **5**, 141–53.

Claridge, M.F. and Dawah, H.A. (1993) Assemblages of herbivorous chalcid wasps and their parasitoids associated with grasses – problems of species and specificity, in *Plant Galls: Organisms, Interactions, Populations* (ed. M.A.J. Williams), Systematics Association Special Volume, 49. Clarendon Press, Oxford, pp. 313–29.

Claridge, M.F. and den Hollander, J. (1983) The biotype concept and its application to insect pests of agriculture. *Crop Protection*, **2**, 85–95.

Claridge, M.F., den Hollander, J. and Morgan, J.C. (1984) Specificity of acoustic signals and mate choice in the brown planthopper *Nilaparvata lugens*. *Entomologia Experimentalis et Applicata* **35**, 221–26.

Claridge, M.F., Morgan, J.C., Steenkiste, A.E., Iman, M. and Damayanti, D. (1999) Seasonal patterns of egg parasitism and natural biological control of rice brown planthopper in Indonesia. *Agricultural and Forest Entomology*, **1**, 297–304.

Claridge, M.F., Morgan, J.C., Steenkiste, A.E., Iman, M. and Damayanti, D. (2000) Experimental field studies on predation and egg parasitism of rice brown planthopper in Indonesia. *Agricultural and Forest Entomology*, **4**, 203–10.

Clark, A.M. and Egen, R.C. (1975) Behavior of gynandromorphs of the wasp *Habrobracon juglandis*. *Developmental Biology*, **45**, 251–59.

Clark, D.L. and Uetz, G.W. (1990) Video image recognition by the jumping spider, *Maevia inclemens* (Araneae: Salticidae). *Animal Behaviour*, **40**, 884–90.

Clark, M.E., Veneti, Z., Bourtzis, K. and Karr, T.L. (2003) *Wolbachia* distribution and cytoplasmic incompatibility during sperm development: the cyst as the basic cellular unit of CI expression. *Mechanisms in Development*, **120**, 185–98.

Clarke, B.C. (1962) Balanced polymorphism and the diversity of sympatric species, in *Taxonomy and Geography* (ed. D. Nichols), Systematics Association Publication No. 4, Systematics Association, Oxford, pp. 47–70.

Clausen, C.P. (1940) *Entomophagous Insects*. McGraw-Hill Co., New York.

Clausen, C.P., Jaynes, H.A. and Gardner, T.R. (1933) Further investigations of the parasites of *Popillia japonica* in the Far East. *United States Department of Agriculture Technical Bulletin*, **366**, 1–58.

de Clercq, R. (1985) Study of the soil fauna in winter wheat fields and experiments on the influence of this fauna on the aphid populations. *Bulletin IOBC/WPRS*, **7**, 133–5.

de Clercq, R. and Pietraszko, R. (1982) Epigeal arthropods in relation to predation of cereal aphids, in *Aphid Antagonists* (ed. R. Cavalloro), Proceedings of the EC Experts' Meeting, Portici. A.A. Balkema, Rotterdam, pp. 88–92.

Cloarec, A. (1977) Alimentation des Larves d'*Anax imperator* Leach dans un milieu naturel (Anisoptera, Aeschnidae). *Odontologica*, **6**, 227–43.

Cloutier, C. (1997) Facilitated predation through interaction between life stages in the stinkbug predator *Perillus bioculatus* (Hemiptera: Pentatomidae). *Journal of Insect Behavior*, **10**, 581–98.

Cloutier, C. and Mackauer, M. (1979) The effect of parasitism by *Aphidius smithi* (Hymenoptera: Aphidiidae) on the food budget of the pea *aphid, Acyrthosiphon pisum. Canadian Journal of Zoology*, **57**, 1605–11.

Cloutier, C. and Mackauer, M. (1980) The effect of superparasitism by *Aphidius smithi* (Hymenoptera: Aphidiidae) on the food budget of the pea aphid, *Acyrthosiphon pisum* (Homoptera: Aphidiidae). *Canadian Journal of Zoology*, **58**, 241–4.

Cloutier, C., Levesque, C.A., Eaves, D.M. and Mackauer, M. (1991) Maternal adjustment of sex–ratio in response to host size in the aphid parasitoid *Ephedrus californicus. Canadian Journal of Zoology*, **69**, 1489–95.

Cloutier, C., Duperron, J., Tertuliano, M. and McNeil, J.N. (2000) Host instar and fitness in the koinobiotic parasitoid *Aphidius nigripes. Entomologia Experimentalis et Applicata*, **97**, 29–40.

Coaker, T.H. (1965) Further experiments on the effect of beetle predators on the numbers of the cabbage root fly, *Erioischia brassicae* (Bouche), attacking crops. *Annals of Applied Biology*, **56**, 7–20.

Cochran, W.G. (1983) *Planning and Analysis of Observational Studies*. John Wiley and Sons, New York.

Cock, M.J.W. (1978) The assessment of preference. *Journal of Animal Ecology*, **47**, 805–16.

Cocuzza, G.E., De Clercq, P., Van de Veire, M., De Cock, A., Degheele, D. and Vacante, V. (1997a) Reproduction of *Orius albidipennis* on pollen and *Ephestia kuehniella* eggs. *Entomologia Experimentalis et Applicata*, **82**, 101–4.

Cocuzza, G.E., De Clercq, P., Lizzio, S., Van de Veire, M., Tirry, L., Degheele, D. and Vacante, V. (1997b) Life tables and predation activity of *Orius laevigatus* and *O. albidipennis* at three constant temperatures. *Entomologia Experimentalis et Applicata*, **85**, 189–98.

Coddington, J.A., Young, L.H. and Coyle, F.A. (1996) Estimating spider species richness in a southern Appalachian cove hardwood forest. *The Journal of Arachnology*, **24**, 111–28.

Coe, R.L. (1966) Diptera: Pipunculidae. *Handbooks for the Identification of British Insects, 10(2c)*. Royal Entomological Society of London, London.

Cohen, A.C. (1984) Food consumption, food utilization, and metabolic rates of *Geocoris punctipes* (Het.: Lygaeidae) fed *Heliothis virescens* (Lep.: Noctuidae) eggs. *Entomophaga*, **29**, 361–7.

Cohen, A.C. (1989) Ingestion efficiency and protein consumption by a heteropteran predator. *Annals of the Entomological Society of America*, **82**, 495–9.

Cohen, A.C. (1993) Organisation of digestion and preliminary characterization of salivary trypsin-like enzymes in a predaceous heteropteran, *Zelus renardii. Journal of Insect Physiology*, **39**, 823–9.

Cohen, A.C. (1995) Extra-oral digestion in predaceous terrestrial Arthropoda. *Annual Review of Entomology*, **40**, 85–103.

Cohen, A.C. and Tang, R. (1997) Relative prey weight influences handling time and biomass extraction in *Sinea confusa* and *Zelus renardii* (Heteroptera: Reduviidae). *Environmental Entomology*, **26**, 559–65.

Cohen, J.E. and Briand, F. (1984) Trophic links of community food webs. *Proceedings of the National Academy of Sciences*, **81**, 4105–9.

Cohen, J.E., Schoenly, K., Heong, K.L., Justo, H., Arida, G., Barrion, A.T. and Litsinger, J.A. (1994) A food web approach to evaluating the effect of insecticide spraying on insect pest population dynamics in a Philippine irrigated rice ecosystem. *Journal of Applied Ecology*, **31**, 747–63.

Cole, L.R. (1967) A study of the life-cycles and hosts of some Ichneumonidae attacking pupae of the green oak-leaf roller moth, *Tortrix viridana* (L.) (Lepidoptera: Tortricidae) in *England. Transactions of the Royal Entomological Society of London*, **119**, 267–81.

Cole, L.R. (1970) Observations on the finding of mates by *Phaeogenes invisor* and *Apanteles medicaginis* (Hymenoptera: Ichneumonidae). *Animal Behaviour*, **18**, 184–192.

Cole, L.R. (1981) A visible sign of a fertilization act during oviposition by an ichneumonid wasp, *Itoplectis maculator. Animal Behaviour*, **29**, 299–300.

Colfer, R.G. and Rosenheim, J.A. (2001) Predation on immature parasitoids and its impact on aphid suppression. *Oecologia*, **126**, 292–304.

Coll, M. (1998) Parasitoid activity and plant species composition in intercropped systems, in *Enhancing Biological Control* (eds C.H. Pickett and R.L. Bugg). University of California Press, Berkeley, pp. 85–120.

Coll, M. (1998) Feeding and living on plants in predatory heteroptera: in *Predatory Heteroptera in Agroecosystems: Their Ecology and Use in Biological Control* (eds M. Coll and J. Ruberson). Thomas Say

Publications in Entomology, Lanham, MD, pp. 89–129.

Coll, M. and Bottrell, D.G. (1996) Movement of an insect parasitoid in simple and diverse plant assemblages. *Ecological Entomology*, **21**, 141–9.

Coll, M. and Guershon, M. (2002) Omnivory in terrestrial arthropods: mixing plant and prey diets. *Annual Review of Entomology*, **47**, 85–103.

Colley, M.R. and Luna, J.M. (2000) Relative attractiveness of potential beneficial insectary plants to aphidophagous hoverflies (Diptera: Syrphidae). *Environmental Entomology*, **29**, 1054–9.

Collier, T.R. and Hunter, M.S. (2001) Lethal interference competition in the whitefly parasitoids *Eretmocerus eremicus* and *Encarsia sophia*. *Oecologia*, **129**, 147–54.

Collier, T., Kelly, S. and Hunter, M. (2002) Egg size, intrinsic competition, and lethal interference in the parasitoids *Encarsia pergandiella* and *Encarsia formosa*. *Biological Control*, **23**, 254–61.

Collins, M.D. and Dixon, A.F.G. (1986) The effect of egg depletion on the foraging behaviour of an aphid parasitoid. *Journal of Applied Entomology*, **102**, 342–52.

Collins, M.D., Ward, S.A and Dixon, A.F.G. (1981) Handling time and the functional response of *Aphelinus thomsoni*, a predator and parasite of the aphid *Drepanosiphum platanoidis*. *Journal of Animal Ecology*, **50**, 479–87.

Colunga-Garcia, M., Gage, S.H. and Landis, D.A. (1997) Response of an assemblage of Coccinellidae (Coleoptera) to a diverse agricultural landscape. *Environmental Entomology*, **26**, 797–804.

Comins, H.N. and Fletcher, B.S. (1988) Simulation of fruit fly population dynamics, with particular reference to the olive fly, *Dacus oleae*. *Ecological Modelling*, **40**, 213–31.

Comins, H.N. and Hassell, M.P. (1976) Predation in multi-prey communities. *Journal of Theoretical Biology*, **62**, 93–114.

Comins, H.N. and Hassell, M.P. (1979) The dynamics of optimally foraging predators and parasitoids. *Journal of Animal Ecology*, **48**, 335–51.

Comins, H.N. and Hassell, M.P. (1996) Persistence in multi-species host-parasitoid interactions in spatially distributed models with local dispersal. *Journal of Theoretical Biology*, **183**, 19–28.

Comins, H.N. and Wellings, P.W. (1985) Density related parasitoid sex ratio: influence on host-parasitoid dynamics. *Journal of Animal Ecology*, **54**, 583–94.

Commonwealth Scientific and Industrial Research Organisation (CSIRO) (1991) *The Insects of Australia, Vol. 1.* Melbourne University Press, Carlton.

Conn, D.L.T. (1976) Estimates of population size and longevity of adult narcissus bulb fly *Merodon equestris* Fab. (Diptera: Syrphidae). *Journal of Applied Ecology*, **13**, 429–34.

Conrad, K.F. and Herman, T.B. (1990) Seasonal dynamics, movements and the effects of experimentally increased female densities on a population of imaginal *Calopteryx aequabilis* (Odonata: Calopterygidae). *Ecological Entomology*, **15**, 119–29.

Conrad, K.F. and Pritchard, G. (1990) Preoviposition mate-guarding and mating behaviour of *Argia vivida* (Odonata: Coenagrionidae). *Ecological Entomology*, **15**, 363–70.

Consoli, E.L., Kitajima, E.W. and Parra, J.R.P. (1999) Sensilla on the antenna and ovipositor of the parasitic wasps *Trichograma galloi* Zucchi and *T. pretiosum* Riley (Hymenoptera., Trichogrammatidae). *Microscopy Research and Technique*, **45**, 313–24.

Conti, E., Salerno, G., Bin, F., Williams, H.J. and Vinson, S.B. (2003) Chemical cues from *Murgantia histrionica* eliciting host location and recognition in the egg parasitoid *Trissolcus brochymenae*. *Journal of Chemical Ecology*, **29**, 115–30.

Cook, D. and Stoltz, D.B. (1983) Comparative serology of viruses isolated from ichneumonid parasitoids. *Virology*, **130**, 215–20.

Cook, J.M. (1993a) Sex determination in the Hymenoptera: a review of models and evidence. *Heredity*, **71**, 421–35.

Cook, J.M. (1993b) Inbred lines as reservoirs of sex alleles in parasitoid rearing programs. *Environmental Entomology*, **22**, 1213–16.

Cook, J.M. (1993c) Experimental tests of sex determination in *Goniozus nephantidis* (Hymenoptera: Bethylidae). *Heredity*, **71**, 130–7.

Cook, J.M. (1996) A beginner's guide to molecular markers for entomologists. *Antenna*, **20**, 53–62.

Cook, J.M. (2002) Sex determination in invertebrates, in *Sex Ratios: Concepts and Research Methods* (ed. I.C.W. Hardy). Cambridge University Press, Cambridge, pp. 178–94.

Cook, J.M. and Butcher, R.D.J. (1999) The transmission and effects of *Wolbachia* bacteria in parasitoids. *Researches on Population Ecology*, **41**, 15–28.

Cook, J.M. and Crozier, R.H. (1995) Sex determination and population biology in the Hymenoptera. *Trends in Ecology and Evolution*, **10**, 281–6.

Cook, J.M., Rivero Lynch, A.P. and Godfray, H.C.J. (1994) Sex ratio and foundress number in the parasitoid wasp *Bracon hebetor*. *Animal Behaviour*, **47**, 687–96.

Cook, J.M., Compton, S.G., Herre, E.A. and West, S.A. (1997) Alternative mating tactics and extreme male dimorphism in fig wasps. *Proceedings of the Royal Society of London B.*, **264**, 747–54.

Cook, J.M., Bean, D. and Power, S. (1999) Fatal fighting in fig wasps – GBH in time and space. *Trends in Ecology and Evolution*, **14**, 257–9.

Cook, L.M., Bower, P.P. and Croze, H.J. (1967) The accuracy of a population estimation from multiple recapture data. *Journal of Animal Ecology*, **36**, 57–60.

Cook, R.M. and Cockrell, B.J. (1978) Predator ingestion rate and its bearing on feeding time and the theory of optimal diets. *Journal of Animal Ecology*, **47**, 529–48.

Cook, R.M. and Hubbard, S.F. (1977) Adaptive searching strategies in insect parasitoids. *Journal of Animal Ecology*, **46**, 115–25.

Cook, S.P., Hain, F.P. and Smith, H.R. (1994) Oviposition and pupal survival of gypsy moth (Lepidoptera: Lymantriidae) in Virginia and North Carolina pine-hardwood forests. *Environmental Entomology*, **23**, 360–66.

Coombes, D.S. and Sotherton, N.W. (1986) The dispersal and distribution of polyphagous predatory Coleoptera in cereals. *Annals of Applied Biology*, **108**, 461–74.

Coombs, M.T. (1997) Influence of adult food deprivation and body size on fecundity and longevity of *Trichopoda giacomellii*: a South American parasitoid of *Nezara viridula*. *Biological Control*, **8**, 119–23.

Cooper, B.A. (1945) Hymenopterist's Handbook. *The Amateur Entomologist*, **7**, 1–160.

Copland, M.J.W. (1976) Female reproductive system of the Aphelinidae (Hymenoptera: Chalcidoidea). *International Journal of Insect Morphology and Embryology*, **5**, 151–66.

Copland, M.J.W. and King, P.E. (1971) The structure and possible function of the reproductive system in some Eulophidae and Tetracampidae. *Entomologist*, **104**, 4–28.

Copland, M.J.W. and King, P.E. (1972a) The structure of the female reproductive system in the Pteromalidae (Chalcidoidea: Hymenoptera). *Entomologist*, **105**, 77–96.

Copland, M.J.W. and King, P.E. (1972b) The structure of the female reproductive system in the Eurytomidae (Chalcidoidea: Hymenoptera). *Journal of Zoology*, **166**, 185–212.

Copland, M.J.W. and King, P.E. (1972c) The structure of the female reproductive system in the Torymidae (Hymenoptera: Chalcidoidea). *Transactions of the Royal Entomological Society of London*, **124**, 191–212.

Copland, M.J.W. and King, P.E. (1972d) The structure of the female reproductive system in the Chalcididae (Hym.). *Entomologist's Monthly Magazine*, **107**, 230–39.

Copland, M.J.W., King, P.E. and Hill, D.S. (1973) The structure of the female reproductive system in the Agaonidae (Chalcidoidea, Hymenoptera). *Journal of Entomology* (A), **48**, 25–35.

Corbet, S.A. (1978) Bee visits and the nectar of *Echium vulgare* L. and *Sinapis alba* L.. *Ecological Entomology*, **3**, 25–37.

Corbet, S.A. (2000) Butterfly nectaring flowers: butterfly morphology and flower form. *Entomologia Experimentalis et Applicata*, **96**, 289–98.

Corbet, S.A., Unwin, D.M. and Prŷs-Jones, O.E. (1979) Humidity, nectar and insect visitors to flowers, with special reference to *Crataegus*, *Tilia* and *Echium*. *Ecological Entomology*, 4, 9–22.

Corbet, S.A., Saville, N.M., Fussell, M. Prŷs-Jones, O.E. and Unwin, D.M. (1995) The competition box: a graphical aid to forecasting pollinator performance. *Journal of Applied Ecology*, **32**, 707–19.

Corbett, A. and Plant, R.E. (1993) Role of movement in the response of natural enemies to agroecosystem diversification: a theoretical evaluation. *Environmental Entomology*, **22**, 519–31.

Corbett, A. and Rosenheim, J.A. (1996a) Quantifying movement of a minute parasitoid, *Anagrus epos* (Hymenoptera, Mymaridae) using fluorescent dust marking and recapture. *Biological Control*, **6**, 35–44.

Corbett, A. and Rosenheim, J.A. (1996b) Impact of a natural enemy overwintering refuge and its interactions with the surrounding landscape. *Ecological Entomology*, **21**, 155–64.

Corbett, A., Murphy, B.C., Rosenheim, J.A. and Bruins, P. (1996) Labelling an egg parasitoid, *Anagrus epos* (Hymenoptera: Mymaridae), with rubidium within an overwintering refuge. *Environmental Entomology*, **25**, 29–38.

Cordoba-Aguilar, A. (2002) Sensory trap as the mechanism of sexual selection in a damselfly genitalic trait (Insecta: Calopterygidae). *American Naturalist*, **160**, 594–601.

Corey, D., Kambhampati, S. and Wilde, G. (1998) Electrophoretic analysis of *Orius insidiosus* (Hemiptera: Anthocoridae) feeding habits in field corn. *Journal of the Kansas Entomological Society*, **71**, 11–17.

Cornelius, M. and Barlow, C.A. (1980) Effect of aphid consumption by larvae on development and reproductive efficiency of a flower-fly, *Syrphus corollae* (Diptera: Syrphidae). *Canadian Entomologist*, **112**, 989–92.

Cornelius, M.L., Duan, J.J. and Messing, R.H. (1999) Visual stimuli and response of female Oriental fruit flies (Diptera: Tephritidae) to fruit-mimicking traps. *Journal of Economic Entomology*, **92**, 121–9.

Cornell, H. and Pimentel, D. (1978) Switching in the parasitoid *Nasonia vitripennis* and its effect on host competition. *Ecology*, **59**, 297–308.

Cornell, H.V. (1988) Solitary and gregarious brooding, sex ratios and the incidence of thelytoky in the parasitic Hymenoptera. *American Midland Naturalist*, **119**, 63–70.

Cornell, H.V., Hawkins, B.A. and Hochberg, M.E. (1998) Towards an empirically–based theory of herbivore demography. *Ecological Entomology*, **23**, 340–9.

Corrigan, J.E. and Lashomb, J.H. (1990) Host influences on the bionomics of *Edovum puttleri* (Hymenoptera: Eulophidae): effects on size and reproduction. *Environmental Entomology*, **19**, 1496–502.

Cortesero, A.M., Stapel, J.O. and Lewis, W.J. (2000) Understanding and manipulating plant attributes to enhance biological control. *Biological Control*, **17**, 35–49.

Cottrell, T.E. and Yeargan, K.V. (1998) Influence of native weed, *Acalypha ostryaefolia* (Euphorbiaceae) on *Coleomegilla maculata* (Coleoptera: Coccinellidae) population density, predation and cannibalism in sweet corn. *Environmental Entomology*, **27**, 1375–85.

Coulson, R.M.R., Curtis, C.F., Ready, P.D., Hill, N. and Smith, D.F. (1990) Amplification and analysis of human DNA present in mosquito bloodmeals. *Medical and Veterinary Entomology*, **4**, 357–66.

Cousin, G. (1933) Étude biologique d'un Chalcidien, *Mormoniella vitripennis*. *Bulletin Biologique de France et Belgique*, **67**, 371–400.

Couty, A., Kaiser, L., Huet, D. and Pham-Delegue, M.H. (1999) The attractiveness of different odour sources from the fruit-host complex on *Leptopilina boulardi*, a larval parasitoid of frugivorous *Drosophila* spp. *Physiological Entomology*, **24**, 76–82.

Cowgill, S.E., Wratten, S.D. and Sotherton, N.W. (1993) The selective use of floral resources by the hoverfly *Episyrphus balteatus* (Diptera: Syrphidae) on farmland. *Annals of Applied Biology*, **122**, 223–31.

Cox, D.R. (1972) Regression models and life tables. *Biometrics*, **38**, 67–77.

Cox, D.R. and Oakes, D. (1984) *Analysis of Survival Data*. Chapman and Hall, London.

Craig, C.C. (1953) On the utilisation of marked specimens in estimating populations of flying insects. *Biometrika*, **40**, 170–76.

Crawley, M.J. (1993) *GLIM for Ecologists*. Blackwell, Oxford.

Crawley, M.J. (2002) *Statistical Computing: An Introduction to Data Analysis Using S-Plus*. John Wiley, New York.

Creswell, J.E., Osborne, J.L. and Goulson, D. (2000) An economic model of the limits to foraging range in central place foragers with numerical solutions for bumblebees. *Ecological Entomology*, **25**, 249–55.

Cresswell, M.J. (1995) Malaise trap: collection attachment modification and collection fluid. *Weta*, **18**, 10–11.

Crocker, R.L., Rodriguez-del-Bosque, L.A., Nailon, W.T. and Wei, X. (1996) Flight periods in Texas of three parasites (Diptera: Pyrgotidae) of adult *Phyllophaga* spp. (Coleoptera: Scarabaeidae), and egg production by *Pyrgota undata*. *Southwestern Entomologist*, **21**, 317–24.

Croft, P. and Copland, M. (1993) Size and fecundity in *Dacnusa sibirica* Telenga. *Bulletin OILB/SROP*, **16**, 53–6.

Cronin, J.T. (2003a) Matrix heterogeneity and host-parasitoid interactions in space. *Ecology*, **84**, 1506–16.

Cronin, J.T. (2003b) Movement and spatial population structure of a prairie planthopper. *Ecology*, **84**, 1179–88.

Cronin, J.T. and Strong, D.R. (1996) Genetics of oviposition success of a thelytokous fairyfly parasitoid, *Anagrus delicatus*. *Heredity*, **76**, 43–54.

Cronin, J.T. and Strong, D.R. (1999) Dispersal-dependent oviposition and the aggregation of parasitism. *American Naturalist*, **154**, 23–36.

Crook, A.M.E. (1996) A monoclonal antibody technique for investigating predation on vine weevil in soft fruit plantations. *Proceedings of the Brighton Crop Protection Conference – Pests and Diseases 1996, BCPC, Farnham, Surrey UK*, 435–6.

Crook, N.E. and Payne, C.C. (1980) Comparison of three methods of ELISA for baculoviruses. *Journal of General Virology*, **46**, 29–37.

Crook, N.E. and Sunderland, K.D. (1984) Detection of aphid remains in predatory insects and spiders by ELISA. *Annals of Applied Biology*, **105**, 413–22.

Crowle, A.J. (1980) Precipitin and microprecipitin reactions in fluid medium and in gels, in *Manual of Clinical Immunology* (eds N.R. Rose and H. Friedman). American Society for Microbiology, Washington DC, pp. 3–14.

Crowley, P.H. and Linton, M.C. (1999) Antlion foraging: tracking prey across space and time. *Ecology*, **80**, 2271–82.

Crowley, P.H. and Martin, E.K. (1989) Functional responses and interference within and between year classes of a dragonfly population, *Journal of the North American Benthological Society*, **8**, 211–21.

Crozier, R.H. (1971) Heterozygosity and sex determination in haplo-diploidy. *American Naturalist*, **105**, 399–412.

Crozier, R.H. (1975) Hymenoptera, in *Animal Cytogenetics 3: Insecta 7* (ed. B. John). Gebruder Borntraeger, Berlin and Stuttgart, pp. 1–95.

Crozier, R.H. (1977) Evolutionary genetics of the Hymenoptera. *Annual Review of Entomology*, **22**, 263–288.

Crudgington, H. and Siva-Jothy, M.T. (2000) Genital damage, kicking and early death. *Nature*, **407**, 855–6.

Crum, D.A., Weiser, L.A. and Stamp, N.E. (1998) Effects of prey scarcity and plant material as a dietary supplement on an insect predator. *Oikos*, **81**, 549–57.

Crumb, S.E., Eide, P.M. and Bonn, A.E. (1941) The European earwig. *USDA Technical Bulletin*, **766**, 1–76.

Culin, J.D. and Rust, R.W. (1980) Comparison of the ground surface and foliage dwelling spider communities in a soybean habitat. *Environmental Entomology*, **9**, 577–82.

Culin, J.D. and Yeargan, K.V. (1983) Comparative study of spider communities in alfalfa and soybean ecosystems: ground-surface spiders. *Annals of the Entomological Society of America*, **76**, 832–8.

Cumming, J.M. (1994) Sexual selection and the evolution of dance fly mating systems (Diptera: Empididae; Empidinae). *Canadian Entomologist*, **126**, 907–20.

Cunha, A.B. and Kerr, W.E. (1957) A genetical theory to explain sex determination by arrhenotokous parthenogenesis. *Forma et Functio*, **1**, 33–36.

Dafni, A. (1992) *Pollination Ecology: A Practical Approach*. IRL Press, Oxford.

Dahms, E.C. (1984a) Revision of the genus *Melittobia* (Chalc. Eulophidae) with descriptions of seven new species. *Memoirs of the Queensland Museum*, **21**, 271–336.

Dahms, E.C. (1984b) A review of the biology of species in the genus *Melittobia* (Hym., Eulophidae) with interpretations and additions using observations on *Melittobia australica*. *Memoirs of the Queensland Museum*, **21**, 337–60.

van Dam, N.M., Hadwich, K. and Baldwin, I.T. (2000) Induced responses in *Nicotiana attenuata* affect behavior and growth of the specialist herbivore *Manduca sexta*. *Oecologia* **122**, 371–9.

Danilevskii, A.S. (1965) *Photoperiodism and Seasonal Development in Insects*. Oliver and Boyd, Edinburgh and London.

Danks, H.V. (1971) Biology of some stem-nesting Aculeate Hymenoptera. *Transactions of the Royal Entomological Society of London*, **122,** 323–99.

Darling, D.C. and Werren, J.H. (1990) Biosystematics of two new species of *Nasonia* (Hym. Pteromalidae) reared from birds' nests in North America. *Annals of the Entomological Society of America*, **83**, 352–70.

Davidson, J. (1944) On the relationship between temperature and the rate of development of insects at constant temperatures. *Journal of Animal Ecology*, **13**, 26–38.

Davies, D.H., Burghardt, R.L. and Vinson, S.B. (1986) Oögenesis of *Cardiochiles nigriceps* Viereck (Hymenoptera: Braconidae): histochemistry and development of the chorion with special reference to the fibrous layer. *International Journal of Insect Morphology and Embryology*, **15**, 363–74.

Davies, J. (1974) The effect of age and diet on the ultrastructure of Hymenopteran flight muscle. *Experimental Gerontology*, **9**, 215–19.

Davies, K.F. and Margules, C.R. (1998) Effects of habitat fragmentation on carabid beetles: experimental evidence. *Journal of Animal Ecology*, **67**, 460–71.

Davies, N.B. (1991) Mating systems, in *Behavioural Ecology 3rd edn*. (eds J.R. Krebs and N.B. Davies). Blackwell Scientific Publishing, Oxford, pp. 263–94.

Davies, N.B. (1992) *Dunnock Behaviour and Social Evolution*. Oxford Series in Ecology and Behaviour. Oxford University Press, Oxford.

Davis, A.J. (1993) The Observer: an integrated software package for behavioural research. *Journal of Animal Ecology*, **62**, 218–9.

Davis, J.R. and Kirkland, R.L. (1982) Physiological and environmental factors related to the dispersal flight of the convergent lady beetle *Hippodamia convergens*. *Journal of the Kansas Entomological Society*, **55**, 187–96.

Dawah, H.A., Hawkins, B.A. and Claridge, M.F. (1995) Structure of the parasitoid communities of grass-feeding chalcid wasps. *Journal of Animal Ecology*, **64**, 708–20.

Dawkins, R. (1980) Good strategy or Evolutionary Stable Strategy?, in *Sociobiology: Beyond Nature/Nurture?* (eds Barlow, G.W. and J. Silverberg). Westview Press, Boulder, pp. 331–67.

Dawson, N. (1965) A comparative study of the ecology of eight species of fenland Carabidae (Coleoptera). *Journal of Animal Ecology*, **34**, 299–314.

Day, S.E. and Jeanne, R.L. (2001) Food volatiles as attractants for yellowjackets (Hymenoptera: Vespidae). *Environmental Entomology*, **30**, 157–65.

Day, W.H. (1994) Estimating mortality caused by parasites and diseases of insects: comparisons of the dissection and rearing methods. *Environmental Entomology*, **23**, 543–50.

Dean, D.A. and Sterling, W.L. (1990) Seasonal patterns of spiders captured in suction traps in Eastern Texas. *Southwestern Entomologist*, **15**, 399–412.

Dean, G.J. and Satasook, C. (1983) Response of *Chrysoperla carnea* (Stephens) (Neuroptera: Chrysopidae) to some potential attractants. *Bulletin of Entomological Research*, **73**, 619–24.

Dean, G.J., Jones, M.G. and Powell, W. (1981) The relative abundance of the hymenopterous parasites attacking *Metopolophium dirhodum* (Walker) and *Sitobion avenae* (F.) (Hemiptera. Aphididae) on cereals during 1973–79 in southern England. *Bulletin of Entomological Research*, **71**, 307–15.

DeAngelis, D.L. and Waterhouse, J.C. (1987) Equilibrium and nonequilibrium concepts in ecological models. *Ecological Monographs*, **57**, 1–21.

DeAngelis, D.L., Goldstein, R.A. and O'Neill, R.V. (1975) A model for trophic interactions. *Ecology*, **56**, 881–92.

DeBach, P. (1943) The importance of host-feeding by adult parasites in the reduction of host populations. *Journal of Economic Entomology*, **36**, 647–58.

DeBach, P. (1946) An insecticidal check method for measuring the efficacy of entomophagous insects. *Journal of Economic Entomology*, **39**, 695–7.

DeBach, P. (1955) Validity of insecticidal check method as measure of the effectiveness of natural enemies of diaspine scale insects. *Journal of Economic Entomology*, **48**, 584–8.

DeBach, P. (1971) The use of natural enemies in insect pest management ecology. *Proceedings of the Tall Timbers Conference on Ecological Animal Control by Habitat Management*, **3**, 211–33.

DeBach, P. and Huffaker, C.B. (1971) Experimental techniques for evaluation of the effectiveness of natural enemies, in *Biological Control* (ed. C.B. Huffaker). Plenum, New York, pp. 113–40.

DeBach, P. and Rosen, D. (1991) *Biological Control by Natural Enemies, 2nd edn.* Cambridge University Press, Cambridge.

De Barro, P.J. (1991) A cheap lightweight efficient vacuum sampler. *Journal of the Australian Entomological Society*, **30**, 207–8.

Debout, G., Fauvergue, X. and Fleury, F. (2002) The effect of foundress number on sex ratio under partial local mate competition. *Ecological Entomology*, **27**, 242–6.

Decker, U.M. (1988) Evidence for semiochemicals affecting the reproductive behaviour of the aphid parasitoids *Aphidius rhopalosiphi* De Stefani-Perez and *Praon volucre* Haliday (Hymenoptera: Aphidiidae) – a contribution towards integrated pest management in cereals. University of Hohenheim, Ph.D. Thesis.

Decker, U.M., Powell, W. and Clark, S.J. (1993) Sex pheromones in the cereal aphid parasitoids *Praon volucre* and *Aphidius rhopalosiphi*. *Entomologia Experimentalis et Applicata*, **69**, 33–9.

De Clercq, P. and Degheele, D. (1997) Effects of mating status on body weight, oviposition, egg load, and predation in the predatory stinkbug *Podisus maculiventris* (Heteroptera: Pentatomidae). *Annals of the Entomological Society of America*, **90**, 121–7.

De Clercq, R. and Pietraszko, R. (1983) Epigeal arthropods in relation to predation of cereal aphids, in *Aphid Antagonists* (ed. R. Cavalloro). A.A. Balkema, Rotterdam, pp. 88–92.

Dedeine, F., Vavre, F., Fleury, F., Loppin, B., Hochberg, M.E. and Bouletreau, M. (2001). Removing symbiotic *Wolbachia* bacteria specifically inhibits oogenesis in a parasitic wasp. *Proceedings of the National Academy of Sciences*, **98**, 6247–52.

Deevey, E.S. (1947) Life tables for natural populations of animals. *Quarterly Review of Biology*, **22**, 283–314.

DeJong, G. (1979) The influence of the distribution of juveniles over patches of food on the dynamics of a population. *Netherlands Journal of Zoology*, **29**, 33–51.

De Keer, R. and Maelfait, J.P. (1987) Life-history of *Oedothorax fuscus* (Blackwall, 1834) (Araneae, Linyphiidae) in a heavily grazed pasture. *Revue d'Ecologie et de Biologie du Sol*, **24**, 171–85.

De Keer, R. and Maelfait, J.P. (1988) Observations on the life cycle of *Erigone atra* (Araneae, Erigoninae) in a heavily grazed pasture. *Pedobiologia*, **32**, 201–12.

De Kraker, J., Van Huis, A., van Lenteren, J.C., Heong, K.L. and Rabbinge, R. (1999) Egg mortality of rice leaffolders *Cnaphalocrocis medinalis* and *Marasmia patnalis* in irrigated rice fields. *BioControl*, **44**, 449–71.

De Kraker, J., Van Huis, A., van Lenteren, J.C., Heong, K.L. and Rabbinge, R. (2001) Effect of prey and predator density on predation of rice leaffolder eggs by the cricket *Metioche villaticollis*. *Biocontrol Science and Technology*, **11**, 67–80.

Delettre, Y.R., Morvan, N., Tréhen, P. and Grootaert, P. (1998) Local biodiversity and multi-habitat use in empidoid flies (Insecta: Diptera, Empidoidea). *Biodiversity and Conservation*, **7**, 9–25.

DeLong, D.M. (1932) Some problems encountered in the estimation of insect populations by the sweeping method. *Annals of the Entomological Society of America*, **25**, 13–17.

Delves, P.J. (1997) *Antibody Production: Essential Techniques.* John Wiley and Sons, Chichester.

Demichelis, S. and Manino, A. (1995) Electrophoretic detection of dryinid parasitoids in *Empoasca* leafhoppers. *Journal of Applied Entomology*, **119**, 543–5.

Demichelis, S. and Manino, A. (1998) Electrophoretic detection of parasitism by Dryinidae in Typhlocybinae leafhoppers (Homoptera: Auchenorrhyncha). *Canadian Entomologist*, **130**, 407–14.

De Moraes, R.R., Funderburk, J.E. and Maruniak, J.E. (1998) Polymerase Chain Reaction techniques to detect multiple nucleopolyhedrovirus in *Anticarsia gemmatalis* (Lepidoptera: Noctuidae) and predator populations in soybean. *Environmental Entomology*, **27**, 968–75.

Dempster, J.P. (1960) A quantitative study of the predators on the eggs and larvae of the broom beetle, *Phytodecta olivacea* Forster, using the precipitin test. *Journal of Animal Ecology*, **29**, 149–67.

Dempster, J.P. (1967) The control of *Pieris rapae* with DDT. 1. The natural mortality of the young stages of *Pieris*. *Journal of Applied Ecology*, **4**, 485–500.

Dempster, J.P. (1975) *Animal Population Ecology*. Academic Press, London.

Dempster, J.P. (1983) The natural control of populations of butterflies and moths. *Biological Reviews*, **58**, 461–81.

Dempster, J.P. (1989) Insect introductions: natural dispersal and population persistence in insects. *The Entomologist*, **108**, 5–13.

Dempster, J.P and Lakhani, K.H. (1979) A population model for cinnabar moth and its food plant ragwort. *Journal of Animal Ecology*, **48**, 143–65.

Dempster, J.P. and Pollard, E. (1986) Spatial heterogeneity, stochasticity and the detection of density dependence in animal populations. *Oikos*, **46**, 413–16.

Dempster, J.P., Lakhani, K.H. and Coward, P.A. (1986) The use of chemical composition as a population marker in insects: a study of the Brimstone butterfly. *Ecological Entomology*, **11**, 51–65.

Deng, D.A. and Li, B.Q. (1981). Collecting ground beetles (Carabidae) in baited pitfall traps. *Insect Knowledge*, **18**, 205–7.

Dennis, P. (1991) Temporal and spatial distribution of arthropod predators of the aphids *Rhopalosiphum padi* (L.) and *Sitobion avenae* (F.) in cereals next to field margin habitats. *Norwegian Journal of Agricultural Science*, **5**, 79–88.

Dennis, P. and Wratten, S.D. (1991) Field manipulation of individual staphylinid species in cereals and their impact on aphid populations. *Ecological Entomology*, **16**, 17–24.

Dennis, P., Wratten, S.D. and Sotherton, N.W (1991) Mycophagy as a factor limiting predation of aphids (Hemiptera: Aphididae) by staphylinid beetles (Coleoptera: Staphylinidae) in cereals. *Bulletin of Entomological Research*, **81**, 25–31.

Dennison, D.F. and Hodkinson, I.D. (1983) Structure of the predatory beetle community in woodland soil ecosystem. 1. Prey selection. *Pedobiologia*, **25**, 109–15.

Dennison, D.F. and Hodkinson, I.D. (1984) Structure of the predatory beetle community in a woodland soil ecosystem. IV. Population densities and community composition. *Pedobiologia*, **26**, 157–70.

Denno, R.F., McClure, M.S. and Ott, J.R. (1995) Interspecific interactions in phytophagous insects: competition rexamined and resurrected. *Annual Review of Entomology*, **40**, 297–331.

Denno, R.F., Gratton, C., Peterson, M.A., Langellotto, G.A., Finke, D.L. and Huberty, A.F. (2002) Bottom-up forces mediate natural-enemy impact in a phytophagous insect community. *Ecology*, **83**, 1443–58.

Denoth, M., Frid, L. and Myers, J.H. (2002) Multiple agents in biological control: improving the odds? *Biological Control*, **24**, 20–31

Dent, D. (1991) *Insect Pest Management*. CAB International, Wallingford.

Denver, K. and Taylor, P.D. (1995) An inclusive fitness model for the sex ratio in a partially sibmating population with inbreeding cost. *Evolutionary Ecology*, **9**, 318–327.

Desender, K. and Maelfait, J.P. (1983) Population restoration by means of dispersal, studied for different carabid beetles (Coleoptera, Carabidae) in a pasture ecosystem, in *New Trends in Soil Biology* (eds P., Lebrun, H.M. Andre, A. De Medts, C. Grégoire-Wibo and G. Wauthy). Dieu-Brichart, Ottignies-Louvain-la Neuve, pp. 541–50.

Desender, K. and Maelfait, J.P. (1986) Pitfall trapping within enclosures: a method for estimating the relationship between the abundances of coexisting carabid species (Coleoptera: Carabidae). *Holarctic Ecology*, **9**, 245–50.

Desender, K. and Segers, R. (1985) A simple device and technique for quantitative sampling of riparian beetle populations with some carabid and staphylinid abundance estimates on different riparian habitats (Coleoptera). *Revue d'Écologie et de Biologie du Sol*, **22**, 497–506.

Desender, K., Maelfait, J.P., D'Hulster, M. and Vanhereke, L. (1981) Ecological and faunal studies on Coleoptera in agricultural land I. Seasonal occurrence of Carabidae in the grassy edge of a pasture. *Pedobiologia*, 22, 379–84.

Desender, K., Mertens, J. D'Hulster, M. and Berbiers, P. (1984) Diel activity patterns of Carabidae (Coleoptera), Staphylinidae (Coleoptera) and

Collembola in a heavily grazed pasture. *Revue d'Écologie et de Biologie du Sol*, **21**, 347–61.

Desender, K., van den Broeck, D. and Maelfait, J.P. (1985) Population biology and reproduction in *Pterostichus melanarius* III. (Coleoptera, Carabidae) from a heavily grazed pasture ecosystem. *Mededelingen van de Fakulteit Landbouwwetenschappen Rijksuniversiteit Gent*, **50**, 567–75.

Desharnais, R.A., Costantino, R.F., Cushing, J.M., Henson, S.M. and Dennis, B. (2001) Chaos and population control of insect outbreaks. *Ecology Letters*, **4**, 229–35.

Desouhant, E., Driessen, G., Lapchin, L., Wielaard, S. and Bernstein, C. (2003) Dispersal between host populations in field conditions: navigation rules in the parasitoid *Venturia canescens*. *Ecological Entomology*, **28**, 257–67.

De Vis, R.M.J., Fuentes, L.E. and van Lenteren, J.C. (2002) Life history of *Amitus fuscipennis* (Hym., Platygastridae) as parasitoid of the greenhouse whitefly *Trialeurodes vaporariorum* (Hom., Aleyrodidae) on tomato as function of temperature. *Journal of Applied Entomology*, **126**, 24–33.

De Vis, R.M.J., Mendez, H. and van Lenteren, J.C. (2003) Comparison of foraging behavior, interspecific host discrimination, and competition of *Encarsia formosa* and *Amitus fuscipennis*. *Journal of Insect Behavior*, **16**, 117–52.

Dewar, A.M., Dean G.J. and Cannon, R. (1982) Assessment of methods for estimating the numbers of aphids (Hemiptera: Aphididae) in cereals. *Bulletin of Entomological Research*, **72**, 675–85.

Dicke, M. and De Jong, M. (1988) Prey preference of the phytoseiid mite *Typhlodromus pyri*. 2. Electrophoretic diet analysis. *Experimental and Applied Acarology*, **4**, 15–25.

Dicke, M. and Sabelis, M.W. (1988) Infochemical terminology: based on a cost benefit analysis rather than origin of compounds? *Functional Ecology*, **2**, 131–9.

Dicke, M., Sabelis, M.W., Dejong, M. and Alers, M.P.T. (1990) Do Phytoseiid mites select the best prey species in terms of reproductive success. *Experimental and Applied Acarology*, **8**, 161–73.

Dickens, J.C. (1999) Predator-prey interactions: olfactory adaptations of generalist and specialist predators. *Agricultural and Forest Entomology*, **1**, 47–54.

Diefenbach, C.W. and Dveksler, G.S. (1995) *PCR Primer. A Laboratory Manual*. Cold Spring Harbor Laboratory Press, New York.

Dietrick, E.J. (1961) An improved back pack motor fan for suction sampling of insect populations. *Journal of Economic Entomology*, **54**, 394–5.

Dietrick, E.J., Schlinger, E.I. and van den Bosch, R. (1959) A new method for sampling arthropods using a suction collecting machine and modified Berlese funnel separator. *Journal of Economic Entomology*, **52**, 1085–91.

Digby, P.G.N. and Kempton, R.A. (1987) *Multivariate Analysis of Ecological Communities*. Chapman and Hall, London.

Digweed, S.C. (1993) Selection of terrestrial gastropod prey by Cychrine and Pterostichine ground beetles (Coleoptera: Carabidae). *Canadian Entomologist*, **125**, 463–72.

Digweed, S.C., Currie, C.R., Cárcamo, H.A. and Spence, J.R. (1995) Digging out the "digging-in effect" of pitfall traps; influences of depletion and disturbance on catches of ground beetles (Coleoptera; Carabidae). *Pedobiologia*, **39**, 561–76.

van Dijken, M.J. (1991) A cytological method to determine primary sex ratio in the solitary parasitoid *Epidinocarsis lopezi*. *Entomologica experimentalis et Applicata*, **60**, 301–4.

van Dijken, M.J. and van Alphen, J.J.M. (1991) Mutual interference and superparasitism in the solitary parasitoid *Epidinocarsis lopezi*. *Mededelingen van de Landbouwwetenschappen Rijksuniversiteit Gent*, **56**, 1003–10.

van Dijken, M.J. and van Alphen, J.J.M. (1998) The ecological significance of differences in host detection behaviour in coexisting parasitoid species. *Ecological Entomology*, **23**, 265–70.

van Dijken, M.J., van Alphen, J.J.M. and van Stratum, P. (1989) Sex allocation in *Epidinocarsis lopezi*: local mate competition. *Entomologia Experimentalis et Applicata*, **52**, 249–55.

Dijkerman, H.J. (1988) Notes on the parasitation behaviour and larval development of *Trieces tricarinatus* and *Triclistus yponomeutae* (Hymenoptera, Ichneumonidae), endoparasitoids of the genus *Yponomeuta* (Lepidoptera, Yponomeutidae). *Proceedings Koninklijke Nederlandse Akademie van Wetenschappen*, **91**, 19–30.

Dijkerman, H.J. (1990) Suitability of eight *Yponomeuta* species as hosts of *Diadegma armillata*. *Entomologia Experimentalis et Applicata*, 54, 173–80.

Dijkerman, H.J. and Koenders, J.T.H. (1988) Competition between *Trieces tricarinatus* and *Triclistus yponomeutae* in multiparasitized hosts, *Entomologia Experimentalis et Applicata*, **47**, 289–95.

Dijkstra, L.J. (1986) Optimal selection and exploitation of hosts in the parasitic wasp *Colpoclypeus florus* (Hym., Eulophidae). *Netherlands Journal of Zoology*, **36, 1** 77–301.

Dimitry, N.Z. (1974) The consequences of egg cannibalism in *Adalia bipunctata* (Coleoptera: Coccinellidae). *Entomophaga*, **19**, 445–51.

Dinter, A. (1995) Estimation of epigeic spider population densities using an intensive D-vac sampling technique and comparison with pitfall trap catches in winter wheat. *Acta Jutlandica*, **70**, 23–32.

Dinter, A. and Poehling, H.M. (1992) Spider populations in winter wheat fields and the side-effects of insecticides. *Aspects of Applied Biology*, **31**, 77–85.

Dixon, A.F.G. (1958) The escape responses shown by certain aphids to the presence of the coccinellid *Adalia decempunctata* (L.). *Transactions of the Royal Entomological Society of London*, **110**, 319–34.

Dixon, A.F.G. (1959a) An experimental study of the searching behaviour of the predatory coccinellid beetle *Adalia decempunctata* (L.). *Journal of Animal Ecology*, **28**, 259–81.

Dixon, A.F.G. (1970) Factors limiting the effectiveness of the coccinellid beetle, *Adalia bipunctata* (L.), as a predator of the sycamore aphid, *Drepanosiphum platanoides* (Schr.). *Journal of Animal Ecology*, **39**, 739–51.

Dixon, A.F.G. (2000) *Insect Predator-prey Dynamics: Ladybird Beetles and Biological Control*. Cambridge University Press, Cambridge.

Dixon, A.F.G. and Barlow, N.D. (1979) Population regulation in the lime aphid. *Zoological Journal of the Linnean Society*, **67**, 225–37.

Dixon, A.F.G. and Guo, Y. (1993) Egg and cluster size in ladybird beetles (Coleoptera: Coccinellidae): the direct and indirect effects of aphid abundance. *European Journal of Entomology*, **90**, 457–63.

Dixon, P.L. and McKinlay, R.G. (1989) Aphid predation by harvestmen in potato fields in Scotland. *Journal of Arachnology*, **17**, 253–5.

Dixon, P.L. and McKinlay, R.G. (1992) Pitfall trap of and predation by *Pterostichus melanarius* and *Pterostichus madidus* in insecticide treated and untreated potatoes. *Entomologia Experimentalis et Applicata*, **64**, 63–72.

Dixon, T.J. (1959b) Studies on oviposition behaviour of Syrphidae. *Transactions of the Royal Entomological Society of London*, **111**, 57–80.

Djemai, I., Casas, J. and Magal, C. (2001) Matching host reactions to parasitoid wasp vibrations. *Proceedings of the Royal Society of London B*, **268**, 2403–8.

Djieto-Lordon, C., Orivel, J. and Dejean, A. (2001) Predatory behavior of the African ponerine ant *Platythyrea modesta* (Hymenoptera : Formicidae). *Sociobiology*, **38**, 303–15.

Doak, P. (2000) The effects of plant dispersion and prey density on parasitism rates in a naturally patchy habitat. *Oecologia*, **122**, 556–67.

Doane, J.F., Scotti, P.D., Sutherland, O.R.W. and Pottinger, R.P. (1985) Serological identification of wireworm and staphylinid predators of the Australian soldier fly (*Inopus rubriceps) and* wireworm feeding on plant and animal food, *Entomologia Experimentalis et Applicata*, **38**, 65–72.

Dobson, S.L. and Tanouye, MA. (1998) Evidence for a genomic imprinting sex determination mechanism in *Nasonia vitripennis* (Hymenoptera; Chalcidoidea). *Genetics*, **149**, 233–42.

Dobson, S.L., Bourtzis, K., Braig, H.R., Jones, B.F., Zhou, W.G., Rousset, F. and O'Neill, S.L. (1999) *Wolbachia* infections are distributed throughout insect somatic and germ line tissues. *Insect Biochemistry and Molecular Biology*, **29**, 153–60.

Dobson, S.L., Marsland, E.J., Veneti, Z., Bourtzis, K. and O'Neill, S.L. (2002) Characterisation of *Wolbachia* host cell range via the *in vitro* establishment of infections. *Applied and Environmental Microbiology*, **68**, 656–60.

Docherty, M. and Leather, S.R. (1997) Structure and abundance of arachnid communities in Scots and lodgepole pine plantations. *Forest Ecology and Management*, **95**, 197–207.

Doerge, R.W. and Churchill, G.A. (1996) Permutation tests for multiple loci affecting a quantitative trait. *Genetics*, **142**, 285–94.

Doerge, R.W. and Rebaï, A. (1996) Significance thresholds for QTL interval mapping test. *Heredity*, **76**, 459–64.

Domenichini, G. (1953) Studio sulla morfologia dell'addome degli Hymenoptera Chalcidoidea. *Bolletino di Zoologie Agraria e Bachicoltura*, **19**, 1–117.

Domenichini, G. (1967) Contributo alla conoscenza biologica e tassinomica dei Tetrastichinae paleartici (Hym., Eulophidae) con particolare riguardo ai materiali dell'Istituto di Entomologia dell'università di Torino. *Bollettino di Zoologia Agraria e Bachicoltura*, **2**, 75–110.

Dominey, W.J. (1984) Alternative mating tactics and evolutionarily stable strategies. *American Zoologist*, **24**, 385–96.

Donaldson, J.S. and Walter, G.H. (1988) Effects of egg availability and egg maturity on the ovipositional activity of the parasitic wasp, *Coccophagus atratus*. *Physiological Entomology*, **13**, 407–17.

Dondale, C.D., Redner, J.H., Farrell, E., Semple, R.B. and Turnbull, A.L. (1970) Wandering of hunting spiders in a meadow. *Bulletin du Musée Nationale d'Histoire Naturelle (Canada)*, **41**, 61–4.

Dondale, C.D., Parent, B. and Pitre, D. (1979) A 6-year study of spiders (Araneae) in a Quebec apple orchard. *Canadian Entomologist*, **111**, 377–80.

Donisthorpe, H. (1936) The dancing habits of some Braconidae. *Entomologist's Record*, **78**, 84.

Dormann, W. (2000) A new pitfall trap for use in periodically inundated habitats, in *Natural History and Applied Ecology of Carabid Beetles* (eds P. Brandmayr, G.L. Lövei, T.Z. Brandmayr, A. Casale and A.V. Taglianti). Pensoft Publishers, Moscow, pp. 247–50.

Dostalkova, I., Kindlmann, P. and Dixon, A.F.G. (2002) Are classical predator-prey models relevant to the real world? *Journal of Theoretical Biology*, **218**, 323–30.

Doutt, R.L. (1959) The biology of parasitic Hymenoptera. *Annual Review of Entomology* **4**, 161–182.

Doutt, R.L., Annecke, D.P. and Tremblay, E. (eds) (1976) Biology and host relationships of parasitoids, in *Theory and Practice of Biological Control* (eds C.B. Huffaker and P.S. Messenger). Academic Press, New York, pp. 143–168.

Dover, B.A. and Vinson, S.B. (1990) Stage-specific effects of *Campoletis sonorensis* parasitism on *Heliothis virescens* development and prothoracic glands. *Physiological Entomology*, **15**, 405–14.

Dowell, F.E., Broce, A.B., Xie, F., Throne, J.E. and Baker, J.E. (2000) Detection of parasitised fly puparia using near infrared spectroscopy. *Journal of Near Infrared Spectroscopy*, **8**, 259–65.

Dowell, R. (1978) Ovary structure and reproductive biologies of larval parasitoids of the alfalfa weevil (Coleoptera: Curculionidae). *Canadian Entomologist*, **110**, 507–12.

Downes, J.A. (1970) The feeding and mating behaviour of the specialized Empidinae (Diptera); observations on four species of *Rhamphomyia* in the high arctic and a general discussion. *Canadian Entomologist*, **102**, 769–91.

Downes, W.L. and Dahlem, G.A. (1987) Keys to the evolution of Diptera: role of Homoptera. *Environmental Entomology*, **16**, 847–54.

Drabble, E. and Drabble, H. (1917) The syrphid visitors to certain flowers, *New Phytologist*, **16**, 105–9.

Drabble, E. and Drabble, H. (1927) Some flowers and their dipteran visitors. *New Phytologist*, **26**, 115–23.

Dransfield, R.D. (1979) Aspects of host-parasitoid interactions of two aphid parasitoids, *Aphidius urticae* (Haliday) and *Aphidius uzbeckistanicus* (Luzhetski) (Hymenoptera, Aphidiidae). *Ecological Entomology*, **4**, 307–16.

Drapeau, M.D. (1999) Local mating and sex ratios. *Trends in Ecology and Evolution*, **14**, 235.

Drapeau, M.D. and Werren J.H. (1999) Differences in mating behaviour and sex ratio between three sibling species of *Nasonia*. *Evolutionary Ecology Research*, **1**, 223–34.

Dreisig, H. (1981) The rate of predation and its temperature dependence in a tiger beetle, *Cicindela hybrida*. *Oikos*, **36**, 196–202.

Dreisig, H. (1988) Foraging rate of ants collecting honeydew or extrafloral nectar, and some possible constraints. *Ecological Entomology*, **13**, 143–54.

Drent, R.H. and Daan, S. (1980) The prudent parent: energetic adjustments in avian breeding. *Ardea*, **68**, 225–52.

Driessen, G. and Hemerik, L. (1992) The time and egg budget of *Leptopilina clavipes*, a parasitoid of larval *Drosophila*. *Ecological Entomology*, **17**, 17–27.

Driessen, G.J., Bernstein, C., van Alphen, J.J.M. and Kacelnik, A., (1995) A count-down mechanism for host search in the parasitoid *Venturia canescens*. *Journal of Animal Ecology*, **64**, 117–25.

Drost, Y.C. and Carde, R.T. (1992) Influence of host deprivation on egg load and oviposition behaviour of *Brachymeria intermedia*, a parasitoid of gypsy moth. *Physiological Entomology*, **17**, 230–34.

Drukker, B., Scutareanu, P. and Sabelis, M. W. (1995) Do anthocorid predators respond to synomones from psylla- infested pear trees under field conditions. *Entomologia Experimentalis et Applicata*, **77**, 193–203.

Drukker, B., Bruin, J. and Sabelis, M.W. (2000) Anthocorid predators learn to associate herbivore-induced plant volatiles with presence or absence of prey. *Physiological Entomology*, **25**, 260–5.

Drumtra, D.E.W. and Stephen, F.M. (1999) Incidence of wildflower visitation by hymenopterous parasitoids of southern pine beetle, *Dendroctonus frontalis* Zimmermann. *Journal of Entomological Science*, **34**, 484–8.

Du, Y.-J., Poppy, G.M. and Powell, W. (1996) Relative importance of semiochemicals from first and second trophic levels in host foraging behavior of *Aphidius ervi*. *Journal of Chemical Ecology*, **22**, 1591–1605.

Duan, J.J. and Messing, R.H. (1999) Effects of origin and experience of patterns of host acceptance by the opiine parasitoid *Diachasmimorpha tryoni*. *Ecological Entomology*, **24**, 284–91.

Dubrovskaya, N.A. (1970) Field carabid beetles (Coleoptera, Carabidae) of Byelorussia. *Entomological Review*, **49**, 476–83.

Duchateau, M.J., Hoshiba H. and Velthuis, H.H.W. (1994) Diploid males in the bumble bee *Bombus terrestris*. *Entomologia Experimentalis et Applicata*, **71**, 263–9.

DuDevoir, D.S. and Reeves, R.M. (1990) Feeding activity of carabid beetles and spiders on gypsy moth larvae (Lepidoptera: Lymantriidae) at high density prey populations. *Journal of Entomological Science*, **25**, 341–56.

Dudgeon, D. (1990) Feeding by the aquatic heteropteran, *Diplonychus rusticum* (Belostomatidae): an effect of prey density on meal size. *Hydrobiologia*, **190**, 93–6.

Dudich, E. (1923) Über einen somatischen Zwitter des Hirschkäfers. *Entomologische Blätter*, **19**, 129–33.

Dudley, R. and Ellington, C.P. (1990) Mechanics of forward flight in bumblebees. 1. Kinematics and morphology. *Journal of Experimental Biology*, **148**, 19–52.

Duelli, P. (1980) Adaptive dispersal and appetitive flight in the green lacewing, *Chrysopa carnea*. *Ecological Entomology*, **5**, 213–20.

Duelli, P. (1981) Is larval cannibalism in lacewings adaptive? *Researches on Population Ecology*, **23**, 193–209.

Duelli, P., Studer, M., Marchand, I. and Jakob, S. (1990) Population movements of arthropods between natural and cultivated areas. *Biological Conservation*, **54**, 193–207.

Duelli, P., Obrist, M.K. and Schmatz, D.R. (1999) Biodiversity evaluation in agricultural landscapes: above-ground insects. *Agriculture, Ecosystems and Environment*, **74**, 33–64.

Duffey, E. (1962) A population study of spiders in limestone grassland. *Journal of Animal Ecology*, **31**, 571–99.

Duffey, E. (1974) Comparative sampling methods for grassland spiders. *Bulletin of the British Arachnological Society*, **3**, 34–7.

Duffey, E. (1978) Ecological strategies in spiders including some characteristics of species in pioneer and mature habitats. *Symposia of the Zoological Society of London*, **42**, 109–23.

Duffey, E. (1980) The efficiency of the Dietrick Vacuum Sampler (D-VAC) for invertebrate population studies in different types of grassland. *Bulletin d'Ecologie*, **11**, 421–31.

Duffey, S.S., K.A. Bloem and Campbell, B.C. (1986) Consequences of sequestration of plant natural products in plant–insect–parasitoid interactions, in *Interactions of Plant Resistance and Parasitoids and Predators of Insects* (eds D.J. Boethel and R.D. Eikenbary). Horwood, Chichester, pp. 31–60.

Dugatkin, L. A. and Alfieri, M. S. (2003) Boldness, behavioral inhibition and learning. *Ethology, Ecology and Evolution* **15**, 43–9.

Dunlap-Pianka, H., Boggs, C.L. and Gilbert, L.E. (1977) Ovarian dynamics in heliconiine butterflies: programmed senescence versus eternal youth. *Science*, **197**, 487–90.

Dupas, S., Fery, F. and Carton, Y. (1998) A single parasitoid segregating factor controls immune suppression in *Drosophila*. *Journal of Heredity*, **89**, 306–11.

Durkis, T.J. and Reeves, R.M. (1982) Barriers increase efficiency of pitfall traps. *Entomological News*, **93**, 8–12.

Dushay, M.S. and Beckage, N.E. (1993) Dose- dependent separation of *Cotesia congregata*-associated polydnavirus effects on *Manduca sexta* larval development and immunity. *Journal of Insect Physiology*, **39**, 1029–40.

Dutton, A. and Bigler, F. (1995) Flight activity assessment of the egg parasitoid *Trichogramma brassicae* (Hym.: Trichogrammatidae) in laboratory and field conditions. *Entomophaga*, **40**, 223–33.

Dwumfour, E.F. (1992) Volatile substances evoking orientation in the predatory flowerbug *Anthocoris nemorum* (Heteroptera, Anthocoridae). *Bulletin of Entomological Research*, **82**, 465–9.

Eady, P. (1994) Intraspecific variation in sperm precedence in the bruchid beetle *Callosobruchus maculatus*. *Ecological Entomology*, **19**, 11–16.

East, R. (1974) Predation on the soil dwelling stages of the winter moth at Wytham Wood, Berkshire. *Journal of Animal Ecology*, **43**, 611–26.

Eastop, V.F. and Pope, R.D. (1969) Notes on the biology of some British Coccinellidae. *The Entomologist*, **102**, 162–4.

Ebbert, M.A. (1993) Endosymbiotic sex ratio distorters in insects and mites, in *Evolution and Diversity of Sex Ratio in Insects and Mites* (eds D.L. Wrensch and M.A. Ebbert). Chapman and Hall, New York, pp. 150–91.

Eberhard, W.G. (1985) *Sexual Selection and Animal Genitalia*. Harvard University Press, Cambridge.

Eberhard, W.G. (1991) Copulatory courtship and cryptic female choice in insects. *Biological Reviews*, **66**, 1–31.

Eberhard, W.G. (1996) *Female control: Sexual Selection by Cryptic Female Choice*. Princeton University Press, Princeton.

Eberhard, W.G. (1997) Sexual selection by cryptic female choice in insects and arachnids, in *The Evolution of Mating Systems in Insects and Arachnids* (eds J.C. Choe and B.J. Crespi). Cambridge University Press, Cambridge, pp. 32–57.

Eberhard, W.G. (2000) Spider manipulation by a wasp larva. *Nature* **406**, 255–6.

Eberhardt, L.L. and Thomas, J.M. (1991) Designing environmental field studies. *Ecological Monographs*, **61**, 53–73.

Edgar, W.D. (1970) Prey and feeding behaviour of adult females of the wolf spider *Pardosa amentata* (Clerk.). *Netherlands Journal of Zoology*, **20**, 487–91.

Edwards, C.A. (1991) The assessment of populations of soil-inhabiting invertebrates. *Agriculture, Ecosystems and Environment*, **34**, 145–76.

Edwards, C.A. and Fletcher, K.E. (1971) A comparison of extraction methods for terrestrial arthropods, in

Methods of Study in Quantitative Soil Ecology, Production and Energy Flow (ed. J. Phillipson), IBP Handbooks, **18**, 150–85.

Edwards, C.A., Sunderland, K.D. and George, K.S. (1979) Studies on polyphagous predators of cereal aphids. *Journal of Applied Ecology*, **16**, 811–23.

Edwards, O.R. and Hopper, K.R. (1999) Using superparasitism by a stem borer to infer a host refuge. *Ecological Entomology*, **24**, 7–12.

Edwards, O.R. and Hoy, M.A. (1995) Random Amplified Polymorphic DNA markers to monitor laboratory-selected, pesticide-resistant *Trioxys pallidus* (Hymenoptera: Aphidiidae) after release into three California walnut orchards. *Environmental Entomology*, **24**, 487–96.

Edwards, R.L. (1954) The effect of diet on egg maturation and resorption in *Mormoniella vitripennis* (Hymenoptera, Pteromalidae). *Quarterly Journal of Microscopical Science*, **95**, 459–68.

Eggers, T. and Jones, T.H. (2000) You are what you eat... or are you? *Trends in Ecology and Evolution*, **15**, 265–6.

Eggleton, P. (1990) Male reproductive behaviour of the parasitoid wasp *Lytarmes maculipennis* (Hymenoptera: Ichneumonidae). *Ecological Entomology*, **15**, 357–60.

Eggleton, P. (1991) Patterns in male mating strategies of the Rhyssini: a holophyletic group of parasitoid wasps (Hymenoptera: Ichneumonidae). *Animal Behaviour*, **41**, 829–38.

Ehler, L.E. and Hall, R.W. (1982) Evidence for competitive exclusion of introduced natural enemies in biological control. *Environmental Entomology*, **11**, 1–4.

Ehler, L.E., Eveleens, K.G. and van den Bosch, R. (1973) An evaluation of some natural enemies of cabbage looper on cotton in California, *Environmental Entomology*, **2**, 1009–15.

Eijs, I., Ellers, J. and van Duinen, G.-J. (1998) Feeding strategies in drosophilid parasitoids: the impact of natural food resources on energy reserves in females. *Ecological Entomology*, **23**, 133–8.

Eisenberg, R.M., Hurd, L.E., Fagan, W.F., Tilmon, K.J., Snyder, W.E., Vandersall, K.S., Datz, S.G. and Welch, J.D. (1992) Adult dispersal of *Tenodera aridifolia sinensis* (Mantodea: Mantidae). *Environmental Entomology*, **21**, 350–3.

Eisner, T., Eisner, M., Rossini, C., Iyengar, V.K., Roach, B.L., Benedikt, E. and Meinwald, J. (2000) Chemical defense against predation in an insect egg. *Proceedings of the National Academy of Sciences*, **97**, 1634–9.

Ekbom, B.S. (1982) Diurnal activity patterns of the greenhouse whitefly, *Trialeurodes vaporariorum* (Homoptera: Aleyrodidae) and its parasitoid *Encarsia formosa* (Hymenoptera: Aphelinidae). *Protection Ecology*, **4**, 141–50.

Elagoze, M., Poirie, M. and Periquet, G. (1995) Precidence of the first male sperm in successive matings in the Hymenoptera *Diadromus pulchellus*. *Entomologia Experimentalis et Applicata*, **75**, 251–55.

Elgar, M.A. and Crespi, B.J. (1992) *Cannibalism: Ecology and Evolution Among Diverse Taxa*. Oxford University Press, Oxford.

Elkinton, J.S., Buonaccossi, J.P., Bellows Jr, T.S. and Van Driesche, R.G. (1992) Marginal attack rate, *k*-values and density dependence in the analysis of contemporaneous mortality factors. *Researches on Population Ecology*, **34**, 29–44.

Eller, F.J., Bartlet, R.J., Jones, R.L. and Kulman, H.M. (1984) Ethyl(Z)-9-hexadecenoate: a sex pheromone of *Syndipnus rubiginosus*, a sawfly parasitoid. *Journal of Chemical Ecology*, **10**, 291–300.

Eller, F.J., Tumlinson, J.H. and Lewis, W.J. (1990) Intraspecific competition in *Microplitis croceipes* (Hymenoptera: Braconidae), a parasitoid of *Heliothis* species (Lepidoptera: Noctuidae). *Annals of the Entomological Society of America*, **83**, 504–8.

Ellers, J. (1996) Fat and eggs: an alternative method to measure the trade-off between survival and reproduction in insect parasitoids. *Netherlands Journal of Zoology*, **46**, 227–35.

Ellers, J. and van Alphen, J.J.M. (1997) Life history evolution in *Asobara tabida*: plasticity in allocation of fat reserves to survival and reproduction. *Journal of Evolutionary Biology*, **10**, 771–85.

Ellers, J. and van Alphen, J.J.M. (2002) A trade-off between diapause duration and fitness in female parasitoids. *Ecological Entomology*, **27**, 279–84.

Ellers, J. and Jervis, M. (2003) Body size and the timing of egg production in parasitoid wasps. *Oikos*, **102**, 164–72.

Ellers, J. van Alphen, J.J.M. and Sevenster J.G (1998) A field study of size-fitness relationships in the parasitoid *Asobara tabida*. *Journal of Animal Ecology*, **67**, 318–24.

Ellers, J., Sevenster, J.G. and Driessen, G. (2000a) Egg load evolution in parasitoids. *American Naturalist*, **156**, 650–5.

Ellers, J., Driessen, G. and Sevenster, J.G. (2000b). The shape of the trade-off function between egg production and life-span in the parasitoid *Asobara tabida*. *Netherlands Journal of Zoology*, **50**, 29–36.

Ellers, J., Bax, M. and van Alphen, J.J.M. (2001) Seasonal changes in female size and its relation to reproduction in the parasitoid *Asobara tabida*. *Oikos*, **92**, 309–14.

Ellington, C.P., Machin, K.E. and Casey, T.M. (1990) Oxygen consumption of bumblebees in forward flight. *Nature*, **347**, 472–3.

Ellington, J., Kiser, K., Ferguson, G. and Cardenas, M. (1984) A comparison of sweepnet, absolute, and Insectavac sampling methods in cotton ecosystems. *Journal of Economic Entomology*, **77**, 599–605.

Elliot, S.L., Blanford, S. and Thomas, M.B. (2002) Host-pathogen interactions in a varying environment: temperature, behavioural fever and fitness. *Proceedings of the Royal Society of London B*, **269**, 1599–1607.

Elliott, J.M. (1967) Invertebrate drift in a Dartmoor stream. *Archives of Hydrobiology*, **63**, 202–37.

Elliott, J.M. (1970) Methods of sampling invertebrate drift in running water. *Annals of Limnology*, **6**, 133–59.

Elliott, J.M. (2003) A comparative study of the functional response of four species of carnivorous stoneflies. *Freshwater Biology*, **48**, 191–202.

Elliott, N.C. and Michels, G.J. (1997) Estimating aphidophagous coccinellid populations in alfalfa. *Biological Control*, **8**, 43–51.

Elton, R.A. and Greenwood, J.J.D. (1970) Exploring apostatic selection. *Heredity*, **25**, 629–33.

Elton, R.A. and Greenwood, J.J.D. (1987) Frequency-dependent selection by predators: comparison of parameter estimates. *Oikos*, **48**, 268–72.

Embree, D.G. (1966) The role of introduced parasites in the control of the winter moth in Nova Scotia. *Canadian Entomologist*, **98**, 1159–68.

Embree, D.G. (1971) The biological control of the winter moth in eastern Canada by introduced parasites, in *Biological Control* (ed. C.B. Huffaker). Plenum, New York, pp. 217–26.

van Emden, H.F. (1965) The role of uncultivated land in the biology of crop pests and beneficial insects. *Scientific Horticulture*, **17**, 121–36.

van Emden, H.F. (1988) The potential for managing indigenous natural enemies of aphids on field crops. *Philosophical Transactions of the Royal Society of London B*, **318**, 183–201.

van Emden, H.F. and Hagen, K.S. (1976) Olfactory reactions of the green lacewing, *Chrysopa carnea* to tryptophan and certain breakdown products. "*Environmental Entomology*, **5**, 469–73.

Emlen, J.M. (1966) The role of time and energy in food preference. *American Naturalist*, **100**, 611–17.

Emlen, S.T. and Oring, L.W. (1977) Ecology, sexual selection, and the evolution of mating systems. *Science*, **197**, 215–23.

Engelmann, F. (1970) *The Physiology of Insect Reproduction*. Pergamon Press, Oxford.

Epstein, M.E. and Kulman, H.M. (1984) Effects of aprons on pitfall trap catches of carabid beetles in forests and fields. *Great Lakes Entomologist* **17**, 215–21.

Erdtmann, C. (1969) *Handbook of Palynology*. Munksgaard, Copenhagen.

Ericson, D, (1977) Estimating population parameters of *Pterostichus cupreus* and *P. melanarius* (Carabidae) in arable fields by means of capture-recapture. *Oikos*, **29**, 407–17.

Ericson, D. (1978) Distribution, activity and density of some Carabidae (Coleoptera) in winter wheat fields. *Pedobiologia*, **18**, 202–17.

Ericson, D. (1979) The interpretation of pitfall catches of *Pterostichus cupreus* and *P. melanarius* (Coleoptera, Carabidae) in cereal fields. *Pedobiologia*, **19**, 320–8.

Ernsting, G. and Huyer, F.A. (1984) A laboratory study on temperature relations of egg production and development in two related species of carabid beetle. *Oecologia*, **62**, 361–7.

Ernsting, G. and Isaaks, J.A. (1988) Reproduction, metabolic rate and survival in a carabid beetle. *Netherlands Journal of Zoology*, **38**, 46–60.

Ernsting, G. and Isaaks, J.A. (1991) Accelerated ageing: a cost of reproduction in the carabid beetle *Notiophilus biguttatus* F. . *Functional Ecology*, **5**, 299–303.

Ernsting, G. and Joosse, E.N.G. (1974) Predation on two species of surface dwelling Collembola. A study with radio-isotope labelled prey. *Pedobiologia*, **14**, 222–31.

Escalante, G. and Rabinovich, J.E. (1979) Population dynamics of *Telenomus fariai* (Hymenoptera; Scelionidae), a parasite of Chagas' disease vectors, IX. Larval competition and population size regulation under laboratory conditions. *Researches on Population Ecology*, **20**, 235–46.

Estoup, A, Solignac, M. and Cornuet, J-M (1994) Precise assessment of the number of matings and relatedness in honey bee colonies. *Proceedings of the Royal Society of London B*, **258**, 1–7.

Ettifouri, M. and Ferran, A. (1993) Influence of larval rearing diet on the intensive searching behaviour of *Harmonia axyridis* (Col.: Coccinellidae) larvae. *Entomophaga*, **38**, 51–9.

Eubank, W.P., Atmar, J.W. and Ellington, J.J. (1973) The significance and thermodynamics of fluctuating versus static thermal environments on *Heliothis zea* egg development rates. *Environmental Entomology*, **2**, 491–6.

Eubanks, M.D. (2001) Estimates of the direct and indirect effects of red imported fire ants on biological control in field crops. *Biological Control*, **21**, 35–43.

Evans, D.A., Miller, B.R. and Bartlett, C.B. (1973) Host searching range of *Dasymutilla nigripes* (Fabricius) as investigated by tagging (Hymenoptera: Mutillidae). *Journal of the Kansas Entomological Society*, **46**, 343–6.

Evans, E.W. (1993) Indirect interactions among phytophagous insects: aphids, honeydew and natural enemies, in *Individuals, Populations and Patterns in Ecology* (eds A.D. Watt, S. Leather, N.J. Mills and K.F.A. Walters). Intercept, Andover, pp. 287–298.

Evans, E.W. (2000) Egg production in response to combined alternative foods by the predator *Coccinella transversalis*. *Entomologia Experimentalis et Applicata*, **94**, 141–7.

Evans, E.W. and Dixon, A.F.G. (1986) Cues for oviposition by ladybird beetles (Coccinellidae); response to aphids. *Journal of Animal Ecology*, **55**, 1027–34.

Evans, E.W. and England, S. (1996) Indirect interactions in biological control of insects: pests and "natural enemies in alfalfa. *Ecological Applications*, **6**, 920–30.

Evans, E.W. and Swallow, J.G. (1993) Numerical responses of natural enemies to artificial honeydew in Utah alfalfa. *Environmental Entomology*, **22**, 1392–401.

Evans, E.W., Stevenson, A.T and Richards, D.R. (1999) Essential versus alternative foods of insect predators: benefits of a mixed diet. *Oecologia*, **121**, 107–12.

Evans, H.F. (1976) Mutual interference between predatory arthropods. *Ecological Entomology*, **1**, 283–6.

Eveleens, K.G., van den Bosch, R. and Ehler, L.E. (1973) Secondary outbreak induction of beet armyworm by experimental insecticide applications in cotton in California. *Environmental Entomology*, **2**, 497–503.

Everts, J.W., Willemsen, I., Stulp, M., Simons, L., Aukema, B. and Kammenga, J. (1991) The toxic effect of deltamethrin on linyphiid and erigonid spiders in connection with ambient temperature, humidity and predation. *Archives of Environmental Contamination and Toxicology*, **20**, 20–4.

Ewing, A.W. (1983) Functional aspects of *Drosophila* courtship. *Biological Reviews*, **58**, 275–92.

Eyre, M.D. and Luff, M.L. (1990) A preliminary classification of European grassland habitats using carabid beetles, in *The Role of Ground Beetles in Ecological and Environmental Studies* (ed. N.E. Stork). Intercept, Andover, pp. 227–36.

Eyre, M.D., Luff, M.L. and Rushton, S.P. (1990) The ground beetle (Coleoptera, Carabidae) fauna of intensively managed agricultural grasslands in northern England and southern Scotland. *Pedobiologia*, **34**, 11–18.

Fabre, J.H. (1907) *Souvenirs Entomologiques. Études sur L'Instinct et les Moeurs des Insectes, 5ieme Séries.* Delagrave, Paris.

Fadamiro, H.Y. and Heimpel, G.E. (2001) Effects of partial sugar deprivation on lifespan and carbohydrate mobilization in the parasitoid *Macrocentrus grandii* (Hymenoptera: Braconidae). *Annals of the Entomological Society of America*, **94**, 909–16.

Faegri, K. and Iversen, J. (1989) *A Textbook of Pollen Analysis, 4th edn* (revised by K. Faegri, P.E. Kaland and K. Kryzwinski). John Wiley and Sons, Chichester.

Fagan, W.F. (1997) Introducing a boundary-flux approach to quantifying insect diffusion rates. *Ecology*, **78**, 579–87.

Fagan, W.F., Hakim, A.L., Ariawan, H., and Yuliyantiningsih, S. (1998) Interactions between biological control efforts and insecticide applications in tropical rice agroecosystems: the potential role of intraguild predation. *Biological Control*, **13**, 121–6.

Fagan, W.F., Lewis, M.A., Neubert, M.G. and van den Driessche, P. (2002) Invasion theory and biological control. *Ecology Letters*, **5**, 148–57.

Falconer, D.S. and Mackay, T.F.C. (1996) *Introduction to Quantitative Genetics*. Longman Group Limited, Harlow.

Fantle, M.S., Dittel, A.I., Schwalm, S.M., Epifanio, C.E. and Fogel, M.L. (1999) A food web analysis of the juvenile blue crab, *Callinectes sapidus*, using stable isotopes in whole animals and individual amino acids. *Oecologia*, **120**, 416–26.

FAO (Food and Agriculture Organization of the United Nations) (1996) Code of conduct for the import and release of exotic biological control Agents, *International Standards for Phytosanitary Measures (ISPM) (Publication No. 3)*. FAO, Rome.

Faria, C.A., Torres, J.B and Farias, A.M.I. (2000) Resposta functional de *Trichogramma pretiosum* Riley (Hymenoptera: Trichogrammatidae) parasitando ovos de *Tuta absoluta* (Meyrick) (Lepidoptera: Gelechiidae): efeito da idade do hospedeiro. *Anais da Entomologica Sociadade do Brasil*, **29**, 85–93.

Farrar, R.R., Barbour, J.D. and Kennedy, G.C. (1989) Quantifying food consumption and growth in insects. *Annals of the Entomological Society of America*, **82**, 593–98.

Faulds, W. and Crabtree, R. (1995) A system for using a Malaise trap in the forest canopy. *New Zealand "Entomologist*, **18**, 97–9.

Fauvergue, X., Hopper, K.R. and Antolin, M.F. (1995) Mate finding via a trail pheromone by a parasitoid wasp. *Proceeding of the National Academy of Sciences* **92**, 900–5.

Fauvergue, X., Fouillet, P, Mesquita, A.L.M. and Boulétreau, M. (1998a) Male orientation to trail sex pheromones in parasitoid wasps: does the spatial distribution of virgin females matter? *Journal of Insect Physiology*, **44**, 667–75.

Fauvergue, X., Hopper K.R., Antolin M.F. and Kazmer, D.J. (1998b) Does time until mating affect progeny sex ratio? A manipulative experiment with the parasitoid wasp *Aphelinus asychis*. *Journal of Evolutionary Biology*, **11**, 611–22.

Fauvergue, X., Fleury, F., Lemaitre, C. and Allemand, R. (1999) Parasitoid mating structures when hosts are patchily distributed: field and laboratory experiments with *Leptopilina boulardi* and *L. heterotoma*. *Oikos*, **86**, 344–56.

Feddersen, I., Sander, K. and Schmidt, O. (1986) Virus-like particles with host protein-like antigenic determinants protect an insect parasitoid from encapsulation. *Experientia*, **42**, 1278–81.

Fedorenko, A.Y. (1975) Instar and species-specific diets in two species of *Chaoborus*. *Limnology and Oceanography*, **20**, 238–42.

Feener, D.H. and Brown, B.V. (1997) Diptera as parasitoids. *Annual Review of Entomology*, **42**, 73–97.

Feinsinger, P. and Swarm, L.A. (1978) How common are ant-repellant nectars? *Biotropica*, **10**, 238–9.

Fellowes, M.D.E. (1998) Do non-social insects get the (kin) recognition they deserve? *Ecological Entomology*, **23**, 223–7.

Fellowes, M.D.E. and Godfray, H.C.J. (2000) The evolutionary ecology of resistance to parasitoids by *Drosophila*. *Heredity*, **84**, 1–8.

Fellowes, M. and Hutcheson, K. (2001) Flies in the face of adversity. *Biologist*, **48**, 75–8.

Fellowes, M.D.E., Masnatta, P., Kraaijeveld, A.R. and Godfray, H.C.J. (1998) Pupal parasitoid attack reduces the relative fitness of *Drosophila* that have previously encapsulated larval parasitoids. *Ecological Entomology*, **23**, 281–4.

Fellowes, M.D.E., Compton, S. and Cook, J.M. (1999a) Sex allocation and local mate competition in Old World non-pollinating fig wasps. *Behavioral Ecology and Sociobiology*, **46**, 95–102.

Fellowes, M.D.E., Kraaijeveld, A.R. and Godfray, H.C.J. (1999b) The relative fitness of *Drosophila melanogaster* (Diptera: Drosophilidae) that have survived attack by the parasitoid *Asobara tabida* (Hymenoptera: Braconidae). *Journal of Evolutionary Biology*, **12**, 123–8.

Felsenstein, J. (1985) Phylogenies and the comparative method. *American Naturalist*, **125**, 1–15.

Feng, J.G., Zhang, Y, Tao, X. and Chen, X.L. (1988) Use of radioisotope ^{32}P to evaluate the parasitization of *Adoxophyes orana* (Lep.: Tortricidae) by mass released *Trichogramma dendrolimi* (Hym.: Trichogrammatidae) in an apple orchard. *Chinese Journal of Biological Control*, **4**, 152–4.

Feng, M.G. and Nowierski, R.M. (1992) Spatial patterns and sampling plans for cereal aphids (Hem.: Aphididae) killed by entomophthoralean fungi and hymenopterous parasitoids in spring wheat. *Entomophaga*, **37**, 265–75.

Fenlon, J.S. and Sopp, P.I. (1991) Some statistical considerations in the determination of thresholds in ELISA. *Annals of Applied Biology*, **119**, 177–89.

Ferkovich, S.M. and Dillard, C.R. (1987) A study of uptake of radiolabelled host proteins and protein synthesis during development of eggs of the endoparasitoid, *Microplitis croceipes* (Cresson) (Braconidae). *Insect Biochemistry*, **16**, 337–45.

Fernández-Arhex, V. and Corley, J.C. (2003) The functional response of parasitoids and its implications for biological control. *Biocontrol Science and Technology*, **13**, 403–13.

Ferran, A., Buscarlet, A. and Larroque, M.M. (1981) The use of $HT^{18}O$ for measuring the food consumption in aged larvae of the aphidophagous ladybeetle, *Semiadalia 11notata* (Col: Coccinellidae), *Entomophaga*, **26**, 71–7 (In French).

Fichter, B.L. and Stephen, W.P. (1979) Selection and use of host-specific antigens. *Entomological Society of America, Miscellaneous Publication*, **11**, 25–33.

Fidgen, J.G., Eveleight, E.S. and Quiring, D.T. (2000) Influence of host size on oviposition behaviour and fitness of *Elachertus cacoeciae* attacking a low-density population of spruce budworm *Choristoneura fumiferana* larvae. *Ecological Entomology*, **25**, 156–64.

Field, J. (1992) Guild structure in solitary spider-hunting wasps (Hymenoptera: Pompilidae) compared with null model predictions. *Ecological Entomology*, **17**, 198–208.

Field, K.G., Olsen, G.J., Lane, D.J., Giovannoni, S.J., Ghiselin, M. Raff, E.C., Pace, N.R. and Raff, R.A (1992) Guild structure in solitary spider-hunting wasps (Hymenoptera: Pompilidae) compared with null model predictions. *Ecological Entomology*, **17**, 198–208.

Field, K.G. Olsen, G.J., Lane, D.J., Giovannoni, S.J., Ghiselin, M.J., Raff, E.C., Pace, N.R. and Raff, R.A. (1998) Molecular phylogeny of the animal kingdom. *Science*, **239**, 748–53.

Field, S.A. (1998) Patch exploitation, patch-leaving, and pre-emptive patch defence in the parasitoid wasp *Trissolocus basalis* (Insecta: Scelionidae). *Ethology*, **104**, 323–38.

Field, S.A., and Austin, A.D. (1994) Anatomy and mechanics of the telescopic ovipositor system of *Scelio* Latreille (Hymenoptera, Scelionidae) and related genera. *International Journal of Insect Morphology and Embryology*, **23**, 135–58.

Field, S.A. and Calbert, G. (1998) Patch defence in the parasitoid wasp *Trissolcus basalis*: when to begin fighting? *Behaviour*, **135**, 629–4.

Field, S.A. and Calbert, G. (1999) Don't count your eggs before they're parasitized: contest resolution and the trade offs during patch defense in a parasitoid wasp. *Behavioral Ecology*, **10**, 122–7.

Field, S.A. and Keller, M.A. (1993a) Alternative mating tactics and female mimicry as post-copulatory mate-guarding behaviour in the parasitic wasp *Cotesia rubecula*. *Animal Behaviour*, **46**, 1183–9.

Field, S.A. and Keller, M.A. (1993b) Courtship and intersexual signalling in the parasitic wasp *Cotesia rubecula* (Hymenoptera: Braconidae). *Journal of Insect Behaviour*, **6**, 737–50.

Field, S.A. and Keller, M.A. (1994) Localization of the female sex pheromone gland in *Cotesia rubecula* Marshall (Hymenoptera: Braconidae). *Journal of Hymenopteran Research*, **3**, 151–6.

Field, S.A. and Keller, M.A. (1999) Short-term host discrimination in the parasitoid wasp *Trissolcus basalis* Wollaston (Hymenoptera: Scelionidae). *Australian Journal of Zoology*, **47**, 19–28.

Field, S.A., Keller, M.A. and Calbert, G. (1997) The pay-off from superparasitism in the egg parasitoid *Trissolcus basalis*, in relation to patch defence. *Ecological Entomology*, **22**, 142–9.

Field, S.A., Calbert, G. and Keller, M.A. (1998) Patch defence in the parasitoid wasp *Trissolcus basalis* (Insecta: Scelionidae): the time structure of pairwise contests, and the 'waiting game'. *Ethology*, **104**, 821–40.

Finch, S. (1992) Improving the selectivity of water traps for monitoring populations of the cabbage root fly. *Annals of Applied Biology*, **120**, 1–7.

Finch, S. (1996) Effect of beetle size on predation of cabbage root fly eggs by ground beetles. *Entomologia Experimentalis et Applicata*. **81**, 199–206.

Finch, S. and Coaker, T.H. (1969) Comparison of the nutritive values of carbohydrates and related compounds to *Erioischia brassicae*. *Entomologia Experimentalis et Applicata*, **12**, 441–53.

Fincke, O.M. (1984) Sperm competition in the damselfly *Enallagma hageni* Walsh (Odon., Coenagrionidae): benefits of multiple mating to males and females. *Behavioral Ecology and Sociobiology*, **14**, 235–40.

Fincke, O.M., Waage, Jonathan K. and Koenig, W.D. (1987) Natural and sexual selection components of odonate mating patterns, in *The Evolution of Mating Systems in Insects and Arachnids* (eds J.C. Choe and B.J. Crespi). Cambridge University Press, Cambridge, pp. 58–74.

Fink, U. and Völkl, W. (1995) The effect of biotic factors on foraging and oviposition success of the aphid parasitoid, *Aphidius rosae*. *Oecologia*, **103**, 371–8.

Finke, D.L. and Denno, R.F. (2002) Intraguild predation diminished in complex-structured vegetation: implications for prey suppression. *Ecology*, **83**, 643–52.

Finke, D.L. and Denno, R.F. (2003) Intra-guild predation relaxes natural enemy impacts on herbivore populations. *Ecological Entomology*, **28**, 67–73.

Finlayson, T. (1967) A classification of the subfamily Pimplinae (Hymenoptera: Ichneumonidae) based on final-instar larval characteristics. *Canadian Entomologist*, **99**, 1–8.

Finlayson, T. (1990) The systematics and taxonomy of final-instar larvae of the family Aphidiidae (Hymenoptera). *Memoirs of the Entomological Society of Canada*, **152**, 1–74.

Finlayson, T. and Hagen, K.S. (1977) Final-instar larvae of parasitic Hymenoptera. *Pest Management Papers, Simon Fraser University*, **10**, 1–111.

Fischer, M.K. and Shingleton, A.W. (2001) Host plant and ants influence the honeydew sugar composition of aphids. *Functional Ecology*, **15**, 544–50.

Fisher, R.A. (1930) *The Genetical Theory of Natural Selection*. Oxford University Press, Oxford.

Fisher, R.A. and Ford, E.B. (1947) The spread of a gene in natural conditions in a colony of the moth *Panaxia dominulua* L.. *Heredity*, **1**, 143–74.

Fisher, R.C. (1961) A study in insect multiparasitism. II. The mechanism and control of competition for the host. *Journal of Experimental Biology*, **38**, 605–28.

Fisher, R.C. (1971) Aspects of the physiology of endoparasitic Hymenoptera. *Biological Reviews*, **46**, 243–78.

Fitt, G.P. (1990) Comparative fecundity, clutch size, ovariole number and egg size of *Dacus tryoni* and *D. jarvisi*, and their relationship to body size. *Entomologia Experimentalis et Applicata*, **55**, 11–21.

Fitzgerald, J.D., Solomon, M.G. and Murray, R.A. (1986) The quantitative assessment of arthropod predation rates by electrophoresis. *Annals of Applied Biology*, **109**, 491–8.

Fjerdingstad, E.J. and Boomsma, J.J. (1998) Multiple mating increases sperm stores of *Atta colombica* leafcutter ant queens. *Behavioral Ecology and Sociobiology*, **42**, 257–61.

Flanders, S.E. (1934) The secretion of the colleterial glands in the parasitic chalcids. *Journal of Economic Entomology*, **27**, 861–2.

Flanders, S.E. (1942) Oösorption and ovulation in relation to oviposition in the parasitic Hymenoptera. *Annals of the Entomological Society of America*, **35**, 251–66.

Flanders, S.E. (1950) Regulation of ovulation and egg disposal in the parasitic Hymenoptera. *Canadian Entomologist*, **82:** 134–40.

Flanders, S.E. (1953) Predatism by the adult hymenopterous parasite and its role in biological control. *Journal of Economic Entomology*, **46**, 541–4.

Flanders, S.E. (1962) The parasitic Hymenoptera: specialists in population regulation, *Canadian Entomologist*, **94**, 1133–47.

Fleischer, S.J., Gaylor, M.J. and Edelson, J.V. (1985) Estimating absolute density from relative sampling of *Lygus lineolaris* (Heteroptera: Miridae) and selected predators in early to mid-season cotton. *Environmental Entomology*, **14**, 709–17.

Fleischer, S.J., Gaylor, M.J., Hue, N.V. and Graham, L.C. (1986) Uptake and elimination of rubidium, a physiological marker, in adult *Lygus lineolaris* (Hemiptera: Miridae). *Annals of the Entomological Society of America*, **79**, 19–25.

Fleming, J.-A.G.W. (1992) Polydnaviruses: mutualists and pathogens. *Annual Review of Entomology*, **37**, 401–25.

Fletcher, B.S. and Kapatos, E.T. (1983) An evaluation of different temperature-development rate models for predicting the phenology of the olive fly *Dacus oleae*, in *Fruit Flies of Economic Importance* (ed. R. Cavalloro), CEC/IOBC Symposium, Athens, November, 1982. Balkema, Rotterdam, pp. 321–30.

Fletcher, B.S., Kapatos, E. and Southwood, T.R.E. (1981) A modification of the Lincoln Index for estimating the population densities of mobile insects. *Ecological Entomology*, **6**, 397–400.

Fleury, F., Allemand, R., Fouillet, P. and Bouletreau, M. (1995) Genetic variation in locomotor activity rhythm among populations of *Leptopilina heterotoma* (Hymenoptera: Eucoilidae), a larval parasitoid of *Drosophila* species. *Behavior Genetics*, **25**, 81–9.

Fleury, F., Vavre, F., Ris, N., Fouillet, P. and Bouletreau, M. (2000) Physiological cost induced by the maternally-transmitted *Drosophila* parasitoid *Leptopilina heterotoma*. *Parasitology*, **121**, 493–500.

Flint, H.M., Salter, S.S. and Walters, S. (1979) Caryophylene: an attractant for the green lacewing. *Environmental Entomology*, **8**, 1123–5.

Flint, H.M., Merkle, J.R. and Sledge, M. (1981) Attraction of male *Collops vittatus* in the field by caryophyllene alcohol. *Environmental Entomology*, **10**, 301–4.

Floren, A. and Linsenmair, K.E. (1998) Non-equilibrium communities of Coleoptera in trees in a lowland rain forest of Borneo. *Ecotropica*, **4**, 55–67.

Follett, P. A., Johnson, M.T. and Jones, V.P. (2000) Parasitoid drift in Hawaiian pentatomoids, in *Nontarget Effects of Biological Control* (eds P.A. Follett and J.J. Duan). Kluwer Academic Publishers, Dordrecht. pp. 78–93.

Folsom, T.C. and Collins, N.C. (1984) The diet and foraging behavior of the larval dragonfly *Anax junius* (Aeshnidae), with an assessment of the role of refuges and prey activity. *Oikos*, **42**, 105–13.

Force, D.C. and Messenger, P.S. (1964) Fecundity, reproductive rates, and innate capacity for increase in three parasites of *Therioaphis maculata* (Buckton). *Ecology*, **45**, 706–15.

Formanowicz D.R. Jr., (1982) Foraging tactics of larvae of *Dytiscus verticalis* (Coleoptera, Dytiscidae): the assessment of prey density. *Journal of Animal Ecology*, **51**, 757–67.

Forsse, E., Smith, S.M. and Bourchier, R.S. (1992) Flight initiation in the egg parasitoid *Trichogramma minutum*: effects of ambient temperature, mates, food, and host eggs. *Entomologia Experimentalis et Applicata*, **62**, 147–54.

Forsythe, T.G. (1987) *Common Ground Beetles, Naturalists' Handbooks No. 8*. Richmond, Slough.

Fournier, E. and Loreau, M. (2000) Movement of *Pterostichus melanarius* in agricultural field margins in relation to hunger state, in *Natural History and Applied Ecology of Carabid Beetles* (eds P. Brandmayr, G.L. Lövei, T.Z. Brandmayr, A. Casale and A.V. Taglianti). Pensoft Publishers, Moscow, pp. 207–19.

Fournier, E. and Loreau, M. (2002) Foraging activity of the carabid beetle *Pterostichus melanarius* III. In field margin habitats. *Agriculture, Ecosystems and Environment*, **89**, 253–9.

Fowler, H.G. (1987) Field behaviour of *Euphasiopteryx depleta* (Diptera: Tachinidae): phonotactically orienting parasitoids of mole crickets (Orthoptera: Cryllotalpidae: *Scapteriscus*). *Journal of the New York Entomological Society*, **95**, 474–80.

Fowler, H.G. (1989) Optimization of nectar foraging in a solitary wasp (Hymenoptera: Sphecidae: *Larra* sp.). *Naturalia*, **14**, 13–17.

Fowler, H.G. and Kochalka, J.N. (1985) New records of *Euphasiopteryx depleta* (Diptera: Tachinidae) from Paraguay: attraction to broadcast calls of *Scapteriscus acletus* (Orthoptera: Gryllotalpidae). *Florida Entomologist*, **68**, 225–6.

Fowler, S.V. (1988) Field studies on the impact of natural enemies on brown planthopper populations on rice in Sri Lanka. *Proceedings of the 6th Auchenorrhyncha Meeting, Turin, Italy, 7–11 September 1987*, 567–74.

Fowler, S.V., Claridge, M.F., Morgan, J.C., Peries, I.D.R. and Nugaliyadde, L. (1991) Egg mortality of'the brown planthopper, *Nilaparvata lugens* (Homoptera: Delphacidae) and green leafhoppers, *Nephotettix* spp. (Homoptera: Cicadellidae), on rice in Sri Lanka. *Bulletin of Entomological Research*, **81**, 161–7.

Fox, L.R. and Murdoch, W.W. (1978) Effects of feeding history on short term and long term functional responses in *Notonecta hoffmani*. *Journal of Animal Ecology*, **47**, 945–59.

Fox, P.M., Pass, B.C. and Thurston, R. (1967) Laboratory studies on the rearing of *Aphidius smithi* (Hymenoptera: Braconidae) and its parasitism of *Acyrthosiphon pisum* (Homoptera: Aphididae). *Annals of the Entomological Society of America*, **60**, 1083–7.

Frank, J.H. (1967a) The insect predators of the pupal stage of the winter moth, *Operophtera brumata* (L.) (Lepidoptera: Hydriomenidae). *Journal of Animal Ecology*, **36**, 375–89.

Frank, J.H. (1967b) The effect of pupal predators on a population of winter moth, *Operophtera brumata* (L.) (Lepidoptera: Hydriomenidae). *Journal of Animal Ecology*, **36**, 611–21.

Frank, J.H. (1968) Notes on the biology of *Philonthus decorus* (Grav.) (Col., Staphylinidae). *Entomologist's Monthly Magazine*, **103**, 273–7.

Frank, J.H. (1979) The use of the precipitin technique in predator-prey studies to 1975. *Entomological Society of America, Miscellaneous Publication*, **11**, 1–15.

Frank, S.A. (1984) The behavior and morphology of the fig wasps *Pegoscapus assuetus* and *P. jimenezi:* descriptions and suggested behavioral characters for phylogenetic studies. *Psyche*, **91**, 289–308.

Frank, S.A. (1985a) Hierarchical selection theory and sex ratios. II. On applying the theory and a test with fig wasps. *Evolution*, **39**, 949–964.

Frank, S.A. (1985b) Are mating and mate competition by the fig wasp *Pegoscapus assuetus* (Agaonidae) random within a fig? *Biotropica*, **17**, 170–2.

Frank, S.A. (1986) Hierarchical selection theory and sex ratios. I. General solutions for structured populations. *Theoretical Population Biology*, **29**, 312–342.

Franke, U., Friebe, B. and Beck, L. (1988) Methodisches zur Ermittlung der Siedlungsdichte von Bodentieren aus Quadratproben und Barberfallen. *Pedobiologia*, **32**, 253–64.

Frazer, B.D. (1988) Coccinellidae, in *World Crop Pests. Aphids, their Biology, Natural Enemies and Control, Vol. 2B* (eds A.K. Minks and P. Harrewijn). Elsevier, Amsterdam, pp. 231–48.

Frazer, B.D. and Gilbert, N. (1976) Coccinelliids and aphids: a quantitative study of the impact of adult ladybirds (Coleoptera: Coccinellidae) preying on field populations of pea aphids (Homoptera: Aphididae). *Journal of the Entomological Society of British Columbia*, **73**, 33–56.

Frazer, B.D. and Gill, B. (1981) Hunger, movement and predation of *Coccinella californica* on pea aphids in the laboratory and in the field. *Canadian Entomologist*, **113**, 1025–33.

Frazer, B.D. and McGregor, R.R. (1992) Temperature-dependent survival and hatching rate of eggs of seven species of Coccinellidae. *Canadian Entomologist*, **124**, 305–12.

Frazer, B.D. and Raworth, D.A. (1985) Sampling for adult coccinellids and their numerical response to strawberry aphids (Coleoptera: Coccinellidae: Homoptera: Aphididae). *Canadian Entomologist*, **117**, 153–61.

Frazer, B.D., Gilbert, N., Ives, P.M. and Raworth, D.A. (1981a) Predator reproduction and the overall predator-prey relationship. *Canadian Entomologist*, **113**, 1015–24.

Frazer, B.D., Gilbert, N., Nealis, V. and Raworth, D.A. (1981b) Control of aphid density by a complex of predators. *Canadian Entomologist*, **113**, 1035–41.

Freckleton, R.P. (2000) Biological control as a learning process. *Trends in Ecology and Evolution*, **15**, 263–4.

Free, C.A., Beddington, J.R. and Lawton, J.H. (1977) On the inadequacy of simple models of mutual interference for parasitism and predation. *Journal of Animal Ecology*, **46**, 543–4.

Freeman, J.A. (1946) The distribution of spiders and mites up to 300 feet in the air. *Journal of Animal Ecology*, **15**, 69–74.

Freeman-Long, R., A. Corbett, C. Lamb, C. Reberg-Horton, J. Chandler and M. Stimmann 1998. Beneficial insects move from flowering plants to nearby crops. *California Agriculture, September-October*, 23–6.

French, B.W., Elliott, N.C., Berberet, R.C. and Burd, J.D. (2001) Effects of riparian and grassland habitats on ground beetle (Coleoptera: Carabidae) assemblages in adjacent wheat fields. *Environmental Entomology*, **30**, 225–34.

Fuester, R.W. and Taylor, P.B. (1996) Differential mortality in male and female gypsy moth (Lepidoptera: Lymantriidae) pupae by invertebrate natural enemies and other factors. *Environmental Entomology*, **25**, 536–47.

Fuester, R.W., Drea, J.J., Cruber, F., Hoyer, H. and Mercardier, G. (1983) Larval parasites and other natural enemies of *Lymantria dispar* (Lepidoptera, Lymantriidae) in Burgenland, Austria, and Wurzburg, Germany. *Environmental Entomology*, **12**, 724–37.

Fukushima, J., Kainoh, Y., Honda, H. and Takabayashi, J. (2002) Learning of herbivore-induced and nonspecific plant volatiles by a parasitoid, *Cotesia kariyai*. *Journal of Chemical Ecology*, **28**, 579–86.

Funke, W. (1971) Food and energy turnover of leaf-eating insects and their influence on primary production. *Ecological Studies*, **2**, 81–93.

Funke, W., Jans, W. and Manz, W. (1995) Temporal and spatial niche differentiation of predatory arthropods of the soil surface in two forest ecosystems. *Acta Zoologica Fennica*, **196**, 111–14.

Fye, R.E. (1965) Methods for placing wasp trap nests in elevated locations. *Journal of Economic Entomology*, **58**, 803–4.

Gadau J., Page R.E. and Werren J.H. (1999) Mapping of hybrid incompatibility loci in *Nasonia*. *Genetics*, **153**, 1731–41.

Gagné, I., Coderre, D. and Maufette, Y. (2002). Egg cannibalism by *Coleomegilla maculata lengi* neonates: preference even in the presence of essential prey. *Ecological Entomology*, **27**, 285–91.

Galis, F. and van Alphen, J.J.M. (1981) Patch time allocation and search intensity of *Asobara tabida* Nees (Hym.: Braconidae). *Journal of Animal Ecology*, **31**, 701–12.

Gallego, V.C., Baltazar, C.R., Cadapan, E.P. and Abad, R.G. (1983) Some ecological studies on the coconut leafminer, *Promecotheca cumingii* Baly (Coleoptera: Hispidae) and its hymenopterous parasitoids in the Philippines. *Philippine Entomologist*, **6**, 471–94.

Gandolfi, M., Mattiacci, L. and Dorn, S. (2003) Mechanisms of behavioral alterations of parasitoids reared in artificial systems. *Journal of Chemical Ecology*, **29**, 1871–87.

Garcia-Salazar, C. and Landis, D.A. (1997) Marking *Trichogramma brassicae* (Hymenoptera: Trichogrammatidae) with fluorescent marker dust and its effect on survival and flight behavior. *Journal of Economic Entomology*, **90**, 1546–50.

Gardener, M.C. and Gillman, M.P. (2001) Analyzing variability in nectar amino acids: composition is less variable than concentration. *Journal of Chemical Ecology*, **27**, 2545–58.

Gardener, M.C. and Gillman, M.P. (2002) The taste of nectar – a neglected area of pollination ecology. *Oikos*, **98**, 552–7.

Gardener, W.A., Shepard, M. and Noblet, T.R. (1981) Precipitin test for examining predator-prey interactions in soybean fields. *Canadian Entomologist*, **113**, 365–9.

Garland, T. Jr., Harvey, P.H. and Ives, A.R. (1992) Procedures for the analysis of comparative data using independent contrasts. *Systematic Biology*, **41**, 18–32.

Gaston, K.J. (1988) The intrinsic rates of increase of insects of different sizes. *Ecological Entomology*, **14**, 399–409.

Gaston, K.J. and Gauld, I.D. (1993) How many species of pimplines (Hymenoptera: Ichneumonidae) are there in Costa Rica? *Journal of Tropical Ecology*, **9**, 491–9.

Gauld, I. and Bolton, B. (1988) *The Hymenoptera*. Oxford University Press, Oxford.

Gauld, I.D. and Huddleston, T. (1976) The nocturnal Ichneumonidae of the British Isles, including a key to genera. *Entomologist's Gazette*, **27**, 35–49.

Gauthier, N. and Monge, J.P. (1999) Could the egg itself be the source of the oviposition deterrent in the ectoparasitoid *Dinarmus basalis*? *Journal of Insect Physiology*, **45**, 393–400.

Geden, C.J., Bernier, U.R., Carlson, D.A. and Sutton, B.D. (1998) Identification of *Muscidifurax* spp., parasitoids of muscoid flies, by composition patterns of cuticular hydrocarbons. *Biological Control*, **12**, 200–7.

van der Geest, L.P.S. and Overmeer, W.P.J. (1985) Experiences with polyacrylamide gradient gel electrophoresis for the detection of gut contents of phytoseiid mites. *Mededelingen van de Factilteit voor Landbouwwetenschappen Rijksuniversiteit Gent*, **50**, 469–71.

Gehan, E.A. and Siddiqui, M.M. (1973) Simple regression methods for survival time studies. *Journal of American Statistical Association*, **68**, 848–56.

Geoghegan, I.E., Chudek, J.A., Mackay, R.L., Lowe, C., Moritz, S., McNicol, R.J., Birch, N.E., Hunter, G. and Majerus, M.E.N. (2000) Study of anatomical changes in *Coccinella septempunctata* (Coleoptera: Coccinellidae) induced by diet and by infection with the larva of *Dinocampus coccinellae* (Hymenoptera: Braconidae) using magnetic resonance microimaging. *European Journal of Entomology*, **97**, 457–61.

Gerber, G.H., Walkof, J. and Juskiw, D. (1992) Portable, solar-powered charging system for blacklight traps. *Canadian Entomologist*, **124**, 553–4.

Gerking, S.D. (1957) A method of sampling the littoral macrofauna and its application. *Ecology*, **38**, 219–26.

Gerling, D. and Legner, E.F. (1968) Developmental history and reproduction of *Spalangia cameroni*, parasite of synanthropic flies. *Annals of the Entomological Society of America*, **61**, 1436–43.

Gerling, D., Quicke, D.L.J. and Orion, T. (1998) Oviposition mechanisms in the whitefly parasitoids *Encarsia transvana* and *Eretmocerus mundus*. *Biocontrol*, **43**, 289–97.

Getz, W.M. (1984) Population dynamics: a resource per capita approach. *Journal of Theoretical Biology*, **108**, 623–44.

Getz, W.M. and Gutierrez, A.P. (1982) A perspective on systems analysis in crop production and insect pest management. *Annual Review of Entomology*, **27**, 447–66.

Geusen-Pfister, H. (1987) Studies on the biology and reproductive capacity of *Episyrphus balteatus* Deg. (Dipt., Syrphidae) under greenhouse conditions. *Journal of Applied Entomology*, **104**, 261–70.

Gherna, R.L., Werren, J.H. and Weisburg, W. (1991) *Arsenophonus nasoniae* gen. nov., sp. nov., the causitive agent of the son-killer trait in the parasitic wasp *Nasonia vitripennis*. *International Journal of Systematic Bacteriology*, **41**, 563–65.

Giamoustaris, A., and Mithen, R. (1995) The effect of modifying the glucosinolate content of leaves of oilseed rape (*Brassica napus* s.sp. *oleifera*) on its interaction with specialist and generalist pests. *Annals of Applied Biology*, **126**, 347–63.

Gibbons, D.W. and Pain, D. (1992) The influence of river flow-rate on the breeding behaviour of *Calopteryx* damselflies. *Journal of Animal Ecology*, **61**, 283–9.

Gilbert, F.S. (1980) Flower visiting by hoverflies. *Journal of Biological Education*, **14**, 70–4.

Gilbert, F.S. (1981) Foraging ecology of hoverfies: morphology of the mouthparts in relation to feeding on nectar and pollen in some common urban species. *Ecological Entomology*, **6**, 245–62.

Gilbert, F.S. (1983) The foraging ecology of hoverflies (Diptera: Syrphidae): circular movements on composite flowers. *Behavioural Ecology and Sociobiology*, **13**, 253–7.

Gilbert, F.S. (1985a) Diurnal activity patterns in hoverflies (Diptera, Syrphidae). *Ecological Entomology*, **10**, 385–92.

Gilbert, F.S. (1985b) Ecomorphological relationships in hoverflies (Diptera, Syrphidae), *Proceedings of the Royal Society of London B*, **224**, 91–105.

Gilbert, F.S. (1986) *Hoverflies, Naturalists' Handbooks No. 5*. Cambridge University Press, Cambridge.

Gilbert, F.S. (1990) Size, phylogeny and life-history in the evolution of feeding specialisation in insect predators, in *Insect Life Cycles: Genetics, Evolution and Co-ordination* (ed. F.S. Gilbert). Springer-Verlag, London, pp. 101–24.

Gilbert, F.S. (1991) Feeding in adult hoverflies. *Hoverfly Newsletter*, **13**, 5–11.

Gilbert, F.S. and Jervis, M.A. (1998) Functional, evolutionary and ecological aspects of feeding-related mouthpart specializations in parasitoid flies. *Biological Journal of the Linnean Society*, **63**, 495–535.

Gilbert, F.S. and Owen, J. (1990) Size, shape, competition, and community structure in hoverflies (Diptera: Syrphidae). *Journal of Animal Ecology*, **59**, 21–39.

Gilbert, L.E. (1976) Postmating female odour in *Heliconius* butterflies: a male-contributed antiaphrodisiac? *Science*, **193**, 419–420.

Gilbert, N. (1984) Control of fecundity in *Pieris rapae*. II. Differential effects of temperature. *Journal of Animal Ecology*, **53**, 589–97.

Gilbert, N., Gutierrez, A.P., Frazer, B.D. and Jones, R.E. (1976) *Ecological Relationships*. W.H. Freeman and Co., Reading.

Giller, P.S. (1980) The control of handling time and its effects on the foraging strategy of a heteropteran predator, *Notonecta*. *Journal of Animal Ecology*, **49**, 699–712.

Giller, P.S. (1982) The natural diet of waterbugs (Hemiptera: Heteroptera): electrophoresis as a potential method of analysis. *Ecological Entomology*, **7**, 233–7.

Giller, P.S. (1984) Predator gut state and prey detectability using electrophoretic analysis of gut contents. *Ecological Entomology*, **9**, 157–62.

Giller, P.S. (1986) The natural diet of the Notonectidae: field trials using electrophoresis. *Ecological Entomology*, **11**, 163–72.

Gillespie, D.R. and Finlayson, T. (1983) Classification of final-instar larvae of the Ichneumoninae (Hymenoptera: lchneumonidae). *Memoirs of the Entomological Society of Canada*, **124**, 1–81.

Gillespie, D.R. and McGregor, R.R. (2000). The functions of plant feeding in the omnivorous predator *Dicyphus hesperus*: water places limits on predation. *Ecological Entomology* **25**, 380–86.

Gillespie, J.H. and Turelli, M. (1989) Genotype-environment interactions and the maintenance of polygenic variation. *Genetics*, **121**, 129–38.

Gilpin, M. and Hanski, I. (1991) *Metapopulation Dynamics – Empirical and Theoretical Investigations*. Academic Press, London.

Gimeno, C., Belshaw, R. and Quicke, D.L.J. (1997) Phylogenetic relationships of the Alysiinae/Opiinae (Hymenoptera: Braconidae) and the utility of cytochrome b, 16S and 28S D2 rRNA. *Insect Molecular Biology*, **6**, 273–84.

Ginzburg, L.R. (1986) The theory of population dynamics: I. Back to first principles. *Journal of Theoretical Biology*, **122**, 385–99.

Giorgini, M. (2001) Induction of males in thelytokous populations of *Encarsia meritoria* and *Encarsia protransvena:* A systematic tool. *BioControl*, **46**, 427–38.

Giron, D. and Casas, J. (2003a) Lipogenesis in an adult parasitic wasp. *Journal of Insect Physiology*, **49**, 141–7.

Giron, D. and Casas, J. (2003b) Mothers reduce egg provisioning with age. *Ecology Letters*, **6**, 271–7.

Giron, D., Rivero, A., Mandon, N., Darrouzet, E. and Casas, J. (2002) The physiology of host feeding in parasitic wasps: implications for survival. *Functional Ecology*, **16**, 750–7.

Gist, C.S. and Crossley, D.A. (1973) A method of quantifying pitfall trapping. *Environmental Entomology*, **2**, 951–2.

Gleick, J. (1987) *Chaos: Making a New Science*. Penguin, New York.

Glen, D.M. (1973) The food requirements of *Blepharidopterus angulatus* (Heteroptera: Miridae) as a predator of the lime aphid, *Eucallipterus tiliae. Entomologia Experimentalis et Applicata*, **16**, 255–67.

Glen, D.M. (1975) Searching behaviour and prey-density requirements of *Blepharidopterus angulatus* (Fall.) (Heteroptera: Miridae) as a predator of the lime aphid, *Eucallipterus tiliae* (L.), and leafhopper, *Alnetoidea alneti* (Dahlbom). *Journal of Animal Ecology*, **44**, 116–35.

Glen, D.M., Milsom, N.F. and Wiltshire, C.W. (1989) Effects of seed-bed conditions on slug numbers and damage to winter wheat in a clay soil. *Annals of Applied Biology*, **115**, 177–90.

Godfray, H.C.J. (1987a) The evolution of clutch size in invertebrates. *Oxford Surveys in Evolutionary Biology*, **4**, 117–54.

Godfray, H.C.J. (1987b) The evolution of clutch size in parasitic wasps. *American Naturalist*, **129**, 221–33.

Godfray, H.C.J. (1988) Virginity in haplodiploid populations: a study on fig wasps. *Ecological Entomology*, **13**, 283–91.

Godfray, H.C.J. (1990) The causes and consequences of constrained sex allocation in haplodiploid animals. *Journal of Evolutionary Biology*, **3**, 3–17.

Godfray, H.C.J. (1994) *Parasitoids: Behavioral and Evolutionary Ecology*. Princeton University Press, Princeton.

Godfray, H.C.J. and Cook, J.M. (1997) Mating systems of parasitoid wasps, in *The Evolution of Mating Systems in Insects and Arachnids* (eds J.C. Choe and B.J. Crespi.). Cambridge University Press, Cambridge, pp. 211–25.

Godfray, H.C.J. and Grenfell, B.T. (1993) The continuing quest for chaos. *Trends in Ecology and Evolution*, **8**, 43–4.

Godfray, H.C.J. and Hardy, I.C.W. (1993) Sex ratio and virginity in haplodiploid insects, in *Evolution and Diversity of Sex Ratio in Insects and Mites* (eds D.L. Wrensch and M.A. Ebbert). Chapman and Hall, New York, pp. 402–17.

Godfray, H.C.J. and Hassell, M.P. (1987) Natural enemies may be a cause of discrete generations in tropical systems. *Nature*, **327**, 144–7.

Godfray, H.C.J. and Hassell, M.P. (1988) The population biology of insect parasitoids. *Science Progress, Oxford*, **72**, 531–48.

Godfray, H.C.J. and Hassell, M.P. (1989) Discrete and continuous insect populations in tropical environments. *Journal of Animal Ecology*, **58**, 153–74.

Godfray, H.C.J. and Hassell, M.P. (1991) Encapsulation and host-parasitoid population biology, in *Parasite-Host Associations* (eds C.A. Toft, A. Aeschlimann and L. Bolis). Oxford University Press, Oxford, pp. 131–47.

Godfray, H.C.J. and Hassell, M.P. (1992) Long time series reveal density dependence. *Nature*, **359**, 673–4.

Godfray, H.C.J. and Ives, A.R. (1988) Stochasticity in invertebrate clutch-size models. *Theoretical Population Biology*, **33**, 79–101.

Godfray, H.C.J. and Müller, C.B. (1999) Host-parasitoid dynamics, in *Insect Populations: in Theory and Practice* (eds J.P. Dempster and I. McLean). Kluwer, London, pp. 135–65.

Godfray, H.C.J. and Pacala, S.W. (1992) Aggregation and the population dynamics of parasitoids and predators. *American Naturalist*, **140**, 30–40.

Godfray, H.C.J. and Shimada, M. (1999) Parasitoids as model systems for ecologists. *Researches on Population Ecology*, **41**, 3–10.

Godfray, H.C.J. and Waage, J.K. (1991) Predictive modelling in biological control: the Mango mealy bug *(Rastrococcus invadens)* and its parasitoids. *Journal of Applied Ecology*, **28**, 434–53.

Godfray, H.C.J., Hassell, M.P. and Holt, R.D. (1994) The population dynamic consequences of phenological asynchrony between parasitoids and their hosts. *Journal of Animal Ecology*, **63**, 1–10.

Godfray, H.C.J., Agassiz, D.J.L., Nash, D.R. and Lawton, J.H. (1995) The recruitment of parasitoid species to two invading herbivores. *Journal of Animal Ecology*, **64**, 393–402.

Goding, J.W. (1986) *Monoclonal Antibodies: Principles and Practice*. Academic Press, San Diego.

Goff, A.M. and Nault, L.R. (1984) Response of the pea aphid parasite *Aphidius ervi* Haliday (Hymenoptera:

Aphidiidae) to transmitted light. *Environmental Entomology*, **13**, 595–8.

Gohole, L.S., Overholt, W.A., Khan, Z.R. and Vet, L.E.M. (2003) Role of volatiles emitted by host and non-host plants in the foraging behaviour of *Dentichasmias busseolae*, a pupal parasitoid of the spotted stemborer *Chilo partellus*. *Entomologia Experimentalis et Applicata*, **107**, 1–9.

Gokool, S., Curtis, C.F. and Smith, D.F. (1993) Analysis of mosquito bloodmeals by DNA profiling. *Medical and Veterinary Entomology*, **7**, 208–16.

Goldschmidt, R. (1915) Vorläufige Mitteilung über weitere Versuche zur Vererbung und Bestimmung des Geschlechts. *Biologisches Zentralblatt*, **35**, 565–70.

Goldson, S.L. and McNeill, M.R. (1992) Variation in the critical photoperiod for diapause induction in *Microctonus hyperodae*, a parasitoid of Argentine stem weevil. *Proceedings of the 45th New Zealand Plant Protection Conference, 1992*, pp. 205–9.

Goldson, S.L., McNeill, M.R., Phillips, C.B. and Proffitt, J.R. (1992) Host specificity testing and suitability of the parasitoid *Microctonus hyperodae* (Hym: Braconidae, Euphorinae) as a biological control agent of *Listronotus bonariensis* (Col.: Curculionidae) in New Zealand. *Entomophaga*, **37**, 483–98.

Goldson, S.L., Proffitt, J.B. and Baird, D.B. (1998) Establishment and phenology of the parasitoid *Microctonus hyperodae* (Hymenoptera: Braconidae) in New Zealand. *Environmental Entomology*, **27**, 1386–92.

Gonzalez, D., Gordh, G., Thompson, S.N. and Adler, J. (1979) Biotype discrimination and its importance to biological control, in *Genetics in Relation to Insect Management* (eds M.A. Hoy and J.J. McKelvey). The Rockefeller Foundation, New York, pp. 129–39.

Gonzalez, J.M., Matthews, R.W. and Matthews, J.R. (1985) A sex pheromone in males of *Melittobia australica* and *M. femorata* (Hym., Eulophidae). *Florida Entomologist*, **68**, 279–86.

Goodenough, J.L., Hartsack, A.W. and King, E.G. (1983) Developmental models for *Trichogramma pretiosum* (Hymenoptera: Trichogrammatidae) reared on four hosts. *Journal of Economic Entomology*, **76**, 1095–102.

Goodhardt, G.J., Ehrenberg, A.S.C. and Chatfield, C. (1984) The Dirichlet: a comprehensive model of buying behaviour. *Journal of the Royal Statistical Society A*, **147**, 621–55.

Goodman, C.L., Greenstone, M.H. and Stuart, M.K. (1997) Monoclonal antibodies to vitellins of bollworm and tobacco budworm (Lepidoptera: Noctuidae): biochemical and ecological implications. *Annals of the Entomological Society of America*, **90**, 83–90.

Goodnight, K.F. and Queller, D.C. (1999) Computer software for performing likelihood tests of pedigree relationship using genetic markers. *Molecular Ecology*, **8**, 1231–4.

Goodpasture, C. (1975) Comparative courtship and karyology in *Monodontomerus* (Hym., Torymidae). *Annals of the Entomological Society of America*, **68**, 391–7.

van der Goot, V.S. and Grabandt, R.A.J. (1970) Some species of the genera *Melanostoma*, *Platycheirus* and *Pyrophaena* (Diptera, Syrphidae) and their relation to flowers *Entomologische Berichten*, **30**, 135–43.

Gordh, G. and DeBach, P. (1976) Male inseminative potential in *Aphytis lignanensis* (Hymenoptera: Aphelinidae). *Canadian Entomologist*, **108**, 583–89.

Gordh, G. and DeBach, P. (1978) Courtship behavior in the *Aphytis lingnanensis* group, its potential usefulness in taxonomy, and a review of sexual behavior in the parasitic Hymenoptera (Chalcidoidea: Aphelinidae). *Hilgardia*, **46**, 37–75.

Gordh, G., Woolley, J.B. and Medeved, R.A. (1983) Biological studies on *Goniozus legneri* Gordh (Hymenoptera: Bethylidae), primary external parasite of the navel orangeworm *Amyelois transitella* and pink bollworm *Pectinophora gossypiella* (Lepidoptera: Pyralidae, Gelechiidae). *Contributions of the American Entomological Institute*, **20**, 433–68.

Gordon, A.H. (1975) Electrophoresis of proteins in polyacrylamide and starch gels, in *Laboratory Techniques in Biochemistry and Molecular Biology* (eds T.S. Work and E. Work). North-Holland, Amsterdam, London.

Gordon, D.M., Nisbet, R.M., de Roos, A., Gurney, W.S.L. and Stewart, R.K. (1991) Discrete generations in host-parasitoid models with contrasting life cycles. *Journal of Animal Ecology*, **60**, 295–308.

Gordon, P.L. and McKinlay, R.G. (1986) Dispersal of ground beetles in a potato crop; a mark-release study. *Entomologia Experimentalis et Applicata*, **40**, 104–5.

Gouinguene, S., Alborn, H. and Turlings, T.C.J. (2003) Induction of volatile emissions in maize by different larval instars of *Spodoptera littoralis*. *Journal of Chemical Ecology*, **29**, 145–62.

Gould, J.R., Elkinton, J.S. and Wallner, W.E. (1990) Density dependent suppression of experimentally created gypsy moth, *Lymantria dispar* (Lepidoptera: Lymantriidae) populations by natural enemies. *Journal of Animal Ecology*, **59**, 213–33.

Gould, J.R., Bellows Jr, T.S. and Paine, T.D.(1992a) Population dynamics of *Siphoninus phillyreae* in California in the presence and absence of a parasitoid, *Encarsia partenopea*. *Ecological Entomology*, **17**,127–34.

Gould, J.R., Bellows, T.S. and Paine, T.D. (1992b) Evaluation of biological control of *Siphoninus phillyreae* (Haliday) by the parasitoid *Encarsia partenopea* (Walker) using life-table analysis. *Biological Control*, **2**, 257–65.

Goulson, D. (2000) Are insects flower constant because they use search images to find flowers? *Oikos*, **88**, 547–52.

Goulson, D. and Wright, N.P. (1998) Flower constancy in the hoverflies *Episyrphus balteatus* (Degeer) and *Syrphus ribesii* (L.) (Syrphidae). *Behavioral Ecology*, **9**, 213–9.

Gowaty, P.A. (1994) Architects of sperm competition. *Trends in Ecology and Evolution*, **9**, 160–62.

Gowling, G.R. and van Emden, H.F. (1994) Falling aphids enhance impact of biological control by parasitoids on partially aphid-resistant plant varieties. *Annals of Applied Biology*, **125**, 233–42.

Gozlan, S., Millot, P., Rousset, A. and Fournier, D. (1997) Test of the RAPD-PCR method to evaluate the efficacy of augmentative biological control with *Orius* (Het., Anthocoridae). *Entomophaga*, **42**, 593–604.

Grafen, A. (1991) Modelling in behavioural ecology, in *Behavioural Ecology, 3rd edn* (eds J.R. Krebs and N.B. Davies). Blackwell Scientific Publishing, Oxford, pp. 5–31.

Graham, H.M., Wolfenbarger, D.A. and Nosky, J.B. (1978) Labeling plants and their insect fauna with rubidium. *Environmental Entomology*, **7**, 379–83.

Graham, H.M., Jackson, C.G. and Lakin, K.R. (1984) Comparison of two methods of using the D-vac to sample mymarids and their hosts in alfalfa. *Southwestern Entomologist*, **9**, 249–52.

Graham, M.W.R. de V. (1989) A remarkable secondary sexual character in the legs of male *Nesolynx glossinae* (Waterston). *Entomologist's Monthly Magazine*, **125**, 231–32.

Graham, M.W.R. de V. (1993) Swarming in Chalcidoidea (Hym.) with the description of a new species of *Torymus* (Hym., Torymidae) involved. *Entomologist's Monthly Magazine*, **129**, 15–22.

Grasswitz, T.R. and Paine, T.D. (1993) Effect of experience on in-flight orientation to host-associated cues in the generalist parasitoid *Lysiphlebus testaceipes*. *Entomologia Experimentalis et Applicata*, **68**, 219–29.

Gravena, S. and Sterling, W.L. (1983) Natural predation on the cotton leafworm (Lepidoptera: Noctuidae). *Journal of Economic Entomology*, **76**, 779–84.

Greany, P.D. and Oatman, E.R. (1972) Analysis of host discrimination in the parasite *Orgilus lepidus* (Hymenoptera: Braconidae). *Annals of the Entomological Society of America*, **65**, 377–83.

Greany, P.D., Hawke, S.D., Carlysle, T.C. and Anthony, D.W. (1977) Sense organs in the ovipositor of *Biosteres* (*Opius*) *longicaudatus*, a parasite of the Caribbean fruit fly *Anastrepha suspensa*. *Annals of the Entomological Society of America*, **70**, 319–21.

Greathead, D.J. (1986) Parasitoids in classical biological control, in *Insect Parasitoids* (eds J. Waage and D. Greathead). Academic Press, London, pp. 289–318.

Greathead, D.J. and Greathead, A.H. (1992) Biological control of insect pests by insect parasitoids and predators: the BIOCAT database. *Biocontrol News and Information*, **13**, 61N–68N.

Greathead, D.J. and Waage, J.K. (1983) Opportunities for biological control of agricultural pests in developing countries. *World Bank Technical Paper*, **11**, 1–44.

Greatorex, E.C. (1996). A molecular technique for examining the gut content of predatory mites. *Proceedings of the Brighton Crop Protection Conference – Pests and Diseases 1996, BCPC, Farnham, Surrey UK*, 437–8.

Greb, R.J. (1933) Effects of temperature on production of mosaics in *Habrobracon*. *Biological Bulletin*, **65**, 179–86.

Greeff, J.M. (1995) Offspring allocation in structured populations with dimorphic males. *Evolutionary Ecology*, **9**, 550–85.

Greeff, J.M. (1996) Alternative mating strategies, partial sibmating and split sex ratios in haplodiploid species. *Journal of Evolutionary Biology*, **9**, 855–69.

Greeff, J.M. (1998) Local mate competition, sperm usage and alternative mating strategies. *Evolutionary Ecology*, **12**, 627–8.

Greeff, J.M. and Ferguson, J.W.H. (1999) Mating ecology of nonpollinating fig wasps of *Ficus ingens*. *Animal Behaviour*, **57**, 215–22.

Greeff, J.M. and Taylor, P.D. (1997) Effects of inbreeding depression on relatedness and optimal sex ratios. *Evolutionary Ecology*, **11**, 245–47.

Greeff, J.M., van Noort, S., Rasplus, J-Y. and Kjellberg, F. (2003) Dispersal and fighting in male pollinating fig wasps. *Comptes Rendus Biologies*, **326**, 121–30.

Green, J. (1999) Sampling method and time determines composition of spider collections. *Journal of Arachnology*, **27**, 176–82.

Green, R.F., Gordh, G. and Hawkins, B.A. (1982) Precise sex ratios in highly inbred parasitic wasps. *American Naturalist*, **120**, 653–65.

Greenblatt, J.A., Barbosa, P. and Montgomery, M.E. (1982) Host's diet effects on nitrogen utilization efficiency for two parasitoid species: *Brachymeria intermedia* and *Coccygomimus turionellae*. *Physiological Entomology*, **7**, 263–7.

Greene, A. (1975) Biology of five species of Cychrini (Coleoptera: Carabidae) in the steppe region of south-eastern Washington. *Melanderia*, **19**, 1–43.

Greenfield, M.D. (1981) Moth sex pheromones: an evolutionary perspective. *Florida Entomologist*, **64**, 4–17.

Greenfield, M.D. and Karandinos, M.G. (1976) Fecundity and longevity of *Synanthedon pictipes* under constant and fluctuating temperatures. *Environmental Entomology*, **5**, 883–7.

Greenstone, M.H. (1977) A passive haemagglutination inhibition assay for the identification of stomach contents of invertebrate predators. *Journal of Applied Ecology*, **14**, 457–64.

Greenstone, M.H. (1979a) A line transect density index for wolf spiders (*Pardosa* spp.) and a note on the applicability of catch per unit effort methods to entomological studies. *Ecological Entomology*, **4**, 23–9.

Greenstone, M.H. (1979b) Passive haemagglutination inhibition: a powerful new tool for field studies of entomophagous predators. *Entomological Society of America, Miscellaneous Publications*, **11**, 69–78.

Greenstone, M.H. (1983) Site-specificity and site tenacity in a wolf-spider: a serological dietary analysis. *Oecologia*, **56**, 79–83.

Greenstone, M.H. (1990) Meteorological determinants of spider ballooning: the roles of thermals vs. the vertical windspeed gradient in becoming airborne. *Oecologia*, **84**, 164–8.

Greenstone, M.H. (1996) Serological analysis of arthropod predation: past, present and future, in *The Ecology of Agricultural Pests: Biochemical Approaches* (eds W.O.C. Symondson and J.E. Liddell), Systematics Association Special Volume 53. Chapman and Hall, London, pp. 265–300.

Greenstone, M.H. (1999) Spider predation: how and why we study it. *Journal of Arachnology*, **27**, 333–42.

Greenstone, M.H. and Edwards, M.J. (1998) DNA hybridization probe for endoparasitism by *Microplitis croceipes* (Hymenoptera: Braconidae). *Annals of the Entomological Society of America*, **91**, 415–21.

Greenstone, M.H. and Hunt, J.H. (1993) Determination of prey-antigen half-life in *Polistes metricus* using a monoclonal antibody-based immunodot assay. *Entomologia Experimentalis et Applicata*, **68**, 1–7.

Greenstone, M.H. and Morgan, C.E. (1989) Predation on *Heliothis zea* (Lepidoptera, Noctuidae) – an instar-specific ELISA assay for stomach analysis. *Annals of the Entomological Society of America*, **82**, 45–9.

Greenstone, M.H. and Trowell, S.C. (1994) Arthropod predation: a simplified immunodot format for predator gut analysis. *Annals of the Entomological Society of America*, **87**, 214–7.

Greenstone, M.H., Morgan, C.E. and Hultsch, A.L. (1985) Spider ballooning: development and evaluation of field trapping methods (Araneae). *Journal of Arachnology*, **13**, 337–45.

Greenstone, M.H., Eaton, R.R. and Morgan, C.E. (1991) Sampling aerially dispersing arthropods: A high-volume, inexpensive, automobile- and aircraft-borne system. *Journal of Economic Entomology*, **84**, 1717–24.

Greenwood, J.J.D. and Elton, R.A. (1979) Analysing experiments on frequency dependent selection by predators. *Journal of Animal Ecology*, **48**, 721–37.

Grégoire-Wibo, C. (1983a) Incidences écologiques des traitments phytosanitaires en culture de betterave sucrière, essais expérimentaux en champ. 1. Les Collemboles épigés. *Pedobiologia*, **25**, 37–48.

Grégoire-Wibo, C. (1983b) Incidences écologiques des traitments phytosanitaires en culture de betterave sucrière, essais expérimentaux en champ. II. Acariens, Polydesmes, Staphylins, Cryptophagides et Carabides. *Pedobiologia*, **25**, 93–108.

Grenier, S., Pintureau, B., Heddi, A., Lassabliere, F., Jager, C., Louis, C. and Khatchadourian, C. (1998) Successful horizontal transfer of *Wolbachia* symbionts between *Trichogramma* wasps. *Proceedings of the Royal Society of London*, **265**, 1441–5.

Grenier, S., Gomes, S.M., Pintureau, B., Lasslbiere, F. and Bolland, P. (2002) Use of tetracycline in larval diet to study the effect of *Wolbachia* on host fecundity and clarify taxonomic status of *Trichogramma* species in cured bisexual lines. *Journal of Invertebrate Pathology*, **80**, 13–21.

Gresens, S.E., Cothran, M.L. and Thorp, J.H. (1982) The influence of temperature on the functional response of the dragonfly *Celithemis fasciata* (Odonata: Libellulidae). *Oecologia*, **53**, 281–4.

Gressitt, J.L. and Gressitt, M.K. (1962) An improved Malaise trap. *Pacific Insects*, **4**, 87–90.

Grether, G.F. (1996) Sexual selection and survival selection on wing colouration and body size in the rubyspot damselfly *Haeterina americana*. *Evolution*, **50**, 1939–48.

Gribbin, S.D. and Thompson, D.J. (1990) Asymmetric intraspecific competition among larvae of the damselfly *Ischnura elegans* (Zygoptera: Coenagrionidae). *Ecological Entomology*, **15**, 37–42.

Griffith, D.M. and Poulson, T.L. (1993) Mechanisms and consequences of intraspecific competition in a carabid cave beetle. *Ecology*, **74**, 1373–83.

Griffiths, D. (1980) The feeding biology of ant-lion larvae: growth and survival in *Morter obscitrus. Oikos*, **34**, 364–70.

Griffiths, D. (1982) Tests of alternative models of prey consumption by predators, using ant-lion larvae. *Journal of Animal Ecology*, **52**, 363–73.

Griffiths, D. (1992) Interference competition in ant-lion (*Macroleon quinquemaculatus*) larvae. *Ecological Entomology*, **17**, 219–26.

Griffiths, E., Wratten, S.D. and Vickerman, G.P. (1985) Foraging behaviour by the carabid *Agonum dorsale* in the field. *Ecological Entomology*, **10**, 181–9.

Griffiths, G., Winder, L., Bean, D., Preston, R., Moate, R., Neal, R., Williams, E., Holland, J. and Thomas G. (2001) Laser marking the carabid *Pterostichus melanarius* for mark-release-recapture. *Ecological Entomology*, **26**, 662–3.

Griffiths, N.T. and Godfray, H.C.J. (1988) Local mate competition, sex ratio and clutch size in bethylid wasps. *Behavioral Ecology and Sociobiology*, **22**, 211–17.

Grigliatti, T. (1986) Mutagenesis, in *Drosophila, a Practical Approach* (ed D.B. Roberts). IRL Press, Oxford, pp. 39–45.

Grissell, E.E. and Goodpasture, C.E. (1981) A review of nearctic Podagrionini, with description of sexual behavior of *Podagrion mantis* (Hym., Torymidae). *Annals of the Entomological Society of America*, **74**, 226–41.

Grissell, E.E. and Schauff, M.E. (1990) *A Handbook of the Families of Nearctic Chalcidoidea (Hymenoptera)*. The Entomological Society of Washington, Washington.

Gromadzka, J. and Trojan, P. (1967) Comparison of the usefulness of an entomological net, photoeclector and biocenometer for investigation of entomocenosis. *Ekologia Polska*, **15**, 505–29.

Grosch, D.S. (1945) The relation of cell size and organ size to mortality in *Habrobracon. Growth*, **9**, 1–17.

Grosch, D.S. (1950) Starvation studies with the parasitic wasp *Habrobracon. Biological Bulletin, Marine Biological Laboratory, Wood's Hole*, **99**,65–73.

Gross, H.R., Lewis, W.J., Jones, R.L. and Nordlund, D.A. (1975) Kairomones and their use for management of entomophagous insects. III. Stimulation of *Trichogramma achaeae, T. pretiosum* and *Microplitis croceipes* with host seeking stimuli at the time of "release to improve efficiency. *Journal of Chemical Ecology*, **1**, 431–8.

Gross, M.R. (1996) Alternative reproductive strategies and tactics: diversity within sexes. *Trends in Ecology and Evolution*, **11**, 92–98.

Gross, P. (1993) Insect behavioral and morphological defenses against parasitoids. *Annual Review of Entomology*, **38**, 251–73.

Grostal P. and Dicke, M. (2000) Recognising one's enemies: a functional approach to risk assessment by prey. *Behavioral Ecology and Sociobiology*, **47**, 258–64.

Gu, H., Wäckers, F., Steindl, P., Günther, D. and Dorn, S. (2001) Different approaches to labelling parasitoids using strontium. *Entomologia Experimentalis et Applicata*, **99**, 173–81.

Guerra, M., Fuentes-Contreras, E. and Niemeyer, H.M. (1998) Differences in behavioural responses of *Sitobion avenae* (Hemiptera: Aphididae) to volatile compounds, following parasitism by *Aphidius ervi* (Hymenoptera: Braconidae). *Ecoscience*, **5**, 334–7.

Guerrieri, E., Pennacchio, F. and Tremblay, E. (1997) Effect of adult experience on in-flight orientation to plant and plant-host complex volatiles in *Aphidius ervi* Haliday (Hymenoptera: Braconidae). *Biological Control*, **10**, 159–65.

Guertin, D.S., Ode, P.J., Strand, M.R. and Antolin, M.F. (1996) Host-searching and mating in an outbreeding parasitoid wasp. *Ecological Entomology*, **21**, 27–33.

Guido, A.S. and Fowler, H.G. (1988) *Megacephala fulgida* (Coleoptera: Cicindelidae): a phonotactically orienting predator of *Scapteriscus* mole crickets (Orthoptera: Gryllotalpidae). *Cicindela*, **20**, 51–2.

Guilford, T. and Dawkins, M.S. (1987) Search images not proven: a reappraisal of recent evidence. *Animal Behaviour*, **35**,1838–45.

Gülel, A. (1988) Effects of mating on the longevity of males and sex ratio of *Dibrachys boarmiae* (Hym., Pteromalidae). *Doga Tu Zooloji*, **12**, 225–30 (In Turkish, with English summary).

Gunasena, G.H., Vinson, S.B. and Williams, H.J. (1990) Effects of nicotine on growth, development, and survival of the tobacco budworm (Lepidoptera: Noctuidae) and the parasitoid *Campoletis sonorensis* (Hymenoptera: Ichneumonidae). *Journal of Economic Entomology*, **83**, 1777–82.

Gurdebeke, S. and Maelfait, J.P. (2002) Pitfall trapping in population genetics studies: finding the right "solution". *Journal of Arachnology*, **30**, 255–61.

Gurney, W.S.C. and Nisbet, R.M. (1985) Fluctuation periodicity, generation separation, and the expression of larval competition. *Theoretical Population Biology*, **28**, 886–923.

Gurr, G.M., Thwaite, W.G. and Nicol, H.I. (1999a) Field evaluation of the effects of the insect growth regulator tebufenoxide on entomophagous arthropods and pests of apples. *Australian Journal of Entomology*, **38**, 135–40.

Gurr, G.M., Wratten, S.D., Irvin, N.A., Hossain, Z., Baggen, L.R., Mensah, R.K. and Walker, P.W. (1999b). Habitat manipulation in Australasia: recent biological control progress and prospects for adoption. *Proceedings of the Sixth Australasian Applied Entomological Research Conference, Volume 2*, 225–34.

Gurr, G.M., Wratten, S.D. and Barbosa, P. (2000) Success in conservation biological control of arthropods, in *Biological Control: Measures of Success* (eds G. Gurr, and S. Wratten). Kluwer Academic Publishers, Dordrecht, pp. 105–32.

Gurr, G.M., Wratten, S.D., Tylianakis, J., Kean, J. and Keller, M. (2004) Providing plant foods for insect natural enemies in farming systems: balancing practicalities and theory, in *Plant-provided Food and Plant-carnivore Mutualism* (eds F. Wäckers, P. van Rijn, and J. Bruin). Cambridge University Press, Cambridge.

Gutierrez, A.P. (1970) Studies on host selection and host specificity of the aphid hyperparasite *Charips victrix* (Hymenoptera: Cynipidae). 6. Description of sensory structures and a synopsis of host selection and host specificity. *Annals of the Entomological Society of America*, **63**, 1705–9.

Gutierrez, A.P., Baumgaertner, J.U. and Hagen, K.S. (1981) A conceptual model for growth, development and reproduction in the ladybird beetle, *Hippodamia convergens* (Coleoptera: Coccinellidae). *Canadian Entomologist*, **113**, 21–33.

Gutierrez, A.P., Baumgartner, J.U. and Summers, C.G. (1984) Multitrophic level models of predator-prey energetics: III. A case study in an alfalfa ecosystem. *The Canadian Entomologist*, **116**, 950–63.

Gutierrez, A.P., Neuenschwander, P., Schulthess, F. Herren, H.R., Baumgartner, J.U., Wermelinger, B., Löhr, B. and Ellis, C.K. (1988a) Analysis of biological control of cassava pests in Africa. II. Cassava mealybug *Phenacoccus manihoti*. *Journal of Applied Ecology*, **25**, 921–40.

Gutierrez, A.P., Yaninek, J.S., Wermelinger, B., Herren, H.R. and Ellis, C.K. (1988b) Analysis of biological control of cassava pests in Africa. III. Cassava green mite *Mononychellus tanajoa*. *Journal of Applied Ecology*, **251**, 941–51.

Gutierrez, A.P., Hagen, K.S. and Ellis, C.K. (1990) Evaluating the impact of natural enemies: a multitrophic perspective, in *Critical Issues in Biological Control* (ed. M. Mackauer L.E. Ehler and J. Roland). Intercept, Andover, pp. 81–107.

Gutierrez, A.P., Neuenschwander, P. and van Alphen, J.J.M. (1993) Factors affecting biological control of cassava mealybug by exotic parasitoids: a ratio-dependent supply-demand driven model. *Journal of Applied Ecology*, **30**, 706–21.

Gutierrez, A.P., Mills, N.J., Schreiber, S.J. and Ellis, C.K. (1994) A physiologically based tritrophic perspective on bottom-up top-down regulation of populations. *Ecology*, **75(8)**, 2227–42.

Györfi, J. (1945) Beobachtungen über die ernahrung der schlupfwespenimagos. *Erdjszeti Kisérletek*, **45**, 100–12.

Haas, V. (1980) Methoden zur Erfassung der Arthropodenfauna in der Vegetationssicht von Grasland-Ökosystemen. *Zoologischer Anzeiger Jena*, **204**, 319–30.

Haber, W.A., Frankie, G.W., Baker, H.G. *et al.* (1981) Ants like flower nectar. *Biotropica*, **13**, 211–14.

Haccou, P. and Meelis, E. (1992) *Statistical Analysis of Behavioural Data, an Approach Based on Time-structured Models*. Oxford University Press, Oxford.

Haccou, P., de Vlas, S.J., van Alphen, J.J.M. and Visser, M.E. (1991) Information processing by foragers: effects of intra-patch experience on the leaving tendency of *Leptopilina heterotoma*. *Journal of Animal Ecology*, **60**, 93–106.

Hackman, W. (1957) Studies on the ecology of the wolf spider *Trochosa terricola* Deg. *Societas Scientiarum Fennica Commentationes Biologicae*, **16**, 1–34.

Hadrys, H. and Siva-Jothy, M.T. (1994) Unravelling the components that underlie insect reproductive traits using a simple molecular approach, in *Molecular Ecology and Evolution: Approaches and Applications* (eds B. Schierwater, B. Streit, G.P. Wagner and R. DeSalle). Birkhäuser, Basel, pp. 75–90.

Hadrys, H., Balick, M. and Schierwater, B. (1992) Applications of random amplified polymorphic DNA (RAPD) in molecular ecology. *Molecular Ecology*, **1**, 55–63.

Hadrys, H., Schierwater, B., Dellaporta, S.L., DeSalle, R. and Buss, L.W. (1993) Determination of paternities in dragonflies by random amplified polymorphic DNA fingerprinting. *Molecular Ecology* **2**, 79–87.

Haeck, J. (1971) The immigration and settlement of carabids in the new Ijsselmeer-polders. *Miscellaneous Papers Landbouwwetenschappen Hogeschule Wageningen*, **8**, 33–52.

Hag Ahmed, S.E.M.K. (1989) Biological control of glasshouse *Myzus persicae* (Sulzer) using *Aphidius matricariae* Haliday. Wye College, University of London, Ph.D. Thesis.

Hagen, K. (1964a) Developmental stages of parasites, in *Biological Control of Insect Pests and Weeds* (ed. P. DeBach). Chapman and Hall, London, pp. 168–246.

Hagen, K. (1964b) Nutrition of entomophagous insects and their hosts, in *Biological Control of Insect Pests and Weeds* (ed. P. DeBach). Chapman and Hall, London, pp. 356–80.

Hagen, K.S. (1986) Ecosystem analysis: plant cultivars (HPR), entomophagous species and food supplements, in *Interactions of Plant Resistance and Parasitoids and Predators of Insects* (eds D.J. Boethel and R.D. Eikenbary). Ellis Horwood, Chichester/John Wiley and Sons, New York, pp. 151–97.

Hagen, K.S. and Tassan, R.L. (1966) The influence of protein hydrolysates of yeasts and chemically defined diets upon the fecundity of *Chrysopa carnea* Stephens (Neuroptera). *Acta Societatis Zoologicae Bohemoslovacae*, **30**, 219–27.

Hagler, J.R. (1998) Variation in the efficacy of several predator gut content immunoassays. *Biological Control*, **12**, 25–32.

Hagler, J.R. and Durand, C.M. (1994) A new method for immunologically marking prey and its use in predation studies. *Entomophaga*, **39**, 257–65.

Hagler, J.R. and Jackson, C.G. (2001) Methods for marking insects: current techniques and future prospects. *Annual Review of Entomology*, **46**, 511–43.

Hagler, J.R. and Naranjo, S.E. (1994a) Qualitative survey of two coleopteran predators of *Bemisia tabaci* (Homoptera: Aleyrodidae) and *Pectinophora gossypiella* (Lepidoptera: Gelechiidae) using a multiple prey gut content ELISA. *Environmental Entomology*, **23**, 193–7.

Hagler, J.R. and Naranjo, S.E. (1994b) Determining the frequency of heteropteran predation on sweet "potato whitefly and pink bollworm using multiple ELISAs. *Entomologia Experimentalis et Applicata*, **72**, 63–70.

Hagler, J.R., Cohen, A.C., Enriquez, F.J. and Bradley-Dunlop, D. (1991) An egg-specific monoclonal antibody to *Lygus hesperus*. *Biological Control*, **1**, 75–80.

Hagler, J.R., Cohen, A.C., Bradley-Dunlop, D. and Enriquez, F.J. (1992a) New approach to mark insects for feeding and dispersal studies. *Environmental Entomology*, **21**, 20–5.

Hagler, J.R., Cohen, A.C., Bradley-Dunlop, D. and Enriquez, F.J. (1992b) Field evaluation of predation on *Lygus hesperus* (Hemiptera: Miridae) using species- and stage-specific monoclonal antibody. *Environmental Entomology*, **21**, 896–900.

Hagler, J.R., Brower, A.G., Tu, Z., Byrne, D.N., Bradley-Dunlop, D. and Enriquez, F.J. (1993) Development of a monoclonal antibody to detect predation of the sweetpotato whitefly, *Bemisia tabaci*. *Entomologia Experimentalis et Applicata*, **68**, 231–6.

Hagler, J.R., Naranjo, S.E., Dunlop-Bradley, D., Enriquez, F.J. and Henneberry, T.J. (1994) A monoclonal antibody to pink bollworm (Lepidoptera: Gelechiidae) egg antigen: a tool for predator gut analysis. *Annals of the Entomological Society of America*, **87**, 85–90.

Hagler, J.R., Buchmann, S.L. and Hagler, D.A. (1995) A simple method to quantify dot blots for predator gut content analyses. *Journal of Entomological Science*, **30**, 95–8.

Hagler, J.A., Machtley, S.A. and Leggett, J.E. (2002a) Parasitoid mark-release-recapture techniques – I. Development of a battery-operated suction trap for collecting minute insects. *Biocontrol Science and Technology*, **12**, 653–9.

Hagler, J.R., Jackson, C.G., Henneberry, T.J. and Gould, J.R. (2002b) Parasitoid mark-release-recapture techniques – II. Development and application of a protein marking technique for *Eretmocerus* spp., parasitoids of *Bemisia argentifolii*. *Biocontrol Science and Technology*, **12**, 661–75.

Hagstrum, D.W. and Smittle, B.J. (1977) Host-finding ability of *Bracon hebetor* and its influence upon adult parasite survival and fecundity. *Environmental Entomology*, **6**, 437–9.

Hagstrum, D.W. and Smittle, B.J. (1978) Host utilization by *Bracon hebetor*. *Environmental Entomology*, **7**, 596–600.

Hågvar, E.B. (1972) The effect of intra- and interspecific larval competition for food (*Myzus persicae*) on the development at 20°C of *Syrphus ribesii* and *Syrphus corollae* (Diptera: Syrphidae). *Entomophaga*, **17**, 71–7.

Hågvar, E.B. (1973) Food consumption in larvae of *Syrphus ribesii* (L.) and *Syrphus corollae* (Fabr.) (Dipt., Syrphidae). *Norsk Entomologisk Tiddskrift*, **201**, 315–21.

Hågvar, E.B. and Hofsvang, T. (1991) Effect of honeydew on the searching behaviour of the aphid parasitoid *Ephedrus cerasicola* (Hymenoptera, Aphidiidae). *Redia*, **74**, 259–64.

Hågvar, E.B., Hofsvang, T., Trandem, N. and Saeterbø, K.G. (1998) Six-year Malaise trapping of the leaf miner *Chromatomyia fuscula* (Diptera: Agromyzidae) and its chalcidoid parasitoid complex in a barley field and its boundary. *European Journal of Entomology*, **95**, 529–43.

Hails, R.S. and Crawley, M.J. (1992) Spatial density dependence in populations of a cynipid gall-former *Andricus quercuscalicis*. *Journal of Animal Ecology*, **61**, 567–83.

Hairston, N.G., Smith, F.E. and Slobodkin, L.B. (1960) Community structure, population control and competition. *American Naturalist*, **44**, 421–5.

Halaj, J. and Cady, A.B. (2000) Diet composition and significance of earthworms as food of harvestmen (Arachnida: Opiliones). *American Midland Naturalist*, **143**, 487–91.

Halaj, J. and Wise, D.H. (2001) Terrestrial trophic cascades: how much do they trickle? *American Naturalist*, **157**, 262–81.

Halaj, J., Ross, D.W. and Moldenke, A.R. (1998) Habitat structure and prey availability as predictors of the abundance and community organisation of spiders in western Oregon forest canopies. *Journal of Arachnology*, **26**, 203–20.

Haldane, J.B.S. (1919) The combination of linkage values, and the calculation of distance between the loci of linked factors. *Journal of Genetics*, **8**, 299–309.

Haley, C.S. and Knott, S.A. (1992) A simple regression method for mapping quantitative trait loci in line crosses using flanking markers. *Heredity*, **69**, 315–24.

Hall, J.C., Siegel, R.W., Tompkins, L. and Kyriacou, C.P. (1980) Neurogenetics of courtship in *Drosophila*. *Stadler Symposia, University of Missouri*, **12**, 43–82.

Hall, R.W. (1993) Alteration of sex ratios of parasitoids for use in biological control, in *Evolution and Diversity of Sex Ratio in Insects and Mites* (eds L. Wrensch and M.A. Ebbert). Chapman and Hall, New York, pp. 542–7.

Hall, S.J. and Raffaelli, D. (1993) Food webs: theory and reality. *Advances in Ecological Research*, **24**, 187–239.

Halsall, N.B. and Wratten, S.D. (1988a) The efficiency of pitfall trapping for polyphagous Carabidae. *Ecological Entomology*, **13**, 293–9.

Halsall, N.B. and Wratten, S.D. (1988b) Video recordings of aphid predation in a wheat crop. *Aspects of Applied Biology*, **17**, 277–80.

Halstead, J.A. (1988) A gynandromorph of *Hockeria rubra* (Ashmead) (Hymenoptera: Chalcididae). *Proceedings of the Entomological Society of Washington*, **90**, 258–9.

Hämäläinen, M., Markkula, M. and Raij, T. (1975) Fecundity and larval voracity of four ladybeetle species (Col., Coccinellidae). *Annales Entomologici Fennici*, **41**, 124–7.

van Hamburg, H. and Hassell, M.P. (1984) Density dependence and the augmentative release of egg parasitoids against graminaceous stalkborers. *Ecological Entomology*, **9**, 101–8.

Hamerski, M.R. and Hall, R.W. (1989) Adult emergence, courtship, mating, and ovipositional behavior of *Tetrastichus gallerucae* (Hymenoptera: Eulophidae), a parasitoid of the elm leaf beetle (Coleoptera: Chrysomelidae). *Environmental Entomology*, **18**, 791–4.

Hames, B.D. and Rickwood, D. (1981) *Gel Electrophoresis of Proteins: A Practical Approach*. IRL Press Ltd., Oxford.

Hamilton, R.M., Dogan, E.B., Schaalje, G.B. and Booth, G.M. (1999) Olfactory response of the lady beetle *Hippodamia convergens* (Coleoptera: Coccinellidae) to prey related odors, including a scanning electron microscopy study of the antennal sensilla. *Environmental Entomology*, **28**, 812–22.

Hamilton, W.D. (1967) Extraordinary sex ratios. *Science*, **156**, 477–88.

Hamilton, W.D. (1979) Wingless and fighting males in fig wasps and other insects, in *Sexual Selection and Reproductive Competition in Insects* (eds M.S. Blum and N.A. Blum). Academic Press, New York, pp. 167–220.

Hamm, A.H. (1934) Syrphidae (Dipt.) associated with flowers. *Journal of the Society for British Entomology*, **1**, 8–9.

Hamon, N., Bardner, R., Allen-Williams, L. and Lee, J.B. (1990) Carabid populations in field beans and their effect on the population dynamics of *Sitona lineatus* (L.). *Annals of Applied Biology*, **117**, 51–62.

Hance, T. (1986) Experiments on the population control of *Aphis fabae* by different densities of Carabidae (Coleoptera: Carabidae). *Annals of the Royal Society of Zoology, Belgium*, **116**, 15–24 (In French).

Hance, T. (1995) Relationships between aphid phenology and predator and parasitoid abundances in maize fields. *Acta Jutlandica*, **70**, 113–23.

Hance, T. and Grégoire-Wibo, C. (1983) Étude du régime alimentaire des Carabidae par voie sérologique, in *New Trends in Soil Biology* (eds P. Lebrun, H. André, A. DeMedts, C. Grégoire-Wibo and G. Wauthy). Imprimerie Dieu-Brichart, Ottignies-Louvain, pp. 620–22.

Hance, T. and Rossignol, P. (1983) Essai de quantification de la prédation des Carabidae par la test ELISA. *Mededelingen van de Fakulteit Landbouwwetenschappen Rijksuniversiteit Gent*, **48**, 475–85.

Hand, L.F. and Keaster, A.J. (1967) The environment of an insect field cage, *Journal of Economic Entomology*, **60**, 910–15.

Hansen, L.S. and Jensen, K.M.V. (2002) Effect of temperature on parasitism and host-feeding of *Trichogramma turkestanica* (Hymenoptera: Trichogrammatidae) on *Ephestia kuehniella* (Lepidoptera: Pyralidae). *Journal of Economic Entomology*, **95**, 50–6.

Hansen, T.F. (1997) Stabilizing selection and the comparative analysis of adaptation. *Evolution*, **51**, 1341–51.

Hanski, I. (1990) Density dependence, regulation and variability in animal populations. *Philosophical Transactions of the Royal Society of London B*, **330**, 141–50.

Hanski, I. (1994) A practical model of metapopulation dynamics. *Journal of Animal Ecology*, **63**, 151–62.

Hanski, I. (1999) *Metapopulation Ecology*. Oxford University Press, Oxford.

Hanski, I., Foley, P. and Hassell, M.P. (1996) Random walks in a metapopulation: how much density-dependence is necessary for long-term persistence? *Journal of Animal Ecology*, **65**, 274–82.

Hanula, J.L. and Franzreb, K. (1998) Source, distribution and abundance of macroarthropods on the bark of longleaf pine: potential prey of the red-cockaded woodpecker. *Forest Ecology and Management*, **102**, 89–102.

Harborne, J.B. (1988) *Introduction to Ecological Biochemistry*. Academic Press, London.

Harder, L.D. (1986) Effects of nectar concentration and flower depth on flower handling efficiency of bumblebees. *Oecologia*, **69**, 309–15.

Hardie, J., Nottingham, S.F., Powell, W. and Wadhams, L.J. (1991) Synthetic aphid sex pheromone lures female parasitoids. *Entomologia Experimentalis et Applicata*, **61**, 97–9.

Hardie, J., Hick, A.J., Höller, C., Mann, J., Merritt, L., Nottingham, S.F., Powell, W., Wadhams, L.J., Witthinrich, J. and Wright, A.F. (1994) The responses of *Praon* spp. parasitoids to aphid sex pheromone components in the field. *Entomologia Experimentalis et Applicata*, **71**, 95–9.

Hardy, I.C.W. (1992) Non-binomial sex allocation and brood sex ratio variances in the parasitoid Hymenoptera. *Oikos*, **65**, 143–58.

Hardy, I.C.W. (1994a) Sex ratio and mating structure in the parasitoid Hymenoptera. *Oikos*, **69**, 3–20.

Hardy, I.C.W. (1994b) Polyandrous parasitoids: multiple-mating for variety's sake? *Trends in Ecology and Evolution*, **9**, 202–3.

Hardy, I.C.W. (2002) The birds and the wasps: a sex ratio meta-analysis. *Trends in Ecology and Evolution*, **17**, 207.

Hardy, I.C.W. and Blackburn, T.M. (1991) Brood guarding in a bethylid wasp. *Ecological Entomology*, **16**, 55–62.

Hardy, I.C.W. and Cook, J.M. (1995) Brood sex ratio variance, developmental mortality and virginity in a gregarious parasitoid wasp. *Oecologia*, **103**, 162–9.

Hardy, I.C.W. and Godfray, H.C.J. (1990) Estimating the frequency of constrained sex allocation in field populations of Hymenoptera, *Behaviour*, **114**, 137–47.

Hardy, I.C.W. and Mayhew, P.J. (1998a) Sex ratio, sexual dimorphism and mating structure in bethylid wasps. *Behavioral Ecology and Sociobiology*, **42**, 383–95.

Hardy, I.C.W. and Mayhew, P.J. (1998b) Partial local mating and the sex ratio: indirect comparative evidence. *Trends in Ecology and Evolution*, **13**, 431–2.

Hardy, I.C.W. and Mayhew, P.J. (1999) Local mating and sex ratios: reply from I.C.W. Hardy and P.J. Mayhew. *Trends in Ecology and Evolution*, **14**, 235.

Hardy, I.C.W., van Alphen, J.J.M. and van Dijken, M.J. (1992a) First record of *Leptopilina longipes* in the Netherlands, and its hosts identified. *Entomologische Berichten*, **52**, 128–30.

Hardy, I.C.W., Griffiths, N.T. and Godfray, H.C.J. (1992b) Clutch size in a parasitoid wasp: a manipulation experiment. *Journal of Animal Ecology*, **61**, 121–9.

Hardy, I.C.W., Dijkstra, L.J., Gillis, J.E.M. and Luft, P.A. (1998) Patterns of sex ratio, virginity and developmental mortality in gregarious parasitoids. *Biological Journal of the Linnean Society*, **64**, 239–70.

Hardy, I.C.W., Pedersen, J.B, Sejr, M.K. and Linderoth, U.H. (1999) Local mating, dispersal and sex ratio in a gregarious parasitoid wasp. *Ethology*, **105**, 57–72.

Hardy, I.C.W., Stokkebo, S., Bønløkke-Pedersen, J. and Sejr, M.K. (2000) Insemination capacity and dispersal in relation to sex allocation decisions in *Goniozus legneri* (Hymenoptera: Bethylidae): why are there more males in larger broods? *Ethology*, **106**, 1021–32.

Hare, J.D. (1992) Effects of plant variation on herbivore-natural enemy interactions, in *Plant Resistance to Herbivores and Pathogens* (eds R.S. Fritz and E.L. Simms). University of Chicago Press, Chicago, pp. 278–300.

Hare, J.D. and Morgan, D.J.W. (2000) Chemical conspicuousness of an herbivore to its natural enemy: effect of feeding site selection. *Ecology*, **81**, 509–19.

Hariri, A.R., Werren, J.H. and Wilkinson, G.S. (1998) Distribution and reproductive effects of *Wolbachia* in stalk eyed flies (Diptera: Diopsidae). *Heredity*, **81**, 254–60.

Hariri, G.E. (1966) Laboratory studies on the reproduction of *Adalia bipunctata* (Coleoptera: Coccinellidae). *Entomologia Experimentalis et Applicata*, **9**, 200–4.

Harmon, J.P., Losey, J.E. and Ives, A.R. (1998) The role of vision and color in the close proximity foraging behaviour of four coccinellid species. *Oecologia*, **115**, 287–92.

Harris, H. and Hopkinson, D.A. (1977) *Handbook of Enzyme Electrophoresis in Human Genetics*. Amsterdam, North-Holland.

Harris, K.M. (1973) Aphidophagous Cecidomyiidae (Diptera): taxonomy, biology and assessments of field populations. *Bulletin of Entomological Research*, **63**, 305–25.

Harris, P. (1985) Biological control of weeds, in *Proceedings of the 6th International Symposium on the Biological Control of Weeds* (ed. E.S. Delfosse). Agriculture Canada, Ottowa, pp. 3–12.

Harris, R.J. (1989). An entrance trap to sample foods of social wasps (Hymenoptera: Vespidae). *New Zealand Journal of Zoology*, **16**, 369–71.

Harris, R.J. and Oliver, E.H. (1993) Prey diets and population densities of the wasps *Vespula vulgaris* and *V. germanica* in scrubland-pasture. *New Zealand Journal of Ecology*, **17**, 5–12.

Harrison, J.F. and Fewell, J.H. (2002) Environmental and genetic influences on flight metabolic rate in the honey bee, *Apis mellifera*. *Comparative Biochemistry and Physiology A*, **133**, 323–33.

Harrison, S. and Taylor, A.D. (1997) Empirical evidence for metapopulation dynamics, in *Metapopulation Biology, Ecology, Genetics and Evolution* (ed. I. Hanski and M.E. Gilpin). Academic Press, San Diego, pp. 27–42.

Harshman, L. and Prout, T. (1994) Sperm displacement without sperm transfer in *Drosophila melanogaster*. *Evolution*, **48**, 758–66.

Hartl, D.L. and Clark, A.G. (1989) *Principles of Population Genetics*. Sinauer Associates, Sunderland.

Harvey, B.A., Davidson, J.M., Linforth, R.S.T. and Taylor, A.J. (2000). Real time flavour release from chewing gum during eating, in *Frontiers of Flavour Science* (eds. P. Schieberle and K-H. Engel). Deutsche Forschunganstalt fuer Lebensmittelchemie, Garching, pp. 271–4.

Harvey, J.A. (1996) *Venturia canescens* parasitizing *Galleria mellonella* and *Anagasta kuehniella*: is the parasitoid a conformer or regulator? *Journal of Insect Physiology*, **42**, 1017–25.

Harvey, J.A. and Gols, G.J.Z. (1998) The influence of host quality on progeny and sex allocation in the pupal ectoparasitoid, *Muscidifurax raptorellus* (Hymenoptera: Pteromalidae). *Bulletin of Entomological Research*, **88**, 299–304.

Harvey, J.A. and Strand, M.R. (2002) The developmental strategies of endoparasitoid wasps vary with host feeding ecology. *Ecology*, **83**, 2439–51.

Harvey, J.A. and Thompson, D.J. (1995) Developmental interactions between the solitary endoparasitoid *Venturia canescens* (Hymenoptera, Ichneumonidae) and 2 of its hosts, *Plodia interpunctella* and *Corcyria cephalonica* (Lepidoptera: Pyralidae). *European Journal of Entomology*, **92**, 427–35.

Harvey, J.A. and Vet, L.E.M. (1997) *Venturia canescens* parasitizing *Galleria mellonella* and *Anagasta kuehniella*: differing suitability of two hosts with highly variable growth potential. *Entomologia Experimentalis et Applicata*, **84**, 93–100.

Harvey, J.A., Harvey, I.F. and Thompson, D.J. (1993) The effect of superparasitism on development of the solitary parasitoid wasp, *Venturia canescens* (Hymenoptera: Ichneumonidae). *Ecological Entomology*, **18**, 203–8.

Harvey, J.A., Harvey, I.F. and Thompson, D.J. (1994) Flexible larval feeding allows use of a range of host sizes by a parasitoid wasp. *Ecology*, **75**, 1420–8.

Harvey, J.A., Jervis, M.A., Gols, R., Jiang, N. and Vet, L.E.M. (1999) Development of the parasitoid, *Cotesia rubecula* (Hymenoptera: Braconidae) in *Pieris rapae* and *Pieris brassicae* (Lepidoptera: Pieridae): evidence for host regulation. *Journal of Insect Physiology*, **45**, 173–82.

Harvey, J.A., Kadash, K. and Strand, M.R. (2000) Differences in larval feeding behavior correlate with altered developmental strategies in two parasitic wasps: implications for the size-fitness hypothesis. *Oikos*, **88**, 621–9.

Harvey, J.A., Harvey, I.F. and Thompson, D.J. (2001) Lifetime reproductive success in the solitary endoparasitoid, *Venturia canescens*. *Journal of Insect Behavior*, **14**, 573–93.

Harvey, J.A., van Dam, N.M and Gols, R. (2003) Interactions over four trophic levels: foodplant quality affects development of a hyperparasitoid as mediated through a herbivore and its primary parasitoid. *Journal of Animal Ecology*, **72**, 520–31.

Harvey, P. (1985) Intrademic group selection and the sex ratio, in *Behavioural Ecology: Ecological consequences of Adaptive Behaviour* (eds R.M. Sibly and R.H. Smith). Blackwell Scientific Publications, Oxford, pp. 59–73.

Harvey, P.H. (1996) Phylogenies for ecologists. *Journal of Animal Ecology*, **65**, 255–63.

Harvey, P.H. and Nee, S. (1997). The phylogenetic foundations of behavioural ecology, in *Behavioural Ecology, 4th edn* (eds J.R. Krebs and N.B. Davies). Blackwell Scientific Publishing, Oxford, pp. 334–49.

Harvey, P.H. and Pagel, M.D. (1991) *The Comparative Method in Evolutionary Biology*. Oxford University Press, Oxford.

Harwood, J.D., Sunderland, K.D. and Symondson, W.O.C. (2001a) Living where the food is: web location by linyphiid spiders in relation to prey availability in winter wheat. *Journal of Applied Ecology*, **38**, 88–99.

Harwood, J.D., Phillips, S.W., Sunderland, K.D. and Symondson, W.O.C. (2001b) Secondary predation: quantification of food chain errors in an aphid-spider-carabid system using monoclonal antibodies. *Molecular Ecology*, **10**, 2049–57.

Harwood, J.D., Sunderland, K.D. and Symondson, W.O.C. (2003) Web location by linyphiid spiders: prey-specific aggregation and foraging strategies. *Journal of Animal Ecology*, **72**, 745–56.

Harwood, J.D., Sunderland, K.D. and Symondson, W.O.C. (2004). Web location and foraging strategies amongst spiders: competition and coexistence. *Journal of Animal Ecology*, in press.

Hase, A. (1925) Beiträge zur Lebensgeschichte der Schlupfwespe *Trichogramma evanescens* Westwood. *Arbuten aus der biologischen Reichsanstalt für Land und Forstwirtschaft*, **14**, 171–224.

Haslett, J.R. (1983) A photographic account of pollen digestion by adult hoverflies. *Physiological Entomology*, **8**, 167–71.

Haslett, J.R. (1989a) Interpreting patterns of resource utilization: randomness and selectivity in pollen feeding by adult hoverflies. *Oecologia*, **78**, 433–42.

Haslett, J.R. (1989b) Adult feeding by holometabolous insects: pollen and nectar as complementary nutrient sources for *Rhingia campestris* (Diptera: Syrphidae). *Oecologia*, **81**, 361–3.

Haslett, J.R. and Entwistle, P.F. (1980) Further notes on *Eriozona syrphoides* (Fall.) (Dipt., Syrphidae) in Hafren Forest, mid-Wales. *Entomologist's Monthly Magazine*, **116**, 36.

Hassall, M., Dangerfield, J.M., Manning, T.P. and Robinson, F.G. (1988) A modified high-gradient extractor for multiple samples of soil macro-arthropods. *Pedobiologia*, **32**, 21–30.

Hassan, E. (1967) Untersuchungen über die bedeutung der kraut- und strauchschicht als nahrungsquelle für imagines entomophager Hymenopteren. *Zeitschrift für Angewandte Entomologie*, **60**, 238–65.

Hassell, M.P. (1968) The behavioural response of a tachinid fly *(Cyzenis albicans* [Fall.]) to its host, the winter moth *(Operophtera brumata* [L.]). *Journal of Animal Ecology*, **37**, 627–39.

Hassell, M.P. (1969) A population model for the interaction between *Cyzenis albicans* (Fall.) (Tachinidae) and *Operophtera brumata* (L.) (Geometridae) at Wytham, Berkshire. *Journal of Animal Ecology*, **38**, 567–76.

Hassell, M.P. (1971a) Mutual interference between searching insect parasites. *Journal of Animal Ecology*, **40**, 473–86.

Hassell, M.P. (1971b) Parasite behaviour as a factor contributing to the stability of insect host-parasite interactions, in *Dynamics of Populations, Proceedings of the Advanced Study Institute on 'Dynamics of Numbers in Populations', Oosterbeck, 1970* (eds P.J. den Boer and C.R. Gradwell). Centre for Agricultural Publishing and Documentation, Wageningen, pp. 366–79.

Hassell, M.P. (1978) *The Dynamics of Arthropod Predator-prey Systems. Monographs in Population Biology, 13.* Princeton University Press, Princeton.

Hassell, M.P. (1980a) Foraging strategies, population models and biological control: a case study. *Journal of Animal Ecology*, **49**, 603–28.

Hassell, M.P. (1980b) Some consequences of habitat heterogeneity for population dynamics. *Oikos*, **35**, 150–60.

Hassell, M.P. (1982a) Patterns of parasitism by insect parasites in patchy environments. *Ecological Entomology*, **7**, 365–77.

Hassell, M.P. (1982b) What is searching efficiency? *Annals of Applied Biology*, **101**, 170–75.

Hassell, M.P. (1984) Parasitism in patchy environments: inverse density dependence can be stabilising. *IMA Journal of Mathematics Applied in Medicine and Biology*, **1**, 123–33.

Hassell, M.P. (1985) Insect natural enemies as regulating factors. *Journal of Animal Ecology*, **54**, 323–34.

Hassell, M.P. (1986) Parasitoids and host population regulation, in *Insect Parasitoids* (eds J. Waage and D. Greathead). Academic Press London, pp. 201–24.

Hassell, M.P. (1987) Detecting regulation in patchily distributed animal populations. *Journal of Animal Ecology*, **56**, 705–13.

Hassell, M.P. (2000a) Host-parasitoid population dynamics. *Journal of Animal Ecology*, **69**, 543–66.

Hassell, M.P. (2000b) *The Spatial and Temporal Dynamics of Host-parasitoid Interactions*. Oxford University Press, Oxford.

Hassell, M.P. and Anderson, R.M. (1984) Host susceptibility as a component in host-parasitoid systems. *Journal of Animal Ecology*, **53**, 611–21.

Hassell, M.P. and Comins, H.N. (1978) Sigmoid functional responses and population stability. *Theoretical Population Biology*, **14**, 62–7.

Hassell, M.P. and Godfray, H.C.J. (1992) The population biology of insect parasitoids, in *Natural Enemies* (ed. M.J. Crawley). Blackwell, Oxford, pp. 265–92.

Hassell, M.P. and May, R.M. (1973) Stability in insect host-parasite models. *Journal of Animal Ecology*, **42**, 693–736.

Hassell, M.P. and May, R.M. (1986) Generalist and specialist natural enemies in insect predator-prey interactions. *Journal of Animal Ecology*, **55**, 923–40.

Hassell, M.P. and May, R.M. (1988) Spatial heterogeneity and the dynamics of parasitoid-host systems. *Annales Zoologici Fennici*, **25**, 55–61.

Hassell, M.P. and Moran, V.C. (1976) Equilibrium levels and biological control. *Journal of the Entomological Society of South Africa*, **39**, 357–66.

Hassell, M.P. and Pacala, S.W. (1990) Heterogeneity and the dynamics of host-parasitoid interactions. *Philosophical Transactions of the Royal Society of London B*, **330**, 203–20.

Hassell, M.P. and Varley, G.C. (1969) New inductive population model for insect parasites and its bearing on biological control. *Nature*, **223**, 1113–37.

Hassell, M.P. and Waage, J.K. (1984) Host-parasitoid population interactions. *Annual Review of Entomology*, **29**, 89–114.

Hassell, M.P., Lawton, J.H. and Beddington, J.R. (1976) The components of arthropod predation. I. The prey death rate. *Journal of Animal Ecology*, **45**, 135–64.

Hassell, M.P., Lawton, J.H. and Beddington, J.R. (1977) Sigmoid functional responses by invertebrate predators and parasitoids. *Journal of Animal Ecology*, **46**, 249–62.

Hassell, M.P., Waage, J.K. and May, R.M. (1983) Variable parasitoid sex ratios and their effect on host-parasitoid dynamics. *Journal of Animal Ecology*, **52**, 889–904.

Hassell, M.P., Lessells, C.M. and McGavin, G.C. (1985) Inverse density dependent parasitism in a patchy environment: a laboratory system. *Ecological Entomology*, **10**, 393–402.

Hassell, M.P., Southwood, T.R.E. and Reader, P.M. (1987) The dynamics of the viburnum whitefly (*Aleurotrachelus jelinekii* (Fraunf)): a case study on population regulation. *Journal of Animal Ecology*, **56**, 283–300.

Hassell, M.P., Latto, J. and May, R.M. (1989a) Seeing the wood for the trees: detecting density dependence from existing life table studies. *Journal of Animal Ecology*, **58**, 883–92.

Hassell, M.P., Taylor, V.A. and Reader, P.M. (1989b) The dynamics of laboratory populations of *Callosobruchus chinensis* and *C. maculatus* (Coleoptera: Bruchidae) in patchy environments. *Researches on Population Ecology*, **31**, 35–51.

Hassell, M.P., May, R.M., Pacala, S.W. and Chesson, P.L. (1991) The persistence of host-parasitoid associations in patchy environments. 1. A general criterion. *American Naturalist*, **138**, 568–83.

Hassell, M.P., Comins, H.N. and May, R.M. (1994) Species coexistence and self-organising spatial dynamics. *Nature*, **353**, 255–8.

Hassell, M.P., Crawley, M.J., Godfray, H.C.J. and Lawton, J.H. (1998) Top-down versus bottom-up and Ruritanian bean bug. *Proceedings of the National Academy of Sciences*, **95**, 10661–664.

Hastings, A., Hom, C.L., Ellner, S., Turchin, P. and Godfray, H.C.J. (1993) Chaos in ecology: is mother nature a strange attractor? *Annual Review of Ecology and Systematics*, **24**, 1–33.

Hattingh, V. and Samways, M.J. (1990) Absence of interference during feeding by the predatory ladybirds *Chilocorus* spp. (Coleoptera: Coccinellidae). *Ecological Entomology*, **151**, 385–90.

Hattingh, V. and Samways, M.J. (1995) Visual and olfactory location of biotopes, prey patches, and individual prey by the ladybeetle *Chilocorus nigritus*. *Entomologia Experimentalis et Applicata*, **75**, 87–98.

Hauber, M.E. (1999) Variation in pit size of antlion (*Myrmeleon carolinus*) larvae: the importance of pit construction. *Physiological Entomology*, **24**, 37–40.

Hauber, M.E. (2002) Conspicuous colouration attracts prey to a stationary predator. *Ecological Entomology*, **27**, 686–91.

Hausmann, C., Wäckers, F.L., and Dorn, S. (2003) Establishing preferences for nectar constituents in the parasitoid *Cotesia glomerata*. *Entomologia Experimentalis et Applicata*, in press.

Havill, N.P. and Raffa, K.F. (2000) Compound effects of induced plant responses on insect herbivores and parasitoids: implications for tritrophic interactions. *Ecological Entomology*, **25**, 171–9.

Hawke, S.D., Farley, R.D. and Greany, P.D. (1973) The fine structure of sense organs in the ovipositor of the parasitic wasp, *Orgilus lepidus* Muesebeck. *Tissue and Cell*, **5**, 171–84.

Hawkes, R.B. (1972) A fluorescent dye technique for marking insect eggs in predation studies. *Journal of Economic Entomology*, **65**, 1477–8.

Hawkins, B. (1992) Parasitoid-host food webs and donor control. *Oikos*, **65**, 159–62.

Hawkins, B.A. (1993) Refuges, host population dynamics and the genesis of parasitoid diversity, in *Hymenoptera and Biodiversity* (eds J. LaSalle and I.D. Gauld). CAB International, Wallingford, pp. 235–56.

Hawkins, B.A. (1994) *Pattern and Process in Host-parasitoid Interactions*. Cambridge University Press, Cambridge.

Hawkins, B.A. and Cornell, H.V. (1994) Maximum parasitism rates and successful biological control. *Science*, **266**, 1886.

Hawkins, B.A. and Cornell, H.V. (1999) *Theoretical Approaches to Biological Control*. Cambridge University Press, Cambridge.

Hawkins, B.A. and Marino, P.C. (1997) The colonization of native phytophagous insects in North America by exotic parasitoids. *Oecologia*, **112**, 566–1.

Hawkins, B.A., Thomas, M.B. and Hochberg, M.E. (1993) Refuge theory and biological control. *Science*, **262**, 1429–32.

Hawkins, B.A., Martinez, N.D. and Gilbert, F. (1997) Source food webs as estimators of community web structure. *Acta Oecologica*, **18**, 575–86.

Hawkins, B.A., Mills, N.J., Jervis, M.A. and Price, P.W. (1999) Is the biological control of insects a natural phenomenon? *Oikos*, **86**, 493–506.

Hay, F. and Westwood, O. (2002) *Practical Immunology*. Blackwell Science, London.

Headrick, D.H. and Goeden, R.D. (2001) Biological control as a tool for ecosystem management. *Biological Control*, **21**, 249–7.

Heads, P.A. (1986) The costs of reduced feeding due to predator avoidance: potential effects on growth and fitness in *Ischnura elegans* larvae (Odonata: Zygoptera). *Ecological Entomology*, **11**, 369–77.

Heads, P.A. and Lawton, J.H. (1983) Studies on the natural enemy complex of the holly leaf-miner: the effects of scale on the detection of aggregative responses and the implications for biological control. *Oikos*, **40**, 267–76.

Heap, M.A. (1988) The pit-light, a new trap for soil-dwelling insects. *Journal of the Australian Entomological Society*, **27**, 239–40.

Heath, B.D., Butcher, R.D.J., Whitfield, W.G.F. and Hubbard, S.F. (1999) Horizontal transfer of *Wolbachia* between phylogenetically distant insect species by a naturally occurring mechanism. *Current Biology*, **9**, 313–6.

Heaversedge, R.C. (1967) Variation in the size of insect parasites of puparia of *Glossina* spp.. *Bulletin of Entomological Research*, **58**, 153–8.

Hedrick, A.V. (2000) Crickets with extravagant mating songs compensate for predation risk with extra caution. *Proceedings of the Royal Society of London B*, **267**, 671–5.

Heidari, M. (1989) Biological Control of Glasshouse Mealybugs Using Coccinellid Predators. Wye College, University of London, Ph.D. Thesis.

Heidari, M. and Copland, M.J.W. (1992) Host finding by *Cryptolaemus montrouzieri (Col.*, Coccinellidae) a predator of mealybugs (Horn., Pseudococcidae). *Entomophaga*, **37**, 621–5.

Heidari, M. and Copland, M.J.W. (1993) Honeydew – a food resource or arrestant for the mealybug predator *Cryptolaemus montrouzieri*. *Entomophaga*, **38**, 63–8.

Heikinheimo, O. and Raatikainen, M. (1962) Comparison of suction and netting methods in population investigations concerning the fauna of grass leys and cereal fields, particularly in those concerning the leafhopper, *Calligypona pellucida* (F.). *Publications of the Finnish State Agricultural Research Board*, **191**, 5–29.

Heimpel, G.E. (1994) Virginity and the cost of insurance in highly inbred Hymenoptera. *Ecological Entomology*, **19**, 299–302.

Heimpel, G.E. and Collier, T.R. (1996) The evolution of host-feeding behaviour in insect parasitoids. *Biological Reviews*, **71**, 373–400.

Heimpel, G.E. and Jervis, M.A. (2004) An evaluation of the hypothesis that floral nectar improves biological control by parasitoids, in *Plant-provided Food and Plant-carnivore Mutualisms* (eds F. Wäckers, P. van Rijn and J. Bruin). Cambridge University Press, Cambridge.

Heimpel, G.E. and Lundgren, J.G. (2000) Sex ratios of commercially reared biological control agents. *Biological Control*, **19**, 77–93.

Heimpel, G.E. and Rosenheim, J.A. (1995) Dynamic host feeding by the parasitoid *Aphytis melinus*: the balance between current and future reproduction. *Journal of Animal Ecology*, **64**, 153–67.

Heimpel, G.E. and Rosenheim, J.A. (1998) Egg limitation in parasitoids: a review of the evidence and a case study. *Biological Control*, **11**, 160–8.

Heimpel, G.E., Rosenheim, J.A. and Adams, J.M. (1994) Behavioral ecology of host feeding in *Aphytis* parasitoids. *Norwegian Journal of Agricultural Sciences Supplement*, **16**, 101–15.

Heimpel, G.E., Rosenheim, J.A. Kattari, D. (1997a) Adult feeding and lifetime reproductive success in the parasitoid *Aphytis melinus*. *Entomologia Experimentalis et Applicata*, **83**, 305–15.

Heimpel, G.E., Rosenheim, J.A. and Mangel, M. (1997b) Predation on adult *Aphytis* parasitoids in the field. *Oecologia*, **110**, 346–52.

Heimpel, G.E., Antolin, M.F., Franqui, R.A. and Strand, M.F. (1997c) Reproductive isolation and genetic variation between two 'strains' of *Bracon hebetor* (Hymenoptera: Braconidae). *Biological Control*, **9**, 149–56.

Heimpel, G.E., Mangel, M. and Rosenheim, J.A. (1998) Effects of time limitation and egg limitation on lifetime reproductive success of a parasitoid in the field. *American Naturalist*, **152**, 273–89.

Heimpel, G.E., Antolin, M.F. and Strand, M.R. (1999) Diversity of sex-determining alleles in *Bracon hebetor*. *Heredity*, **82**, 282–91.

Heimpel, G.E., Neuhauser, C. and Hoogendoorn, M. (2003) Effects of parasitoid fecundity and host resistance on indirect interactions among hosts sharing a parasitoid. *Ecological Letters*, **6**, 556–66.

Heimpel, G.E., Lee, J.C., Wu. Z., Weiser, L., Wäckers, F. and Jervis, M.A. (2004) Gut sugar analysis in field-caught parasitoids: adapting methods originally developed for biting flies. *International Journal of Pest Management*, **50**, 193–198.

Hein, I., Mach, R.L., Farnleitner, A.H. and Wagner, M. (2003) Application of single-strand conformation polymorphism and denaturing gradient gel electrophoresis for *fla* sequence typing of *Campylobacter jejuni*. *Journal of Microbiological Methods*, **52**, 305–13.

Heinrich, B. (1975) Energetics of pollination. *Annual Review of Ecology and Systematics*, **6**, 139–70.

Heinrich, B. (1985) Insect foraging energetics, in *Handbook of Experimental Pollination Biology* (eds C.E. Jones and R.J. Little). Van Nostrand Reinhold, New York, pp. 187–214.

Heinrichs, E.A., Aquino, G.B., Chelliah, S., Valencia, S.L. and Reisseg, W.H. (1982) Resurgence of *Nilaparvata lugens* (Stål) populations as influenced by method and timing of insecticide applications in lowland rice. *Environmental Entomology*, **11**, 78–84.

Heinz, K.M. (1991) Sex-specific reproductive consequences of body size in the solitary ectoparasitoid, *Diglyphus begini*. *Evolution*, **45**, 1511–15.

Heinz, K.M. (1998) Host size-dependent sex allocation behaviour in a parasitoid: implications for *Catolaccus grandis* (Hymenoptera: Pteromalidae) mass rearing programmes. *Bulletin of Entomological Research*, **88**, 37–45.

Heinz, K.M. and Parrella, M.P. (1990) Holarctic distribution of the leafminer parasitoid *Diglyphus begini* (Hymenoptera: Eulophidae) and notes on its life history attacking *Liriomyza trifolii* (Diptera: Agromyzidae) in chrysanthemum. *Annals of the Entomological Society of America*, **83**, 916–24.

Heinze, J, Hölldobler, B. and Yamauchi, K. (1998) Male competition in *Cardiocondyla* ants. *Behavioral Ecology and Sociobiology*, **42**, 239–46.

Heitmans, W.R.B., Overmeer, W.P.J. and Van der Geest, L.P.S. (1986) The role of *Orius vicinus* Ribaut (Heteroptera; Anthocoridae) as a predator of phytophagous and predacious mites in a Dutch orchard. *Journal of Applied Entomology*, **102**, 391–402.

Held, J. and Krueger, H. (1999) FIT on the top of an organizer: a new techniques for multi-purpose event recording. *Proceedings of the Eighth International Conference on Human-Computer Interaction (Munich, Germany, August 22–27, 1999)*, p. 117–8.

Held, J., Brüesch, M., Krueger, H. and Pasch, T. (1999) The FIT-system: a new hand-held computer tool for ergonomic assessement. *Medical and Biological Engineering and Computing, Suppl.* 2, **37**, 862–3.

Helenius, J. (1990) Incidence of specialist natural enemies of *Rhopalosiphum padi* (L.) (Hom., Aphididae) on oats in monocrops and mixed intercrops with faba beans. *Journal of Applied Entomology*, **109**, 136–43.

Helenius, J. (1995) Rate and local scale spatial pattern of adult emergence of the generalist predator *Bembidion guttula* in an agricultural field. *Acta Jutlandica*, **70**, 101–11.

Helenius, J. and Tolonen, T. (1994) Enhancement of generalist aphid predators in cereals: effect of green manuring on recruitment of ground beetles (Col., Carabidae). *Bulletin SROP/WPRS*, **17**, 201–10.

Helenius, J., Holopainen, J., Muhojoki, M., Pokki, P., Tolonen, T. and Venäläinen, A. (1995) Effect of undersowing and green manuring on abundance of ground beetles (Coleoptera, Carabidae) in cereals. *Acta Zoologica Fennica*, **196**, 156–9.

Hemerik, L. and Harvey, J.A. (1999) Flexible larval development and the timing of destructive feeding by a solitary endoparasitoid: an optimal foraging problem in evolutionary perspective. *Ecological Entomology*, **24**, 308–15.

Hemerik, L. and van der Hoeven, N. (2003) Egg distributions of solitary parasitoids revisited. *Entomologia Experimentalis et Applicata*, **107**, 81–6.

Hemerik, L., Driessen, G.J. and Haccou, P. (1993) Effects of intra-patch experiences on patch time, search time and searching efficiency of the parasitoid *Leptopilina clavipes*. *Journal of Animal Ecology*, **62**, 33–44.

Hemerik, L., van der Hoeven, N. and van Alphen, J.J.M. (2002) Egg distributions and the information a solitary parasitoid has and uses for its oviposition decisions. *Acta Biotheoretica*, **50**, 167–88.

Hemingway, J., Small, G.J., Lindsay, S.W. and Collins, F.H. (1996) Combined use of biochemical, immunological and molecular assays for infection, species identification and resistance detection in field populations of *Anopheles* (Diptera: Culicidae), in *The Ecology of Agricultural Pests: Biochemical Approaches* (eds W.O.C. Symondson and J.E. Liddell). Chapman and Hall, London, pp. 31–49.

Hemptinne, J.L. and Desprets, A. (1986) Pollen as a spring food for *Adalia bipunctata*, in *Ecology of Aphidophaga* (ed. I. Hodek). Academia, Prague and W. Junk, Dordrecht, pp. 29–35.

Hemptinne, J.-L., Dixon, A.F.G. and Gauthier, C. (2000) Nutritive cost of intraguild predation on eggs of *Coccinella septempuncata* and *Adalia bipunctata* (Coleoptera: Coccinellidae). *European Journal of Entomology*, **97**, 559–62.

Henaut, Y., Pablo, J., Ibarra-Nuñez, G. and Williams, T. (2001) Retention, capture and consumption of experimental prey by orb-web weaving spiders in coffee plantations of Southern Mexico. *Entomologia Experimentalis et Applicata*, **98**, 1–8.

Henderson, I.F. and Whitaker, T.M. (1976) The efficiency of an insect suction sampler in grassland. *Ecological Entomology*, **2**, 57–60.

Hengeveld, R. (1980) Polyphagy, oligophagy and food specialization in ground beetles (Coleoptera, Carabidae). *Netherlands Journal of Zoology*, **30**, 564–84.

Heng-Moss, T., Baxendale, F. and Riordan, T. (1998) Beneficial arthropods associated with buffalograss. *Journal of Economic Entomology*, **91**, 1167–72.

Henneicke, K., Dawah, H.A. and Jervis, M.A. (1992) The taxonomy and biology of final instar larvae of some Eurytomidae (Hymenoptera: Chalcidoidea) associated with grasses in Britain. *Journal of Natural History*, **26**, 1047–87.

Henneman, M.L. and Memmott, J. (2002) Infiltration of a Hawaiian community by introduced biological control agents. *Science*, **293**, 1314–16.

Henter, H.J. (1995) The potential coevolution in a host-parasitoid system. 2. Genetic variation within a population of wasps in the ability to parasitise an aphid host. *Evolution*, **49**, 427–38.

Hentz, M.G., Ellsworth, P.C., Naranjo, S.E. and Watson, T.F. (1998) Development, longevity, and fecundity of *Chelonus* sp. nr *curvimaculatus* (Hymenoptera: Braconidae), an egg-larval parasitoid of pink bollworm (Lepidoptera: Gelechiidae). *Environmental Entomology*, **27**, 443–9.

Heong, K.L., Aquino, C.B. and Barrion, A.T. (1991) Arthropod community structures of rice ecosystems in the Philippines. *Bulletin of Entomological Research*, **81**, 407–16.

Herard, F. and Prevost, G. (1997) Suitability of *Yponomeuta malinellus* and *Y. cagnagellus* (Lepidoptera: Yponomeutidae) as hosts of *Diadegma armillata* (Hymenoptera: Ichneumonidae). *Environmental Entomology*, **26**, 933–38.

Herard, F., Keller, M.A. and Lewis, W.J. (1988) Rearing *Microplitis demolitor* Wilkinson (Hymenoptera: Braconidae) in the laboratory for use in studies of semiochemical mediated searching behaviour. *Journal of Entomological Science*, **23**, 105–11.

Heraty, J.M. and Quicke, D.L.J. (2003) Phylogenetic implications of ovipositor structure in Eucharitidae and Perilampidae (Hymenoptera: Chalcidoidea). *Journal of Natural History*, **37**, 1751–64.

Herre, E.A. (1985) Sex ratio adjustment in fig wasps. *Science*, **228**, 896–98.

Herre, E.A. (1987) Optimality, plasticity and selective regime in fig wasp sex ratios. *Nature*, **329**, 627–29.

Herre, E.A., Machado, C.A., Bermingham, E, Nason, J.D., Windsor, D.M., MacCafferty, S.S., van Houten, W. and Bachmann, K. (1996) Molecular phylogenies of figs and their pollinator wasps. *Journal of Biogeography*, **23**, 521–30.

Herre, E.A., West, S.A., Cook, J.M., Compton, S.G. and Kjellberg, F. (1997) Fig-associated wasps: pollinators and parasites, sex-ratio adjustment and male polymorphism, population structure and its consequences, in *The Evolution of Mating Systems in Insects and Arachnids* (eds J.C. Choe and B.J. Crespi). Cambridge University Press, Cambridge, pp. 226–39.

Herrera, C.J., van Driesche, R.G. and Bellotti, A.C. (1989) Temperature-dependent growth rates for the cassava mealybug, *Phenacoccus herreni*, and two of its encyrtid parasitoids, *Epidinocarsis diversicornis* and *Acerophagus coccois* in Colombia. *Entomologia Experimentalis et Applicata*, **50**, 21–7.

Herting, B. (1960) Biologie der wespaldarktischen Raupenfliegen (Diptera:Tachnidae). *Monographien zur Angewandte Entomologie*, **16**, 1–202.

Hertlein, M.B. and Thorarinsson, K. (1987) Variable patch times and the functional response of *Leptopilina boulardi* (Hymenoptera: Eucoilidae). *Environmental Entomology*, **16**, 593–8.

Hespenheide, H.A. (1958) Insect visitors to extrafloral nectaries of *Byttneria aculeata* (Sterculaceaceae): relative importance and roles. *Ecological Entomology*, **10** *191–204*.

Hess, A.D. (1941) New limnological sampling equipment. *Limnological Society of America, Special Publication*, **6**, 1–15.

Hewitt, G.M. and Butlin, R.F. (1997) Causes and consequences of population structure, in *Behavioural Ecology, 4th edition* (eds J.R. Krebs and N.B. Davies). Blackwell Scientific Publishing, Oxford, pp. 350–72.

Heydemann, B. (1953) Agrarökologische Problematik. University of Kiel, Ph.D. Thesis.

Heydemann, B. (1962) Untersuchungen über die Aktivitäts- und Besiedlungsdichte bei epigäischen Spinnen, *Zoologischer Anzeiger Supplement*, **25**, 538–56.

Hickman, J.M. and Wratten, S.D. (1996) Use of *Phacelia tanacetifolia* (Hydrophyllaceae) as a pollen source to enhance hoverfly (Diptera: Syrphidae) populations in cereal fields. *Journal of Economic Entomology*, **89**, 832–40.

Hickman, J.M., Wratten, S.D., Jepson, P.C. and Frampton, C.M. (2001) Effect of hunger on yellow water trap catches of hoverfly (Diptera: Syrphidae) adults. *Agricultural and Forest Entomology*, **3**, 35–40.

Hilborn, R. and Stearns, S.C. (1982) On inference in ecology and evolutionary biology: the problem of multiple causes. *Acta Biotheoretica*, **31**, 145–64.

Hildrew, A.G. and Townsend, C.R. (1976) The distribution of two predators and their prey in an iron rich stream. *Journal of Animal Ecology*, **45**, 41–57.

Hildrew, A.G. and Townsend, C.R. (1982) Predators and prey in a patchy environment: a freshwater study. *Journal of Animal Ecology*, **51**, 797–815.

Hill, M.G. (1988) Analysis of the biological control of *Mythimna separata* (Lepidoptera: Noctuidae) by *Apanteles ruficrus* (Braconidae: Hymenoptera) in New Zealand. *Journal of Applied Ecology*, **25**, 197–208.

Hill, M.O. (1973) Reciprocal averaging: an eigenvector method of ordination. *Journal of Ecology*, **61**, 237–49.

Hillis, D.M. and Dixon, S.K. (1991) Ribosomal DNA: molecular evolution and phylogenetic inference. *Quarterly Review of Biology*, **66**, 411–53.

Hillis, D.M. and Moritz, C. (eds) (1990) *Molecular Systematics*. Sinauer, Sunderland.

Hills, O.A. (1933) A new method for collecting samples of insect populations. *Journal of Economic Entomology*, **26**, 906–10.

Hiroki, M., Kato, Y., Kamito, T. and Miura, K. (2002) Feminization of genetic males by a symbiotic bacterium in a butterfly, *Eurema hecabe* (Lepidoptera: Pieridae). *Naturwissenschaften*, **89**, 167–70.

Hirose, Y., Vinson, S.B. and Hirose, Y. (1988) Protandry in the parasitoid *Cardiochiles nigriceps*, as related to its mating system. *Ecological Research*, **3**, 217–26.

Hobson, K.A. (1999) Tracing origins and migration of wildlife using stable isotopes: a review. *Oecologia*, **120**, 314–26.

Hobson, K.A., Wassenaar, L.I. and Taylor, O.R. (1999) Stable isotopes (delta D and delta C-13) are geographic indicators of natal origins of monarch butterflies in eastern North America. *Oecologia*, **120**, 397–404.

Hochberg, M.E. and Hawkins, B.A. (1994) The implications of population dynamics theory to parasitoid diversity and biological control, in *Parasitoid Community Ecology* (eds B.A. Hawkins and W. Sheehan). Oxford University Press, Oxford, pp. 451–71.

Hochberg, M.E. and Holt, R.D. (1999) The uniformity and density of pest exploitation as guides to success in biological control, in *Theoretical Approaches to Biological Control* (eds H.V. Cornell and B.A. Hawkins). Cambridge University Press, Cambridge, UK. pp. 71–85.

Hochberg, M.E. and Ives, A.R. (2000) *Parasitoid Population Biology*. Princeton University Press, Princeton.

Hochberg, M.E., Hassell, M.P. and May, R.M. (1990) The dynamics of host-parasitoid-pathogen interactions. *American Naturalist*, **135**, 74–94.

Hochuli, A., Pfister-Wilhelm, R. and Lanzrein, B. (1999) Analysis of endoparasitoid-released proteins and their effects on host development in the system *Chelonus inanitus* (Braconidae)-*Spodoptera littoralis* (Noctuidae). *Journal of Insect Physiology*, **45**, 823–33.

Hocking, B. (1953) The intrinsic range and speed of flight in insects. *Transactions of the Royal Entomological Society of London*, **104**, 223–345.

Hocking, H. (1967) The influence of food on longevity and oviposition in *Rhyssa persuasoria* (L.) (Hymenoptera: Ichneumonidae), *Journal of the Australian Entomological Society*, **6**, 83–8.

Hoddle, M.S., Van Driesche, R.G., Elkinton, J.S. and Sanderson, J.P. (1998) Discovery and utilization of *Bemisia argentifolii* by *Eretmocerus eremicus* and *Encarsia formosa* (Beltsville strain) in greenhouses. *Entomologia Experimentalis et Applicata*, **87**, 15–28.

Hodek, I. (1973) *Biology of Coccinellidae*. W. Junk, The Hague/Czechoslovakian Academy of Sciences, Prague.

Hodek, I. and Hodková, M. (1988) Multiple role of temperature during insect diapause: a review. *Entomologia Experimentalis et Applicata*, **49**, 153–65.

Hodek, I. and Honěk, A. (1996) *Ecology of Coccinellidae*, Kluwer Academic Publishers, Dordrecht, Holland.

Hodges, C.M. (1985a) Bumblebee foraging: the threshold departure rule. *Ecology*, **66**, 179–87.

Hodges, C.M. (1985b) Bumblebee foraging: energetic consequences of using a threshold departure rule. *Ecology*, **66**, 188–97.

Hodgson, C.J. and Mustafa, T.M. (1984) Aspects of chemical and biological control of *Psylla pyricola* Forster in England. *Bulletin IOBC/WPRS Working Group Integrated Control of Pear Psyllids*, **7**, 330–53.

Hoelmer, K.A., Roltsch, W.J., Chu, C.C. and Henneberry, T.J. (1998) Selectivity of whitefly traps in cotton for *Eretmocerus eremicus* (Hymenoptera: Aphelinidae), a native parasitoid of *Bemisia argentifolii* (Homoptera: Aleyrodidae). *Environmental Entomology*, **27**, 1039–44.

Hoelscher, C.E. and Vinson, S.B. (1971) The sex ratio of a hymenopterous parasitoid, *Campoletis perdistinctus*, as affected by photoperiod, mating and temperature. *Annals of the Entomological Society of America*, **64**, 1373–76.

Hoelzel, A.R. (1992) *Molecular Genetic Analysis of Populations*. Oxford University Press, Oxford.

Hoelzel, A.R. and Bancroft (1992) Statistical analysis of variation, in *Molecular Genetic Analysis of Populations. A Practical Approach* (ed. A.R. Hoelzel). Oxford University Press, Oxford, pp. 297–305.

van der Hoeven, N. and Hemerik, L. (1990) Superparasitism as an ESS: to reject or not to reject, that is the question. *Journal of Theoretical Biology*, **146**, 467–82.

Hoffmann, A.A. and Parsons, P.A. (1988) The analysis of quantitative variation in natural populations with isofemale strains. *Genetics Selection Evolution*, **20**, 87–98.

Hoffmann, A.A. and Parsons, P.A. (1989) An integrated approach to environmental stress tolerance and life-history variation: dessication tolerance in *Drosophila*. *Biological Journal of the Linnean Society*, **37**, 117–36.

Hoffmann, A.A. and Schiffer, M. (1998) Changes in the heritability of five morphological traits under combined environmental stresses in *Drosophila melanogaster*. *Evolution*, **52**, 1207–12.

Hoffmann, A.A. and Turelli, M. (1997) Cytoplasmic incompatability in insects, in *Influential Passengers: Inherited Microorganisms and Arthropod Reproduction* (eds S.L. O'Neill, A.A. Hoffmann and J.H. Werren). Oxford University Press, Oxford, pp. 42–80.

Hoffmann, A.A., Hercus, M. and Dagher, H. (1998) Population dynamics of the *Wolbachia* infection causing cytoplasmic incompatibility in *Drosophila melanogaster*. *Genetics*, **148**, 221–31.

Hoffmann, M.P., Wright, M.G., Pitcher, S.A. and Gardner, J. (2002) Inoculative releases of *Tricho-*

gramma ostriniae for suppression of *Ostrinia nubilalis* (European corn-borer) in sweet corn: field biology and population dynamics. *Biological Control*, **25**, 249–58.

Hoffmeister, T.S. (2000) Marking decisions and host discrimination in a parasitoid attacking concealed hosts. *Canadian Journal of Zoology-Revue Canadienne de Zoologie*, **78**, 1494–9.

Hoffmeister, T.S. and Gienapp, P. (2001) Discrimination against previously searched, host-free patches by a parasitoid foraging for concealed hosts. *Ecological Entomology*, **26**, 487–94.

Hoffmeister, T.S. and Vidal, S. (1994) The diversity of fruit fly (Diptera: Tephritidae) parasitoids, in *Parasitoid Community Ecology* (eds B.A. Hawkins and W. Sheehan). Oxford University Press, Oxford, pp. 47–76.

Hofsvang, T. and Hågvar, E.B. (1975a) Duration of development and longevity in *Aphidius ervi* and *Aphidius platensis* (Hymenoptera, Aphidiidae), two parasites of *Myzus persicae* (Homoptera, Aphididae). *Entomophaga*, **20**, 11–22.

Hofsvang, T. and Hågvar, E.B. (1975b) Developmental rate, longevity, fecundity, and oviposition period of *Ephedrus cerasicola* Starý (Hym.: Aphidiidae) parasitizing *Myzus persicae* Sulz. (Hom.: Aphididae) on paprika. *Norwegian Journal of Entomology*, **22**, 15–22.

Hohmann, C.L., Luck, R.F., Oatman, E.R. and Platner, C.R. (1989) Effects of different biological factors on longevity and fecundity of *Trichogramma platneri* Nagarkatti (Hymenoptera: Trichogrammatidae). *Anais da Sociedade Entomologica do Brasil*, **18**, 61–70.

Hokkanen, H. and Pimentel, D. (1984) New approach for selecting biological control agents. *Canadian Entomologist*, **116**, 1109–21.

Holden, P.R., Brookfield, J.F.Y. and Jones, P. (1993) Cloning and characterization of an *fts-Z* homolog from a bacterial symbiont of *Drosophila melanogaster*. *Molecular and General Genetics*, **240**, 213–20.

Holland, J.M. and Reynolds, C.J.M. (2003) The impact of soil cultivation on arthropod (Coleoptera and Araneae) emergence on arable land. *Pedobiologia*, **47**, 181–91.

Holland, J.M. and Smith, S. (1997) Capture efficiency of fenced pitfall traps for predatory arthropods within a cereal crop, in *New Studies in Ecotoxicology* (eds P.T. Haskell and P.K. McEwen). The Welsh Pest Management Forum, Cardiff, pp. 34–36.

Holland, J.M. and Smith, S. (1999) Sampling epigeal arthropods: an evaluation of fenced pitfall traps using mark-release-recapture and comparisons to unfenced pitfall traps in arable crops. *Entomologia Experimentalis et Applicata*, **91**, 347–57.

Holland, J.M., Thomas, S.R. and Hewett, A. (1996) Some effects of polyphagous predators on an outbreak of cereal aphid (*Sitobion avenae* F.) and orange wheat blossom midge (*Sitodiplosis mosellana* Gehin). *Agriculture, Ecosystems and Environment*, **59**, 181–90.

Hölldobler, B. and Wilson, E.O. (1983) The evolution of communal nestweaving in ants. *American Scientist*, **71**, 489–99.

Höller, C. (1991) Evidence for the existence of a species closely related to the cereal aphid parasitoid *Aphidius rhopalosiphi* De Stefani-Perez based on host ranges, morphological characters, isolectric focusing banding patterns, cross-breeding experiments and sex pheromone specificities (Hymenoptera, Braconidae, Aphidiinae). *Systematic Entomology*, **16**, 15–28.

Höller, C. and Braune, H.J. (1988) The use of isoelectric focusing to assess percentage hymenopterous parasitism in aphid populations. *Entomologia Experimentalis et Applicata*, **47**, 105–14.

Höller, C., Christiansen-Weniger, P., Micha, S.G., Siri, N. and Borgemeister, C. (1991) Hyperparasitoid-aphid and hyperparasitoid-primary parasitoid relationships. *Redia*, **74**, 153–61.

Holling, C.S. (1959a) The components of predation as revealed by a study of small mammal predation of the European sawfly. *Canadian Entomologist*, **91**, 293–320.

Holling, C.S. (1959b) Some characteristics of simple types of predation and parasitism. *Canadian Entomologist*, **91**, 385–98.

Holling, C.S. (1961) Principles of insect predation. *Annual Review of Entomology*, **6**, 163–82.

Holling, C.S. (1965) The functional response of predators to prey density and its role in mimicry and population regulation. *Memoirs of the Entomological Society of Canada*, **45**, 3–60.

Holling, C.S. (1966) The functional response of invertebrate predators to prey density. *Memoirs of the Entomological Society of Canada*, **48**, 1–86.

Holloway, A.K., Heimpel, G.E. Strand, M.R. and Antolin, M.F. (1999) Survival of diploid males in *Bracon* sp. near *hebetor* (Hymenoptera: Braconidae). *Annals of the Entomological Society of America*, **92**, 110–6.

Holloway, B.A. (1976) Pollen-feeding in hover-flies (Diptera: Syrphidae). *New Zealand Journal of Zoology*, **3**, 339–50.

Holloway, G.J., de Jong, P.W. and Ottenheim, M. (1993) The genetics and costs of chemical defense in the two-spot ladybird (*Adalia bipunctata* L.). *Evolution*, **47**, 1229–39.

Holmes, H.B. (1976) *Mormoniella Publications Supplement*. Mimeographed edn, circulated by the author.

Holmes, P.R. (1984) A field study of the predators of the grain aphid, *Sitobion avenae* (F.) (Hemiptera: Aphididae), in winter wheat. *Bulletin of Entomological Research*, **74**, 623–31.

Holopäinen, J.K. and Helenius, J. (1992) Gut contents of ground beetles (Col., Carabidae), and activity of these and other epigeal predators during an outbreak of *Rhopalosiphum padi* (Hom., Aphididae). *Acta Agriculturae Scandinavica B*, **42**, 57–61.

Holopäinen, J.K. and Varis, A.L. (1986) Effects of a mechanical barrier and formalin preservative on pitfall catches of carabid beetles (Coleoptera, Carabidae) in arable fields. *Journal of Applied Entomology*, **102**, 440–45.

Holt, J., Cook, A.G., Perfect, T.J. and Norton, G.A. (1987) Simulation analysis of brown planthopper *(Nilaparvata lugens)* population dynamics on rice in the Phillipines. *Journal of Applied Ecology*, **24**, 87–103.

Holt, R.D. (1977) Predation, apparent competition, and the structure of prey communities. *Theoretical Population Biology*, **12**, 197–229.

Holt, R.D. and Hassell, M.P. (1993) Environmental heterogeneity and the stability of host–parasitoid interactions. *Journal of Animal Ecology*, **62**, 89–100.

Holt, R.D. and Hochberg, M.E. (2001) Indirect interactions, community modules and biological control: a theoretical perspective, in *Evaluating Indirect Ecological Effects of Biological Control* (eds E. Wajnberg, J.K. Scott and P.C. Quimby). CAB International, Wallingford, pp. 13–37.

Holt, R.D. and Kotler, B.P. (1987) Short-term apparent competition. *American Naturalist*, **130**, 412–30.

Holt, R.D. and Lawton, J.H. (1994) The ecological consequences of shared natural enemies. *Annual Review of Ecology and Systematics*, **25**, 495–520.

Holt, R.D. and Polis, G.A. (1997) A theoretical framework for intraguild predation. *American Naturalist*, **149**, 745–64.

Holtkamp, R.H. and Thompson, J.I. (1985) A lightweight, self-contained insect suction sampler. *Journal of the Australian Entomological Society*, **24**, 301–2.

Holyoak, M. (1994) Identifying delayed density dependence in time series data. *Oikos*, **70**, 296–04.

Holyoak, M. (2000) Effects of nutrient enrichment on predator-prey metapopulation dynamics. *Journal of Animal Ecology*, **67**, 985–97.

Holyoak, M. and Lawler, S.P. (1996) The role of dispersal in predator-prey metapopulation dynamics. *Journal of Animal Ecology*, **65**, 640–52.

Holzapfel, E.P. and Perkins, B.D. (1969) Trapping of air-borne insects on ships in the Pacific, Part 7. *Pacific Insects*, **11**, 455–76.

Honěk, A. (1986) Production of faeces in natural populations of aphidophagous coccinellids (Col.) and estimation of predation rates. *Zeitschrift für angewandte Entomologie*, **102**, 467–76.

Honěk, A. (1988) The effect of crop density and microclimate on pitfall trap catches of Carabidae, Staphylinidae (Coleoptera) and Lycosidae (Araneae) in cereal fields. *Pedobiologia*, **32**, 233–42.

Honěk, A. (1997) The effect of temperature on the activity of Carabidae (Coleoptera) in a fallow field. *European Journal of Entomology*, **94**, 97–104.

Honěk, A. and Kocourek, F. (1986) The flight of aphid predators to a light trap: possible interpretations, in *Ecology of Aphidophaga* (ed. I. Hodek). Academia, Prague, pp. 333–37.

Honěk, A., Jarošík, V., Lapchin, L. and Rabasse, J.-M. (1998) Host choice and offspring sex allocation in the aphid parasitoid *Aphelinus abdominalis* (Hymenoptera: Aphelinidae). *Journal of Agricultural Entomology*, **15**, 209–21.

Honeycut, R.L., Nelson, K., Schilter, D.A. and Sherman, P.W. (1991) Genetic variation within and among populations of the naked mole-rat: evidence from nuclear and mitochondrial genomes. in *The Biology of the Naked Mole-rat* (eds P.W. Sherman, J.U.M. Jarvis and R.D. Alexander). Princeton University Press, Princeton, pp. 195–208.

Hoogendoorn, M. and Heimpel, G.E. (2001) PCR-based gut content analysis of insect predators: using ribosomal ITS-1 fragments from prey to estimate predation frequency. *Molecular Ecology*, **10**, 2059–67.

Hoogendoorn, M. and Heimpel, G.E. (2002) Indirect interactions between an introduced and a native ladybird beetle species mediated by a shared parasitoid. *Biological Control*, **25**, 224–30.

Hooker, M.E., Barrows, E.M. and Ahmed, S.W. (1987) Adult longevity as affected by size, sex, and maintenance in isolation or groups in the parasite *Pediobius foveolatus* (Hymenoptera: Eulophidae). *Annals of the Entomological Society of America*, **80**, 655–9.

Hooper, R.E. and Siva-Jothy, M.T. (1996) Last male sperm precedence in a damselfly demonstrated by RAPD profiling. *Molecular Ecology*, **5**, 449–52.

Hooper, R.E. and Siva-Jothy, M.T. (1997) "Flybys": a remote assessment behaviour in female *Calopteryx splendens xanthostoma*. *Journal of Insect Behaviour*, **10**, 165–75.

Hopper, K.R. (1991) Ecological applications of elemental labeling: analysis of dispersal, density, mortality and feeding. *Southwestern Entomologist Supplement*, **14**, 71–83.

Hopper, K.R. (1996) Making biological control introductions more effective, in *Biological Control Introductions: Opportunities for Improved Crop Production* (chaired by J.K. Waage), British Crop Protection Council, Farnham, pp. 61–76.

Hopper, K.R. and King, E.G. (1986) Linear functional response of *Microplitis croceipes* (Hymenoptera: Braconidae) to variation in *Heliothis* spp. (Lepidoptera: Noctuidae) density in the field. *Environmental Entomology*, **15**, 476–80.

Hopper, K.R. and Roush, R.T. (1993) Mate finding, dispersal, number released, and the success of biological control introductions. *Ecological Entomology*, **18**, 321–31.

Hopper, K.R. and Woolson, E.A. (1991) Labeling a parasitic wasp, *Microplitis croceipes* (Hymenoptera: Braconidae), with trace elements for mark-recapture studies. *Annals of the Entomological Society of America*, **84**, 255–62.

Hopper, K.R., Powell, J.E. and King, E.G. (1991) Spatial density dependence in parasitism of *Heliothis virescens* (Lepidoptera: Noctuidae) by *Microplitis croceipes* (Hymenoptera: Braconidae) in the field. *Environmental Entomology*, **20**, 292–302.

Hopper, K.R., Roush, R.T. and Powell, W. (1993) Management of genetics of biological control introductions. *Annual Review of Entomology*, **38**, 27–51.

Hopper, K.R., Aidara, S., Agret, S., Cabal, J., Coutinot, D., Dabire, R., Lesieux, C., Kirk, G., Reichert, S., Tronchetti, F. and Vidal, J. (1995) Natural enemy impact on the abundance of *Diuraphis noxia* (Homoptera: Aphididae) in wheat in southern France. *Environmental Entomology*, **24**, 402–8.

Hopper, K.R. , Crowley, P.H. and Kielman, D. (1996) Density dependence, hatching synchrony, and within-cohort cannibalism in young dragonfly larvae. *Ecology*, **77**, 191–200.

Horn, D.J. (1981) Effect of weedy backgrounds on colonization of collards by green peach aphid, *Myzus persicae*, and its major predators. *Environmental Entomology*, **10**, 285–9.

Horne, P.A. and Horne, J.A. (1991) The effects of temperature and host density on the development and survival of *Copidosoma koehleri*. *Entomologia Experimentalis et Applicata*, **59**, 289–92.

Horton, D.R., Hinojosa, T. and Olson, S.R. (1998) Effects of photoperiod and prey type on diapause tendency and preoviposition period in *Perillus bioculatus* (Hemiptera:Pentatomidae) *Canadian Entomologist*, **130**, 315–20.

Hossain, Z., Gurr, G.M. and Wratten, S.D. (1999) Capture efficiency of insect natural enemies from tall and short vegetation using vacuum sampling. *Annals of Applied Biology*, **135**, 463–7.

Hossain, Z., Gurr, G.M., Wratten, S.D. and Raman, A. (2002) Habitat manipulation in lucerne *Medicago sativa*: arthropod population dynamics in harvested and 'refuge' crop strips. *Journal of Applied Ecology*, **39**, 445–54.

Houle, D. (1992) Comparing evolvability and variability of quantitative traits. *Genetics*, **130**, 195–204.

Hövemeyer, K. (1995) Trophic links, nutrient fluxes and natural history of the *Allium ursinum* food web, with particular reference to life history traits of two hoverfly herbivores (Diptera: Syrphidae). *Oecologia*, **102**, 86–94.

Howard, L.O. (1886) The excessive voracity of the female mantis. *Science*, **8**, 326.

Howarth, F.G. (1991) Environmental impacts of classical biological control. *Annual Review of Entomology*, **36**, 485–509.

Howe, R.W. (1967) Temperature effects on embryonic development in insects. *Annual Review of Entomology*, **12**, 15–42.

Howling, G.C. (1991) Slug foraging behaviour: attraction to food items from a distance. *Annals of Applied Biology*, **119**, 147–53.

Howling, G.G. and Port, G.R. (1989) Time-lapse video assessment of molluscicide baits, in *Slugs and Snails in World Agriculture*. *BCPC* Monograph No. 41, pp. 161–6.

Hoy, M.A. (1994) *Insect Molecular Genetics*. Academic Press, San Diego.

Hoy, M.A., Jeyaprakash, A., Morakote, R., Lo, P.K.C. and Nguyen, R. (2000) Genomic analyses of two populations of *Ageniaspis citricola* (Hymenoptera: Encyrtidae) suggest that a cryptic species may exist. *Biological Control*, **17**, 1–10.

Hradetzky, R. and Kromp, B. (1997) Spatial distribution of flying insects in an organic rye field and an adjacent hedge and forest edge. *Biological Agriculture and Horticulture*, **15**, 353–7.

Hsiao, T.H. (1996) Studies of interactions between alfalfa weevil strains, *Wolbachia* endosymbionts and parasitoids, in *The Ecology of Agricultural Pests* (eds W.O.C. Symondson and J.E. Liddell). Chapman and Hall, London, pp. 51–71.

Hu, G.Y. and Frank, J.H. (1996) Effect of the red imported fire ant (Hymenoptera: Formicidae) on dung-inhabiting arthropods in Florida. *Environmental Entomology*, **25**, 1290–6.

Hubbard, S.F., Marris, G., Reynolds, A. and Rowe, G.W. (1987) Adaptive patterns in the avoidance of superparasitism by solitary parasitic wasps. *Journal of Animal Ecology*, **56**, 387–401.

Hubbard, S.F., Harvey, I.F. and Fletcher, J.P. (1999) Avoidance of superparasitism: a matter of learning? *Animal Behaviour*, **57**, 1193–7.

Hudson, L. and Hay, F.C. (1989) *Practical Immunology, 3rd edn*. Blackwell, London.

Huelsenbeck, J.P. and Ronquist, F. (2001) MRBAYES: Bayesian inference of phylogenetic trees. *Bioinformatics*, **17**, 754–5.

Hufbauer, R.A. (2001) Pea aphid-parasitoid interactions: have parasitoids adapted to differential resistance? *Ecology*, **82**, 717–25.

Huffaker, C.B. (1958) Experimental studies on predation: dispersion factors and predator-prey oscillations. *Hilgardia*, **27**, 343–83.

Huffaker, C.B. and Kennett, C.E. (1966) Biological control of *Parlatoria oleae* (Colvee) through the compensatory action of two introduced parasites. *Hilgardia*, **37**, 283–335.

Huffaker, C.B., Kennett, C.E. and Finney, G.L. (1962) Biological control of olive scale, *Parlatoria oleae* (Colvee) in California by imported *Aphytis maculicornis* (Masi) (Hymenoptera: Aphelinidae). *Hilgardia*, **32**, 541–636.

Huffaker, C.B., Messenger, P.S. and DeBach, P. (1971) The natural enemy component in natural control and the theory of biological control, in *Biological Control* (ed. C.B. Huffaker). Plenum, New York, pp. 16–67.

Huffaker, C.B., Luck, R.F. and Messenger, P.S. (1977) The ecological basis of biological control. *Proceedings of the 15th International Congress of Entomology, Washington, 1976*, 560–86.

Huger, A.M., Skinner, S.W. and Werren, J.H. (1985) Bacterial infections associated with the son-killer trait in the parasitoid wasp *Nasonia* (=*Mormoniella*) *vitripennis* (Hymenoptera: Pteromalidae). *Journal of Invertebrate Pathology*, **46**, 272–80.

Hughes, R.D. (1962) A method for estimating the effects of mortality on aphid populations. *Journal of Animal Ecology*, **31**, 389–96.

Hughes, R.D. (1963) Population dynamics of the cabbage aphid, *Brevicoryne brassicae* (L.). *Journal of Animal Ecology*, **32**, 393–424.

Hughes, R.D. (1972) Population dynamics, in *Aphid Technology* (ed. H.F. van Emden). Academic Press, London, pp. 275–93.

Hughes, R.D. (1989) Biological control in the open field, in *World Crop Pests, Vol. 2C: Aphids, Their Biology, Natural Enemies and Control* (eds A.K. Minks and P. Harrewijn). Elsevier, Amsterdam, pp. 167–98.

Hughes, R.D. and Gilbert, N. (1968) A model of an aphid population – a general statement. *Journal of Animal Ecology*, **37**, 553–63.

Hughes, R.D. and Sands, P. (1979) Modelling bushfly populations. *Journal of Applied Ecology*, **16**, 117–39.

Hughes, R.D., Woolcock, L.T. and Hughes, M.A. (1992) Laboratory evaluation of parasitic Hymenoptera used in attempts to biologically control aphid pests of crops in Australia. *Entomologia Experimentalis et Applicata*, **63**, 177–85.

Hughes-Schrader, S. (1948) Cytology of coccids (Coccoidea). *Advances in Genetics*, **2**, 127–203.

Huigens, M.E., Luck, R.F., Klasen, R.H.G., Maas, M.F.P.M., Timmermans, M.J.T.N. and Stouthamer, R. (2000) Infectious parthenogenesis. *Nature*, **405**, 178–9.

van Huizen, T.H.P. (1990) 'Gone with the Wind': flight activity of carabid beetles in relation to wind direction and to the reproductive state of females in flight, in *The Role of Ground Beetles in Ecological and Environmental Studies* (ed. N.E. Stork). Intercept, Andover, pp. 289–93.

Hulspas-Jordaan, P.M. and van Lenteren, J.C. (1978) The relationship between host-plant leaf structure and parasitization efficiency of the parasitic wasp *Encarsia formosa* Gahan (Hymenoptera: Aphelinidae). *Mededelingen van de Faculteit Landbouwwetenschappen Rijksuniversiteit Gent*, **43**, 431–40

Hunt, G.J., Page, R.E., Fondrk, M.K. and Dullem, C.J. (1995) Major quantitative trait loci affecting honeybee foraging behavior. *Genetics*, **141**, 1537–45.

Hunt, J.H., Brown, P.A., Sago, K.M. and Kerker, J.A. (1991) Vespid wasps eat pollen. *Journal of the Kansas Entomological Society*, **64**, 127–30.

Hunter, M.D. (2001) Multiple approaches to estimating the relative importance of top-down and bottom-up forces on insect populations: experiments, life tables, and time-series analysis. *Basic and Applied Ecology*, **2**, 295–309.

Hunter, M.D, and Price, P. (1992) Playing chutes and ladders: heterogeneity and the relative roles of bottom-up and top-down forces in natural communities. *Ecology*, **73**, 724–32.

Hunter, M.D. and Price, P. (1998) Cycles in insect populations: delayed density dependence or exogenous driving variables? *Ecological Entomology*, **23**, 216–22.

Hunter, M.D., Varley, G.C. and Gradwell, G.R. (1997). Estimating the relative roles of top-down and bottom-up forces on insect herbivore populations: A classic study revisited. *Proceeding of the National Academy of Sciences*, **94**, 9176–81.

Hunter, M.S., Nur, U. and Werren, J.H. (1993) Origin of males by genome loss in an autoparasitoid wasp. *Heredity*, **70**, 162–171.

Hunter, M.S., Antolin, M.F. and Rose, M. (1996) Courtship behavior, reproductive relationships, and allozyme patterns of three North American populations of *Eretmocerus* nr. *californicus* (Hymenoptera: Aphelinidae) parasitizing the whitefly *Beimia* sp. *tabaci* complex (Homoptera: Aleyrodidae). *Proceedings of the Entomological Society of Washington*, **98**, 126–137.

Hunter, M.S., Collier, T.R. and Kelly, S.E. (2002) Does an autoparasitoid disrupt host suppression provided by a primary parasitoid? *Ecology*, **83**, 1459–69.

Hurlbutt, B. (1987) Sexual size dimorphism in parasitoid wasps. *Biological Journal of the Linnean Society*, **30**, 63–89.

Hurst, G., Majerus, M. and Walker, L. (1992) Cytoplasmic male killing elements in the two spot ladybird, *Adalia bipunctata* (Linnaeus) (Coleoptera: Coccinellidae). *Heredity*, **71**, 84–91.

Hurst, G.D.D., Hurst, L.D. and Majerus, M.E.N. (1993a) Altering sex-ratios – the games microbes play. *Bioessays*, **15**, 695–7.

Hurst, G.D.D., Majerus, M.E.N. and Walker, L.E. (1993b) The importance of cytoplasmic male killing elements in natural populations of the two spot ladybird, *Adalia bipunctata* (Linnaeus) (Coleoptera: Coccinellidae). *Biological Journal of The Linnean Society*, **49**, 195–202.

Hurst, G.G.D., Hurst, L.D. and Majerus, M.E.N. (1996) Cytoplasmic sex-ratio distorters, in *Influential Passengers: Microbes and Invertebrate reproduction* (eds S.L. O'Neill, A.A. Hoffmann and J.H. Werren). Oxford University Press, Oxford, pp. 125–54.

Hurst, G.D.D., Jiggins, F.M. and Robinson, S.J.W. (2001) What causes inefficient transmission of male-killing *Wolbachia* in *Drosophila*? *Heredity*, **87**, 220–6.

Hurst, L.D. (1991) The incidences and evolution of cytoplasmic male killers. *Proceedings of the Royal Society of London B*, **244**, 91–99.

Hurst, L.D. (1992) Intragenomic conflicts as an evolutionary force. *Proceedings of the Royal Society of London B*, **248**, 135–140.

Hurst, L.D. (1993) The incidences, mechanisms and evolution of cytoplasmic sex-ratio distorters in animals. *Biological Reviews*, **68**, 121–194.

Ibarra-Núñez, G., Garcia, J.A., López, J.A. and Lachaud, J.P. (2001) Prey analysis in the diet of some ponerine ants (Hymenoptera: Formicidae) and web-building spiders (Araneae) in coffee plantations in Chiapas, Mexico. *Sociobiology*, **37**, 723–55.

Ibrahim, M.M. (1955) Studies on *Coccinella undecimpunctata aegyptiaca* Reiche 2. Biology and life-history. *Bulletin de la Societé Entomologique d'Egypte*, **39**, 395–423.

Ichikawa, T. (1976) Mutual communication by substrate vibration in the mating behaviour of planthoppers (Homoptera: Delphacidae). *Applied Entomology and Zoology*, **11**, 8–23.

Idris, A.B. and Grafius, E. (1995) Wildflowers as nectar sources for *Diadegma insulare* (Hymenoptera: Ichneumonidae), a parasitoid of diamondback moth (Lepidoptera: Yponomeutidae). *Environmental Entomology*, **24**, 1726–35.

Idris, A.B. and Grafius, E. (1998) Diurnal flight activity of *Diadegma insulare* (Hymenoptera: Ichneumonidae), a parasitoid of the diamondback moth (Lepidoptera: Plutellidae), in the field. *Environmental Entomology*, **27**, 406–14.

Iizuka, T. and Takasu, K. (1998) Olfactory associative learning of the pupal parasitoid *Pimpla luctuosa* Smith (Hymenoptera: Ichneumonidae). *Journal of Insect Behavior*, **11**, 743–60.

Ijichi, N., Kondo, N. Matsumoto, R., Shimada, M., Ishikawa, H. and Fukatsu, T. (2002). Internal spatiotemporal population dynamics of infection with three *Wolbachia* strains in the azuki bean beetle, *Callosobruchus chinensis* (Coleoptera: Bruchidae). *Applied and Environmental Microbiology*, **68**, 4074–80.

Ikawa, T., Shimada, M., Matsuda, H. and Okabe, H. (1993) Sex allocation of parasitic wasps: local mate competition, dispersal before mating and host quality variation. *Journal of Evolutionary Biology*, **6**, 79–94.

Ilovai, Z. and van Lenteren, J.C. (1997) Development of a method for testing adult-fly capacity of *Aphidius colemani* Vierck (Hymenoptera: Braconidae). *Bulletin OILB/SROP*, **20**, 207–14.

Inouye, D.W. (1978) Resource partitioning in bumblebees: experimental studies of foraging behaviour. *Ecology*, **59**, 672–8.

Iperti, G. (1966) Some components of efficiency in aphidophagous coccinellids, in *Ecology of Aphidophagous Insects* (ed. I. Hodek). Academia, Prague and W. Junk, The Hague, p. 253.

Iperti, G. and Buscarlet, L.A. (1986) Seasonal migration of the ladybird *Semiadalia undecimnotata*, in *Ecology of Aphidophaga* (ed. I. Hodek). Academia, Prague and Dr. W. Junk, Dordrecht, pp. 199–204.

Irvin, N.A., Wratten, S.D. and Frampton, C.M. (2000). Understorey management for the enhancement of the leafroller parasitoid *Dolichogenidea tasmanica* (Cameron) in orchards at Canterbury, New Zealand, in *Hymenoptera: Evolution, Biodiversity and Biological Control* (eds A.D. Austin and M. Dowton). CSIRO, Melbourne, pp. 396–402.

Irwin, M.E., Gill, R.W. and Gonzales, D. (1974) Field-cage studies of native egg predators of the Pink Bollworm in southern California cotton. *Journal of Economic Entomology*, **67**, 193–6.

Ishii, M., Sato, Y. and Tagawa, J. (2000). Diapause in the braconid wasp, *Cotesia glomerata* (L.) II. Factors inducing and terminating diapause. *Entomological Science*, 3, 201–6.

Itioka, T., Inoue, T., Matsumoto, T. and Ishida, N. (1997) Biological control by two exotic parasitoids: eight-year population dynamics and life-tables of the arrowhead scale. *Entomologia Experimentalis et Applicata*, **85**, 65–74.

Ives, A.R. (1992) Continuous-time models of host-parasitoid interactions. *American Naturalist*, **140**, 1–29.

Ives, A.R. and Jansen, V.A.A. (1998) Complex dynamics in stochastic tritrophic models. *Ecology*, **79**, 1039–52.

Ives, A.R. and Settle, W.H. (1997) Metapopulation dynamics and pest control in agricultural systems. *American Naturalist*, **149**, 220–46.

Ives, A.R., Schooler, S.S., Jagar, V.J., Knuteson, S.E., Grbic, M. and Settle, W.H. (1999) Variability and parasitoid foraging efficiency: a case study of pea aphids and *Aphidius ervi*. *American Naturalist*, **154**, 652–73.

Ives, P.M. (1981a) Feeding and egg production of two species of coccinellids in the laboratory. *Canadian Entomologist*, **113**, 999–1005.

Ives, P.M. (1981b) Estimation of coccinellid numbers and movement in the field. *Canadian Entomology*, **113**, 981–97.

Iwao, K. and Ohsaki, N. (1996) Inter- and intraspecific interactions among larvae of specialist and generalist parasitoids. *Researches on Population Ecology*, **38**, 265–73.

Iwasa, Y., Higashi, M. and Matsuda, H. (1984) Theory of oviposition strategy of parasitoids. 1. Effect of mortality and limited egg number. *Theoretical Population Biology*, **26**, 205–27.

Iwata, K. (1959) The comparative anatomy of the ovary in Hymenoptera. Part 3. Braconidae (including Aphidiidae) with descriptions of ovarian eggs. *Kontyû*, **27**, 231–8.

Iwata, K. (1960) The comparative anatomy of the ovary in Hymenoptera. Part 5. Ichneumonidae. *Acta Hymenopterologica*, **1**, 115–69.

Iwata, K. (1962) The comparative anatomy of the ovary in Hymenoptera. Part 6. Chalcidoidea with descriptions of ovarian eggs. *Acta Hymenopterologica*, **1**, 383–91.

Jachmann, F. and van den Assem (1993) The interaction of external and internal factors in the courtship of parasitic wasps (Hym., Pteromalidae). *Behaviour*, **125**, 1–19.

Jachmann, F. and van den Assem (1995) A causal ethological analysis of the courtship behaviour of an insect (the parasitic wasp *Nasonia vitripennis*, Chalc., Pteromalidae). *Behaviour*, **133** 1051–75.

Jackson, C.G. (1986) Effects of cold storage of adult *Anaphes ovijentatus* on survival, longevity, and oviposition. *Southwestern Entomologist*, **11**, 149–53.

Jackson, C.G. (1991) Elemental markers for entomophagous insects. *Southwestern Entomologist, Supplement* **14**, 65–70.

Jackson, C.G. and Debolt, J.W. (1990) Labeling of *Leiophron uniformis*, a parasitoid of *Lygus* spp., with rubidium. *Southwest Entomologist*, **15**, 239–44.

Jackson, C.G., Cohen, A.C. and Verdugo, C.L. (1988) Labeling *Anaphes ovijentatus* (Hymenoptera: Mymaridae), an egg parasite of *Lygus* spp. (Hemiptera: Miridae), with rubidium. *Annals of the Entomological Society of America*, **81**, 919–22.

Jackson, D.J. (1928) The biology of *Dinocampus (Perilitus) rutilus* Nees, a braconid parasite of *Sitona lineata* L. Part 1. *Proceedings of the Zoological Society of London*, **1928**, 597–630.

James, D.G., Warren, G.N. and Whitney, J. (1992) Phytoseiid mite populations on dormant grapevines: extraction using a microwave oven. *Experimental and Applied Acarology*, **14**, 175–8.

James, H.G. and Nicholls, C.F. (1961) A sampling cage for aquatic insects. *Canadian Entomologist*, **93**, 1053–5.

James, H.G. and Redner, R.L. (1965) An aquatic trap for sampling mosquito predators. *Mosquito News*, **25**, 35–7.

Jamnongluk, W., Kittayapong, P., Baimai, V. and O'Neill, S.L. (2002) *Wolbachia* infections of tephritid fruit flies: molecular evidence for five distinct strains in a single host species. *Current Microbiology*, **45**, 255–60.

Janowski-Bell, M.E. and Horner, N.V. (1999) Movement of the male brown tarantula, *Aphonopelma hentzi* (Araneae, Theraphosidae), using radio telemetry. *Journal of Arachnology*, **27**, 503–12.

Jansen, R.C. and Stam, P. (1994) High resolution of quantitative traits into multiple loci via interval mapping. *Genetics*, **136**, 1447–55.

Janssen, A. (1989) Optimal host selection by *Drosophila* parasitoids in the field. *Functional Ecology*, **3**, 469–79.

Janssen, A., van Alphen, J., Sabelis, M. and Bakker, K. (1991) Microhabitat selection behaviour of *Leptopilina heterotoma* changes when odour of competitor is present. *Redia*, **74**, 302–10.

Janssen, A., Pallini, A., Venzon, M. and Sabelis, M.W. (1998) Behaviour and indirect interactions in food webs of plant-inhabiting arthropods. *Experimental and Applied Acarology*, **22**, 497–521.

Japyassu, H.F. and Viera, C. (2002) Predatory plasticity in *Nephilengys cruentata* (Araneae: Tetragnathidae): relevance for phylogeny reconstruction. *Behaviour*, **139**, 529–44.

Jarne, P. and Lagoda, P.J.L. (1996) Microsatellites, from molecules to populations and back. *Trends in Ecology and Evolution*, **11**, 424–9.

Jarošik, V. (1992) Pitfall trapping and species-abundance relationships: a value for carabid beetles (Coleoptera, Carabidae). *Acta Entomologica Bohemoslovica*, **89**, 1–12.

Jayanth K.P, and Bali, G. (1993) Diapause behavior of *Zygogramma bicolorata* (Coleoptera: Chrysomelidae), a biological control agent for *Parthenium hysterophorus* (Asteraceae), in Bangalore, India. *Bulletin of Entomological Research*, **83**, 383–8.

Jedlicková, J. (1997) Modification of a leaf-washing apparatus for the recovery of mites. *Experimental and Applied Acarology*, **21**, 273–7.

Jennings, D.T., Houseweart, M.W. and Cokendolpher, J.C. (1984) Phalangids (Arachnida: Opiliones) associated with strip clearcut and dense spruce-fir forests of Maine. *Environmental Entomology*, **13**, 1306–11.

Jensen, P.B. (1997) The influence of unspraying on diversity of soil-related hymenopteran parasitoids in cereal fields. *Journal of Applied Entomology*, **121**, 417–24.

Jensen, T.S., Dyring, L., Kristensen, B., Nilesen, B.O. and Rasmussen, E.R. (1989) Spring dispersal and summer habitat distribution of *Agonum dorsale* (Coleoptera: Carabidae). *Pedobiologia*, **33**, 155–65.

Jepson, P., Cuthbertson, P., Downham, M., Northey, D., O'Malley, S., Peters, A., Pullen, A., Thacker, R., Thackray, D., Thomas, C. and Smith, C. (1987) A quantitative ecotoxicological investigation of the impact of synthetic pyrethroids on beneficial insects in winter cereals. *Bulletin SROP/WPRS, 1987/X/1*, 194–205.

Jervis, M.A. (1978) Homopteran Bugs, in *A Dipterist's Handbook* (eds A. Stubbs and P.J. Chandler). Amateur Entomologist's Society, Hanworth, pp. 173–6.

Jervis, M.A. (1979) Courtship, mating and 'swarming' in *Aphelopus melaleucus* (Hym., Dryinidae). *Entomologist's Gazette*, **30**, 191–3.

Jervis, M.A. (1980a) Life history studies of *Aphelopus* species (Hymenoptera: Dryinidae) and *Chalarus* species (Diptera: Pipunculidae), primary parasites of typhlocybine leafhoppers (Homoptera: Cicadellidae). *Journal of Natural History*, **14**, 769–80.

Jervis, M.A. (1980b) Ecological studies on the parasite complex associated with typhlocybine leafhoppers (Homoptera: Cicadellidae), *Ecological Entomology*, **5**, 123–36.

Jervis, M.A. (1992) A taxonomic revision of the pipunculid fly genus *Chalarus* Walker, with particular reference to the European fauna. *Zoological Journal of the Linnean Society*, **105**, 243–352.

Jervis, M.A. (1998) Functional and evolutionary aspects of mouthpart structure in parasitoid wasps. *Biological Journal of the Linnean Society*, **63**, 461–93.

Jervis, M.A. and Kidd, N.A.C. (1986) Host-feeding strategies in hymenopteran parasitoids. *Biological Reviews*, **61**, 395–434.

Jervis, M.A. and Kidd, N.A.C. (1991) The dynamic significance of host-feeding by insect parasitoids–what modellers ought to consider. *Oikos*, **62**, 97–9.

Jervis, M.A. & Kidd, N.A.C. (1993) Integrated pest management in European olives – new developments. *Antenna*, **17**, 108–14.

Jervis, M.A. and Kidd, N.A.C. (1999) Parasitoid nutritional ecology: implications for biological control, in *Theoretical Approaches to Biological Control* (eds B.A. Hawkins and H.V. Cornell). Cambridge University Press, Cambridge, pp. 131–51.

Jervis, M.A., Kidd, N.A.C. and Walton, M. (1992a) A review of methods for determining dietary range in adult parasitoids. *Entomophaga*, **37**, 565–74.

Jervis, M.A., Kidd, N.A.C., McEwen, P., Campos, M. and Lozano, C. (1992b) Biological control strategies in olive pest management, in *Research Collaboration in European IPM Systems* (ed. P.T. Haskell). British Crop Protection Council, Farnham, pp. 31–9.

Jervis, M.A., Kidd, N.A.C., Fitton, M.G., Huddleston, T. and Dawah, H.A. (1993) Flower-visiting by hymenopteran parasitoids. *Journal of Natural History*, **27**, 67–105.

Jervis, M.A., Kidd, N.A.C. and Almey, H.A. (1994) Post-reproductive life in the parasitoid *Bracon hebetor* (Say) (Hym., Braconidae). *Journal of Applied Entomology*, **117**, 72–7.

Jervis, M.A., Hawkins, B.A. and Kidd, N.A.C. (1996a) The usefulness of destructive host feeding parasitoids in classical biological control: theory and observation conflict. *Ecological Entomology*, **21**, 41–6.

Jervis, M.A., Kidd, N.A.C. and Heimpel, G.E. (1996b) Parasitoid adult feeding and biological control – a review. *Biocontrol News and Information*, **17**, 1N-22N.

Jervis, M.A., Heimpel, G.E., Ferns, P.N., Harvey, J.A. and Kidd, N.A.C. (2001) Life-history strategies in parasitoid wasps: a comparative analysis of 'ovigeny'. *Journal of Animal Ecology*, **70**, 442–58.

Jervis, M.A., Ferns, P.N. and Heimpel, G.E. (2003) Body size and the timing of egg production in parasitoid wasps: a comparative analysis. *Functional Ecology*, **17**, 375–83.

Jervis, M.A., Lee, J.C. and Heimpel, G.E. (2004) Use of behavioural and life-history studies to understand the effects of habitat manipulation, in *Ecological Engineering for Pest Management* (eds G. Gurr, S.D. Wratten and M. Altieri). CSIRO Press, Melbourne, in press.

Jeyaprakash, A. and Hoy, M.A. (2000) Long PCR improves *Wolbachia* DNA amplification: *wsp* sequences found in 76% of sixty-three arthropod species. *Insect Molecular Biology*, **9**, 393–405.

Jia, F., Margolies, D.C., Boyer, J.E. and Charlton, R.E. (2002) Genetic variation in foraging traits among inbred lines of a predatory mite. *Heredity*, **89**, 371–9.

Jiang, C. and Zeng, Z.-B. (1995) Multiple trait analysis of genetic mapping for quatitative trait loci. *Genetics*, **140**, 1111–27.

Jiggins, F.M. (2002a) Widespread "hilltopping" in *Acraea* butterflies and the origin of sex-role-reversed swarming in *Acraea encedon* and *A. encedana*. *African Journal of Ecology*, **40**, 228–31.

Jiggins, F.M. (2002b) The rate of recombination in *Wolbachia* bacteria. *Molecular Biology Evolution*, **19**, 1640–3.

Jiggins, F.M., Hurst, G.D.D., Jiggins, C.D., Von der Schulenburg, J.H.G. and Majerus. M.E.N. (2000) The butterfly *Danaus chrysippus* is infected by a male-killing *Spiroplasma* bacterium. *Parasitology*, **120**, 439–46.

Jiggins, F.M., Hurst, G.D.D., Hinrich, J., von der Schulenburg, J.H.G. and Majerus, M.E.N. (2001a) Two male-killing *Wolbachia* strains coexist within a population of the butterfly *Acraea encedon*. *Heredity*, **86**, 161–6.

Jiggins, F.M., von der Schulenburg, J.H.G., Hurst, G.D.D. and Majerus, M.E.N. (2001b) Recombination confounds interpretations of *Wolbachia* evolution. *Proceedings of the Royal Society of London B*, **268**, 1423–7.

Jiggins, F.M., Hurst, G.D.D. and Yang, Z.H. (2002) Host-symbiont conflicts: positive selection on an outer membrane protein of parasitic but not mutualistic Rickettsiaceae. *Molecular Biology Evolution*, **19**, 1341–9.

Jmhasly, P. and Nentwig, W. (1995) Habitat management in winter wheat and evaluation of subsequent spider predation on insect pests. *Acta Oecologica*, **16**, 389–403.

Johanowicz, D.L. and Hoy, M.A. (1999) *Wolbachia* infection dynamics in experimental laboratory populations of *Metaseiulus occidentalis*. *Entomologia Experimentalis et Applicata*, **93**, 259–68.

Johnson, C.G. and Taylor, L.R. (1955) The development of large suction traps for airborne insects. *Annals of Applied Biology*, **43**, 51–61.

Johnson, C.G., Southwood, T.R.E. and Entwistle, H.M. (1957) A new method of extracting arthropods and molluscs from grassland and herbage with a suction apparatus. *Bulletin of Entomological Research*, **48**, 211–8.

Johnson, D.M., Akre, B.C. and Crowley, P.H. (1975) Modeling arthropod predation: wasteful killing by damselfly naiads. *Ecology*, **36**, 1081–93.

Johnson, D.M., Bohanan, R.E., Watson, C.N. and Martin, T.H. (1984) Coexistence of *Enallagma divagans* and *Enallagma traviatum* (Zygoptera: Coenagrionidae) in Bays Mountain Lake: an in situ enclosure experiment. *Advances in Odonatology*, **2**, 57–70.

Johnson, J.W., Eikenbary, R.D. and Holbert, D. (1979) Parasites of the greenbug and other graminaceous aphids: identity based on larval meconia and features of the empty aphid mummy. *Annals of the Entomological Society of America*, **72**, 759–66.

Johnson, L.K. (1982) Sexual selection in a tropical brentid weevil. *Evolution*, **36**, 251–62.

Johnson, P.C. and Reeves, R.M. (1995) Incorporation of the biological marker rubidium in gypsy moth (Lepidoptera: Lymantriidae) and its transfer to the predator *Carabus nemoralis* (Coleoptera: Carabidae). *Environmental Entomology*, **24**, 46–51.

Johnston, H. (1998) Automating behaviour measurement. *Scientific Computing World*, **35**, 29–30.

Jolly, C.M. (1965) Explicit estimates from capture-recapture data with both death and immigration-stochastic model. *Biometrika*, **52**, 225–47.

Jones, M.G. (1979) The abundance and reproductive activity of common Carabidae in a winter wheat crop. *Ecological Entomology*, **4**, 31–43.

Jones, R.L. and Lewis, W.J. (1971) Physiology of the host-parasite relationship between *Heliothis zea* and *Microplitis croceipes*. *Journal of Insect Physiology*, **17**, 921–7.

Jones, S.A. and Morse, J.G. (1995) Use of isoelectric focusing electrophoresis to evaluate citrus thrips (Thysanoptera: Thripidae) predation by *Euseius tularensis* (Acari: Phytoseiidae). *Environmental Entomology*, **24**, 1040–51.

Jones, T.H. (1986). Patterns of Parasitism by *Trybliographa rapae* (Westw.), a cynipid parasitoid of the Cabbage Root Fly. Imperial College, University of London, Ph.D. Thesis.

Jones, T.H. and Hassell, M.P. (1988) Patterns of parasitism by *Trybliographa rapae*, a cynipid parasitoid of the cabbage root fly, under laboratory and field conditions. *Ecological Entomology*, **13**, 65–92.

Jones, T.H., Godfray, H.C.J. and Hassell, M.P. (1996) Relative movement patterns of a tephritid fly and its parasitoid wasps. *Oecologia*, **106**, 317–24.

Jones, T.M., Balmford, A. and Quinnell, R.J. (1998) Fisherian flies: the benefits of female choice in a lekking sandfly. *Proceedings of the Royal Society of London B*, **265**, 1651–57.

Jones, W.A., Greenberg, S.M. and Legaspi Jr, B. (1999) The effect of varying *Bemisia argentifolii* and *Eretmocerus mundus* ratios on parasitism. *BioControl*, **44**, 13–28.

Jones, W.T. (1982) Sex ratio and host size in a parasitic wasp. *Behavioral Ecology and Sociobiology*, **101**, 207–10.

de Jong, P.W. and van Alphen, J.J.M. (1989) Host size selection and sex allocation in *Leptomastix dactylopii*, a parasitoid of *Planococcus citri*. *Entomologia Experimentalis et Applicata*, **50**, 161–9.

Joseph, S.B., Snyder, W.E. and Moore, J. (1999) Cannibalizing *Harmonia axyridis* (Coleoptera: Coccinellidae) larvae use endogenous cues to avoid eating relatives. *Journal of Evolutionary Biology*, **12**, 792–7.

Joyce, K.A., Jepson, P.C., Doncaster, C.P. and Holland, J.M. (1997) Arthropod distribution patterns and dispersal processes within the hedgerow, in *Species Dispersal and Land Use Processes* (eds A. Cooper and J. Power), Proceedings of the 6th Annual Conference of the International Association for Landscape Ecology, Coleraine, Northern Ireland pp. 103–10.

Judd, W.W. (1970) Insects associated with flowering wild carrot, *Daucus carota* L., in southern Ontario. *Proceedings of the Entomological Society of Ontario*, **100**, 176–81.

Juliano, S.A. (1985) The effects of body size on mating and reproduction in *Brachinus lateralis* (Coleoptera: Carabidae). *Ecological Entomology*, **10**, 271–80.

Juliano, S.A. (2001) Non-linear curve-fitting: predation and functional response curves, in *Design and Analysis of Ecological Experiments, 2nd edn* (eds S.M. Scheives and J. Gurevitch). Chapman and Hall, New York, pp. 198–19.

Juvonen-Lettington, A. and Pullen, A. (2001) The assessment of the saproxylic Coleoptera fauna of lowland parkland and wood pasture: an evaluation of sampling techniques. *Antenna*, **25**, 97–9.

Kádár, F. and Lövei, G.L. (1992) Light trapping of carabids (Coleoptera: Carabidae) in an apple orchard in Hungary. *Acta Phytopathologica et Entomologica Hungarica*, **27**, 343–8.

Kádár, F. and Szentkirályi, F. (1998) Seasonal flight pattern of *Harpalus rufipes* (De Geer) captured by light traps in Hungary (Coleoptera: Carabidae). *Acta Phytopathologica et Entomologica Hungarica*, **33**, 367–77.

Kageyama, D., Hoshizaki, S. and Ishikawa, Y. (1998) Female-biased sex ratio in the Asian corn borer, *Ostrinia furnacalis*: evidence for the occurrence of feminizing bacteria in an insect. *Heredity*, **81**, 311–16.

Kageyama, D., Nishimura, G., Hoshizaki, S. and Ishikawa, Y. (2002) Feminizing *Wolbachia* in an insect, *Ostrinia furnacalis* (Lepidoptera: Crambidae). *Heredity*, **88**, 444–9.

Kairo, M.T.K., Cock, M.J.W. and Quinlan, M.M. (2003). An assessment of the use of the Code of Conduct for the Import and Release of Exotic Biological Control Agents (ISPM No. 3) since its endorsement as an international standard. *Biocontrol News and Information*, **24**, 15N–27N.

Kaitala, A. (1988) Wing muscle dimorphism: two reproductive pathways of the waterstrider *Gerris thoracicus* in relation to habitat instability. *Oikos*, **53**, 222–8.

Kaitala, A. (1991) Phentotypic plasticity in reproductive behaviour of waterstriders: trade-offs between reproduction and longevity during food stress. *Functional Ecology*, **5**, 12–18.

Kaitala, A. and Huldén, L. (1990) Significance of spring migration and flexibility in flight muscle histolysis in waterstriders (Heteroptera, Gerridae). *Ecological Entomology*, **15**, 409–18.

Kajak, A. and Lukasiewicz, J. (1994) Do semi-natural patches enrich crop fields with predatory epigean arthropods? *Agriculture Ecosystems and Environment*, **49**, 149–61.

Kajita, H. (1989) Mating activity of the aphelinid wasp, *Encarsia* sp. in the field (Hymenoptera, Aphelinidae). *Applied Entomology and Zoology*, **24**, 313–16.

Kajita, H. and van Lenteren, J.C. (1982) The parasite-host relationship between *Encarsia formosa* (Hymenoptera: Aphelinidae) and *Trialeurodes vaporariorum* (Homoptera: Aleyrodidae), XIII. Effect of low temperatures on egg maturation of *Encarsia formosa*. *Zeitschrift für Angewandte Entomologie*, **93**, 430–9.

Kakehashi, N., Suzuki, Y. and Iwasa, Y. (1984) Niche overlap of parasitoids in host-parasitoid systems: its consequences to single versus multiple introduction controversy in biological control. *Journal of Applied Ecology*, **21**, 115–31.

Kamano, Y., Shimizu, K., Kanioh, Y. and Tatsuki, S. (1989) Mating behaviour of *Ascogaster reticulatus* Watanabe (Hymenoptera: Braconidae), an egg-larval parasitoid of the smaller tea tortrix, *Adoxophyes* sp. (Lepidoptera: Tortricidae). II Behavioral sequence and the role of sex pheromone. *Applied Entomology and Zoology*, **24**, 372–8.

Karban, R. (1998) Caterpillar basking behaviour and nonlethal parasitism by tachinid flies. *Journal of Insect Behavior*, **11**, 713–23.

Karban, R. and English-Loeb, G. (1997) Tachinid parasitoids affect host plant choice by caterpillars to increase caterpillar survival. *Ecology*, **78**, 603–11.

Karczewski, J. (1967) The observations on flower-visiting species of Tachinidae and Calliphoridae (Diptera). *Fragmenta Faunistica*, **13**, 407–84.

Kareiva, P. (1990) Establishing a foothold for theory in biological control practice: using models to guide experimental design and release protocols, in *New Directions in Biological Control* (eds R.R. Baker and P.E. Dunn). Alan R. Liss, New York, pp. 65–81.

Karlsson, B. and Wickman, P.-O. (1989) The cost of prolonged life: an experiment on a nymphalid butterfly. *Functional Ecology*, **3**, 399–405.

Karowe, D.N. and Martin, M.M. (1989) The effects of quantity and quality of diet nitrogen on the growth, efficiency of food utilisation, nitrogen budget, and metabolic rate of fifth-instar *Spodoptera eridania* larvae (Lepidoptera: Noctuidae). *Journal of Insect Physiology*, **35**, 699–708.

Kathuria, P., Greeff, J.M., Compton, S.G. and Ganeshaiah, K.N. (1999) What fig wasp sex ratios may or may not tell us about sex allocation strategies. *Oikos*, **87**, 520–30.

Kato, M. (1994) Alternation of bottom-up and top-down regulation in a natural population of an agromyzid leafminer, *Chromatomyia suikazurae*. *Oecologia*, **97**, 9–16.

Katsoyannos, P. (1984) The establishment of *Rhizobius forestieri* (Col, Coccinellidae) in Greece and its efficiency as an auxiliary control agent against a heavy infestation of *Saissetia oleae* (Hom, Coccidae). *Entomophaga*, **29**, 387–97.

Katsoyannos, P., Kontodimas, D.C. and Stathas, G.J. (1997) Summer diapause and winter quiescence of *Coccinella septempunctata* (Col., Coccinellidae) in central Greece. *Entomophaga*, **42**, 483–91.

Kawauchi, S.E. (1985) Effects of photoperiod on the induction of diapause, the live weight of emerging adult and the duration of development of three species of aphidophagous coccinellids (Coleoptera: Coccinellidae). *Kontyû*, **53**, 536–46.

Kayhart, M. (1956) A comparative study of dose-action curves for visible eye-color mutations induced by X-rays, thermal neutrons, and fast neutrons in *Mormoniella vitripennis*. *Radiation Research*, **4**, 65–76.

Kazmer, D.J. (1990) Isoelectric focussing procedures for the analysis of allozymic variation in minute arthropods. *Annals of the Entomological Society of America*, **84**, 332–9.

Kazmer, D.J. and Luck, R.F. (1991) The genetic-mating structure of natural and agricultural populations of *Trichogramma*, in *Proceedings of the 3rd International Symposium on* Trichogramma *and other egg parasitoids, San Antonio (Tx, USA)* (eds E. Wajnberg and S.B. Vinson). INRA, Paris, pp. 107–10.

Kazmer, D.J. and Luck, R.F. (1995) Field tests of the size-fitness hypothesis in the egg parasitoid *Trichogramma pretiosum*. *Ecology*, **76**, 412–5.

Kazmer, D.J., Maiden, K., Ramualde, N., Coutinot, D. and Hopper, K.R. (1996) Reproductive compatibility, mating behaviour, and random amplified polymorphic DNA variability in some *Aphelinus asychis* (Hymenoptera: Aphelinidae) derived from the Old World. *Annals of the Entomological Society of America*, 89, 212–300.

Kean, J.M. and Barlow, N.D. (2000) The effects of density-dependence and local dispersal in individual-based stochastic metapopulations. *Oikos*, **88**, 282–90.

Kearns, C.A. and Inouye, D.W. (1995) *Techniques for Pollination Biologists*. University Press of Colorado, Niwot.

Kearsey, M.J. and Farquhar, A.G.L. (1998) QTL analysis; where are we now? *Heredity*, **80**, 137–142.

Kearsey, M.J. and Pooni, H.S. (1996) *The Genetical Analysis of Quantitative Traits*. Chapman and Hall, London.

Keeler, K.H. (1978) Insects feeding at extrafloral nectaries of *Ipomoea carnea* (Convolvulaceae). *Entomological News*, **89**, 163–8.

Keen, D.P., Keen, J.E., He, Y. and Jones, C.J. (2001) Development of an enzyme-linked immunosorbent assay for detection of the gregarious hymenopteran parasitoid *Muscidifurax raptorellus* in house fly pupae. *Biological Control*, **21**, 140–51.

Kegel, B. (1990) Diurnal activity of carabid beetles living on arable land, in *The Role of Ground Beetles in Ecological and Environmental Studies* (ed. N.E. Stork). Intercept, Andover, pp. 65–76.

Kempson, D., Lloyd, M. and Ghelardi, R. (1963) A new extractor for woodland litter. *Pedobiologia*, **8**, 1–21.

Kenmore, P.E., Carino, F.O., Perez, C.A., Dyck, V.A. and Gutierrez, A.P. (1985) Population regulation of the rice brown planthopper *(Nilaparvata lugens* Stål) within rice fields in the Philippines. *Journal of Plant Protection in the Tropics*, **1**, 19–37.

Kennedy, B.H. (1979) The effect of multilure on parasites of the European elm bark beetle, *Scolytus multistriatus*. *ESA Bulletin*, **25**, 116–18.

Kennedy, J.S. (1978) The concepts of olfactory 'arrestment' and 'attraction'. *Physiological Entomology*, **3**, 91–8.

Kennedy, P.J. and Randall, N.P. (1997) A semi-field method to assess the effect of dimethoate on the density and mobility of ground beetles (Carabidae), in *Arthropod Natural Enemies in Arable Land III. The Individual, the Population and the Community* (ed. W. Powell). *Acta Jutlandica*, **72**, 21–37.

Kennedy, P.J., Randall, N.P. and Hackett, B. (1996) Effects of dimethoate on ground beetles in semi-field enclosures: a mark-recapture study. *Brighton Crop Protection Conference – Pests and Diseases 1996*, **3**, 1199–204.

Kennedy, T.F., Evans, G.O. and Feeney, A.M. (1986) Studies on the biology of *Tachyporus hypnorum* F. (Col. Staphylinidae), associated with cereal fields in Ireland. *Irish Journal of Agricultural Research*, **25**, 81–95.

Kerkut, G.A. and Gilbert, L.I. (1985) *Comprehensive Insect Physiology, Biochemistry and Pharmacology. Vol 1. Embryogenesis and Reproduction*. Pergamon, Oxford.

Kessel, E.L. (1955) The mating activities of balloon flies. *Systematic Zoology*, **4**, 97–104.

Kessel, E.L. (1959) Introducing *Hilara wheeleri* Melander as a balloon maker, and notes on other North American balloon flies. *Wasmann Journal of Biology*, **17**, 221–30.

Kevan, P.G. (1973) Parasitoid wasps as flower visitors in the Canadian high arctic. *Anzeiger Schädlungskunde*, **46**, 3–7.

Kfir, R. (1981) Fertility of the polyembryonic parasite *Copidosoma koehleri*, effect of humidities on life length and relative abundance as compared with that of *Apanteles subandinus* in potato tuber moth. *Annals of Applied Biology*, **99**, 225–30.

Kfir, R. (2002) Increase in cereal stem borer populations through partial elimination of natural enemies. *Entomologia Experimentalis et Applicata*, **104**, 299–306.

Kfir, R. and van Hamburg, H. (1988) Interspecific competition between *Telenomus ullyetti* (Hymenoptera: Scelionidae) and *Trichogramma lutea* (Hymenoptera: Trichogrammatidae) parasitizing eggs of the cotton bollworm *Heliothis armigera* in the laboratory. *Environmental Entomology*, **17**, 664–70.

Kfir, R. and Luck, R.F. (1979) Effects of constant and variable temperature extremes on sex ratio and progeny production by *Aphytis melinus* and *A. lingnanensis* (Hymenoptera: Aphelinidae). *Ecological Entomology*, **4**, 335–44.

Kharboutli, M.S. and Mack, T.P. (1993) Comparison of three methods for sampling arthropod pests and their natural enemies in peanut fields. *Journal of Economic Entomology*, **86**, 1802–10.

Kidd, N.A.C. (1979) Simulation of population processes with a programmable pocket calculator. *Journal of Biological Education*, **13**, 284–90.

Kidd, N.A.C. (1984) A BASIC program for use in teaching population dynamics. *Journal of Biological Education*, **18**, 227–8.

Kidd, N.A.C. (1990a) The population dynamics of the large pine aphid, *Cinara pinea* (Mordv.). I. Simulation of laboratory populations. *Researches on Population Ecology*, **32**, 189–208.

Kidd, N.A.C. (1990b) The population dynamics of the large pine aphid, *Cinara pinea* (Mordv.). II. Simulation of field populations. *Researches on Population Ecology*, **32**, 209–26.

Kidd, N.A.C. (1990c) A synoptic model to explain long-term population changes in the large pine aphid, in *Population Dynamics of Forest Insects* (eds A.D. Watt, S.R. Leather, M.D. Hunter and N.A.C. Kidd). Intercept, Andover, pp. 317–27.

Kidd, N.A.C. and Jervis, M.A. (1989) The effects of host-feeding behaviour on the dynamics of parasitoid-host interactions, and the implications for biological control. *Researches on Population Ecology*, **31**, 235–74.

Kidd, N.A.C. and Jervis, M.A. (1991) Host-feeding and oviposition strategies of parasitoids in relation to host stage. *Researches on Population Ecology*, **33**, 13–28.

Kidd, N.A.C. and Jervis, M.A. (1997) The impact of parasitoids and predators on forest insect populations. in Forests and Insects. (eds A.D. Watt, N.E. Stork and M.D. Hunter). Chapman and Hall, London. pp. 49–68.

Kidd, N.A.C. and Mayer, A.D. (1983) The effect of escape responses on the stability of insect host-parasite models. *Journal of Theoretical Biology*, **104**, 275–87.

Kiman, Z.B. and Yeargan, K.V. (1985) Development and reproduction of the predator *Orius insidiosus* (Hemiptera: Anthocoridae) reared on diets of selected plant material and arthropod prey. *Annals of the Entomological Society of America*, **78**, 464–7.

Kimsey, L.S. (1980) The behaviour of male orchid bees (Apidae, Hymenoptera, Insecta) and the question of leks. *Animal Behaviour*, **28**, 996–1004.

Kindlmann, P. and Dixon, A.F.G. (1999) Generation time ratios – determinants of prey abundance in insect predator-prey interactions. *Biological Control*, **16**, 133–8.

Kindlmann, P. and Dixon, A.F.G. (2001) When and why top–down regulation fails in arthropod predator–prey systems. *Basic and Applied Ecology*, **2**, 33–40.

Kindlmann, P., Dixon, A.F.G. and Dostálková, I. (2001) Role of ageing and temperature in shaping reaction norms and fecundity functions in insects. *Journal of Evolutionary Biology*, **14**, 835–40.

King, B.H. (1987) Offspring sex ratios in parasitoid wasps. *Quarterly Review of Biology*, **62**, 367–96.

King, B.H. (1989) Host-size dependent sex ratios among parasitoid wasps: does host growth matter. *Oecologia*, **78**, 420–6.

King, B.H. (1993) Sex ratio manipulation by parasitoid wasps, in *Evolution and Diversity of Sex Ratio in Insects and Mites* (eds D.L. Wrensch and M.A. Ebbert). Chapman and Hall, New York, pp. 418–41.

King, B.H. (1998) Host age response in the parasitoid wasp *Spalangia cameroni* (Hymenoptera: Pteromalidae). *Journal of Insect Behaviour*, **11**, 103–17.

King, B.H. and King, R.B. (1994) Sex-ratio manipulation in response to host size in the parasitoid wasp *Spalangia cameroni* – is it adaptive? *Behavioral Ecology*, **5**, 448–54.

King, B.H. and Skinner, S.W. (1991) Sex ratio in a new species of *Nasonia* with fully-winged males. *Evolution*, **45**, 225–8.

King, B.H., Grimm, K.M. and Reno, H.E. (2000). Effects of mating on female locomotor activity in the pteromalid wasps *Nasonia vitripennis* (Hymenoptera; Pteromalidae). *Environmental Entomology* **29**, 927–33.

King, P.E. (1961) The passage of sperm to the spermatheca during mating in *Nasonia vitripennis*. *Entomologist's Monthly Magazine*, **97**, 136.

King, P.E. (1962) The structure and action of the spermatheca in *Nasonia vitripennis*. *Proceedings of the Royal Entomological Society of London A*, **37**, 3–5.

King, P.E. (1963) The rate of egg resorption in *Nasonia vitripennis* (Walker) (Hymenoptera: Pteromalidae) deprived of hosts. *Proceedings of the Royal Entomological Society of London A* **38**, 98–100.

King, P.E. and Copland, M.J.W. (1969) The structure of the female reproductive system in the Mymaridae (Chalcidoidea: Hymenoptera). *Journal of Natural History*, **3**, 349–65.

King, P.E. and Fordy, M.R. (1970) The external morphology of the 'pore' structures on the tip of the ovipositor in Hymenoptera. *Entomologist's Monthly Magazine*, **106**, 65–6.

King, P.E. and Ratcliffe, N.A. (1969) The structure and possible mode of functioning of the female reproductive system in *Nasonia vitripennis* (Hymenoptera: Pteromalidae). *Journal of Zoology*, **157**, 319–44.

King, P.E. and Richards, J.G. (1968) Oösorption in *Nasonia vitriperinis* (Hymenoptera: Pteromalidae). *Journal of Zoology*, **54**, 495–516.

King, P.E. and Richards, J.G. (1969) Oögenesis in *Nasonia vitripennis* (Walker) (Hymenoptera: Pteromalidae). *Proceedings of the Royal Entomological Society, London A*, **44**, 143–57.

King, P.E., Askew, R.R. and Sanger, C. (1969a) The detection of parasitised hosts by males of *Nasonia vitripennis* (Hym., Pteromalidae) and some possible implications. *Proceedings of The Royal Entomological Society of London A*, **44**, 85–90.

King, P.E., Ratcliffe, N.A. and Copland, M.J.W. (1969b) The structure of the egg membranes in *Apanteles glomeratus* (L.) (Hymenoptera: Braconidae). *Proceedings of the Royal Entomological Society of London A*, **44**, 137–42.

King, P.E., Ratcliffe, N.A. and Fordy, M.R. (1971) Oögenesis in a Braconid, *Apanteles glomeratus* (L.) possessing an hydropic type of egg. *Zeitschrift für Zellforschung und Mikroskopische Anatomie*, **119**, 43–57.

Kinoshita, M., Kasuya, E. and Yahara, Y. (2002) Effects of time-dependent competition for oviposition sites on clutch sizes and offspring sex ratios in a fig wasp. *Oikos*, **96**, 31–35.

Kirby, R.D. and Ehler, L.E. (1977) Survival of *Hippodamia convergens* in grain sorghum. *Environmental Entomology*, **6**, 777–80.

Kiritani, K., Kawahara, S., Sasaba, T. and Nakasuji, F. (1972) Quantitative evaluation of predation by spiders on the green rice leafhopper, *Nephotettix cincticeps* Uhler, by a sight-count method. *Researches on Population Ecology*, **13**, 187–200.

Kirk, W.D.J. (1984) Ecologically selective coloured traps. *Ecological Entomology*, **9**, 35–41.

Kishimoto, H. and Takagi, K. (2001) Evaluation of predation on *Panonychus citri* (McGregor) (Acari: Tetranychidae) from feeding traces on eggs. *Applied Entomology and Zoology*, **36**, 91–5.

Kiss, B. and Samu, F. (2000) Evaluation of population densities of the common wolf spider *Pardosa agrestis* (Araneae: Lycosidae) in Hungarian alfalfa fields using mark-recapture. *European Journal of Entomology*, **97**, 191–5.

Kitano, H. (1969) Experimental studies on the parasitism of *Apanteles glomeratus* L. with special reference to its encapsulation-inhibiting capacity. *Bulletin Tokyo Gakugei University*, **21**, 95–136.

Kitano, H. (1975) Studies on the courtship behavior of *Apanteles glomeratus*, 2. Role of the male wings during courtship and the release of mounting and copulatory behavior in the males. *Kontyû*, **43**, 513–21. (In Japanese with English summary).

Kitano, H. (1976) Studies on the courtship behaviour of *Apanteles glomeratus* L. 3. On the behaviour of males and females after their emergence from cocoons. *Physiological Ecology Japan* **17**, 383–93.

Kitano, H. (1986) The role of *Apanteles glomeratus* venom on the defensive response of its host, *Pieris rapae crucivora*. *Journal of Insect Physiology*, **32**, 369–75.

Kitching, R.C. (1977) Time resources and population dynamics in insects. *Australian Journal of Ecology*, **2**, 31–42.

Kitching, R.L., Bergelson, J.M., Lowman, M.D., McIntyre, S. and Carruthers, G. (1993) The biodiversity of arthropods from Australian rainforest canopies: General introduction, methods, sites and ordinal results. *Australian Journal of Ecology*, **18**, 181–91.

Kittayapong, P., Jamnongluk, W., Thipaksorn, A., Milne, J.R. and Sindhusake, C. (2003) *Wolbachia* infection complexity among insects in the tropical rice-field community. *Molecular Ecology*, **12**, 1049–60.

Klazenga, N. and De Vries, H.H. (1994) Walking distances of five differently sized ground beetle species. *Proceedings of the Section Experimental and Applied Entomology of the Netherlands Entomological Society*, **5**, 99–100.

Kleinhenz, A. and Büchs, W. (1993) Einfluss verschiedener landwirtschaftlicher Produktionsintensitäten auf die Spinnenfauna in der Kultur in Zuckerrübe. *Verhandlungen der Gesellschaft für Ökologie*, **22**, 81–88.

Klomp, H. and Teerink, B.J. (1962) Host selection and number of eggs per oviposition in the egg parasite. *Trichogramma embryophagum* Htg.. *Nature*, **195**, 1020–21.

Klomp, H., Teerink, B.J. and Ma, W.C. (1980) Discrimination between parasitized and unparasitized hosts in the egg parasite *Trichogramma embryophagum* (Hym: Trichogrammatidae): a matter of learning and forgetting. *Journal of Animal Ecology*, **30**, 254–77.

Knuth, P. (1906) *Handbook of Flower Pollination*, Vol I. Clarendon Press, Oxford.

Knuth, P. (1908) *Handbook of Flower Pollination*, Vol II. Clarendon Press, Oxford.

Knuth, P. (1909) *Handbook of Flower Pollination*, Vol III. Clarendon Press, Oxford.

Kobayashi, M. and Ishikawa, H. (1993) Breakdown of indirect flight muscles of alate aphids (*Acyrthosiphon pisum*) in relation to their flight, feeding and reproductive behaviour. *Journal of Insect Physiology*, **39**, 549–54.

Kogan, M. and Herzog, D.C. (eds) (1980) *Sampling Methods in Soybean Entomology*. Springer-Verlag, New York.

Kogan, M. and Legner, E.F. (1970) A biosystematic revision of the genus *Muscidifurax* (Hymn. Pteromalidae) with descriptions of four new species. *Canadian Entomologist*, **102**, 1268–90.

Kogan, M. and Parra, J.R.P. (1981) Techniques and applications of measurements of consumption and utilization of food by phytophagous insects, in *Current Topics in Insect Endocrinology and Nutrition* (eds G. Bhaskaran, S. Friedman and J.G. Rodriguez). Plenum Press, New York, pp. 337–52.

Kogan, M. and Pitre, H.N. (1980) General sampling methods for above-ground populations of soybean arthropods, in *Sampling Methods in Soybean Entomology* (eds M. Kogan and D.C. Herzog). Springer-Verlag, New York, pp. 30–60.

Kokuba, H. and Duelli, P. (1980) Aerial population movement and vertical distribution of aphidophagous insects in cornfields (Chrysopidae, Coccinellidae and Syrphidae), in *Ecology of Aphidophaga* (ed. I. Hodek). Academia, Prague and W. Junk, Dordrecht, pp. 279–84.

Kondo, N., Shimada, M. and Fukatsu, T. (1999) High prevalence of *Wolbachia* in the azuki bean beetle *Callosubruchus chinensis* (Coleoptera, Bruchidae). *Zoological Science*, **16**, 955–62.

Kondo, N., Nikoh, N., Ijichi, N., Shimada, M. and Fukatsu, T. (2002) Genome fragment of *Wolbachia* endosymbiont transferred to X chromosome of host insect. *Proceedings of the National Academy of Sciences*, **99**, 14280–85.

Koomen, P. (1998) Winter activity of *Anyphaena accentuata* (Walckenaer, 1802) (Araneae: Anyphaenidae). *Proceedings of the 17th European Colloquium of Arachnology*. British Arachnological Society, Buckinghamshire, UK, pp. 222–5.

Kopelman, A.H. and Chabora, P.C. (1992) Resource availability and life-history parameters of *Leptopilina boulardi* (Hymenoptera: Eucoilidae). *Annals of the Entomological Society of America*, **85**, 195–9.

Korol, A.B., Ronin, Y.I. and Kirzher, V.M. (1995) Interval mapping of quantitative trait loci employing correlated trait complexes. *Genetics*, **140**, 1137–47.

Kosambi, D.D. (1944) The estimation of map distances from recombination values. *Annals of Eugenetics*, **12**, 172–5.

Kose, H. and Karr, T.L. (1995) Organization of *Wolbachia pipientis* in the *Drosophila* fertilized egg and embryo revealed by an anti-*Wolbachia* monoclonal antibody. *Mechanisms of Development*, **51**, 275–88.

Kouame, K.L. and Mackauer, M. (1991) Influence of aphid size, age and behaviour on host choice by the parasitoid wasp *Ephedrus californicus:* a test of host-size models. *Oecologia*, **88**, 197–203.

Kowalski, R. (1976) Obtaining valid population indices from pitfall trapping data. *Bulletin de l'Académie Polonaise des Sciences*, **23**, 799–803.

Kowalski, R. (1977) Further elaboration of the winter moth population models. *Journal of Animal Ecology*, **46**, 471–82.

Kozanec, M. and Belcari, A. (1977) Structure of the ovipositor, associated sensilla and spermathecal system of entomoparasitic pipunculid flies (Diptera: Pipunculidae). *Journal of Natural History*, **31**, 1273–88.

Kraaijeveld, A.R. (1999) Kleptoparasitism as an explanation for paradoxical oviposition decisions of the parasitoid *Asobara tabida*. *Journal of Evolutionary Biology*, **12**, 129–133.

Kraaijeveld, A.R. and van Alphen, J.J.M. (1994) Geographical variation in resistance of the parasitoid *Asobara tabida* against encapsulation by *Drosophila melanogaster* larvae: the mechanism explored. *Physiological Entomology*, **19**, 9–14.

Kraaijeveld, A.R. and van Alphen, J.J.M. (1995a) Geographical variation in encapsulation ability of *Drosophila melanogaster* larvae and evidence for parasitoid-specific components. *Evolutionary Ecology*, **9**, 10–17.

Kraaijeveld, A.R. and van Alphen, J.J.M. (1995b) Variation in diapause and sex ratio in the parasitoid *Asobara tabida*. *Entomologia Experimentalis et Applicata*, **74**, 259–65.

Kraaijeveld, A.R. and Godfray, H.C.J. (1997) Trade-off between parasitoid resistance and larval competitive ability in *Drosophila melanogaster*. *Nature*, **389**, 278–80.

Kraaijeveld, A.R., van Alphen, J.J.M. and Godfray, H.C.J. (1998) The coevolution of host resistance and parasitoid virulence. *Parasitology*, **116**, 29–45.

Kraaijeveld, A.R., Adriaanse, I.C.T., and van den Bergh, B. (1999) Parasitoid size as a function of host sex: potential for different sex allocation strategies. *Entomologia Experimentalis et Applicata*, **92**, 289–94.

Kraaijeveld, A.R., Hutcheson, K.A., Limentani, E.C. and Godfray, H.C.J. (2001) Costs of counterdefences to host resistance in a parasitoid of *Drosophila*. *Evolution*, **55**, 1815–21.

Krafsur, E.S., Kring, T.J., Miller, J.C., Nariboli, P., Obrycki, J.J., Ruberson, J.R. and Schaefer, P.W. (1997) Gene flow in the exotic colonizing ladybeetle *Harmonia axyridis* in North America. *Biological Control*, **8**, 207–14.

Kramer, D.A., Stinner, R.E. and Hain, F.P. (1991) Time versus rate in parameter estimation of nonlinear temperature-dependent development models. *Environmental Entomology*, **20**, 484–8.

Krebs, C.J. (1999) *Ecological Methodology, 2nd edition*, Benjamin Cummins, Harlow.

Krebs, J.R. and Davies, N.B. (eds) (1978) *Behavioural Ecology*, Blackwell, Oxford.

Krebs, J.R. and Davies, N.B. (1993) *An Introduction to Behavioural Ecology*. Blackwell, Oxford.

Kring, T.J., Gilstrap, F.E. and Michels, G.J. (1985) Role of indigenous coccinellids in regulating greenbugs (Homoptera: Aphididae) on texas grain sorghum. *Journal of Economic Entomology*, **78**, 269–73.

Krishnamoorthy, A. (1984) Influence of adult diet on the fecundity and survival of the predator, *Chrysopa scelestes* (Neur.: Chrysopidae). *Entomophaga*, **29**, 445–50.

Krishnamoorthy, A. (1989) Effect of cold storage on the emergence and survival of the adult exotic parasitoid, *Leptomastix dactylopii* How. (Hym., Encyrtidae). *Entomon*, **14**, 313–18.

Kristoffersen, A.B., Lingjaerde, O.C., Stenseth, N.C. and Shimada, M. (2001) Non-parametric modelling of non-linear density dependence: a three-species host-parasitoid system. *Journal of Animal Ecology*, **70**, 808–19.

Krombein, K.V. (1967) *Trap-nesting Wasps and Bees: Life-histories, Nests and Associates*. Smithsonian Press, Washington, DC.

Kromp, B. and Nitzlader, M. (1995) Dispersal of ground beetles in a rye field in Vienna, Eastern Austria. *Acta Jutlandica*, **70**, 269–77.

Kromp, B., Pflügl, C., Hradetzky, R. and Idinger, J. (1995) Estimating beneficial arthropod densities using emergence traps, pitfall traps and the flooding method in organic fields (Vienna, Austria). *Acta Jutlandica*, **70**, 87–100.

Kruse, K.C. (1983) Optimal foraging by predaceous diving beetle larvae on toad tadpoles. *Oecologia*, **581**, 383–8.

Kullenberg, B. and Bergström, G. (1976) Hymenoptera Aculeata males as pollinators of *Ophrys* orchids. *Zoologica Scripta*, **5**, 13–23.

Kummer, H. (1960) Experimentelle Untersuchungen zur Wirkung von Fortpflanzungsfaktoren auf die Lebensdauer von *Drosophila melanogaster* Weibchen. *Zeitschrift für Vergleichende Physiologie*, **43**, 642–79.

Kunkel, J.G. and Nordin, J.H. (1985) Yolk proteins, in *Comparative Insect Physiology, Biochemistry and Pharmacology* (eds G.A. Kerkut and L.J. Gilbert). Pergamon Press, Oxford, pp. 83–111.

Kuno, E. (1971) Sampling and analysis of insect populations. *Annual Review of Entomology*, **36**, 285–304.

Kuno, E. and Dyck, V.A. (1985) Dynamics of Philippine and Japanese populations of the brown planthopper: comparison of basic characteristics, in *Proceedings of the ROC-Japan Seminar on the Ecology and the Control of the Brown Planthopper*, Republic of China, National Science Council, pp. 1–9.

Kuperstein, M.L. (1974) Utilisation of the precipitin test for the quantitative estimation of the influence of *Pterostichus crenuliger* (Coleoptera: Carabidae) upon the population dynamics of *Eurygaster integriceps* (Hemiptera: Scutelleridae). *Zoologicheski Zhurnal*, **53**, 557–62.

Kuperstein, M.L. (1979) Estimating carabid effectiveness in reducing the sunn pest, *Eurygaster integriceps* Puton (Heteroptera: Scutelleridae) in the USSR. Serology in insect predator-prey studies. *Entomological Society of America, Miscellaneous Publications*, **11**, 80–4.

Kuschel, G. (1991) A pitfall trap for hypogean fauna. *Curculio*, **31**, 5.

Kuschka, V., Lehmann, G. and Meyer, U. (1987) Zur Arbeit mit Bodenfallen. *Beitrage zur Entomologie*, **37**, 3–27.

Kutsch, W. (1999) Telemetry in insects: the "intact animal approach". *Theory in Biosciences*, **118**, 29–53.

Kvarnemo, C. and Ahnestjö, I. (2002) Operational sex ratios and mating competition. in *Sex Ratios: Concepts and Research Methods* (ed. I.C.W. Hardy). Cambridge University Press, Cambridge, pp. 366–82.

Labine, P.A. (1966) The population biology of the butterfly, *Euphydryas editha*. IV. Sperm precedence - a preliminary report. *Evolution*, **20**, 580–6.

Lack, D. (1947) The significance of clutch size. *Ibis*, **89**, 309–52.

Laing, D.R. and Caltagirone, L.E. (1969) Biology of *Habrobracon lineatellae* (Hymenoptera: Braconidae). *Canadian Entomologist*, **101**, 135–42.

Lajtha, K. and Michener, R.H. (1994) *Stable Isotopes in Ecology and Environmental Science*. Blackwell Scientific Publications, Oxford, UK.

Lancaster, J., Hildrew, A.C. and Townsend, C.R. (1991) Invertebrate predation on patchy and mobile prey in streams. *Journal of Animal Ecology*, **601**, 625–41.

Lander, E., Green, P., Abrahamson, J., Barlow, A., Daly, M., Lincoln, S. and Newbury, L. (1987) MAPMAKER: an interactive computer package for constructing primary genetic linkage maps of experimental and natural populations. *Genomics*, **1**, 174–81.

Lander, E.S. and Botstein, D. (1989) Mapping Mendelian factors underlying quantitative traits using RFLP linkage maps. *Genetics*, **121**, 185–99.

Landis, D.A. and Van der Werf, W. (1997) Early-season predation impacts the establishment of aphids and spread of beet yellows virus in sugar beet. *Entomophaga*, **42**, 499–516.

Landis, D.A., Wratten, S.D. and Gurr, G.M. (2000) Habitat management to conserve natural enemies of arthropod pests in agriculture. *Annual Review Entomology*, **45**, 175–201.

Landolt, P.J. (1998) Chemical attractants for trapping yellowjackets *Vespula germanica* and *Vespula pensylvanica* (Hymenoptera: Vespidae). *Environmental Entomology*, **27**, 1229–34.

Lane, S.D., Mills, N.J. and Getz, W.M. (1999) The effects of parasitoid fecundity and host taxon on the biological control of insect pests: the relationship between theory and data. *Ecological Entomology*, **24**, 181–90.

Lang, A. (2000) The pitfalls of pitfalls: a comparison of pitfall trap catches and absolute density estimates of epigeal invertebrate predators in arable land. *Journal of Pest Science*, **73**, 99–106.

Lang, A. (2003) Intraguild interference and biocontrol effects of generalist predators in a winter wheat field. *Oecologia*, **134**, 144–53.

Lange, K. and Boehnke, M. (1982) How many polymorphic genes will it take to span the human genome? *American Journal of Human Genetics*, **34**, 842–45.

Lanza, J. (1991) Response of fire ants (Formicidae: *Solenopsis invicta* and *S. geminata*) to artificial nectars with amino acids. *Ecological Entomology*, **16**, 203–10.

Lapchin, L., Ferran, A., Iperti, G., Rabasse, J.M. and Lyon, J.P. (1987) Coccinellids (Coleoptera: Coccinellidae) and syrphids (Diptera: Syrphidae) as predators of aphids in cereal crops: a comparison of sampling methods. *Canadian Entomologist*, **119**, 815–22.

Lapchin, L., Boll, R., Rochat, J., Geria, A.M. and Franco, E. (1997) Projection pursuit nonparametric regression used for predicting insect densities from visual abundance classes. *Environmental Entomology*, **26**, 736–44.

Larget, B. and Simon, D.L. (1999) Markov chain Monte Carlo algorithms for Bayesian analysis of phylogenetic trees. *Molecular Biology Evolution*, **16**, 750–9.

Larsson, F.K. and Kustvall, V. (1990). Temperature reverses size-dependent mating success of a cerambycid beetle. *Functional Ecology*, **4**, 85–90.

Larsson, S., Bjorkman, C. and Kidd, N.A.C. (1993) Outbreaks in Diprionid Sawflies: Why some species and not others? in *Sawfly Life History Adaptations to Woody Plants*. (eds M. Wagner and K.F. Raffa). Academic Press, San Diego, pp. 453–83.

LaSalle, J. and Gauld, I.D. (1994) Hymenoptera: their diversity, and their impact on the diversity of other organisms, in *Hymenoptera and Biodiversity* (eds J. LaSalle and I.D. Gauld). C.A.B. International, Wallingford, pp. 1–26.

Laska, M.S. and Wootton, J.T. (1998) Theoretical concepts and empirical approaches to measuring interaction strength. *Ecology*, **79**, 461–76.

Laughlin, R. (1965) Capacity for increase: a useful population statistic. *Journal of Animal Ecology*, **34**, 77–91.

Laurent, M. and Lamarque, P. (1974) Utilisation de la methode des captures successives (De Lury) pour L'evaluation des peuplements piscicoles. *Annals of Hydrobiology*, **5**, 121–32.

Laurent, V., Wajnberg, E., Mangin, B., Schiex, T., Gaspin, C. and Vanlerberghe-Masutti, F. (1998) A composite genetic map of the parasitoid wasp *Trichogramma brassicae* based on RAPD markers. *Genetics*, **150**, 275–82.

Lauzière, I., Pérez–Lachaud, G. and Brodeur, J. (2000) Effect of female body size and adult feeding on the fecundity and longevity of the parasitoid *Cephalonomia stephanoderis* Betrem (Hymenoptera: Bethylidae). *Annals of the Entomological Society of America*, **93**, 103–9.

Lavigne, R.J. (1992) Ethology of *Neoaratus abludo* Daniels (Diptera: Asilidae) in South Australia, with notes on *N. pelago* (Walker) and *N. rufiventris* (Macquart). *Proceedings of the Entomological Society of Washington*, **94**, 253–62.

Lavine, M.D. and Beckage, N.E. (1996) Temporal pattern of parasitism-induced immunosuppression in *Manduca sexta* larvae by parasitized *Cotesia congregata*. *Journal of Insect Physiology*, **42**, 41–51.

Lawrence, E.S. and Allen, J.A. (1983) On the term 'search image'. *Oikos*, **40**, 313–14.

Lawrence, P.O. (1990) The biochemical and physiological effects of insect hosts on the development and ecology of their insect parasites: An overview. *Archives of Insect Biochemistry and Physiology* **13**: 217–28.

Lawton, J.H. (1970) Feeding and food energy assimilation in larvae of the damselfly *Pyrrhosoma nymphula* (Sulz.) (Odonata: Zygoptera). *Journal of Animal Ecology*, **39**, 669–89.

Lawton, J.H. (1986) The effect of parasitoids on phytophagous insect communities, in *Insect Parasitoids* (eds J. Waage and D. Greathead). Academic Press, London, pp. 265–87.

Lawton, J.H., Beddington, J.R. and Bonser, R. (1974) Switching in invertebrate predators, in *Ecological Stability* (eds M.B. Usher and M.H. Williamson). Chapman and Hall, London, pp. 141–58.

Lawton, J.H., Hassell, M.P. and Beddington, J.R. (1975) Prey death rates and rates of increase of arthropod predator populations. *Nature*, **255**, 60–62.

Lawton, J.H., Thompson, B.A. and Thompson, D.J. (1980) The effects of prey density on survival and growth of damsel fly larvae. *Ecological Entomology*, **5**, 39–51.

Leatemia, J.A, Laing, J.E. and Corrigan, J.E. (1995) Production of exclusively male progeny by mated, honey-fed *Trichogramma minutum* Riley (Hym., Trichogrammatidae). *Journal of Applied Entomology*, **119**, 561–6.

Leather, S.R. (1988) Size, reproductive potential and fecundity in insects: things aren't as simple as they seem. *Oikos*, **51**, 386–9.

Leather, S.R., Walters, K.F.A. and Bale, J.S. (1993) *The Ecology of Insect Overwintering*. Cambridge University Press, Cambridge.

Ledieu, M.S. (1977) Ecological aspects of parasite use under glass, in *Pest Management in Protected Culture Crops* (eds F.F. Smith and R.E. Webb), Proceedings of the 15th International Congress of Entomology, Washington, 1976, USDA, pp. 75–80.

Lee, J.C. and Heimpel, G.E. (2003). Sugar feeding by parasitoids in cabbage fields and the consequences for pest control. *Proceedings of the 1st International Symposium on Biological Control of Arthropods. Honolulu, Hawaii, 2002*. In press.

Lee, J.H., Johnson, S.J. and Wright, V.L. (1990). Quantitative survivorship analysis of the velvetbean caterpillar (Lepidoptera: Noctuidae) pupae in soybean fields in Louisiana. *Environmental Entomology*, **19**, 978–86.

Lee, K.Y., Barr, R.O., Cage, S.H. and Kharkar, A.N. (1976) Formulation of a mathematical model for insect pest ecosystems – the cereal leaf beetle problem. *Journal of Theoretical Biology*, **59**, 33–76.

Leeper J.R. (1974) Adult feeding behavior of *Lixophaga spenophori*, a tachinid parasite of the New Guinea sugarcane weevil. *Proceedings of the Hawaiian Entomological Society*, **21**, 403–12.

Leereveld, H. (1982) Anthecological relations between reputedly anemophilous flowers and syrphid flies III. Worldwide survey of crop and intestine contents of certain anthophilous syrphid flies. *Tijdschrift voor Entomologie*, **125**, 25–35.

Lees, A.D. (1955) *The Physiology of Diapause in Arthropods*. Cambridge University Press, Cambridge.

Legaspi Jr. B.A.C. Shepard, B.M. and Almazan, L.P. (1987) Oviposition behaviour and development of *Goniozus triangulifer* (Hymenoptera: Bethylidae). *Environmental Entomology*, **16**, 1284–86.

Legaspi, J.C., O'Neill, R.J. and Legaspi, B.C. (1996) Trade-offs in body weights, eggs loads and fat reserves of field-collected *Podisus maculiventris* (Heteroptera: Pentatomidae). *Environmental Entomology*, **25**, 155–64.

Legner, E.F. (1968) Adult emergence interval and reproduction in parasitic Hymenoptera influenced by host size and density. *Annals of the Entomological Society of America*, **62**, 220–6.

Legner, E.F. (1969) Reproductive isolation and size variations in the *Muscidifurax raptor* complex. *Annals of the Entomological Society of America*, **62**, 382–5.

Legner, E.F. (1985) Effects of scheduled high temperature on male production in thelytokous *Muscidifurax uniraptor* (Hymenoptera: Pteromalidae). *Canadian Entomology*, **117**, 383–9.

Legner, E.F. (1987) Transfer of thelytoky to arrhenotokous *Muscidifurax raptor* Girault and Sanders (Hymenoptera: Pteromalidae). *Canadian Entomology*, **119**, 265–71.

Lei, G. and Hanski, I. (1997) Metapopulation structure of *Cotesia melitaearum*, a specialist parasitoid of the butterfly, *Melitaea cinxia*. *Oikos*, **78**, 91–100.

Lei, G. and Hanski, I. (1998) Spatial dynamics of two competing specialist parasitoids in a host metapopulation. *Journal of Animal Ecology*, **67**, 422–33.

Lei, G.C. and Camara, M.D. (1999) Behaviour of the generalist parasitoid, *Cotesia melitaearum*: from individual behaviour to metapopulation processes. *Ecological Entomology*, **24**, 59–72.

Leius, K. (1960) Attractiveness of different foods and flowers to the adults of some hymenopterous parasites. *Canadian Entomologist*, **92**, 369–76.

Leius, K. (1962) Effects of the body fluids of various host larvae on fecundity of females of *Scambus buolianae* (Htg.) (Hymenoptera: Ichneumonidae). *Canadian Entomologist*, **94**, 1078–82.

Leius, K. (1963) Effects of pollens on fecundity and longevity of adult *Scambus buolianae* (Htg.) (Hymenoptera: Ichneumonidae). *Canadian Entomologist*, **95**, 202–7.

Le Lannic, J. and Nenon, J.P. (1999) Functional morphology of the ovipositor in *Megarhyssa atrata* (Hymenoptera, Ichneumonidae) and its penetration into wood. *Zoomorphology*, **119**, 73–79.

Lemieux, J.P. and Lindgren, B.S. (1999) A pitfall trap for large-scale trapping of Carabidae: comparison against conventional design, using two different preservatives. *Pedobiologia*, **43**, 245–53.

van Lenteren, J.C. (1976) The development of host discrimination and the prevention of superparasitism in the parasite *Pseudeucoila bochei* Weld (Hym.: Cynipidae). *Netherlands Journal of Zoology*, **26**, 1–83.

van Lenteren, J.C. (1980) Evaluation of control capabilities of natural enemies: does art have to become science? *Netherlands Journal of Zoology*, **30**, 369–81.

van Lenteren, J.C. (1986) Parasitoids in the greenhouse: successes with seasonal inoculative release systems, in *Insect Parasitoids* (eds J. Waage and D. Greathead). Academic Press, London, pp. 341–74.

van Lenteren, J.C. (1991) Encounters with parasitised hosts: to leave or not to leave the patch. *Netherlands Journal of Zoology*, **41**, 144–57.

van Lenteren, J.C. and Bakker, K. (1978) Behavioural aspects of the functional responses of a parasite (*Pseudeucoila bochei* Weld) to its host (*Drosophila melanogaster*). *Netherlands Journal of Zoology*, **28**, 213–33.

van Lenteren, J.C. and van Roermund, H.J.W. (1999) Why is the parasitoid *Encarsia formosa* so successful in controlling whiteflies? *Theoretical Approaches to Biological Control* (eds B.A. Hawkins and H.V. Cornell). Cambridge University Press, Cambridge, pp. 116–30.

van Lenteren, J.C., Bakker, K. and van Alphen, J.J.M. (1978) How to analyse host discrimination. *Ecological Entomology*, **3**, 71–5.

van Lenteren. J.C., van Vianen, A., Gast, H.F. and Kortenhoff, A. (1987) The parasite-host relationship between *Encarsia formosa* Gahan (Hymenoptera: Aphelinidae) and *Trialeurodes vaporariorum* (Westwood) (Homoptera: Aleyrodidae), XVI. Food effects on oögenesis, life span and fecundity of *Encarsia formosa* and other hymenopterous parasites. *Zeitschrift für Angewandte Entomologie*, **103**, 69–84.

van Lenteren, J.C., Isidoro, N. and Bin, F. (1998) Functional anatomy of the ovipositor clip in the parasitoid *Leptopilina heterotoma* (Thompson) (Hymenoptera: Eucolidae), a structure to grip escaping host larvae. *International Journal of Insect Morphology and Embryology*, **27**, 263–8.

van Lenteren, J.C., Babendreier, D., Bigler, F., Burgio, G., Hokkanen, H.M.T., Kuske, S., Loomans, A.J.M., Menzler-Hokkanen, I., van Rijn, P.C.J., Thomas, M.B., Tommasini, M.G and Zeng, G.-G. (2003). Environmental risk assessment of exotic natural enemies used in inundative biological control. *BioControl*, **48**, 3–38.

Leonard, S.H. and Ringo, J.M. (1978) Analysis of male courtship patterns and mating behavior of *Brachymeria intermedia*. *Annals of the Entomological Society of America*, **71**, 817–26.

Leopold, R.A. (1976) The role of male accessory glands in insect reproduction. *Annual Review of Entomology*, **21**, 199–221.

Le Ralec, A. (1995) Egg contents in relation to host-feeding in some parasitic Hymenoptera. *Entomophaga*, **40**, 87–93.

Le Ralec, A., Rabasse, J.M. and Wajnberg, E. (1996) Comparative morphology of the ovipositor of some parasitic Hymenoptera in relation to characteristics of their hosts. *Canadian Entomologist*, **128**, 413–33.

Le Ru, B. and Makosso, J.P.M. (2001) Prey habitat location by the cassava mealybug predator *Exochomus flaviventris*: olfactory responses to odor of plant, mealybug, plant mealybug complex, and plant-mealybug-natural enemy complex. *Journal of Insect Behavior*, **14**, 557–72.

LeSar, C.D. and Unzicker, J.D. (1978) Soybean spiders: species composition, population densities, and vertical distribution. *Illinois Natural History Survey Biological Notes*, **107**, 1–14.

Lesiewicz, D.S., Lesiewicz, J.L., Bradley, J.R. and van Duyn, I.W. (1982) Serological determination of carabid (Coleoptera: Adephaga) predation on corn earworm (Lepidoptera, Noctuidae) in field corn. *Environmental Entomology*, **11**, 1183–6.

Leslie, G.W. and Boreham, P.F.L. (1981) Identification of arthropod predators of *Eldana saccharina* Walker (Lepidoptera: Pyralidae) by cross-over electrophoresis. *Journal of the Entomological Society of South Africa*, **44**, 381–8.

Leslie, P.H. (1945) On the use of matrices in certain population mathematics. *Biometrika*, **33**, 183–212.

Leslie, P.H. (1948) Some further notes on the use of matrices in population mathematics. *Biometrika*, **351**, 213–45.

Lessard, E.J., Martin, M.P. and Montagnes, D.J.S. (1996) A new method for live-staining protists with DAPI and its application as a tracer of ingestion by walleye pollock (*Theragra chalcogramma* (Pallas)) larvae. *Journal of Experimental Marine Biology and Ecology*, **204**, 43–57.

Lessells, C.M. (1985) Parasitoid foraging: should parasitism be density dependent? *Journal of Animal Ecology*, **54**, 27–41.

Letourneau, D.K. (1987) The enemies hypothesis: tritrophic interactions and vegetational diversity in tropical agroecosystems. *Ecology*, **68**, 1616–22.

Levesque, C. and Levesque, G.Y. (1996) Seasonal dynamics of rove beetles (Coleoptera: Staphylinidae) in a raspberry plantation and adjacent sites in eastern Canada. *Journal of the Kansas Entomological Society*, **69**, 285–301.

Lewis, O.T., Memmott, J., LaSalle, J., Lyal, C.H.C., Whitefoord, C. and Godfray, H.C.J. (2002) Structure of a diverse tropical forest insect-parasitoid community. *Journal of Animal Ecology*, **71**, 855–73.

Lewis, S.M. and Austad, S.N. (1994) Sexual selection in flour beetles: the relationship between sperm precedence and male olfactory attractiveness. *Behavioral Ecology*, **5**, 219–24.

Lewis, W.H., Vinay, P. and Zenger, V.E. (1983) *Airborne and Allergenic Pollen of North America*. Johns Hopkins Press, Baltimore.

Lewis, W.J. and Takasu, K. (1990) Use of learned odors by a parasitic wasp in accordance with host and food needs. *Nature*, **348**, 635–6.

Liddell, J.E. and Cryer, A. (1991) *A Practical Guide to Monoclonal Antibodies*. Wiley and Sons Ltd., Chichester.

Liddell, J.E. and Symondson, W.O.C. (1996). The potential of combinatorial gene libraries in pest-predator relationship studies, in *The Ecology of Agricultural Pests: Biochemical Approaches* (eds W.O.C. Symondson and J.E. Liddell), Systematics Association Special Volume No. 53. Chapman and Hall, London, pp 347–66.

Liddell, J.E. and Weeks, I. (1995) *Antibody Technology*. BIOS, Oxford.

Lighton, J.R.B. and Duncan, F.D. (2002) Energy cost of locomotion: validation of laboratory data by *in situ* respirometry. *Ecology*, **83**, 3517–22.

Lill, J.T. (1999) Structure and dynamics of a parasitoid community attacking larvae of *Psilocorsis quercicella* (Lepidoptera: Oecophoridae). *Environmental Entomology*, **28**, 1114–23.

Lim, U.T. and Lee, J.H. (1999) Enzyme-linked immunosorbent assay used to analyze predation of *Nilaparvata lugens* (Homoptera: Delphacidae) by *Pirata subpiraticus* (Araneae: Lycosidae). *Environmental Entomology*, 28, 1177–82.

Limburg, D.D. and Rosenheim, J.A. (2001) Extrafloral nectar consumption and its influence on survival and development of an omnivorous predator, larval *Chrysoperla plorabunda* (Neuroptera: Chrysopidae). *Environmental Entomology*, **30**, 595–604.

Lincoln, F.C. (1930) Calculating waterfowl abundance on the basis of banding returns. *USDA Circular*, **118**, 1–4.

Lindsley, D.L. and Zimm, G.G. (1992) *The Genome of Drosophila melanogaster*. Academic Press, San Diego.

Linforth, R.S.T. and Taylor, A.J. (1998) Apparatus and methods for the analysis of trace constituents in gases. *European Patent Application EP 0819 937 A2*.

Linforth, R.S.T., Ingham, K.E. and Taylor, A.J. (1996) Time course profiling of volatile release from foods during the eating process, in *Flavour Science: Recent Developments* (eds D.S. Mottram and A.J. Taylor). Royal Society of Chemistry, pp. 361–8.

Lingren, P.D., Ridgway, R.L. and Jones, S.L. (1968) Consumption by several common arthropod predators of eggs and larvae of two *Heliothis* species that attack cotton. *Annals of the Entomological Society of America*, **61**, 613–18.

Liske, E. and Davis, W.J. (1984) Sexual behaviour in the Chinese praying mantis. *Animal Behaviour*, **32**, 916–18.

Liske, E. and Davis, W.J. (1987) Courtship and mating behaviour of the Chinese praying mantis, *Tenodera aridifolia sinensis*. *Animal Behaviour*, **35**, 1524–37.

Lister, A., Usher, M.B. and Block, W. (1987) Description and quantification of field attack rates by predatory mites: an example using electrophoresis method with a species of Antarctic mite. *Oecologia*, **72**, 185–91.

Liu, H. and Beckenbach, A.T. (1992). Evolution of the mitochondrial cytochrome oxidase II gene among 10 insect orders. *Molecular Phylogenetics and Evolution*, **1**, 41–52.

Liu, S.-S. (1985a) Development, adult size and fecundity of *Aphidius sonchi* reared in two instars of its aphid host, *Hyperomyzus lactucae*. *Entomologia Experimentalis et Applicata*, **37**, 41–8.

Liu, S.-S. (1985b) Aspects of the numerical and functional responses of the aphid parasite, *Aphidius sonchi*, in the laboratory. *Entomologia Experimentalis et Applicata*, **37**, 247–56.

Liu, Y.H. and Tsai, J.H. (2002) Effect of temperature on development, survivorship, and fecundity of *Lysiphlebia mirzai* (Hymenoptera: Aphidiidae), a parasitoid of *Toxoptera citricida* (Homoptera: Aphididae). *Environmental Entomology*, **31**, 418–24.

Livdahl, T.P. and Sugihara, G. (1984) Non-linear interactions of populations and the importance of estimating *per capita* rates of change. *Journal of Animal Ecology*, **53**, 573–80.

Lloyd, J.E. (1966) Studies on the flash communication system in *Photinus* fireflies. *University of Michigan*

Museum of Zoology Miscellaneous Publications, **130**, 1–95.

Lloyd, J.E. (1971) Bioluminiscent communication in insects. *Annual Review of Entomology*, **16**, 97–122.

Lloyd, J.E. (1975) Aggressive mimicry in *Photuris*: signal repertoires by femmes fatales. *Science*, **187**, 452–3.

Löbner, U. and Fuchs, E. (1994) Tissue print-immunoblotting, a rapid technique for detecting proteins of aphids in the digestive tract of predatory arthropods. *Archives of Phytopathology and Plant Protection*, **29**, 179–84.

Löbner, U. and Hartwig, O. (1994) Soldier beetles (Col., Cantharidae) and nabid bugs (Het., Nabidae) – occurrence and importance as aphidophagous predators in winter wheat fields in the surroundings of Halle/Saale (Sachsen-Anhalt). *Bulletin IOBC/WPRS Working Group Integrated Control of Cereal Pests*, **17**, 179–87.

Lobry de Bruyn, L.A. (1999) Ants as bioindicators of soil function in rural environments. *Agriculture, Ecosystems and Environment*, **74**, 425–41.

Loch, A.D. and Walter, G.H. (1999) Does the mating system of *Trissolcus basalis* (Wollaston) (Hymenoptera: Scelionidae) allow outbreeding? *Journal of Hymenoptera Research*, **8**, 238–50.

Loch, A.D. and Walter, G.H. (2002) Mating behaviour of *Trissolcus basalis* (Wollaston) (Hymenoptera: Scelionidae): potential for outbreeding in a predominantly inbreeding species *Journal of Insect Behavior*, **15**, 13–23.

Logan, J.A. and Allen, J.C. (1992) Nonlinear dynamics and chaos in insect populations. *Annual Review of Entomology*, **37**, 455–77.

Logan, J.A., Wollkind, D.J., Hoyte, S.C. and Tanigoshi L.K. (1976) An analytical model for description of temperature dependent rate phenomena in arthropods. *Environmental Entomology*, **5**, 1133–40.

Lohr, B., Varela, A.M. and Santos, B. (1989) Life-table studies on *Epidinocarsis lopezi* (DeSantis) (Hym., Encyrtidae), a parasitoid of the cassava mealybug, *Phenacoccus manihoti* Mat.-Ferr. (Hom., Pseudococcidae). *Journal of Applied Entomology*, **107**, 425–34.

Lohse, G.A. (1981) Bodenfallenfänge im Naturpark Wilseder Berg mit einer kritischen Beurteilung ihrer Aussagekraft. *Jahrbücher der naturwissenschaftlichen Vereinigung Wuppertal*, **34**, 43–7.

Long, A.D., Mullaney, S.L., Reid, L.A., Fry, J.D., Langley, C.H. and Mackay, T.F.C. (1995) High resolution mapping of genetic factors affecting abdominal bristle number in *Drosophila melanogaster*. *Genetics*, **139**, 1273–91.

Longino, J.T. and Colwell, R.K. (1997) Biodiversity assessment using structured inventory: capturing the ant fauna of a tropical rain forest. *Ecological Applications*, **7**, 1263–77.

de Loof, A. (1987) The impact of the discovery of vertebrate-type steroids and peptide hormone-like substances in insects. *Entomologia Experimentalis et Applicata*, **45**, 105–13.

Lopez, R., Ferro, D.N. and Van Driesche, R.C. (1995) Two tachinid species discriminate between parasitised and non-parasitised hosts. *Entomologia Experimentalis et Applicata*, **74**, 37–45.

Lord, W.D., DiZinno, J.A., Wilson, M.R., Budowle, B., Taplin, D. and Meinking, T.L. (1998) Isolation, amplification and sequencing of human mitochondrial DNA obtained from human crab louse, *Pthirus pubis* (L.) blood meals. *Journal of Forensic Sciences*, **43**, 1097–100.

Loreau, M. (1984a) Population density and biomass of Carabidae (Coleoptera) in a forest community. *Pedobiologia*, **27**, 269–78.

Loreau, M. (1984b) Étude experimentale de l'alimentation de *Abax ater* Villers, *Carabus problematicus* Herbst et *Cychrus attenuatus* Fabricius (Coleoptera: Carabidae). *Annales de la Société Royale Zoologique de Belgique*, **114**, 227–40.

Loreau, M. (1987) Vertical distribution of activity of carabid beetles in a beech forest floor. *Pedobiologia*, **30**, 173–8.

Loreau, M. and Nolf, C.L. (1993) Occupation of space by the carabid beetle *Abax ater*. *Acta Oecologica*, **14**, 247–58.

Losey, J.E. and Denno, R.F. (1998a) The escape response of pea-aphids to foliar foraging predators: factors affecting dropping behaviour. *Ecological Entomology*, **23**, 53–61.

Losey, J.E. and Denno, R.F. (1998b) Interspecific variation in the escape responses of aphids: effect on risk of predation from foliar-foraging and ground-foraging predators. *Oecologia*, **115**, 245–52.

Losey, J.E. and Denno, R.F. (1998c) Positive predator-predator interactions: enhanced predation rates and synergistic suppression of aphid populations. *Ecology*, **79**, 2143–52.

Lotka, A.J. (1922) The stability of the normal age distribution. *Proceedings of the National Academy of Sciences*, **8**, 339–45.

Lotka, A.J. (1925) *Elements of Physical Biology*. Williams and Wilkins, Baltimore.

Louda, S.M., Kendall, D., Connor, J. and Simberloff, D. (1997) Ecological effects of an insect introduced for the biological control of weeds. *Science*, **277**, 1088–90.

Louda, S.M., Pemberton, R.W., Johnson, M.T. and Follett, P.A. (2003) Nontarget effects – the Achilles' heel of biological control? Retrospective analyses to reduce risk associated with biocontrol introductions. *Annual Review of Entomology*, **48**, 365–96.

Loughton, B.G., Derry, C. and West, A.S. (1963) Spiders and the spruce budworm. *Memoirs of the Entomological Society of Canada*, **31**, 249–68.

Lovallo, N., McPheron, B.A. and Cox-Foster, D.L (2002) Effects of the polydnavirus of *Cotesia congregata* on the immune system and the development of non-habitual hosts of the parasitoid. *Journal of Insect Physiology*, **48**, 517–26.

Lövei, G.L. (1986) The use of biochemical methods in the study of carabid feeding: the potential of isozyme analysis and ELISA, in *Feeding Behaviour and Accessibility of Food for Carabid Beetles* (eds P.J. Den Boer, P.J. Grum and J. Szyszko). Warsaw Agricultural University Press, Warsaw, pp. 21–7.

Lövei, G.L. and Szentkirályi, F. (1984) Carabids climbing maize plants. *Zeitschrift für Angewandte Entomologie*, **97**, 107–10.

Lövei, G.L., Monostori, E. and Andó, I. (1985) Digestion rate in relation to starvation in the larva of a carabid predator, *Poecilus cupreus*. *Entomologia Experimentalis et Applicata*, **37**, 123–7.

Lövei, C.L., McDougall, D., Bramley, G., Hodgson, D.J. and Wratten, S.D. (1992) Floral resources for natural enemies: the effect of *Phacelia tanacetifolia* (Hydrophylaceae) on within-field distribution of hoverflies (Diptera: Syrphidae). *Proceedings of the 45th New Zealand Plant Protection Conference*, 1992, pp. 60–61.

Lövei, G.L., Stringer, I.A.N., Devine, C.D. and Cartellieri, M. (1997) Harmonic radar – a method using inexpensive tags to study invertebrate movement on land. *New Zealand Journal of Ecology*, **21**, 187–93.

Lowe, A.D. (1968) The incidence of parasitism and disease in some populations of the cabbage aphid (*Brevicoryne brassicae* L.) in New Zealand. *New Zealand Journal of Agricultural Research*, **11**, 821–8.

Loxdale, H.D. (1994) Isozyme and protein profiles of insects of agricultural and horticultural importance, in *The Identification and Characterisation of Pest Organisms* (ed. D.L. Hawksworth). CAB International, Wallingford, pp. 337–75.

Loxdale, H.D. and den Hollander, J. (eds) (1989) *Electrophoretic studies on Agricultural Pests*, Systematics Association Special Volume No. 39. Clarendon Press, Oxford.

Loxdale, H.D. and Lushai, G. (1998) Molecular markers in entomology. *Bulletin of Entomological Research*, **88**, 577–600.

Loxdale, H.D., Castañera, P. and Brookes, C.P. (1983) Electrophoretic study of enzymes from cereal aphid populations. 1. Electrophoretic techniques and staining systems for characterising isoenzymes from six species of cereal aphid (Hemiptera: Aphididae). *Bulletin of Entomological Research*, **73**, 645–57.

Loxdale, H.D., Brookes, C.P. and De Barro, P.J. (1996) Application of novel molecular markers (DNA) in agricultural entomology, in *Ecology of Agricultural Pests* (eds W.O.C. Symondson and J.E. Liddell). Chapman and Hall, London, pp. 149–98.

Lucas, J.R. (1985) Antlion pit structure and the behaviour of prey. *American Zoologist*, **25**, A26–A26.

Luck, R.F. (1990) Evaluation of natural enemies for biological control: a behavioural approach. *Trends in Ecology and Evolution*, **5**, 196–9.

Luck, R.F., Podoler, H. and Kfir, R. (1982) Host selection and egg allocation behaviour by *Aphytis melinus* and *A. lingnanensis*: a comparison of two facultatively gregarious parasitoids. *Ecological Entomology*, **7**, 397–408.

Luck, R.F., Shepard, B.M. and Kenmore, P.E. (1988) Experimental methods for evaluating arthropod natural enemies. *Annual Review of Entomology*, **33**, 367–91.

Luck, R.F., Stouthamer, R. and Nunney, L.P. (1993) Sex determination and sex ratio patterns in parasitic Hymenoptera, in *Evolution and Diversity of Sex Ratio in Insects and Mites* (eds D.L. Wrensch and M.A. Ebbert). Chapman and Hall, New York, pp. 442–76.

Luck, R.F., Jiang, G. and Houck, I.A. (1999) A laboratory evaluation of the astigmatid mite *Hemisarcoptes cooremani* Thomas (Acari: Hemisarcoptidae) as a potential biological control agent for an armored scale, *Aonidiella aurantii* (Maskell) (Homoptera : Diaspididae). *Biological Control*, **15**, 173–83.

Luff, M.L. (1968) Some effects of formalin on the numbers of Coleoptera caught in pitfall traps. *Entomologist's Monthly Magazine*, **104**, 115–16.

Luff, M.L. (1975) Some features influencing the efficiency of pitfall traps. *Oecologia*, **19**, 345–57.

Luff, M.L. (1978) Diel activity patterns of some field Carabidae. *Ecological Entomology*, **3**, 53–62.

Luff, M.L. (1986) Aggregation of some Carabidae in pitfall traps, in *Carabid Beetles – Their Adaptations and Dynamics* (eds P.J. den Boer, M.L. Luff, D. Mossakowski and F. Weber). Fischer, Stuttgart, pp. 385–97.

Luff, M.L. (1987) Biology of polyphagous ground beetles in agriculture. *Agricultural Zoology Reviews*, **2**, 237–78.

Luff, M.L. (1996) Use of carabids as environmental indicators in grasslands and cereals. *Annales Zoologici Fennici*, **33**, 185–95.

Luft, P.A. (1996) Fecundity, longevity, and sex ratio of *Goniozus nigrifemur* (Hymenoptera: Bethylidae). *Biological Control*, **17**, 17–23.

Lum, P.T. (1961) The reproductive system of some Florida mosquitoes II. The male accessory glands and their roles. *Annals of the Entomological Society of America*, **54**, 430–3.

Lum, P.T.M. and Flaherty, B.R. (1973) Influence of continuous light on oocyte maturation in *Bracon hebetor*. *Annals of the Entomological Society of America*, **66**, 355–7.

Lund, R.D. and Turpin, F.T. (1977) Serological investigation of black cutworm larval consumption by ground beetles. *Annals of the Entomological Society of America*, **70**, 322–4.

Lundgren, J.G. and Heimpel, G.E. (2003) Quality assessment of three species of commercially produced *Trichogramma* and the first report of thelytoky in commercially produced *Trichogramma*. *Biological Control*, **26**, 68–73.

Lunt, D.H., Zhang, D.-X., Szymura, J.M. and Hewitt, G.M. (1996) The insect cytochrome oxidase I gene: evolutionary patterns and conserved primers for phylogenetic studies. *Insect Molecular Biology*, **5**, 153–65.

Lynch, L.D. (1998) Indirect mutual interference and the $CV^2 > 1$ rule. *Oikos*, **83**, 318–26.

Lynch, L.D. and Thomas, M.B. (2000) Nontarget effects in the biocontrol of insects with insects, nematodes and microbial agents: the evidence. *Biocontrol News and Information*, **21**, 117N–30N.

Lynch, L.D., Ives, A.R., Waage, J.K., Hochberg, M.E. and Thomas, M.B. (2002) The risks of biocontrol: transient impacts and minimum nontarget densities. *Ecological Applications*, **12**, 1872–82.

Lynch, M. and Walsh, B. (1998) *Genetics and Analysis of Quantitative Traits*. Sinauer Associates, Inc., Sunderland.

Lys, J.A. (1995) Observation of epigeic predators and predation on artificial prey in a cereal field. *Entomologia Experimentalis et Applicata*, **75**, 265–72.

Lys, J.A. and Nentwig, W. (1991) Surface activity of carabid beetles inhabiting cereal fields. Seasonal phonology and the influence of farming operations on five abundant species. *Pedobiologia*, **35**, 129–38.

MacArthur, R.H. and Pianka, E.R. (1966) On optimal use of a patchy environment. *American Naturalist*, **100**, 603–9.

MacGregor, H.C. (1993) *An Introduction to Animal Cytogenetics*. Chapman and Hall, London, UK.

Mack, T.P. and Smilowitz, Z. (1982) CMACSIM, a temperature-dependent predator-prey model simulating the impact of *Coleomegilla maculata* (DeGeer) on green peach aphids on potato plants. *Environmental Entomology*, **11**, 1193–201.

Mackauer, M. (1976) An upper boundary for the sex ratio in a haploid insect. *Canadian Entomologist*, **108**, 1399–1402

Mackauer, M. (1983) Quantitative assessment of *Aphidius smithi* (Hymenoptera: Aphidiidae): fecundity, intrinsic rate of increase, and functional response. *Canadian Entomologist*, **115**, 399–415.

Mackauer, M. (1986) Growth and developmental interactions in some aphids and their hymenopteran parasites. *Journal of Insect Physiology*, **32**, 275–80.

Mackauer, M. (1990) Host discrimination and larval competition in solitary endoparasitoids, in *Critical Issues in Biological Control* (eds M. Mackauer, L.E. Ehler and J. Roland). Intercept, Andover, pp. 14–62.

Mackauer, M. and Henkelman, D.H. (1975) Effect of light-dark cycles on adult emergence in the aphid parasite *Aphidius smithi*. *Canadian Journal of Zoology*, **53**, 1201–206.

Mackauer, M. and Sequeira, R. (1993) Patterns of development in insect parasites, in *Parasites and Pathogens of Insects, Vol. 1* (eds N.E. Beckage, S.N. Thompson and BA. Frederici). Academic Press, London, pp. 1–23.

Mackauer, M. and Way, M.J. (1976) *Myzus persicae* (Sulz.) an aphid of world importance, in *I.B.P. Vol. 9, Studies in Biological Control* (ed. V.L. Delucchi). Cambridge University Press, Cambridge, pp. 51–120.

Mackauer, M., Sequeira, R. and Otto, M. (1997) Growth and development in parasitoid wasps: adaptation to variable host resources. Vertical Food Web Interactions, in *Evolutionary Patterns and Driving Forces. Ecological Studies 130* (eds K. Dettmer, G. Bauer and W. Volkl). Springer-Verlag, Berlin, pp. 191–203.

Mackay, A.I. and Kring, T.J. (1998) Acceptance and utilization of diapausing *Helicoverpa zea* (Lepidoptera: Noctuidae) pupae by *Ichneumon promissorius* (Hymenoptera: Ichneumonidae). *Environmental Entomology*, **27**, 1006–9.

Macleod, A. (1999) Attraction and retention of *Episyrphus balteatus* DeGeer (Dipera: Syrphidae) at an arable field margin with rich and poor flora resource. *Agriculture Ecosystems Environment*, **73**, 237–44.

Macleod, A., Wratten, S.D. and Harwood, R.W.J. (1994) The efficiency of a new lightweight suction sampler for sampling aphids and their predators in arable land. *Annals of Applied Biology*, **124**, 11–17.

Macleod, A., Sotherton, N.W., Harwood, R.W.J. and Wratten, S.D. (1995) An improved suction sampling device to collect aphids and their predators in agroecosystems. *Acta Jutlandica*, **70**, 125–31.

Mader, H.J., Schell, C. and Kornacher, P. (1990) Linear barriers to arthropod movements in the landscape. *Biological Conservation*, **54**, 209–22.

Maes, D. and Pollet, M. (1997) Dolichopodid communities (Diptera: Dolichopodidae) in "De Kempen" (eastern Belgium): biodiversity, faunsistics and ecology. *Bulletin and Annales de la Société Royale Belge d'Entomologie*, **133**, 419–38.

Maia, A.D.N., Luiz, A.J.B. and Campanhola, C. (2000) Statistical inference on associated fertility life table parameters using jacknife technique: computational aspects. *Journal of Economic Entomology*, **93**, 511–18.

Maier, C.T. and Waldbauer, G.P. (1979) Diurnal activity patterns of flower flies (Diptera, Syrphidae) in an Illinois sand area. *Annals of the Entomological Society of America*, **72**, 237–45.

Maingay, H.D., Bugg, R.L., Carlson, R.W. and Davidson, N.A. (1991) Predatory and parasitic wasps (Hymenoptera) feeding at flowers of sweet fennel *(Foeniculum vulgare* Miller var. *dulce* Battandier & Trabut, Apiaceae) and Spearmint *(Mentha spicata* L., Lamiaceae) in Massachusetts. *Biological Agriculture and Horticulture*, **7**, 363–83.

Mainx, F. (1964) The genetics of *Megaselia scalaris* Loew (Phoridae): a new type of sex determination in Diptera. *American Naturalist*, **98**, 415–30.

Majer, J.D., Recher, H. and Keals, N. (1996) Branchlet shaking: a method for sampling tree canopy arthropods under windy conditions. *Australian Journal of Ecology*, **21**, 229–34.

Majerus, M.E.N. (1994) *Ladybirds*. Harper Collins, London.

Majerus, M.E.N. and Hurst, G.D.D. (1997) Ladybirds as a model system for the study of male-killing symbionts. *Entomophaga*, **42**, 13–20.

Majerus, T.M.O., Majerus, M.E.N., Knowles, B., Wheeler, J., Bertrand, D., Kuznetzov, V.N., Ueno, H. and Hurst, G.D.D. (1998) Extreme variation in the prevalence of inherited male-killing microorganisms between three populations of *Harmonia axyridis* (Coleoptera: Coccinellidae). *Heredity*, **81**, 683–91.

Malaise, R. (1937) A new insect trap. *Entomologisk Tidskrift*, **58**, 148–60.

Malloch, G., Fenton, B. and Butcher, R.D.J. (2000) Molecular evidence for multiple infections of a new group of *Wolbachia* in the European raspberry beetle *Byturus tomentosus*. *Molecular Ecology*, **9**, 77–90.

Manga, N. (1972) Population metabolism of *Nebria brevicollis* (F.) (Coleoptera: Carabidae). *Oecologia*, **10**, 223–42.

Mangel, M. (1989a) Evolution of host selection in parasitoids: does the state of the parasitoid matter? *American Naturalist*, **133**, 157–72.

Mangel, M. (1989b) An evolutionary explanation of the motivation to oviposit. *Journal of Evolutionary Biology*, **2**, 157–72.

Mangel, M., and Clark, C.W. (1988) *Dynamic Modeling in Behavioural Ecology*. Princeton University Press, Princeton, New Jersey.

Mani, M. and Nagarkatti, S. (1983) Relationship between size of *Eucelatoria bryani* Sabrosky females and their longevity and fecundity. *Entomon*, **8**, 83–6.

Manly, B.F.J. (1971) Estimates of a marking effect with capture-recapture sampling. *Journal of Applied Ecology*, **8**, 181–89.

Manly, B.F.J. (1972) Tables for the analysis of selective predation experiments. *Researches on Population Ecology*, **14**, 74–81.

Manly, B.F.J. (1973) A linear model for frequency dependent selection by predators. *Researches on Population Ecology*, **14**, 137–50.

Manly, B.F.J. (1974) A model for certain types of selection experiments. *Biometrics*, **30**, 281–94.

Manly, B.F.J. (1977) The determination of key factors from life table data. *Oecologia*, **31**, 111–17.

Manly, B.F.J. (1988) A review of methods for key factor analysis, in *Estimation and Analysis of Insect Populations* (eds L. McDonald, B. Manly, J. Lockwood and J. Logan). Springer-Verlag, Berlin, pp. 169–90.

Manly, B.F.J. (1990) *Stage-structured Populations*. Chapman and Hall, London.

Manly, B.F.J. and Parr, M.J. (1968) A new method of estimating population size, survivorship, and birth rate from capture-recapture data. *Transactions of the Society for British Entomology*, **18**, 81–9.

Manly, B.F.J., Miller, P. and Cook, L.M. (1972) Analysis of a selective predation experiment. *American Naturalist*, **106**, 719–36.

Mardulyn, P. and Cameron, S.A. (1999) The major opsin in bees (Insecta: Hymenoptera); a promising nuclear gene for higher level phylogenetics. *Molecular Phylogenetics and Evolution*, **12**, 168–76.

Margolies, D.C. and Cox, T.S. (1993) Quantitative genetics applied to haplodiploid insects and mites, in *Evolution and Diversity of Sex Ratio in Insects and Mites* (eds D.L. Wrensch and M.A. Ebbert). Chapman and Hall, New York, pp. 548–559.

Markgraf, A. and Basedow, T. (2000) Carabid assemblages associated with fields of sugar beet and their margins in Germany, shown by different methods of trapping and sampling, in *Natural History and Applied Ecology of Carabid Beetles* (eds P. Brandmayr, G.L. Lövei, T.Z. Brandmayr, A. Casale and A.V. Taglianti). Pensoft Publishers, Moscow, pp. 295–305.

Markgraf, A. and Basedow, T. (2002) Flight activity of predatory Staphylinidae in agriculture in central Germany. *Journal of Applied Entomology* **126**, 79–81.

Marks, R.J. (1977) Laboratory studies of plant searching behaviour by *Coccinella septempunctata* L. larvae. *Bulletin of Entomological Research*, **67**, 235–41.

van Marle, J. and Piek, T. (1986) Morphology of the venom apparatus, in *Venoms of the Hymenoptera, Biochemical, Pharmacological and Behavioural Aspects*. (ed. T.O. Piek). Academic Press, London, pp. 17–44.

Marples, N.M., de Jong, P.W., Ottenheim, M.M.V.M.D. and Brakefield, P.M. (1993) The inheritance of a wingless character in the 2-spot ladybird (*Adalia bipunctata*). *Entomologica Experimentalis et Applicata*, **69**, 69–73.

Marris, G.C. and Casperd, J. (1996) The relationship between conspecific superparasitism and the outcome of in vitro contests staged between different larval instars of the solitary endoparasitoid *Venturia canescens*. *Behavioral Ecology and Sociobiology*, **39**, 61–69.

Marris, G.C., Hubbard, S.F. and Scrimgeour, C. (1996) The perception of genetic similarity by the solitary parthenogenetic parasitoid *Venturia canescens*, and its effects on the occurrence of superparasitism. *Entomologia Experimentalis et Applicata*, **78**, 167–74.

Marston, N.L. (1980) Sampling parasitoids of soybean insect pests, in *Sampling Methods in Soybean Entomology* (eds M. Kogan and D.C. Herzog). Springer-Verlag, New York, pp. 481–504.

Marston, N., Davis, D.G. and Gebhardt, M. (1982) Ratios for predicting field populations of soybean insects and spiders from sweep-net samples. *Journal of Economic Entomology*, **75**, 976–81.

Martin, P. and Bateson, P. (1996) *Measuring Behaviour: an Introductory Guide, 2nd edn*. Cambridge University Press, Cambridge.

Martinez, N.D., Hawkins, B.A., Dawah, H.A. and Feifarek, B.P. (1999) Effects of sampling effort on characterization of food-web structure. *Ecology* **80**, 1044–55.

Martínez-Martínez, L. and Bernal, J.S. (2002) *Ephestia kuehniella* Zeller as a factitious host for *Telenomus remus* Nixon: host acceptance and suitability. *Journal of Entomological Science*, **37**, 10–26.

Martins, E.P. (ed.) (1996) *Phylogenetics and the Comparative Method in Animal Behaviour*. Oxford University Press, Oxford.

Mascanzoni, D. and Wallin, H. (1986) The harmonic radar: a new method of tracing insects in the field. *Ecological Entomology*, **11**, 387–90.

Masner, L. (1976) Yellow pan traps (Moreicke traps, Assiettes jaunes). *Proctos*, **2**, 2.

Masner, L. and Goulet, H. (1981) A new model of flight-interception trap for some hymenopterous insects. *Entomology News*, **92**, 199–202.

Mason, C.E. and Blocker, H.D. (1973) A stabilised drop trap for unit-area sampling of insects in short vegetation. *Environmental Entomology*, **2**, 214–6.

Masui, S., Sasaki, T. and Ishikawa, H. (1997) *Gro*-E-homologous operon of *Wolbachia*, an intracellular symbiont of arthropods: a new approach for their phylogeny. *Zoological Science*, **14**, 701–6.

Masui, S., Kamoda, S., Sasaki, T. and Ishikawa, H. (1999) The first detection of the insert sequence ISW1 in the intracellular reproductive parasite *Wolbachia*. *Plasmid*, **42**, 13–19.

Masui, S., Kamoda, S., Sasaki, T. and Ishikawa, H. (2000) Distribution and evolution of bacteriophage WO in *Wolbachia*, the endosymbiont causing sexual alterations in arthropods. *Journal of Molecular Evolution*, **51**, 491–7.

le Masurier, A.D. (1991) Effect of host size on clutch size in *Cotesia glomerata*. *Journal of Animal Ecology*, **60**, 107–18.

Matos, M., Rose, M.R., Pite, M.T.R., Rego, C., and Avelar, T. (2000) Adaptation to the laboratory environment in *Drosophila subobscura*. *Journal of Evolutionary Biology*, **13**, 9–19.

Matsura, T. (1986) The feeding ecology of the pit-making ant-lion larva, *Myrmeleon bore*: feeding rate and species composition of prey in a habitat. *Ecological Research*, **1**, 15–24.

Matsura, T. and Morooka, K. (1983) Influences of prey density on fecundity in a mantis, *Paratenodera angustipennis* (S.). *Oecologia*, **56**, 306–12.

Matthews, R.W. (1975) Courtship in parasitic wasps, in *Evolutionary Strategies of Parasitic Insects and Mites* (ed. P.W. Price). Plenum Press, New York, pp. 66–86.

Matthews, R.W. and Matthews, J.R. (1971) The Malaise trap: its utility and potential for sampling insect populations. *The Michigan Entomologist*, **4**, 117–22.

Matthews, R.W. and Matthews, J.R. (1983) Malaise traps: the Townes model catches more insects. *Contributions of the American Entomological Institute*, **20**, 428–32.

Maund, C.M. and Hsiao, T.H. (1991) Differential encapsulation of two *Bathyplectes* parasitoids among alfalfa weevil strains, *Hypera postica* (Gyllenhal). *Canadian Entomologist*, **123**, 197–203.

Maurizio, A. (1975) How bees make honey, in *Honey: A Comprehensive Survey* (ed. E. Crane). Heinemann, London, pp. 77–105.

Maxwell, M.R. (1998) Lifetime mating opportunities and male mating behaviour in sexually cannibalistic praying mantids. *Animal Behaviour*, **55**, 1011–28.

Maxwell, M.R. (2000) Does a single meal affect female reproductive output in the sexually cannibalistic praying mantid *Iris oratoria*? *Ecological Entomology*, **25**, 54–62.

May, B. (1992) Starch gel electrophoresis of allozymes, in *Molecular Genetic Analysis of Populations. A Practical Approach* (ed. A.R. Hoelzel). Oxford University Press, Oxford, pp. 1–27.

May, P. (1988) Determinants of foraging profitability in two nectarivorous butterflies. *Ecological Entomology*, **13**, 171–84.

May, R.M. (1974a) Biological populations with nonoverlapping generations: stable points, stable cycles and chaos. *Science*, **186**, 645–7.

May, R.M. (1974b) *Stability and Complexity in Model Ecosystems*. Princeton University Press, Princeton.

May, R.M. (1976) Estimating *r*: a pedagogical note. *American Naturalist*, **110**, 496–9.

May, R.M. (1978) Host-parasitoid systems in patchy environments: a phenomenological model. *Journal of Animal Ecology*, **47**, 833–43.

May, R.M. (1986) When two and two do not make four: nonlinear phenomena in ecology. *Proceedings of the Royal Society of London B*, **228**, 241–66.

May, R.M. and Hassell, M.P. (1981) The dynamics of multiparasitoid-host interactions. *American Naturalist*, **117**, 234–61.

May, R.M. and Hassell, M.P. (1988) Population dynamics and biological control. *Philosophical Transactions of the Royal Society of London B*, **318**, 129–69.

May, R.M., Hassell, M.P., Anderson, R.M. and Tonkyn, D.W. (1981) Density dependence in host-parasitoid models. *Journal of Animal Ecology*, **50**, 855–65.

Mayhew, P.J. and van Alphen, J.J.M. (1999) Gregarious development in alysiine parasitoids evolved through a reduction in larval aggression. *Animal Behaviour*, **58**, 131–41.

Mayhew, P.J. and Blackburn, T.M. (1999) Does development mode organize life-history traits in parasitoid Hymenoptera? *Journal of Animal Ecology*, **68**, 906–16.

Mayhew, P.J. and Glaizot, O. (2001) Integrating theory of clutch size and body size evolution for parasitoids. *Oikos*, **92**, 372–76.

Mayhew, P.J. and Pen, I. (2002) Comparative analysis of sex ratios, in *Sex Ratios: Concepts and Research Methods* (ed. I.C.W. Hardy). Cambridge University Press, Cambridge, pp. 132–56.

Maynard Smith, J. (1972) *On Evolution*. Edinburgh University Press, Edinburgh.

Maynard Smith, J. (1974) The theory of games and the evolution of animal conflicts. *Journal of Theoretical Biology*, **47**, 209–21.

Maynard Smith, J. (1982) *Evolution and the Theory of Games*. Cambridge University Press, Cambridge.

Mayse, M.A., Price, P.W. and Kogan, M. (1978) Sampling methods for arthropod colonization studies in soybean. *Canadian Entomologist*, **110**, 265–74.

McArdle, B.H. (1977) An investigation of a *Notonecta glauca-Daphnia magna* predator-prey system. University of York, PhD. Thesis.

McBrien, H. and Mackauer, M. (1990) Heterospecific larval competition and host discrimination in two species of aphid parasitoids: *Aphidius ervi* and *Aphidius smithi*. *Entomologia Experimentalis et Applicata*, **56**, 145–53.

McCann, K., Hastings, A. and Huxel, G.R. (1998) Weak trophic interactions and the balance of nature. *Nature*, **395**, 794–8.

McCarty, M.T., Shepard, M. and Turnipseed, S.C. (1980) Identification of predacious arthropods in soybeans by using autoradiography. *Environmental Entomology*, **9**, 199–203.

McCauley, E. and Murdoch, W.W. (1987) Cyclic and stable populations: plankton as paradigm. *American Naturalist*, **129**, 97–121.

McCauley, V.J.E. (1975) Two new quantitative samplers for aquatic phytomacrofauna. *Hydrobiologia*, **47**, 81–9.

McClain, D.C., Rock, G.C. and Stinner, R.E. (1990a) Thermal requirements for development and simulation of the seasonal phonology of *Encarsia perniciosi* (Hymenoptera: Aphelinidae), a parasitoid of the San Jose Scale (Homoptera: Diaspididae) in North Carolina orchards. *Environmental Entomology*, **19**, 1396–1402.

McClain, D.C., Rock, G.C. and Woolley, J.B. (1990b) Influence of trap color and San jose scale (Homoptera: Diaspididae) pheromone on sticky trap catches of 10 aphelinid parasitoids (Hymenoptera). *Environmental Entomology*, **19**, 926–31.

McDaniel, S.C. and Sterling, W.L. (1979) Predator determination and efficiency on *Heliothis virescens* eggs in cotton using ^{32}P. *Environmental Entomology*, **8**, 1083–87.

McDonald, I.C. (1976) Ecological genetics and the sampling of insect populations for laboratory colonization. *Environmental Entomology*, **5**, 815–20.

McDonald, L., Manly, B., Lockwood, J. and Logan, J. (1988) *Estimation and Analysis of Insect Populations*. Springer-Verlag, Berlin.

McDougall, S.J. and Mills, N.J. (1997) The influence of hosts, temperature and food sources on the longevity of *Trichogramma platneri*. *Entomologia Experimentalis et Applicata*, **83**, 195–203.

McEwen, P. (1997) Sampling, handling and rearing insects, in *Methods in Ecological and Agricultural Entomology* (eds D.R. Dent and M.P. Walton). CAB International, Wallingford, UK, pp. 5–26.

McEwen, P., Clow, Jervis, M.A. and Kidd, N.A.C. (1993a) Alteration in searching behaviour of adult female green lacewings (*Chrysoperla carnea*) (Neur: Chrysopidae) following contact with honeydew of the black scale (*Saissetia oleae*) (Hom: Coccidae) and solutions containing L-tryptophan. *Entomophaga*, **38**, 347–54.

McEwen, P., Jervis, M.A. and Kidd, N.A.C. (1993b) Influence of artificial honeydew on larval development and survival in *Chrysoperla carnea* (Neur.: Chrysopidae). *Entomophaga*, **38**, 241–44.

McEwen, P., Jervis, M.A. and Kidd, N.A.C. (1993c) The effect on olive moth (*Prays oleae*) population levels, of applying artificial food to olive trees. *Proceedings A.N.P.P. 3rd International Conference on Pests in Agriculture, Montpellier, December, 1993*, 361–8.

McEwen, P.K., Jervis, M.A. and Kidd, N.A.C. (1994) Use of a sprayed L-tryptophan solution to concentrate numbers of the green lacewing *Chrysoperla carnea* in olive tree canopy. *Entomologia Experimentalis et Applicata*, **70**, 97–9.

McEwen, P.K., Jervis, M.A. and Kidd, N.A.C. (1996) The influence of an artificial food supplement on larval and adult performance in the green lacewing *Chrysoperla carnea* (Stephens). *International Journal of Pest Management*, 42, 25–27.

McEwen, P., New, T.R. and Whittington, A.E. (2001) *Lacewings in the Crop Environment*. Cambridge Universitry Press, Cambridge.

McGregor, R. (1997) Host-feeding and oviposition by parasitoids on hosts of different fitness value: influence of egg load and encounter rate. *Journal of Insect Behavior*, **10**, 451–62.

McGuire, A.D. and Armbruster, W.S. (1991) An experimental test for reproductive interactions between 2 sequentially blooming *Saxifraga* species (Saxifragaceae). *American Journal of Botany*, **78**, 214–19.

McInnis, D.O., Wang, T.T.Y. and Nishimoto, J. (1986) The inheritance of a black body mutant in *Biosteres longicaudatus* (Hymenoptera; Braconidae) from Hawaii. *Proceedings of the Hawaii Entomological Society*, **27**, 37–40.

McIver, I.D. (1981) An examination of the utility of the precipitin test for evaluation of arthropod predator-prey relationships. *Canadian Entomologist*, **113**, 213–22.

McKemey, A.R., Symondson, W.O.C. and Glen, D.M. (2001) Effect of slug size on predation by *Pterostichus melanarius*. *Biocontrol Science and Technology*, **11**, 83–93.

McKemey, A.R., Symondson, W.O.C. and Glen, D.M. (2003). Predation and prey size choice by the carabid beetle *Pterostichus melanarius*: the dangers of extrapolating from laboratory to field. *Bulletin of Entomological Research*, **93**, 227–34.

McKenna, M.A. and Thomson, J.D. (1988) A technique for sampling and measuring small amounts of floral nectar. *Ecology*, **69**, 1306–7.

McMurtry, J.A. (1992) Dynamics and potential impact of "generalist" phytoseiids in agroecosystems and possibilities for establishment of exotic species. *Experimental and Applied Acarology*, **14**, 371–82.

McNabb, D.M., Halaj, J. and Wise, D.H. (2001) Inferring trophic positions of generalist predators and their linkage to the detrital food web in agroecosystems: a stable isotope analysis. *Pedobiologia*, **45**, 289–97.

McNair, J.N. (1986) The effects of refuges on predator-prey interactions: a reconsideration. *Theoretical Population Biology*, **29**, 38–63.

McNeil, J.N. and Brodeur, J. (1995) Pheromone-mediated mating in the aphid parasitoid, *Aphidius nigripes* (Hymenoptera: Aphididae). *Journal of Chemical Ecology*, **21**, 959–72.

McNeil, J.N. and Rabb, R.L. (1973) Physical and physiological factors in diapause initiation of two hyperparasites of the tobacco hornworm, *Manduca sexta*. *Journal of Insect Physiology*, **19**, 2107–18.

McNeill, M.R., Vittum, P.J. and Baird, D.B. (1999) Suitability of *Listronotus maculicollis* (Coleoptera: Curculionidae) as a host for *Microctonus hyperodae* (Hymenoptera: Braconidae). *Journal of Economic Entomology*, **92**, 1292–1300.

McNeill, M.R., Vittum, P.J. and Jackson, T.A. (2000) *Serratia marcescens* as a rapid indicator of *Microctonus hyperodae* oviposition activity in *Listronotus maculicollis*. *Entomologia Experimentalis et Applicata*, **95**, 193–200.

McPeek, M.A. and Crowley, P.H. (1987) The effects of density and relative size on the aggressive behaviour, movement, and feeding of damselfly larvae (Odonata: Coenagrionidae). *Animal Behaviour*, **35**, 1051–61.

McPeek, M.A., Grace, M. and Richardson, J.M.L. (2001) Physiological and behavioural responses to predators shape the growth/predation risk trade-off in damselflies. *Ecology*, **82**, 1535–45.

McPhail, M. (1937) Relation of time of day, temperature and evaporation to attractiveness of fermenting sugar solution to Mexican fruitfly. *Journal of Economic Entomology*, **30**, 793–9.

McPhail, M. (1939) Protein lures for fruit flies. *Journal of Entomology*, **32**, 758–61.

McVey, M.E. and Smittle, B.J. (1984) Sperm precedence in the dragonfly *Erythemis simplicicollis*. *Journal of Insect Physiology*, **30**, 619–28.

Meagher, R.L. and Mitchell, E.R. (1999) Nontarget Hymenoptera collected in pheromone- and synthetic floral volatile-baited traps. *Environmental Entomology*, **28**, 367–71.

Meelis, E. (1982) Egg distribution of insect parasitoids: a survey of models. *Acta Biotheoretica*, **31**, 109–26.

van Meer, M.M.M., Can Kan, F.J.P.M. and Stouthamer, R. (1999a) Spacer 2 region and 5S rDNA variation of *Wolbachia* strains involved in cytoplasmic incompatibility or sex-ratio distortion in arthropods. *Letters in Applied Microbiology*, **28**, 17–22.

van Meer, M.M.M., Witteveldt, J. and Stouthamer, R. (1999b) Development of a microinjection protocol for the parasitoid *Nasonia vitripennis*. *Entomologia Experimentalis et Applicata*, **93**, 325–9.

van der Meijden, E. and Klinkhamer, P.G.L. (2000) Conflicting interests of plants and the natural enemies of herbivores. *Oikos*, **89**, 202–8.

Meiners, T., Wäckers, F. and Lewis, W.J. (2003) Associative learning of complex odours in parasitoid host location. *Chemical Senses*, **28**, 231–6.

van den Meiracker, R.A.F. (1994) Induction and termination of diapause in *Orius* predatory bugs. *Entomologia Experimentalis et Applicata*, **73**, 127–37.

van den Meiracker, R.A.F., Hammond, W.N.O. and van Alphen, J.J.M. (1990) The role of kairomone in prey finding by *Diomus* sp. and *Exochomus* sp., two coccinellid predators of the cassava mealybug, *Phenacoccus manihoti*. *Entomologia Experimentalis et Applicata*, **56**, 209–17.

Melbourne, B.A. (1999) Bias in the effect of habitat structure on pitfall traps: An experimental evaluation. *Australian Journal of Ecology*, **24**, 228–39.

Mellini, E. (1972) Studi sui Detteri Larvevoridi. XXV. Sul determinismo ormonale delle influenze esercitate dagli ospiti sui loro parassiti. *Bolletino di Istituto di Bologna*. **31**, 165–203.

Memmott, J. (1999) The structure of a plant-pollinator food web. *Ecology Letters*, **2**, 276–80.

Memmott, J. (2000) Food webs as a tool for studying nontarget effects in biological control, in *Nontarget Effects of Biological Control* (eds P.A. Follett and J.J. Duan). Kluwer Academic Publishers, Dordrecht, pp. 147–63.

Memmott, J. and Godfray, H.C.J. (1993) Parasitoid webs, in *Hymenoptera and Biodiversity* (eds J. LaSalle and I.D. Gauld). CAB International, Wallingford, UK, pp. 217–34.

Memmott, J. and Godfray, H.C.J. (1994) The use and'construction of parasitoid webs, in *Parasitoid Community Ecology* (eds B. A. Hawkins and W. Sheehan). Oxford University Press, Oxford, pp. 300–18.

Memmott, J., Godfray, H.C.J. and Gauld, I.D. (1994) The structure of a tropical host-parasitoid community. *Journal of Animal Ecology*, **63**, 521–40.

Memmott, J., Martinez, N.D. and Cohen, J.E. (2000) Predators, parasitoids and pathogens: species richness, trophic generality and body sizes in a natural food web. *Journal of Animal Ecology*, **69**, 1–15.

Mendel, M.J., Shaw, P.B. and Owens, J.C. (1987) Life history characteristics of *Anastatus semiflavidus* (Hymenoptera: Eupelmidae), an egg parasitoid of the range caterpillar, *Hemileuca oliviae* (Lepidoptera: Saturniidae) over a range of temperatures. *Environmental Entomology*, **16**, 1035–41.

Mendes, S.M., Bueno, V.H.P., Argolo, V.M. and Silveira, L.C.P. (2002) Type of prey influences biology and consumption rate of *Orius insidiosus* (Say) (Hemiptera, Anthocoridae). *Revista Brasileira de Entomologia*, 46, 99–103.

Mendis, V.W. (1997) A study of slug egg predation using immunological techniques. University of Cardiff, Ph.D. Thesis.

Mendis, V.W., Bowen, I.D., Liddell, J.E. and Symondson, W.O.C. (1996). Monoclonal antibodies against *Deroceras reticulatum* and *Arion ater* eggs for use in predation studies, in *Slug and Snail Pests in Agriculture*, BCPC Symposium Proceedings No. 66, 99–106.

Menken, S.B.J. (1982) Enzyme characterisation of nine endoparasite species of small ermine moths (Yponomeutidae). *Experimentia*, **38**, 1461–2.

Menken, S.B.J. and Raijmann, L.E.L. (1996). Biochemical systematics: principles and perspectives for pest management, in *The Ecology of Agricultural Pests: Biochemical Approaches* (eds W.O.C. Symondson and J.E. Liddell). Systematics Association Special Volume 53, Chapman and Hall, London, pp. 7–29.

Menken, S.B.J. and Ulenberg, S.A. (1987) Biochemical characters in agricultural entomology. *Agricultural Zoology Reviews*, **2**, 305–60.

Mensah, R.K. (1997) Yellow traps can be used to monitor populations of *Coccinella transversalis* (F.) and *Adalia bipunctata* (L.) (Coleoptera: Coccinellidae) in cotton crops. *Australian Journal of Entomology*, **36**, 377–81.

de Menten, L., Niculita, H., Gilbert, M., Delneste, D. and Aron, S. (2003) Fluorescence in situ hybridization: a new method for determining primary sex ratio in ants. *Molecular Ecology*, **12**, 1637–48.

Mercot, H., Llorente, B., Jacques, M., Atlan, A. and Montchamp-Moreau, C. (1995) Variability within the Seychelles cytoplasmic incompatibility system in *Drosophila simulans*. *Genetics*, **141**, 1015–23.

Messenger, P.S. (1964a) The influence of rhythmically fluctuating temperature on the development and reproduction of the spotted alfalfa aphid *Therioaphis maculata*. *Journal of Economic Entomology*, **57**, 71–6.

Messenger, P.S. (1964b) Use of life tables in a bioclimatic study of an experimental aphid-braconid wasp host-parasite system. *Ecology*, **45**, 119–31.

Messenger, P.S. (1970) Bioclimatic inputs to biological control and pest management programs, in *Concepts of Pest Management* (eds R.L. Rabb and F.E. Guthrie). North Carolina State University, Raleigh, pp. 84–99.

Messenger, P.S. and van den Bosch, R. (1971) The adaptability of introduced biological control agents. In *Biological Control* (ed. C.F. Huffaker). Plenum, New York, pp. 68–92.

Messing, R.H. (2001) Centrufugal phylogeny as a basis for non-target host testing in biological control: is it relevant for parasitoids? *Phytoparasitica*, **29**, 187–9.

Messing, R.H. and Aliniazee, M.T. (1988) Hybridization and host suitability of two biotypes of *Trioxys pallidus* (Hymenoptera: Aphidiidae). *Annals of the Entomological Society of America*, **81**, 6–9.

Messing, R.H. and Aliniazee, M.T. (1989) Introduction and establishment of *Trioxys pallidus* (Hym.: Aphidiidae) in Oregon, USA for control of filbert aphid *Myzocallis coryli* (Hom.: Aphididae). *Entomophaga*, **34**, 153–63.

Messing, R.H. and Wong, T.T.Y. (1992) An effective trapping method for field studies of opiine braconid parasitoids of tephritid fruit flies. *Entomophaga*, **37**, 391–6.

Messing, R.H., Klungness, L.M., Purcell, M. and Wong, T.T.Y. (1993) Quality control parameters of mass-reared opiine parasitoids used in augmentative biological control of tephritid fruit flies in Hawaii. *Biological Control*, **3**, 140–7.

Metz, C.W. (1938) Chromosome behavior, inheritance and sex determination in *Sciara*. *American Naturalist*, **72**, 485–520.

Meunier, J. and Bernstein, C. (2002) The influence of local mate competition on host-parasitoid dynamics. *Ecological Modelling*, **152**, 77–88.

Meyhöfer, R. (2001) Intraguild predation by aphidophagous predators on parasitised aphids: the use of multiple video cameras. *Entomologia Experimentalis et Applicata*, **100**, 77–87.

Meyhöfer, R. and Casas, J. (1999) Vibratory stimuli in host location by parasitic wasps. *Journal of Insect Physiology*, **45**, 967–71.

Meyhöfer, R. and Klug, T. (2002) Intraguild predation on the aphid parasitoid *Lysiphlebus fabarum* (Marshall) (Hymenoptera: Aphidiidae): mortality risks and behavioral decisions made under threats of predation. *Biological Control*, **25**, 239–48.

Meyhöfer, R., Casas, J. and Dorn, S. (1994) Host location by a parasitoid using leafminer vibrations: characterising the vibrational signals produced by the leafmining host. *Physiological Entomology*, **19**, 349–59.

Meyhöfer, R., Casas, J. and Dorn, S. (1997) Vibration-mediated interactions in a host-parasitoid system. *Proceedings of the Royal Society of London B*, **264**, 261–6.

Michaud, J.P. (2003) A comparative study of larval cannibalism in three species of ladybird. *Ecological Entomology*, **28**, 92–101.

Micheli, M.R., Beva, R., Pascale, E. and D'Ambrosio, E. (1994) Reproducible DNA fingerprinting with the random amplified polymorphic DNA (RAPD) method. *Nucleic Acids Research*, **22**, 1921–2.

Michels Jr, G.J. and Behe, R.W. (1991) Effects of two prey species on the development of *Hippodamia sinuata* (Coleoptera: Coccinellidae) larvae at constant temperatures. *Journal of Economic Entomology*, **84**, 1480–84.

Michelsen, A. (1983) Biophysical basis of sound communication, in *Bioacoustics: A Comparative Approach* (ed. B. Lewis). Academic Press, London, pp. 3–38.

Michelsen, A., Fink, F., Gogala, M. and Traue, D. (1982) Plants as transmission channels for insect vibrational songs. *Behavioral Ecology and Sociobiology*, **11**, 269–81.

Michiels, N.K. (1992) Consequences and adaptive significance of variation in copulation duration in the dragonfly *Sympetrum danae*. *Behavioral Ecology and Sociobiology*, **29**, 429–35.

Michiels, N.K. and Dhondt, A.A. (1988) Direct and indirect estimates of sperm precedence and displacement in the dragonfly *Sympetrum danae* (Odonata: Libellulidae). *Behavioral Ecology and Sociobiology*, **23**, 257–63.

Middleton, R.J. (1984) The distribution and feeding ecology of web-spinning spiders living in the canopy of Scots Pine (*Pinus sylvestris* L.). University College, Cardiff, Ph.D. Thesis.

Mikheev, A.V. and Kreslavskii, A.G. (2000) Marking insects by burning. *Entomological Review*, **80**, 367–9.

Miles, L.R. and King, E.G. (1975) Development of the tachinid parasite, *Lixophaga diatraeae*, on various developmental stages of the sugar cane borer in the laboratory. *Environmental Entomology*, **4**, 811–14.

Miller, M.C. (1981) Evaluation of enzyme-linked immunosorbent assay of narrow- and broad-spectrum anti-adult Southern pine beetle serum. *Annals of the Entomological Society of America*, **74**, 279–82.

Miller, M.C., Chappell, W.A., Gamble, W. and Bridges, J.R. (1979) Evaluation of immunodiffusion and immunoelectrophoretic tests using a broad-spectrum anti-adult Southern Pine Beetle serum. *Annals of the Entomological Society of America*, **72**, 99–104.

Miller, P.L. (1982) Genital structure, sperm competition and reproductive behaviour in some African libellulid dragonflies. *Advances in Odonatology*, **1**, 175–92.

Miller, P.L. (1984) The structure of the genitalia and the volumes of sperm stored in male and female *Nesciothemis farinosa* and *Orthetrum chrysostigma*. *Odonatologica*, **13**, 415–28.

Miller, P.L. (1987) *Dragonflies, Naturalists' Handbooks No 7*. Cambridge University Press, Cambridge.

Miller, P.L. (1991) The structure and function of the genitalia in the Libellulidae (Odonata). *Zoological Journal of the Linnean Society*, **102**, 43–74.

Miller, R.S., Passoa, S., Waltz, R.D. and Mastro, V. (1993). Insect removal from sticky traps using a citrus oil solvent. *Entomological News*, **104**, 209–13.

Mills, N.J. (1981) Some aspects of the rate of increase of a coccinellid. *Ecological Entomology*, **6**, 293–9.

Mills, N.J. (1982a) Satiation and the functional response: test of a new model. *Ecological Entomology*, **7**, 305–15.

Mills, N.J. (1982b) Voracity, cannibalism and coccinellid predation. *Annals of Applied Biology*, **101**, 144–8.

Mills, N.J. (1994) Biological control: some emerging trends, in *Individuals, Populations and Patterns in Ecology* (eds S.R. Leather, A.D. Watt, N.J. Mills and K.F.A.Walters). Intercept, Andover, pp. 213–22.

Mills, N.J. (1997) Techniques to evaluate the efficacy of natural enemies, in *Methods in Ecological and Agricultural Entomology* (eds D.R. Dent and M.P. Walton). CAB International, Wallingford, pp. 271–91.

Mills, N.J. (2000) Biological control: the need for realistic models and experimental approaches to parasitoid introductions, in *Parasitoid Population Biology* (eds M.E. Hochberg and A.R. Ives). Princeton University Press, Princeton, pp. 217–34.

Mills, N.J. (2001) Factors influencing top-down control of insect pest populations in biological control systems. *Basic and Applied Ecology*, **2**, 323–32.

Mills, N.J. and Getz, W.M. (1996) Modelling the biological control of insect pests: a review of host-parasitoid models. *Ecological Modelling*, **92**, 121–43.

Mills, N.J. and Gutierrez, A.P. (1999) Biological control of insect pests: a tritrophic perspective, in *Theoretical Approaches to Biological Control* (eds B.A. Hawkins and H.V. Cornell). Cambridge University Press, Cambridge, pp. 89–102.

Mills, N.J. and Kuhlmann, U. (2000) The relationship between egg load and fecundity among *Trichogramma* parasitoids. *Ecological Entomology*, **25**, 315–24.

Mills, N.J. and Lacan, I. (2004) Ratio-dependence in the functional response of insect parasitoids: evidence from *Trichogramma minutum* foraging for eggs in small host patches. *Ecological Entomology*, **29**, 208–16.

Minari, O. and Zilversmit, D.B. (1963) Use of KCN for stabilisation of colour in direct Nesslerisation of Kjeldahl digests. *Analytical Biochemistry*, **6**, 320–7.

Minkenberg, O. (1989) Temperature effects on the life history of the eulophid wasp *Diglyphus isaea*, an ectoparasitoid of leafminers (*Liriomyza* spp.), on tomatoes. *Annals of Applied Biology*, **115**, 381–97.

Mitchell, P., Arthur, W. and Farrow, M. (1992) An investigation of population limitation using factorial experiments. *Journal of Animal Ecology*, **611**, 591–8.

Mitter, C., Farrell, B. and Futuyma, D.J. (1991) Phylogenetic studies of insect-plant interactions: insights into the genesis of diversity. *Trends in Ecology and Evolution*, **6**, 290–3.

Miura, K. (1990) Life-history parameters of *Gonatocerus cincticipitis* Sahad (Hym., Mymaridae), an egg parasitoid of the green rice leafhopper, *Nephotettix cincticeps* Uhler (Hem., Cicadellidae). *Journal of Applied Entomology*, **110**, 353–7.

Miyashita, T. (1999) Life-history variation in closely related generalist predators living in the same habitat: a case study with three *Cyclosa* spiders. *Functional Ecology*, **13**, 307–14.

Molbo, D. and Parker, E.D. Jr. (1996) Mating structure and sex ratio variation in a natural population of *Nasonia vitripennis*. *Proceedings of the Royal Society of London B*, **263**, 1703–9.

Molbo, D., Krieger, M.J.B., Herre, E.A. and Keller, L. (2002) Species-diagnostic microsatellite loci for the fig wasp genus *Pegoscaptus*. *Molecular Ecology Notes*, **2**, 440–2.

Mollema, C. (1988) Genetical aspects of resistance in a host-parasitoid interaction. University of Leiden, Ph.D thesis.

Mollema, C. (1991) Heritability estimates of host selection behaviour by the *Drosophila* parasitoid *Asobara tabida*. *Netherlands Journal of Zoology*, **41**, 174–83.

Møller, A.P. and Swaddle, J. (1997) *Asymmetry, Developmental Stability and Evolution*. Oxford University Press, Oxford.

Mols, P.J.M. (1987) Hunger in relation to searching behaviour, predation and egg production of the carabid beetle *Pterostichus coerulescens* L.: results of simulation. *Acta Phytopathologica et Entomologica Hungarica*, **22**, 187–205.

Mommertz, S., Schauer, C., Kösters, N., Lang, A. and Filser, J. (1996) A comparison of D-Vac suction, fenced and unfenced pitfall trap sampling of epigeal arthropods in agroecosystems. *Annales Zoologici Fennici*, **33**, 117–24.

Monconduit, H. and Prevost, G. (1994) Avoidance of encapsulation by *Asobara tabida*, a larval parasitoid of *Drosophila* species. *Norwegian Journal of Agricultural Science, Supplement*, **16**, 301–9.

Mondor, E.B. and Roitberg, B.D. (2003) Age-dependent fitness costs of alarm signaling in aphids. *Canadian Journal of Zoology*, **81**, 757–62.

Moore, P.D., Webb, J.A. and Colinson, M.E. (1991) *Pollen Analysis*. Blackwell, Oxford.

Moore, R. (2001) Emergence trap developed to capture adult large pine weevil *Hylobius abietis* (Coleoptera: Curculionidae) and its parasite *Bracon hylobii* (Hymenoptera: Braconidae). *Bulletin of Entomological Research*, **91**, 109–15.

Moore, W.S. (1995) Inferring phylogenies from mtDNA variation: mitochondrial-gene trees versus nuclear-gene trees. *Evolution*, **49**, 718–26.

Morales, J. and Hower, A.A. (1981) Thermal requirements for development of the parasite *Microctonus aethiopoides*. *Environmental Entomology*, **10**, 279–84.

Moran, M.D. and Hurd, L.E. (1998) A trophic cascade in a diverse arthropod community caused by a generalist arthropod predator. *Oecologia*, **113**, 126–32.

Moran, M.D. and Scheidler, A.R. (2002) Effects of nutrients and predators on an old-field food chain: interactions of top-down and bottom-up processes. *Oikos*, **98**, 116–24.

Moratorio, M.S. (1987) Effect of host species on the parasitoids *Anagrus mutans* and *Anagrus silwoodensis* Walker (Hymenoptera: Mymaridae). *Environmental Entomology*, **16**, 825–7.

Moreby, S. (1991) A simple time-saving improvement to the motorized insect suction sampler. *Entomologist*, **110**, 2–4.

Moreno, D.S., Gregory, W.A. and Tanigoshi, L.K. (1984) Flight response of *Aphytis melinus* (Hymenoptera: Aphelinidae) and *Scirtothrips citri* (Thysanoptera: Thripidae) to trap color, size and shape. *Environmental Entomology*, **13**, 935–40.

Morgan, D.J.W. and Cook, J.M. (1994) Extremely precise sex ratios in small clutches of a bethylid wasp. *Oikos*, **71**, 423–30.

Morin, P.J. and Lawler, S.P. (1995) Food web architecture and population dynamics: theory and empirical evidence. *Annual Review of Ecology and Systematics*, **26**, 505–29.

Moritz, C., Dowling, T.E. and Brown, W.M. (1987) Evolution of animal mitochondrial DNA: relevance for population biology and systematics. *Annual Review of Entomology and Systematics*, **18**, 269–92.

Morris, D.L. (1948) Quantitative determination of carbohydrates with Dreywood's anthrone reagent. *Science*, **107**, 254.

Morris, R.F. (1959) Single factor analysis in population dynamics. *Ecology*, **40**, 580–88.

Morris, R.F. and Fulton, W.C. (1970) Models for the development and survival of *Hyphantria cunea* in relation to temperature and humidity. *Memoires de la Société Entomologique du Canada*, **70**, 1–60.

Morris, R.J. and Fellowes, M.D.E. (2002) Learning and natal host influence host preference, handling time and sex allocation behaviour in a pupal parasitoid. *Behavioral Ecology and Sociobiology*, **51**, 386–93.

Morris, R.J., Müller, C.B. and Godfray, H.C.J. (2001) Field experiments testing for apparent competition between primary parasitoids mediated by secondary parasitoids. *Journal of Animal Ecology*, **70**, 301–9.

Morris, T. and Campos, M. (1996) A hybrid beating tray. *Entomologist*, **115**, 20–2.

Morris, T.I., Symondson, W.O.C., Kidd, N.A.C., Jervis, M.A. and Campos, M. (1998) Are ants significant predators of the olive moth, *Prays oleae*? *Crop Protection* **17**, 365–6.

Morris, T.I., Campos, M., Kidd, N.A.C., Jervis, M.A. and Symondson, W.O.C. (1999a) Dynamics of the predatory arthropod community in Spanish olive groves. *Agricultural and Forest Entomology*, **1**, 219–28.

Morris, T.I., Campos, M., Kidd, N.A.C. and Symondson, W.O.C. (1999b) What is consuming *Prays oleae* (Bernard) (Lep.: Yponomeutidae) and when: a serological solution. *Crop Protection* **18**, 17–22.

Morris, T.I., Symondson, W.O.C., Kidd, N.A.C. and Campos, M. (2002) The effect of different ant species on the olive moth, *Prays oleae* (Bern.), in a Spanish olive orchard. *Journal of Applied Entomology*, **126**, 224–30.

Morrison, G. and Strong Jr, D.R. (1980) Spatial variations in host density and the intensity of parasitism: some empirical examples. *Environmental Entomology*, **9**, 149–52.

Morrison, G. and Strong Jr, D.R. (1981) Spatial variation in egg density and the intensity of parasitism in a neotropical chrysomelid (*Cephaloleia consanguinea*). *Ecological Entomology*, **6**, 55–61.

Morrison, L.W. (2002) Long-term impacts of an arthropod-community invasion by the imported fire ant, *Solenopsis invicta*. *Ecology*, **83**, 2337–45.

Mountford, M.D. (1988) Population regulation, density dependence, and heterogeneity. *Journal of Animal Ecology*, **57**, 845–58.

Mousseau, T.A. and Roff, D.A. (1987) Natural selection and the heritability of fitness components. *Heredity*, **59**, 181–97.

Mowes, M., Freier, B., Kreuter, T. and Triltsch, H. (1997) Tiller counting or total plot harvest – how exact are predator counts in winter wheat? *Anzeiger für Schadlingskunde, Pflanzenschutz, Umweltschutz*, **70**, 121–6.

Mueke, J.M., Manglitz, G.R. and Kerr, W.R. (1978) Pea aphid: interaction of insecticides and alfalfa varieties. *Journal of Economic Entomology*, **71**, 61–5.

Mueller, U.G. and LaReesa Wolfenberger, L. (1999) AFLP genotyping and fingerprinting. *Trends in Ecology and Evolution*, **14**, 389–94.

Mühlenberg, M. (1993) *Freilandökologie*. Quelle and Meyer Verlag, Heidelberg.

Mukerji, M.K. and LeRoux, E.J. (1969) The effect of predator age on the functional response of *Podisus maculiventris* to the prey size of *Galleria mellonella*. *Canadian Entomologist*, **101**, 314–27.

Müller, C.B. and Brodeur, J. (2002) Intraguild predation in biological control and conservation biology. *Biological Control*, **25**, 216–23.

Müller, C.B. and Godfray, H.C.J. (1997) Apparent competition between two aphid species. *Journal of Animal Ecology*, **66**, 57–64.

Müller, C.B. and Godfray, H.C.J. (1999) Predators and mutualists influence the exclusion of aphid species from natural communities. *Oecologia*, **119**, 120–5.

Müller, C.B., Adriaanse, I.C.T., Belshaw, R. and Godfray, H.C.J. (1999) The structure of an aphid-parasitoid community. *Journal of Animal Ecology*, **68**, 346–70.

Müller, F.P. (1983) Differential alarm pheromone responses between strains of the aphid *Acyrthosiphon pisum*. *Entomologia Experimentalis et Applicata*, **34**, 347–8.

Müller, H. (1883) *Fertilisation of Flowers* (translated by W. D'Arcy Thompson). Macmillan and Co., London.

Müller, H.J. (1970) Formen der Dormanz bei Insekten. *Nova Acta Leopoldina*, **35**, 1–27.

Müller, J.K. (1984) Die Bedeutung der Fallenfang-Methode für die Lösung ökologischer Fragestellungen. *Zoologischer Jahrbucher, Abteilung für Systematik Okologie und Geographie der Tiere*, **111**, 281–305.

Münster-Swendsen, M. and Nachman, G. (1978) Asynchrony in insect host-parasite interaction and its effect on stability, studied by a simulation model. *Journal of Animal Ecology*, **47**, 159–71.

Murakami, Y. and Tsubaki, Y. (1984) Searching efficiency of the lady beetle *Coccinella septempunctata* larvae in uniform and patchy environments. *Journal of Ethology*, **2**, 1–6.

Murchie, A.K., Burn, D.J., Kirk, W.D.J. and Williams, I.H. (2001) A novel mechanism for time-sorting insect catches, and its use to derive diel flight periodicity of brassica pod midge *Dasineura brassicae* (Diptera: Cecidomyiidae). *Bulletin of Entomological Research*, **91**, 199–203.

Murdoch, W.W. (1969) Switching in general predators: experiments on predator specificity and stability of prey populations. *Ecological Monographs*, **39**, 335–54.

Murdoch, W.W. (1990) The relevance of pest-enemy models to biological control, in *Critical Issues in Biological Control* (eds M. Mackauer, L.E. Ehler and J. Roland). Intercept, Andover, pp. 1–24.

Murdoch, W.W. (1994) Population regulation in theory and practice – the Robert H. MacArthur award lecture presented August 1991 in San Antonio, Texas, USA. *Ecology*, **75**, 271–87.

Murdoch W.W. and Briggs, C.J. (1996) Theory for biological control: recent developments. *Ecology*, **77**, 2001–13.

Murdoch, W.W. and McCauley, E. (1985) Three distinct types of dynamic behaviour shown by a single planktonic system. *Nature*, **316**, 628–30.

Murdoch, W.W. and Oaten, A.A. (1975) Predation and population stability. *Advances in Ecological Research*, **9**, 1–131.

Murdoch, W.W. and Reeve, J.D. (1987) Aggregation of parasitoids and the detection of density dependence in field populations. *Oikos*, **50**, 137–41.

Murdoch, W.W. and Sih, A. (1978) Age-dependent interference in a predatory insect. *Journal of Animal Ecology*, **47**, 581–92.

Murdoch, W.W. and Stewart-Oaten, A. (1989) Aggregation by parasitoids and predators: effects on equilibrium and stability. *American Naturalist*, **134**, 288–310.

Murdoch, W.W. and Walde, S.J. (1989) Analysis of insect population dynamics, in *Towards a More Exact Ecology* (eds P.J. Grubb, and J.P. Whittaker). Blackwell, Oxford, pp. 113–40.

Murdoch, W.W., Chesson, J. and Chesson, P.L (1985) Biological control in theory and practice. *American Naturalist*, **125**, 344–66.

Murdoch, W.W., Nisbet, R.M., Blythe, S. Gurney, W.S.C. and Reeve, J.D. (1987) An invunerable age-class and stability in delay-differential parasitoid-host models. *American Naturalist*, **129**, 263–82.

Murdoch, W.W., McCauley, E., Nisbet, R.M Curney, W.S.C. and de Roos, A.M. (1992a) Individual-based models: combining testability and generality, in *Individual-based Models and Approaches in Ecology* (eds D.L. DeAngeles an L.J. Gross). Chapman and Hall, New York, pp. 18–35.

Murdoch, W.W., Nisbet, R.M., Luck, R.F., Godfray, H.C.J. and Gurney, W.S.C. (1992b) Size-selective sex-allocation and host-feeding in a parasitoid–host model. *Journal of Animal Ecology*, **61**, 533–41.

Murdoch, W.W., Briggs, C.J. and Nisbet, R.M. (1997) Dynamical effects of host size- and parasitoid state-dependent attacks by parasitoids. *Journal of Animal Ecology*, **66**, 542–56.

Murdoch, W.W., Briggs, C.J. and Collier, T.R. (1998) Biological control of insects: implication for theory in population ecology, in *Insect Populations in Theory and Practice* (eds J.P. Dempster and I.F.G. McLean). Kluwer Academic Publishers, Dordrecht, pp. 167–86.

Murphy, B.C., Rosenheim, J.A., Granett, J., Pickett, C.H. and Dowell, R.V. (1998) Measuring the impact of a natural enemy refuge: the prune tree/vineyard example, in *Enhancing Biological Control* (eds C.H. Pickett and R.L. Bugg). University of California Press, Berkeley, pp. 297–309.

Murphy, W.L. (1985) Procedure for the removal of insect specimens from sticky-trap material. *Annals of the Entomological Society of America*, **78**, 881.

Murray, R.A. and Solomon, M.G. (1978) A rapid technique for analysing diets of invertebrate predators by electrophoresis. *Annals of Applied Biology*, **90**, 7–10.

Murray, R.A., Solomon, M.G. and Fitzgerald, J.D. (1989) The use of electrophoresis for determining patterns of predation in arthropods, in *Electrophoretic Studies on Agricultural Pests* (eds H.D. Loxdale and J. den Hollander). Systematics Association Special Volume No. 39, Clarendon Press, Oxford, pp. 467–83.

Myers, J.H. (1985) How many insect species are necessary for successful biological control?, in *Proceedings of the VI International Symposium on Biological Control of Weeds, August 19–25, 1984, Vancouver* (ed. E.S. Delfosse). Agriculture Canada, Ottowa.

Myers, J.H., Higgins, C. and Kovacs, E. (1989) How many insect species are necessary for the biological control of insects? *Environmental Entomology* **18**, 541–47.

Myers, J.H., Smith, J.N.M. and Elkinton, J.S. (1994) Biological control and refuge theory. *Science*, **265**, 811.

Myint, W.W. and Walter, G.H. (1990) Behaviour of *Spalangia cameroni* males and sex ratio theory. *Oikos*, **59**,163–74.

Nabli, H., Bailey, W.C. and Necibi, S. (1999) Beneficial insect attraction to light traps with different wavelengths. *Biological Control*, **16**, 185–8.

Nadel, H. (1987) Male swarms discovered in Chalcidoidea (Hymenoptera: Encyrtidae, Pteromalidae). *Pan-Pacific Entomologist*, **63**, 242–6.

Nadel, H. and Luck, R.F. (1985) Span of female emergence and male sperm depletion in the female biased quasi gregarious parasitoid *Pachycrepoideus vindemmiae* (Hym., Pteromalidae). *Annals of the Entomological Society of America*, **78**, 410–4.

Nadel, H. and Luck, R.F. (1992) Dispersal and mating structure of a parasitoid with a female-biased sex ratio: implications for theory. *Evolutionary Ecology*, **6**, 270–8.

Nagai, K. (1990) Suppressive effect of *Orius* sp. (Hemiptera: Anthocoridae) on the population density of *Thrips palmi* (Thysanoptera: Thripidae) in eggplant in an open field. *Japanese Journal of Applied Entomology and Zoology*, **34**, 109–14.

Nagarkatti, S. (1970) The production of a thelytokous hybrid in an interspecific cross between two species of *Trichogramma*. *Current Science*, **39**, 76–8.

Nagarkatti, S. and Fazaluddin, M. (1973) Biosystematic studies on *Trichogramma* species (Hymenoptera: Trichogrammatidae). II. Experimental hybridization between some *Trichogramma* species from the New World. *Systematic Zoology*, **22**, 103–17.

Nagelkerke, C.J. and Hardy, I.C.W. (1994) The influence of developmental mortality on optimal sex allocation under local mate competition. *Behavioral Ecology*, **5**, 401–11.

Nahrung, H.F. and Merritt, D.J. (1999) Moisture is required for the termination of egg diapause in the chrysomelid beetle, *Homichloda barkeri*. *Entomologia Experimentalis et Applicata*, **93**, 201–7.

Nakamura, M. and Nakamura, K. (1977) Population dynamics of the chestnut gall wasp. *Dryocosmus kuriphilus* Yasumatsu (Hymenoptera: Cynipidae). *Oecologia*, **27**, 97–116.

Nakamuta, K. (1982) Switchover in searching behaviour of *Coccinella septempunctata* L. (Coleoptera: Coccinellidae) caused by prey consumption. *Applied Entomology and Zoology*, **17**, 501–6.

Nakashima, Y. and Hirose, Y. (1999) Trail sex pheromone as a cue for searching mates in an insect predator *Orius sauteri*. *Ecological Entomology*, **24**, 115–7.

Nakashima, Y., Teshiba, M. and Hirose, Y. (2002) Flexible use of patch marks in an insect predator: effect of sex, hunger state, and patch quality. *Ecological Entomology*, **27**, 581–7.

Naranjo, S.E. and Gibson, R.L. (1996). Phytophagy in predaceous heteroptera: effects on life history and population dynamics, in *Zoophytophagous Heteroptera: Implications for Life History and Integrated Pest Management*, (eds O. Alomar and R.N. Wiedenmann). Thomas Say Publications in Entomology, Lanham, MD, pp. 57–93.

Naranjo, S.E. and Hagler, J.R. (1998) Characterizing and estimating the effect of heteropteran predation, in *Predatory Heteroptera: Their Ecology and Use in Biological Control* (eds M. Coll and J.R. Ruberson). Entomological Society of America, Maryland, pp. 171–97.

Naranjo, S.E. and Hagler, J.R. (2001) Toward the quantification of predation with predator gut immunoassays: a new approach integrating functional response behavior. *Biological Control*, **20**, 175–89.

Narisu, Lockwood, J.A. and Schell, S.P. (1999) A novel mark-recapture technique and its application to monitoring the direction and distance of local movements of rangeland grasshoppers (Orthoptera: Acrididae) in the context of pest management. *Journal of Applied Ecology*, **36**, 604–17.

Naseer, M. and Abdurahman, U.C. (1990) Reproductive biology and predatory behaviour of the anthocorid bugs (Anthocoridae: Hemiptera) associated with the coconut caterpillar *Opisina arenosella* (Walker). *Entomon*, **15**, 149–58.

Navajas, M., Lagnel, J., Fauvel, G. and De Moraes, G. (1999) Sequence variation of ribosomal internal transcribed spacers (ITS) in commercially important Phytoseiidae mites. *Experimental and Applied Acarology*, **23**, 851–9.

Naylor, T.H. (1971) *Computer Simulation Experiments with Models of Economic Systems*. John Wiley, New York.

Nealis, V.G. (1988) Weather and the ecology of *Apanteles fumiferanae* Vier. (Hymenoptera: Braconidae). *Memoirs of the Entomological Society of Canada*, **146**, 57–70.

Nealis, V.G. and Fraser, S. (1988) Rate of development. reproduction, and mass-rearing of *Apanteles fumiferanae* Vier. (Hymenoptera: Braconidae) under controlled conditions. *Canadian Entomologist*, **120**, 197–204.

Nealis, V.G., Jones, R.E. and Wellington, W.G. (1984) Temperature and development in host-parasite relationships. *Oecologia*, **61**, 224–9.

Nechols, J.R. and Tauber, M.J. (1977) Age specific interaction between the greenhouse whitefly and *Encarsia formosa:* influence of host on the parasite's oviposition and development. *Environmental Entomology*, **6**, 143–49.

Nechols, J.R., Tauber, M.J. and Helgesen, R.G. (1980) Environmental control of diapause and postdiapause development in *Tetrastichus julis* (Hymenoptera: Eulophidae), a parasite of the cereal leaf beetle, *Oulema melanopus* (Coleoptera: Chrysomelidae). *Canadian Entomologist*, **112**, 1277–84.

Neigel, J.E. (1997) A comparison of alternative strategies for estimating gene flow from genetic markers. *Annual Review of Ecology and Systematics*, **28**, 105–28.

Neilsen, T. (1969) Population studies on *Helophilus hybridus* Loew and *Sericomyia silentis* (Harris) (Dipt., Syrphidae) on Jaeren, S.W. Norway. *Norsk Entomologisk Tidsskrift*, **16**, 33–8.

Nelemans, M.N.E. (1986) Marking techniques for surface-dwelling Coleoptera larvae. *Pedobiologia*, **29**, 143–6.

Nelemans, M.N.E. (1988) Surface activity and growth of larvae of *Nebria brevicollis* (F.) (Coleoptera, Carabidae). *Netherlands Journal of Zoology*, **38**, 74–95.

Nelemans, M.N.E., den Boer, P.J. and Spee, A. (1989) Recruitment and summer diapause in the dynamics of a population of *Nebria brevicollis* (Coleoptera: Carabidae). *Oikos*, **56**, 157–69.

Němec, V. and Starý, F. (1983) Electromorph differentiation in *Aphidius ervi* Hal. biotype on *Microlophium carnosum* (Bkt.) related to parasitization on *Acyrthosiphon pisum* (Harr.) (Hym., Aphidiidae). *Zeitschrift für Angewandte Entomologie*, **95**, 524–30.

Němec, V. and Starý, P. (1984) Population diversity of *Diaeretiella rapae* (Mclnt.) (Hym., Aphidiidae), an aphid parasitoid in agroecosystems. *Zeitschrift für Angewandte Entomologie*, **97**, 223–33.

Němec, V. and Starý, P. (1985) Genetic diversity and host alternation in aphid parasitoids (Hymenoptera: Aphidiidae). *Entomologia Generalis*, **10**, 253–58.

Němec, V. and Starý, P. (1986) Population diversity centers of aphid parasitoids (Hym: Aphidiidae): a new strategy in integrated pest management, in *Ecology of Aphidophaga* (ed. I. Hodek). Academia, Prague and W. Junk, Dordrecht, pp. 485–8.

Nentwig, W. (1998) Weedy plant species and their beneficial arthropods: potential for manipulation in field crops, in *Enhancing Biological Control* (eds C.H. Pickett and R.L. Bugg). University of California Press, Berkeley, pp. 49–72.

Netting, J.F. and Hunter, M.S. (2000) Ovicide in the whitefly parasitoid, *Encarsia formosa*. *Animal Behaviour*, **60**, 217–26.

Neuenschwander, P. (1982) Beneficial insects caught by yellow traps used in mass-trapping of the olive fly, *Dacus oleae*. *Entomologia Experimentalis et Applicata*, **32**, 286–96.

Neuenschwander, P. (2001) Biological control of the cassava mealybug in Africa: a review. *Biological Control*, **21**, 214–29.

Neuenschwander, P. and Gutierrez, A.P. (1989) Evaluating the impact of biological control measures, in *Biological Control: A Sustainable Solution to Crop Pest Problems in Africa* (eds J.S. Yaninek and H.R.

Herren), Proceedings of The Inaugral Conference and Workshop of the IITA Biological Control Program Center for Africa, 5–9 December 1988, Cotonou IITA, lbadan, pp. 147–54.

Neuenschwander, F. and Herren, H. (1988) Biological control of the cassava mealybug, *Phenacoccus manihoti*, by the exotic parasitoid *Epidinocarsis lopezi* in Africa, *Philosophical Transactions of the Royal Society of London B*, **318**, 319–33.

Neuenschwander, P. and Michelakis, S. (1980) The seasonal and spatial distribution of adult and larval chrysopids on olive-trees in Crete. *Acta Oecologia/ Oecologia Applicata*, **1**, 93–102.

Neuenschwander, P., Canard, M. and Michelakis, S. (1981) The attractivity of protein hydrolysate baited McPhail traps to different chrysopid and hemerobiid species (Neuroptera) in a Cretan olive orchard. *Annals of the Entomological Society of France*, **17**, 213–20.

New, T.R. (1969) The biology of some species of *Alaptus* (Mymaridae) parasitizing eggs of Psocoptera. *Transactions of the Society for British Entomology*, **18**, 181–93.

New, T.R. (1991) *Insects as Predators*. New South Wales University Press, Kensington.

Newton, B.L. and Yeargan, K.V. (2002) Population characteristics of *Phalangium opilio* (Opiliones: Phalangiidae) in Kentucky agroecosystems. *Environmental Entomology*, **31**, 92–8.

Ngumbi, P.M., Lawyer, P.G., Johnson, R.N., Kiilu, G. and Asiago, C. (1992) Identification of phlebotomine sandfly bloodmeals from Baringo district, Kenya, by direct enzyme-linked immunosorbent assay (ELISA). *Medical and Veterinary Entomology*, **6**, 385–8.

Nichols, W.S. and Nakamura, R.M. (1980) Agglutination and agglutination inhibition assays, in *Manual of Clinical Immunology* (eds N.R. Rose and H. Friedman). American Society for Microbiology, Washington DC., pp. 15–22.

Nicholls, C.F. and Bérubé, J.A.C. (1965) An expendable cage for feeding tests of coccinellid predators of aphids. *Journal of Economic Entomology*, **58**, 1169–70.

Nicholson, A.J. (1957) Comments on paper of T.B. Reynoldson. *Cold Spring Harbor Symposium on Quantitative Biology*, 22 (ed. K. Brehme Warren). Cold Spring Harbor, New York, pp. 313–27.

Nicholson, A.J. and Bailey, V.A. (1935) The balance of animal populations. Part I. *Proceedings of the Zoological Society of London*, **3**, 551–98.

Nicol, C.M.Y. and Mackauer, M. (1999) The scaling of body size and mass in a host parasitoid association: influence of host species and stage. *Entomologia Experimentalis et Applicata*, **90**, 83–92.

Niemelä, J. (1990) Spatial distribution of carabid beetles in the southern Finnish taiga: the question of scale, in *The Role of Ground Beetles in Ecological and Environmental Studies* (ed. N.E. Stork). Intercept, Andover, pp. 143–55.

Niemelä, J., Halme, E., Pajunen, T. and Haila, Y. (1986) Sampling spiders and carabid beetles with pitfall traps: the effect of increased sampling effort. *Annales Entomologici Fennici*, **52**, 109–11.

Nienstedt, K. and Poehling, H.M. (1995) Markierung von Aphiden mit ^{15}N-eine geignete Methode zur Quantifizierung der Prädationsleistung polyphager Prädatoren? *Mitteilungen der Deutschen Gesselchaft für allgemeine und angewandte Entomologie*, **10**, 227–30.

Nienstedt, K. and Poehling, H.M. (2000) ^{15}N-marked aphids for predation studies under field conditions. *Entomologia Experimentalis et Applicata*, **94**, 319–23.

Nisbet, R.M. and Gurney, W.S.C. (1983) The systematic formulation of population models for insects with dynamically varying instar duration. *Theoretical Population Biology*, **23**, 114–35.

Nishida, T. (1958) Extrafloral glandular secretions, a food source for certain insects. *Proceedings of the Hawaiian Entomological Society*, **26**, 379–86.

Nishimura, K. (1997) Host selection by virgin and inseminated females of the parasitic wasp, *Dinarmus basalis* (Pteromalidae, Hymenoptera). *Functional Ecology*, **11**, 336–41.

Noda, H., Miyoshi, T., Zhang, Q., Watanabe, K., Deng, K. and Hoshizaki, S. (2001) *Wolbachia* infection shared among planthoppers (Homoptera : Delphacidae) and their endoparasite (Strepsiptera: Elenchidae): a probable case of interspecies transmission. *Molecular Ecology*, **10**, 2101–106.

Noda, H., Miyoshi, T. and Koizumi, Y. (2002) *In vitro* cultivation of *Wolbachia* in insect and mammalian cell lines. *In vitro Cellular and Developmental Biology – Animal*, **38**, 423–7.

Noldus, L.P.J.J. (1991). *The Observer*: a software system for collection and analysis of observational data. *Behavior Research Methods, Instruments and Computers*, **23**, 415–29.

Noldus, L.P.J.J., Lewis, W.J., Tumlinson, J.H. and van Lenteren, J.C. (1988) Olfactometer and wind tunnel experiments on the role of sex pheromones of noctuid moths in the foraging behaviour of *Trichogramma* spp., in *Trichogramma and Other Egg Parasites* (eds J. Voegele, J.K. Waage and J.C. van Lenteren). *Colloques d'INRA, Paris*, **43**, 223–38.

Noldus, L.P.J.J., Buma, M.O.S., Jansen, R.G. and Takken, W. (1995). *EthoVision*: a video tracking and analysis system for the study of locomotory and flight behaviour of insects. Abstract for the 9th

International Symposium on Insect-Plant relationships, 24–30 June 1995, Gwatt, Switzerland.

Noldus, L.P.J.J., Trienes, R.J.H., Hendriksen, A.H.M., Jansen, H. and Jansen, R.G. (2000). The Observer Video-Pro: new software for the collection, management, and presentation of time-structured data from videotapes and digital media files. *Behavior Research Methods, Instruments and Computers*, **32**, 197–206.

Nordlander, G. (1987) A method for trapping *Hylobius abietis* (L.) with a standardized bait and its potential for forecasting seedling damage. *Scandinavian Journal of Forest Research*, **2**, 199–213.

Norling, U. (1971) The life history and seasonal regulation of *Aeschna viridis* Eversm. in southern Sweden (Odonata). *Entomologica Scandinavica*, **2**, 170–90.

Norling, U. (1984) Photoperiodic control of larval development in *Leucorrhinia dubia*: A comparison between populations from northern and southern Sweden (Anisoptera: Libellulidae). *Odonatologica*, **13**, 529–50.

van Nouhuys, S. and Hanski, I. (2000) Apparent competition between parasitoids mediated by a shared hyperparasitoid. *Ecology Letters*, **3**, 82–4.

van Nouhuys, S. and Via, S. (1999) Natural selection and genetic differentiation of behaviour between parasitoids from wild and cultivated habitats. *Heredity*, **83**, 127–37.

Noyes, J.S. (1982) Collecting and preserving chalcid wasps (Hymenoptera: Chalcidoidea). *Journal of Natural History*, **16**, 315–34.

Noyes, J.S. (1984) In a fog. *Chalcid Forum*, **3**, 4–5.

Noyes, J.S. (1989) A study of five methods of sampling Hymenoptera (Insecta) in a tropical forest, with special reference to the Parasitica. *Journal of Natural History*, **23**, 285–98.

Nufio, C.R. and Papaj, D.R. (2001) Host marking behavior in phytophagous insects and parasitoids. *Entomologia Experimentalis et Applicata*, **99**, 273–93.

Nunney, L. and Luck, R.F. (1988) Factors influencing the optimum sex ratio in a structured population. *Theoretical Population Biology*, **33**, 1–30.

Nur, U. (1989) Reproductive biology and genetics. Chromosomes, sex ratios and sex determination, in *Armoured Scale Insects, Their Biology, Natural Enemies and Control, Vol. A* (ed. D. Rosen). Elsevier Science Publishers, Amsterdam, the Netherlands, pp. 179–190.

Nur, U., Werren, J.H., Eickbush, D.G., Burke, W.D. and Eickbush, T.H. (1988) A Selfish B chromosome that enhances its transmission by eliminating the paternal genome. *Science*, **240**, 512–4.

Nyffeler, M. (1982) Field studies on the ecological role of the spiders as insect predators in agroecosystems (abandoned grassland, meadows and cereal fields). Swiss Federal Institute of Technology, Zurich, Ph.D. Thesis.

Nyffeler, M. (1999) Prey selection of spiders in the field. *Journal of Arachnology*, **27**, 317–24.

Nyffeler, M. and Benz, G. (1988a) Feeding ecology and predatory importance of wolf spiders (*Pardosa* spp.) (Araneae, Lycosidae) in winter wheat fields. *Journal of Applied Ecology*, **106**, 123–34.

Nyffeler, M. and Benz, G. (1988b) Prey and predatory importance of micryphantid spiders in winter wheat fields and hay meadows. *Journal of Applied Entomology*, **105**, 190–7.

Nyffeler, M., Sterling, W.L. and Dean, D.A. (1992) Impact of the striped lynx spider (Araneae: Oxyopidae) and other natural enemies on the cotton fleahopper (Hemiptera: Miridae) in Texas cotton. *Environmental Entomology*, **21**, 1178–88.

Obara, M. and Kitano, H. (1974) Studies on the courtship behavior of *Apanteles glomeratus* L. I. Experimental studies on releaser of wing vibrating behavior in the male. *Kontyû*, **42**, 208–4.

Obrist, M.K. and Duelli, P. (1996) Trapping efficiency of funnel- and cup traps for epigeal arthropods. *Mitteilungen der Schweizerischen Entomologischen Gesellschaft*, **69**, 361–9.

Ode, P.J. and Strand, M.R. (1995) Progeny and sex allocation decisions of the polyembryonic wasp *Copidosoma floridanum*. *Journal of Animal Ecology*, **64**, 213–24.

Ode, P.J., Antolin, M.F, and Strand, M.R. (1995) Brood-mate avoidance in the parasitic wasp *Bracon hebetor* Say. *Animal Behaviour*, **49**, 1239–48.

Ode, P.J., Antolin, M.F, and Strand, M.R. (1996) Sex allocation and sexual asymmetries in intra-brood competition in the parasitic wasp *Bracon hebetor*. *Journal of Animal Ecology*, **65**, 690–700.

Ode, P.J., Antolin, M.F, and Strand, M.R. (1997) Constrained oviposition and female-biased sex allocation in a parasitic wasp. *Oecologia*, **109**, 447–55.

Ode, P.J., Antolin, M.F. and Strand, M.R. (1998) Differential dispersal and female-biased sex allocation in a parasitic wasp. *Ecological Entomology*, **23**, 314–18.

O'Donnell, D.J. (1982) Taxonomy of the immature stages of parasitic Hymenoptera associated with aphids. Imperial College, University of London, Ph.D. Thesis.

O'Donnell, D.J. and Mackauer, M. (1989) A morphological and taxonomic study of first instar larvae of Aphidiinae (Hymenoptera: Braconidae). *Systematic Entomology*, **14**, 197–219.

Oelbermann, K. and Scheu, S. (2002) Stable isotope enrichment (δ^{15}N and δ^{13}C) in a generalist predator (*Pardosa lugubris*, Araneae: Lycosidae): effects of prey quality. *Oecologia*, **130**, 337–44.

Ogloblin, A.A. (1924) The role of extra-embryonic blastoderm of *Dinocampus terminatus* Nees during larval development. *Memoires de la Société Royale des Sciences de Bohème*, **3**, 1–27 (In French).

Oh, H.W., Kim, M.I.G., Shin, S.W., Bae, K.S., Ahn, Y.J. and Park, H.Y. (2000) Ultrastructural and molecular identification of a *Wolbachia* endosymbiont in a spider, *Nephila clavata. Insect Molecular Biology*, **9**, 539–43.

Ohgushi, T. (1992) Resource limitation on insect herbivore populations, in *Effects of Resource Distribution on Animal-plant Interactions*, (eds M.D. Hunter, T. Ohgushi, and P.W. Price). Academic Press, San Diego, U.S.A. pp. 199–241.

Ohiagu, C.E. and Boreham, P.F.L. (1978) A simple field test for evaluating insect predator-prey relationships. *Entomologia Experimentalis et Applicata*, **23**, 40–7.

Olive, C.W. (1982) Behavioral response of a sit-and-wait predator to spatial variation in foraging gain. *Ecology*, **63**, 912–20.

Olson, D.M. and Andow, D.A. (1998) Larval crowding and adult nutrition effects on longevity and fecundity of female *Trichogramma nubilale* Ertle & Davis (Hymenoptera: Trichogrammatidae). *Environmental Entomology*, **27**, 508–14.

Olson, D.M. and Andow, D.A. (2002) Inheritance of an oviposition behavior by an egg parasitoid. *Heredity*, **88**, 437–43.

Olson, D.M., Fadamiro, H., Lundgren, J.G. and Heimpel, G.E. (2000) Effects of sugar feeding on carbohydrate and lipid metabolism in a parasitoid wasp. *Physiological Entomology*, **25**, 17–26.

Omacini, M., Chaneton, E.J., Ghersa, C.M. and Müller, C.B. (2001) Symbiotic fungal endophytes control insect host-parasite interaction webs. *Nature*, **409**, 78–81.

O'Neill, K.M. and Skinner, S.W. (1990) Ovarian egg size and number in relation to female size in five species of parasitoid wasps. *Journal of Zoology*, **220**, 115–22.

O'Neill, S.L. and Karr, T.L. (1990) Bi-directional incompatability between conspecific populations of *Drosophila simulans*. *Nature*, **348**, 178–80.

O'Neill, S.L., Giordano, R., Colbert, A.M.E., Karr, T.L. and Robertson, H.M. (1992) 16S ribosomal RNA phylogenetic analysis of the bacterial endosymbionts associated with cytoplasmic incompatibility in insects. *Proceedings of the National Academy of Sciences*, **89**, 2699–702.

O'Neill, S.L., Hoffmann, A.A. and Werren, J.H. (1997) *Influential Passengers; Inherited Microorganisms and Arthropod Reproduction*. Oxford University Press, Oxford.

Ongagna, P. and Iperti, G. (1994) Influence of temperature and photoperiod in *Harmonia axyridis* Pall. (Col., Coccinellidae): Obtaining rapidly fecund adults or dormancy. *Journal of Applied Entomology*, **117**, 314–17.

Ooi, P.A.C. (1986) Insecticides disrupt natural biological control of *Nilaparvata lugens* in Sekinchani Malaysia, in *Biological Control in the Tropics* (eds M.Y. Hussein and A.G. Ibrahim). Universiti Pertanian Malaysia, Serdang, pp. 109–20.

Opp, S.B. and Luck, R.F. (1986) Effects of host size on selected fitness components of *Aphytis melinus* and *A. lingnanensis* (Hymenoptera: Aphelinidae). *Annals of the Entomological Society of America*, **79**, 700–04.

Orr, B.K., Murdoch, W.W. and Bence, J.R. (1990) Population regulation, convergence and cannibalism in *Notonecta* (Hemiptera). *Ecology*, **71**, 68–82.

Orr, D.B. and Boethel, D.J. (1986) Influence of plant antibiosis through four trophic levels. *Oecologia*, **70**, 242–9.

Orr, D.B. and Borden, J.H. (1983) Courtship and mating behavior of *Megastigmus pinus* (Hym., Torymidae). *Journal of the Entomological Society of British Columbia*, **80**, 20–24.

Orr, D.B. and Pleasants, J.M. (1996) The potential of native prairie plant species to enhance the effectiveness of the *Ostrinia nubilalis* parasitoid *Macrocentrus grandii*. *Journal of the Kansas Entomological Society*, **69**, 133–43.

Orr, D.B., Garcia-Salazar, C. and Landis, D.A. (2000) *Trichogramma* nontarget impacts: a method for biological control risk assessment, in *Nontarget Effects of Biological Control* (eds P.A. Follett and J.J. Duan). Kluwer Academic Publishers, Dordrecht, pp. 111–25.

Orr, M.R., De Camargo, R.X. and Benson, W.W. (2003) Interactions between ant species increase arrival rates of an ant parasitoid. *Animal Behaviour*, **65**, 1187–93.

Orzack, S.H. (1993) Sex ratio evolution in parasitic wasps, in *Evolution and Diversity of Sex Ratio in Insects and Mites* (eds D.L. Wrensch and M.A. Ebbert). Chapman and Hall, New York, pp. 477–511.

Orzack, S.H. and Gladstone, J. (1994) Quantitative genetics of sex-ratio traits in the parasitic wasp, *Nasonia vitripennis*. *Genetics*, **137**, 211–20.

Orzack, S.H. and Parker, E.D. (1986) Sex-ratio control in a parasitic wasp, *Nasonia vitripennis*. I. Genetic variation in facultative sex-ratio adjustment. *Evolution*, **40**, 331–40.

Orzack, S.H. and Parker, E.D. (1990) Genetic variation for sex ratio traits within a natural population of a parasitic wasp, *Nasonia vitripennis*. *Genetics*, **124**, 373–84.

Orzack, S.H., Parker, E.D. and Gladstone, J. (1991) The comparative biology of genetic variation for conditional sex ratio behavior in a parasitic wasp, *Nasonia vitripennis*. *Genetics*, **127**, 583–99.

Osborne, J.L., Clark, S.J., Morris, R.J., Williams, I.H., Riley, J.R., Smith, A.D., Reynolds, D.R. and Edwards, A.S. (1999) A landscape-scale study of bumble bee foraging range and constancy, using harmonic radar. *Journal of Applied Ecology*, 36, 519–33.

Osborne, K.A., Robichon, A., Burgess, E., Butland, S., Shaw, R.A., Coulthard, A., Pereira, H.S., Greenspan, R.J. and Sokolowski, M.B. (1997) Natural behaviour polymorphism due to a cGMP-dependent protein kinase of *Drosophila*. *Science*, **277**, 834–6.

Osborne, K.H. and Allen, W.W. (1999) Allen-vac: an internal collection bag retainer allows for snag-free arthropod sampling in woody scrub. *Environmental Entomology*, **28**, 594–6.

Osborne, L.S. (1982) Temperature-dependent development of greenhouse whitefly and its parasite *Encarsia formosa*. *Environmental Entomology*, **111**, 483–85.

Ostrom, P.H., Colunga-Garcia, M. and Gage, S.H. (1997) Establishing pathways of energy flow for insect predators using stable isotope ratios: field and laboratory evidence. *Oecologia*, **109**, 108–13.

Ôtake, A. (1967) Studies on the egg parasites of the smaller brown planthopper, *Laodelphax striatellus* (Fallén) (Hemiptera: Delphacidae), I. A device for assessing the parasitic activity, and the results obtained in 1966. *Bulletin of the Shikoku Agricultural Experiment Station*, **17**, 91–103.

Ôtake, A. (1970) Estimation of parasitism by *Anagrus* nr. *flaveolus* Waterhouse (Hymenoptera, Mymaridae). *Entomophaga*, **15**, 83–92.

Otis, D.L., Burnham, G.C., White, G.C. and Anderson, D.R. (1978) Statistical inference from capture data on closed populations. *Wildlife Monographs*, **62**, 1–135.

Otronen, M. (1995) Energy reserves and mating success in males of the yellow dung fly, *Scathophaga stercoraria*. *Functional Ecology*, **9**, 683–8.

Ottenheim, M., Holloway, G.J. and de Jong, P.W. (1992) Sex ratio in ladybirds (Coccinellidae). *Ecological Entomology*, **17**, 366–8.

Ottenheim, M., Volmer, A.D. and Holloway, G.J. (1996) The genetics of phenotypic plasticity in adult abdominal colour pattern of *Eristalis arbustorum* (Diptera: Syrphidae). *Heredity*, **77**, 493–9.

Otto, M. and Mackauer, M. (1998) The developmental strategy of an idiobiont ectoparasitoid, *Dendrocerus carpenteri*: influence of variations in host quality on offspring growth and fitness. *Oecologia*, **117**, 353–64.

Ouedraogo, R.M., Cusson, M., Goetell, M.S. and Brodeur, J. (2003) Inhibition of fungal growth in thermoregulating locusts, *Locusta migratoria*, infected by the fungus *Metarhizium anisopliae* var. *acridum*. *Journal of Invertebrate Pathology*, **82**, 103–9.

Outreman, Y., Le Ralec, A., Plantegenest, M., Chaubet, B., and Pierre, J.S. (2001) Superparasitism limitation in an aphid parasitoid: cornicle secretion avoidance and host discrimination ability. *Journal of Insect Physiology*, **47**, 339–48.

Owen, J.A. (1981) Trophic variety and abundance of hoverflies (Diptera, Syrphidae) in an English suburban garden. *Holarctic Ecology*, **4**, 221–8.

Owen, J.A. (1995) A pitfall trap for repetitive sampling of hypogean arthropod faunas. *Entomologist's Record*, **107**, 225–8.

Owen, J.A., Townes, H. and Townes, M. (1981) Species diversity of Ichneumonidae and Serphidae (Hymenoptera) in an English suburban garden. *Biological Journal of the Linnean Society*, **16**, 315–36.

Owen, R.E. and Packer, L. (1994) Estimation of the proportion of diploid males in populations of Hymenoptera. *Heredity*, **72**, 219–27.

Özder, N. and Sağlam, O. (2003) Effects of aphid prey on larval development and mortality of *Adalia bipunctata* and *Coccinella septempuncata* (Coleoptera: Coccinellidae). *Biocontrol Science and Technology*, **13**, 449–53.

Pacala, S.W. and Hassell, M.P. (1991) The persistence of host-parasitoid associations in patchy environments. II. Evaluation of field data. *American Naturalist*, **138**, 584–605.

Pacala, S.W., Hassell, M.P. and May, R.M. (1990) Host-parasitoid associations in patchy environments. *Nature*, **344**, 150–3.

Page, R.D.M. (1993) *Tangled Trees*. University of Chicago Press, Chicago and London.

Page, R.F. (1980) The evolution of multiple mating behaviour by honey bee queens (*Apis mellifera* L.). *Genetics*, **96**, 263–73.

Paill, W. (2000) Slugs as prey for larvae and imagines of *Carabus violaceus* (Coleoptera: Carabidae), in *Natural History and Applied Ecology of Carabid Beetles* (eds P. Brandmayr, G.L. Lövei, T.Z. Brandmayr, A. Casale and A.V. Taglianti). Pensoft Publishers, Moscow, pp. 221–7.

Pajunen, V.I. (1990) A note on the connection between swarming and territorial behaviour in insects. *Annales Entomologici Fennici*, **46**, 53–55.

Pakarinen, E. (1994) Autotomy in arionid and limacid slugs. *Journal of Molluscan Studies*, **60**, 19–23.

Pallewatta, P.K.T.N.S. (1986) Factors Affecting Progeny and Sex Allocation by the Egg Parasitoid, *Trichogramma evanescens* Westwood. Imperial College, University of London, Ph.D. Thesis.

Palmer, D.F. (1980) Complement fixation test, in *Manual of Clinical Immunology* (eds N.R. Rose and H. Friedman). American Society for Microbiology, Washington, pp. 35–47.

Palumbi, S.R., Cipriano, F. and Hare, M.P. (2001) Predicting nuclear gene coalescence from mitochondrial data: the three-times rule. *Evolution*, **55**, 859–68.

Pampel, W. (1914) Die weiblischen Geschlechtsorgane der Ichneumoniden. *Zeitschrift für Wissenschaftliche Zoologie*, **108**, 290–357.

Pannebakker, B.A., Pijnacker, L.P., Zwaan, B.J. and Beukeboom, L.W. (2004) Cytology of *Wolbachia*-induced parthenogenesis in *Leptopilina clavipes* (Hymenoptera: Figitidae). *Genome*, **47**, 299–303.

Papaj, D.R. and Vet, L.E.M. (1990) Odor learning and foraging success in the parasitoid, *Leptopilina heterotoma*. *Journal of Chemical Ecology*, **16**, 3137–50.

Paradise, C.J. and Stamp, N.E. (1990) Variable quantities of toxic diet cause different degrees of compensatory and inhibitory responses by juvenile praying mantids. *Entomologia Experimentalis et Applicata*, **55**, 213–22.

Paradise, C.J. and Stamp, N.E. (1991) Abundant prey can alleviate previous adverse effects on growth of juvenile praying mantids (Orthoptera: Mantidae). *Annals of the Entomological Society of America*, **84**, 396–406.

Paradise, C.J. and Stamp, N.E. (1993) Episodes of unpalatable prey reduce consumption and growth of juvenile praying mantids. *Journal of Insect Behavior*, **6**, 155–66.

Paris, O.H. and Sikora, A. (1967) Radiotracer analysis of the trophic dynamics of natural isopod populations, in *Secondary Productivity of Terrestrial Ecosystems* (ed. K. Petrusewicz). Panstwowe Wydawnictwo Naukoe, Warsaw, pp. 741–71.

Parker, E.D. and Orzack, S.H. (1985) Genetic variation for the sex ratio in *Nasonia vitripennis*. *Genetics*, **110**, 93–105.

Parker, F.D. and Bohart, R.M. (1966) Host-parasite associations in some twig-nesting Hymenoptera from western North America. *Pan-Pacific Entomologist*, **42**, 91–8.

Parker, F.D. and Pinnell, R.E. (1973) Effect on food consumption of the imported cabbageworm when parasitized by two species of *Apanteles*. *Environmental Entomology*, **2**, 216–19.

Parker, G.A. (1970a) Sperm competition and its evolutionary consequences in the insects. *Biological Reviews*, **45**, 525–67.

Parker, G.A. (1970b) The reproductive behaviour and the nature of sexual selection in *Scatophaga stercoraria* L. I. Diurnal and seasonal changes in population density around the site of mating and oviposition. *Journal of Animal Ecology*, **39**, 185–204.

Parker, G.A. (1970c) The reproductive behaviour and the nature of sexual selection in *Scatophaga stercoraria* L. (Diptera: Scatophagidae). II. The fertilization rate and spatial and temporal relationships of each sex around the site of mating and oviposition. *Journal of Animal Ecology*, **39**, 205–28.

Parker, G.A. (1970d) The reproductive behaviour and the nature of sexual selection in *Scatophaga stercoraria* L. V. The female's behaviour at the oviposition site. *Behaviour*, **38**, 140–68.

Parker, G.A. (1971) The reproductive behaviour and the nature of sexual selection in *Scatophaga stercoraria* L. VI. The adaptive significance of emigration from the oviposition site during the phase of genital contact. *Evolution*, **40**, 215–33.

Parker, G.A. (1978) Searching for mates, in *Behavioural Ecology, an Evolutionary Approach* (eds J.R. Krebs and N.B. Davies). Blackwell, Oxford, pp. 214–44.

Parker, G.A. (1984) Evolutionarily stable strategies, in *Behavioural Ecology: An Evolutionary Approach* (eds J.R. Krebs and N.B. Davies). Blackwell, Oxford, pp. 30–61.

Parker, G.A. and Courtney, S.P. (1984) Models of clutch size in insect oviposition. *Theoretical Population Biology*, **26**, 27–48.

Parker, G.A., Baker, R.R. and Smith, V.G.F. (1972) The origin and evolution of gamete dimorphism and the male-female phenomenon. *Journal of Theoretical Biology*, **36**, 529–53.

Parker, G.A., Simmons, L.W. and Kirk, H. (1990) Analysing sperm competition data: simple models for predicting mechanisms. *Behavioral Ecology and Sociobiology*, **27**, 55–65.

Parker, G.G., Smith, A.P. and Hogan, K.P. (1992) Access to the upper forest canopy with a large tower crane. *BioScience*, **42**, 664–70.

Parker, H.L. and Thompson, W.R. (1925) Contribution à la biologie des Chalcidiens entomophages. *Annals de la Société Entomologique de France*, **97**, 425–65.

Parmenter, L. (1956) Flies and their selection of the flowers they visit. *Entomologist's Record and Journal of Variation*, **68**, 242–3.

Parmenter, L. (1961) Flies visiting the flowers of wood spurge, *Euphorbia amygdaliodes* L. (Euphorbiaceae). *Entomologist's Record and Journal of Variation*, **73**, 48–9.

Parmenter, R.R. and MacMahon, J.A. (1989) Animal density estimation using a trapping web design: field validation experiments. *Ecology*, **70**, 169–79.

Parr, M.J. (1965) A population study of a colony of imaginal *Ischnura elegans* (van der Linden) (Odonata: Coenagriidae) at Dale, Pembrokeshire. *Field Studies*, **2**, 237–82.

Parsons, P.A. (1980) Isofemale strains and evolutionary strategies in natural populations, in *Evolutionary Biology, Vol. 13* (eds M. Hecht, W. Steere and B.Wallace). Plenum Publishing Corporation, NewYork.

Partridge, L. (1989) An experimentalist's approach to the role of costs in the evolution of life-histories, in *Towards a More Exact Ecology* (eds P.J. Grubb and J.B. Whittaker). Blackwell Scientific Publications, Oxford, pp. 231–46.

Partridge, L. and Farquhar, M. (1981) Sexual activity reduces lifespan of male fruitflies. *Nature*, **294**, 580–82.

Pashley, D.P., McPheron, B.A. and Zimmer, E.A. (1993) Systematics of holometabolous insect orders based on 18S rRNA. *Molecular Phylogenetics and Evolution*, **2**, 131–142.

Pasteur, N., Pasteur, G., Bonhomme, F., Catalan, J. and Britton-Davidian, J. (1988) *Practical Isozyme Genetics*. Ellis Horwood Ltd., Chichester.

Pastorok, R.A. (1981) Prey vulnerability and size selection by *Chaoborus* larvae. *Ecology*, **62**, 1311–24.

Patel, K.J. and Schuster, D.J. (1983) Influence of temperature on the rate of development of *Diglyphus intermedius* (Hymenoptera: Eulophidae) Girault, a parasite of *Lyriomyza* spp. (Diptera: Agromyzidae). *Environmental Entomology*, **12**, 885–7.

Patil, N.G., Baker, P.S., Groot, W. and Waage, J.K. (1994) Competition between *Psyllaephagus yaseeni* and *Tamarixia leucaenae*, two parasitoids of the leucaena psyllid (*Heteropsylla cubana*). *International Journal of Pest Management*, **40**, 211–15.

Patt, J.M., Hamilton, G.C. and Lashomb, J.H. (1997) Foraging success of parasitoid wasps on flowers: interplay of floral architecture and searching behavior. *Entomologia Experimentalis et Applicata*, **83**, 21–30.

Patt, J.M., Hamilton, G.C. and Lashomb, J.H. (1999) Response of two parasitoid wasps to nectar odors as a function of experience. *Entomologia Experimentalis et Applicata*, **90**, 1–8.

Pausch, R.D., Roberts, S.J., Bamey, R.J. and Armbrust, E.J. (1979) Linear pitfall traps, a modification of an established trapping method. *The Great Lakes Entomologist*, **12**, 149–51.

Payne, J.A. and Wood, B.W. (1984) Rubidium as a marking agent for the hickory shuckworm (Lepidoptera: Tortricidae). *Environmental Entomology*, **13**, 1519–21.

Pearl, R. and Miner, J.R. (1935) Experimental studies on the duration of life, XIV. The comparative mortality of certain lower organisms. *Quarterly Review of Biology*, **10**, 60–79.

Pearl, R. and Parker, S.L. (1921) Experimental studies on the duration of life. I. Introductory discussion of the duration of life in *Drosophila*. *American Naturalist*, **55**, 481–509.

Pekár, S. (1999) Some observations on overwintering of spiders (Araneae) in two contrasting orchards in the Czech Republic. *Agriculture, Ecosystems and Environment*, **73**, 205–10.

Pekár, S. (2000) Webs, diet, and fecundity of *Theridion impressum* (Araneae: Theridiidae). *European Journal of Entomology*, **97**, 47–50.

Pekár, S. (2002) Differential effects of formaldehyde concentration and detergent on the catching efficiency of surface active arthropods by pitfall traps. *Pedobiologia*, **46**, 539–47.

Penagos, D.I., Magallanes, R., Valle, J., Cisneros, J., Martinez, A.M., Goulson, D., Chapman, J.W., Caballero, P., Cave, R.D. and Williams, T. (2003) Effect of weeds on insect pests of maize and their natural enemies in Southern Mexico. *International Journal of Pest Management*, **49**, 155–61.

Pendleton, R.C. and Grundmann, A.W. (1954) Use of ^{32}P in tracing some insect-plant relationships of the thistle, *Cirsium undulatum*. *Ecology*, **35**, 187–91.

Peng, R., Christian, K. and Gibb, K. (1999) The effect of levels of green ant, *Oecophylla smaragdina* (F.) colonisation on cashew yield in northern Australia, in *Symposium on Biological Control in the Tropics* (eds L.W. Hong and S.S. Sastroumo). CABI Publishing, Wallingford UK, pp. 24–8.

Percival, M.S. (1961) Types of nectar in angiosperms. *New Phytologist*, **60**, 235–81.

Percival, M.S. (1965) *Floral Biology*. Pergamon Press, Oxford.

Perera, H.A.S. (1990) Effect of Host Plant on Mealybugs and Their Parasitoids. Wye College, University of London, Ph.D. Thesis.

Pérez-Lachaud, G., Hardy, I.C.W. and Lachaud, J.P. (2002) Insect gladiators: competitive interactions between three species of bethylid wasps attacking the coffee berry borer, *Hypothenemus hampei* (Coleoptera: Scolytidae). *Biological Control*, **25**, 231–8.

Pérez-Maluf, R. and Kaiser, L. (1998) Mating and oviposition experience influence odor learning in *Leptopilina boulardi* (Hymenoptera: Eucoilidae), a parasitoid of *Drosophila*. *Biological Control*, **11**, 154–9.

Pérez-Maluf, R., Kaiser, L., Wajnberg, E. and Carton, Y. (1998) Genetic variability of conditioned probing responses to a fruit odor in *Leptopilina boulardi* (Hymenoptera: Eucoilidae), a *Drosophila* parasitoid. *Behavior Genetics*, **28**, 67–73.

Perfecto, I., Horwith, B., van der Meer, J., Schultz, B., McCuinness, H. and Dos Santos, A. (1986) Effects of plant diversity and density on the emigration rate of two ground beetles, *Harpalus pennsylvanicus* and *Evarthrus sodalis* (Coleoptera: Carabidae), in a system of tomatoes and beans. *Environmental Entomology*, **15**, 1028–31.

Periquet, G., Hedderwick, M.P., El Agoze, M. and Poirié, M. (1993) Sex determination in the hymenopteran *Diadromus pulchellus* (Ichneumonidae): validation of the one-locus multi-allele model. *Heredity*, **70**, 420–7.

Perrot-Minnot, M.J. and Werren, J.H. (1999) *Wolbachia* infection and incompatibility dynamics in experimental selection lines. *Journal of Evolutionary Biology*, **12**, 272–82.

Perry, J.N. (1989) Review: population variation in entomology 1935–1950. *The Entomologist*, **108**, 184–98.

Perry, J.N. (1998) Measures of spatial pattern for counts. *Ecology*, **79**, 1008–17.

Perry, J.N. and Bowden, J. (1983) A comparative analysis of *Chrysoperla carnea* catches in light- and suction-traps. *Ecological Entomology*, **8**, 383–94.

Persons, M.H. (1999) Hunger effects on foraging responses to perceptual cues in immature and adult wolf spiders (Lycosidae). *Animal Behaviour*, **57**, 81–8.

Peters, R.H. (1983) *The Ecological Implications of Body Size*. Cambridge University Press, New York.

Petersen, M.K. (1999) The timing of dispersal of the predatory beetles *Bembidion lampros* and *Tachyporus hypnorum* from hibernating sites into arable fields. *Entomologia Experimentalis et Applicata*, **90**, 221–4.

Petersen, M.K and Hunter, M.S. (2002) Ovipositional preference and larval-early adult performance of two generalist lacewing predators of aphids in pecans. *Biological Control*, **25**, 101–9.

Peterson, B.J. and Fry, B. (1987) Stable isotopes in ecosystem studies. *Annual Review of Ecology and Systematics*. **18**, 293–320.

Petters, R.M. and Grosch, D.S. (1976) Increased production of genetic mosaics in *Habrobracon juglandis* by cold shock of newly oviposited eggs. *Journal of Embryology and Experimental Morphology*, **36**, 127–31.

Petters, R.M. and Mettus, R.V. (1980) Decreased diploid male viability in the parasitic wasp, *Bracon hebetor*. *Journal of Heredity*, **71**, 353–6.

Pettersson, J. (1970) An aphid sex attractant. I. Biological Studies. *Entomologia Scandinavica*, **1**, 63–73.

Pettersson, J. (1972) Technical data of a serological method for quantitative predator efficiency studies on *Rhopalosiphum padi* (L.). *Swedish Journal of Agricultural Research*, **2**, 65–69.

Pettersson, R.B. (1996) Effect of forestry on the abundance and diversity of arboreal spiders in the boreal spruce forest. *Ecography*, **19**, 221–8.

Pexton, J. and Mayhew, P.J. (2002) Siblicide and life-history evolution in parasitoids. *Behavioral Ecology*, **13**, 690–5.

Pfiffner, L. and Luka, H. (2000) Overwintering of arthropods in soils of arable fields and adjacent semi-natural habitats. *Agriculture, Ecosystems and Environment*, **78**, 215–22.

Phillips, A., Milligan, P.J.M., Broomfield, G. and Molyneux, D.H. (1988) Identification of medically important Diptera by analysis of cuticular hydrocarbons, in *Biosystematics of Haematophagous Insects Systematics Association Special Volume No.* 37 (ed. M.W. Service). Clarendon Press, Oxford, Chapter 4.

Phillips, C.B. and Baird, D.B. (1996) A morphometric method to assist in defining the South American origins of *Microctonus hyperodae* Loan (Hymenoptera: Braconidae) established in New Zealand. *Biocontrol Science and Technology*, **6**, 189–205.

Phillipson, J. (1960) A contribution to the feeding biology of *Mitopus morio* (F.) (Phalangida). *Journal of Animal Ecology*, **29**, 35–43.

Pianka, E.R. (1973) The structure of lizard communities. *Annual Review of Ecology and Systematics*, **4**, 53–74.

Pickard, R.S. (1975) Relative abundance of syrphid species in a nest of the wasp *Ectemnius cavifrons* compared with that in the surrounding habitat. *Entomophaga*, **20**, 143–51.

Pickett, J.A. (1988) Integrating use of beneficial organisms with chemical crop protection. *Philosophical Transactions of the Royal Society of London B*, **318**, 203–11.

Pickett, J.A., Wadhams, L.J., Woodcock, C.M., and Hardie, J. (1992) The chemical ecology of aphids. *Annual Review of Entomology*, **37**, 67–90.

Pickup, J. and Thompson, D.J. (1990) The effects of temperature and prey density on the development rates and growth of damselfly larvae (Odonata: Zygoptera). *Ecological Entomology*, **15**, 187–200.

Piek, T. (1986) *Venoms of the Hymenoptera. Biochemical, Pharmacological and Behavioural Aspects*. Academic Press, London.

Pielou, E.C. (1984) *The Interpretation of Ecological Data*. Wiley, New York.

Pierce, C.L., Crowley, P.H. and Johnson, D.M. (1985) Behaviour and ecological interactions of larval Odonata. *Ecology*, **66**, 1504–12.

van der Pijl, L. and Dodson, C.H. (1966) *Orchid Flowers: Their Pollination and Evolution*. University of Miami Press, Coral Gables.

Pijls, J.W.A.M., Steenbergen, H.J. and van Alphen, J.J.M. (1996) Asexuality cured. The relations and differences between sexual and asexual *Apoanagyrus diversicornis*. *Heredity*, **76**, 506–13.

Pimentel, D. (1963) Introducing parasites and predators to control native pests. *Canadian Entomologist*, **95**, 785–92.

Pimentel, D., Nagel, W.P. and Madden, J.L. (1963) Space-time structure of the environment and the survival of parasite-host systems. *American Naturalist*, **97**, 141–66.

Pimentel, D., Glenister, C., Fast, S. and Gallahan, D. (1984) Environmental risks of biological pest controls. *Oikos*, **42**, 283–90.

Pinnegar, J.K. and Polunin, N.V.C. (2000) Contributions of stable-isotope data to elucidating food webs of Mediterranean rocky littoral fishes. *Oecologia*, **122**, 399–409.

Pintureau, B. and Babault, M. (1981) Enzymatic characterisation of *Trichogramma evanescens* and *T. maidis* (Hym.: Trichogrammatidae); study of hybrids. *Entomophaga*, **26**, 11–22.

Pintureau, B., Grenier, S., Boleat, B., Lassabliere, F., Heddi, A. and Khatchadourian, C. (2000) Dynamics of *Wolbachia* populations in transfected lines of *Trichogramma*. *Journal of Invertebrate Pathology*, **76**, 20–5.

Piper, R.W. and Compton, S.G. (2002) A novel technique for relocating concealed insects. *Ecological Entomology*, **27**, 251–3.

Pivnick, K.A. (1993) Diapause initiation and pupation site selection of the braconid parasitoid *Microplitis mediator* (Haliday): A case of manipulation of host behaviour. *Canadian Entomologist*, **125**, 825–30.

Platt, J., Caldwell, J.S. and Kok, L.T. (1999) An easily replicated, inexpensive Malaise-type trap design. *Journal of Entomological Science*, **34**, 154–7.

Plaut, H.N. (1965) On the phenology and control value of *Stethorus punctillum* Weise as a predator of *Tetranychus cinnabarinus* Boid. in Israel. *Entomophaga*, **10**, 133–7.

Pleasants, J.M. (1989) Optimal foraging by nectarivores: a test of the marginal value theorem. *American Naturalist*, **134**, 51–71.

Plepys, D., Ibarra, F., Francke, W. and Löfstedt, C. (2002) Odour-mediated nectar foraging in the silver Y moth, *Autographa gamma* (Lepidoptera; Noctuidae): behavioural and electrophysiological responses to floral volatiles. *Oikos*, **99**, 75–82.

Plowright, C.M.S. and Plowright, R.C. (1997) The advantage of short tongues in bumblebees (*Bombus*) – analyses of species distributions according to flower corolla depth, and of working speeds on white clover. *Canadian Entomologist*, **129**, 51–9.

Podoler, H. and Rogers, D. (1975) A new method for the identification of key factors from life-table data. *Journal of Animal Ecology*, **44**, 85–114.

Poehling, H.M. (1987) Effect of reduced dose rates of pirimicarb and fenvalerate on aphids and beneficial arthropods in winter wheat. *Bulletin IOBC/WPRS Working Group Integrated Control of Cereal Pests*, **10**, 184–93.

Poirié, M., Périquet, G. and Beukeboom, L. (1993) The hymenopteran way of determining sex. *Seminars in Developmental Biology*, **3**, 357–61.

Poirié, M., Hita, M., Frey, F., Huguet, E., Lemeunier, F., Periquet, G. and Carton, Y. (1999) *Drosophila* resistance genes against parasitoids: a gene-for-species'concept. *Abstract of the 7th European Society for Evolutionary Biology Congress, Barcelona*, **II**, p. 235.

Polgar, L.A. and Hardie, J. (2000) Diapause induction in aphid parasitoids. *Entomologia Experimentalis et Applicata*, **97**, 21–7.

Polgar, L.A., Mackauer, M. and Völkl, W. (1991) Diapause induction in two species of aphid parasitoids: The influence of aphid morph. *Journal of Insect Physiology*, **37**, 699–702.

Polgar, L.A., Darvas, B. and Völkl, W. (1995) Induction of dormancy in aphid parasitoids: Implications for enhancing their field effectiveness. *Agriculture, Ecosystems and Environment*, **52**, 19–23.

Polis, G.A. (1981) The evolution and dynamics of intraspecific predation. *Annual Review of Ecology and Systematics*, **12**, 225–32.

Polis, G.A. (1991) Complex trophic interactions in deserts: an empirical critique of food-web theory. *American Naturalist*, **138**, 123–55.

Polis, G.A. (1999) Why are parts of the world green? Multiple factors control productivity and the distribution of biomass. *Oikos*, **86**, 3–15.

Polis, G.A. and Holt, R.D. (1992) Intraguild predation: The dynamics of complex trophic interactions. *Trends in Ecology and Evolution*, **7**, 151–4.

Polis, G.A. and Strong, D.R. (1996) Food web complexity and community dynamics. *American Naturalist*, **147**, 813–46.

Polis, G.A., Myers, C.A. and Holt, R.D. (1989) The ecology and evolution of intraguild predation: potential competitors that eat each other. *Annual Review of Ecology and Systematics*, **20**, 297–330.

Polis, G.A., Holt, R.D., Menge, B.A. and Winemiller, K.O. (1996) Time, space, and life history: influences on food webs, in *Food Webs: Integration of Patterns and Dynamics* (eds G.A. Polis and K.O. Winemiller). Chapman and Hall, New York, pp. 435–60.

Pollard, E. (1968) Hedges. III. The effect of removal of the bottom flora of a hawthorn hedgerow on the

Carabidae of the hedge bottom. *Journal of Applied Ecology*, **5**, 125–39.

Pollard, E. (1971) Hedges, IV. Habitat diversity and crop pests: a study of *Brevicoryne brassicae* and its syrphid predators. *Journal of Applied Ecology*, **8**, 751–80.

Pollard, E. (1977) A method for assessing changes in the abundance of butterflies. *Biological Conservation*, **12**, 115–34.

Pollard, E., Lakhani, K.H. and Rothery, P. (1987) The detection of density-dependence from a series of annual censuses. *Ecology*, **68**, 2046–55.

Pollard, S. (1988) Partial consumption of prey: the significance of prey water loss on estimates of biomass intake. *Oecologia*, **76**, 475–6.

Pollard, S.D. (1990) A methodological constraint influencing measurement of food intake rates in sucking predators. *Oecologia*, **82**, 569–71.

Pollet, M. and Desender, K. (1985) Adult and larval feeding ecology in *Pterostichus melanarius* Ill. (Coleoptera, Carabidae). *Mededelingen van de Fakulteit Landbouwwetenschappen Rijksuniversiteit Gent*, **50**, 581–94.

Pollet, M. and Desender, K. (1986) Prey selection in carabid beetles (Col: Carabidae): are diel activity patterns of predators and prey synchronised? *Mededelingen van de Fakulteit Landbouwwetenschappen Rijksuniversiteit Gent*, **51**, 957–71.

Pollet, M. and Desender, K. (1988) Quantification of prey uptake in pasture inhabiting carabid beetles. *Mededelingen van de Fakulteit Landbouwwetenschappen Rijksuniversiteit Gent*, **53**, 1119–29.

Pollet, M. and Desender, K. (1990) Investigating the food passage in *Pterostichus melanarius* (Coleoptera, Carabidae): an attempt to explain its feeding behaviour. *Mededelingen van de Fakulteit Landbouwwetenschappen Rijksuniversiteit Gent*, **55**, 527–40.

Pollet, M. and Grootaert, P. (1987) Ecological data on Dolichopodidae (Diptera) from a woodland ecosystem: I. Colour preference, detailed distribution and comparison of different sampling techniques. *Bulletin de l'Institut Royal des Sciences Naturelles de Belgique, Entomologie*, **57**, 173–86.

Pollet, M. and Grootaert, P. (1993) Factors affecting the responses of Empidoidea (Insecta, Diptera) to coloured traps. *Belgian Journal of Zoology*, **123**, *Supplement* **1**, 58–9.

Pollet, M. and Grootaert, P. (1994) Optimizing the water trap technique to collect Empidoidea (Diptera). *Studia Dipterologica*, **1**, 33–48.

Pollet, M. and Grootaert, P. (1996) An estimation of the natural value of dune habitats using Empidoidea (Diptera). *Biodiversity and Conservation*, **5**, 859–80.

Pollock, K.H., Nichols, J.D., Brownie, C. and Hines, J.E. (1990) Statistical inference for capture-recapture experiments. *Wildlife Monographs*, **107**, 1–97.

Pompanon, F., Fouillet, P. and Boulétreau, M. (1995) Emergence patterns and protandry in relation to daily patterns of locomotor activity in *Trichogramma* species. *Evolutionary Ecology*, **9**, 467–77.

Pompanon, F., De Schepper, B., Mourer, Y., Fouillet, P. and Boulétreau, M. (1997) Evidence for a substrate-borne sex pheromone in the parasitoid wasp *Trichogramma brassicae*. *Journal of Chemical Ecology*, **23**, 1349–60.

Ponsard, S. and Arditi, R. (2000) What can stable isotopes (δ^{15}N and δ^{13}C) tell about the food web of soil macro-invertebrates? *Ecology*, **81**, 852–64.

Porres, M.A., McMurty, J.A. and March, R.B. (1975) Investigations of leaf sap feeding by three species of phytoseiid mites by labelling with radioactive phosphoric acid ($H_3^{32}PO_4$). *Annals of the Entomological Society of America*, **68**, 871–2.

Poser, T. (1988) Chilopoden als Prädatoren in einem Laubwald. *Pedobiologia*, **31**, 261–81.

Post, D.M. (2002) Using stable isotopes to estimate trophic position: models, methods and assumptions. *Ecology*, **83**, 703–18.

Post, D.M., Pace, M.L. and Hairston, N.G. (2000) Ecosystem size determines food-chain length in lakes. *Nature*, **405**, 1047–9.

Postek, M.T., Howard, K.S., Johnson, A.H. and McMichael, K.L. (1980) *Scanning Electron Microscopy – A Student's Handbook*. Ladd Research Industries, Louisiana.

Potter, C.F., and Bertin, R.I. (1988) Amino acids in artificial nectar: feeding preferences of the flesh fly *Sarcophaga bullata*. *American Midland Naturalist*, **120**, 156–62.

Potting, R.P.J., Vet, L.E.M. and Dicke, M. (1995) Host microhabitat location by stem-borer parasitoid *Cotesia flavipes* – The role of herbivore volatiles and locally and systemically induced plant volatiles. *Journal of Chemical Ecology*, **21**, 525–39.

Potting, R.P.J., Snelling, H.M. and Vet, L.E.M. (1997) Fitness consequences of superparasitism and mechanism of host discrimination in the stemborer parasitoid *Cotesia flavipes*. *Entomologia Experimentalis et Applicata*, **82**, 341–8.

Powell, J.F. and King, E.G. (1984) Behaviour of adult *Microplitis croceipes* (Hymenoptera: Braconidae) and parasitism of *Heliothis* spp. (Lepidoptera: Noctuidae) host larvae in cotton. *Environmental Entomology*, **13**, 272–7.

Powell, W. (1980) *Toxares deltiger* (Haliday) (Hymenoptera: Aphidiidae) parasitising the cereal aphid, *Metopolophium dirhodum* (Walker) (Hemiptera:

Aphididae), in southern England: a new host-parasitoidrecord. *Bulletin of Entomological Research*, **70**, 407–9.

Powell, W. (1982) The identification of hymenopterous parasitoids attacking cereal aphids in Britain. *Systematic Entomology*, **7**, 465–73.

Powell, W. (1986) Enhancing parasitoid activity in crops, in *Insect Parasitoids* (eds J. Waage and D. Greathead). Academic Press, London, pp. 319–40.

Powell, W. (1999) Parasitoid hosts, in *Pheromones of Non-Lepidopteran Insects Associated with Agricultural Plants* (eds J. Hardie and A.K. Minks). CABI Publishing, Wallingford, pp. 405–27.

Powell, W. and Bardner, R. (1984) Effects of polyethylene barriers on the numbers of epigeal predators caught in pitfall traps in plots of winter wheat with and without soil-surface treatments of fonofos. *Bulletin IOBC/WPRS Working Group Integrated Control of Cereal Pests*, **8**, 136–8.

Powell, W. and Walton, M.P. (1989) The use of electrophoresis in the study of hymenopteran parasitoids of agricultural pests, in *Electrophoretic Studies on Agricultural Pests* (eds H.D. Loxdale and J. den Hollander). Clarendon Press, Oxford, pp. 443–66.

Powell, W. and Wright, A.F. (1992) The influence of host food plants on host recognition by four aphidiine parasitoids (Hymenoptera: Braconidae). *Bulletin of Entomological Research*, **81**, 449–53.

Powell, W., Dean, G.J. and Bardner, R. (1985) Effects of pirimicarb, dimethoate and benomyl on natural enemies of cereal aphids in winter wheat. *Annals of Applied Biology*, **106**, 235–42.

Powell, W., Hardie, J., Hick, A.J., Höller, C., Mann, J., Merritt, L., Nottingham, S.F., Wadhams, L.J., Witthinrich, J. and Wright, A.F. (1993) Responses of the parasitoid *Praon volucre* (Hymenoptera: Braconidae) to aphid sex pheromone lures in cereal fields in autumn: Implications for parasitoid manipulation. *European Journal of Entomology*, **90**, 435–8.

Powell, W., Hawthorne, A., Hemptinne, J.L., Holopainen, J.K., Den Nijs, L.J.F.M., Riedel, W. and Ruggle, P. (1995) Within-field spatial heterogeneity of arthropod predators and parasitoids. *Acta Jutlandica*, **70**, 235–42.

Powell, W., Pennacchio, F., Poppy, G.M. and Tremblay, E. (1998) Strategies involved in the location of hosts by the parasitoid *Aphidius ervi* Haliday (Hymenoptera: Braconidae: Aphidiinae). *Biological Control*, **11**, 104–12.

Prasifka, J.R., Krauter, P.C., Heinz, K.M., Sansone, C.G. and Minzenmayer, R.R. (1999) Predator conservation in cotton: using grain sorghum as a source for insect predators. *Biological Control*, **16**, 223–9.

Prasifka, J.R., Heinz, K.M. and Sansone, C.G. (2001) Field testing rubidium marking for quantifying intercrop movement of predatory arthropods. *Environmental Entomology*, **30**, 711–19.

Preston-Mafham, K.G. (1999) Courtship and mating in *Empis (Xanthempis) trigramma* Meig., *E. tessellata* F. and *E. (Polyblepharis) opaca* F. (Diptera: Empididae) and the possible implications of 'cheating behaviour'. *Journal of Zoology*, **247**, 239–46.

Prevost, G. and Lewis, W.J. (1990) Heritable differences in the response of the braconid wasp *Microplitis croceipes* to volatile allelochemicals. *Journal of Insect Behaviour*, **3**, 277–87.

Price, P.W. (1970) Trail odours: recognition by insects parasitic in cocoons. *Science*, **170**, 546–7.

Price, P.W. (1975) Reproductive strategies of parasitoids, in *Evolutionary Strategies of Parasitic Insects and Mites* (ed. P.W. Price). Plenum, New York, pp. 87–111.

Price, P.W. (1981) Semiochemicals in evolutionary time, in *Semiochemicals their Role in Pest Control* (eds D.A. Nordlund, R.L. Jones and W.J. Lewis). Wiley and Sons, New York, pp. 251–79.

Price, P.W. (1987) The role of natural enemies in insect populations, in *Insect Outbreaks* (eds P. Barbosa and J.C. Schultz). Academic Press, New York, pp. 287–312.

Price, P.W. (1988) Inversely density-dependent parasitism: the role of plant refuges for hosts. *Journal of Animal Ecology*, **57**, 89–96.

Price, P.W. (1992) Plant resources as the mechanistic basis for insect-herbivore population dynamics, in *Effects of Resource Distribution on Animal-plant Interactions* (eds M.D. Hunter, T. Ohgushi and P.W. Price). Academic Press, San Diego. U.S.A. pp. 139–3.

Price, T. (1997a) Review of '*Phylogenies and the Comparative method in Animal Behaviour*' E.P. Martins (ed) (1996). Oxford University Press, Oxford. *Animal Behaviour*, **54**, 235–8.

Price, T. (1997b) Correlated evolution and independent contrasts. *Philosophical Transactions of the Royal Society of London B*, **352**, 519–29.

Primack, R.B. (1983) Insect pollination in the New Zealand mountain flora. *New Zealand Journal of Botany*, **21**, 317–33.

Primrose, S.B. (1995) *Principles of Genome Analysis*. Blackwell Science, Oxford, UK.

Principi, M.M. (1949) Contributi allo studio dei neurotteri italiani. 8. Morfologia, anafomia e funzionamento degli apparati genitali nel gen. *Chrysopa* Leach (*Chrysopa septempunctata* Wesm. e *Chrysopa formosa* Brauer). *Bolletino di Istituto Enomologia di Bologna*, **17**, 316–62.

Principi, M.M. and Canard, M. (1984) Feeding habits, in *Biology of the Chrysopidae* (eds M. Canard, Y. Semeria and T. New). W. Junk, The Hague, pp. 76–92.

Pritchard, G. (1989) The roles of temperature and diapause in the life history of a temperate-zone dragonfly: *Argia vivida*. *Ecological Entomolology*, **14**, 99–108.

Proft, J., Maier, W.A. and Kampen, H. (1999) Identification of six sibling species of the *Anopheles maculipennis* complex (Diptera: Culicidae) by a polymerase chain reaction assay. *Parasitology Research*, **85**, 837–43.

Pruess, K.P. (1983) Day-degree methods for pest management. *Environmental Entomology*, **12**, 613–9.

Pruess, K.P., Lal Saxena, K.M. and Koinzan, S. (1977) Quantitative estimation of alfalfa insect populations by removal sweeping. *Environmental Entomology*, **6**, 705–8.

Pschorn-Walcher, H. (1977) Biological control of forest insects. *Annual Review of Entomology*, **22**, 1–22.

Purvis, A. and Rambaut, A. (1995) Comparative analysis by independent contrasts (CAIC): an Apple Macintosh application for analysing comparative data. *Computer Applications in the Biosciences*, **11**, 247–51.

Purvis, G. and Fadl, A. (1996) Emergence of Carabidae (Coleoptera) from pupation: a technique for studying the "productivity" of carabid habitats. *Annales Zoologici Fennici*, **33**, 215–23.

Putman, R.J. and Wratten, S.D. (1984) *Principles of Ecology*. Croom Helm, London.

Putman, W. (1963) Nectar of peach leaf glands as insect food. *Canadian Entomologist*, **95**, 108–9.

Putman, W.L. (1965) Paper chromatography to detect predation on mites. *Canadian Entomologist*, **97**, 435–41.

Putters, F.A. and van den Assem, J. (1988) The analysis of preference in a parasitic wasp. *Animal Behaviour*, **36**, 933–5.

Putters, F.A. and Vonk, M. (1991) The structure oriented approach in ethology: network models and sex ratio adjustments in parasitic wasps. *Behaviour*, **114**, 148–60.

Pyke, G.H. (1983) Animal movements: an optimal foraging approach, in *The Ecology of Animal Movement* (eds I.R. Swingland and P.J. Greenwood). Clarendon Press, Oxford, pp. 7–31.

Pyke, G.H. (1984) Optimal foraging theory: a critical review. *Annual Review of Ecology and Systematics*, **15**, 523–75.

Pyke, G.H. (1991) What does it cost a plant to produce floral nectar? *Nature*, **350**, 58–9.

Queller, D.C., Strassmann, J.E. and Hughes, C. (1993) Microsatellites and kinship. *Trends in Ecology and Evolution*, **8**, 285–8.

Quicke, D.L.J. (1993) *Principles and Techniques of Contemporary Taxonomy*. Chapman and Hall, London.

Quicke, D.L.J. (1997) *Parasitic Wasps*. Chapman and Hall, London.

Quicke, D.L.J., Fitton, M.G. and Ingram S. (1992) Phylogenetic implications of the structure and distribution of ovipositor valvilli in the Hymenoptera (Insecta). *Journal of Natural History* **26**, 587–608.

Quicke, D.L.J., Fitton, M.G., Tunstead, J.R., Ingram, S.N. and Gaitens, P.V. (1994) Ovipositor structure and relationships within the Hymenoptera, with special reference to the Ichneumonoidea. *Journal of Natural History*, **28**, 635–82.

Quimio, G.M. and Walter, G.H. (2000) Swarming, delayed sexual maturation of males, and mating behaviour of *Fopius arisanus* (Sonan) (Hymenoptera: Braconidae). *Journal of Insect Behavior*, **13**, 797–813.

Quinn, M.A., Kepner, R.L., Walgenbach, D.D., Foster, R.N., Bohls, R.A., Pooler, P.D., Reuter, K.C. and Swain, J.L. (1993) Grasshopper stages of development as indicators of nontarget arthropod activity: implications for grasshopper management programs on mixed-grass rangeland. *Environmental Emtomology*, **22**, 532–40.

Rácz, V. (1983) Populations of predatory Heteroptera in apple orchards under different types of management. *Proceedings of the International Conference of Integrated Plant Protection*, **2**, 34–9.

Radwan, Z. and Lövei, C.L. (1982) Distribution and bionomics of ladybird beetles (Col., Coccinellidae) living in an apple orchard near Budapest, Hungary. *Zeitschrift für Angewandte Entomologie*, **94**, 169–75.

Raffa, K.F., Havill, N.P., and Nordheim, E.V. (2002) How many choices can your test animal compare effectively? Evaluating a critical assumption of behavioral preference tests. *Oecologia*, **133**, 422–9.

Raffaelli, D. and Hall, S.J. (1992) Compartments and predation in an estuarine food web. *Journal of Animal Ecology*, **61**, 551–60.

Ragsdale, D.W., Larson, A.D. and Newsom, L.D. (1981) Quantitative assessement of the predators of *Nezara viridula* eggs and nymphs within a soybean agroecosystem using ELISA. *Environmental Entomology*, **10**, 402–5.

Ragusa, S. (1974) Influence of temperature on the oviposition rate and longevity of *Opius concolor siculus* (Hymenoptera: Braconidae). *Entomophaga*, **19**, 61–6.

Rahman, M.H., Fitton, M.G. and Quicke, D.L.J. (1998) Ovipositor internal microsculpture in the Braconidae (Insecta, Hymenoptera). *Zoologica Scripta*, **27**, 319–31.

Ram, A. and Sharma, A.K. (1977) Selective breeding for improving the fecundity and sex-ratio of *Tricho-*

gramma fasciatum (Perkins) (Trichogrammatidae: Hymenoptera), an egg parasite of Lepidopterous hosts. *Entomon*, **2**, 133–7.

Ramadan, M.M., Wong, T.T. and Wong, M.A. (1991) Influence of parasitoid size and age on male mating success of Opiinae (Hymenoptera: Braconidae), larval parasitoids of fruit flies (Diptera: Tephritidae). *Biological Control*, **1**, 248–55.

Ratcliffe, N.A. (1982) Cellular defence reactions of insects. *Fortschritte Zoologie*. **27**, 223–44.

Ratcliffe, N.A. and King, P.E. (1969) Morphological, ultrastructural, histochemical and electrophoretic studies on the venom system *of Nasonia vitripennis* Walker (Hymenoptera: Pteromalidae). *Journal of Morphology*, **127**, 177–204.

Ratcliffe, N.A. and Rowley, A.F. (1987) Insect responses to parasites and other pathogens, in *Immune Responses in Parasitic Infections: Immunology, Immunopathology, and Immunoprophylaxis* (ed. E.J.L. Soulsby). CRC Press, Boca Raton, pp. 271–332.

Rathay, E. (1883) Untersuchungen über die Spermogonien der Rostpilze. *Denkschriften der Kiaserlichen Akademie der Wissenschafen zu Wien*, **46**, 1–51.

Ratner, S. and Vinson, S.B. (1983) Encapsulation reactions in vitro by haemocytes of *Heliothis virescens*. *Journal of Insect Physiology*, **29**, 855–63.

Ravlin, F.W. and Haynes, D.L. (1987) Simulation of interactions and management of parasitoids in a multiple host system. *Environmental Entomology*, **16**, 1255–65.

Raworth, D.A. and Choi, M.Y. (2001) Determining numbers of active carabid beetles per unit area from pitfall-trap data. *Entomologia Experimentalis et Applicata*, **98**, 95–108.

Raymond, B., Darby, A.C. and Douglas, A.E. (2000) Intraguild predators and the spatial distribution of a parasitoid. *Oecologia*, **124**, 367–72.

Read, A.F., Narara, A., Nee, S., Keymer, A.E. and Day, K.P. (1992) Gametocyte sex ratios as indirect measures of outcrossing rates in malaria. *Parasitology*, **104**, 387–95.

Read, A.F., Smith, T.G., Nee, S. and West, S.A (2002) Sex ratios of malaria parasites and related protozoa. in *Sex Ratios: Concepts and Research Methods* (ed. I.C.W. Hardy). Cambridge University Press, Cambridge, pp. 314–32.

Readshaw, J.L. (1973) The numerical response of predators to prey density. *Journal of Applied Ecology*, **10**, 342–51.

Redfern, M. and Askew, R.R. (1992) *Plant Galls, Naturalists' Handbooks No 17*. Richmond Publishing Co. Ltd, Slough.

Redfern, M., Jones, T.H. and Hassell, M.P. (1992) Heterogeneity and density dependence in a field study of a tephritid-parasitoid interaction. *Ecological Entomology*, **17**, 255–62.

Reeve, J.D. (1990) Stability, variability and persistence in host-parasitoid systems. *Ecology*, **71**, 422–6.

Reeve, J.D., Kerans, B.L. and Chesson, P.L. (1989) Combining different forms of parasitoid aggregation: effects on stability and patterns of parasitism. *Oikos*, **56**, 233–9.

Reeve, J.D., Simpson, J.A. and Fryar, J.S. (1996) Extended development in *Thanasimus dubius* (F.) (Coleoptera: Cleridae), a predator of the southern pine beetle. *Journal of Entomological Science*, **31**, 123–31.

Reimann, J.G., Moen, D.O. and Thorson, B.J. (1967) Female monogamy and its control in the house fly *Musca domestica* L.. *Journal of Insect Physiology*, **13**, 407–18.

Reitsma, C. (1966) Pollen morphology of some European Rosaceae. *Acta Botanica Neerlandensis*, **15**, 290–307.

Reitz, S.R. and Trumble, J.T. (1997) Effects of linear furanocoumarins on the herbivore *Spodoptera exigua* and the parasitoid *Archytas marmoratus*: host quality and parasitoid success. *Entomologia Experimentalis et Applicata*, **84**, 9–16.

Ren, S.X., Stansly, P.A. and Liu, T.X. (2002) Life history of the whitefly predator *Nephaspis oculatus* (Coleoptera: Coccinellidae) at six constant temperatures. *Biological Control*, **23**, 262–8.

Renner, F. (1986) Zur Nischendifferenzierung bei Pirata-Arten (Araneida, Lycosidae). *Verhandlungen des Naturwissenschaftlichen Vereins im Hamburg (NF)*, **28**, 75–90.

Reuter, M. and Keller, L. (2003) High levels of multiple *Wolbachia* infection and recombination in the ant *Formica exsecta*. *Molecular Biology Evolution*, **20**, 748–53.

Reynolds, D.R., Riley, J.R., Armes, N.J., Cooter, R.J., Tucker, M.R. and Colvin, J. (1997). Techniques for quantifying insect migration, in *Methods in Ecological and Agricultural Entomology* (eds D.R. Dent and M.P. Walton). CAB International, Wallingford, pp. 111–45.

Reynolds, J.D. (1996) Animal breeding systems. *Trends in Ecology and Evolution*, **11**, 68–72.

Ricci, C. (1986) Seasonal food preferences and behaviour of *Rhyzobius litura*, in *Ecology of Aphidophaga* (ed. I. Hodek). Academia, Prague and W. Junk, Dordrecht, pp. 119–23.

Rice, R.E. and Jones, R.A. (1982) Collections of *Prospaltella perniciosi* Tower (Hymenoptera: Aphelinidae) on San Jose scale (Homoptera: Diaspididae) pheromone traps. *Environmental Entomology*, **11**, 876–80.

Rice, W.R. (1996) Sexually antagonistic male adaptation triggered by experimental arrest of female evolution. *Nature*, **381**, 232–4.

Richards, O.W. and Davies, R.C. (1977) *Imms' General Textbook of Entomology, Vols 1 and 2*. Chapman and Hall, London.

Richardson, B.J., Baverstock, P.R. and Adams, M. (1986) *Allozyme Electrophoresis*. Academic Press, London.

Riddick, E.W. (2003) Factors affecting progeny production of *Anaphes iole*. *Biocontrol*, **48**, 177–89.

Riddick, E.W. and Mills, N.J. (1994) Potential of adult carabids (Coleoptera: Carabidae) as predators of fifth-instar codling moth (Lepidoptera: Tortricidae) in apple orchards in California. *Environmental Entomology*, **23**, 1338–45.

Ridley, M. (1988) Mating frequency and fecundity in insects, *Biological Reviews*, **63**, 509–49.

Ridley, M. (1989) Why not to use species in comparative tests. *Journal of Theoretical Biology*, **136**, 361–4.

Ridley, M. (1993a) *Evolution*. Blackwell Scientific Publishing, Oxford.

Ridley, M. (1993b) Clutch size and mating frequency in the parasitic Hymenoptera. *American Naturalist*, **142**, 893–910.

Riecken, U. (1999) Effects of short-term sampling on ecological characterization and evaluation of epigeic spider communities and their habitats for site assessment studies. *Journal of Arachnology*, **27**, 189–95.

Riecken, U. and Raths, U. (1996) Use of radio telemetry for studying dispersal and habitat use of *Carabus coriaceus* L.. *Annales Zoologici Fennici*, **33**, 109–16.

Rieux, R., Simon, S. and Defrance, H. (1999) Role of hedgerows and ground cover management on arthropod populations in pear orchards. *Agriculture, Ecosystems and Environment*, **73**, 119–27.

Rigaud, T. (1997) Inherited microorganisms and sex determination of arthropod hosts, in *Influential Passengers: Inherited Microorganisms and Arthropod Reproduction* (eds S.L. O'Neill, A.A. Hoffmann and J.H. Werren). Oxford University Press, Oxford, pp. 81–101.

van Rijn, P.C.J. (2002) The Impact of Supplementary Food on a Prey-predator Interaction. University of Amsterdam, Ph.D. Thesis.

van Rijn, P.C.J. and Tanigoshi, L.K. (1999) The contribution of extrafloral nectar to survival and reproduction of the predatory mite *Iphiseius degenerans* on *Ricinus communis*. *Experimental and Applied Acarology*, **23**, 281–96.

van Rijn, P.C.J., van Houten, Y.M. and Sabelis, M.W. (2002) How plants benefit from providing food to predators even when it is also edible to herbivores. *Ecology*, **83**, 2664–79.

Riley, J.R., Smith, A.D., Reynolds, D.R., Edwards, A.S., Osborne, J.L., Williams, I.H., Carreck, N.L. and Poppy, G.M. (1996) Tracking bees with harmonic radar. *Nature*, **379**, 29–30.

Ringel, M.S., Rees, M. and Godfray, H.C.J. (1998) The evolution of diapause in a coupled host-parasitoid system. *Journal of Theoretical Biology*, **194**, 195–204.

Risch, S.J., Andow, D.A. and Altieri, M. (1983) Agroecosystems diversity and pest control: data, tentative conclusions, and new research directions. *Environmental Entomology*, 12, 625–9.

Rivero, A. and Casas, J. (1999) Rate of nutrient allocation to egg production in a parasitic wasp. *Proceedings of the Royal Society of London B*, **266**, 1169–74.

Rivero, A. and West, S.A. (2002) The physiological costs of being small in a parasitic wasp. *Evolutionary Ecology Research*, **4**, 407–20.

Rizki, R.M. and Rizki, T.M. (1984) Selective destruction of a host blood type by a parasitoid wasp. *Proceedings of the National Academy of Sciences*, **81**, 6154–8.

Rizki, R.M. and Rizki, T.M. (1990) Parasitoid virus-like particles destroy *Drosophila* cellular immunity. *Proceedings of the National Academy of Sciences*, **87**, 8388–92.

Roberts, H.R. (1973) Arboreal Orthoptera in the rain forests of Costa Rica collected with insecticide: A report on the grasshoppers (Acrididae), including new species. *Proceedings of the Academy of Natural Sciences, Philadelphia*, **125**, 46–66.

Robertson, C. (1928) *Flowers and Insects: Lists of Visitors of Four Hundred and Fifty Three Flowers*. The Science Press, Lancaster.

Robertson P.L. (1968) A morphological and functional study of the venom apparatus in representatives of some major groups of Hymenoptera. *Australian Journal of Zoology*, **16**, 133–66.

Robinson, J.V. and Novak, K.L. (1997) The relationship between mating system and penis morphology in ischnuran damselflies (Odonata: Coenagrionidae). *Biological Journal of the Linnean Society*, **60**, 187–200.

Roderick, G.K. (1992) Post-colonization evolution of natural enemies, in *Selection Criteria and Ecological Consequences of Importing Natural Enemies* (eds W.C. Kaufmann and J.R. Nechols), Thomas Say, Entomological Society of America, pp. 71–86.

Roderick, G.K. (1996) Geographic structure of insect populations: gene flow, phylogeography, and their uses. *Annual Review of Entomology*, **41**, 325–52.

Rodriguez-Saona, C. and Miller, J.C. (1999) Temperature-dependent effects on development, mortality, and growth of *Hippodamia convergens* (Coleoptera: Coccinellidae). *Environmental Entomology*, **28**, 518–22.

Roeder, K.D. (1935) An experimental analysis of the sexual behavior of the praying mantis (*Mantis religiose* L.). *Biological Bulletin*, **69**, 203–20.

Roff, D.A. (1992) *The Evolution of Life Histories: Theory and Analysis*. Chapman and Hall, New York.

Roff, D.A. (1996) The evolution of genetic correlations: an analysis of patterns. *Evolution*, **50**, 1392–1403.

Roff, D.A. (1997) *Evolutionary Quantitative Genetics*. Chapman and Hall, New York.

Roff, D.A. (2002) *Life History Evolution*. Sinauer, Sunderland.

Rogers, D. (1972) Random search and insect population models. *Journal of Animal Ecology*, **41**, 369–83.

Rogers, D.J. and Hassell, M.P. (1974) General models for insect parasite and predator searching behaviour: interference. *Journal of Animal Ecology*, **43**, 239–53.

Roitberg, B.D., Mangel, M., Lalonde, R.C., Roitberg, C.A., van Alphen, J.J.M. and Vet, L. (1992) Seasonal dynamic shifts in patch exploitation by parasitic wasps. *Behavioral Ecology*, **3**, 156–65.

Roitberg, B., Sircom, J., van Alphen, J.J.M. and Mangel, M. (1993) Life expectancy and reproduction. *Nature*, **364**, 108.

Roitberg, B.D., Boivin, G. and Vet, L.E.M. (2001) Fitness, parasitoids, and biological control: an opinion. *Canadian Entomologist*, **133**, 429–38.

Roitt, I.M., Brostoff, J. and Male, D. (2001) *Immunology*. Mosby, London.

Rokas, A. (2000) Selfish element wars in a wasp battleground. *Trends in Ecology and Evolution*, **16**, 277.

Rokas, A., Atkinson, R.J., Nieves-Aldrey, J.-L., West, S.A., and Stone, G.N. (2002) The incidence and diversity of *Wolbachia* in gallwasps (Hymenoptera: Cynipidae) on oak. *Molecular Ecology*, **11**, 1815–29.

Roland, J. (1990) Interaction of parasitism and predation in the decline of winter moth in Canada, in *Population Dynamics of Forest Insects* (eds A.D. Watt, S.R. Leather, M.D. Hunter and N.A.C. Kidd). Intercept, Andover, pp. 289–302.

Roland, J. (1994) After the decline: what maintains low winter moth density after successful biological control? *Journal of Animal Ecology*, **63**, 392–98.

Roland, J., McKinnon, G., Backhouse, C. and Taylor, P.D. (1996) Even smaller radar tags on insects. *Nature*, **381**, 120.

Rolff, J. and Kraaijeveld, A.R. (2001) Host preference and survival in selected lines of a *Drosophila* parasitoid, *Asobara tabida*. *Journal of Evolutionary Biology*, **14**, 742–5.

Rollinson, D. and Stothard, J.R. (1994) Identification of pests and pathogens by random amplification of polymorphic DNA (RAPDs), in *Identification and Characterization of Pest Organisms* (ed. D.L. Hawksworth). CAB International, Wallingford, UK.

Roltsch, W., Hanna, R., Zalom, F., Shorey, H. and Mayse, M. (1998) Spiders and vineyard habitat relationships in central California., in *Enhancing Biological Control* (eds C.H. Pickett and R.L. Bugg). University of California Press, Berkeley, California, USA, pp. 311–38.

Romeis, J. and Wäckers, F.L. (2000) Feeding responses by female *Pieris brassicae* butterflies to carbohydrates and amino acids. *Physiological Entomology*, **25**, 247–53.

Romeis, J., Shanower, T.G. and Zebitz, C.P.W. (1998) Response of *Trichogramma* egg parasitoids to coloured sticky traps. *BioControl*, **43**, 17–27.

Room, R.M. (1977) ^{32}P labelling of immature stages *of Heliothis armigera* (Hubner) and *H. punctigera* Wallengren (Lepidoptera: Noctuidae): relationships of doses to radioactivity, mortality and label half-life. *Journal of the Australian Entomological Society*, **16**, 245–51.

Root, R.B. (1973) Organization of a plant-arthropod association in simple and diverse habitats: the fauna of collards (*Brassica oleracea*). *Ecological Monographs*, **43**, 95–124.

Rose, N.R. and Friedman, H. (eds) (1980) *Manual of Clinical Immunology*. American Society for Microbiology, Washington DC.

Röse, U.S.R., Lewis, W.J. and Tumlinson, J.H. (1998) Specificity of systemically released cotton volatiles as attractants for specialist and generalist parasitic wasps. *Journal of Chemical Ecology*, **24**, 303–19.

Rosengren, R., Vepsaläinen, K. and Wuorenrinne, H. (1979) Distribution, nest densities and ecological significance of wood ants (the *Formica rufa* group) in Finland. *Bulletin SROP, 1979*, **2**, 181–213.

Rosenheim, J.A. (1996) An evolutionary argument for egg limitation. *Evolution*, **50**, 2089–94.

Rosenheim, J.A. (1999) The relative contributions of time and eggs to the cost of reproduction. *Evolution*, **53**, 376–85.

Rosenheim, J.A. (2001) Source-sink dynamics for a generalist insect predator in habitats with strong higher-order predation. *Ecological Monographs*, **71**, 93–116.

Rosenheim, J.A. and Brodeur, J. (2002) A simple trap to study small-scale movement by walking arthropods. *Entomologia Experimentalis et Applicata*, **103**, 283–5.

Rosenheim, J.A. and Rosen, D. (1992) Influence of egg load and host size on host-feeding behaviour of the parasitoid *Aphytis lingnanensis*. *Ecological Entomology*, **17**, 263–72.

Rosenheim, J.A., Wilhoit, L.R. and Armer, C.A. (1993) Influence of intraguild predation among generalist insect predators on the suppression of a herbivore population. *Oecologia*, **96**, 439–49.

Rosenheim, J.A., Kaya, H.K., Ehler, L.E., Marois, J.J. and Jaffee, B.A. (1995) Intraguild predation among biocontrol agents: Theory and Evidence. *Biological Control*, **5**, 303–35.

Rosenheim, J.A., Nonacs, P. and Mangel, M. (1996) Sex ratios and multifaceted parental investment. *American Naturalist*, **148**, 501–35.

Rosenheim, J.A., Limburg, D.D. and Colfer, R.G. (1999) Impact of generalist predators on a biological control agent, *Chrysoperla carnea*: direct observations. *Ecological Applications*, **9**, 409–17.

Rosenheim, J.A., Heimpel, G.E. and Mangel, M. (2000) Egg maturation, egg resorption and the costliness of transient egg limitation in insects. *Proceedings of the Royal Society of London B*, **267**, 1565–73.

Roskam, J.C. (1982) Larval characters of some eurytomid species (Hymenoptera, Chalcidoidea). *Proceedings Koninklijke Nederlandse Akademie van Wetenschappen*, **85**, 293–305.

Ross, K.G., Vargo, E.L., Keller, L. and Trager, J.C. (1993) Effect of a founder event on variation in the genetic sex-determining system of the fire ant *Solenopsis invicta*. *Genetics*, **135**, 843–54.

Rossi, M.N. and Fowler, H.G. (2000) Ant predation of larval *Diatraea saccharalis* Fab. (Lep., Crambidae) in new sugarcane in Brazil. *Journal of Applied Entomology*, **124**, 245–7.

Rössler, Y. and DeBach, P. (1973) Genetic variability in a thelytokous form of *Aphytis mytilaspidis* (Le Baron) (Hymenoptera: Aphelinidae). *Hilgardia*, **42**, 149–75.

Rotary, N. and Gerling, D. (1973) The influence of some external factors upon the sex ratio of *Bracon hebetor* Say (Hymenoptera: Braconidae). *Environmental Entomology*, **2**, 134–8.

Rotheram, S.M. (1967) Immune surface of eggs of a parasitic insect. *Nature*, **214**, 700.

Rotheram, S. (1973a) The surface of the egg of a parasitic insect. 1. The surface of the egg and first-instar larva of *Nemeritis*. *Proceedings of the Royal Society of London B*, **183**, 179–94.

Rotheram, S. (1973b) The surface of the egg of a parasitic insect. 2. The ultrastructure of the particulate coat on the egg of *Nemeritis*. *Proceedings of the Royal Society of London B*, **183**, 195–204.

Rotheray, G.E. (1979) The biology and host searching behaviour of a cynipoid parasite of aphidophagous Syrphidae. *Ecological Entomology*, **4**, 75–82.

Rotheray, G.E. (1981) Courtship, male swarming and a sex pheromone of *Diplazon pectoratorius* (Thunberg) (Hym., Ichneumonidae). *Entomologist's Gazette*, **32**, 193–6.

Rotheray, G.E. and Barbosa, P. (1984) Host related factors affecting oviposition behaviour in *Brachymeria intermedia*. *Entomologia Experimentalis et Applicata*, **35**, 141–5.

Rothery, P., Newton, I., Dale, L. and Wesolowski, T. (1997) Testing for density dependence allowing for weather effects. *Oecologia*, **112**, 518–23.

Rothschild, G.H.L. (1966) A study of a natural population of *Conomelus anceps* (Germar) (Homoptera: Delphacidae) including observations on predation using the precipitin test. *Journal of Animal Ecology*, **35**, 413–34.

Rott, A.S. and Godfray, H.C.J. (2000) The structure of a leafminer-parasitoid community. *Journal of Animal Ecology*, **69**, 274–89.

Roush, R.T. (1989) Genetic variation in natural enemies: critical issues for colonization in biological control, in *Critical Issues in Biological Control* (eds M. Mackauer, L.E. Ehler and J. Rolands). Intercept, Andover, pp. 263–88.

Roush, R.T. (1990) Genetic considerations in the propagation of entomophagous species, in *New Directions in Biological Control: Alternatives for Suppressing Agricultural Pests and Diseases* (eds R.R. Baker and P.E. Dunn). A. R. Liss, New York, pp. 373–87.

Rousset, F., Bouchon, D., Pintureau, B., Juchault, P. and Solignac, M. (1992) *Wolbachia* endosymbionts responsible for various alterations of sexuality in arthropods. *Proceedings of the Royal Society of London B*, **250**, 91–8.

Roux, V. and Raoult, D. (1995) Phylogenetic analysis of the genus Rickettsia by 16S rDNA sequencing. *Research in Microbiology*, **146**, 385–96.

Rowe, L., Arnqvist, G. Sih, A. and Krupa, J.J. (1994) Sexual conflict and the evolutionary ecology of mating patterns: water striders as a model system. *Trends in Ecology and Evolution*, **9**, 289–93.

Roy, M., Brodeur, J. and Cloutier, C. (2003) Effect of temperature on intrinsic rates of natural increase (r_m) of a coccinellid and its spider mite prey. *BioControl*, **48**, 57–72.

Royama, T. (1971) A comparative study of models for predation and parasitism. *Researches on Population Ecology, Supplement*, **1**, 1–91.

Royama, T. (1977) Population persistence and density dependence. *Ecological Monographs*, **47**, 1–35.

Royer, L., Fournet, S., Brunel, E. and Boivin, G. (1999) Intra- and interspecific host discrimination by host-seeking larvae of coleopteran parasitoids. *Oecologia*, **118**, 59–68.

Ruberson, J. and Greenstone, M.H. (1998) Predators of bollworm/budworm eggs in cotton: an immunological study. *Proceedings of the 1998 Beltwide Cotton Conference*, **2**, 1095–8.

Ruberson, J.R. and Kring, T.J. (1993) Parasitism of developing eggs by *Trichogramma pretiosum* (Hymenoptera: Trichogrammatidae): host age preference and suitability. *Biological Control*, **3**, 39–46.

Ruberson, J.R., Tauber, M.J. and Tauber, C.A. (1988) Reproductive biology of two biotypes of *Edovum puttleri*, a parasitoid of Colorado potato beetle eggs. *Entomologia Experimentalis et Applicata*, **46**, 211–19.

Ruberson, J.R., Tauber, C.A., and Tauber, M.J. (1989) Development and survival of *Telenomus lobatus*, a parasitoid of chrysopid eggs: effect of host species. *Entomologia Experimentalis et Applicata*, **51**, 101–6.

Ruberson, J.R., Tauber, M.J., Tauber, C.A. and Gollands, B. (1991) Parasitization by *Edovum puttleri* (Hymenoptern: Eulophidae) in relation to host density in the field. *Ecological Entomology*, **161**, 81–9.

Ruberson, J.R., Shen, Y.J., and Kring, T.J. (2000) Photoperiodic sensitivity and diapause in the predator *Orius insidiosus* (Heteroptera: Anthocoridae). *Annals of the Entomological Society of America*, **93**, 1123–30.

Rubia, E.G. and Shepard, B.M. (1987) Biology of *Metioche vittaticollis* (Stål) (Orthoptera: Gryllidae), a predator of rice pests. *Bulletin of Entomological Research*, **77**, 669–76.

Rubia, E.G., Ferrer, E.R. and Shepard, B.M. (1990) Biology and predatory behaviour of *Conocephalus longipennis* (de Haan) (Orthoptera: Tettigoniidae). *Journal of Plant Protection in the Tropics*, **7**, 205–11.

Rugman-Jones, P. and Eady, P.E. (2001) The sterile male technique: a source of error in estimating levels of sperm precedence? *Entomologia Experimentalis et Applicata*, **98**, 241–44.

Ruiz-Narvaez, E. and Castro-Webb, N. (2002) A simple method to estimate percentage parasitism when the host and parasitoid phenologies are unknown: A statistical approach. *BioControl*, **48**, 87–100.

Russel, R.J. (1970) The effectiveness of *Anthocoris nemorum* and *A. confusus* (Hemiptera: Anthocoridae) as predators of the sycamore aphid, *Drepanosiphum platanoides*. I. The number of aphids consumed during development. *Entomologia Experimentalis et Applicata*, **13**, 194–207.

Russell, E.P. (1989) Enemies hypothesis: a review of the effect of vegetational diversity on predatory insects and parasitoids. *Environmental Entomology*, **18**, 590–9.

Ruther, J., Homann, M. and Steidle, J.L.M. (2000) Female-derived sex pheromone mediates courtship behaviour in the parasitoid *Lariophagus distinguendus*. *Entomologia Experimentalis et Applicata*, **96**, 265–74.

Rutledge, C.E. and Wiedenmann, R.N. (2003) An attempt to change habitat preference of a parasitoid, *Cotesia sesamiae* (Hymenoptera: Braconidae), through artificial selection. *Journal of Entomological Science*, **38**, 93–103.

Růžička, V. (2000) Spiders in rocky habitats in central Bohemia. *Journal of Arachnology*, **28**, 217–22.

Růžička, V. and Antuš, P. (1997) Collecting spiders from rocky habitats. *Newsletter of the British Arachnological Society*, **80**, 4–5.

Ryan, R.B. (1990) Evaluation of biological control: introduced parasites of larch casebearer (Lepidoptera: Coleophoridae) in Oregon. *Environmental Entomology*, **19**, 1873–81.

Ryan, R.B. (1997) Before and after evaluation of biological control of the larch casebearer (Lepidoptera: Coleophoridae) in the Blue Mountains of Oregon and Washington, 1912–1995. *Environmental Entomology*, **26**, 703–15.

Ryoo, M.I., Hong, Y.S. and Yoo, C. K. (1991) Relationship between temperature and development of *Lariophagus distinguendus* (Hymenoptera: Pteromalidae), an ectoparasitoid of *Sitophilus oryzae* (Coleoptera: Curculionidae). *Journal of Economic Entomology*, **84**, 825–9.

Sabelis, M.W. (1990) How to analyse prey preference when prey density varies? A new method to discriminate between gut fullness and prey type composition. *Oecologia*, **82**, 289–98.

Sabelis, M.W. (1992) Predatory arthropods, in *Natural Enemies: The Population Biology of Predators, Parasites and Diseases* (ed. M.J. Crawley). Blackwell, Oxford, pp. 225–64.

Sabelis, M.W. and van de Baan, H.E. (1983) Location of distant spider-mite colonies by phytoseiid predators. Demonstration of specific kairomones emitted by *Tetranychus urticae and Panonychus ulmi* (Acari: Tetranychidae, Phytoseiidae). *Entomologia Experimentalis et Applicata*, **33**, 303–14.

Sabelis, M.W., Afman, B.P. and Slim, P.J. (1984) Location of distant spider mite colonies by *Phytoseiulus persimilis:* Localization and extraction of a kairomone. *Acarology*, **6**, 431–40.

Sadeghi, H. and Gilbert, F.S. (1999) Individual variation in oviposition preference, and its interaction with larval performance in an insect predator. *Oecologia*, **118**, 405–11.

Sadeghi, H. and Gilbert, F. (2000) The effect of egg load and host deprivation on oviposition behaviour in aphidophagous hoverflies. *Ecological Entomology*, **25**, 101–8.

Sagarra, L.A. and Vincent, C. (1999) Influence of host stage on oviposition, development, sex ratio and survival of *Anagyrus kamali* Moursi (Hymenoptera:

Encyrtidae), a parasitoid of the Hibiscus Mealybug, *Maconellicoccus hirsutus* Green (Homoptera: Pseudococcidae). *Biological Control*, **15**, 51–6.

Sagarra, L.A., Peterkin, D.D., Vincent, C. and Stewart, R.K. (2000a) Immune response of the hibiscus mealybug, *Maconellicoccus hirsutus* (Homoptera: Pseudococcidae), to oviposition of the parasitoid *Anagyrus kamali* Moursi (Hymenoptera: Encyrtidae). *Journal of Insect Physiology*, **46**, 647–53.

Sagarra, L.A., Vincent, C., Peters, N.F. and Stewart, R.K. (2000b) Effect of host density, temperature, and photoperiod on the fitness of *Anagyrus kamali*, a parasitoid of the hibiscus mealybug *Maconellicoccus hirsutus*. *Entomologia Experimentalis et Applicata*, **96**, 141–7.

Sahad, K.A. (1982) Biology and morphology of *Gonatocerus* sp. (Hymenoptera, Mymaridae), an egg parasitoid of the green rice leafhopper, *Nephotettix cincticeps* Uhler (Homoptera: Deltocephalidae). I. Biology. *Kontyû*, **50**, 246–60.

Sahad, K.A. (1984) Biology of *Anagrus optabilis* (Perkins) (Hymenoptera, Mymaridae), an egg parasitoid of delphacid leafhoppers. *Esakia*, **22**, 129–44.

Sahragard, A., Jervis, M.A. and Kidd, N.A.C. (1991) Influence of host availability on rates of oviposition and host-feeding, and on longevity in *Dicondylus indianus* Olmi (Hym., Dryinidae), a parasitoid of the Rice Brown Planthopper, *Nilaparvata lugens* Stål (Hem., Delphacidae). *Journal of Applied Entomology*, **112**, 153–62.

Sakuma, M. (1998) Probit analysis of preference data. *Applied Entomology and Zoology*, **33**, 339–47.

Sakurai, H., Hirano, T. and Takeda, S. (1986) Physiological distinction between aestivation and hibernation in the lady beetle, *Coccinella septempunctata bruckii* (Coleoptera: Coccinellidae). *Applied Entomology and Zoology*, **21**, 424–29.

Salmon, J.T. and Horner, N.V. (1977) Aerial dispersion of spiders in north central Texas. *Journal of Arachnology*, **5**, 153–7.

Salt, G. (1932) Superparasitism by *Collyria calcitrator* Grav. *Bulletin of Entomological Research*, **23**, 211–6.

Salt, G. (1934) Experimental studies in insect parasitism, II. Superparasitism. *Proceedings of the Royal Society of London B*, **114**, 455–76.

Salt, G. (1936) Experimental studies in insect parasitism. IV. The effect of superparasitism on populations of *Trichogramma evanescens*. *Journal of Experimental Biology*, **13**, 363–75.

Salt, G. (1940) Experimental studies in insect parasitism. VII. The effects of different hosts on the parasite *Trichogramma evanescens* Westw. (Hym.-Chalcidoidea). *Proceedings of the Royal Entomological Society of London A*, **15**, 81–124.

Salt, G. (1941) The effects of hosts upon their insect parasites. *Biological Reviews*, **16**, 239–64.

Salt, G. (1958) Parasite behaviour and the control of insect pests. *Endeavour*, **17**, 145–8.

Salt, G. (1961) Competition among insect parasitoids. Mechanisms in biological competition. *Symposium of the Society for Experimental Biology*, **15**, 96–119.

Salt, G. (1968) The resistance of insect parasitoids to the defence reactions of their hosts. *Biological Reviews*, **43**, 200–32.

Salt, G. (1970) *The Cellular Defence Reactions of Insects*. Cambridge University Press, Cambridge.

Sambrook, J., Fritsch, E.F. and Maniatis, T. (1989) *Molecular Cloning: a Laboratory Manual. Volumes I-III*. (ed. C. Nolan). Cold Spring Harbor Laboratory Press, USA.

Samson-Boshuizen, M., van Lenteren, J.C. and Bakker, K. (1974) Success of parasitization of *Pseudeucoila bochei* Weld (Hym. Cynip.): a matter of experience. *Netherlands Journal of Zoology*, **24**, 67–85.

Samu, F. and Kiss, B. (1997) Mark-recapture study to establish population density of the dominant wolf spider in Hungarian agricultural fields. *Bulletin of the British Ecological Society*, **28**, 265–9.

Samu, F. and Lövei, G.L. (1995) Species richness of a spider community (Araneae): Extrapolation from simulated increasing sampling effort. *European Journal of Entomology*, **92**, 633–8.

Samu, F. and Sárospataki, M. (1995a) Design and use of a hand-hold suction sampler, and its comparison with sweep net and pitfall trap sampling. *Folia Entomologica Hungarica*, **56**, 195–203.

Samu, F. and Sárospataki, M. (1995b) Estimation of population sizes and "home ranges" of polyphagous predators in alfalfa using mark-recapture: an exploratory study. *Acta Jutlandica*, **70**, 47–55.

Samu, F., Sunderland, K.D., Topping, C.J. and Fenlon, J.S. (1996) A spider population in flux: selection and abandonment of artificial web-sites and the importance of intraspecific interactions in *Lepthyphantes tenuis* (Araneae: Linyphiidae) in wheat. *Oecologia*, **106**, 228–39.

Samu, F., Németh, J. and Kiss, B. (1997) Assessment of the efficiency of a hand-held suction device for sampling spiders: improved density estimation or oversampling? *Annals of Applied Biology*, **130**, 371–8.

Samways, M.J. (1983) Community structure of ants (Hymenoptera: Formicidae) in a series of habitats associated with citrus. *Journal of Applied Ecology*, **20**, 833–47.

Samways, M.J. (1986) Spatial and temporal population patterns of *Aonidiella aurantii* (Hemiptera; Diaspididae) parasitoids (Hymenoptera: Aphelinidae and

Encyrtidae) caught on yellow sticky traps in citrus. *Bulletin of Entomological Research*, **76**, 265–74.

Samways, M.J. (1989) Climate diagrams and biological control: an example from the areaography of the ladybird *Chilocorus nigritus* (Fabricius, 1798) (Insecta, Coleoptera, Coccinellidae). *Journal of Biogeography*, **16**, 345–51.

Samways, M.J. (1997) Classical biological control and biodiversity conservation: what risks are we prepared to accept? *Biodiversity and Conservation*, **6**, 1309–16.

Sandlan, K. (1979a) Sex ratio regulation in *Coccygomimus turionellae* Linnaeus (Hymenoptera: Ichneumonidae) and its ecological implications. *Ecological Entomology*, **4**, 365–78.

Sandlan, K. (1979b) Host feeding and its effects on the physiology and behaviour of the ichneumonid parasite *Coccygomimus turionellae*. *Physiological Entomology*, **4**, 383–92.

Sanger, C. and King, P.E. (1971) Structure and function of the male genitalia in *Nasonia vitripennis* (Walker) (Hym.: Pteromalidae). *Entomologist*, **104**, 137–49.

Sanon, A., Ouedraogo, A.F., Tricault, Y., Credland, P.F. and Huignard, J. (1998) Biological control of bruchids in cowpea stores by release of *Dinarmus basilis* (Hymenoptera: Pteromalidae). *Environmental Entomology*, **27**, 717–25.

Sargent, J.R. and George, S.G. (1975) *Methods in Zone Electrophoresis, 3rd edn*. BDH Chemicals Ltd., Poole.

Sasaba, T. and Kiritani, K. (1975) A systems model and computer simulation of the green rice leafhopper populations in control programmes. *Researches on Population Ecology*, **16**, 231–44.

Sasaki, T., Kubo, T. and Ishikawa, H. (2002) Interspecific transfer of *Wolbachia* between two lepidopteran insects expressing cytoplasmic incompatibility: a *Wolbachia* variant naturally infecting *Cadra cautella* causes male killing in *Ephestia kuehniella*. *Genetics*, **162**, 1313–9.

Sato, Y., Tagawa, J. and Hidaka, T. (1986) Effects of the gregarious parasitoids *Apanteles rufricus* and *A. kariyai* on host growth and development. *Journal of Insect Physiology*, **32**, 281–6.

Saugstad, E.S., Bram, R.A. and Nyquist, W.E. (1967) Factors influencing sweep net sampling of alfalfa. *Journal of Economic Entomology*, **60**, 421–6.

Saul, C.B., Whiting, P.W., Saul, S.W. and Heidner, C.A. (1965) Wild-type and mutant stocks of *Mormoniella*. *Genetics*, **52**, 1317–27.

Sauphanor, B., Chabrol, L., Faivre d'Arcier, F., Sureau, F. and Lenfant, C. (1993) Side effects of diflubenzuron on a pear psylla predator: *Forficula auricularia*. *Entomophaga*, **38**, 163–74.

Sawoniewicz, J. (1973) Ichneumonidae (Hymenoptera) visiting the flowers of *Peucedanum oreoselinum* L. (Umbelliferae). *Folia Forestalia Polonica*, **21**, 43–78 (In Polish).

Sawyer, R. (1981) *Pollen Identification for Beekeepers*. University College, Cardiff Press, Cardiff.

Sawyer, R. (1988) *Honey Identification*. Cardiff Academic Press, Cardiff.

Schaupp, W.C. and Kulman, H.M. (1992) Attack behaviour and host utilization of *Coccygomimus disparis* (Hymenoptera: Ichneumonidae) in the laboratory. *Environmental Entomology*. **21**, 401–8.

Schausberger, P. (1999) Predation preference of *Typhlodromus pyri* and *Kampimodromus aberrans* (Acari: Phytoseiidae) when offered con- and heterospecific immature life stages. *Experimental and Applied Acarology*, **23**, 389–98.

Scheller, H.V. (1984) Pitfall trapping as the basis for studying ground beetle (Carabidae) predation in spring barley. *Tiddskrift voor Plantearl*, **88**, 317–24.

Schellhorn, N.A., Kuhman, T.R., Olson, A.C. and Ives, A.R. (2002) Competition between native and introduced parasitoids of aphids: nontarget effects and biological control. *Ecology*, **83**, 2745–57.

Schelvis, J. and Siepel, H. (1988) Larval food spectra of *Pterostichus oblongopunctatus* and *P. rhaeticus* in the field (Coleoptera: Carabidae). *Entomologia Generalis*, **13**, 61–6.

Schenk, D. and Bacher, S. (2002) Functional response of a generalist insect predator to one of its prey species in the field. *Journal of Animal Ecology*, **71**, 524–31.

Scheu, S. and Falca, M. (2000) The soil food web of two beech forests (*Fagus sylvatica*) of contrasting humus type: stable isotope analysis of a macro- and a mesofauna-dominated community. *Oecologia*, **123**, 285–96.

Schilthuizen, M. and Stouthamer, R. (1998) Distribution of *Wolbachia* among the guild associated with the parthenogenetic gall wasp *Diplolepis rosae*. *Heredity*, **81**, 270–4.

Schilthuizen, M., Nordlander, G., Stouthamer, R. and van Alphen, J.J.M. (1998) Morphological and molecular phylogenetics in the genus *Leptopilina* (Hymenoptera: Cynipoidea: Eucoilidae). *Systematic Entomology*, **23**, 253–64.

Schlinger, E.I. and Hall, J.C. (1960) The biology, behaviour and morphology of *Praon palitans* Muesebeck, an internal parasite of the spotted alfalfa aphid, *Therioaphis maculata* (Buckton) (Hymenoptera; Braconidae, Aphidiinae). *Annals of the Entomological Society of America*, **53**, 144–60.

Schmaedick, M.A., Ling, K.S., Gonsalves, D. and Shelton, A.M. (2001) Development and evaluation of an enzyme-linked immunosorbent assay to detect *Pieris rapae* remains in guts of arthropod predators. *Entomologia Experimentalis et Applicata*, **99**, 1–12.

Schmid-Araya, J.M., Schmid, P.E., Robertson, A., Winterbottom, J., Gjerlov, C. and Hildrew, A.G. (2002) Connectance in stream food webs. *Journal of Animal Ecology*, **71**, 1056–62.

Schmidt, J.M. and Smith, J.J.B. (1985) Host volume measurement by the parasitoid wasp *Trichogramma minutum*: the roles of curvature and surface area. *Entomologia Experimentalis et Applicata*, **39**, 213–21.

Schmidt, J.M. and Smith, J.J.B. (1987a) The effect of host spacing on the clutch size and parasitization rate of *Trichogramma minutum*. *Entomologia Experimentalis et Applicata*, **43**, 125–31.

Schmidt, J.M. and Smith, J.J.B. (1987b) Short interval time measurement by a parasitoid wasp. *Science*, **237**, 903–5.

Schmidt, J.M. and Smith, J.J.B. (1989) Host examination walk and oviposition site selection of *Trichogramma minutum:* studies on spherical hosts. *Journal of Insect Behavior*, **2**, 143–71.

Schmidt-Nielsen, K. (1984) *Scaling: Why is Animal Size So Important?* Cambridge University Press, New York.

Schmitz, O.J. (2003) Top predator control of plant biodiversity and productivity in an old-field ecosystem. *Ecology Letters*, **6**, 156–63.

Schmitz, O.J. and Suttle, K.B. (2001) Effects of top predator species on direct and indirect interactions in a food web. *Ecology*, **82**, 2072–81.

Schmitz, O.J., Hambäck, P.A. and Beckerman, A.P. (2000) Trophic cascades in terrestrial systems: a review of the effects of carnivore removals on plants. *American Naturalist*, **155**, 141–53.

Schneider, F. (1969) Bionomics and physiology of aphidophagous Syrphidae. *Annual Review of Entomology*, **14**, 103–24.

Schneider, K. and Duelli, P. (1997) A comparison of trap efficiency of window and Malaise traps. *Mitteilungen der Deutschen Gesellschaft für Allgemeine und Angewandte Entomologie*, **11**, 843–6.

Schneider, M.V., Beukeboom, L.W., Driessen, G., Lapchin, L., Bernstein, C. and van Alphen. J.J.M. (2002) Geographical distribution and genetic relatedness of sympatrical thelytokous and arrhenotokous populations of the parasitoid *Venturia canescens* (Hymenoptera). *Journal of Evolutionary Biology*, **15**, 191–200.

Schoemaker, D.D., Machado, C.A., Molbo, D., Werren, J.H., Windsor, D.M. and Herre, E.A. (2002) The distribution of *Wolbachia* in figwasps; correlation with host phylogeny, ecology and population structure. *Proceedings of the Royal Society of London B*, **269**, 2257–67.

Schoenly, K., Beaver, R.A. and Heumier, T.A. (1991) On the trophic relations of insects: a food-web approach. *American Naturalist*, **137**, 597–638.

Schoenly, K.G., Cohen, J.E., Heong, K.L., Litsinger, J.A., Aquino, G.B., Barrion, A.T. and Arida, G. (1996a) Food web dynamics of irrigated rice fields at five elevations in Luzon, Philippines. *Bulletin of Entomological Research*, **86**, 451–466.

Schoenly, K.G., Cohen, J.E., Heong, K.L., Arida, G.S., Barrion, A.T. and Litsinger, J.A. (1996b) Quantifying the impact of insecticides on food web structure of rice-arthropod populations in a Philippine farmer's irrigated field: a case study, in *Food Webs: Integration of Patterns and Dynamics* (eds G.A. Polis and K.O. Winemiller). Chapman and Hall, New York, pp. 343–51.

Schönrogge, K. and Crawley, M.J. (2000) Quantitative webs as a means of assessing the impact of alien insects. *Journal of Animal Ecology*, **69**, 841–68.

Schoof, D.D., Palchick, S. and Tempelis, C.H. (1986) Evaluation of predator-prey relationships using an enzyme immunoassay. *Annals of the Entomological Society of America*, **79**, 91–5.

Schoonhoven, L.M., Jermy, T and van Loon, J.J.A. (1998) *Insect–plant Biology: From Physiology to Evolution*. Chapman and Hall, London.

Schulten, G.G.M. (1997) The FAO Code of Conduct for the import and release of exotic biological control agents. *Bulletin OEPP/EPPO*, **27**, 29–36.

Schultz, D.R. and Clarke, M.E. (1995) An immunological study of predation on hatchery-reared, juvenile red drum (*Sciaenops ocellatus*): preparation and assays of a "red drum-specific" protein for predator-prey experiments. *Journal of Experimental Marine Biology and Ecology*, **189**, 233–49.

Schulz, K.L. and Yurista, P.M. (1995) Diet composition from allozyme analysis in the predatory cladoceran *Bythotrephes cederstroemi*. *Limnology and Oceanography*, **40**, 821–6.

Scott, J.A. (1973) Convergence of population biology and adult behaviour in two sympatric butterflies, *Neominois ridingsii* (Papilionoidea: Nymphalidae) and *Amblyscirtes simius* (Hesperoidea: Hesperiidae). *Journal of Animal Ecology*, **42**, 663–72.

Scott, M. Berrigan, D. and Hoffmann, A.A. (1997) Costs and benefits of acclimation to elevated temperature in *Trichogramma carverae*. *Entomologia Experimentalis et Applicata*, **85**, 211–9.

Scott, M.P. and Williams, S.M. (1994) Measuring reproductive success in insects, in *Molecular Ecology and Evolution: Approaches and Applications* (eds B. Schierwater, B. Streit, G.P. Wagner and R. DeSalle, eds). Birkhäuser, Basel, pp. 61–74.

Scott, S.M. and Barlow, C.A. (1984) Effect of prey availability during development on the reproductive output of *Metasyrphus corollae* (Diptera: Syrphidae). *Environmental Entomology*, **13**, 669–74.

Scott, T.A. and Melvin, E.H. (1953) Determination of dextran with anthrone. *Analytical Chemistry*, **25**, 1656–61.

Scriber, J.M. and Slansky Jr, F. (1981) The nutritional ecology of immature insects. *Annual Review of Entomology*, **26**, 183–211.

Scribner, K.T. and Bowman, T.D. (1998) Microsatellites identify depredated waterfowl remains from glaucous gull stomachs. *Molecular Ecology*, **7**, 1401–5.

Scudder, G.G.E. (1971) Comparative morphology of insect genitalia. *Annual Review of Entomology*, **16**, 379–406.

Scutareanu, P., Drukker, B., Bruin, J., Posthumus, M.A. and Sabelis, M.W. (1997) Volatiles from *Psylla*-infested pear trees and their possible involvement in attraction of anthocorid predators. *Journal of Chemical Ecology*, **23**, 2241–60.

Seal, D.R., Stansly, P.A. and Schuster, D.J. (2002) Influence of temperature and host on life history parameters of *Catolaccus hunteri* (Hymenoptera: Pteromalidae). *Environmental Entomology*, **31**, 354–60.

Seber, G.A.F. (1965) A note on the multiple-recapture census. *Biometrika*, **52**, 249.

Seber, G.A.F. (1973) *The Estimation of Animal Abundance and Related Parameters*. Griffin, London.

Secord, D. and Kareiva, P. (1996) Perils and pitfalls in the host specificity paradigm. *Bioscience*, **46**, 448–53.

Sedivy, J. and Kocourek, F. (1988) Comparative studies on two methods for sampling insects in an alfalfa seed stand in consideration of a chemical control treatment. *Journal of Applied Entomology*, **106**, 312–8.

Seifter, S.S., Dayton, S., Novic, B. and Muntwyler, E. (1950) The estimation of glycogen with the anthrone reagent. *Archives of Biochemistry and Biophysics*, **25**, 191–200.

Sekamatte, B., Latigo, M. and Russell-Smith, A. (2001) The potential of protein- and sugar-based baits to enhance predatory ant activity and reduce termite damage to maize in Uganda. *Crop Protection*, **20**, 653–62.

Sem'yanov, V.P. (1970) Biological properties of *Adalia bipunctata* L. (Coleoptera, Coccinellidae) in conditions of Leningrad region. *Zashchchita Rastenii Vreditelet' i Boleznii* **127**, 105–12.

Sengonca, Ç. and Kranz, J. (2001) A modified, four-armed olfactometer for determining olfactory reactions of beneficial arthropods. *Anzeiger für Schadlingskunde*, **74**, 127–32.

Sequeira, R. and Mackauer, M. (1992a) Covariance of adult size and development time in the parasitoid wasp *Aphidius ervi* in relation to the size of its host, *Acyrthosiphon pisum*. *Evolutionary Ecology*, **6**, 34–44.

Sequeira, R. and Mackauer, M. (1992b) Nutritional ecology of an insect host-parasitoid association: the pea aphid-*Aphidius ervi* system. *Ecology*, **73**, 183–9.

Sequeira, R. and Mackauer, M. (1994) Variation in selected life-history parameters of the parasitoid wasp, *Aphidius ervi*: influence of host developmental stage. *Entomologia Experimentalis et Applicata*, **71**, 15–22.

Sergeeva, T.K. (1975) Use of cross immunoelectrophoresis for increasing sensitivity of serological reactions in studying trophic relations in insects. *Zoologicheski Zhurnal*, **54**, 1014–19.

Service, M.W. (1973) Study of the natural predators of *Aedes cantans* (Meigen) using the precipitin test. *Journal of Medical Entomology*, **10**, 503–10.

Service, M.W., Voller, A. and Bidwell, E. (1986) The enzyme-linked immnunosorbent assay (ELISA) test for the identification of blood-meals of haematophagous insects. *Bulletin of Entomological Research*, **76**, 321–30.

Settle, W.H. and Wilson, L.T. (1990) Behavioral-factors affecting differential parasitism by *Anagrus epos* (Hymenoptera, Mymaridae), of two species of erythroneuran leafhoppers (Homoptera, Cicadellidae). *Journal of Animal Ecology*, **59**, 877–91.

Settle, W.H., Ariawan H., Astuti, E.T., Cahyana, W., Hakima, A.L., Hindayana, D., Lestari, A.S., and Pajarningsih, S. (1996) Managing tropical rice pests through conservation of generalist natural enemies and alternative prey. *Ecology*, **77**, 1975–88.

Sevenster, J.G., Ellers, J. and Driessen, G. (1998) An evolutionary argument for time-limitation. *Evolution*, **52**, 1241–4.

Severinghaus, L., Kurtak, B.H. and Eickwort, G.C. (1981) The reproductive behaviour of *Anthidium manicatum* and the significance of size for territorial males. *Behavioral Ecology and Sociobiology*, **9**, 51–58.

Shahjahan, M. (1974) *Erigeron* flowers as a food and attractive odour source for *Peristenus pseudopallipes*, a braconid parasitoid of the tarnished plant bug. *Environmental Entomology*, **3**, 69–72.

Shahjahan, M. and Streams, F.A. (1973) Plant effects on host-finding by *Leiophron pseudopallipes* (Hymenoptera: Braconidae), a parasitoid of the tarnished plant bug. *Environmental Entomology*, **2**, 921–5.

Shaltiel, L. and Ayal, Y. (1998) The use of kairomones for foraging decisions by an aphid parasitoid in small host aggregations. *Ecological Entomology*, **23**, 319–29.

Shannon, R.E. (1975) *Systems Simulation: The Art and the Science.* Prentice-Hall, New Jersey.

Sharpe, P.J.H., Curry, C.L., DeMichele, D.W. and Cole, C.L. (1977) Distribution model of organism development times. *Journal of Theoretical Biology*, **661**, 21–38.

Shaw, C.R. and Prasad, R. (1970) Starch gel electrophoresis of enzymes – a compilation of recipes. *Biochemical Genetics*, **4**, 297–320.

Shaw, M.R. (1997) Rearing parasitic Hymenoptera. *The Amateur Entomologist*, **25**, 1–46.

Shaw, M.R. and Huddleston, T. (1991) Classification and biology of braconid wasps. *Handbooks for the Identification of British Insects, Royal Entomological Society of London* **7(11)**, 1–126.

Shea, K. and Possingham, H.P. (2000) Optimal release strategies for biological control agents: an application of stochastic dynamic programming. *Journal of Applied Ecology*, **37**, 77–86.

Shea, K., Nisbet, R.M., Murdoch, W.W. and Yoo, H.J.S. (1996) The effect of egg limitation on stability in insect host–parasitoid population models. *Journal of Animal Ecology*, **65**, 743–55.

Shea, K., Possingham, H.P., Murdoch, W.W. and Roush, R. (2002) Active adaptive management in insect pest and weed control: intervention with a plan for learning. *Ecological Applications*, **12**, 927–36.

Sheehan, W., Wäckers, F.L. and Lewis, W.J. (1993) Discrimination of previously searched, host-free sites by *Microplitis croceipes* (Hymenoptera: Braconidae). *Journal of Insect Behavior*, **6**, 323–31.

Sheldon, J.K. and MacLeod, E.G. (1971) Studies on the biology of Chrysopidae. 2. The feeding behaviour of the adult *Chrysopa carnea* (Neuroptera). *Psyche*, **78**, 107–21.

Shepard, B.M. and Ooi, P.A.C. (1991) Techniques for evaluating predators and parasitoids in rice, in *Rice Insects: Management Strategies* (eds E.A. Heinrichs and T.A. Miller). Springer-Verlag, New York, pp. 197–214.

Shepard, M. and Waddill, V.H. (1976) Rubidium as a marker for Mexican bean beetles, *Epilachana varivestis* (Coleoptera: Coccinellidae). *Canadian Entomologist*, **108**, 337–9.

Sherlock, P.L., Bowden, J. and Digby, P.G.N. (1986) Studies of elemental composition as a biological marker in insects. V. The elemental composition of *Rhopalosiphum padi* (L.) (Hemiptera: Aphididae) from *Prunus padus* at different localities. *Bulletin of Entomological Research*, **76**, 621–32.

Sherratt, T.N. and Harvey, I.F. (1993) Frequency dependent food selection by arthropods: a review. *Biological Journal of the Linnean Society*, **48**, 167–86.

Shonouda, M.L., Bombosch, S., Shalaby, A.M. and Osman, S.I. (1998) Biological and chemical characterization of a kairomone excreted by the bean aphids, *Aphis fabae* Scop. (Hom.: Aphididae), and its effect on the predator *Metasyrphus corollae*. Fabr. II–Behavioural response of the predator *M. corollae* to the aphid kairomone. *Journal of Applied Entomology*, **122**, 25–8.

Short, J.R.T. (1952) The morphology of the head of larval Hymenoptera with special reference to the head of the Ichneumonoidea, including a classification of the final instar larvae of the Braconidae. *Transactions of the Royal Entomological Society of London*, **103**, 27–84.

Short, J.R.T. (1959) A description and classification of the final instar larvae of the Ichneumonidae (Insecta, Hymenoptera). *Proceedings of the United States National Museum*, **110**, 391–511.

Short, J.R.T. (1970) On the classification of the final instar larvae of the Ichneumonidae (Hymenoptera). *Transactions of the Royal Entomological Society of London, Supplement*, **122**, 185–210.

Short, J.R.T. (1978) The final larval instars of the Ichneumonidae. *Memoirs of the American Entomology Institute*, **25**, 1–508.

Siddiqui, W.H., Barlow, C.A. and Randolph, P.A. (1973) Effect of some constant and alternating temperature on population growth of the pea aphid *Acyrthosiphon pisum* (Hom: Aphididae). *Canadian Entomologist*, **105**, 145–56.

Siekmann, G., Tenhumberg, B. and Keller, M.A. (2001) Feeding and survival in parasitic wasps: sugar concentration and timing matter. *Oikos*, **95**, 425–30.

Sigsgaard, L. (1996) Serological analysis of predators of *Helicoverpa armigera* Hübner (Lepidoptera: Noctuidae) eggs in sorghum-pigeonpea intercropping at ICRISAT, India: a preliminary field study, in *The Ecology of Agricultural Pests: Biochemical Approaches* (eds W.O.C. Symondson and J.E. Liddell), Systematics Association Special Volume 53. Chapman and Hall, London, pp. 367–81.

Sigsgaard, L., Villareal, S., Gapud, V. and Rajotte, E. (1999) Directional movement of predators between the irrigated rice field and its surroundings, in *Symposium on Biological Control in the Tropics* (eds L.W. Hong and S.S. Sastroumo). CABI Publishing, Wallingford UK, pp. 43–7.

Sih, A. (1980) Optimal foraging: partial consumption of prey. *American Naturalist*, **116**, 281–90.

Sih, A. (1982) Foraging strategies and avoidance of predation by an aquatic insect, *Notonecta hoffmani*. *Ecology*, **63**, 786–96.

Sih, A. (1987) Nutritional ecology of aquatic insect predators, in *Nutritional Ecology of Insects, Mites, Spiders and Related Invertebrates* (eds F. Slansky Jr and

J.G. Rodriguez). Wiley-Interscience, New York, pp. 579–607.

Sikorowski, P.P., Powell, J.E. and Lawrence, A.M. (1992) Effects of bacterial contamination on development of *Microplitis croceipes* (Hym, Braconidae). *Entomophaga*, **37**, 475–81.

Silva, I.M.M.S., Honda, J., Van Kan, F., Hu, J., Neto, L., Pintureau, B. and Stouthamer, R. (1999) Molecular differentiation of five *Trichogramma* species occurring in Portugal. *Biological Control*, **16**, 177–84.

Simberloff, D. and Stiling, P. (1996) How risky is biological control? *Ecology*, **77**, 1965–74.

Simmonds, F.J. (1943) The occurrence of superparasitism in *Nemeritis canescens* Grav.. *Revue Canadienne de Biologie*, **2**, 15–58.

Simmons, L.W. and Siva-Jothy, M.T. (1998) Insect sperm competition: mechanisms and the potential for selection, in *Sperm Competition and Sexual Selection* (eds T. Birkhead and A.P. Møller). Academic Press, London, pp. 341–434.

Simon, C., Frati, F., Beckenbach, A., Crespi, B., Liu, H. and Flook, P. (1994) Evolution, weighting and phylogenetic utility of mitochondrial gene sequences and a compilation of conserved polymerase chain reaction primers. *Annals of the Entomological Society of America*, **87**, 651–701.

Simon, U., Pfütze, J. and Thömen, D. (2001) A timesorting stem-eclector. *Ecological Entomology*, **26**, 325–9.

Simons, A.M. and Roff, D.A. (1994) The effect of environmental variablity on the heritabilities of traits of a field cricket. *Evolution*, **48**, 1637–49.

Singh, R. and Sinha, T.B. (1980) Bionomics of *Trioxys (Binodoxys) indicus* Sibba Rao and Sharma, an aphidiid parasitoid of *Aphis craccivora* Koch.. *Zeitschrift für Angewandte Entomologie*, **90**, 233–7.

Sinkins, S.P., Braig, H.R. and O'Neill, S.L. (1995) *Wolbachia pipientis*: bacterial density and unidirectional cytoplasmic incompatibility between infected populations of *Aedes albopictus*. *Experimental Parasitology*, **81**, 284–91.

Sinkins, S.P., Curtis, C.F. and O'Neill, S.L. (1997) The potential of inherited symbiont systems to pest control, in *Influential Passengers: Inherited Microorganisms and Arthropod Reproduction* (eds S.L. O'Neill, A.A. Hoffmann and J.H. Werren). Oxford University Press, Oxford, pp. 155–275.

Sirot, E. and Bernstein, C. (1996) Time sharing between host searching and food searching in parasitoids: state-dependent optimal strategies. *Behavioral Ecology*, **7**, 189–94.

Sirot, E. and Bernstein, C. (1997) Food searching and superparasitism in solitary parasitoids. *Acta Oecologia*, **18**, 63–72.

Sirot, E., Ploye, E. and Bernstein, C. (1997) State dependent superparasitism in a solitary parasitoid: egg load and survival. *Behavioral Ecology*, **8**, 226–32.

Sisterton, M.S. and Averill, A.L. (2002) Costs and benefits of food foraging for a braconid parasitoid. *Journal of Insect Behavior*, **15**, 571–88.

Siva-Jothy, M.T. (1984) Sperm competition in the Libellulidae (Anisoptera) with special reference to *Crocothemis erythraea* (Brulle) and *Orthetrum cancellatum* (L.). *Advances in Odonatology*, **2**, 195–207.

Siva-Jothy, M.T. (1987) Variation in copulation duration and the resultant degree of sperm removal in *Orthetrum cancellatum* (L.) (Libellulidae: Odonata). *Behavioral Ecology and Sociobiology*, **20**, 147–151.

Siva-Jothy, M.T. and Hadrys, H. (1998) A role for molecular biology in testing ideas about cryptic female choice, in *Molecular Approaches to Ecology and Evolution* (eds R. DeSalle and B. Schierwater). Birkhäuser, Basel, pp. 37–53.

Siva-Jothy, M.T. and Hooper, R. (1996) Differential use of sperm during oviposition in *Calopteryx splendens xanthostoma*. *Behavioral Ecology and Sociobiology*, **39**, 389–93.

Siva-Jothy, M.T. and Skarstein, F. (1998) Towards a functional understanding of "good genes". *Ecology Letters*, **1**, 178–85.

Siva-Jothy, M.T. and Tsubaki, Y. (1989) Variation in copulation duration in *Mnais pruinosa pruinosa* (Calopterygidae; Odonata). 1. Alternative matesecuring tactics and sperm precedence. *Behavioral Ecology and Sociobiology*, **24**, 39–45.

Siva-Jothy, M.T. and Tsubaki, Y. (1994) Sperm competition and sperm precedence in the Dragonfly *Nannophya pygmaea*. *Physiological Entomology*, **19**, 363–6.

Siva-Jothy, M.T., Gibbons, D.W. and Pain, D. (1995) Female oviposition-site preference and egg hatching success in the damselfly *Calopteryx splendens xanthostoma*. *Behavioral Ecology and Sociobiology*, **37**, 39–44.

Siva-Jothy, M.T., Tsubaki, Y. and Hooper, R.E. (1998) Decreased immune response as a proximate cost of copulation and oviposition in a damselfly. *Physiological Entomology*, **23**, 274–7.

Sivasubramaniam, W., Wratten, S.D. and Klimaszewski J. (1997) Species composition, abundance, and activity of predatory arthropods in carrot fields, Canterbury, New Zealand. *New Zealand Journal of Zoology*, **24**, 205–12.

Sivinksi, J. and Webb, J.C. (1989) Acoustic signals produced during courtship in *Diachasmimorpha* (= *Biosteres*) *longicaudata* (Hymenoptera, Braconidae) and other Braconidae. *Annals of the Entomological Society of America*, **82**, 116–20.

Skinner, G.J. (1980) The feeding habits of the woodant, *Formica rufa* (Hymenoptera: Formicidae), in limestone woodland in north-west England. *Journal of Animal Ecology*, **49**, 417–33.

Skinner, S.W. (1982) Maternally inherited sex ratio in the parasitoid wasp, *Nasonia vitripennis*. *Science*, **215**, 1133–4.

Skinner, S.W. (1985a) Clutch size as an optimal foraging problem for insects. *Behavioral Ecology and Sociobiology*, **17**, 231–8.

Skinner, S.W. (1985b) Son-killer: a third extrachromosomal factor affecting sex-ratio in the parasitoid wasp, *Nasonia vitripennis*. *Genetics*, **109**, 745–59.

Skinner, S.W. and Werren, J.W. (1980) The genetics of sex determination in *Nasonia vitripennis* (Hymenoptera: Pteromalidae). *Genetics*, **94**, 98–106.

Skovgård, H. (2002) Dispersal of the filth fly parasitoid *Spalangia cameroni* (Hymenoptera: Pteromalidae) in a swine facility using fluorescent dust marking and sentinel pupal bags. *Environmental Entomology*, **31**, 425–31.

Skuhravý, V. (1960) Die Nahrung des Ohrwurmes (*Forficula auricularia* L.) in den Feldkulturen. *Acta Societatis Entomologicae Čechoslovenicae*, **57**, 329–39.

Skuhravý, V. (1970) Zur Anlockungsfahigkeit von Formalin für Carabiden in Bodenfallen. *Beitrage zur Entomologie*, **20**, 371–4.

Slansky Jr, F. (1986) Nutritional ecology of endoparasitic insects and their hosts: an overview. *Journal of Insect Physiology*, **32**, 255–61.

Slansky Jr, F. and Rodriguez, J.G. (1987) Nutritional ecology of insects, mites, spiders and related invertebrates: an overview, in *Nutritional Ecology of Insects, Mites, Spiders and Related Invertebrates* (eds F. Slansky Jr and J.G. Rodriguez). Wiley Interscience, New York, pp. 1–69.

Slansky Jr, F. and Scriber, J.M. (1982) Selected bibliography and summary of quantitative food utilization by immature insects. *Bulletin of the Entomological Society of America*, **28**, 43–55.

Slansky Jr, F. and Scriber, J.M. (1985) Food consumption and utilization, in *Comprehensive Insect Physiology, Biochemistry and Pharmacology* (eds G.A. Kerkut and L.I. Gilbert). Pergamon Press, Oxford, pp. 87–163.

Smilowitz, Z. and Iwantsch, G.F. (1973) Relationships between the parasitoid *Hyposoter exiguae* and the cabbage looper *Trichoplusia ni*: effects of host age on developmental rate of the parasitoid. *Environmental Entomology*, **2**, 759–63.

Smith, A.D. and Riley, J.R. (1996) Signal processing in a novel radar system for monitoring insect migration. *Computers and Electronics in Agriculture*, **15**, 267–78.

Smith, A.P. and Alcock, J. (1980). A comparative study of the mating systems of Australian eumenid wasps (Hymenoptera). *Zeitschrift für Tierspsychologie*, **53**, 41–60.

Smith, B.C. (1961) Results of rearing some coccinellid (Coleoptera: Coccinellidae) larvae on various pollens. *Proceedings of the Entomological Society of Ontario*, **91**, 270–1.

Smith, B.C. (1965) Growth and development of coccinellid larvae on dry foods (Coleoptera, Coccinellidae). *Canadian Entomologist*, **97**, 760–8.

Smith, C.C. and Fretwell, S.D. (1974) The optimal balance between size and number of offspring. *American Naturalist*, **108**, 499–506.

Smith, C.R., Heinz, K.M., Sansone, C.G. and Flexner, J.L. (2000) Impact of recombinant baculovirus applications on target heliothines and nontarget predators in cotton. *Biological Control*, **19**, 201–14.

Smith, H.S. and DeBach, P. (1942) The measurement of the effect of entomophagous insects on population densities of their hosts. *Journal of Economic Entomology*, **35**, 845–9.

Smith, J.G. (1976) Influence of crop background on natural enemies of aphids on Brussels sprouts, *Annals of Applied Biology*, **83**, 15–29.

Smith, J.W., Stadelbacher, E.A. and Gantt, C.W. (1976) A comparison of techniques for sampling beneficial arthropod populations associated with cotton. *Environmental Entomology*, **5**, 435–44.

Smith, K.G.V. (1974) Rearing the Hymenoptera Parasitica. *Leaflets of the Amateur Entomologist's Society*, **35**, 1–15.

Smith, L. and Rutz, D.A. (1987) Reproduction, adult survival and intrinsic rate of growth of *Urolepis rufipes* (Hymenoptera: Pteromalidae), a pupal parasitoid of house flies, *Musca domestica*. *Entomophaga*, **32**, 315–27.

Smith, R.H. (1991) Genetic and phenotypic aspects of life-history evolution in animals. *Advances in Ecological Research*, **21**, 63–120.

Smith, R.L. (1984) *Sperm Competition and the Evolution of Animal Mating Systems*. Academic Press, London.

Smith, S.M. (1988) Pattern of attack on spruce budworm egg masses by *Trichogramma minutum* (Hymenoptera: Trichogrammatidae) released in forest stands. *Environmental Entomology*, **17**, 1009–15.

Snider, R.M. and Snider, R.J. (1997) Efficiency of arthropod extraction from soil cores. *Entomological News*, **108**, 203–8.

Snodgrass, G.L. (2002) Characteristics of a red-eye mutant of the tarnished plant bug (Heteroptera: Miridae). *Annals of the Entomological Society of America*, **95**, 366–9.

Snodgrass, R.E. (1935) *Principles of Insect Morphology*. McGraw-Hill, London.

Snow, R. (1963) Alcoholic hydrochloric acid-carmine as a stain for chromosomes in squash preparations. *Stain Technology*, **38**, 9–13.

Snyder, W.E. and Ives, A.R. (2001) Generalist predators disrupt biological control by a specialist parasitoid. *Ecology*, **82**, 705–16.

Snyder, W.E. and Ives, A.R. (2003) Interactions between specialist and generalist natural enemies: parasitoids, predators, and pea aphid biocontrol. *Ecology*, **84**, 91–107.

Snyder, W.E. and Wise, D.H. (2001) Contrasting trophic cascades generated by a community of generalist predators. *Ecology*, **82**, 1571–83.

Soenjaro, E. (1979) Effect of labelling with the radioisotope ^{65}Zn on the performance of the eulophid, *Colpoclypeus florus*, a parasite of Tortricidae. *Entomologia Experimentalis et Applicata*, **25**, 304–10.

Sokolovska, N., Rowe, L., and Johansson, F. (2002) Fitness and body size in mature odonates. *Ecological Entomology*, **25**, 239–48.

Sokolowski, M.B. (1980) Foraging strategies of *Drosophila melanogaster*: a chromosomal analysis. *Behavioral Genetics*, **10**, 291–302.

Sokolowski, M.B. and Turlings, T.C.J. (1987) *Drosophila* parasitoid-host interactions: vibrotaxis and ovipositor search from the host's perspective. *Canadian Journal of Zoology*, **65**, 461–4.

Sokolowski, M.B., Bauer, S.J., Wai-Ping, V., Rodriguez, L., Wong, J.L. and Kent, C. (1986) Ecological genetics and behaviour of *Drosophila melanogaster* larvae in nature. *Animal Behaviour*, **34**, 403–8.

Sokolowski, M.B., Pereira, H.S. and Hughes, K. (1997) Evolution of foraging behavior in *Drosophila* by density-dependent selection. *Proceedings of the National Academy of Sciences*, **94**, 7373–7.

Sol, R. (1961) Über den Eingriff von Insektiziden in das Wechselspiel von *Aphis fabae* Scop. und einigen ihrer Episiten. *Entomophaga*, **6**, 7–33.

Sol, R. (1966) The occurrence of aphidivorous Syrphids and their larvae on different crops, with the help of coloured water traps, in *Ecology of Aphidophagoüs Insects* (ed. I. Hodek). Academia, Prague and W. Junk, Dordrecht, pp. 181–4.

Solomon, M.E. (1949) The natural control of animal populations. *Journal of Animal Ecology*, **18**, 1–35.

Solomon, M.G., Fitzgerald, J.D. and Murray, R.A. (1996) Electrophoretic approaches to predator-prey interactions, in *The Ecology of Agricultural Pests: Biochemical Approaches* (eds W.O.C. Symondson and J.E. Liddell), Systematics Association Special Volume 53. Chapman and Hall, London, pp. 457–68.

Soper, R.S., Shewell, G.E. and Tyrrell, D. (1976) *Colcondamyia auditrix* nov. sp. (Diptera: Sarcophagidae), a parasite which is attracted by the mating song of its host, *Okanagana rimosa* (Homoptera: Cicadidae). *Canadian Entomologist*, **108**, 61–8.

Sopp, P.I. and Sunderland, K.D. (1989) Some factors affecting the detection period of aphid remains in predators using ELISA. *Entomologia Experimentalis et Applicata*, **51**, 11–20.

Sopp, P. I. and Wratten, S.D. (1986) Rates of consumption of cereal aphids by some polyphagous predators in the laboratory. *Entomologia Experimentalis et Applicata*, **41**, 69–73.

Sopp, P.I., Sunderland, K.D., Fenlon, J.S. and Wratten, S.D. (1992) An improved quantitative method for estimating invertebrate predation in the field using an enzyme-linked immunosorbent assay (ELISA). *Journal of Applied Ecology*, **29**, 295–302.

Sotherton, N.W. (1982) Predation of a chrysomelid beetle (*Gastrophysa polygoni*) in cereals by polyphagous predators. *Annals of Applied Biology*, **101**, 196–9.

Sotherton, N.W. (1984) The distribution and abundance of predatory arthropods overwintering on farmland. *Annals of Applied Biology*, **105**, 423–9.

Sotherton, N.W. (1985) The distribution and abundance of predatory arthropods overwintering in field boundaries. *Annals of Applied Biology*, **106**, 17–21.

Sotherton, N.W., Wratten, S.D. and Vickerman, C.P. (1985) The role of egg predation in the population dynamics of *Gastrophysa polygoni* (Coleoptera) in cereal fields. *Oikos*, **43**, 301–8.

Soussi, R. and Le Ru, B. (1998) Influence of the host plant of the cassava mealybug *Phenacoccus manihoti* (Hemiptera: Pseudoccidae) on biological characteristics of its parasitoid *Apoanagyrus lopezi* (Hymenoptera: Encyrtidae). *Bulletin of Entomological Research*, **88**, 75–82.

Southwood, T.R.E. (1962) Migration of terrestrial arthropods in relation to habitat. *Biological Reviews*, **37**, 171–214.

Southwood, T.R.E. (1978a) *Ecological Methods with Particular Reference to the Study of Insect Populations, 2nd edn*. Chapman and Hall, London.

Southwood, T.R.E. (1978b) The components of diversity, in *Diversity of Insect Faunas* (eds L.A. Mound and N. Waloff). Blackwell, Oxford, pp. 19–40.

Southwood, T.R.E. and Henderson, P.A. (2000) *Ecological Methods, 3rd edn*. Blackwell Science, Oxford.

Southwood, T.R.E. and Jepson, W.F. (1962) Studies on the populations of *Oscinella frit* L. (Diptera: Chloropidae) in the wheat crop. *Journal of Animal Ecology*, **31**, 481–95.

Southwood, T.R.E. and Pleasance, H.J. (1962) A hand-operated suction apparatus for the extraction of arthropods from grassland and similar habitats, with notes on other models. *Bulletin of Entomological Research*, **53**, 125–8.

Southwood, T.R.E., Hassell, M.P., Reader, P.M. and Rogers, D.J. (1989) Population dynamics of the viburnum whitefly (*Aleurotrachelus jelinekii*). *Journal of Animal Ecology*, **58**, 921–42.

Sparks, A.N., Chiang, H.C., Burkhardt, C.C., Fairchild, M.L. and Weekman, G.T. (1966) Evaluation of the influence of predation on corn borer populations. *Journal of Economic Entomology*, **59**,104–7.

Speicher, B.R. (1936) Oogenesis, fertilization and early cleavage in *Habrobracon*. *Journal of Morphology*, **59**, 401–421.

Speicher, B.R., Speicher, K.G. and Roberts, F.L. (1965) Genetic segregation in the unisexual wasp *Devorgilla*. *Genetics*, **52**, 1035–41.

Speicher, K.G. (1934) Impaternate females in *Habrobracon*. *Biological Bulletin*, **67**, p. 277.

Speight, M.R. and Lawton, J.H. (1976) The influence of weed-cover on the mortality imposed on artificial prey by predatory ground beetles in cereal fields. *Oecologia*, **23**, 211–23.

Spence, J.R. (1986) Relative impacts of mortality factors in field populations of the waterstrider *Gerris buenoi* Kirkaldy (Heteroptera: Gerridae). *Oecologia*, **70**, 68–76.

Spence, J.R. and Niemelä, J.K. (1994) Sampling carabid assemblages with pitfall traps: the madness and the method. *Canadian Entomologist*, **126**, 881–94.

Spieles, D.J. and Horn, D.J. (1998) The importance of prey for fecundity and behaviour in the gypsy moth (Lepidoptera: Lymantriidae) predator *Calosoma sycophanta* (Cleoptera: Carabidae). *Environmental Entomology*, 27, 458–62.

Spiller, D.A. and Schoener, T.W. (1990) A terrestrial field experiment showing the impact of eliminating top predators on foliage damage. *Nature*, **347**, 469–72.

Staak, C., Allmang, B., Kampe, U. and Mehlitz, D. (1981) The complement fixation test for the species identification of blood meals from tsetse flies. *Tropenmedizin und Parasitologie*, **32**, 97–8.

Stacey, D.A. and Fellowes, M.D.E. (2002) Influence of temperature on pea aphid (*Acyrthosiphon pisum*) (Hemiptera: Aphididae) resistance to natural enemy attack. *Bulletin of Entomological Research*, **92**, 351–7.

Stack, P.A. and Drummond, F.A. (1997) Reproduction and development of *Orius insidiosus* in a blue light-supplemented short photoperiod. *Biological Control*, **9**, 59–65.

Stadler, B. and Mackauer, M. (1996) Influence of plant quality on interactions between the aphid parasitoid *Ephedrus californicus* Baker (Hymenoptera: Aphidiidae) and its host, *Acyrthosiphon pisum* (Harris) (Homoptera: Aphididae). *Canadian Entomologist*, **128**, 27–39.

Stam, P. (1993) Construction of integrated genetic linkage maps by means of a new computer package: JoinMap. *The Plant Journal*, **3**, 739–44.

Standen, V. (2000) The adequacy of collecting techniques for estimating species richness of grassland invertebrates. *Journal of Applied Ecology*, **37**, 884–93.

Stark, J.D., Vargas, R.I. and Thalman, R.K. (1991). Diversity and abundance of Oriental Fruit Fly parasitoids in guava orchards in Kauai, Hawaii. *Journal of Economic Entomology*, **84**, 1460–7.

Starý, P. (1970) Biology of Aphid Parasites (Hymenoptera: Aphidiidae) with respect to integrated control. *Series Entomologica*, **6**, 1–643.

Stearns, S.C. (1992) *The Evolution of Life Histories*. Oxford University Press, Oxford.

Steffan, S.A., Daane, K.M. and Mahr, D.L. (2001) ^{15}N-enrichment of plant tissue to mark phytophagous insects, associated parasitoids, and flower-visiting entomophaga. *Entomologia Experimentalis et Applicata*, **98**, 173–80.

Steiner, W.W.M. (1988) Electrophoretic techniques for the genetic study of aphids, in *Aphids. Their Biology, Natural Enemies and Control*, (eds A.K. Minks and P. Harrewijn). Elsevier, Amsterdam, pp. 135–43.

Stelleman, P. and Meeuse, A.D.J. (1976) Anthecological relations between reputedly anemophilous flowers and syrphid flies. 1. The possible role of syrphid flies as pollinators of *Plantago*. *Tijdschrift voor Entomologie*, **119**, 15–31.

Stephen, F.M., Lih, M.P. and Browne, L.E. (1996) Biological control of southern pine beetle through enhanced nutrition of its adult parasitoids. *Proceedings of the North American Forest Insect Work Conference, Publication 160 of the Texas Forest Service*, 34–5.

Stephens, D.W. and Krebs, J.R. (1986) *Foraging Theory*. Princeton University Press, Princeton.

Stephens, M.J., France, C.M., Wratten, S.D. and Frampton, C. (1998) Enhancing biological control of leafrollers (Lepidoptera: Tortricidae) by sowing buckwheat (*Fagopyrum esculentum*) in an orchard. *Biocontrol Science and Technology*, **8**, 547–58.

Sterling, W. (1989) Estimating the abundance and impact of predators and parasites on *Heliothis* populations, in *Proceedings of the Workshop on Biological Control of Heliothis: Increasing the Effectiveness of Natural Enemies* (eds E.G. King and R.D. Jackson). U.S. Department of Agriculture, New Delhi, pp. 37–56.

Stern, C. (1968) *Genetic Mosaics and Other Essays*. Harvard University Press, Cambridge, Massachusetts.

Stern, V.M., Schlinger, E.I. and Bowen, W.R. (1965) Dispersal studies of *Trichogramma semifumatum* (Hymenoptera: Trichogrammatidae) tagged with radioactive phosphorus. *Annals of the Entomological Society of America*, **58**, 234–40.

Steward, V.B., Smith, K.G. and Stephen, F.M. (1988) Predation by wasps on lepidopteran larvae in an Ozark forest canopy. *Ecological Entomology*, **13**, 81–6.

Stewart, LA. and Dixon, A.F.G. (1989) Why big species of ladybird beetles are not melanic. *Functional Ecology*, **3**, 165–77.

Stewart, L.A., Hemptinne, J.-L. and Dixon, A.F.G. (1991) Reproductive tactics of ladybird beetles: relationships between egg size, ovariole number and development time. *Functional Ecology*, **5**, 380–85.

Stewart-Oaten, A. and Murdoch, W.W. (1990) Temporal consequences of spatial density dependence. *Journal of Animal Ecology*, **59**, 1027–45.

Steyskal, G.C. (1977) History and use of the McPhail trap. *The Florida Entomologist*, **60**, 11–16.

Steyskal, G.C. (1981) A bibliography of the Malaise *trap. Proceedings of the Entomological Society of Washington*, **83**, 225–9.

Stiling, P.D. (1987) The frequency of density dependence in insect host parasitoid systems. *Ecology*, **68**, 844–56.

Stiling, P. (1988) Density-dependent processes and key factors in insect populations. *Journal of Animal Ecology*, **57**, 581–94.

Stiling, P.D. (1990) Calculating the establishment rates of parasitoids in biological control. *American Entomologist*, **36**, 225–30.

Stiling, P. and Rossi, A.M. (1997) Experimental manipulations of top-down and bottom-up factors in a tritrophic system. *Ecology*, **78**, 1602–6.

Stiling, P. and Simberloff, D. (2000) The frequency and strength of nontarget effects of invertebrate biological control agents of plant pests and weeds, in *Nontarget Effects of Biological Control* (eds P.A. Follett and J.J. Duan). Kluwer Academic Publishers, Dordrecht, pp. 31–43.

Stinner, R.E., Gutierrez, A.P. and Butler, G.D. (1974) An algorithm for temperature dependent growth rate simulation. *Canadian Entomologist*, **106**, 519–24.

Stinner, R.E., Butler Jr, G.D., Bacheler, J.S. and Tuttle, C. (1975) Simulation of temperature-dependent development in population dynamics models. *Canadian Entomologist*, **107**, 1167–74.

Stoewen, J.F. and Ellis, C.R. (1991) Evaluation of a technique for monitoring predation of Western corn rootworm eggs, *Diabrotica virgifera virgifera* (Coleoptera: Chrysomelidae). *Proceedings of the Entomological Society of Ontario*, **122**, 27–33.

Stoffolano, J.G. (1995) Regulation of a carbohydrate meal in adult Diptera, in *Regulatory Mechanisms in Insect Feeding* (eds R.F. Chapman and G. de Boer). Chapman and Hall, London, pp. 210–47.

Stoks, R. (2001) Food stress and predator-induced stress shape developmental performance in a damselfly. *Oecologia*, **127**, 222–9.

Stoks, R. and McPeek, M.A. (2003) Predators and life'histories shape *Lestes* damselfly assemblages along a freshwater habitat gradient. *Ecology*, **84**, 1576–87.

Stoltz, D.B. (1981) A putative baculovirus in the ichneumonid parasitoid *Mesoleius tenthredinis*. *Canadian Journal of Microbiology*, **27**, 116–22.

Stoltz, D.B. (1986) Interactions between parasitoid-derived products and host insects: an overview. *Journal of Insect Physiology*, **32**, 347–50.

Stoltz, D.B. and Vinson, S.B. (1979) Viruses and parasitism in insects. *Advances in Virus Research*, **24**, 125–71.

Stoltz, D.B., Krell, P.J., Summers, M.D. and Vinson, S.B. (1984) Polydnaviridae - a proposed family of insect viruses with segmented, double-stranded, circular DNA genomes. *Intervirology*, **21**, 1–4.

Stone, N.D. and Gutierrez, A.P. (1986) Pink bollworm control in southwestern desert cotton. I. A field-oriented simulation model. *Hilgardia*, **54**, 1–24.

Storck-Weyhermüller, S. (1988) Einfluss naturlicher Feinde auf die Populationsdynamik der Getreideblattlause im Winterweizen Mittelhessens (Homoptera: Aphididae). *Entomologia Generalis*, **13**, 189–206.

Stork, N.E. and Hammond, P.M. (1997) Sampling arthropods from tree-crowns by fogging with knockdown insecticides: lessons from oak tree beetle assemblages in Richmond Park (UK), in *Canopy Arthropods* (eds N.E. Stork, J. Adis and R.K. Didham). Chapman and Hall, London, pp. 3–26.

Stork, N.E., Hammond, P.M., Russell, B.L. and Hadwen, W.L. (2001) The spatial distribution of beetles within the canopies of oak trees in Richmond Park, U.K. *Ecological Entomology*, **26**, 302–11.

Stott, D.I. (1989) Immunoblotting and dot blotting. *Journal of Immunological Methods*, **119**, 153–87.

Stoutamire, W.P. (1974) Australian terrestrial orchids, thynnid wasps, and pseudocopulation. *American Orchid Society Bulletin*, **1974**, 13–18.

Stouthamer, R. (1993) The use of sexual versus asexual wasps in biological control. *Entomophaga*, **38**, 3–6.

Stouthamer, R. (1997) *Wolbachia*-induced thelytoky, in *Influential Passengers: Microbes and Invertebrate Reproduction* (eds S.L. O'Neill, A.A. Hoffmann and J.H. Werren). Oxford University Press, Oxford, pp. 102–24.

Stouthamer, R. and Kazmer, D.J. (1994) Cytogenetics of microbe-associated parthenogenesis and its consequences for gene flow in *Trichogramma* wasps. *Heredity*, **73**, 317–27.

Stouthamer, R. and Mak, F. (2002). Influence on the offspring production of the *Wolbachia*-infected parthenogenetic parasitoid *Encarsia formosa*. *Journal of Invertebrate Pathology*, **80**, 41–5.

Stouthamer, R. and Werren, J.H. (1993) Microbes associated with parthenogenesis in wasps of the genus *Trichogramma*. *Journal of Invertebrate Pathology*, **61**, 6–9.

Stouthamer, R., Luck, R.F. and Hamilton, W.D. (1990a) Antibiotics cause parthenogenetic *Trichogramma* (Hymenoptera, Trichogrammatidae) to revert to sex. *Proceedings of the National Academy of Sciences*, **87**, 2424–7.

Stouthamer, R., Pinto, J.D., Platner, G.R. and Luck, R.F. (1990b) Taxonomic status of thelytokous species of *Trichogramma* (Hymenoptera: Trichogrammatidae). *Annals of the Entomological Society of America*, **83**, 475–81.

Stouthamer, R., Luck, R.F. and Werren, J.H. (1992) Genetics of sex determination and the improvement of biological control using parasitoids. *Environmental Entomology*, **21:** 427–35.

Stouthamer, R. Breeuwer, J.A.J., Luck, R.F. and Werren, J.H. (1993) Molecular identification of microorganisms associated with parthenogenesis. *Nature*, **361**, 66–8.

Stouthamer, R., Breeuwer, J.A.J. and Hurst, G.D.D. (1999a) *Wolbachia pipientis*: Microbial manipulator of arthropod reproduction. *Annual Review of Microbiology*, **53**, 71–102.

Stouthamer, R., Hu, J., Van Kan, F.J.P.M., Platner, G.R. and Pinto, J.D. (1999b) The utility of internally transcribed spacer 2 DNA sequences of the nuclear ribosomal gene for distinguishing sibling species of *Trichogramma*. *BioControl*, **43**, 421–40.

Stouthamer, R., van Tilborg, M., de Jong, H., Nunney, L. and Luck, R.F. (2001). Selfish element maintains sex in natural populations of a parasitoid wasp. *Proceedings of the Royal Society of London B*, **268**, 617–22.

Stouthamer, R., Hurst, G.D.D. and Breeuwer, J.A.J. (2002) Sex ratio distorters and their detection, in *Sex Ratios: Concepts and Research Methods* (ed. I.C.W. Hardy). Cambridge University Press, Cambridge, 195–215.

Stowe, M.K., Turlings, T.C.J., Loughrin, J.H., Lewis, W.J. and Tumlinson, J.H. (1995) The chemistry of eavesdropping, alarm, and deceit. *Proceedings of the National Academy of Sciences*, **92**, 23–8.

Stradling, D.J. (1987) Nutritional ecology of ants, in *Nutritional Ecology of Insects, Mites, Spiders and Related Invertebrates* (eds F. Slansky and J.G. Rodriguez). John Wiley and Sons, New York, pp. 927–69.

Strand, M.R. (1986) The physiological interactions of parasitoids with their hosts and their influence on reproductive strategies, in *Insect Parasitoids* (eds J. Waage and D. Greathead). Academic Press, London, pp. 97–136.

Strand, M.R. (1988) Variable sex ratio strategy of *Telenomus heliothidis* (Hymenoptera: Scelionidae): adaptation to host and conspecific density. *Oecologia*, **77**, 219–24.

Strand, M.R. (1989) Oviposition behavior and progeny allocation of the polyembryonic wasp *Copidosoma floridanum* (Hymenoptera: Encyrtidae). *Journal of Insect Behavior*, **2**, 355–69.

Strand, M.R. (2000) Development traits and life history evolution in parasitoids, in *Parasitoid Population Biology* (eds M.E. Hochberg and A.R. Ives). Princeton University Press, Princeton and Oxford, pp. 139–62.

Strand, M.R. and Godfray, H.C.J. (1989) Superparasitism and ovicide in parasitic Hymenoptera: theory and a case study of the ectoparasitoid *Bracon hebetor*. *Behavioral Ecology and Sociobiology*, **24**, 421–32.

Strand, M.R. and Pech, L.L. (1995) Immunological basis for compatibilty in parasitoid-host relationships. *Annual Review of Entomology*, **40**, 31–56.

Strand, M.R. and Vinson, S.B. (1982) Source and characterization of an egg recognition kairomone of *Telenomus heliothidis*, a parasitoid of *Heliothis virescens*. *Physiological Entomology*, **7**, 83–90.

Strand, M.R. and Vinson, S.B. (1983) Analyses of an egg recognition kairomone of *Telenomus heliothidis* (Hymenoptera: Scelionidae): isolation and function. *Journal of Chemical Ecology*, **9**, 423–32.

Strand, M.R. and Vinson, S.B. (1985) *In vitro* culture of *Trichogramma pretiosum* on an artificial medium. *Entomologia Experimentalis et Applicata*, **391**, 203–9.

Strand, M.R., Meola, S.M. and Vinson, S.B. (1986) Correlating pathological symptoms in *Heliothis virescens* eggs with development of the parasitoid *Telenomus heliothidis*. *Journal of Insect Physiology*, **32**, 389–402.

Strand, M.R., Johnson, J.A. and Culin, J.D. (1988) Developmental interactions between the parasitoid *Microplitis demolitor* (Hymenoptera: Braconidae) and its host *Heliothis virescens* (Lepidoptera: Noctuidae). *Annals of the Entomological Society of America*, **81**, 822–30.

Strand, M.R., Roitberg, B.D. and Papaj, D.R. (1990) Acridine orange: a potentially useful internal marker of Hymenoptera and Diptera. *Journal of the Kansas Entomological Society*, **63**, 634–7.

Strickland, E.H. (1923) Biological notes on parasites of prairie cutworms. *Bulletin of the Department of Agriculture Dominion of Canada, Entomology Branch*, **22**, 1–40.

van Strien-van Liempt, W.T.F.H. (1983) The competition between *Asobara tabida* Nees von Esenbeck, 1834 and *Leptopilina heterotoma* (Thomson, 1862) in multiparasitized hosts. *Netherlands Journal of Zoology*, **33**, 125–63.

Strong, D.R., Lawton, J.H. and Southwood, T.R.E. (1984) *Insects on Plants*. Blackwell Scientific Publications, Oxford.

Stuart, M.K. and Burkholder, W.E. (1991) Monoclonal antibodies specific for *Laelius pedatus* (Bethylidae) and *Bracon hebetor* (Braconidae), two hymenopterous parasitoids of stored product pests. *Biological Control*, **1**, 302–8.

Stuart, M.K. and Greenstone, M.H. (1990) Beyond ELISA: a rapid, sensitive, specific immunodot assay for identification of predator stomach contents. *Annals of the Entomological Society of America*, **83**, 1101–7.

Stuart, M.K. and Greenstone, M.H. (1996) Serological diagnosis of parasitism: a monoclonal antibody-based immunodot assay for *Microplitis croceipes* (Hymenoptera: Braconidae), in *The Ecology of Agricultural Pests: Biochemical Approaches* (eds W.O.C. Symondson and J.E. Liddell), Systematics Association Special Volume 53. Chapman and Hall, London, pp. 300–21.

Stuart, M.K. and Greenstone, M.H. (1997) Immunological detection of hymenopteran parasitism in *Helicoverpa zea* and *Heliothis virescens*. *Biological Control*, **8**, 197–202.

Stubblefield, J.W. and Seger, J. (1990) Local mate competition with variable fecundity: dependence of offspring sex ratios on information utilization and mode of male production. *Behavioral Ecology*, **1**, 68–80.

Stubbs, M. (1980) Another look at prey detection by coccinellids. *Ecological Entomology*, **5**, 179–82.

Stutt, A. and Siva-Jothy, M.T. (2001) The cost of traumatic insemination in the bed-bug. *Proceedings of the National Academy of Sciences*, **98**, 5683–7.

Suckling, D.M., Burnip, G.M., Gibb, A.R., Stavely, F.J.L. and Wratten, S.D. (1996) Comparison of suction and beating tray sampling for apple pests and their natural enemies. *Proceedings of the 49th New Zealand Plant Protection Conference*, New Zealand Plant Protection Society, Rotorua, New Zealand, pp. 41–7.

Suckling, D.M., Gibb, A.R., Burnip, G.M. and Delury, N.C. (2002) Can parasitoid sex pheromones help in insect biocontrol? A case study of codling moth (Lepidoptera; Tortricidae) and its parasitoids *Ascogaster quadridentata* (Hymenoptera: Braconidae). *Environmental Entomology*, **31**, 947–52.

Sugihara, G., Bersier, L.F. and Schoenly, K. (1997) Effects of taxonomic and trophic aggregation on food web properties. *Oecologia*, **112**, 272–84.

Suh, C.P., Orr, D.B. and Van Duyn, J.W. (2000) *Trichogramma* releases in North Carolina cotton: why releases fail to suppress Heliothine pests. *Journal of Economic Entomology*, **93**, 1137–45.

Sullivan, D.J. (1971) Comparative behaviour and competition between two aphid hyperparasites: *Alloxysta victrix* and *Asaphes californicus*. *Environmental Entomology*, **1**, 234–44.

Summers, C.G., Garrett, R.E. and Zalom, F.G. (1984) New suction device for sampling arthropod populations. *Journal of Economic Entomology*, **77**, 817–23.

Sumner, A.T. (1972) A simple technique for demonstrating centromeric heterochromatin. *Experimenal Cell Research*, **75**, 304–6.

Sun, L.V., Babaratsas, A., Savakis, C., O'Neill, S.L. and Bourtzis, K. (1999) Gene organization of the *dna*-A region of *Wolbachia*. *Journal of Bacteriology*, **181**, 4708–10.

Sunderland, K.D. (1975) The diet of some predatory arthropods in cereal crops. *Journal of Applied Ecology*, **12**, 507–15.

Sunderland, K.D. (1987) Spiders and cereal aphids in Europe. *Bulletin IOBC/WPRS Working Group Integrated Control of Cereal Pests*, **10**, 82–102.

Sunderland, K.D. (1988) Quantitative methods for detecting invertebrate predation occurring in the field. *Annals of Applied Biology*, **112**, 201–24.

Sunderland, K.D. (1991) The ecology of spiders in cereals. *Proceedings of the 6th International Symposium on Pests and Diseases of Small Grain Cereals and Maize, Halle/Saale*, **1**, 269–80.

Sunderland, K.D. (1992) Effects of pesticides on the population ecology of polyphagous predators. *Aspects of Applied Biology*, **31**, 19–28.

Sunderland, K.D. (1996) Progress in quantifying predation using antibody techniques, in *The Ecology of Agricultural Pests: Biochemical Approaches* (eds W.O.C. Symondson and J.E. Liddell), Systematics Association Special Volume 53. Chapman and Hall, London, pp. 419–55.

Sunderland, K.D. (2002) Invertebrate pest control by carabids, in *The Agroecology of Carabid Beetles* (ed. J. Holland). Intercept, Andover, pp. 165–214.

Sunderland, K.D. and Sutton, S.L. (1980) A serological study of arthropod predation on woodlice in a dune grassland ecosystem. *Journal of Animal Ecology*, **49**, 987–1004.

Sunderland, K.D. and Topping, C.J. (1995) Estimating population densities of spiders in cereals. *Acta Jutlandica*, **70**, 13–22.

Sunderland, K.D. and Vickerman, G.P. (1980) Aphid feeding by some polyphagous predators in relation to aphid density in cereal fields. *Journal of Applied Ecology*, **17**, 389–96.

Sunderland, K.D., Fraser, A.M. and Dixon, A.F.G. (1986a) Distribution of linyphiid spiders in relation to capture of prey in cereal fields. *Pedobiologia*, **29**, 367–75.

Sunderland, K.D., Fraser, A.M. and Dixon, A.F.G. (1986b) Field and laboratory studies on money spiders (Linyphiidae) as predators of cereal aphids. *Journal of Applied Ecology*, **23**, 433–47.

Sunderland, K.D., Crook, N.E., Stacey, D.L. and Fuller, B.T. (1987a) A study of feeding by polyphagous predators on cereal aphids using ELISA and gut dissection. *Journal of Applied Ecology*, **24**, 907–33.

Sunderland, K.D., Hawkes, C., Stevenson, J.H. McBride, J., Smart, L.E., Sopp, P., Powell, W., Chambers, R.J. and Carter, O.C.R. (1987b) Accurate estimation of invertebrate density in cereals. *Bulletin IOBC/WPRS Working Group Integrated Control of Cereal Pests*, **10**, 71–81.

Sunderland, K.D., Chambers, R.J., Helyer, N.L. and Sopp, P.I. (1992) Integrated pest management of greenhouse crops in Northern Europe. *Horticultural Reviews*, **13**, 1–66.

Sunderland, K.D., De Snoo, G.R., Dinter, A., Hance, T., Helenius, J., Jepson, P., Kromp, B., Lys, J.A., Samu, F., Sotherton, N.W., Toft, S. and Ulber, B. (1995a) Density estimation for invertebrate predators in agroecosystems. *Acta Jutlandica*, **70**, 133–62.

Sunderland, K.D., Lövei, G.L. and Fenlon, J. (1995b) Diets and reproductive phenologies of the introduced ground beetles *Harpalus affinis* and *Clivina australasiae* (Coleoptera: Carabidae) in New Zealand. *Australian Journal of Zoology*, **43**, 39–50.

Sunderland, K.D., Axelsen, J.A., Dromph, K., Freier, B., Hemptinne, J.L., Holst, N.H., Mols, P.J.M., Petersen, M.K., Powell, W., Ruggle, P., Triltsch, H. and Winder, L. (1997) Pest control by a community of natural enemies. *Acta Jutlandica*, **72**, 271–326.

Suomaläinen, E., Saura, A. and Lokki, J. (1987) *Cytology and Evolution in Parthenogenesis*. CRC Press, Boca Raton, Florida.

Surber, E.W. (1936) Rainbow trout and bottom fauna production in one mile of stream. *Transactions of the American Fisheries Society*, **66**, 193–202.

Sutherland, J.P., Sullivan, M.S. and Poppy, G.M. (1999) The influence of floral character on the foraging behaviour of the hoverfly, *Episyrphus balteatus*. *Entomologia Experimentalis et Applicata*, **93**, 157–64.

Sutherland, W.J. (1996) *Ecological Census Techniques*. Cambridge University Press, Cambridge.

Sutherst, R.W. and Maywald, G.F. (1985) A computerised system for matching climates in ecology. *Agriculture, Ecosystems and Environment*, **13**, 281–99.

Suzuki, Y. and Hiehata, K. (1985) Mating systems and sex ratios in the egg parasitoids, *Trichogramma dendrolimi* and *T. papilionis* (Hymenoptera: Trichogrammidae). *Animal Behaviour*, **33**, 1223–7.

Suzuki, Y., Tsuji, H. and Sasakawa, M. (1984) Sex allocation and the effects of superparasitism on secondary sex ratios in the gregarious parasitoid, *Trichogramma chilonis* (Hymenoptera: Trichogrammatidae). *Animal Behaviour*, **32**, 478–84.

Svensson, B.G. and Petersson, E. (2000) Swarm site fidelity in the sex role-reversed dance fly *Empis borealis*. *Journal of Insect Behavior*, **13**, 785–96.

Svensson, B.G., Petersson, E. and Frisk, M. (1990) Nuptial gift size prolongs copulation in the dance fly *Empis borealis*. *Ecological Entomology*, **15**, 225–9.

Swofford, D.L. (1998) PAUP*: phylogenetic analysis using parsimony (*and other methods). Sinauer Associates, Sunderland.

Syme, P.D. (1975) Effects on the longevity and fecundity of two native parasites of the European Pine Shoot Moth in Ontario. *Environmental Entomology*, **4**, 337–46.

Symondson, W.O.C. (1992). Pest control, in *The Complete Manual of Organic Gardening* (ed. B. Caplan). Headline Book Publishing, London, pp. 290–333.

Symondson, W.O.C. (2002a). Diagnostic techniques for determining carabid diets, in *The Agroecology of Carabid Beetles* (ed. J. Holland). Intercept, Andover, UK, pp. 137–64.

Symondson, W.O.C. (2002b). Molecular identification of prey in predator diets. *Molecular Ecology*, **11**, 627–41.

Symondson, W.O.C. and Hemingway, J. (1997) Biochemical and molecular techniques, in *Methods in Ecological and Agricultural Entomology* (eds D.R. Dent and M.P. Walton). CAB International, Wallingford, pp. 293–350.

Symondson, W.O.C. and Liddell, J.E. (1993a) The development and characterisation of an anti-haemolymph antiserum for the detection of mollusc remains within carabid beetles. *Biocontrol Science and Technology*, **3**, 261–75.

Symondson, W.O.C. and Liddell, J.E. (1993b) The detection of predation by *Abax parallelepipedus* and *Pterostichus madidus* (Coleoptera: Carabidae) on Mollusca using a quantitative ELISA. *Bulletin of Entomological Research*, **83**, 641–7.

Symondson, W.O.C. and Liddell, J.E. (1993c) A monoclonal antibody for the detection of arionid slug remains in carabid predators. *Biological Control*, **3**, 207–14.

Symondson, W.O.C. and Liddell, J.E. (1993d) The detection of predator-mollusc interactions using advanced antibody technologies. *A.N.P.P. 3rd International Conference on Pests in Agriculture, Montpellier 7–9 December, 1993*, 417–24.

Symondson, W.O.C. and Liddell, J.E. (1993e) Differential antigen decay rates during digestion of molluscan prey by carabid predators. *Entomologia Experimentalis et Applicata*, **69**, 277–87.

Symondson, W.O.C. and Liddell, J.E. (1995) Decay rates for slug antigens within the carabid predator *Pterostichus melanarius* monitored with a monoclonal antibody. *Entomologia Experimentalis et Applicata*, **75**, 245–50.

Symondson, W.O.C. and Liddell, J.E. (eds) (1996a) *The Ecology of Agricultural Pests: Biochemical Approaches*. Chapman and Hall, London.

Symondson, W.O.C. and Liddell, J.E. (1996b) A species-specific monoclonal antibody system for detecting remains of field slugs, *Deroceras reticulatum* (Miller) (Mollusca: Pulmonata) in carabid beetles (Coleoptera: Carabidae). *Biocontrol Science and Technology*, **6**, 91–9.

Symondson, W.O.C. and Liddell, J.E. (1996c) Polyclonal, monoclonal and engineered antibodies to investigate the role of predation in slug population dynamics, in *The Ecology of Agricultural Pests: Biochemical Approaches* (eds W.O.C. Symondson and J.E. Liddell). Chapman and Hall, London, pp. 323–43.

Symondson, W.O.C., Mendis, V.W. and Liddell, J.E. (1995). Monoclonal antibodies for the identification of slugs and their eggs. *EPPO Bulletin*, **25**, 377–82.

Symondson, W.O.C., Glen, D.M., Wiltshire, C.W., Langdon, C.J. and Liddell, J.E. (1996a) Effects of cultivation techniques and methods of straw disposal on predation by *Pterostichus melanarius* (Coleoptera: Carabidae) upon slugs (Gastropoda: Pulmonata) in an arable field. *Journal of Applied Ecology*, **33**, 741–53.

Symondson, W.O.C., Erickson, M.L. and Liddell, J.E. (1996b) Progress in the development of antibodies to detect predation on slugs – a review, plus new data on a monoclonal antibody against *Tandonia budapestensis*, in *Slug and Snail Pests in Agriculture. Proceedings of a Symposium*. BCPC, Farnham, UK, pp. 263–70.

Symondson, W.O.C., Erickson, M.L. and Liddell, J.E. (1997). Species-specific detection of predation by Coleoptera on the milacid slug *Tandonia budapestensis* (Mollusca: Pulmonata). *Biocontrol Science and Technology*, **7**, 457–65.

Symondson, W.O.C., Erickson, M.L. and Liddell, J.E. (1999a). Development of a monoclonal antibody for the detection and quantification of predation on slugs within the *Arion hortensis* agg. (Mollusca: Pulmonata). *Biological Control*, **16**, 274–82.

Symondson, W.O.C., Erickson, M.L., Liddell, J.E. and Jayawardena, K.G.I. (1999b) Amplified detection, using a monoclonal antibody, of an aphid-specific epitope exposed during digestion in the gut of a predator. *Insect Biochemistry and Molecular Biology*, **29**, 873–82.

Symondson, W.O.C., Gasull, T. and Liddell, J.E. (1999c). Rapid identification of adult whiteflies in plant consignments using monoclonal antibodies. *Annals of Applied Biology* 134, 271–6.

Symondson, W.O.C., Glen, D.M., Erickson, M.L., Liddell, J.E. and Langdon, C.J. (2000) Do earthworms help to sustain the slug predator *Pterostichus melanarius* (Coleoptera: Carabidae) within crops? Investigations using monoclonal antibodies. *Molecular Ecology*, **9**, 1279–92.

Symondson, W.O.C., Sunderland, K.D. and Greenstone, M. (2002) Can generalist predators be effective biocontrol agents? *Annual Review of Entomology*, **47**, 561–94.

Syrett, P. and Penman, D.R. (1981) Developmental threshold temperatures for the brown lacewing, *Micromus tasmaniae* (Neuroptera: Hemerobiidae). *New Zealand Journal of Zoology*, **8**, 281–3.

Syrjämäki, J. (1976) The mystery of the missing females in connection with male swarming of *Blacus rificornis* Nees (Hym: Braconidae). *Annales Entomologici Fennici*, **42**, 66–8.

Szyszko, J., Vermeulen, H.J.W. and den Boer, P.J. (1996) Survival and reproduction in relation to habitat quality and food availability for *Pterostichus oblongopunctatus* F. (Carabidae, Col.). *Acta Jutlandica*, **71**, 25–40.

Tagami, Y., Miura, K. and Stouthamer, R. (2001) How does infection with parthenogenesis-inducing *Wolbachia* reduce the fitness of *Trichogramma*? *Journal of Invertebrate Pathology*, **78**, 267–72.

Tagawa, J. (1977) Localization and histology of the female sex pheromone producing gland in the parasitic wasp *Apanteles glomeratus*. *Journal of Insect Physiology*, **23**, 49–56.

Tagawa, J. (1983) Female sex pheromone glands in the parasitic wasp, genus *Apanteles*. *Applied Entomology and Zoology*, **18**, 416–27.

Tagawa, J. (1987) Post-mating changes in the oviposition tactics of the parasitic wasp *Apanteles glomeratus* L. (Hym., Braconidae). *Applied Entomology and Zoology*, **22**, 537–42.

Tagawa, J. and Hidaka, T. (1982) Mating behaviour of the braconid wasp, *Apanteles glomeratus* (Hym., Braconidae): mating sequence and the factor for correct orientation of male to female. *Applied Entomology and Zoology*, **17**, 32–9.

Tagawa, J. and Kitano, H. (1981) Mating behaviour of the braconid wasp, *Apanteles glomeratus*. (Hymenoptera, Braconidae) in the field. *Applied Entomology and Zoology*, **16**, 345–50.

Tagawa, J., Asano, S., Ohtsubo, T. Kamomae, M. and Gotoh, T. (1985) Influence of age on the mating behaviour of the braconid wasp, *Apanteles glomeratus*. *Applied Entomology and Zoology*, **20**, 227–30.

Takabayashi, J., Takahashi, S., Dicke, M. and Posthumus, M.A. (1995) Developmental stage of herbivore *Pseudaletia separata* affects production of herbivore-induced synomone by corn plants. *Journal of Chemical Ecology*, **21**, 273–87.

Takabayashi, J., Sato, Y., Horikoshi, M., Yamaoka, R., Yano, S., Ohsaki, N. and Dicke, M. (1998) Plant effects on parasitoid foraging: differences between two tritrophic systems. *Biological Control*, **11**, 97–103.

Takagi, M. (1985) The reproductive strategy of the gregarious parasitoid, *Pteromalus puparum* (Hymenoptera: Pteromalidae). 1. Optimal number of eggs in a single host. *Oecologia*, **68**, 1–6.

Takahashi, S. and Sugai, T. (1982) Mating behavior of the parasitoid wasp *Tetrastichus hagenowii* (Hym., Eulophidae). *Entomologia Generalis*, **7**, 287–93.

Takasu, K. and Lewis, W.J. (1993) Host- and food-foraging of the parasitoid *Microplitis croceipes*: learning and physiological state effects. *Biological Control*, **3**, 70–4.

Takasu, K., Takano, S.I., Sasaki, M., Yagi, S. and Nakamura, S. (2003) Host recognition by the tick parasitoid *Ixodiphagus hookeri* (Hymenoptera: Encyrtidae). *Environmental Entomology*, **32**, 614–7.

Tanaka, C., Kainoh, Y. and Honda, H. (1999) Physical factors in host selection of the parasitoid fly, *Exorista japonica* Townsend (Diptera: Tachinidae). *Applied Entomology and Zoology*, **34**, 91–7.

Tanaka, T. (1987) Effect of the venom of the endoparasitoid, *Apanteles kariyai* Watanabe, on the cellular defense reaction of the host, *Pseudaletia separata* Walker. *Journal of Insect Physiology*, **33**, 413–20.

Tanksley, S.D. (1993) Mapping polygenes. *Annual Review of Genetics*, **27**, 205–33.

Tardieux, I. and Rabasse, J.M. (1988) Induction of a thelytokous reproduction in the *Aphidius colemani* (Hym., Aphidiidae) complex. *Journal of Applied Entomology*, **106**, 58–61.

Tassan, R.L., Hagen, K.S. and Sawall, E.F. (1979) The influence of field food sprays on the egg production rate of *Chrysopa carnea*. *Environmental Entomology*, **8**, 81–5.

Tauber, M.J., Tauber, C.A. and Masaki, S. (1986) *Seasonal Adaptations of Insects*. Oxford University Press, New York.

Tauber M.J., C.A. Tauber, J.P. Nyrop and Villani, M.G. (1998) Moisture, a vital but neglected factor in the seasonal ecology of insects: hypotheses and tests of mechanisms. *Environmental Entomology*, **27**, 523–30.

Taubert, S. and Hertl, F. (1985) Eine neue tragbare Insektensaugfalle mit Elektro-Batterie-Betreib. *Mitteilungen Deutsche Gesellschaft für Algemeine und Angewandte Entomologie*, **4**, 433–7.

Taylor, A.D. (1988a) Host effects on larval competition in the gregarious parasitoid *Bracon hebetor*. *Journal of Animal Ecology*, **57**, 163–72.

Taylor, A.D. (1988b) Parasitoid competition and the dynamics of host-parasitoid models. *American Naturalist*. **132**, 417–36.

Taylor, A.D. (1990) Metapopulations, dispersal, and predator-prey dynamics: an overview. *Ecology*, **71**, 429–33.

Taylor, A.D. (1991) Studying metapopulation effects in predator-prey systems. *Biological Journal of the Linnean Society*, **42**, 305–23.

Taylor, A.J., Linforth, R.S.T., Ingham, K.E. and Clawson, A.R. (1995) Methodology for measuring aroma release, in *Bioflavour 95* (eds P.X. Etievant and P. Schreier). INRA, Paris, pp. 23–8.

Taylor, A.J., Müller, C.B. and Godfray, H.C.J. (1998) Effect of aphid predators on oviposition behaviour of aphid parasitoids. *Journal of Insect Behavior*, **11**, 297–302.

Taylor, B.W., Anderson, C.R. and Peckarsky, B.L. (1998). Effects of size at metamorphosis on stonefly fecundity, longevity and reproductive success. *Oecologia*, **114**, 494–502.

Taylor, D.B. and Szalanski, A.L. (1999) Identification of *Muscidifurax* spp. by polymerase chain reaction–restriction fragment length polymorphism. *Biological Control*, **15**, 270–3.

Taylor, L.R. (1955) The standardization of air-flow in insect suction traps. *Annals of Applied Biology*, **43**, 390–408.

Taylor, L.R. (1962) The absolute efficiency of insect suction traps. *Annals of Applied Biology*, **50**, 405–21.

Taylor, P.D. and Bulmer, M.G. (1980) Local mate competition and the sex ratio. *Journal of Theoretical Biology*, **86**, 409–19.

Taylor, R.J. (1984) *Predation*. Chapman and Hall, London.

Teder, T. (1998) Limited variability of genitalia in the genus *Pimpla* (Hymenoptera: Ichneumonidae): inter- or intraspecific causes? *Netherlands Journal of Zoology*, **48**, 335–47.

Tenhumberg, B., Keller, M.A., and Possingham, H.P. (2001a) Using Cox's proportional hazard models to implement optimal strategies: An example from behavioural ecology. *Mathematical and Computer Modelling*, **33**, 597–607.

Tenhumberg, B., Keller, M.A., Tyre, A.J. and Possingham, H.P. (2001b) The effect of resource aggregation at different scales: optimal foraging behavior of *Cotesia rubecula*. *American Naturalist*, **158**, 505–18.

Tepidino, V.J. (1988) Incidence of pre-emergence sib mating in *Monodontomerus obsoletus*, *Pteromalus venustus* and *Testrastichus megachilidis*, three chalcid parasitoids of the alfalfa leafcutting bee *Megachile rotundata* (Hymenoptera, Chalcididae). *Pan-Pacific Entomologist*, **64**, 63–66.

Thaler, J.S. (2002) Effect of jasmonate-induced plant responses on the natural enemies of herbivores. *Journal of Animal Ecology*, **71**, 141–50.

Theopold, U., Krause, E. and Schmidt, O. (1994) Cloning of a VLP-protein coding gene from a parasitoid wasp *Venturia canescens*. *Archives of Insect Biochemistry and Physiology*, **26**, 137–45.

Thibout, E. (1975) Analyse des causes de l'inhibition de la receptivite sexualle et de l'influence d'une eventual seconde copulation sur la reproduction chez la Teigne du poireau, *Acrolepia assectella*. *Entomologia Experimentalis et Applicata*, **18**, 105–16.

Thiede, U. (1981) Über die Verwendung von Acryglasrohrchen zur Untersuchung der Biologie und ökologie solitarer aculeater Hymenopteren. *Deutsche Entomologische Zeitschrift*, **28**, 45–53.

Thiel, A. and Hoffmeister, T.S. (2004) Knowing your habitat: linking patch-encounter rate and patch exploitation in parasitoids. *Behavioral Ecology*, **15**, 419–25.

Thiele, H.V. (1977) *Carabid Beetles in Their Environments*. Springer-Verlag, Berlin.

Thoday, J.M. and Thompson, Jr. J.N. (1976) The number of segregating genes implied by continuous variation. *Genetica*, **46**, 335–44.

Thomas, C.D. (1989) Predator-herbivore interactions and the escape of isolated plants from phytophagous insects. *Oikos*, **55**, 291–98.

Thomas, C.F.G. and Jepson, P.C. (1997) Field-scale effects of farming practices on linyphiid spider populations in grass and cereals. *Entomologia Experimentalis et Applicata*, **84**, 59–69.

Thomas, C.F.G. and Jepson, P.C. (1999) Differential aerial dispersal of linyphiid spiders from a grass and a cereal field. *Journal of Arachnology*, **27**, 294–300.

Thomas, C.F.G., Parkinson, L. and Marshall, E.J.P. (1998) Isolating the components of activity-density for the carabid beetle *Pterostichus melanarius* in farmland. *Oecologia*, **116**, 103–12.

Thomas, D.B. and Sleeper, E.L. (1977) The use of pitfall-traps for estimating the abundance of arthropods, with special reference to the Tenenebrionidae (Coleoptera). *Annals of the Entomological Society of America*, **70**, 242–8.

Thomas, M. and Waage, J. (1996) *Integration of Biological Control and Host-Plant Resistance Breeding*. CTA, Wageningen.

Thomas, M.B. and Willis, A.J. (1998) Biocontrol – risky but necessary? *Trends in Ecology and Evolution*, **13**, 325–9.

Thomas, M.B., Wratten, S.D. and Sotherton, N.W. (1992) The creation of 'island' habitats in farmland to manipulate populations of beneficial arthropods: predator densities and species composition. *Journal of Applied Ecology*, **29**, 524–31.

Thomas, R.S. (2002) An immunological and behavioural study of the role of carabid beetle larvae as slug control agents in cereal crops. Cardiff University, PhD Thesis.

Thompson, D.J. (1975) Towards a predator-prey model incorporating age-structure: the effects of predator and prey size on the predation of *Daphnia magna* by *Ischnura elegans*. *Journal of Animal Ecology*, **44**, 907–16.

Thompson, D.J. (1978) Prey size selection by larvae of the damselfly *Ischnura elegans* (Odonata). *Journal of Animal Ecology*, **47**, 786–96.

Thompson, D.J. and Fincke, O.M. (2002) Body size and fitness in Odonata, stabilising selection and a meta-analysis too far? *Ecological Entomology*, **27**, 378–84.

Thorbek, P., Topping, C. and Sunderland, K. (2002) Validation of a simple method for monitoring aerial activity of spiders. *Journal of Arachnology*, **30**, 57–64.

Thornhill, E.W. (1978) A motorised insect sampler. *PANS*, **24**, 205–7.

Thornhill, R. (1976) Sexual selection and nuptial feeding behaviour in *Bittacus apicalis*. *American Naturalist*. **110**, 529–48.

Thornhill, R. (1980a) Mate choice in *Hylobittacus apicalis* (Insecta: Mecoptera) and its relation to some models of female choice. *Evolution*, **34**, 519–38.

Thornhill, R. (1980b) Sexual selection in the black-tipped hanging fly. *Scientific American*, **242**, 138–45.

Thornhill, R. (1988) Mate choice in *Hylobittacus apicalis*. *Evolution*, **34**, 519–38.

Thornhill, R. (1992a) Female preference for the pheromone of males with low fluctuating asymmetry in the Japanese Scorpionfly (*Panorpa japonica*: Mecoptera). *Behavioral Ecology*, **3**, 277–83.

Thornhill, R. (1992b) Fluctuating asymmetry and the mating system of the Japanese scorpionfly, *Panorpa japonica*. *Animal Behaviour*, **44**, 867–79.

Thornhill, R. and Alcock, J. (1982) *The Evolution of Insect Mating Systems*. Harvard University Press, Cambridge.

Thorpe, W.H. (1939) Further experiments on olfactory conditioning in a parasitic insect. The nature of the conditioning process. *Proceedings of the Royal Society of London B*, **126**, 370–97.

Thorpe, W.H. and Caudle, H.B. (1938) A study of the olfactory responses of insect parasites to the food plant of their host. *Parasitology*, **30**, 523–8.

Thu, G.H.T. and Ueno, T. (2002) Biology of *Hemiptarsenus varicornis* (Hymenoptera: Eulophidae), a parasitoid wasp of the leafminer *Liriomyza trifolii* (Diptera : Agromyzidae). *Journal of the Faculty of Agriculture Kyushu University*, **47**, 45–54.

Ticehurst, M. and Reardon, R. (1977) Malaise trap: a comparison of 2 models for collecting adult stage of gypsy moth parasites. *Melsheimer Entomological Series*, **23**, 17–19.

Tijssen, P. (1985) *Practice and Theory of Enzyme Immunoassays*. Elsevier, Oxford.

Tillman, P.G. and Powell, J.E. (1992) Intraspecific host discrimination and larval competition in *Microplitis croceipes, Microplitis demolitor, Cotesia kazak* (Hym.: Braconidae), and *Hyposoter didymator* (Hym.: Ichneumonidae), parasitoids of *Heliothis virescens* (Lep,: Noctuidae). *Entomophaga*, **37**, 429–37.

Tilmon, K.J., Danforth, B.N., Day, W.H. and Hoffmann, N.P. (2000) Determining parasitoid species composition in a host population: a molecular approach. *Annals of the Entomological Society of America*, **93**, 640–7.

Tinbergen, L. (1960) The dynamics of insect and bird populations in pine woods. *Archives Néerlandaises de Zoologie*, **13**, 259–473.

Tingle, C.C.D. and Copland, M.J.W. (1988) Predicting development of the mealybug parasitoids *Anagyrus pseudococci, Leptomastix dactylopii*, and *Leptomastidea abnormis* under glasshouse conditions. *Entomologia Experimentalis et Applicata*, **46**, 19–28.

Tingle, C.C.D. and Copland, M.J.W. (1989) Progeny production and adult longevity of the mealybug parasitoids *Anagyrus pseudococci, Leptomastix dactylopii*, and *Leptomastidea abnormis* (Hym.: Encyrtidae) in relation to temperature. *Entomophaga*, **34**, 111–20.

Tobin, J.E. (1997) Competition and coexistence of ants in a small patch of rainforest canopy in Peruvian Amazonia. *Journal of the New York Entomological Society*, **105**, 105–12.

Tobolewski, J., Kaliszewski, M.J., Colwell, R.K. and Oliver, J.H. (1992) Detection and identification of mammalian DNA from the gut of museum specimens of ticks. *Journal of Medical Entomology*, **29**, 1049–51.

Tod, M.E. (1973) Notes on beetle predators of molluscs. *Entomologist*, **106**, 196–201.

Todd, F.E. and Vansell, G.H. (1942) Pollen grains in nectar and honey. *Journal of Economic Entomology*, **35**, 728–31.

Toft, C.A. (1983) Community patterns of nectivorous adult parasitoids (Diptera: Bombyliidae) on their resources. *Oecologia*, **57**, 200–15.

Toft, CA. (1984a) Activity budgets in two species of bee flies (*Lordotus*: Bombyliidae, Diptera): a comparison of species and sexes. *Behavioral Ecology and Sociobiology*, **14**, 287–96.

Toft, C.A. (1984b) Resource shifts in bee flies (Bombyliidae): interactions among species determine choice of resources. *Oikos*, **43**, 104–12.

Toft, C.A. (1989a) Population structure and mating system of a desert bee fly (*Lordotus pulchrissimus*) (Diptera: Bombyliidae). 1. Male demography and interactions. *Oikos*, **54**, 345–58.

Toft, C.A. (1989b) Population structure and mating system of a desert bee fly (*Lordotus pulchrissimus*) (Diptera: Bombyliidae). 2. Female demography, copulations and characteristics of swarm sites. *Oikos*, **54**, 359–69.

Toft, S., Vangsgaard, C. and Goldschmidt, H. (1995) The distance method used to measure densities of web spiders in cereal fields. *Acta Jutlandica*, **90**, 33–45.

Tokeshi, M. (1985) Life-cycle and production of the burrowing mayfly, *Ephemera danica:* a new method for estimating degree-days required for growth. *Journal of Animal Ecology*, **54**, 919–30.

Tomiuk, J. and Wöhrmann, K. (1980) Population growth and population structure of natural populations of *Macrosiphum rosae* (L.) (Hemiptera, Aphididae). *Zeitschrift für Angewandte Entomologie*, **90**, 464–73.

Tomlin, A.D., McLeod, D.G.R., Moore, L.V., Whistlecraft, J.W., Miller, J.J. and Tolman, J.H. (1992) Dispersal of *Aleochara bilineata* (Col.: Staphylinidae) following inundative releases in urban gardens. *Entomophaga*, **37**, 55–63.

Tonkyn, D.W. (1980) The formula for the volume sampled by a sweep net. *Annals of the Entomological Society of America*, **73**, 452–4.

Tooker, J.F. and Hanks, L.M. (2000). Flowering plant hosts of adult hymenopteran parasitoids of central Illinois. *Annals of the Entomological Society of America*, **93**, 580–8.

Topham, M., and Beardsley, J.W. (1975) Influence of nectar source plants on the New Guinea sugarcane weevil parasite, *Lixophaga sphenophori* (Villeneuve). *Proceedings of the Hawaiian Entomological Society*, **22**, 145–54.

Topping, C.J. (1993) Behavioural responses of three linyphiid spiders to pitfall traps. *Entomologia Experimentalis et Applicata*, **68**, 287–93.

Topping, C.J. and Luff, M.L. (1995) Three factors affecting the pitfall trap catch of linyphiid spiders (Araneae: Linyphiidae). *Bulletin of the British Arachnological Society*, **10**, 35–8.

Topping, C.J. and Sunderland, K.D. (1992) Limitations to the use of pitfall traps in ecological studies exemplified by a study of spiders in a field of winter wheat. *Journal of Applied Ecology*, **29**, 485–91.

Topping, C.J. and Sunderland, K.D. (1994) Methods for quantifying spider density and migration in cereal crops. *Bulletin of the British Arachnological Society*, **9**, 209–13.

Topping, C.J. and Sunderland, K.D. (1995) Method for monitoring aerial dispersal by spiders. *Acta Jutlandica*, **70**, 245–56.

Topping, C.J. and Sunderland, K.D. (1998) Population dynamics and dispersal of *Lepthyphantes tenuis* in an ephemeral habitat. *Entomologia Experimentalis et Applicata*, **87**, 29–41.

Topping, C.J., Sunderland, K.D. and Bewsey, J. (1992) A large improved rotary trap for sampling aerial invertebrates. *Annals of Applied Biology*, **121**, 707–14.

Törmälä, T. (1982) Evaluation of five methods of sampling field layer arthropods, particularly the leafhopper community, in grassland. *Annales Entomologici Fennici* **48**, 1–16.

Toth, R.S. and Chew, R.M. (1972) Development and energetics of *Notonecta undulata* during predation on *Culex tarsalis*. *Annals of the Entomological Society of America*, **65**, 1270–9.

Tourtellot, M.K. (1992) Software review: *The Observer*. *Journal of Insect Behavior*, **5**, 415–416.

Townes, H. (1939) Protective odors among the Ichneumonidae (Hymenoptera). *Bulletin of the Brooklyn Entomological Society*, **34**, 29–30.

Townes, H. (1958) Some biological characteristics of the Ichneumonidae (Hymenoptera) in relation to biological control. *Journal of Economic Entomology*, **51**, 650–52.

Townes, H. (1962) Design for a Malaise trap. *Proceedings of the Entomological Society of Washington*, **64**, 253–62.

Tran, T.V. and Takasu, K. (2000) Life history of the pupal parasitoid *Diadromus subtilicornis* (Gravenhorst) (Hymenoptera: Ichneumonidae) as influenced by temperature, photoperiod, and availability of food and hosts. *Entomological Science*, **3**, 255–64.

Traugott, M. (2002) Dispersal power, home range and habitat preference of cantharid larvae (Coleoptera: Cantharidae) in arable land. *European Journal of Soil Biology*, **38**, 79–83.

Traugott, M. (2003) The prey spectrum of larval and adult *Cantharis* species in arable land: An electrophoretic approach. *Pedobiologia*, **47**, 161–9.

Traut, W. (1994) Sex determination in the fly *Megaselia scalaris*, a model system for primary steps of sex-chromosome evolution. *Genetics*, **136**, 1097–1104.

Traut, W. and Willhoeft, U. (1990) A jumping sex determining factor in the fly *Megaselia scalaris*. *Chromosoma*, **99**, 407–12.

Traveset, A. (1990) Bruchid egg mortality on *Acacia farnesiana* caused by ants and abiotic factors. *Ecological Entomology*, **15**, 463–7.

Travis, S., Maschinski, J. and Keim, P. (1996) An analysis of genetic variation in *Astragalus cremnophylax* var. *cremnophylax*, a critically endangered plant, using AFLP markers. *Molecular Ecology*, **5**, 735–45.

Treacy, M.F., Benedict, J.H., Walmsley, M.H., Lopez, J.D. and Morrison, R.K. (1987) Parasitism of bollworm (Lepidoptera: Noctuidae) eggs on nectaried and nectariless cotton. *Environmental Entomology*, **16**, 420–3.

Tretzel, E. (1955) Technik und Bedeutung des Fallenfanges für ökologische Untersuchungen. *Zoologische Anzeiger*, **155**, 276–87.

Trexler, J.C., McCulloch, C.E. and Travis, J. (1988) How can the functional response best be determined? *Oecologia*, **76**, 206–14.

Triltsch, H. (1997) Gut contents in field sampled adults of *Coccinella septempunctata* L. (Col.: Coccinellidae). *Entomophaga*, **42**, 125–31.

Trimble, R.M. and Brach, E.J. (1985) Effect of color on sticky-trap catches of *Pholetesor ornigis* (Hymenoptera: Braconidae), a parasite of the spotted tentiform leafminer *Phyllonorycter blancardella* (Lepidoptera: Gracillariidae). *Canadian Entomologist*, **117**, 1559–64.

Trimble, R.M., Blommers, L.H.M. and Helsen, H.H.M. (1990) Diapause termination and thermal requirements for postdiapause development in *Aphelinus mali* at constant and fluctuating temperatures. *Entomologia Experimentalis et Applicata*, **56**, 61–9.

Tripathi, R.N. and Singh, R. (1991) Aspects of life-table studies and functional response of *Lysiphlebia mirzai*. *Entomologia Experimentalis et Applicata*, **59**, 279–87.

True, J.R., Liu, J., Stam, F., Zeng, Z.-B. and Laurie, C.C. (1997) Quantitative genetic analysis of divergence in male secondary sexual traits between *Drosophila simulans* and *Drosophila mauritiana*. *Evolution*, **51**, 816–32.

Tsuchida, T., Koga, R., Shibao, H., Matsumoto, T. and Fukatsu, T. (2002) Diversity and geographic distribution of secondary endosymbiotic bacteria in natural populations of the pea aphid, *Acyrthosiphon pisum*. *Molecular Ecology*, **11**, 2123–35.

Tukey, H.B. (1971) Leaching of substances from plants, in *Ecology of Leaf Surface Micro-organisms* (eds T.F. Preece and C.H. Dickinson). Academic Press, pp. 67–80.

Tullberg, B.S. and Hunter, A.F. (1996) Evolution of larval gregariousness in relation to repellant defences and warning coloration in tree-feeding Macrolepidoptera: a phylogenetic analysis based on independent contrasts. *Biological Journal of the Linnean Society*, **57**, 253–76.

Tullberg, B.S., Gamerale-Stille, G. and Solbreck, C. (2000) Effects of food plant and group size on predator defence: differences between co-occurring aposematic Lygaeinae bugs. *Ecological Entomology*, **25**, 220–5.

Turchin, P. and Berryman, A.A. (2000) Detecting cycles and delayed density dependence: a comment on Hunter and Price (1998). *Ecological Entomology*, **25**, 119–21.

Turelli, M. and Hoffmann, A.A. (1995) Cytoplasmic incompatability in *Drosophila simulans*: dynamics and parameter estimates from natural-populations. *Genetics*, **140**, 1319–38.

Turelli, M. and Hoffmann, A.A. (1999) Microbe-induced cytoplasmic incompatibility as a mechanism for introducing transgenes into arthropod populations. *Insect Molecular Biology*, **8**, 243–55.

Turlings, T.C.J., Bernasconi, M., Bertossa, R., Bigler, F., Caloz, G. and Dorn, S. (1998) The induction of volatile emissions in maize by three herbivore species with different feeding habits: possible consequences for their natural enemies. *Biological Control*, **11**, 122–9.

Turnbow Jr, R.H., Franklin, R.T. and Nagel, W.P. (1978) Prey consumption and longevity of adult *Thanasimus dubius*. *Environmental Entomology*, **7**. 695–7.

Turnbull, A.L. (1962) Quantitative studies of the food of *Linyphia triangularis* Clerk (Araneae: Linyphiidae). *Canadian Entomologist*, **94**, 1233–49.

Turnbull, A.L. and Nicholls, C.F. (1966) A 'quick trap' for area sampling of arthropods in grassland communities. *Journal of Economic Entomology*, **59**, 1100–4.

Udayagiri, S., Mason, C.E. and Pesek, J.D. (1997) *Coleomegilla maculata*, *Coccinella septempunctata* (Coleoptera: Coccinellidae), *Chrysoperla carnea* (Neuroptera: Chrysopidae), and *Macrocentrus grandii* (Hymenoptera: Braconidae) trapped on colored sticky traps in corn habitats. *Environmental Entomology*, **26**, 983–8.

Ueno, T. (1994) Self-recognition by the parasitic wasp *Itoplectis naranyae* (Hymenoptera, Ichneumonidae). *Oikos*, **70**, 333–9.

Ueno, T. (1997) Effects of superparasitism, larval competition, and host feeding on offspring fitness in the parasitoid *Pimpla nipponica* (Hymenoptera: Ichneumonidae). *Annals of the Entomological Society of America*, **90**, 682–8.

Ueno, T. (1998a) Selective host-feeding on parasitized hosts by the parasitoid *Itoplectis naranyae* (Hymenoptera: Ichneumonidae) and its implication for biological control. *Bulletin of Entomological Research*, **88**, 461–6.

Ueno, T. (1998b) Adaptiveness of sex ratio control by the pupal parasitoid *Itoplectis naranyae* (Hymenoptera: Ichneumonidae) in response to host size. *Evolutionary Ecology*, **12**, 643–54.

Ueno, T. (1999a) Reproduction and host-feeding in the solitary parasitoid wasp *Pimpla nipponica* (Hymenoptera: Ichneumonidae). *Invertebrate Reproduction and Development*, **35**, 231–7.

Ueno, T. (1999b) Host-feeding and acceptance by a parasitic wasp (Hymenoptera: Ichneumonidae) as influenced by egg load and experience in a patch. *Evolutionary Ecology*, **13**, 33–44.

Ueno, T. (1999c) Multiparasitism and host feeding by solitary parasitoid wasps (Hymenoptera: Ichneumonidae) based on the pay-off from parasitised hosts. *Annals of the Entomological Society of America*, **99**, 601–8.

Ueno, T. (1999d) Host suitability and sex ratio differences in wild-caught and laboratory-reared parasitoid *Pimpla parnarae* (Hymenoptera: Ichneumonidae). *Annals of the Entomological Society of America*, **99**, 609–14.

Ueno, T. and Tanaka, T. (1996) Self-host discrimination by a parasitic wasp: the role of short-term memory. *Animal Behaviour*, **52**, 875–83.

Ueno, T. and Tanaka, T. (1997) Comparison of primary and secondary sex ratios in parasitoid wsaps using a method for observing chromosomes. *Entomologia Experimentalis et Applicata*, **82**, 105–8.

Uetz, G.W. and Unzicker, J.D. (1976) Pitfall trapping in ecological studies of wandering spiders. *Journal of Arachnology*, **3**, 101–11.

Ulber, B. and Wolf-Schwerin, G. (1995) A comparison of pitfall trap catches and absolute density estimates of carabid beetles in oilseed rape fields. *Acta Jutlandica*, **70**, 77–86.

Unruh, T.R., Gonzalez, D. and Gordh, G. (1984) Electrophoretic studies on parasitic Hymenoptera and implications for biological control. *Proceedings of the XVIIth International Congress of Entomology*, p. 705.

Unruh, T.R., White, W., Gonzalez, D. and Luck, R.F. (1986) Electrophoretic studies of parasitic Hymenoptera and implications for biological control, in *Biological Control of Muscoid Flies* (eds R.S. Patterson and D.A. Rutz). Entomological Society of America, Miscellaneous Publication, pp. 150–63.

Unruh, T.R., White, W., Gonzalez, D. and Woolley, J.B. (1989) Genetic relationships among seventeen *Aphidius* (Hymenoptera: Aphidiidae) populations, including six species. *Annals of the Entomological Society of America*, **82**, 754–68.

Unterman, B.M., Baumann, P. and McLean, D.L. (1989) Pea aphid symbiont relationships established by analysis of 16S rRNAs. *Journal of Bacteriology*, **171**, 2970–4.

Unwin, D.M. and Corbet, S.A., (1991) *Insects, Plants and Microclimate, Naturalists' Handbooks No 15*. Richmond Press, Slough.

Urbaneja, A., Llácer, E., Garrido, A. and Jacas, J.-A. (2001a) Effect of variable photoperiod on development and survival of *Cirrospilus* sp. nr *lyncus* (Hymenoptera: Eulophidae), an ectoparasitoid of *Phyllocnistis citrella* (Lepidoptera: Gracillariidae). *Florida Entomologist*, **84**, 305–7.

Urbaneja, A., Llacer, E., Garrido, A. and Jacas, J.A. (2001b) Effect of temperature on the life history of *Cirrospilus* sp. near *lyncus* (Hymenoptera : Eulophidae), a parasitoid of *Phyllocnistis citrella* (Lepidoptera : Gracillariidae). *Biological Control*, **21**, 293–9.

Valicente, F.H. and O'Neill, R.J. (1995) Effects of host plants and feeding regimes on selected life-history characteristics of *Podisus maculiventris* (Say) (Heter"optera, Pentatomidae). *Biological Control*, **5**, 449–61.

Valladares, G.R. and Salvo, A. (1999) Insect-plant food webs could provide new clues for pest management. *Environmental Entomology*, **28**, 539–44.

Van Baarlen, P., Sunderland, K.D. and Topping, C.J. (1994) Eggsac parasitism of money spiders (Araneae, Linyphiidae) in cereals, with a simple method for estimating percentage parasitism of *Erigone* spp. eggsacs by Hymenoptera. *Journal of Applied Entomology*, **118**, 217–23.

Van Borm, S., Wenseleers, T., Billen, J. and Boomsma, J.J. (2003). Cloning and sequencing of *wsp* encoding gene fragments reveals a diversity of co-infecting *Wolbachia* strains in *Acrmyrmex* leafcutter ants. *Molecular Phylogenetics and Evolution*, **26**, 102–9.

Van den Berg, H., Ankasah, D., Muhammad, A., Rusli, R., Widayanto, H.A., Wirasto, H.B. and Yully, I. (1997) Evaluating the role of predation in population fluctuations of the soybean aphid *Aphis glycines* in farmers' fields in Indonesia. *Journal of Applied Ecology*, **34**, 971–84.

Van Dijk, T.S. (1994) On the relationship between food, reproduction and survival of two carabid beetles: *Calathus melanocephalus* and *Pterostichus versicolor*. *Ecological Entomology*, **19**, 262–70.

Van Driesche, R.G. (1983) Meaning of 'per cent parasitism' in studies of insect parasitoids. *Environmental Entomology*, **12**, 1611–22.

Van Driesche, R.C. (1988) Field levels of encapsulation and superparasitism for *Cotesia glomerata* (L.) (Hymenoptera: Braconidae) in *Pieris rapae* (L.) (Lepidoptera: Pieridae). *Journal of the Kansas Entomological Society*, **61**, 328–31.

Van Driesche, R.G. and Bellows Jr, T.S. (1988) Use of host and parasitoid recruitment in quantifying losses from parasitism in insect populations. *Ecological Entomology*, **13**, 215–22.

Van Driesche, R.G. and Bellows Jr, T.S. (1996) *Biological Control*. Chapman and Hall, London.

Van Driesche, R.C. and Taub, G. (1983) Impact of parasitoids on *Phyllonorycter* leafminers infesting apple in Massachusetts, USA. *Protection Ecology*, **51**, 303–17.

Van Driesche, R.G., Bellotti, A., Herrera, C.J. and Castillo, J.A. (1986) Encapsulation rates of two encyrtid parasitoids by two *Phenacoccus* spp. of cassava mealybugs in Colombia. *Entomologia Experimentalis et Applicata*, **42**, 79–82.

Van Driesche, R.G., Bellows Jr, T.S., Elkinton, J.S., Gould, J.R. and Ferro, D.N. (1991) The meaning of percentage parasitism revisited: solutions to the problem of accurately estimating total losses from parasitism in a host generation. *Environmental Entomology*, **20**, 1–7.

Van Handel, E. (1972), The detection of nectar in mosquitoes. *Mosquito News*, **32**, 458.

Van Handel, E. (1984) Metabolism of nutrients in the adult mosquito. *Mosquito News*, **44**, 573–9.

Van Handel, E. (1985a) Rapid determination of glycogen and sugars in mosquitoes. *Journal of the American Mosquito Association*, **1**, 299–301.

Van Handel, E. (1985b) Rapid determination of lipids in mosquitoes. *Journal of the American Mosquito Association*, **1**, 302–4.

Van Handel, E. and Day, J.F. (1988) Assay of lipids, glycogen and sugars in individual mosquitoes: correlations with wing length in field-collected *Aedes vexans*. *Journal of the American Mosquito Association*, **4**, 549–50.

Van Handel, E., Haeger, J.S. and Hansen, C.W. (1972) The sugars of some Florida nectars. *American Journal of Botany*, **59**, 1030–2.

Van Hezewijk, B.H., Bourchier, R.S. and Smith, S.M. (2000) Searching speed of *Trichogramma minutum* and its potential as a measure of parasitoid quality. *Biological Control*, **17**, 139–46.

Van Kan, F.J.P.M., Silva, I.M.M.S., Schilthuizen, M., Pinto, J.D. and Stouthamer, R. (1996) Use of DNA-based methods for the identification of minute wasps of the genus *Trichogramma*. *Proceedings of Experimental and Applied Entomology*, **7**, 233–7.

Van Laerhoven, S.L. and Stephen, F.M. (2002) Height distribution of adult parasitoids of the southern pine beetle complex. *Environmental Entomology*, **31**, 982–7.

Van Noordwijk, A.J. and De Jong, G. (1986) Acquisition and allocation of resources: their influence on variation in life history tactics. *American Naturalist*, **128**, 137–42.

Van Opijnen, T. and Breeuwer, J.A.J. (1999) High temperatures eliminate *Wolbachia*, a cytoplasmic incompatibility inducing endosymbiont, from the two-spotted spider mite. *Experimental and Applied Acarology*, **23**, 871–81.

Vanlerberghe-Masutti, F. (1994) Molecular identification and phylogeny of parasitic wasp species (Hymenoptera: Trichogrammatidae) by mitochondrial DNA RFLP and RAPD markers. *Insect Molecular Biology*, **3**, 229–37.

Vannini, M., Conti, A., Ferreti, J. and Becciolini, A. (1993) Trophic exchange in *Pardosa hortensis* (Lycosidae, Araneae). *Journal of Zoology*, **231**, 163–6.

Vargas, R.I., Stark, J.D., Prokopy, R.J. and Green, T.I. (1991) Response of Oriental fruit fly and associated parasitoids to coloured balls. *Journal of Economic Entomology*, **84**, 1503–7.

Varley, G.C. and Gradwell, C.R. (1960) Key factors in population studies. *Journal of Animal Ecology*, **29**, 99–401.

Varley, G.C. and Gradwell, G.R. (1965) Interpreting winter moth population changes. *Proceedings of the 12th International Congress of Entomology*, 377–8.

Varley, G.C. and Gradwell, C.R. (1968) Population models for the winter moth, in *Insect Abundance* (ed. T.R.E. Southwood), 4th Symposium of the Royal Entomological Society of London, 132–42.

Varley, G.C. and Gradwell, C.R. (1970) Recent advances in insect population dynamics. *Annual Review of Entomology*, **15**, 1–24.

Varley, G.C., Gradwell, G.R. and Hassell, M.P. (1973) *Insect Population Ecology: an Analytical Approach*. Blackwell, Oxford.

Varley, M.J., Copland, M.J.W., Wratten, S.D. and Bowie, M.H. (1993) Parasites and predators, in *Video Techniques in Animal Ecology and Behaviour* (ed. S.D. Wratten). Chapman and Hall, London, pp. 33–63.

Vaughn, T.T. and Antolin, M.F. (1998) Population genetics of an opportunistic parasitoid in an agricultural landscape. *Heredity*, **80**, 152–62.

Vavra, J. and Larsson, J.I.R. (1999). Structure of the microsporidia, in *The Microsporidia and Microsporidosis* (eds M. Wittner and L.M. Weiss). ASM Press, Washington, pp. 7–84.

Vavre, F., Fleury, F., Lepetit, D., Fouillet, P. and Bouletreau, M. (1999) Phylogenetic evidence for horizontal transmission of *Wolbachia* in host-parasitoid associations. *Molecular Biology and Evolution*, **16**, 1711–23.

Vavre, F., Fleury, F., Varaldi, J., Fouillet, P. and Bouletreau, M. (2000) Evidence for female mortality in *Wolbachia*-mediated cytoplasmic incompatibility in haplodiploid insects: Epidemiologic and evolutionary consequences. *Evolution*, **54**, 191–200.

Vavre, F., Dedeine, F., Quillon, M., Fouillet, P., Fleury, F. and Bouletreau, M. (2001) Within-species diversity of *Wolbachia*-induced cytoplasmic incompatibility in haplodiploid insects. *Evolution*, **55**, 1710–14.

van Veen, J.C. (1981) The biology of *Poecilostictus cothurnatus* (Hymenoptera, Ichneumonidae) an endoparasite of *Bupalus pinarius* (Lepidoptera, Geometridae) *Annales Entomologici Fennici*, **47**, 77–93.

Vega, F.E., Barbosa, P. and Panduro, A.P. (1990) An adjustable water-pan trap for simultaneous sampling of insects at different heights. *Florida Entomologist*, **73**, 656–60.

Vehrencamp, S.L. and Bradbury, J.W. (1984) Mating Systems and Ecology, in *Behavioural Ecology: An Evolutionary Approach, 2nd edn.* (eds J.R. Krebs and N.B. Davies). Blackwell, Oxford, pp. 251–78.

Venzon, M., Janssen, A. and Sabelis, M.W. (2002) Prey preference and reproductive success of the generalist predator *Orius laevigatus*. *Oikos*, **97**, 116–24.

Verkerk, R.H.J., Leather, S.R. and Wright, D.J. (1998) The potential for manipulating crop-pest-natural enemy interactions for improved insect pest management. *Bulletin of Entomological Research*, **88**, 493–501.

Vet, L.E.M. (1983) Host-habitat location through olfactory cues by *Leptopilina clavipes* (Hartig) (Hym.: Eucoilidae), a parasitoid of fungivorous *Drosophila:* the influence of conditioning. *Netherlands Journal of Zoology*, **33**, 225–48.

Vet, L.E.M. (1999) From chemical to population ecology: infochemical use in an evolutionary context. *Journal of Chemical Ecology*, **25**, 31–49.

Vet, L.E.M. and van Alphen, J.J.M. (1985) A comparative functional approach to the host detection behaviour of parasitic wasps. I. A qualitative study on Eucoilidae and Alysiinae. *Oikos*, **44**, 478–86.

Vet, L.E.M. and Dicke, M. (1992) Ecology of infochemical use by natural enemies in a tritrophic context. *Annual Review of Entomology*, **37**, 141–72.

Vet, L.E.M. and Groenewold, A.W. (1990) Semiochemicals and learning in parasitoids. *Journal of Chemical Ecology*, **16**, 3119–35.

Vet, L.E.M., van Lenteren, J.C., Heymans, M. and Meelis, E. (1983) An airflow olfactometer for measuring olfactory responses of hymenopterous parasitoids and other small insects. *Physiological Entomology*, **8**, 97–106.

Vet, L.E.M, Janse, C., van Achterberg, C. and van Alphen, J.J.M. (1984) Microhabitat location and niche segregation in two sibling species of drosophilid parasitoids: *Asobara tabida* (Nees) and *A.'rufescens* (Foerster) (Braconidae: Alysiinae). *Oecologia*, **61**, 182–8.

Vet, L.E.M., Lewis, W.J., Papaj, D.R. and van Lenteren, J.C. (1990) A variable-response model for parasitoid foraging behaviour. *Journal of Insect Behavior*, **3**, 471–90.

van Vianen, A. and van Lenteren, J.C. (1986) The parasite-host relationship between *Encarsia formosa* Cahan (Hym., Aphelinidae) and *Trialeurodes vaporariorum* (Westwood) (Hom., Aleyrodidae), XIV. Genetic and environmental factors influencing body-size and number of ovarioles of *Encarsia formosa*. *Journal of Applied Entomology*, **101**, 321–31.

Vickerman, G.P. and Sunderland, K.D. (1975) Arthropods in cereal crops: nocturnal activity, vertical distribution and aphid predation. *Journal of Applied Ecology*, **12**, 755–66.

Viggiani, G. and Battaglia, D. (1983) Courtship and mating behaviour in a few Aphelinidae (Hym., Chalcidoidea). *Bolletino de Laboratoria Entomologia Agraria Filippo Silvestri, Portici*, **40**, 89–96.

Vigneault, C., Roger, C., Hui, K.P.C. and Boivin, G. (1998) An image analysis system developed for the evaluation of *Coleomegilla maculata* larvae's behavior. *Canadian Agricultural Engineering*, **40**, 55–60.

Viktorov, G.A. (1968) The influence of population density on sex ratio in *Trissolcus grandis* Thoms. (Hymenoptera: Scelionidae). *Zoologichesky Zhurnal*, **47**, 1035–19 (In Russian).

Viktorov, G.A. and Kochetova, N.I. (1971) Significance of population density to the control of sex ratio in *Trissolcus grandis* (Hymenoptera: Scelionidae). *Zoologichesky Zhurnal*, **50**, 1735–55 (In Russian).

Vilhelmsen, L., Isidoro, N., Romani, R., Basibuyuk, H.H. and Quicke, D.L.J. (2001) Host location and oviposition in a basal group of parasitic wasps: the subgenual organ, ovipositor apparatus and associated structures in the Orussidae (Hymenoptera, Insecta). *Zoomorphology*, **121**, 63–84.

Villemant, C. and Ramzi, H. (1995). Predators of *Lymantria dispar* (Lep. Lymantriidae) egg masses: spatio-temporal variation of their impact during the 1988–89 pest generation in the Mamora cork oak forest (Morocco). *Entomophaga*, **40**, 441–56.

Vinson, S.B. (1972) Effect of the parasitoid *Campoletis sonorenesis* on the growth of its host, *Heliothis virescens*. *Journal of Insect Physiology*, **18**, 1509–14.

Vinson, S.B. (1976) Host selection by insect parasitoids. *Annual Review of Entomology*, **21**, 109–33.

Vinson, S.B. (1978) Courtship behavior and source of a sexual pheromone from *Cardiochiles nigriceps*. *Annals of the Entomological Society of America*, **71**, 832–37.

Vinson, S.B. (1985) The behaviour of parasitoids, in *Comprehensive Insect Physiology, Biochemistry and Pharmacology*, (eds G.A. Kerkut and L.I. Gilbert). Pergamon Press, New York, pp. 417–69.

Vinson, S.B. (1988) Physiological studies of parasitoids reveal new approaches to the biological control of insect pests. *ISI Atlas of Animal and Plant Sciences*, **1**, 25–31.

Vinson, S.B. (1990) Physiological interactions between the host genus *Heliothis* and its guild of parasitoids. *Archives of Insect Biochemistry and Physiology*, **13**, 63–81.

Vinson, S.B. (1999) Parasitoid manipulation as a plant defense strategy. *Annals of the Entomological Society of America*, **92**, 812–28.

Vinson, S.B. and Barras, D.J. (1970) Effects of the parasitoid *Cardiochiles nigriceps* on the growth, development and tissues of *Heliothis virescens*. *Journal of Insect Physiology*, **16**, 1329–38.

Vinson, S.B. and Iwantsch, G.F. (1980a) Host regulation by insect parasitoids. *Quarterly Review of Biology* **55**: 143–164.

Vinson, S.B. and Iwantsch, G.F. (1980b) Host suitability for insect parasitoids. *Annual Review of Entomology*, **25**, 397–419.

Vinson, S.B. and Sroka, P. (1978) Effects of superparasitism by a solitary endoparasitoid on the host, parasitoid and field samplings *Southwestern Entomologist*, **3**, 299–303.

Vinson, S.B. and Stolz, D.B. (1986) Cross protection experiments with two parasitoid (Hymenoptera: Ichneumonidae) viruses. *Annals of The Entomological Society of America*, **79**, 216–18.

Vinson, S.B., Harlan, D.P. and Hart, W.G. (1978) Response of the parasitoid *Microterys flavus to* the brown soft scale and its honeydew. *Environmental Entomology*, **7**, 874–8.

Visser, M.E. (1992) Adaptive self- and conspecific superparasitism in the solitary parasitoid *Leptopilina heterotoma*. *Behavioral Ecology*, **4**, 22–8.

Visser, M.E. (1993) *The Observer*: a software package for behavioural observations. *Animal Behaviour*, **45**, 1045–48.

Visser, M. (1994) The importance of being large: the relationship between size and fitness in females of the parasitoid *Aphaereta minuta* (Hymenoptera: Braconidae). *Journal of Animal Ecology*, **63**, 963–78.

Visser, M.E. and Driessen, G. (1991) Indirect mutual interference in parasitoids. *Netherlands Journal of Entomology*, **41**, 214–27.

Visser, M.E., van Alphen, J.J.M. and Nell, H.W. (1990) Adaptive superparasitism and time allocation in solitary parasitoids: the influence of the number of parasitoids depleting the patch. *Behaviour*, **114**, 21–36.

Visser, M.E., van Alphen, J.J.M. and Hemerik, L. (1992a) Adaptive superparasitism and patch time allocation in solitary parasitoids: an ESS model. *Journal of Animal Ecology*, **61**, 93–101.

Visser, M.E., van Alphen, J.J.M. and Nell, H.W. (1992b) Adaptive superparasitism and patch time allocation in solitary parasitoids: The influence of pre-patch experience. *Behavioral Ecology and Sociobiology*, **31**, 163–72.

Visser, M.E., Luyckx, B., Nell, H.W. and Boskamp, G.J.F. (1992c) Adaptive superparasitism in solitary parasitoids: marking of parasitised hosts in relation to the pay-off from superparasitism, *Ecological Entomology*, **17**, 76–82.

Visser, M.E., Jones, T.H. and Driessen, G. (1999) Interference among insect parasitoids: a multi-patch experiment. *Journal of Animal Ecology*, **68**, 108–20.

Vogt, E.A. and Nechols, J.R. (1991) Diel activity patterns of the squash bug egg parasitoid *Gryon pennsylvanicum* (Hymenoptera: Scelionidae). *Annals of the Entomological Society of America*, **84**, 303–8.

Vogt, J.T., Grantham, R.A., Smith, W.A. and Arnold, D.C. (2001) Prey of the red imported fire ant (Hymenoptera: Formicidae) in Oklahoma peanuts. *Environmental Entomology*, **30**, 123–8.

Völkl, W. and Starý, P. (1988) Parasitisation of *Uroleucon* species (Hom., Aphididae) on thistles (Compositae, Cardueae). *Journal of Applied Entomology*, **106**, 500–6.

Voller, A., Bidwell, D.E. and Bartlett, A. (1979) *The Enzyme-Linked Immunosorbent Assay (ELISA)*. Dynatech Europe, Guernsey.

Volterra, V. (1931) Variations and fluctuations of the number of individuals in animal species living together (translation from 1928 version), in *Animal Ecology* (ed. R.N. Chapman). Arno, New York.

Vonk, M., Putters, F. and Velthuis, B.J. (1991) The causal analysis of an adaptive system: sex-ratio decisions as observed in a parasitic wasp and simulated by a network model, in *From Animals to Animals* (eds J.A. Meyer and S.W. Wilson). MIT Press, Cambridge, Massachusetts, pp. 485–91.

Vorley, W.T. (1986) The activity of parasitoids (Hymenoptera: Braconidae) of cereal aphids (Hemiptera: Aphididae) in winter and spring in southern England. *Bulletin of Entomological Research*, **76**, 491–504.

Vorley, W.T. and Wratten, S.D. (1985) A simulation model of the role of parasitoids in the population development of *Sitobion azienae* (Hemiptera: Aphididae) on cereals. *Journal of Animal Ecology*, **22**, 813–23.

Vos, M., Kole, M. and van Alphen, J.J.M. (1993) Sex ratios and mating structure in *Leptomastix dactylopii*, a parasitoid of the citrus mealybug, *Planococcus citri*. *Mededelingen Faculteit Landbouwwetenschap Universiteit Gent*, **53**, 561–8.

Vos, P., Hogers, R., Bleeker, M., Reijans, M., Van de Lee, T., Hornes, M., Frijters, A., Pot, J., Peleman, J., Kuiper, M. and Zabeau, M. (1995) AFLP – A new technique for DNA-fingerprinting. *Nucleic Acids Research*, **23**, 4407–14.

Waage, Jeffrey K. (1978) Arrestment responses of a parasitoid, *Nemeritis canescens*, to a contact chemical produced by its host, *Plodia interpunctella*. *Physiological Entomology*, **3**, 135–46.

Waage, Jeffrey K. (1979) Foraging for patchily distributed hosts by the parasitoid *Nemeritis canescens*. *Journal of Animal Ecology*, **48**, 353–71.

Waage, Jeffrey K. (1982) Sib-mating and sex ratio strategies in scelionid wasps. *Ecological Entomology*, **7**, 102–12.

Waage, Jeffrey K. (1983) Aggregation in field parasitoid populations: foraging time allocation by a population of *Diadegma* (Hymenoptera, Ichneumonidae). *Ecological Entomology*, **8**, 447–53.

Waage, Jeffrey K. (1986) Family planning in parasitoids: adaptive patterns of progeny and sex allocation, in *Insect Parasitoids* (eds J. Waage and D. Greathead). Academic Press, London, pp. 63–95.

Waage, Jeffrey K. (1990) Ecological theory and the selection of biological control agents, in *Critical Issues in Biological Control* (eds M. Mackauer, L.E. Ehler and J. Roland). Intercept, Andover, pp. 135–157.

Waage, Jeffrey K. (1992) Biological control in the year 2000, in *Pest Management and the Environment in 2000* (eds A.-A.S.A. Kadir and H.S. Barlow). CAB International, Wallingford, pp. 329–40.

Waage, Jeffrey K. (2001) Indirect ecological effects in biological control: the challenge and the opportunity, in *Evaluating Indirect Ecological Effects of Biological Control* (eds E. Wajnberg, J.K. Scott and P.C. Quimby). CAB International, Wallingford, pp. 1–12.

Waage, Jeffrey K. and Barlow, N.D. (1993) Decision tools for biological control, in *Decision Tools for Pest Management* (eds G.A. Norton and J.D. Mumford). CAB International, Wallingford, pp. 229–45.

Waage, Jeffrey K. and Godfray, H.C.J. (1985) Reproductive strategies and population ecology of insect parasitoids, in *Behavioural Ecology, Ecological Consequences of Adaptive Behaviour* (eds R.M. Sibly and R.H. Smith). Blackwell Scientific Publications, Oxford, pp. 449–70.

Waage, Jeffrey K. and Greathead, D.J. (1988) Biological control: challenges and opportunities. *Philosophical Transactions of the Royal Society of London B*, **318**, 111–28.

Waage, Jeffrey K. and Hassell, M.P. (1982) Parasitoids as biological control agents – a fundamental approach. *Parasitology*, **84**, 241–68.

Waage, Jeffrey K. and Lane, J.A. (1984) The reproductive strategy of a parasitic wasp. II. Sex allocation and local mate competition in *Trichogramma evanescens*. *Journal of Animal Ecology*, **53**, 417–26.

Waage, Jeffrey K. and Mills, N.J. (1992) Biological Control, in *Natural Enemies: The Population Biology of Predators, Parasites and Diseases* (ed. M.J. Crawley). Blackwell, Oxford, pp. 412–30.

Waage, Jeffrey K. and Ng. S.-M. (1984) The reproductive strategy of a parasitic wasp. I. Optimal progeny and sex allocation in *Trichogramma evanescens*. *Journal of Animal Ecology*, **53**, 401–16.

Waage, Jeffrey K., Carl, K.P., Mills, N.J. and Greathead, D.J. (1985) Rearing entomophagous insects, in *Handbook of Insect Rearing, Vol. 1*, (eds P. Singh and R.F. Moore). Elsevier, Amsterdam, pp. 45–66.

Waage, Jonathan K. (1973) Reproductive behaviour and its relation to territoriality in *Calopteryx maculata* (Beauvois) (Odonata, Calopterygidae). *Behaviour*, **47**, 240–56.

Waage, Jonathan K. (1979a) Dual function of the damselfly penis: sperm removal and transfer. *Science*, **203**, 916–18.

Waage, Jonathan K. (1979b) Adaptive significance of postcopulatory guarding of mates and non-mates by male *Calopteryx maculata*. *Behavioral Ecology and Sociobiology*, **6**, 147–54.

Waage, Jonathan K. (1984) Sperm competition and the evolution of odonate mating systems, in *Sperm Competition and the Evolution of Animal Mating Systems* (ed. R.L. Smith). Academic Press, New York, pp. 257–90.

Waage, Jonathan K. (1996) Parental investment – Minding the kids or keeping control?, in *Feminism and Evolutionary Biology* (ed. P.A. Gowaty). Chapman and Hall, New York, pp. 527–53.

Wäckers, F.L. (1994) The effect of food deprivation on the innate visual and olfactory preferences in the parasitoid *Cotesia rubecula*. *Journal of Insect Physiology*, **40**, 641–9.

Wäckers, F.L. (1999) Gustatory response by the hymenopteran parasitoid *Cotesia glomerata* to a range of nectar- and honeydew-sugars. *Journal of Chemical Ecology*, **12**, 2863–77.

Wäckers, F.L. (2000) Do oligosaccharides reduce the suitability of honeydew for predators and parasitoids? A further facet to the function of insect-synthesized honeydew sugars. *Oikos*, **90**, 197–201.

Wäckers, F. (2001) A comparison of nectar- and honeydew sugars with respect to their utilization by the hymnopteran parasitoid *Cotesia glomerata*. *Journal of Insect Physiology*, **47**, 1077–84.

Wäckers, F.L., Bjornsten, A. and Dorn, S. (1996) A comparison of flowering herbs with respect to their nectar accessibility for the parasitoid *Pimpla turionellae*. *Proceedings of Experimental and Applied Entomology*, **7**, 177–82

Wäckers, F.L., Mitter, E. and Dorn, S. (1998) Vibrational sounding by the pupal parasitoid *Pimpla* (Coccygomimus) *turionellae*: An additional solution to the reliability detectability problem. *Biological Control*, **11**, 141–6.

Waddington, K.D. (1987) Nutritional ecology of bees, in *Nutritional Ecology of Insects, Mites, Spiders and Related Invertebrates* (eds F. Slansky and J.G. Rodriguez). Wiley International, New York, pp. 393–419.

Wagner, J.D., Glover, M.D., Moseley, J.B. and Moore A.J. (1999) Heritability and fitness consequences of cannibalism of *Harmonia axyridis*. *Evolutionary Ecology Research*, **1**, 375–88.

Wagner, T.L., Wu, H.-I., Sharpe, P.J.H. and Coulson, R.N. (1984) Modeling distributions of insect development time: A literature review and application of the Weibull function. *Annals of the Entomological Society of America*, **77**, 475–87.

Wagner, T.L., Hsin-I, W., Feldman, R.M., Sharpe, P.J.H. and Coulson, R.N. (1985) Multiple-cohort approach for simulating development of insect population under variable temperatures. *Annals of the Entomological Society of America*, **78**, 691–9.

Wajnberg, E. and Colazza, S. (1998) Genetic variability in the area searched by a parasitic wasp: analysis from automatic video tracking of the walking path. *Journal of Insect Physiology*, **44**, 437–44.

Wajnberg, E., Bouletreau, M., Prevost, G. and Fouillet, P. (1990) Developmental relationships between *Drosophila* larvae and their endoparasitoid *Leptopilina* (Hymenoptera: Cynipidae) as affected by crowding. *Archives of Insect Biochemistry and Physiology*, **13**, 239–45.

Wajnberg, E., Rosi, M.C. and Colazza, S. (1999) Genetic variation in patch time allocation in a parasitic wasp. *Journal of Animal Ecology*, **68**, 121–33.

Waldbauer, C.P. (1968) The consumption and utilization of food by insects. *Advances in Insect Physiology*, **5**, 229–88.

Walde, S.J. (1994) Immigration and the dynamics of predator-prey interaction in biological control. *Journal of Animal Ecology*, **63**, 337–46.

Walde, S.J. and Murdoch. W.W. (1988) Spatial density dependence in parasitoids. *Annual Review of Entomology*, **33**, 441–66.

Walker, M. and Jones, T.H. (2001) Relative roles of top-down and bottom-up forces in terrestrial tritrophic plant-insect herbivore-natural enemy systems. *Oikos*, **93**, 177–87.

Walker, T.J. (1993) Phonotaxis in female *Ormia ochracea* (Diptera: Tachinidae), a parasitoid of field crickets. *Journal of Insect Behavior*, **6**, 389–410.

Wallin, H. (1985) Spatial and temporal distribution of some abundant carabid beetles (Coleoptera: Carabidae) in cereal fields and adjacent habitats. *Pedobiologia*, **28**, 19–34.

Wallin, H. (1986) Habitat choice of some field-inhabiting carabid beetles (Coleoptera: Carabidae) studied by recapture of marked individuals. *Ecological Entomology*, **11**, 457–66.

Wallin, H. and Ekbom, B.S. (1988) Movements of carabid beetles (Coleoptera: Carabidae) inhabiting cereal fields: a field tracing study. *Oecologia*, **77**, 39–43.

Wallin, H. and Ekbom, B. (1994) Influence of hunger level and prey densities on movement patterns in three species of *Pterostichus* beetles (Coleoptera: Carabidae). *Environmental Entomology*, **23**, 1171–81.

Wallin, H., Chiverton, P.A., Ekbom, B.S. and Borg, A. (1992) Diet, fecundity and egg size in some polyphagous predatory carabid beetles. *Entomologia Experimentalis et Applicata*, **65**, 129–40.

Waloff, N. (1975) The parasitoids of the nymphal and adult stages of leafhoppers (Auchenorrhyncha, Homoptera). *Transactions of the Royal Entomological Society of London*, **126**, 637–86.

Waloff, N. and Jervis, M.A. (1987) Communities of parasitoids associated with leafhoppers and planthoppers in Europe. *Advances in Ecological Research*, **17**, 281–402.

Walrant, A. and Loreau, M. (1995) Comparison of iso-enzyme electrophoresis and gut content examination for determining the natural diet of the ground beetle species *Abax ater* (Coleoptera: Carabidae). *Entomologia Generalis*, **19**, 253–9.

Walter, G.H. (1988) Activity patterns and egg production in *Coccophagus bartletti*. an aphelinid parasitoid of scale insects. *Ecological Entomology*, **13**, 95–105.

Walton, M.P. (1986) The Application of Gel Electrophoresis to the Study of Cereal Aphid Parasitoids. Hatfield Polytechnic, Ph.D. Thesis.

Walton, M.P., Loxdale, H.D. and Allen-Williams, L. (1990a) Electrophoretic 'keys' for the identification of parasitoids (Hymenoptera: Braconidae: Aphelinidae) attacking *Sitobion avenae* (F.) (Hemiptera: Aphididae). *Biological Journal of the Linnean Society*, **40**, 333–46.

Walton, M.P., Powell, W., Loxdale, H.D. and Allen-Williams, L. (1990b) Electrophoresis as a tool for estimating levels of hymenopterous parasitism in field populations of the cereal aphid, *Sitobion avenae*. *Entomologia Experimentalis et Applicata*, **54**, 271–9.

Wang, T. and Laing, J.E. (1989) Diapause termination and morphogenesis of *Holcothorax testaceipes* Ratzeburg (Hymenoptera, Encyrtidae), an introduced parasitoid of the spotted tentiform leafminer, *Phyllonorycter blancardella* (F.). (Lepidoptera, Gracillariidae). *Canadian Entomologist*, **121**, 65–74.

Ware, A.B. and Compton S.G. (1994a) Dispersal of adult female fig wasps 1. Arrivals and departures. *Entomologia Experimentalis et Applicata*, **73**, 221–9.

Ware, A.B. and Compton S.G. (1994b) Dispersal of adult female fig wasps 2. Movements between trees. *Entomologia Experimentalis et Applicata*, **73**, 231–8.

Warren, P.H. (1994) Making connections in food webs. *Trends in Ecology and Evolution*, **9**, 136–41.

Washino, R.K. and Tempelis, C.H. (1983) Mosquito host bloodmeal identification: methodology and data analysis. *Annual Review of Entomology*, **28**, 179–201.

Watanabe, M. (2002) Cold tolerance and *myo*-inositol accumulation in overwintering adults of a ladybeetle, *Harmonia axyridis* (Coleoptera: Coccinellidae). *European Journal of Entomology*, **99**, 5–9.

Waterhouse, D.F. (1991) *Guidelines for Biological Control Projects in the Pacific*. South Pacific Commission, Noumea.

Waters, T.F. (1962) Diurnal periodicity in the drift of stream invertebrates. *Ecology*, **43**, 316–20.

Watt, A.D., Stork, N.E., McBeath, C. and Lawson, G.L. (1997) Impact of forest management on insect abundance and damage in a lowland tropical forest in southern Cameroon. *Journal of Applied Ecology*, **34**, 985–98.

Way, M.J. and Banks, C.J. (1968) Population studies on the active stages of the black bean aphid, *Aphis fabae* Scop., on its winter host *Euonymus europaeus* L. *Annals of Applied Biology*, **62**, 177–97.

Way, M.J., Javier, G. and Heong, K.L. (2002) The role of ants, especially the fire ant, *Solenopsis geminata*

(Hymenoptera: Formicidae), in the biological control of tropical upland rice pests. *Bulletin of Entomological Research*, **92**, 431–7.

Webb, J.C., Calkins, C.O., Chambers, D.L., Schweinbacher, W. and Russ, K. (1983) Acoustical aspects of behaviour of the Mediterranean fruitfly *(Ceratitis capitata):* analysis and identification of courtship sounds. *Entomologia Experimentalis et Applicata*, **33**, 1–8.

Weekman, G.T. and Ball, H.J. (1963) A portable electrically operated collecting device. *Journal of Economic Entomology*, **56**, 708–9.

Weeks, R.D. and McIntyre, N.E. (1997) A comparison of live versus kill pitfall trapping techniques using various killing agents. *Entomologia Experimentalis et Applicata*, **82**, 267–73.

Wehling, A. and Heimbach, U. (1991) Untersuchungen zur Wirkung von Pflanzenschutzmitteln auf Spinnen (Araneae) am Beispiel einiger Insektizide. *Nachrichtenblatt des Deutschen Pflanzenschutzdienstes*, **43**, 24–30.

Wei, Q. (1986) The foraging behaviour of *Cyrtorhinus lividipennis* (Reusen), as a predator of *Nilaparvata lugens* (Stål). University of Wales, M.Sc. Thesis.

Weigenberg, I. and Roff, D.A. (1996) Natural heritabilities: can they be reliably estimated in the laboratory. *Evolution*, **50**, 2149–57.

Weinbrenner, M. and Völkl, W. (2002) Oviposition behaviour of the aphid parasitoid, *Aphidius ervi*: Are wet aphids recognized as host? *Entomologia Experimentalis et Applicata*, **103**, 51–9.

Weir, B.S. (1990) *Genetic Data Analysis: Methods for Discrete Population Genetic Data*. Sinauer Associates, Sunderland.

Weisberg, W.G., Barns, S.M., Pelletier, D.A. and Lane, D.J. (1991) 16S ribosomal DNA amplification for phylogenetic study. *Journal of Bacteriology*, **173**, 697–703.

Weiss, I.M. and Vossbrinck, C.R. (1998) Microsporidiosis: Molecular and diagnostic aspects. *Advances in Parasitology*, **40**, 351–95.

Weiss, M.R. (2003) Good housekeeping: why do shelter-dwelling caterpillars fling their frass? *Ecology Letters*, **6**, 361–70.

Weisser, L.A. and Stamp, N. (1998) Combined effects of allelochemicals, prey availability, and supplemental food on growth of a generalist predator. *Entomologia Experimentalis et Applicata*, **87**, 181–9.

Weisser, W.W. (1994) Age-dependent foraging behavior and host-instar preference of the aphid parasitoid *Lysiphlebus cardui*. *Entomologia Experimentalis et Applicata*, **70**, 1–10.

Weisser, W.W. and Hassell, M.P. (1996) Animals 'on the move' stabilise host-parasitoid systems. *Proceedings of the Royal Society of London B*, **263**, 749–4.

Weisser, W.W. and Houston, A.I. (1993) Host discrimination in parasitic wasps: when is it advantageous? *Functional Ecology*, **7**, 27–39.

Weisser, W.W. and Völkl, W. (1997) Dispersal in the aphid parasitoid, *Lysiphlebus cardui* (Marshall) (Hym., Aphidiidae). *Journal of Applied Entomology*, **121**, 23–8.

Weisser, W.W., Völkl, W. and Hassell, M.P. (1997) The importance of adverse weather conditions for behaviour and population ecology of an aphid parasitoid. *Journal of Animal Ecology*, **66**, 386–400.

Weisser, W.W., Braendle, C. and Minoretti, N. (1999) Predator-induced morphological shift in the pea aphid. *Proceedings of the Royal Society of London B*, **266**, 1175–81.

Welburn, S.C., Maudlin, I. and Ellis, D.S. (1987) *In vitro* cultivation of Rickettsiae-like organisms from *Glossina* spp. *Annals of Tropical and Medical Parasitology*, **81**, 331–5.

Welch, R.C. (1993) Ovariole development in Staphylinidae (Coleoptera). *Invertebrate Reproduction and Development*, **23**, 225–34.

Wellings, P., Morton, R. and Hart, P.J. (1986) Primary sex-ratio and differential progeny survivorship in solitary haplo-diploid parasitoids. *Ecological Entomology*, **11**, 341–8.

Wellington, W.G. and Fitzpatrick, S.M. (1981) Territoriality in the drone fly, *Eristalis tenax* (Diptera: Syrphidae). *Canadian Entomologist*, **113**, 695–704.

Wenseleers, T., Sundström, L. and Billen, J. (2002) Deleterious *Wolbachia* in the ant *Formica truncorum*. *Proceedings of the Royal Society of London B*, **269**, 623–9.

Went, D.F. (1982) Egg activation and parthenogenetic reproduction in insects. *Biological Reviews*, **57**, 319–44.

Went, D.F. and Krause, G. (1973) Normal development of mechanically activated, unlaid eggs of an endoparasitic hymenopteran. *Nature*, **244**, 454–5.

Werren, J.H. (1980) Sex ratio adaptations to local mate competition in a parasitic wasp. *Science*, **208**, 1157–9.

Werren, J.H. (1983) Sex ratio evolution under local mate competition in a parasitic wasp. *Evolution*, **37**, 116–24.

Werren, J.H. (1984) A model for sex ratio selection in parasitic wasps: local mate competition and host quality effects. *Netherlands Journal of Zoology*, **34**, 81–96.

Werren, J.H. (1987) The coevolution of autosomal and cytoplasmic sex ratio factors. *Journal of Theoretical Biology*, **124**, 317–34.

Werren, J.H. (1991) The paternal-sex-ratio chromosome of *Nasonia*. *American Naturalist*, **137**, 392–402.

Werren, J.H. (1997) Biology of *Wolbachia*. *Annual Review of Entomology*, **42**, 587–609.

Werren, J.H. (1998) *Wolbachia* and speciation, in *Endless Forms: Species and Speciation* (eds D.J. Howard and S.H. Berlocher). Oxford University Press, Oxford, pp. 245–60.

Werren, J.H. (2000) Evolution of *Wolbachia* symbioses in invertebrates. *American Zoologist*, **40**, 1255.

Werren, J.H. and van den Assem, J. (1986) Experimental analysis of a paternally inherited extrachromosomal factor. *Genetics*, **114**, 217–33.

Werren, J.H. and Bartos, J.D. (2001) Recombination in *Wolbachia*. *Current Biology*, **11**, 431–5.

Werren, J.H. and O'Neill, S.L. (1997) The evolution of heritable symbionts, in *Influential Passengers: Microbes and Invertebrate reproduction* (eds. S.L. O'Neill, A.A. Hoffmann and J.H. Werren). Oxford University Press, Oxford, pp. 1–41.

Werren, J.H. and Simbolotti, G. (1989) Combined effects of host quality and local mate competition on sex allocation in *Lariophagus distinguendus*. *Evolutionary Ecology*, **3**, 203–13.

Werren, J.H. and Stouthamer, R. (2003) PSR (paternal sex ratio) chromosomes: the ultimate selfish genetic elements. *Genetica*, **117**, 85–101.

Werren, J.H. and Windsor, D.M. (2000) *Wolbachia* infection frequencies in insects: evidence of a global equilibrium? *Proceedings of the Royal Society of London B*, **267**, 1277–85.

Werren, J.H., Skinner, S.W. and Charnov, E.L. (1981) Paternal inheritance of a daughterless sex ratio factor. *Nature*, **293**, 467–8.

Werren, J.H., Skinner, S.W. and Huger, A.M. (1986) Male-killing bacteria in a parasitic wasp. *Science*, **231**, 990–2.

Werren, J.H., Nur, U. and Eickbush, D. (1987) An extrachromosomal factor causing loss of paternal chromosomes. *Nature*, **327**, 75–6.

Werren, J.H., Zhang, W. and Guo, L.R. (1995a) Evolution and phylogeny of *Wolbachia*: reproductive parasites of arthropods. *Proceedings of the Royal Society of London B*, **261**, 55–71.

Werren, J.H., Windsor, D. and Guo, L.R. (1995b) Distribution of *Wolbachia* among neotropical arthropods. *Proceedings of the Royal Society of London B*, **262**, 197–204.

Weseloh, R.M. (1972) Sense organs of the hyperparasite *Cheiloneurus noxius* (Hymenoptera: Encyrtidae) important in host selection processes. *Annals of the Entomological Society of America*, **65**, 41–6.

Weseloh, R.M. (1980) Sex pheromone gland of the gypsy moth parasitoid, *Apanteles melanoscelus*: reevaluation and ultrastructural survey. *Annals of the Entomological Society of America*, **73**, 576–80.

Weseloh, R.M. (1981) Relationship between colored sticky panel catches and reproductive behavior of forest tachinid parasitoids. *Environmental Entomology*, **10**, 131–5.

Weseloh, R.M. (1986a) Biological control of gypsy moths: help from a beetle. *Frontiers of Plant Science*, **39**, 2–3.

Weseloh, R.M. (1986b) Effect of photoperiod on progeny production and longevity of gypsy moth (Lepidoptera: Lymantriidae) egg parasite *Ooencyrtus kuvanae* (Hymenoptera: Encyrtidae). *Environmental Entomology*, **15**, 1149–53.

Weseloh, R.M. (1989a) Temperature-based models of development for the Gypsy Moth (Lepidoptera: Lymantridae) predator, *Calosoma sycophanta* (Coleoptera: Carabidae). *Environmental Entomology*, **18**, 1105–11.

Weseloh, R.M. (1989b) Simulation of predation by ants based on direct observations of attacks on gypsy moth larvae. *Canadian Entomologist*, **121**, 1069–76.

Weseloh, R.M. (1990) Estimation of predation rates of gypsy moth larvae by exposure of tethered caterpillars. *Environmental Entomology*, **19**, 448–55.

Weseloh, R.M. (1993) Manipulation of forest ant (Hymenoptera: Formicidae) abundance and resulting impact on gypsy moth (Lepidoptera: Lymantriidae) populations. *Environmental Entomology*, **22**, 587–94.

Wesolowska, W. and Russell-Smith, A. (2000) Jumping spiders from Mkomazi Game Reserve in Tan"zania (Araneae: Salticidae). *Tropical Zoology*, **13**, 11–27.

West, S.A. and Herre, E.A. (1998a) Partial local mate competition and the sex ratio: a study on non-pollinating fig wasps. *Journal of Evolutionary Biology*, **11**, 531–48.

West, S.A. and Herre, E.A. (1998b) Stabilizing selection and variance in fig wasp sex ratios, *Evolution*, **52**, 475–85.

West, S.A. and Rivero, A. (2000) Using sex ratios to estimate what limits reproduction in parasitoids. *Ecology Letters*, **3**, 294–9.

West, S.A. and Sheldon, B.C. (2002) Constraints in the evolution of sex ratio adjustment. *Science*, **295**, 1685–8.

West, S.A., Flanagan, K.E. and Godfray, H.C.J. (1996) The relationship between parasitoid size and fitness in the field, a study of *Achrysocharoides zwoelferi* (Hymenoptera: Eulophidae). *Journal of Animal Ecology*, **65**, 631–9.

West, S.A., Herre, E.A., Compton, S.G., Godfray, H.C.J. and Cook, J.M. (1997) A comparative study of virginity in fig wasps. *Animal Behaviour*, **54**, 437–450.

West, S.A., Compton, S.G., Vincent, S.L, Herre, E.A. and Cook, J.M. (1998a) Virginity in haplodiploid populations: a comparison of estimation methods. *Ecological Entomology*, **23**, 207–10.

West, S.A., Cook, J.M., Werren, J.H. and Godfray, H.C.J. (1998b) *Wolbachia* in two insect host-parasitoid communities. *Molecular Ecology*, **7**, 1457–65.

Weston, R.F., Qureshi, I. and Werren, J.H. (1999) Genetics of wing size differences between two *Nasonia* species. *Journal of Evolutionary Biology*, **12**, 586–595.

Weyman, G.S., Jepson, P.C. and Sunderland, K.D. (1995) Do seasonal changes in numbers of aerially dispersing spiders reflect population density on the ground or variation in ballooning motivation? *Oecologia*, **101**, 487–93.

Whalon, M.E. and Parker, B.L. (1978) Immunological identification of tarnished plant bug predators. *Annals of the Entomological Society of America*, **71**, 453–56.

Whalon, M.E. and Smilowitz, Z. (1979) The interaction of temperature and biotype on development of the green peach aphid, *Myzus persicae* (Sulz.). *American Potato Journal*, **56**, 591–6.

Wheeler, W.M. (1910) A gynandromorphous Mutillid. *Psyche*, **17**, 186–190.

Whitcomb, W.H. (1980) Sampling spiders in soybean fields, in *Sampling Methods in Soybean Entomology* (eds M. Kogan and D.C. Herzog). Springer-Verlag, New York, pp. 544–58.

Whitcomb, W.H. and Bell, K. (1964) Predaceous insects, spiders, and mites of Arkansas cotton fields. *University of Arkansas, Agricultural Experiment Station Bulletin*, **690**, 1–84.

White, A.J., Wratten, S.D., Berry, N.A. and Weigmann, U. (1995) Habitat manipulation to enhance biological control of *Brassica* pests by hover flies. *Journal of Economic Entomology*, **88**, 1171–6.

White, G.C., Anderson, D.R., Burnham, K.P. and Otis, D.L. (1982) *Capture-recapture and Removal Methods.* Los Alamos National Laboratory, Los Alamos.

White, M.J.D. (1954) *Animal Cytology and Evolution.* Cambridge University Press, Cambridge.

Whitfield, J.B. and Cameron, S.A. (1998) Hierarchical analysis of variation in the mitochondrial 16S rRNA gene among Hymenoptera. *Molecular Biology and Evolution*, **15**, 1728–43.

Whiting, A.R. (1961) Genetics of *Habrobracon*. *Advances in Genetics*, **10**, 295–348.

Whiting, A.R. (1967) The biology of the parasitic wasp *Mormoniella vitripennis* (= *Nasonia brevicornis*) (Walker). *Quarterly Review of Biology*, **42**, 333–406.

Whiting, P.W. (1932) Mutants in *Habrobracon*. *Genetics*, **17**, 1–30.

Whiting, P.W. (1934) Mutants in *Habrobracon* II. *Genetics*, **19**, 268–91.

Whiting, P.W. (1935) Recent X-ray mutations in *Habrobracon*. *Proceedings of the Pacific Academy of Science*, **9**, 60–3.

Whiting, P.W. (1939) Sex determination and reproductive economy in *Habrobracon*. *Genetics*, **24**, 110–1.

Whiting, P.W. (1940) Multiple alleles in sex determination of *Habrobracon*. *Journal of Morphology*, **66**, 323–55.

Whiting, P.W. (1943) Multiple alleles in complementary sex determination of *Habrobracon*. *Genetics*, **28**, 365–82.

Whiting, P.W. (1954) Comparable mutant eye colors in *Mormoniella* and *Pachycrepoideus* (Hymenoptera: Pteromalidae). *Evolution*, **8**, 135–47.

Whiting, P.W. (1961) Genetics of *Habrobracon*. *Advances in Genetics*, **10**, 295–348.

Whiting, P.W. and Whiting, A.R. (1925). Diploid males from fertilized eggs in Hymenoptera. *Science*, **62**, 437–8.

Wickman, P.-O. and Karlsson, B. (1989) Abdomen size, body size and the reproductive effort of insects. *Oikos*, **56**, 209–14.

Wiebe, A.P. and Obrycki, J.J. (2002) Prey suitability of *Galerucella pusilla* eggs for two generalist predators, *Coleomegilla maculata* and *Chrysoperla carnea*. *Biological Control*, **23**, 143–8.

Wiedenmann, R.N. & R.J. O'Neil 1992. Searching strategy of the predator *Podisus maculiventris*. *Environmental Entomology*, **21**, 1–9.

Wigglesworth V.B. (1972) *The Principles of Insect Physiology*. Methuen, London.

Wilbert, H. and G. Lauenstein (1974) Die eignung von *Megoura viciae* (Buckt.) (Aphid) für larven und erwachsene weibehen von *Aphelinus asychus* Walker (Aphelinidae). *Oecologia*, **16**, 311–22.

de Wilde, J. and de Loof, A. (1973) Reproduction, in *The Physiology of the Insecta, Vol. I, 2nd edn* (ed. M. Rockstein). Academic Press, New York, pp. 11–95.

Wilkes, A. (1963) Environmental causes of variation in the sex ratio of an arrhenotokous insect, *Dahlbominus fuliginosus* (Nees) (Hymenoptera: Eulophidae), *Canadian Entomologist*, **95**, 183–202.

Wilkes, A. (1964) Inherited male-producing factor in an insect that produces its males from unfertilized eggs. *Science*, **144**, 305–7.

Wilkes, A. (1965) Sperm transfer and utilization following insemination in the wasp *Dahlbominus fuscipennis* (Zett.) (Hymenoptera: Eulophidae). *Canadian Entomologist*, **97**, 647–57.

Wilkes, A. (1966) Sperm utilization following multiple insemination in the wasp *Dahlbominus fuscipennis*. *Canadian Journal of Genetics and Cytology*, **8**, 451–61.

Wilkes, A. and Lee, P.E. (1965) The ultrastructure of dimorphic spermatozoa in the hymenopteran *Dahl-*

bomius fuscipennis (Zett.) (Eulophidae). *Canadian Journal of Genetics and Cytology*, **7**, 609–19.

Wilkinson, J.D., Schmidt, G.T. and Biever, K.D. (1980) Comparative efficiency of sticky and water traps for sampling beneficial arthropods in red clover and the attraction of clover head caterpillar adults to anisyl acetone. *Journal of the Georgia Entomological Society*, **15**, 124–31.

Williams, D.D and Feltmate, B.W. (1992) *Aquatic Insects*. CAB International, Wallingford.

Williams, G. (1958) Mechanical time-sorting of pitfall captures. *Journal of Animal Ecology*, **27**, 27–35.

Williams, H.J., Elzen, G.W. and Vinson, S.B. (1988) Parasitoid-host plant interactions emphasizing cotton (*Gossypium*), in *Novel Aspects of Insect-Plant Allelochemicals and Host Specificity* (eds P. Barbosa and D. Letourneau). John Wiley, New York, pp. 171–200.

Williams, J.G.K., Kubelik, A.R., Livak, K.J., Rafalski, J.A. and Tingey, S.V. (1990) DNA polymorphisms amplified by arbitrary primers are useful as genetic markers. *Nucleic Acids Research*, **18**, 6531–5.

Williams, R.J. and Martinez, N.D. (2000) Simple rules yield complex food webs. *Nature*, **404**, 180–3.

Williamson, M. (1972) *The Analysis of Biological Populations*. Edward Arnold, London.

Willis, J.C. and Burkill, I.H. (1895) Flowers and insects in Great Britain. I. *Annals of Botany*, **9**, 227–73.

Willmer, P.C. (1980) The effects of insect visitors on nectar constituents in temperate plants. *Oecologia*, **47**, 270–77.

Willmer, P.G. (1983) Thermal constraints on activity patterns in nectar-feeding insects. *Ecological Entomology*, **8**, 455–69.

Wilson, E.O. (1971) *The Insect Societies*. Belknap Press, Harvard.

Wilson, E.O. (1974) *Sociobiology: The Abridged Edition*. Belknap Press, Harvard.

Wilson, F. and Woolcock, L.T. (1960) Environmental determination of sex in a parthenogenetic parasite. *Nature*, **186**, 99–100.

Wilson, H.B. and Hassell, M.P. (1997) Host-parasitoid spatial models: the interplay of demographic stochasticity and dynamics. *Proceedings of the Royal Society of London B*, **264**, 1189–95.

Wilson, K. and Goulding, K.H. (1986) *A Biologist's Guide to the Principles and Techniques of Practical Biochemistry, 3rd edn*. Edward Arnold, London.

Wilson, L.T. and Gutierrez, A.P. (1980) Within-plant distribution of predators on cotton: comments on sampling and predator efficiences. *Hilgardia*, **48**, 1–11.

Wilson, N., Tubman, S., Eady, P. and Robertson, G.W. (1997) Female genotype affects male success in sperm competition. *Proceedings of The Royal Society of London B*, **264**, 1491–95.

Winder, L., Hirst, D.J., Carter, N., Wratten, S.D. and Sopp, P.I. (1994) Estimating predation of the grain aphid *Sitobion avenae* by polyphagous predators. *Journal of Applied Ecology*, **31**, 1–12.

Winder, L., Holland, J.M., Perry, J.N., Woolley, C. and Alexander, C.J. (2001) The use of barriered pitfall trapping for sampling beetles and spiders. *Entomologia Experimentalis et Applicata*, **98**, 249–58.

Wink, M., Grimm, C., Koschmieder, C., Sporer, F. and Bergeot, O. (2000) Sequestration of phorbolesters by the aposematically coloured bug *Pachycoris klugii* (Heteroptera : Scutelleridae) feeding on *Jatropha curcas* (Euphorbiaceae). *Chemoecology*, **10**, 179–84.

Wittmann, E.J. and Leather, S.R. (1997) Compatibility of *Orius laevigatus* Fieber (Hemiptera: Anthocoridae) with *Neoseiulus* (*Amblyseius*) *cucumeris* Oudemmans (Acari: Phytoseiidae) and *Iphiseius* (*Amblyseius*) *degenerans* Berlese (Acari: Phytoseiidae) in the biocontrol of *Frankliniella occidentalis* Pergande (Thysanoptera: Thripidae). *Experimental and Applied Acarology*, **21**, 523–38.

Wolf, R. and Wolf, D. (1988) Activation by calcium ionophore injected into unfertilized ovarian eggs explanted from *Pimpla turionellae* (Hymenoptera). *Zoologische Jahrbucher, Abteilung für Allgemeine Zoologie und Physiologie der Tiere*, **92**, 501–12.

Woodward, G. and Hildrew, A.G. (2001) Invasion of a stream food web by a new top predator. *Journal of Animal Ecology*, **70**, 273–88.

Woodward, G. and Hildrew, A.G. (2002) Body-size determinants of niche overlap and intraguild predation within a complex food web. *Journal of Animal Ecology*, **71**, 1063–74.

Wool, D., van Emden, H.F. and Bunting, S.D. (1978) Electrophoretic detection of the internal parasite *Aphidius matricariae* in *Myzus persicae*. *Annals of Applied Biology*, **90**, 21–6.

Wool, D., Gerling, D. and Cohen, I. (1984) Electrophoretic detection of two endoparasite species, *Encarsia lutea* and *Eretmocerus mundus* in the whitefly, *Bemesia tabaci* (Genn.) (Hom., Aleurodidae). *Zeitschrift für Angewandte Entomologie*, **98**, 276–9.

Work, T.T., Buddle, C.M., Korinus, L.M. and Spence, J.R. (2002) Pitfall trap size and capture of three taxa of litter-dwelling arthropods: implications for biodiversity studies. *Environmental Entomology*, **31**, 438–48.

Workman, C. (1978) Life cycle and population dynamics of *Trochosa terricola* Thorell (Araneae: Lycosidae) in a Norfolk grass heath. *Ecological Entomology*, **3**, 329–40.

Worner, S.P., Goldson, S.B. and Frampton, E.R. (1989) Comparative ecoclimatic assessments of *Anaphes diana* (Hymenoptera: Mymaridae) and its intended host, *Sitona discoideus* (Coleoptera: Curculionidae), in New Zealand. *Journal of Economic Entomology*, **82**, 1085–90.

Wratten, S.D. (1973) The effectiveness of the coccinellid beetle, *Adalia bipunctata* (L.), as a predator of the lime aphid, *Eucallipterus tiliae* L. *Journal of Animal Ecology*, **42**, 785–802.

Wratten, S. (1992) Population regulation in insect herbivores – top-down or bottom-up? *New Zealand Journal of Ecology*, **16**, 145–7.

Wratten, S.D. (1994) *Video Techniques in Animal Ecology and Behaviour*. Chapman and Hall, London.

Wratten, S.D. and Pearson, J. (1982) Predation of sugar beet aphids in New Zealand. *Annals of Applied Biology*, **101**, 178–81.

Wratten, S.D., White, A.J., Bowie, M.H., Berry, N.A. and Weigmann, U. (1995) Phenology and ecology of hoverflies (Diptera: Syrphidae) in New Zealand. *Environmental Entomology*, **24**, 595–600.

Wratten, S.D., Bowie, M.H., Hickman, J.M., Evans, A.M., Sedcole, J.R. and Tylianakis, J.M. (2003) Field boundaries as barriers to movement of hover flies (Diptera: Syrphidae) in cultivated land. *Oecologia*, **134**, 605–11.

Wrensch, D.L. and Ebbert, M.A. (1993) *Evolution and Diversity of Sex Ratio*. Chapman and Hall, New York and London.

Wright, A.F. and Stewart, A.J.A. (1992) A study of the efficacy of a new inexpensive type of suction apparatus in quantitative sampling of grassland invertebrate populations. *Bulletin of the British Ecological Society*, **23**, 116–20.

Wright, D.W., Hughes, R.D. and Worrall, J. (1960) The effect of certain predators on the numbers of cabbage root fly (*Erioischia brassicae*) (Bouche) and on the subsequent damage caused by the pest. *Annals of Applied Biology*, **48**, 756–63.

Wylie, H.C. (1983) Delayed development of *Microctonus vittatae* (Hymenoptera: Braconidae) in superparasitised adults of *Phyllotreta cruciferae* (Coleoptera: Chrysomelidae). *Canadian Entomologist*, **115**, 441–2.

Wysoki, M., de Long, M., Rene, S. (1988) *Trichogramma platneri* Nagarkatti (Hymenoptera: Trichogrammatidae), its biology and ability to search for eggs of two lepidopterous avocado pests, *Boarmia (Ascotis) selenaria* (Schiffermuller) (Geometridae) and *Cryptoblabes gnidiella* (Milliere) (Phycitidae) in Israel. *Colloques de L' INRA*, **43**, 295–301.

Wyss, E. (1995) The effects of weed strips on aphids and aphidophagous predators in an apple orchard. *Entomologia Experimentalis et Applicata*, **75**, 43–9.

Yahiro, K. (1997) Ground beetles (Coleoptera, Caraboidea) caught by a light trap during ten years. *Esakia*, **37**, 57–69.

Yamada, Y.Y. and Kitashiro, S. (2002) Infanticide in a dryinid parasitoid, *Haplogonatopus atratus*. *Journal of Insect Behavior*, **15**, 415–27.

Yamaguchi, T. and Hasegawa, M. (1996) An experiment on ant predation in soil using a new bait trap method. *Ecological Research*, **11**, 11–16.

Yamamura, N. and Yano, E. (1988) A simple model of host-parasitoid interaction with host-feeding. *Researches on Population Ecology*, **30**, 353–69.

Yano, E. (1989a) A simulation study of population interaction between the greenhouse whitefly, *Trialeurodes vaporariorum* Westwood (Homoptera: Aleyrodidae) and the parasitoid *Encarsia formosa* Gahan (Hymenoptera: Aphelinidae). I. Description of the model. *Researches on Population Ecology*, **31**, 73–88.

Yano, E. (1989b) A simulation study of population interaction between the greenhouse whitefly, *Trialeurodes vaporariorum* Westwood (Homoptera: Aleyrodidae) and the parasitoid *Encarsia formosa* Cahan (Hymenoptera: Aphelinidae). II. Simulation analysis of population dynamics and strategy of biological control. *Researches on Population Ecology*, **31**, 89–104.

Yasuda, H. (1995) Effect of prey density on behaviour and development of the predatory mosquito, *Toxorhynchites towadensis*. *Entomologia Experimentalis et Applicata*, **76**, 97–103.

Yeargan, K.V. and Barney, W.E. (1996) Photoperiodic induction of reproductive diapause in the predators *Nabis americoferus* and (Heteroptera: Nabidae). *Annals of the Entomological Society of America*, **89**, 70–4.

Yeargan, K.V. and Cothran, W.R. (1974) An escape barrier for improved suction sampling of *Pardosa ramulosa* and *Nabis* spp. populations in alfalfa. *Environmental Entomology*, **3**, 189–91.

Yodzis, P. and Winemiller, K.O. (1999) In search of operational trophospecies in a tropical aquatic food web. *Oikos*, **87**, 327–40.

Yoshida, S. (1978) Behaviour of males in relation to the female sex pheromone in the parasitoid wasp, *Anisopteromalus calandrae* (Hym., Pteromalidae). *Entomologia Experimentalis et Applicata*, **23**, 152–62.

Yoshida, S. and Hidaka, T. (1979) Determination of the position of courtship display of the young unmated male *Anisopteromalus calandrae*. *Entomologia Experimentalis et Applicata*, **26**, 115–20.

Young, L.C. (1980) Field estimation of the functional response of *Itoplectis behrensi*, a parasite of the California oakworm, *Phryganidia californica*. *Environmental Entomology*, **9**, 49–50.

Young, O.P. and Hamm, J.J. (1986) Rate of food passage and fecal production in *Calosoma sayi* (Coleoptera: Carabidae). *Entomological News*, **97**, 21–7.

Yu, D.S. and Luck, R.F. (1988) Temperature dependent size and development of California Red Scale (Homoptera: Diaspididae) and its effect on host availability for the ecto-parasitoid *Aphytis melinus* De Bach (Hymenoptera: Aphelinidae). *Environmental Entomology*, **17**, 154–61.

Yu, D.S., Luck, R.F. and Murdoch, W.W. (1990) Competition, resource partitioning and coexistence of an endoparasitoid *Encarsia perniciosi* and an ectoparasitoid *Aphytis melinus* of the California red scale. *Ecological Entomology*, **15**, 469–80.

Zacaro, A.A. and Porter, S.D. (2003) Female reproductive system of the decapitating fly *Pseudacteon wasmanni* Schmitz (Diptera: Phoridae). *Arthropod Structure and Development*, **31**, 329–37.

Zaidi, R.H., Jaal, Z., Hawkes, N.J., Hemingway, J. and Symondson, W.O.C. (1999) Can the detection of prey DNA amongst the gut contents of invertebrate predators provide a new technique for quantifying predation in the field? *Molecular Ecology*, **8**, 2081–8.

Zandstra, B.H. and Motooka, P.S. (1978) Beneficial effects of weeds in pest mangement. *PANS*, **24**, 333–8.

Zaslavsky V.A. and Vagina, N.P. (1996) Joint and separate effects of photoperiodic and alimentary induction of diapause in *Coccinella septempunctata* (Coleoptera, Coccinellidae). *Zoologichesky Zhurnal*, **75**, 1474–82.

Zaviezo, T. and Mills, N. (2000) Factors influencing the evolution of clutch size in a gregarious insect parasitoid. *Journal of Animal Ecology*, **69**, 1047–57.

Zchori-Fein, E., Faktor, O., Zeidan, M., Gottlieb, Y., Czosnek, H. and Rosen, D. (1995) Parthenogenesis-inducing microorganisms in *Aphytis* (Hymenoptera: Aphelinidae). *Insect Molecular Biology*, **4**, 173–8.

Zchori-Fein, E., Gottlieb, Y., Kelly, S.E., Brown, JK., Wilson, J.M., Karr, T.L. and Hunter, M.S. (2001) A newly discovered bacterium associated with parthenogenesis and a change in host selection behaviour in parasitoid wasps. *Proceedings of the National Academy of Sciences*, **98**, 12555–60.

Zeng, Z.-B. (1994) Precision mapping of quantitative trait loci. *Genetics*, **136**, 1457–68.

Zeng, Z.-B., Kao, C.-H. and Basten, C. J. (1999) Estimating the genetic architecture of quantitative traits. *Genetics Research, Cambridge*, **74**: 279–89.

Zenger, J.T. and Gibb, T.J. (2001) Identification and impact of egg predators of *Cyclocephala lurida* and *Popillia japonica* (Coleoptera: Scarabaeidae) in turfgrass. *Environmental Entomology*, **30**, 425–30.

Zhang, J., Drummond, F.A., Liebman, M. and Hartke, A. (1997) Phenology and dispersal of *Harpalus rufipes* DeGeer (Coleoptera: Carabidae) in agroecosystems in Maine. *Journal of Agricultural Entomology*, **14**, 171–86.

Zheng, Y., Hagen, K.S., Daane, K.M. and Mittler, T.E. (1993a) Influence of larval dietary supply on the food consumption, food utilisation efficiency, growth and development of the lacewing *Chrysoperla carnea*. *Entomologia Experimentalis et Applicata*, **67**, 1–7.

Zheng, Y., Hagen, K.S., Daane, K.M. and Mittler, T.E. (1993b) Influence of larval food consumption on the fecundity of the lacewing *Chrysoperla carnea*. *Entomologia Experimentalis et Applicata*, **67**, 9–14.

Zhou, W.G., Rousset, F. and O'Neill, S.L. (1998) Phylogeny and PCR-based classification of *Wolbachia* strains using *wsp* gene sequences. *Proceedings of the Royal Society of London B* **265**, 509–15.

Zhou, X., Carter, N. and Powell, W. (1994) Seasonal distribution and aerial movement of adult coccinellids on farmland. *Biocontrol Science and Technology*, **4**, 167–75.

Zhu, Y.C. and Greenstone, M.H. (1999) Polymerase chain reaction techniques for distinguishing three species and two strains of *Aphelinus* (Hymenoptera: Aphelinidae) from *Diuraphis noxia* and *Schizaphis graminum* (Homoptera: Aphididae). *Annals of the Entomological Society of America*, **92**, 71–9.

Zhu, Y.C., Burd, J.D., Elliott, N.C. and Greenstone, M.H. (2000) Specific ribosomal DNA marker for early PCR detection of *Aphelinus hordei* (Hymenoptera: Aphelinidae) and *Aphidius colemani* (Hymenoptera: Aphidiidae) from *Diuraphis noxia* (Homoptera: Aphididae). *Annals of the Entomological Society of America*, **93**, 486–91.

Zoebelein, G. (1955) Der honigtau als nahrung der Insekten. I. *Zeitschrift für Angewandte Entomologie*, **38**, 369–416.

Zuk, M. (1988) Parasite load, body size and age of wild caught male field crickets: effects on sexual selection. *Evolution*, **42**, 969–76.

Zwölfer, H. (1971) The structure and effect of parasite complexes attacking phytophagous host insects, in *Dynamics of Populations* (eds P.J. den Boer and G.R. Gradwell), Proceedings of the Advanced Study Institute on 'Dynamics of Numbers in Populations', Oosterbeck, 1970. Centre for Agricultural Publishing and Documentation, Wageningen, pp. 405–18.

Zwölfer, H. (1979) Strategies and counterstrategies in insect population systems competing for space and food in flower heads and plant galls. *Fortschritte für Zoologie*, **25**, 331–53.

van Zyl, A., van der Linde, T.C.D. and Grimbeek, R.J. (1997) Metabolic rates of pit-building and non-pit-building antlion larvae (Neuroptera: Myrmeleontidae) from southern Africa. *Journal of Arid Environments*, **37**, 355–65.

AUTHOR INDEX

GENUS AND SPECIES INDEX

SUBJECT INDEX